Autodesk Inventor 2024
Black Book

By
Gaurav Verma
Matt Weber
(CADCAMCAE Works)

Edited by
Kristen

ISBN # 978-1-77459-105-5

DEDICATION

To teachers, who make it possible to disseminate knowledge
to enlighten the young and curious minds
of our future generations

To students, who are the future of the world

THANKS

To my friends and colleagues

To my family for their love and support

Training and Consultant Services

At CADCAMCAE Works, we provide effective and affordable one to one online training on various software packages in Computer Aided Design(CAD), Computer Aided Manufacturing(CAM), Computer Aided Engineering (CAE), Computer programming languages (C/C++, Java, .NET, Android, Javascript, HTML, and so on). The training is delivered through remote access to your system and voice chat via Internet at any time, any place, and at any pace to individuals, groups, students of colleges/universities, and CAD/CAM/CAE training centers. The main features of this program are:

Training as per your need

Highly experienced Engineers and Technician conduct the classes on the software applications used in the industries. The methodology adopted to teach the software is totally practical based, so that the learner can adapt to the design and development industries in almost no time. The efforts are to make the training process cost effective and time saving while you have the comfort of your time and place, thereby relieving you from the hassles of traveling to training centers or rearranging your time table.

Software Packages on which we provide basic and advanced training are:

CAD/CAM/CAE: CATIA, Creo Parametric, Creo Direct, SolidWorks, Autodesk Inventor, Solid Edge, UG NX, AutoCAD, AutoCAD LT, EdgeCAM, MasterCAM, SolidCAM, DelCAM, BOBCAM, UG NX Manufacturing, UG Mold Wizard, UG Progressive Die, UG Die Design, SolidWorks Mold, Creo Manufacturing, Creo Expert Machinist, NX Nastran, Hypermesh, SolidWorks Simulation, Autodesk Simulation Mechanical, Creo Simulate, Gambit, ANSYS, and many others.

Computer Programming Languages: C++, VB.NET, HTML, Android, Javascript and so on.

Game Designing: Unity.

Civil Engineering: AutoCAD MEP, Revit Structure, Revit Architecture, AutoCAD Map 3D and so on.

We also provide consultant services for design and development on the above mentioned software packages

For more information you can mail us at:
cadcamcaeworks@gmail.com

Table of Contents

Chapter 3 : Dimensioning and Constraining

Chapter 4 : Creating Solid Models

Chapter 5 : Advanced Solid Modeling Tools

Chapter 6 : Advanced Modeling Tools and Practical

Chapter 7 : Assembly Design

Chapter 8 : Advanced Assembly and Design

Chapter 9 : Advanced Assembly and Design-II

Chapter 13 : Surface Design and Freeform Creation

Chapter 14 : Drawing Creation

Chapter 15 : Analyses and Simulation

Chapter 16 : Model Based Annotations

Chapter 17 : Application Management

Preface

Autodesk Inventor is a product of Autodesk Inc. Autodesk Inventor 2024 is a parametric, feature-based solid modeling tool that not only unites the three-dimensional (3D) parametric features with two-dimensional (2D) tools, but also addresses every design-through-manufacturing process. The continuous enhancements in the software has made it a complete PLM software. The software is capable of performing analysis with an ease. Its compatibility with CAM software is remarkable. Based mainly on the user feedback, this solid modeling tool is remarkably user-friendly and it allows you to be productive from day one.

The **Autodesk Inventor 2024 Black Book** is the 5th edition of our series on Autodesk Inventor. With lots of features and thorough review, we present a book to help professionals as well as beginners in creating some of the most complex solid models. The book follows a step by step methodology. In this book, we have tried to give real-world examples with real challenges in designing. We have tried to reduce the gap between university use of Autodesk Inventor and industrial use of Autodesk Inventor. In this edition of book, we have included the enhancements made in latest version of the software. The book covers almost all the information required by a learner to master the Autodesk Inventor. The book starts with sketching and ends at advanced topics like Mold Design, Sheetmetal, Weldment, and MBD. Some of the salient features of this book are :

In-Depth explanation of concepts

Every new topic of this book starts with the explanation of the basic concepts. In this way, the user becomes capable of relating the things with real world.

Topics Covered

Every chapter starts with a list of topics being covered in that chapter. In this way, the user can easy find the topic of his/her interest easily.

Instruction through illustration

The instructions to perform any action are provided by maximum number of illustrations so that the user can perform the actions discussed in the book easily and effectively. There are about 2000 small and large illustrations that make the learning process effective.

Tutorial point of view

At the end of concept's explanation, the tutorial make the understanding of users firm and long lasting. Almost each chapter of the book has tutorials that are real world projects. Moreover most of the tools in this book are discussed in the form of tutorials.

Project

Projects and exercises are provided to students for practicing on demand.

For Faculty

If you are a faculty member, then you can ask for video tutorials on any of the topic, exercise, tutorial, or concept. As faculty, you can register on our website to get electronic desk copies of our latest books, self-assessment, and solution of practical. Faculty resources are available in the **Faculty Member** page of our website (**www. cadcamcaeworks.com**) once you login. Note that faculty registration approval is manual and it may take two days for approval before you can access the faculty website.

Formatting Conventions Used in the Text

All the key terms like name of button, tool, drop-down, etc. are kept bold.

Free Resources

Link to the resources used in this book are provided to the users via email. To get the resources, mail us at ***cadcamcaeworks@gmail.com*** with your contact information. With your contact record with us, you will be provided latest updates and informations regarding various technologies. The format to write us mail for resources is as follows:

Subject of E-mail as ***Application for resources of _____ book***.
Also, given your information like
Name:
Course pursuing/Profession:
E-mail ID:

Note: We respect your privacy and value it. If you do not want to give your personal informations then you can ask for resources without giving your information.

About Authors

The author of this book, Gaurav Verma, has authored and assisted in more than 17 titles in CAD/CAM/CAE which are already available in market. He has authored **AutoCAD Electrical Black Books** which are available in both **English** and **Russian** language. He has also authored books on various modules of Creo Parametric and SolidWorks. He has provided consultant services to many industries in US, Greece, Canada, and UK. He has assisted in preparing many Government aided skill development programs. He has been speaker for Autodesk University, Russia 2014. He has assisted in preparing AutoCAD Electrical course for Autodesk Design Academy. He has worked on Sheetmetal, Forging, Machining, and Casting designs in Design and Development department.

The author of this book, Matt Weber, has authored many books on CAD/CAM/CAE available already in market. **SolidWorks Simulation Black Books** are one of the most selling books in SolidWorks Simulation field. The author has hands on experience on almost all the CAD/CAM/CAE packages. If you have any query/doubt in any CAD/CAM/CAE package, then you can contact the author by writing at cadcamcaeworks@gmail.com

For Any query or suggestion

If you have any query or suggestion, please let us know by mailing us on *cadcamcaeworks@gmail.com*. Your valuable constructive suggestions will be incorporated in our books and your name will be addressed in special thanks area of our books on your confirmation.

Page left blank intentionally

Chapter 0

Basics of CAD, CAM, and CAE

Topics Covered

The major topics covered in this chapter are:

- *Introduction to CAD*
- *Introduction to CAM*
- *Introduction to CAE*

INTRODUCTION TO CAD

In earlier days of Mechanical industry, designer engineers had to draw every mechanical component on paper or cloth using drafter and geometry tools like pencils, markers, scale, erasers, and so on. But the age of manually drawing is gone and now a days, we use CAD (Computer Aided Design) software to create engineering drawings. There is a long list of CAD software available in market like Autodesk Inventor, SolidWorks, Creo Parametric, and so on. Broadly there are two ways in which CAD software perform 3D modeling:

- Parametric Modeling
- Direct Modeling

Parametric Modeling V/S Direct Modeling

In Parametric Modeling, the model is create based on parameters. All the parameters that you specify while creating the model are recorded and can be changed any point of time while working on the model. Like, if you are creating a box in parametric modeling then its length, width, and height will be recorded with model and can be changed anytime. AutoCAD, Autodesk Inventor, SolidWorks, Creo Parametric are name of some of the software capable of performing Parametric modeling.

In Direct Modeling, the model is created by direct approach rather than specifying parameters for model. To create a model with direct modeling approach, you place primitive shapes and them drag-drop the key points to change the shape of model. Although Direct Modeling is a nice approach to create models for animators but for Mechanical Engineers, Parametric modeling is an important requirement.

2D Drawing

2D Drawings are used to represent 3D objects on paper for manufacturing. 2D drawings are still the requirement of manufacturers for manufacturing any engineering product. There are various symbols and standards established to created 2D drawing for engineers. These drawings can be furthers divided into different categories based on their application areas like mechanical drawing, electrical drawing, electronic drawing, civil drawing and so on. Our concern for this book is mechanical drawings. For representing objects in mechanical 2D drawings, we use two type of projects of objects on paper: First Angle Projection and Third Angle Projects.

First Angle Projection

In First Angle Projection, the object is imagined to be in first quadrant; refer to Figure-1. In projection system, the vertical plane is used to generate Front view and horizontal plane is used to generate Top view. Now, assume these planes to be hinged at the center and if you move the horizontal plane clockwise then in First Angle projection, the Top view is placed below Front view while placing orthographic views and Left view is placed on the right of Front view; refer to Figure-2.

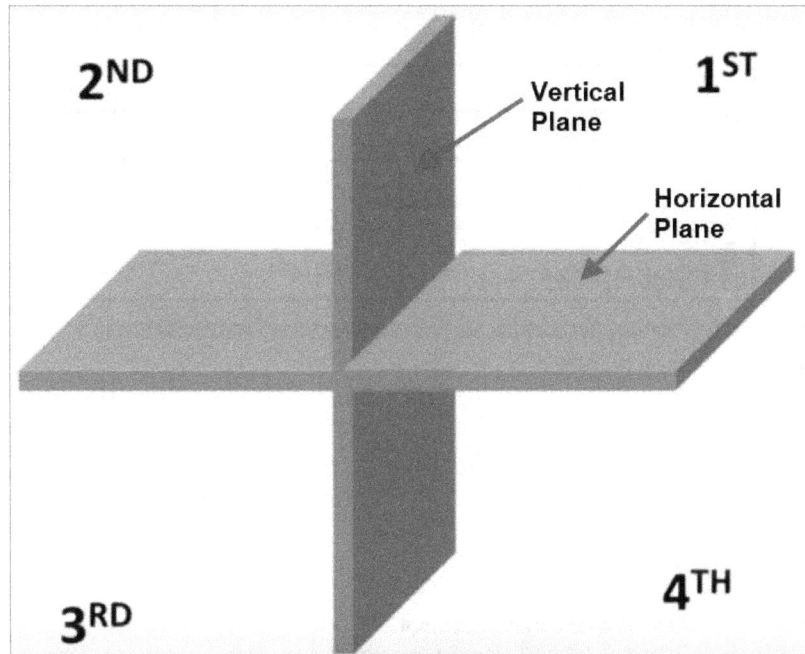

Figure-1. Quadrant system for orthographic view projections

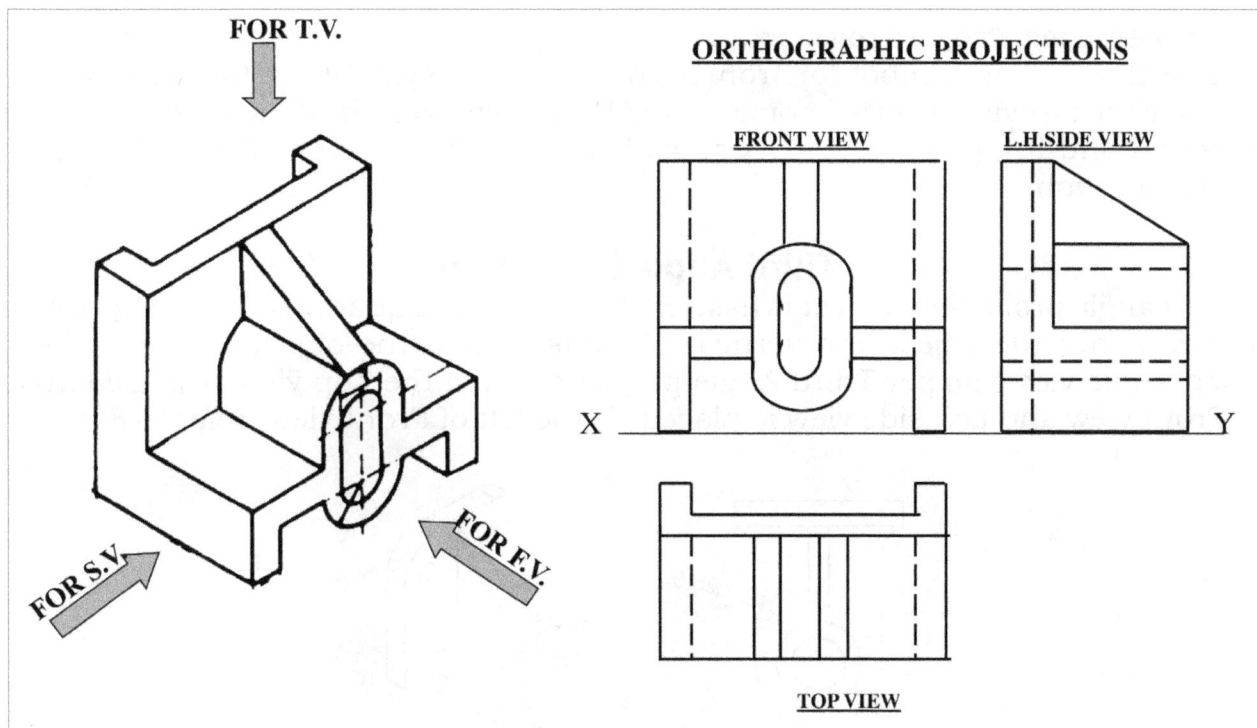

Figure-2. First Angle Projection Views placement

The symbol for projection is always given in Title box of manufacturing drawings; refer to Figure-3. The symbol for Third Angle projection is given in Figure-3. For First Angle projection and Third Angle projection, the symbols are given in Figure-4.

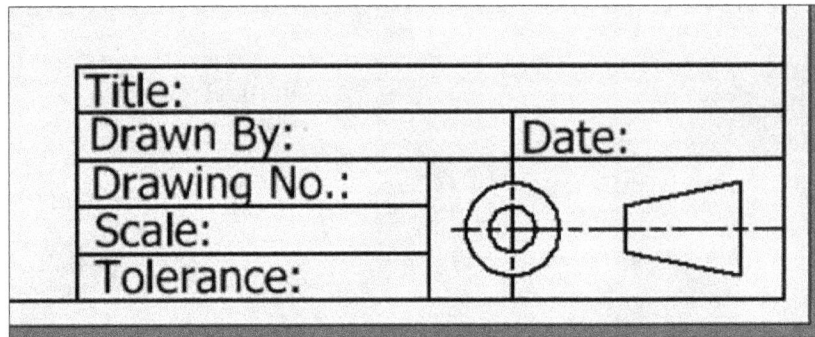

Figure-3. Title block sample

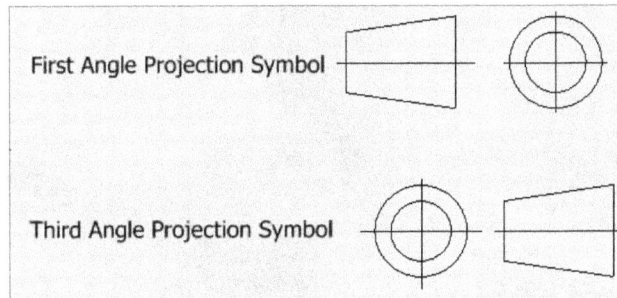

Figure-4. Projection symbols

How to remember Projection Symbols

Assume that ▱ is symbol for front view and ⊕ is symbol for top view. Always remember if top view symbol comes after front view symbol then it is First Angle projection and if top view symbol comes before front view symbol then it is Third Angle projection.

Third Angle Projection

In Third angle projection, object is assumed to be in third quadrant so, the horizontal plane is above the object and vertical plane is behind the object. When we place orthographic views as per Third Angle projection then the Top view is placed above the Front view and Left side view is placed on the left of Front view; refer to Figure-5.

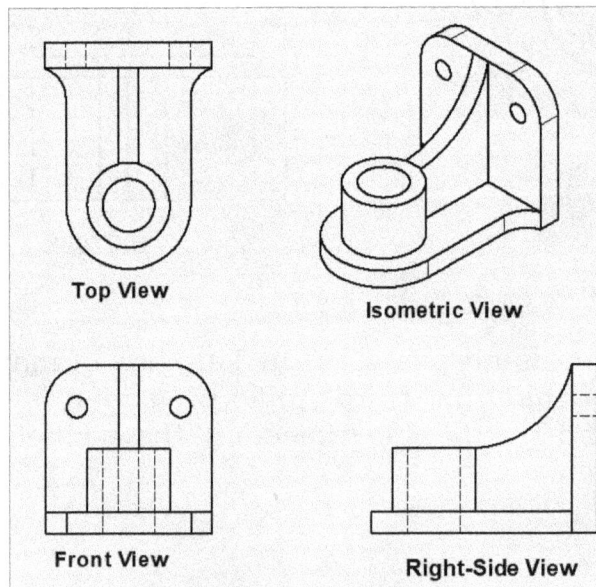

Figure-5. Third Angle projection view placement

The projections discussed earlier are used for Orthographic views. Apart from orthographic views, we also use Isometric and Trimetric views to represent 3D objects in 2D drawings. These views are discussed next.

Axonometric Projections

There are three types of axonometric projections; Isometric, Dimetric and Trimetric. These projections are discussed next.

Isometric means equal measures. Isometric drawing is way of presenting designs/drawings in three dimensions. In order for a design to appear three dimensional, a 30 degree angle is applied to sides object. The cube shown in Figure-6 is as per isometric projection.

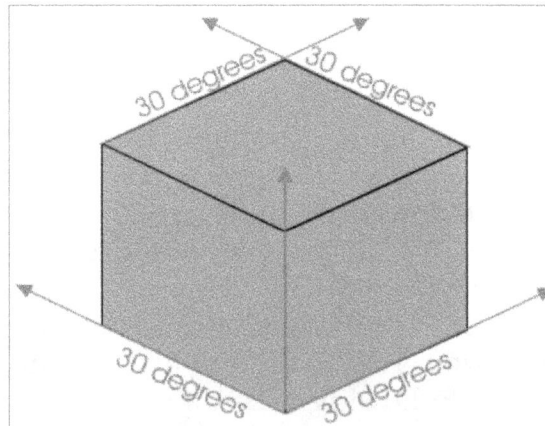

Figure-6. Isometric view projection

In Trimetric projection, the projection of the three angles between the axes are unequal. Thus, three separate scales are needed to generate a trimetric projection of an object. Figure-7 shows an example of different projections.

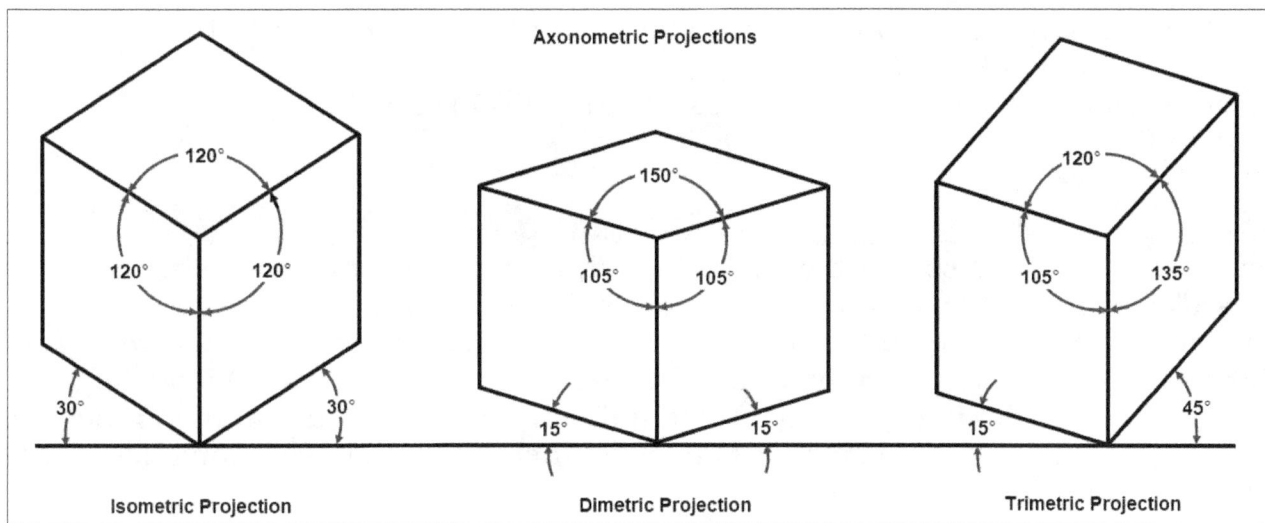

Figure-7. Axonometric Projections

In Dimetric projection, two of an objects axes make equal angles with the plane of projection and the third angle is larger or smaller than the other two; refer to Figure-7.

Drafting Standards

Drafting Standards are the collection of rules defined for creating 2D drawings. In CAD software 2D drawings, following parameters are controlled by Drafting Standards:

- Mechanical object behavior.
- What layers Mechanical Objects are created on.
- The properties of these layers.
- Text heights and colors.
- Projection settings for use with Power View.
- Dimension styles.
- Hole chart settings and formats.
- Centerline format.
- Section line format.
- Thread line format.
- Note text and leader formats.
- Symbology formats.
- Bill of Materials (BOM), parts list and balloon formats.

There are various standards followed by different countries for drafting like, ANSI, BSI, CSN, DIN, GB, ISO JIS, GOST, and so on. ANSI drafting standard was developed by American National Standards Institute. This drafting standard is widely used by American manufacturers. ISO drafting standard was developed by International Organization for Standardization. ANSI and ISO are two most popular standards for drafting engineering drawing. Following are some of the major differences between the two standards:

- ANSI dimensions are read horizontally. ISO dimensions are parallel to the dimension line.
- ANSI dimensions are centered on the dimension line. ISO dimension are placed above the dimension line.
- ANSI tends to use abbreviations. ISO uses symbols. (example: RAD, DIAM, 3 PLACES versus R, Ø, 3X)
- Dimensions have a different syntax. ANSI: 1.000 DIAM 3 PLACES and ISO: 3X Ø 1.000

3D Modeling

Before the first CAD software was developed, manufacturers were using geometry tools like pencil, scale, drafter, and so on to create drawings. Since then the CAD software have developed a lot and so has the requirement of manufacturers. Now, 3D replication of object is created in the computer using various 3D Modeling tools and later the model is used to generate different views with annotations, perform analysis, or generate programs for CNC machines. There is a very long list of CAD software available in market.

Following are some of the functions that can be performed using latest CAD software:
- 3D Modeling
- Drafting (2D Drawing Creation and generation
- Assembly (Top-Down Approach and Bottom-Up Approach)
- 3D Printing

- Computer Aided Manufacturing (CAM)
- Computer Aided Engineering (CAE)

You will learn about various aspects of CAD software later in this book. Now, we will discuss the role of CAD Engineer in mechanical industry.

Role of CAD Engineer in Industry

Following are some of the roles and responsibilities of a CAD engineer:

- Configure, deploy, maintain, and upgrade CAD models as per the client requirement.
- Design, develop, and engineer high quality models using 3D and 2D CAD tools for manufacturing and analysis.
- Produce designs that meet targets for feasibility, performance, costs, quality, safety, legislation and timing.
- Ensure that all work carried out is in compliance with company design, safety, quality, environmental compliance, and procedural standards.
- Interact with architect and client, as necessary to obtain critical design information necessary to complete project within intended time frame.
- Update and maintain product design files.
- Assist in improving daily processes to ensure that the CAD systems meet customer requirements.
- Train and guide Production Engineers on engineered design.
- Determine limitations, assumptions and solutions in the design and development of CAD models.
- Assist in implementation of CAD engineering applications.
- Determine design specifications and parameters for CAD models.

Documents Prepared by CAD Engineers in Automobile Industry

A CAD Engineer is involved in designing of new parts and very soon gets involved in Design Engineer's work. There are various documents that are prepared by CAD/Design Engineers in mechanical industries for development of new parts and processes. Automotive Industry Action Group (AIAG) has developed a standard packages of documentation for Automotive industries world-wide called PPAP. Production Part Approval Process (PPAP) is used in automotive industry supply chains to establish confidence between supplier. Various documents that are prepared in PPAP package are given next.

Design Records

Design records means printed copy of engineering drawings of components to be manufactured. If the customer is responsible for designing, this is a copy of customer drawing that is sent together with the Purchase Order (PO). If supplier is responsible for designing then these drawings are released in supplier's release system. "Each and every feature must be "ballooned" or "road mapped" to correspond with the inspection results (including print notes, standard tolerance notes and specifications, and anything else relevant to the design of the part)."

Authorized Engineering Change (note) Documents

The Authorized Engineering Change Documents (notes) are used to convey changes in original design. The detailed description of changes is noted in this document. Usually this document is called "Engineering Change Notice", but it may be covered by the customer PO or any other engineering authorization.

Engineering Approval

This approval is usually the Engineering trial with production parts performed at the customer plant. A "temporary deviation" usually is required to send parts to customer before PPAP. Customer may require other "Engineering Approvals".

DFMEA

A copy of the Design Failure Mode and Effect Analysis (DFMEA) is reviewed and signed-off by supplier and customer. If customers are design responsible then customers may not share this document with the supplier. However, the list of all critical or high impact product characteristics should be shared with the supplier, so they can be addressed on the PFMEA and Control Plan.

Process Flow Diagram

A copy of the Process Flow, indicating all steps and sequence in the manufacturing process including incoming components.

PFMEA

A copy of the Process Failure Mode and Effect Analysis (PFMEA), reviewed and signed-off by supplier and customer. The PFMEA follows the Process Flow steps, and indicates "what could go wrong" during the manufacturing of each component.

Control Plan

A copy of the Control Plan, reviewed and signed-off by supplier and customer. The Control Plan follows the PFMEA steps, and provides more details on how the "potential issues" are checked in the incoming quality, assembly process or during inspections of finished products.

Measurement System Analysis Studies (MSA)

MSA usually contains lists of Gauges and other measuring instruments required to measure critical or high impact characteristics, and a confirmation that gauges used to measure these characteristics are calibrated.

Dimensional Results

A list of every dimension noted on the ballooned drawing. This list shows the product characteristic, specification, the measurement results and the assessment showing if this dimension is "OK" or "not OK". Usually a minimum of 6 pieces are reported per product/process combination.

Records of Material / Performance Tests

A summary of every test performed on the part. This summary is usually on a form of DVP&R (Design Verification Plan and Report), which lists each individual test, when it was performed, the specification, results and the assessment pass/fail. If there is an Engineering Specification, usually it is noted on the print. The DVP&R shall be reviewed and signed off by both customer and supplier engineering groups. The quality engineer will look for a customer signature on this document. In addition, this section lists all material certifications (steel, plastics, plating, etc.), as specified on the print. The material certification shall show compliance to the specific call on the print.

Initial Sample Inspection Report

The report for material samples which is initially inspected before prototype made.

Initial Process Studies

Usually this section shows all Statistical Process Control charts affecting the most critical characteristics. The intent is to demonstrate that critical processes have stable variability and that is running near the intended nominal value.

Qualified Laboratory Documentation

Copy of all laboratory certifications (e.g. A2LA, TS, NABL) of the laboratories that performed the tests reported in this section.

Appearance Approval Report

A copy of the AAI (Appearance Approval Inspection) form signed by the customer. Applicable for components affecting appearance only.

Sample Production Parts

A sample from the same lot of initial production run. The PPAP package usually shows a picture of the sample and where it is kept (customer or supplier).

Master Sample

A sample signed off by customer and supplier, that usually is used to train operators on subjective inspections such as visual or for noise.

Checking Aids

When there are special tools for checking parts, this section shows a picture of the tool and calibration records, including dimensional report of the tool.

Customer-Specific Requirements

Each customer may have specific requirements to be included on the PPAP package. It is a good practice to ask the customer for PPAP expectations before even quoting for a job. North America auto makers OEM (Original Equipment Manufacturer) requirements are listed on the IATF website.

Part Submission Warrant (PSW)

This is the form that summarizes the whole PPAP package. This form shows the reason for submission (design change, annual revalidation, etc.) and the level of documents submitted to the customer. There is a section that asks for "results meeting all drawing and specification requirements: yes/no" refers to the whole package. If there is any deviations the supplier should note on the warrant or inform that PPAP cannot be submitted.

Augmented Reality and Virtual Reality
Augmented Reality

Augmented Reality is a way to project information on different displays. Augmented Reality has vast application area from social media and entertainment industry to surgical procedures in hospitals; refer to Figure-8. The game Pokemon Go is an example of AR.

Figure-8. AR information displayed from map

Augmented Reality also finds applications in CAD. Although for mechanical engineers it is applicable to civil engineers as well. Civil engineers can project the image of a whole building designed in computer to the customers while there is no building at all. This way they can collect funding for a building project.

Mechanical Engineers can show their final CAD design of a product to their customer without even starting a manufacturing step; refer to Figure-9. If customer approves the design then they can start manufacturing.

Figure-9. Augmented Reality CAD model

Virtual Reality

Virtual Reality is a computer simulated environment to show different types of objects and project real experience through our sensory system. Virtual Reality shuts you from real world and keeps you inside a computer simulated environment. VR is very common with smart phones these days. For CAD engineer, VR can be a life saver sometimes. Using Virtual Reality, you can assemble different components of a large machine virtually and then find any shortcoming based on the experience.

INTRODUCTION TO CAM

The story of CAM starts with CNC machines. CNC represent Computer Numeric Control. CNC machines used numeric codes generated by CAM software to perform various operations. A CAM software takes the input from user and based on specified parameters, it generates CNC programs with G-codes and M-codes. There are various software available for CAM like MasterCAM, BobCAM, EdgeCAM and so on. These software are specialized for CAM. Now a days, most of the CAD software also come with CAM modules like SolidWorks, Creo Parametric, and so on. The NC codes generated by these CAM software depend on the controller hardware installed on your machine. There are various controllers available in the market like Fanuc controller, Siemens controller, Heidenhain controller, and so on. The numeric codes change according to the controller used in the machine. These numeric codes are compiled in the form of a program, which is fed in the machine controller via a storage media. The numeric codes are generally in the form of G-codes and M-codes. For understanding purpose, some of the G-codes and M-codes are discussed next with their functions for a Fanuc controller.

Code		Function
G00	-	Rapid movement of tool.
G01	-	Linear movement while creating cut.
G02	-	Clockwise circular cut.
G03	-	Counter-clockwise circular cut.
G20	-	Starts inch mode.
G21	-	Starts mm mode.
G96	-	Provides constant surface speed.
G97	-	Constant RPM.
G98	-	Feed per minute

G99 -   Feed per revolution

M00 -   Program stop
M02 -   End of program
M03 -   Spindle rotation Clockwise.
M04 -   Spindle rotation Counter Clockwise.
M05 -   Spindle stop
M08 -   Coolant on
M09 -   Coolant off
M98 -   Subprogram call
M99 -   Subprogram exit

Once you have created an NC program in CAM software, you can simulate the cutting operations in software to check the toolpaths; refer to Figure-10.

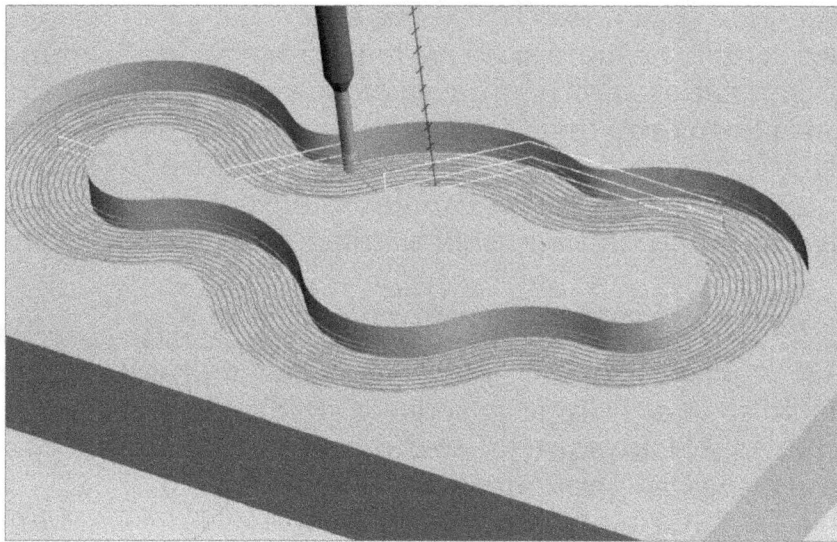

Figure-10. Toolpath Simulation

Role of CAM Engineer

A CAM engineer works closely with CAD engineer and in most of the small industries, CAD engineer and CAM engineer is the same person. Various tasks that a CAM engineer perform in industry are given next.

*   Modifying model as per the customer requirement.
*   Deciding machining strategy and tools required for machining the part.
*   Creating CNC programs depending on NC controller for the machine.

INTRODUCTION TO CAE

CAE means Computer Aided Engineering. Software like Ansys, Cosmol, SolidWorks Simulation, and so on are dedicated to perform different types of analyses. The types of analyses that can be performed using CAE software are given next.

*   Structural Analysis
*   Thermal Analysis
*   Computational Flow Analysis
*   Mold Flow Analysis

- Electronic Circuit Analysis
- Topology Optimization and many others.

Static Analysis

This is the most common type of analysis we perform. In this analysis, loads are applied to a body due to which the body deforms and the effects of the loads are transmitted throughout the body. To absorb the effect of loads, the body generates internal forces and reactions at the supports to balance the applied external loads. These internal forces and reactions cause stress and strain in the body. Static analysis refers to the calculation of displacements, strains, and stresses under the effect of external loads, based on some assumptions. The assumptions are as follows.

- All loads are applied slowly and gradually until they reach their full magnitudes. After reaching their full magnitudes, load will remain constant (i.e. load will not vary against time).
- Linearity assumption: The relationship between loads and resulting responses is linear. For example, if you double the magnitude of loads, the response of the model (displacements, strains and stresses) will also double. You can make linearity assumption if:

1. All materials in the model comply with Hooke's Law that is stress is directly proportional to strain.
2. The induced displacements are small enough to ignore the change is stiffness caused by loading.
3. Boundary conditions do not vary during the application of loads. Loads must be constant in magnitude, direction, and distribution. They should not change while the model is deforming.

If the above assumptions are valid for your analysis, then you can perform **Linear Static Analysis**. For example, a cantilever beam fixed at one end and force applied on other end; refer to Figure-11.

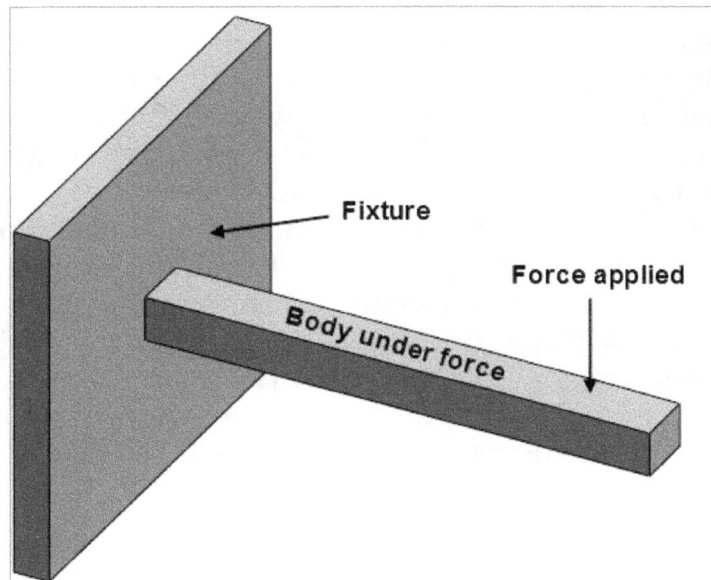

Figure-11. Linear static analysis example

If the above assumptions are not valid, then you need to perform the **Non-Linear Static analysis**. For example, force applied on an object attached with a spring; refer to Figure-12.

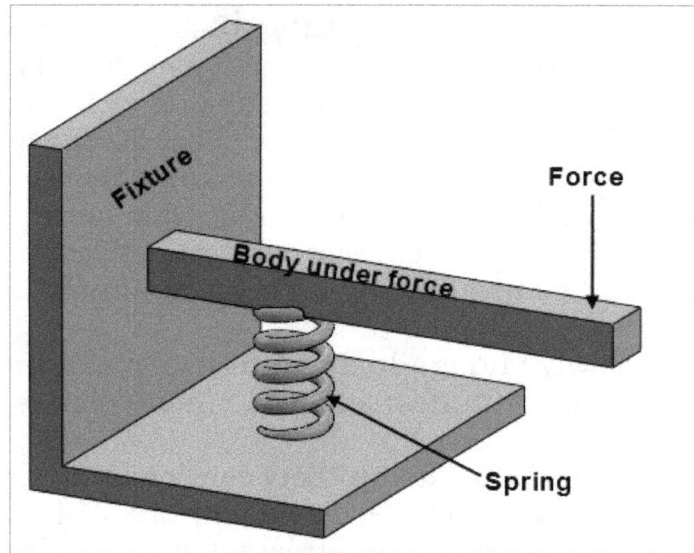

Figure-12. Non-linear static analysis example

Modal Analysis (Vibration Analysis)

By its very nature, vibration involves repetitive motion. Each occurrence of a complete motion sequence is called a "cycle." Frequency is defined as so many cycles in a given time period. "Cycles per seconds" or "Hertz". Individual parts have what engineers call "natural" frequencies. For example, a violin string at a certain tension will vibrate only at a set number of frequencies, that's why you can produce specific musical tones. There is a base frequency in which the entire string is going back and forth in a simple bow shape.

Harmonics and overtones occur because individual sections of the string can vibrate independently within the larger vibration. These various shapes are called "modes". The base frequency is said to vibrate in the first mode, and so on up the ladder. Each mode shape will have an associated frequency. Higher mode shapes have higher frequencies. The most disastrous kinds of consequences occur when a power-driven device such as a motor, produces a frequency at which an attached structure naturally vibrates. This event is called "resonance." If sufficient power is applied, the attached structure will be destroyed. Note that armies, which normally marched "in step," were taken out of step when crossing bridges. Should the beat of the marching feet align with a natural frequency of the bridge, it could fall down. Engineers must design in such a way that resonance does not occur during regular operation of machines. This is a major purpose of Modal Analysis. Ideally, the first mode has a frequency higher than any potential driving frequency. Frequently, resonance cannot be avoided, especially for short periods of time. For example, when a motor comes up to speed it produces a variety of frequencies. So, it may pass through a resonant frequency.

Thermal analysis

There are three mechanisms of heat transfer. These mechanisms are Conduction, Convection, and Radiation. Thermal analysis calculates the temperature distribution in a body due to some or all of these mechanisms. In all three mechanisms, heat flows from a higher-temperature medium to a lower temperature one. Heat transfer by conduction and convection requires the presence of an intervening medium while heat transfer by radiation does not.

There are two modes of heat transfer analysis.

Steady State Thermal Analysis

In this type of analysis, we are only interested in the thermal conditions of the body when it reaches thermal equilibrium, but we are not interested in the time it takes to reach this status. The temperature of each point in the model will remain unchanged until a change occurs in the system. At equilibrium, the thermal energy entering the system is equal to the thermal energy leaving it. Generally, the only material property that is needed for steady state analysis is the thermal conductivity.

Transient Thermal Analysis

In this type of analysis, we are interested in knowing the thermal status of the model at different instances of time. A thermos designer, for example, knows that the temperature of the fluid inside will eventually be equal to the room temperature(steady state), but designer is interested in finding out the temperature of the fluid as a function of time. In addition to the thermal conductivity, we also need to specify density, specific heat, initial temperature profile, and the period of time for which solutions are desired.

Thermal Stress Analysis

The Thermal Stress Analysis is performed to check the stresses induced in part when thermal and structural loads act on the part simultaneously. Thermal Stress Analysis is important in cases where material expands or contracts due to heating or cooling of the part to certain temperature in irregular way. One example where thermal stress analysis finds its importance is two material bonded strip working in a high temperature environment.

Event Simulation

The Event Simulation analysis is used to study the effect of object velocity, initial velocity, acceleration, time dependent loads, and constraints in the design. The results of this analysis include displacements, stresses, strains, and other measurements throughout a specified time period. You can perform this analysis when you need to check the effect of throwing a phone from some height or similar cases where motion is involved.

Shape Optimization

The Shape Optimization is not an analysis but a study to find the shape of part which utilizes minimum material but sustains the applied load up to required factor of safety.

There are various equations and parameters involved in analysis by CAE software and results are displayed in the form of tables & graphical representations.

Role of CAE Engineer

The CAE engineer performs many tasks related to analyses like checking and deciding material of product, defining real-world scenario for the analysis in software, preparing mesh model of product for analysis, running different types of analyses, and analyzing the results.

FOR STUDENT NOTES

Chapter 1

Starting with Autodesk Inventor

Topics Covered

The major topics covered in this chapter are:

- *Starting Autodesk Inventor 2024.*
- *Starting a new document.*
- *Autodesk Inventor Interface.*
- *Opening a document.*
- *Closing documents.*
- *Basic Settings for Autodesk Inventor Professional*

DOWNLOADING AND INSTALLING AUTODESK INVENTOR STUDENT EDITION

Autodesk gives a free license of 1 Year for students to practice on Autodesk Inventor and many other software. You cannot use a student edition in manufacturing products but you can use this edition to learn software. The procedure to download and install latest Autodesk Inventor educational version is given next.

- Open the link **https://www.autodesk.com/education/edu-software/overview** in your Web Browser. The web page for Autodesk software will be displayed; refer to Figure-1.

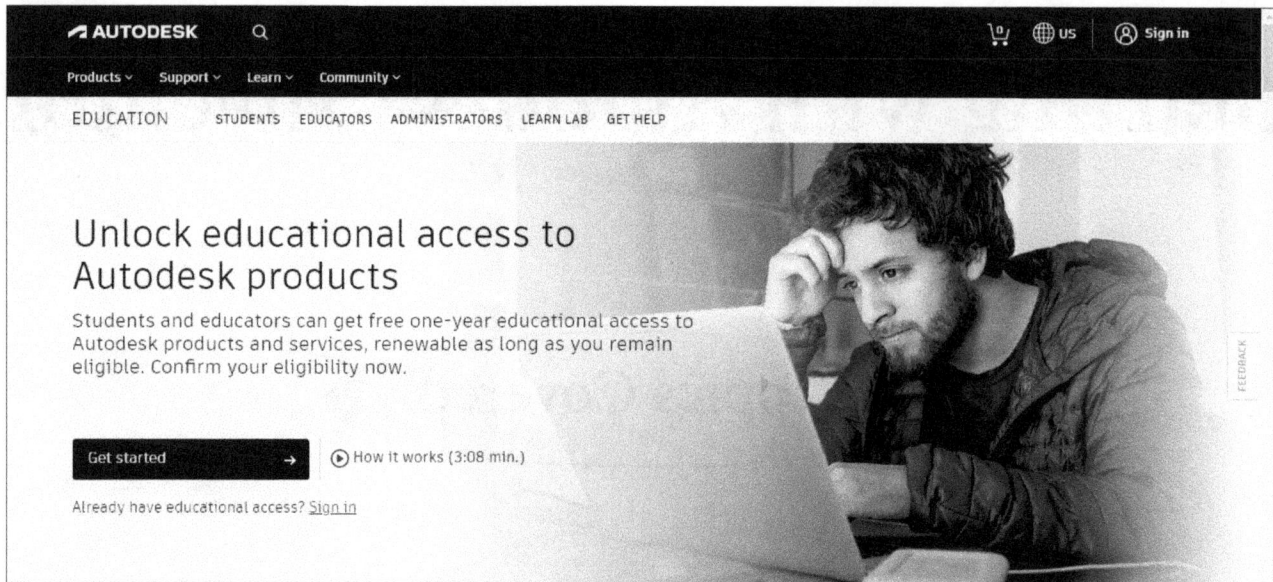

Figure-1. Autodesk page of education softwares

- If you do not have an Autodesk Student account then click on the **Get started** button on this page and create your account.
- If you have an Autodesk account then click on **Sign In** link button on this page and sign in with your student account details.
- After signing in, the web page of Autodesk educational software will be displayed. Scroll-down the page and click on **Get product** link button of **Inventor Professional** section. You will be asked to select the Operating system, Version, and Language of the software.
- Set the parameters as required and click on **INSTALL** drop-down button, the two options will be displayed; refer to Figure-2. Click on the **DOWNLOAD** option and the software will begin to download. After downloading the software, install the software by accepting the license terms and following the instructions as displayed.

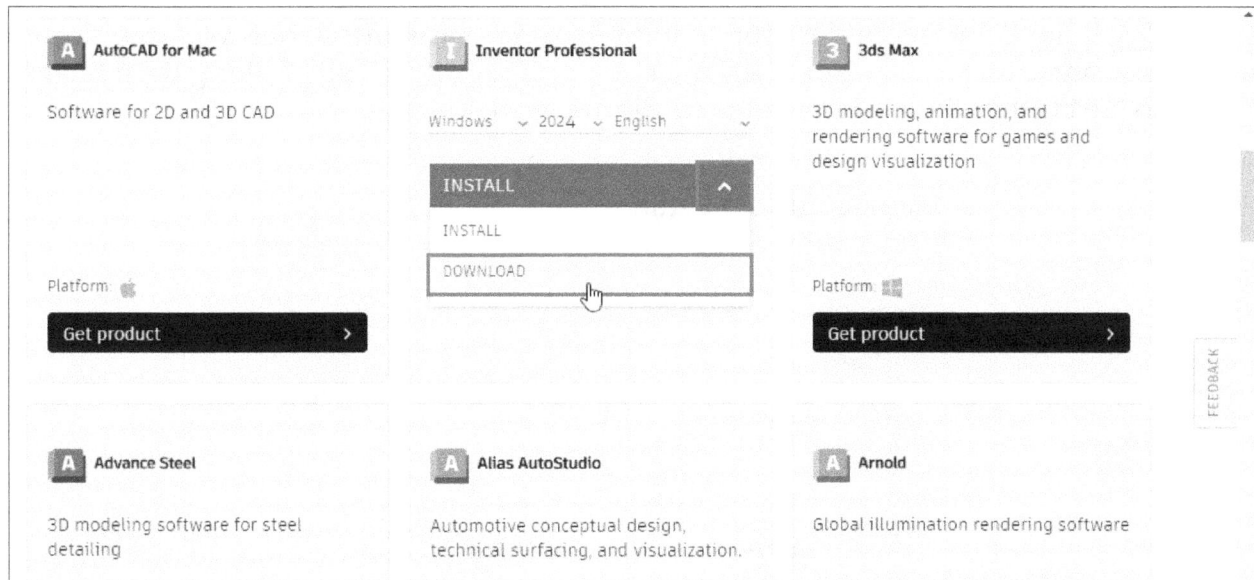

Figure-2. INSTALL drop down options

STARTING AUTODESK INVENTOR

I hope you have installed Autodesk Inventor 2024 in your system, so that you can follow instructions given in this book!!

There are various ways to start Autodesk Inventor but we will use the fastest general method to start Autodesk Inventor in Microsoft Windows.

- Click on the **Start** button at the **Taskbar**. The menu of application shortcuts will be displayed.
- Type **Autodesk Inventor** (in Microsoft Windows 10). The applications with the name Autodesk Inventor will be displayed which is only one in our case; refer to Figure-3.

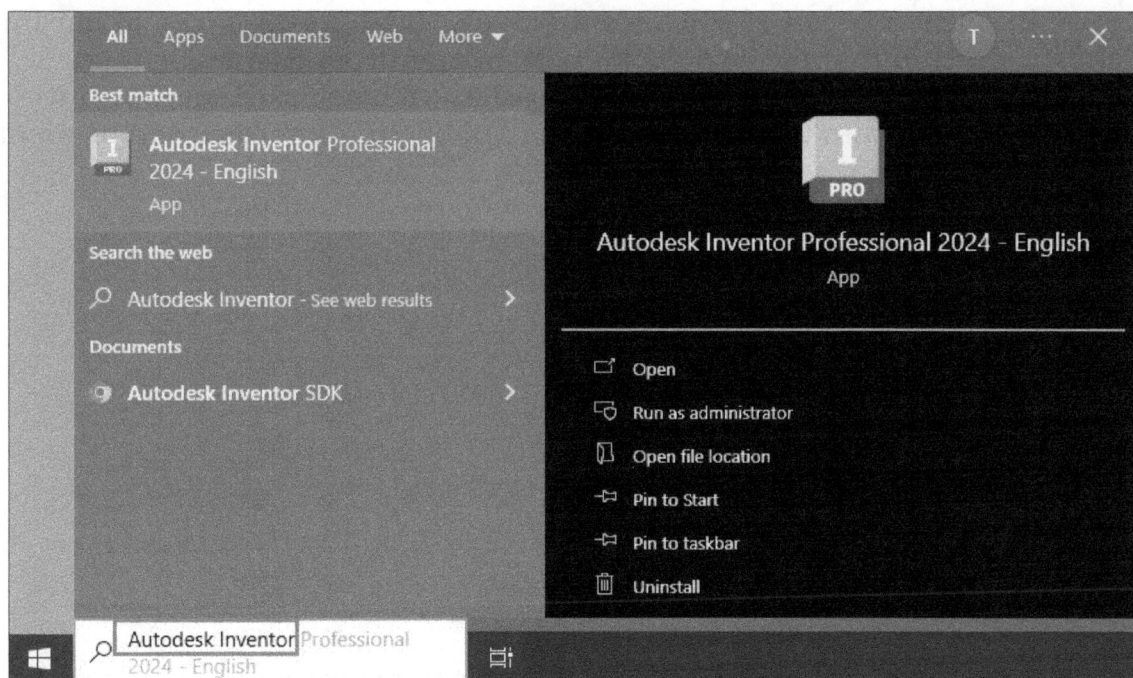

Figure-3. Start menu

- Click on the **Autodesk Inventor Professional 2024-English** link button. Note that if you have created a desktop icon of Autodesk Inventor then you can double-click on that icon to start application. The application will start along with **Migrate Custom Settings** dialog box; refer to Figure-4.

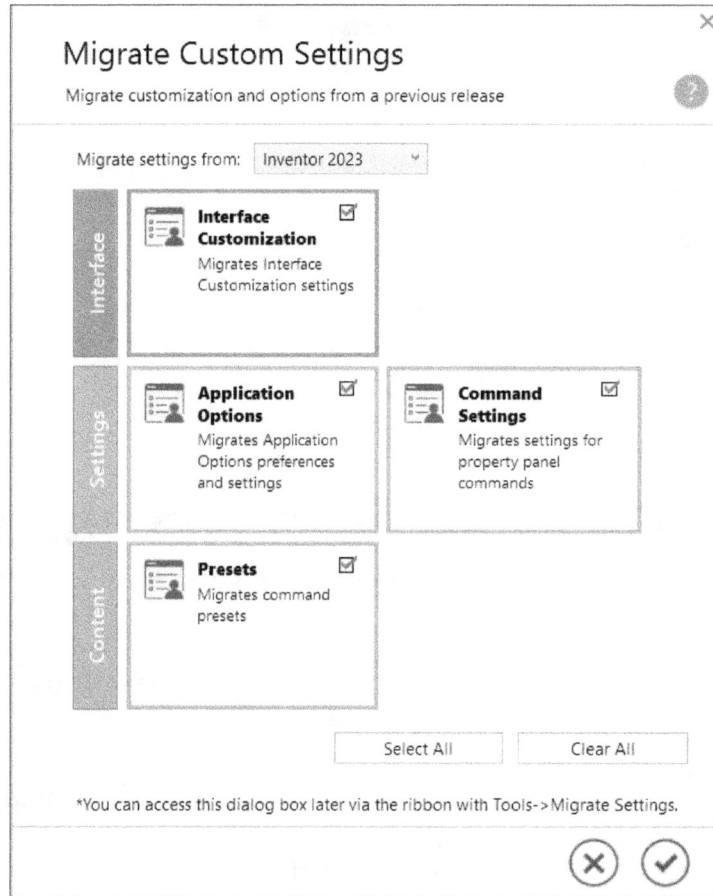

Figure-4. Migrate Custom Settings dialog box

- If you want to migrate the customization and options from the previous version of software into this version of software then select the check boxes of parameters which you want to migrate and click on **OK** button from the dialog box.
- If you do not want to migrate the customization then select **Clear All** button and click on **OK** button from the dialog box. The application interface will be displayed as shown in Figure-5.

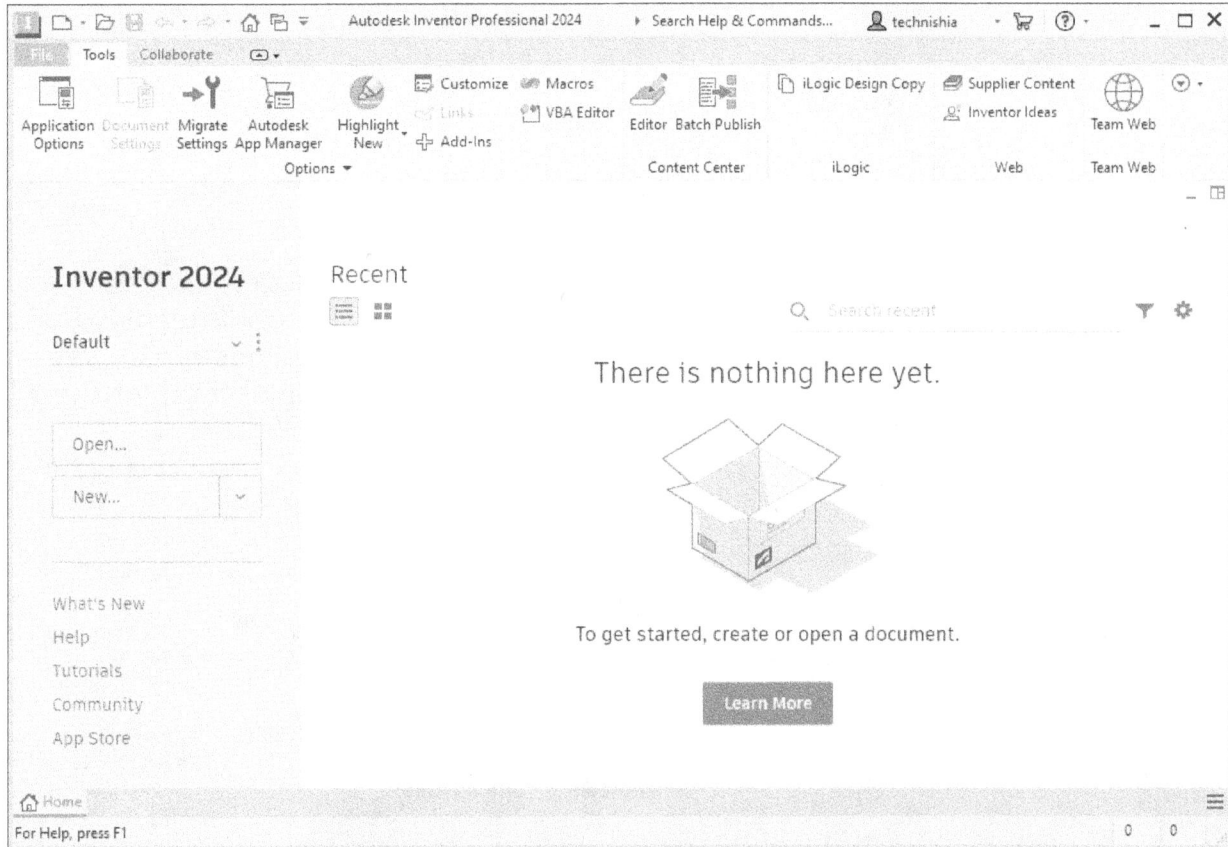

Figure-5. Autodesk Inventor interface

Autodesk Inventor INTERFACE

Autodesk always keeps on improving the interface of Inventor for better usability. The interface of Autodesk Inventor 2024 is displayed as shown in Figure-5. Various components of interface are displayed in Figure-6.

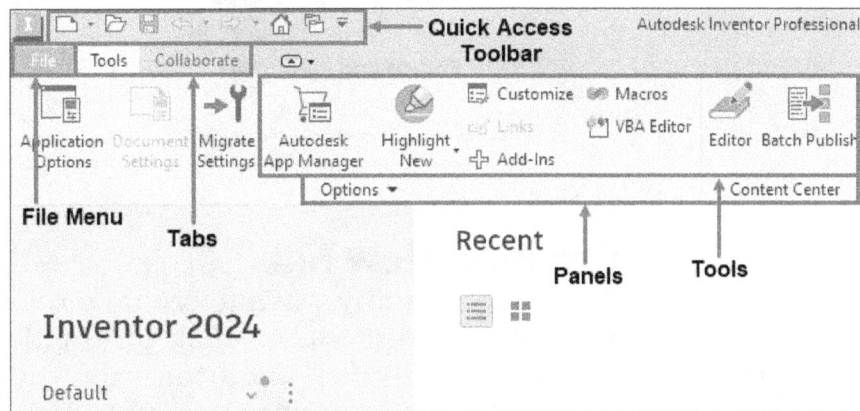

Figure-6. Components of Autodesk Inventor interface

Ribbon

Ribbon is the area of the application window that holds all the tools for designing and editing; refer to Figure-7. **Ribbon** is divided into **Tabs** which are further divided into **Panels**. Each panel is collection of tools dedicated to similar operations. The tools in these panels will be discussed in this chapter and subsequent chapters.

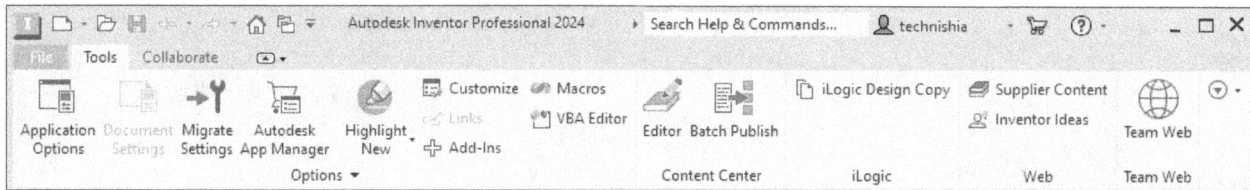

Figure-7. Ribbon

FILE MENU

The options in the **File** menu are used to manage the overall functioning of Autodesk Inventor. Once, you click on the **File** button at the top-left corner of the window, the **File** menu will be displayed as shown in Figure-8. Various options of the **File** menu are discussed next.

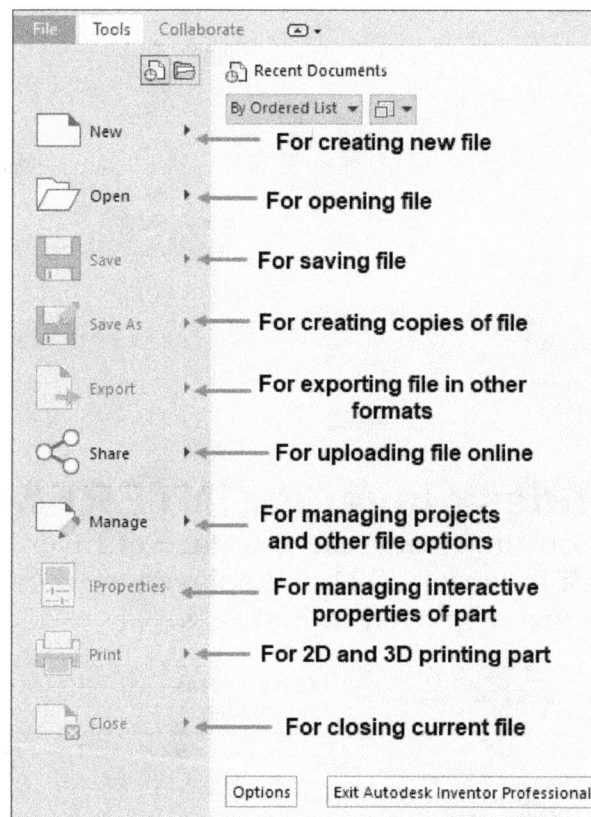

Figure-8. File menu

Creating New File

Like other products of Autodesk, there are many ways to create a new file. To create a new file, use the **New** option from the **File** menu or click on the **New** button from **Home** tab; refer to Figure-9. You can also use the **New** button from the **Quick Access Toolbar** at the top-left of the application window; refer to Figure-10 or press **CTRL+N** from the keyboard.

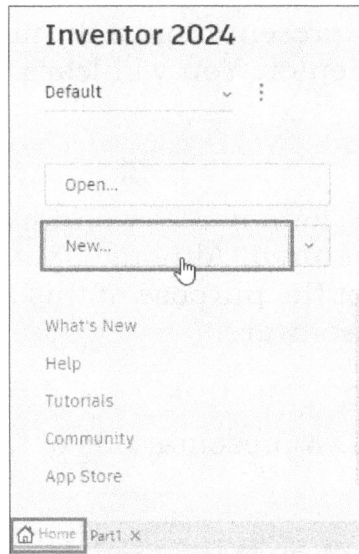

Figure-9. New button in Home tab

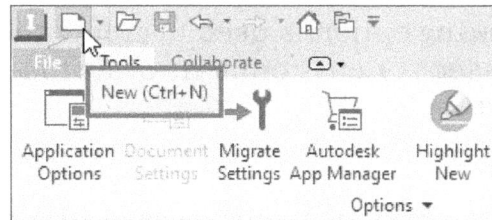

Figure-10. New button in Quick Access Toolbar

We will discuss the method of starting new file using **File** menu here. You will learn about the other methods later in the book.

Starting a New File using File menu

- Click on the **File** button at the top in the **Ribbon**. The **File** menu will be displayed.
- Hover the cursor on the **New** option in the menu. The options for creating new file will be displayed in the **New** cascading menu; refer to Figure-11.

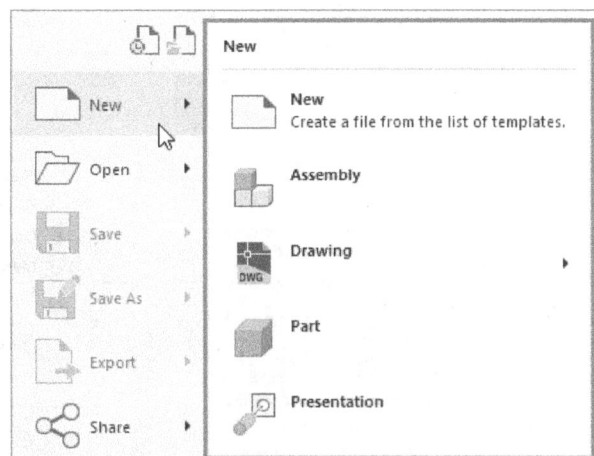

Figure-11. New cascading menu

There are five options in the **New** cascading menu viz. **New**, **Assembly**, **Drawing**, **Part**, and **Presentation**. The functions of these options are discussed next.

The **New** option is used to create new file by using the templates saved in the Autodesk directory.

The **Assembly** option is used to create assembly file. On clicking the **Assembly** option, the Assembly environment is displayed. An assembly file contains the assembly of various parts created in Inventor. For example, you can save the assembly model of motorbike in the form of assembly file in Autodesk Inventor. You will learn more about the assembly files and Assembly environment, later in the book.

The **Drawing** option is used to create the 2D representation of the model created in **Part** or **Assembly** environment of Autodesk Inventor. You will learn more in the later chapters.

The **Part** option is used to create the part file for any real-world model. A real-world model means a model that can be manufactured. Although, you can create unreal objects in Autodesk Inventor but that is not the purpose of this software, for creating those objects you should use animation software.

The **Presentation** option is used to create 3D representations of the model for presentation to the client.

We will start with the **Part** environment and then discuss the predefined templates.

- Click on the **New** option from the **New** cascading menu in the **File** menu. The **Create New File** dialog box will be displayed; refer to Figure-12.

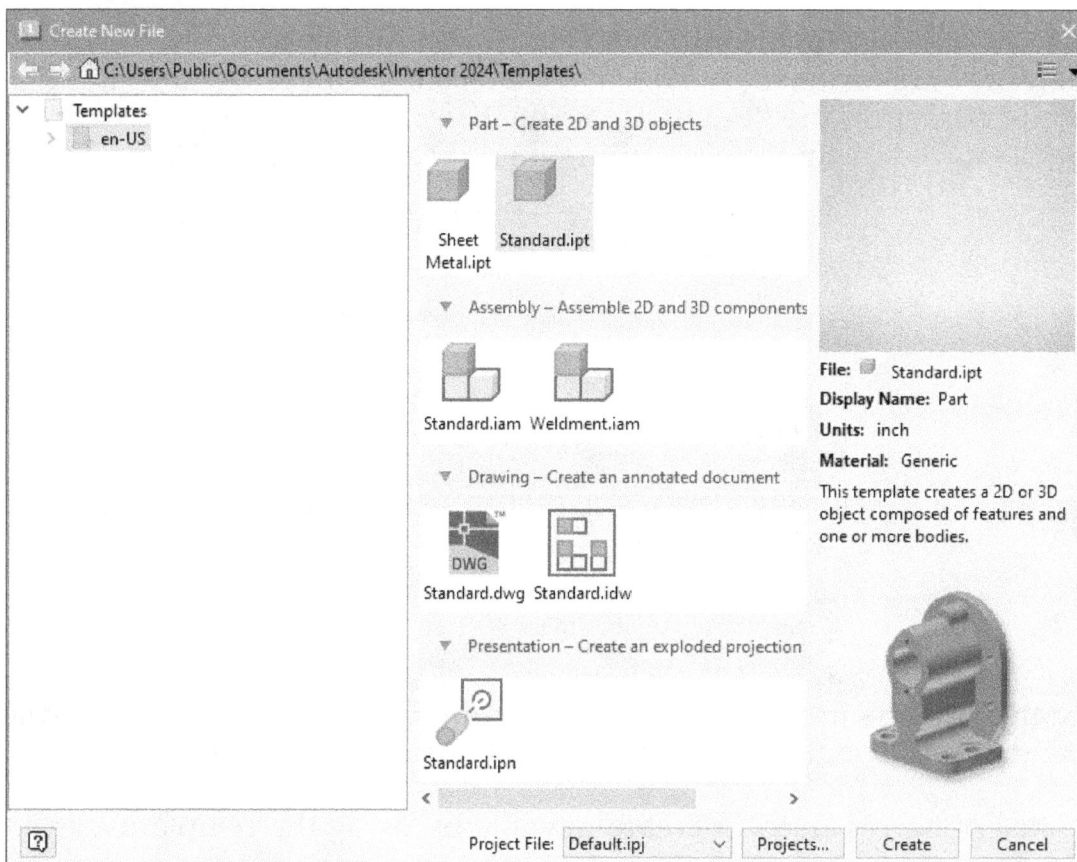

Figure-12. Create New File dialog box

- Expand the **en-US** node from the **Templates** category and select desired folder from **English** or **Metric** in the **Templates** area at the left of dialog box. If you select the **English** folder then the templates with **English** units (Feet and Inches) will be displayed in the right of the dialog box. If you select the **Metric** folder from the left of the dialog box then the templates with **Metric** units (Meters, Millimeters) will be displayed in the right of the dialog box.

- To start a new file with unit as mm, click on the **Metric** folder and then click on the **Standard (mm).ipt** icon from the right of the dialog box; refer to Figure-13.

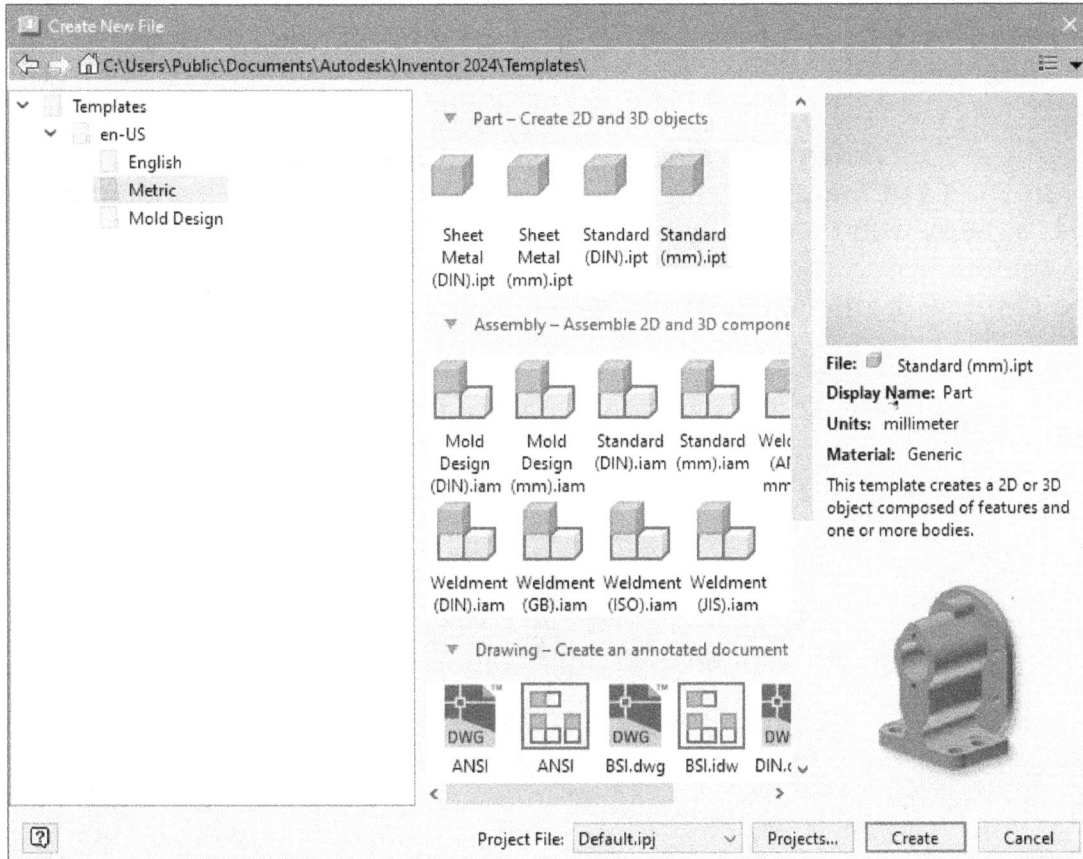

Figure-13. Creating part file in metric

- Click on the **Create** button from the dialog box. The new file will open and part environment will be displayed; refer to Figure-14.

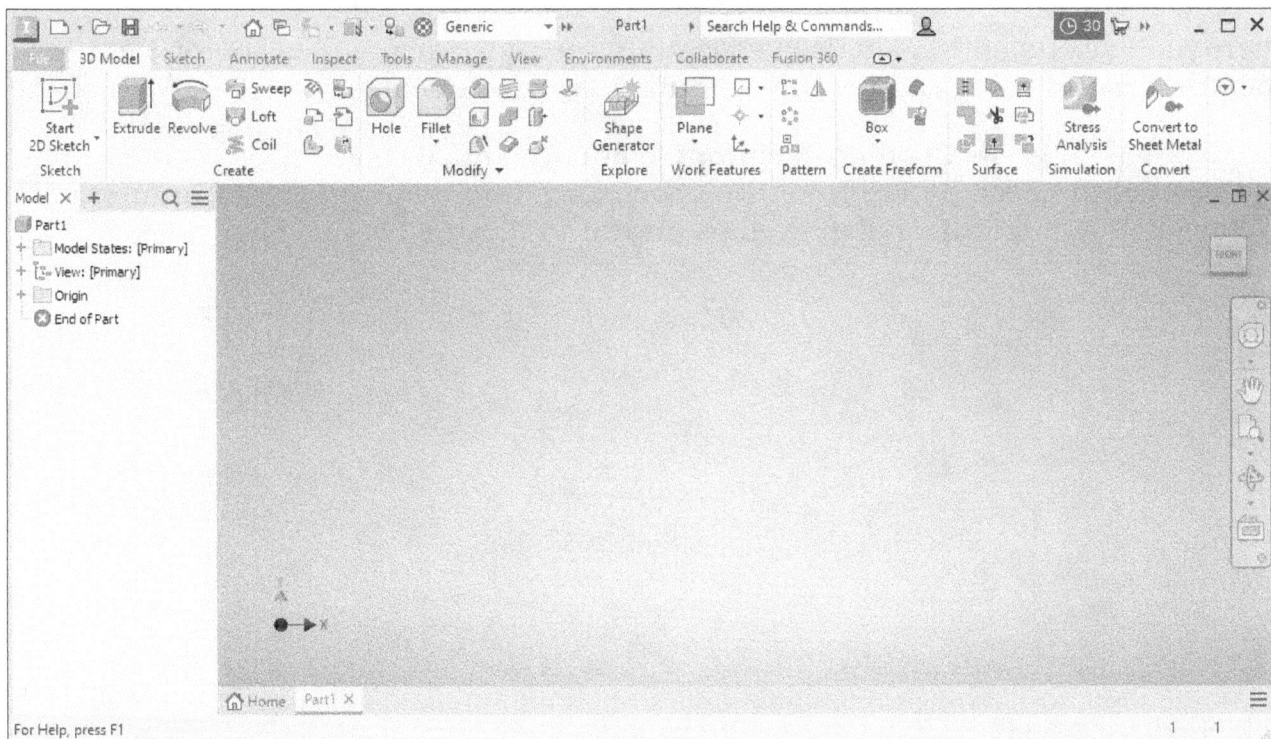

Figure-14. Part environment

Opening a file

Similar to creating new file, there are many ways to open files in Autodesk Inventor like, using the **Open** option from the **File** menu, using the **Open** button from the **Quick Access Toolbar**, or pressing the **CTRL+O** key from the keyboard. Here, we will discuss the procedure to open a file by using the **Open** option from the **File** menu.

Opening a File using File Menu Options

• Click on the **File** menu button at the top-left corner of the application window. The **File** menu will be displayed.

• Hover the cursor on the **Open** option in the **File** menu. The **Open** cascading menu will be displayed; refer to Figure-15.

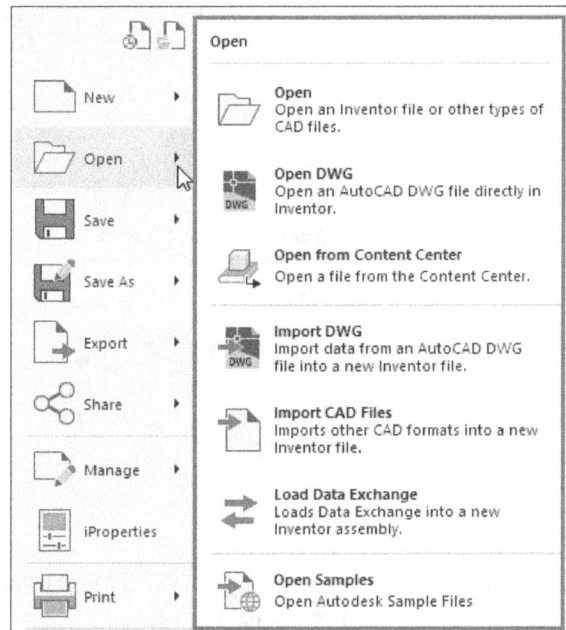

Figure-15. Open cascading menu

There are various options in **Open** cascading menu to open a file. The procedures of opening files by using each of these options are given next.

Opening Inventor file and other CAD files

• Click on the **Open** option from the **Open** cascading menu in the **File** menu. The **Open** dialog box will be displayed as shown in Figure-16.

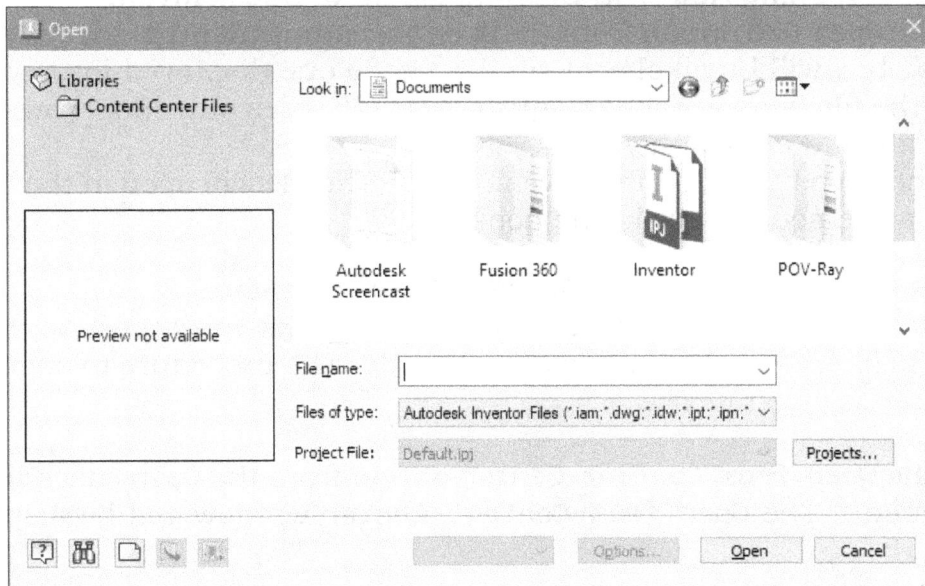

Figure-16. Open dialog box

- Click on the **Files of type** drop-down and select the format of file you want to open; refer to Figure-17.

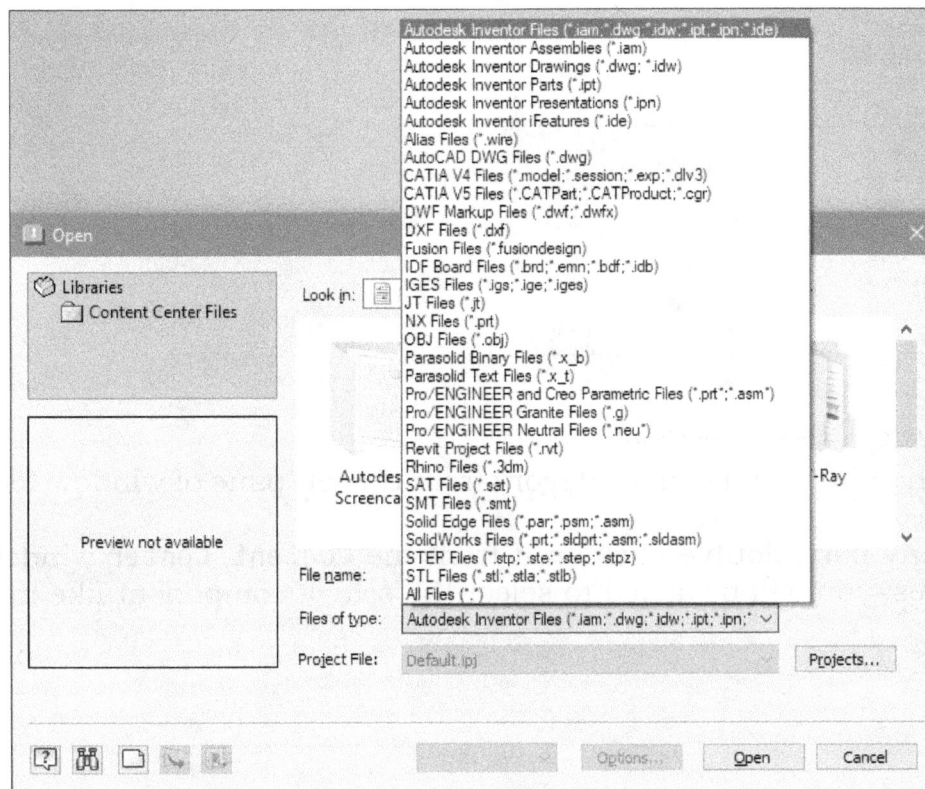

Figure-17. Files of type drop down

- Browse to the location of file by using general Microsoft Windows functions and double-click on the file to open it.

Note that you will learn more options of **Open** dialog box later in the book.

Opening AutoCAD Drawing file in Autodesk Inventor

- Click on the **Open DWG** option from the **Open** cascading menu in the **File** menu. The **Open** dialog box will be displayed similar to the one displayed in previous topic.
- Note that the file type is locked to AutoCAD DWG files only. So, browse to desired AutoCAD file and click on it.
- Click on the **Open** button from the dialog box. The file will open in the layout mode which we will discuss later in the book.

Opening file from Content Center

Like other Autodesk products, Autodesk Inventor gives you access to the library of standard parts like, Gear, bearing, connector, etc. The procedure to open parts from **Content Center** is given next.

- Click on the **Open from Content Center** option from the **Open** cascading menu in the **File** menu. The **Open from Content Center** window will be displayed; refer to Figure-18.

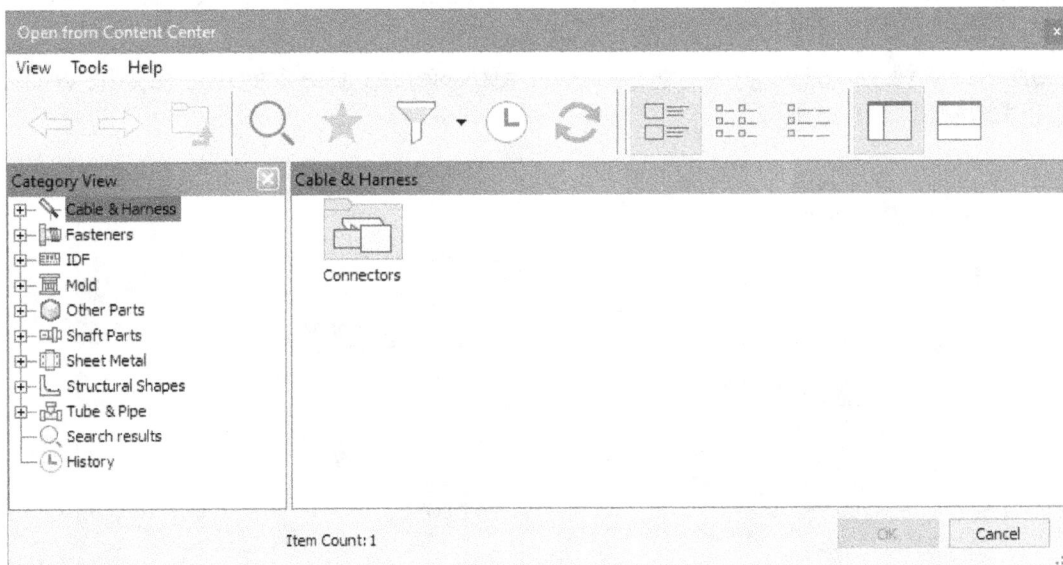

Figure-18. Open from Content Center window

- Click on the plus sign (**+**) of a category from the left pane of window to check subcategories.
- To open any part, double-click on it from the **Content Center** window. In some of the cases, you will be asked to select the size of component like in Figure-19.

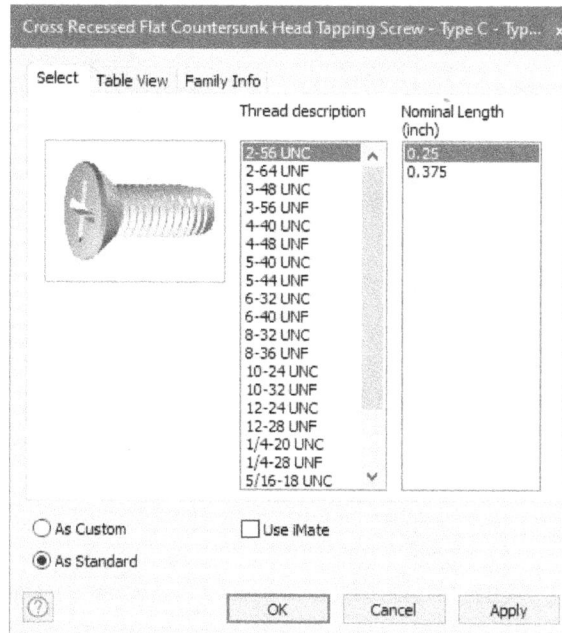

Figure-19. Specifying size for tapping screw

- After selecting the size, click on the **OK** button. The part will open.

Importing AutoCAD DWG files

Importing models of other software can reduces lots of extra work of rebuilding the base sketch for Inventor models. In Autodesk Inventor, we can directly use the AutoCAD drawing files for creating or manipulating the model. The procedure to import the AutoCAD Drawing files is given next.

- Click on the **Import DWG** option from the **Open** cascading menu in the **File** menu. The **Import** dialog box will be displayed as shown in Figure-20.

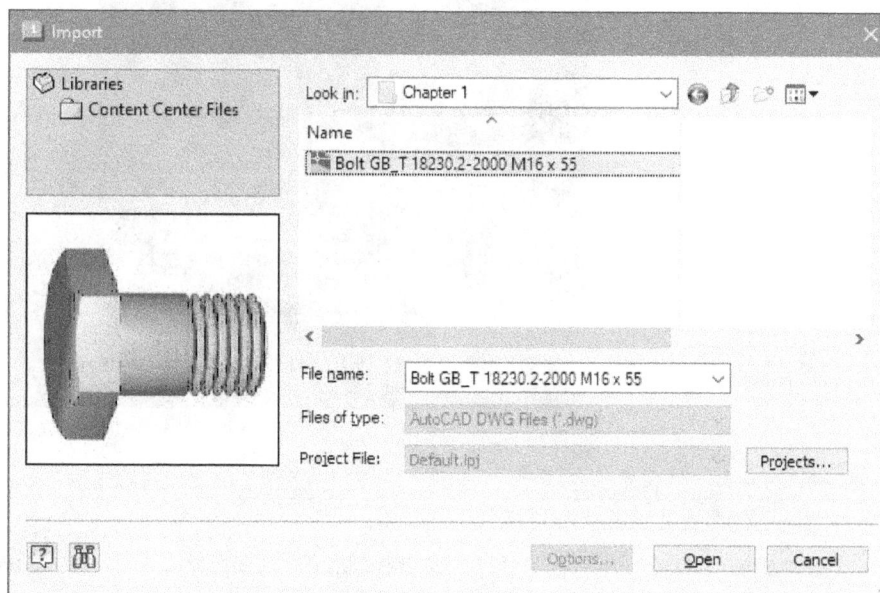

Figure-20. Import dialog box

- Double-click on desired file from the dialog box. The **DWG/DXF File Wizard** dialog box will be displayed; refer to Figure-21.

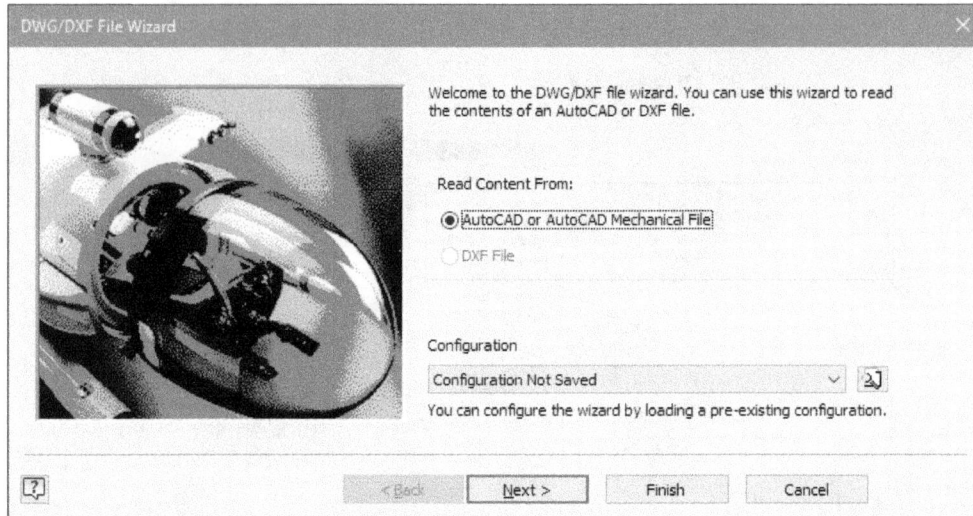

Figure-21. DWG DXF File Wizard dialog box

- The format of file will be recognized automatically and the relevant radio buttons will become active on this page. The file we are using is a dwg file, so it will show only **AutoCAD or AutoCAD Mechanical File** radio button as active. If there is any configuration saved with the file then it will appear in the **Configuration** drop-down. Select desired options and click on the **Next** button. The **Layers and Objects Import Options** dialog box will be displayed; refer to Figure-22.

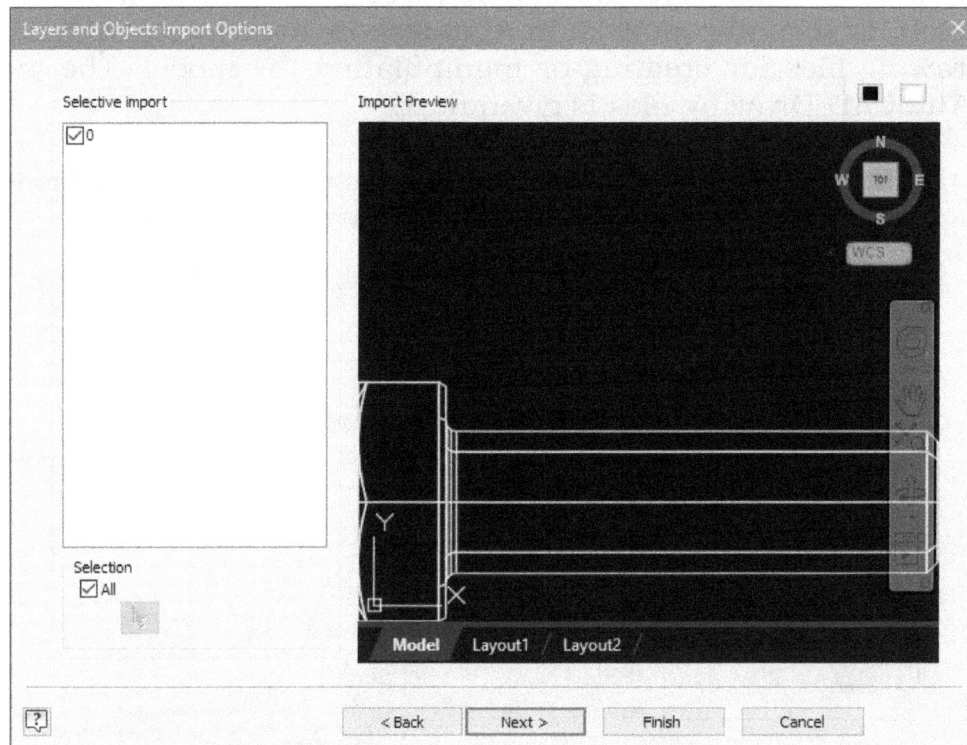

Figure-22. Layers and Objects Import Options dialog box

- Select desired layers/objects from the **Selective import** list box and check the preview in the right of the dialog box. Once the preview is as per your requirement, click on the **Next** button from the dialog box. The **Import Destination Options** dialog box will be displayed; refer to Figure-23.

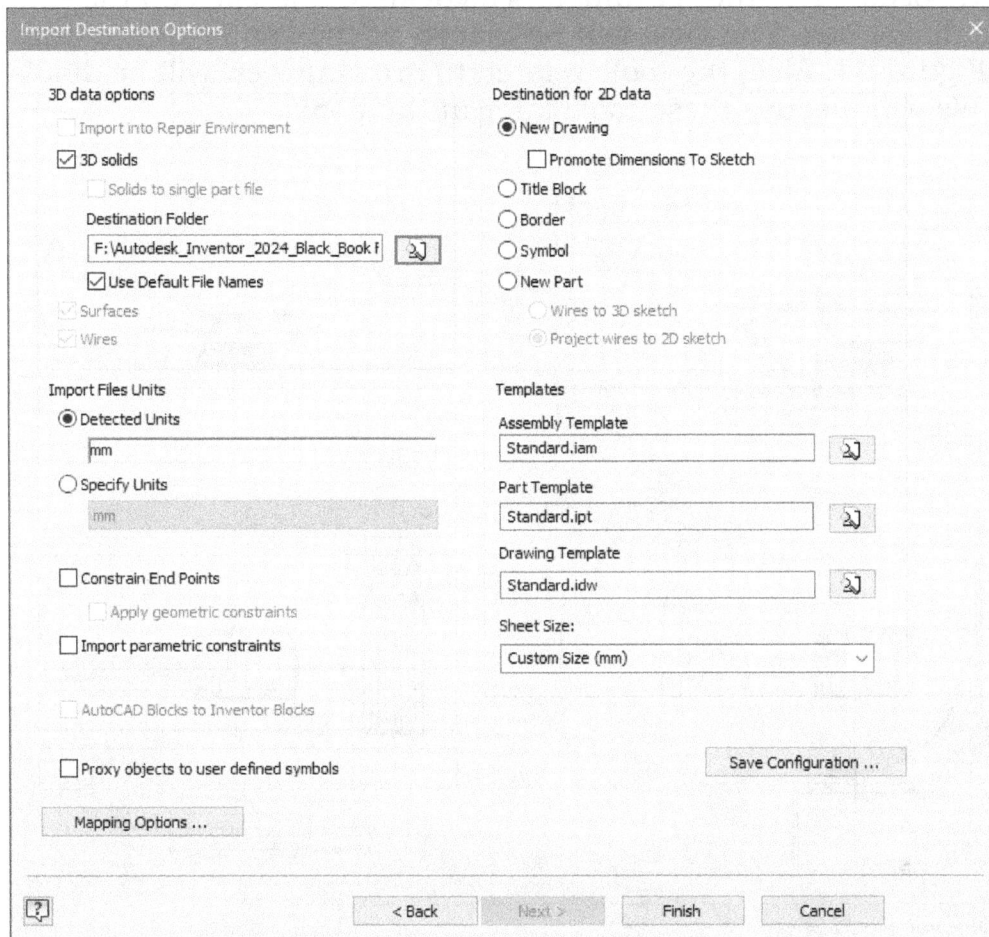

Figure-23. Import Destination Options dialog box

- Since, we do not have a 3D model in our drawing, so we are not selecting the **3D Solids** check box from the **3D data options** area of the dialog box. If you need a 3D model to be imported from the AutoCAD drawing file then select this check box and specify the relevant parameters. Similarly, you can select other check boxes in the 3D data options area to import respective components of file.
- By default, the **Detected Units** radio button is selected in the **Import Files Units** area of the dialog box, so the units of imported files are used. If you want to import the file in different unit then select the **Specify Units** radio button from the area and select desired unit from the drop-down below the radio button.
- Select the **Constrain End Points** check box if you want to constrain all the open end points of the imported drawing model. After selecting this check box, you can select the **Apply geometric constraints** check box for applying geometric constraints of Autodesk Inventor. Select the **Import parametric constraints** check box if you want to import the geometric constraints applied in AutoCAD.
- Select the **AutoCAD Blocks to Inventor Blocks** check box if you want to import blocks from the AutoCAD and convert them into Inventor blocks.
- Select the **Proxy objects to user defined symbols** check box to copy all the objects as user defined symbols in Autodesk Inventor.
- In the **Destination for 2D Data** area, select desired radio button to define location where imported 2D data will be placed in Inventor file.

- Similarly, specify the other parameters as desired and click on the **Finish** button from the dialog box. The view will be placed at the center of the drawing sheet; refer to Figure-24. Also, the tools related to drawing view will be displayed in the **Ribbon**. We will discuss these tools later in the book.

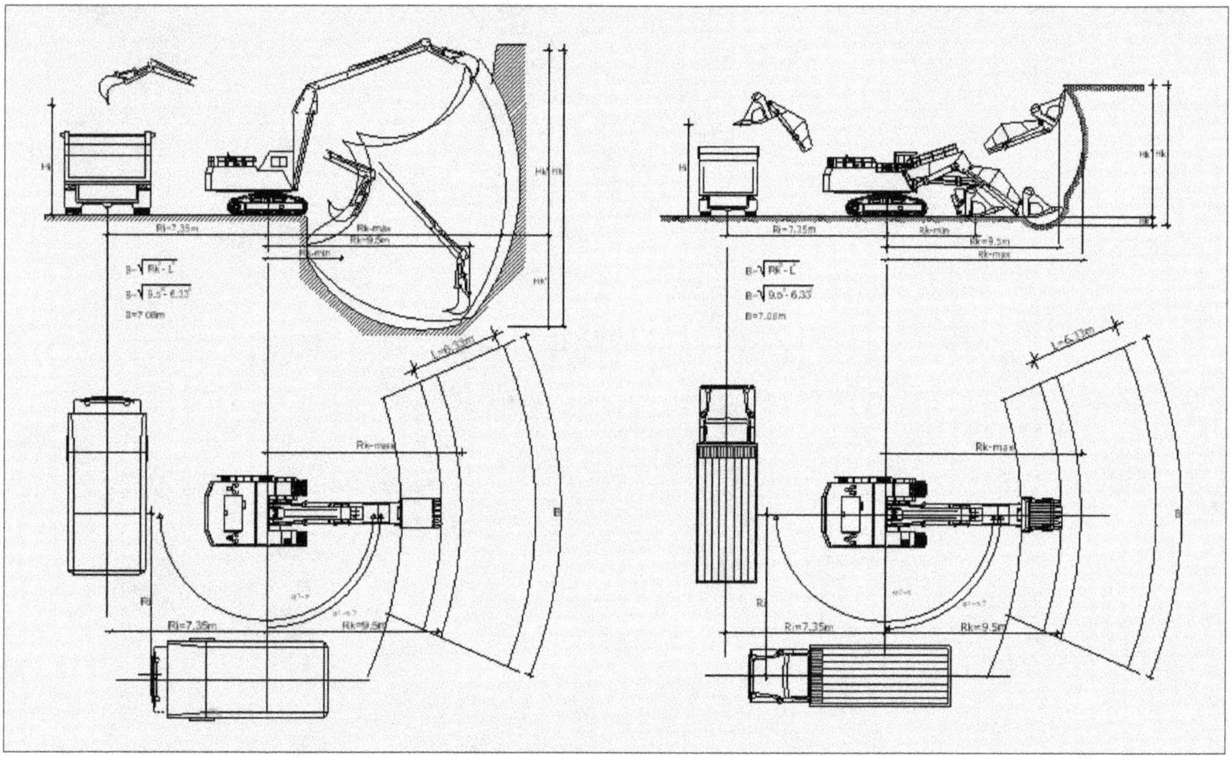

Figure-24. Drawing view placed

Importing CAD files

- Click on the **Import CAD Files** tool from the **Open** cascading menu in the **File** menu. The **Open Document** dialog box will be displayed as shown in Figure-25.

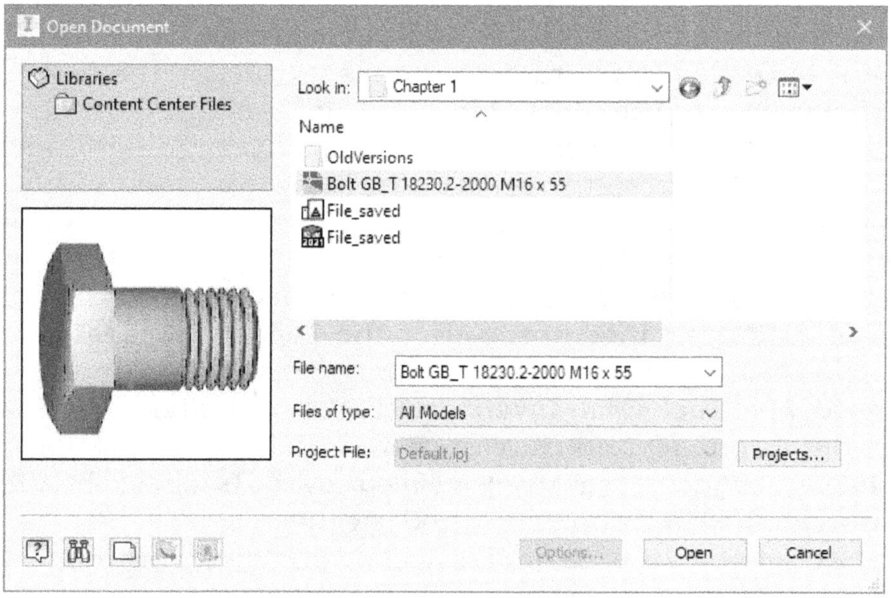

Figure-25. Open Document dialog box

- Double-click on desired CAD file from the dialog box. The **Import** dialog box will be displayed; refer to Figure-26. Note that depending on the type of document, you might get different dialog box for specifying parameters. We get this dialog box when you are opening part file created by Creo Parametric.

Figure-26. Import dialog box

- There are two radio buttons in **Import Type** area of the dialog box; **Reference Model** and **Convert Model**. If you select the **Reference Model** radio button then you cannot make changes in the model imported. You can use it in your inventor model as referenced, make more features on it, and can use it in assembly and drawing as well. On selecting the **Reference Model** radio button, the options related to **Object Filters** will be displayed with **Inventor Length Units** drop-down. Select the check boxes for objects to be imported from the **Object Filters** area and select desired unit from the **Inventor Length Units** drop-down.

- If you select the **Convert Model** radio button then the imported model will be converted to inventor format. After selecting this radio button, specify the name of converted file in the **Name** edit box and set the location of the new file.

- Select the **Individual** or **Composite** option from the **Surfaces** drop-down in the **Part Options** area of the dialog box. The **Individual** option is used to create individual surfaces of the model whereas selecting the **Composite** option will create a combined surface feature.

- Click on the **Select** tab from the **Import** dialog box. The options in the dialog box will be displayed as shown in Figure-27.

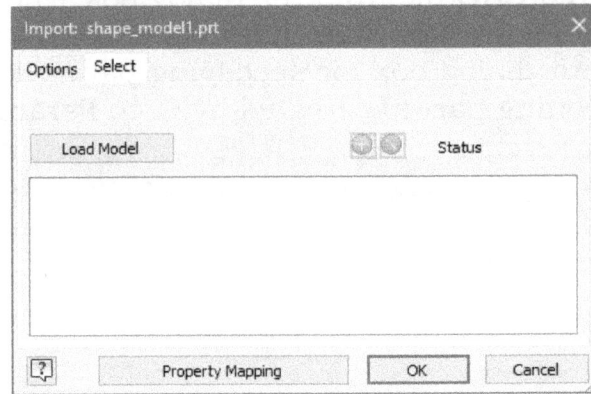

Figure-27. Select tab of Import dialog box

- Click on the **Load Model** button from the dialog box. The objects of the selected model will be displayed; refer to Figure-28.

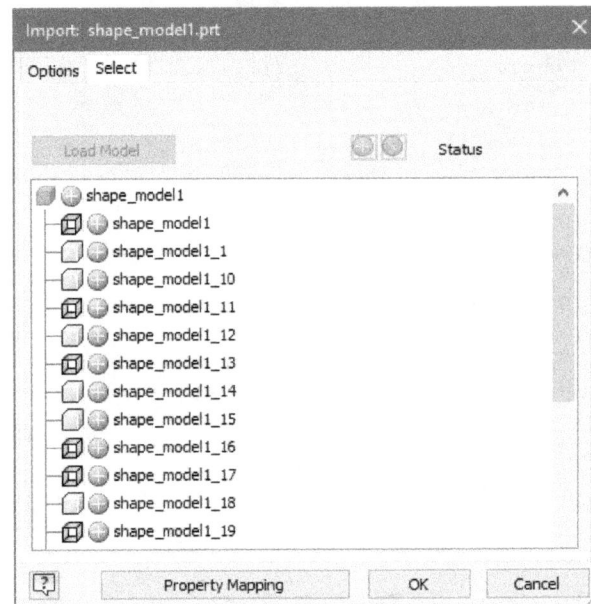

Figure-28. Objects in the Select tab

- Objects with plus sign will be imported and the other objects will be skipped.
- Click on the plus sign against the part to skip it from importing and click on the **OK** button from the dialog box. The object will be displayed in the modeling area.

Load Data Exchange

The **Load Data Exchange** option is used to import part of a shared revit project into Inventor assembly for creating suitable models. For using this option, there must be a common project shared with your team members and an administrator of your project can share view of the model with you via data exchange. To use this option, you must have paid subscription of Autodesk Docs.

Opening Sample Files

The **Open Samples** tool in **Open** cascading menu of the **File** menu is used to download and open the sample files provided by Autodesk for Inventor. On clicking this tool, a web page will be displayed in your default Web Browser. Download and open the file as desired.

Saving File

Saving file is as important as planting trees in backyard!! Jokes apart, its very important to save the model created, so that you can reuse it later. There are many options in Autodesk Inventor to save files. These options are discussed next.

Save option

The **Save** option is used to save the active file. This option is available in the **Save** cascading menu of the **File** menu. The procedure to use this option is given next.

* After creating the model, click on the **Save** option from the **Save** cascading menu in the **File** menu; refer to Figure-29 or press **CTRL+S** from keyboard.

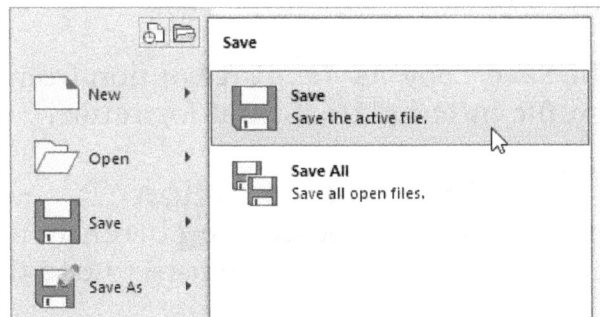

Figure-29. Save option

* If you are saving the file for the first time then the **Save As** dialog box will be displayed; refer to Figure-30.

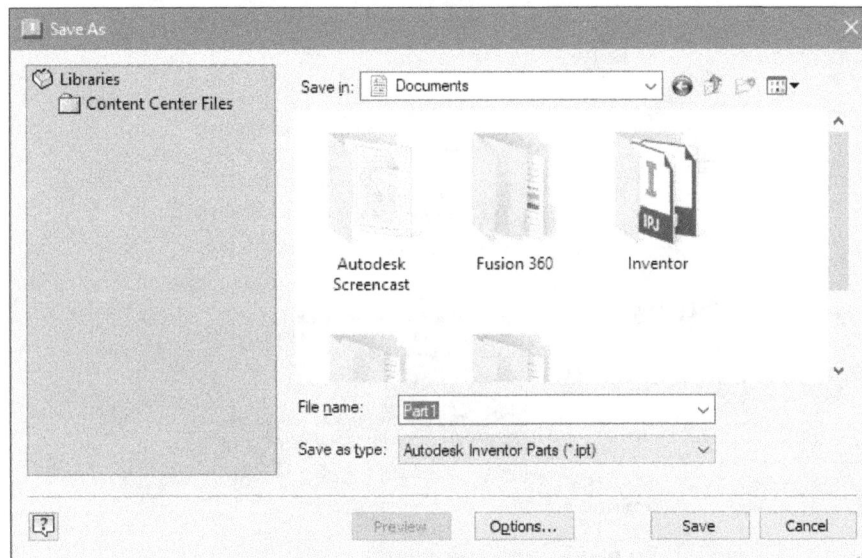

Figure-30. Save As dialog box

* Click on the **Save in** drop-down at the top in the dialog box and select desired location for saving file.
* Specify desired name in the **File name** edit box and click on the **Save** button. The file will be saved by specified name in the selected location.
* If you use the **Save** option after first time then the file will be over-written on the previous session of file.

Save All option

The **Save All** option does the same as the **Save** option do but it saves all the open files.

Save As option

The **Save As** option available in the **Save As** cascading menu is used to save the file with new name and at different location. The operation of this tool is same as the operation of **Save** option for the first time.

Save Copy As option

The **Save Copy As** option is used to save another copy of the current file. Note that if you select the **Save Copy As** option then the file will be saved with another name but it will not open in Inventor automatically.

Similarly, you can use the **Save Copy As Template** option from the **Save As** cascading menu to save the current file in template format for reuse.

Pack and Go option

The **Pack and Go** option is used to package the currently active file and all of its referenced files into a single location. The procedure to use this option is discussed next.

*   After creating and saving the model, click on the **Pack and Go** option from **Save As** cascading menu in the **File** menu; refer to Figure-31. The **Pack and Go** dialog box will be displayed; refer to Figure-32.

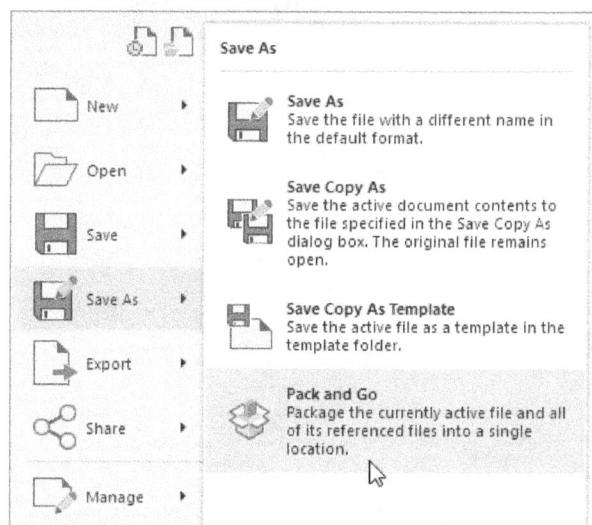

Figure-31. Pack and Go option

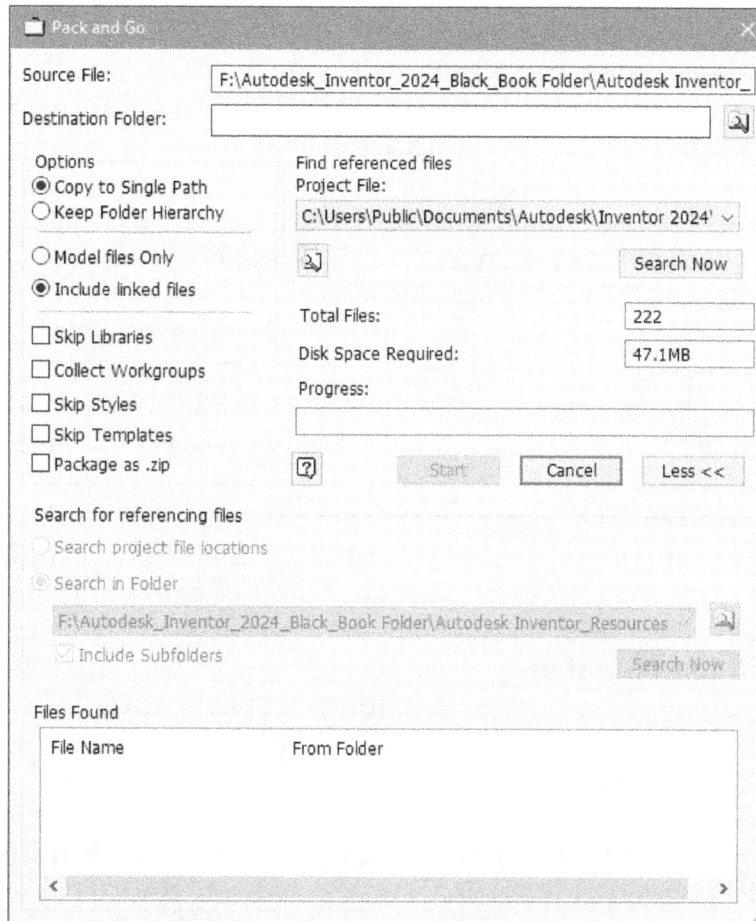

Figure-32. Pack and Go dialog box

- The **Source File** box displays the path and the file name of the file to package.
- Click on **Browse** button from **Destination Folder** box to find the appropriate destination for the packaged files or specify the path and folder name. If the folder does not exist, a prompt to create the folder will be displayed.
- Select **Copy to Single Path** radio button from **Options** area to copy the referenced files to a single folder with the packaged file.
- Select **Keep Folder Hierarchy** radio button from **Options** area to build a folder hierarchy under the destination folder that preserves the folder hierarchy under the original project locations and copies the selected file and its referenced and referencing files to the appropriate subfolders.
- Select **Model files Only** radio button from **Options** area to copy only Autodesk Inventor model files (.ipt, .idw, .ide, .dwg) to the destination folder.
- Select **Include linked files** radio button from **Options** area to copy all referenced files to the destination folder including spreadsheets, text files, and other files.
- Select **Skip Libraries** check box from **Options** area to not copy the library files with the packaged file.
- Select **Collect Workgroups** check box from **Options** area to collect workgroups and the workspace into a single root folder.
- Select **Skip Styles** check box from **Options** area to not copy the styles with the packaged file.
- Select **Skip Templates** check box from **Options** area to not copy the templates with the packaged file.

- Select **Package as .zip** check box from **Options** area to package the files in .zip format.
- The **Project Files** box in the **Find referenced files** area displays the default active project file. Click on the **Browse** button to search for a different project file.
- Click on the **Search Now** button from **Find referenced files** area to search for files that are referenced from the selected file.
- The **Total Files** box displays the total number of files to package.
- The **Disk Space Required** box displays the disk space needed on destination media.
- Select **Search project file locations** radio button from **Search for referencing files** area to search for referencing files in the workspaces, workgroups, and library locations specified in the project file.
- Select **Search in Folder** radio button from **Search for referencing files** area to search for referencing files in the folder identified by the path in the field immediately below this option.
- Select **Include Subfolders** check box from **Search for referencing files** area to search the selected folder and all subfolders for referencing files.
- Click on **Search Now** button from **Search for referencing files** area to search for files that reference the selected file and its transitively referenced files.
- The **Files Found** list box displays the list of files to be packaged.
- After specifying desired parameters, click on **Start** button to begin the process of packaging.
- The **Progress** box displays the progress of the packaging.

Pack and Go creates a log file of the operation and places it in the same directory as the packaged file. See the log file for information about the packaging operation and the names with source paths of all referenced files. The log file is overwritten each time you package an Autodesk Inventor file to the same destination.

Exporting Files

The options in the **Export** cascading menu of **File** menu are used to create alien matter. (No!! I mean it.) Okay, getting back to software terms, the options in the **Export** cascading menu are used to save the inventor files in native format of the other software like, you can save the file for Pro-E, CATIA, AutoCAD, and so on; refer to Figure-33.

Figure-33. Export cascading menu

The options of **Export** cascading menu are discussed next.

Image option

The **Image** option in the **Export** cascading menu is used to save images of the model. After making the model and export it to image format by using this option, you can get beautiful print of the model (which can be your company product) and paste it to the wall. The procedure to use the **Image** option is given next.

* Click on the **Image** option from the **Export** cascading menu in the **File** menu. A very familiar **Save As** dialog box will be displayed; refer to Figure-34.

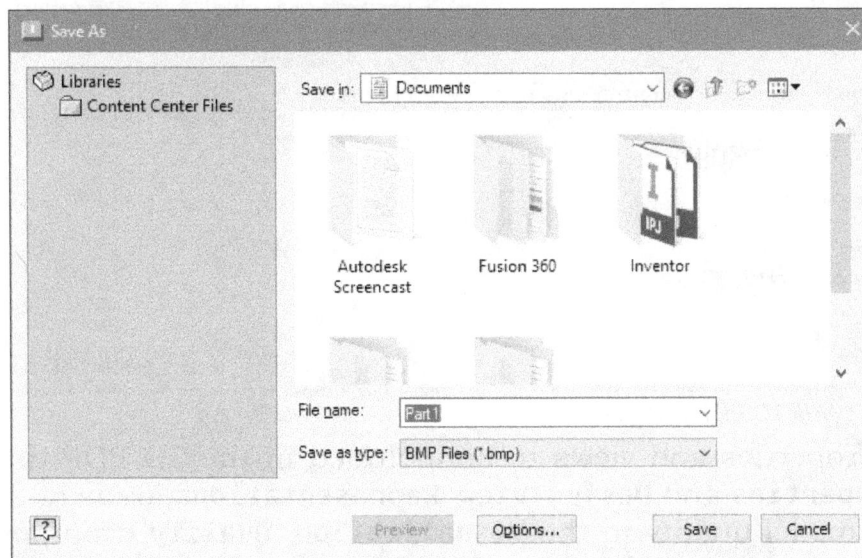

Figure-34. Save As dialog box

- Select desired image format from the **Save as type** drop-down and click on the **Save** button from the dialog box. The current model in the modeling area will be saved in the image with background.

PDF option

PDF is one of the most popular document format. Due to its compact size and easy handling, this format can also be used in presentations of models. Using the **PDF** option in the **Export** cascading menu, you can create the pdf file of Inventor model. Note that the pdf created will be just like image that we created in previous topic.

3D PDF option

Although, iges and other CAD portability formats are useful data for engineers and designers but they are not useful data for marketing. For market executives, there is requirement of a format by which they can display important features of model with some dimensions. 3D PDF is one of the important format for this purpose. The procedure to create 3D PDF is given next.

- Click on the **3D PDF** tool from the **Export** cascading menu in the **File** menu after creating the model. The **Publish 3D PDF** dialog box will be displayed; refer to Figure-35.

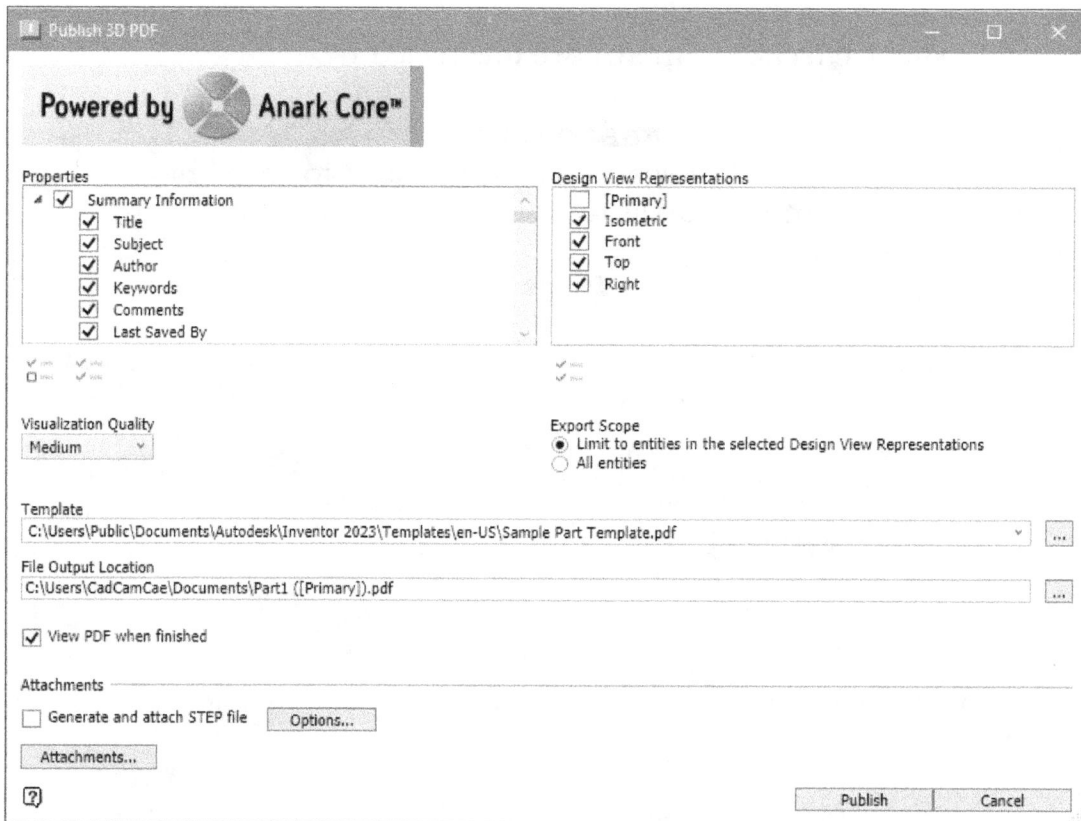

Figure-35. Publish 3D PDF dialog box

- Select the properties and views to be included in the 3D PDF using the check boxes in **Properties** and **Design View Representations** areas of the dialog box.
- Set desired model quality in the **Visualization Quality** drop-down. Note that increasing the quality will increase the PDF size.
- Set desired template and location using the respective fields in the dialog box.

- Select the **Generate and attach STEP file** check box to attach a STEP in PDF. To modify options related to STEP file, click on the **Options** button next to the check box.
- You can add more attachments to the PDF by using the **Attachments** button.
- Click on the **Publish** button after specify desired parameters. The 3D PDF will be generated and displayed in your default PDF Reader Application; refer to Figure-36.

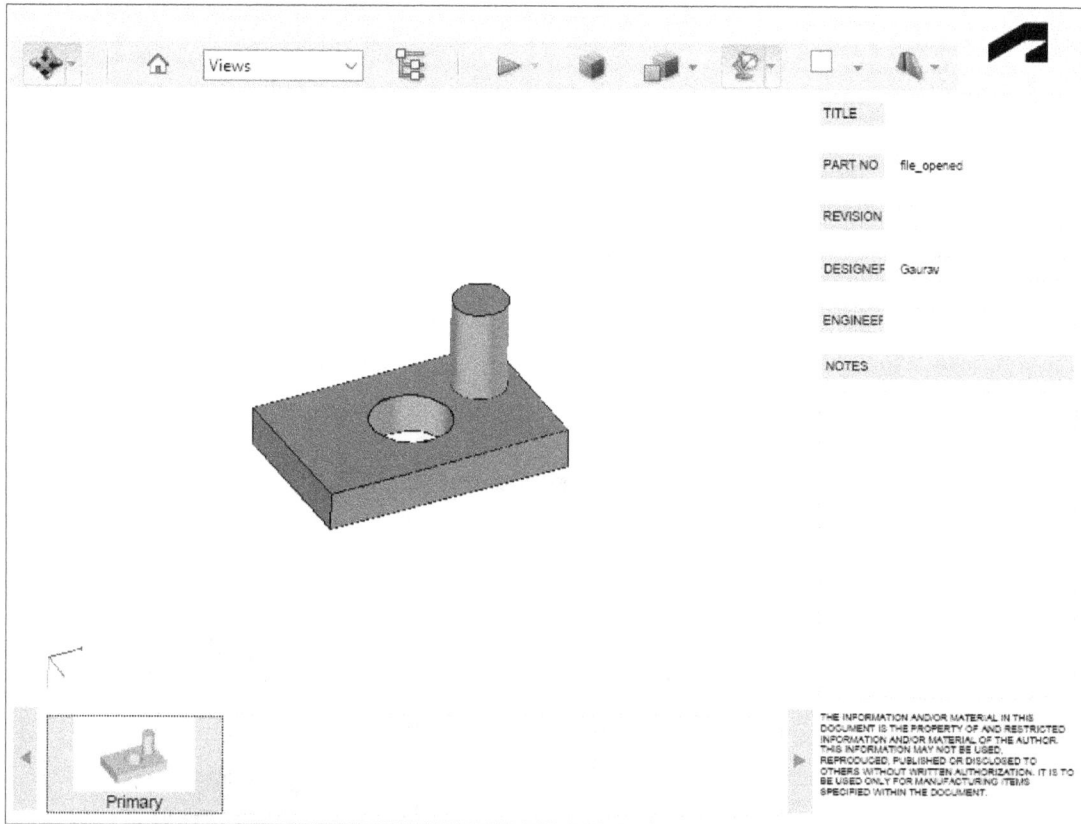

Figure-36. 3D PDF generated

CAD Format option

The **CAD Format** option in the **Export** cascading menu is used to export the Inventor model in other CAD formats like iges, stp, jt, stl, etc. Out of all the export formats, IGES and STP are the most popular formats. So, we will discuss the procedure to export file in these formats.

- Click on the **CAD Format** option from the **Export** cascading menu in the **File** menu. The well known **Save As** dialog box will be displayed. You know what to do now.

Exporting File to IGES Format

- Select the **IGES Files** option from the **Save as type** drop-down; refer to Figure-37.

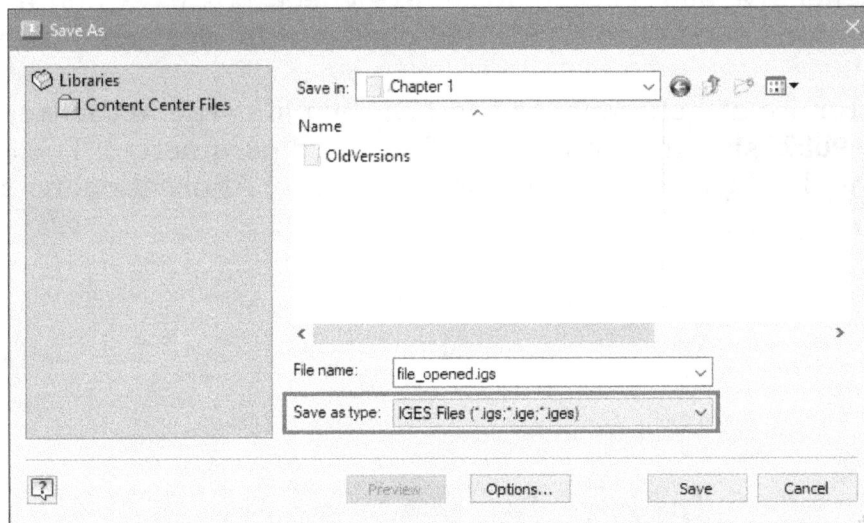

Figure-37. Save as type drop down

- Click on the **Options** button from the dialog box. The **IGES File Save As Options** dialog box will be displayed as shown in Figure-38.

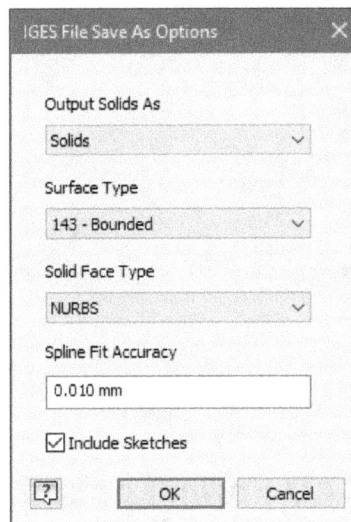

Figure-38. IGES File Save As Options dialog box

- Here, we are exporting a solid part so the options related to solid part are displayed. If you are exporting a surface part or assembly file then the options in the dialog box will change accordingly.
- In the current scenario, select desired option from the **Output Solids As** drop-down. You can export the part as surfaces, solids, or wireframe. On selecting the **Surfaces** option, the **Solid Face Type** drop-down will become inactive and on selecting the **Wireframe** option, the **Surface Type** drop-down will also become inactive.
- Select desired surface type from the **Surface Type** drop-down and desired solid face type from the **Solid Face Type** drop-down.
- Using the **Spline Fit Accuracy** edit box, you can specify the accuracy for conversion.
- Select the **Include Sketches** check box to export the sketches along with the solids/surfaces/wireframe.
- Click on the **OK** button from the dialog box to apply the selected parameters.

STEP Format

- Select the **STEP Files** option from the **Save as type** drop-down.
- Click on the **Options** button from the **Save As** dialog box. The **STEP File Save As Options** dialog box will be displayed as shown in Figure-39.

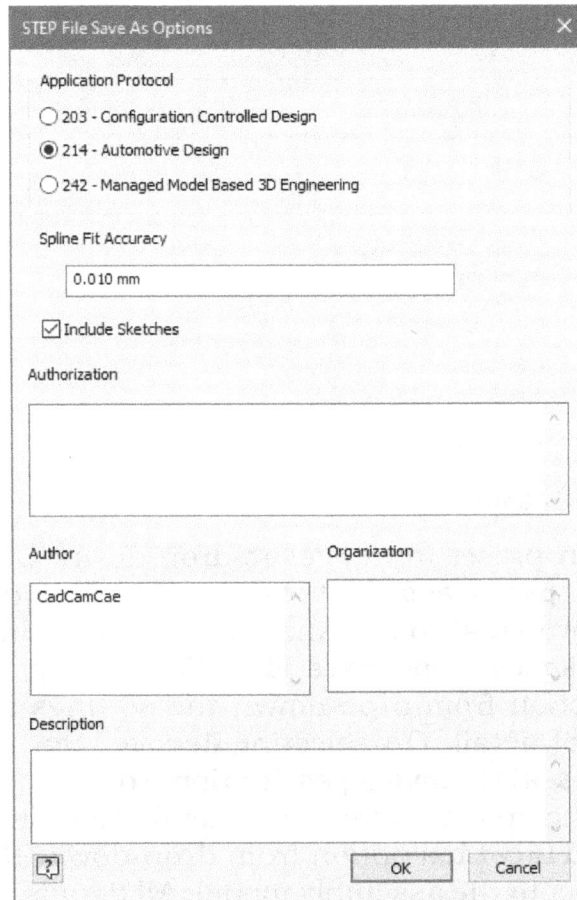

Figure-39. STEP File Save As Options dialog box

- Select desired radio button from the **Application Protocol** area to define the applications of current file.
- As discussed earlier, you can set the accuracy of model by using the **Spline Fit Accuracy** edit box and you can include the sketches in exported file by using the **Include Sketches** check box.
- Specify the other user parameters by using the edit boxes in the dialog box and click on the **OK** button to apply the parameters.
- Click on the **Save** button from the **Save As** dialog box to export the file in selected format.

Similarly, you can export the current model to **DWG** or **DWF** format by using the respective option from the **Export** cascading menu.

RVT option

The **RVT** option in the **Export** cascading menu is used to export a version of an assembly model as RVT file. Revit model do not require manufacturing level detail in models to use for designs. The procedure to use this tool is given next.

- Click on the **RVT** option from **Export** cascading menu in the **File** menu after creating the assembly model. The **Simplify Property** panel will be displayed; refer to Figure-40.

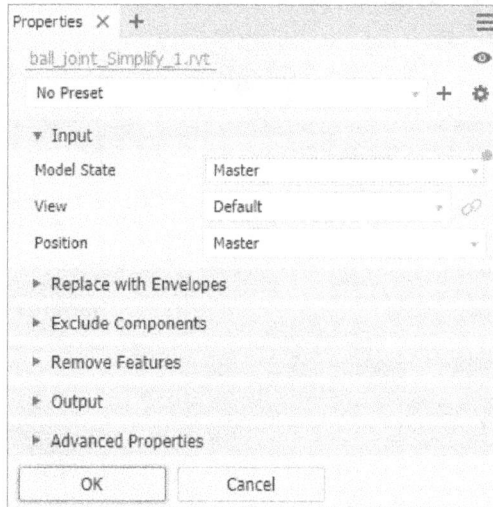

Figure-40. Simplify Property panel

- Select desired built-in preset from **Preset** drop-down. On selecting **Remove the least detail (small parts and features)** option from drop-down, the settings are designed to remove small parts and features resulting in the most detailed model. On selecting **Remove moderate detail (medium sized parts and all listed features)** option from drop-down, the settings are designed to remove a moderate amount of detail. On selecting **Remove the most detail (replace top-level components with envelopes)** option from drop-down, the settings are designed to remove the greatest amount of detail resulting in very basic shapes. On selecting **No Simplification** option from drop-down, the simplified version is full fidelity with respect to the assembly model. All details are visible in the model.
- Click on **Create new preset** ➕ button to create your own preset.
- Click on the **Preset Settings** ⚙ button to create desired settings for the preset.
- The options in the **Model State** drop-down of **Input** rollout define the envelope size. Changes in a model state affect the envelope size; refer to Figure-40.
- Select desired option from **View** drop-down of **Input** rollout to display active design view. Changes in design views do not affect envelope size. Click on the **Associative** 🔗 button to link the design view with the simplified model.
- The options in the **Position** drop-down of **Input** rollout define the envelope size. Changes in position view can affect the envelope size.
- Select **None** option from **Replace** drop-down of **Replace with Envelopes** rollout if you are not using envelopes for the simplication. Components and Features groups are available for use; refer to Figure-41.

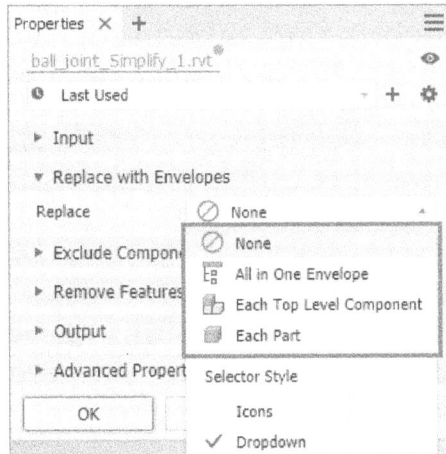

Figure-41. Replace drop down of Replace with Envelopes rollout

- Select **All in One Envelope** option from **Replace** drop-down of **Replace with Envelopes** rollout to create an envelope around the top-level assembly. The result is imprecise and useful for developing a "keep out" area for the design. Components and Features groups are hidden.

- Select **Each Top Level Component** option from **Replace** drop-down of **Replace with Envelopes** rollout to create an envelope around assembly first-level components. Component selection tools display. Feature selection tools are hidden.

- Select **Each Part** option from **Replace** drop-down of **Replace with Envelopes** rollout to create envelopes for all parts in the assembly. Feature selection tools are hidden.

- Select **Exclude parts by size** check box from **Exclude Components** rollout and specify the maximum diagonal value for a bounding box in the **Max. Diagonal** edit box; refer to Figure-42. Click on the arrow ▸ button at the end of edit box to select a recently used value. Click on the **Pick a part to get its bounding box diagonal length** button and select a part/component to supply the diagonal value.

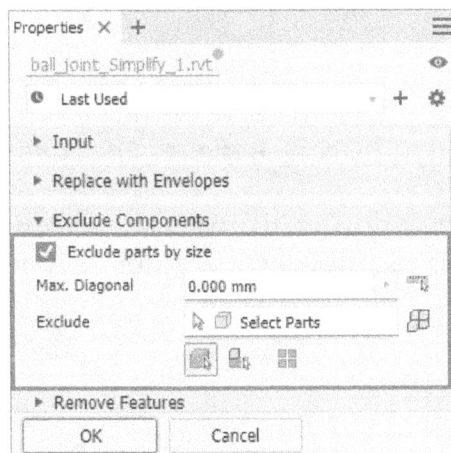

Figure-42. Exclude Components rollout options

- Click on desired selection priority you want to use from **Exclude Components** rollout. Select **Parts** button to select only parts of selected item. Select **Components** button to select only components of selected item. Select **All Occurrences** button to select all occurrences of the selected item.

- Click on the **Show excluded parts** ⊞ button from **Exclude Components** rollout to display the components being removed in the **Exclude** box. Click on the **Show included parts** ◥ button to display the components that are included in the simplified model in the **Include** box.

- The **Remove Features** rollout provides options for selecting features by type or size to include in or exclude from the simplified part. The list of features include holes, fillets, chamfers, pockets (subtractive), embosses (additive), and tunnels (single and multiple entry); refer to Figure-43. On selecting **None** option from drop-down, no features of this type are removed. On selecting **All** option from drop-down, all features of this type are removed. On selecting **Range** option from drop-down, all features equal to or smaller than the specified parameter are removed. Specify the value that defines the maximum size to remove.

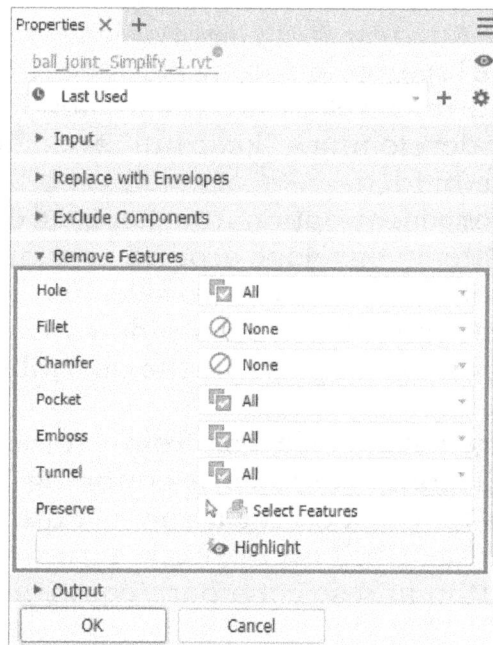

Figure-43. Remove Features rollout options

- Click on the **Select Features** button from **Preserve** section of **Remove Features** rollout to select the features you want to retain from those being removed.

- Click on the **Highlight** button from **Remove Features** rollout to apply a color to the features to be removed.

- The **Type** section of **Output** rollout outputs your inventor design as a Revit model, so you can provide it to building designers or other downstream uses; refer to Figure-44.

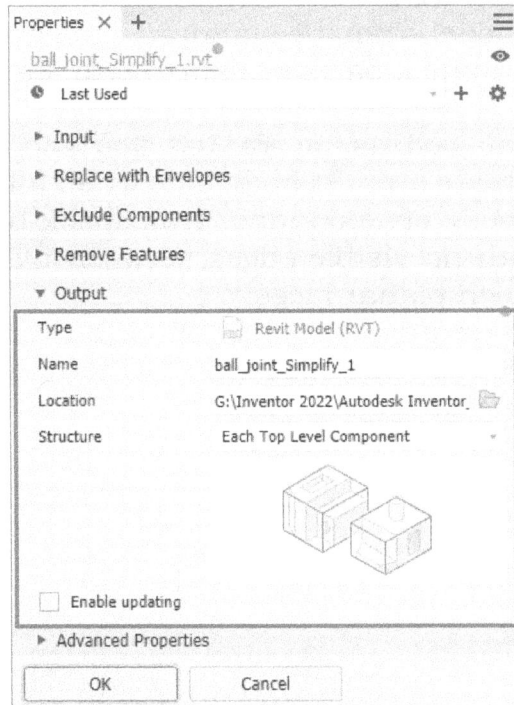

Figure-44. Output rollout options

- Specify desired file name in the **Name** edit box of **Output** rollout or use the default assembly name.
- Specify desired location of the new file in the **Location** edit box of **Output** rollout.
- Select **All in One Element** option from **Structure** drop-down of **Output** rollout to specify a Revit model with one element representing the simplified assembly is output. Select **Each Top Level Component** option from drop-down to specify a Revit model with multiple elements representing the first level components of the simplified model.
- Select **Enable updating** check box from **Output** rollout to create a browser node that enables you to edit the simplification and update the exported RVT.
- Select **Fill internal voids** check box from **Advanced Properties** rollout to automatically filled all internal voids; refer to Figure-45.

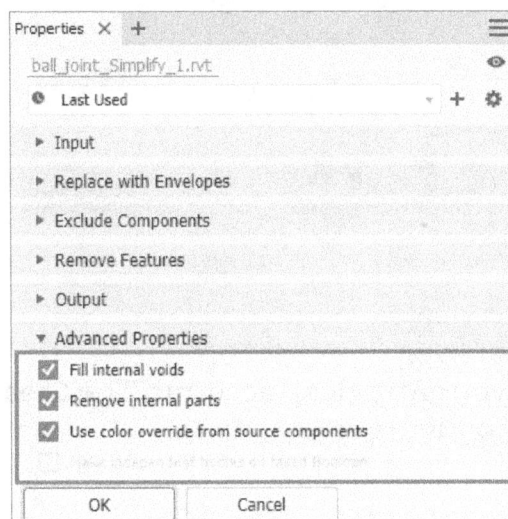

Figure-45. Advanced Properties rollout options

- On selecting **Remove internal parts** check box from **Advanced Properties** rollout, Inventor looks at the model from 14 standard directions (six orthographic and eight isometric) to determine a parts visibility state. Parts deemed not visible are removed.

- Select **Make independent bodies on failed Boolean** check box from **Advanced Properties** rollout to create a multi-body part when a boolean operation fails on one of the single solid body style options. This check box is available only when Style is Single body with no visible edges between planar faces or Single body with visible edges between planar faces.

- After specifying desired parameters, click on **OK** button from the panel or press **ENTER**. The **Exporting New Revit Model (RVT)** window will be displayed exporting the model; refer to Figure-46.

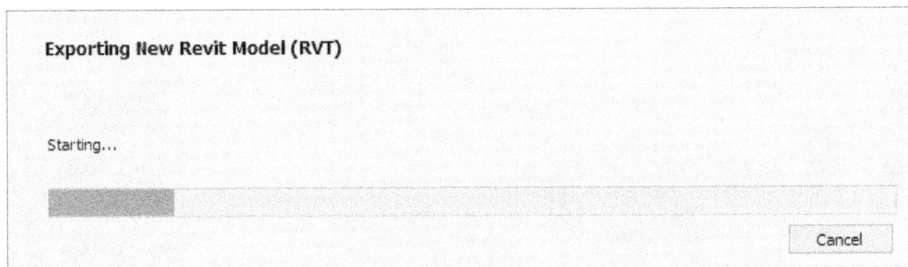

Exporting New Revit Model (RVT)

Starting...

Cancel

Figure-46. Exporting New Revit Model RVT window

Sharing View

The **Share View** tool in **Share** cascading menu of the **File** menu is used to share views of current model with other. The procedure to use this tool is given next.

- Click on the **Share View** tool from the **Share** cascading menu of the **File** menu. The **Sign in** page will be displayed to sign into your Autodesk account; refer to Figure-47.

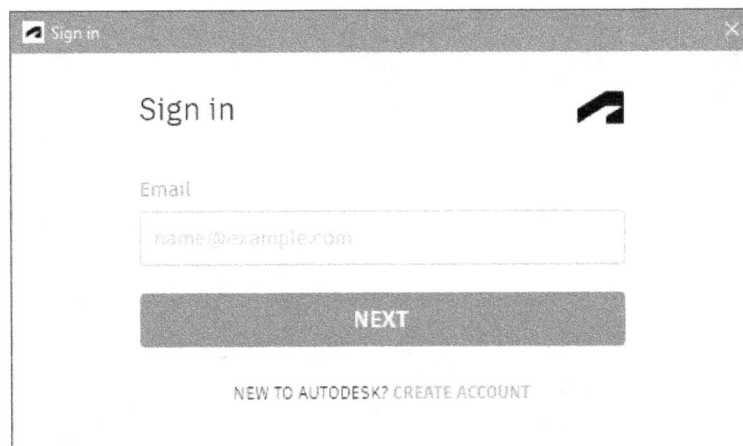

Sign in

Email

name@example.com

NEXT

NEW TO AUTODESK? CREATE ACCOUNT

Figure-47. Sign in page of Autodesk account

- Enter the credentials of your Autodesk account. The **Create a Shared View** window will be displayed; refer to Figure-48.

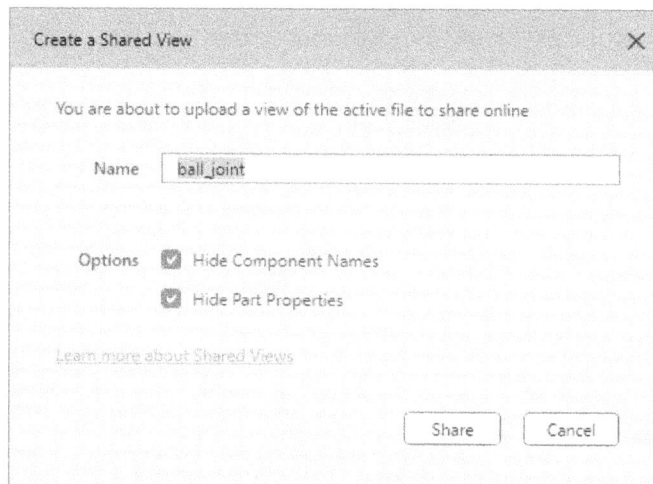

Figure-48. Create a Shared View dialog box

- Specify desired name of view in the **Name** edit box. Select the `Hide Component Names` check box from **Options** area of the dialog box if you want to hide component name from shared view. Similarly, select the `Hide Part Properties` check box if you want to hide properties of part from the view.
- After setting desired parameters, click on the **Share** button from the dialog box.
- Once the processing is complete. The view will be available for sharing.

Managing Project files and features

The options to manage project file and other related features are available in the **Manage** cascading menu; refer to Figure-49. Various options of this menu are discussed next.

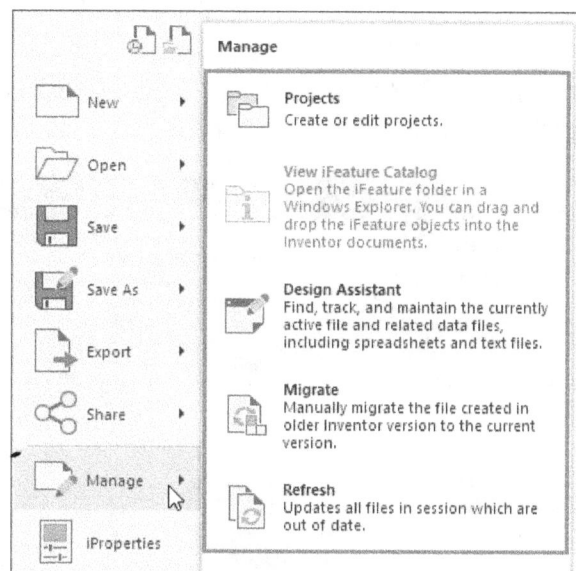

Figure-49. Manage cascading menu

Projects

The **Projects** option is used to create and manage the project files. The procedure to use this tool is given next.

- Click on the **Projects** option from **Manage** cascading menu in the **File** menu. The **Projects** dialog box will be displayed as shown in Figure-50.

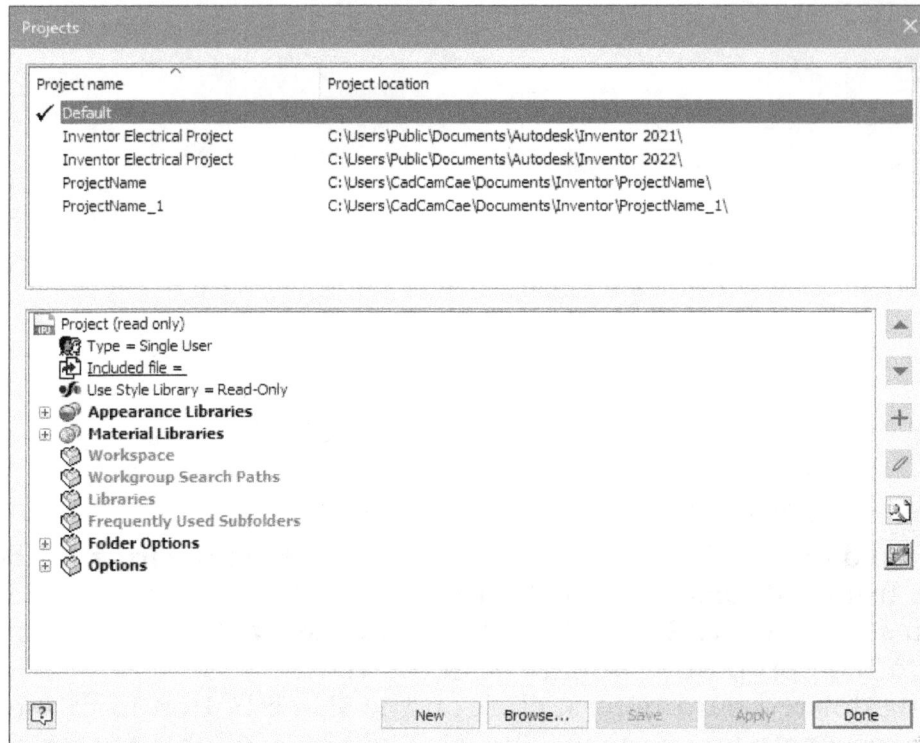

Figure-50. Projects dialog box

- Select the project file that you want to edit from the upper area of the dialog box. The related options will be displayed in the lower area of the dialog box.
- To change any parameter in the dialog box, right-click on it and select the **Edit** option from the shortcut menu.

Adding New Project

- To add new project, click on the **New** button from the **Projects** dialog box. The **Inventor project wizard** dialog box will be displayed; refer to Figure-51.

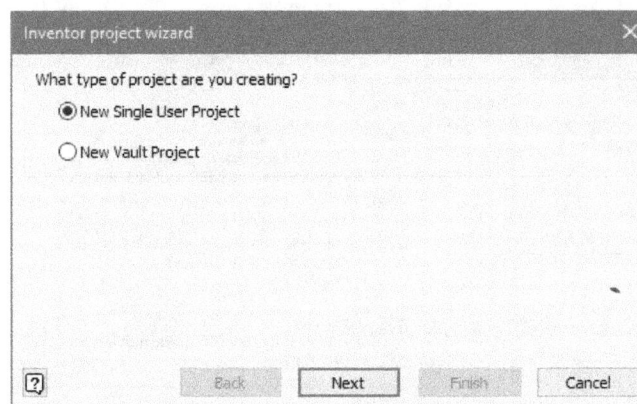

Figure-51. Inventor project wizard dialog box

- If you are using Vault server and want to create a vault project then select the **New Vault Project** radio button. Otherwise, select the **New Single User Project** radio button and click on the **Next** button. The updated **Inventor project wizard** dialog box will be displayed as shown in Figure-52. (Note that we have selected the **New Single User Project** radio button in this case.)

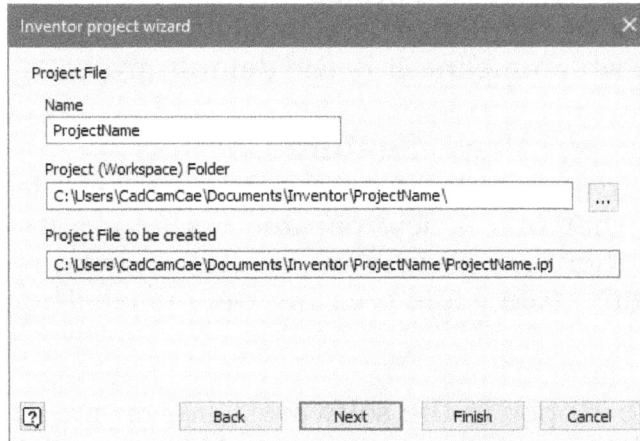

Figure-52. Updated Inventor project wizard dialog box

- Specify desired name of project file in the **Name** edit box. Similarly, you can set the project folder in the **Project (Workspace) Folder** edit box.
- Click on the **Next** button after specifying desired values. The **Select Libraries** page of **Inventor project wizard** dialog box will be displayed; refer to Figure-53.

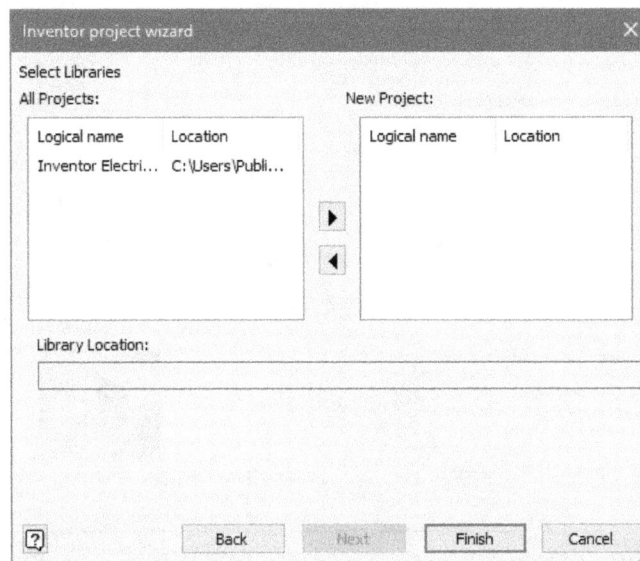

Figure-53. Select libraries page

- Click on the arrow button in the middle of the dialog box to add or remove the library from the new project.
- Click on the **Finish** button from the dialog box to create the project file.
- Click on the **Done** button from the **Projects** dialog box to exit. Now, you can use the newly created file as project while starting new file.
- To make the newly created project as active project, click on the **Projects** button from the **Quick Access Toolbar**; refer to Figure-54. The **Projects** dialog box will be displayed as discussed earlier.

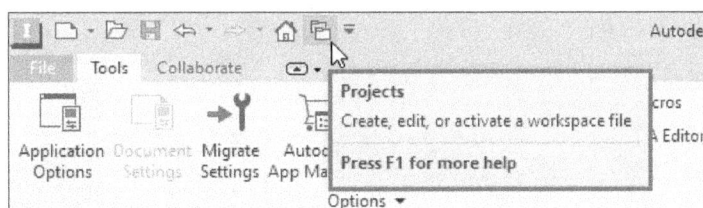

Figure-54. Projects button

- Double-click on the newly created project file that you want to make active project from the upper area of the dialog box. A tick mark will be displayed with the selected project file which means it is the default project.

View iFeature Catalog

In terms of Autodesk Inventor, iFeature means intelligent features. iFeatures are pre-designed features that take a few parameters as input and create real-world designs. For example, there are shapes of punches, cones, rectangular tubes, and so on that can be directly used while creating the model. The procedure to use this tool is given next.

- Make sure a new file is open in the software. Click on the **View iFeature Catalog** tool from the **Manage** cascading menu in the **File** menu. The **Catalog** folder will be displayed in the Windows Explorer; refer to Figure-55.

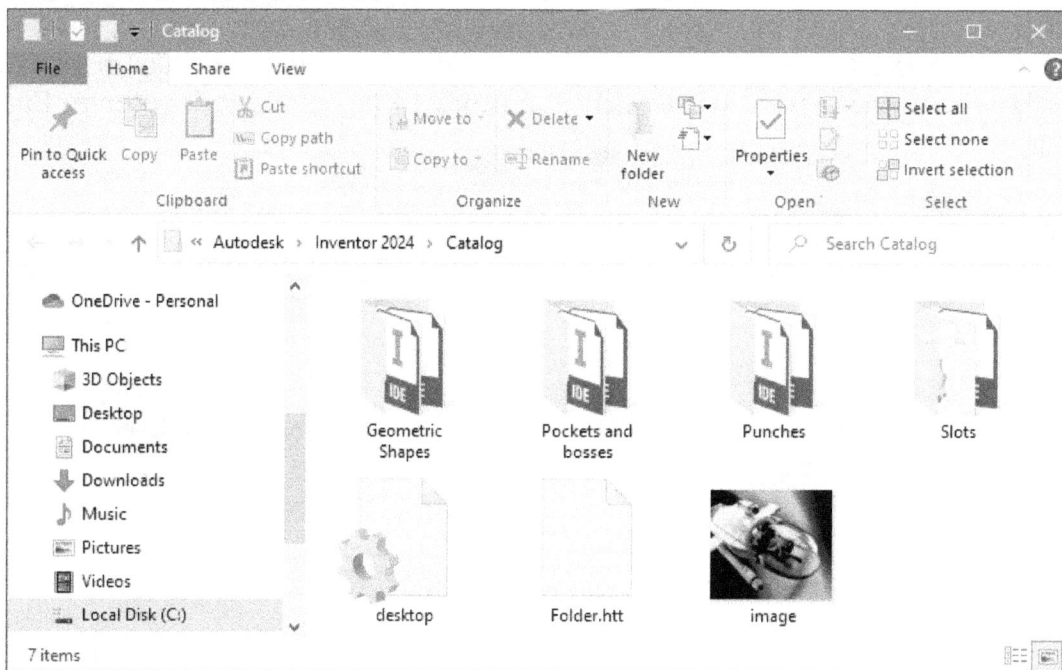

Figure-55. Catalog folder

- Open desired category from the **Windows Explorer** and drag desired feature to modeling area of Autodesk Inventor; refer to Figure-56.

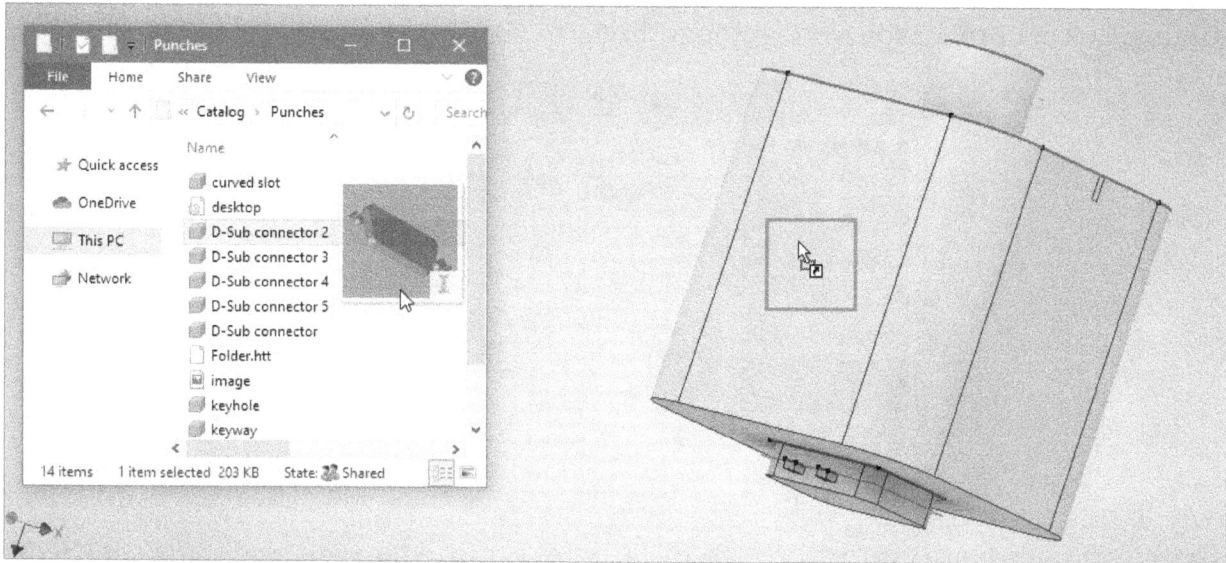

Figure-56. Dragging iFeatures in Inventor

- Once you release the mouse button after dragging the feature on model, the preview of iFeature will be displayed with **Insert iFeature** dialog box; refer to Figure-57.

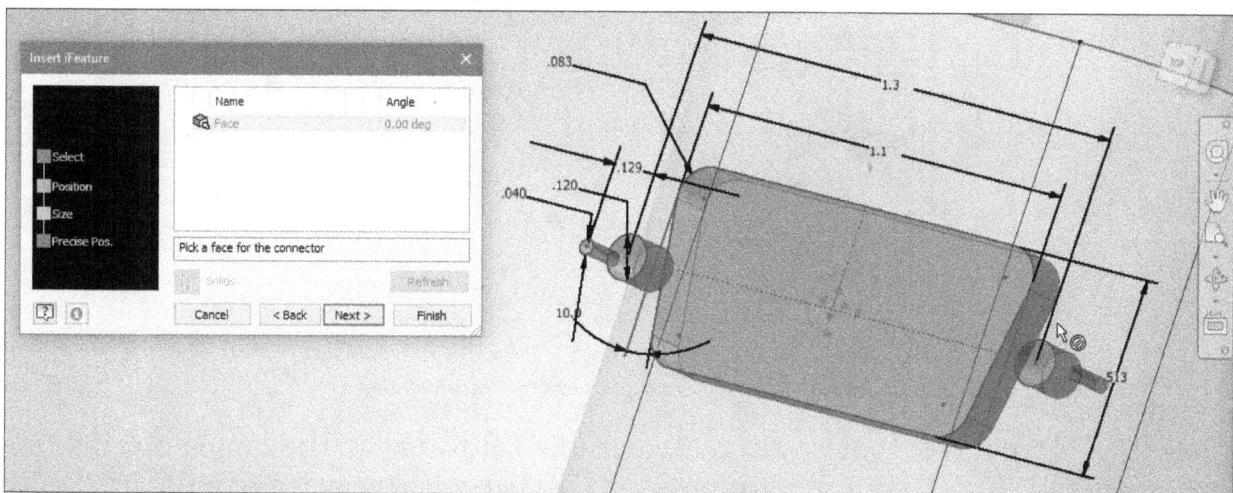

Figure-57. Insert iFeature dialog box

- Click on desired face of the model to place the feature.
- If you want to rotate the feature then click in the field under **Angle** column and specify the value of rotation angle; refer to Figure-58.

Figure-58. Angle edit box

- Click on the **Next** button from the dialog box. The **Size** page of **Insert iFeature** dialog box will be displayed; refer to Figure-59.

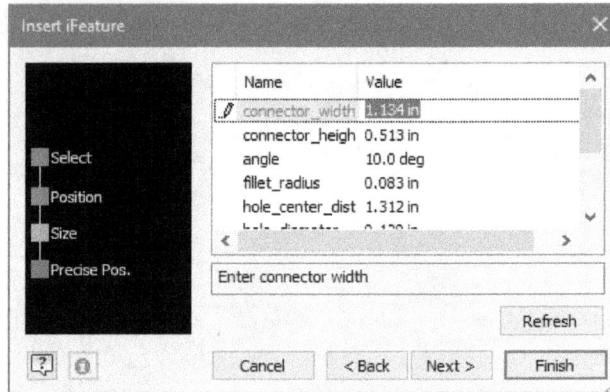

Figure-59. Size page of Insert iFeature dialog box

- Specify desired values for the size parameters of the feature and click on the **Next** button from the dialog box. The **Precise Position** page of **Insert iFeature** dialog box will be displayed; refer to Figure-60.

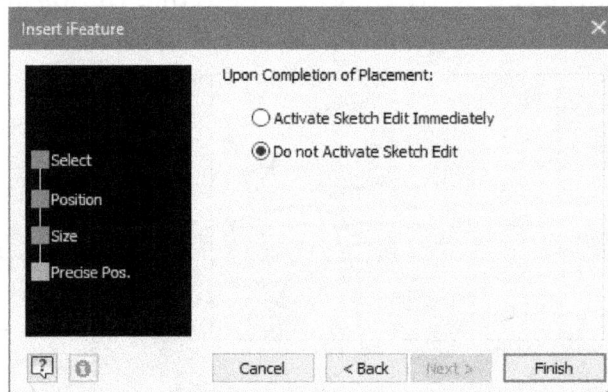

Figure-60. Precise Position page of Insert iFeature dialog box

- Select the **Do not Activate Sketch Edit** radio button from the dialog box if you are satisfied with the current position of the iFeature. If you want to edit the position of the iFeature precisely then select the **Activate Sketch Edit Immediately** radio button and click on the **Finish** button. The sketching environment will be displayed and using the **Dimension** tool, you can place the feature precisely. We will learn about the sketching tools later in the book.
- After specifying the position, click on the **Finish Sketch** button.

Design Assistant

The **Design Assistant** option is used to find, track, and maintain Autodesk Inventor files and related word processing, spreadsheet, or text files. The procedure to use this tool is discussed next.

- Click on the **Design Assistant** option from **Manage** cascading menu in the **File** menu. The **Design Assistant** window will be displayed with the properties and preview of current opened model; refer to Figure-61.

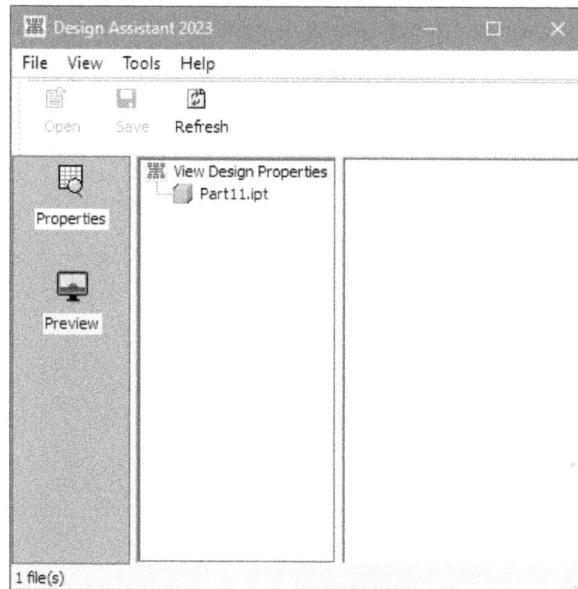

Figure-61. Design Assistant window

- Click on the **Properties** button from left side of the window and select the model from **View Design Properties** box of the window to display the properties of selected model on the right side of the window.
- Click on the **Preview** button from left side of the window and select the model from **View Design Properties** box of the window to display the preview of selected model on the right side of the window; refer to Figure-62.

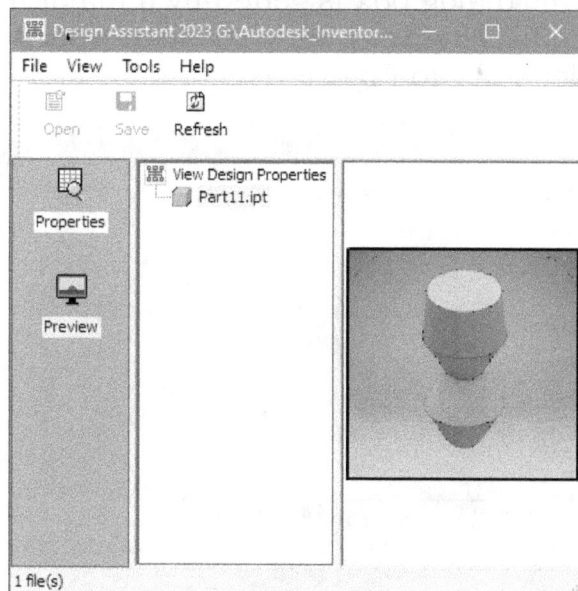

Figure-62. Preview of model

iProperties

The **iProperties** option is used to change general properties of the model. The procedure to use this option is given next.

- Click on the **iProperties** option from the **File** menu. The **iProperties** dialog box will be displayed as shown in Figure-63.

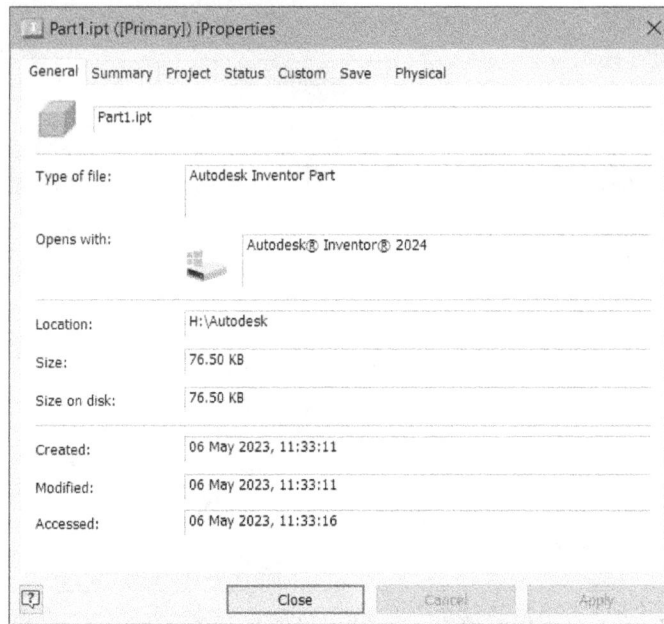

Figure-63. iProperties dialog box

- The options in **General** tab display the general information of the part. Note that you cannot modify any of the values in this tab.
- Click on the **Summary** tab to specify the user details for the part. Similarly, you can specify the project details, status, and other custom details by using the respective tabs.
- The **Physical** tab in this dialog box is generally a major attraction for engineers. Using the options in the **Physical** tab, you can specify and check the mass properties of the model; refer to Figure-64.

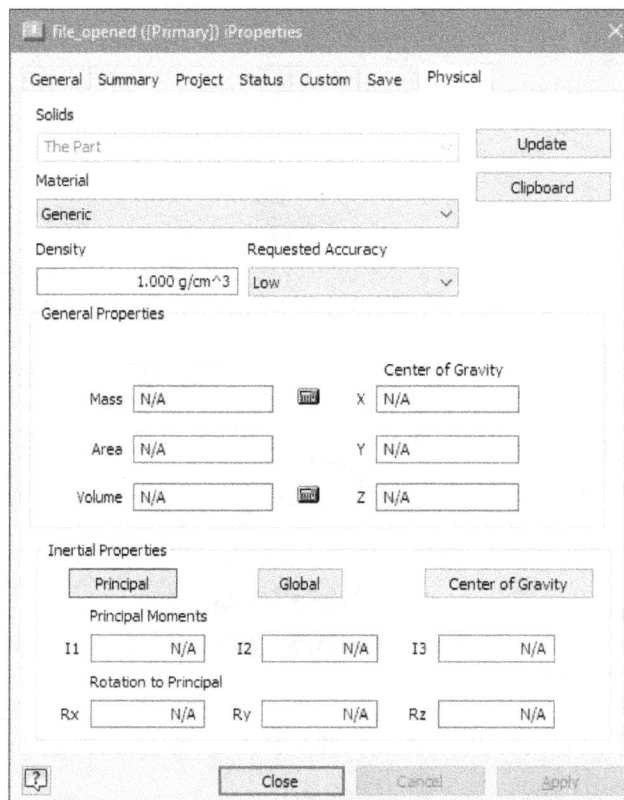

Figure-64. Physical tab of iProperties dialog box

- Select desired material from the **Material** drop-down. The value of density will change accordingly and hence the other property parameters like mass, center of gravity, and so on will change.

Printing Document

Printout is an important requirement in manufacturing area. You can not reach to a boiler/furnace with a tablet/laptop/e-gadgets in your hand as temperature near boiler/furnace is very high. In those cases, we need a copy of engineering drawing as printout that can be used by manufacturers for reference. Also, it is not feasible to hang 100 e-gadgets to display 100 of your products in a presentation room but you can take printouts of your products and display them alongside the products in the presentation room. There are various tools related to printing available in the **Print** cascading menu; refer to Figure-65. Various options in the menu are discussed next.

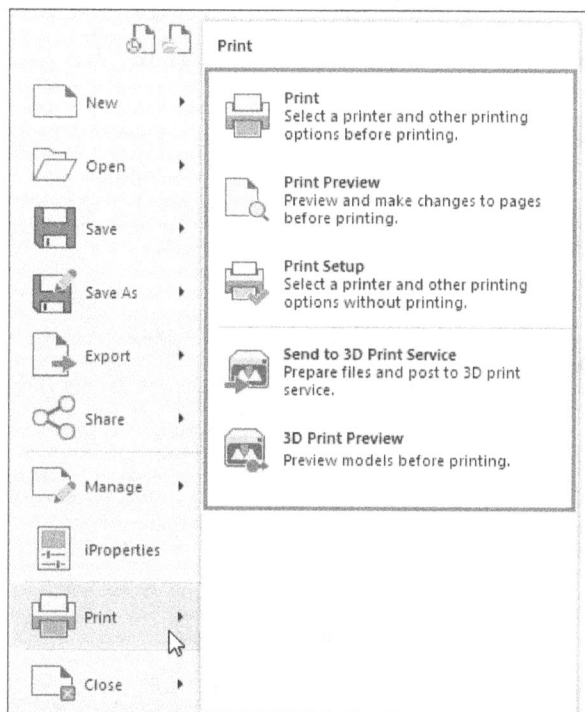

Figure-65. Print cascading menu

Print Tool

As the name suggests, the **Print** tool is used to get print out of the model/drawing on paper. The procedure to use this tool is given next.

- Click on the **Print** tool from the **Print** cascading menu in the **File** menu. The **Print** dialog box will be displayed as shown in Figure-66.

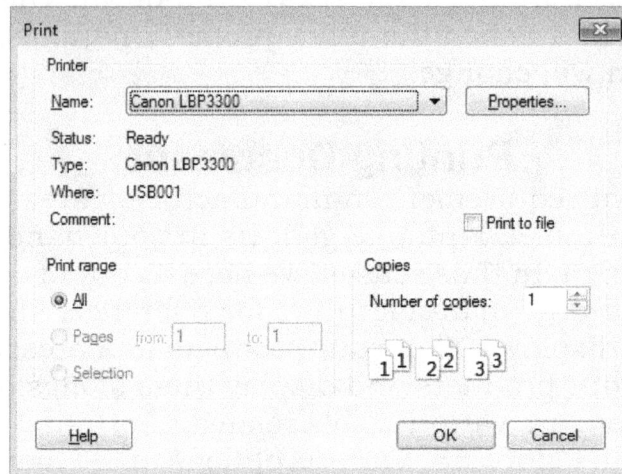

Figure-66. Print dialog box

- Select desired printer from the **Name** drop-down in the **Printer** area of the dialog box.
- Click on the **Properties** button next to **Name** drop-down. The **Printer Properties** dialog box will be displayed as shown in Figure-67.

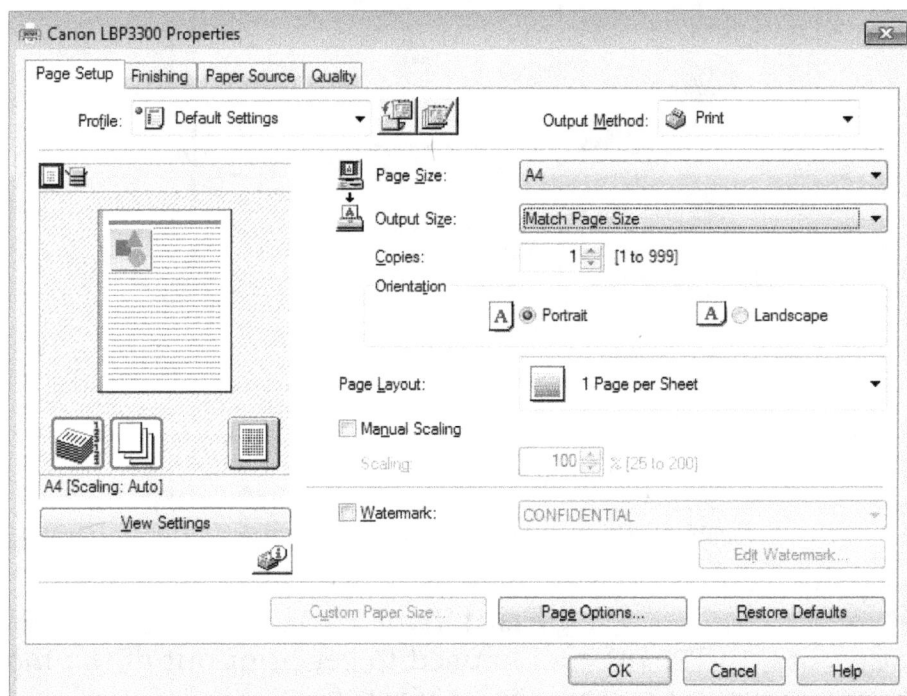

Figure-67. Printer Properties dialog box

- Select desired paper size and orientation from the **Page Size** drop-down and **Orientation** area of the dialog box, respectively.
- To apply watermark on the printout, select the **Watermark** check box and enter desired text in the edit box next to it.
- Click on the **OK** button from the **Printer Properties** dialog box to apply the changes.
- Specify desired number of copies of drawing using the **Number of copies** spinner in the **Copies** area of the dialog box.
- Click on the **OK** button from the dialog box to print the document.

Print Preview

The **Print Preview** tool is used to check the print before sending command to printer/plotter. To check the preview, click on the **Print Preview** tool from the **Print** cascading menu in the **File** menu. If the preview is satisfactory then click on the **Print** button, otherwise click on the **Close** button from the toolbar displayed at the top of the modeling area; refer to Figure-68.

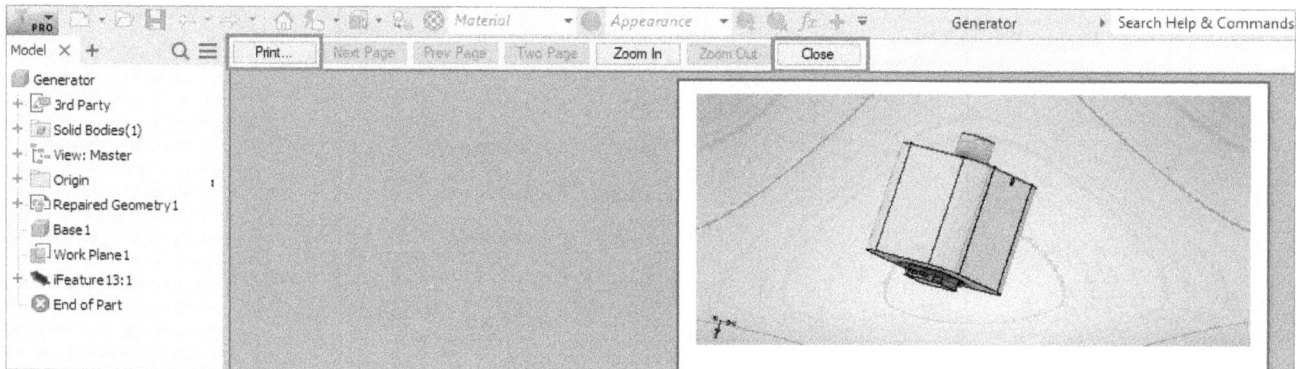

Figure-68. Print preview of the model

Print Setup

The **Print Setup** tool is used to specify settings for the printing. The procedure to use this tool is given next.

- Click on the **Print Setup** tool from the **Print** cascading menu in the **File** menu. The **Print Setup** dialog box will be displayed as shown in Figure-69.

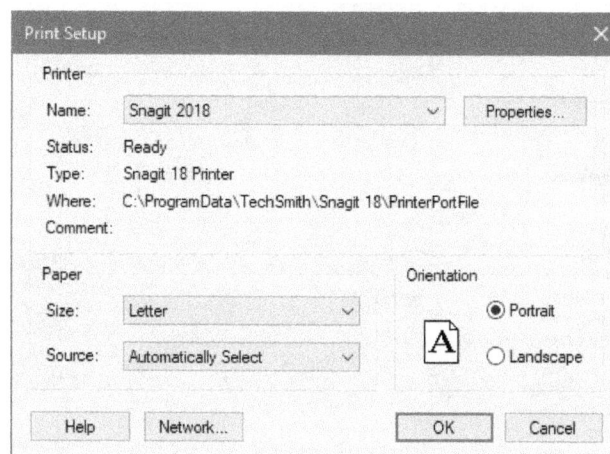

Figure-69. Print Setup dialog box

- The options in this dialog box are same as discussed earlier for the **Printer Properties** dialog box. Specify the size and orientation for the print and click on the **OK** button.

Send to 3D Print service

One of the great invention of this century, 3D Printing facilitate prototype creation at blazing speed. 3D Printing is finding scope in almost every area of engineering whether it is biotechnology or it is Construction engineering. Now, every good CAD package has the option to send file for 3D print service. The procedure to send file for 3D Print service from Autodesk Inventor is given next.

- After creating a solid model, click on the **Send to 3D Print Service** tool from the **Print** cascading menu in the **File** menu. The **Send to 3D Print Service** dialog box will be displayed as shown in Figure-70.

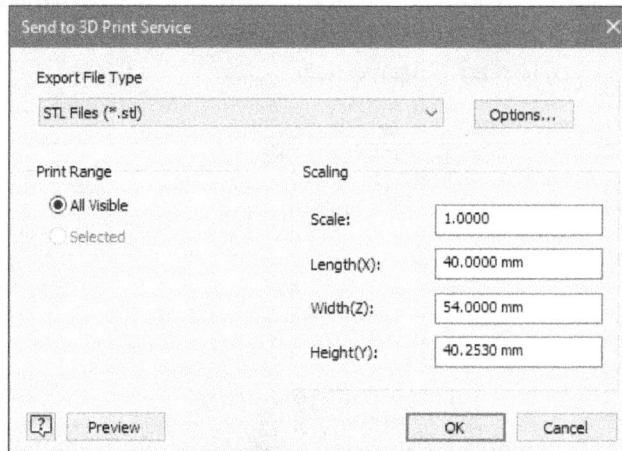

Figure-70. Send to 3D Print Service dialog box

- Experienced CAD users can easily find that the export format in this dialog box, **STL Files** is a very old CAD format and is in use for many years. So, we can understand that the dialog box actually apply some scaling changes and then convert the model to STL file which can be directly used by 3D Printer.
- From the **Scaling** area, specify desired parameters to enlarge or diminish the model.
- Click on the **Options** button next to **Export File Type** drop-down. The **STL File Save As Options** dialog box will be displayed; refer to Figure-71.

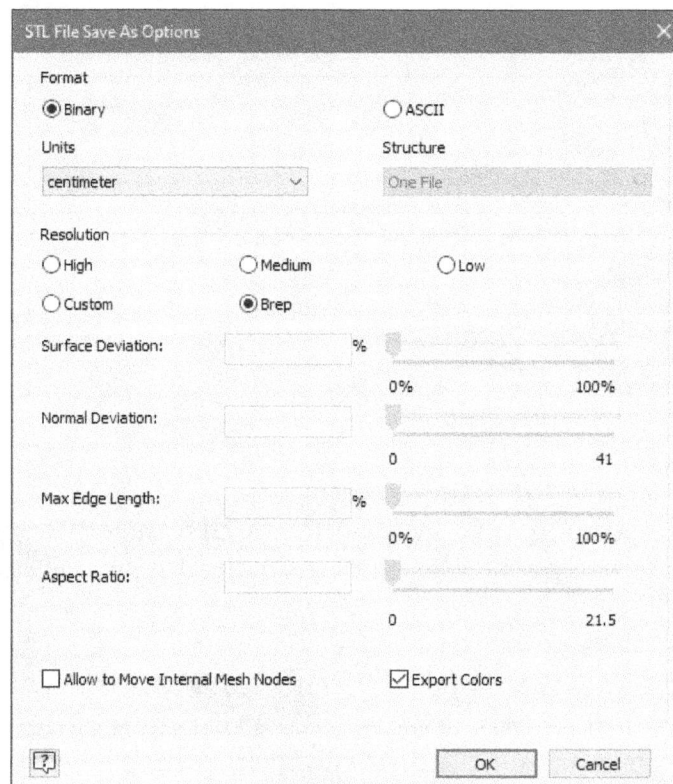

Figure-71. STL File Save As Options dialog box

- Set the quality of export file by using the options in this dialog box and click on the **OK** button.

- Click on the **OK** button from the **Send to 3D Print Service** dialog box to export the file. The **Save Copy As** dialog box will be displayed; refer to Figure-72.

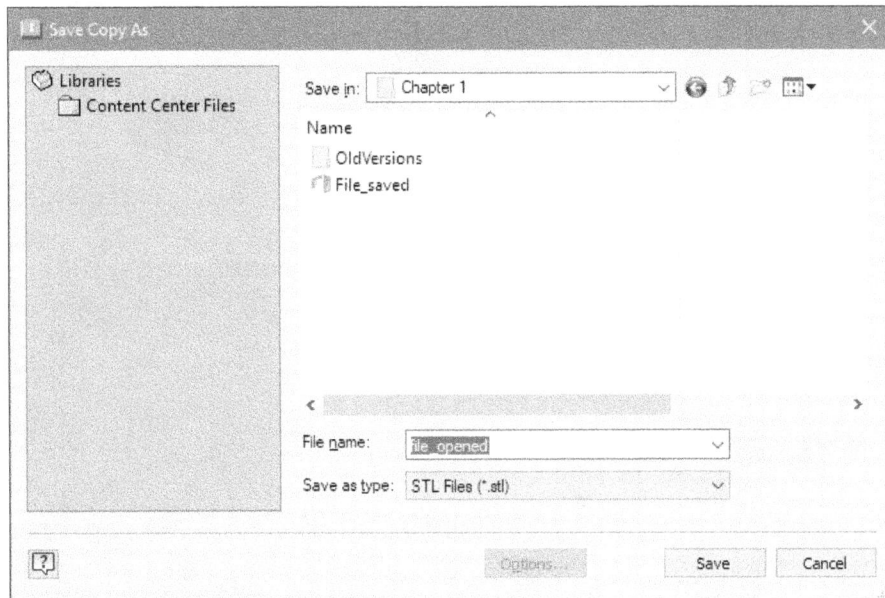

Figure-72. Save Copy As dialog box

- Click on the **Save** button from the dialog box to save the file.

3D Print Preview

The **3D Print Preview** tool works in the same way as the **Print Preview** tool works. Click on this tool and check the quality of 3D print to be created.

Till this point, we have learned about various settings and basic file operations of Autodesk Inventor. This chapter is a reference material and will be utilized in one or another way in all the upcoming chapters. In the next chapter, we will learn about sketching tools and their applications.

SELF ASSESSMENT

Q1. Which of the following options manages the overall document management of Autodesk Inventor?

a) Tools
b) File menu
c) Get Started
d) Collaborate

Q2. By which of the following options, you cannot create a new file?

a) File menu
b) Get Started
c) Collaborate
d) Quick Access Toolbar

Q3. Which of the following options is used to create the 2D representation of the model?

a) Presentation
b) Drawing
c) Part
d) Assembly

Q4. While importing CAD files in Autodesk Inventor, which option in the dialog box cannot make changes in the imported model?

a) Individual
b) Reference Model
c) Convert Model
d) Composite

Q5. Which of the following options is used to save the file with new name and location?

a) Save As
b) Save Copy As
c) Save All
d) Save

Q6. Which tool in Autodesk Inventor allows you to save the file for other software like Pro-E, Catia, and so on?

a) Save
b) Save Copy As Template
c) Pack and Go
d) Export

Q7. Which of the following options is used to export drawing file in Inventor?

a) Image
b) CAD Format
c) Export to DWG
d) Export to DWF

Q8. By which option the intelligent features like shapes of punches, cones, and rectangular tubes can be directly used in the model?

a) iProperties
b) View iFeature Catalog
c) Design Assistant
d) Projects

Q9. If you are using Vault server and want to create a vault project then select **New Single User Project** in the **Inventor project wizard** dialog box. (True/False)

Q10. The options in the **Manage** cascading menu are used to create and manage the project files. (True/False)

FOR STUDENT NOTES

Chapter 2

Creating Sketches

Topics Covered

The major topics covered in this chapter are:

- *Introduction to Sketching*
- *Sketch creation tools*
- *Modification tools*

INTRODUCTION TO SKETCHING

After a bang-bang of settings and file handling options, we are going to start with sketching. Sketch is the base bone (should I say spine!!) of 3D models. If you are creating a wrong sketch then you cannot expect desired results from the 3D model. So, it is very important to understand the tools of sketching as well as method of sketching. The tools related to sketching are available in the **Sketch** tab of the **Ribbon**; refer to Figure-1. In this chapter, we will discuss these tools one by one. But before that let's understand the concept of sketching planes.

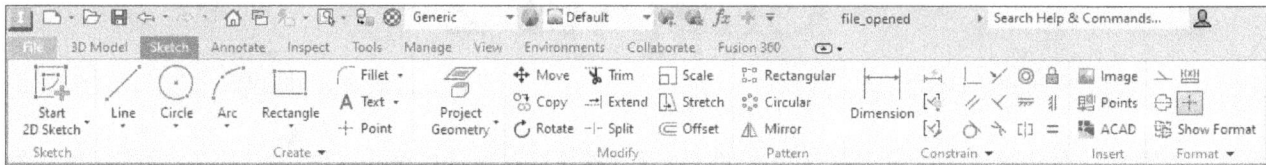

Figure-1. Sketch tab

SKETCHING PLANE

In a CAD software, everything is referenced to other entity like a line created must be referenced to any other geometry, so that you can clearly define the position of line with respect to the other geometry. But, what if there is no geometry in the sketch to reference from. In these cases, we have tools to create reference geometries like reference planes, axes, points, curves, and so on. Out of these reference geometries, the sketching plane acts as foundation for other geometries. By default, there are three planes perpendicular to each other in Autodesk Inventor; refer to Figure-2.

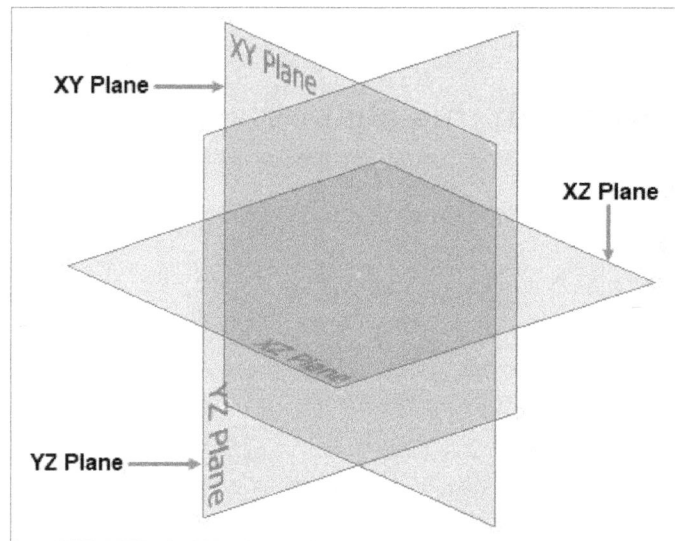

Figure-2. Sketching planes in Inventor

Relation between sketch, plane, and 3D model

Sketch has a direct relationship with planes and the outcome which is generally a 3D model. Refer to Figure-3. In this figure, rectangle is created on the **XY** Plane which is also called **Front** plane. A circle is created on the **YZ** Plane which is also called **Right** plane. A polygon is created on the **XZ** plane which is also called **Top** plane. In a 3D model, the geometry seen from the Front view should be drawn on the **Front** plane. Similarly, geometry seen from the Right view should be drawn at **Right** plane and geometry seen from the Top view should be drawn at the **Top** plane.

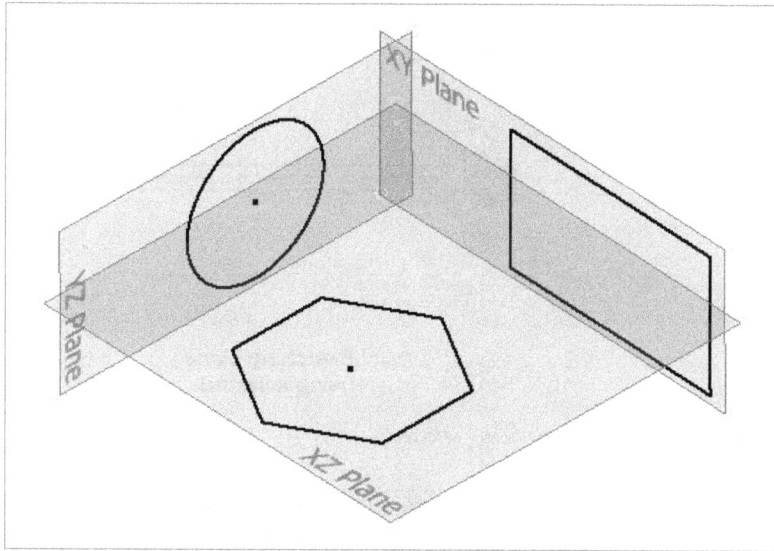

Figure-3. Sketches created on different planes

We will learn more about planes at the beginning of 3D Modeling. We will now start with Sketching tools.

START 2D SKETCH

The **Start 2D Sketch** tool is used to start sketching environment. Once we have started 2D sketch, we will be able to use the sketching tools on the selected plane. The procedure to start 2D sketch is given next.

- Click on the **Start 2D Sketch** tool from the **Start 2D Sketch** drop-down in the **Sketch** tab of the **Ribbon**; refer to Figure-4. You will be asked to select a plane for sketching.

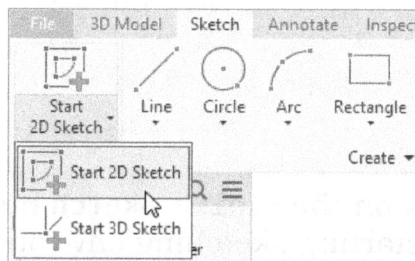

Figure-4. Start 2D Sketch tool

- Click on desired plane from the highlighted planes in modeling area; refer to Figure-5. The sketching environment will be displayed with horizontal and vertical reference lines; refer to Figure-6. Note that these reference lines as actually other two planes.

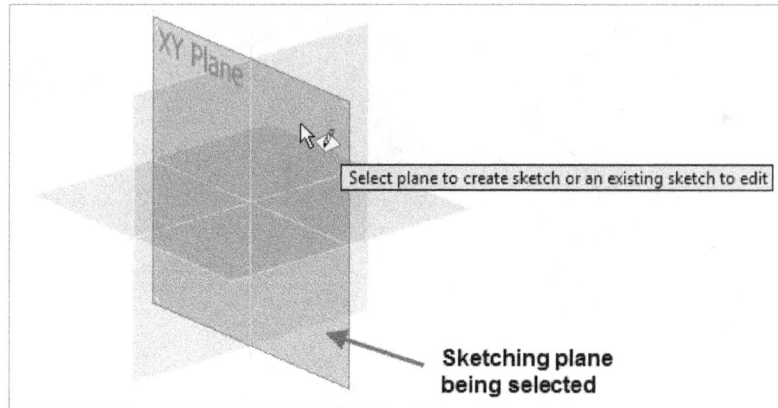

Figure-5. Sketching plane being selected

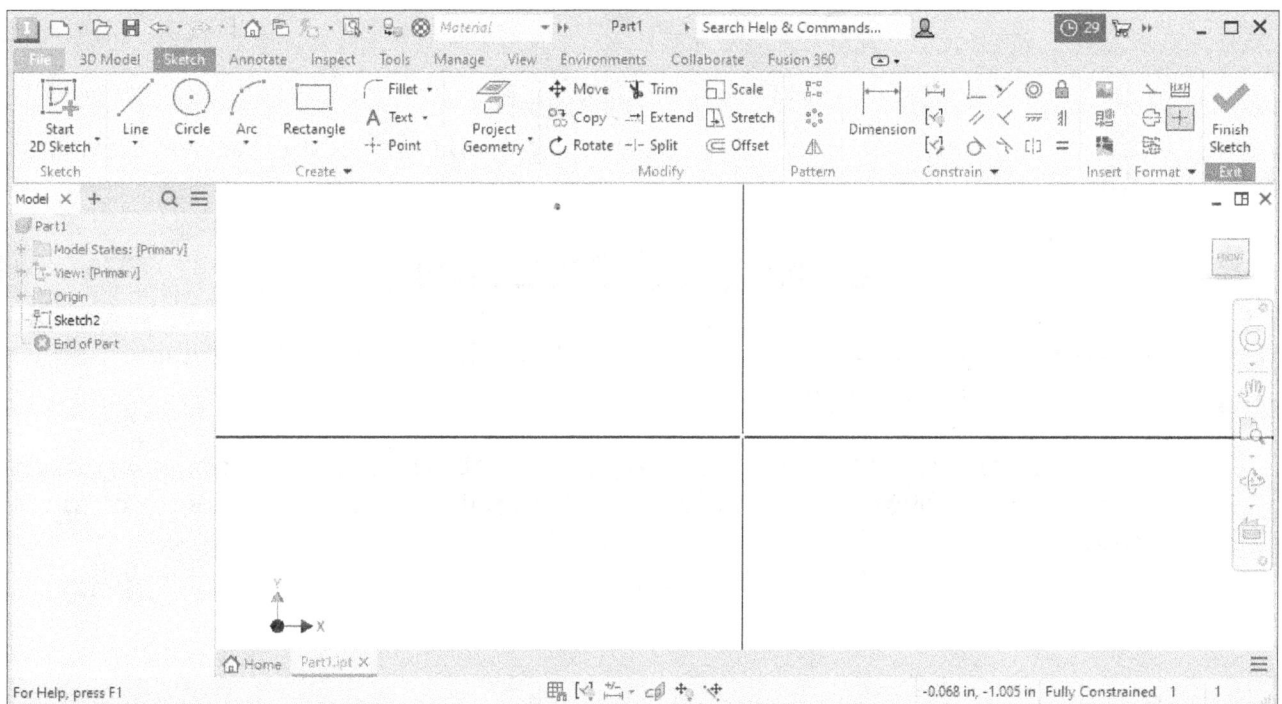

Figure-6. Sketching environment

- After creating sketch, click on the **Finish Sketch** button. Note that we generally use sketching tools after starting sketching environment and before clicking on the **Finish Sketch** button.

SKETCH CREATION TOOLS

The tools that are available in the **Create** panel are named as sketch creation tools. The tools that are most commonly used for sketch creation are given next.

Line tool

As everyone can guess, the **Line** tool is used to create a line in the sketch. The procedure to create a line is given next.

- Click on the **Line** tool from **Create** panel in the **Ribbon**; refer to Figure-7. You will be asked to specify start point of the line. Also, the **Coordinate Input box** will be displayed; refer to Figure-8.

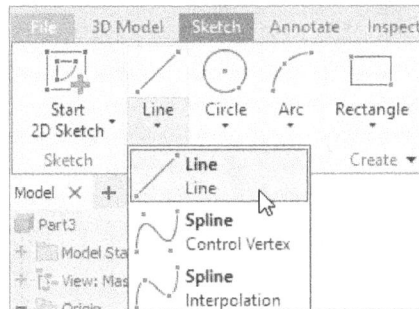

Figure-7. Line tool

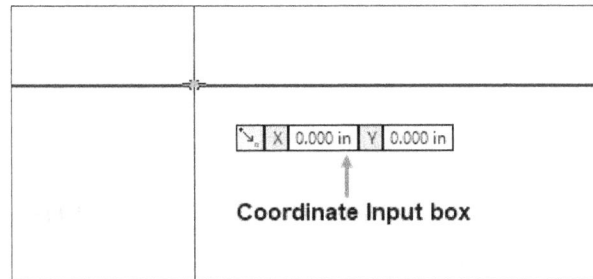

Figure-8. Coordinate Input box

- Specify desired coordinates in the Input box (Note that **X** edit box is used to specify X coordinate and **Y** edit box is used to specify Y coordinate in the Input box) or click in the modeling area to specify the starting point. On doing so, you will be asked to specify the end point of the line; refer to Figure-9.

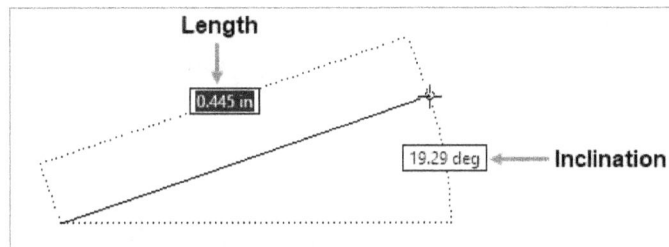

Figure-9. Specifying end point of line

- Specify desired length and angle in the input boxes or click at desired location in the modeling area to specify the end point of the line. Note that if you want to specify the values in the input boxes then press **TAB** to toggle between the two input boxes and press **ENTER** after specifying the values. The specified values will be displayed as dimension along the line; refer to Figure-10.

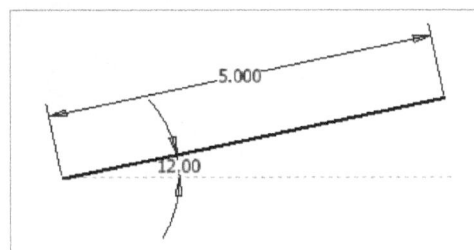

Figure-10. Dimension assigned to line

- Note that you can keep on specifying the end points of the consecutive lines as per your requirement. To stop creation of consecutive lines, press **ESC** button.

- To create an arc using the **Line** tool, click and hold the left mouse button (LMB) after creating a line and drag the cursor to create an arc tangent to previous line; refer to Figure-11.

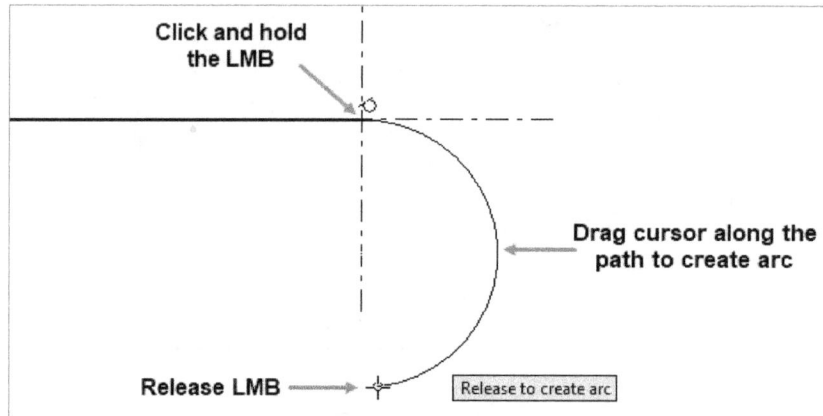

Figure-11. Creating arc with Line tool

Note that if you want to change the unit system for model then click on the **Document Settings** tool from the **Options** panel in the **Tools** tab of **Ribbon**. The **Document Settings** dialog box will be displayed. Click on the **Units** tab and set desired unit system in the dialog box; refer to Figure-12. You will learn more about **Document Settings** dialog box later in the book.

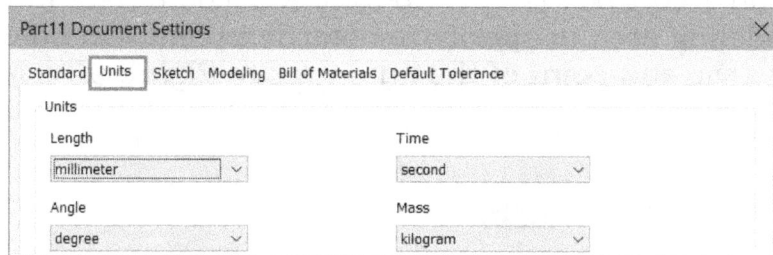

Figure-12. Units tab

Control Vertex Spline

The **Control Vertex Spline** tool is used to create spline with the help of control vertices. A spline is a dynamic curve passing through specified points or controlled by specified vertices. Splines are generally used to create models of artistic objects. The procedure to use this tool is given next.

- Click on the **Control Vertex Spline** tool from **Line** drop-down in the **Create** panel of the **Ribbon**; refer to Figure-13. You will be asked to specify the first point of the spline.

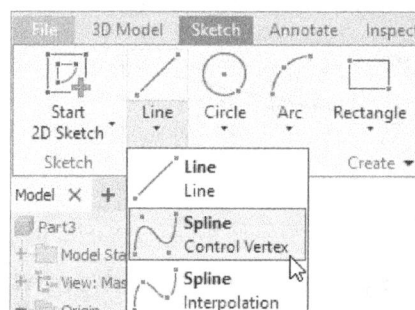

Figure-13. Control Vertex Spline tool

- Click to specify the first point of spline. You are asked to specify the first control vertex of the spline.
- Click to specify the vertex point. You are asked to specify the next control vertices; refer to Figure-14.

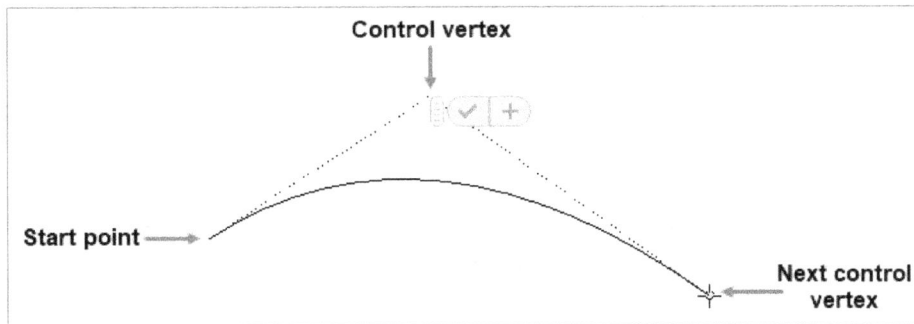

Figure-14. Specifying control vertex

- Keep on specifying the control vertices as per your requirement and press **ENTER** when you have specified control vertices as per your requirement. The spline will be created with the specified control vertices. Press **ESC** to exit the tool.
- Note that to modify the shape of spline, you can drag any of the control vertex to desired position.

Interpolation Spline

The **Interpolation Spline** tool works in the same way as the **Control Vertex Spline** tool works but in case of interpolation spline, we specify the points through which the spline passes. The procedure to use this tool is given next.

- Click on the **Interpolation Spline** tool from **Line** drop-down in the **Create** panel of the **Ribbon**; refer to Figure-15. You will be asked to specify starting point of the spline.

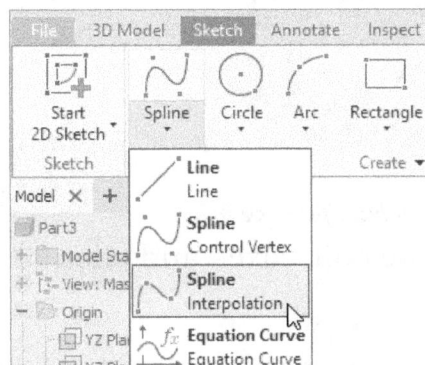

Figure-15. Interpolation Spline tool

- Click to specify the starting point. You are asked to specify the next points of spline.
- Keep on specifying the points through which you want the spline to pass.
- Press **ENTER** to exit the spline creation mode. Press **ESC** to exit the tool.

Equation Curve

The **Equation Curve** tool is used to create curve with the help of an equation. There are three components of equation:

x(t), y(t), and values of t_{min} & t_{max}. Here, x(t) and y(t) are functions of **t** like,

x(t) = t^2
y(t) = t^3+t

t varies from **t_{min}** to **t_{max}**.

To create curve based on equation, follow the steps given next.

- Click on the **Equation Curve** tool from **Line** drop-down in the **Create** panel of the **Ribbon**; refer to Figure-16. The edit boxes related to curve equation will be displayed; refer to Figure-17.

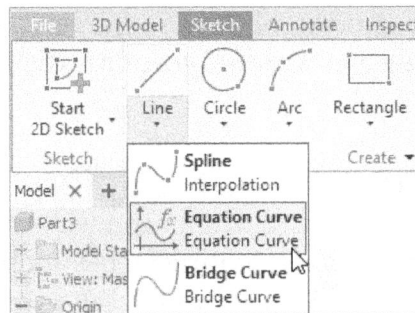

Figure-16. Equation Curve tool

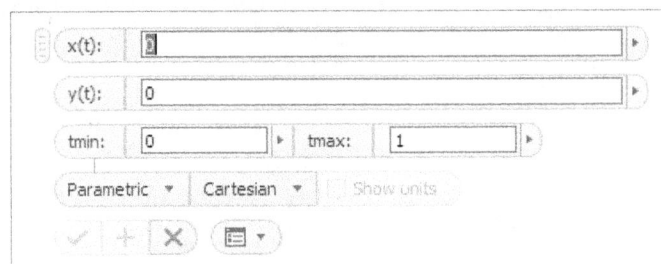

Figure-17. Edit boxes for curve equation

- Specify desired values in the edit boxes and click on the **OK** button to create the curve.

Bridge Curve

As the name suggests, the **Bridge Curve** tool is used to create a curve that connects two curves. The procedure to use this tool is given next.

- Click on the **Bridge Curve** tool from **Line** drop-down in the **Create** panel of the **Ribbon**; refer to Figure-18. You will be asked to select first curve.

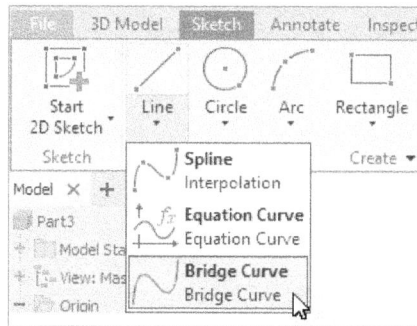

Figure-18. Bridge Curve tool

- Click on the first curve. You will be asked to select the second curve.
- Click on the second curve. A bridge curve will be created connecting both the curves. Note that your selection region of curve will decide the start point/end point of the bridge curve; refer to Figure-19. So, click on the curve near the point which you want to use as start or end point.

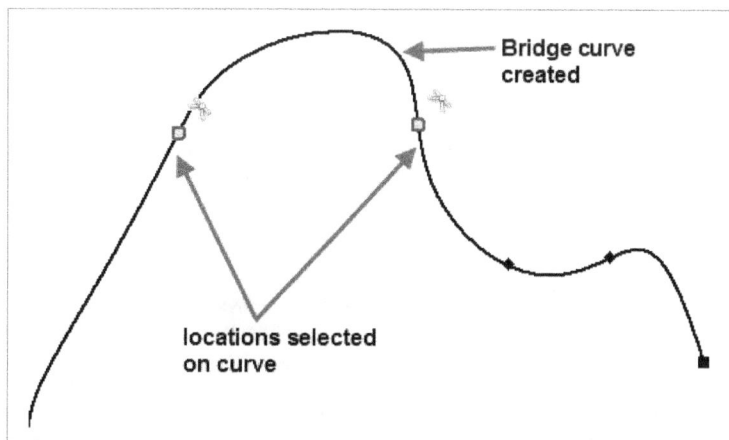

Figure-19. Bridge curve created

Circle

Circle has a great history and it is useful in various designs (No! we will not discuss history here). There are numerous examples where we will be using the circle tool to create designs like wheels, gears, shafts, and so on. The procedure to use this tool is given next.

- Click on the **Center Point Circle** tool from **Create** panel in the **Ribbon**; refer to Figure-20. You will be asked to specify the center point of the circle.

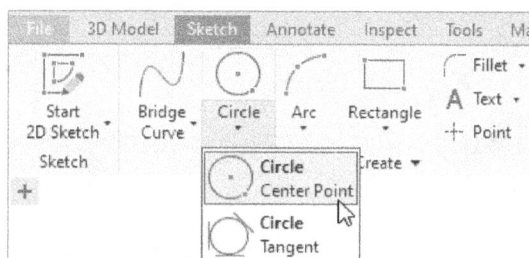

Figure-20. Center Point Circle tool

- Click to specify the center point of the circle. You will be asked to specify diameter of the circle; refer to Figure-21.

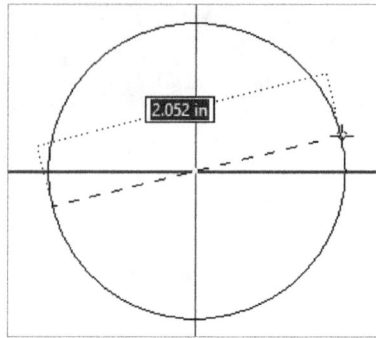

Figure-21. Specifying diameter of circle

- Specify desired diameter value in the edit box. If you want to specify the radius value in place of diameter then right-click in the modeling area before specifying value in edit box. A shortcut menu will be displayed; refer to Figure-22.

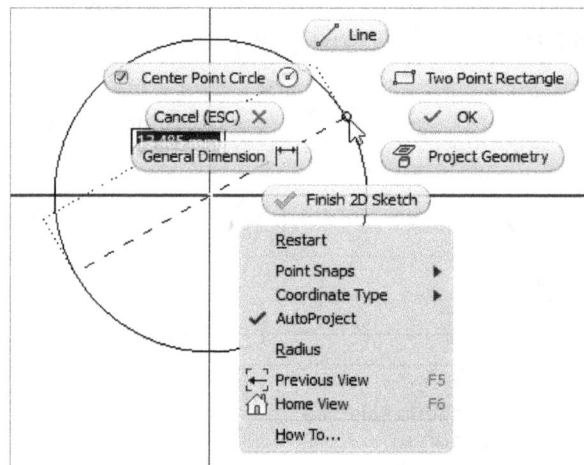

Figure-22. Shortcut menu for circle

- Select the **Radius** option from the shortcut menu. You will be asked to specify radius of the circle; refer to Figure-23.

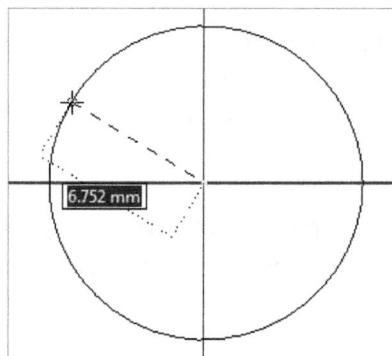

Figure-23. Specifying radius of circle

- After specifying desired value, press **ENTER** from the keyboard to create the circle. Press **ESC** to exit the tool.

Tangent Circle

The **Tangent Circle** tool is used to create circle tangent to three selected lines. The procedure to use this tool is given next.

- Click on the **Tangent Circle** tool from **Circle** drop-down in the **Create** panel of the **Ribbon**; refer to Figure-24. You will be asked to select the first line.

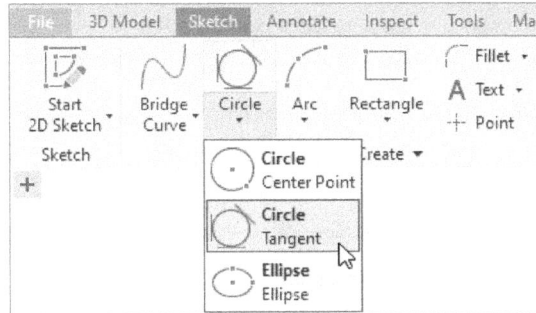

Figure-24. Tangent Circle tool

- Click on the first line in the modeling area; refer to Figure-25. You will be asked to select the second line.

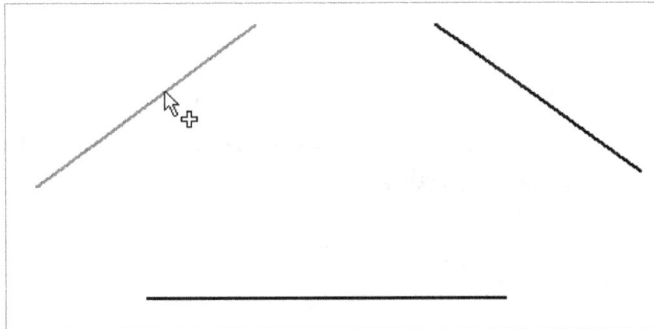

Figure-25. Selecting first line for tangent circle

- Click on the second line and then on the third line. A circle tangent to the three lines will be displayed; refer to Figure-26.

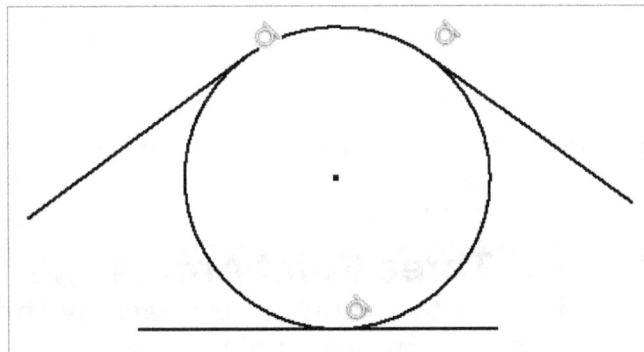

Figure-26. Circle tangent to the lines

Note that it is not compulsory for the circle to touch all the line to be tangent to them.

Ellipse

As the name suggests, the **Ellipse** tool is used to create ellipse in the sketch. The procedure to use this tool is given next.

- Click on the **Ellipse** tool from **Circle** drop-down in the **Create** panel of the **Ribbon**; refer to Figure-27. You will be asked to specify center point of the ellipse.

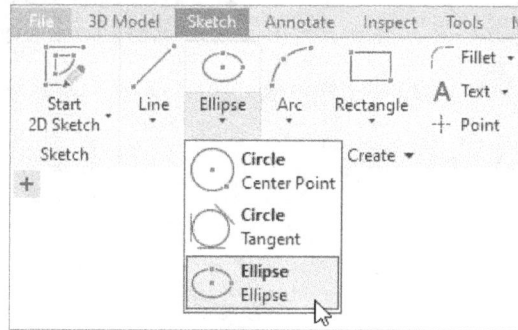

Figure-27. Ellipse tool

• Click to specify the center point of the ellipse. You will be asked to specify the end point of the first axis of ellipse; refer to Figure-28.

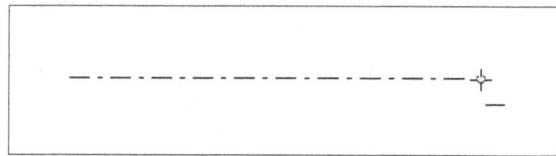

Figure-28. Specifying endpoint of first axis

• Click to specify the end point of axis. You will be asked to specify a circumferential point of the ellipse; refer to Figure-29.

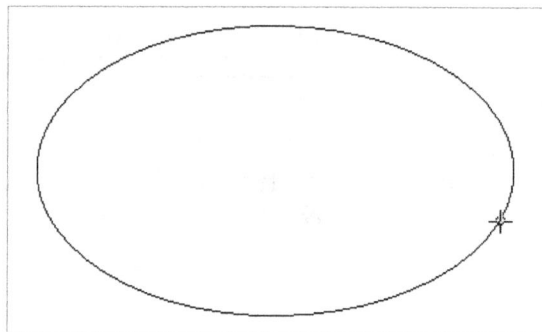

Figure-29. Specifying circumferential point of ellipse

• Click to specify circumferential point of ellipse. The ellipse will be created. Press **ESC** to exit the tool.

Three Point Arc

The **Three Point Arc** tool is used to create an arc passing through specified three points. The procedure to create 3 point arc is given next.

• Click on the **Three Point Arc** tool from **Create** panel in the **Ribbon**; refer to Figure-30. You will be asked to specify the start point of the arc.

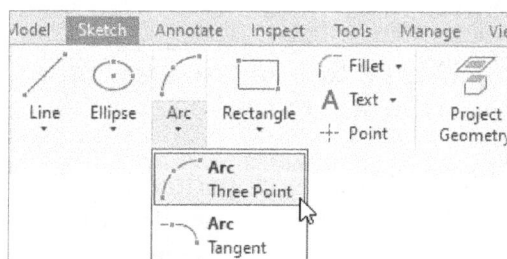

Figure-30. Three Point Arc tool

- Click in the modeling area to specify the start point. You will be asked to specify the end point of the arc; refer to Figure-31.

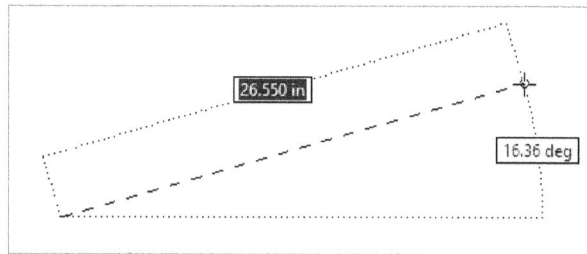

Figure-31. Specifying end point of the arc

- Click to specify the end point of the arc. You will be asked to specify a circumferential point of the arc; refer to Figure-32.

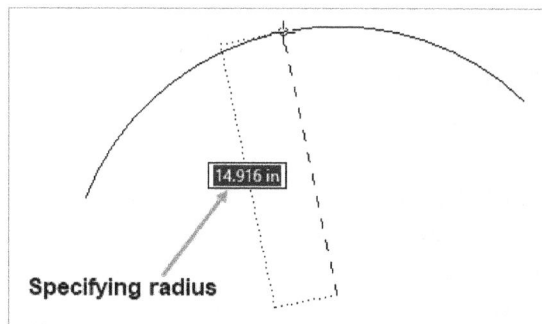

Figure-32. Specifying circumferential point of arc or radius of arc

- Enter desired radius value or click to specify the circumferential point. The arc will be created.

Tangent Arc

As the name suggests, the **Tangent Arc** tool is used to create arc tangent to the selected entity. The procedure to use this tool is given next.

- Click on the **Tangent Arc** tool from **Arc** drop-down in the **Create** panel of the **Ribbon**; refer to Figure-33. You will be asked to select start point of the arc.

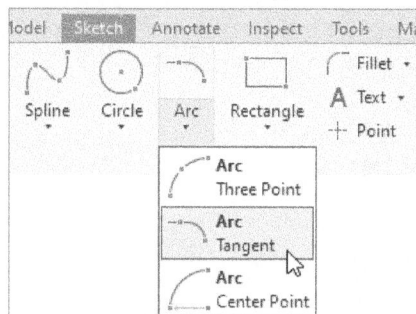

Figure-33. Tangent Arc tool

- Hover the cursor on an entity. The nearest endpoint of the entity will be highlighted in green color.
- Click on the entity when you get desired endpoint. You will be asked to specify the end point of the arc; refer to Figure-34.

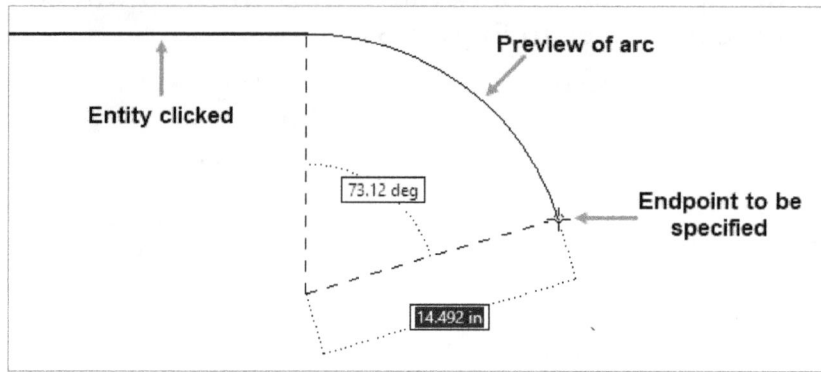

Figure-34. Creating tangent arc

• Click to specify the endpoint of the arc. The arc will be created tangent to the selected entity.

Center Point Arc

The **Center Point Arc** tool is used to create an arc with the help of center point and radius (or circumferential point). If you do not have reference points (for 3 Point Arc) or an entity to make tangent then this is the tool to be used. The procedure to use the **Center Point Arc** tool is given next.

• Click on the **Center Point Arc** tool from **Arc** drop-down in the **Create** panel of the **Ribbon**; refer to Figure-35. You will be asked to specify the center of the arc.

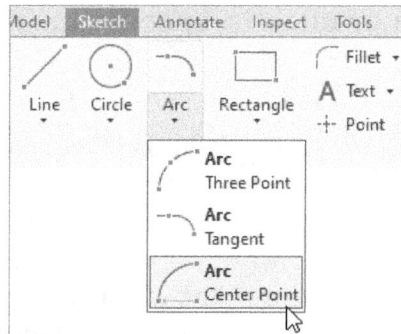

Figure-35. Center Point Arc tool

• Click to specify the center point of the arc. You will be asked to specify the start point of the arc; refer to Figure-36.

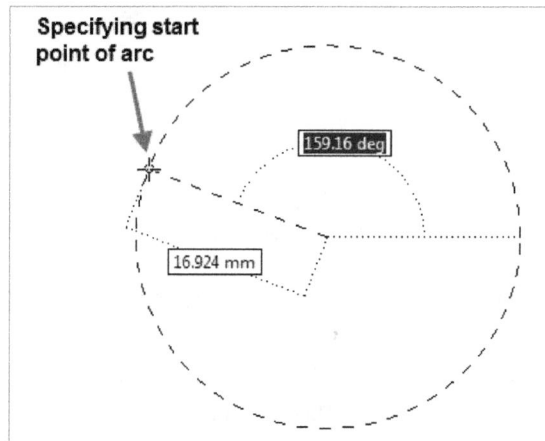

Figure-36. Specifying start point of arc

- Click to specify the start point. You will be asked to specify the span angle of the arc; refer to Figure-37.

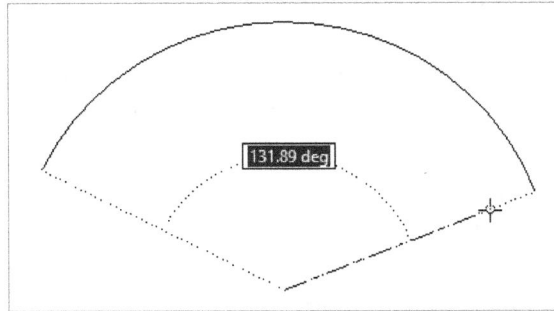

Figure-37. Specifying the angle span value

- Enter the angle span value or click to specify the end point of the arc.

Two Point Rectangle

The **Two Point Rectangle** tool is used to create a rectangle with specified two points. The procedure to use this tool is given next.

- Click on the **Two Point Rectangle** tool from **Create** panel in the **Ribbon**; refer to Figure-38. You will be asked to specify the first corner point of the rectangle.

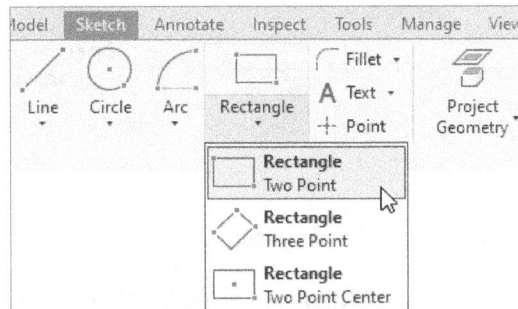

Figure-38. Two Point Rectangle tool

- Click to specify the first corner point. You will be asked to specify the opposite corner point; refer to Figure-39.

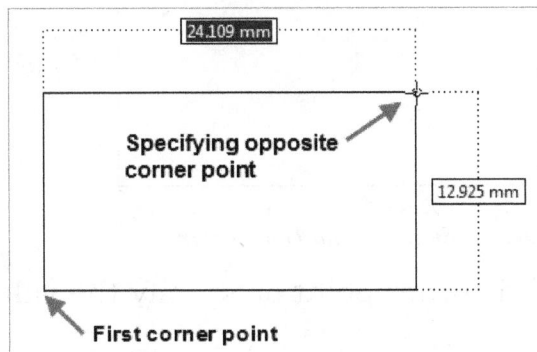

Figure-39. Specifying corner points of rectangle

- Click to specify the other corner point of rectangle. The rectangle will be created. Press **ESC** to exit the tool.

Three Point Rectangle

The **Three Point Rectangle** tool is used to create rectangle by using three points. The procedure to use this tool is given next.

- Click on the **Three Point Rectangle** tool from **Rectangle** drop-down in the **Create** panel of the **Ribbon**; refer to Figure-40. You will be asked to specify a corner point of the rectangle.

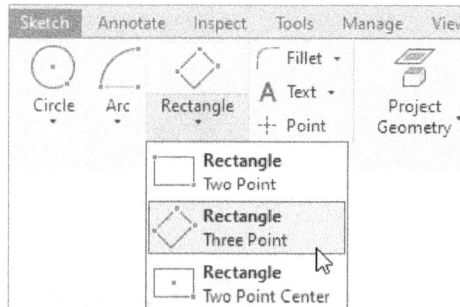

Figure-40. Three Point Rectangle tool

- Click to specify the first corner point. You will be asked to specify second corner point; refer to Figure-41.

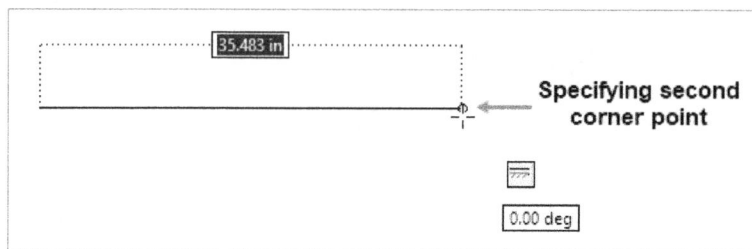

Figure-41. Specifying second corner point

- Click to specify the second corner point. You will be asked to specify the third corner point; refer to Figure-42.

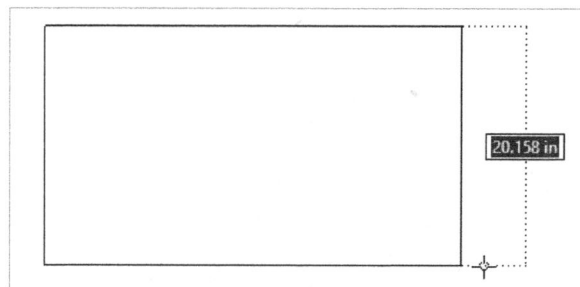

Figure-42. Specifying third corner point

- Click to specify the third corner point or specify the value in the edit box. Press **ESC** to exit the tool.

Two Point Center Rectangle

The **Two Point Center Rectangle** tool is used to create rectangle by specifying center point and corner point of the rectangle. The procedure to use this tool is given next.

- Click on the **Two Point Center Rectangle** tool from **Rectangle** drop-down in the **Create** panel of the **Ribbon**; refer to Figure-43. You will be asked to specify the center point for the rectangle.

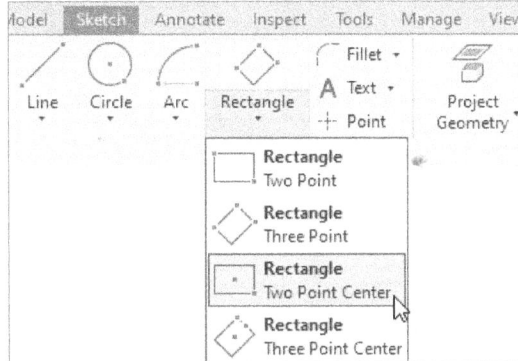

Figure-43. Two Point Center Rectangle tool

- Click to specify the center point for the rectangle. You will be asked to specify one corner point of the rectangle; refer to Figure-44.

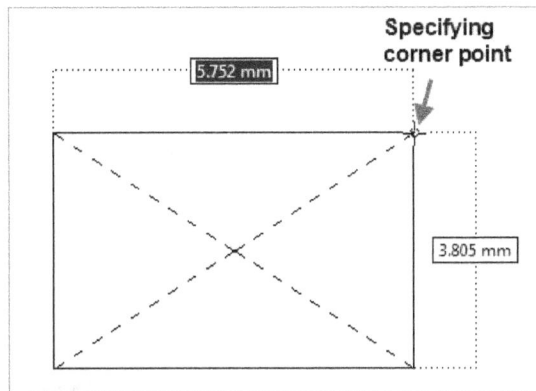

Figure-44. Specifying corner point of center rectangle

- Click to specify the corner point or enter the dimensions of rectangle in the edit boxes. Note that to switch between the edit boxes, you need to press **TAB** from keyboard.

Three Point Center Rectangle

Using the **Three Point Center Rectangle** tool, you can create a rectangle with the help of center point, center line, and corner point. The procedure to use this tool is given next.

- Click on the **Three Point Center Rectangle** tool from **Rectangle** drop-down in the **Create** panel of the **Ribbon**; refer to Figure-45. You will be asked to specify the center point of the rectangle.

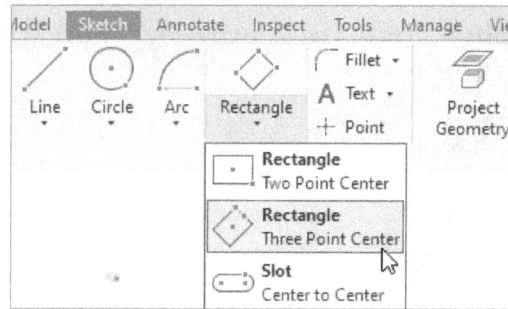

Figure-45. Three Point Center Rectangle tool

- Click to specify the center point of the rectangle. You will be asked to specify second point for the centerline of rectangle; refer to Figure-46.

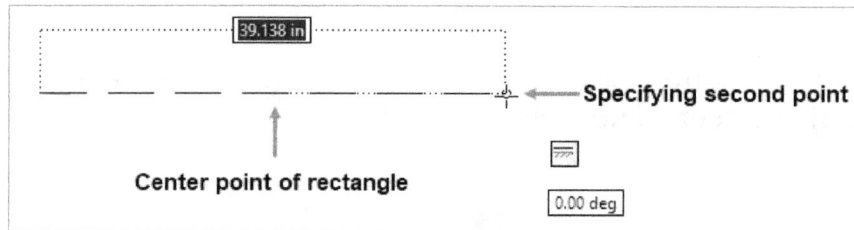

Figure-46. Specifying centerline second point of rectangle

- Click to specify the second point of centerline or enter desired value in the edit box. You will be asked to specify the third point of the rectangle; refer to Figure-47.

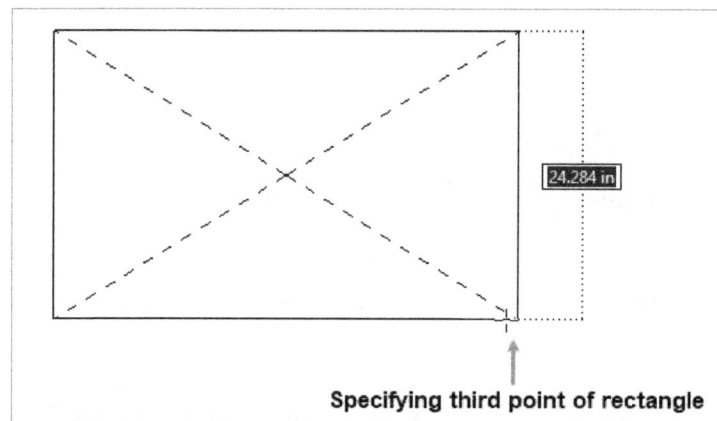

Figure-47. Specifying third point of rectangle

- Click to specify the third point of the rectangle or enter desired dimension in edit box. Press **ESC** to exit the tool.

Center to Center Slot

The **Center to Center Slot** tool is used to create slots in the sketch. In engineering, slots are used to make way for keys, bolts, and other assembly objects. The procedure to use this tool is given next.

- Click on the **Center to Center Slot** tool from **Rectangle** drop-down in the **Create** panel of the **Ribbon**; refer to Figure-48. You will be asked to specify starting center point.

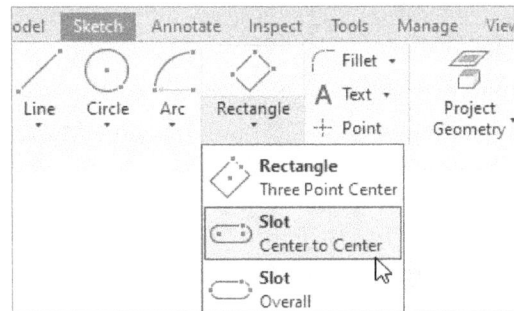
Figure-48. Center to Center Slot tool

- Click to specify the first center point. You will be asked to specify end center point; refer to Figure-49.

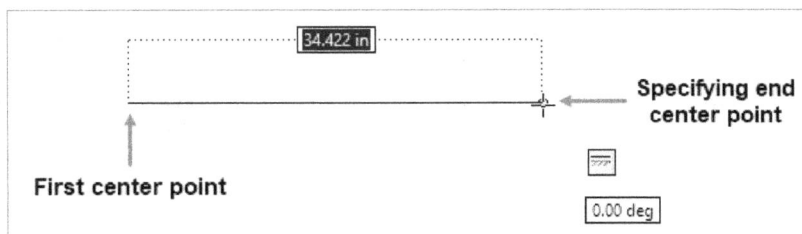
Figure-49. Specifying end center point of slot

- Click to specify the end center point. You will be asked to specify the width of the slot.
- Click at desired location to specify the width of slot or enter desired value in the edit box; refer to Figure-50. The slot will be created.

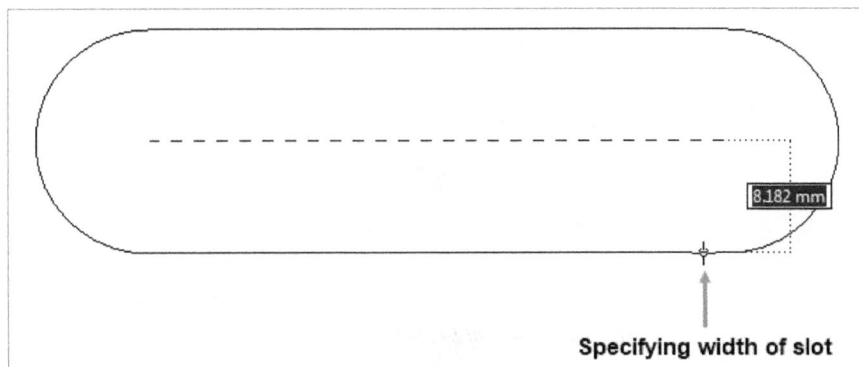
Figure-50. Specifying width of slot

Overall Slot

The **Overall Slot** tool is used to create slot by specifying total length of slot and width of slot. The procedure to use this tool is given next.

- Click on the **Overall Slot** tool from **Rectangle** drop-down in the **Create** panel of the **Ribbon**; refer to Figure-51. You will be asked to specify the start point of the slot.

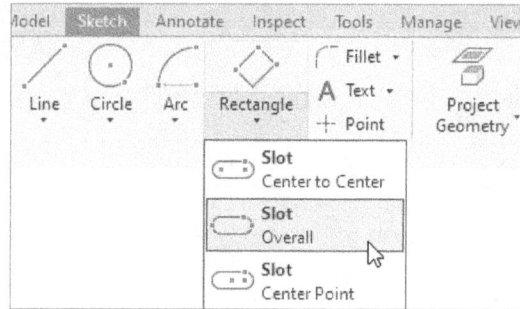

Figure-51. Overall Slot tool

- Click to specify the start point. You will be asked to specify the end point of the slot; refer to Figure-52.

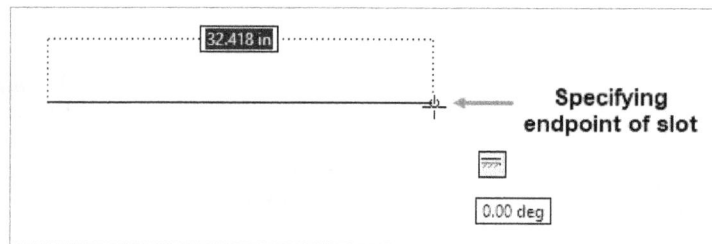

Figure-52. Specifying end point of slot

- Click to specify the end point of the slot. You will be asked to specify the width of the slot.
- Click to specify the width of the slot or enter desired value in the edit box; refer to Figure-53. The slot will be created.

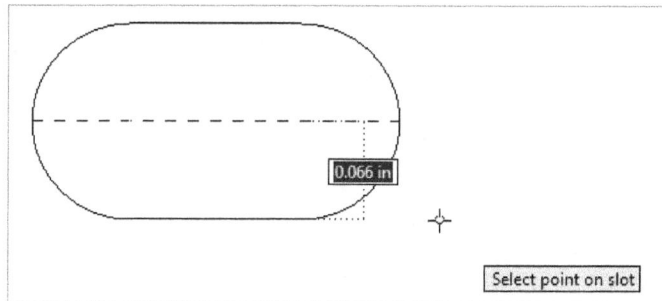

Figure-53. Specifying width of Overall slot

Center Point Slot

The **Center Point Slot** tool is used to create linear slot defined by a center point. The procedure to create slot using this tool is given next.

- Click on the **Center Point Slot** tool from **Rectangle** drop-down in the **Create** panel of the **Ribbon**; refer to Figure-54. You will be asked to specify the center point of the slot.

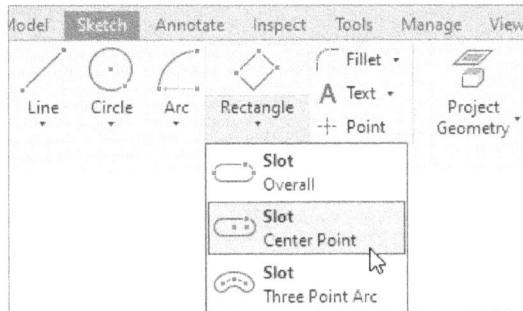

Figure-54. Center Point Slot tool

- Click to specify the center point. You will be asked to specify the second point of the slot; refer to Figure-55.

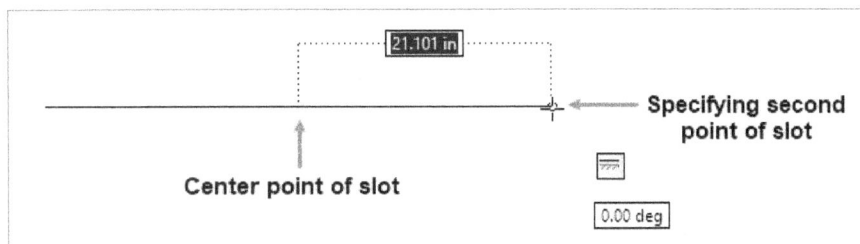

Figure-55. Specifying second point of slot

- Click to specify the second point of the slot. You will be asked to specify the width of the slot.
- Click to specify the width of the slot or enter desired value in the edit box; refer to Figure-56. The slot will be created.

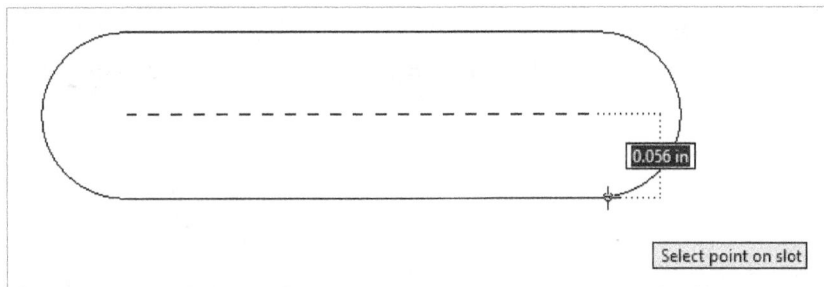

Figure-56. Specifying width of center point slot

Three Point Arc Slot

The **Three Point Arc Slot** tool is used to create slot along the specified arc. The procedure to create slot using this tool is given next.

- Click on the **Three Point Arc Slot** tool from **Rectangle** drop-down in the **Create** panel of the **Ribbon**; refer to Figure-57. You will be asked to specify the start point of the center arc.

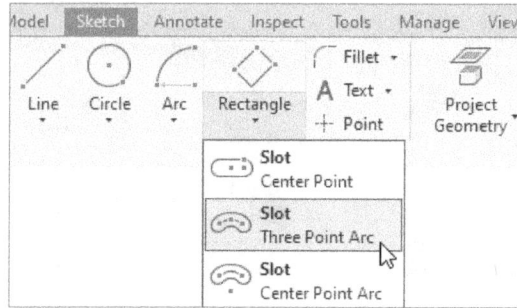

Figure-57. Three Point Arc Slot tool

- Click to specify the start point of the arc. You will be asked to specify the end point of the arc.
- Click to specify the end point of arc or enter desired distance value in the edit box displayed. You will be asked to specify a point on the arc; refer to Figure-58.

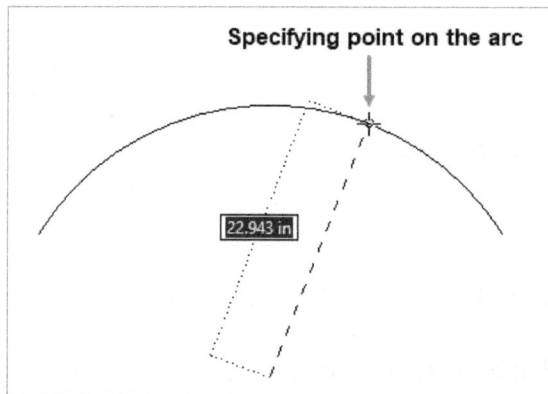

Figure-58. Specifying point on the arc

- Click to specify the arc point or enter desired value in the edit box displayed. You will be asked to specify the width of the slot; refer to Figure-59.

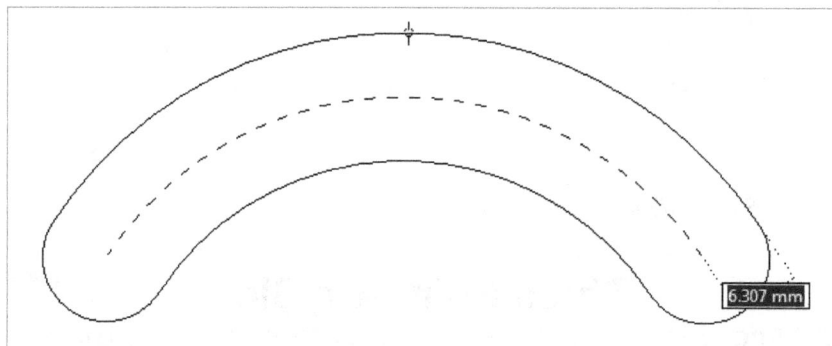

Figure-59. Specifying width of arc slot

- Click at desired location to specify the width or enter desired value of width in the edit box displayed. The slot will be created. Press **ESC** to exit the tool.

Center Point Arc Slot

The **Center Point Arc Slot** tool is used to create slot along the specified arc. The procedure to create slot using this tool is given next.

- Click on the **Center Point Arc Slot** tool from **Rectangle** drop-down in the **Create** panel of the **Ribbon**; refer to Figure-60. You will be asked to specify the center point for the arc.

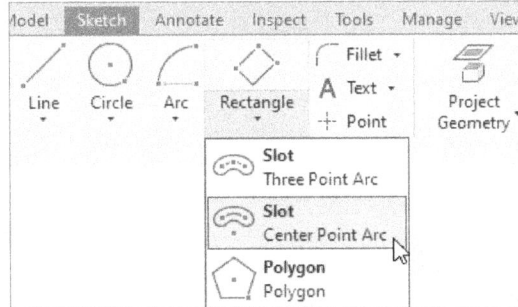

Figure-60. Center Point Arc Slot tool

- Click to specify the center point of the arc. You will be asked to specify the starting point of the arc; refer to Figure-61.

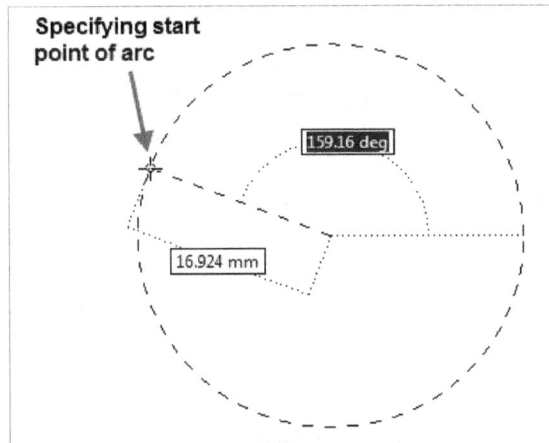

Figure-61. Specifying start point of arc

- Click to specify the starting point. You will be asked to specify end point of the arc; refer to Figure-62.

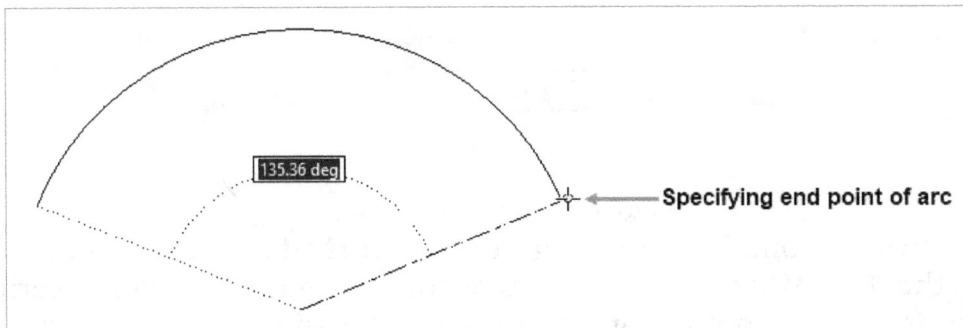

Figure-62. Specifying end point of arc

- Click to specify the end point or enter desired angle value in the edit box displayed. You will be asked to specify the width of slot.
- Enter desired value in the edit box. The slot will be created; refer to Figure-63.

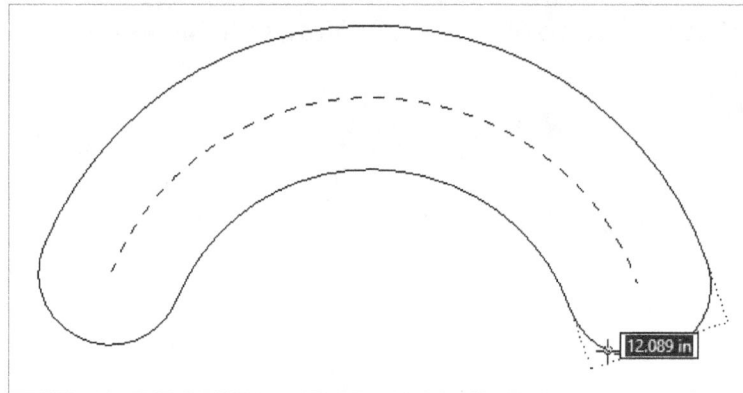

Figure-63. Specifying width of slot

Polygon

As the name suggests, the **Polygon** tool is used to create polygons in the sketch. The procedure to use this tool is given next.

*   Click on the **Polygon** tool from **Rectangle** drop-down in the **Create** panel of the **Ribbon**; refer to Figure-64. The **Polygon** dialog box will be displayed; refer to Figure-65.

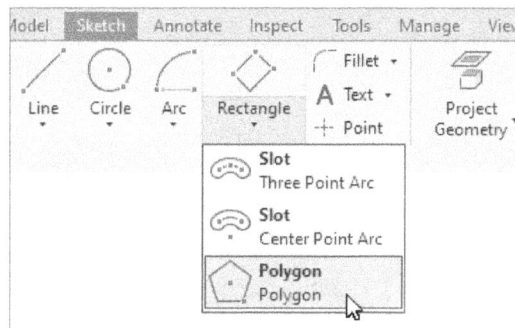

Figure-64. Polygon tool

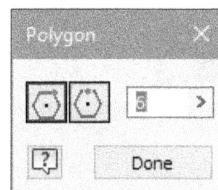

Figure-65. Polygon dialog box

*   There are two buttons in the dialog box. **Inscribed** and **Circumscribed**. Click on the **Inscribed** button if you want to create polygon inscribed in the construction circle and click on the **Circumscribed** button if you want to create the polygon circumscribed to the construction circle.
*   Click in the edit box and specify the number of sides of polygon.
*   Click in the drawing area to specify center of construction circle for polygon. You will be asked to specify a point on the polygon circumference; refer to Figure-66.

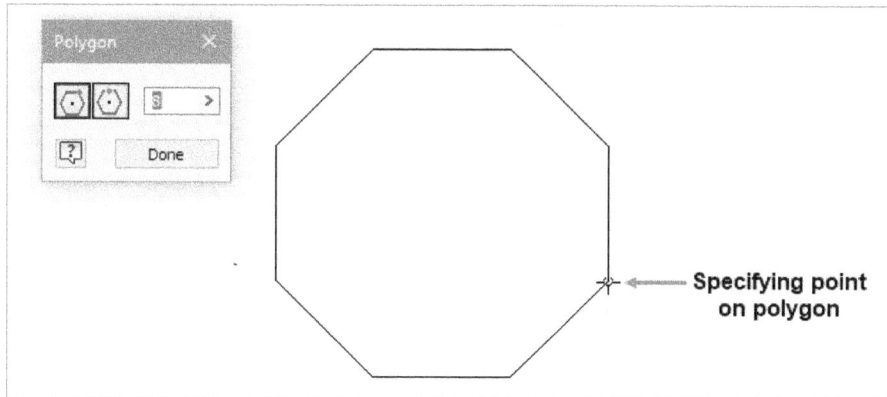

Figure-66. Specifying point on polygon

- Click to specify the point. The polygon will be created. Click on the **Done** button from the dialog box to exit.

Fillet

The **Fillet** tool is used to apply round at the sharp corners. Procedure to use this tool is given next.

- Click on the **Fillet** tool from **Create** panel in the **Ribbon**; refer to Figure-67. The **2D Fillet** dialog box will be displayed as shown in Figure-68.

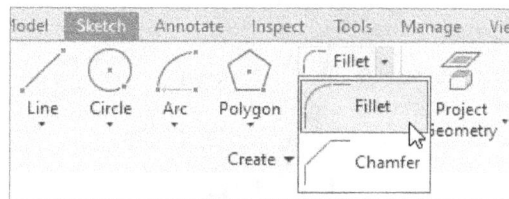

Figure-67. Fillet tool

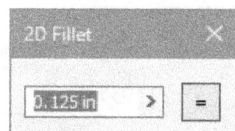

Figure-68. 2D Fillet dialog box

- Specify desired value of radius for fillet in the edit box of dialog box.
- Click on the first edge in sketch and then hover the cursor on other intersecting edge. Preview of the fillet will be displayed; refer to Figure-69.

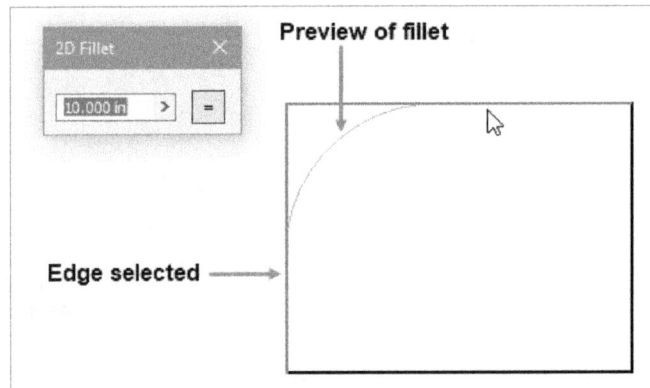

Figure-69. Preview of fillet

- Click on the other edge if preview is as per your requirement. The fillet will be created. Press **ESC** to exit the tool.

Chamfer

The **Chamfer** tool is used to chisel the sharp edges of the model. The procedure to use this tool is given next.

- Click on the **Chamfer** tool from **Fillet** drop-down in the **Create** panel of the **Ribbon**; refer to Figure-70. The **2D Chamfer** dialog box will be displayed; refer to Figure-71.

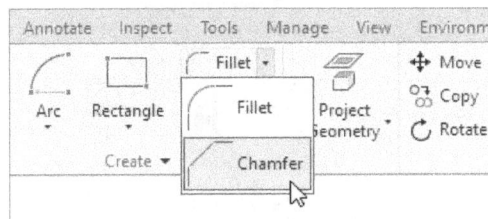

Figure-70. Chamfer tool

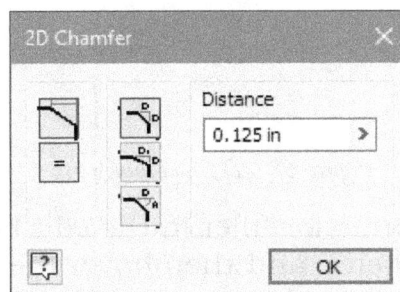

Figure-71. 2D Chamfer dialog box

- There are three buttons for specifying parameters of chamfer, **Distance** , **Distance 1 x Distance 2** , and **Distance x Angle** . If you want to specify equal length of chamfer on both sides to the corner point then select the **Distance** button. If you want to specify different distance on both sides of the corner point then select the **Distance 1 x Distance 2** button. If you want to specify angle and distance for the chamfer then select the **Distance x Angle** button.
- Select desired button from the dialog box and specify the related parameters.
- One by one click on the two intersecting lines. The chamfer will be created; refer to Figure-72. Repeat this step to create more chamfers and press **ESC** to exit the tool.

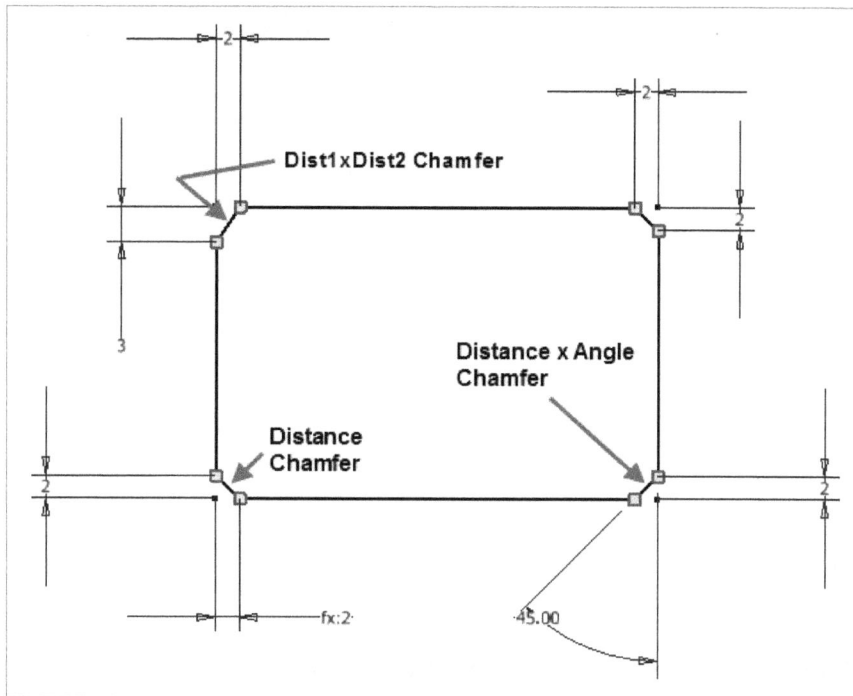

Figure-72. Chamfers in drawing

Text

The **Text** tool is used to write text in the sketch. This text can later be used to engrave marking on the component with the help of CAM software. The procedure to create text is given next.

- Click on the **Text** tool from **Create** panel in the **Ribbon**; refer to Figure-73. You will be asked to create a boundary box by specifying corner points.

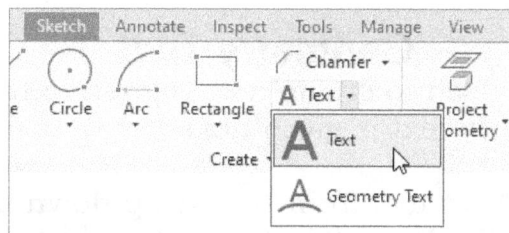

Figure-73. Text tool

- Click and drag the cursor to create boundary box; refer to Figure-74. On creating the box, the **Format Text** dialog box is displayed; refer to Figure-75.

Figure-74. Creating boundary box

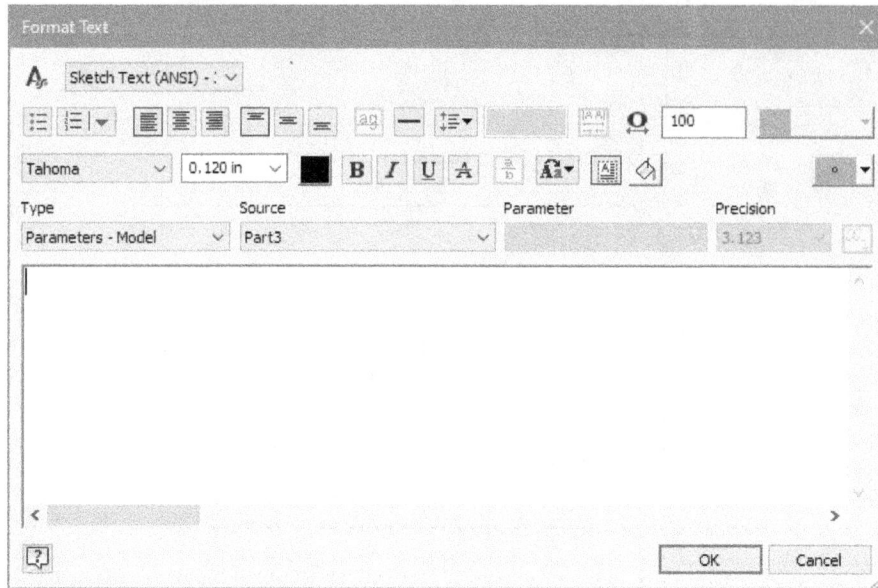

Figure-75. Format Text dialog box

- Type desired text in the input area of the dialog box.
- Set desired format for the text by using buttons in the dialog box like changing color, making text boldface, applying numbering or bullet, and so on.
- Click on the **OK** button from the dialog box. The text will be created; refer to Figure-76.

Figure-76. Text created

Geometry Text

The **Geometry Text** tool is used to create text aligned to the selected geometry. The procedure to create geometry text is given next.

- Click on the **Geometry Text** tool from **Text** drop-down in the **Create** panel of the **Ribbon**; refer to Figure-77. You will be asked to select the geometry to which you want the text to be aligned.

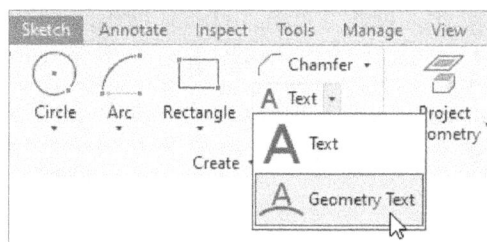

Figure-77. Geometry Text tool

- Click on the geometry; refer to Figure-78. The **Geometry-Text** dialog box will be displayed; refer to Figure-79.

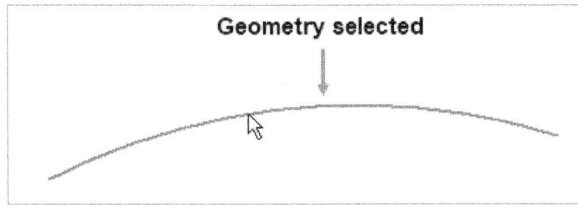

Figure-78. Geometry selected for text

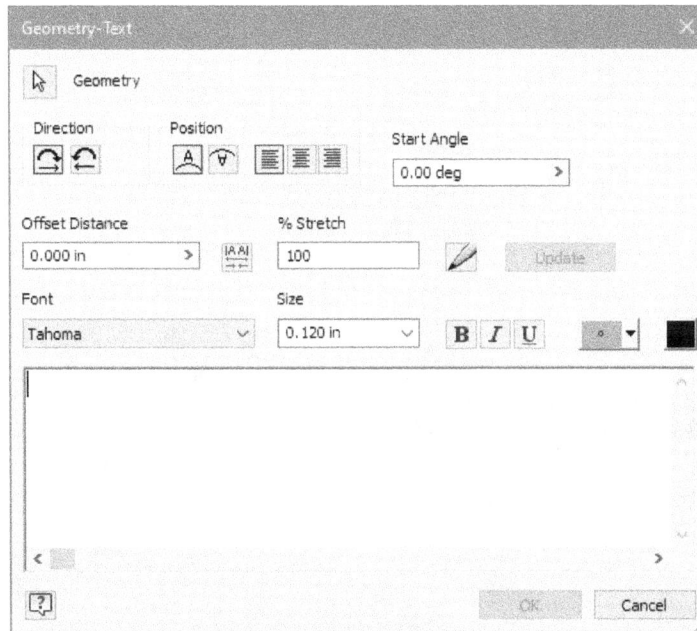

Figure-79. Geometry Text dialog box

- Set the direction and position of the text by using the options in the dialog box.
- Type desired text in the input box and click on the **OK** button from the dialog box. The text will be created; refer to Figure-80.

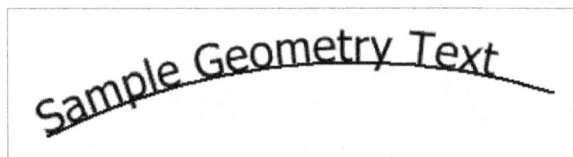

Figure-80. Geometry text created

Point

Point is smallest geometric entity available in any CAD package. Point is mainly used to provide reference for other geometric entities. The procedure to create point is given next.

- Click on the **Point** tool from **Create** panel in the **Ribbon**; refer to Figure-81. You will be asked to specify position of the point; refer to Figure-82.

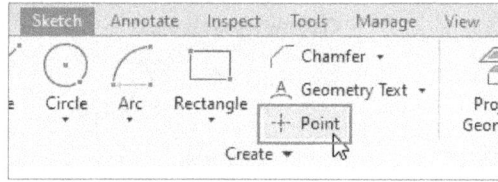

Figure-81. Point tool

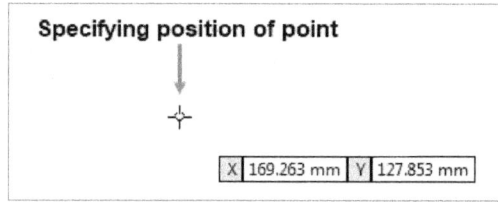

Figure-82. Specifying position of point

- Click to specify the position of point or type desired coordinates for **X** and **Y** directions. Note that you need to press **TAB** to switch between **X** edit box and **Y** edit box.

SKETCH MODIFICATION TOOLS

The sketch modification tools are used to modify the sketch entities. These tools are available in the **Modify** panel of the **Sketch** tab in the **Ribbon**; refer to Figure-83. These tools are discussed next.

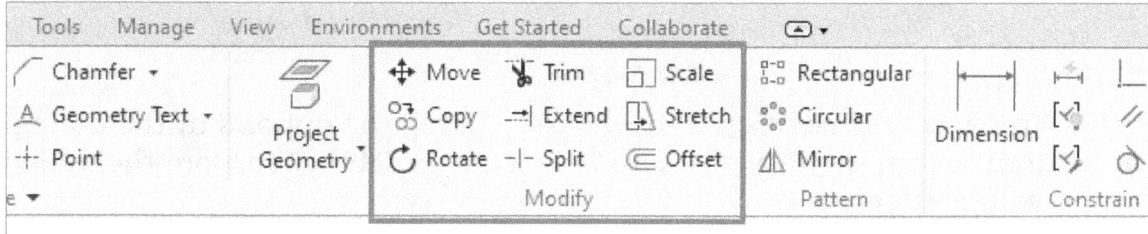

Figure-83. Modify panel

Move tool

The **Move** tool is used to move sketched entities. The procedure to use this tool is given next.

- Click on the **Move** tool from **Modify** panel in the **Ribbon**. The **Move** dialog box will be displayed; refer to Figure-84. Also, you will be asked to select objects.

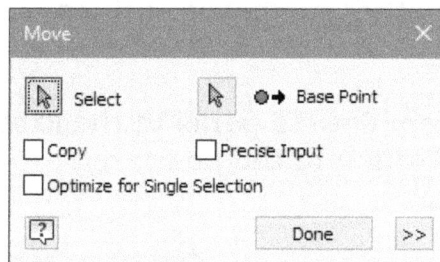

Figure-84. Move dialog box

- Select the object that you want to move; refer to Figure-85. Press **ENTER** from keyboard, you will be asked to select base point for moving the object.

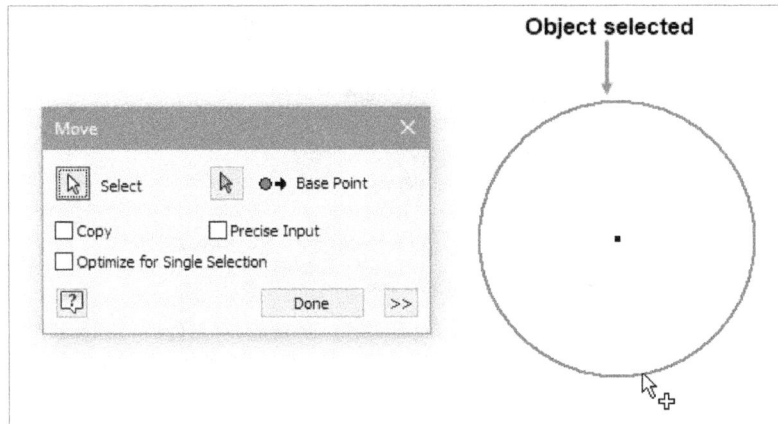

Figure-85. Object selected

- Click on desired location on the object to make it base point. You will be asked to specify the placement position for the object.
- Click to specify the new position. The object will move to the new location as specified; refer to Figure-86.

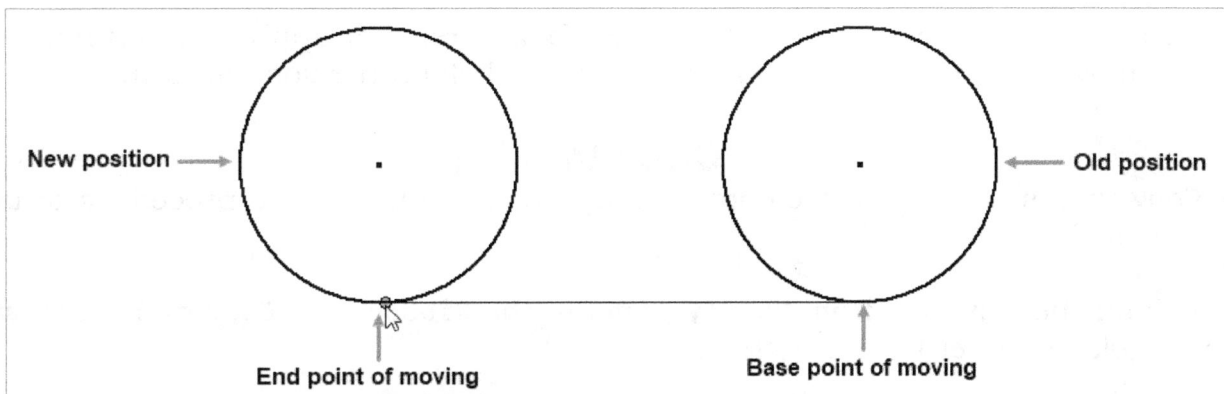

Figure-86. Moving object

- You can create a new copy of the selected object at desired location by selecting the **Copy** check box.
- You can specify the precise positions in the **Inventor Precise Input** box by selecting the **Precise Input** check box.
- Click on the **>>** button from the dialog box to display advanced options of the dialog box; refer to Figure-87.

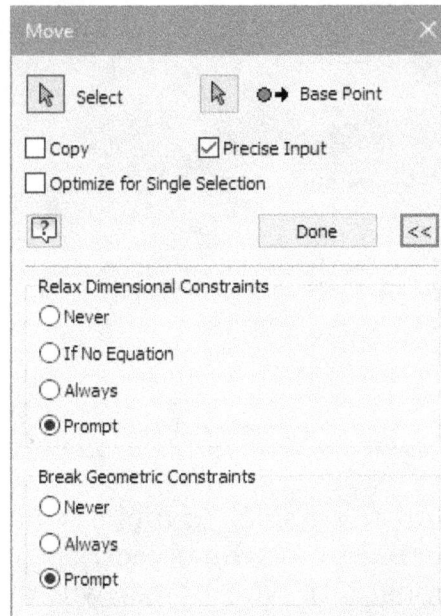

Figure-87. Move dialog box with advanced options

- Select desired radio buttons in the dialog box to relax or apply constraints.
- After moving objects or creating their copies, click on the **Done** button.

Copy tool

The **Copy** tool is used to create copy of the selected entities. The procedure to use this tool is given next.

- Click on the **Copy** tool from **Modify** panel in the **Ribbon**. The **Copy** dialog box will be displayed; refer to Figure-88.

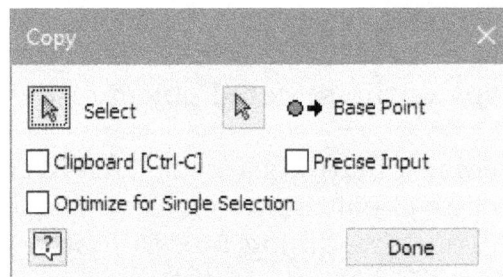

Figure-88. Copy dialog box

- Select the entities that you want to copy. Note that you can select multiple entities to create copy. Press **ENTER** to complete selection of entities. You will be asked to select base point on the entities.
- Click to select the base point. You will be asked to specify the placement location for copied entities.
- Click at desired location. The copies of selected entities will be placed at the specified location. Note that you can create multiple copies by specifying more than one placement points; refer to Figure-89.
- Click on the **Done** button from the dialog box to exit.

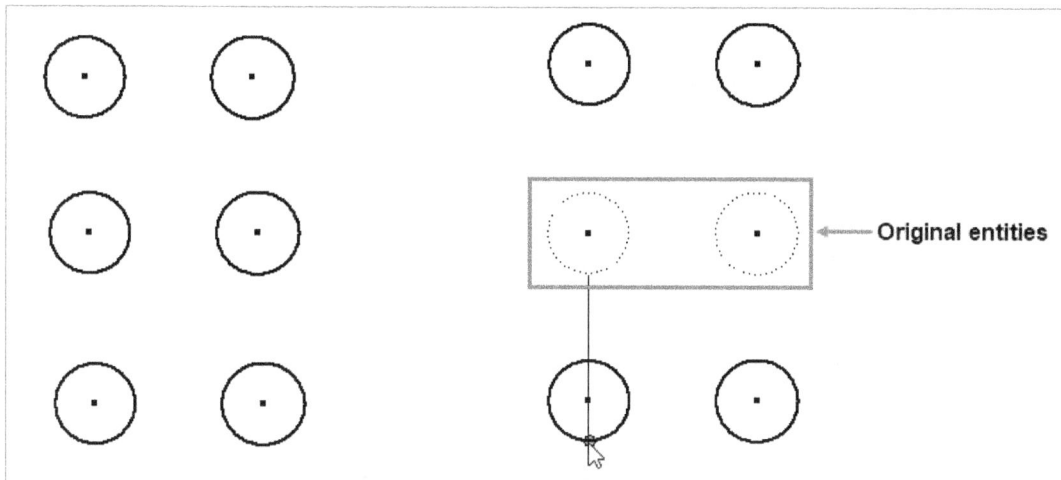

Figure-89. Copies created

Rotate tool

The **Rotate** tool is used to rotate selected entities. Procedure to rotate the entities is given next.

- Click on the **Rotate** tool from **Modify** panel in the **Ribbon**. The **Rotate** dialog box will be displayed; refer to Figure-90.

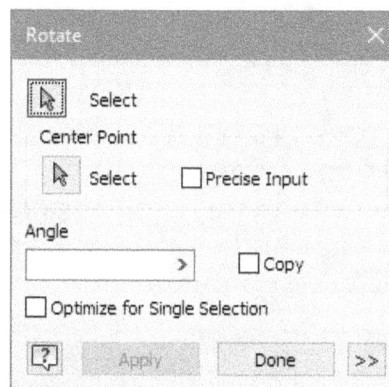

Figure-90. Rotate dialog box

- Select the entities that you want to rotate.

Note that till now we have selected the entities individually. We can also select the entities by windows selection. There are two ways for windows selection; Cross-window selection and Window selection. To select all the entities that intersect with our window is called Cross-window selection. To do so, click in the drawing area and drag the cursor towards left; refer to Figure-91. To make window selection, click in the drawing area and drag the cursor towards right; refer to Figure-92. All the entities that completely come inside the window will be selected.

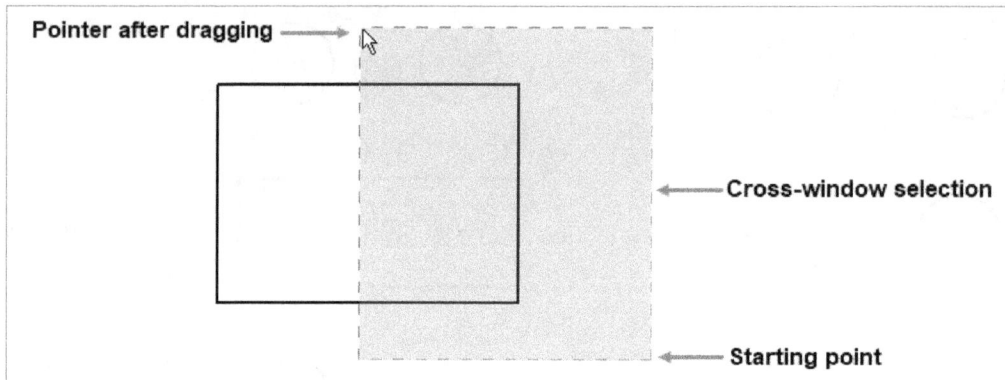

Figure-91. Cross window selection

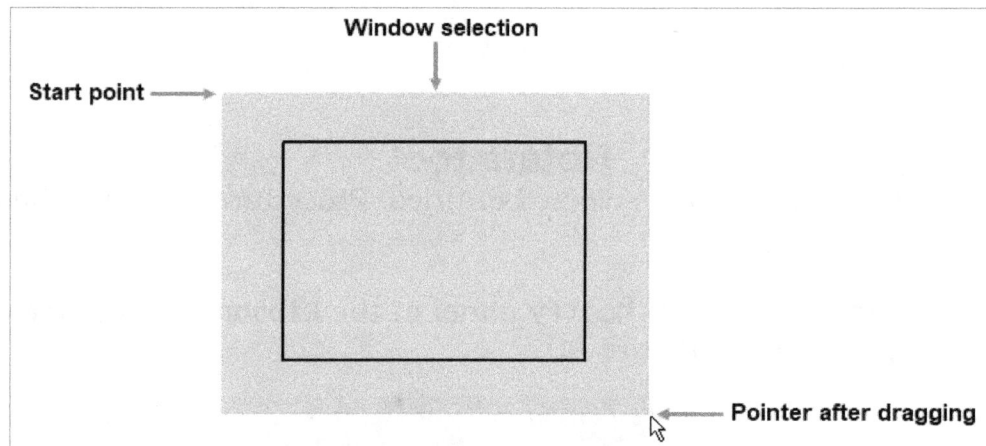

Figure-92. Window selection

- After selecting the entities, press **ENTER** from the keyboard. You will be asked to specify center point for rotation of entities.
- Click at desired location/point to make it pivot point for rotation. If your entities are constrained then a confirmation box will be displayed as shown in Figure-93. Click on **Yes** button to accept the changes.

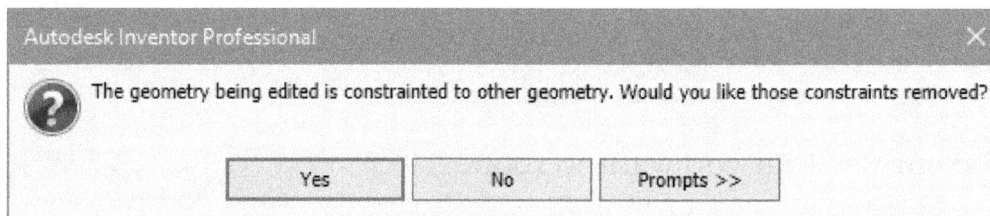

Figure-93. Confirmation box

- You will be prompted to specify the angle value for the rotation; refer to Figure-94. Enter desired angle value in the **Angle** edit box in dialog box and click on the **Apply** button. The entity will be rotated.

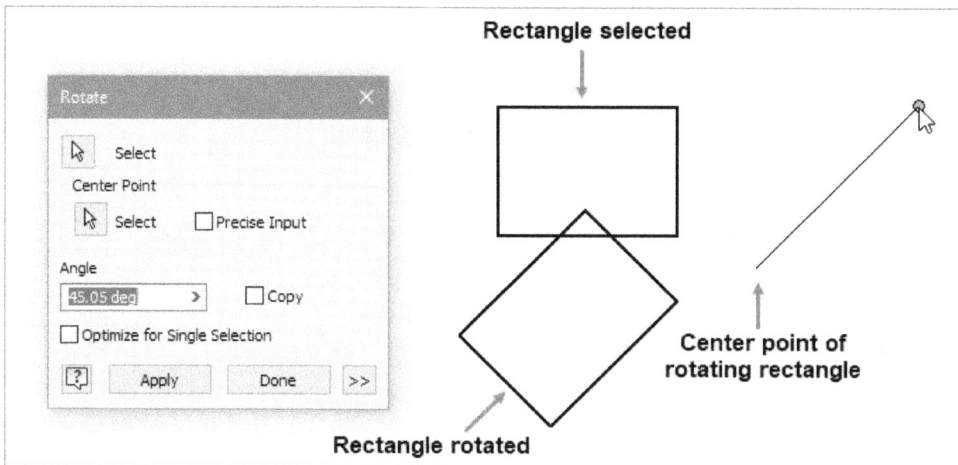

Figure-94. Rotating entities

- You can create a rotated copy of the selected entity by selecting the **Copy** check box. Click on the **Done** button from the dialog box to exit the tool.

Trim tool

As the name suggests, the **Trim** tool is used to remove unwanted section of sketch. The procedure to use this tool is given next.

- Click on the **Trim** tool from **Modify** panel in the **Ribbon**. Cursor is ready with a scissor to remove sketch entity.
- Hover the cursor on the entity that you want to remove. It will be displayed in dashed line type; refer to Figure-95.

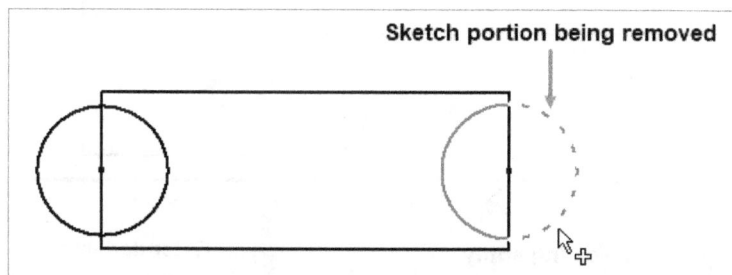

Figure-95. Sketched portion being trimmed

- Click on it to remove the portion. You can keep on selecting the entities until you have removed all the unwanted entities. In place of selecting entities individually, you can remove multiple entities by clicking **LMB** and dragging the cursor over the unwanted entities; refer to Figure-96.

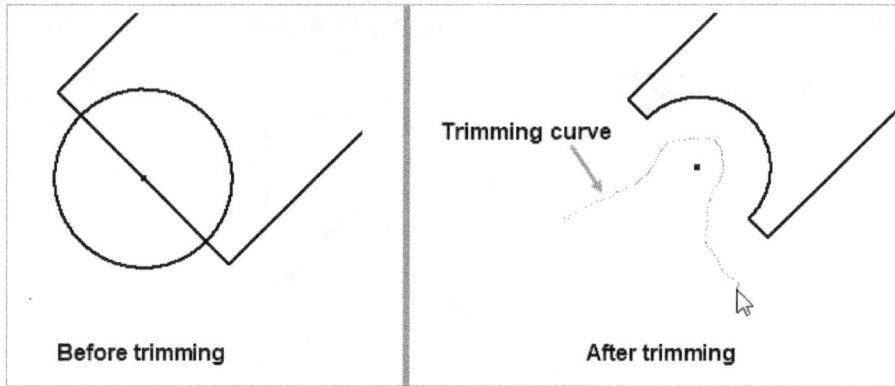

Figure-96. Trimming entities

- Press **ESC** from the keyboard to exit the tool.

Extend tool

The **Extend** tool does reverse of the **Trim** tool. This tool creates link between two entities by extending the selected entity. The procedure to use this tool is given next.

- Click on the **Extend** tool from **Modify** panel in the **Ribbon**. You will be asked to select entities to extend.
- Hover the cursor on the entity. Preview of extension will be displayed; refer to Figure-97.

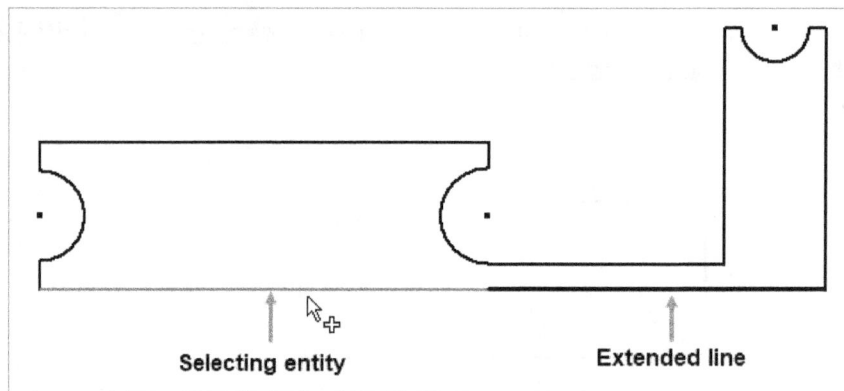

Figure-97. Extending entities

- Click on desired entities to extend them. Press **ESC** to exit the tool.

Split tool

The **Split** tool is used to break the entity into two parts. The procedure to use this tool is given next.

- Click on the **Split** tool from **Modify** panel in the **Ribbon**. You will be asked to select entity to be split.
- Click on the entity near desired snap point from where you want to split the entity; refer to Figure-98. The selected entity will split at the snap point.
- Press **ESC** to exit the tool.

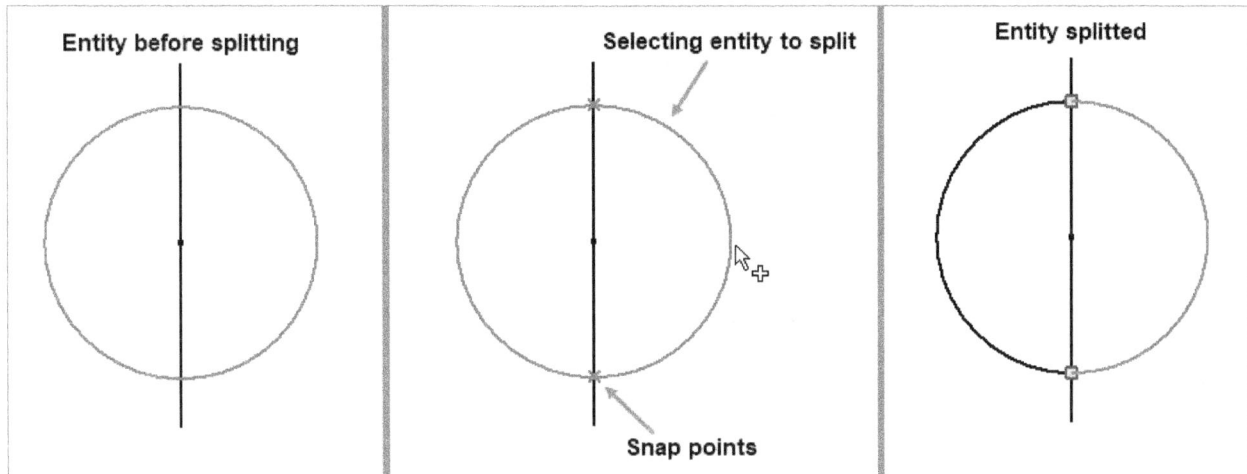

Figure-98. Splitting entity

Scale tool

The **Scale** tool is used to enlarge or diminish the selected entities by same proportion. The procedure to use this tool is given next.

* Click on the **Scale** tool from **Modify** panel in the **Ribbon**. The **Scale** dialog box will be displayed and you will be asked to select the entities; refer to Figure-99.

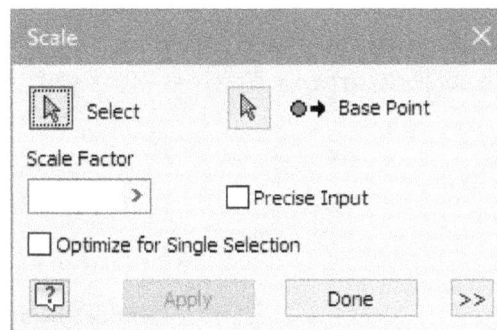

Figure-99. Scale dialog box

* Select the sketch entities that you want to enlarge or diminish and press **ENTER**. You will be asked to specify base point for scaling.
* Click at desired point to make it base point. You will be asked to specify the value of scale factor; refer to Figure-100.

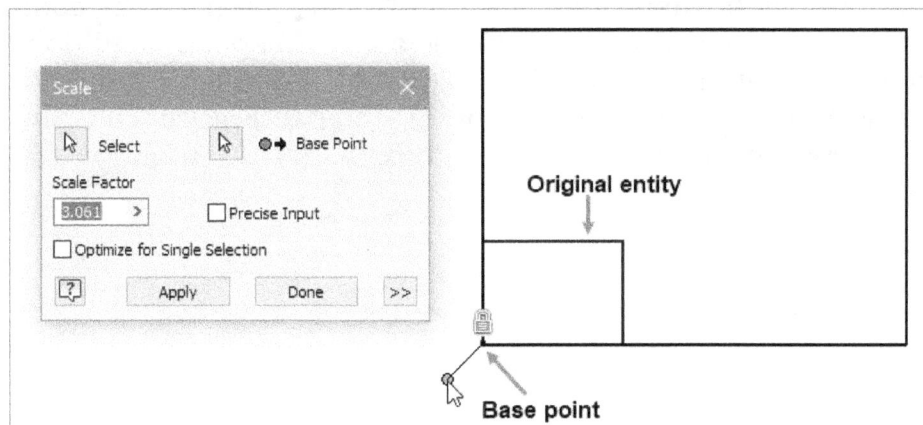

Figure-100. Specifying scale factor

- Specify desired value in the **Scale Factor** edit box in the dialog box. Note that negative values are not applicable for scale factor. Values less than 1 make the selected objects diminish and values greater than 1 enlarge the selected objects.
- Click on the **Apply** button to scale the objects.
- Click on the **Done** button from the dialog box to exit the tool.

Stretch tool

The **Stretch** tool is used to stretch the component by using the selected references. The procedure to use this tool is given next.

- Click on the **Stretch** tool from **Modify** panel in the **Ribbon**. The **Stretch** dialog box will be displayed; refer to Figure-101. Also, you will be asked to select the entities to stretch.

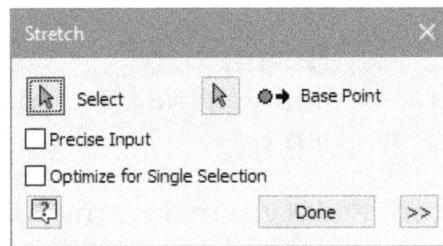

Figure-101. Stretch dialog box

- Select the entities that you want to stretch; refer to Figure-102. Press **ENTER** from the keyboard to complete selection of entities. You will be asked to select the base point.

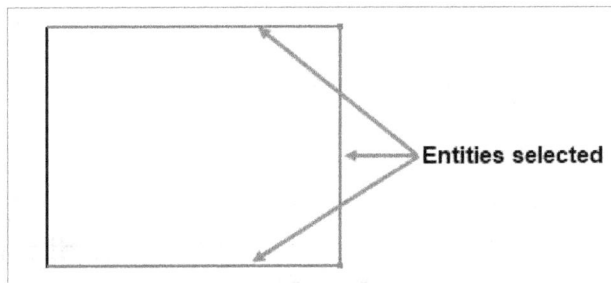

Figure-102. Entities selected for stretching

- Click at desired location to specify base point. If your sketch is constrained then a confirmation box will be displayed. Click on **Yes** button from the confirmation box to allow removal of constraints. You will be asked to specify end point for stretch.
- Click at desired location to specify end point. The object will be stretched by distance between base point and end point; refer to Figure-103.
- Click on the **Done** button from the dialog box to exit the tool.

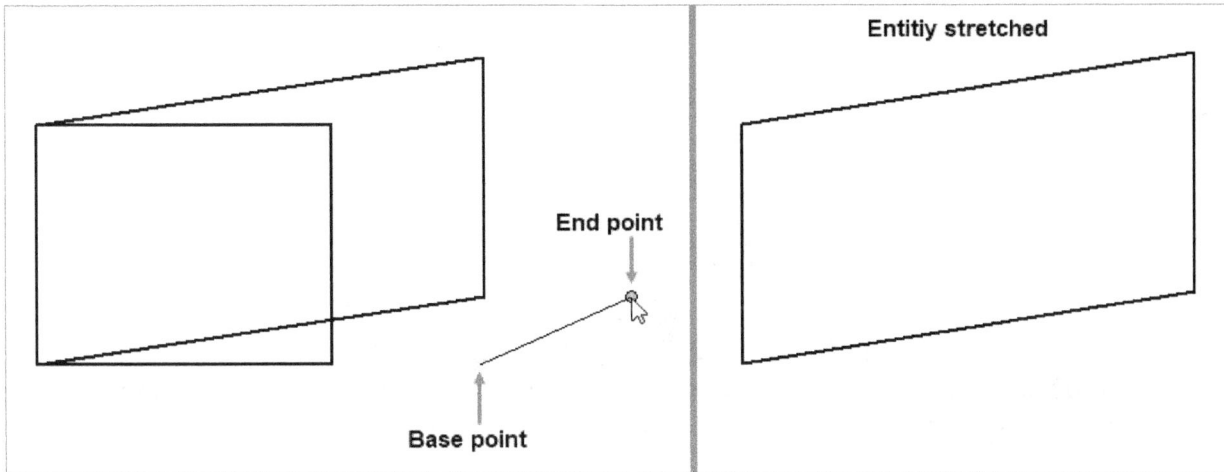

Figure-103. Entitiy stretched

Offset tool

The **Offset** tool is used to create enlarged or diminished copy of the selected entity by specified distance. The procedure to use this tool is given next.

*   Click on the **Offset** tool from **Modify** panel in the **Ribbon**. You will be asked to select the curve/entity to offset.
*   Select the curve/entity. You will be asked to specify the offset distance; refer to Figure-104.

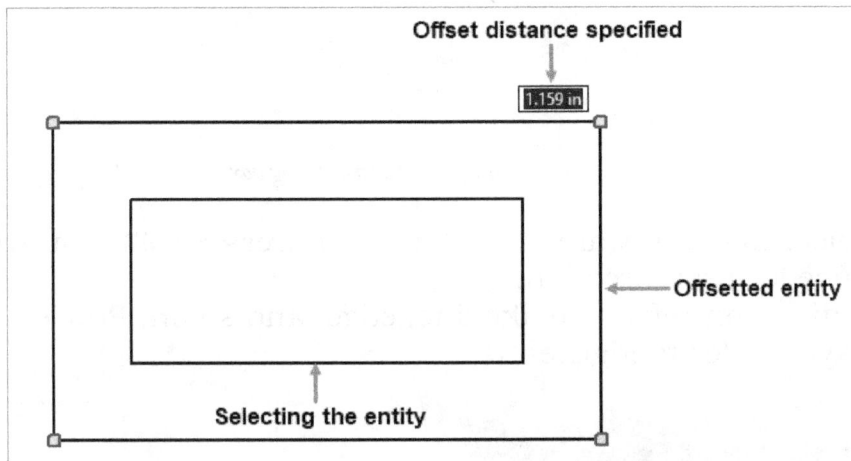

Figure-104. Offset entity created

*   Click to specify the offset distance or enter desired value in the edit box. The offset copy of selected curve will be created. Press **ESC** to exit the tool.

CREATING PATTERNS OF SKETCH ENTITIES

Patterns are arrangement of selected entities in symmetric ways. There are three tools in the **Pattern** panel of the **Ribbon**; refer to Figure-105. These tools are discussed next.

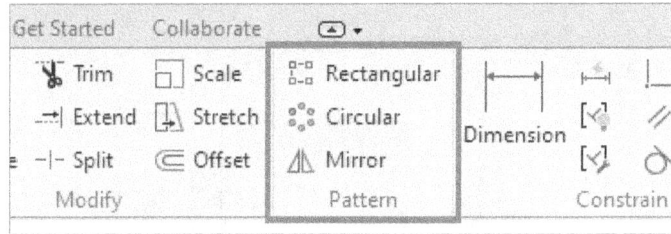

Figure-105. Pattern panel

Rectangular Pattern tool

The **Rectangular Pattern** tool is used to create multiple copies of the selected object in linear symmetry. The procedure to use this tool is given next.

*   Click on the **Rectangular Pattern** tool from **Pattern** panel in the **Ribbon**. The **Rectangular Pattern** dialog box will be displayed; refer to Figure-106. Also, you will be asked to select the entity to be patterned.

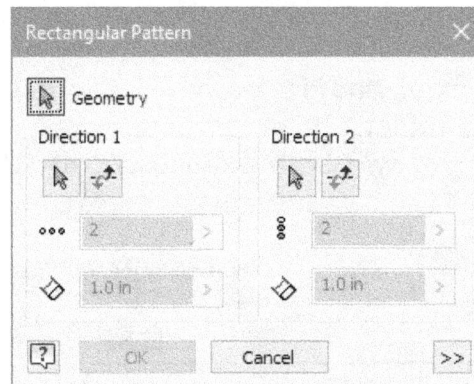

Figure-106. Rectangular Pattern dialog box

*   Select the object that you want to pattern and press **ENTER**. You will be asked to select reference for first direction.
*   Click on the direction reference like line, edge, and so on. Preview of the pattern will be displayed; refer to Figure-107.

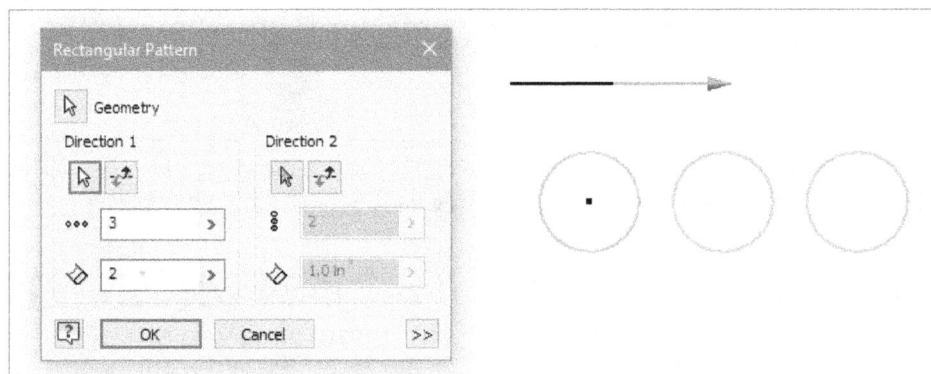

Figure-107. Preview of rectangular pattern

*   Click in the **Count** edit box in the dialog box and specify the number of entities.
*   Click in the **Spacing** edit box and specify the distance between the two consecutive entities of the pattern.

- To create the pattern in the other direction, click on the **Select** button in the **Direction 2** area of the dialog box; refer to Figure-108. You will be asked to select reference for the other direction.

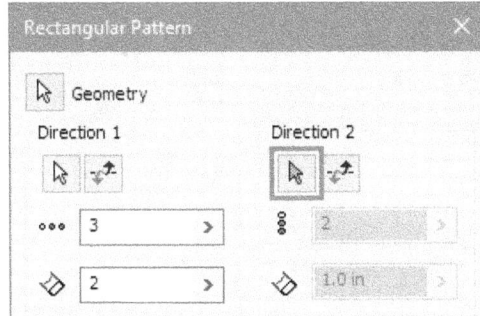

Figure-108. Selection button in Direction 2

- Click on the reference for direction 2, preview of the pattern will be displayed; refer to Figure-109.

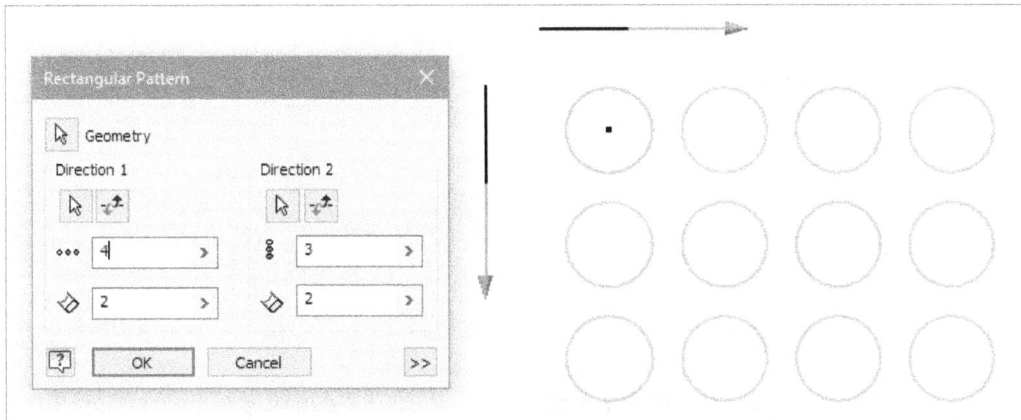

Figure-109. Preview of rectangular pattern in both directions

- Click on the **OK** button from the dialog box. The pattern will be created; refer to Figure-110.

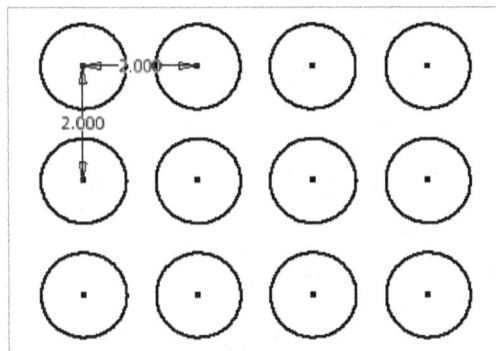

Figure-110. Rectangular pattern created

Circular Pattern

The **Circular Pattern** tool is used to create multiple copies of the selected entity in circular symmetry with respect to the selected axis. The procedure to create circular pattern is given next.

- Click on the **Circular Pattern** tool from **Pattern** panel in the **Ribbon**. The **Circular Pattern** dialog box will be displayed; refer to Figure-111. Also, you will be asked to select the entities for creating pattern.

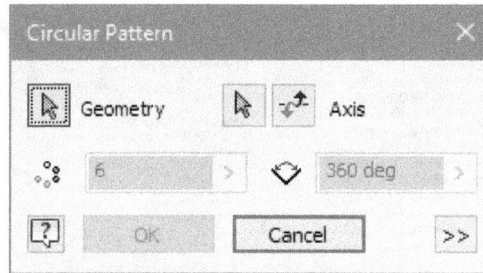

Figure-111. Circular Pattern dialog box

- Select the entity and press **ENTER**. You will be asked to select a reference for axis.
- Click on the reference, preview of the pattern will be displayed; refer to Figure-112.

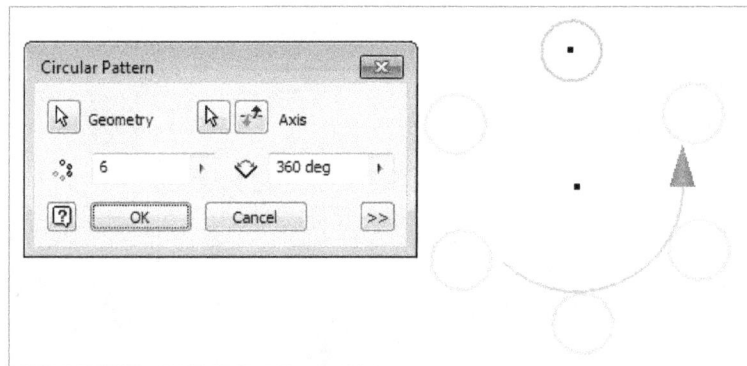

Figure-112. Preview of circular pattern

- Specify the number of entities of pattern in the **Count** edit box and specify the total angle span in the **Angle** edit box. Note that if you clear the **Fitted** check box from the expanded dialog box then you need to specify the angle between two consecutive instances of the object in pattern; refer to Figure-113.

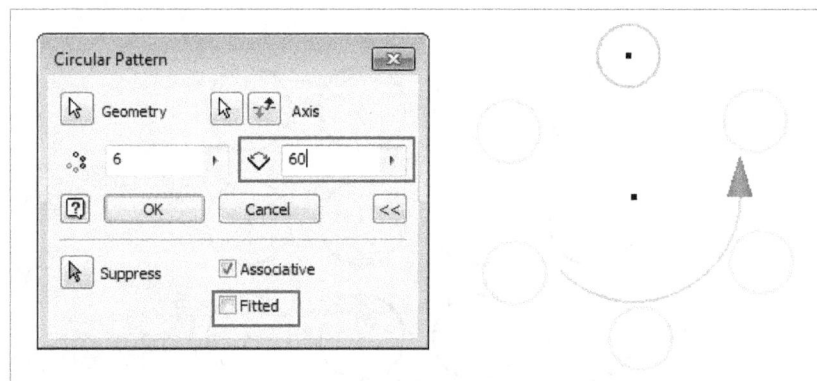

Figure-113. Expanded Circular Pattern dialog box

- You can skip desired instances of pattern by using the **Suppress** button in the expanded dialog box. To do so, click on the **Suppress** button and select the instances from the preview that you do not want to created. The line type of selected entities will be changed to dashed.
- Click on the **OK** button from the dialog box. The circular pattern will be created; refer to Figure-114.

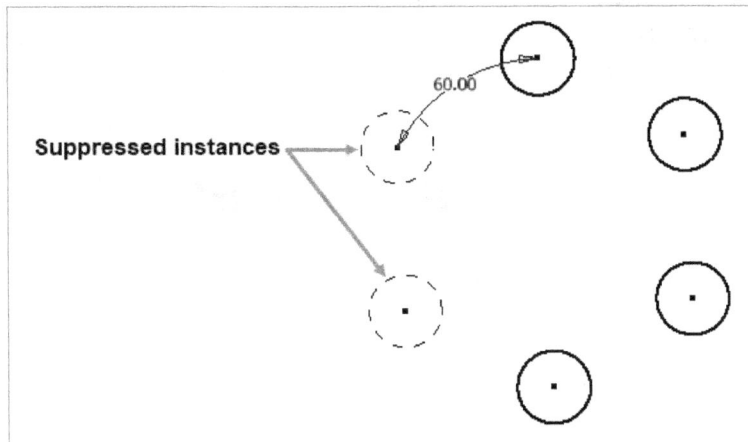

Figure-114. Circular pattern created

Mirror tool

The **Mirror** tool is used to create a mirror copy of the selected entities. The procedure to use this tool is given next.

- Click on the **Mirror** tool from **Pattern** panel in the **Ribbon**. The **Mirror** dialog box will be displayed; refer to Figure-115. Also, you will be asked to select the entities.

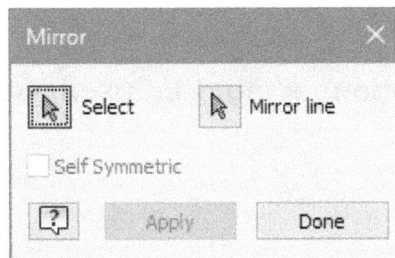

Figure-115. Mirror dialog box

- Select the entity/entities that you want to mirror copy and press **ENTER**. You will be asked to select a mirror line.
- Click on the line that you want to use as mirror line and click on the **Apply** button from the dialog box. The mirror copy will be created; refer to Figure-116.

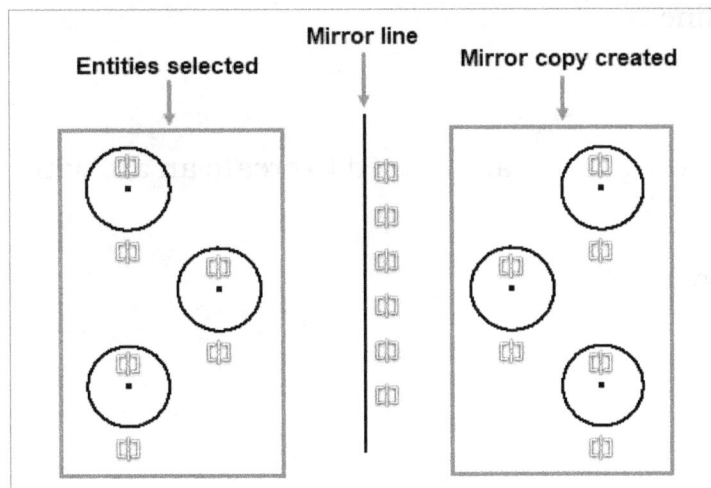

Figure-116. Creating mirror copy

- Click on the **Done** button from the dialog box to exit.

Till this point, we have discussed the tools related to sketching. But, sketches without dimensions and constraints are useless for engineers and designers. In next chapter, we will discuss tools related to dimensions and constraints and then we will work on some practical exercises.

SELF ASSESSMENT

Q1. Which of the following options should be selected first to start sketching environment?

a) Start 2D Sketch
b) Select Plane
c) Finish Sketch
d) Create line

Q2. While creating line, which button should be pressed to toggle between the two input boxes for specifying coordinates?

a) Enter
b) Shift
c) Ctrl
d) Tab

Q3. Which of the following options should be used to create an arc using the **Line** tool?

a) Click and hold the MMB
b) Click and hold the RMB
c) Click and hold the LMB
d) Press the LMB

Q4. By using which of the following options, you can create curve using an equation?

a) Line
b) Interpolation Spline
c) Equation Curve
d) Bridge Curve

Q5. Which of the following tools can be used to create an arc without reference points or an entity?

a) Control Vertex Spline
b) Center Point Arc
c) Tangent Arc
d) Three Point Arc

Q6. Which of the following tools can be used in creating a slot along the specified arc?

a) Overall Slot
b) Center to Center Slot
c) Center Point Slot
d) Center Point Arc Slot

Q7. By the help of which of the following tools, you can apply round at the sharp corners?

a) Chamfer
b) Fillet
c) Polygon
d) Ellipse

Q8. Which of the following is the smallest geometric entity available in any CAD package?

a) Point
b) Text
c) Geometry Text
d) Line

Q9. Which of the following tools is used to enlarge or diminish the entity by same proportion?

a) Extend
b) Stretch
c) Scale
d) Trim

Q10. Which of the following tools can be used to create enlarged or diminished copy of the selected entity?

a) Rectangular Pattern
b) Circular Pattern
c) Offset
d) Mirror

FOR STUDENT NOTES

Chapter 3

Dimensioning and Constraining

Topics Covered

The major topics covered in this chapter are:

- *Introduction to Dimensioning*
- *Dimensioning Tools*
- *Introduction to Constraints*
- *Constrain tools*
- *Formatting tools*
- *Inserting objects in sketch*

INTRODUCTION TO DIMENSIONING

Dimensions are used to specify shape and size of entities. Dimensions in sketch environment should not be confused with dimensions in engineering drawing as they can be represented in same way or in a different way. Purpose of dimensions in sketch is to constrain the size and shape of object whereas purpose of dimensions in drawing is to represent engineering intent of the component for manufacturer. You will learn about drawing dimensions later in the book. Here, we will discuss the use of dimensioning tools in sketch.

DIMENSIONING TOOLS

The tools to apply dimensions are given in **Constrain** panel of the **Sketch** tab; refer to Figure-1. The dimensioning tools in this panel are discussed next.

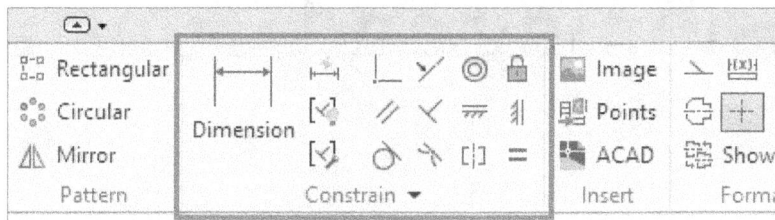

Figure-1. Constrain panel

General Dimension Tool

The **General Dimension** tool is used to apply almost all type of dimensions possible in sketching. Since this single tool is applicable for dimensioning various entities, we will discuss the tool as per its applications.

Dimensioning Line

- Click on the **General Dimension** tool from **Constrain** panel in the **Ribbon**. You will be asked to select entities to be dimensioned.
- Click anywhere in the middle of line, a dimension will get attached to the cursor; refer to Figure-2.

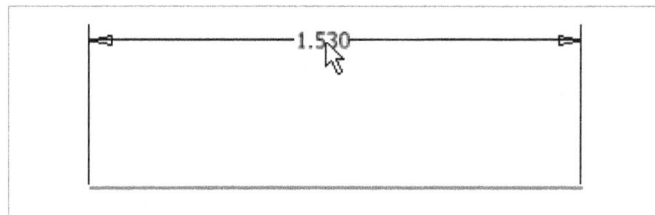

Figure-2. Dimension attached to cursor

- Click at desired location to place the dimension. The **Edit Dimension** dialog box will be displayed; refer to Figure-3.

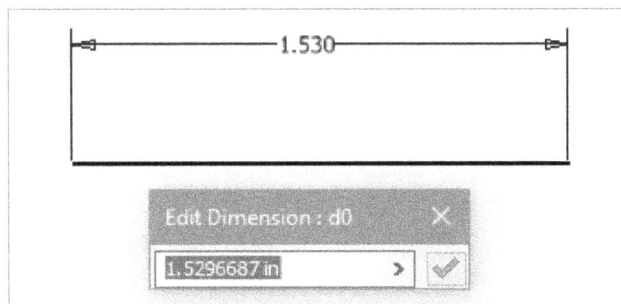

Figure-3. Edit Dimension dialog box

- Specify desired value of dimension and press **ENTER** to apply dimension.

Or

- After clicking on the **General Dimension** tool, one by one click on the end points of the line. Preview of dimension will be displayed; refer to Figure-4.

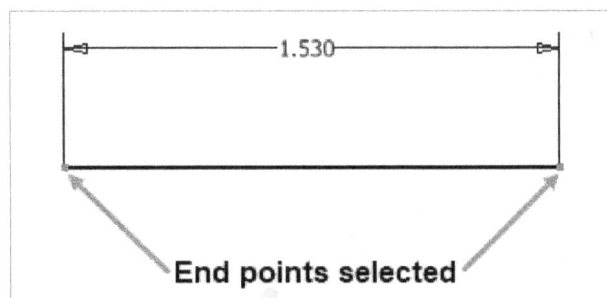

Figure-4. End points of line

- Click to place the dimension and enter desired value in the dialog box displayed.

Aligned Dimension

- After selecting the **General Dimension** tool, click anywhere in the middle of the inclined line. Preview of dimension will be displayed as either horizontal dimension or vertical dimension; refer to Figure-5.

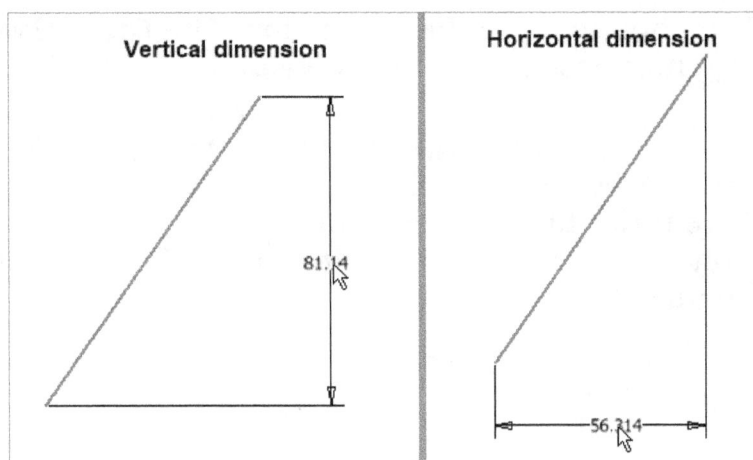

Figure-5. Preview of dimension for inclined line

- Right-click in the drawing area when dimension is attached to the cursor. A shortcut menu will be displayed; refer to Figure-6.

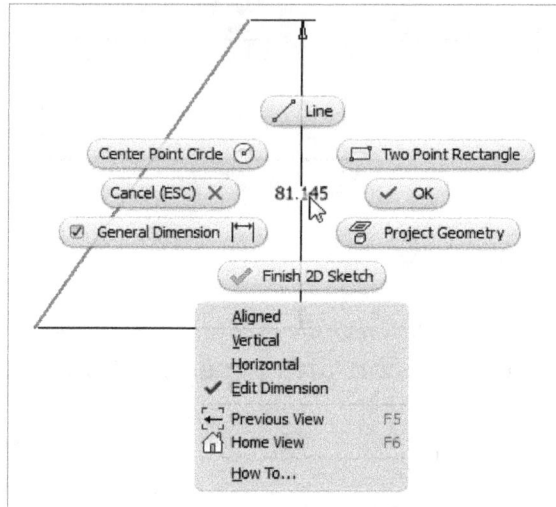

Figure-6. Shortcut menu for dimension

- Click on the **Aligned** option from the shortcut menu. The aligned dimension will get attached to the cursor; refer to Figure-7.

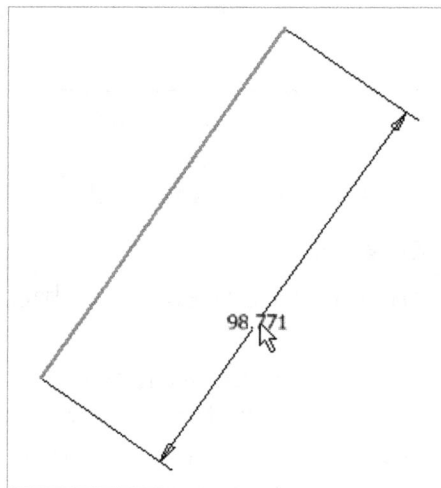

Figure-7. Aligned dimension attached to cursor

- Click at desired location to place the dimension. The **Edit Dimension** dialog box will be displayed. Enter desired dimension value to assign the dimension.

Dimensioning Circle

- Click on the **General Dimension** tool from **Constrain** panel in the **Ribbon**. You will be asked to select entities to be dimensioned.
- Click anywhere on the circle. The diameter dimension will get attached to the cursor; refer to Figure-8.

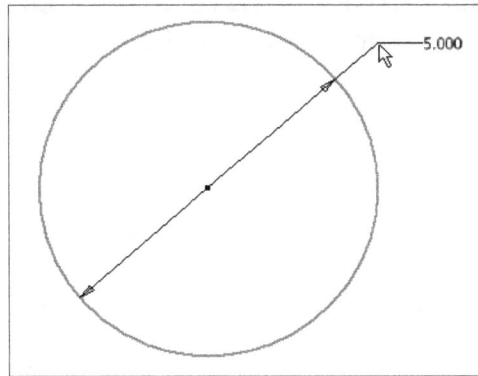
Figure-8. Diameter dimension attached to cursor

- Click at desired location to place the dimension. The **Edit Dimension** dialog box will be displayed; refer to Figure-9.

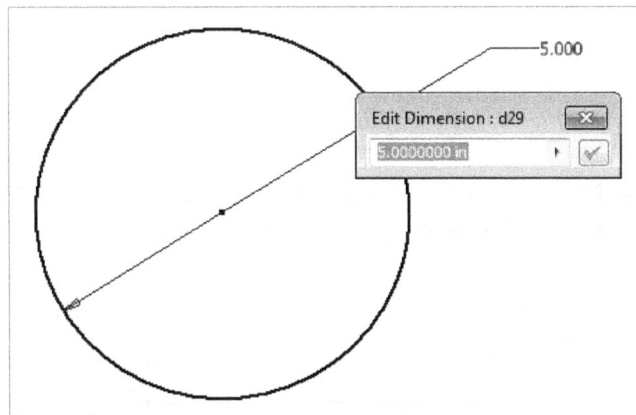
Figure-9. Edit Dimension dialog box for circle

- Enter desired value in the edit box. The dimension will be assigned.

Dimensioning Arc

- Click on the **General Dimension** tool from **Constrain** panel in the **Ribbon**. You will be asked to select entities to be dimensioned.
- Click anywhere on the arc. The radial dimension will get attached to the cursor; refer to Figure-10.

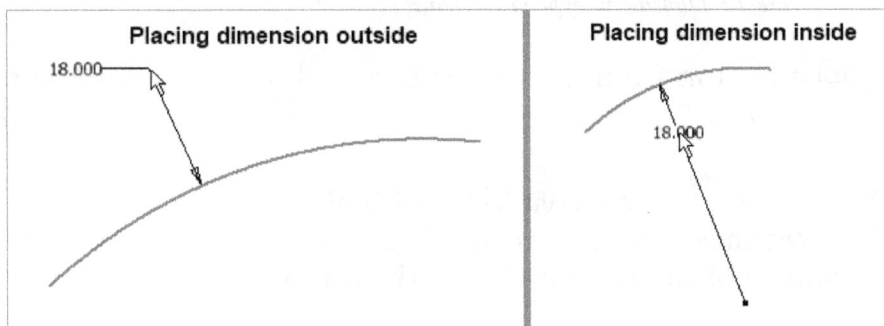
Figure-10. Preview of radial dimension

- Click at desired location to place the dimension. The **Edit Dimension** dialog box will be displayed.
- Enter desired value of dimension. The dimension will be applied to arc.

In the same way, you can assign dimensions to the other sketch entities.

Automatic Dimensions and Constraints

The **Automatic Dimensions and Constraints** tool, as the name suggests, is used to automatically apply dimensions and constraints to all the entities in the sketch. (No! This is not a tool of lazy people, this tool helps to make dimensioning process faster.) The procedure to use this tool is given next.

- Click on the **Automatic Dimensions and Constraints** tool from **Constrain** panel in the **Ribbon**. The **Auto Dimension** dialog box will be displayed; refer to Figure-11. Also, you will be asked to select the entities which are to be dimensioned.

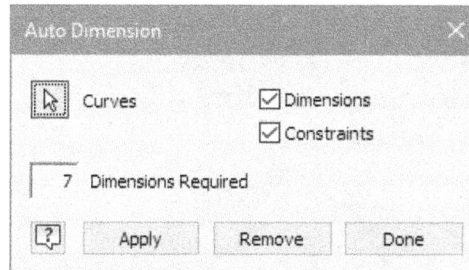

Figure-11. Auto Dimension dialog box

- Select all the entities that you want to dimension and click on the **Apply** button from the dialog box. The dimensions will be applied to the sketch; refer to Figure-12.

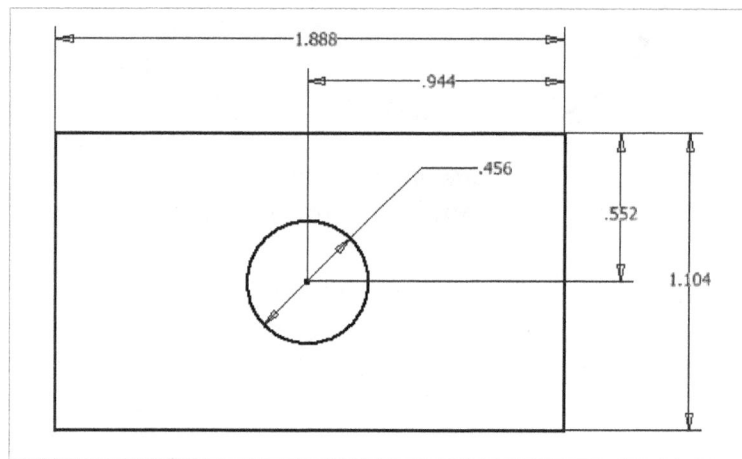

Figure-12. Dimensions applied automatically

- Click on the **Done** button from the dialog box and drag the dimensions to desired locations.

Show Constraints

The **Show Constraints** tool is used to display constraints of the selected sketch entities. The procedure of using this tool is given next.

- Click on the **Show Constraints** tool from **Constrain** panel in the **Ribbon**. You will be asked to select curve/point to display its constraint.
- Click on desired sketch entity. The constraints applied to the selected entity will be displayed; refer to Figure-13.

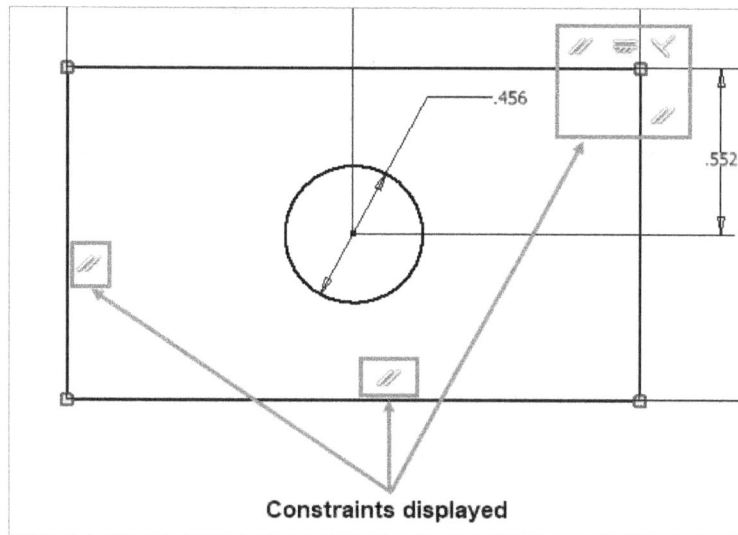

Figure-13. Constraints displayed

Since, we have displayed the constraints and will be using them now onwards, so let's have a brief introduction about them.

CONSTRAINTS

Constraints are used to compel the sketch to retain its shape and position. There are twelve type of constraints available in Autodesk Inventor. These constraints with their respective tools in Inventor are discussed next.

Coincident Constraint

The **Coincident** constraint makes the selected entity coincide with other line, arc, circle, ellipse, point, etc. The procedure to apply this constraint is given next.

- Click on the **Coincident Constraint** button from **Constrain** panel in the **Ribbon**. You will be asked to select first curve/point.
- Select the first curve/point. You will be asked to select second curve/point.
- Select the second curve/point. The two curves/points will become coincident; refer to Figure-14.

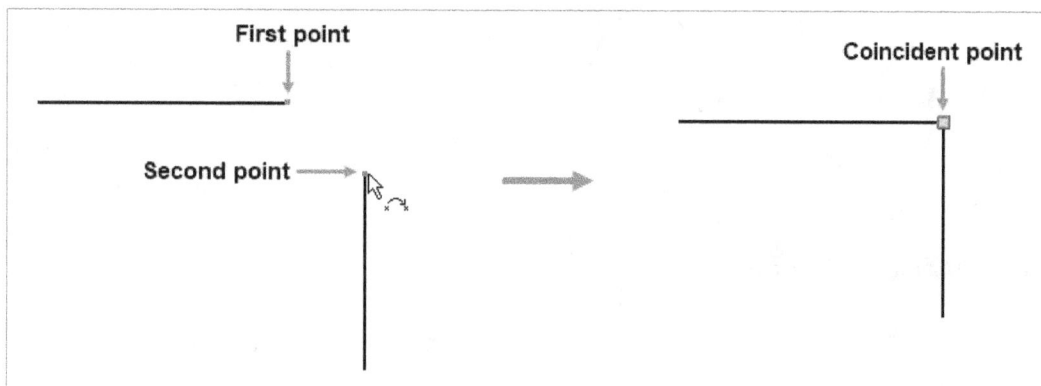

Figure-14. Making points coincident

Collinear Constraint

The **Collinear Constraint** tool is used to make the two or more selected entities collinear. The procedure to use this tool is given next.

- Click on the **Collinear Constraint** tool from **Constrain** panel in the **Ribbon**. You will be asked to select the first line segment or axis.
- Click on the first line segment or axis. You will be asked to select the second line segment or axis.
- Click on the second line segment/axis. The selected entities will become collinear; refer to Figure-15.

Figure-15. Making lines collinear

Concentric Constraint

The **Concentric Constraint** tool is used to make any two of circles, arcs, or ellipses share the same center point. The procedure to use this tool is given next.

- Click on the **Concentric Constraint** tool from **Constrain** panel in the **Ribbon**. You will be asked to select first circle, arc, or ellipse.
- Click on the first circle, arc, or ellipse. You will be asked to select the second circle/arc/ellipse.
- Click on the second circle/arc/ellipse. Both the entities will become concentric; refer to Figure-16.

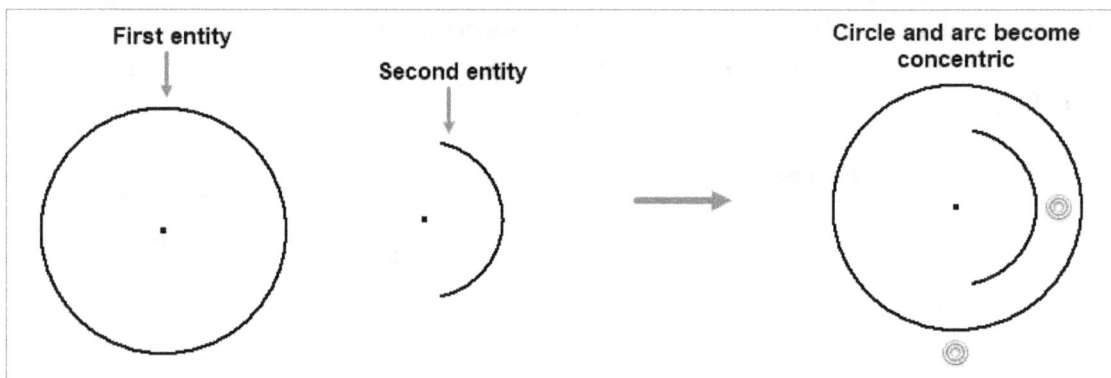

Figure-16. Making entities concentric

Fix Constraint

The **Fix Constraint** tool is used to fix the selected curves/points at their current position with respect to system coordinate system. The procedure to use this constraint is given next.

- Click on the **Fix** tool from **Constrain** panel in the **Ribbon**. You will be asked to select the curve/point which you want to fix.
- Select the curve/point, a lock icon will be attached to the curve/point marking it as fixed; refer to Figure-17.

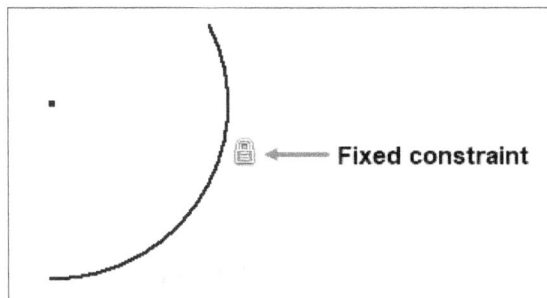

Figure-17. Applying fixed constraint

Parallel Constraint

The **Parallel Constraint** tool is used to make linear geometries parallel to each other. The procedure to make the entities parallel is given next.

- Click on the **Parallel Constraint** tool from **Constrain** panel in the **Ribbon**. You will be asked to select the first line/axis.
- Click on the first line/axis. You will be asked to select the second line/axis.
- Click on the second line/axis. The two lines/axes will become parallel to each other; refer to Figure-18.

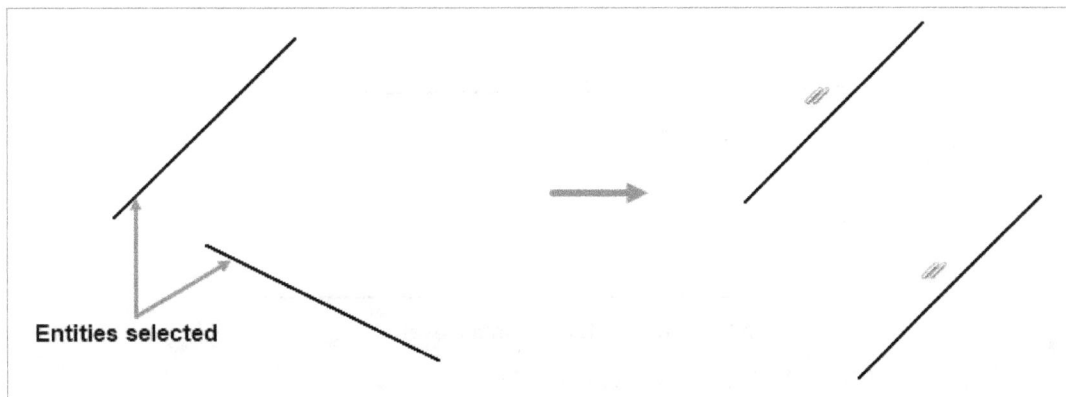

Figure-18. Applying parallel constraint

Perpendicular Constraint

The **Perpendicular Constraint** tool is used to make geometries perpendicular to each other. The procedure to make the entities perpendicular is given next.

- Click on the **Perpendicular Constraint** tool from **Constrain** panel in the **Ribbon**. You will be asked to select the first sketch entity.
- Click on the first sketch entity. You will be asked to select the second sketch entity.
- Click on the second sketch entity. The two entities will become perpendicular to each other; refer to Figure-19. Note that you can make arc, line, circle, ellipse, and spline handles perpendicular to each other.

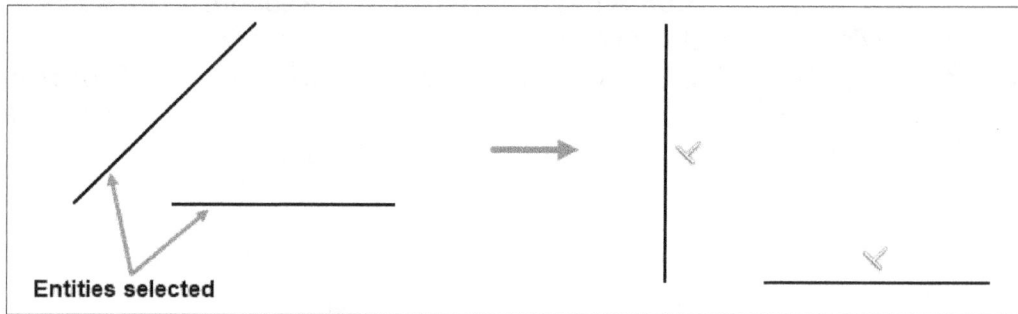

Figure-19. Applying perpendicular constraint

Horizontal Constraint

The **Horizontal Constraint** tool is used to make geometries horizontal or two points at the same horizontal level. The procedure to apply Horizontal constraint is given next.

- Click on the **Horizontal Constraint** tool from **Constrain** panel in the **Ribbon**. You will be asked to select the line/axis or point of first entity.
- Click on the line/axis. It will become horizontal.
- To make the two points at same level, click on the first point and then the second point. Both the points will become at same level; refer to Figure-20.

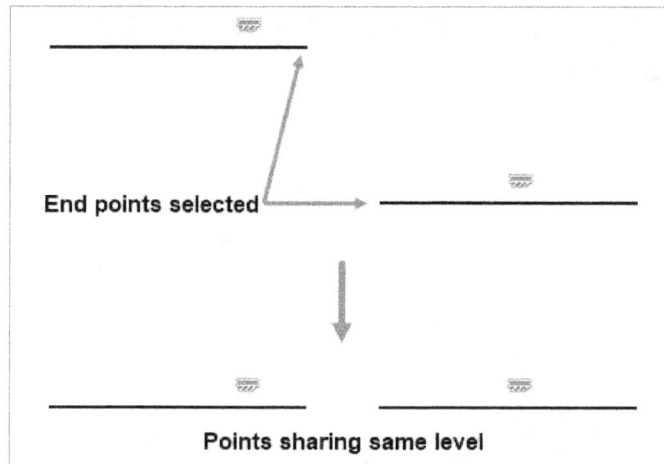

Figure-20. Making points at same level

Vertical Constraint

The **Vertical Constraint** tool is used to make geometries vertical or two points at the same vertical level. The procedure to apply Vertical constraint is given next.

- Click on the **Vertical Constraint** tool from **Constrain** panel in the **Ribbon**. You will be asked to select the line/axis or point of first entity.
- Click on the line/axis. It will become vertical.
- To make the two points at same level, click on the first point and then the second point. Both the points will become at same level; refer to Figure-21.

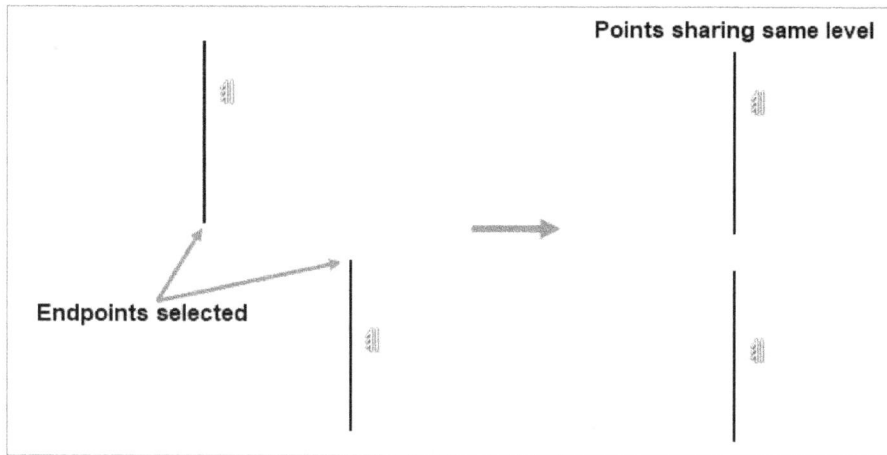

Figure-21. Making points at same level

Tangent Constraint

The **Tangent Constraint** tool is used to make selected curves tangent to the other curve. The procedure to apply this constraint is given next.

- Click on the **Tangent** tool from **Constrain** panel in the **Ribbon**. You will be asked to select the first curve.
- Click on the first curve. You will be asked to select the second curve.
- Click on the second curve. The curves will become tangent to each other; refer to Figure-22.

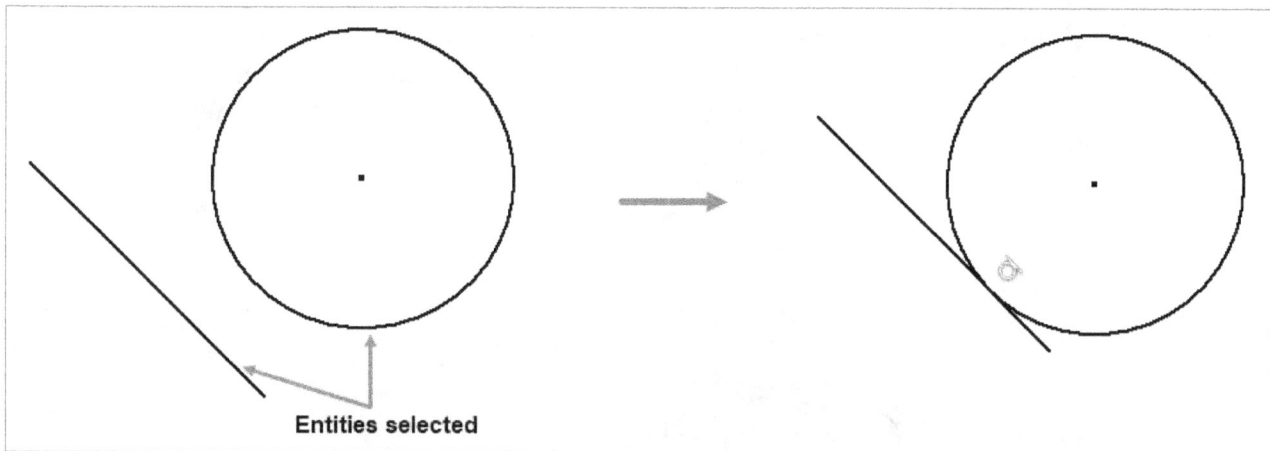

Figure-22. Applying tangent constraint

Smooth (G2) Constraint

The **Smooth (G2) Constraint** tool is used to smoothen the spline by applying curvature continuous condition (G2) on it. The procedure to apply this constraint is given next.

- Click on the **Smooth (G2)** tool from **Constrain** panel in the **Ribbon**. You will be asked to select the first curve.
- Select the first curve. You will be asked to select the second curve.
- Select the spline that you want to smoothen. The Smooth constraint will be applied to the spline; refer to Figure-23.

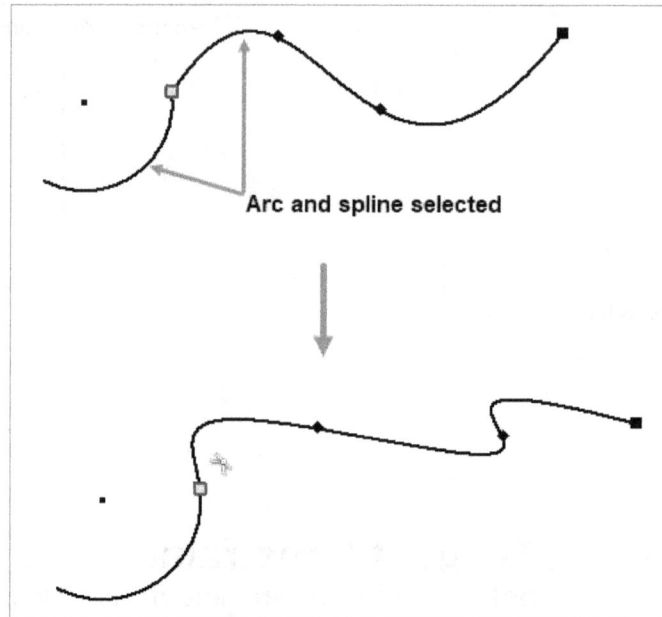

Figure-23. Applying smooth constraint

Symmetric Constraint

The **Symmetric Constraint** is used to make two entities symmetric about a reference line. The procedure to apply this constraint is given next.

- Click on the **Symmetric** tool from **Constrain** panel in the **Ribbon**. You will be asked to select the first curve.
- Select the first curve. You will be asked to select the second curve.
- Select the second curve. You will be asked to select the centerline.
- Click on the centerline. Both the curves will become symmetric with respect to the centerline; refer to Figure-24.

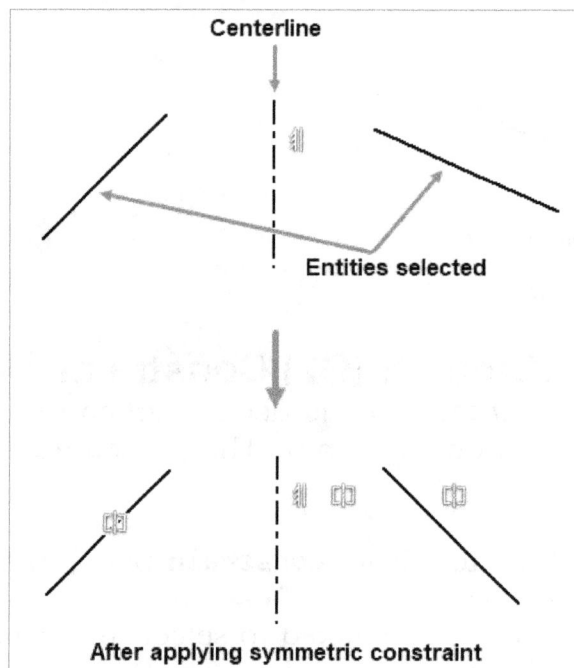

Figure-24. Applying symmetric constraint

Equal Constraint

= The **Equal Constraint** is used to make the two selected entities equal in size. The procedure to apply this constraint is given next.

- Click on the **Equal** tool from **Constrain** panel in the **Ribbon**. You will be asked to select the first curve.
- Select the first curve. You will be asked to select the second curve.
- Select the second curve. Both the curves will become equal in size; refer to Figure-25.

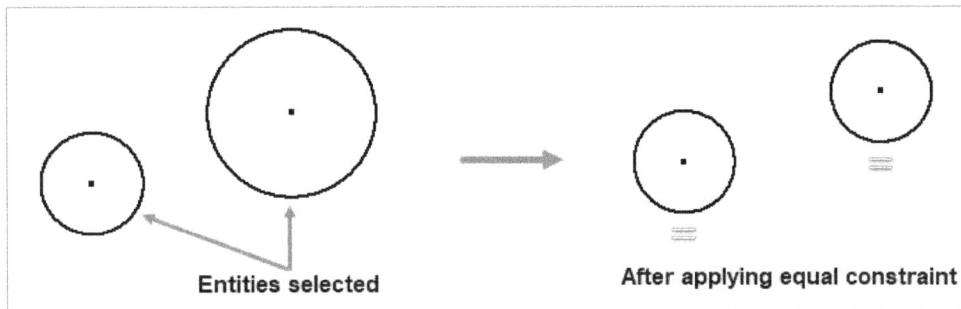

Figure-25. Applying equal constraint

There are a few settings that should be taken care of while applying constraints in sketch. The tool to change these settings is discussed next.

Constraint Settings

The **Constraint Settings** tool is used to change the settings and parameters related to constraint. The procedure to use this tool is given next.

- Click on the **Constraint Settings** tool from **Constrain** panel in the **Ribbon**. The **Constraint Settings** dialog box will be displayed; refer to Figure-26.

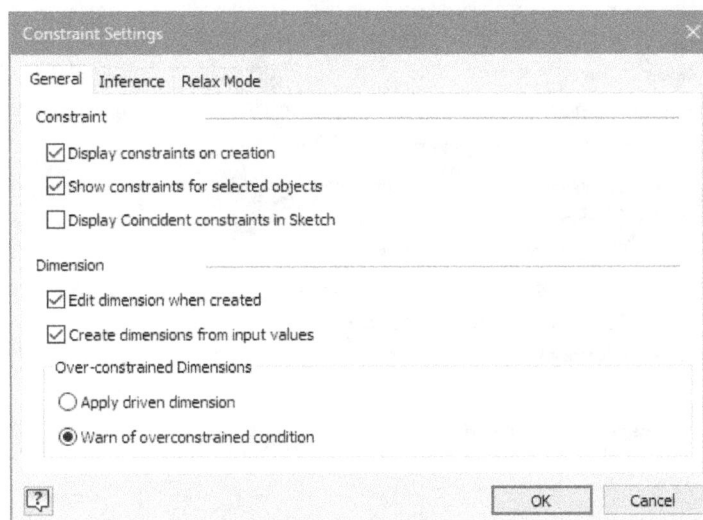

Figure-26. Constraint Settings dialog box

- Select desired check boxes to display respective constraints.
- From the **Dimension** area, select the check boxes to enable editing and assignment of dimension of sketch entity during creation.

- Similarly, select desired radio button from the **Over-constrained Dimensions** area of the dialog box to set solution in case of over-constraining of dimensions.

Inference tab

- Click on the **Inference** tab from the dialog box to display options related to inferencing of constraints (automatic constraining of the entity based on its position and orientation); refer to Figure-27.

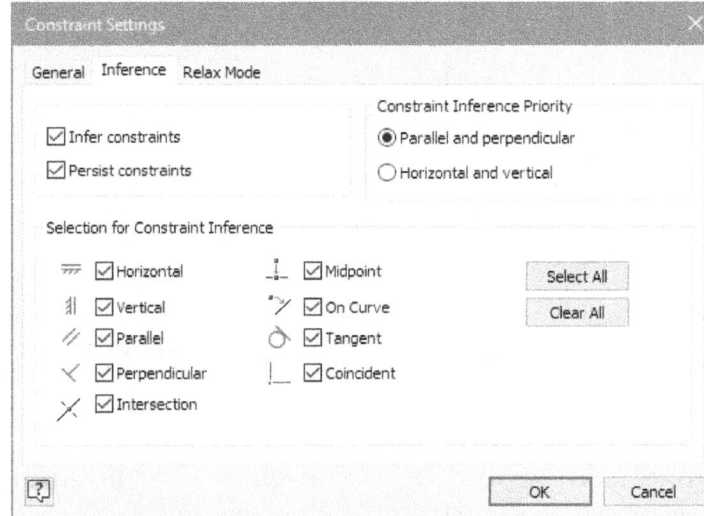

Figure-27. Inference tab in Constraint Settings dialog box

- Select only those check boxes from the dialog box that you want to apply to the sketch entities while creating them.

Relax Mode

- Click on the **Relax Mode** tab from the dialog box to modify settings related to relax mode; refer to Figure-28.

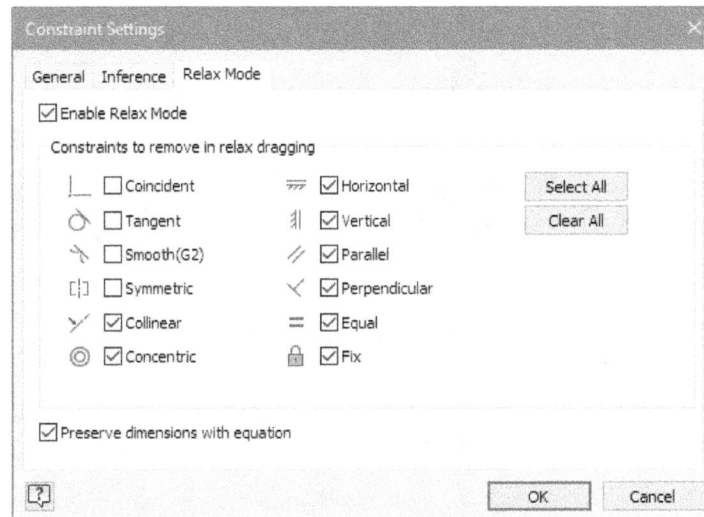

Figure-28. Relax Mode tab in the Constraint Settings dialog box

- Click on the **Enable Relax Mode** check box to enable the relax mode. In relax mode, constraints are automatically removed when you drag the sketch entity by using its key points.
- Select the check boxes against the constraints that you want to be removed in relax mode editing.

- Click on the **OK** button from the dialog box to apply the settings related to constraints.

Concept of Over-constrained, Fully constrained, and Under-constrained

Ah! A new topic, all of a sudden!! Actually, the topic is related to constraining. There are three ways a sketch is constrained; Fully constrained, Over-constrained, and Under-constrained.

Fully Constrained

The Fully constrained sketch is the one in which all the degrees of freedom of each sketch entity are completely defined. Every sketch entity has six degree of freedom denoting rotation and translation in each direction (X direction, Y direction, and Z direction). Once you fix all the degrees of freedom of the entity then it can neither be rotated nor translated in any of the direction.

Under-Constrained

The Under constrained sketch is the one in which some degrees of freedom of the entities are not defined. Which means they are free to move or rotate in directions that are not frozen by constraining their degrees of freedom.

Over-Constrained

The Over constrained sketch is the one in which more constraints are applied on the sketch entity than the required one to fully constrain the sketch.

After understanding the details above, you should always try to fully constrain the sketch, so that it always retains the shape and size which is required as per the drawing or client. Now, question that arises here in some curious minds is, How do I know that my sketch is fully constrained or not? Answer to the question is: check the Bottom Bar of Inventor; refer to Figure-29. In this figure, the sketch is fully dimensioned and constrained, so fully constrained is displayed in the Bottom Bar. If there is any dimension or constrained required then the number of dimensions required to fully define the sketch are displayed in the Bottom Bar; refer to Figure-30.

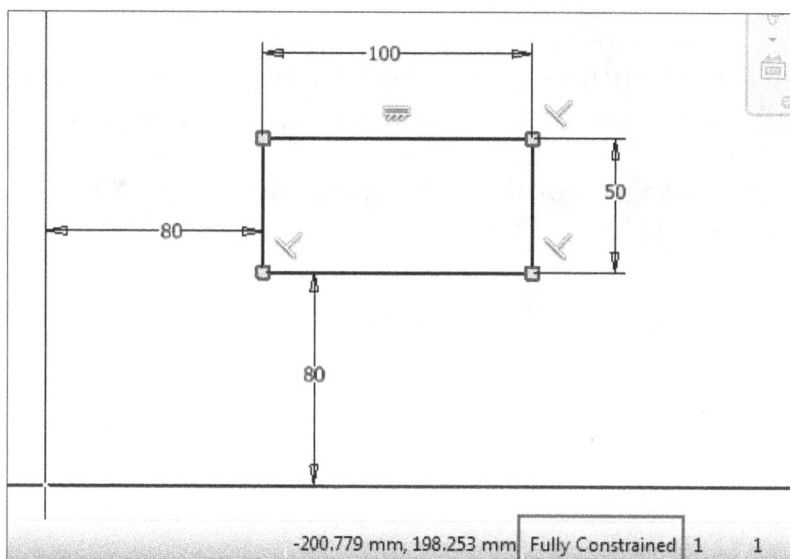

Figure-29. Identifying constrain condition

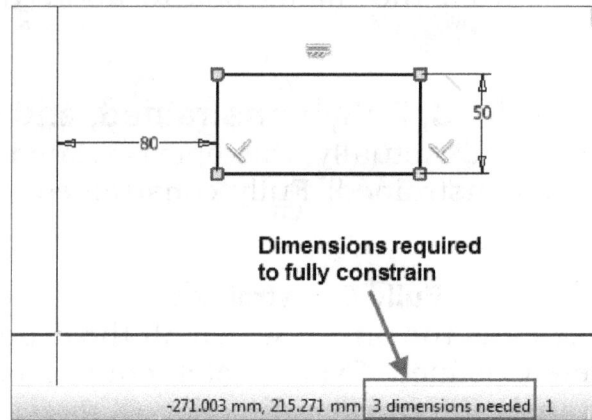

Figure-30. Dimensions required to fully constrain

There are a few miscellaneous tools in the **Insert** panel of **Sketch** tab that are helpful in sketching. These tools are discussed next.

INSERTING OBJECTS IN SKETCH

The tools to insert objects of other software in current sketch are available in the **Insert** panel of **Sketch** tab in the **Ribbon**; refer to Figure-31. There are three tools in this panel which are discussed next.

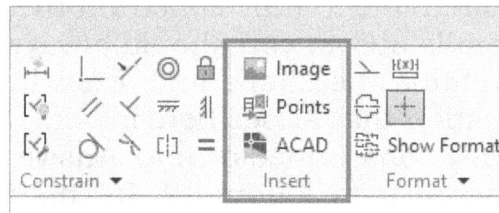

Figure-31. Insert panel

Insert Image Tool

The **Insert Image** tool is used to insert image in the current sketch. Using the image, you can create outline of the sketch using the edges of component in the image (photograph). The procedure to use this tool is given next.

• Click on the **Insert Image** tool from **Insert** panel in the **Ribbon**. The **Open** dialog box will be displayed; refer to Figure-32.

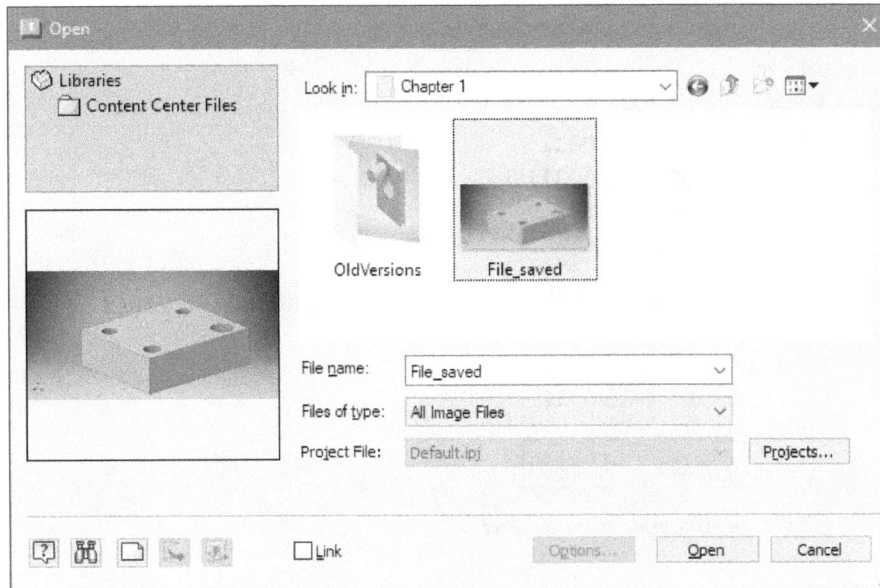

Figure-32. Open dialog box for inserting image

- Double-click on the image file that you want to use in the sketch. The image box will get attached to the cursor; refer to Figure-33.

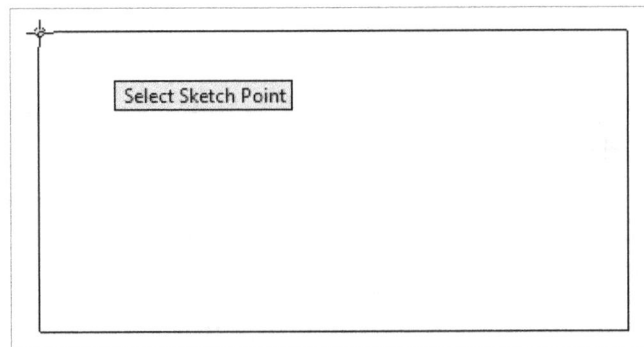

Figure-33. Image box attached to cursor

- Click at desired location to place the image. If the file location is not in folder of active project then a warning box will be displayed asking you to confirm that you want to open the file from its current location.
- Click on the **OK** button from the warning box. The image file will be placed at the specified location. Press **ESC** to exit the tool.

Import Points tool

The **Import Points** tool is used to insert points with the help of excel spreadsheet. The procedure to use this tool is given next.

- Click on the **Import Points** tool from **Insert** panel in the **Ribbon**. The **Open** dialog box will be displayed for selecting the Excel file.
- Select the file which has coordinates of points defined in it in the format as shown in Figure-34. The points will be created in the sketch; refer to Figure-35.

	A	B	C	D	E	F
1	m					
2	x	y				
3	1	0.00126				
4	0.9928	0.00322				
5	0.97989	0.00668				
6	0.96352	0.01098				
7	0.94455	0.01584		x and y coordinates defined		
8	0.9235	0.02106				
9	0.90075	0.02652				
10	0.87658	0.03212				
11	0.85123	0.03776				
12	0.82489	0.04337				

Figure-34. Example of spreadsheet

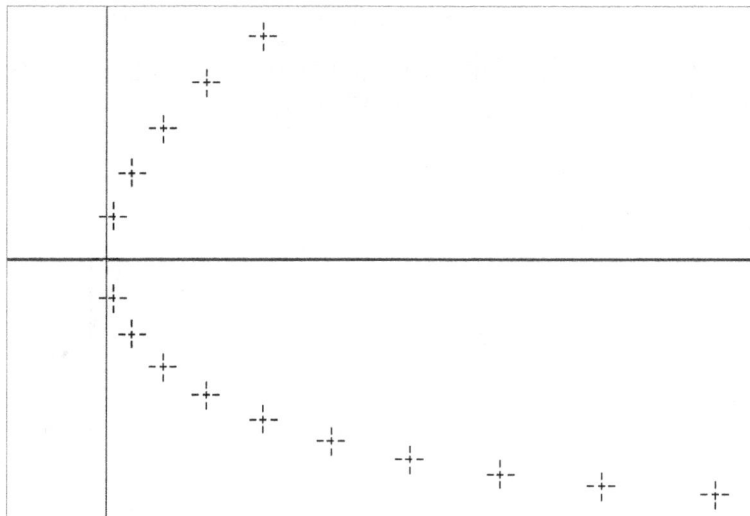

Figure-35. Points created in sketch

Insert AutoCAD File

ACAD Using the **Insert AutoCAD File** tool, you can insert a sketch made in AutoCAD into Inventor sketch. The procedure to use this tool is given next.

*   Click on the **Insert AutoCAD File** tool from **Insert** panel in the **Ribbon**. The **Open** dialog box will be displayed prompting you to select AutoCAD drawing file.
*   Double-click on desired AutoCAD file. The **Layers and Objects Import Options** dialog box will be displayed; refer to Figure-36.

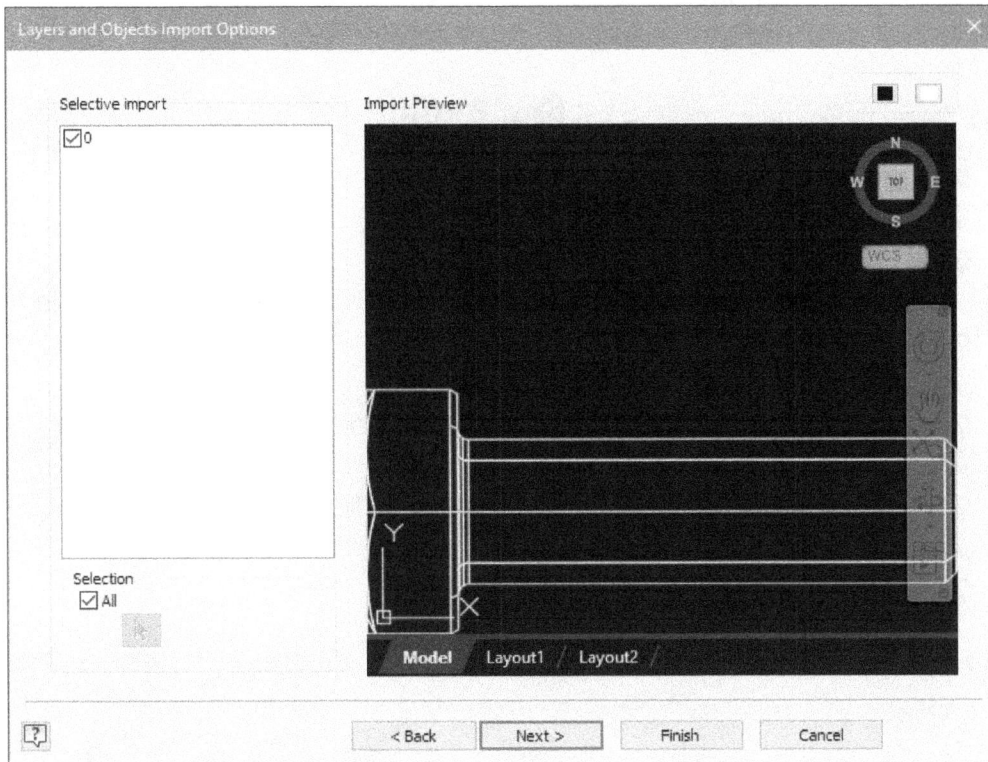

Figure-36. Layers and Objects Import Options dialog box

- Select the layers that you want to import in the sketch and click on the **Finish** button. The objects on selected layers will be inserted in the sketch automatically.

Till this point, we have covered all the important tools required for sketching. I want the users of this book to revise all the tools we have discussed so far. Now, we will work on some practical problems.

PRACTICAL 1
In this practical, we will create a sketch as shown in Figure-37.

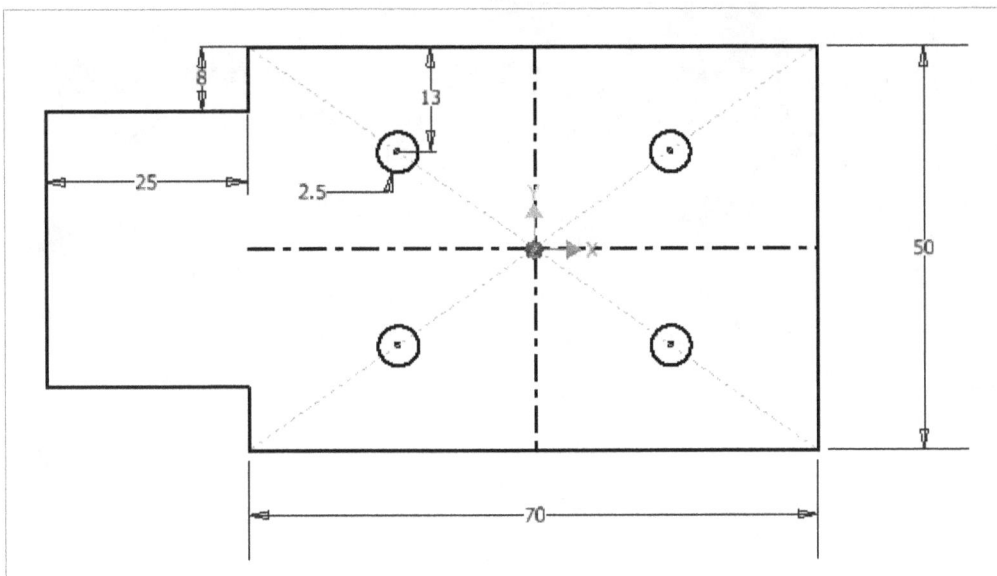

Figure-37. Practical 1

The sketch shown in this figure is not actual drawing that you will get in industry. This sketch is just to help you start in sketching.

Starting Sketch

- Start Autodesk Inventor by double-clicking on the Autodesk Inventor Professional 2024 icon from the desktop.
- Click on the **New** button from the **Quick Access Toolbar**. The **Create New File** dialog box will be displayed.
- Double-click on **Standard(mm).ipt** icon from the **Metric** template; refer to Figure-38. The Part environment of Autodesk Inventor will be displayed.

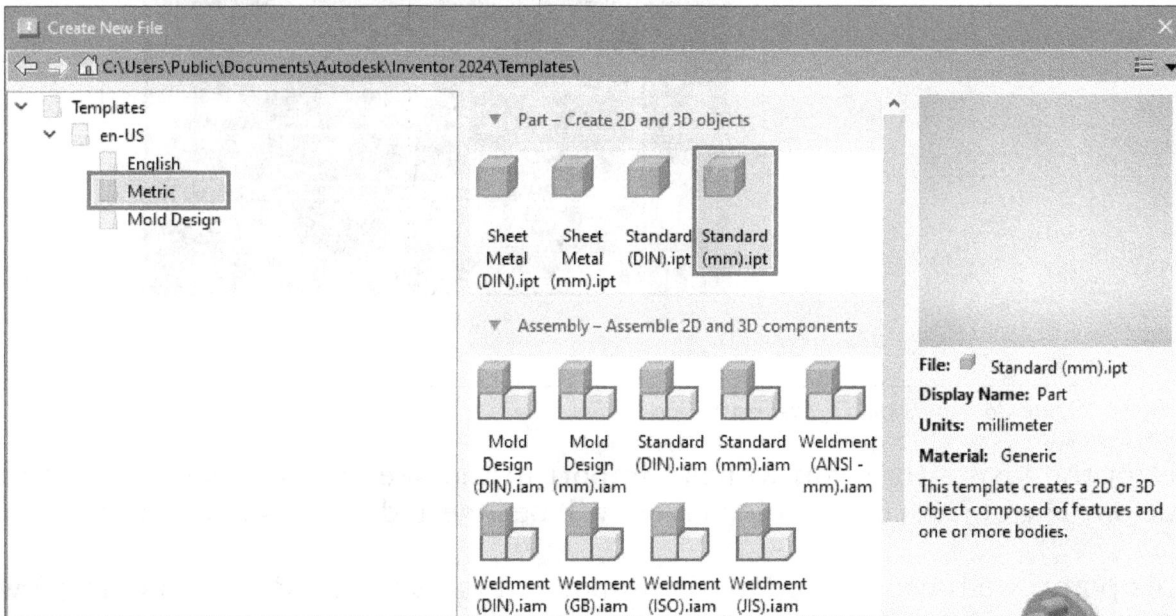

Figure-38. Standard(mm) icon in Create New File dialog box

- Click on the **Start 2D Sketch** button from **Sketch** panel in the **3D Model** tab of the **Ribbon**; refer to Figure-39. You will be asked to select a sketching plane; refer to Figure-40.

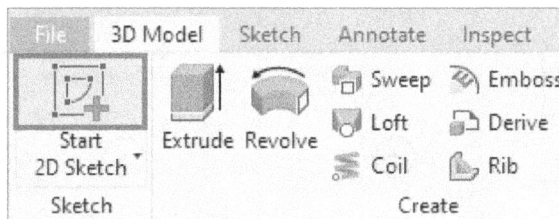

Figure-39. Start 2D Sketch button

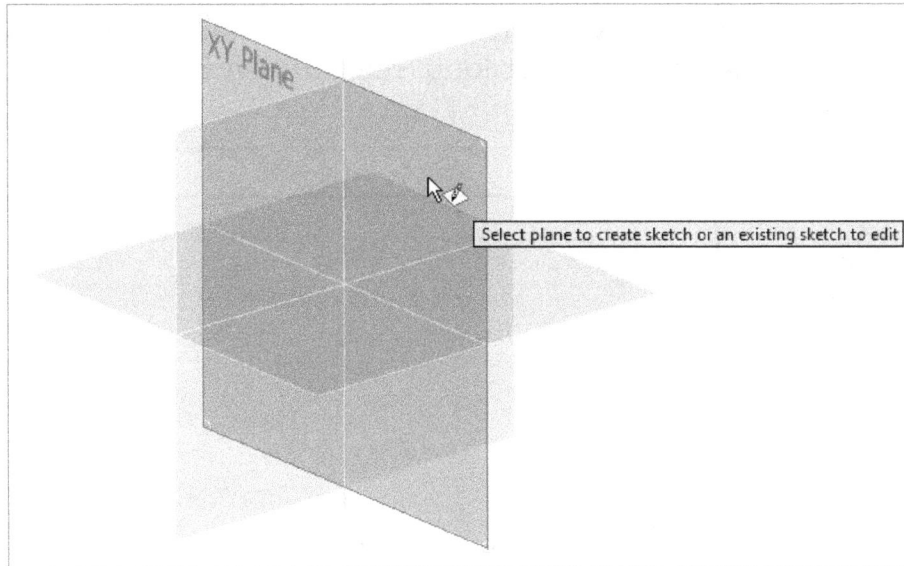

Figure-40. Selecting sketching plane

- Click on desired plane to create sketch on it.

Creating outer loop

- Click on the **Line** tool from **Create** panel in the **Sketch** tab of the **Ribbon**. You will be asked to specify the starting point of the line.
- Click at the origin to specify the starting point of the line. You will be asked to specify the end point of the line.
- Move the cursor straight towards right, type **70** in the input box displayed; refer to Figure-41 and press **ENTER**.

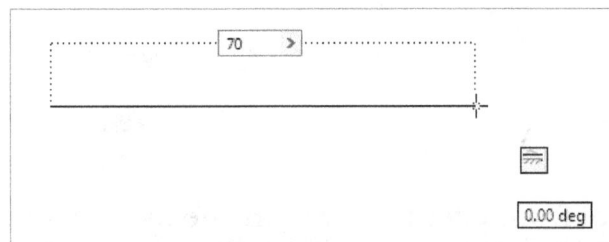

Figure-41. Specifying length of line

- Move the cursor vertically upwards and enter **50** in the input box displayed; refer to Figure-42.

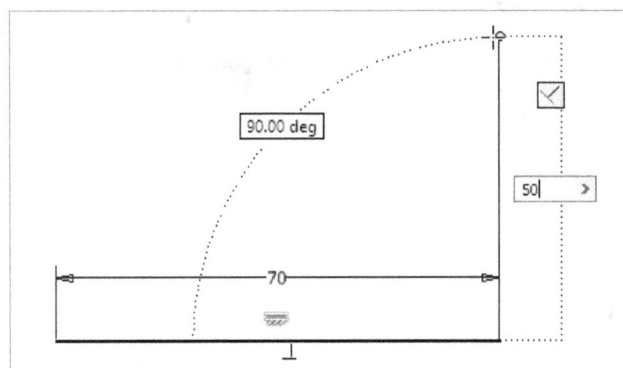

Figure-42. Specifying length of vertical line

- Move the cursor straight towards left and enter the value as **70** in the input box.
- Move the cursor vertically downward and enter the value as **8** in the input box.
- Similarly, create other lines of outer loop; refer to Figure-43.

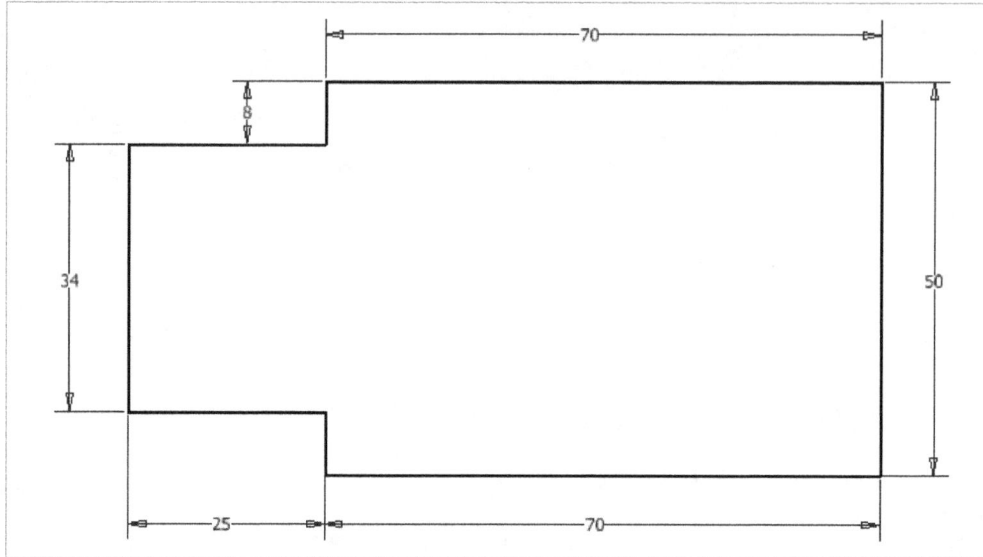

Figure-43. Outer loop

Creating Circles

- Click on the **Construction** button from **Format** panel in the **Sketch** tab of the **Ribbon**; refer to Figure-44. The sketch will be created in construction mode.

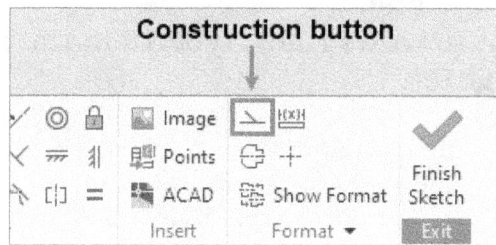

Figure-44. Construction button

- Click on the **Line** tool from **Create** panel in the **Sketch** tab of the **Ribbon**. Create a line as shown in Figure-45, joining the two corners of outer loop diagonally.

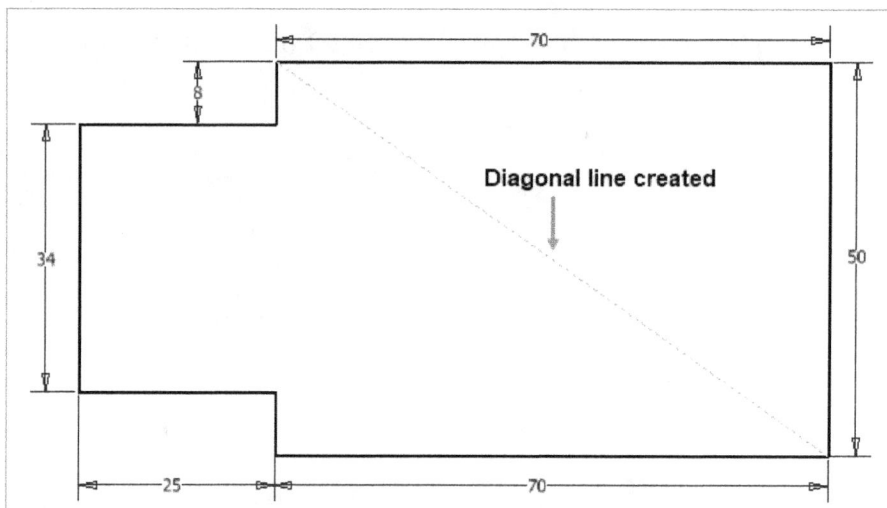

Figure-45. Creating diagonal line

- Similarly, create the other diagonal line; refer to Figure-46.

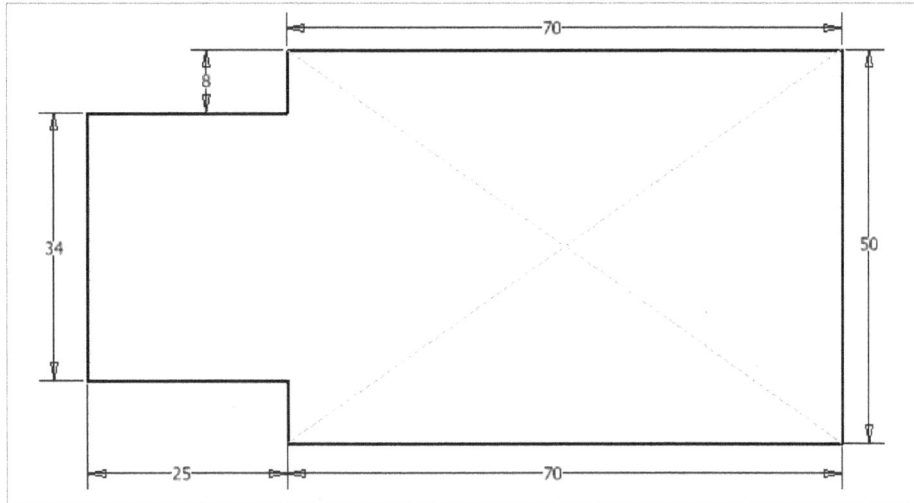

Figure-46. Creating diagonal construction lines

- Click again on the **Construction** button from **Format** panel in the **Sketch** tab to stop creating construction geometries.
- Click on the **Center Point Circle** tool from **Create** panel in the **Sketch** tab of the **Ribbon**. You will be asked to specify the center point of the circle.
- Click on the diagonal line to specify the center point of the circle; refer to Figure-47. You will be asked to specify the radius of circle.

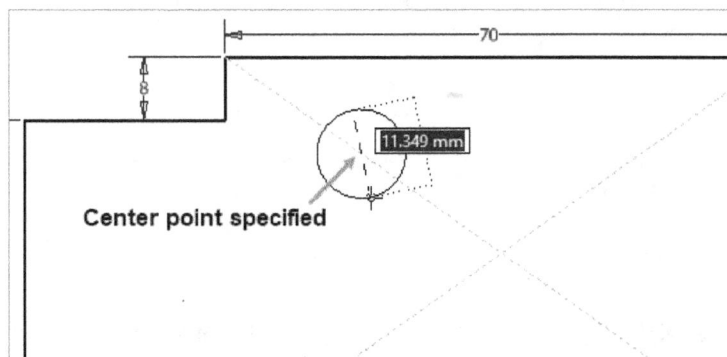

Figure-47. Specifying center point of circle

- Enter the value of radius as **2.5**.
- Click on the **Centerline** button from **Format** panel in the **Sketch** tab of the **Ribbon**; refer to Figure-48. Now, you can create center lines for the sketch.

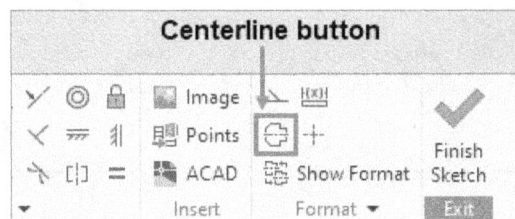

Figure-48. Centerline button

- Click on the **Line** tool from **Create** panel in the **Ribbon** and create the lines as shown in Figure-49. We will use these lines as mirror references.

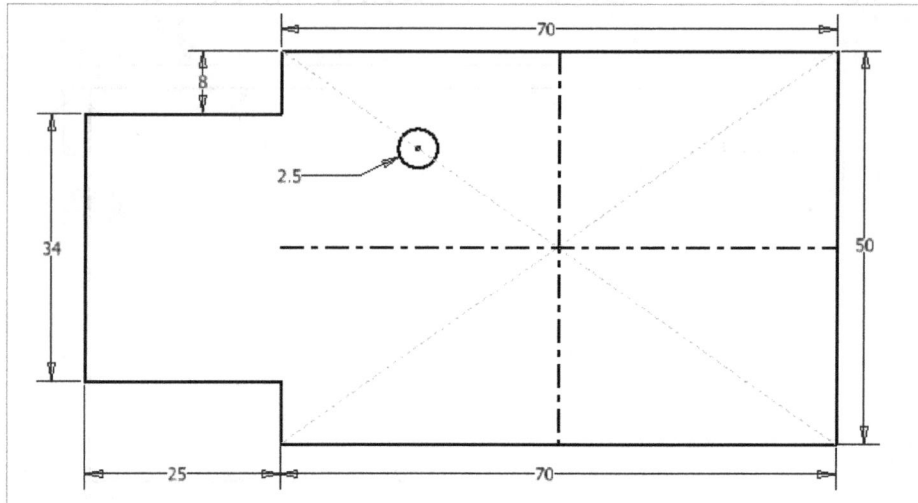

Figure-49. Centerlines created

- Click on the **Mirror** tool from the **Pattern** panel in the **Sketch** tab of **Ribbon**. The **Mirror** dialog box will be displayed and you will be prompted to select entity to mirror.
- Click on the circle created earlier; refer to Figure-50.

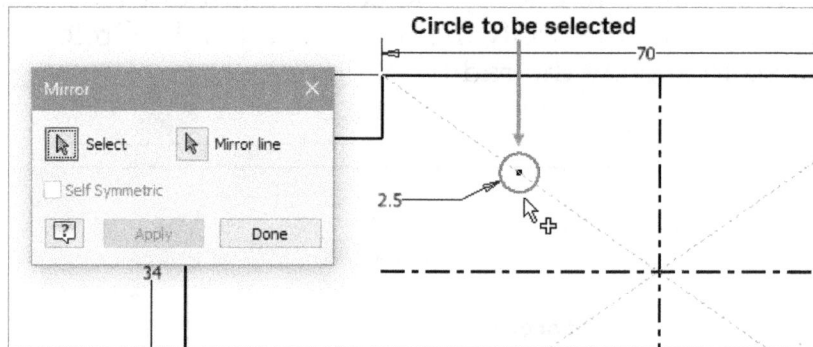

Figure-50. Circle to be selected

- Press **ENTER** and select the vertical centerline. The **Apply** button will become active in the **Mirror** dialog box.
- Click on the **Apply** button to create the mirror copy; refer to Figure-51.

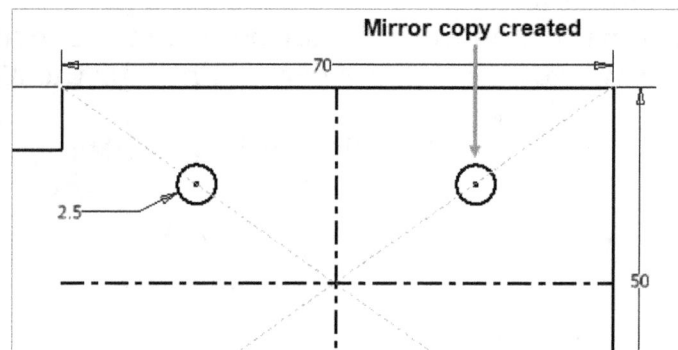

Figure-51. Mirror copy of circle

- Now, select both the circles and press **ENTER**. You will be prompted again to select a mirror line.

- Select the horizontal center line and click on the **Apply** button from the dialog box. The mirror copy of both the circles will be created as shown in Figure-52.

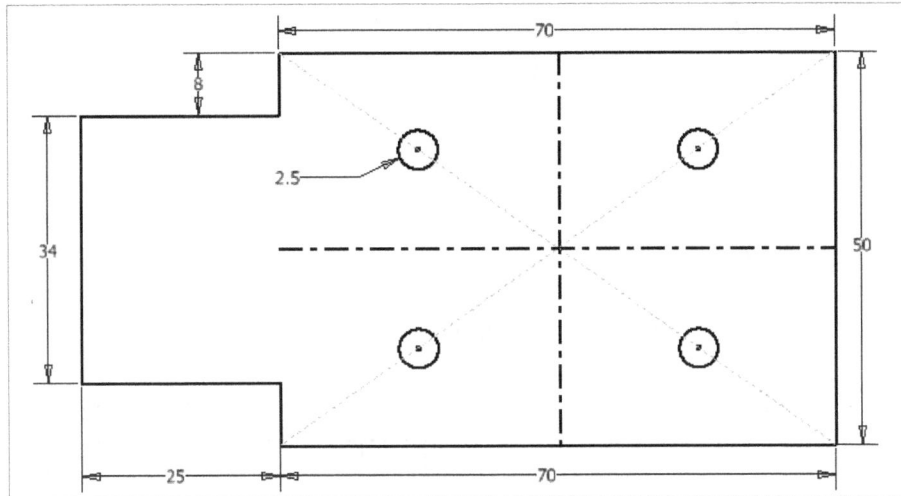

Figure-52. Mirror copies of circle

- Click on the **Done** button from the dialog box to exit the tool.

Applying Dimension to Circle

- Click on the **General Dimension** tool from **Constrain** panel in the **Sketch** tab of the **Ribbon**. You will be asked to select entities to dimension.
- Click on the center point of the circle created at first and then click on the horizontal line above the circle in outer loop; refer to Figure-53. The dimension will get attached to cursor.

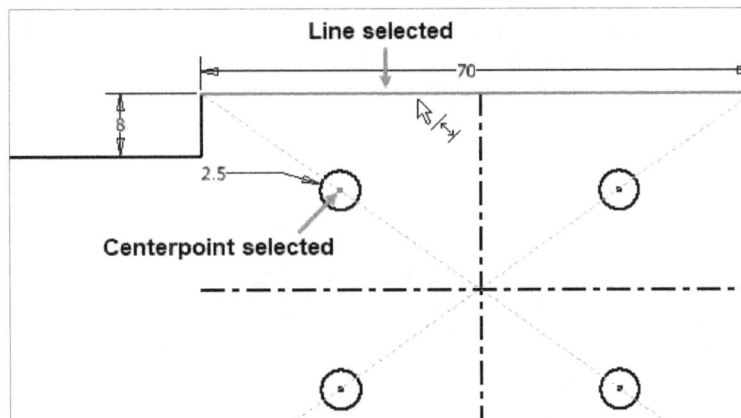

Figure-53. Entities selected for dimensioning

- Click at desired location to place the dimension. The **Edit Dimension** box will be displayed; refer to Figure-54.

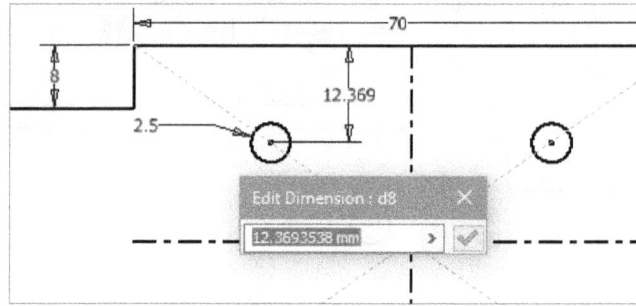

Figure-54. Dimension to edit

• Type the value as **13** and press **ENTER**. The selected circle and all the mirror copies will be dimensioned accordingly; refer to Figure-55.

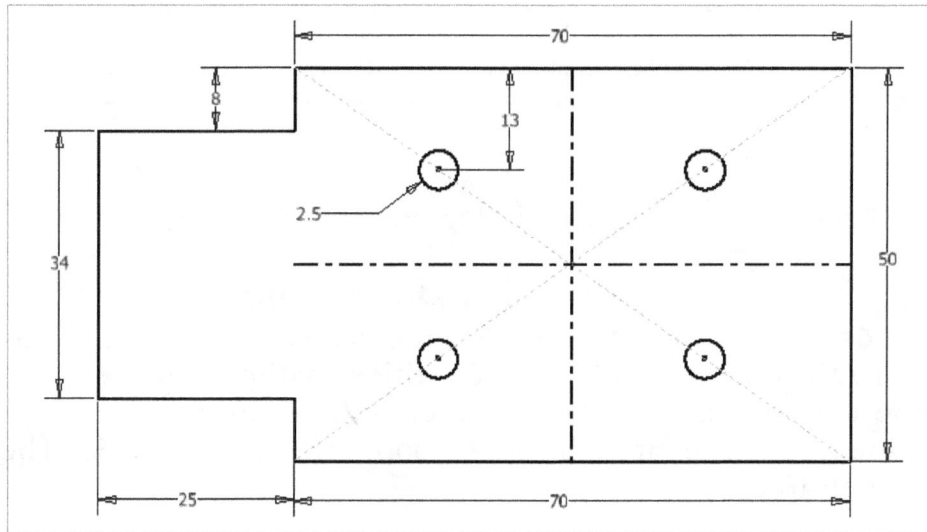

Figure-55. Final sketch of Practical 1

PRACTICAL 2

Create a sketch as shown in Figure-56.

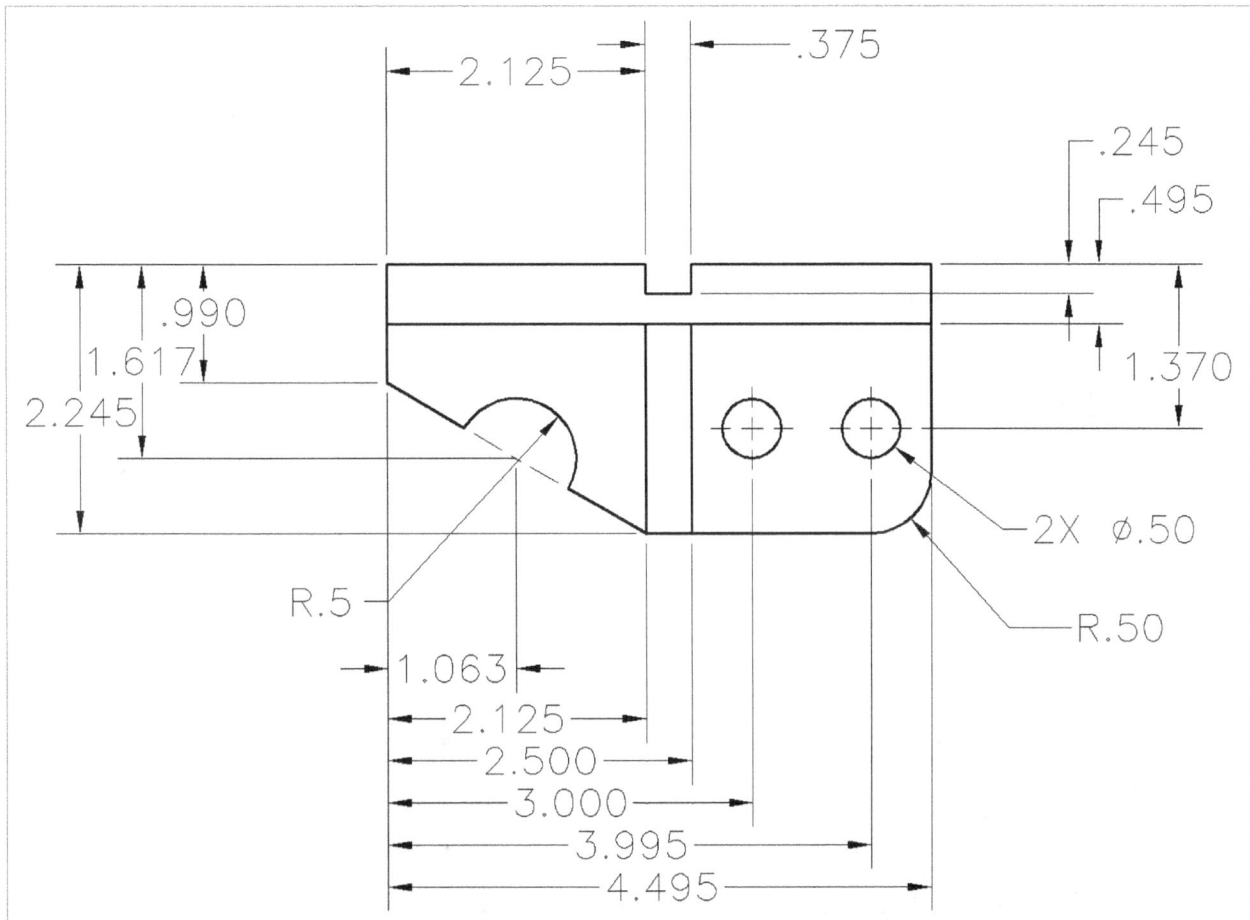

Figure-56. Practical 2

Starting Sketch

- Start Autodesk Inventor by double-clicking on the Autodesk Inventor Professional 2024 icon from the desktop. (If not started yet.)
- Click on the **New** button from the **Quick Access Toolbar**. The **Create New File** dialog box will be displayed.
- Double-click on **Standard(in).ipt** icon from the **English** templates. The Part environment of Autodesk Inventor will be displayed.
- Click on the **Start 2D Sketch** button from the **Sketch** panel in the **3D Model** tab of the **Ribbon**. You will be asked to select a sketching plane.
- Click on desired plane to create sketch on it.

Creating Line sketch

- Click on the **Two Point Rectangle** tool from **Create** panel in the **Sketch** tab of the **Ribbon**. You will be asked to specify the starting corner point of the rectangle.
- Click on the origin to specify the starting corner point. You will be asked to specify the end corner point of the rectangle. Type the length and width of rectangle as **4.495** and **2.245**, respectively in the input boxes on screen; refer to Figure-57. Note that you need to press **TAB** from keyboard to switch between the two dimensions.

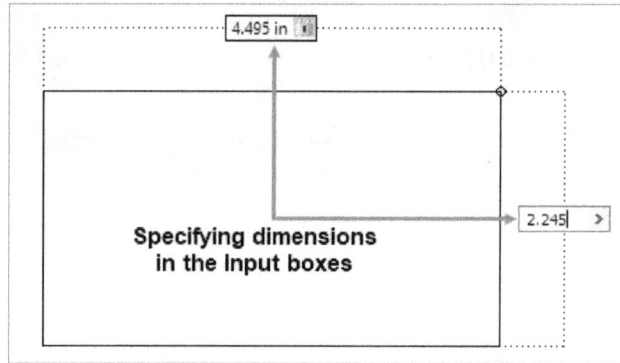

Figure-57. Specifying dimensions of rectangle

- Press **ENTER** to apply the dimensions.
- Click on the **Line** tool from **Create** panel in the **Sketch** tab of the **Ribbon** and create a line intersecting the vertical and horizontal side of rectangle as shown in Figure-58.

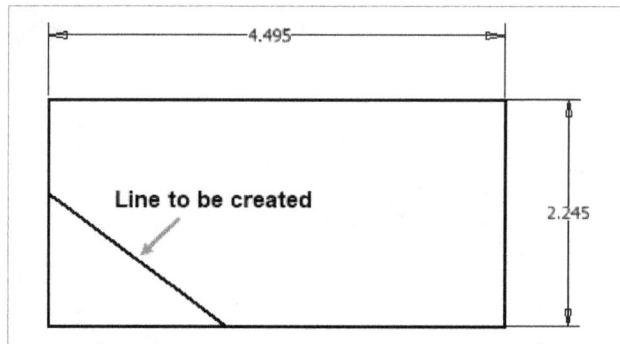

Figure-58. Line to be created

- Click on the **General Dimension** tool from **Constrain** panel in the **Sketch** tab of the **Ribbon** and specify distance of end points of the line as shown in Figure-59.

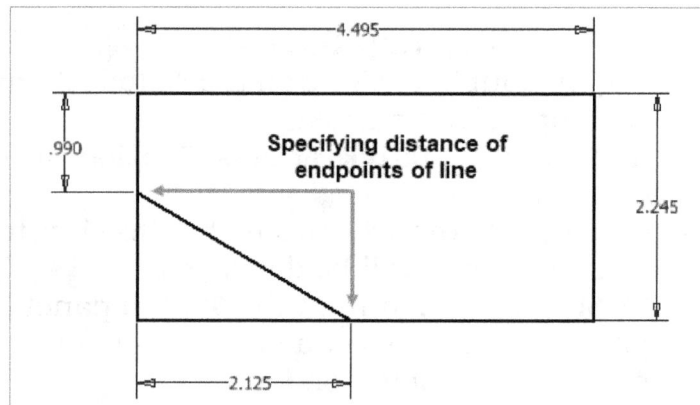

Figure-59. Dimensioning the line

- Similarly, create the other lines of the sketch; refer to Figure-60.

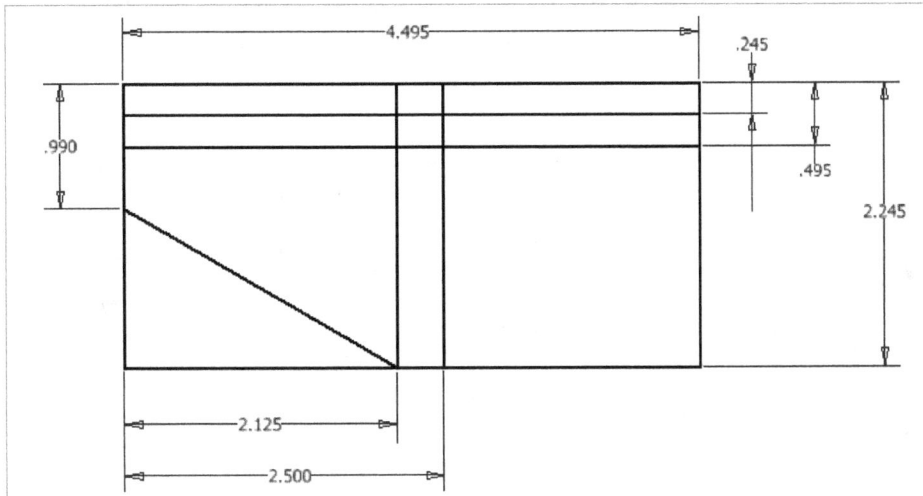
Figure-60. Sketch after creating lines

Creating Circles

- Click on the **Center Point Circle** tool from **Create** panel in the **Sketch** tab of the **Ribbon**. You will be asked to specify the center of the circle.
- Click on the midpoint of the inclined line to specify the center; refer to Figure-61. You will be asked to specify the radius of the circle.

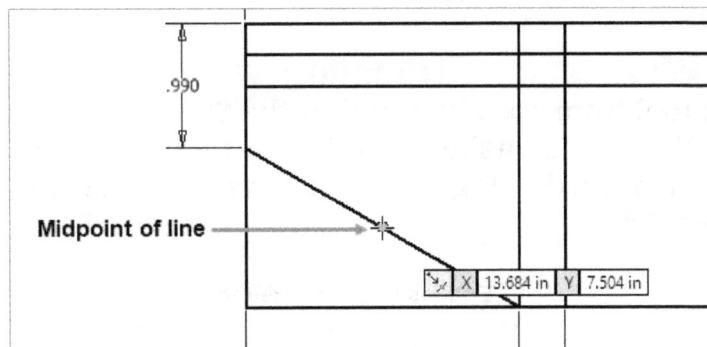
Figure-61. Midpoint of inclined line

- Enter the radius as **0.5** in the input box. The circle will be created.
- Similarly, create the other two circles of diameter **0.5** at random locations as displayed in Figure-62.

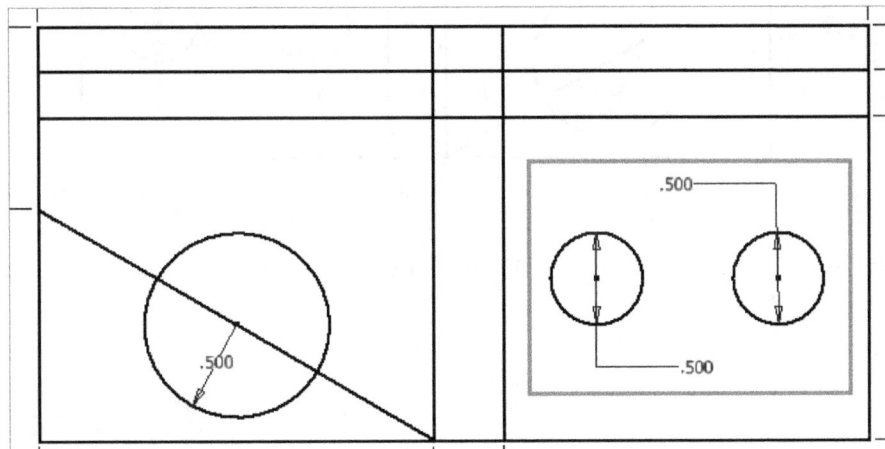
Figure-62. Circles created

- Click on the **General Dimension** tool from **Constrain** panel in the **Sketch** tab of the **Ribbon** and set the dimensions of circles as shown in Figure-63.

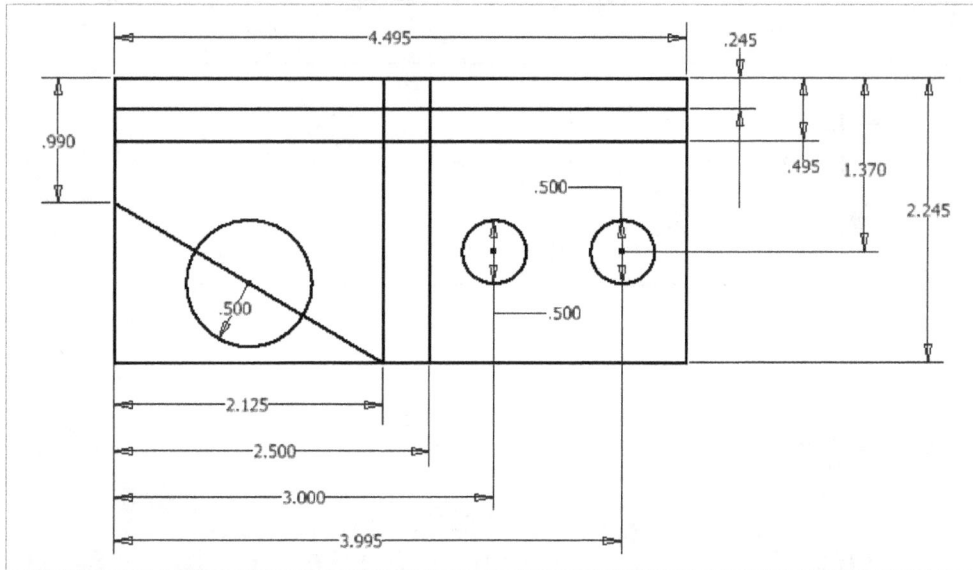

Figure-63. Dimensioning the circles

Now, we are done with creating the entities. There are some extra portions of sketch that need to be trimmed. The procedure is given next.

Trimming

- Click on the **Trim** tool from **Modify** panel in the **Sketch** tab of the **Ribbon** or press **X** from keyboard. You will be asked to select the entities to trim.
- Click on the entities marked in Figure-64 with red dot. The sketch will be displayed as shown in Figure-65.

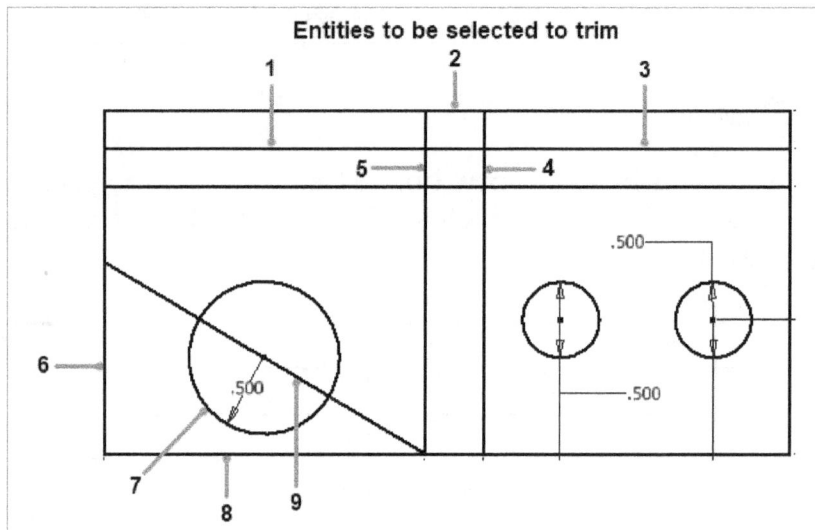

Figure-64. Entities selected for trimming

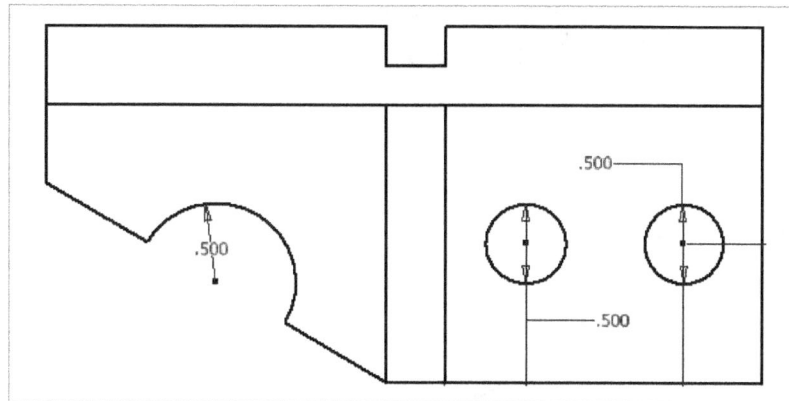

Figure-65. Sketch after trimming

- Create a fillet at the corner by using **Fillet** tool from **Create** panel in the **Sketch** tab of the **Ribbon** and apply rest of the dimensions as given in the drawing; refer to Figure-66.

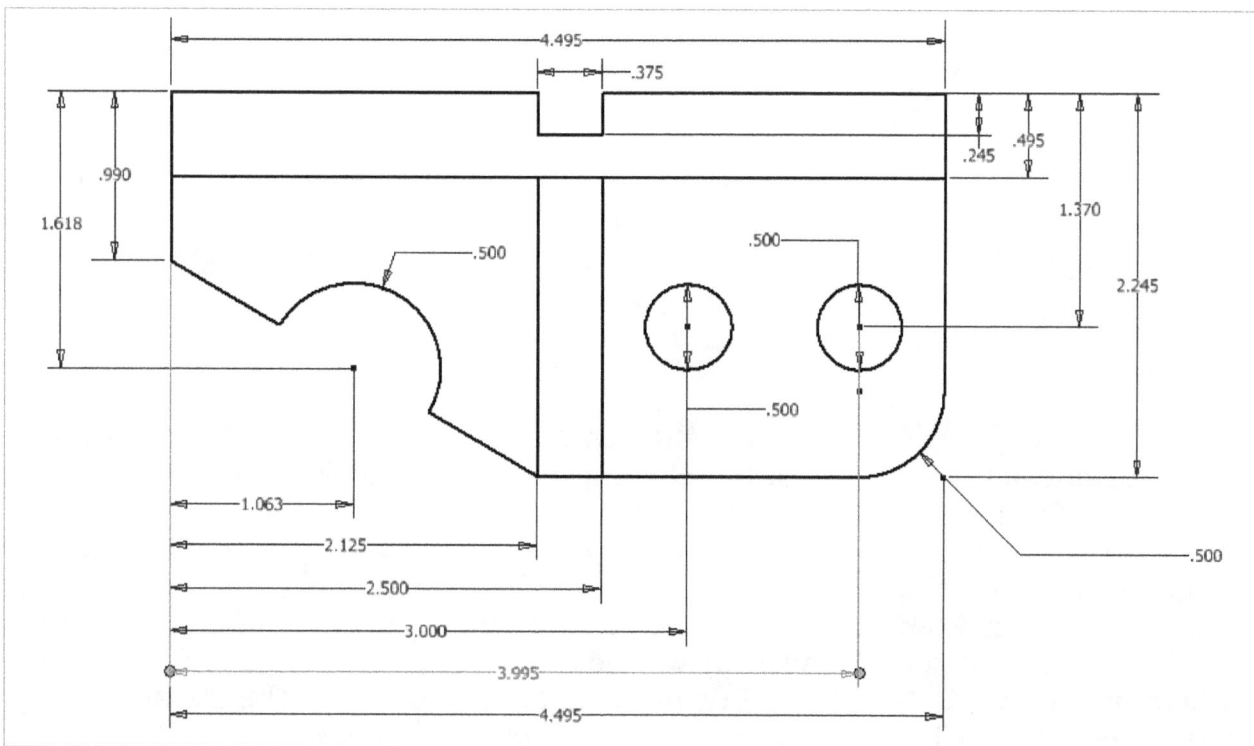

Figure-66. After applying dimensions and creating fillet

PRACTICAL 3

Create the sketch as shown in Figure-67.

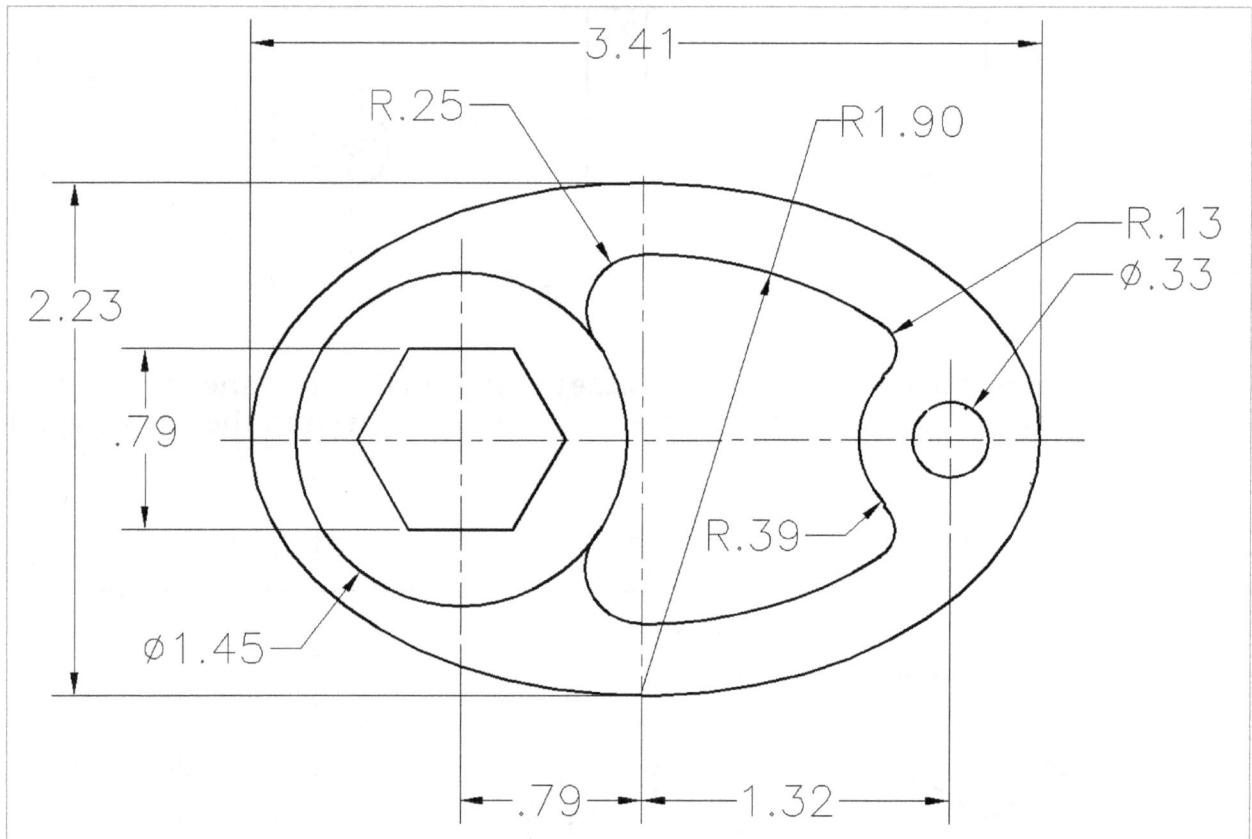

Figure-67. Practical 3

Starting Sketch

- Start Autodesk Inventor by double-clicking on the Autodesk Inventor Professional 2024 icon from the desktop. (If not started yet.)
- Click on the **New** button from the **Quick Access Toolbar**. The **Create New File** dialog box will be displayed.
- Double-click on **Standard(in).ipt** icon from the **English** templates. The Part environment of Autodesk Inventor will be displayed.
- Click on the **Start 2D Sketch** button from **Sketch** panel in the **3D Model** tab of the **Ribbon**. You will be asked to select a sketching plane.
- Click on desired plane to create sketch on it.

Creating Ellipse

- Click on the **Ellipse** tool from **Circle** drop-down in the **Create** panel of the **Ribbon**; refer to Figure-68. You will be asked to specify center of the ellipse.

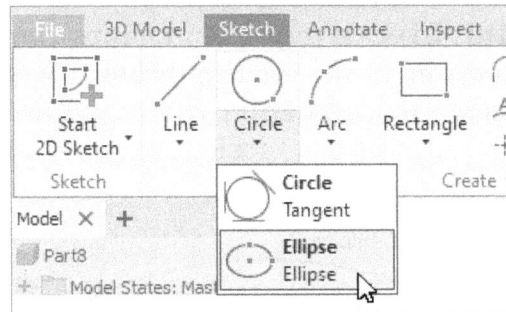

Figure-68. Ellipse tool

- Click on the origin to make it center point of the ellipse. You will be asked to specify the first axis end point.
- Click to specify the first axis end point; refer to Figure-69. You will be asked to specify a point on ellipse; refer to Figure-70.

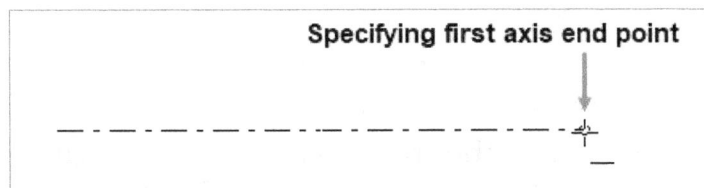

Figure-69. Specifying end point of ellipse axis

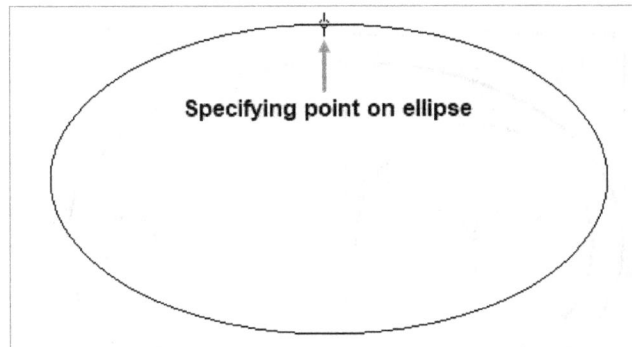

Figure-70. Specifying point on ellipse

- Click at desired location to specify the point on ellipse.
- Click on the **General Dimension** tool and specify the length and width of ellipse as **1.705** and **1.115**, respectively at its axes; refer to Figure-71.

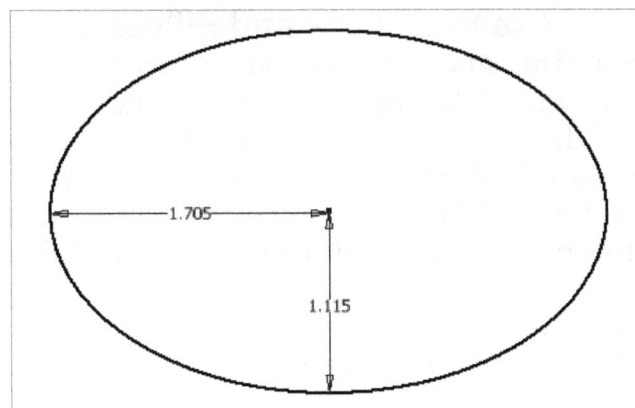

Figure-71. Ellipse after specifying dimensions

Creating Circles

- Click on the **Center Point Circle** tool from **Create** panel in the **Ribbon**. You will be asked to specify the center point for the circle.
- Click at a location collinear to the center of ellipse as shown in Figure-72. You will be asked to specify radius of the circle.

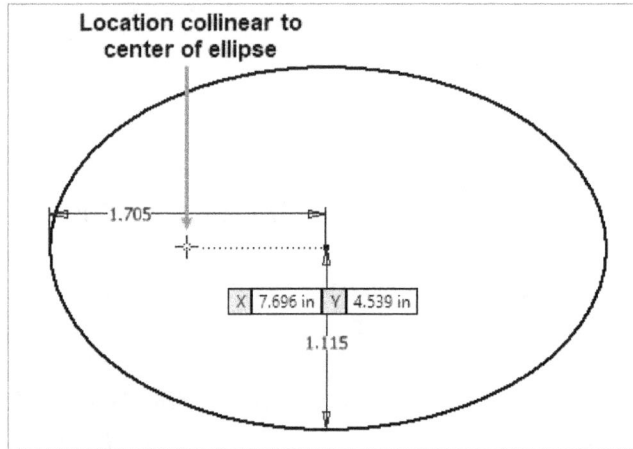

Figure-72. Specifying location for center of circle

- Enter the diameter as **1.45** in the input box. The circle will be created.
- Similarly, create the other circle of diameter **0.33** and specify the distances by using the **General Dimension** tool; refer to Figure-73.

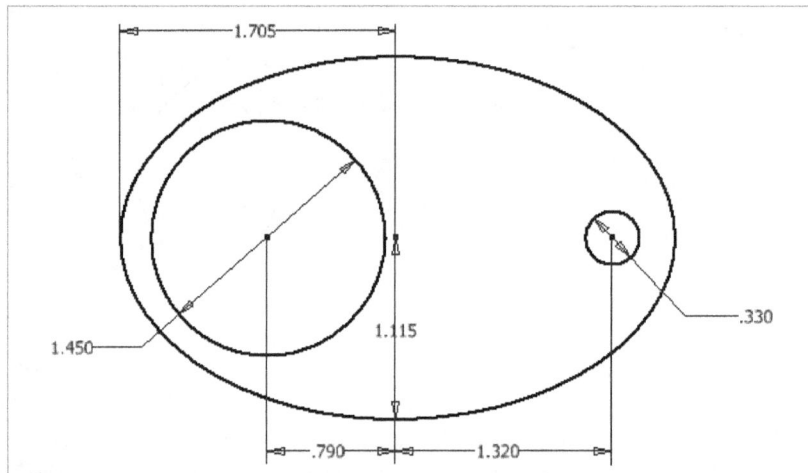

Figure-73. Sketch after creating circles

Creating Arcs and Fillets

- Click on the **Center Point Arc** tool from **Arc** drop-down in the **Create** panel of the **Ribbon**. You will be asked to specify center of the arc.
- Click on the center of the circle with diameter **0.330** to specify center of the arc. You will be asked to specify the starting point of the arc.
- Type the radius of arc as **0.390** in the input box and press **TAB** from keyboard. You will be asked to specify the location of starting point on the construction circle of radius **0.390**.
- Click at the location where angle value is approximately **120** degree; refer to Figure-74. You will be asked to specify end point of the arc.

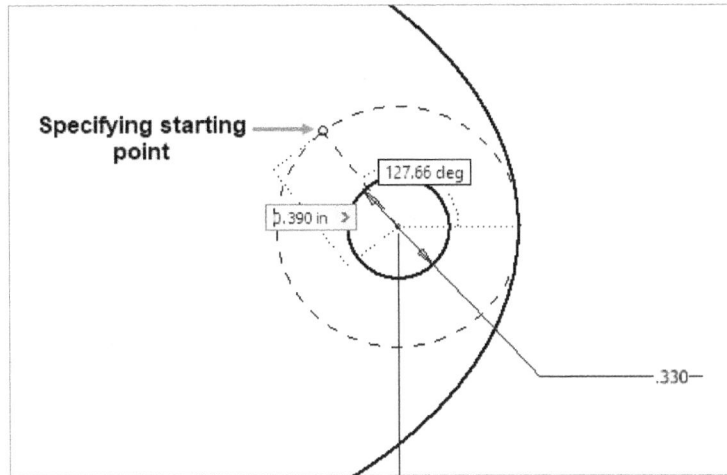

Figure-74. Specifying starting point of arc

- Click at the location where arc angle is approximately **120** degree. The arc will be created.
- Click again on the **Center Point Arc** tool and specify the center point of arc along vertical line of center and at the circle as shown in Figure-75. You will be asked to specify the starting point of the arc.

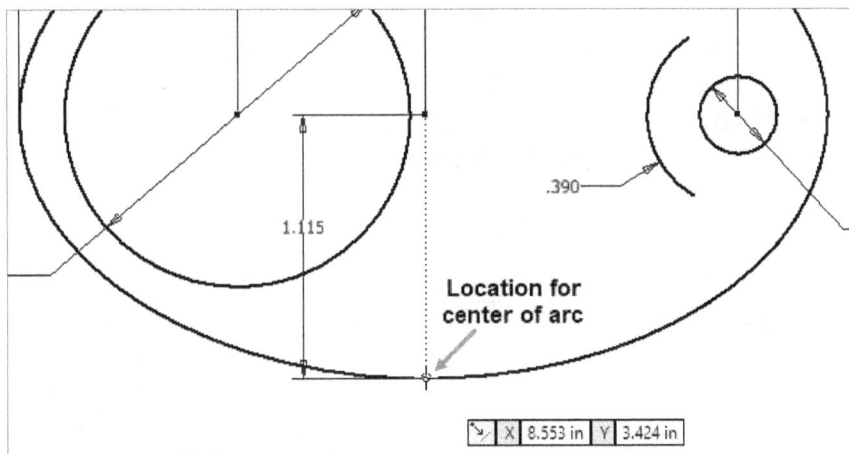

Figure-75. Location for center of arc

- Enter **1.90** in the **Radius** input box and specify the starting point and end point of the arc as shown in Figure-76.

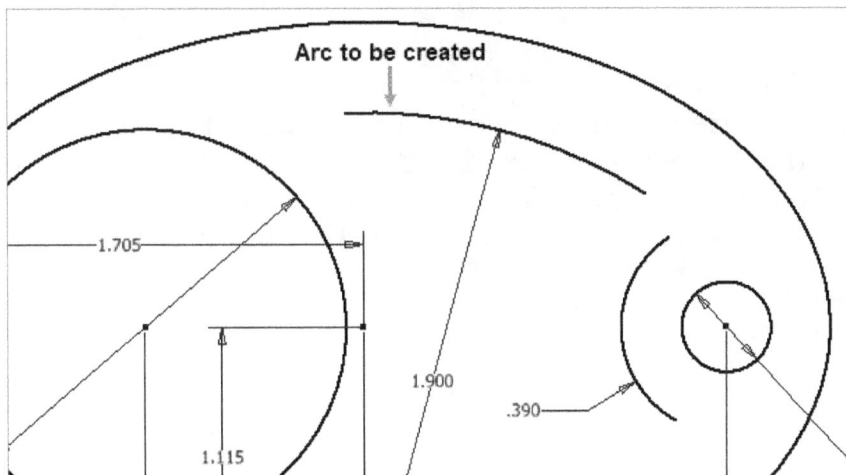

Figure-76. Arc to be created

Applying Constraint

- Click on the **Coincident Constraint** button from **Constrain** panel in the **Ribbon**. You will be asked to select curves or points.
- Select the curves or points to be coincident as shown in Figure-77. The entities will become coincident to each other.

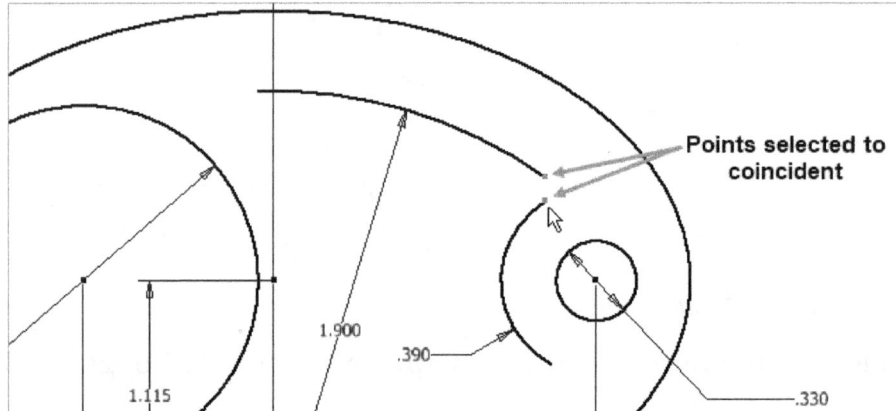

Figure-77. Points selected to coincident

- Press **ESC** from keyboard to exit the tool.

Creating Fillets

- Click on the **Fillet** tool from **Create** panel in the **Ribbon**. You will be asked to select the entities to apply the fillet.
- Type the value of radius for fillet as **0.13** in the edit box displayed in the **2D Fillet** dialog box and click on the arcs as shown in Figure-78. The fillet will be created.

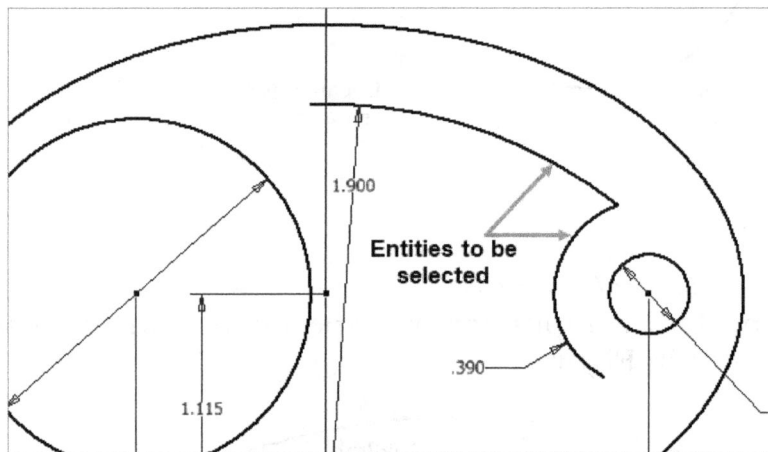

Figure-78. Entities to select for fillet

- If you have earlier tried then you would be knowing that we can not create fillet between arc and circle by using the **Fillet** tool in Inventor. Yes!! We have solution! Click on the **Tangent Arc** tool from **Arc** drop-down in the **Create** panel of the **Ribbon**. You will be asked to specify the starting point of the arc.
- Click on the arc as shown in Figure-79. You will be asked to specify the end point of the arc.

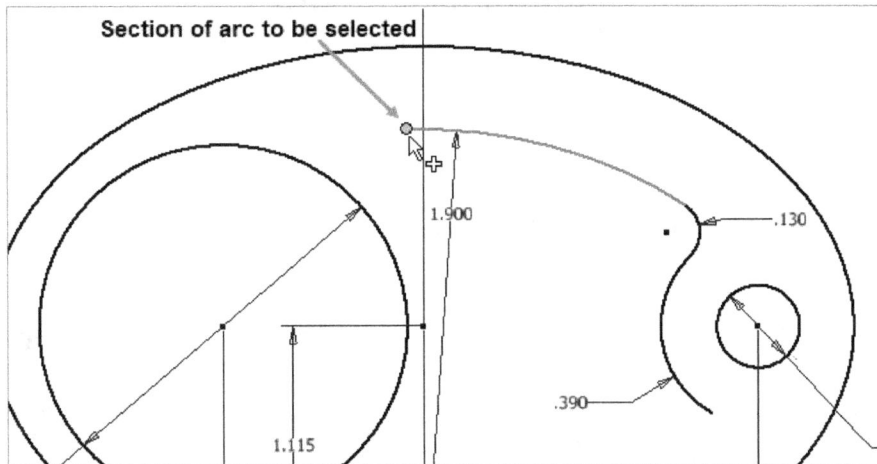

Figure-79. Section of arc to be selected

- Click on the circle at any random location to specify the end point of the arc; refer to Figure-80.

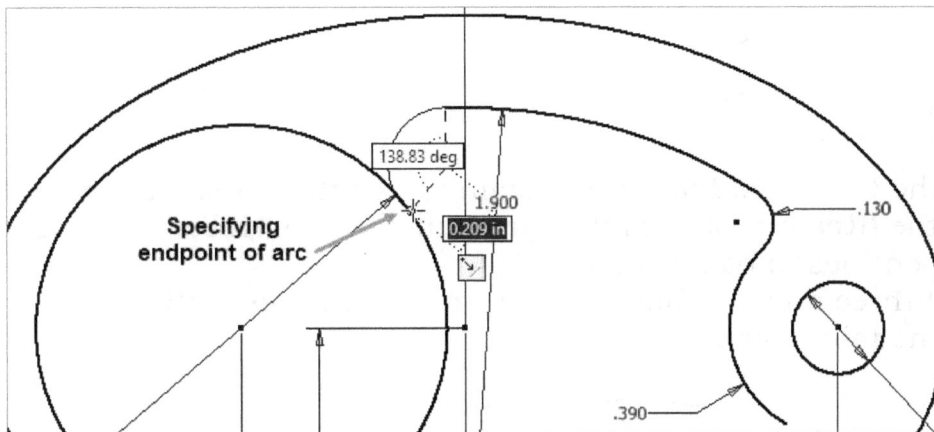

Figure-80. Specifying end point of arc

- Click on the **Tangent** button from **Constrain** panel in the **Ribbon**. You will be asked to select the first curve.
- One by one select the newly created arc and circle as shown in Figure-81. The entities will become tangent to each other. Specify the dimension of the arc as **0.25** using the **General Dimension** tool.

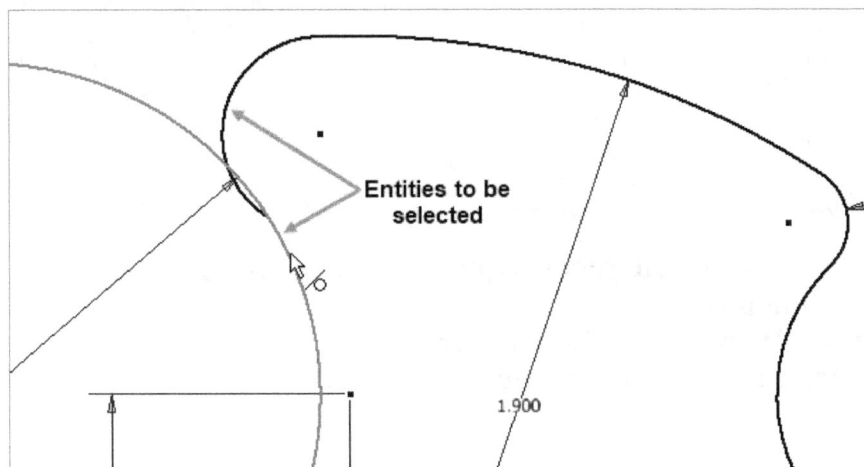

Figure-81. Entities selected for tangency

Creating Mirror copy

- Click on the **Construction** button from **Format** panel in the **Ribbon**. The construction mode will become active.
- Click on the **Line** tool from **Create** panel in the **Ribbon**. You will be asked to specify the starting point of the line.
- Click on the centers of the circles one by one to create the line; refer to Figure-82. Press **ESC** from keyboard to exit the tool.

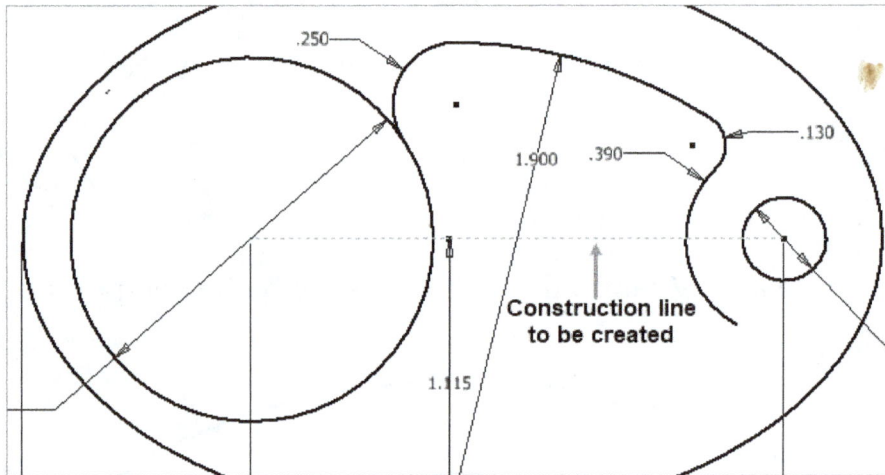

Figure-82. Construction line created

- Click on the **Construction** button again to exit the construction mode.
- Click on the **Mirror** tool from **Pattern** panel in the **Ribbon**. You will be asked to select the entities to be mirrored.
- Select the three arcs as shown in Figure-83 and press **ENTER**. You will be asked to select the mirror line.

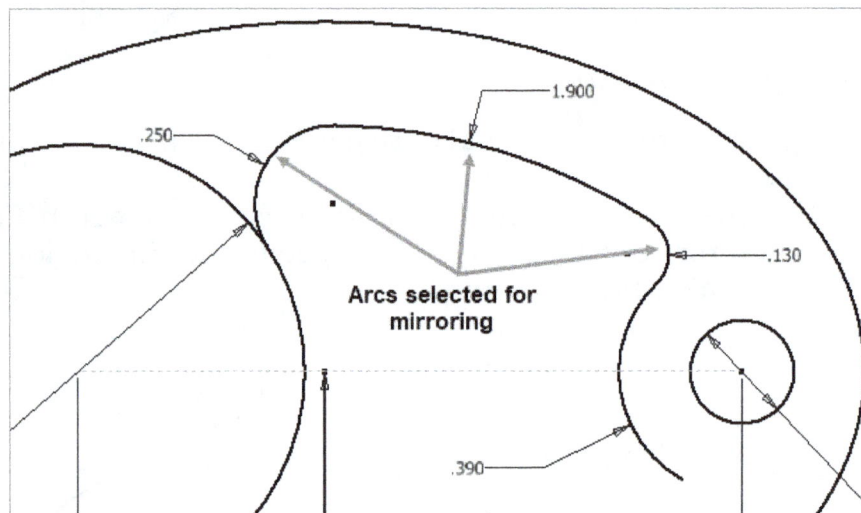

Figure-83. Selecting arcs for mirroring

- Select the construction line recently created. The **Apply** button will become active in the **Mirror** dialog box.
- Click on the **Apply** button. The mirror copy will be created; refer to Figure-84. Click on the **Done** button to exit the dialog box.

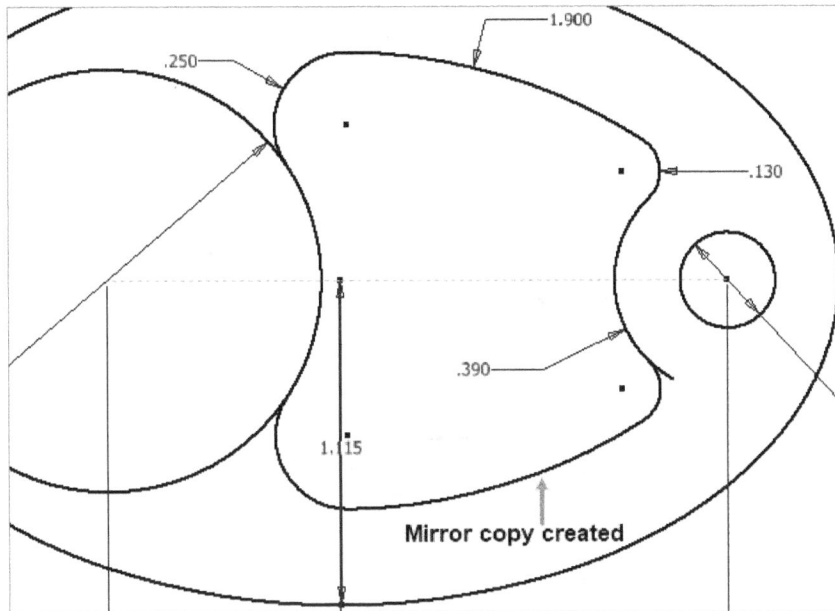

Figure-84. Mirror copy created

Creating Polygon

- Click on the **Polygon** tool from **Rectangle** drop-down in the **Create** panel of the **Ribbon**. You will be asked to specify the number of sides of the polygon in the **Polygon** dialog box; refer to Figure-85.

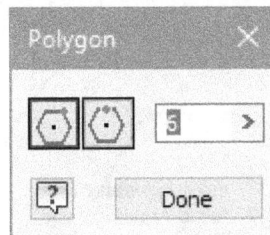

Figure-85. Polygon dialog box

- Specify the value as **6** in the edit box and click at the center of the left circle to specify the center of the polygon. You will be asked to specify a point on the polygon; refer to Figure-86.

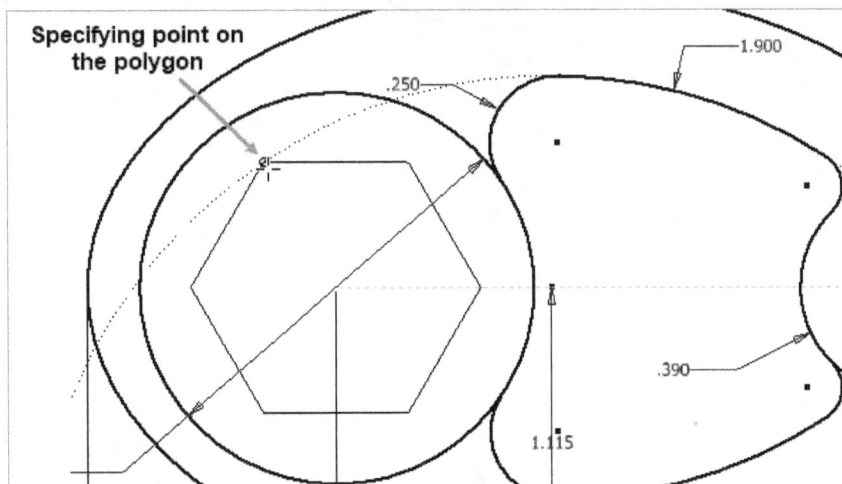

Figure-86. Point to be specified for polygon

- Click randomly at any location inside the circle and press **ESC** from keyboard. The polygon will be created; refer to Figure-87.

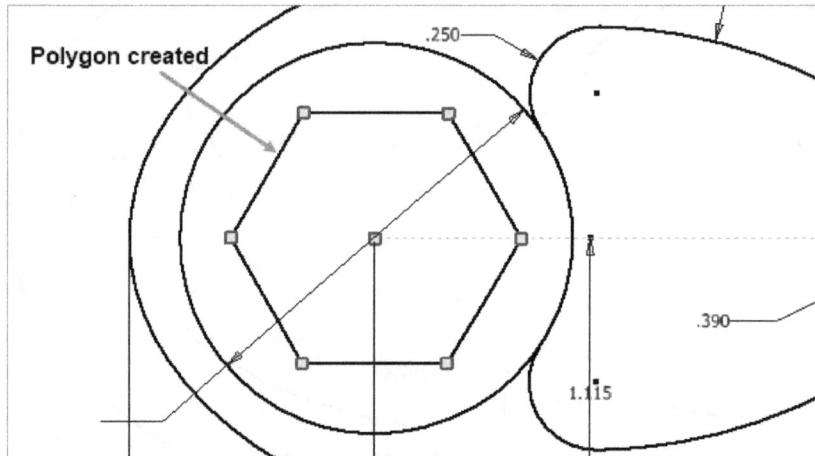

Figure-87. Polygon created

- Click on the **Horizontal Constraint** button from **Constrain** panel in the **Ribbon**. You will be asked to select the entity.
- Click on any line of the polygon to orient it as it is in the drawing.
- Click on the **General Dimension** tool from **Constrain** panel and specify the dimension as given in Figure-88. The final sketch will be created as shown in Figure-89.

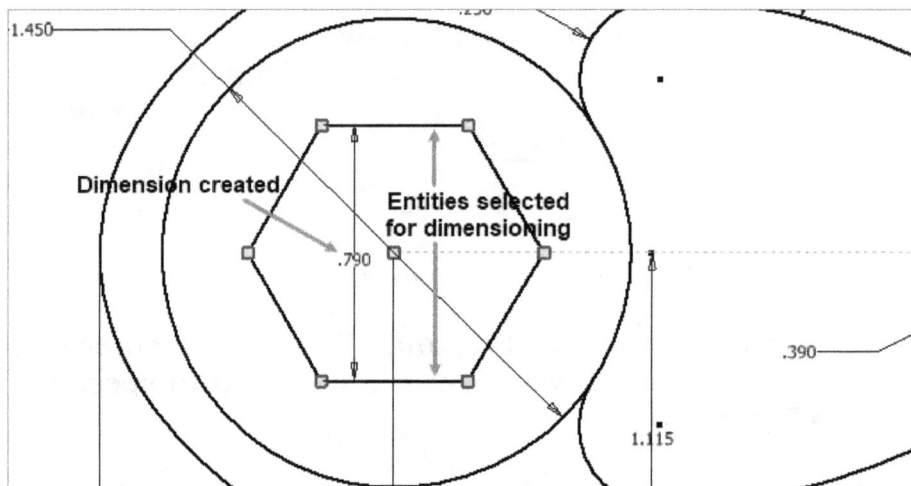

Figure-88. Applying dimension to polygon

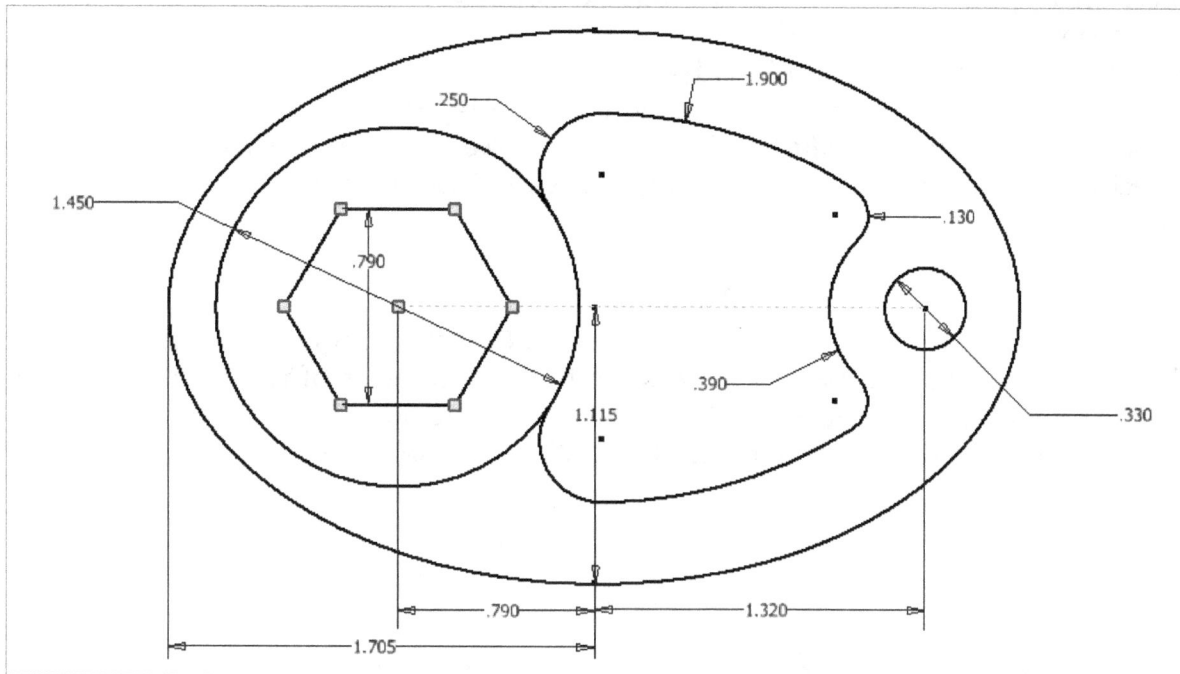

Figure-89. Final sketch for practical 3

PRACTICAL 4

Create the sketch as shown in Figure-90.

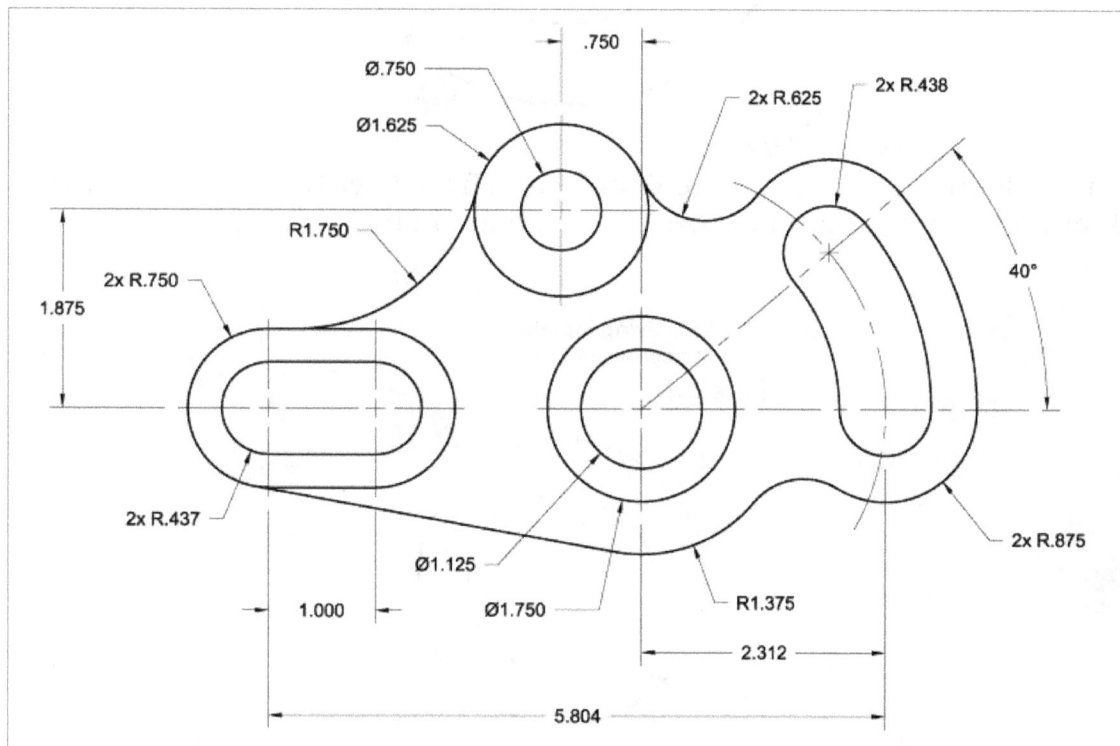

Figure-90. Practical 4

Starting Sketch

- Start Autodesk Inventor by double-clicking on the Autodesk Inventor Professional 2024 icon from the desktop. (If not started yet)

- Click on the **New** button from the **Quick Access Toolbar**. The **Create New File** dialog box will be displayed.
- Double-click on **Standard(in).ipt** icon from the **English** templates. The Part environment of Autodesk Inventor will be displayed.
- Click on the **Start 2D Sketch** button from **Sketch** panel in the **3D Model** tab of the **Ribbon**. You will be asked to select a sketching plane.
- Click on desired plane to create sketch on it.

Creating Circles

- Click on the **Center Point Circle** tool from **Create** panel in the **Ribbon** and click on the origin. You will be asked to specify the diameter of the circle.
- Enter the value of diameter as **1.125** in the input box. The circle will be created.
- Similarly, create the circle of diameter **1.75** at the same center; refer to Figure-91.

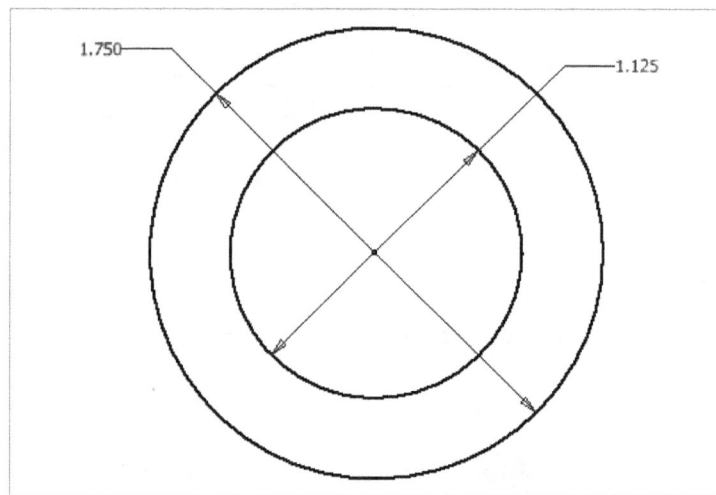

Figure-91. Circles created at origin

- Again, click on the **Center Point Circle** tool from **Create** panel in the **Ribbon** and click at a random location above earlier created circles to specify center; refer to Figure-92.

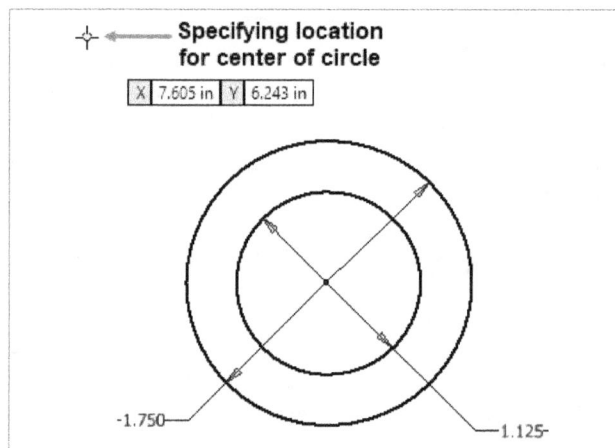

Figure-92. Specifying location for center of circle

- Enter the value of diameter as **0.750** in the Input box. The circle will be created.
- Create a circle of diameter **1.625** at the same center; refer to Figure-93.

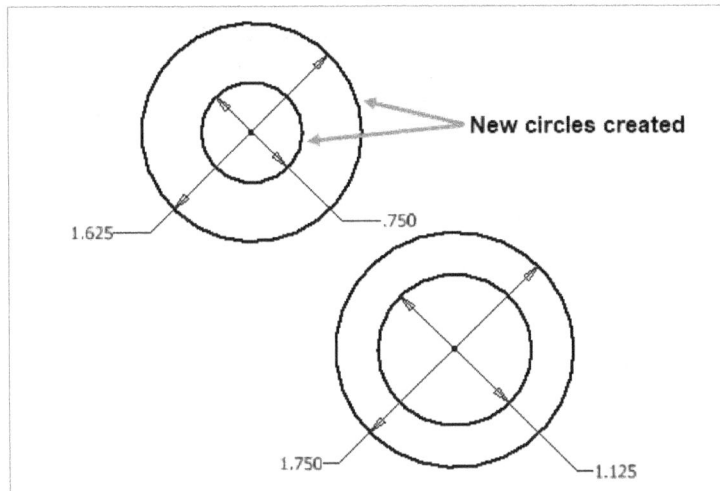

Figure-93. New circles created

- Click on the **General Dimension** tool from **Constrain** panel in the **Ribbon** and specify the position of newly created circles with respect to the origin as shown in Figure-94.

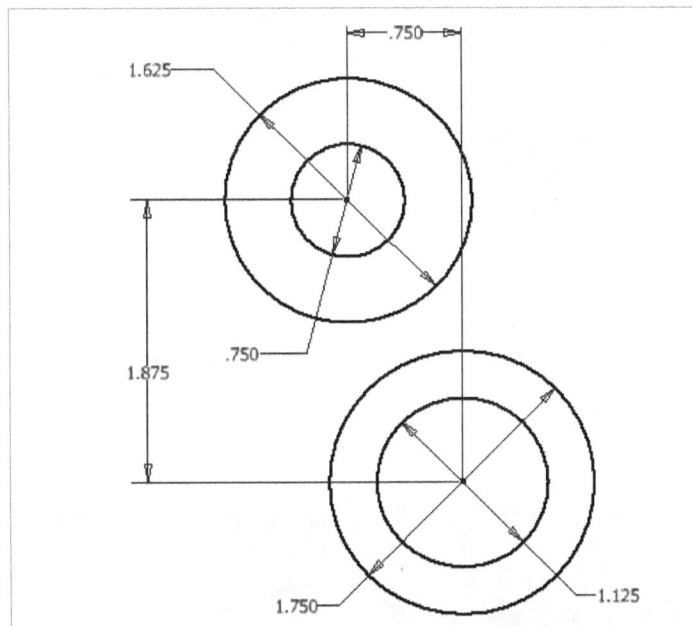

Figure-94. Positioning newly created circle

Creating Slots

- Click on the **Center Point Arc Slot** tool from **Rectangle** drop-down in the **Create** panel of the **Ribbon**. You will be asked to specify center point for center arc.
- Click on the origin to specify center. You will be asked to specify starting point of center arc.
- Type the radius as **2.312** in the Input box and press **TAB**. You will be asked to specify angle for starting point.
- Enter **0** in the Input box. You will be asked to specify end point of the arc.
- Move the cursor above and enter **40** in the Input box. You will be asked to specify diameter of the slot; refer to Figure-95.

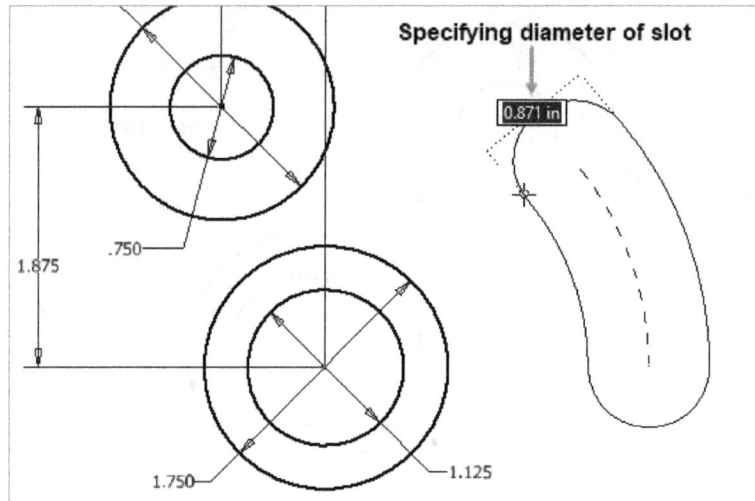

Figure-95. Specifying diameter of slot

- Enter **0.876** (0.438 x 2) in the Input box. The slot will be created; refer to Figure-96.

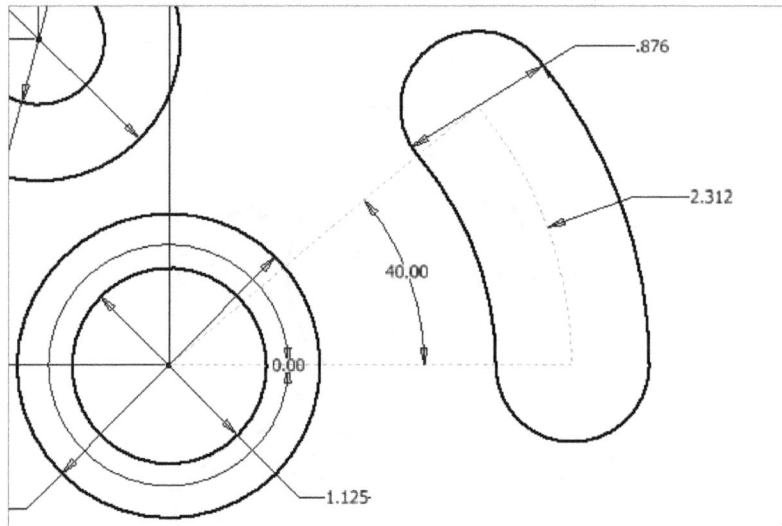

Figure-96. Slot created

- Similarly, create a slot of width **1.75** at the same location; refer to Figure-97.

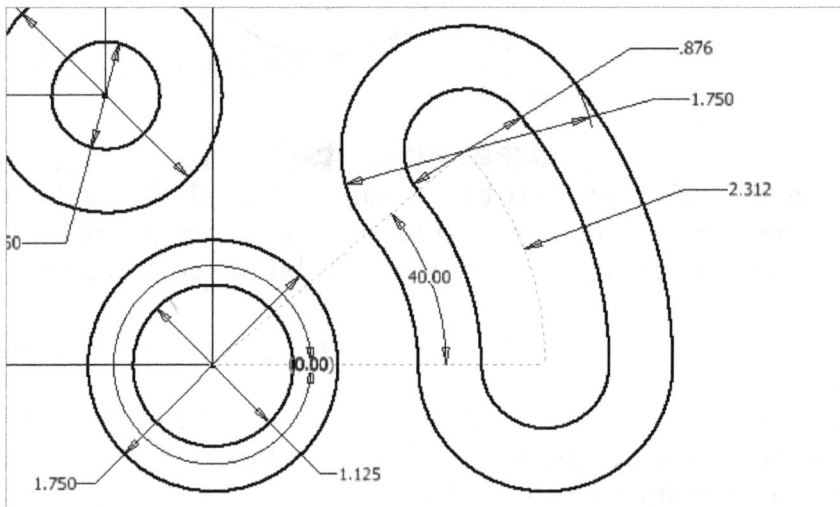

Figure-97. Second slot created

- Click on the **Center to Center Slot** tool from **Rectangle** drop-down in the **Create** panel of the **Ribbon**. You will be asked to specify starting center point of the slot; refer to Figure-98.

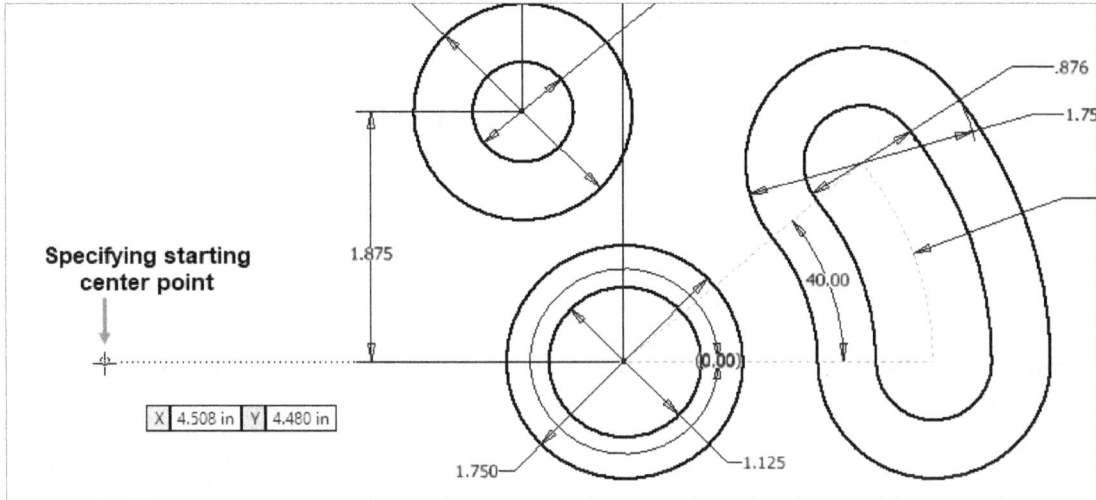

Figure-98. Specifying starting center point of slot

- Click at the location straight left to the origin. You will be asked to specify the end center point for the slot.
- Move cursor towards right and enter **1** in the Input box. You will be asked to specify width of the slot; refer to Figure-99.

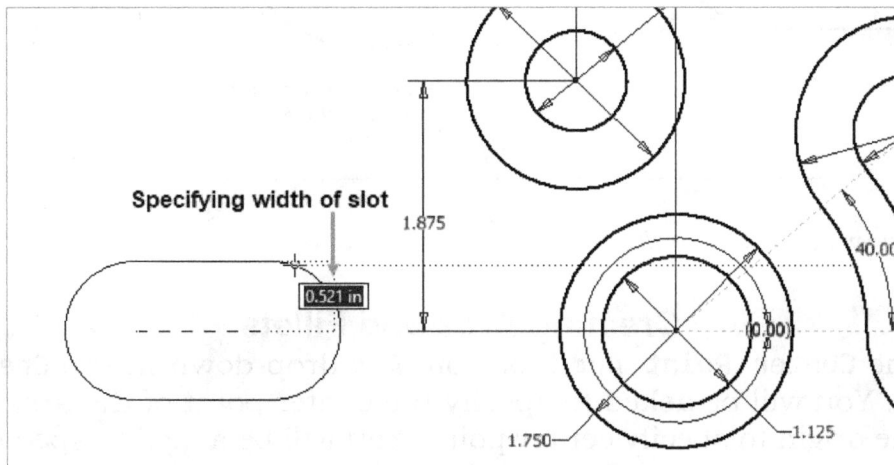

Figure-99. Specifying width of slot

- Right-click and select **Radius** option from the shortcut menu displayed. You will be asked to specify radius for the slot.
- Enter **0.437** in the Input box. The slot will be created.
- Similarly, create a slot of radius **0.75** at the same location; refer to Figure-100.

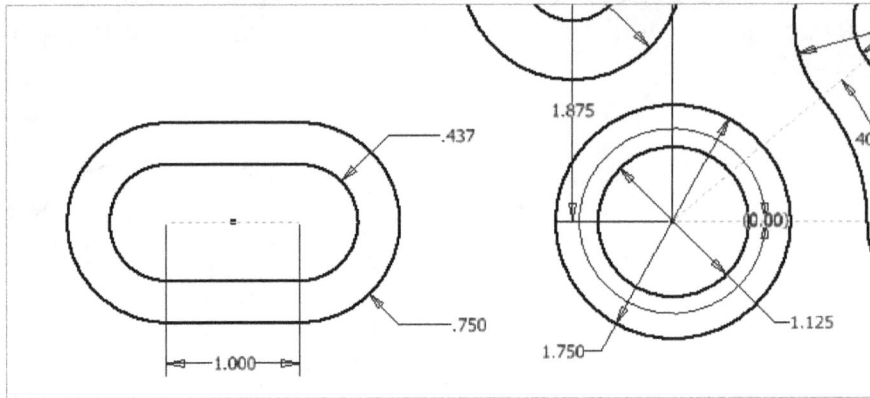

Figure-100. Straight slots created

- Click on the **General Dimension** tool from **Constrain** panel in the **Ribbon** and specify the location of the slot as shown in Figure-101.

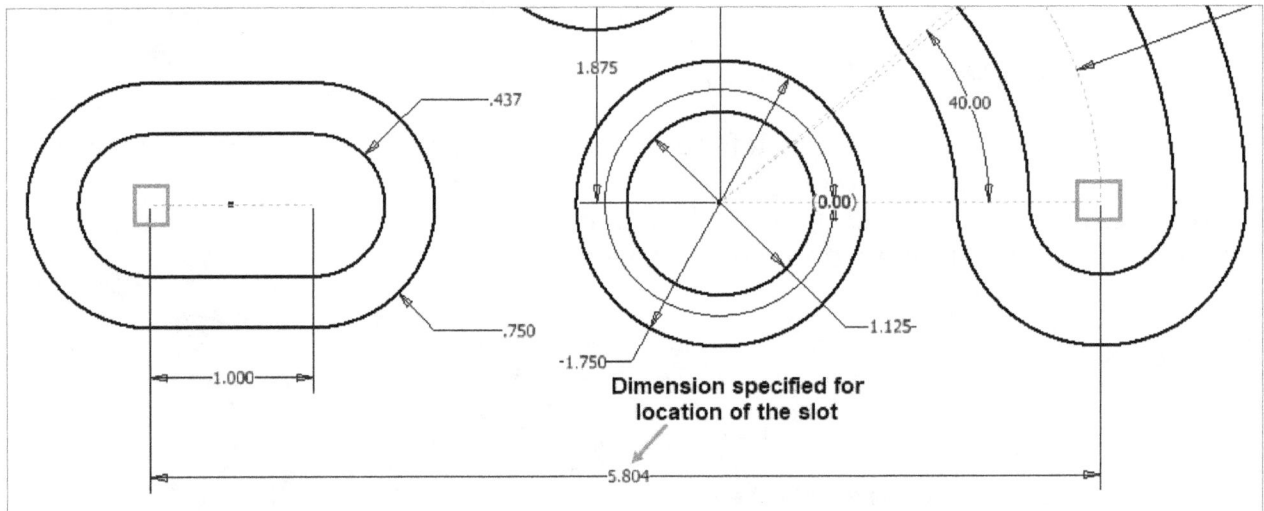

Figure-101. Dimensioning location of the slot

Creating Arcs and Fillets

- Click on the **Center Point Arc** tool from **Arc** drop-down in the **Create** panel of the **Ribbon**. You will be asked to specify the center point of the arc.
- Click on the origin to specify center point. You will be asked to specify start point of the arc.
- Type **1.375** in the Input box and press **TAB** from the keyboard. You will be asked to specify angle of the arc.
- Enter **30** in the Input box. You will be asked to specify angle for end point of the arc; refer to Figure-102.

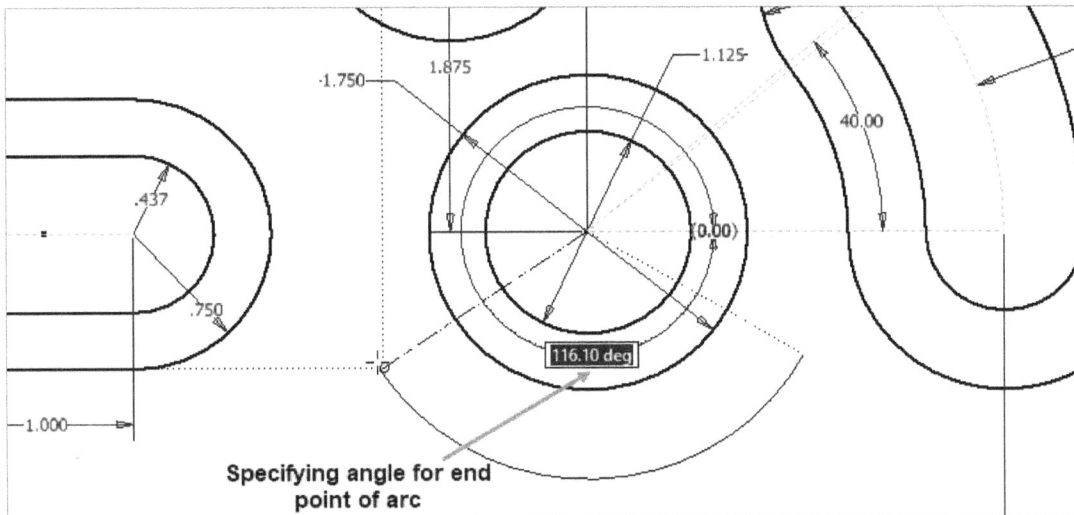

Figure-102. Specifying angle for end point

- Enter **120** in the Input box. The arc will be created.
- Click on the **Three Point Arc** tool from **Create** panel in the **Ribbon**. You will be asked to specify start point of the arc.
- Click on the circle and then on the line to specify start and end points of the arc, respectively; refer to Figure-103. You will be asked to specify radius of the arc.

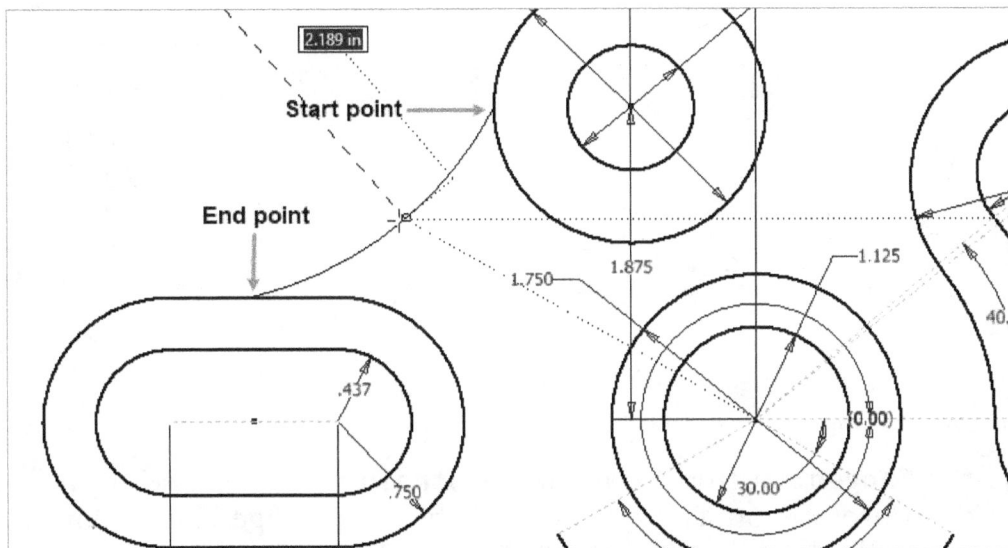

Figure-103. Start and end points of arc

- Enter **1.750** in the Input box. The arc will be created.
- Click on the **Tangent** button from **Constrain** panel in the **Ribbon** and click on newly created arc and then on the connected circle. Similarly, select the newly created arc and then line of slot to make the arc tangent to circle and line of slot.
- Similarly, you can create the other arcs; refer to Figure-104.

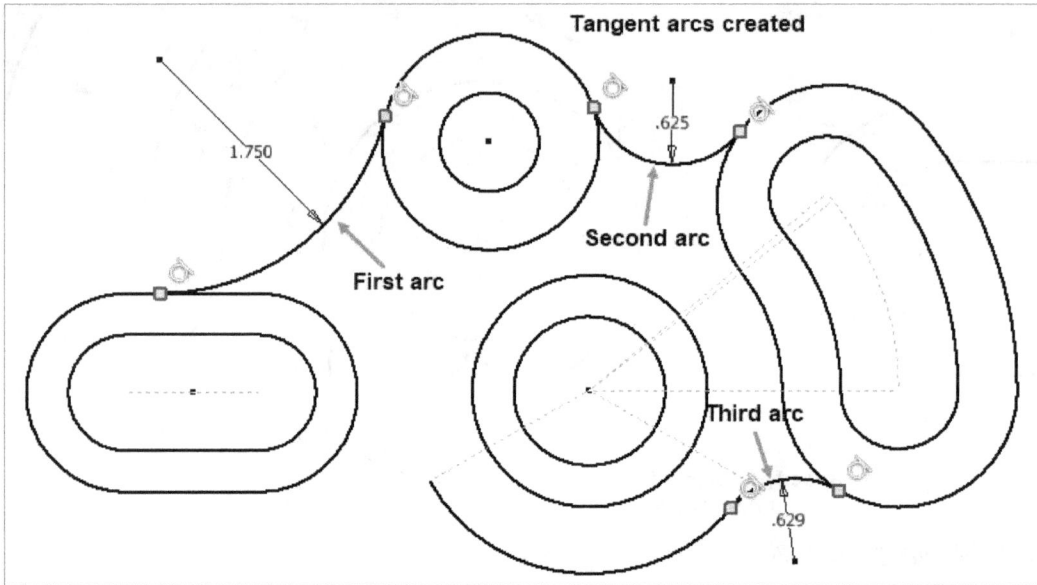

Figure-104. Tangent arcs created

Creating Tangent Line

- Click on the **Line** tool and connect arc of slot to the bottom arc; refer to Figure-105.

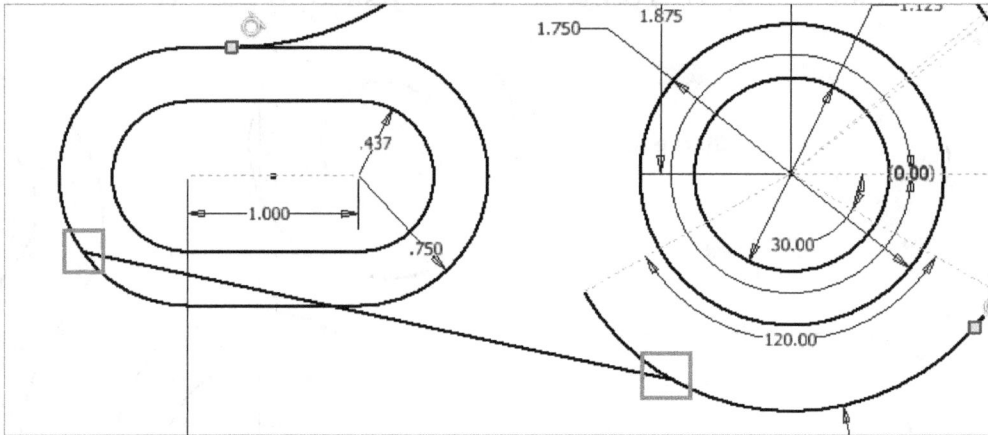

Figure-105. Connecting arc and slot

- Click on the **Tangent** button from the **Constrain** panel and select newly created line and then arc of slot. The entities will become tangent. Similarly, make the line tangent to the bottom arc; refer to Figure-106.

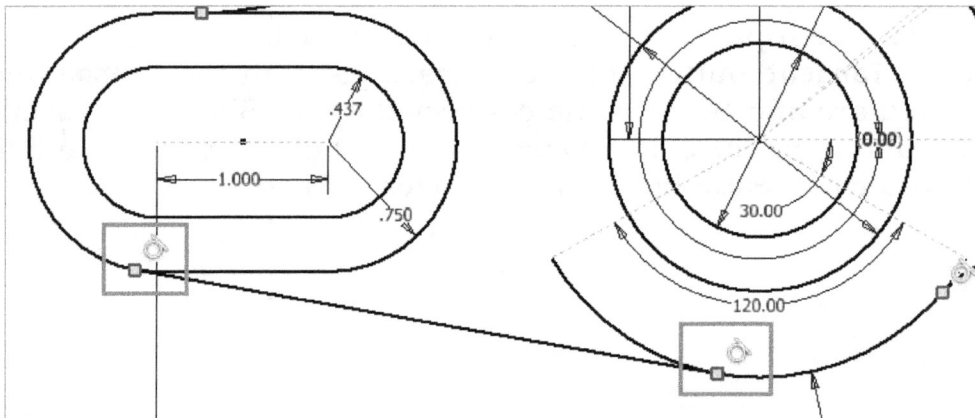

Figure-106. Line tangent to bottom arc and arc of slot

Trimming Extra Sections

- Click on the **Trim** tool from **Modify** panel in the **Ribbon**. You will be asked to select the entities to be removed.
- Click on all the portions of sketch that are not required; refer to Figure-107. The final sketch will be displayed as shown in Figure-108.

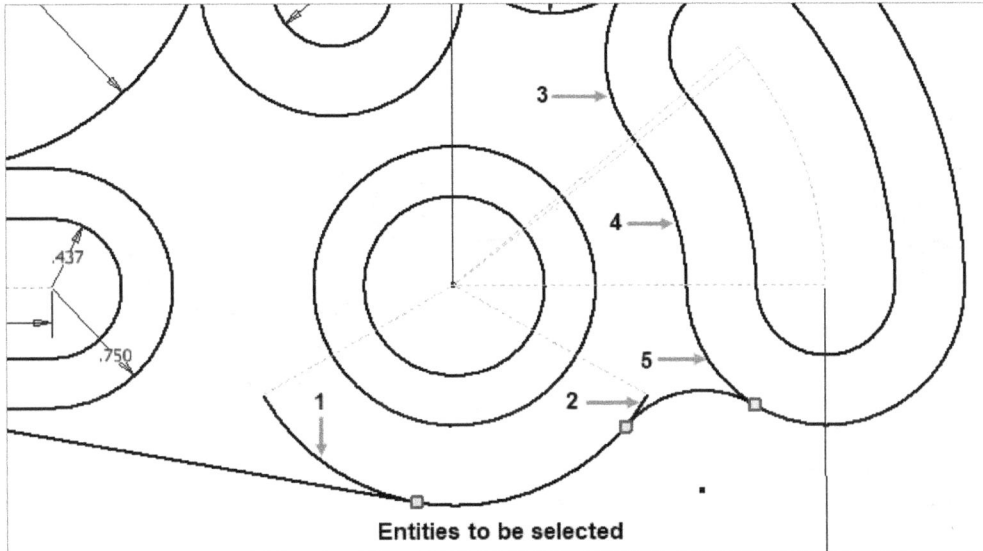

Figure-107. Entities selected for trim

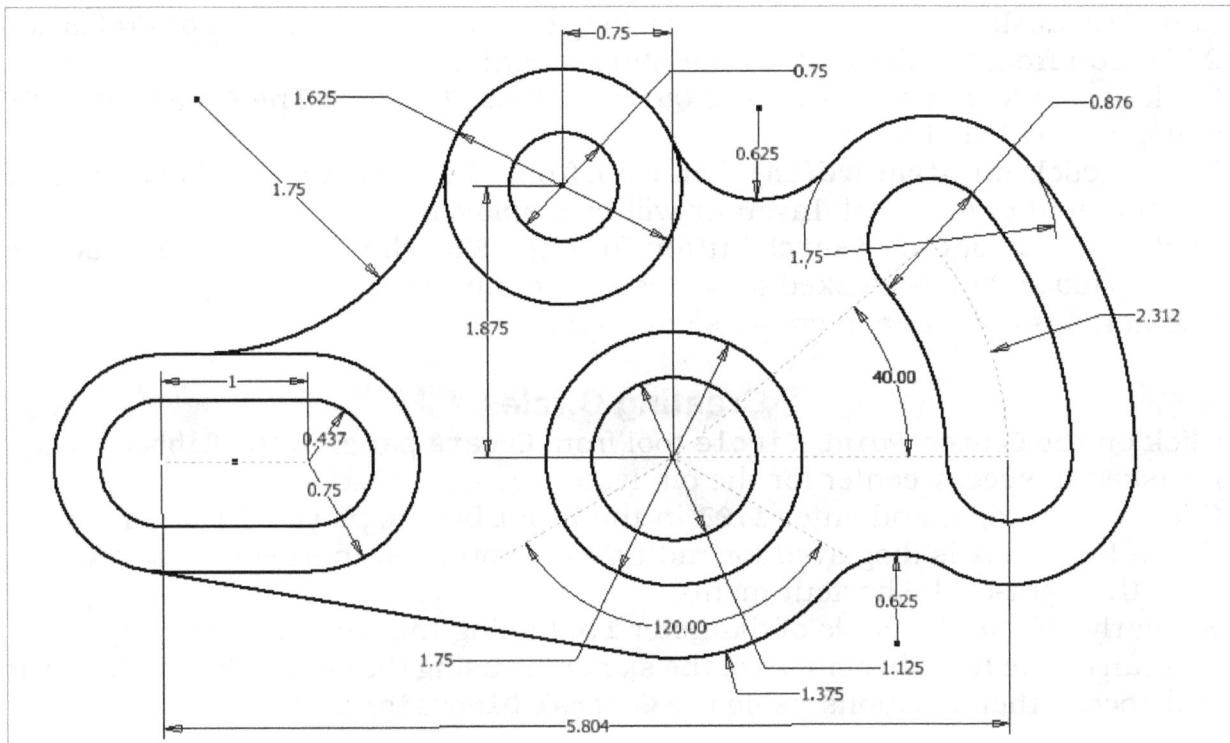

Figure-108. Final sketch for practical 4

Practical 5

In this practical, we will create a sketch for the drawing given in Figure-109.

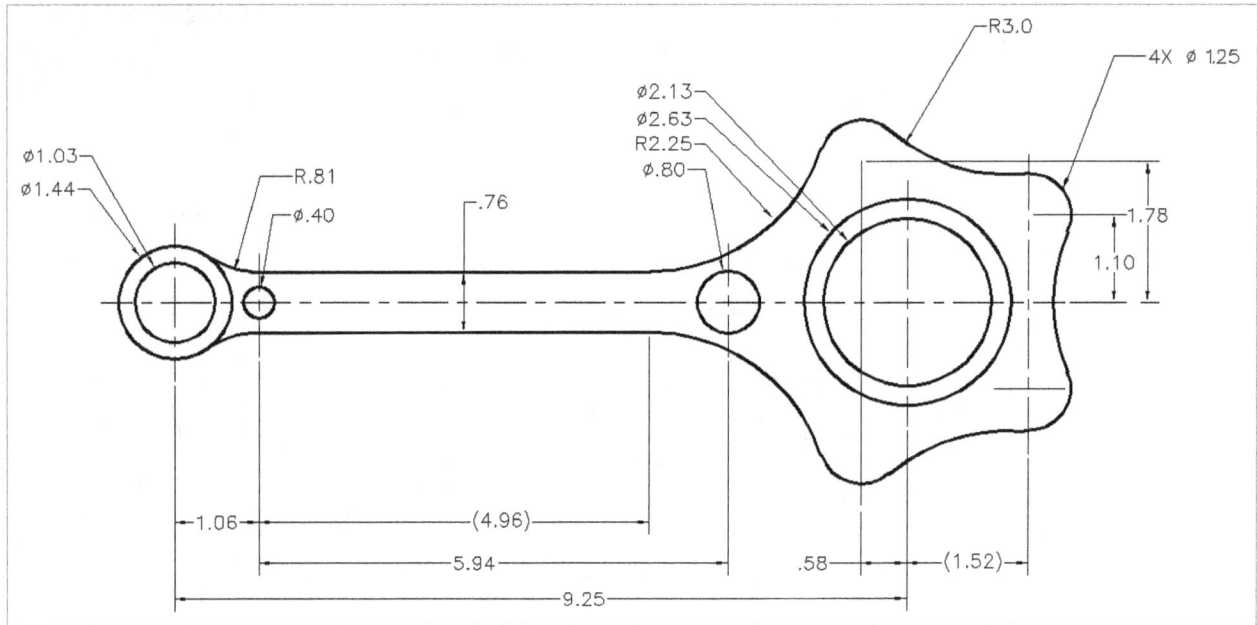

Figure-109. Practical 5

Starting Sketch

- Start Autodesk Inventor by double-clicking on the Autodesk Inventor Professional 2024 icon from the desktop. (If not started yet.)
- Click on the **New** button from the **Quick Access Toolbar**. The **Create New File** dialog box will be displayed.
- Double-click on **Standard(in).ipt** icon from the **English** templates. The Part environment of Autodesk Inventor will be displayed.
- Click on the **Start 2D Sketch** button from the **Sketch** panel in the **3D Model** tab of the **Ribbon**. You are asked to select a sketching plane.
- Click on desired plane to create sketch on it.

Creating Circles

- Click on the **Center Point Circle** tool from **Create** panel in the **Ribbon**. You will be asked to specify center for the circle.
- Click at the origin and enter **1.03** in the Input box displayed for diameter. Note that if Input box is displayed for radius then you need to select **Diameter** option from the right-click shortcut menu.
- Similarly, create the circle of diameter **1.44** using the same center point.
- You can create the other circles of the sketch by using the **Center Point Circle** tool and specify their locations using the **General Dimension** tool; refer to Figure-110.

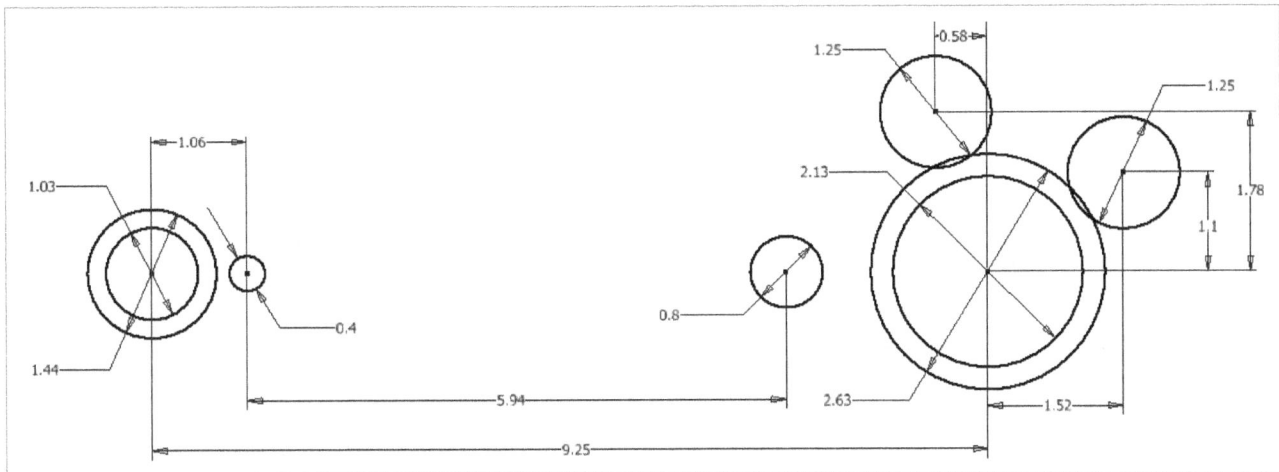

Figure-110. Sketch on creating circles

Creating Mirror Copy

- Click on the **Centerline** button from **Format** panel in the **Ribbon**. The centerline mode will become active.
- Click on the **Line** tool from **Create** panel in the **Ribbon**. You will be asked to specify the starting point of the line.
- Click on the centers of the circles one by one to create the line; refer to Figure-111. Press **ESC** from keyboard to exit the tool.

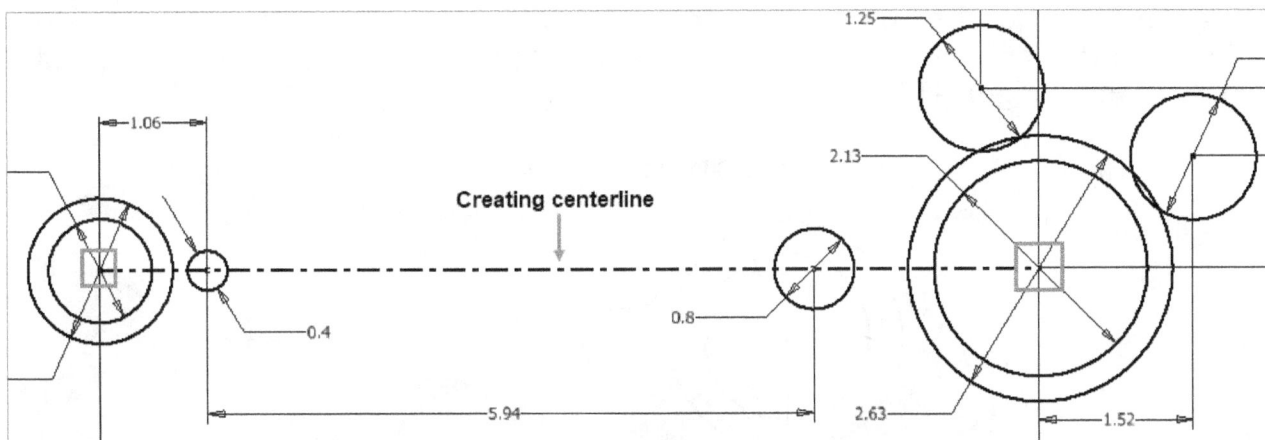

Figure-111. Centerline created

- Click on the **Centerline** button again to exit the centerline mode.
- Click on the **Mirror** tool from **Pattern** panel in the **Ribbon**. You will be asked to select the entities to be mirror copied.
- Select the two circles as shown in Figure-112 and press **ENTER**. You will be asked to select the centerline.

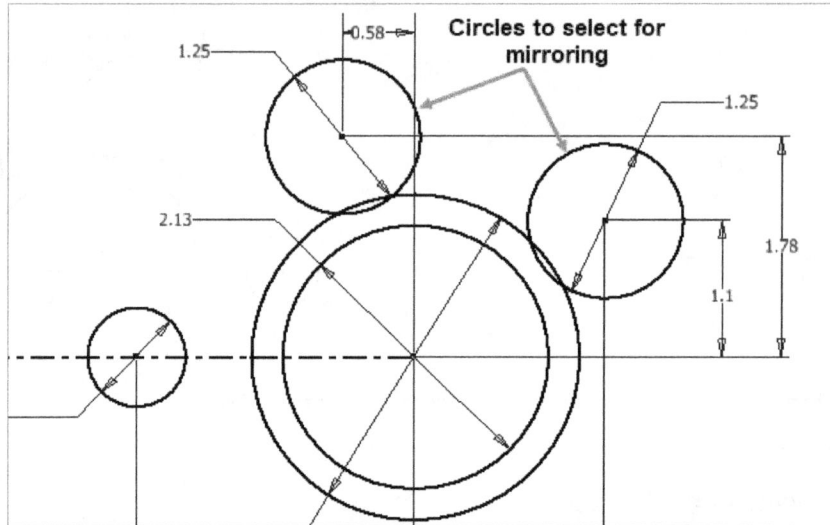

Figure-112. Circles to be selected for mirroring

- Click on the centerline passing through center of circle with diameter **2.130**. The **Apply** button will become active in the **Mirror** dialog box.
- Click on the **Apply** button. The mirror copy of circles will be created. Click on the **Done** button to exit the tool.

Creating Lines

- Click on the **Line** tool from **Create** panel in the **Ribbon**. You will be asked to specify starting point of the line.
- Click on the location which is in vertical line to the center of circle having radius **0.4**; refer to Figure-113.

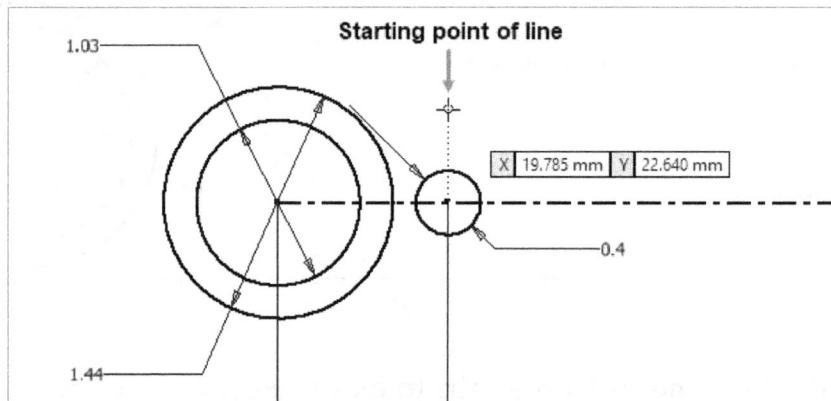

Figure-113. Specifying starting point of line

- Move the cursor straight towards right and enter the distance value as **4.96** in the input box; refer to Figure-114.

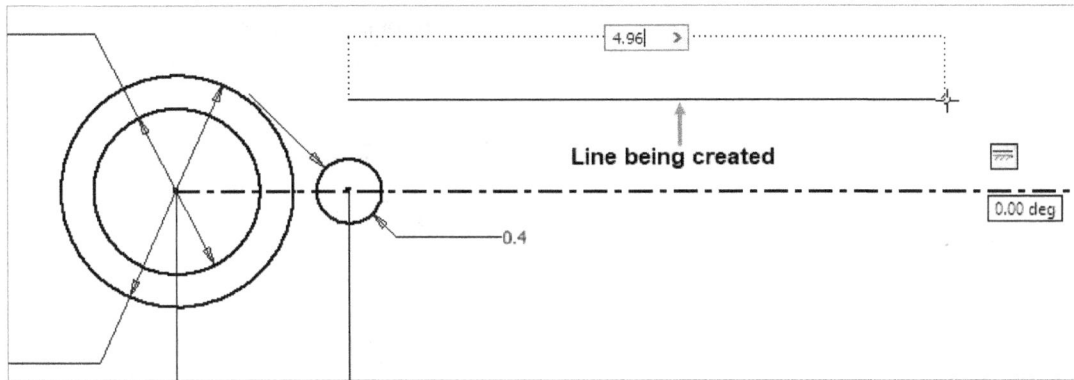

Figure-114. Specifying length of line

- Click on the **General dimension** tool and specify the distance of line from the centerline as **0.38**.
- Create a mirror copy of the line on the other side of centerline; refer to Figure-115.

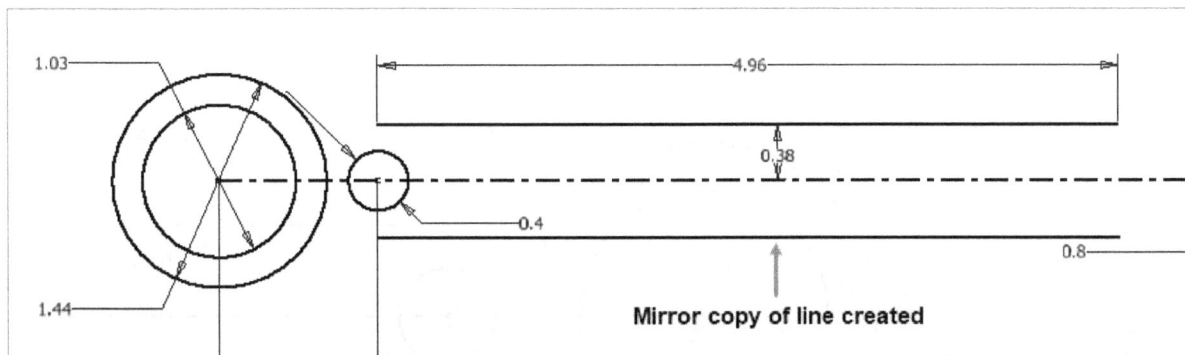

Figure-115. Mirror copy of line created

Creating Arcs

- Click on the **Three Point Arc** tool and click on a point on the circle of diameter **1.44**; refer to Figure-116. You will be asked to specify end point of the arc.

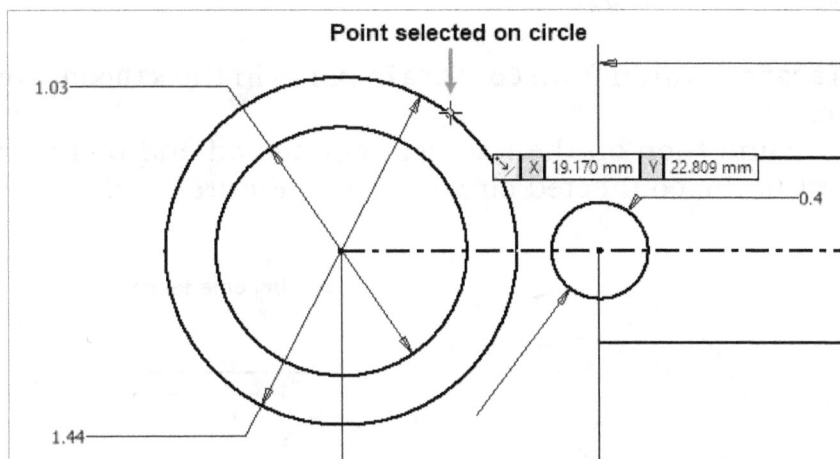

Figure-116. Point selected on circle

- Click at the end point of the line; refer to Figure-117. You will be asked to specify radius of the arc.

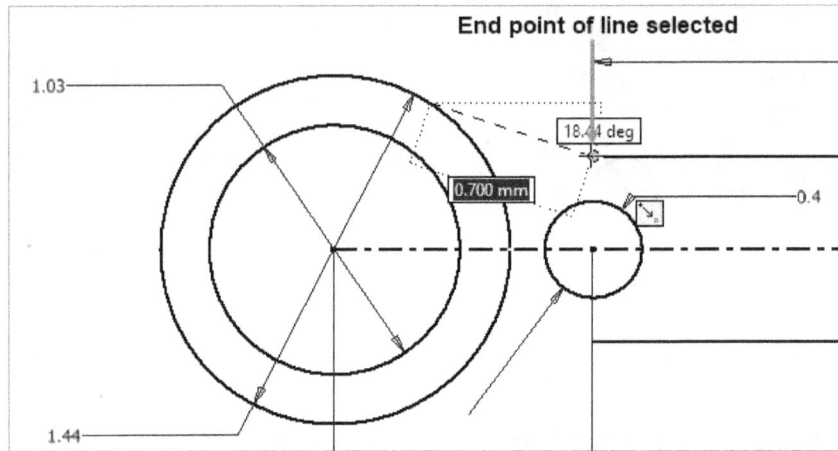

Figure-117. End point of arc specified

- Enter the radius value as **0.81** in the Input box; refer to Figure-118. The arc will be created but it is not tangent to circle and line.

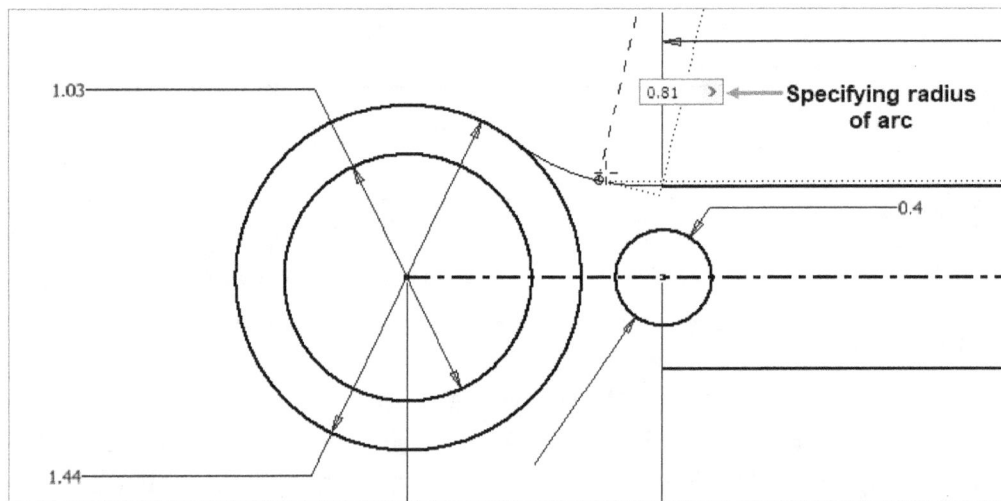

Figure-118. Specifying radius of arc

- Click on the **Tangent** button from **Constrain** panel in the **Ribbon**. You will be asked to select the entities.
- Click on the arc and then on the line near connected end point. Similarly, make the arc tangent to the connected circle; refer to Figure-119.

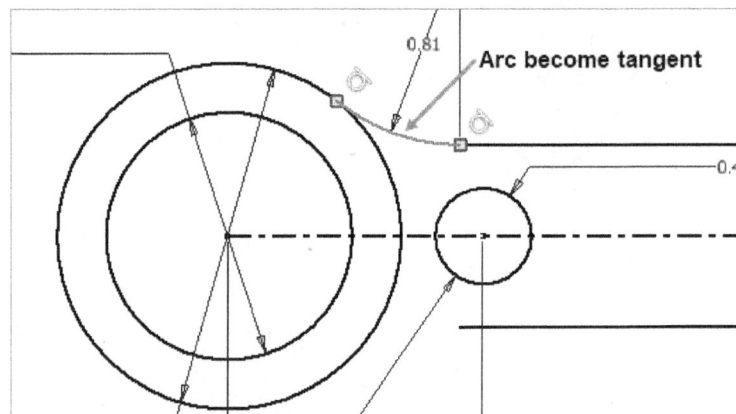

Figure-119. Making arc tangent

- Similarly, create the other arcs and trim the extra sections; refer to Figure-120. The final sketch of Practical 5 will be created; refer to Figure-121.

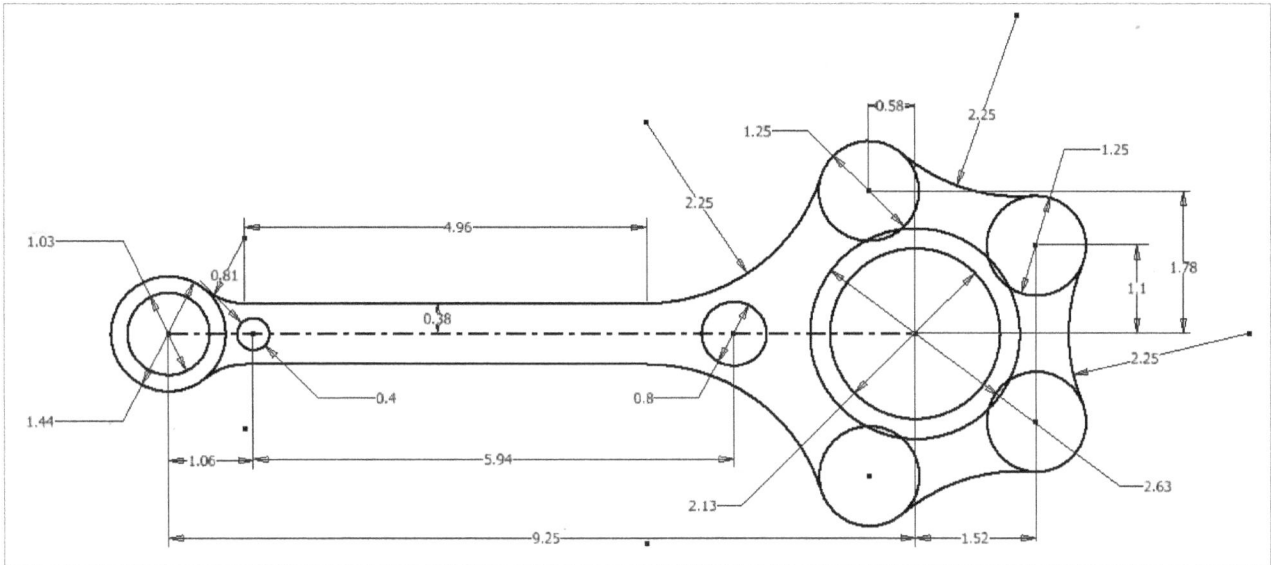

Figure-120. Sketch after creating arcs

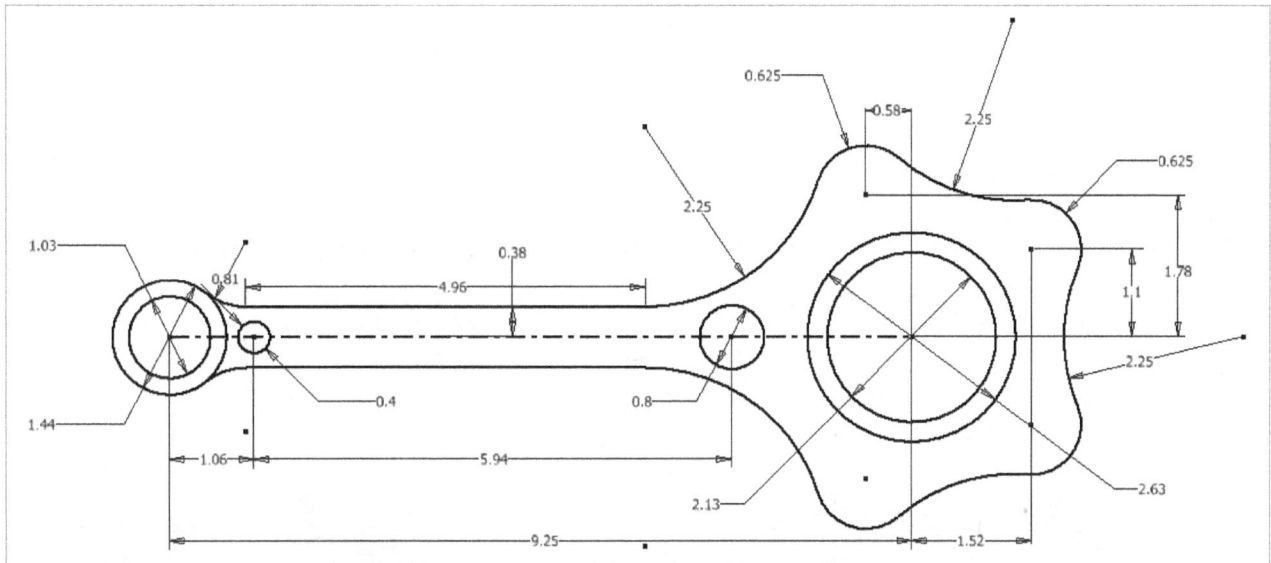

Figure-121. Final sketch for Practical 5

PRACTICE 1

In this practice session, we will create a sketch for the drawing given in Figure-122.

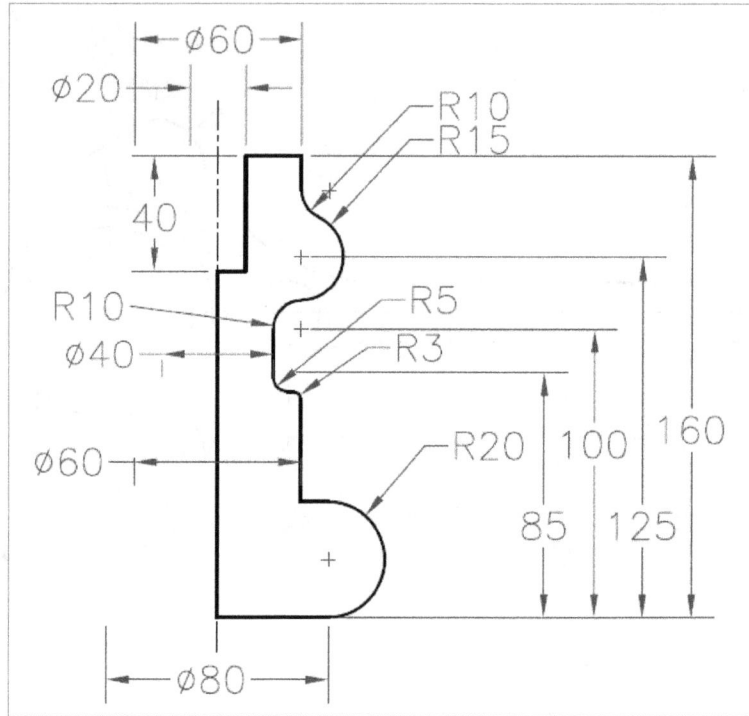

Figure-122. Practice 1

PRACTICE 2

In this practice session, we will create a sketch for the drawing given in Figure-123.

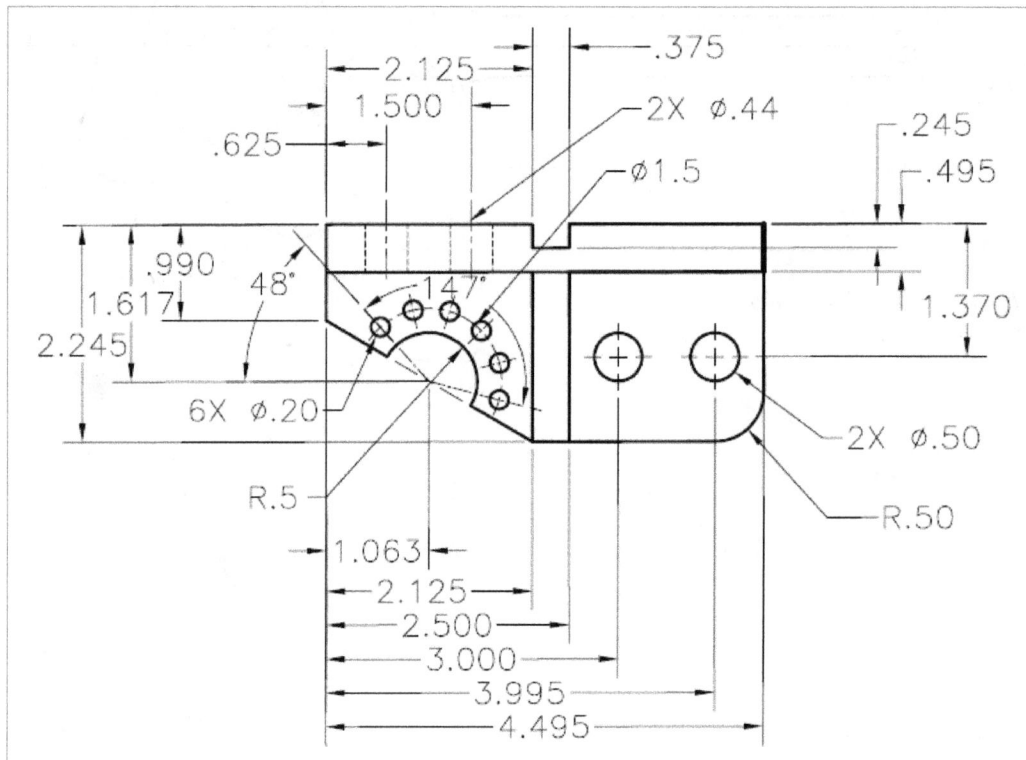

Figure-123. Practice 2

PRACTICE 3

In this practice session, we will create a sketch for the drawing given in Figure-124.

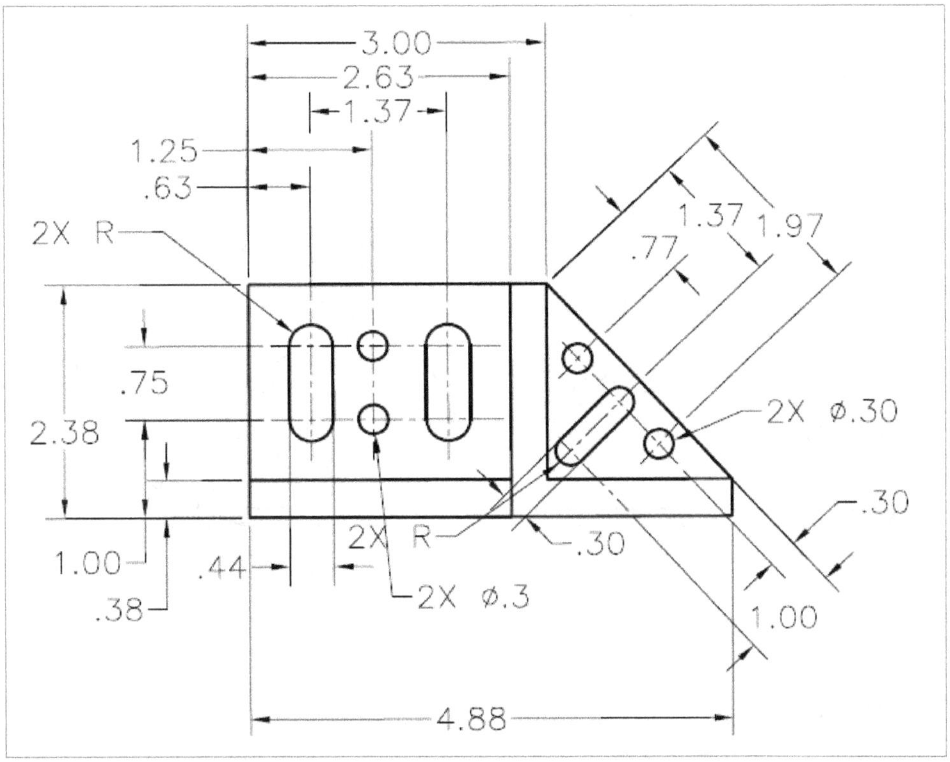

Figure-124. Practice 3

PRACTICE 4

In this practice session, we will create a sketch for the drawing given in Figure-125.

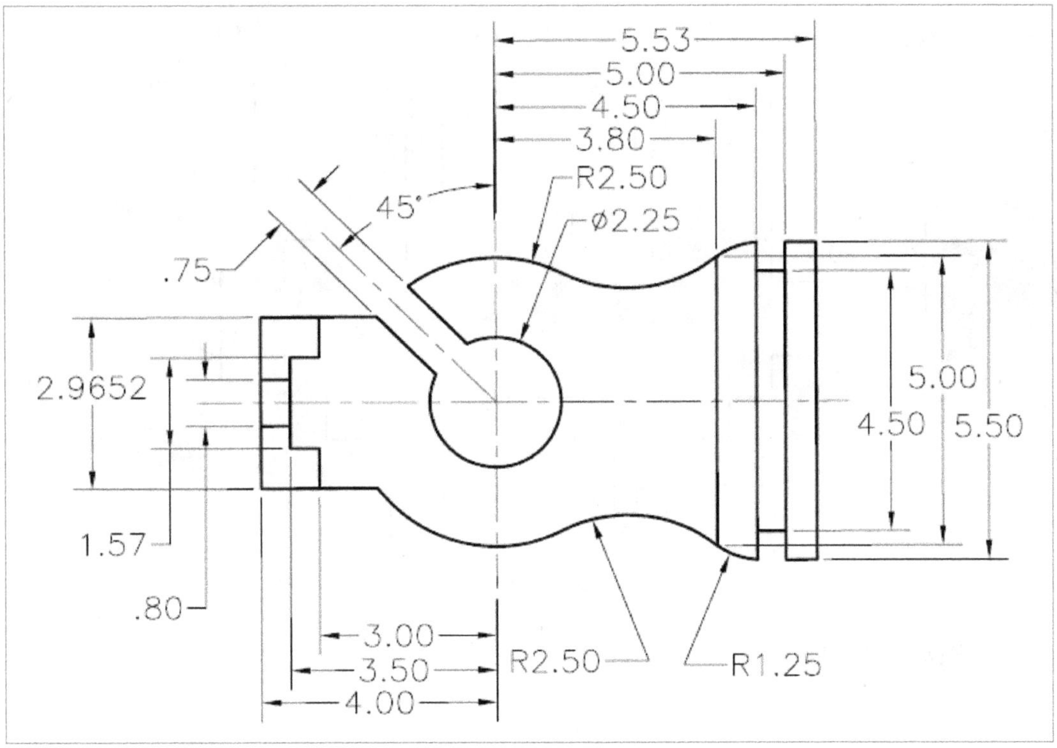

Figure-125. Practice 4

PRACTICE 5

In this practice session, we will create a sketch for the drawing given in Figure-126.

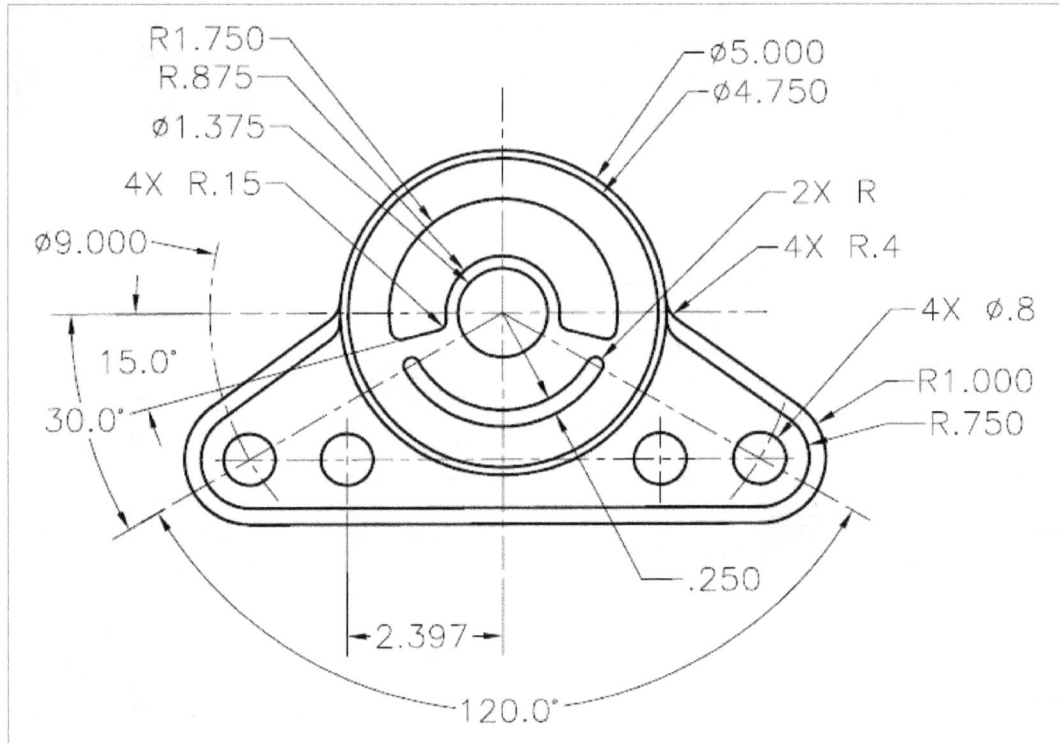

Figure-126. Practice 5

PRACTICE 6

In this practice session, we will create a sketch for the drawing given in Figure-127.

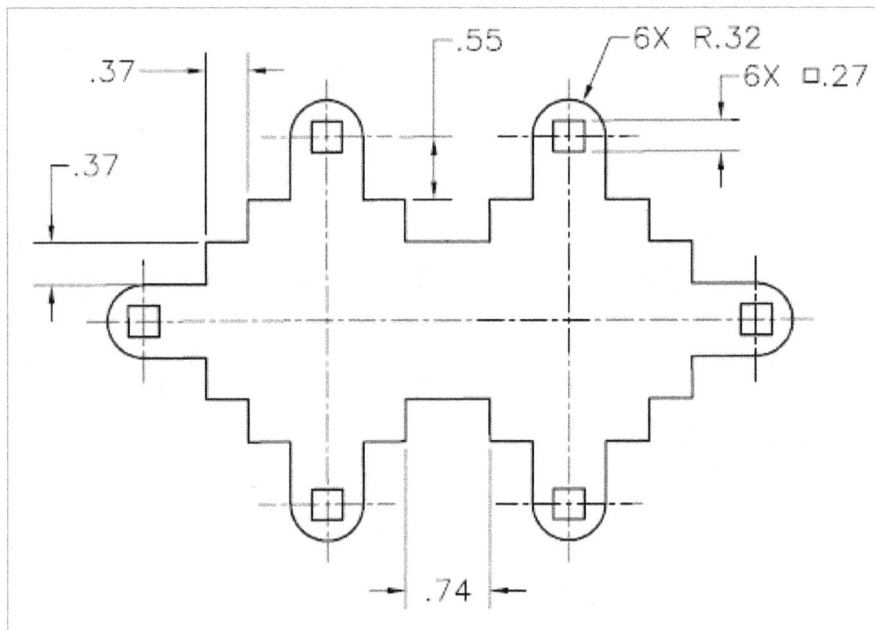

Figure-127. Practice 6

SELF ASSESSMENT

Q1. Which of the following geometries can be dimensioned with **General Dimension** tool?

a) Line
b) Circle
c) Arc
d) All of the above

Q2. Which of the following constraints is used to make two circles, arcs, or ellipses share the same center point?

a) Coincident
b) Collinear
c) Concentric
d) Parallel

Q3. Which of the following constraints is used to make two entities similar about a reference line?

a) Smooth
b) Symmetric
c) Equal
d) Fix

Q4. By enabling which of the following options, the constraints are automatically removed when the sketch entity is selected?

a) Infer constraint
b) Persist constraint
c) Relax mode
d) Fix constraint

Q5. Which of the following objects can be inserted with the help of excel spreadsheet?

a) Text
b) Drawing
c) Image
d) Point

Q6. The **Automatic Dimensions and Constraints** tool is used to automatically apply dimensions and constraints to all the entities in the sketch. (True/False)

Q7. The **Perpendicular** constraint tool is used to make geometries perpendicular to each other. (True/False)

Q8. The **Symmetric** constraint is used to make two selected entities equal in size. (True/False)

Q9. The .... tool is used to display constraints of the selected sketch entities.

Q10. Every sketch entity has .... degree of freedom denoting rotation and translation in each direction.

FOR STUDENT NOTES

Chapter 4

Creating Solid Models

Topics Covered

The major topics covered in this chapter are:

- *Introduction to 3D Models.*
- *Work Planes, axes, and geometric point creation.*
- *Extrude Feature.*
- *Brief note on Drawings.*
- *Sketching plane selection.*
- *Practical and Practice*

INTRODUCTION

We welcome you to the 3D world of Autodesk Inventor. Like any other 3D CAD modeling software, Autodesk Inventor has full set of tools to create 3D models on the basis of specified dimensions. Till this point, we have learned about the 2D sketching tools. Now, we will be using those 2D sketches to create 3D models which can later be used for CAM (Computer Aided Manufacturing) or CAE (Computer Aided Engineering) applications. The tools to create 3D models are available in the **3D Model** tab of the **Ribbon**; refer to Figure-1. We will discuss these tools one by one in this chapter.

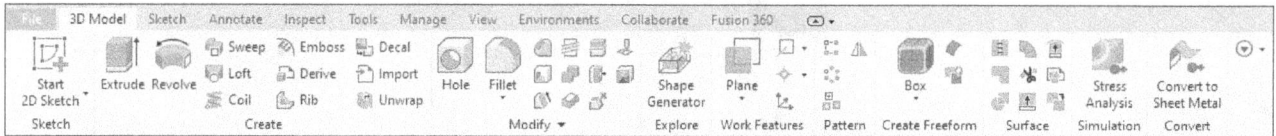

Figure-1. 3D Model tab

Before we starting using the modeling tools, we need to understand the construction geometries like plane, axis, points, and coordinate system. Various construction geometries and their respective tools are discussed next.

WORK PLANE

The Work planes act as floor for 3D model features. By default, three planes are available in Inventor named, XY Plane (Front Plane), YZ Plane (Right Plane), and XZ Plane (Top Plane); refer to Figure-2. To display the default planes, select them from the **Model Browse Bar** in the left of the dialog box while holding the **CTRL** key and right-click on them. A shortcut menu will be displayed; refer to Figure-3. Click on the **Visibility** option from the shortcut menu. The planes will be displayed permanently. We can also create planes whenever required by using the tools available in the **Plane** drop-down of **Work Features** panel in the **Ribbon**; refer to Figure-4. We will learn to use the plane creation tools one by one. Note that there might be some features used for creating plane that you will learn to create later in the book.

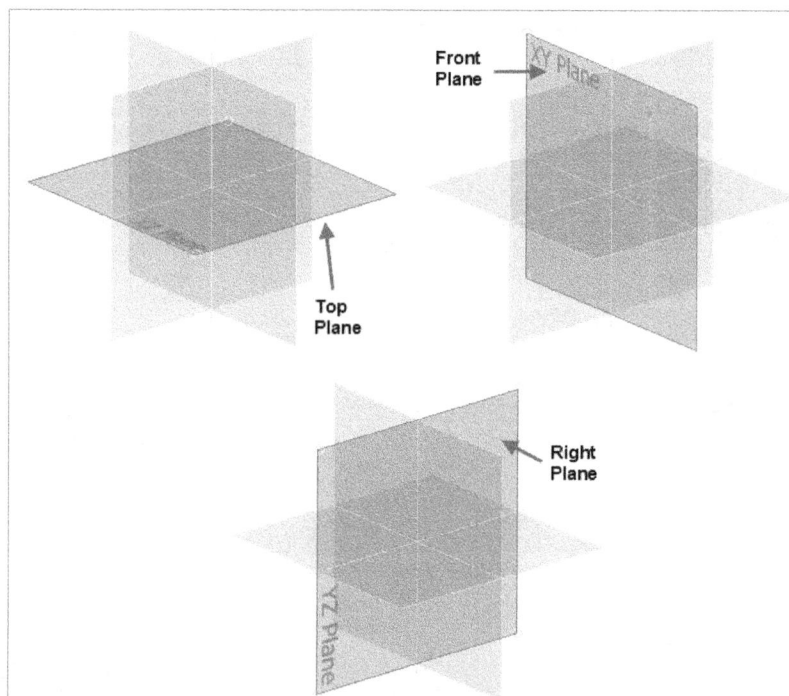

Figure-2. Default Planes

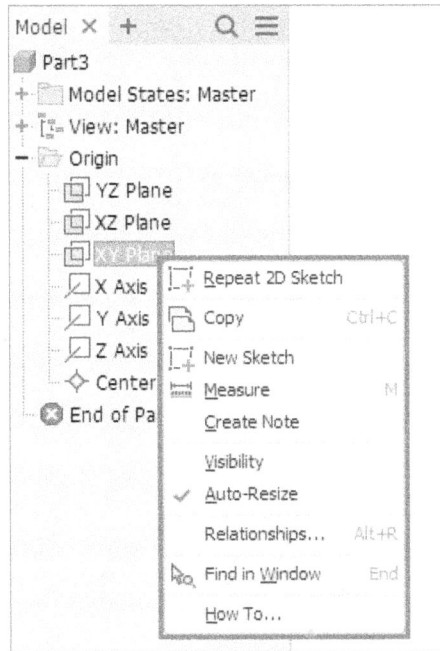

Figure-3. Shortcut menu for planes

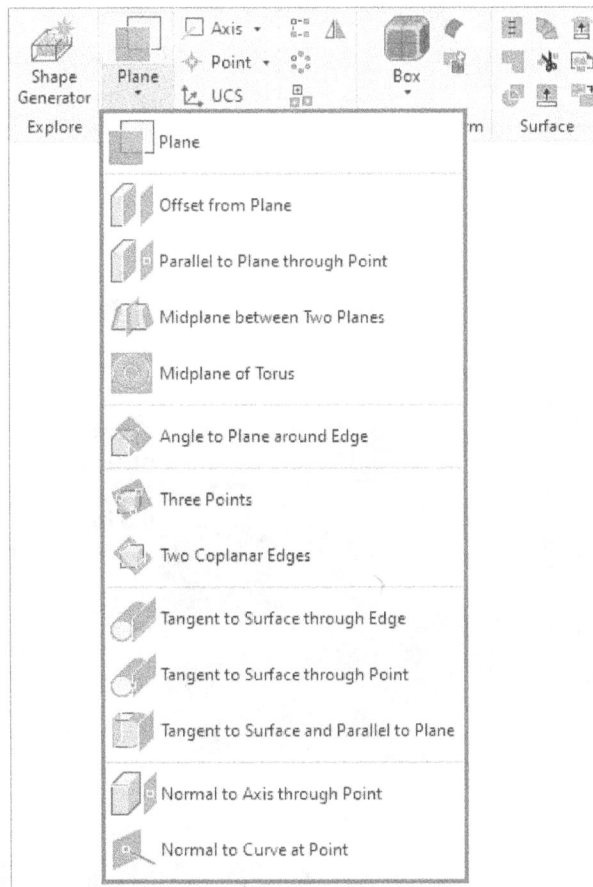

Figure-4. Plane drop-down

Offset from Plane

The **Offset from Plane** tool is used to create a plane at specified offset distance from the selected plane/flat face. The procedure to use this tool is given next.

- Click on the **Offset from Plane** tool from **Plane** drop-down in the **Work Features** panel of the **Ribbon**. You will be asked to select a plane or planar face.
- Select the plane or flat face. You will be asked to specify the offset distance; refer to Figure-5.

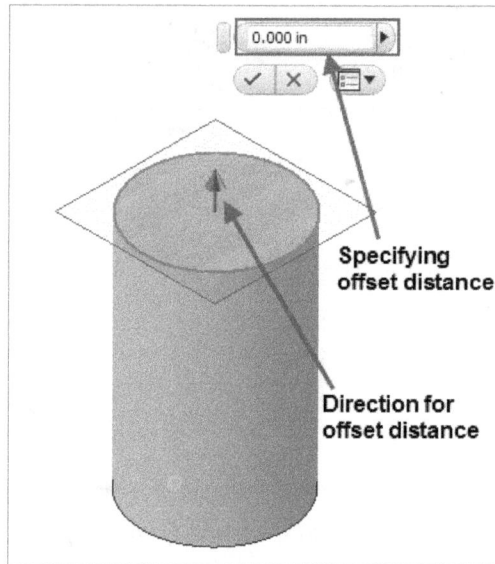

Figure-5. Specifying offset distance for plane

- Enter desired value of distance in the Input box. The plane will be created; refer to Figure-6. Note that you can specify a negative value to create plane in reverse direction.

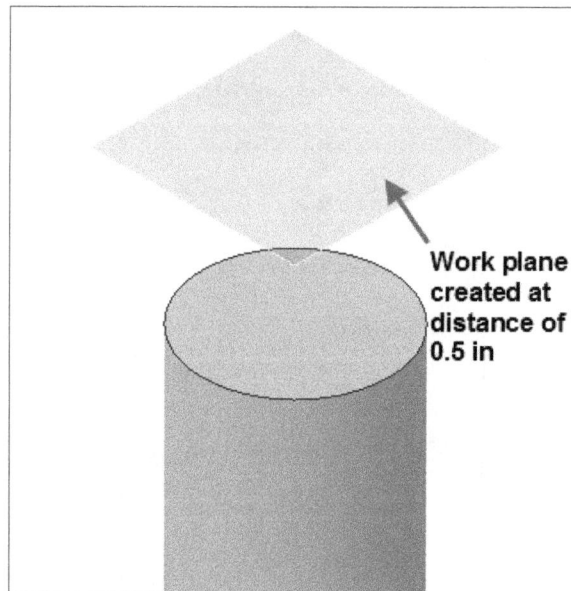

Figure-6. Workplane created

Parallel to Plane through Point

The **Parallel to Plane through Point** tool, as the name suggests, is used to create plane parallel to the selected plane and passing through selected point. The procedure to create plane using this tool is given next.

- Click on the **Parallel to Plane through Point** tool from **Plane** drop-down in the **Work Features** panel of the **Ribbon**. You will be asked to select a point or plane.

- Click on the plane or face to which you want the new plane to be parallel. You will be asked to select a point through which you want the plane to pass.
- Click on the point. The new work plane will be created; refer to Figure-7.

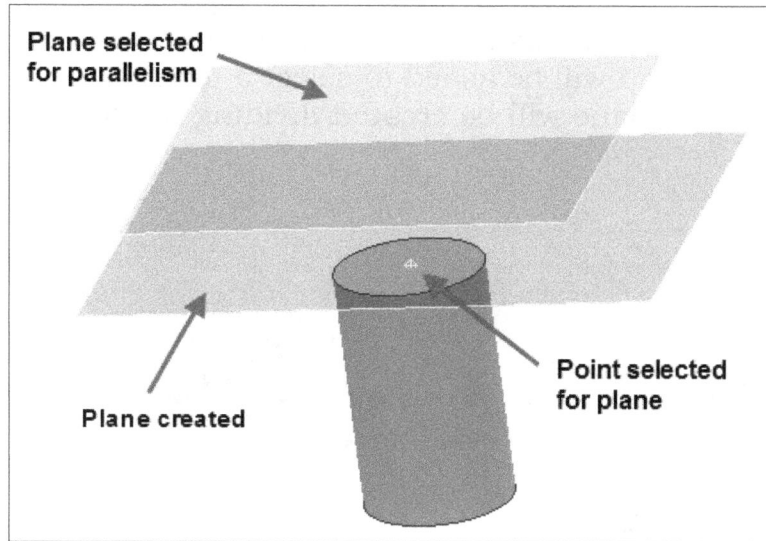

Figure-7. Plane created through point

Midplane between Two Planes

The **Midplane between Two Planes** tool is used to create a plane through the median of selected planes. The procedure to create midplane is given next.

- Click on the **Midplane between Two Planes** tool from **Plane** drop-down in the **Work Features** panel of the **Ribbon**. You will be asked to select the first plane.
- Click on the first plane/face. You will be asked to select the second plane.
- As soon as you hover the cursor on other plane/face, preview of the new plane will be displayed; refer to Figure-8.

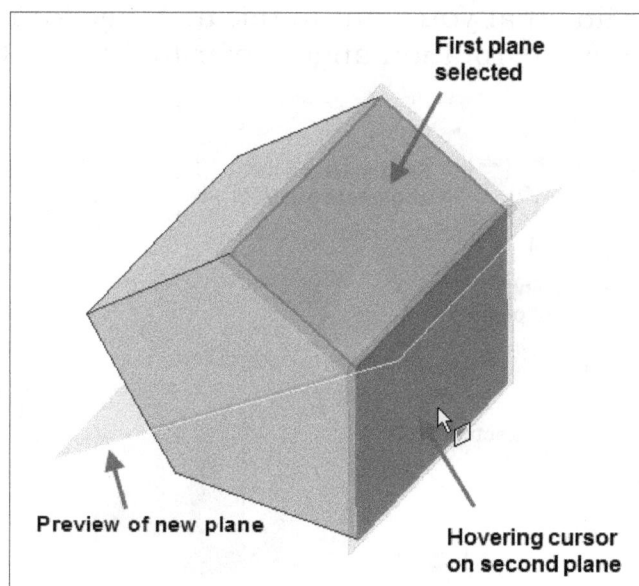

Figure-8. Creating midplane

- Click on the second plane/face to create the mid plane.

Midplane of Torus Tool

The **Midplane of Torus** tool is specific tool to create plane at the middle of torus. The procedure to use this tool is given next.

- Click on the **Midplane of Torus** tool from **Plane** drop-down in the **Work Features** panel of the **Ribbon**. You will be asked to select a torus.
- Click on the torus. A plane will be created dividing the torus at middle; refer to Figure-9.

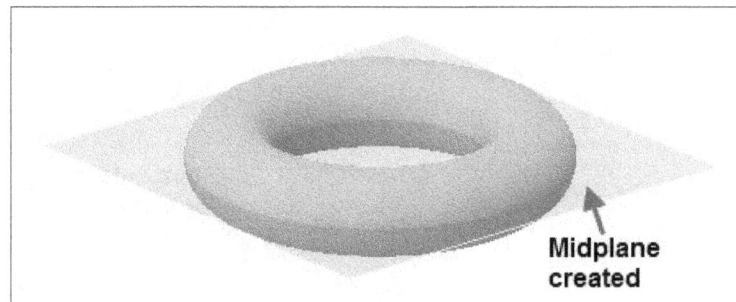

Figure-9. Midplane of torus

Angle to Plane around Edge

The **Angle to Plane around Edge** tool is used to create a plane at an angle to the selected plane around the selected edge. This option is very helpful when you are going to create features at an angle to the other features. The procedure to create plane by using this tool is given next.

- Click on the **Angle to Plane around Edge** tool from **Plane** drop-down in the **Work Features** panel of the **Ribbon**. You will be asked to select a line or a plane.
- Click on the plane which you want to set as reference for angle. In simple words, select the plane to which the new plane will be at angle. You will be asked to select a line or edge.
- Click on the edge or line that you want to use as hinge for rotation of plane. You will be asked to specify the rotation angle; refer to Figure-10.

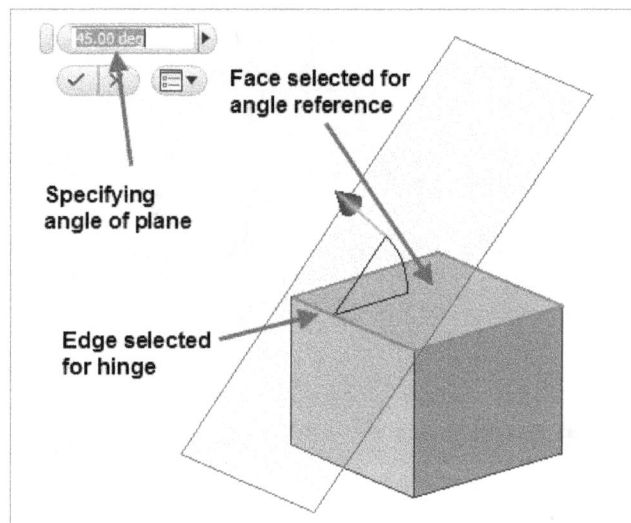

Figure-10. Creating plane at angle

- Enter desired value of angle in the Input box. The plane will be created at specified angle; refer to Figure-11.

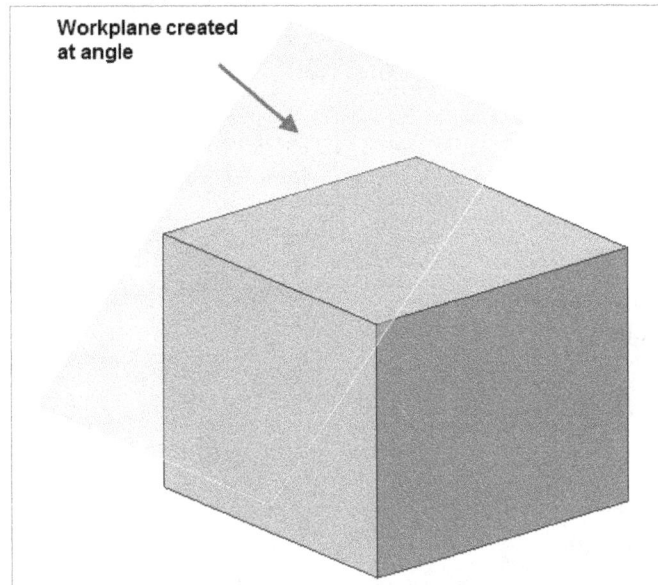

Figure-11. Workplane created at angle

Three Points

The **Three Points** tool is used to create plane passing through the specified three points. The procedure to use this tool is given next.

- Click on the **Three Points** tool from the **Plane** drop-down. You will be asked to select the reference points.
- Click on desired three points one by one. The plane will be created; refer to Figure-12.

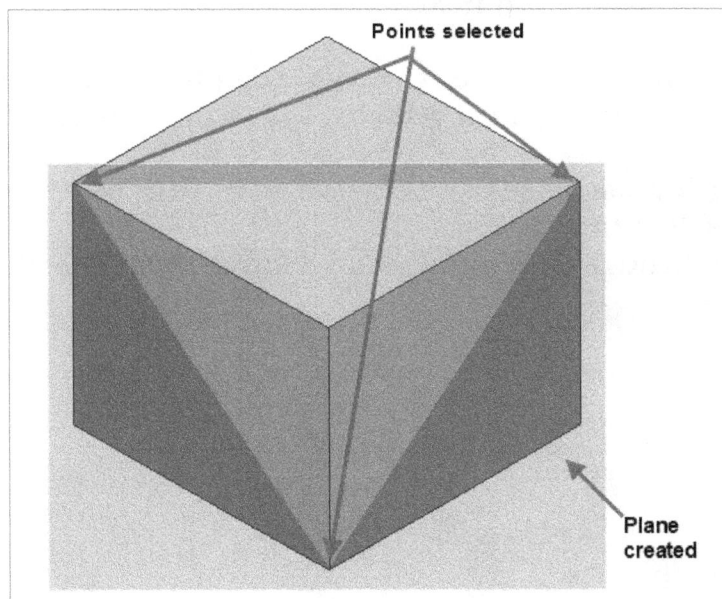

Figure-12. Three point plane created

Two Coplanar Edges

The **Two Coplanar Edges** tool is used to create a plane passing through the two selected coplanar edges. The procedure to create plane by using this tool is given next.

- Click on the **Two Coplanar Edges** tool from **Plane** drop-down in the **Work Features** panel of the **Ribbon**. You will be asked to select an edge, axis, or line.
- Click on desired edge/axis/line. You will be asked to select the second edge/axis/line.
- Hover the cursor on desired edge/axis/line. Preview of the plane will be displayed; refer to Figure-13.
- Click on the edge/line/axis to create the plane.

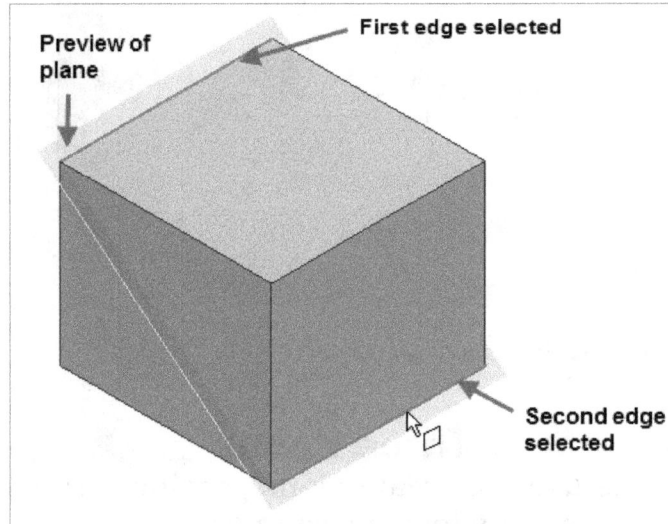

Figure-13. Preview of plane

Tangent to Surface through Edge

The **Tangent to Surface through Edge** tool, as the name suggests, is used to create a plane tangent to the selected surface and passing through the selected edge. The procedure to use this tool is given next.

- Click on the **Tangent to Surface through Edge** tool from the **Plane** drop-down in the **Work Features** panel of the **Ribbon**. You will be asked to select an edge or surface.
- Click on desired round surface to which you want to make the plane tangent. You will be asked to select the edge.
- Click on the edge through which the plane should pass. The plane will be created; refer to Figure-14.

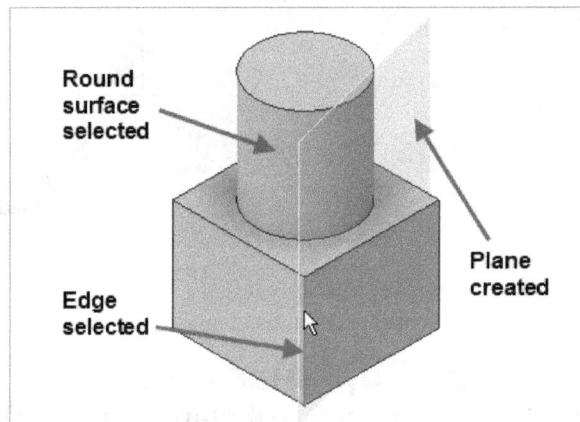

Figure-14. Tangent plane through edge

Tangent to Surface through Point

The **Tangent to Surface through Point** tool is used to create a plane tangent to the selected surface and passing through the selected point. The procedure to create plane by using this tool is given next.

- Click on the **Tangent to Surface through Point** tool from **Plane** drop-down in the **Work Features** panel of the **Ribbon**. You will be asked to select a point or surface.
- Click on the point through which you want the plane to pass. You will be asked to select a round surface.
- Click on the surface to which you want the plane to be tangent. The plane will be created; refer to Figure-15. Note that to create this plane, the point must be on the curved surface.

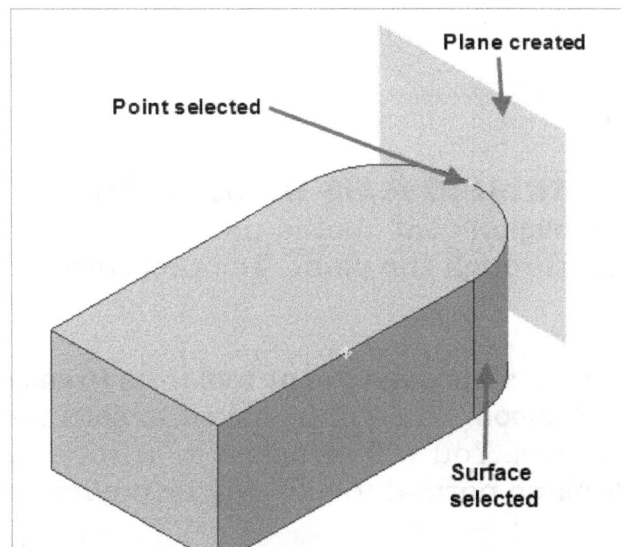

Figure-15. Tangent plane through point

Tangent to Surface and Parallel to Plane

The **Tangent to Surface and Parallel to Plane** tool is used to create a plane tangent to the selected surface and parallel to the selected plane. The procedure to use this tool is given next.

- Click on the **Tangent to Surface and Parallel to Plane** tool from **Plane** drop-down in the **Work Features** panel of the **Ribbon**. You will be asked to select a curved surface or a plane.
- Select the curved surface that you want to be tangent to the plane. You will be asked to select a plane.
- Click on the plane to which you want the new plane to be parallel. The plane will be created; refer to Figure-16.

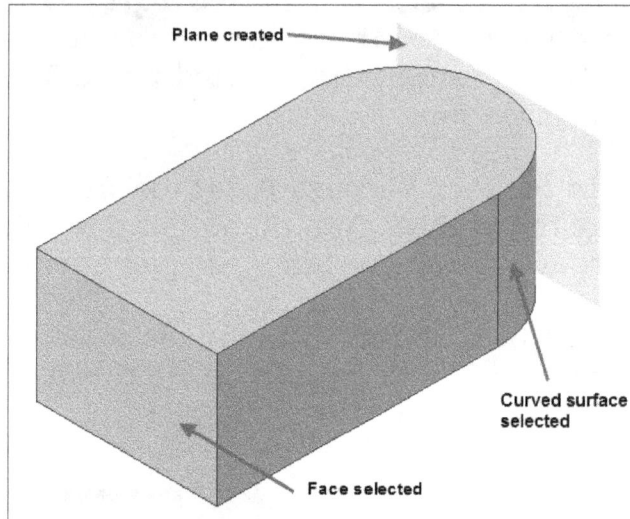

Figure-16. Tangent plane parallel to selected face

Normal to Axis through Point

The **Normal to Axis through Point** tool is used to create a plane normal to the selected axis and passing through the point. The procedure to use this tool is given next.

- Click on the **Normal to Axis through Point** tool from **Plane** drop-down in the **Work Features** panel of the **Ribbon**. You will be asked to select an edge/axis or point.
- Click on desired edge/axis. You will be asked to select a point.
- Click on the point. A plane normal to axis will be created; refer to Figure-17.

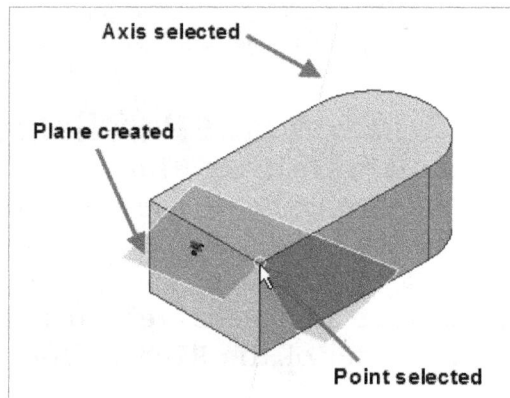

Figure-17. Plane created normal to axis

Normal to Curve at Point

The **Normal to Curve at Point** tool is used to create a plane normal to the selected curve and passing through the selected point. The procedure to use this tool is given next.

- Click on the **Normal to Curve at Point** tool from **Plane** drop-down in the **Work Features** panel of the **Ribbon**. You will be asked to select a curve or point.
- Click on desired point of the curve to which the plane should be perpendicular. You will be asked to select a curve.

- As soon as you hover the cursor on desired curve, a preview of plane will be displayed; refer to Figure-18. Click on the curve to create the plane.

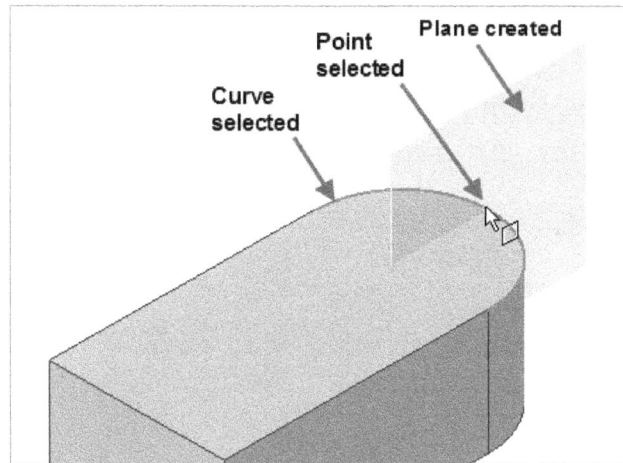

Figure-18. Plane created normal to curve

Plane

Yes!! This was the first tool in the **Plane** drop-down and we are discussing it at the last. This is because the **Plane** tool can create all the type of planes discussed earlier except the offsetted planes. The procedure to use this tool is given next.

- Click on the **Plane** tool from **Plane** drop-down in the **Work Features** panel of the **Ribbon**. You will be asked to select a geometry.
- Select desired geometries one by one. The plane will be created based on the selection.

AXIS

Axis is generally used to support creation of round features and creation of planes. It can also be used as reference for directions. The tools to create axes are available in the **Axis** drop-down in the **Work Features** panel of the **Ribbon**; refer to Figure-19. Tools in this drop-down are discussed next.

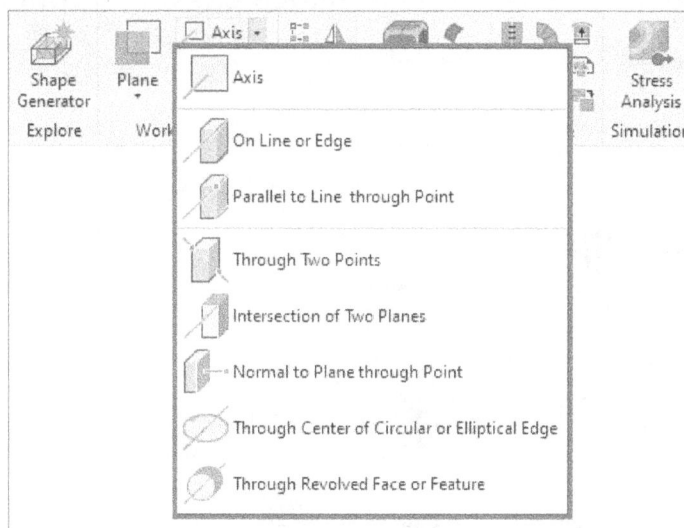

Figure-19. Axis drop-down

On Line or Edge

The **On Line or Edge** tool is used to create an axis on the selected line/edge. The procedure to use this tool is given next.

- Click on the **On Line or Edge** tool from **Axis** drop-down in the **Work Features** panel of the **Ribbon**. You will be asked to select an edge or line.
- Click on the edge or line. An axis will be created along the selected edge/line; refer to Figure-20.

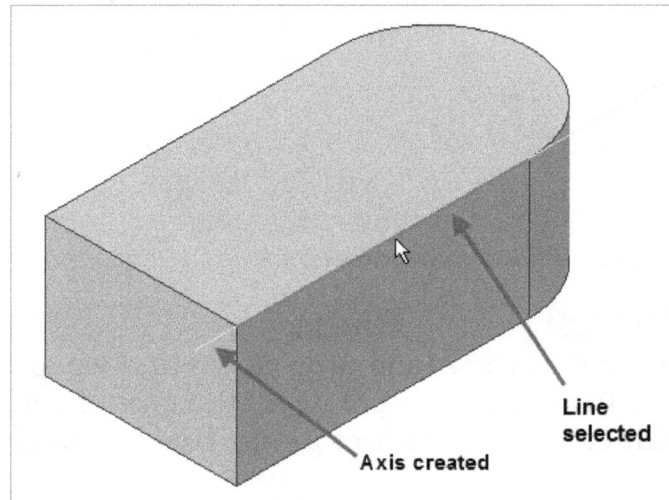

Figure-20. Axis created along selected edge or line

Parallel to Line through Point

The **Parallel to Line through Point** tool, as the name suggests, is used to create an axis parallel to selected edge/line and passing through the selected point. The procedure to create axis by this tool is given next.

- Click on the **Parallel to Line through Point** tool from **Axis** drop-down in the **Work Features** panel of the **Ribbon**. You will be asked to select a line/edge or point.
- Click on desired line/edge. You will be asked to select a point.
- Click on desired point. The axis will be created passing through the selected point; refer to Figure-21.

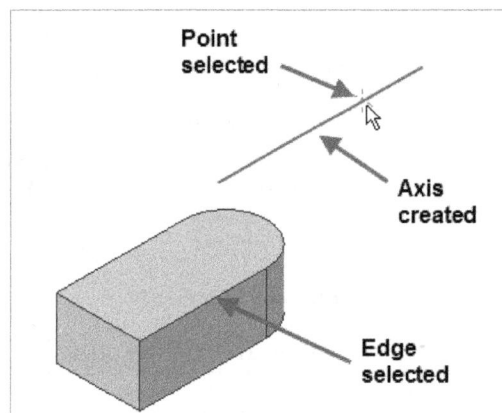

Figure-21. Axis created passing through point

Through Two Points

The **Through Two Points** tool is used to create an axis passing through selected two points. The procedure to use this tool is given next.

* Click on the **Through Two Points** tool from **Axis** drop-down in the **Work Features** panel of the **Ribbon**. You will be asked to select a point.
* Click on desired two points one by one. The axis will be created; refer to Figure-22.

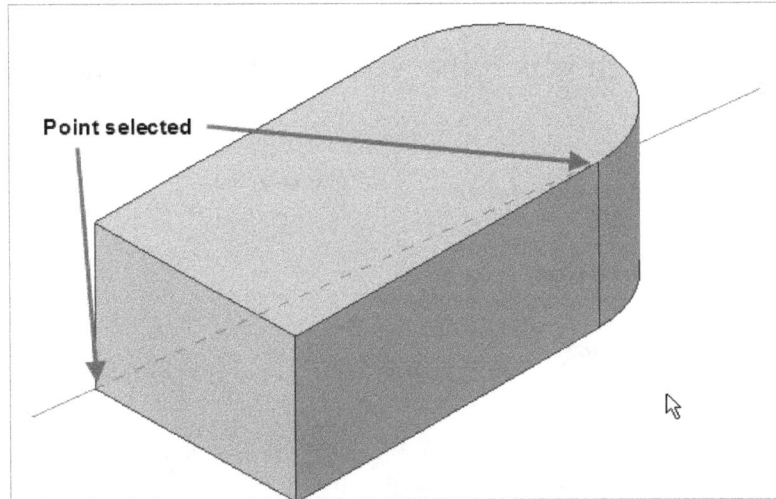

Figure-22. Axis created through points

Intersection of Two Planes

The **Intersection of Two Planes** tool is used to create an axis at the intersection of two planes. The procedure to use this tool is given next.

* Click on the **Intersection of Two Planes** tool from **Axis** drop-down in the **Work Features** panel of the **Ribbon**. You will be asked to select a plane.
* Click on the two intersecting planes/faces. The axis at the intersection of planes/faces will be created; refer to Figure-23.

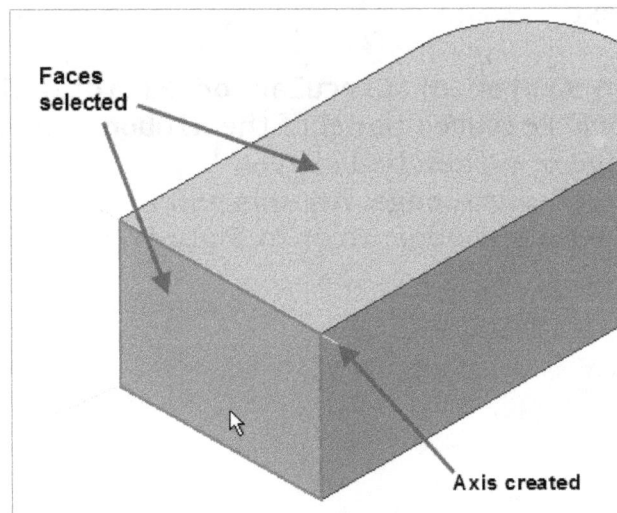

Figure-23. Axis created at intersection

Normal to Plane through Point

The **Normal to Plane through Point** tool is used to create an axis perpendicular to selected plane and passing through the selected point. The procedure to use this tool is given next.

- Click on the **Normal to Plane through Point** tool from **Axis** drop-down in the **Work Features** panel of the **Ribbon**. You will be asked to select a plane or point.
- Select the plane to which the axis should be normal. You will be asked to select a point.
- Click on the point through which the axis should pass. The axis will be created; refer to Figure-24.

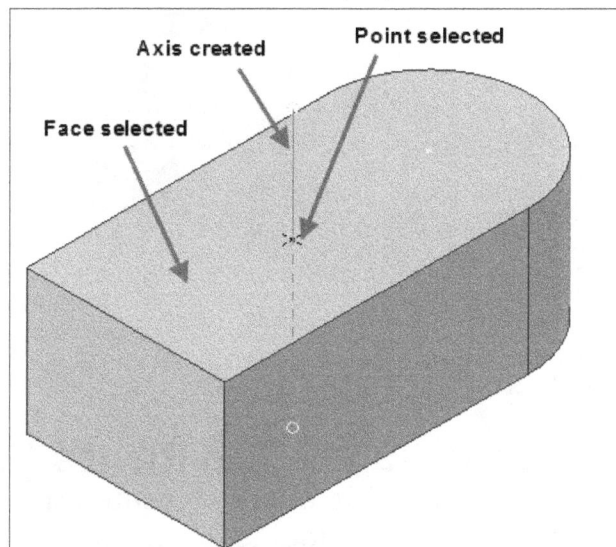

Figure-24. Axis passing through point

Through Center of Circular or Elliptical Edge

The **Through Center of Circular or Elliptical Edge** tool is used to create an axis passing through the center of circular or elliptical edge selected. The procedure to use this tool is given next.

- Click on the **Through Center of Circular or Elliptical Edge** tool from **Axis** drop-down in the **Work Features** panel of the **Ribbon**. You will be asked to select circular/elliptical edge or a sketched curve.
- Click on the circular/elliptical edge. An axis will be created passing through the center of circular or elliptical edge; refer to Figure-25.

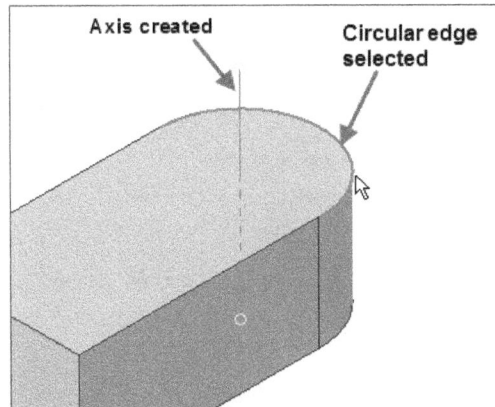

Figure-25. Axis created at center of circular edge

Through Revolved Face or Feature

The **Through Revolved Face or Feature** tool is used to create an axis passing through the centerline of revolved face/feature. The procedure to use this tool is given next.

*   Click on the **Through Revolved Face or Feature** tool from **Axis** drop-down in the **Work Features** panel of the **Ribbon**. You will be asked to select a cylindrical or revolved surface.
*   Select the cylindrical/revolved surface. An axis will be created; refer to Figure-26.

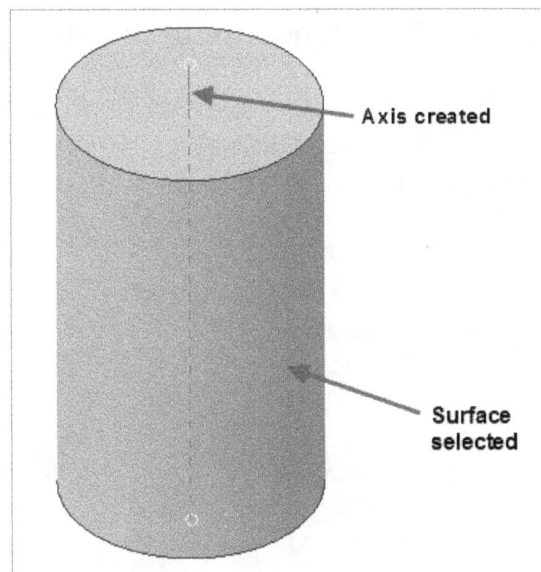

Figure-26. Axis through cylindrical surface

Axis

The **Axis** tool can be used to create any type of axis we have discussed before. After selecting this tool, make the selections based on axis requirement.

POINT

Point is an imaginary smallest geometric entity. Theoretically, point has zero size. Points are used to provide reference for other geometric features. The tools to create point are available in **Point** drop-down in the **Work Features** panel of the **Ribbon**; refer to Figure-27.

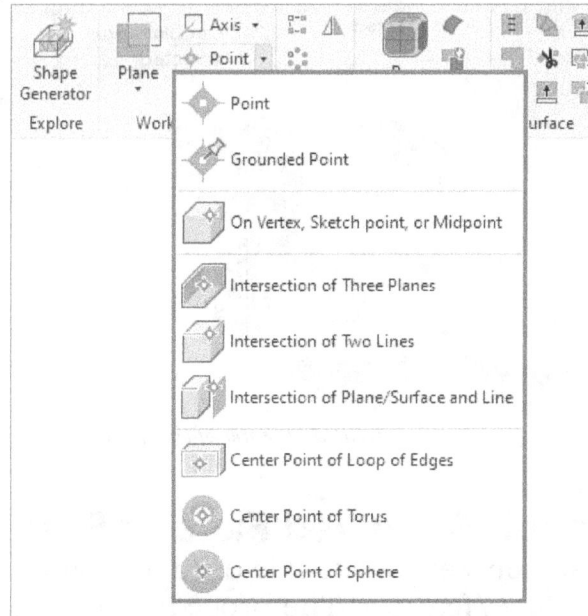

Figure-27. Point drop-down

The tools in this drop-down are discussed next.

Grounded Point

The **Grounded Point** tool is used to create points in reference to selected key point (like end point, mid point, etc.). The procedure to use this tool is given next.

- Click on the **Grounded Point** tool from the **Point** drop-down in the **Work Features** panel of the **Ribbon**. You will be asked to select the vertex or work point.
- Click on desired vertex or work point. A triad will be displayed at the selected vertex; refer to Figure-28.

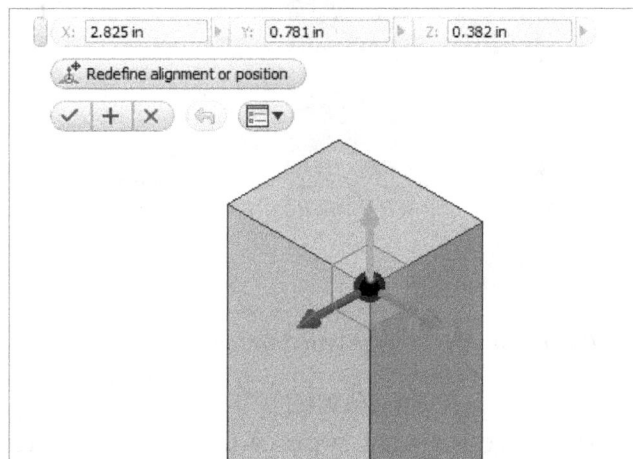

Figure-28. Triad displayed on point

- Click on the arrow to move the point in respective direction. The edit box for selected direction will be active in the Input boxes; refer to Figure-29.

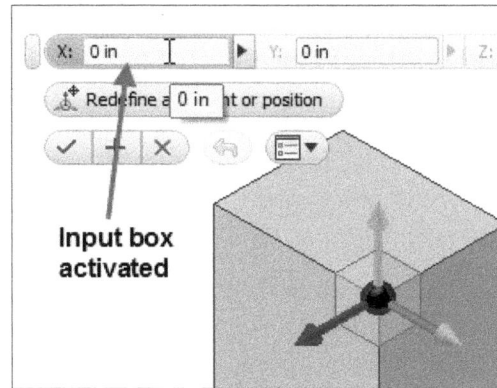

Figure-29. Input box active

- Enter desired distance value in the Input box to specify location of the point in respective direction. Similarly, you can set the location of point in other directions.
- If you want to add more points then click on the **Apply** button + from the Input boxes. You will be asked to specify location of the next point. Repeat the procedure until you get desired number of points.
- Click on the **OK** ✓ button from the Input boxes to create the point.

Similarly, you can use the other tools in the **Point** drop-down to create points at desired locations.

Since, you have learned the basic understanding of planes, axes, and points; it is now time to create some basic solid features. The tools to create solid features are available in the **Create** panel of the **3D Model** tab in the **Ribbon**; refer to Figure-30. These tools are discussed next.

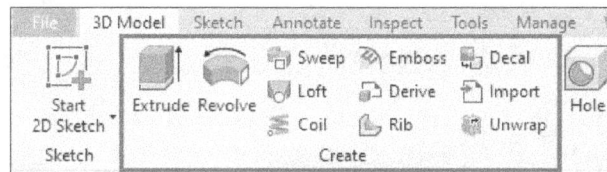

Figure-30. Create panel of 3D Model tab

EXTRUDE TOOL

The **Extrude** tool is used to create a solid volume by adding height to the selected sketch. In other words, this tool adds material by using the boundaries of sketch in the direction perpendicular to the plane of sketch. The procedure to use this tool is given next.

- Click on the **Extrude** tool from **Create** panel in the **3D Model** tab of the **Ribbon**. You will be asked to select a sketching plane.
- Select a plane and create a desired sketch for extrusion.
- Click on the **Finish Sketch** tool to exit the sketching environment. The **Extrude** dialog box will be displayed along with the preview of extrusion; refer to Figure-31.

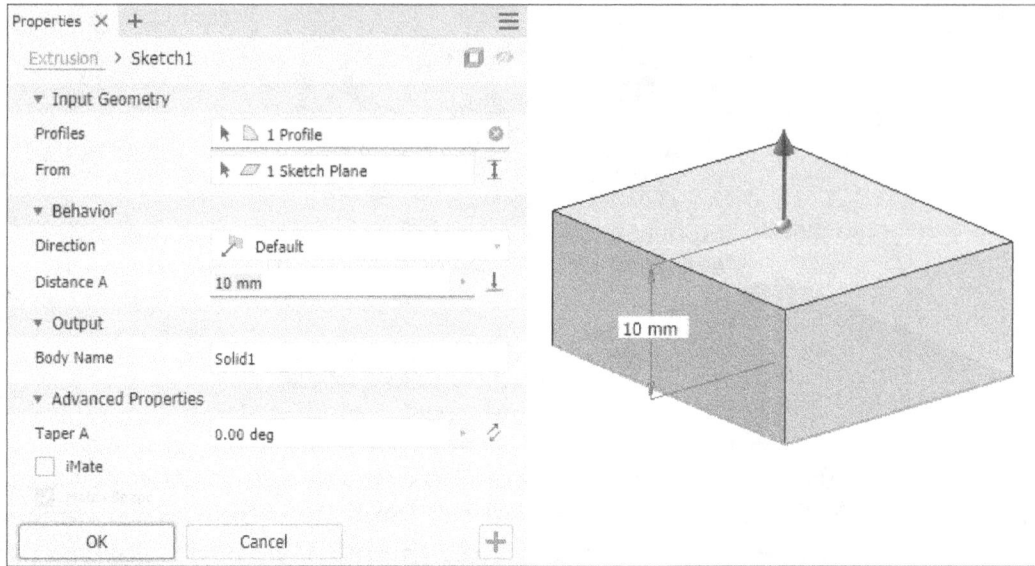

Figure-31. Extrude dialog box with preview of extrusion

Specifying Distance for Extrusion

- To specify the depth of extrusion, enter the desired value in the **Distance A** edit box in the dialog box; refer to Figure-32.

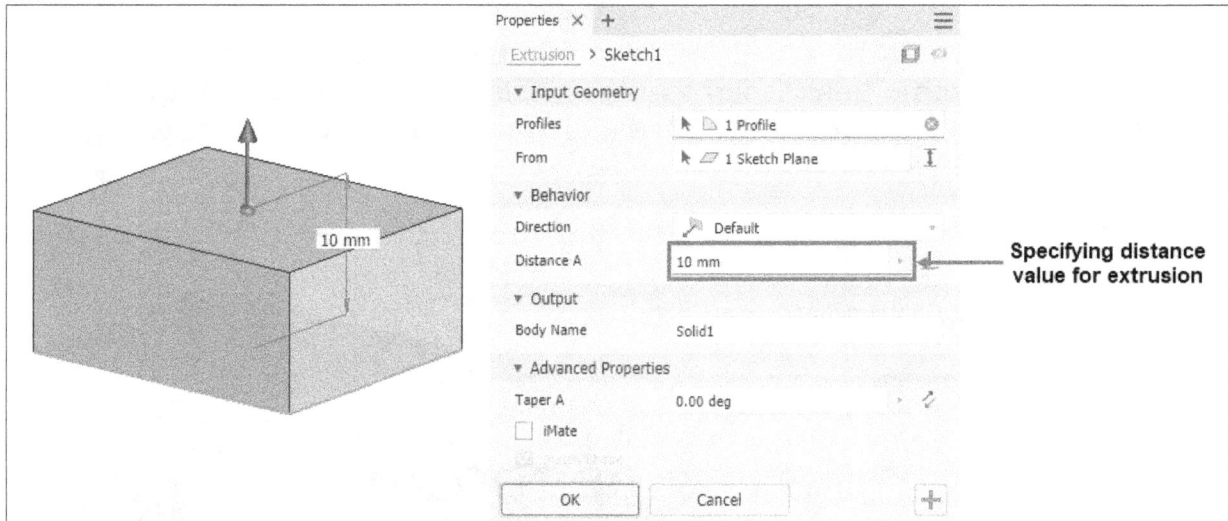

Figure-32. Specifying depth of extrusion

- There are four options to manage direction of extrusion viz. **Default**, **Flipped**, **Symmetric**, and **Asymmetric**; refer to Figure-33.

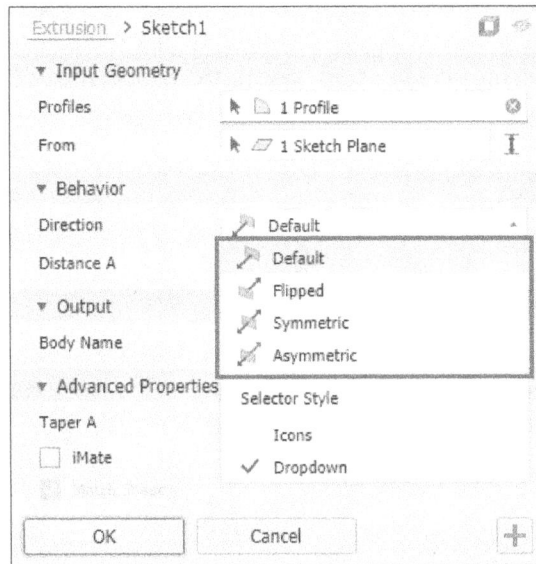

Figure-33. Direction of extrusion

When we were creating planes, you might have noticed two colors of plane, light orange and light blue. If no! Try to rotate the model by holding the **SHIFT** key and dragging the cursor using middle mouse button. The light orange side of plane is positive side and in case of extrusion, Default Direction. The light blue side of plane is negative side and in case of extrusion, Flipped Direction.

- Select the **Default** or **Flipped** button to create the extrusion in respective direction. If you want to create extrusion to both sides of sketching plane with same depth value then select the **Symmetric** button. If you want to create extrusion to both sides of sketching plane but with different depth value in each direction then click on the **Asymmetric** option. On selecting the **Asymmetric** option, two distance edit boxes will be displayed in the dialog box; refer to Figure-34. Specify desired distance for each direction.

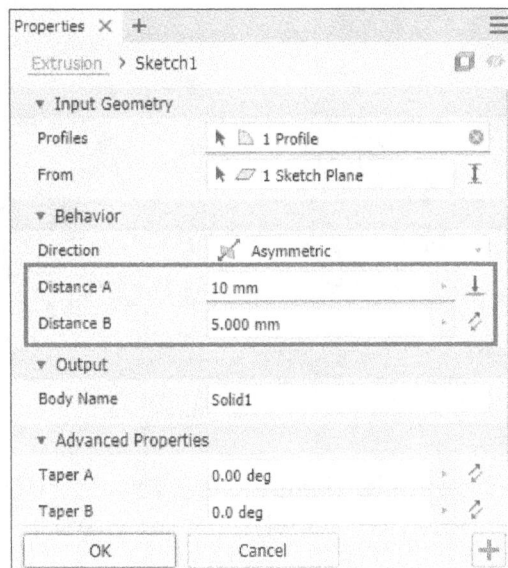

Figure-34. Specifying distances for asymmetric extrusion

Setting Extent of Extrusion

While designing, we might get some reference geometries up to which the extrusion is to be done and we do not have the distance values. To fulfil such conditions, the options are available next to the **Distance A** edit box; refer to Figure-35. These options are discussed next.

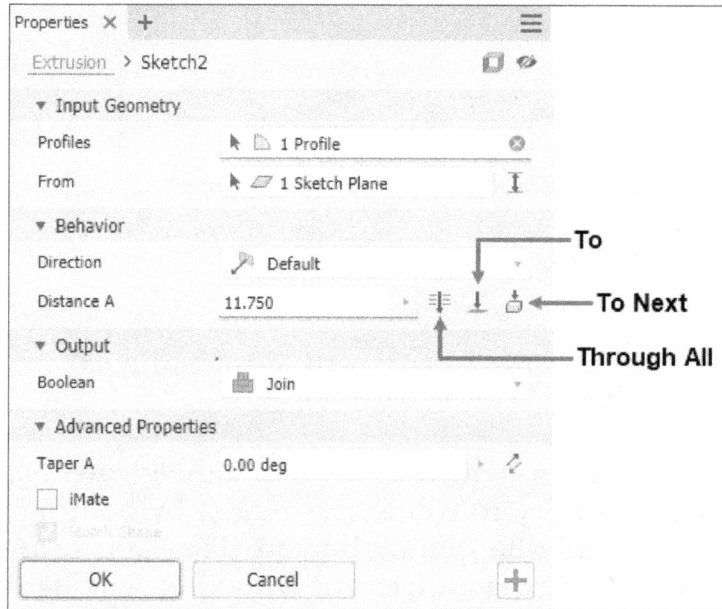

Figure-35. Extent options

- Click on the **Through All** option to create extrusion intersecting with all the solids/ faces coming in the direction of extrusion; refer to Figure-36.

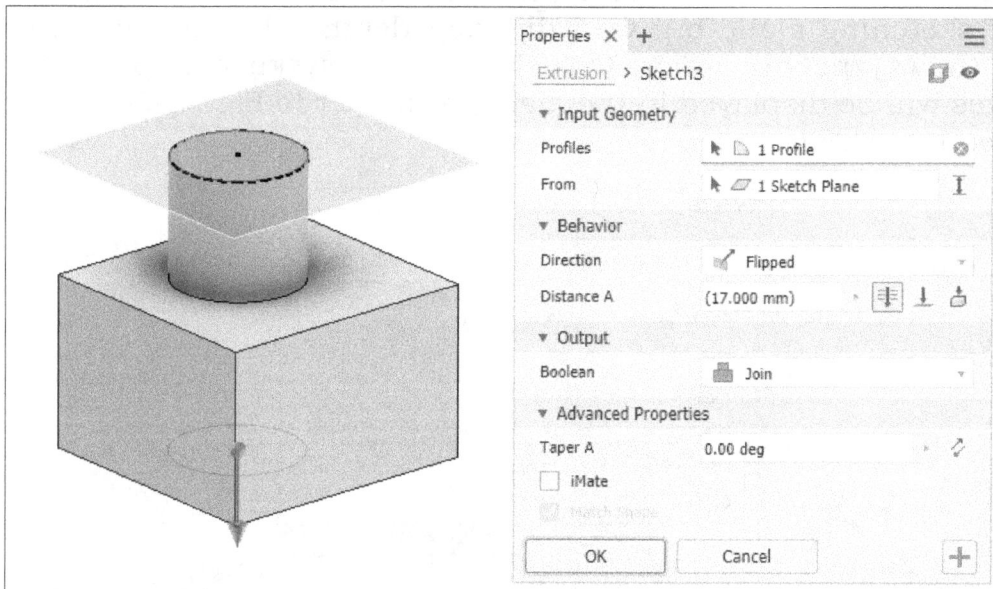

Figure-36. Through All extrusion

- Click on the **To** option and you will be asked to select the terminating geometry. Select the plane, face, or point up to which you want to create extrusion; refer to Figure-37.

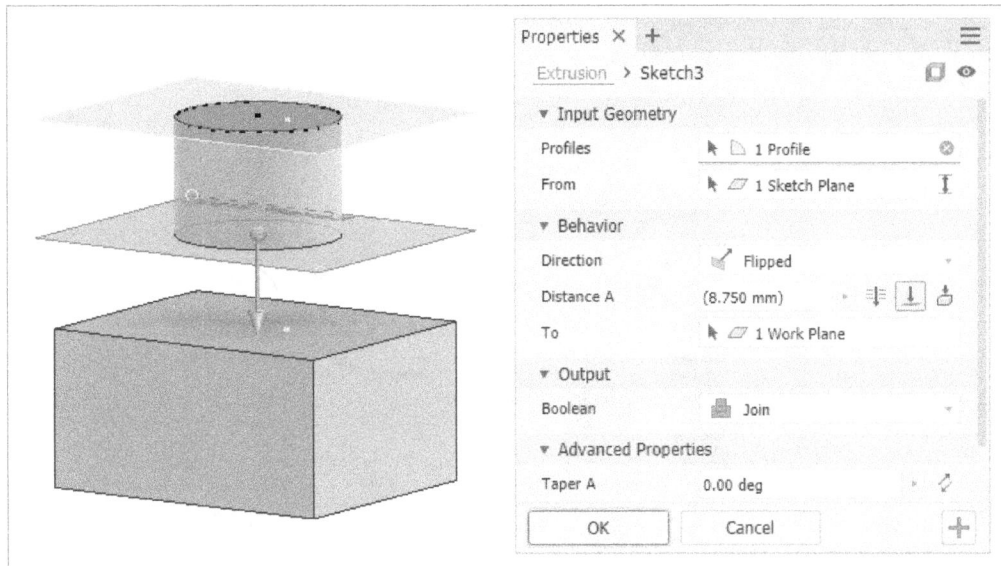

Figure-37. Creating extrusion to plane

- Click on the **To Next** option if there is any solid body/face in the extrusion direction. The extrusion distance will automatically be specified up to the next intersecting body/face; refer to Figure-38. Note that the terminating solid/face must cover the extrusion sketch, completely.

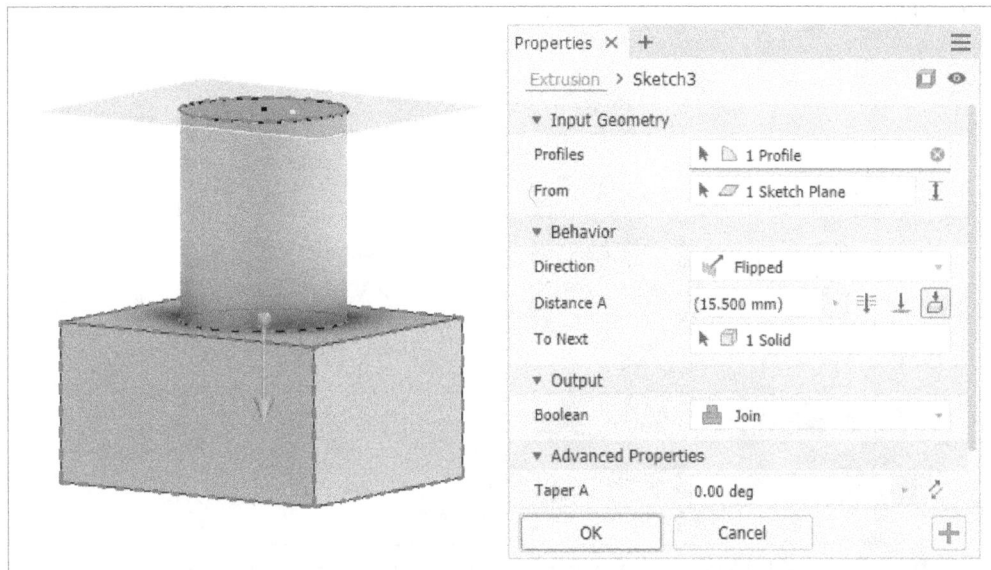

Figure-38. Creating extrusion upto next

Boolean Operations

There are some basic boolean operations that can be performed with solid features like Join (addition), Cut (subtraction), and Intersection. There is one more option by which you can make the newly created solid feature as solid having no relation with other solids in modeling space with the name **New Solid**.

The buttons to apply these options are available from **Boolean** drop-down in **Extrude** dialog box; refer to Figure-39. Note that these buttons will be available only when you have a solid base feature and you are creating another feature on it.

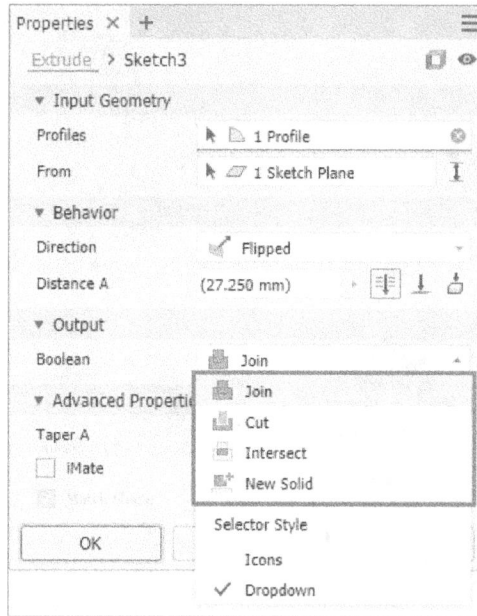

Figure-39. Options to apply Boolean operations

Joining Features

When we are creating complex model, it is not possible to create model by using single operation of tool. In such cases, we may need multiple extrusions joined together like in Figure-40.

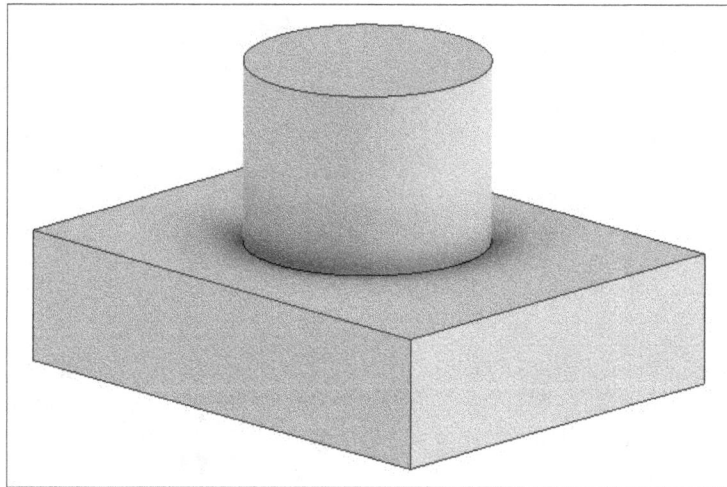

Figure-40. Model created by joining two extrude features

The method to do so is given next.

- Click on the **Extrude** tool from **Create** panel in the **Ribbon** after creating the base extrude feature. You will be asked to select a sketching plane or an existing sketch.
- Select the top face of the base feature as sketching plane; refer to Figure-41. The sketching environment will be activated.

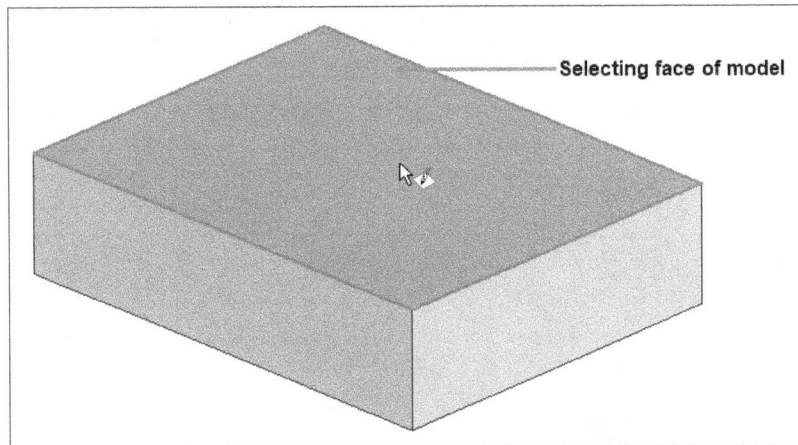

Figure-41. Selecting face of model as sketching plane

- Create the closed sketch and then click on the **Finish Sketch** button from **Exit** panel in the **Ribbon**; refer to Figure-42. Preview of the extrusion will be displayed.

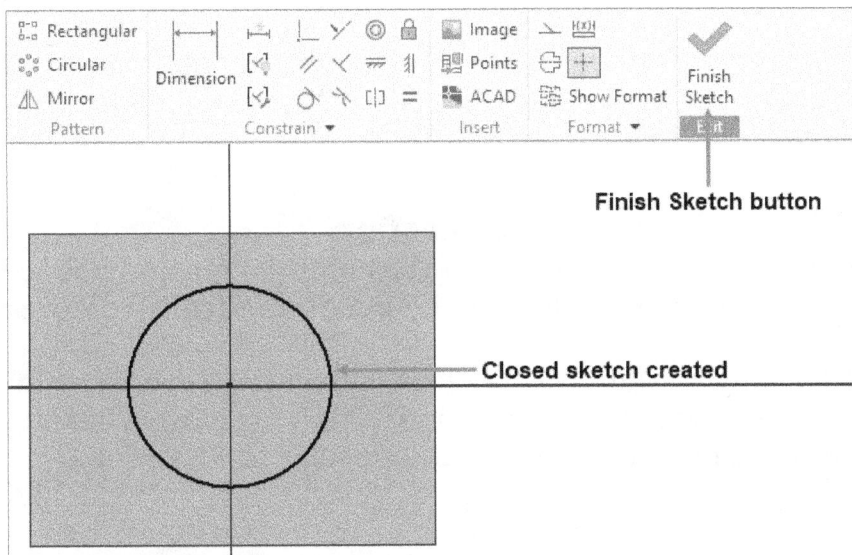

Figure-42. Creating sketch on face of solid feature

- Click on the **Join** button from the **Extrude** dialog box to join the extrusion to base feature; refer to Figure-43.

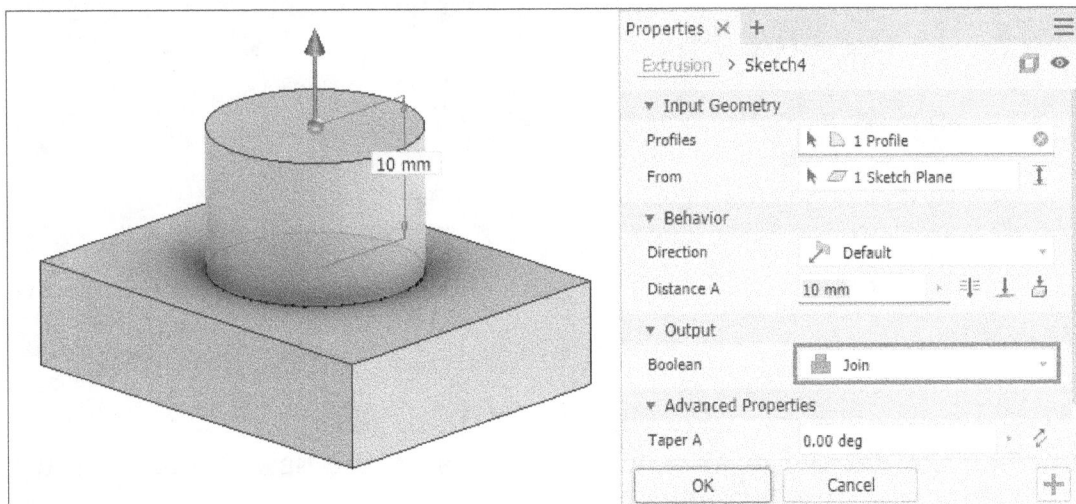

Figure-43. Preview of extrusion joined to base feature

Cutting extrusion from base feature (Cut button)

The **Cut** button is used to remove material from the base feature.

- In the same example, if you reverse the direction of extrusion by clicking on the **Flipped** button ◿ and select the **Cut** button 🖨 then preview of cut feature will be displayed; refer to Figure-44.

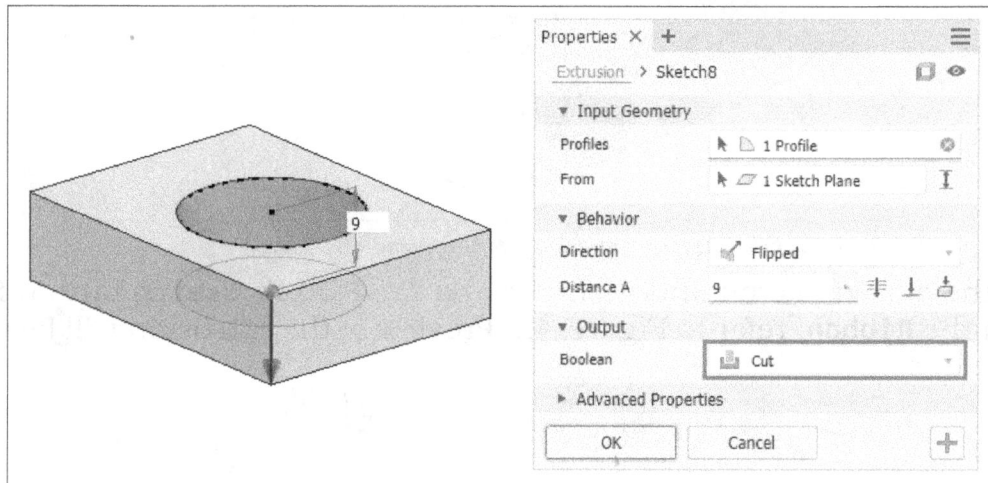

Figure-44. Preview of cut feature

Intersection

The **Intersect** button 🖨 is used to extract the common portion between base feature and extrusion, rest of the material in model is automatically removed.

- In the condition shown in Figure-44, if you select the **Intersect** button then only the cylindrical model will remain and rest will be removed because base feature and extrusion has this portion as common; refer to Figure-45.

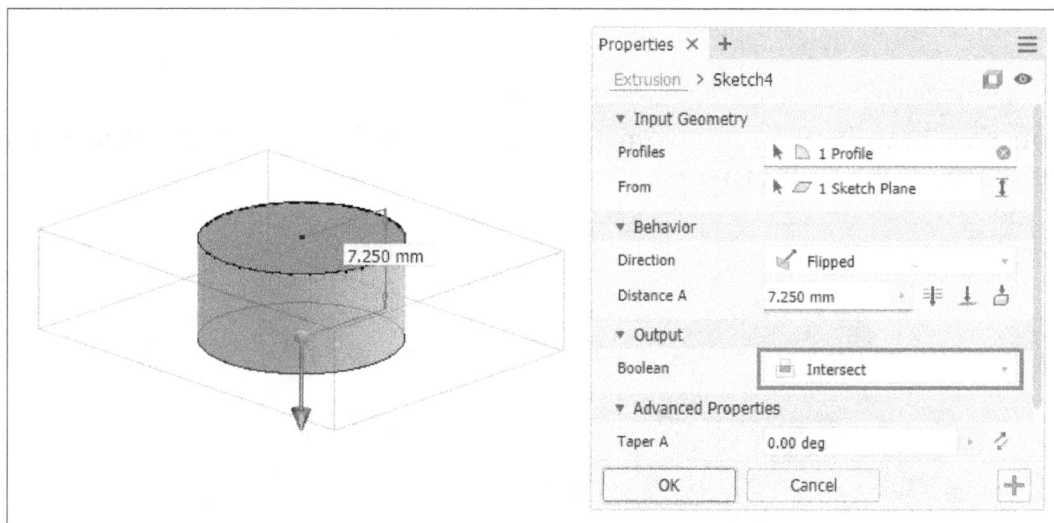

Figure-45. Preview of extrusion with intersect button

New Solid

You can create the extrusion as new solid by using the **New Solid** button ⬜ from **Boolean** drop-down in the **Extrude** dialog box.

Adding Taper to Extrusion

- To add taper to the extrusion, click on the **Advanced Properties** rollout from the dialog box and specify desired angle value in the **Taper A** edit box; refer to Figure-46. The taper angle will be applied to the extrusion keeping the base of extrusion fixed; refer to Figure-47.

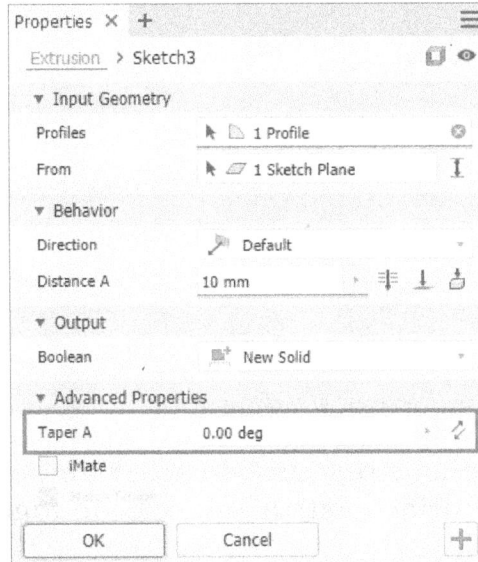

Figure-46. Taper edit box

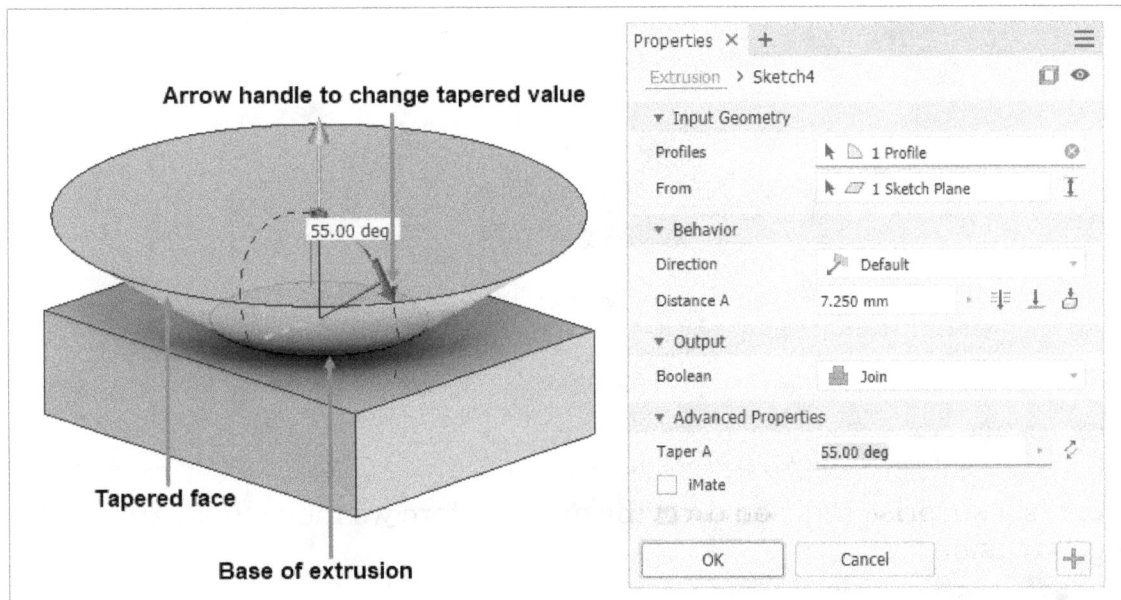

Figure-47. Extrusion with taper angle

And finally, click on the **OK** button from the dialog box to create the extrude feature.

Surface Mode of Extrusion

- Click on the **Surface mode** button from **Extrude** dialog box to create the surface extrusion of feature. The updated **Extrude** dialog box will be displayed with preview of surface extrusion; refer to Figure-48.

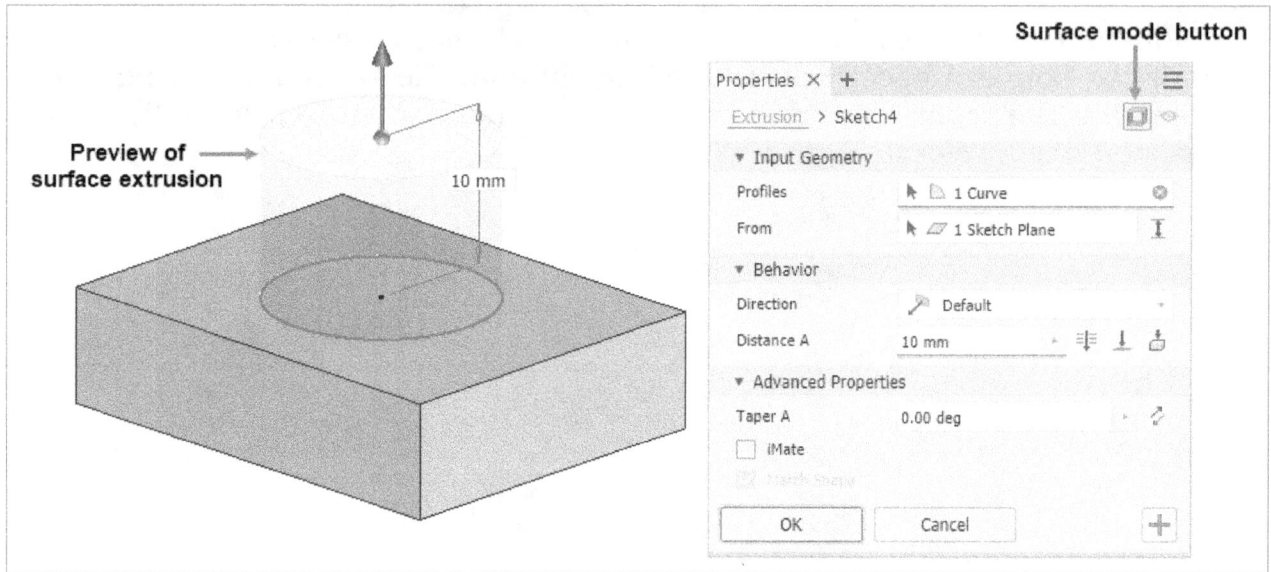

Figure-48. Updated Extrude dialog box with preview of surface extrusion

- Specify desired parameters as discussed earlier and click on the **OK** button from the dialog box to create surface extrusion; refer to Figure-49.

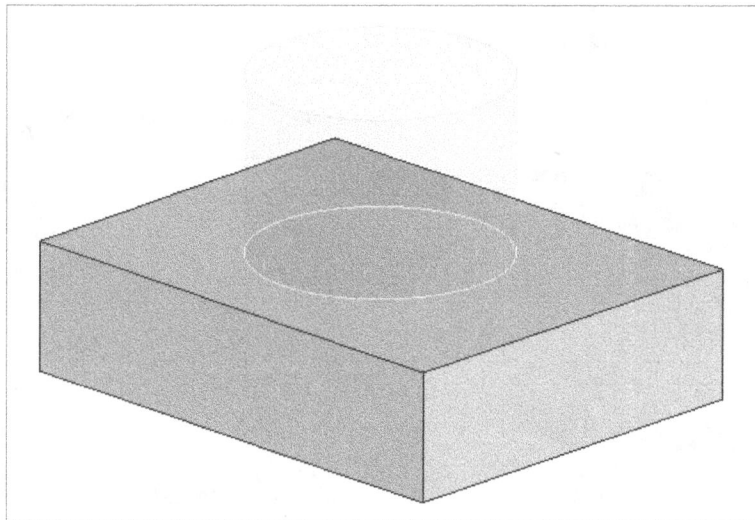

Figure-49. Surface extrusion created

Now, there is a surprise topic on drawing again before we move to create solid models by using extrusion.

BRIEF NOTE ON DRAWINGS

Till this point, we have created sketches and then we have learned to create extrusion with the help of those sketches. But, in real-world conditions, we are not going to have sketches. We will be finding our sketches from the engineering drawings. Being a mechanical engineer, you should be knowing about the engineering drawings but still its good idea to refresh the topic.

Engineering Drawing is the exact representation of an engineering component on the paper. There are a few qualities in Engineering Drawing given as:

- It is a clear and unmistakable representation of engineering component.
- It shows all the shapes and sizes of the engineering component.
- After reading the drawing, we get only one interpretation and there is no scope of confusion.

The Engineering Drawings are broadly classified into four categories:

Machine Drawing

It is pertaining to machine parts or components. It is presented through a number of orthographic views, so that the size and shape of the component is fully understood. Part drawings and assembly drawings belong to this classification. An example of a machine drawing is given in Figure-50.

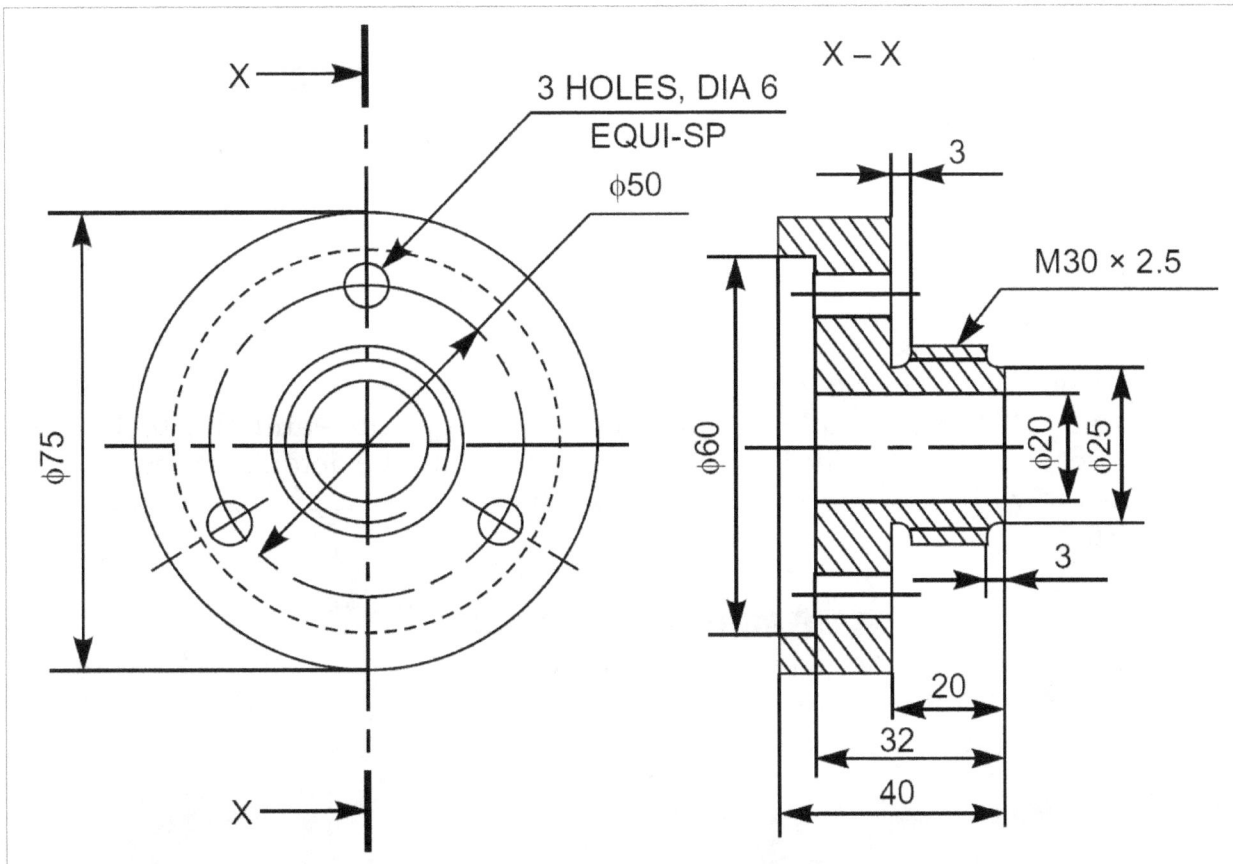

Figure-50. Machine drawing

Production Drawing

A production drawing, also referred to as working drawing, should furnish all the dimensions, limits, and special finishing processes such as heat treatment, honing, lapping, and surface finish to guide the craftsman on the shop floor in producing the component. The title should also mention the material used for the product, number of parts required for the assembled unit, and so on. Figure-51 shows an example of production drawing.

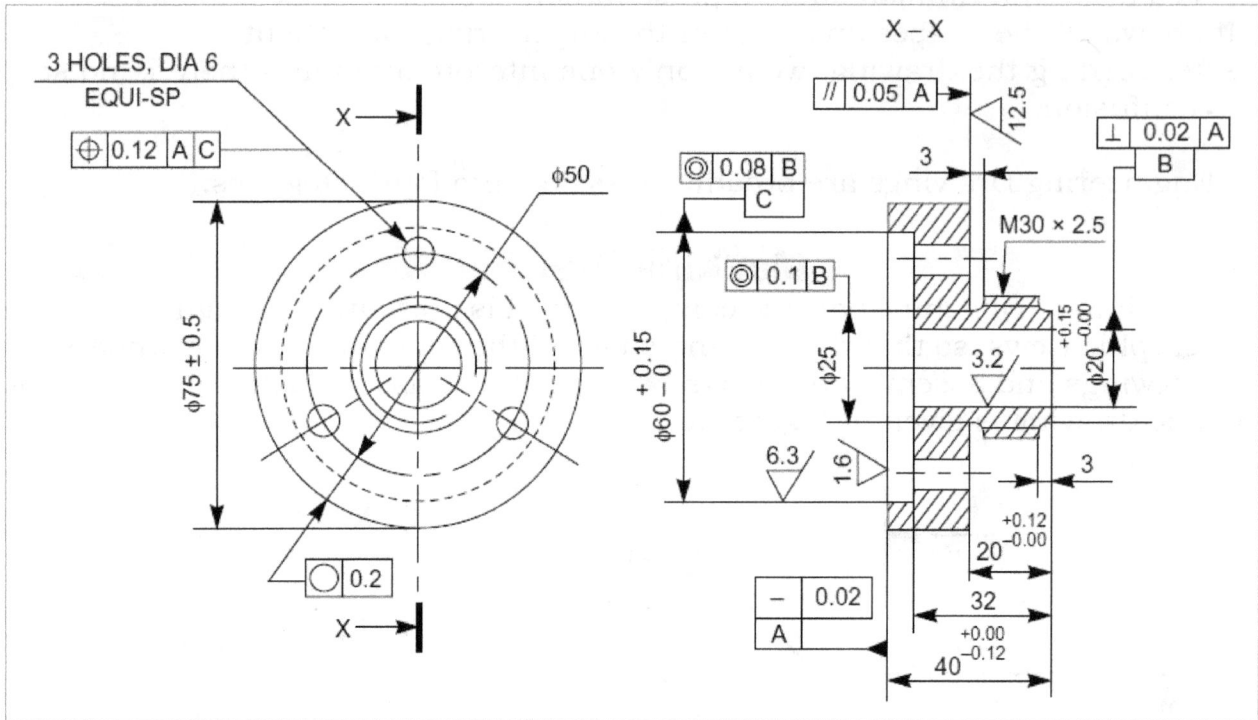

Figure-51. Production drawing

Part Drawing

Component or part drawing is a detailed drawing of a component to facilitate its manufacture. All the principles of orthographic projection and the technique of graphic representation must be followed to communicate the details in a part drawing. A part drawing with production details is rightly called as a production drawing or working drawing.

Assembly Drawing

A drawing that shows various parts of a machine in their correct working locations is an assembly drawing; refer to Figure-52.

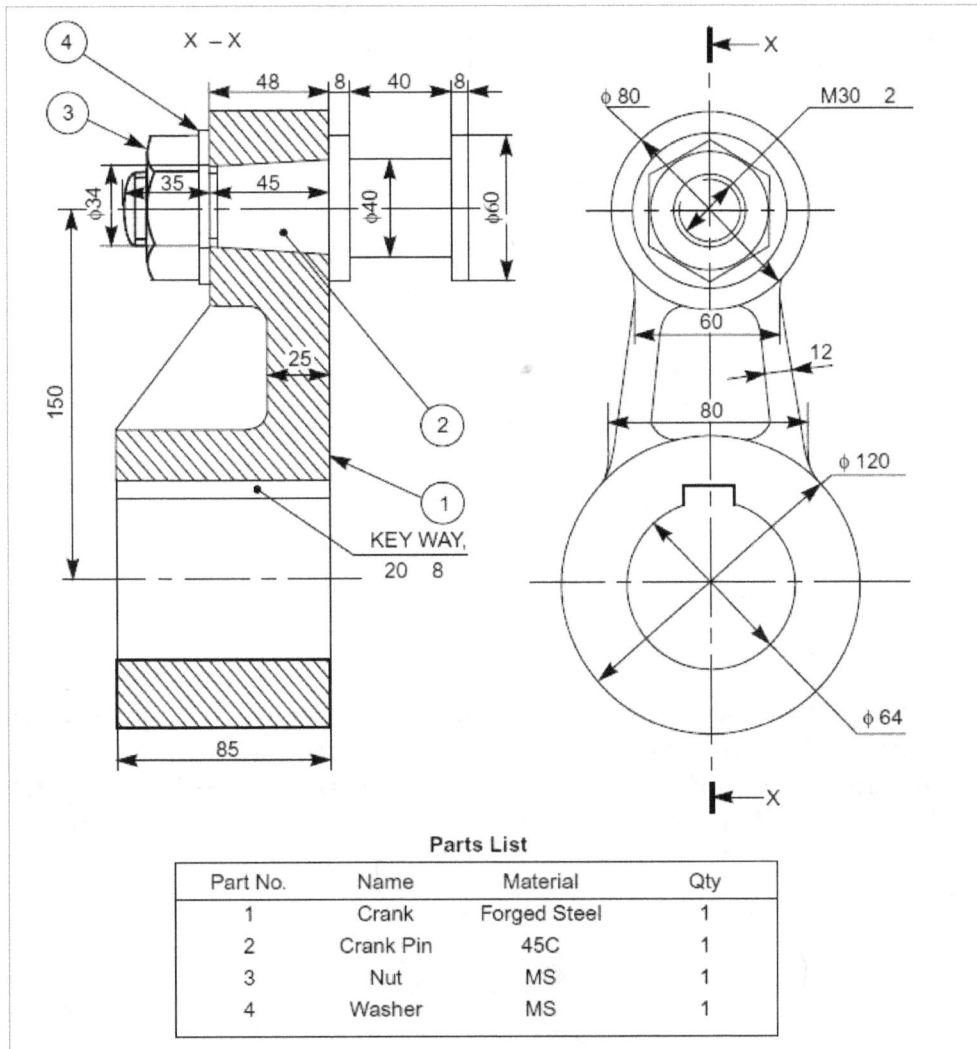

Figure-52. Assembly drawing

Part No.	Name	Material	Qty
1	Crank	Forged Steel	1
2	Crank Pin	45C	1
3	Nut	MS	1
4	Washer	MS	1

Parts List

COMPONENT REPRESENTATION METHODS

Broadly, there are two ways to present a component in engineering drawing; **Orthographic representation** and **Isometric representation**.

The orthographic representation is the method in which component is placed in the form of various views to completely define its shape and size. These orthographic views can be Front view, Right view, Top view, and so on. Refer to Figure-53.

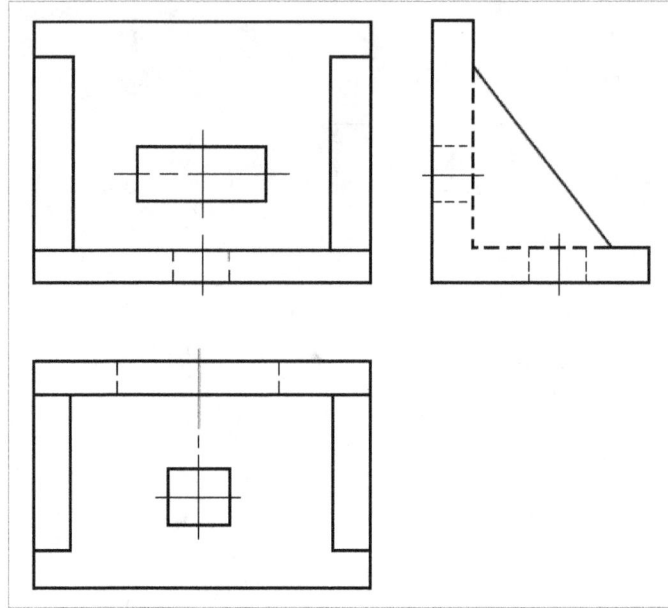

Figure-53. Orthographic views

The isometric representation is the method to display wireframe model of the object with its dimensions at standard angles from base planes. Note that all the objects can not be completely defined by this representation. So, it is useful when we need to represent the shape of the object with some highly needed dimensions; refer to Figure-54.

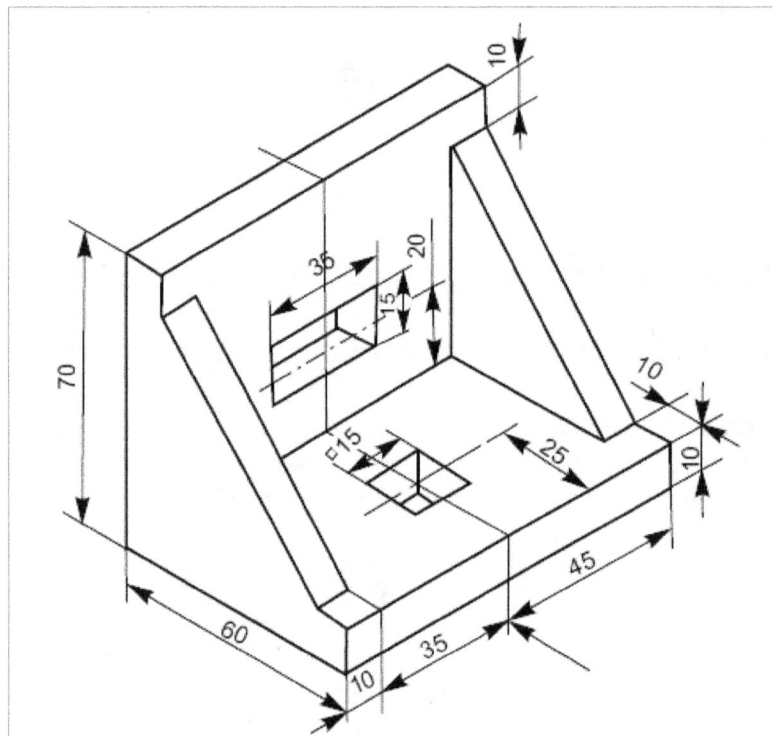

Figure-54. Isometric view

We will be using both the representations to create our models for practice.

SKETCHING PLANE SELECTION

Plane selection is a very important task while creating 3D models because the orientation of the model depends totally on the selection of plane. Since, most of the time we will get orthographic views of the model, so let's check the method of sketching plane selection for orthographic views.

Selecting planes according to orthographic views

Yes!! We know that there are top plane, front plane, right plane, and so on in orthographic views of component. But, how to identify which view is top and which view is front. Answer to this question is projection type. There are two type of projections mainly used for orthographic views; **First Angle Projection** and **Third Angle Projection**. In each engineering drawing, they are shown by their symbols in the Title Block; refer to Figure-55. Let A, B, C, D, and E are the three positions to see a car from front, left, right, top, and bottom. Then car views are placed as shown in Figure-56 for first angle projection and as shown in Figure-57 for third angle projection.

Projection	Symbol
First angle	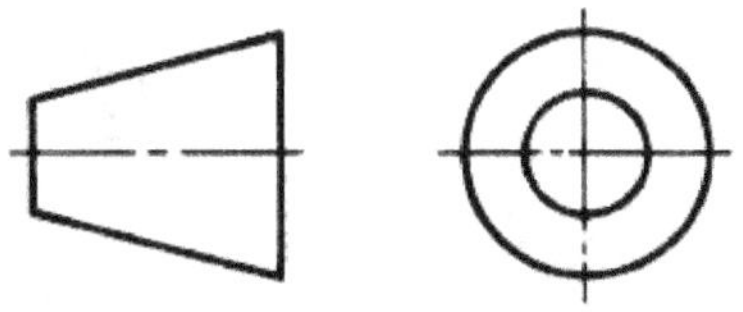
Third angle	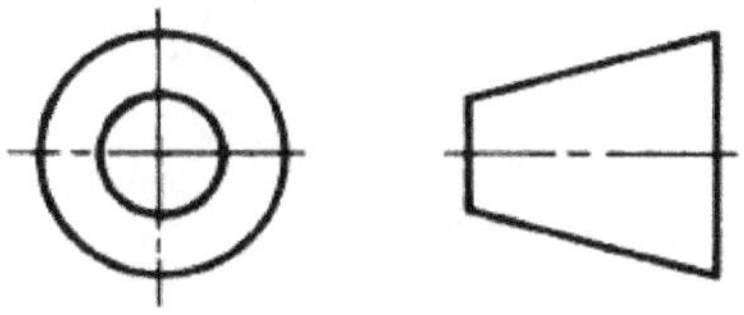

Figure-55. Symbols for projection

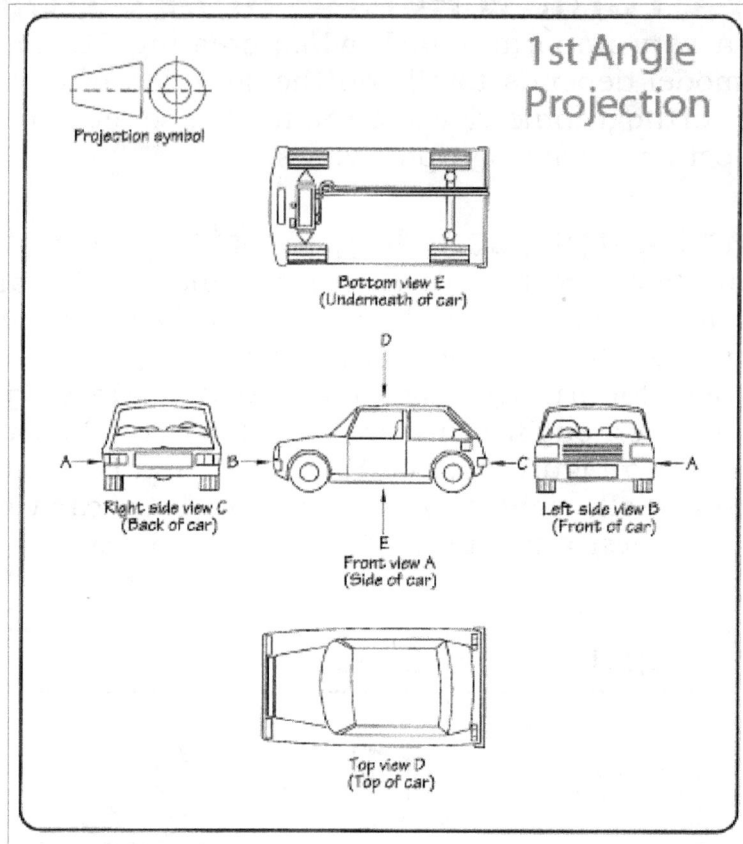

Figure-56. Views in 1st Angle Projection

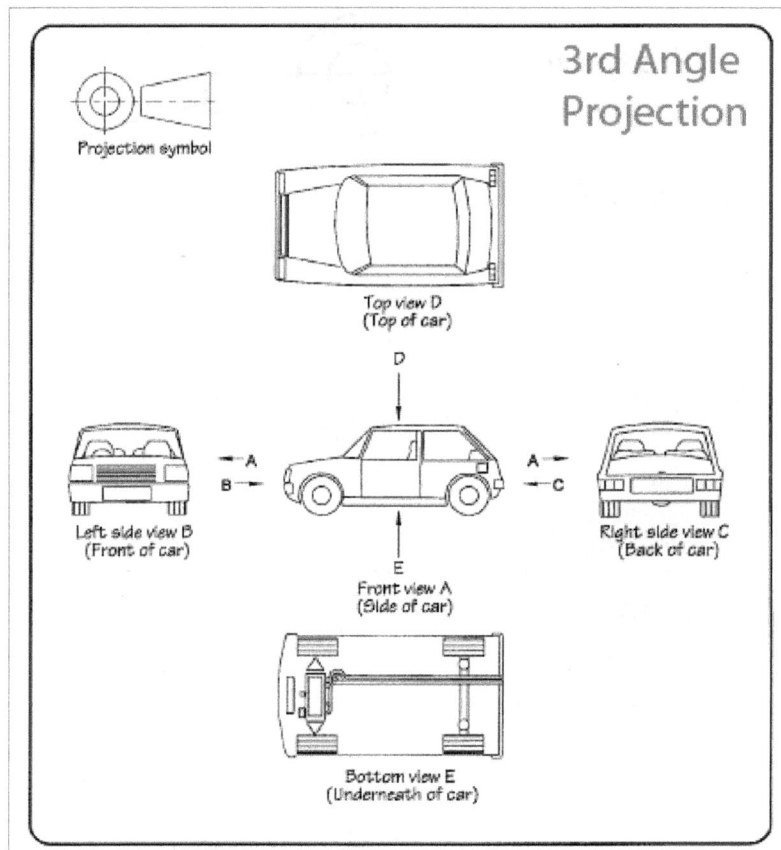

Figure-57. Views in 3rd Angle Projection

PRACTICAL 1

In this practical, we will create a 3D model from the drawing given in Figure-58. The drawing has both isometric as well as orthographic views. You can use any of the two to create the model.

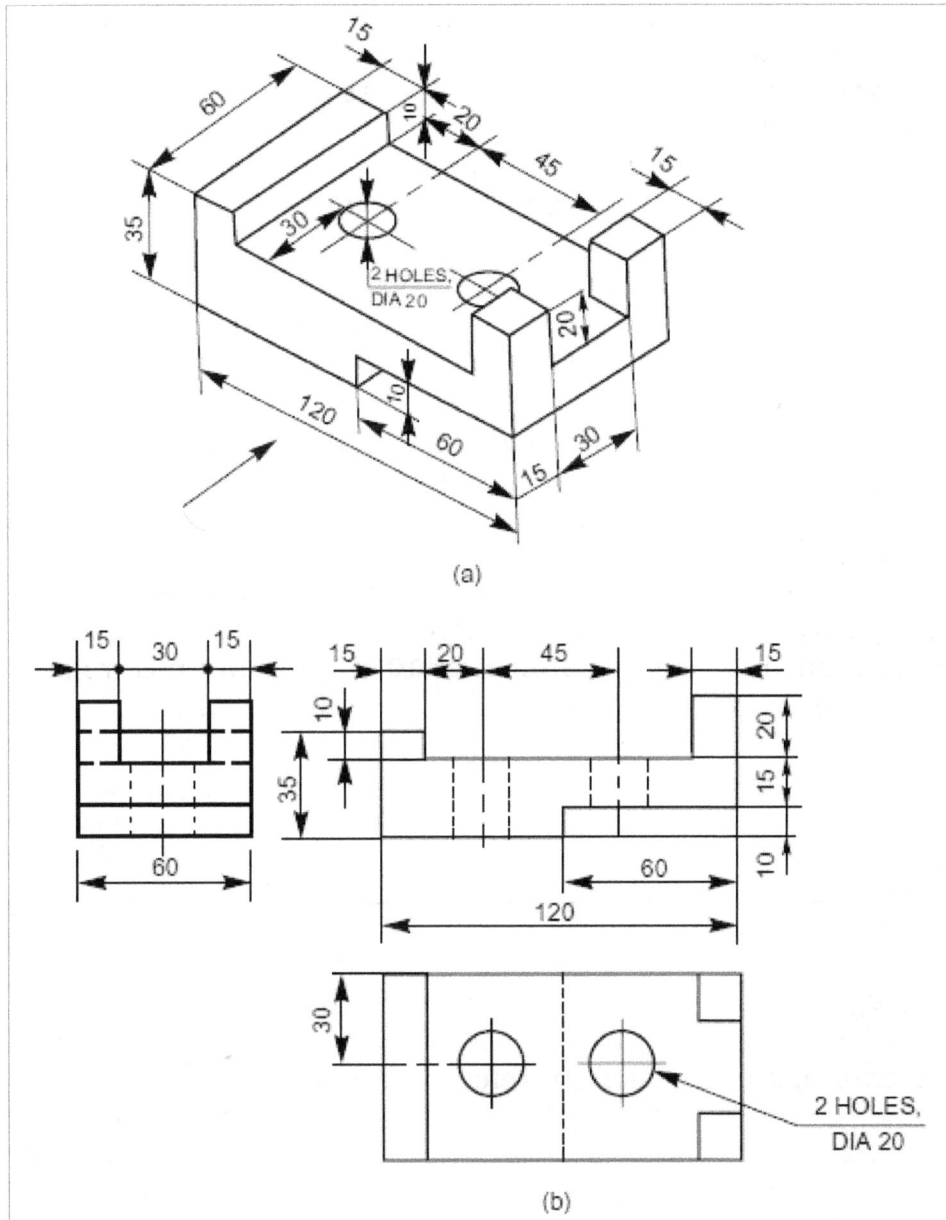

Figure-58. Practical 1

Strategy for Creating model

1. Looking at the isometric view and orthographic views, we can find that the drawing is in **First Angle Projection**. Because keeping the arrow direction in isometric view as Front view, the Right view of model is placed at the left of the Front view in Orthographic view. We are going to use Front view from orthographic views for creating sketch for base feature. Because in this way, we will use the extrusion tool at minimum. (Although, we can create this model using any other view and using same tool multiple times but in the end, we are more time which means more cost to the company.)

2. We will create two circles in a sketch and then create an extrude cut from the base feature by using the sketch to create holes.

3. We will use one more extrude cut to remove extra material from the model.

Starting Part File

- Start Autodesk Inventor by double-clicking on the Autodesk Inventor Professional 2024 icon from the desktop. (If not started yet.)
- Click on the **New** button from the **Quick Access Toolbar**. The **Create New File** dialog box will be displayed.
- Double-click on **Standard(mm).ipt** icon from the **Metric** templates. The Part environment of Autodesk Inventor will be displayed.

Creating sketch

- Click on the **Start 2D Sketch** button from **Sketch** panel in the **3D Model** tab of the **Ribbon**. You will be asked to select a sketching plane.
- Click on the **XY** Plane (Front Plane) to create sketch on it.
- Click on the **Line** tool from **Create** panel in the **Sketch** tab of the **Ribbon**. You will be asked to specify the starting corner point of the line.
- Click on the origin to specify the starting point of line. You will be asked to specify end point of line.
- Move the cursor straight right and enter **60** in the Input box displayed. You will be asked to specify next end point of line.
- Move the cursor straight upward and enter **10** in the Input box; refer to Figure-59.

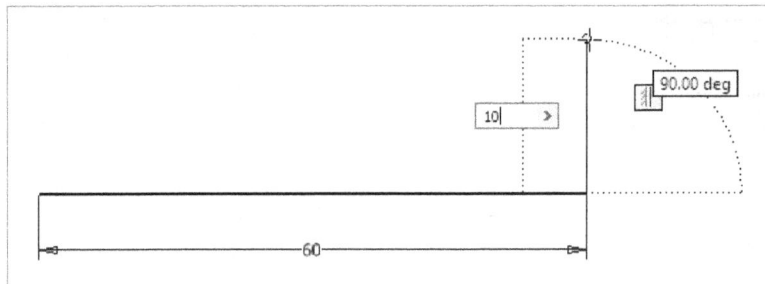

Figure-59. Creating lines

- Similarly, create rest of the sketch; refer to Figure-60.

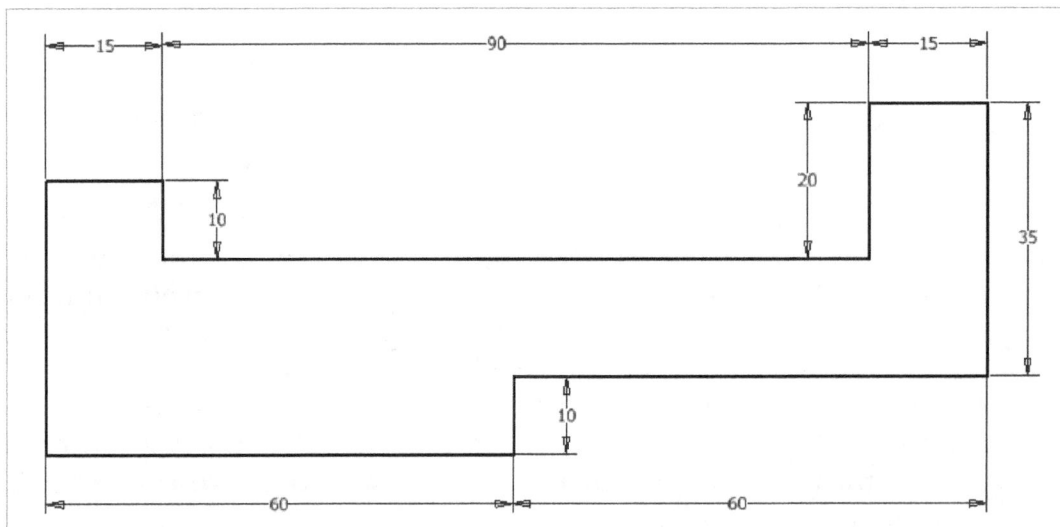

Figure-60. Sketch to be created

- Click on the **Finish Sketch** button from the **Exit** panel to exit sketching environment.

Creating base extrude feature

- Click on the **Extrude** tool from **Create** panel in the **3D Model** tab of the **Ribbon**. The **Extrude** dialog box will be displayed with the preview of extrusion; refer to Figure-61.

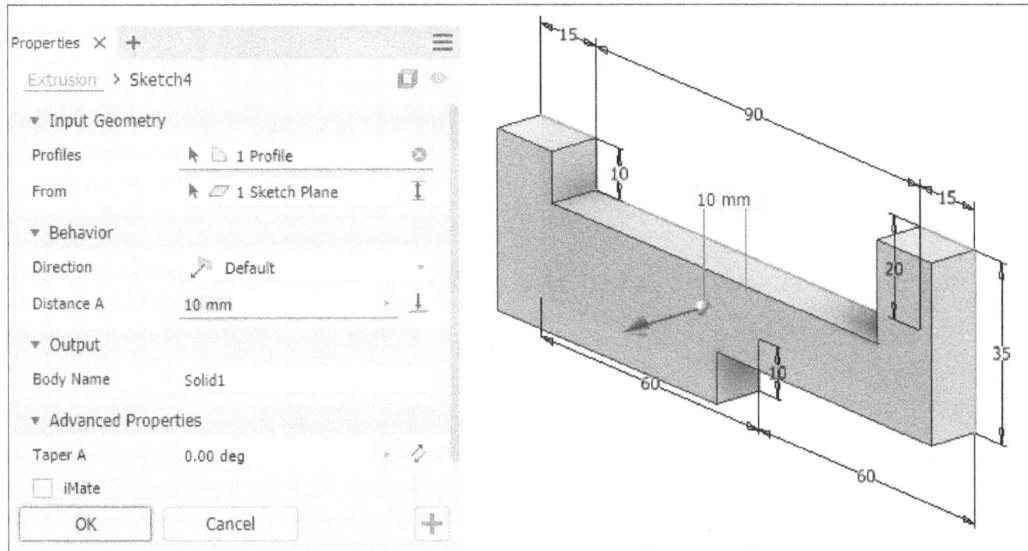

Figure-61. Extrude dialog box with preview of extrusion

- Enter the distance value as **60** in the **Distance A** edit box in the **Extrude** dialog box. The extrusion feature will be created; refer to Figure-62.

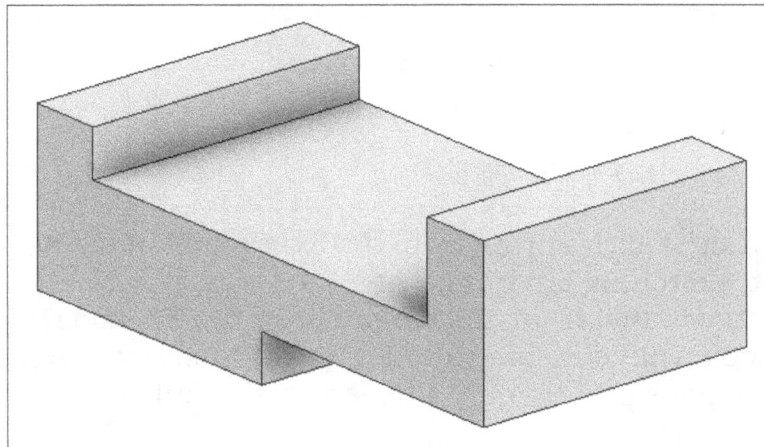

Figure-62. Extrusion created

Creating Holes

- Click on the **Start 2D Sketch** tool from **Sketch** panel in the **3D Model** tab of the **Ribbon**. You will be asked to select a plane/face for sketching.
- Click on the face as shown in Figure-63. The sketching environment will be displayed.

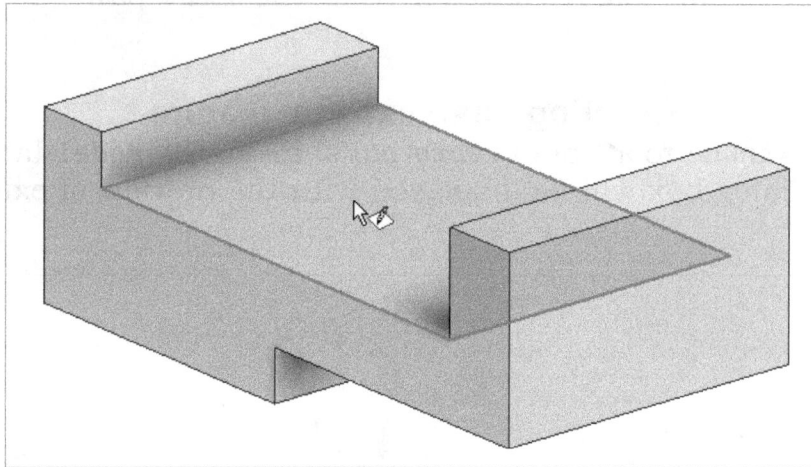

Figure-63. Face to be selected for creating hole

- Create two circles of diameter **20** at the dimensions as shown in Figure-64.

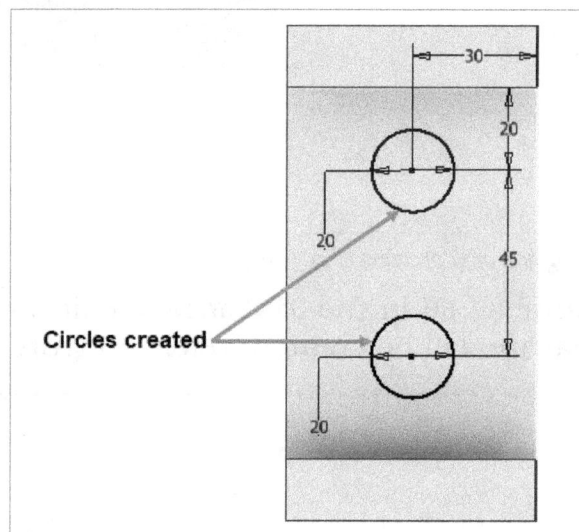

Figure-64. Creating circles

- Click on the **Finish Sketch** button from the **Exit** panel in the **Sketch** tab of **Ribbon**. You will exit the sketching environment.
- Click on the **Extrude** tool from **Create** panel in the **3D Model** tab of **Ribbon**. The **Extrude** dialog box will be displayed asking you to select the profiles.
- Select both the circles one by one. Preview of the cut feature will be displayed; refer to Figure-65.

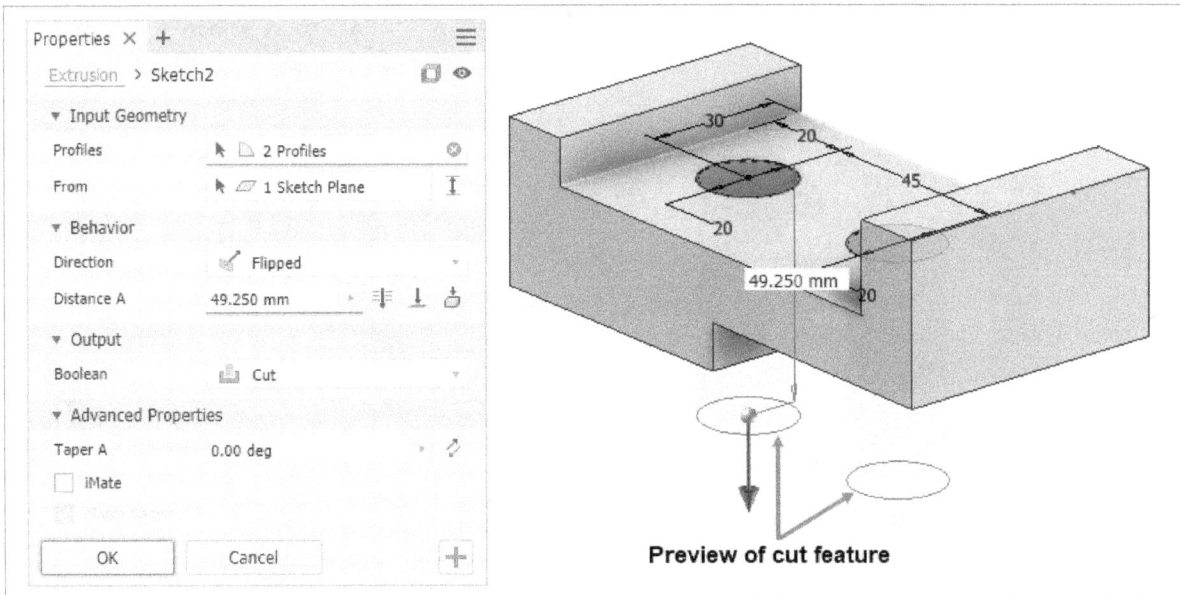

Figure-65. Preview of cut feature

- Make sure the buttons in the dialog box are selected as shown in Figure-65 and enter the distance value as **25** in the **Distance A** edit box. The holes will be created in the base feature; refer to Figure-66.

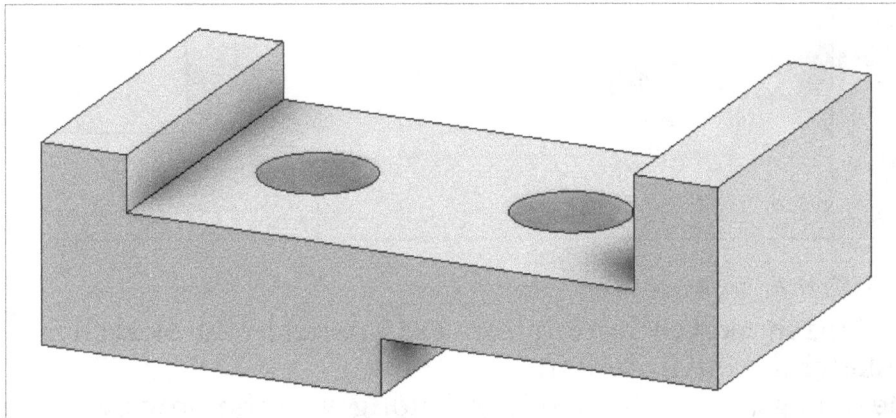

Figure-66. Holes created in the base feature

Creating cut feature

- Click on the **Extrude** tool from **Create** panel in the **3D Model** tab of the **Ribbon**. You will be asked to select a plane/face for sketching.
- Click on the face as shown in Figure-67. The sketching environment will be displayed.

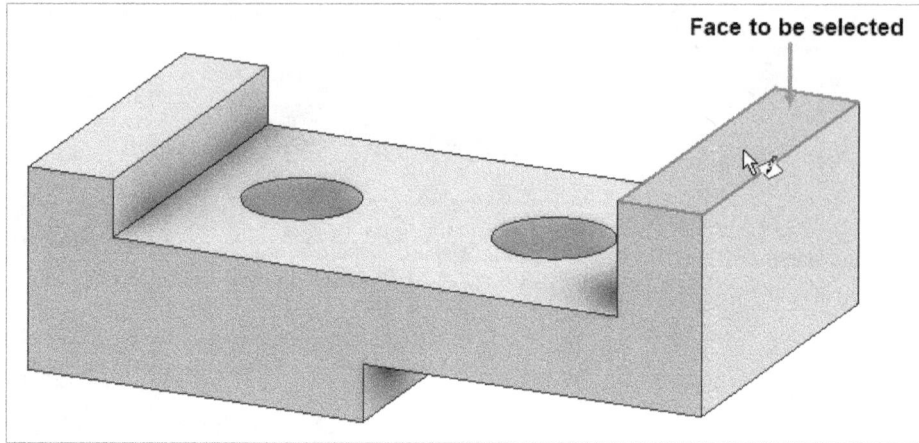

Figure-67. Face to be selected for creating cut feature

- Create the sketch as shown in Figure-68.

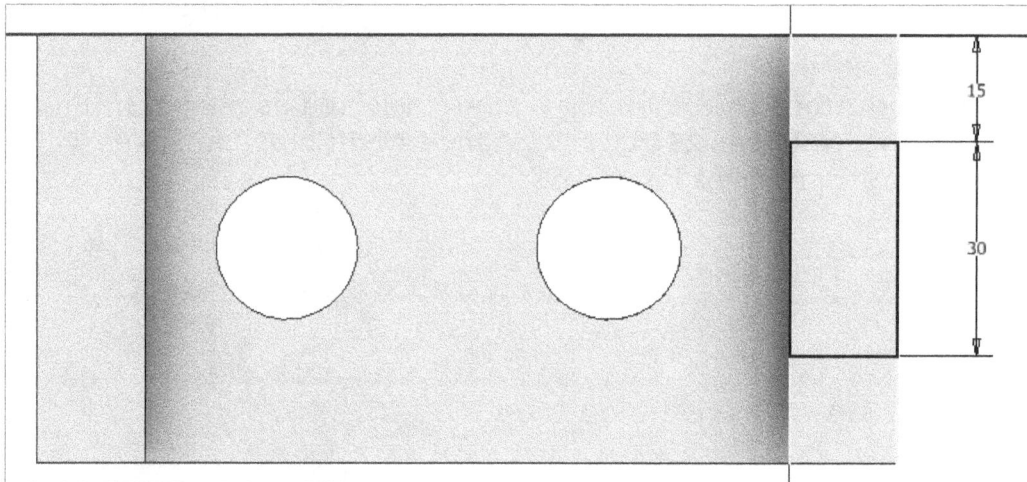

Figure-68. Sketch for cut feature

- Click on the **Finish Sketch** button from **Exit** panel in the **Sketch** tab of the **Ribbon** to exit the sketching environment.
- The **Extrude** dialog box will be displayed along with the preview of extrusion; refer to Figure-69.

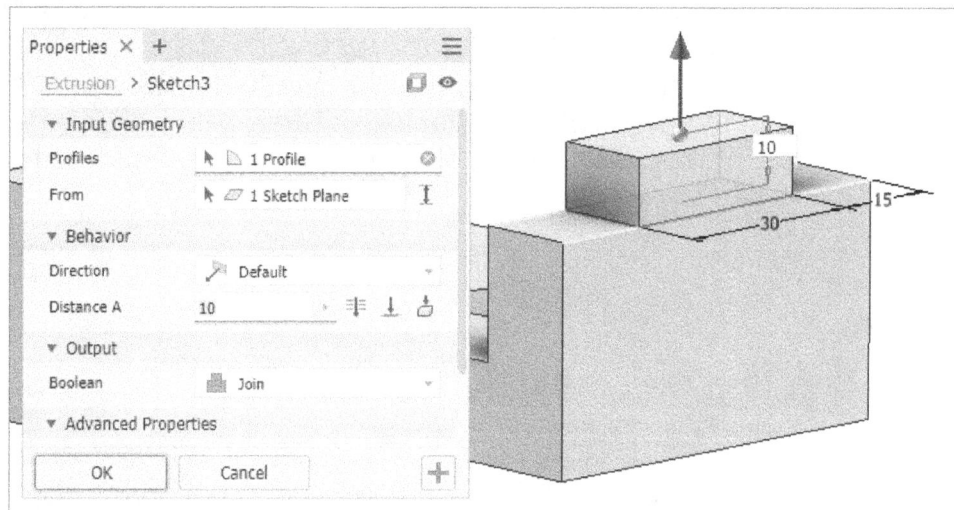

Figure-69. Extrude dialog box with preview of extrusion

- Select the **Flipped** option from **Direction** drop-down and **Cut** option from **Boolean** drop-down from the dialog box and specify the cut depth value as **20** in the **Distance A** edit box; refer to Figure-70.

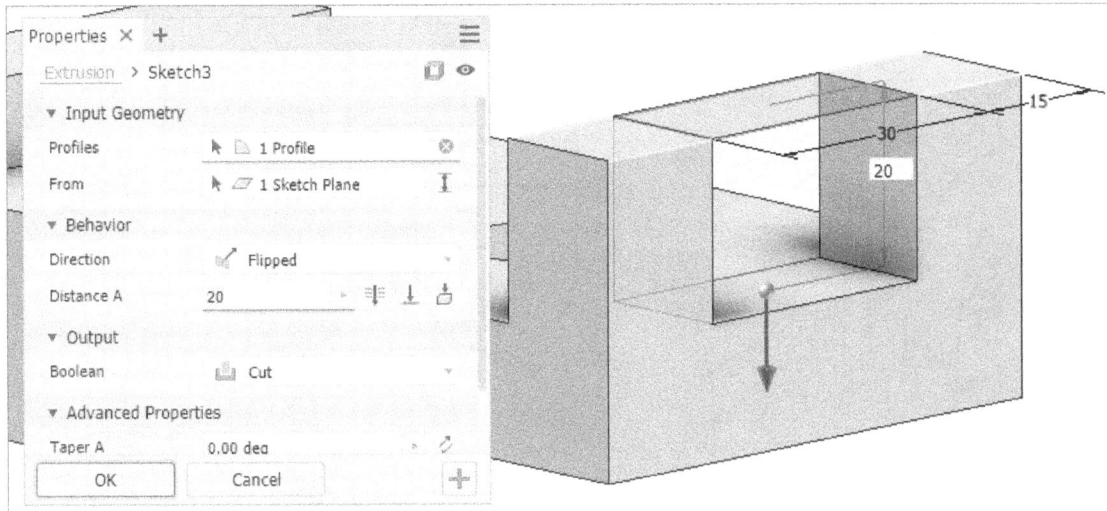

Figure-70. Creating cut feature

- Click on the **OK** button from the dialog box. The final model will be displayed as shown in Figure-71.

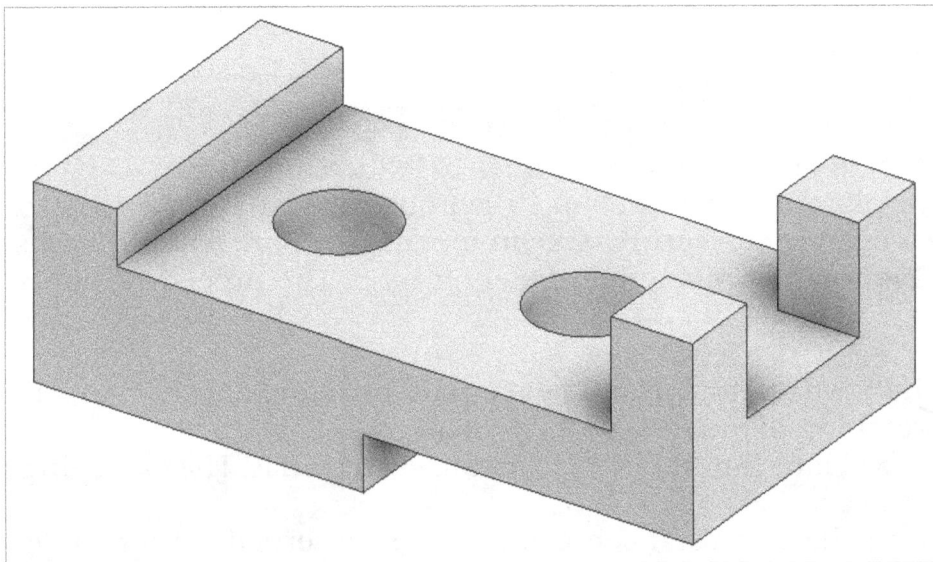

Figure-71. Final Model for Practical 1

PRACTICE 1

Create the model as shown in Figure-72 with the help of extrusion. The orthographic views are as per the Third Angle Projection. Note that you need to first create the base extrusion by using the outlines of Top view.

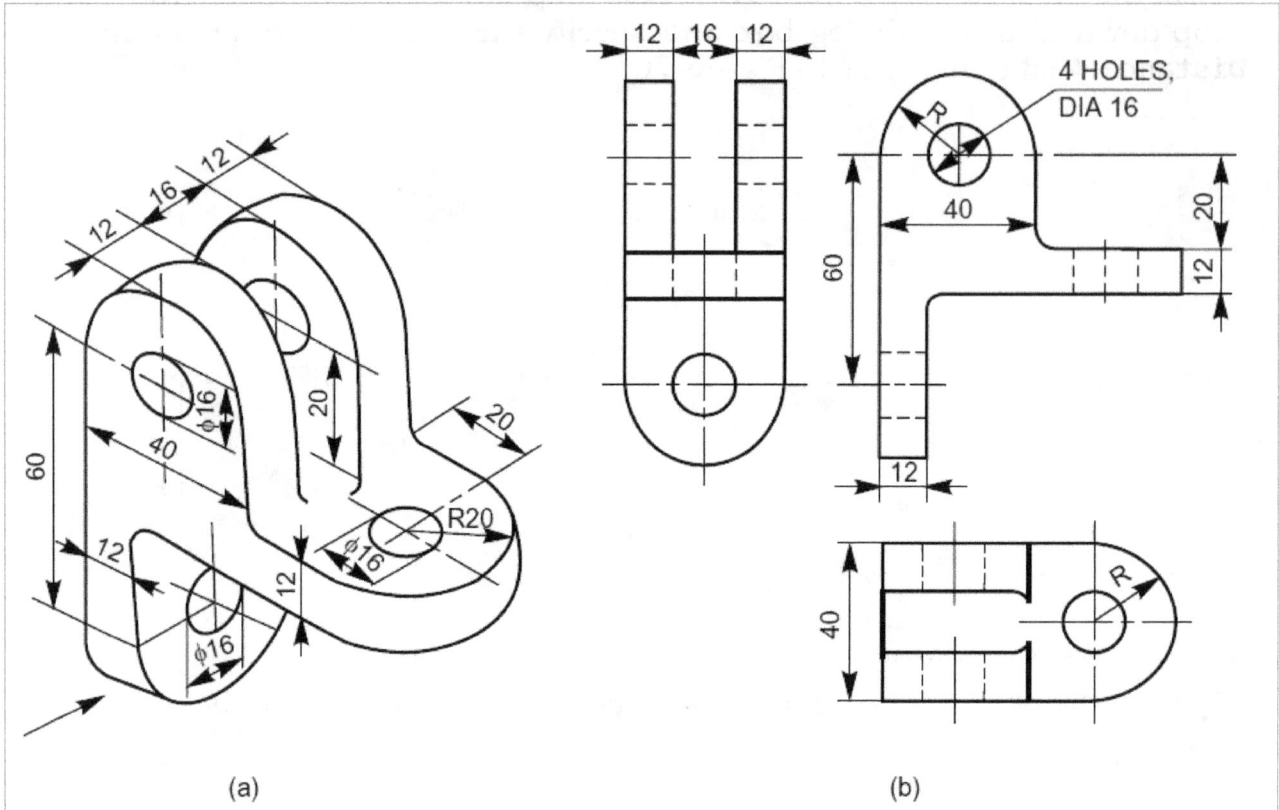

Figure-72. Drawing for Practice 1

REVOLVE TOOL

The **Revolve** tool is used to create cylindrical features. You can also use the **Revolve** tool to remove material from a solid in cylindrical fashion. At most of the places, where we need cylindrical features like in modeling a pen, a shaft, and so on; we can use the **Revolve** tool to reduce the modeling time. The procedure to use this tool is given next.

- Click on the **Revolve** tool from **Create** panel in the **Ribbon**. You will be asked to select a sketching plane.
- Click on the **XY** Plane (Front Plane) to create sketch on it. The sketching environment will be displayed.
- Create closed sketch with a centerline on its either side; refer to Figure-73.

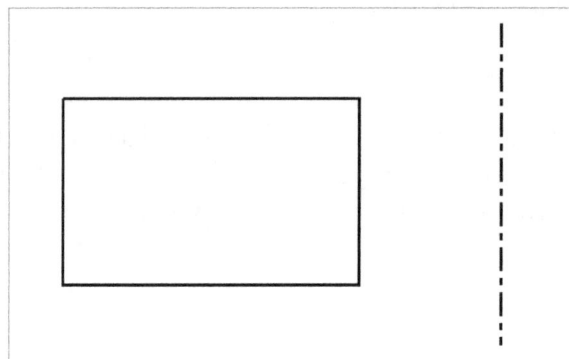

Figure-73. Sketch with centerline

- Click on the **Finish Sketch** button from **Exit** panel in the **Sketch** tab of the **Ribbon**. The **Revolve** dialog box will be displayed along with the preview of revolve feature; refer to Figure-74.

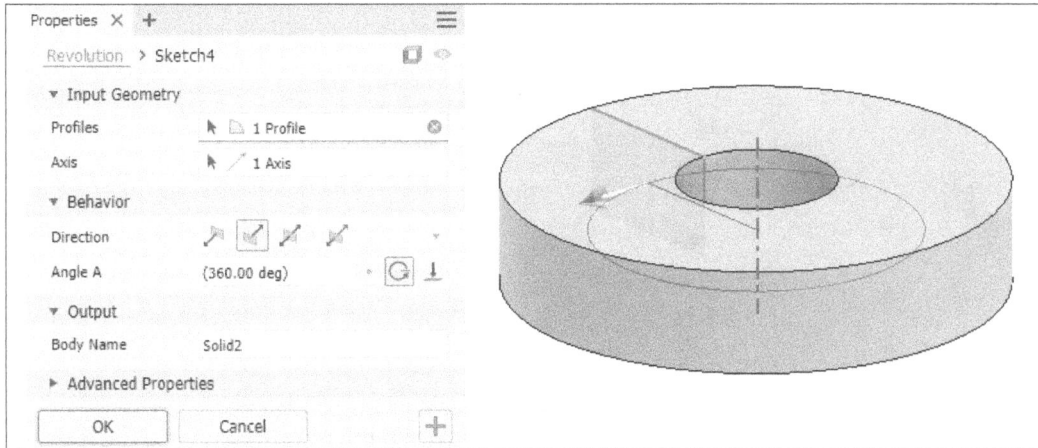

Figure-74. Revolve dialog box with preview of revolve feature

Specifying Angle of Revolution

- Most of the options in this dialog box are same as discussed for **Extrude** tool with the only difference that in place of distance, we will be specifying revolution angle for the revolve feature to define its extents. The options to define angle extent are available next to the **Angle A** edit box; refer to Figure-75.

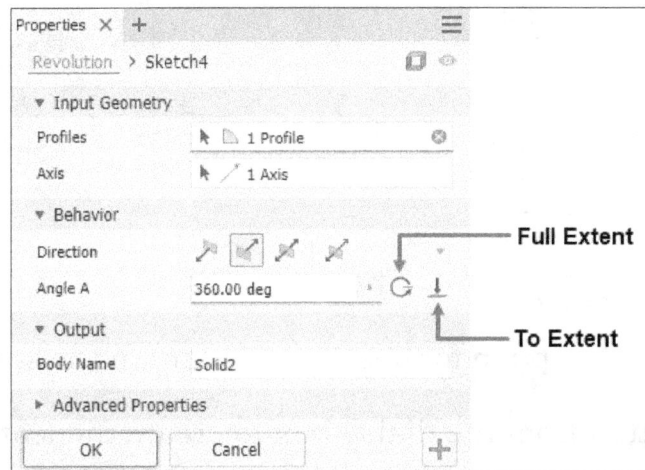

Figure-75. Extent options for revolve feature

- Click on the **Full** option from the **Extent** options. The sketch will be revolved by 360 degree of full revolution.
- Click on the **Angle A** edit box to specify the angle of revolution. Select desired option from **Direction** drop-down to specify the direction of revolve feature. The options will be displayed as shown in Figure-76.

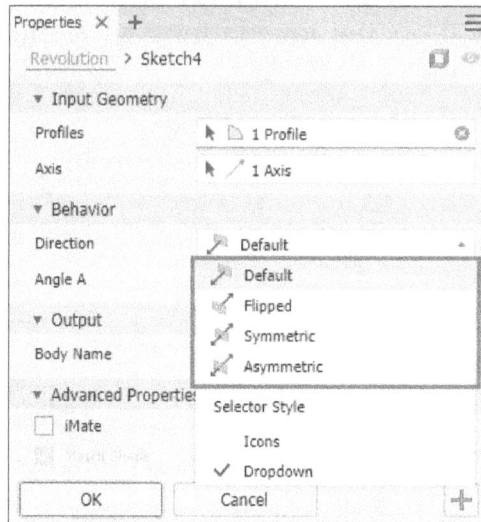

Figure-76. Options to specify angle of revolution

- Specify desired value of angle in the edit box and confirm the revolve feature with preview displayed; refer to Figure-77.

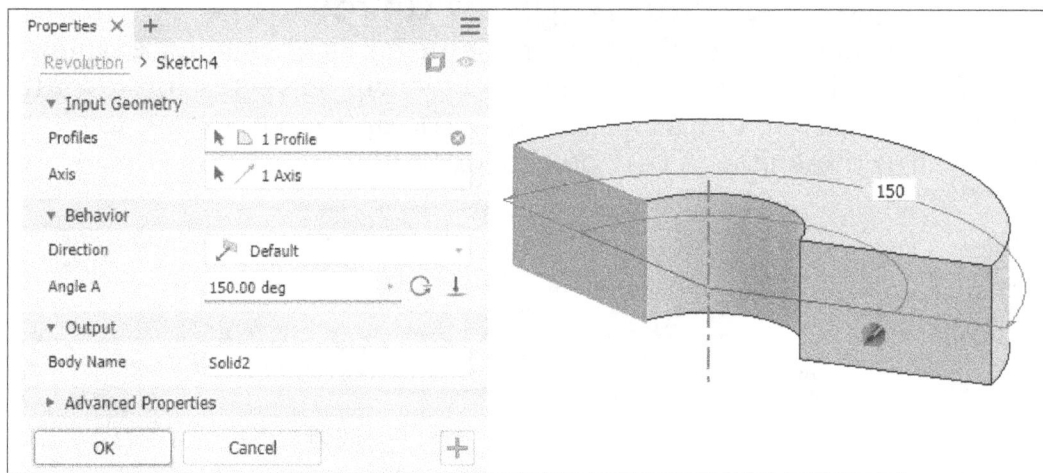

Figure-77. Preview of revolve feature

- You can also use the **To** option for specifying extent in the same way as discussed for **Extrude** tool.
- Click on the **OK** button from the dialog box to create the feature.

PRACTICAL 2

Create the model with dimensions as shown in Figure-78.

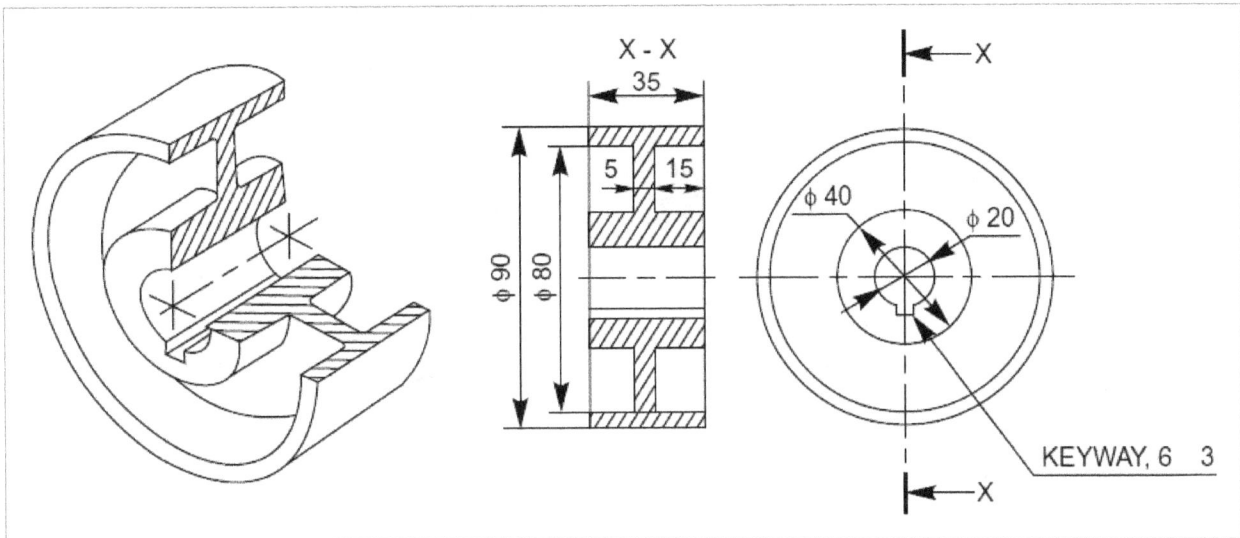

Figure-78. Drawing for Practical 2

Strategy for Creating model

1. Looking at the isometric view and orthographic views, we can find the sketch for revolve feature should be created on the Right plane (YZ Plane).
2. Using the **Revolve** tool, we will create the solid base of the model.
3. Using the **Extrude** tool, we will create the keyway from the base feature.

Starting Part File

- Start Autodesk Inventor by double-clicking on the Autodesk Inventor Professional 2024 icon from the desktop. (If not started yet.)
- Click on the **New** button from the **Quick Access Toolbar**. The **Create New File** dialog box will be displayed.
- Double-click on **Standard(mm).ipt** icon from the **Metric** templates. The Part environment of Autodesk Inventor will be displayed.

Creating sketch

- Click on the **Revolve** tool from **Create** panel in the **3D Model** tab of the **Ribbon**. You will be asked to select a sketching plane.
- Click on the **YZ** Plane (Right Plane) to create sketch on it.
- Click on the **Line** tool from the **Create** panel in the **Sketch** tab of the **Ribbon**. You will be asked to specify the starting point of the line.
- Click on the **Centerline** button from **Format** panel in the **Sketch** contextual tab of **Ribbon** and create a centerline as shown in Figure-79.

Figure-79. Centerline created

- Click again on the **Centerline** button to toggle centerline creation.
- Create the sketch as shown in Figure-80.

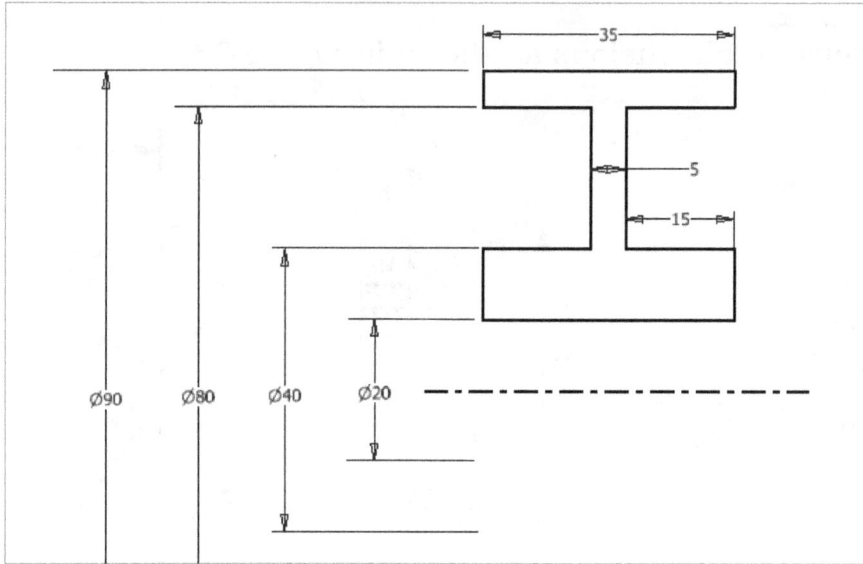

Figure-80. Sketch created for revolve

Creating Revolve Feature

• Click on the **Finish Sketch** button from **Exit** panel in the **Sketch** tab of the **Ribbon**. The **Revolve** dialog box will be displayed along with the preview of revolve feature; refer to Figure-81.

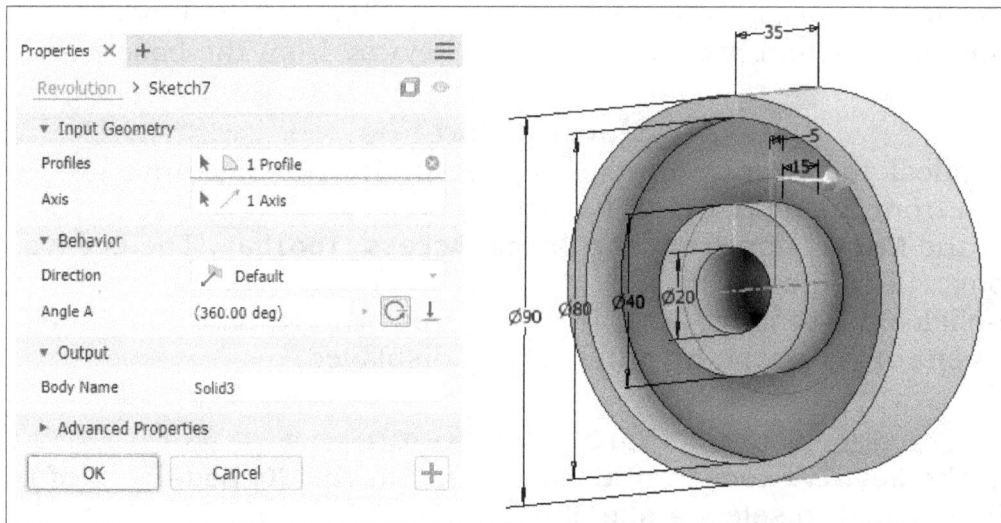

Figure-81. Preview of revolve feature

• Click on the **OK** button from the dialog box to create the feature.

Creating the Cut Feature

• Click on the **Extrude** tool from **Create** panel in the **3D Model** tab of the **Ribbon**. You will be asked to select a plane/face to create the sketch.
• Select the face as shown in Figure-82 and create the sketch as shown in Figure-83.

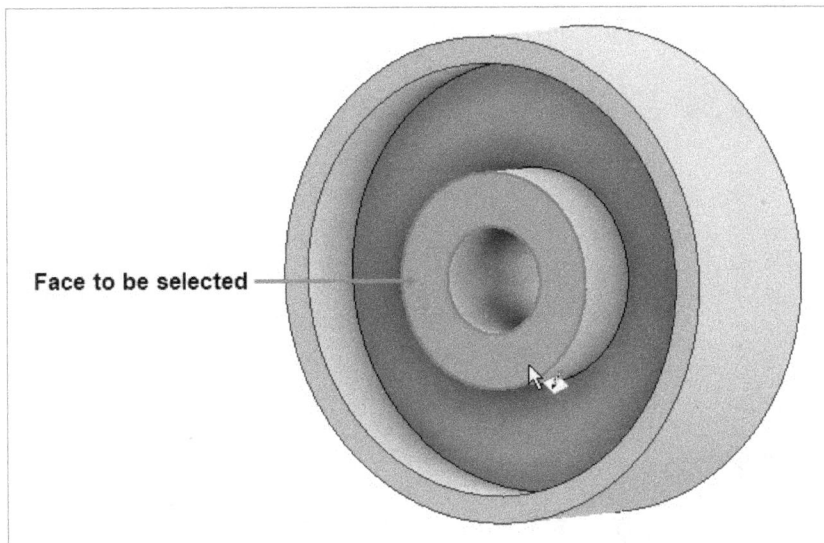

Figure-82. Face selected for extrusion

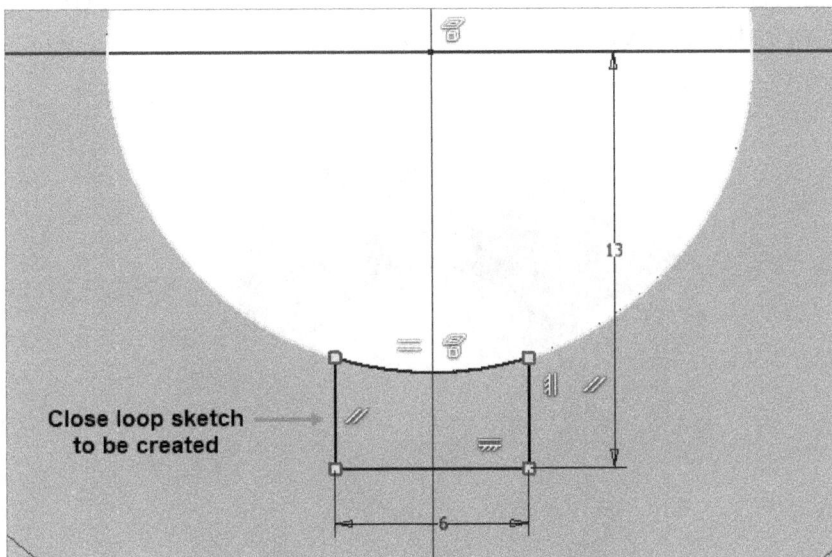

Figure-83. Sketch for extrude cut

- Click on the **Finish Sketch** button from **Exit** panel in the **Sketch** tab of the **Ribbon**. The **Extrude** dialog box will be displayed asking you to select the profiles.
- Select the newly created sketch from the model. The preview of extrusion will be displayed.
- Click on the **To Next** option from the **Extent** options next to the **Distance A** edit box in the dialog box. Preview of the cut feature will be displayed; refer to Figure-84.

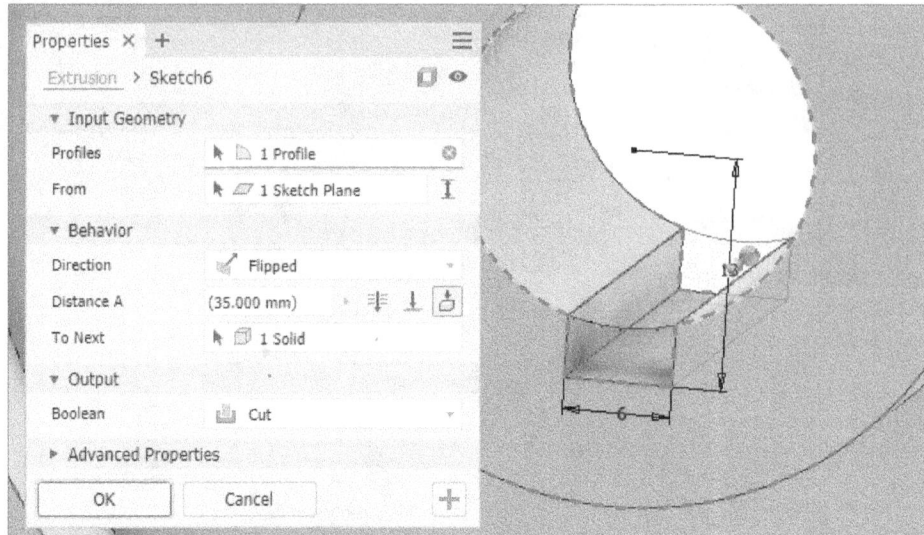

Figure-84. Preview of the cut feature

• Click on the **OK** button from the dialog box to create the cut feature. The final model will be created; refer to Figure-85.

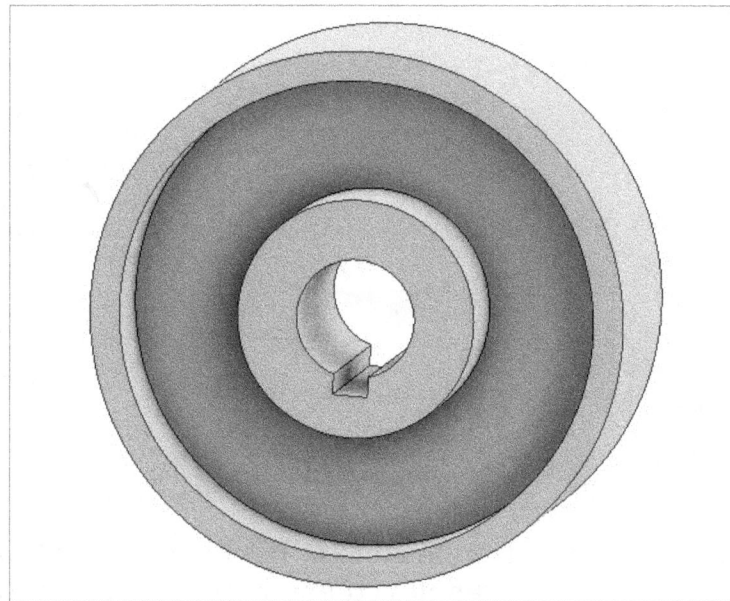

Figure-85. Final model of Practical 2

PRACTICAL 3

Create the model as shown in Figure-86.

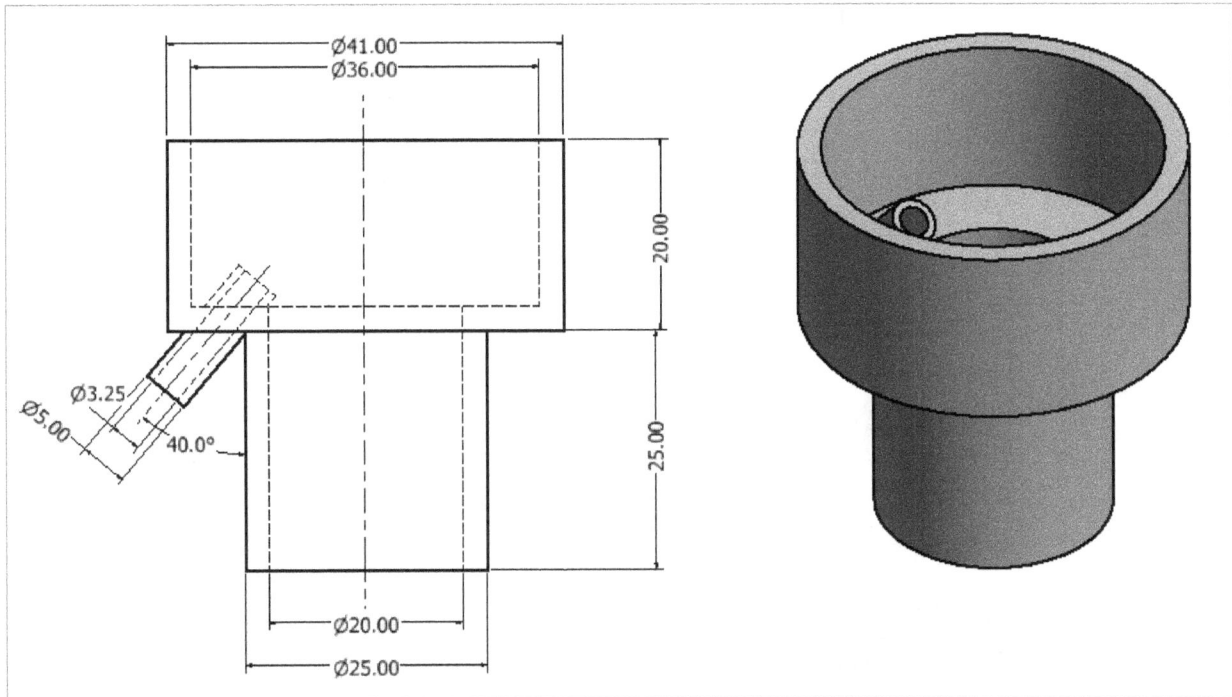

Figure-86. Drawing for Practical 3

Strategy for Creating model

1. Looking at the isometric view and orthographic views, we can find the sketch for revolve feature should be created on the Front plane (YZ Plane).
2. Using the **Revolve** tool, we will create the solid base of the model.
3. Using the Plane tools, we will create a plane at an angle of 40 degree to the vertical wall of base feature.
4. Using the **Extrude** tool, we will create the pipe joined to the base feature.

Starting Part File

- Start Autodesk Inventor by double-clicking on the Autodesk Inventor Professional 2024 icon from the desktop. (If not started yet.)
- Click on the **New** button from the **Quick Access Toolbar**. The **Create New File** dialog box will be displayed.
- Double-click on **Standard(mm).ipt** icon from the **Metric** templates. The Part environment of Autodesk Inventor will be displayed.

Creating sketch

- Click on the **Revolve** button from **Create** panel in the **3D Model** tab of the **Ribbon**. You will be asked to select a sketching plane.
- Select the **XY** Plane (Front Plane) as sketching plane. The sketching environment will be displayed.
- Create the sketch as shown in Figure-87.

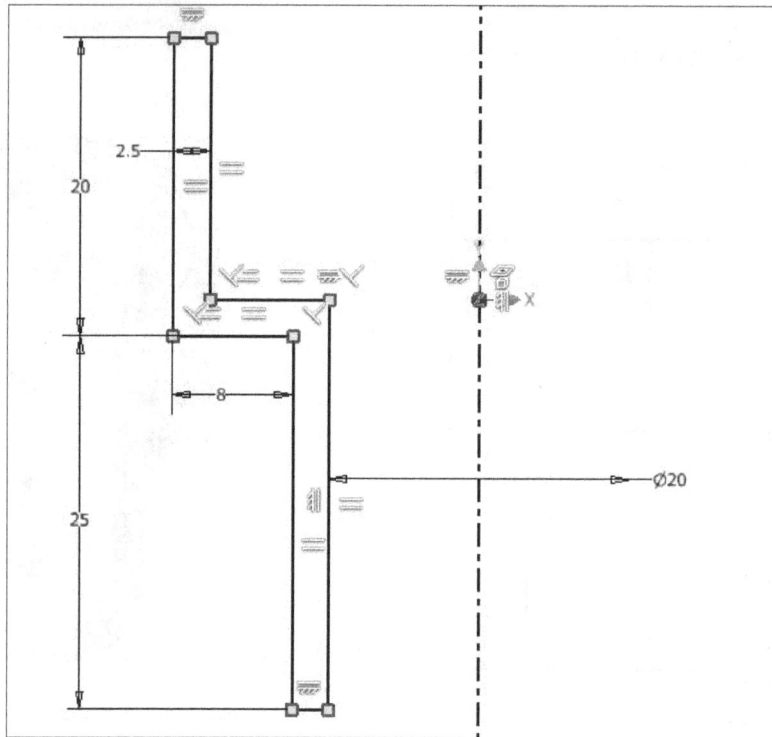

Figure-87. Sketch with constrains

Creating Revolve Feature

- Click on the **Finish Sketch** button from **Exit** panel in the **Sketch** tab of the **Ribbon.** The **Revolve** dialog box will be displayed along with the preview of revolve feature; refer to Figure-88.

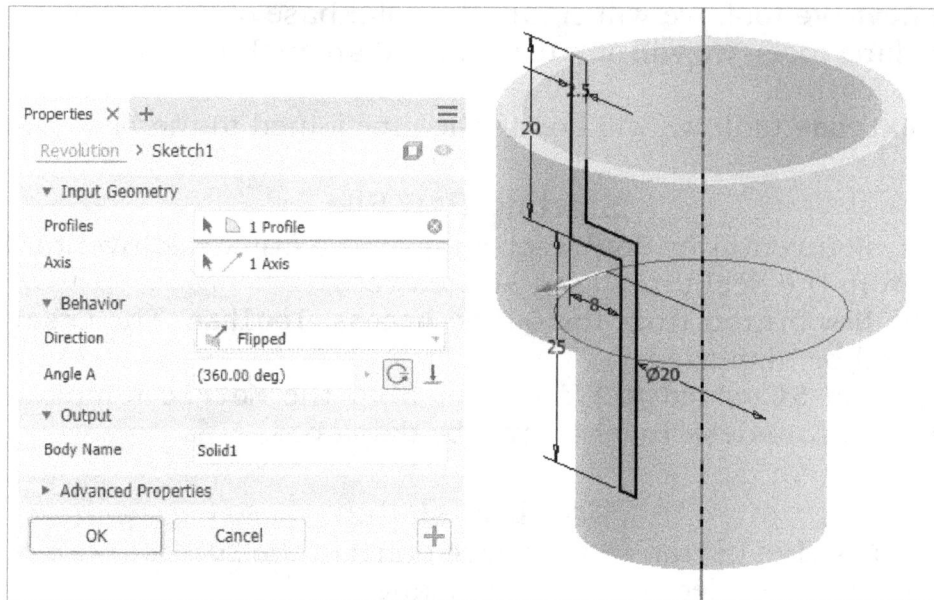

Figure-88. Preview of revolve feature

- Click on the **OK** button from the dialog box to create the revolve feature; refer to Figure-89.

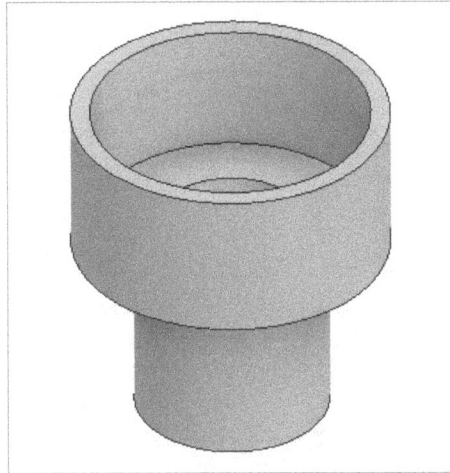

Figure-89. Revolve feature created

Creating Plane at angle

- Click on the **Start 2D Sketch** tool from **Sketch** panel in the **3D Model** tab of the **Ribbon**. You will be asked to select a sketching plane.
- Click on the bottom face of the model as shown in Figure-90 and create a line tangent to the edge of base model as shown in Figure-91.

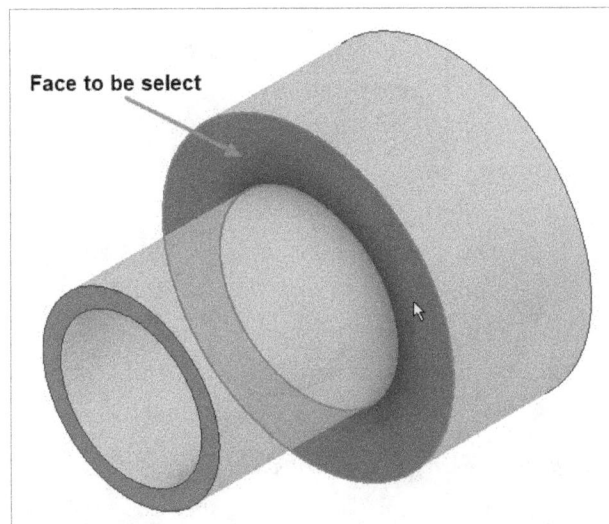

Face to be select

Figure-90. Face to select

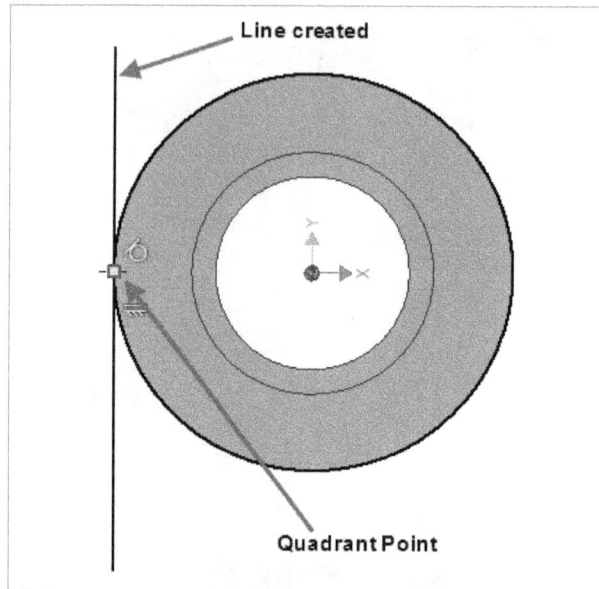

Figure-91. Line to create in sketch

- Click on the **Finish Sketch** button from **Exit** panel in the **Sketch** tab of the **Ribbon**. In isometric view, the sketch should be displayed as shown in Figure-92.

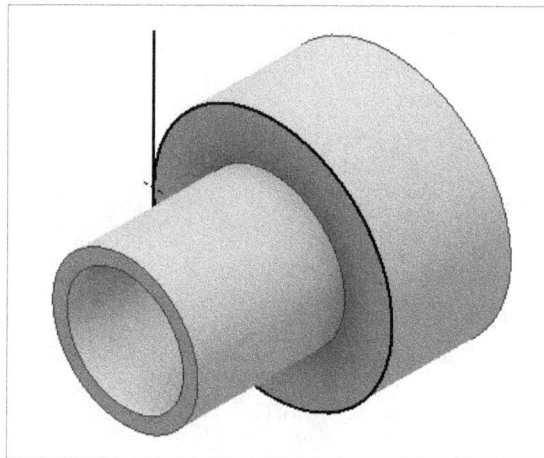

Figure-92. Sketch in isometric view

- Click on the **Angle to Plane around Edge** tool from **Plane** drop-down in the **Work Features** panel of the **3D Model** tab in the **Ribbon**. You will be asked to select a line or plane.
- Click on the line newly created and select the bottom face of the base feature. You will be asked to specify the angle between face and plane; refer to Figure-93.

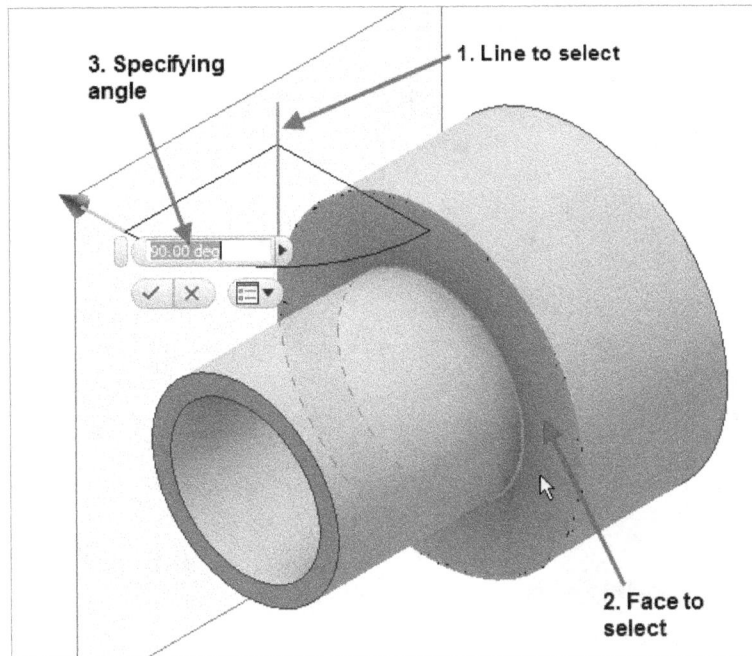

Figure-93. Selecting line and plane

- Specify the angle value as **40** in the edit box and click on the **OK** button to create the plane. The plane will be displayed as shown in Figure-94.

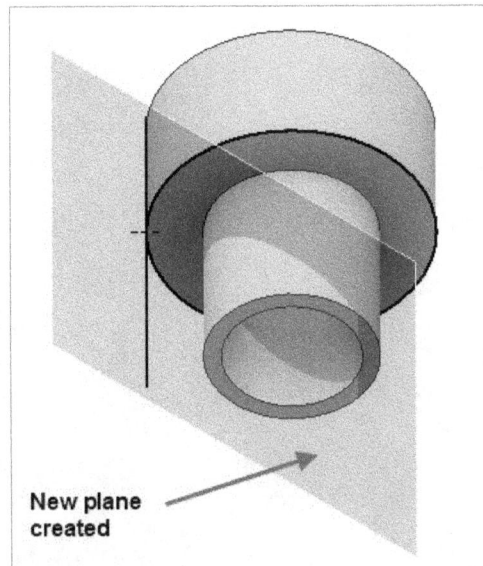

Figure-94. New plane created

Creating Tube

- Click on the **Start 2D Sketch** button from **Sketch** panel in the **3D Model** tab of the **Ribbon**. You will be asked to select the sketching plane.
- Select the newly created plane. The sketching environment will be displayed.

- Create a sketch as shown in Figure-95. Make sure you have applied proper constraint as given in the figure.

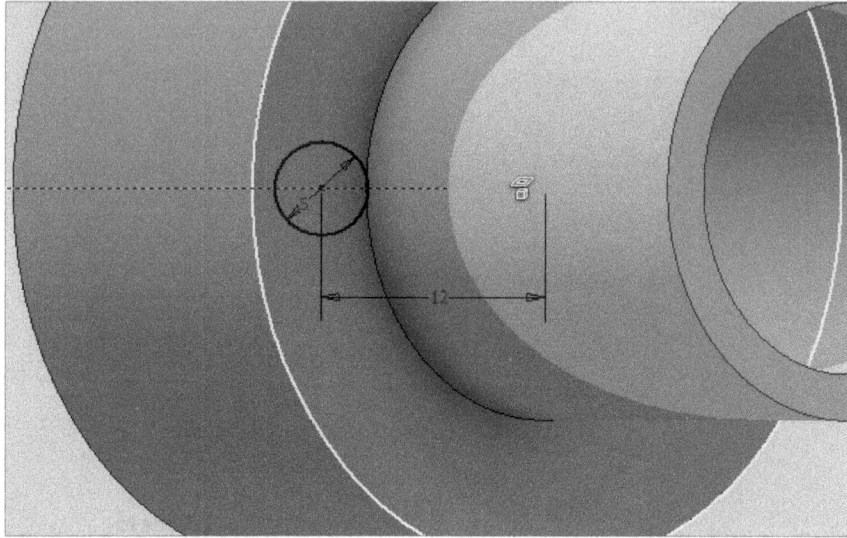

Figure-95. Circle to be created

- Click on the **Finish Sketch** button from **Exit** panel in the **Sketch** tab of the **Ribbon** to exit the sketching environment.
- Click on the **Extrude** tool from **Create** panel in the **3D Model** tab of the **Ribbon**. The **Extrude** dialog box will be displayed along with the preview of extrusion. Specify the distances as 10 inside the base feature and 5 outside the base feature; refer to Figure-96.

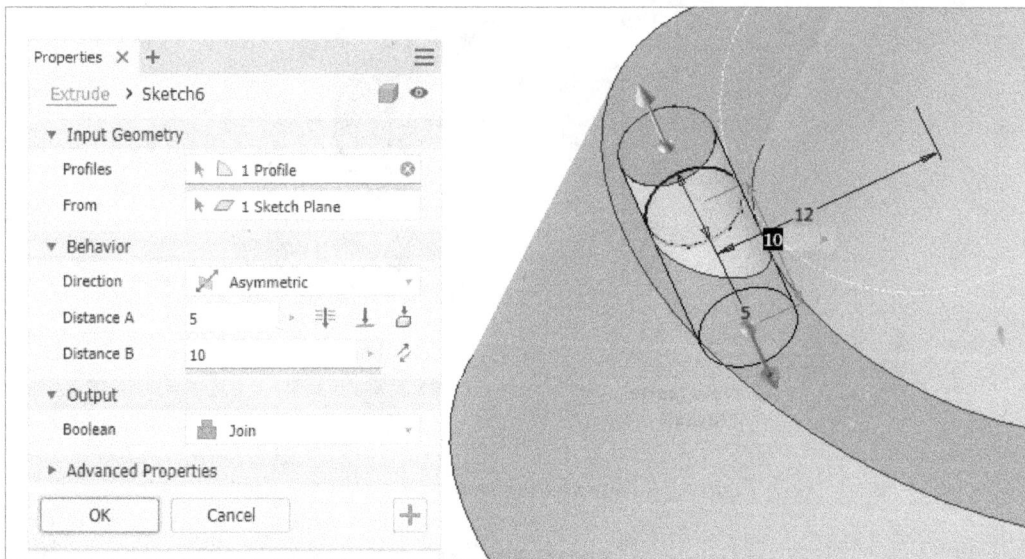

Figure-96. Preview of extrusion

- Click on the **OK** button from the dialog box. The extrusion will be displayed as shown in Figure-97.

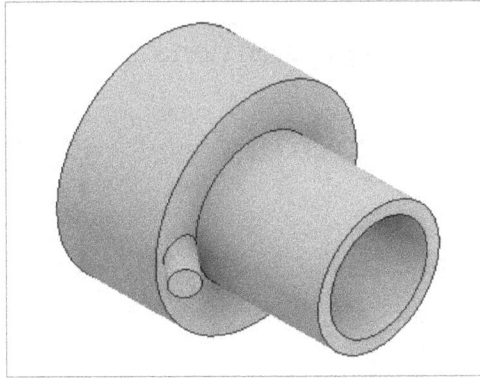

Figure-97. Model after creating extrusion

- Click on the **Extrude** tool again and using the face of newly created extrusion, create a extrude cut for making hole of diameter **3.25**; refer to Figure-98.

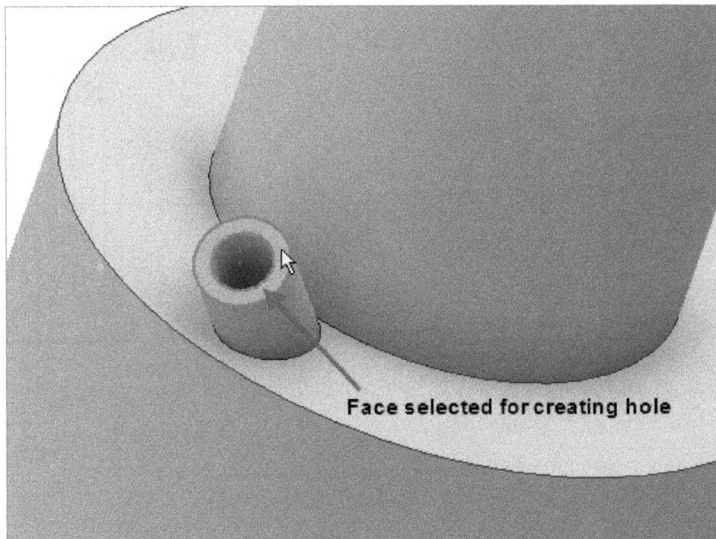

Figure-98. Creating hole

PRACTICE 1

Create the model of pulley by using dimensions given in the Figure-99.

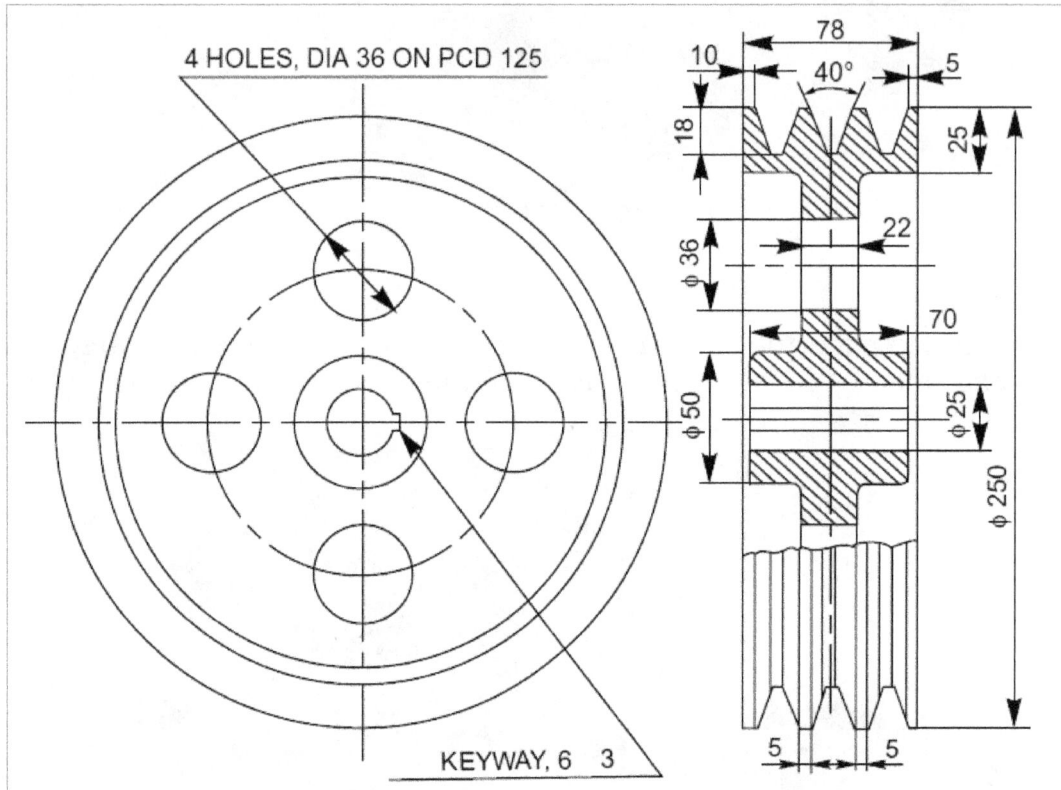

Figure-99. Drawing for Practice 1

PRACTICE 2

Create the model as shown in Figure-100.

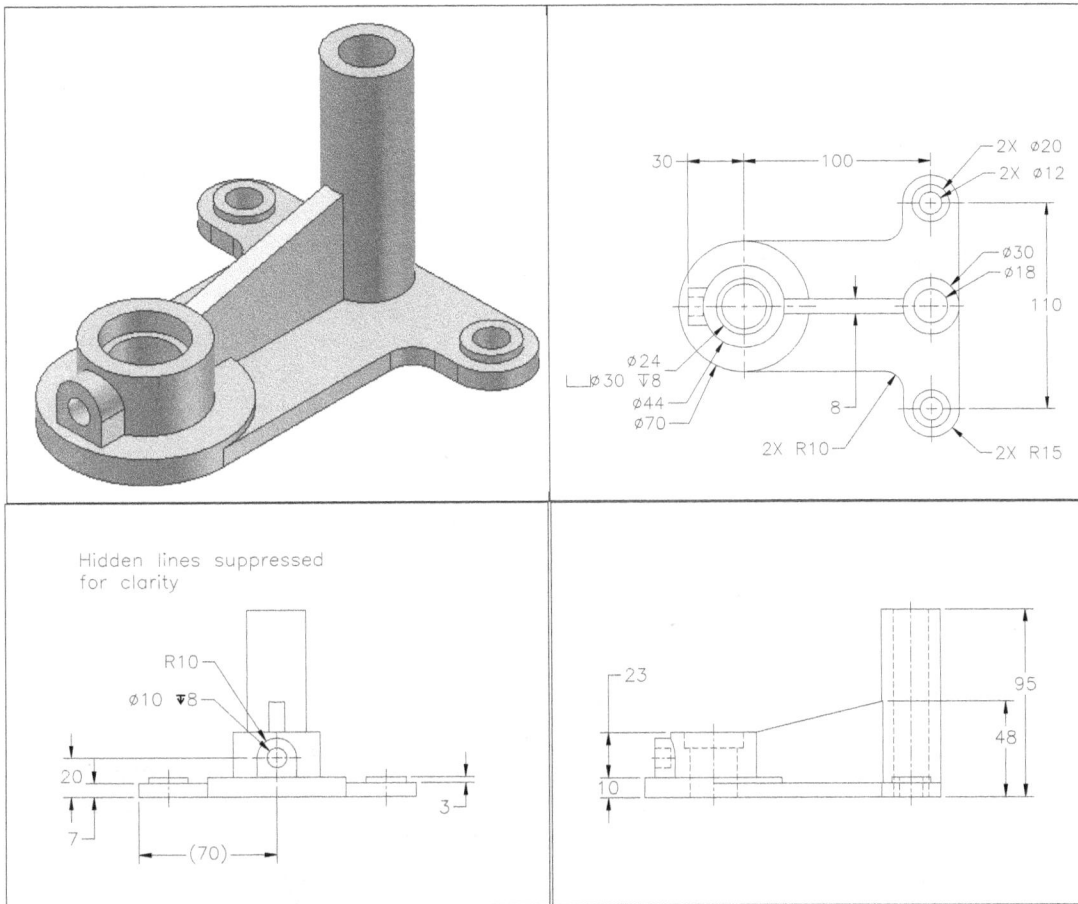

Figure-100. Views and dimensions of the model

PRACTICE 3

Create the model as shown in Figure-101.

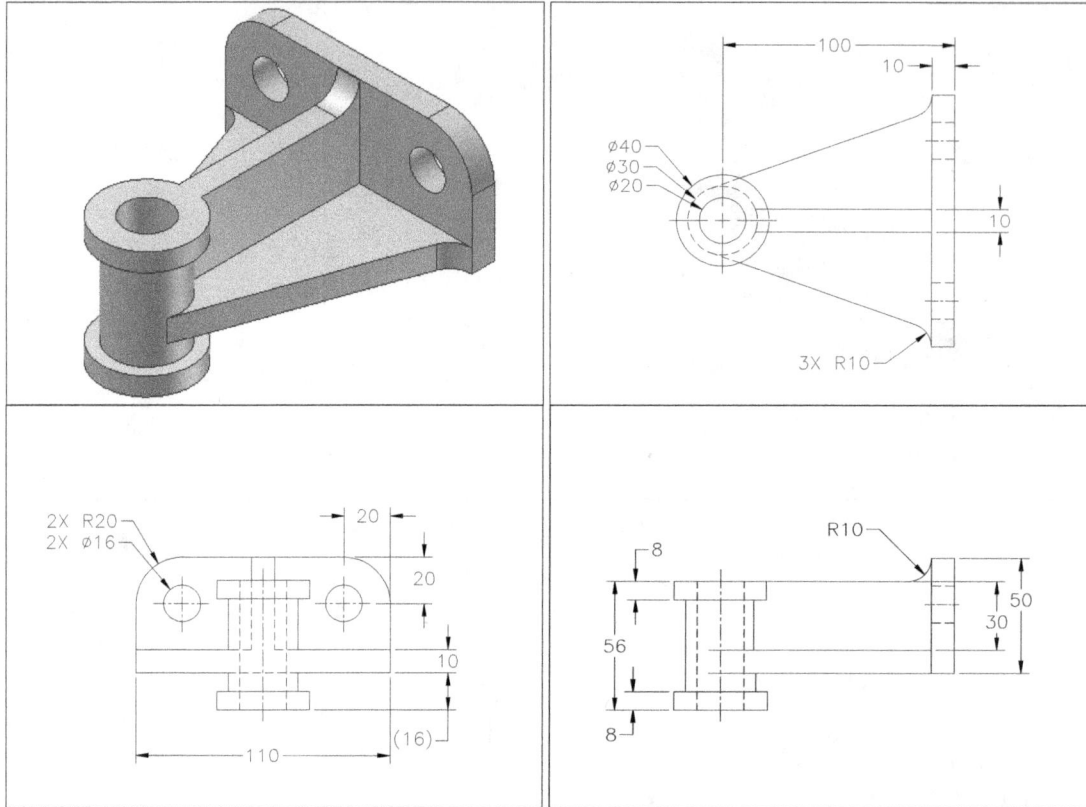

Figure-101. Views and dimensions of the model

PRACTICE 4

Create the model shown in Figure-102.

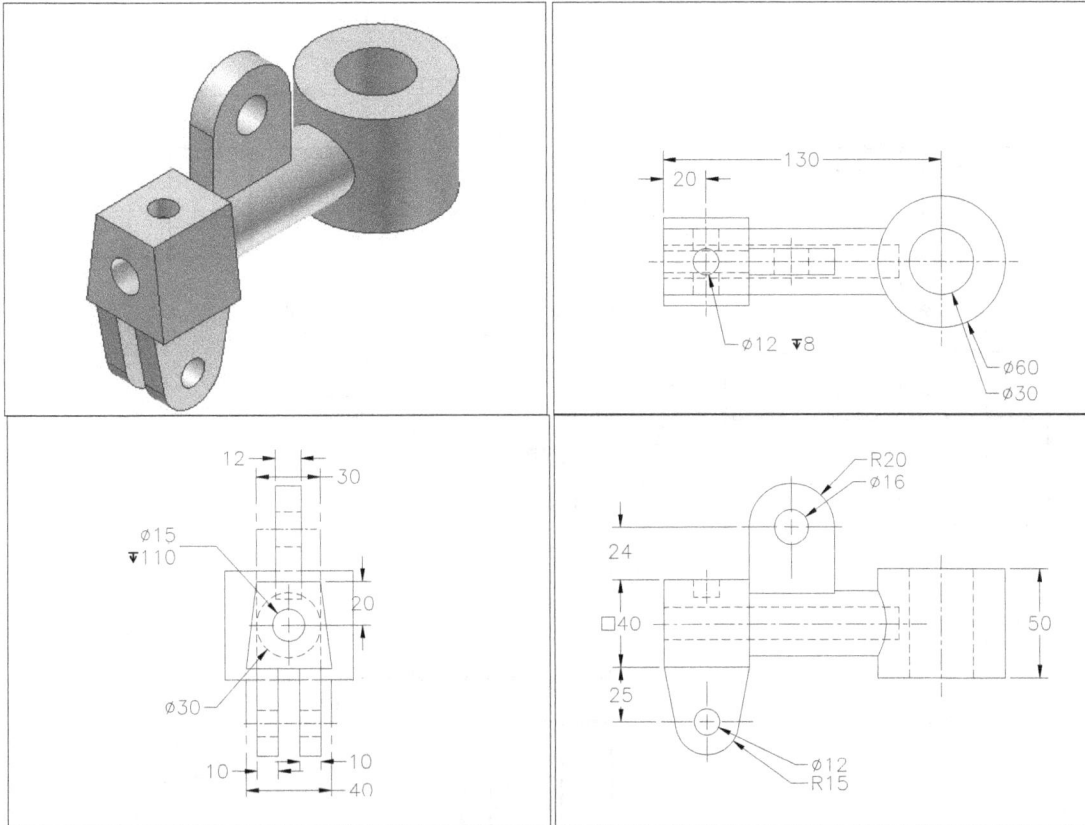

Figure-102. Views and dimensions of the model

PRACTICE 5

Create the model by using the dimensions given in Figure-103.

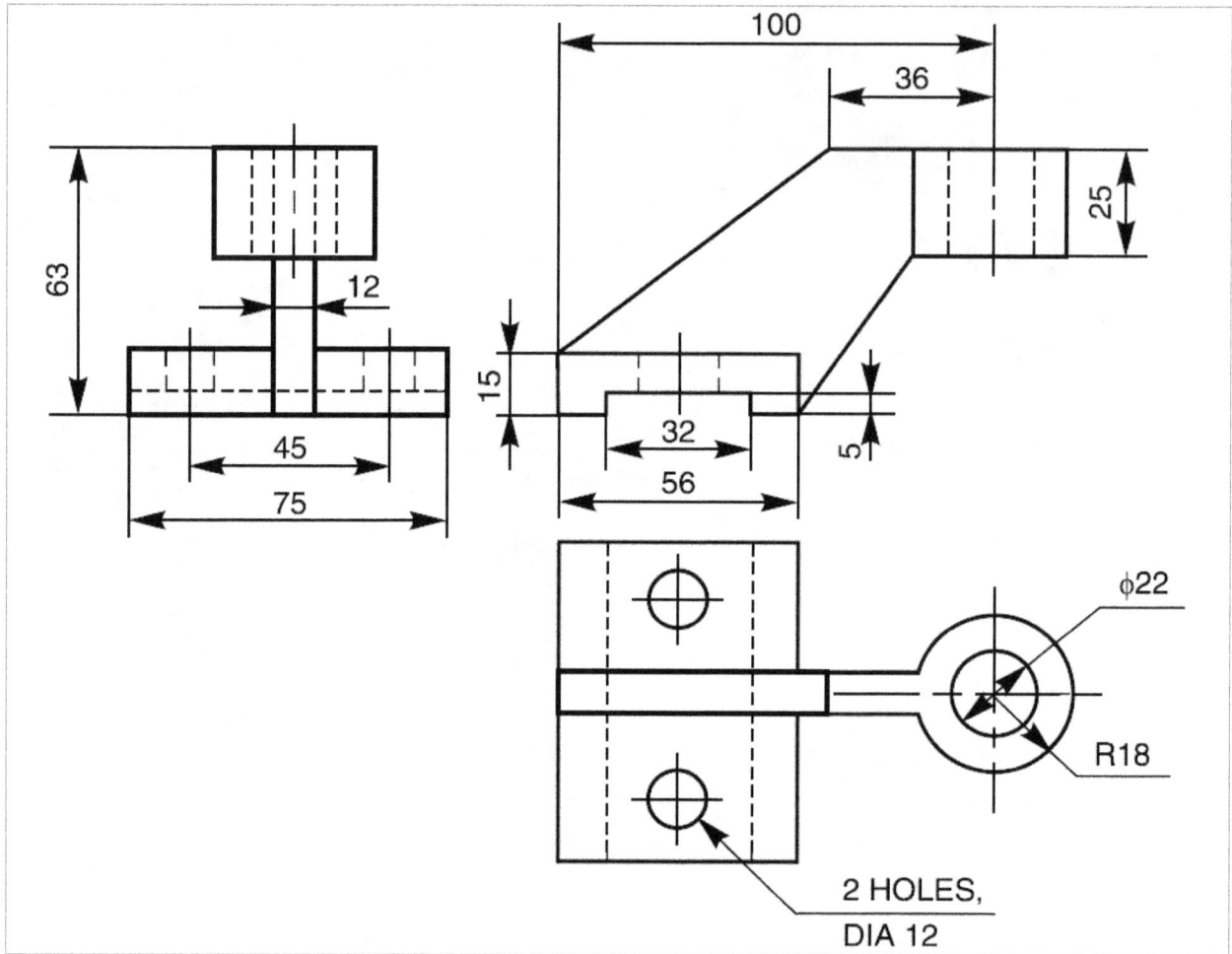

Figure-103. Practice 5

SELF ASSESSMENT

Q1. Which of the following work planes do not exist by default in the software?

a) XY Plane
b) YZ Plane
c) YX Plane
d) XZ Plane

Q2. By which of the following tools can you create a plane through the median of selected planes?

a) Midplane between two Planes
b) Parallel to Plane through Point
c) Midplane of Torus
d) Offset from Plane

Q3. A **Plane** tool can create all the types of plane except one. Which type of plane is it?

a) Normal to Curve at Point
b) Normal to Axis through Point
c) Offset from Plane
d) Two Coplanar Edges

Q4. Which of the following tools can create an axis perpendicular to selected plane and passing through the selected point?

a) On Line or Edge
b) Normal to Plane through Point
c) Through Two Points
d) Axis

Q5. What is the size of Point theoretically?

a) 3
b) 2
c) 1
d) 0

Q6. Which type of Boolean operations should be used to extract the common portion between base feature and extrusion, so that rest of the material in model is automatically removed?

a) Intersect
b) Join
c) Cut
d) New Solid

Q7. Which type of projections are mainly used for orthographic views?

a) First Angle Projection and Second Angle Projection
b) First Angle Projection and Third Angle Projection
c) First Angle Projection and Fourth Angle Projection
d) Third Angle Projection and Fourth Angle Projection

Q8. The **Revolve** tool can be used to remove material from a solid in cylindrical fashion. (True/False)

Q9. If you want to create extrusion to both sides of sketching plane with same depth then select the **Asymmetric** option. (True/False)

Q10. Broadly, there are two ways to present a component in engineering drawing: .... representation and .... representation.

FOR STUDENT NOTES

Chapter 5

Advanced Solid Modeling Tools

Topics Covered

The major topics covered in this chapter are:

- *Sweep tool*
- *Loft tool*
- *Coil tool*
- *Emboss tool*
- *Rib tool*
- *Derive tool*
- *Decal tool*
- *Import tool*

INTRODUCTION

In the previous chapter, we learnt about extrude and revolve features. We also learnt the use of planes, axes, and points. In this chapter, we will learn to use some advanced tools for creating solid models. These tools are also available in the **Create** panel of the **3D Model** tab in the **Ribbon**.

SWEEP TOOL

Sweep The **Sweep** tool is used to create solid/surface feature by sweeping closed loop sketch along the selected trajectory. This tool is generally used when we need to create tubes/bars along a curve. The procedure to use this tool is given next.

- Create a closed loop sketch section and a path in the modeling area; refer to Figure-1.

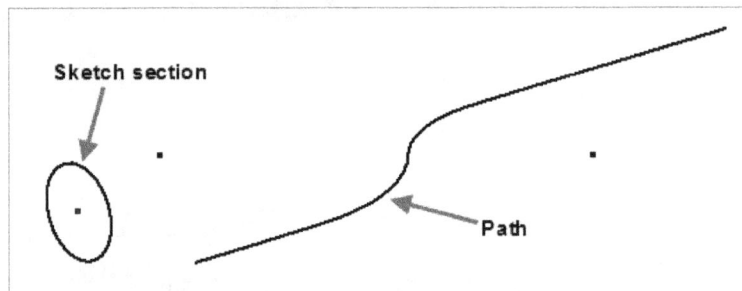

Figure-1. Sketch section and path

- Click on the **Sweep** tool from **Create** panel in the **3D Model** tab of the **Ribbon**. The **Sweep** dialog box will be displayed along with the preview of sweep feature; refer to Figure-2. If there are more than one sections in the modeling space then you will be asked to select a section (Profile).

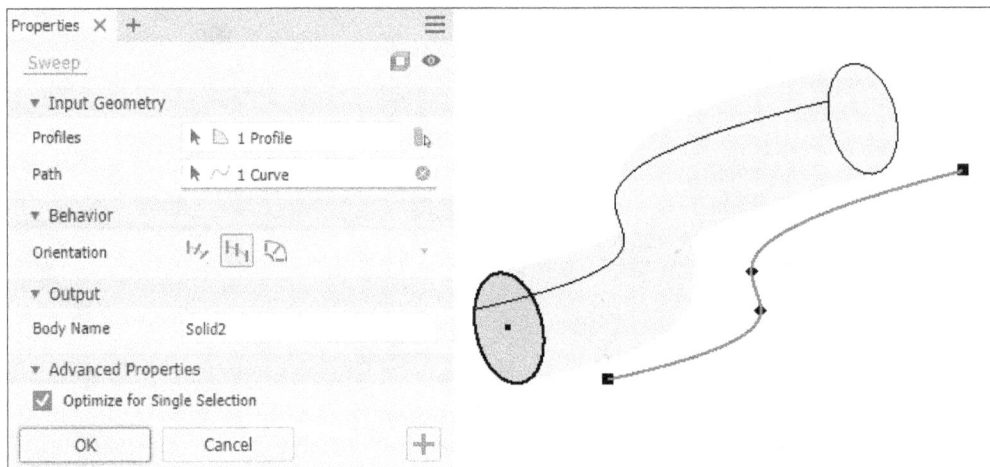

Figure-2. Preview of sweep feature

- There are three types of orientations available for creating sweep feature: **Follow Path**, **Fixed**, and **Guide**; refer to Figure-3.

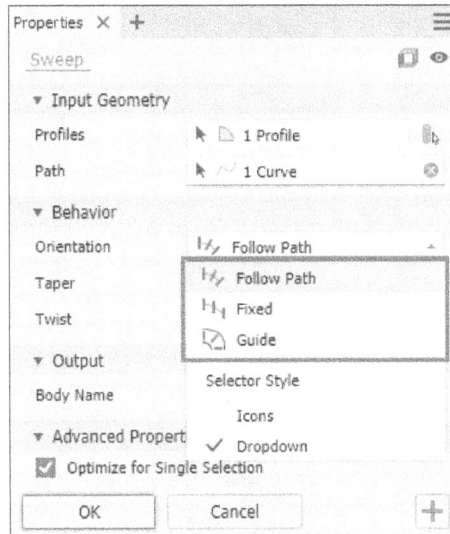

Figure-3. Orientation for sweep

Creating Sweep feature with Follow Path

- If you select the **Follow Path** option from **Orientation** drop-down in the dialog box then profile will be of same shape and size throughout the path followed for sweep; refer to Figure-4.

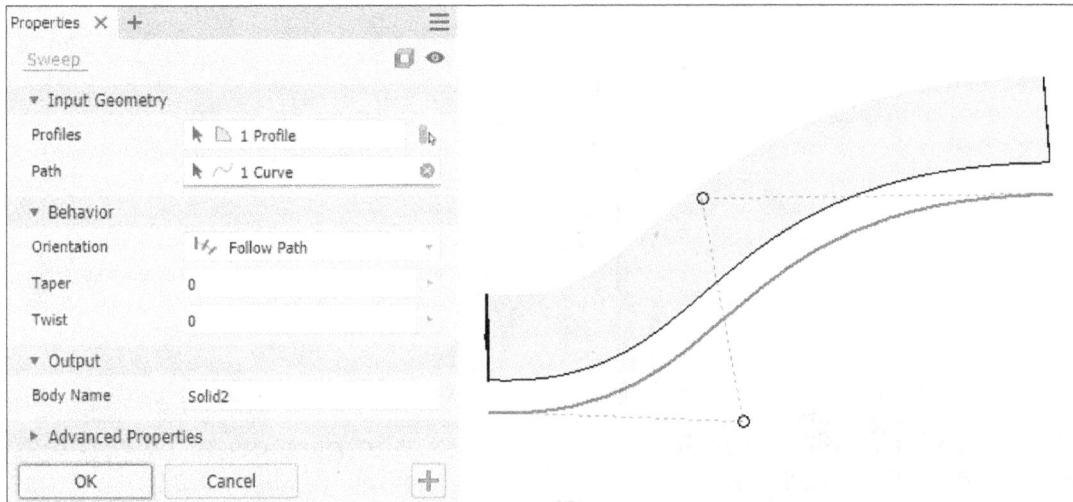

Figure-4. Sweep with Follow Path option

- On selecting the **Follow Path** option, you can specify the taper angle and twist angle in the **Taper** and **Twist** edit boxes of the **Sweep** dialog box, respectively; refer to Figure-5.

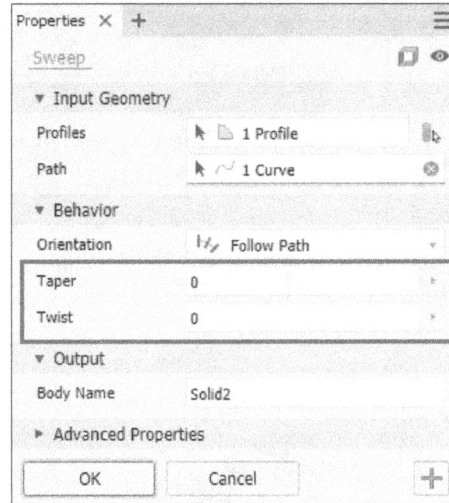

Figure-5. Taper and Twist edit boxes

- Specify the taper angle value in **Taper** edit box, the preview of sweep feature will be displayed; refer to Figure-6.

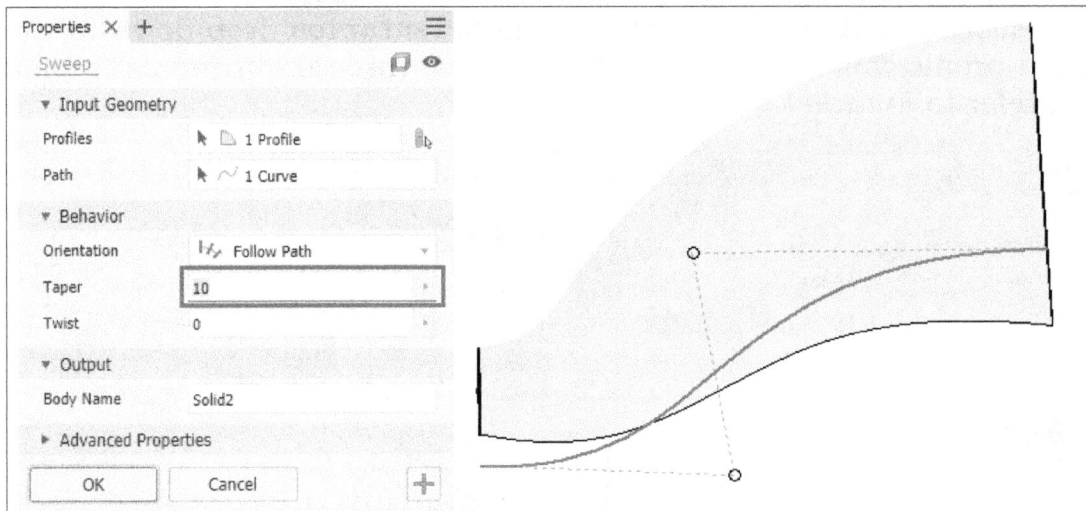

Figure-6. Tapered sweep

- Specify the twist angle value in **Twist** edit box, the preview of sweep feature will be displayed; refer to Figure-7.

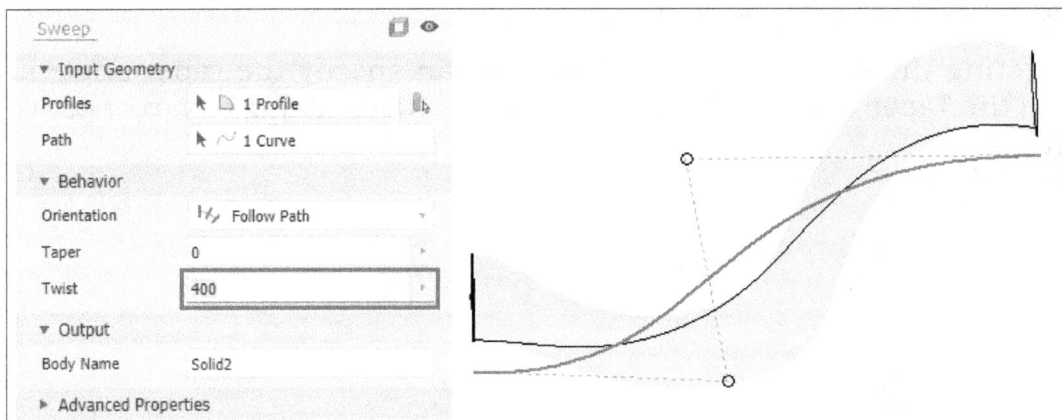

Figure-7. Twisted sweep

Creating Sweep feature with Fixed

- If you select the **Fixed** option from **Orientation** drop-down in the dialog box then shape/size of the profile can be changed to make it follow the exact shape of path; refer to Figure-8.

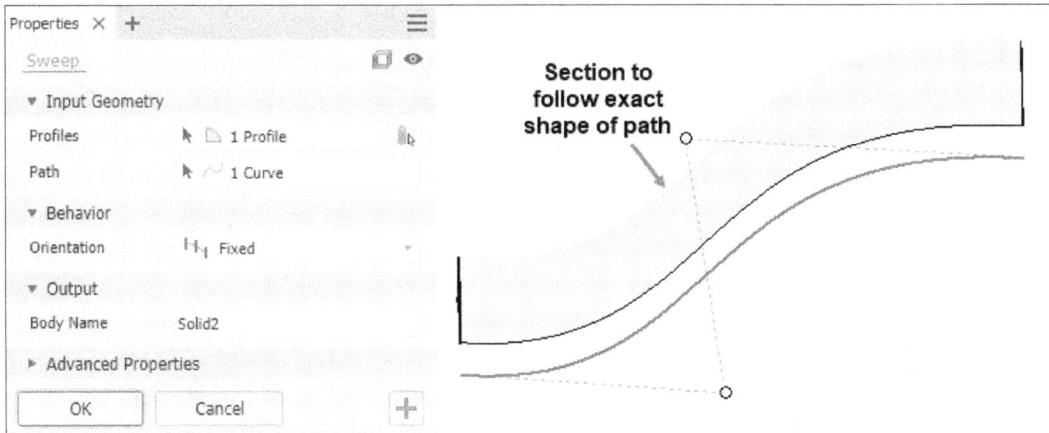

Figure-8. Sweep with Fixed option

Creating Sweep feature with Guide

- Sometimes, it becomes necessary to control the outer shape of the sweep feature. You can create this feature with **Guide** options which is available in **Orientation** drop-down in the dialog box. To use this option, you must have a guide rail curve along with section and path; refer to Figure-9.

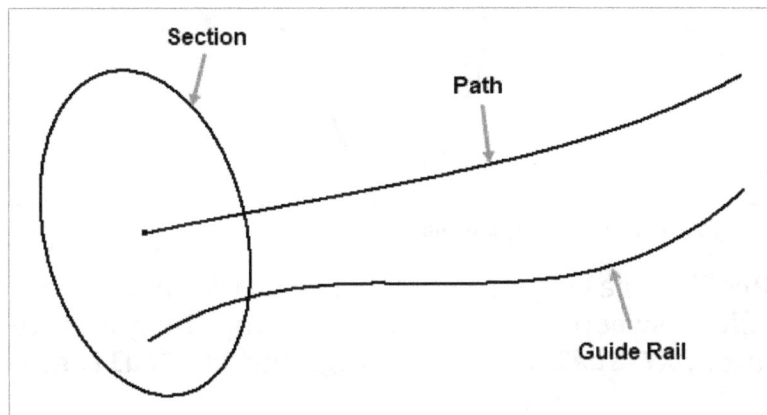

Figure-9. Section with path and guide rail

- Once you have the three sketch entities (section, path, and guide rail), click on the **Sweep** tool from **Create** panel in the **3D Model** tab of the **Ribbon**. The **Sweep** dialog box will be displayed with profile selection and you will be asked to select the path and guide rail; refer to Figure-10.

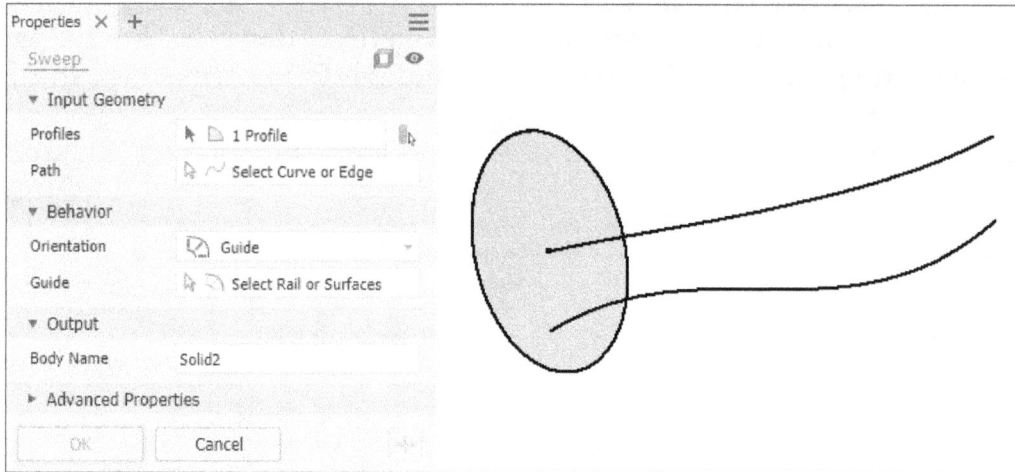

Figure-10. Sweep dialog box with profile selection

*   Click on the **Path** selection button and select the path for sweep.
*   Next, click on the **Guide** selection button and select the guide rail for sweep. Preview of the sweep feature will be displayed; refer to Figure-11.

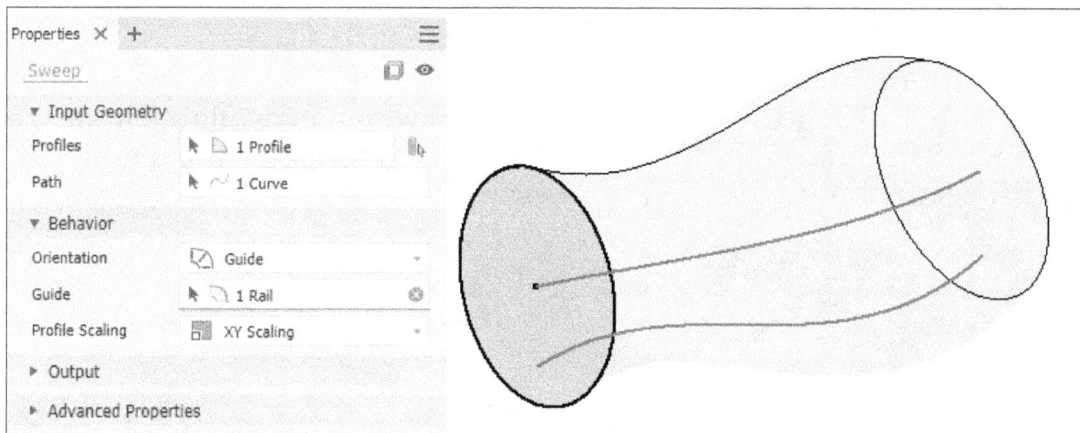

Figure-11. Preview of sweep feature with guide rail

*   Options in the **Profile Scaling** area below the **Guide** selection button are used to manage the profile of sweep feature while following the guide rail. There are three options in this area, **XY Scaling**, **X Scaling**, and **No Scaling**; refer to Figure-12.

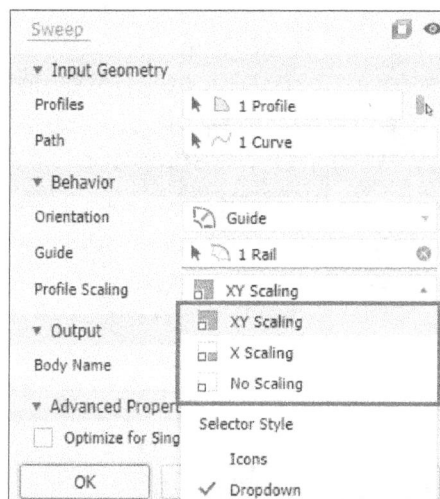

Figure-12. Profile Scaling options

- Select the **XY Scaling** option from the **Profile Scaling** drop-down if you want the profile section to re-size in X and Y both directions as per the guide rail; refer to Figure-13.

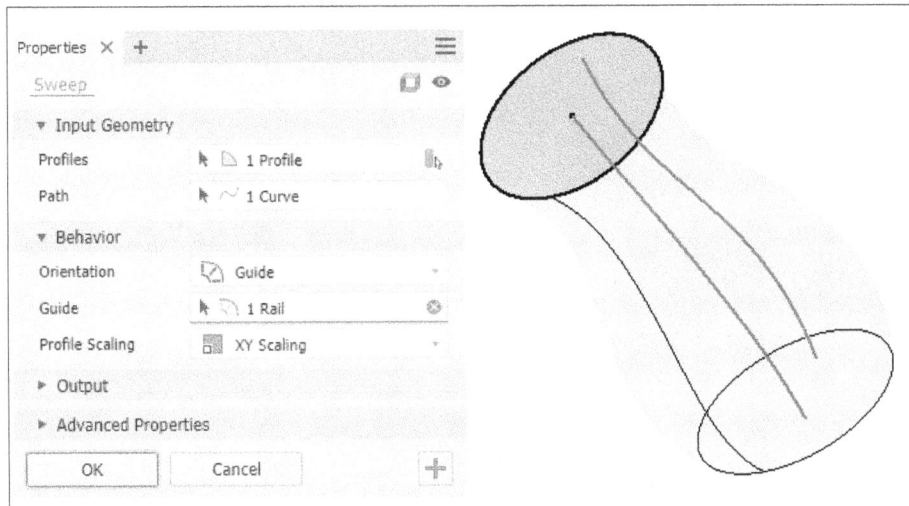

Figure-13. Sweep with XY profile scaling option

- Select the **X Scaling** option if you want the profile section to re-size only in X direction as per the guide rail; refer to Figure-14.

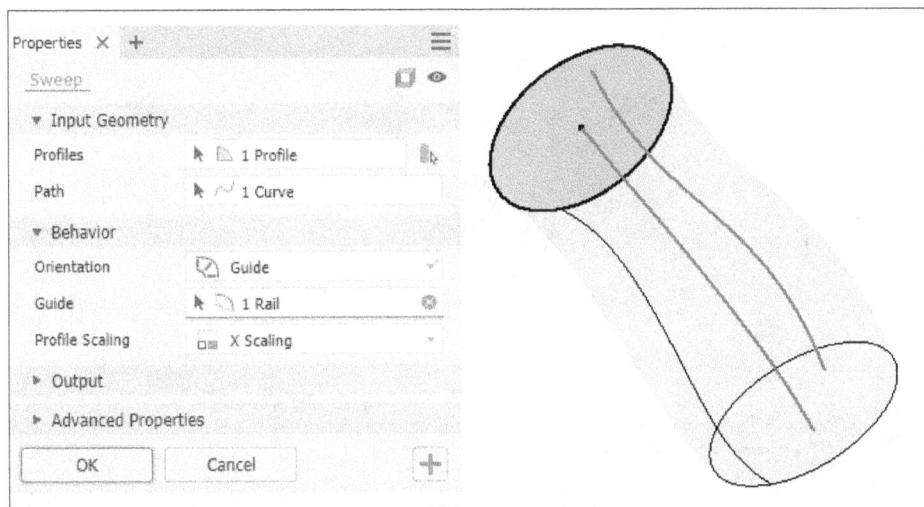

Figure-14. Sweep with X profile scaling option

- If you do not want to scale the profile in any direction then click on the **No Scaling** option from the **Profile Scaling** drop-down; refer to Figure-15.

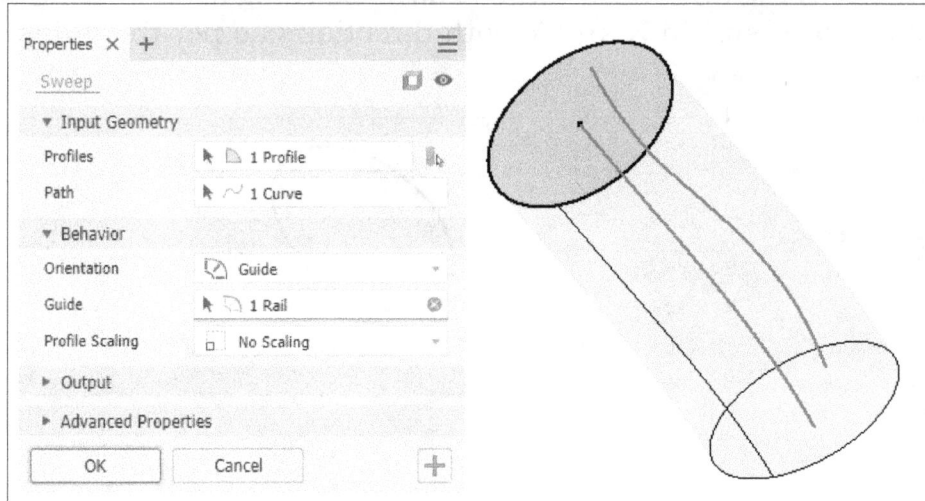

Figure-15. Sweep with No profile scaling option

LOFT TOOL

The **Loft** tool is used to create solid/surface using the transition between two or more sketches; refer to Figure-16. The procedure to use this tool is given next.

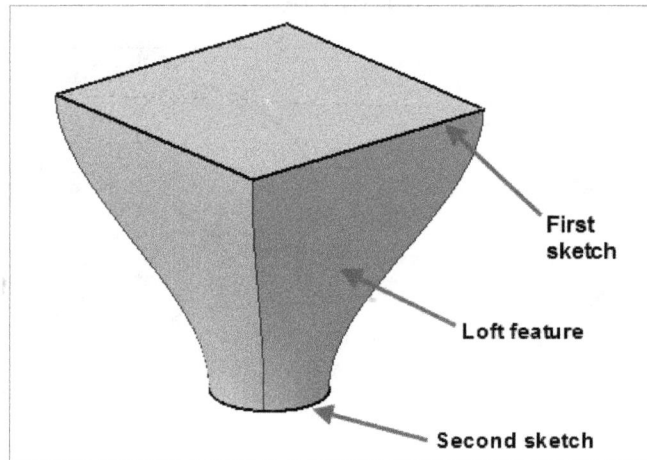

Figure-16. Example of loft feature

- Create desired shape outlines (sketches) on planes parallel to each other; refer to Figure-17.

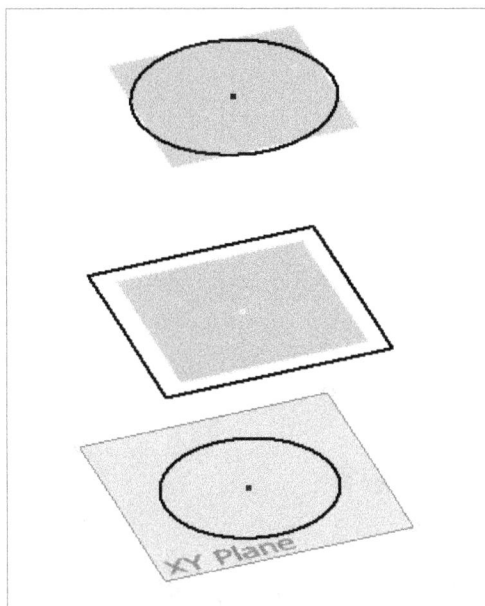

Figure-17. Sketches created on parallel planes

- Click on the **Loft** tool from **Create** panel in the **3D Model** tab of the **Ribbon**. The **Loft** dialog box will be displayed; refer to Figure-18. Also, you will be asked to select the curves.

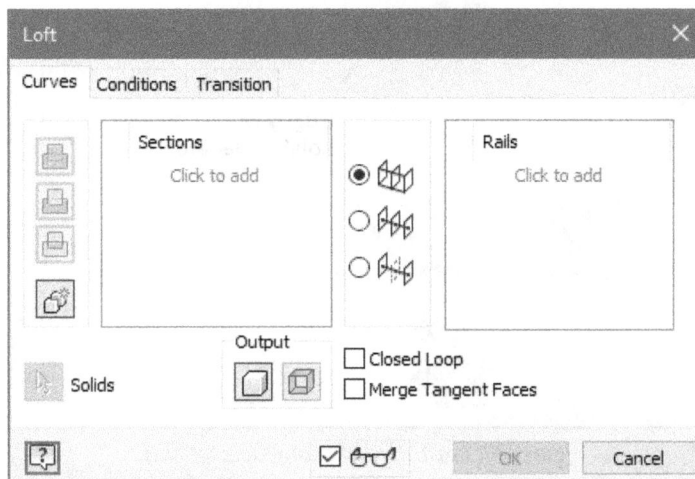

Figure-18. Loft dialog box

- One by one select the curves in a sequence by which you want the transition between curves. Preview of the loft feature will be displayed; refer to Figure-19.

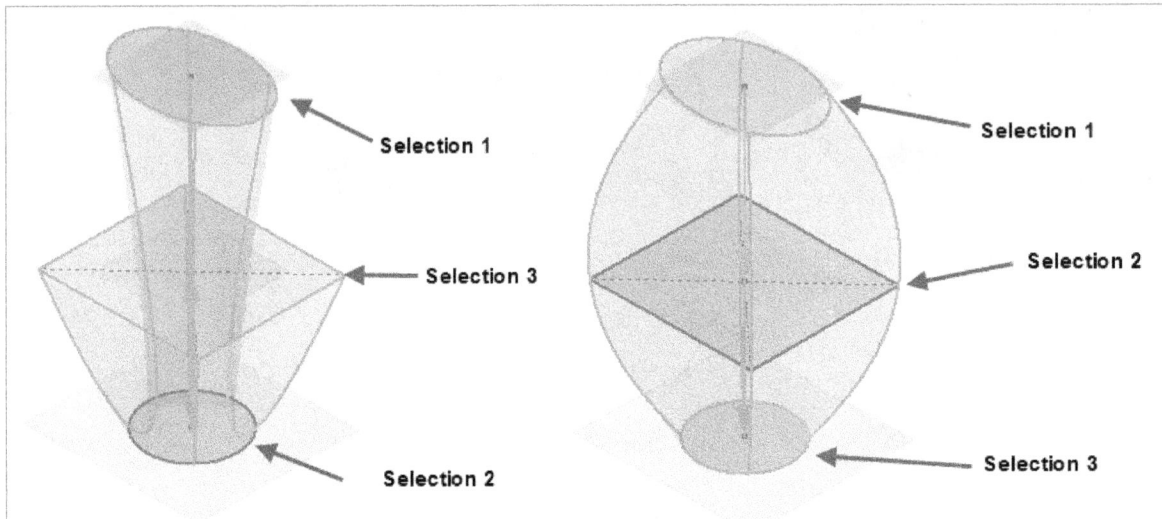

Figure-19. Preview of loft feature

- You can also select the edges of solids to create loft feature; refer to Figure-20.

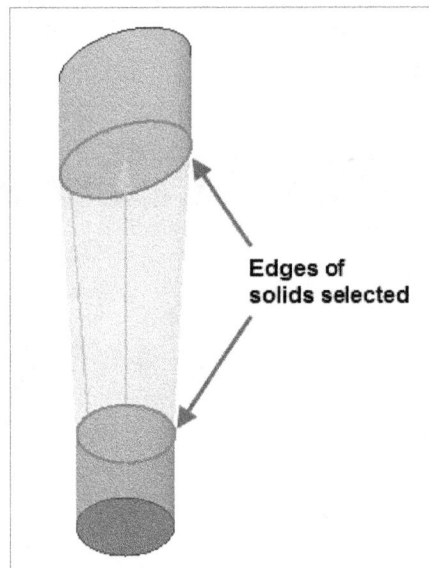

Figure-20. Creating loft using edges of solids

- You can define the transition type for loft feature in dialog box by using the **Rails**, **Center Line**, or **Area Loft** radio button; refer to Figure-21.

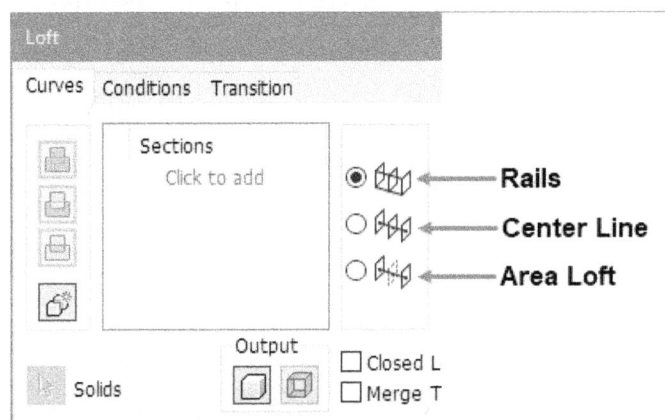

Figure-21. Options to define transition type

Loft using Rail option

- Click on the **Rail** radio button from the dialog box and select the rail for creating loft. Preview of loft following the rail curvature will be displayed; refer to Figure-22.

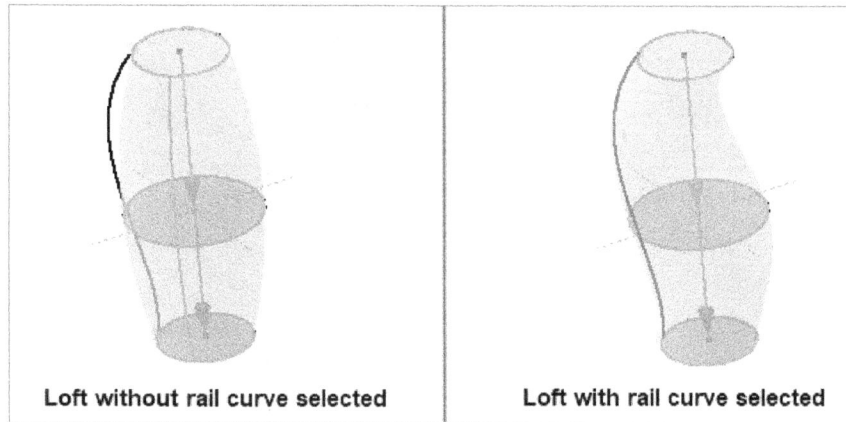

| Loft without rail curve selected | Loft with rail curve selected |

Figure-22. Preview of loft with rail curve

```
Note that the rail curve to be selected must intersect with all the
sections (profile sketches) and that too only one time.
```

Loft with Center Line

Using this option, you can select a center line to refine the shape of loft as per requirement.

- Click on the **Center Line** radio button from the dialog box to activate the options related to center line in the dialog box; refer to Figure-23.

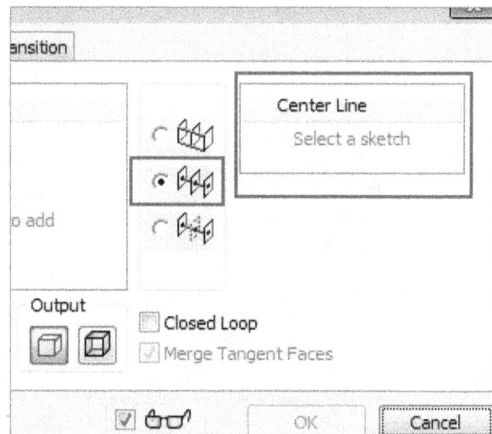

Figure-23. Center Line option in Loft dialog box

- After selecting the section sketches, click in the **Center Line** box in the dialog box and select the curve created for centerline. Preview of the loft will be displayed; refer to Figure-24.

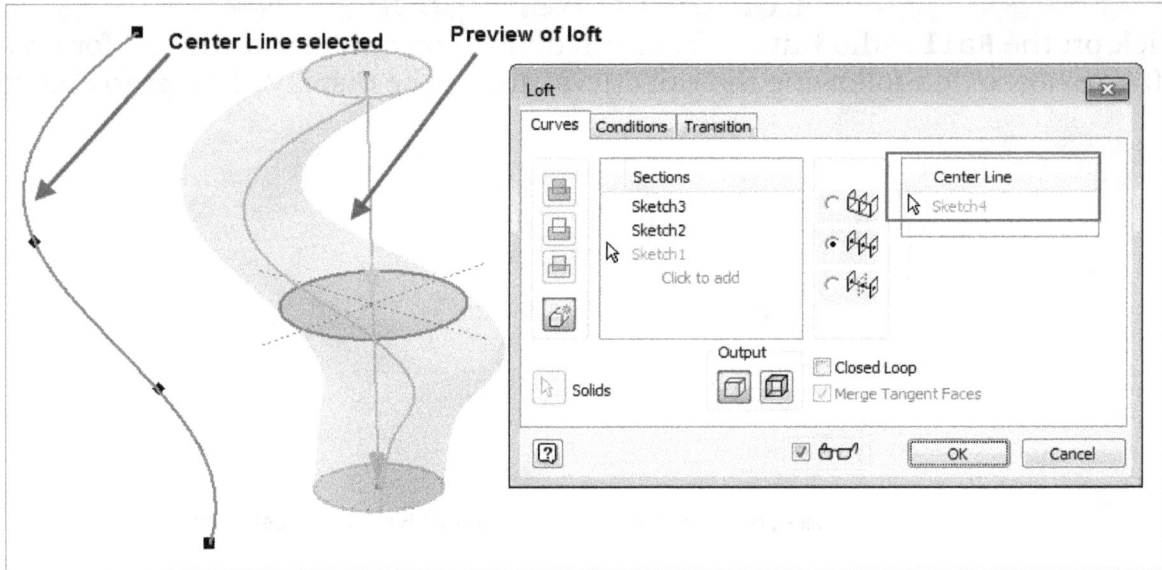

Figure-24. Preview of loft with centerline

Loft with Area Loft option

The **Area Loft** option is used to create loft feature by specifying area of the sections used for loft feature.

- Click on the **Area Loft** radio button from the dialog box. The options in the dialog box will be displayed as shown in Figure-25. Notice the area values for various positions on center line.

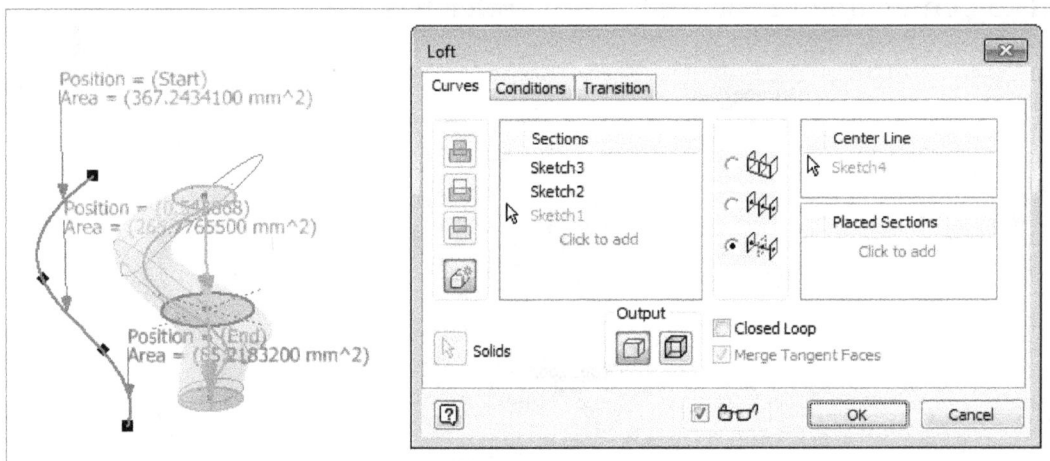

Figure-25. Loft dialog box with Area loft option

- Double-click on the area value that you want to change from the model. The **Section Dimensions** dialog box will be displayed; refer to Figure-26.

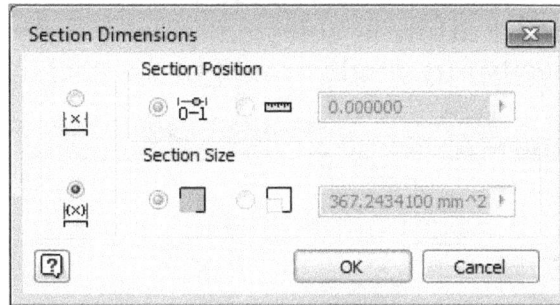

Figure-26. Section Dimensions dialog box

- Click on the **Driving Section** radio button from the dialog box. The **Section Size** area of the dialog box will become active; refer to Figure-27.

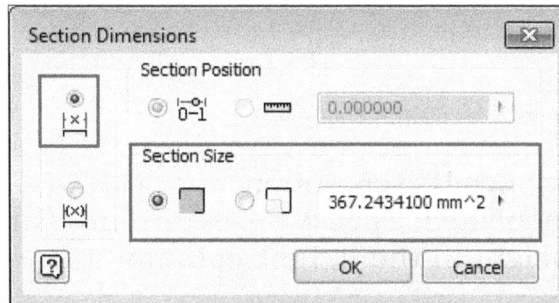

Figure-27. Driving section option of Section Dimensions dialog box

- Select desired radio button from the **Section Size** area of the dialog box and specify desired value in the edit box. If you select the **Area** radio button then you will be asked to specify the value of area of the section. If you select the **Scale Factor** radio button then you need to specify the value for the scale factor for increasing or decreasing the section area.
- Click on the **OK** button from the dialog box to change the area of the section.

Changing Conditions at the starting and end of loft

The options in the **Conditions** tab of the dialog box are used to specify the tangency conditions of the loft feature at the starting and end sections.

- Click on the **Conditions** tab in the **Loft** dialog box. The dialog box will be displayed as shown in Figure-28.

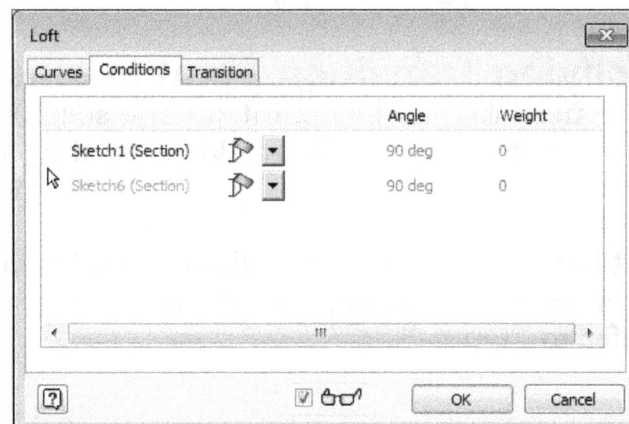

Figure-28. Loft dialog box with Conditions tab selected

- Select desired condition from the **Condition** drop-down; refer to Figure-29. By default, **Free Condition** button is selected from the drop-down and hence you will not be asked to specify any angle or weight for specifying tangency condition.

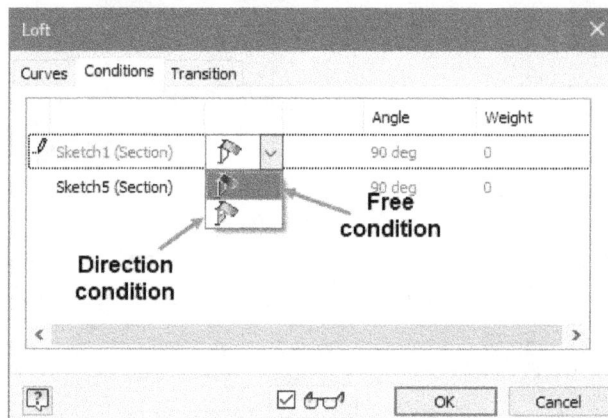

Figure-29. Conditions drop-down

- Click on the **Direction Condition** option from the **Conditions** drop-down. The related options will be activated. Specify desired value of angle in the **Angle** column and desired value of weight in the **Weight** column; refer to Figure-30.

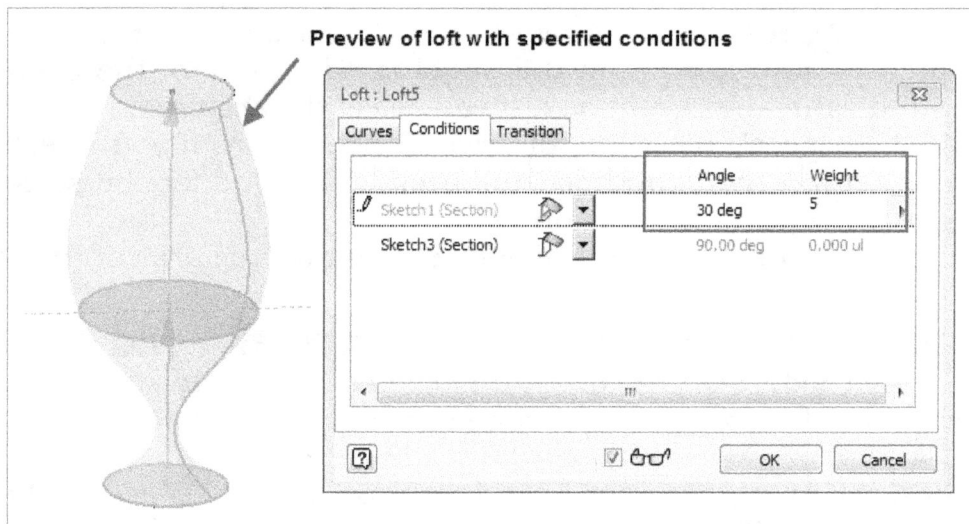

Figure-30. Preview of loft with specified end conditions

Specifying Transition Point Positions

The transition points are used to map key points of one section to the other section. The option to manage transition points is available in the **Transition** tab of the dialog box.

- Click on the **Transition** tab from the **Loft** dialog box. By default, the **Automatic Mapping** check box is selected. Hence, options to manage the transition points are not available; refer to Figure-31.

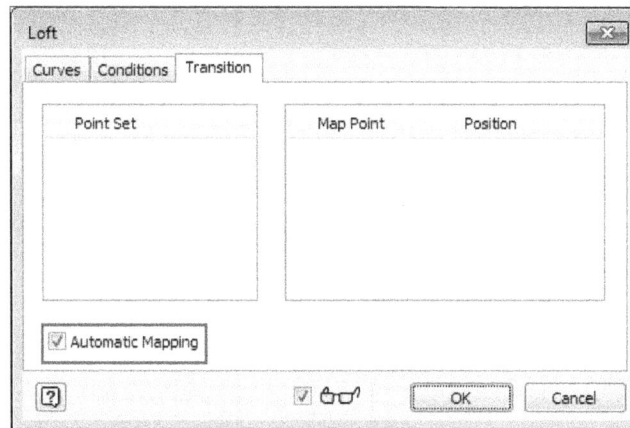

Figure-31. Transition tab of Loft dialog box

- Clear the **Automatic Mapping** check box. The options to change position of transition points will be displayed; refer to Figure-32.

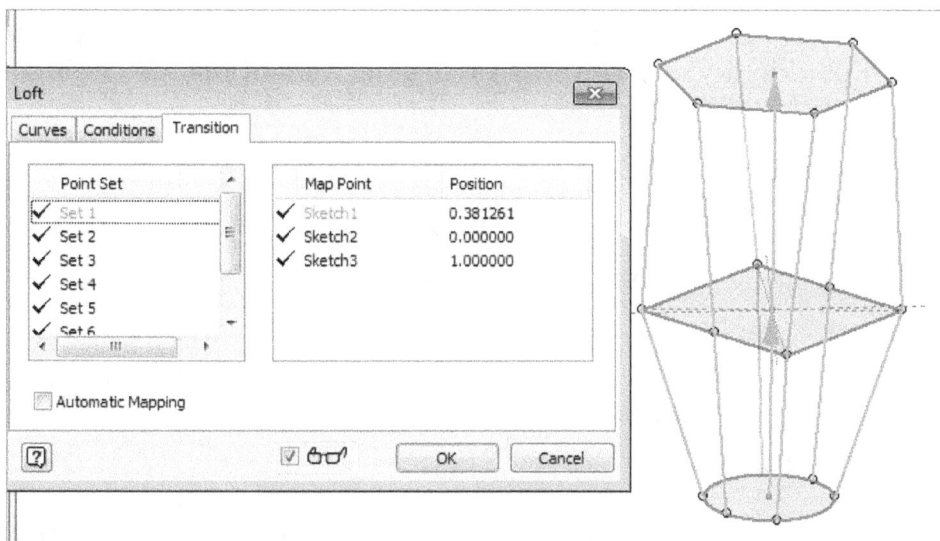

Figure-32. Options to modify position of transition points

- You can drag the key points to change the transition between the sketches.
- Click on the **OK** button from the dialog box to create the loft feature.

COIL TOOL

The **Coil** tool is used to create coil with the help of a profile and an axis. The procedure to create coil is given next.

- Create the profile for coil and an axis; refer to Figure-33.

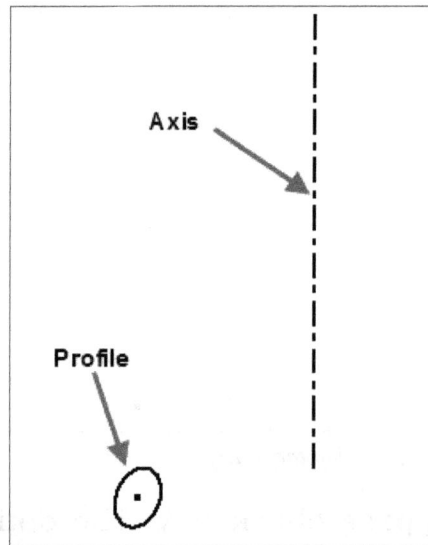

Figure-33. Profile and axis for coil

- Click on the **Coil** tool from **Create** panel in the **3D Model** tab of the **Ribbon**. The **Coil** dialog box will be displayed along with the selection of profile; refer to Figure-34. You will be asked to select the axis.

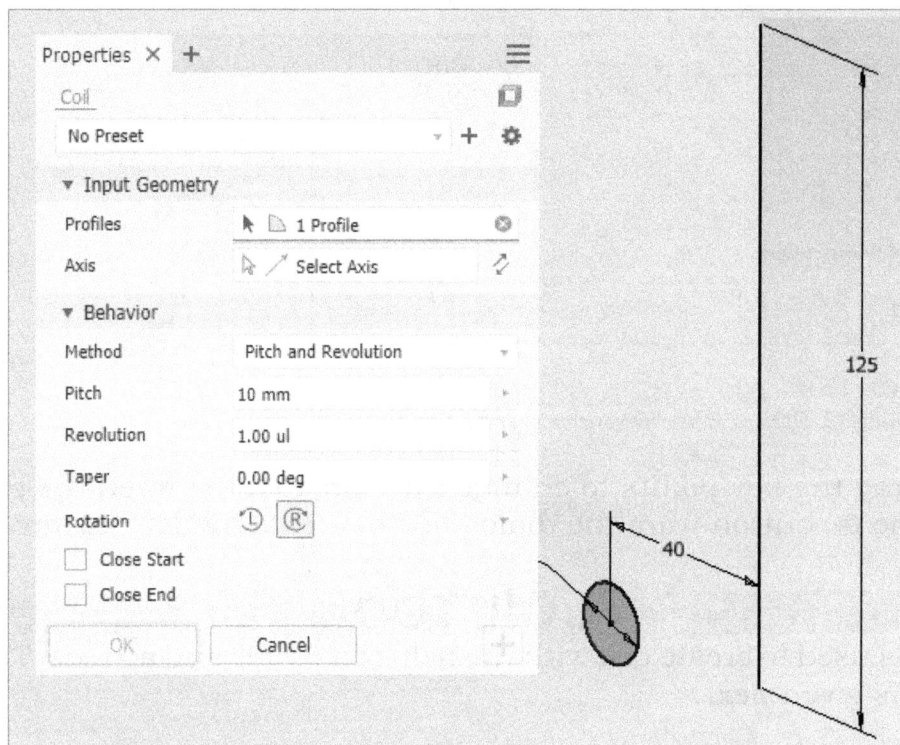

Figure-34. Coil dialog box with selection of profile

- Click on the centerline created for the coil axis. Preview of the coil will be displayed; refer to Figure-35.

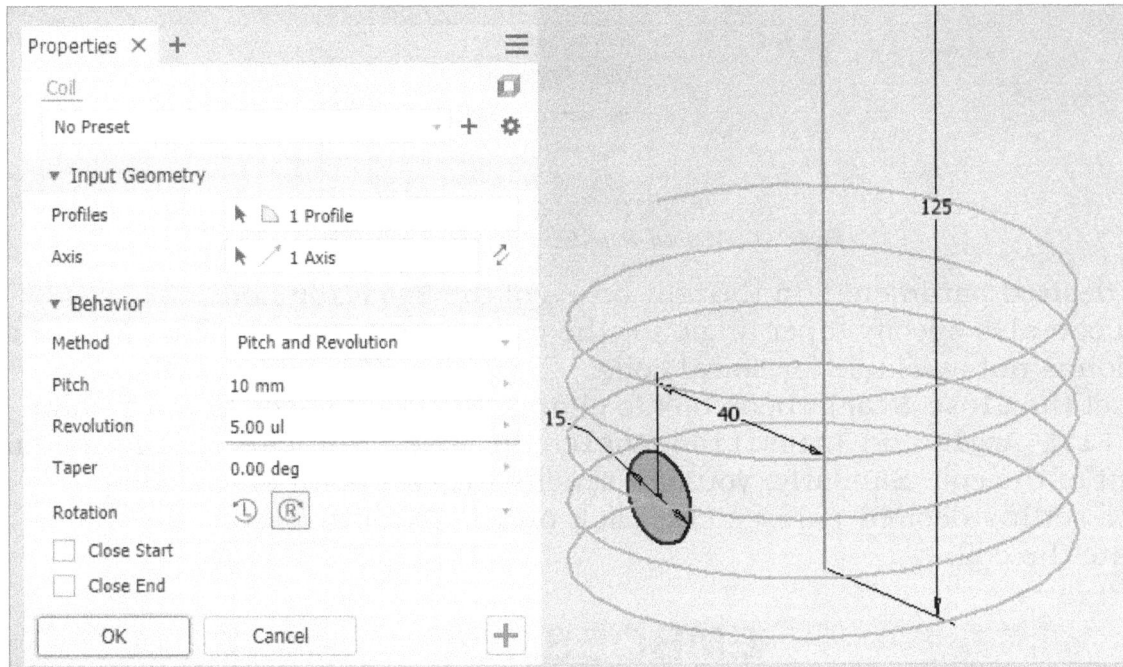

Figure-35. Preview of the coil

- Click on the flip button ⟳ next to **Axis** selection button to change the direction of the coil preview between upward or downward.
- Using the buttons in the **Rotation** area of the dialog box, you can switch between left handed and right handed direction of coil creation.

Changing the coil size and Behavior

The options to change the coil size are available in the **Behavior** node of the dialog box; refer to Figure-36.

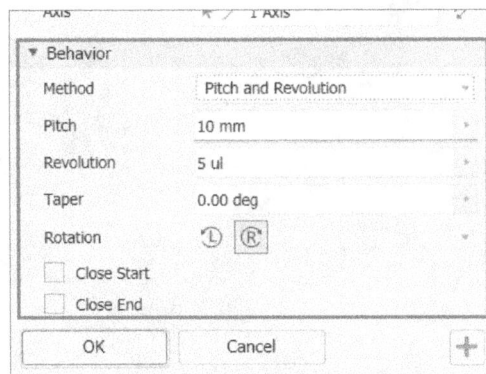

Figure-36. Coil Size tab of Coil dialog box

- There are four option in the **Method** drop-down, **Pitch and Revolution**, **Revolution and Height**, **Pitch and Height**, and **Spiral**; refer to Figure-37. Click on desired option from the drop-down. The related parameters will become active in the dialog box. The **Pitch and Revolution** option is used to specify pitch and number of revolution of the coil. The **Revolution and Height** option is used to specify total number of revolutions and total height of the coil. The **Pitch and Height** option is used to specify the pitch and total height of the coil. The **Spiral** option is used to create a spiral by specifying pitch and number of revolution.

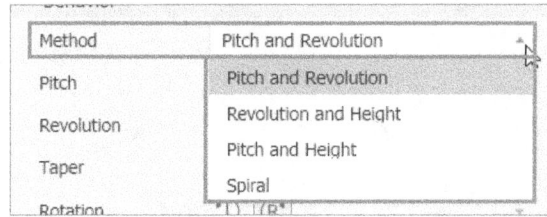

Figure-37. Method drop-down in the Coil dialog box

- Set desired parameters in the edit boxes of the **Behavior** node.
- You can also specify taper angle for the coil in the **Taper** edit box. Note that this option is not available for spiral coil.
- Select the **Close Start** check box to change direction of coil at start point. Using the **Flat Angle** and **Transition Angle** edit boxes, you can specify the starting point of the coil. Similarly, you can specify the end point of the coil.
- After setting desired parameters, click on the **OK** button from the dialog box to create the coil.

EMBOSS TOOL

The **Emboss** tool is used to emboss or engrave a profile on the selected face. Note that you need to have a sketch for embossing. The procedure to use this tool is given next.

- Create desired profile on the face on which you want it to be engraved or embossed; refer to Figure-38.

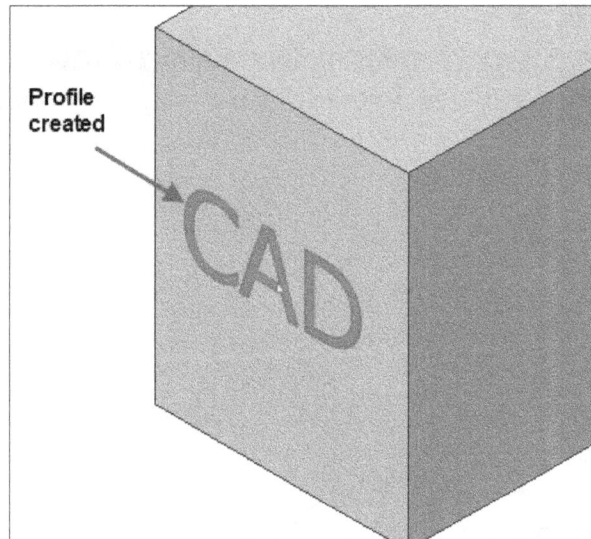

Figure-38. Profile created on face

- Click on the **Emboss** tool from **Create** panel in the **3D Model** tab of the **Ribbon**. The **Emboss** dialog box will be displayed; refer to Figure-39 and you will be asked to select the profile.

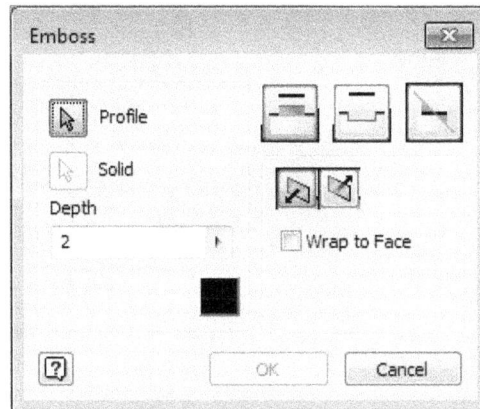

Figure-39. Emboss dialog box

- Click on the profile created earlier and click on desired button from the **Emboss** dialog box; refer to Figure-40.

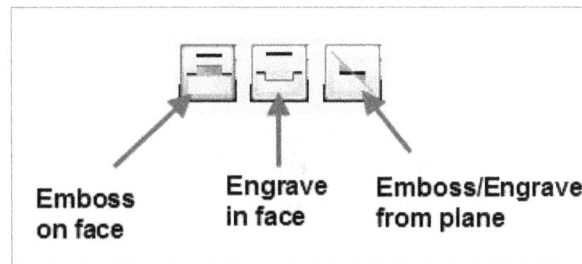

Figure-40. Buttons in Emboss dialog box

- In the **Depth** edit box, specify desired depth. You can also change the appearance of the top face of embossing/engraving by using the **Top Face Appearance** button from the dialog box; refer to Figure-41.

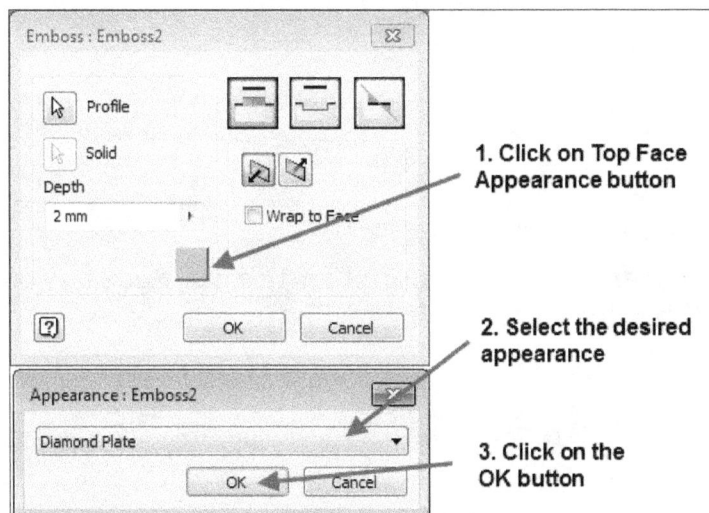

Figure-41. Changing appearance of emboss and engrave

- Click on the **OK** button from the **Emboss** dialog box to create the feature; refer to Figure-42.

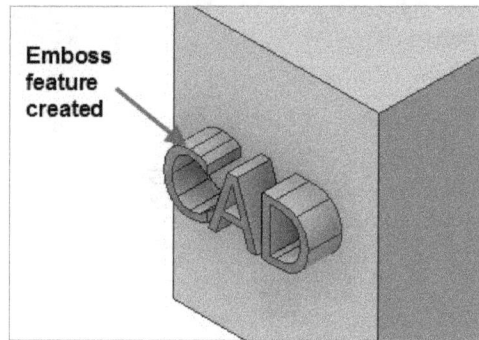

Figure-42. Emboss feature created

RIB TOOL

The **Rib** tool is used to create thin wall support in the structure. A rib can effectively increase the strength of a structure to bear load. The procedure to use this tool is given next.

- Create profile for rib feature in the model space; refer to Figure-43. Note that the profile should be connected to the edges of walls and there should be no extra portion of sketch profile hanging over the edges. In simple words, profile should connect one edge of a wall to edge of the other wall.

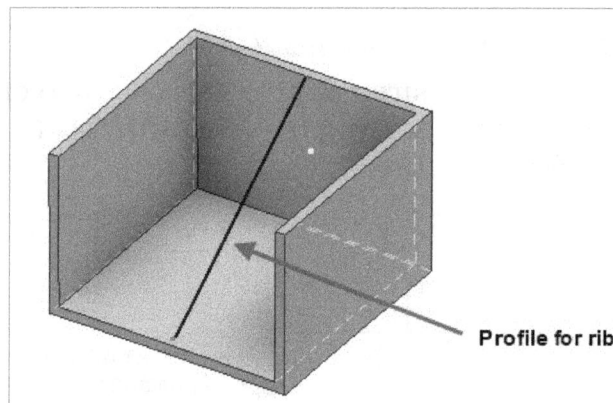

Figure-43. Profile for rib feature

- Click on the **Rib** tool from **Create** panel in the **3D Model** tab of the **Ribbon**. The **Rib** dialog box will be displayed; refer to Figure-44.

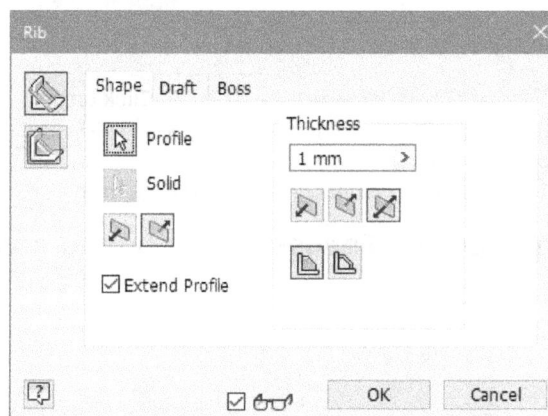

Figure-44. Rib dialog box

- Click on desired button from **Normal to Sketch Plane** ⬛ and **Parallel to Sketch Plane** ⬛ buttons. The **Normal to Sketch Plane** button is used to create rib feature perpendicular to the plane selected for creating profile. The **Parallel to Sketch Plane** button is used to create the rib feature parallel to the sketching plane used for creating profile.
- Select the **Direction 1** ⬛ or **Direction 2** ⬛ button to define the direction for creating rib feature with respect to sketching plane. Note that you will be required to switch the direction of rib feature towards the walls to create it.
- Type desired thickness in the edit box available in **Thickness** area of dialog box.
- Click on desired button from the **Thickness** area to specify the side of rib feature creation; refer to Figure-45.

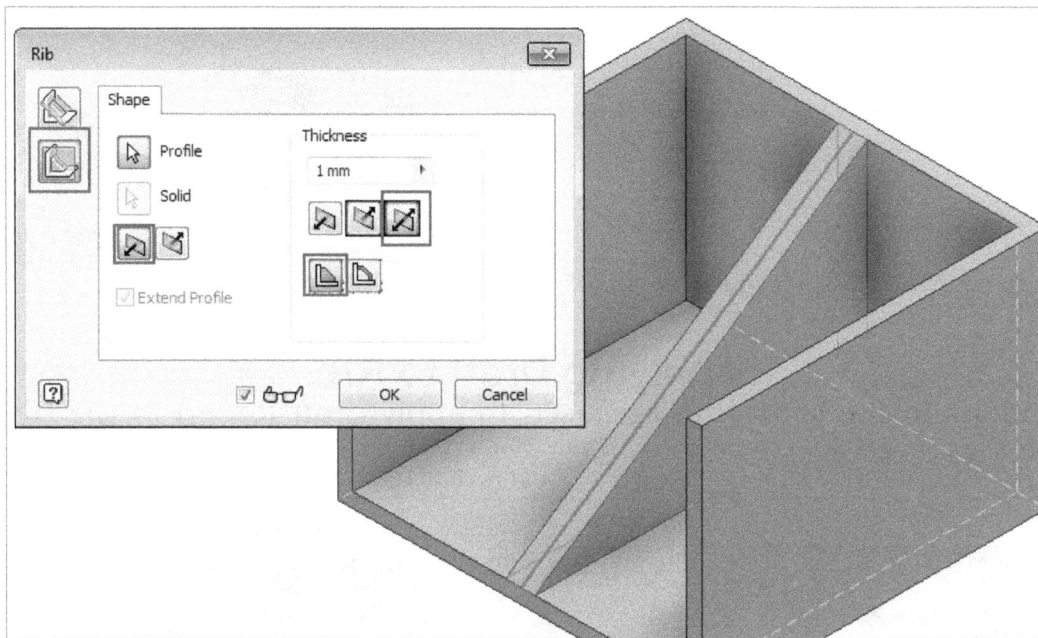

Figure-45. Preview of rib feature

- Click on desired button between **To Next** and **Finite** from the **Thickness** area to define the depth of the rib feature. The **To Next** button is selected by default and it creates the rib feature up to next face. If you select the **Finite** button then you can specify the depth of the rib in the **Extent** edit box; refer to Figure-46.

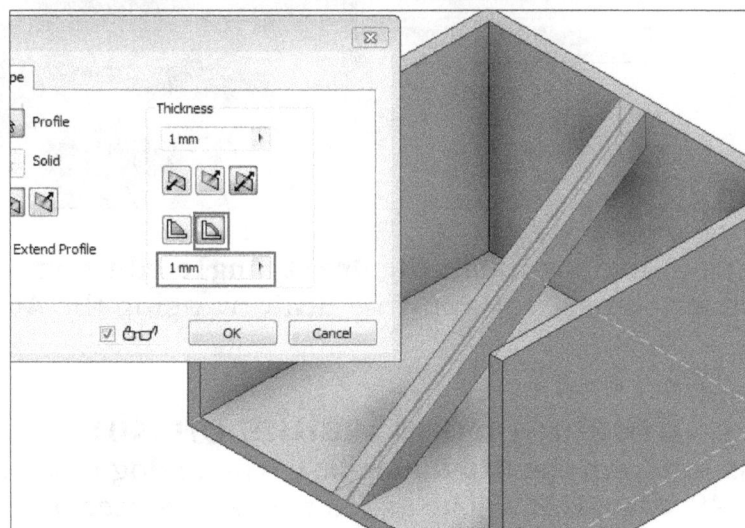

Figure-46. Specifying depth of rib feature

- If you are creating the rib feature perpendicular to the sketching plane by using the **Normal to Sketch Plane** button then **Draft** and **Boss** tabs will be available in the dialog box; refer to Figure-47.

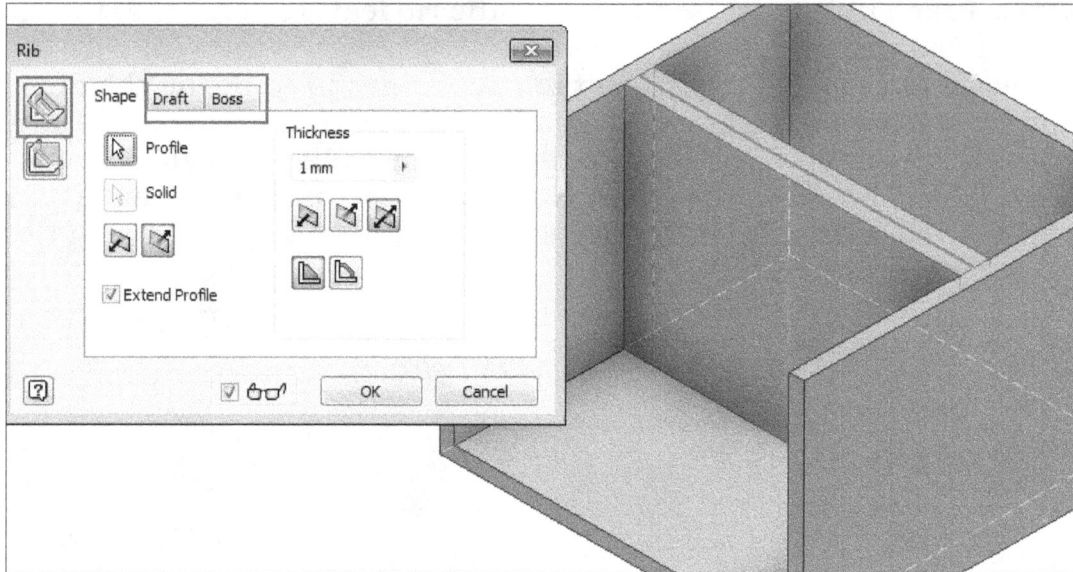

Figure-47. Normal to sketching plane rib feature

Applying Draft to Rib

- The options in the **Draft** tab are used to apply draft (taper) to the rib feature; refer to Figure-48.

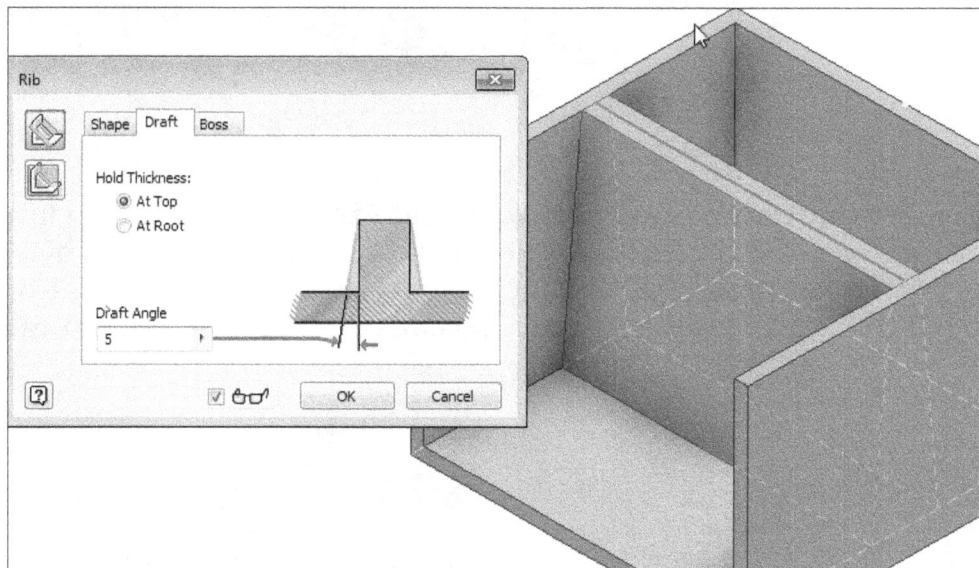

Figure-48. Applying draft to rib feature

- Specify desired draft angle value in the **Draft Angle** edit box.
- You can also define the base for applying draft by using the **At Top** and **At Root** radio buttons.

Creating Boss feature on Rib

- Click on the **Boss** tab from the **Rib** dialog box. The dialog box will be displayed as shown in Figure-49. Also, you will be asked to select center point for boss feature.

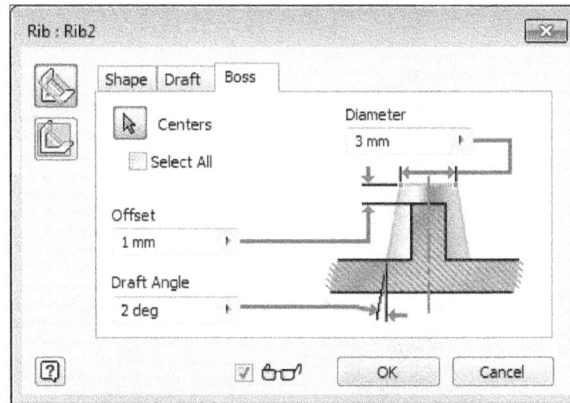

Figure-49. Boss tab of Rib dialog box

- Select the center point for boss feature. Note that the point should be coincident to the profile. You can create one or more points on the profile while creating the sketch for profile. On selecting the point, preview of the boss feature will be displayed; refer to Figure-50.

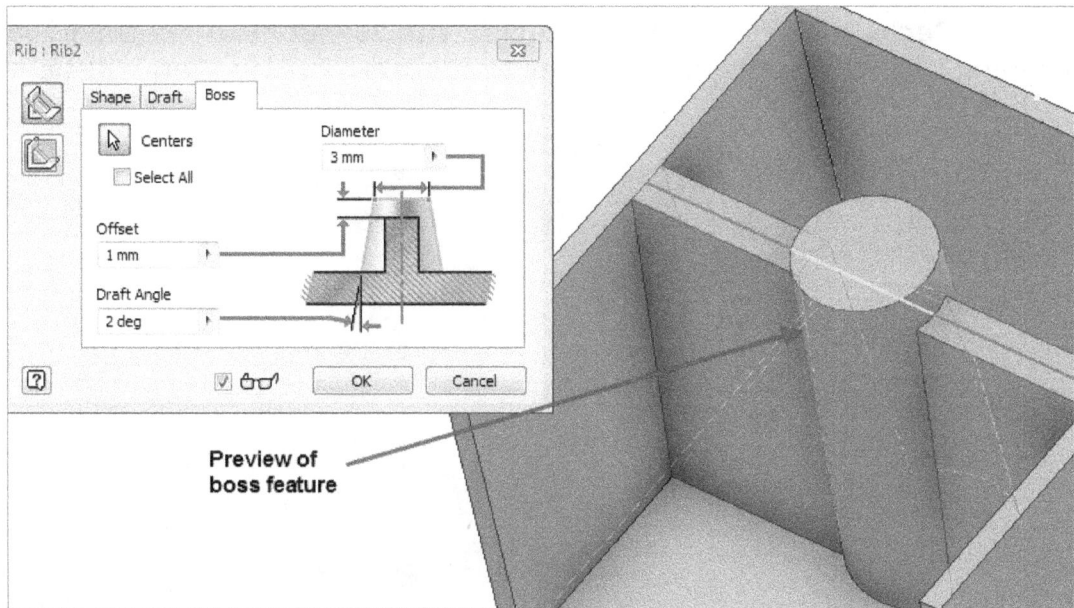

Figure-50. Preview of boss feature on rib

- Click on the **OK** button from the dialog box to create the rib feature with specified settings.

DECAL TOOL

The **Decal** tool is used to apply an image to face of the model. The procedure to use this tool is given next.

- Insert an image on sketching plane parallel to desired face; refer to Figure-51.

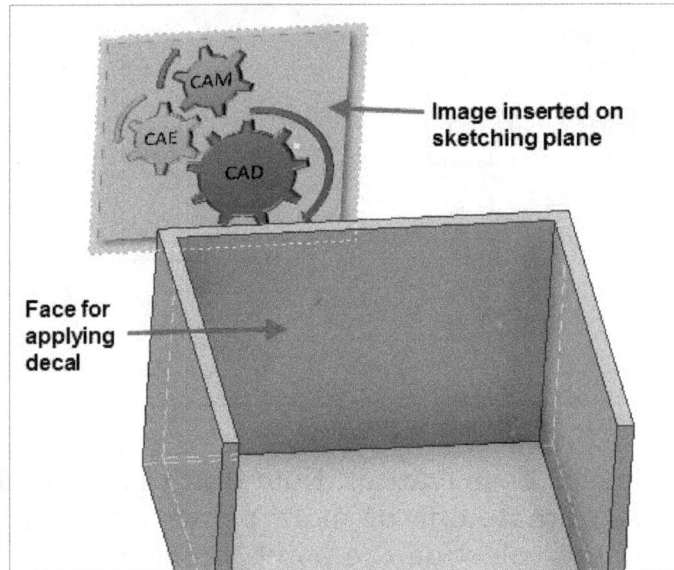

Figure-51. Image and face for decal tool

- Click on the **Decal** tool from **Create** panel in the **3D Model** tab of the **Ribbon**. The **Decal** dialog box will be displayed; refer to Figure-52. Also, you will be asked to select an image file.

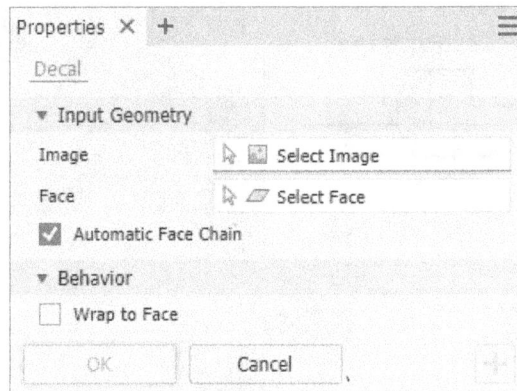

Figure-52. Decal dialog box

- Click on desired image file. You will be asked to select the face.
- Click on the face on which you want to apply decal.
- If you want to wrap the image around the selected face then select the **Wrap to Face** check box from **Behavior** node; refer to Figure-53.

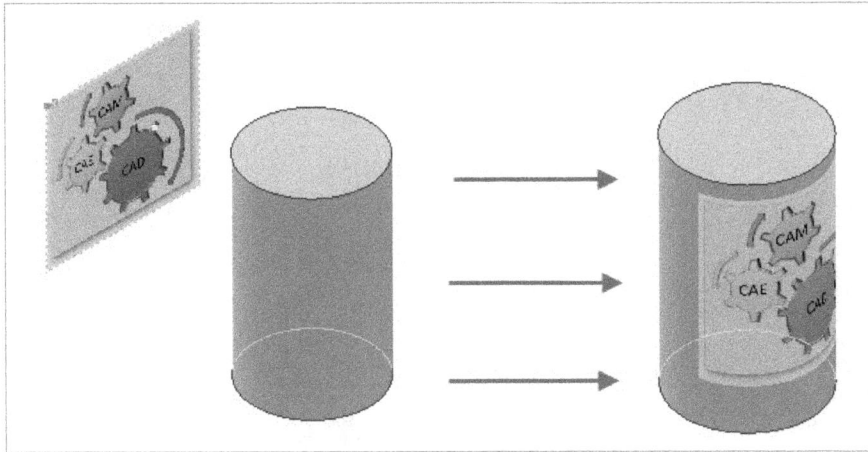

Figure-53. Preview of decal with wrap to face option selected

- If you have selected the **Automatic Face Chain** check box with **Wrap to Face** check box then you can apply decal on face smaller than the image size. The extra decal will be wrapped on the connected faces; refer to Figure-54.

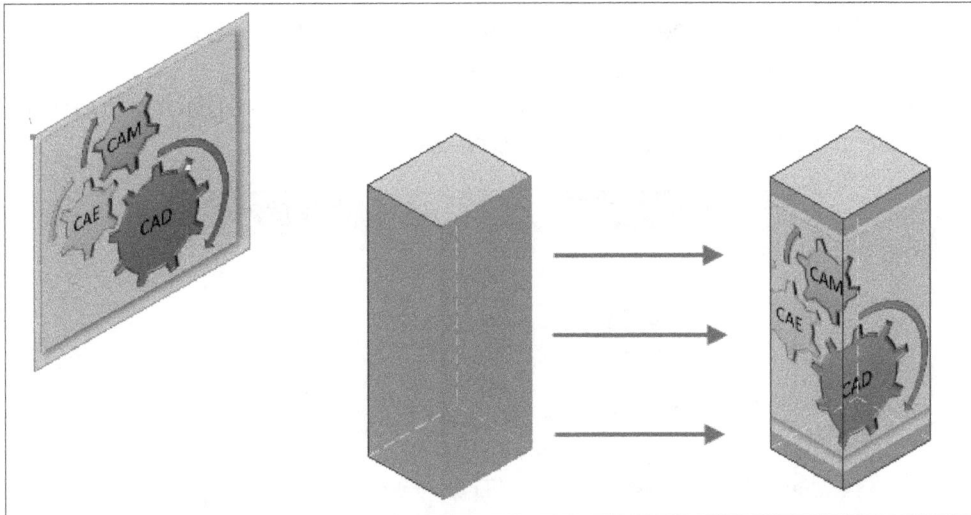

Figure-54. Applying decal on chain faces

- Click on the **OK** button from the dialog box to create the decal feature.

IMPORTING FOREIGN CAD FILES

No! Foreign file does not mean files from different country. In CAD terms, it means CAD files of non-native format. The procedure to import files is given next.

- Click on the **Import** tool from **Create** panel in the **3D Model** tab of the **Ribbon**. The **Import** dialog box will be displayed as shown in Figure-55.

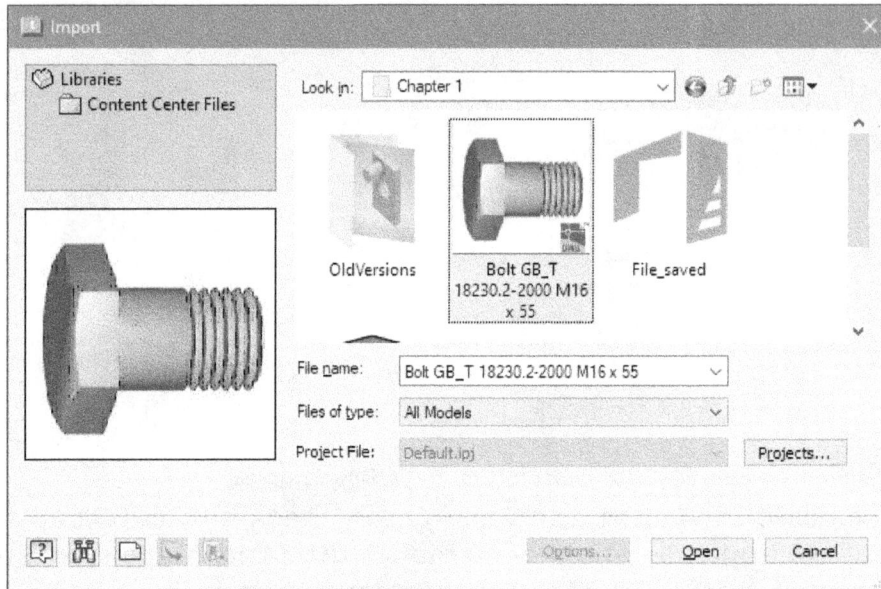

Figure-55. Import dialog box

- Click in the **Files of type** drop-down and select desired format. The files of selected format will be displayed in the file browser.
- Double-click on desired file. The **Import** dialog box with options to include various features will be displayed; refer to Figure-56.

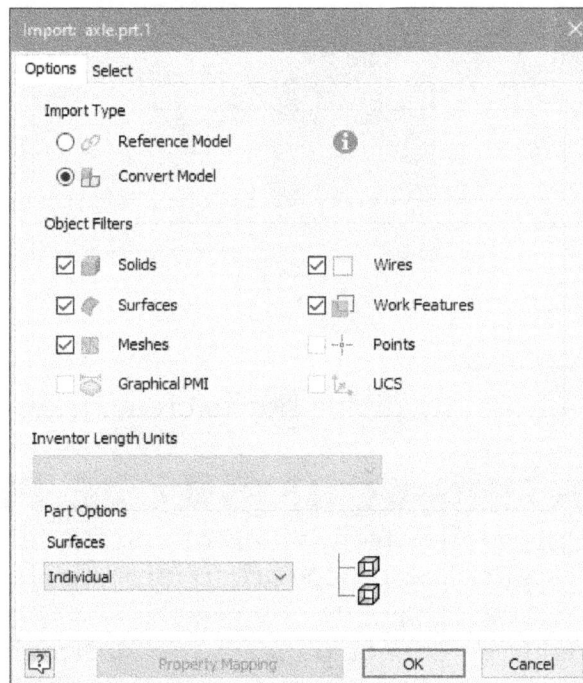

Figure-56. Import dialog box with options to include features

- There are two options in the **Import Type** area of the dialog box. Select the **Reference Model** radio button if you want the imported feature to be updated when the base model is changed in other software. Select the **Convert Model** radio button if you want to create a new model from imported file locally saved in current file.
- Select desired options from the dialog box and click on the **OK** button from the dialog box to import the model; refer to Figure-57.

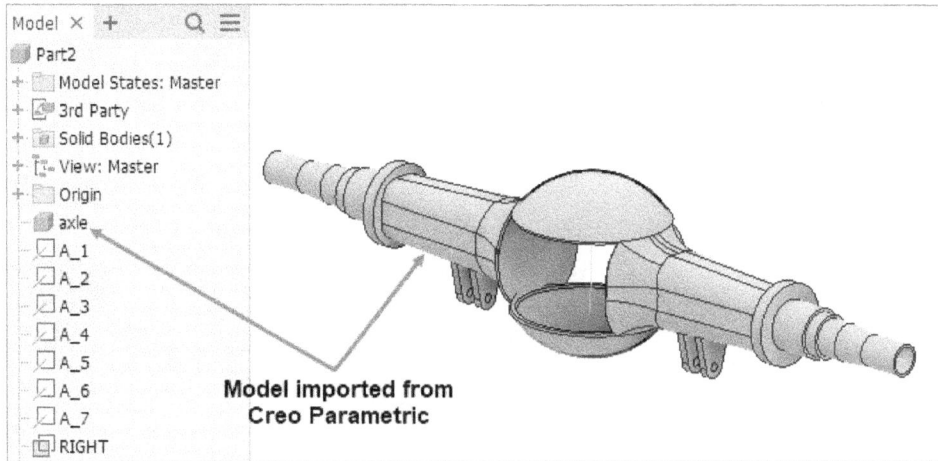

Figure-57. Model imported from Creo Parametric

Now, you can use the imported model as base solid/surface to create other features.

Unwrap Tool

The **Unwrap** tool is used to unwrap faces that cannot be flattened with the Unfold or sheet metal flat pattern command. The procedure to use this tool is given next.

• Click on the **Unwrap** tool from **Create** panel in the **3D Model** tab of the **Ribbon**. The **Unwrap** dialog box will be displayed; refer to Figure-58. You will be asked to select the faces.

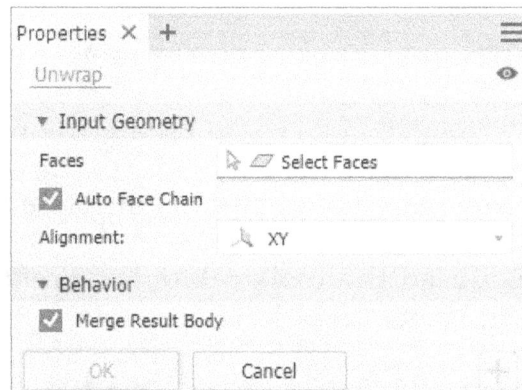

Figure-58. Unwrap dialog box

• Select the contiguous face from the model. The updated **Unwrap** dialog box will be displayed along with the preview of unwrapped faces; refer to Figure-59.

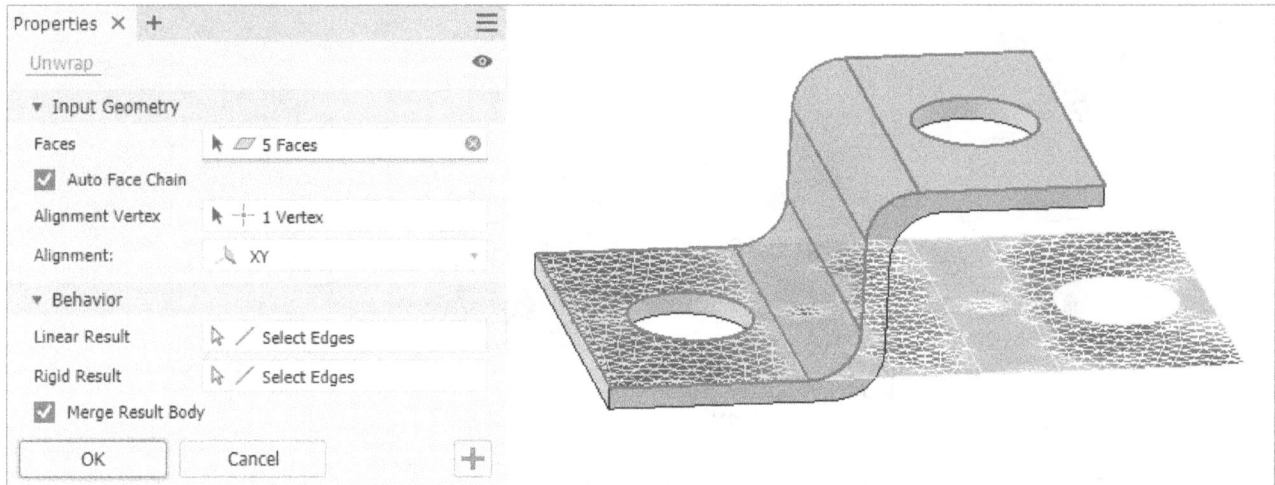

Figure-59. Preview of unwrapped faces

- Select the **Auto Face Chain** check box to select multiple faces with one pick.
- The **Alignment Vertex** area uses the vertex closest to your selection while selecting the face to unwrap.
- Select **Model** option from **Alignment** drop-down to place the unwrapped surface in the same orientation as the model. Select **XY** option from drop-down to place the unwrapped surface at the origin and in the same orientation as the **XY** plane. Select **XZ** option from drop-down to place the unwrapped surface at the origin and in the same orientation as the **XZ** plane. Select **YZ** option from drop-down to place the unwrapped surface at the origin and in the same orientation as the **YZ** plane.
- Click on the **Linear Result** area to select one or more contiguous edges to remain straight.
- Click on the **Rigid Result** area to select one or more contiguous edges to remain rigid.
- Select the **Merge Result Body** check box to create the output as a single surface face.
- Click on the **OK** button from the dialog box to complete the unwrapping process.

Till this point, we have discussed the tools related to creation of features. Now, we will discuss the tools related to modifications of solids.

MODIFICATION TOOLS

The tools to modify features are available in the **Modify** panel of the **3D Model** tab in the **Ribbon**; refer to Figure-60. These tools are discussed next.

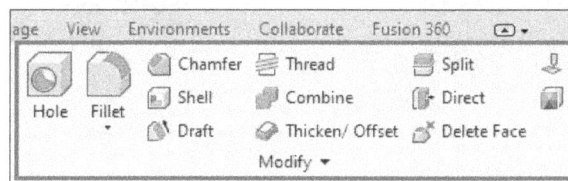

Figure-60. Modify panel

Hole Tool

The **Hole** tool in the **Modify** panel is used to create hole in the solid model. The procedure to create hole is given next.

- Click on the **Hole** tool from **Modify** panel in the **3D Model** tab of the **Ribbon**. The **Hole** dialog box will be displayed; refer to Figure-61. Also, you will be asked to specify the location of the hole.

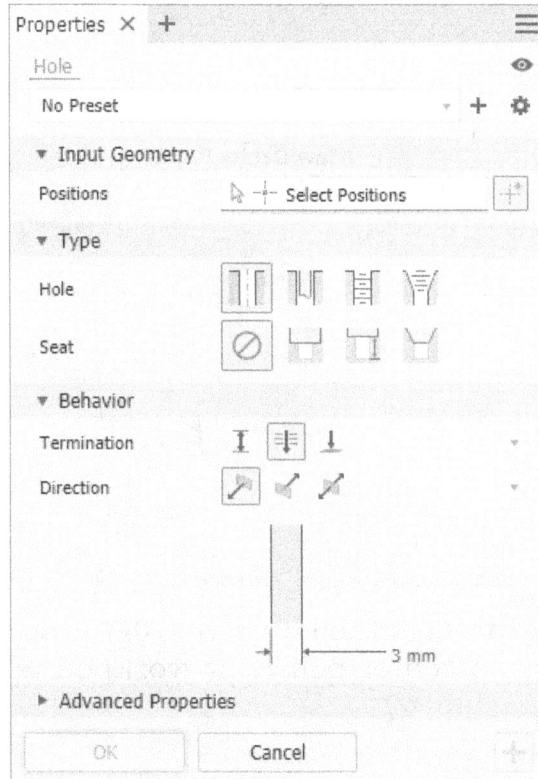

Figure-61. Hole dialog box

- Click at desired location on face of the model. Preview of the hole will be displayed; refer to Figure-62.

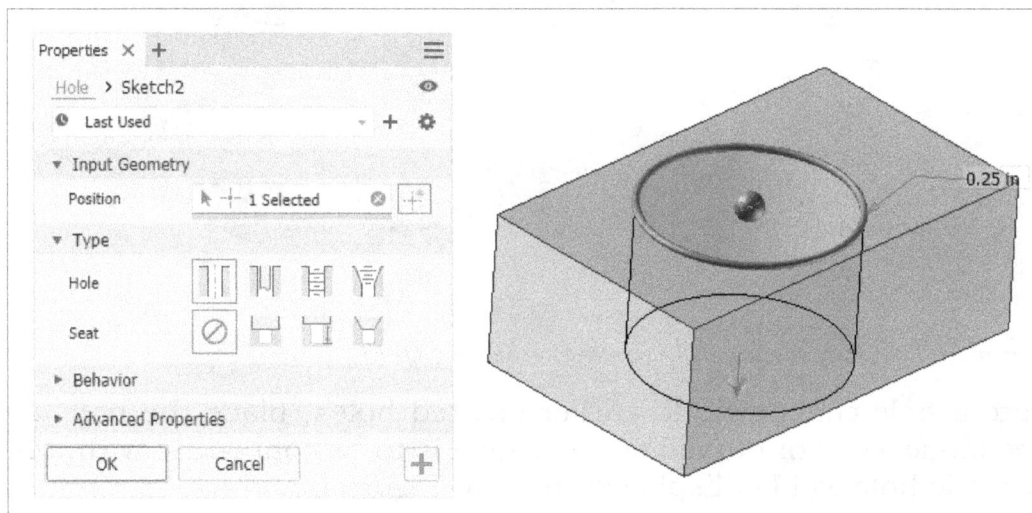

Figure-62. Preview of hole

Specifying location of hole

An existing sketch is not required for specifying the location of hole. The valid inputs for specifying hole locations include sketch point, work point, or a face.

- By default, **Allow Center Point Creation** button is **ON** in the **Position** area of the dialog box. When **Allow Center Point Creation** button is **ON**, you can add center points randomly on the part face; refer to Figure-63. Hence, you can click at any desired location on the face of model to create hole.

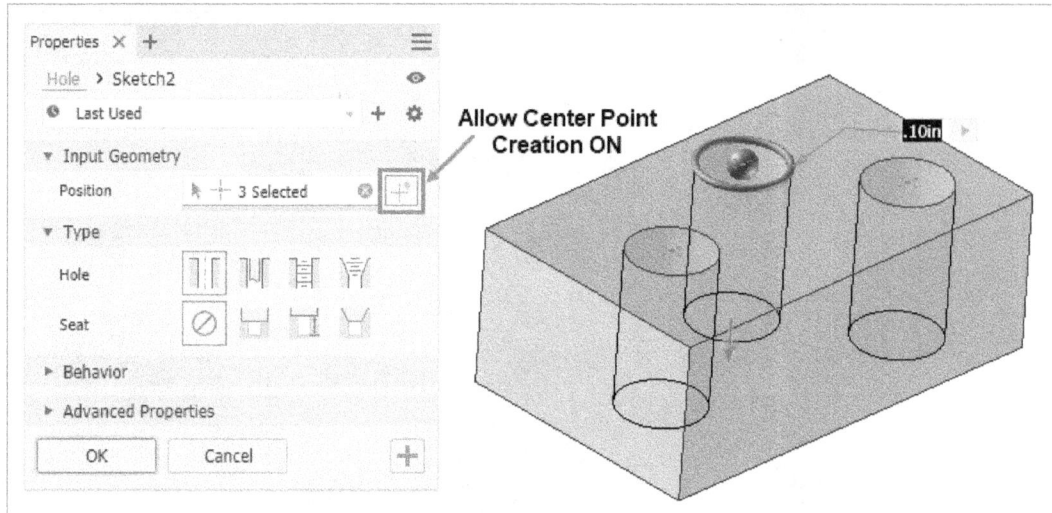

Figure-63. Preview of hole when allow center point button is ON

- When **Allow Center Point Creation** button is **OFF**, you can only select points in an existing sketch, sketch center points or workpoints to create the hole; refer to Figure-64.

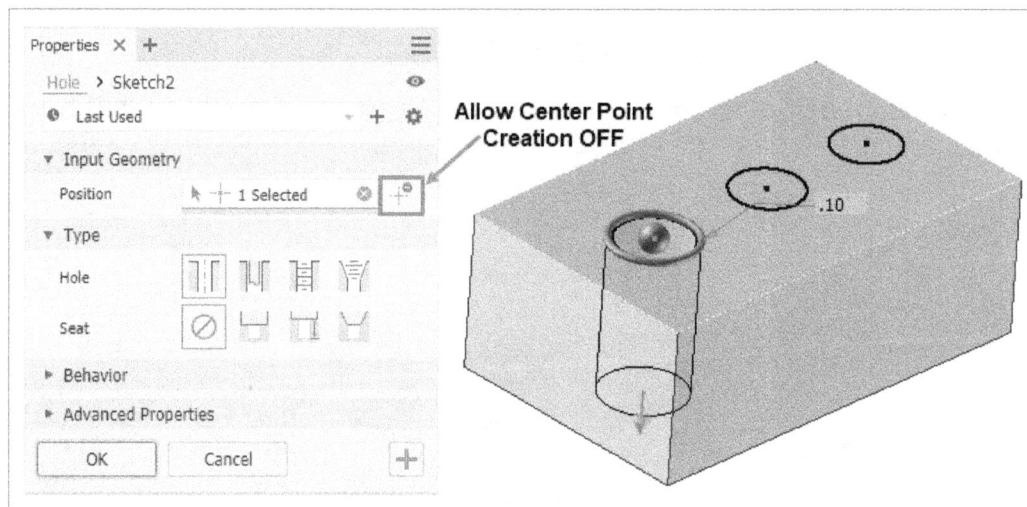

Figure-64. Preview of hole when allow center point button is OFF

- To create a hole concentric to earlier created holes, place the hole center and click the model edge or curved face the hole is to be concentric with, the preview of concentric hole will be displayed; refer to Figure-65.

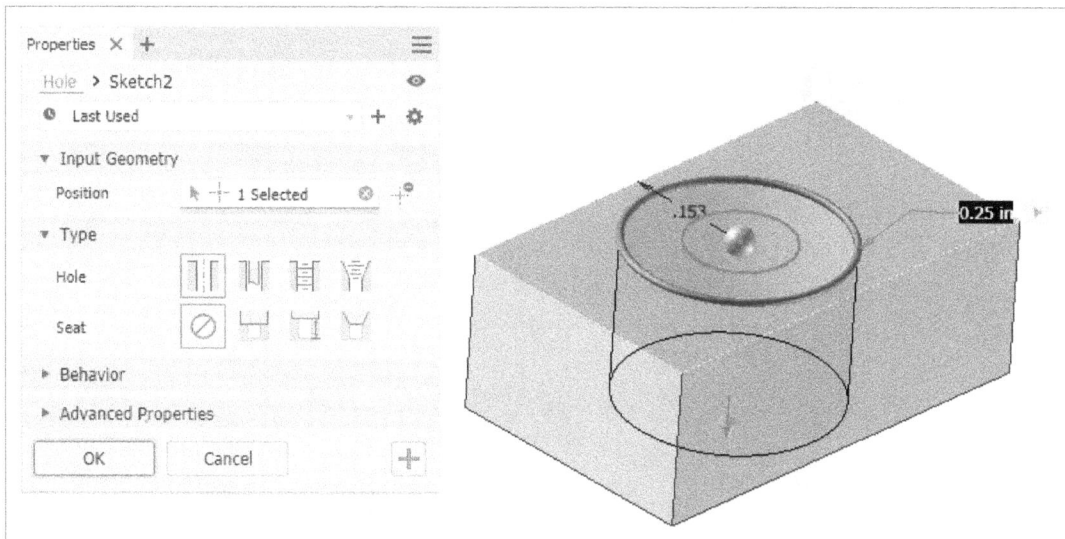

Figure-65. Preview of concentric hole

Setting Shape and Dimension of Hole

- Select desired option from **Hole** section in the **Type** area of the dialog box to define the type of hole as **Simple Hole**, **Clearance Hole**, **Tapped Hole**, and **Taper Tapped Hole**; refer to Figure-66. Options to define parameters of hole will be displayed in the **Behavior** area of the dialog box; refer to Figure-67.

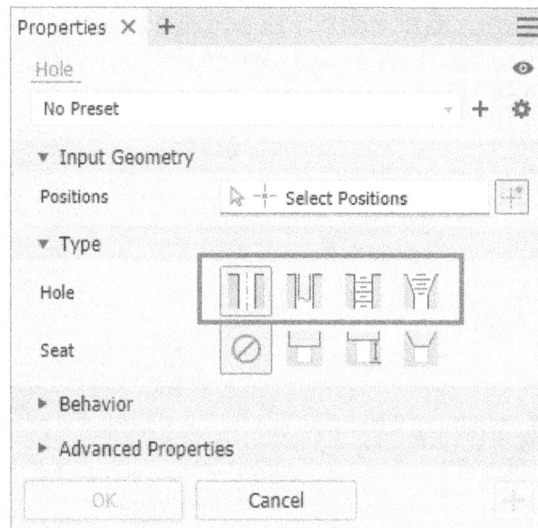

Figure-66. Options to define type of hole

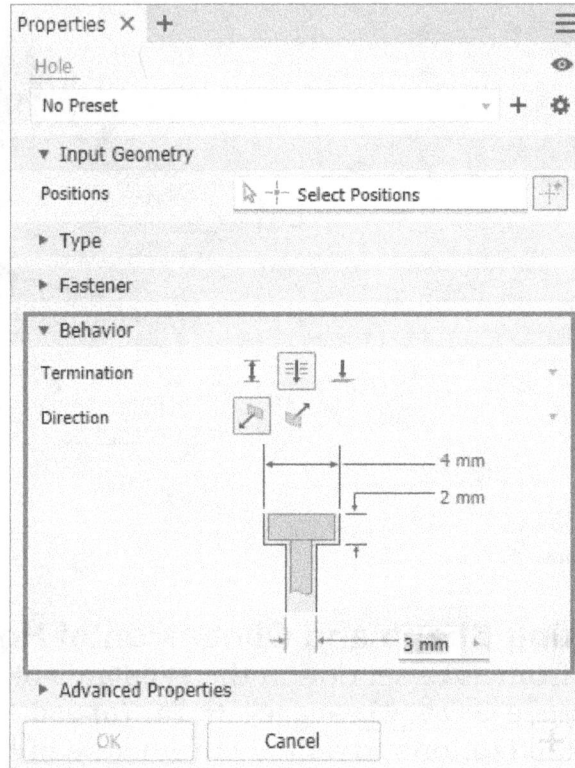

Figure-67. Options to define parameters of hole

- Type desired dimensions in the edit boxes to define size of hole.
- On selecting the **Clearance Hole** option from the type area in the dialog box, options to choose standard holes from the library will be displayed in the **Fastener** area; refer to Figure-68.

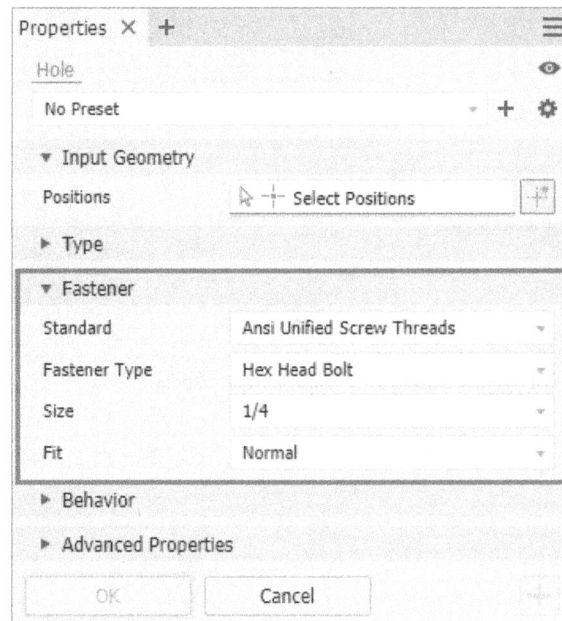

Figure-68. Fastener area

- You can also define the shape of hole by selecting the options as **None**, **Counterbore**, **Spotface**, and **Countersink** from **Seat** section in the **Type** area of the dialog box; refer to Figure-69.

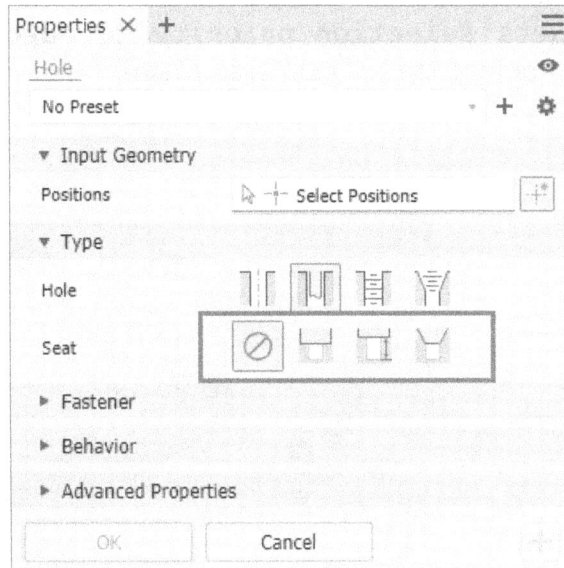

Figure-69. Options to define shape of hole

- Click on the **OK** button from the dialog box to exit the **Hole** tool.

Fillet Tool

The **Fillet** tool is used to apply radius at the sharp edges. The procedure to use this tool is given next.

- Click on the **Fillet** tool from **Modify** panel in the **3D Model** tab of the **Ribbon**. The **Fillet** dialog box will be displayed; refer to Figure-70.

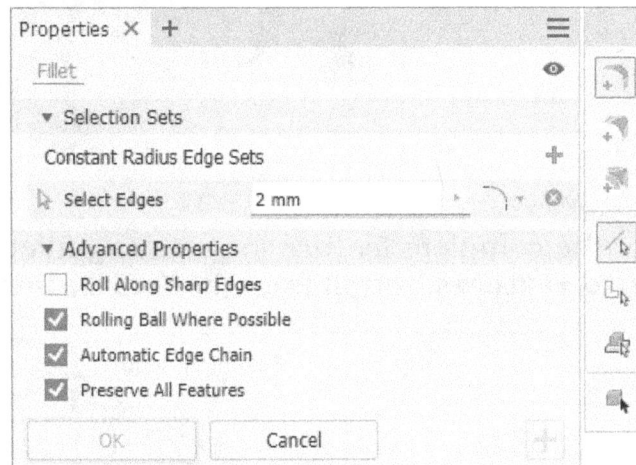

Figure-70. Fillet dialog box

Creating Constant Radius Edge Fillet

- Select the **Add constant radius edge set** button ⬜ from **Tool Palette** in the dialog box, if not selected already. The options related to edge fillet will be displayed; refer to Figure-70.

- Now, there are three modes of selection in the dialog box; **Edges**, **Edge Loops**, and **Features**. Select the **Sets selection priority to edges** button from the **Tool Palette** if you want to select the edges individually; refer to Figure-71.

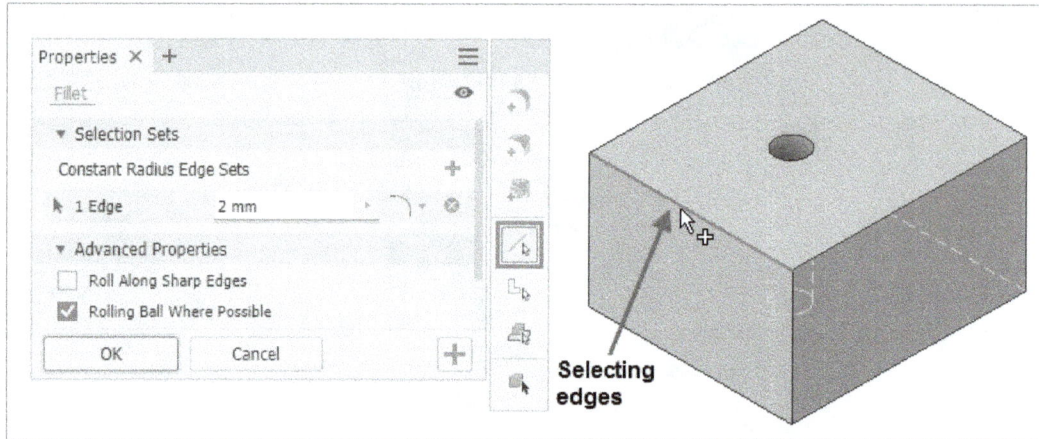

Figure-71. Selecting edges for fillet

- If you want to select a loop of edges then select the **Sets selection priority to edge loops** button from the **Tool Palette**; refer to Figure-72.

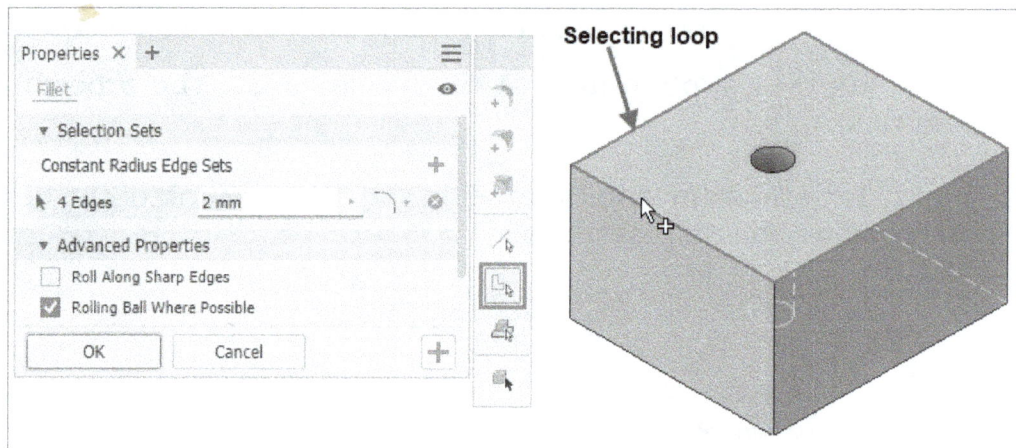

Figure-72. Selecting loop for fillet

- If you want to select the complete feature for creating fillet then select the **Sets selection priority to features** button from the **Tool Palette**; refer to Figure-73.

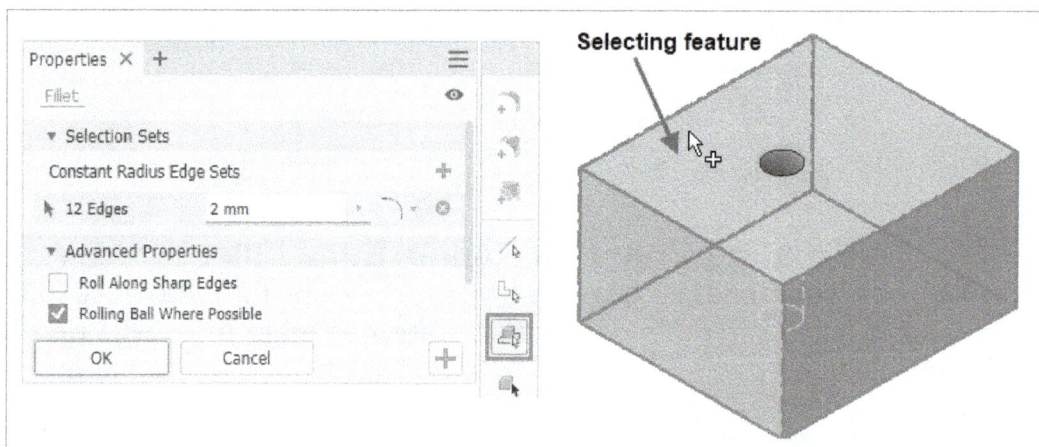

Figure-73. Selecting feature for fillet

- After making desired selection, click on the **Fillet Constant Radius** edit box. You will be asked to define the radius value; refer to Figure-74.

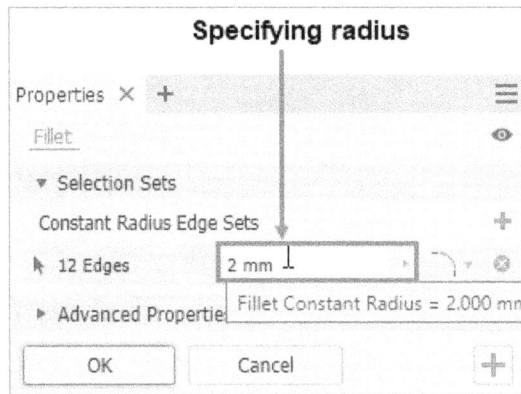

Figure-74. Specifying radius of fillet

- Specify desired radius value in the edit box.
- Select the type of fillet from the drop-down next to radius value in the dialog box. There are three options, **Tangent**, **Smooth (G2)**, and **Inverted**; refer to Figure-75.

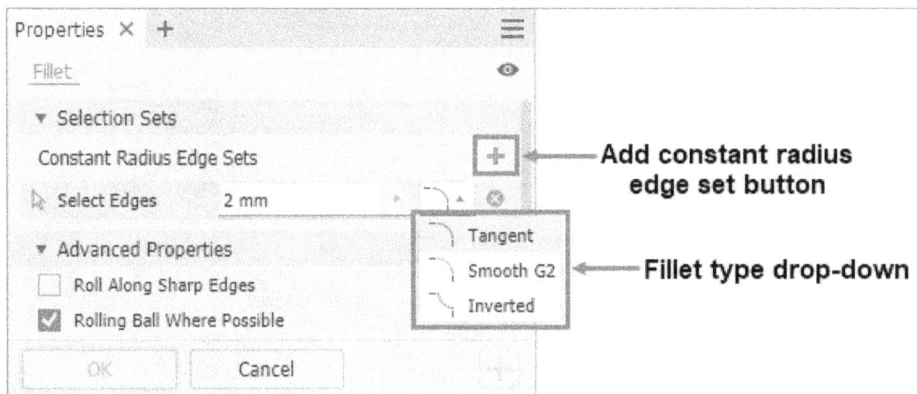

Figure-75. Fillet type drop down

- To create new set of fillet, click on the **Add constant radius edge set** button from **Constant Radius Edge Sets** area of the **Selection Sets** rollout in the dialog box. You will be asked to select a new set of edge.
- Select the new edges and specify desired radius for the new set; refer to Figure-76.

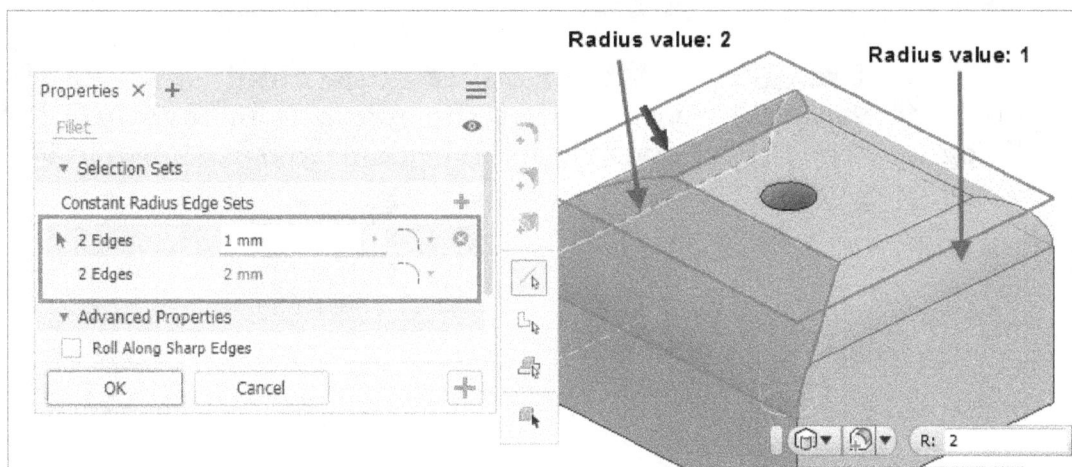

Figure-76. Applying fillet in different sets

- Click on the **Solid selection mode** button from the **Tool Palette** in the dialog box. The options related to solid selection mode will be displayed; refer to Figure-77.

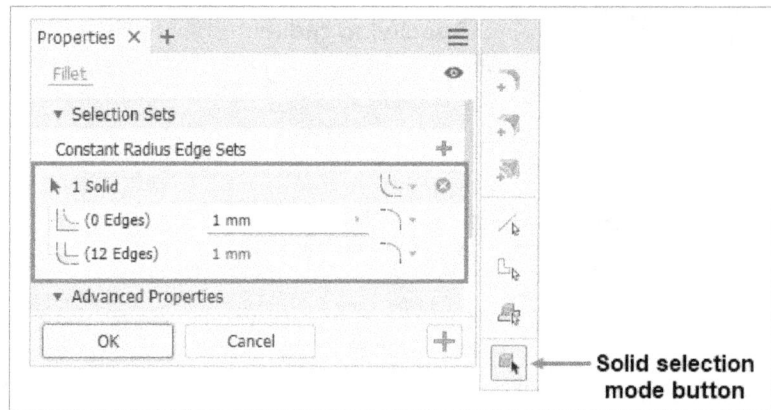

Figure-77. Solid selection mode options

- You can create fillets on all the edges of the solid by selecting the **All Rounds** option from **Solid** drop-down; refer to Figure-78.

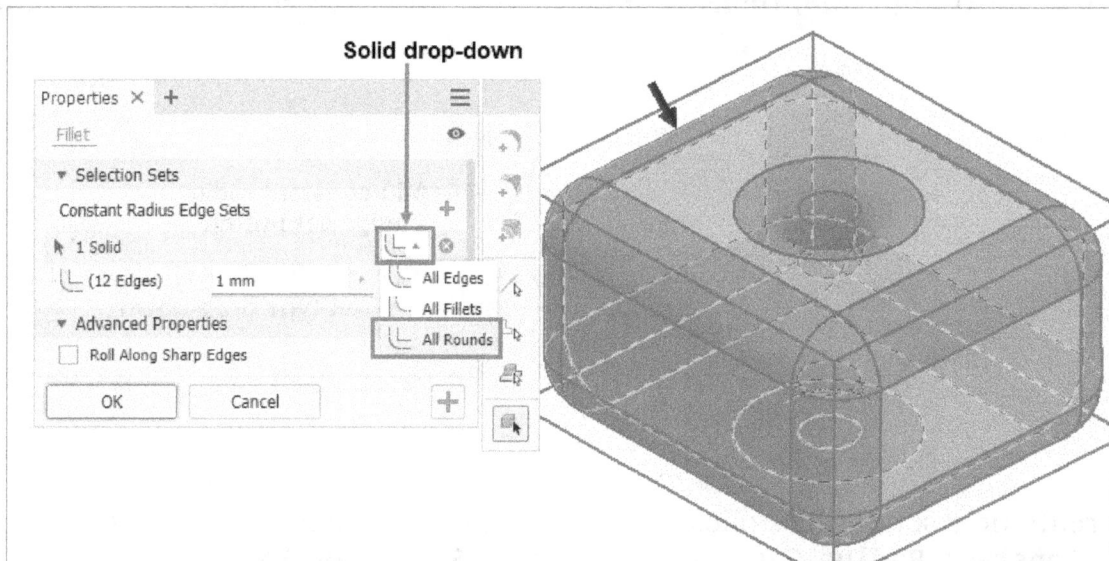

Figure-78. Creating rounds on all the edges

- After specifying the parameters, click on the **OK** button from the dialog box to complete the process.

Creating Variable Radius Edge Fillet

- Click on the **Add variable radius fillet** button from **Tool Palette** in the **Fillet** dialog box. The **Fillet** dialog box will be displayed as shown in Figure-79. Also, you will be asked to select the edges.

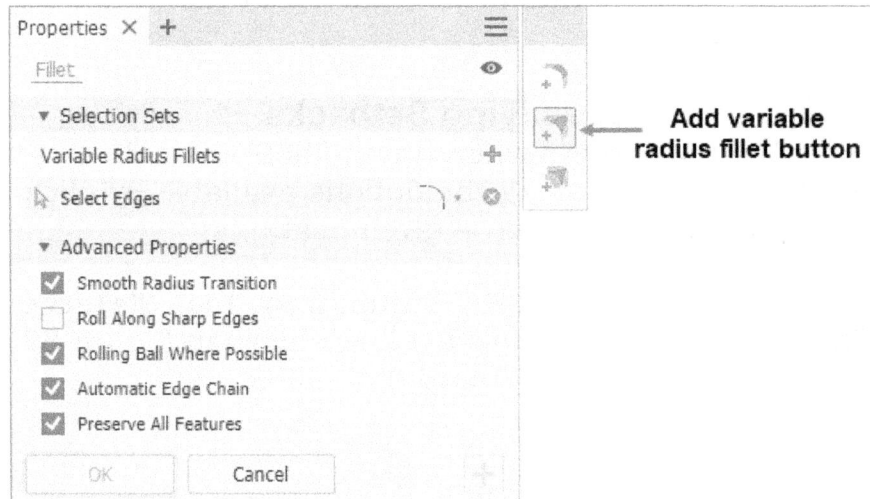

Figure-79. Fillet dialog box with variable radius fillet options

- Select desired edge from the model. You will be asked to select a point on the edge.
- Click on the edge if you want to specify radius at any intermediate point on the edge; refer to Figure-80.

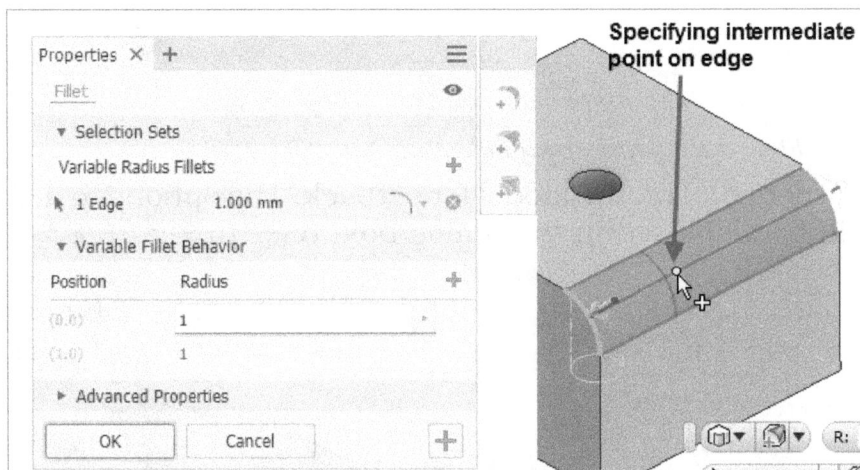

Figure-80. Specifying intermediate point on edge for variable fillet

- Click on the radius value for desired point under the **Radius** column in the **Variable Fillet Behavior** rollout of the dialog box and specify desired radius value; refer to Figure-81.

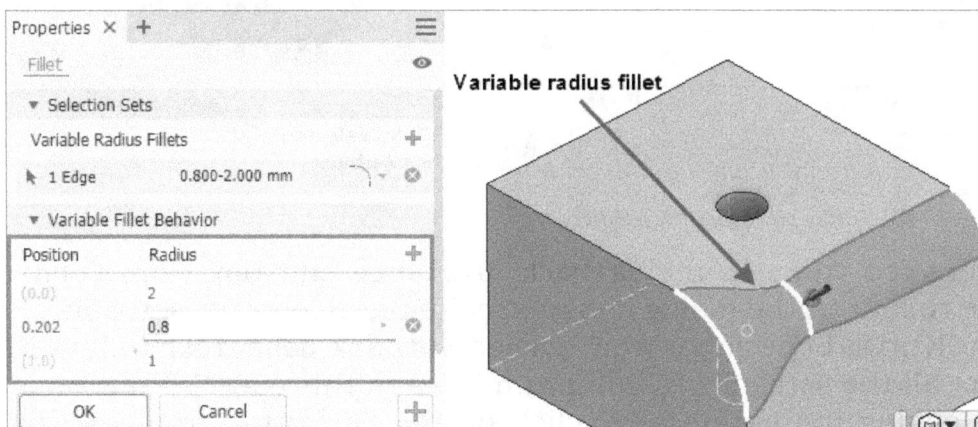

Figure-81. Variable radius fillet

- After specifying the parameters, click on the **OK** button from the dialog box to complete the process.

Specifying Setbacks

Setbacks are created at the corners where three fillets coincide at a point. You can define the shape of setbacks by using the options available by clicking on the **Add corner setback** button. The procedure is given next.

- Click on the **Add corner setback** button from **Tool Palette** in the **Fillet** dialog box. The dialog box will be displayed as shown in Figure-82. Also, you will be asked to select a vertex.

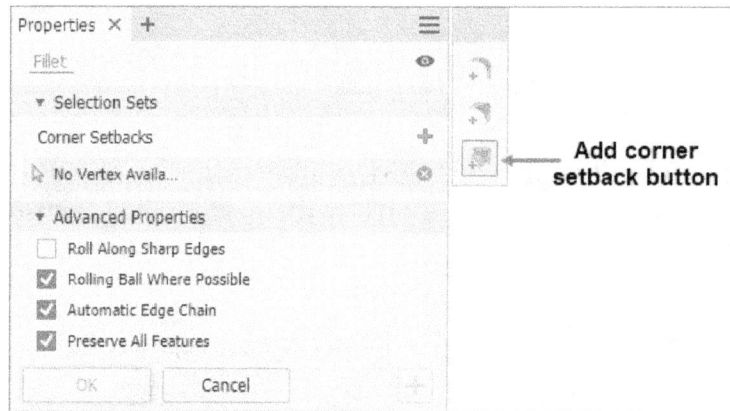

Figure-82. Setback button of Fillet dialog box

- Click at the corner point on the model for setback. The options to define parameters of setback will be displayed in the dialog box; refer to Figure-83.

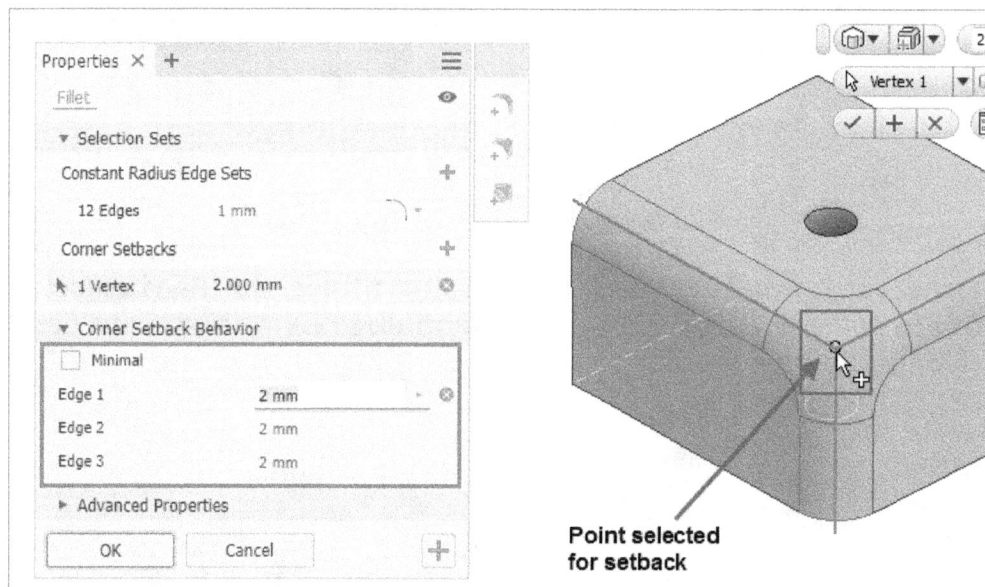

Figure-83. Setting parameters for setbacks

- Specify desired value in the **Setback Distance** edit box from **Corner Setback Behavior** rollout in the dialog box to specify parameter for setback.
- Select the **Minimal** check box from **Corner Setback Behaviour** rollout in the dialog box to set all the setbacks to minimum i.e. **0** in this case.
- After specifying the parameters, click on the **OK** button from the dialog box to complete the process.

Face Fillet Tool

You can create fillets by using faces of the model in place of selecting edges. The procedure to do so is given next.

- Click on the **Face Fillet** tool from **Fillet** drop-down in the **Modify** panel of **3D Model** tab in the **Ribbon**; refer to Figure-84. The **Face Fillet** dialog box will be displayed as shown in Figure-85. Also, you will be asked to select a face.

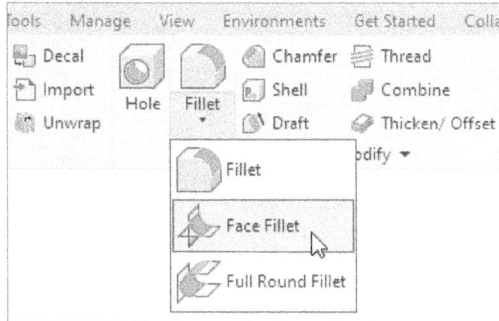

Figure-84. Face Fillet tool

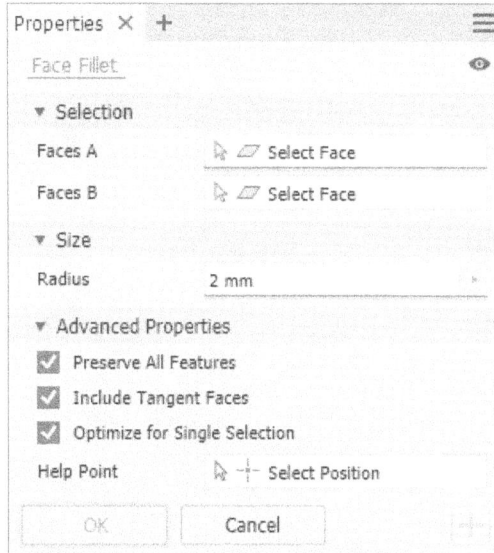

Figure-85. Face Fillet dialog box

- Click on the first face. You will be asked to select second face.
- Select the second face. Preview of the fillet will be displayed; refer to Figure-86.

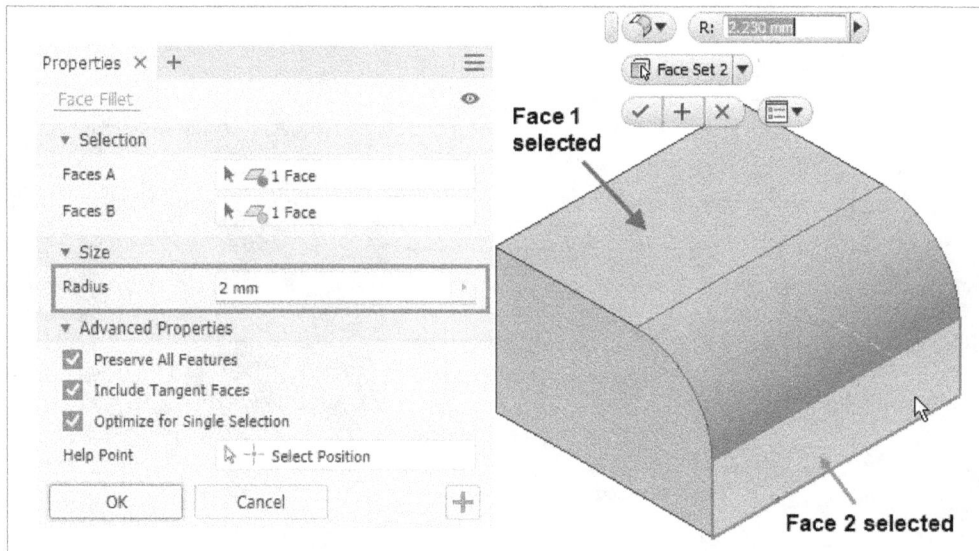

Figure-86. Preview of fillet created using faces

- Set desired radius value in the **Radius** edit box from **Size** rollout of the dialog box.
- If you select the **Include Tangent Faces** check box from **Advanced Properties** rollout then faces that are tangent to the selected faces will also be included while creating the fillet.
- After specifying the parameters, click on the **OK** button from the dialog box to complete the process.

Full Round Fillet Tool

You can create full round fillet by using the **Full Round Fillet** tool. The procedure to create full round fillet is given next.

- Click on the **Full Round Fillet** tool from **Fillet** drop-down in the **Modify** panel of **3D Model** tab in the **Ribbon**; refer to Figure-87. The options in the dialog box will be displayed as shown in Figure-88. Also, you will be asked to select first side face of the model.

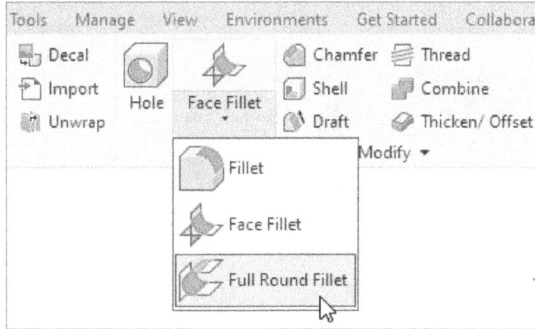

Figure-87. Full Round Fillet tool

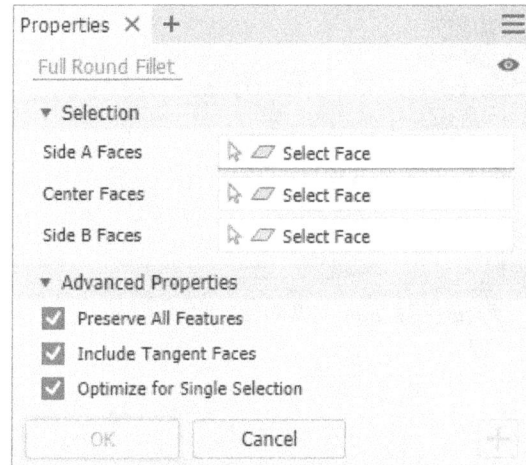

Figure-88. Full Round Fillet dialog box

- Select the first side face. You will be asked to select the center connected face.
- Select the center face. You will be asked to select the second side face.
- Select the second side face. Preview of the full round fillet will be displayed; refer to Figure-89.

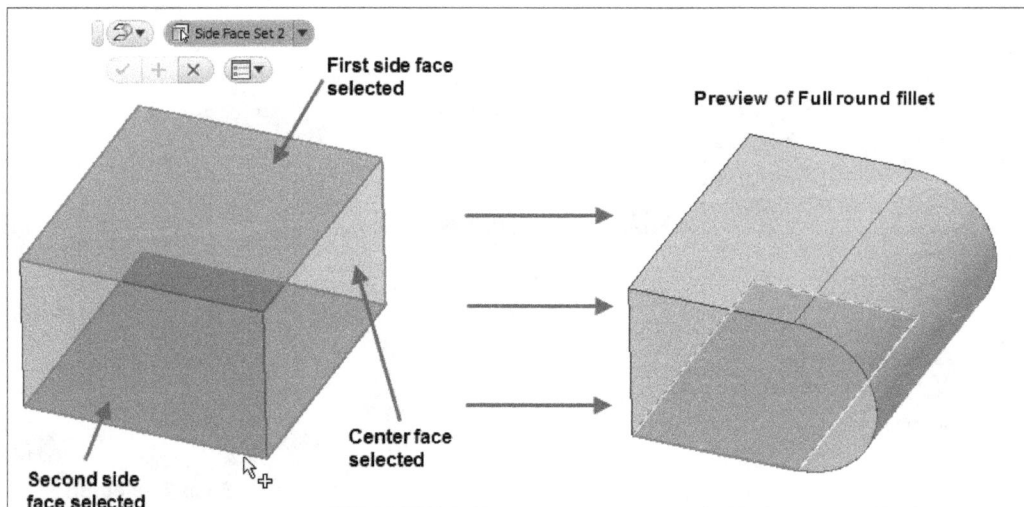

Figure-89. Faces selected for creating full round

- Click on the **OK** button from the dialog box to create the fillet/fillets.

Chamfer Tool

The **Chamfer** tool is used to chisel the sharp edges of the model. The procedure to use the **Chamfer** tool is given next.

- Click on the **Chamfer** tool from **Modify** panel in the **3D Model** tab of the **Ribbon**. The **Chamfer** dialog box will be displayed; refer to Figure-90. Also you will be asked to select the edges.

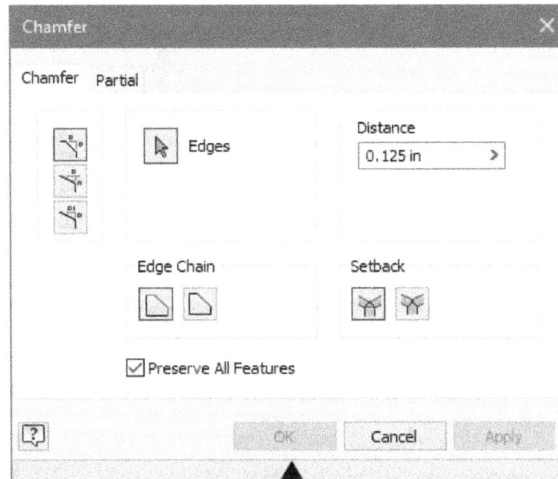

Figure-90. Chamfer dialog box

- Select the edges on which you want to apply chamfer. Preview of the chamfers will be displayed; refer to Figure-91.

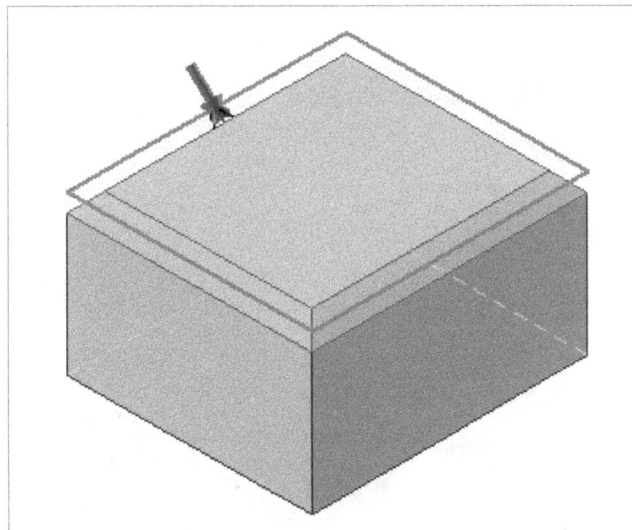

Figure-91. Preview of chamfer

Distance Chamfer

- By default, **Distance** button is selected in the dialog box so you can specify the distance value in both the direction by using the **Distance** edit box in the dialog box.

Distance and Angle Chamfer

- Select the **Distance and Angle** button from the dialog box if you want to specify distance and angle value for chamfer. The **Chamfer** dialog box will be displayed as shown in Figure-92 and you will be asked to select a face.

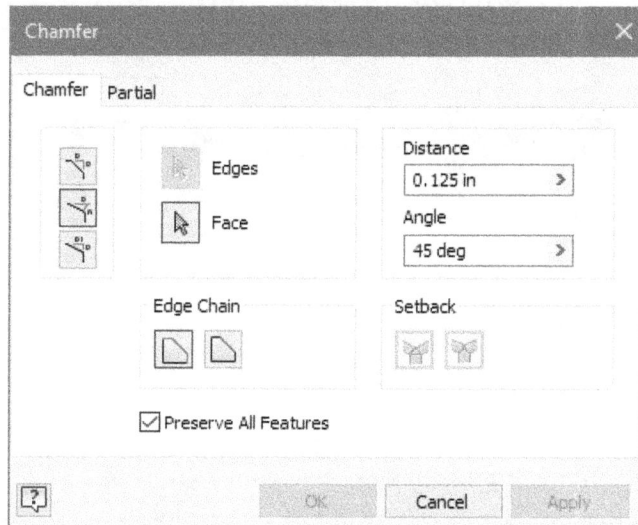

Figure-92. Chamfer dialog box with Distance and Angle button selected

- Select the face with respect to which the chamfer will be at angle. You will be asked to select the edges. Note that you can select only the boundary edges of the selected face to create chamfer.
- Select the edge(s) and specify desired parameters in the **Distance** and **Angle** edit boxes in the dialog box. Preview of the chamfer will be displayed; refer to Figure-93.

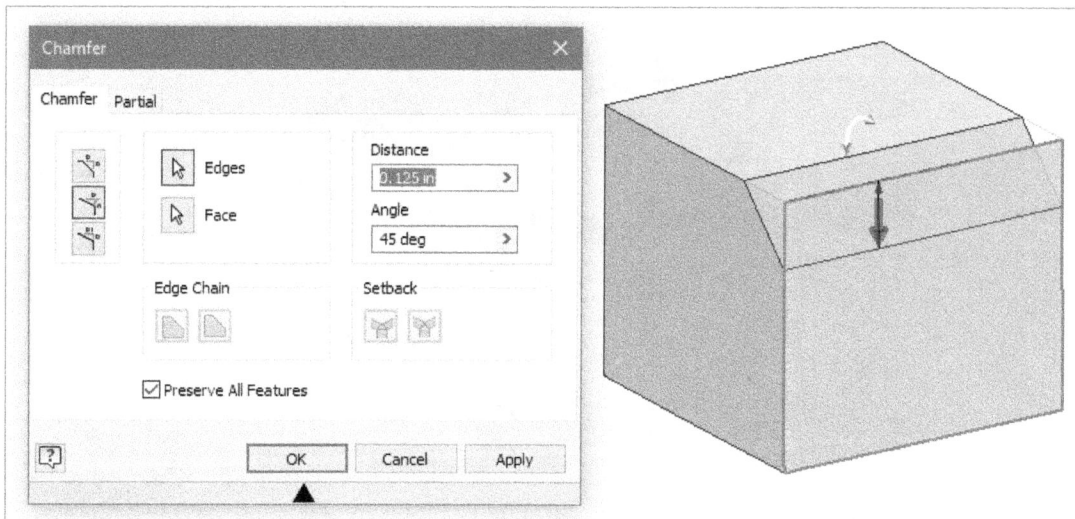

Figure-93. Preview of Distance and Angle chamfer

Two Distances Chamfer

- Select the **Two Distances** button if you want to specify different values for distances in the two directions of the chamfer. The dialog box will be displayed as shown in Figure-94. Also, you will be asked to select the edge(s).

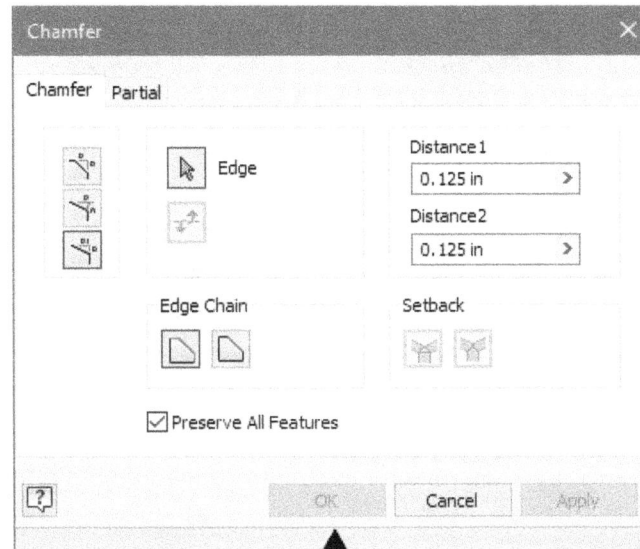

Figure-94. Chamfer dailog box with Two Distances button selected

- Select the edges on which you want to apply chamfer. Preview of the chamfer will be displayed; refer to Figure-95.

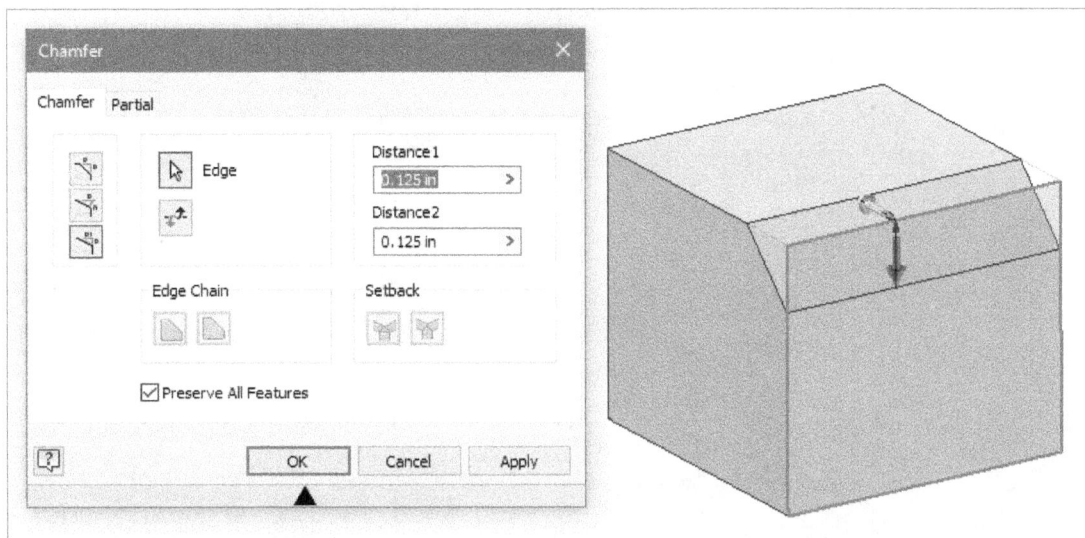

Figure-95. Chamfer created with Two Distance option

- Specify desired distance values in the edit boxes of the dialog box.
- Click on the **OK** button from the dialog box to create the chamfer.

Creating Partial Chamfers

You can create a partial chamfer by defining the location of the start and end vertices along an existing chamfer edge.

- Select the **Partial** tab from the dialog box and you will be asked to locate the end vertex.
- Click on desired chamfer edge to locate the end vertex; the preview of partial chamfer will be displayed; refer to Figure-96.

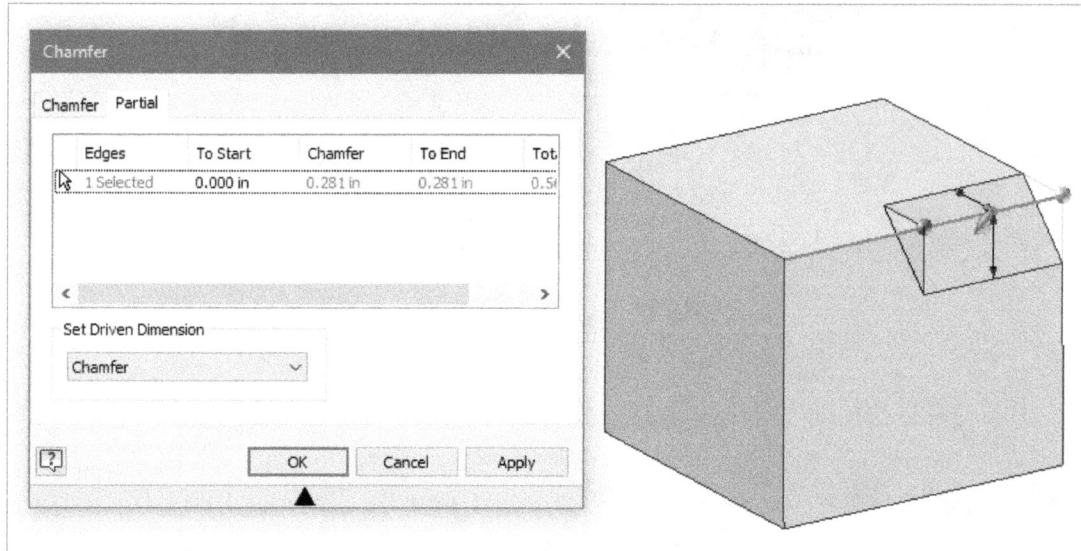

Figure-96. Preview of partial chamfer

- Specify the chamfer distance by dragging the ball or by specifying value in the edit boxes in **Partial** tab of the dialog box; refer to Figure-97.

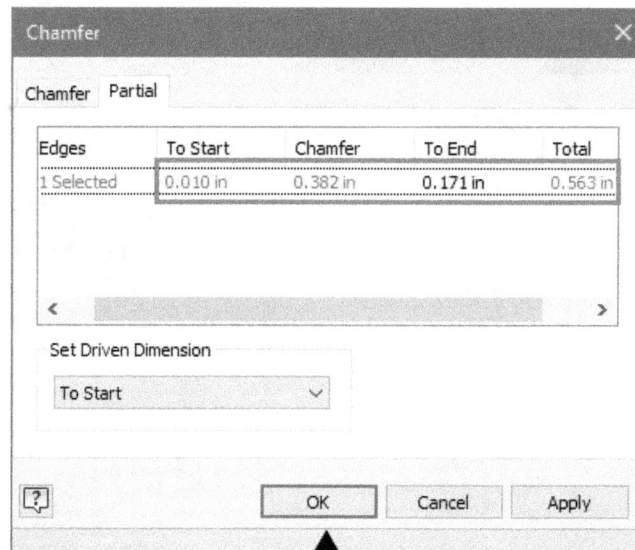

Figure-97. Edit boxes for partial chamfer

- Change the driven dimension type as desired from **Set Driven Dimension** drop-down. There are three options in this drop-down as **To Start**, **Chamfer**, and **To End**.
- Select the **To Start** option to specify the distance from start of edge to start of chamfer. The End and Chamfer length are fixed and do not change when the edge length changes.
- Select the **Chamfer** option to specify the distance as the chamfer length. The Start and End length are fixed and do not change when the edge length changes.
- Select the **To End** option to specify the distance from end of edge to end of chamfer. The Start and Chamfer length are fixed and do not change when the edge length changes.
- Click on the **OK** button from the dialog box to create the chamfer.

Shell Tool

The **Shell** tool is used to scoop out material from the solid base. The procedure to use this tool is given next.

- Click on the **Shell** tool from **Modify** panel in the **3D Model** tab of the **Ribbon**. The **Shell** dialog box will be displayed along with the preview of shell feature; refer to Figure-98. Also, you will be asked to select faces to be removed.

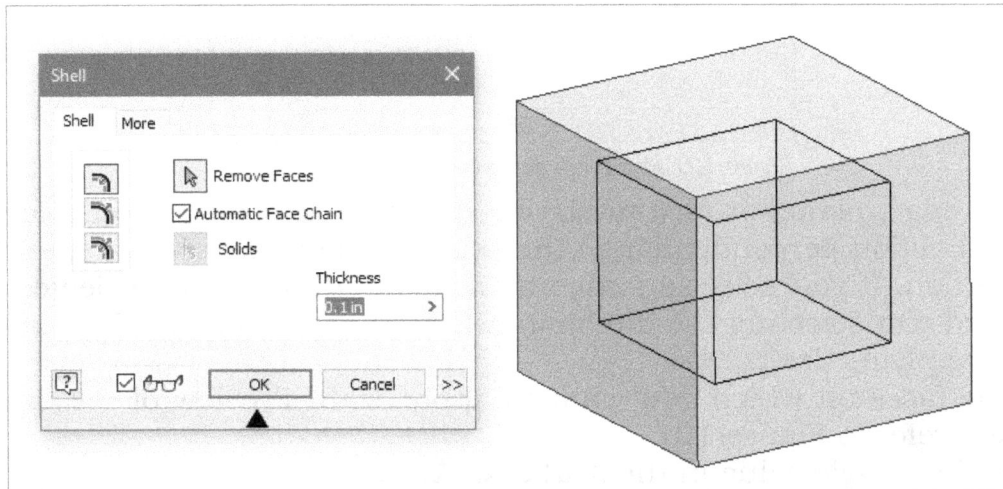

Figure-98. Shell dialog box with preview

- Click on the face(s) that you want to remove from the base model after scooping out the material; refer to Figure-99.

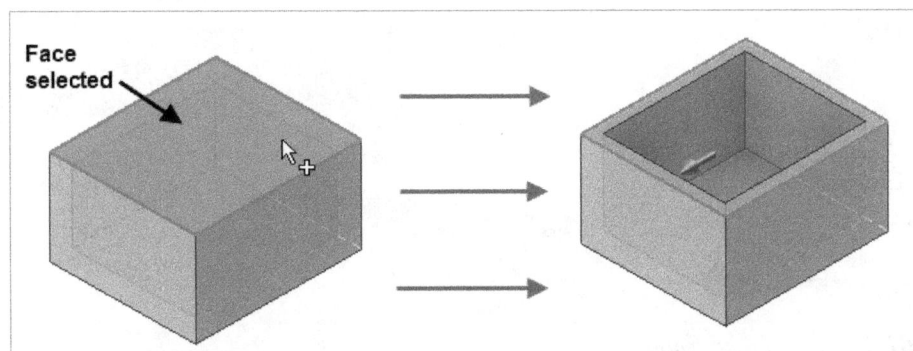

Figure-99. Face selected to remove

- To specify thickness side, click on desired button from **Inside** 🔲, **Outside** 🔲, or **Both** 🔲 from the dialog box.
- Type desired value in the **Thickness** edit box and click on the **OK** button from the dialog box.

Draft Tool

The **Draft** tool is used to apply taper to selected face. The procedure to apply draft is given next.

- Click on the **Draft** tool from **Modify** panel in the **3D Model** tab of the **Ribbon**. The **Face Draft** dialog box will be displayed as shown in Figure-100. Also, you will be asked to select a reference to specify pull direction.

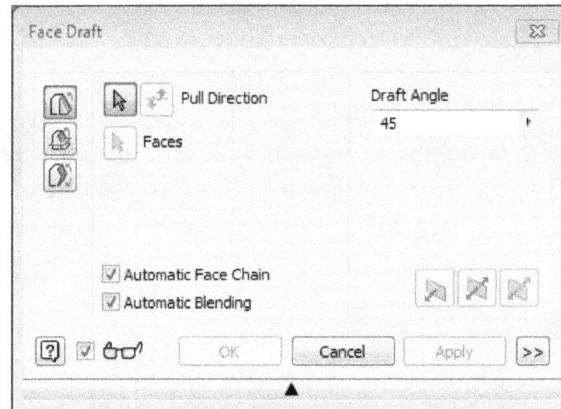

Figure-100. Face Draft dialog box

- Select an edge, plane, or face to specify direction of draft. If you select a plane or face, direction perpendicular to the selected plane or face will be selected as pull direction. If you select an edge then direction along the selected edge will be selected. On selecting the direction reference, you will be asked to select the faces for applying draft.
- Select the faces on which you want to apply draft. Preview of the draft will be displayed; refer to Figure-101.
- Enter desired angle value in the **Draft Angle** edit box.

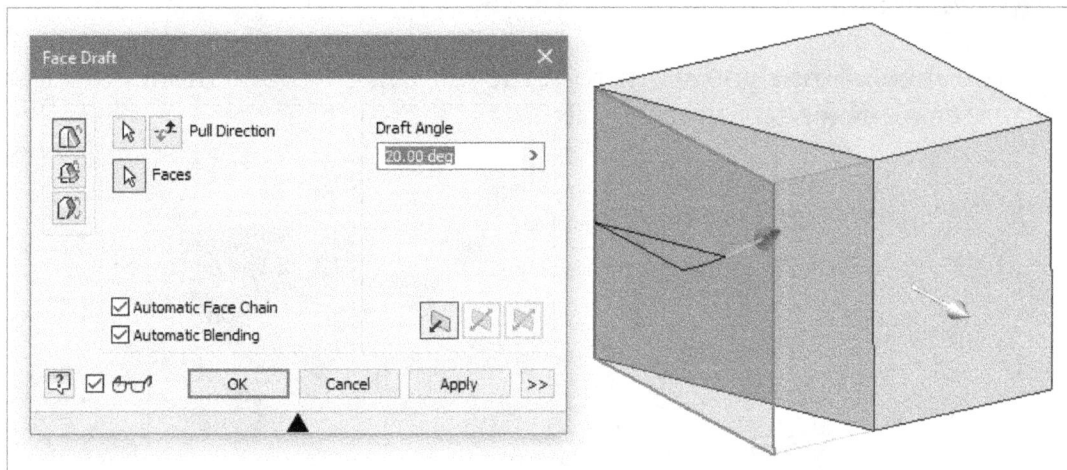

Figure-101. Preview of draft feature

Fixed Plane Draft

- Select the **Fixed Plane** button from the **Face Draft** dialog box. The dialog box will be displayed with **One Way** button selected by default as shown in Figure-102. Also, you will be asked to select a plane to be fixed as angle reference.

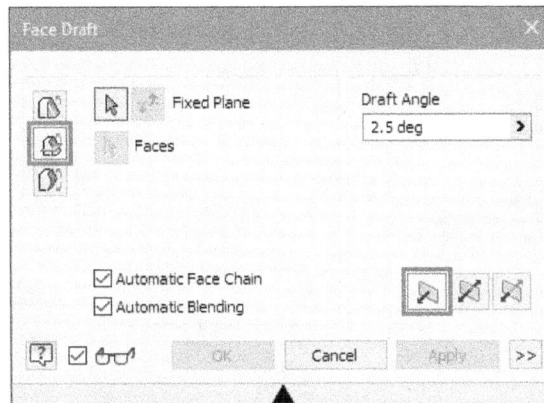

Figure-102. Face Draft dialog box with Fixed Plane button selected

- Select the plane or face that you want as fixed reference for angle; refer to Figure-103. You will be asked to select the faces to apply draft.

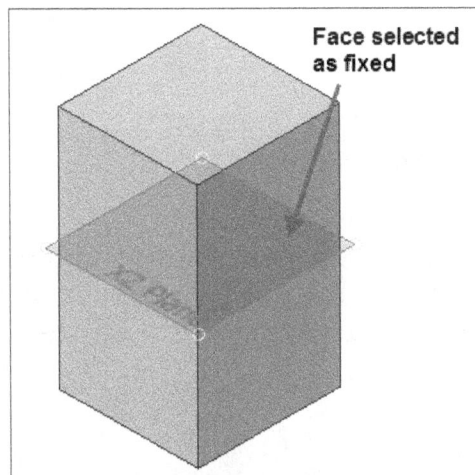

Figure-103. Face selected as fixed

- Select the faces on which you want to apply draft. Preview of draft will be displayed; refer to Figure-104.

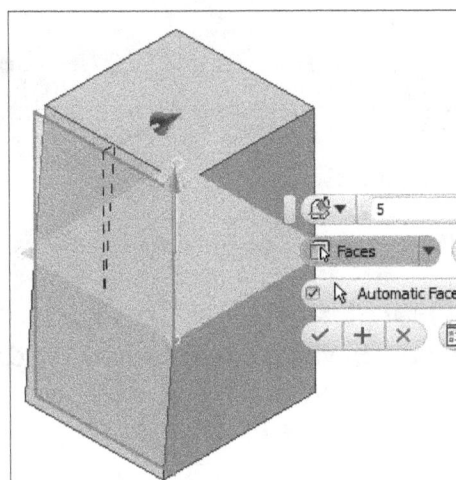

Figure-104. Preview of draft feature with Fixed Plane option

- Specify desired draft angle value in the **Draft Angle** edit box in the dialog box.

- If you want to apply symmetric draft to both sides of face divided by fixed plane then select the **Symmetric** ⊠ button from the dialog box. Preview of the draft will be displayed; refer to Figure-105.

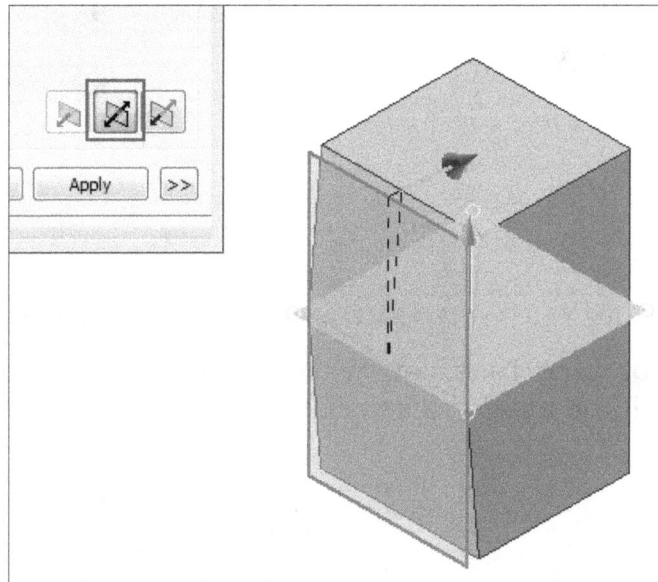

Figure-105. Preview of symmetric draft

- Specify desired angle value in the edit box.
- You can specify different values of draft for two sides of face with respect to fixed plane by using the **Asymmetric** button ⊠. Preview of the draft will be displayed with different draft angles for faces; refer to Figure-106.

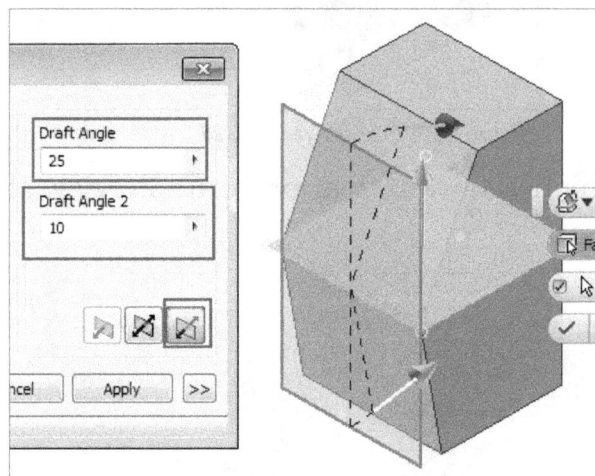

Figure-106. Preview of asymmetric draft

Draft Using Parting Line

Parting Line also works in the same way as fixed plane in draft with the only difference that now, you can also select a curve dividing the face to create different draft on either directions. The procedure to create draft with parting line is given next.

- Click on the **Parting Line** button ⓪ from the **Face Draft** dialog box. The dialog box will be displayed as shown in Figure-107. Also, you will be asked to select a reference for specifying the pull direction.

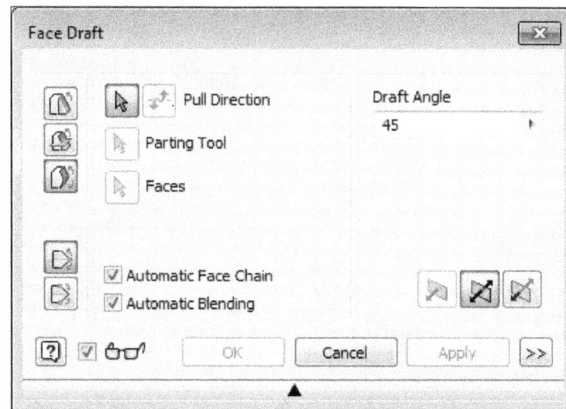

Figure-107. Face Draft dialog box with Parting Line button selected

- Select a face, plane, or edge to specify the pull direction. You will be asked to select the **Parting Tool** reference.
- Select a sketching line, plane, or surface to make it parting tool. You will be asked to select face on which the draft is to be applied.
- Select desired face. Preview of the draft feature will be displayed; refer to Figure-108.

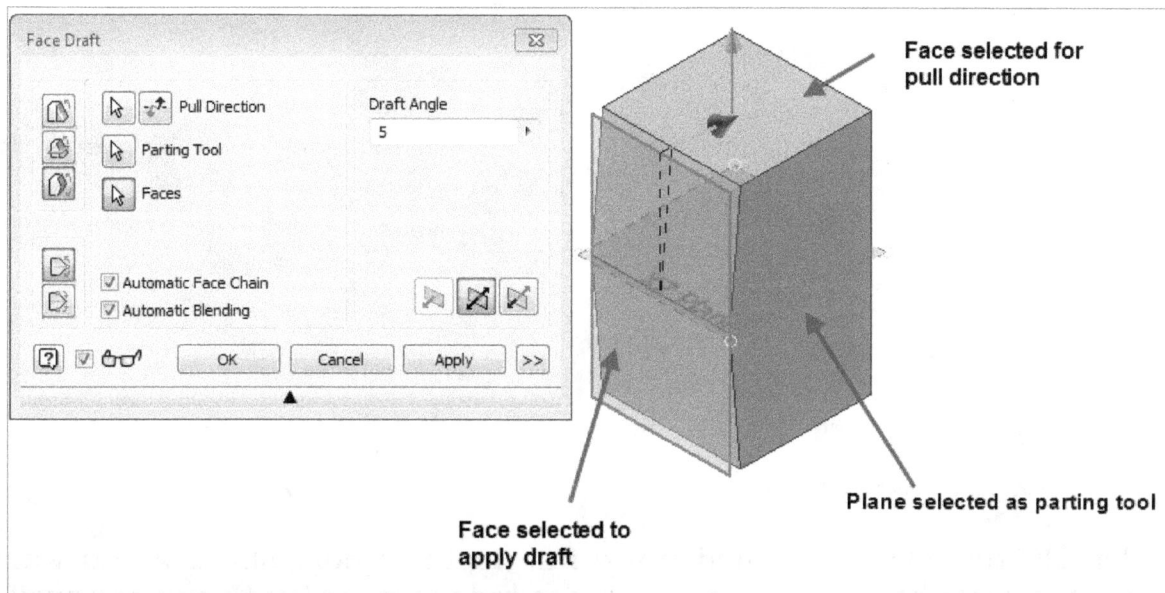

Figure-108. Preview of face draft with parting line selected

- Enter desired angle value and select the **Symmetric** or **Asymmetric** button as per your requirement.
- Click on the **OK** button to apply draft.

Thread Tool

As the name suggests, the **Thread** tool is used to create threads in selected hole or over the selected shaft. The procedure to use this tool is given next.

- Click on the **Thread** tool from **Modify** panel in the **3D Model** tab of the **Ribbon**. The **Thread** dialog box will be displayed; refer to Figure-109. Also, you will be asked to select the face on which threads are to be applied.

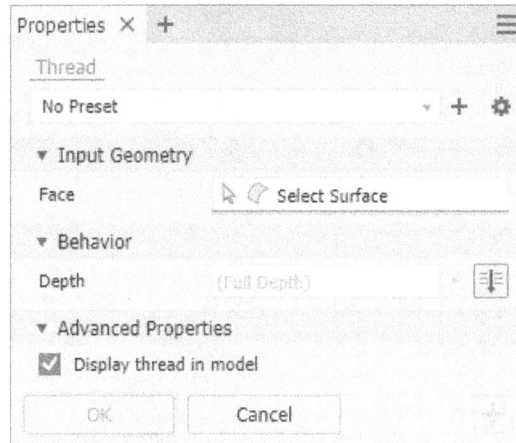

Figure-109. Thread dialog box

- Select the face on which you want to apply thread. The updated **Thread** dialog box will be displayed along with the preview of thread; refer to Figure-110.

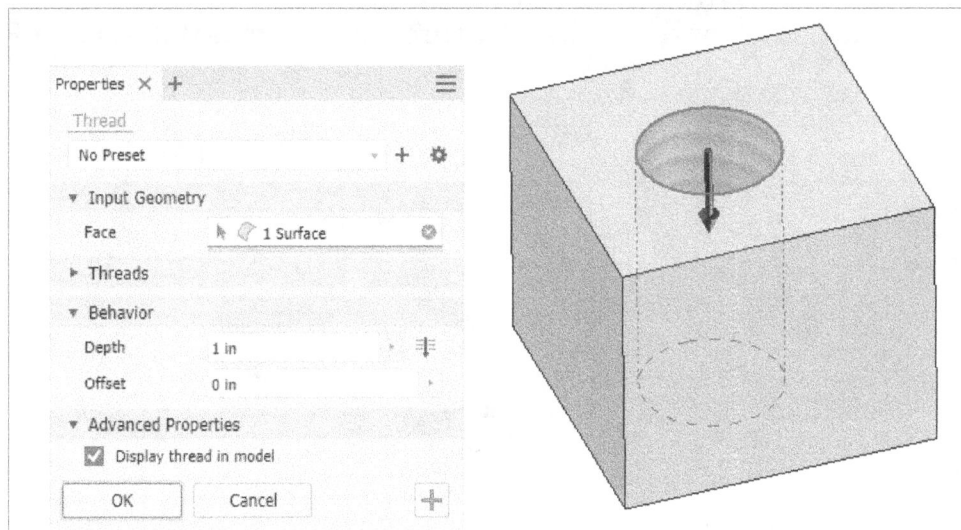

Figure-110. Preview of thread

- By default, threads are created up to full depth of hole/shaft. If you want to specify the depth of thread then specify desired values in **Depth** edit box available in **Behavior** area of the dialog box.
- Click on the **Threads** area of the dialog box to display the options related to size of thread; refer to Figure-111.

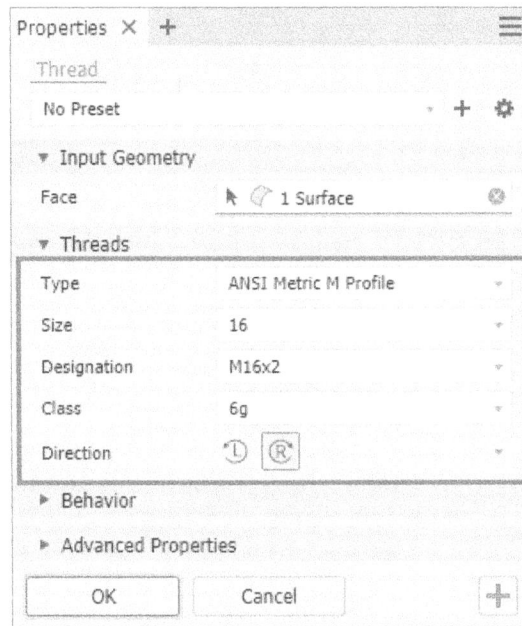

Figure-111. Threads area of the dialog box

- Select desired thread type from the **Type** drop-down and select desired size of the thread.
- Click on the **OK** button from the dialog box. The threads will be created; refer to Figure-112.

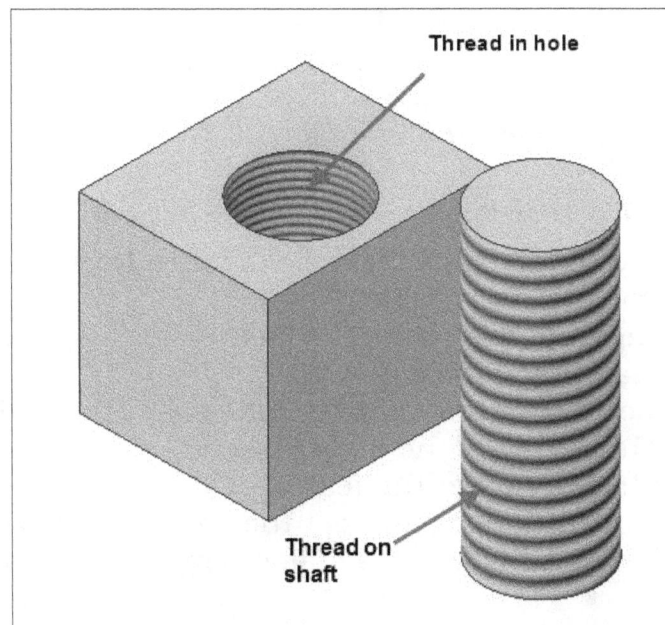

Figure-112. Threads on shaft and hole

Combine Tool

The **Combine** tool is used to perform Boolean operations on the solids. This tool is used to combine two bodies, subtract one body from another, or find out the intersecting portion between selected bodies. The procedure to use this tool is given next.

- Click on the **Combine** tool from **Modify** panel in the **3D Model** tab of the **Ribbon**. The **Combine** dialog box will be displayed; refer to Figure-113. Also, you will be asked to select the base body.

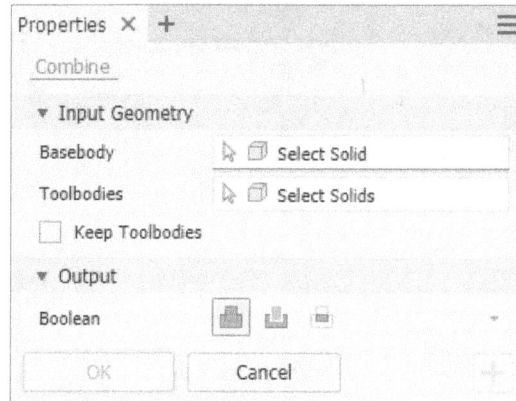

Figure-113. Combine dialog box

- Select the base body. You will be asked to select the tool body.
- Select the second solid body to perform boolean operation; refer to Figure-114.

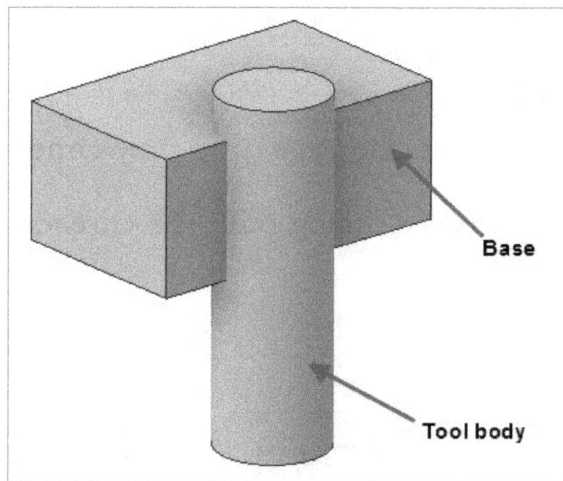

Figure-114. Base and tool body selected for combining

- If you want to join two solid bodies then click on the **Join** button 🔲 from **Boolean** area in the **Output** rollout of the dialog box.
- If you want to subtract tool body from the base solid then click on the **Cut** button 🔲 from **Boolean** area of the dialog box.
- If you want to find the intersecting volume of the solids then click on the **Intersect** button 🔲 from **Boolean** area of the dialog box.
- Click on the **OK** button to create the feature. Note that if you want to keep the tool body even after operation then select the **Keep Toolbodies** check box in the **Input Geometry** rollout of the **Combine** dialog box.

Figure-115 shows the output of the three buttons on the model.

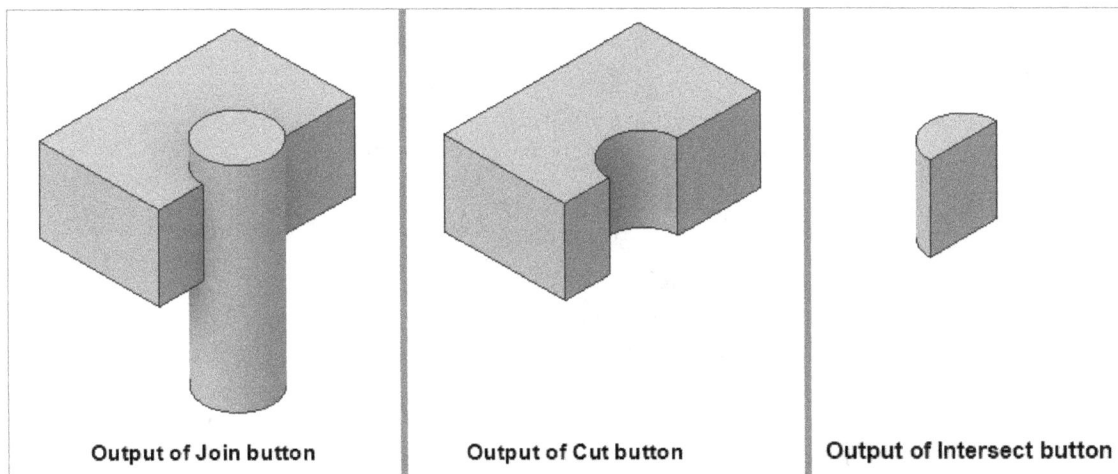

Figure-115. Output of Combining

Thicken/Offset Tool

The **Thicken/Offset** tool, as the name suggests, is used to thicken or offset the solids/surfaces. The procedure to use this tool is given next.

*   Click on the **Thicken/Offset** tool from **Modify** panel in the **3D Model** tab of **Ribbon**. The **Thicken** dialog box will be displayed; refer to Figure-116. Also, you will be asked to select the face.

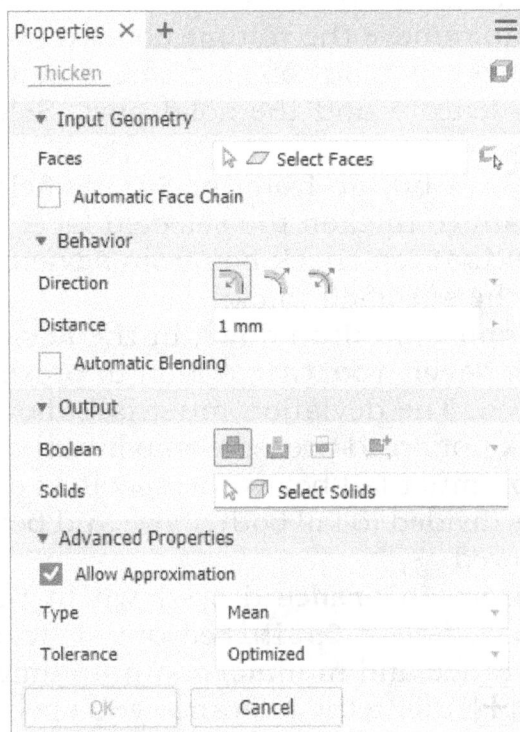

Figure-116. Thicken dialog box

*   Select the face that you want to thicken or offset. Preview of the thicken/offset will be displayed; refer to Figure-117.

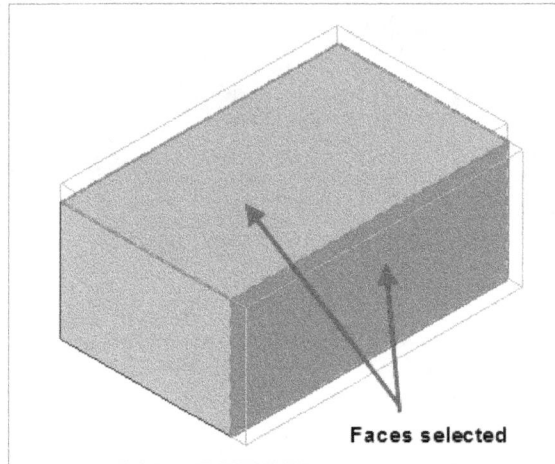

Figure-117. Faces selected for thicken or offset

- Select desired option from **Inside, Outside,** or **Center** option in the **Direction** area of **Behavior** rollout in the dialog box.
- Specify desired thickness or distance of the offset in the **Distance** edit box of **Behavior** rollout.
- Select **Automatic Blending** check box from **Behavior** rollout to automatically move tangential faces and also creates new blends, if required. Note that when using **Automatic Blending**, the **Output** rollout will not be available.
- Select desired option from **Boolean** area in the **Output** rollout of the dialog box. Select **Join** button to add the volume created by the Thicken feature to the solid part. Select **Cut** button to remove the volume created by the Thicken feature from the solid part. Select **Intersect** button to create a new feature from the shared volume of the Thicken feature and the solid part. Select **New Solid** button to create a new solid body.
- Click on the **Select Solids** button from the **Solids** field in the **Output** rollout of the dialog box if you want to thicken the selected face.
- Select the **Allow Approximation** check box from **Advanced Properties** rollout to allow a deviation from the specified thickness.
- Select desired option from **Type** drop-down in the **Advanced Properties** rollout of the dialog box. Select **Never too thin** option to preserve minimum distance if approximation is allowed. The deviation must fall above the specified distance. Select **Never too thick** option to preserve maximum distance if approximation is allowed. The deviation must fall below the specified distance. On selecting the **Mean** option, deviation is divided to fall both above and below the specified distance if approximation is allowed.
- Select desired option from **Tolerance** drop-down in the **Advanced Properties** rollout of the dialog box. Select **Optimized** option to compute approximation using a reasonable tolerance and minimal compute time. Select **Specified** option to compute approximation using the tolerance you specify. Valid tolerance range is from 0-100 %.
- Click on the **OK** button from the dialog box to create the feature.
- If you want to create the surface offset feature then click on the **Surface mode** button from top right corner of the **Thicken** dialog box. The **OffsetSrf** dialog box will be displayed along with the preview of surface offset feature; refer to Figure-118.

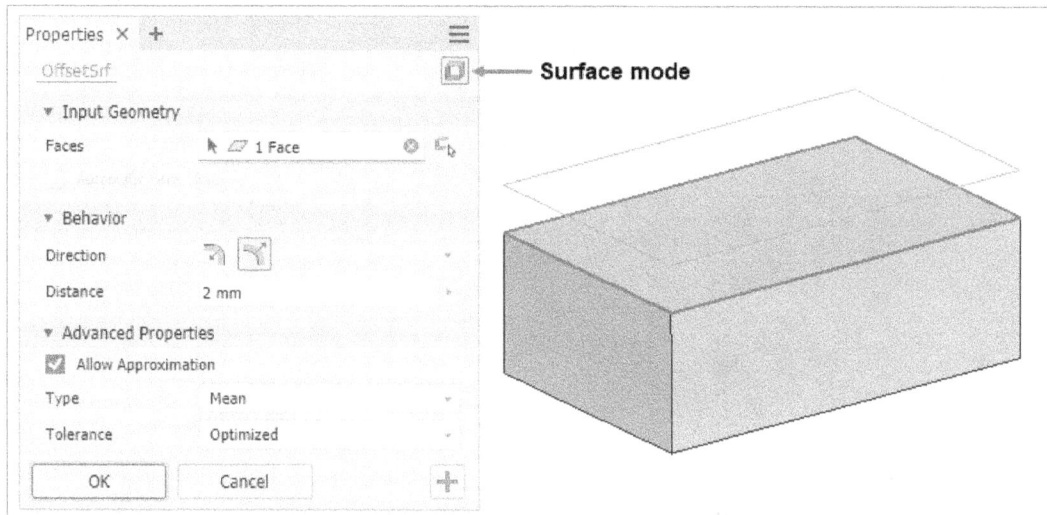

Figure-118. OffsetSrf dialog box with preview of surface offset feature

- Specify desired parameters as discussed earlier and click on **OK** button from **OffsetSrf** dialog box to create the feature.

Split Tool

The **Split** tool is used to divide the selected entity (Face or solid body) with the help of trimming reference (Sketch, Plane, or Surface). The procedure to use this tool is given next.

- Click on the **Split** tool from **Modify** panel in the **Ribbon**. The **Split** dialog box will be displayed as shown in Figure-119. Also, you will be asked to select the sketch profile, plane, or surface as split tool.

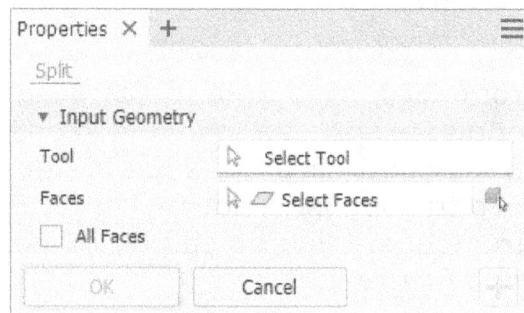

Figure-119. Split dialog box

- Select the object to be used as split tool. You will be asked to select the faces to be split up.
- Select the faces that you want to split. Preview of the splitting will be displayed; refer to Figure-120.

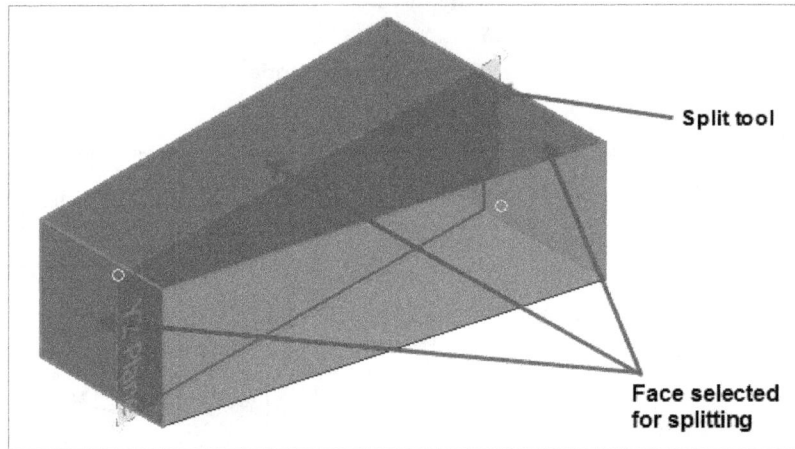

Figure-120. Selecting faces to split

- Select **All Faces** check box from **Input Geometry** rollout to split all the faces of selected body.
- Click on the **Solid selection** button from **Faces** area in the **Input Geometry** rollout. The updated **Split** dialog box will be displayed; refer to Figure-121.

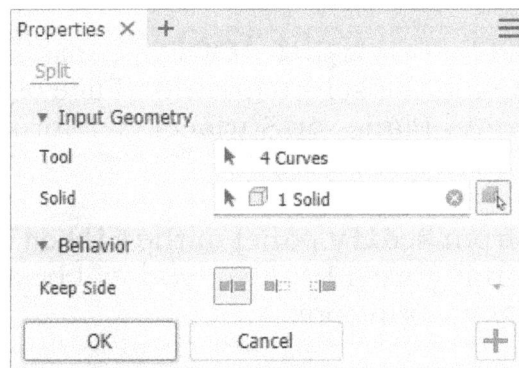

Figure-121. Updated Split dialog box

- Select desired option from **Keep Side** area in the **Behavior** rollout of the dialog box. Select **Both Sides** option to split the solid and keep both sides. Select **Default Side** option to split the body and keep the default side. Select **Flip Side** option to split the solid and keep the opposite side;
- After specifying the parameters, click on the **OK** button from the dialog box to create the feature; refer to Figure-122.

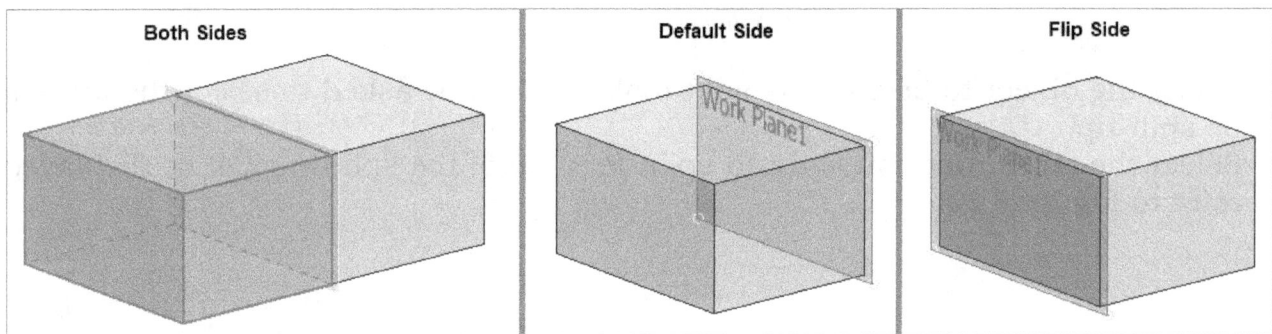

Figure-122. Splitting the body at different types

Delete Face Tool

The **Delete Face** tool is used to convert a part body into a surface body by deleting one or two faces from the solid body. The procedure to use this tool is given next.

- Click on the **Delete Face** tool from **Modify** panel in the **3D Model** tab of the **Ribbon**. The **Delete Face** dialog box will be displayed; refer to Figure-123. Also, you will be asked to select the face to be delete.

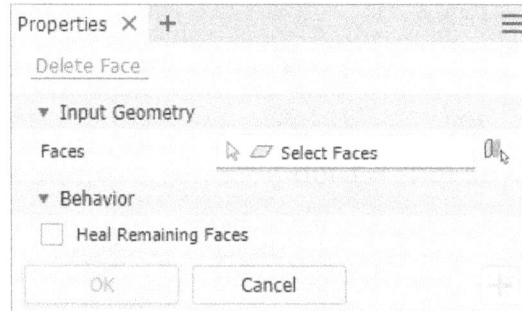

Figure-123. Delete Face dialog box

- Select desired face to be delete from the solid body.
- Select **Heal Remaining Faces** check box from **Behavior** rollout in the dialog box to heal gaps by extending adjacent faces until they intersect.
- Click on the **Lump or Void toggle** button from **Faces** area in the **Input Geometry** rollout of the dialog box and select individual faces, lumps, or voids that you want to delete.
- Click on the **OK** button from the dialog box; the selected face will be deleted; refer to Figure-124.

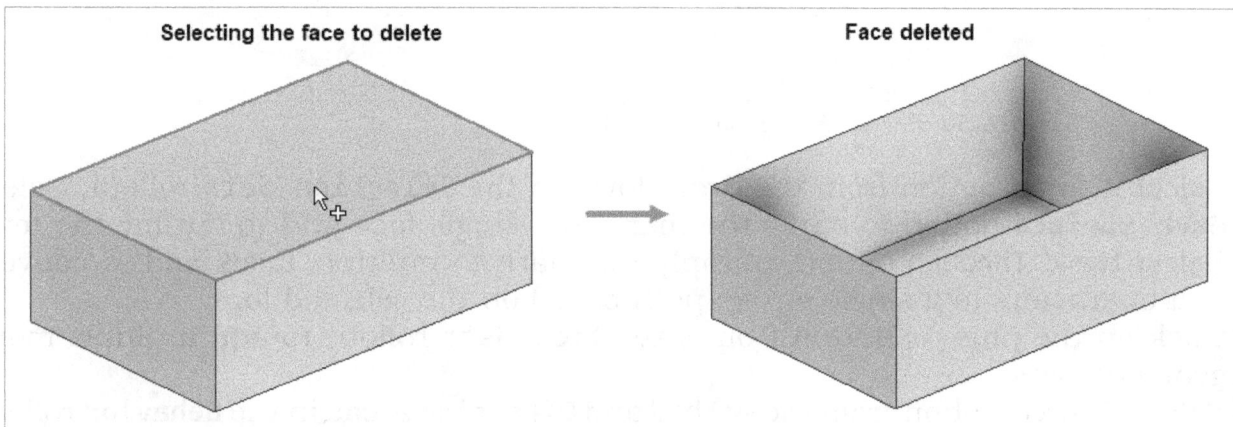

Figure-124. Selected face deleted

Mark

The **Mark** tool is used to prepare the text for machining via processes like laser marking, etching, and engraving. The procedure to use this tool is discussed next.

- Click on the **Mark** tool from **Modify** panel in the **3D Model** tab of the **Ribbon**. The **Mark** dialog box will be displayed; refer to Figure-125. You will be asked to select the object to be marked.

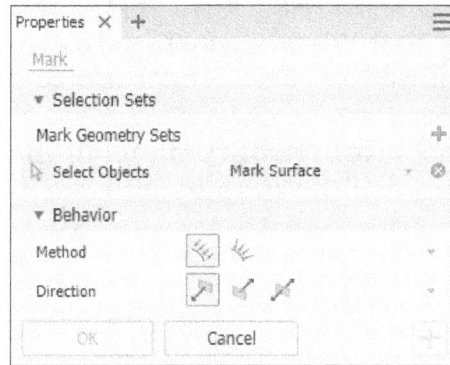

Figure-125. Mark dialog box

- Select desired geometry or text which you want to mark; refer to Figure-126.

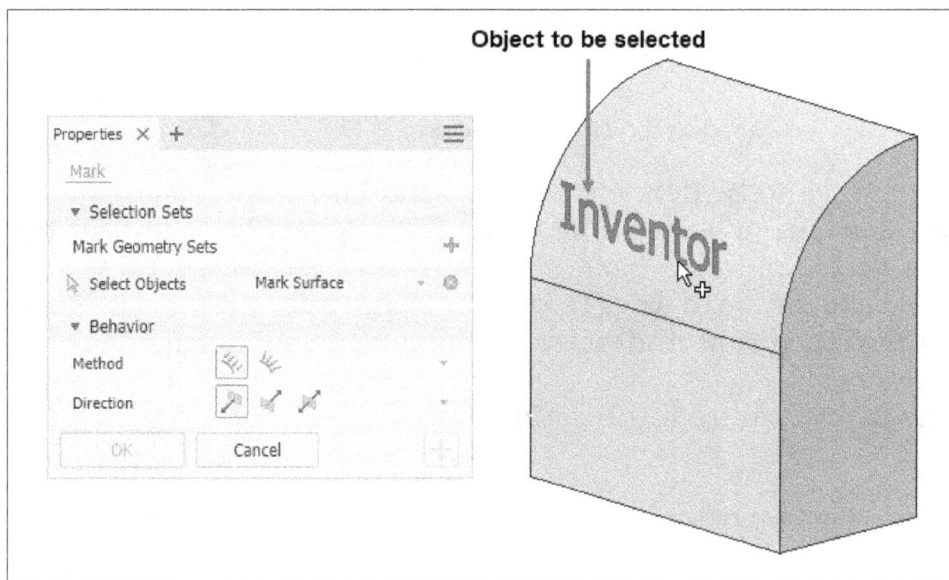

Figure-126. Selecting the object to mark

- Select desired option from the drop-down in the **Selection Sets** rollout. Select **Mark Surface** option to apply the mark to a single face and to outline the text. Select **Mark Through** option to apply the mark to multiple faces and to convert text characters to a single stroke path based on the selected font.
- Click on the plus ⊞ button from **Selection Sets** rollout to add multiple mark geometry sets.
- Select desired option from the **Method** and **Direction** areas in the **Behavior** rollout of the dialog box. Select **Project** option from **Method** area and select desired direction option from **Direction** area to project the sketch geometry to a face. Select **Wrap** option from **Method** area and select desired cylindrical face for the **Face** area to wrap the sketch geometry to a cylindrical face.
- After specifying desired parameters, click on the **OK** button from the dialog box. The selected object will be marked; refer to Figure-127.

Mark Surface with Project option	Mark Through with Project option	Mark Surface with Wrap option

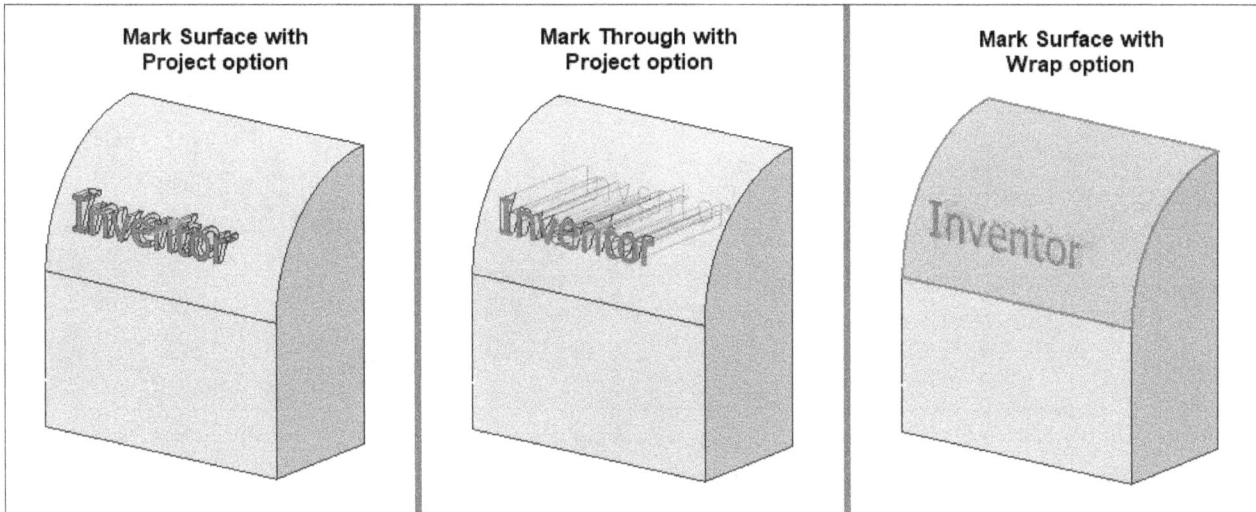

Figure-127. Object marked

Finish

The **Finish** tool is used to apply texture and material appearances to the model in graphics area. The procedure to use this tool is discussed next.

- Click on the **Finish** tool from **Modify** panel in the **3D Model** tab of the **Ribbon**. The **Finish** dialog box will be displayed; refer to Figure-128.

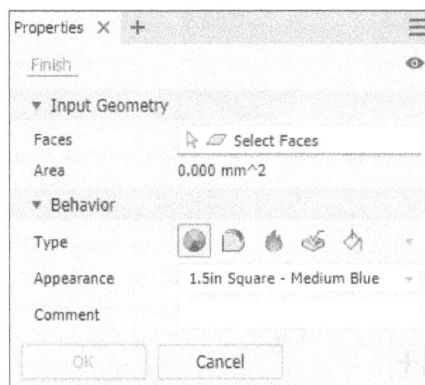

Figure-128. Finish dialog box

- Make sure the **Select Faces** selection button is active in **Faces** section of **Input Geometry** rollout and then select desired faces from the model in graphics area. Preview of appearance will be displayed in graphics area and total area of selected faces will be displayed in the **Area** section of dialog box.
- Select desired type of finish to be applied on the face(s) from **Type** section of **Behavior** rollout. Select **Appearance** option to select an appearance from the Appearance library that you want to apply. Select **Material Coating** option to specify the material coating to be applied to selected faces. Select **Heat Treatment** option to specify the heat treatment to be applied to selected faces. Select **Surface Texture** option to specify the surface texture to be applied to selected faces. Select **Paint** option to specify the paint to be applied to selected faces.
- Specify desired properties of the selected finish type in the dialog box.
- After specifying desired parameters, click on the **OK** button from the dialog box. The finish will be applied to the face(s); refer to Figure-129.

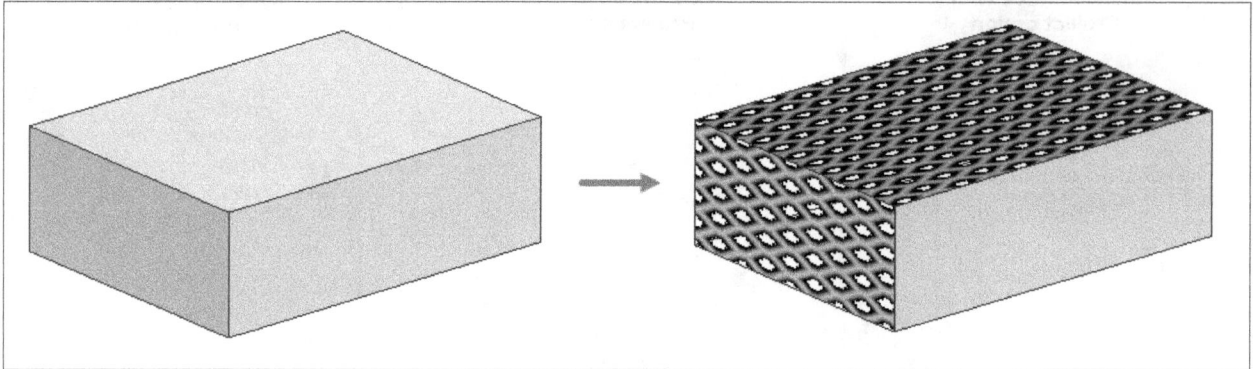

Figure-129. Finish applied to the faces

We have discussed many tools related to advanced 3D Modeling. Let's get some real-world practical on the tools.

PRACTICAL 1

In this practical, we will create a model as shown in Figure-130 using the dimensions given in Figure-131.

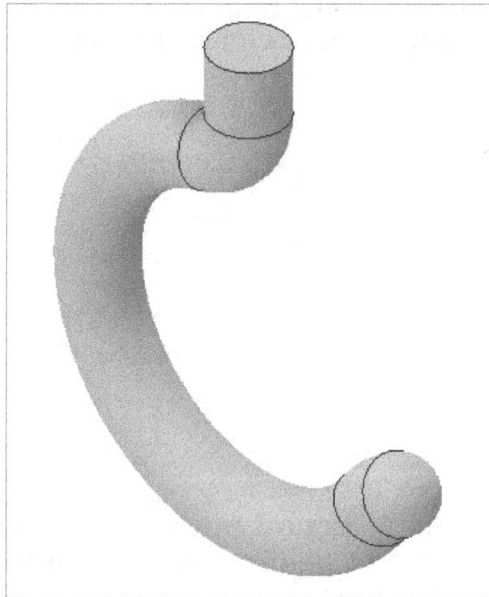

Figure-130. Model for Practical 1

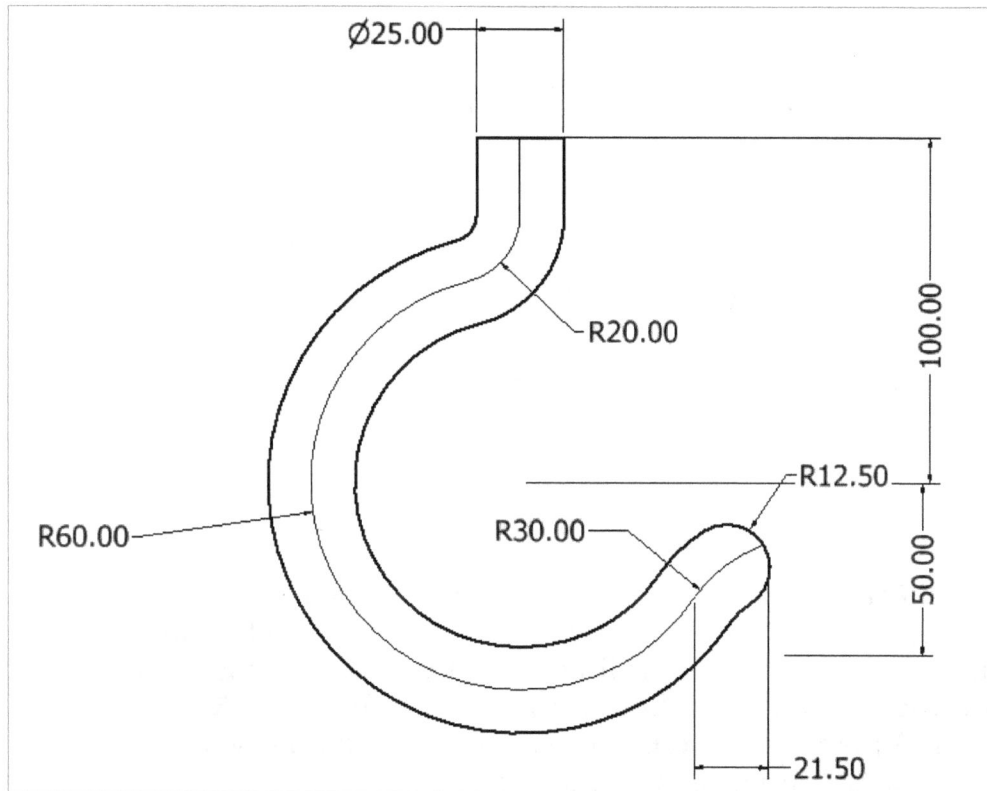

Figure-131. Practical 1 drawing view

Strategy for Creating model

1. Looking at the isometric view and orthographic views, we can find that the sketch for revolve feature should be created on the Front plane (YZ Plane).
2. Using the **Revolve** tool, we will create the solid base of the model.
3. Using the Plane tools, we will create a plane at an angle of 40 degree to the vertical wall of base feature.
4. Using the **Extrude** tool, we will create the pipe joined to the base feature.

Starting Part File

- Start Autodesk Inventor by double-clicking on the Autodesk Inventor Professional 2024 icon from the desktop. (If not started yet.)
- Click on the **New** button from the **Quick Access Toolbar**. The **Create New File** dialog box will be displayed.
- Double-click on **Standard(mm).ipt** icon from the **Metric** templates. The Part environment of Autodesk Inventor will be displayed.

Creating Sketch

- Click on the **Start 2D Sketch** button from **Sketch** panel in the **3D Model** tab of the **Ribbon**. You will be asked to select a sketching plane.
- Select the **XY** Plane (Front Plane) as sketching plane. The sketching environment will become active.
- Create the sketch as shown in Figure-132.

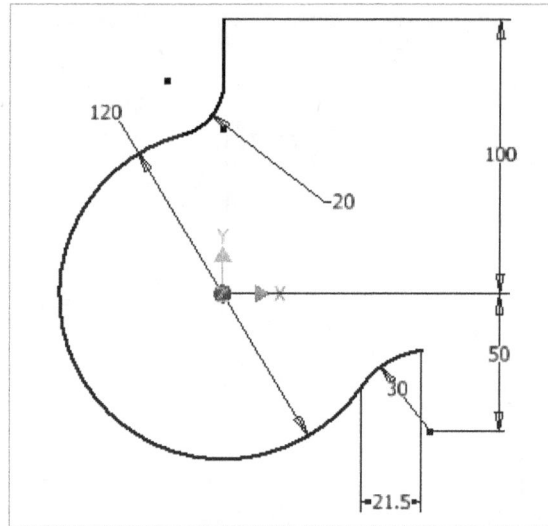

Figure-132. Sketch for practical 1

- Click on the **Finish Sketch** button from **Exit** panel in the **Ribbon**.
- Click on the **Plane** tool from **Plane** drop-down in the **Work Features** panel of the **3D Model** tab in the **Ribbon**. You will be asked to select the geometry for references.
- Select the top point of the sketch and then select the **XZ Plane** from the **Model Browser** bar. A plane will be created passing through the selected point and parallel to the selected **XZ** plane; refer to Figure-133.

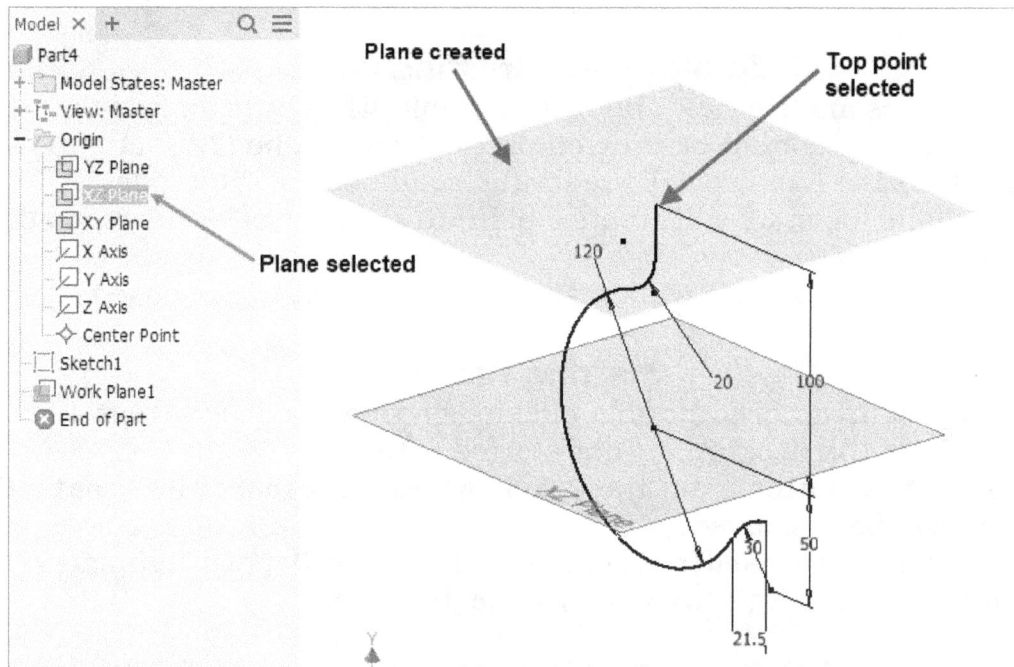

Figure-133. Plane created

- Again, click on the **Start 2D Sketch** button from **Sketch** panel in the **Ribbon**. You will be asked to select a sketching plane.
- Select the work plane recently created. The sketching environment will be displayed.
- Create the circle of diameter **25** with center at the coordinate system as shown in Figure-134.

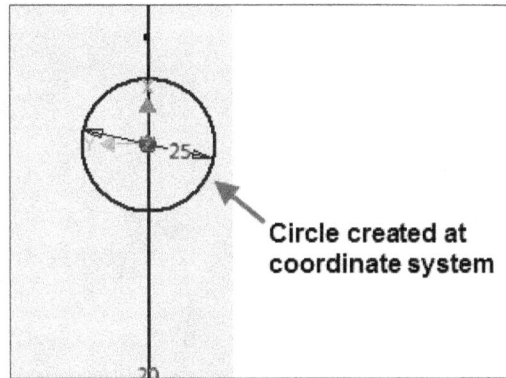

Figure-134. Circle to be created

- Click on the **Finish Sketch** button to exit the sketching environment. In isometric orientation, the sketches will be displayed as shown in Figure-135.

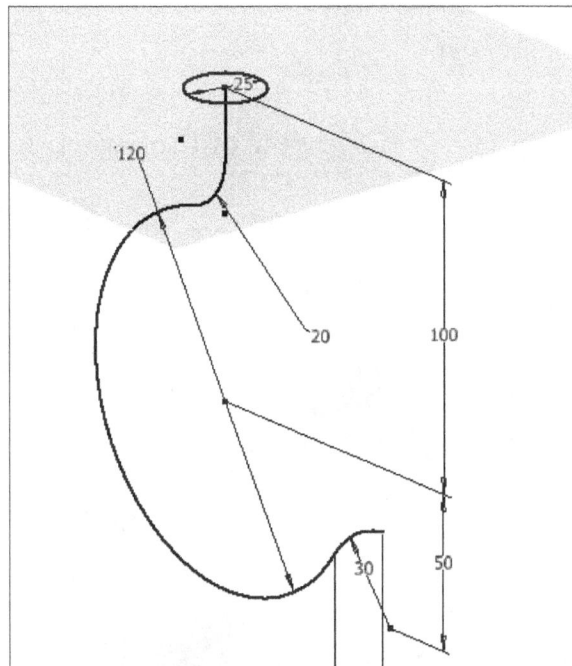

Figure-135. Sketches in isometric orientation

Creating Sweep Feature

- Click on the **Sweep** tool from **Create** panel in the **3D Model** tab of the **Ribbon**. The **Sweep** dialog box will be displayed and the profile will get selected automatically; refer to Figure-136. Also, you will be asked to select the path.

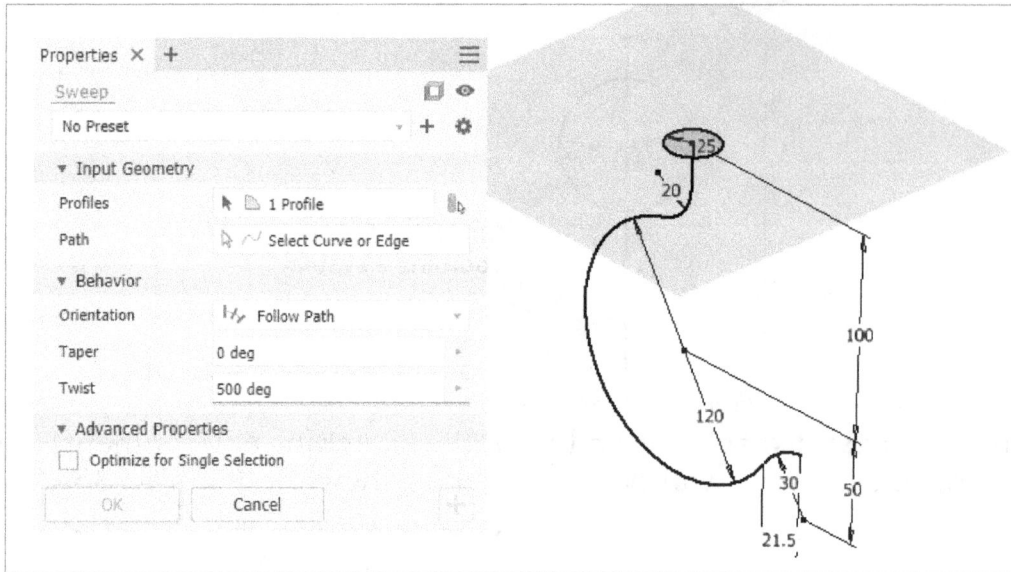

Figure-136. Sweep dialog box with profile selected

- Select the path for sweep feature. Preview of the sweep feature will be displayed; refer to Figure-137.

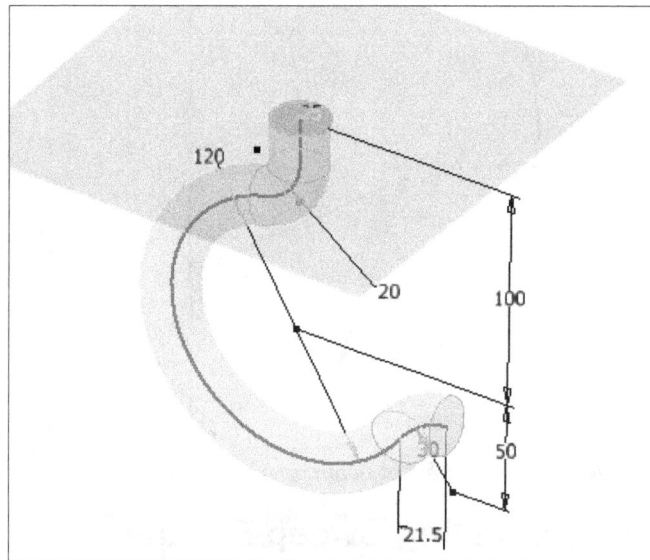

Figure-137. Preview of the sweep feature

- Click on the **OK** button from the dialog box to create the feature.

Applying Fillet

- Click on the **Fillet** tool from **Modify** panel in the **3D Model** tab of the **Ribbon**. The **Fillet** dialog box will be displayed; refer to Figure-138. Also, you will be asked to select the edge on which fillet is to be applied.

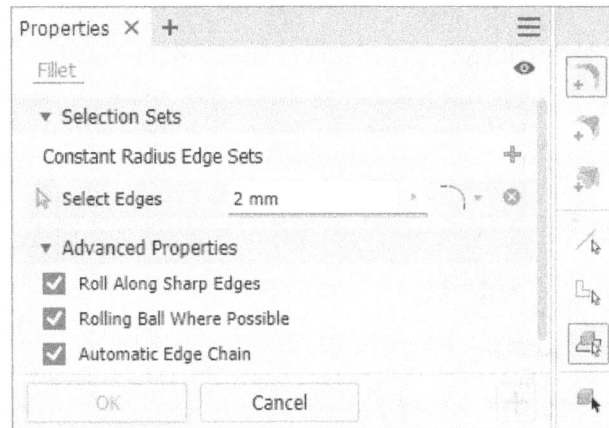

Figure-138. Fillet dialog box displayed

- Select the edge as shown in Figure-139 and specify the radius as **12.5**. Preview of the fillet will be displayed; refer to Figure-140.

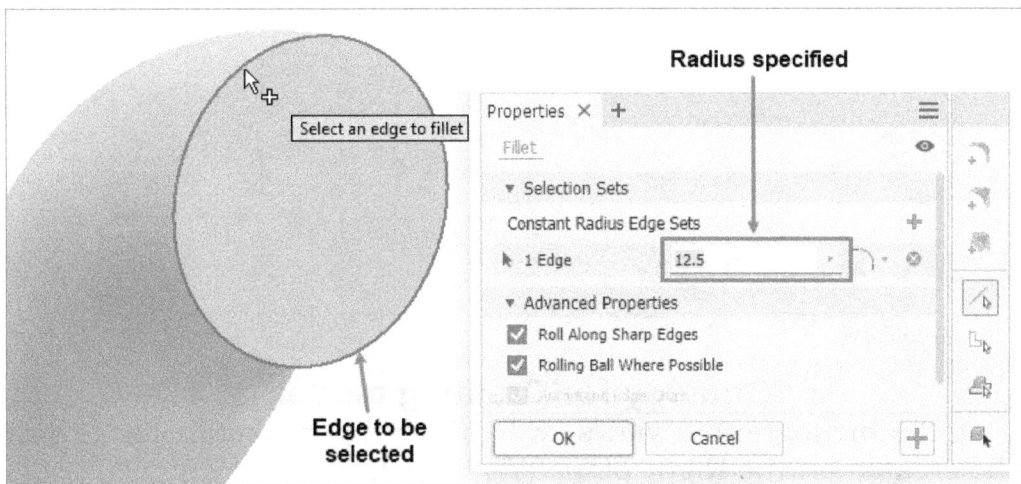

Figure-139. Edge selected for fillet

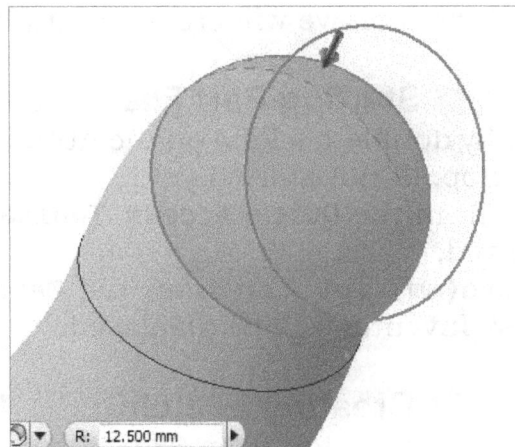

Figure-140. Preview of fillet

- Click on the **OK** button from the dialog box to create fillet.

PRACTICAL 2

Create the model using the drawings shown in Figure-141.

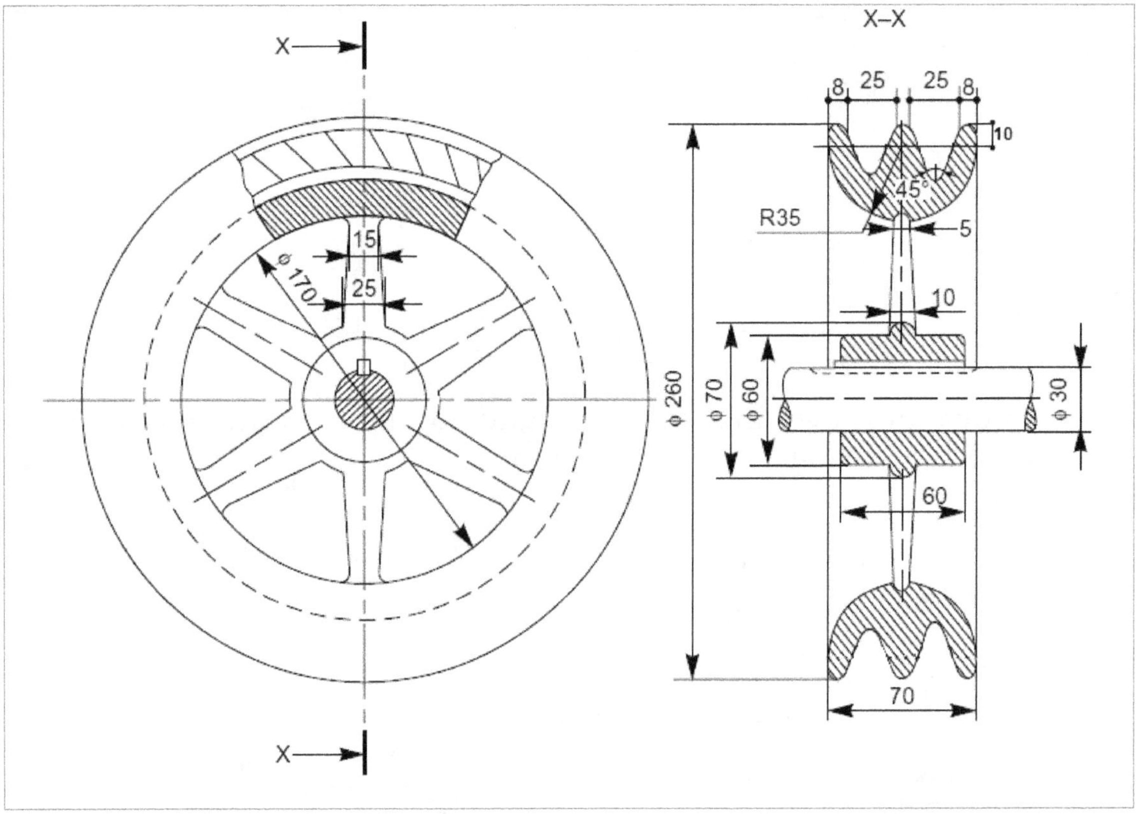

Figure-141. Rope Pulley

Strategy for Creating model

1. Looking at the orthographic views, we can find the sketch for revolve feature should be created on the Right plane (YZ Plane).
2. Using the **Revolve** tool, we will create the solid base of the model.
3. Using the **Loft** tool, we will create a spoke of the rim.
4. Using the **Circular Pattern** tool, we will create multiple instances of spokes.

Starting Part File

- Start Autodesk Inventor by double-clicking on the Autodesk Inventor Professional 2024 icon from the desktop. (If not started yet.)
- Click on the **New** button from the **Quick Access Toolbar**. The **Create New File** dialog box will be displayed.
- Double-click on **Standard(mm).ipt** icon from the **Metric** templates. The Part environment of Autodesk Inventor will be displayed.

Creating Sketch

- Click on the **Start 2D Sketch** button from **Sketch** panel in the **3D Model** tab of the **Ribbon**. You will be asked to select a sketching plane.
- Select the **YZ Plane** (Right Plane) as sketching plane. The sketching environment will become active.
- Create the sketch as shown in Figure-142.

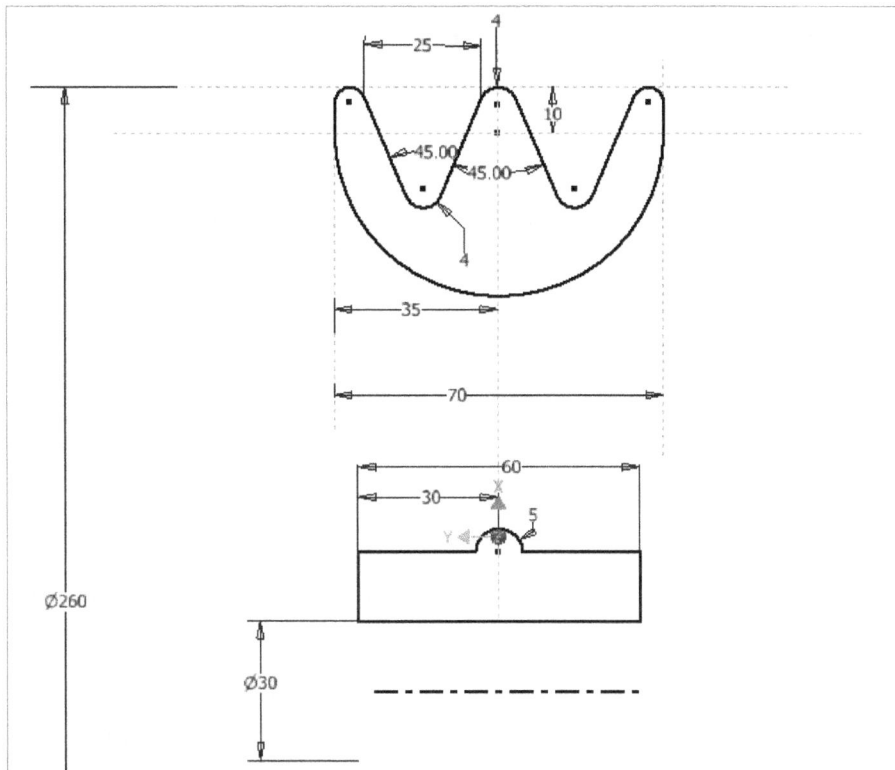

Figure-142. Sketch for practical 2

- Click on the **Finish Sketch** button from the **Ribbon**.

Creating Revolve Feature

- Click on the **Revolve** tool from **Create** panel in the **3D Model** tab of the **Ribbon**. You will be asked to select the sections to be revolved; refer to Figure-143.

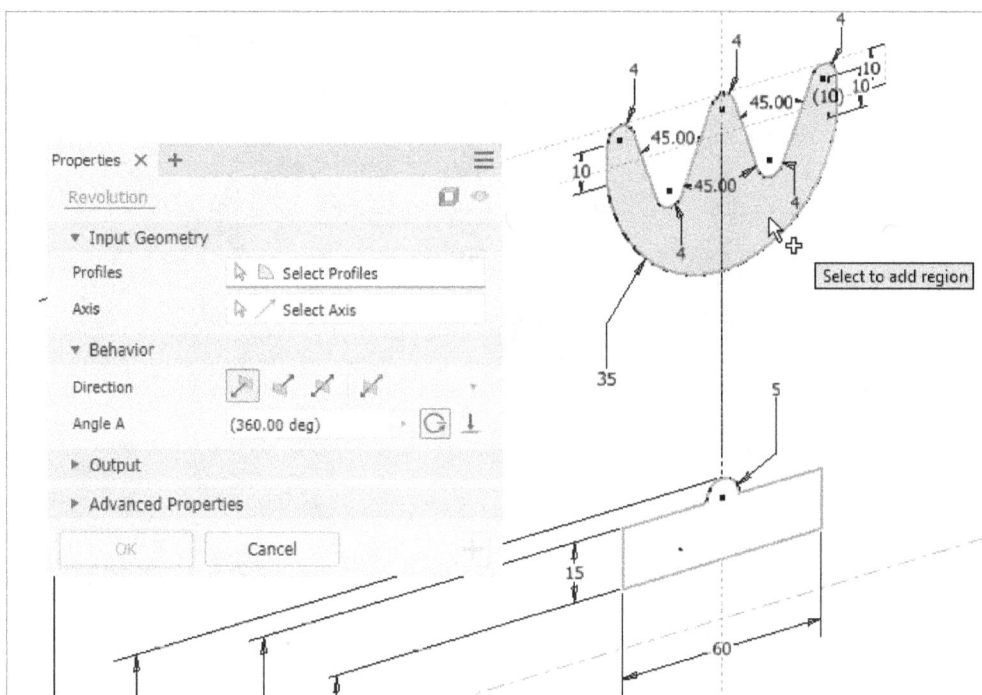

Figure-143. Selecting sections to revolve

- Select the sections as shown in Figure-144 and click on the **Select Axis** button from the **Revolve** dialog box.

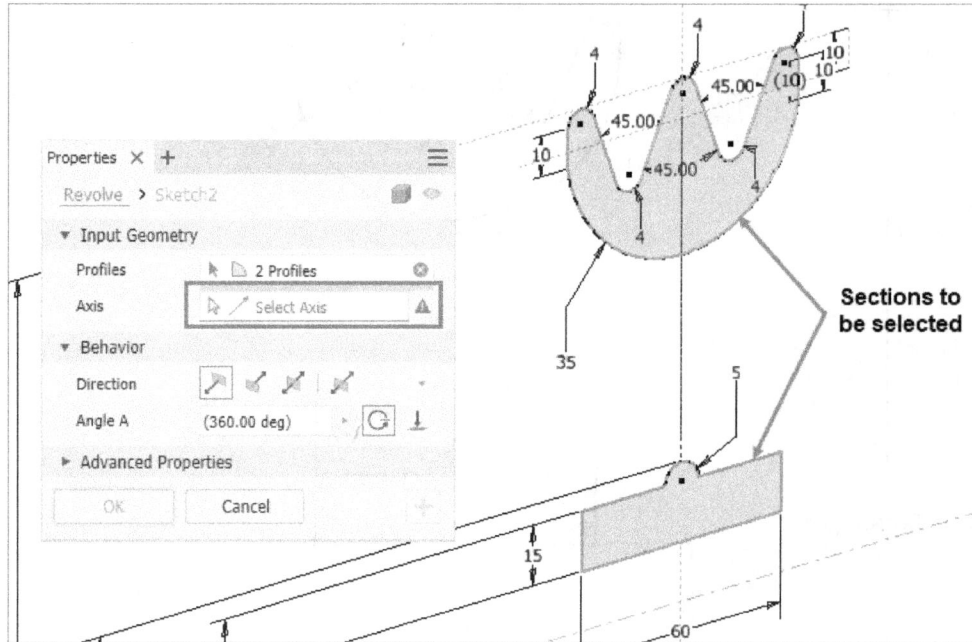

Figure-144. Sections selected

- Select the center line created in the sketch; refer to Figure-145. Preview of the revolve feature will be displayed; refer to Figure-146.

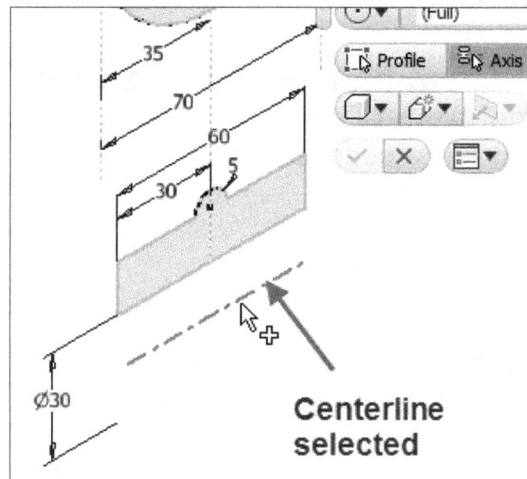

Figure-145. Centerline to select

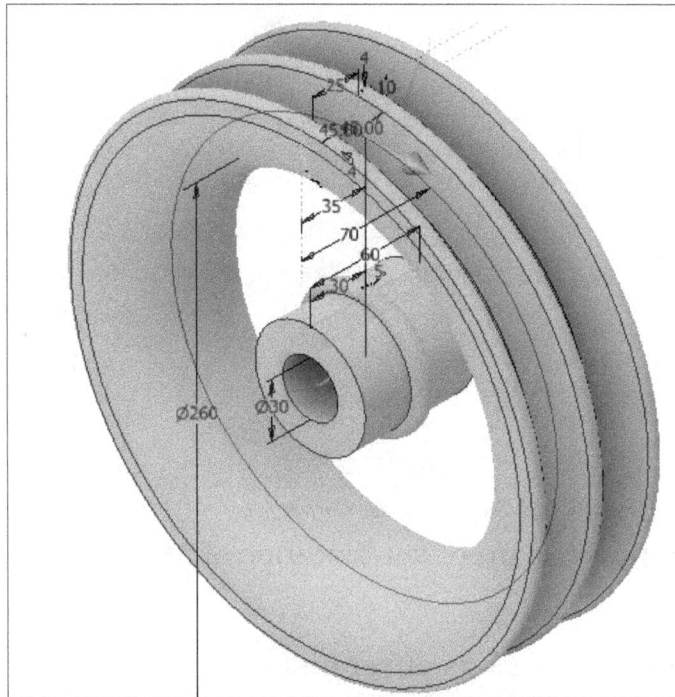

Figure-146. Preview of revolve feature

- Click on the **OK** button from the dialog box to create the feature.

Creating Planes for Sketches of Loft Feature

- Click on the **Tangent to Surface and Parallel to Plane** tool from **Plane** drop-down in the **Work Features** panel of the **Ribbon**. You will be asked to select the surface or plane.
- Select the **XZ** Plane from the **Model Browser** bar in the left of application window; refer to Figure-147. You will be asked to select the surface to be tangent.

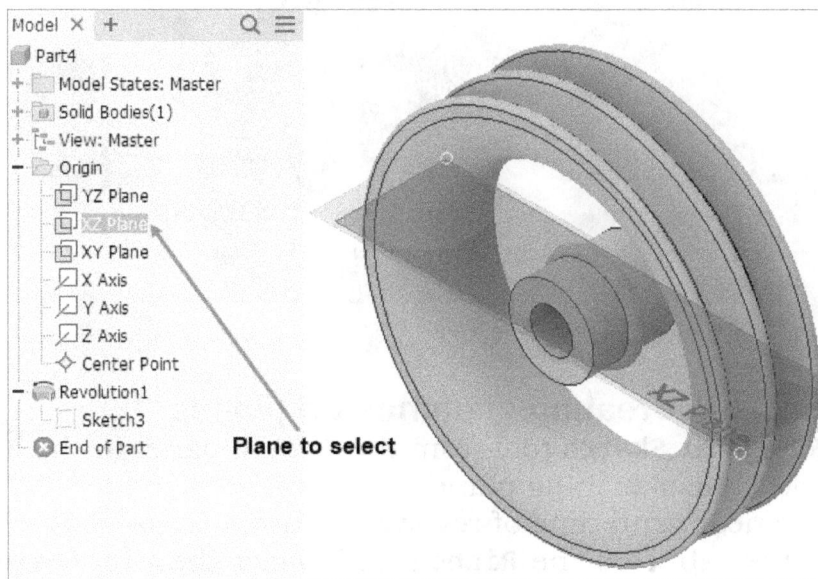

Figure-147. Selecting plane

- Select the surface as shown in Figure-148. The plane will be created.

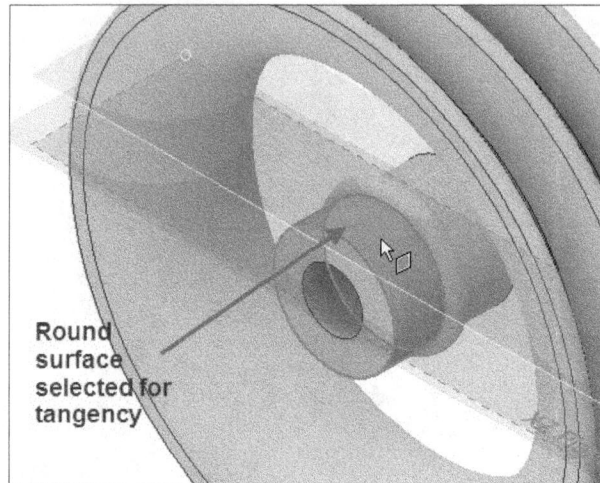

Figure-148. Surface selected for tangency

- Similarly, create a plane at an offset distance of **55** from the newly created plane; refer to Figure-149.

(If you have question like why **55** as offset distance then check the distance between these surfaces in the sketch created for revolve feature; refer to Figure-149.)

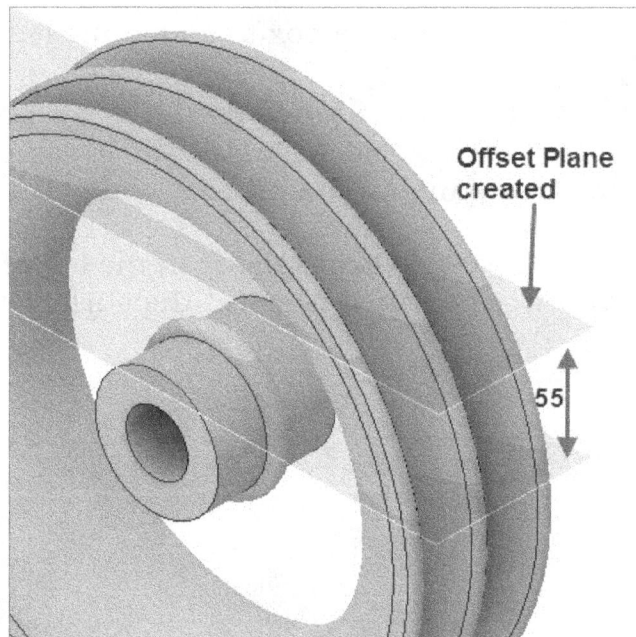

Figure-149. Offset plane created

Creating sketches on planes

- Click on the **Start 2D Sketch** tool from the **Sketch** panel in the **Ribbon**. You will be asked to select the sketching plane.
- Select the plane near centerline of revolve feature.
- Click on the **View** tab from the **Ribbon** and select the **Wireframe with Hidden Edges** option from the **Visual Style** drop-down in the **Appearance** panel of the **Ribbon**; refer to Figure-150.

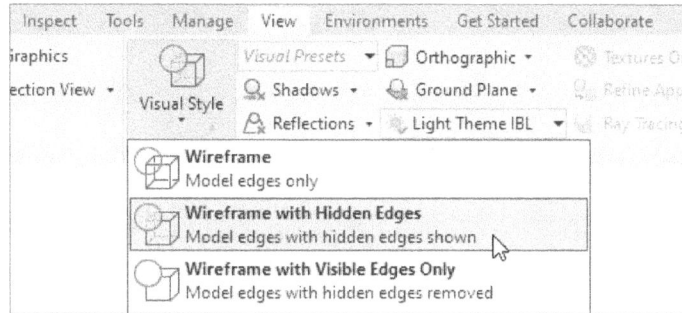

Figure-150. Visual style selected

- Create a rectangle at the center of the sketch by using the parameters as given in Figure-151.

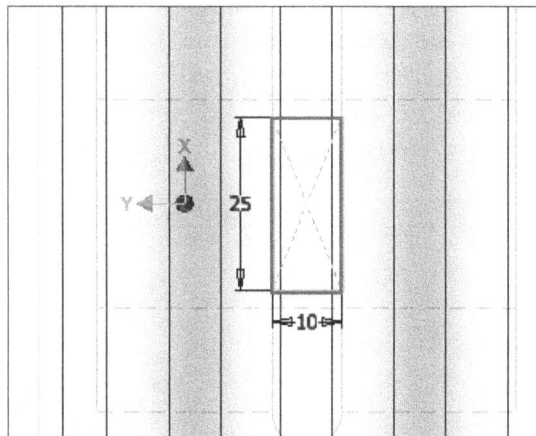

Figure-151. Rectangle created

- Click on the **Finish Sketch** button from the **Ribbon**.
- Similarly, create rectangle of size **15x5** on other plane as shown in Figure-152.

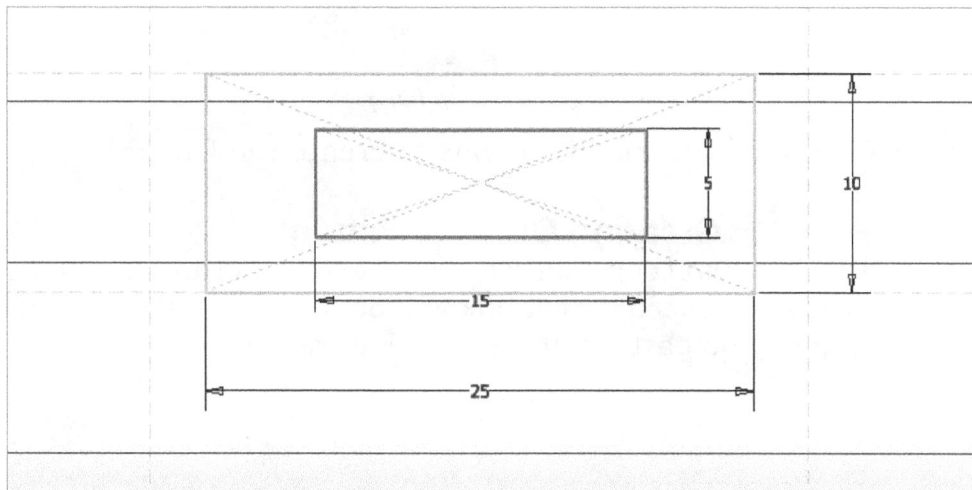

Figure-152. Second rectangle created

- Now, change the visual style to **Shaded with Edges** from the **Visual Style** drop-down in the **View** tab of **Ribbon**; refer to Figure-153.

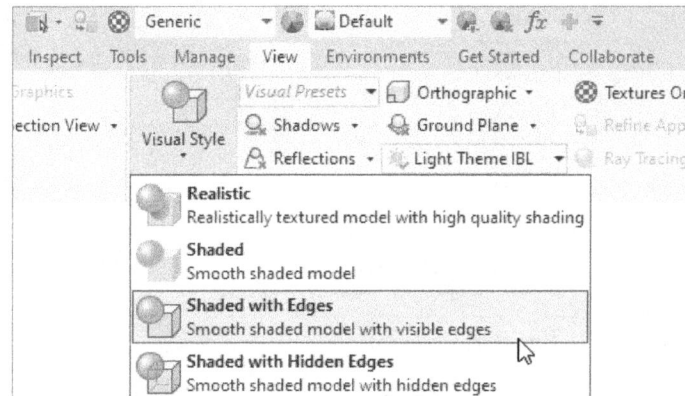

Figure-153. Selecting visual style

Creating the Loft Feature

- Click on the **Loft** tool from **Create** panel in the **3D Model** tab of the **Ribbon**. The **Loft** dialog box will be displayed and you will be asked to select the sketches.
- One by one select the two sketches created recently. Preview of the Loft feature will be displayed; refer to Figure-154.

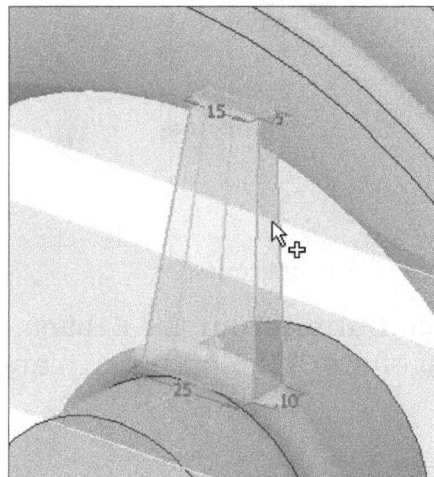

Figure-154. Preview of loft feature

- Click on the **OK** button from the dialog box to create the feature.

Creating Circular Pattern

- Click on the **Circular Pattern** tool from **Pattern** panel in the **3D Model** tab of **Ribbon**. The **Circular Pattern** dialog box will be displayed and you will be asked to select the feature to be patterned; refer to Figure-155.

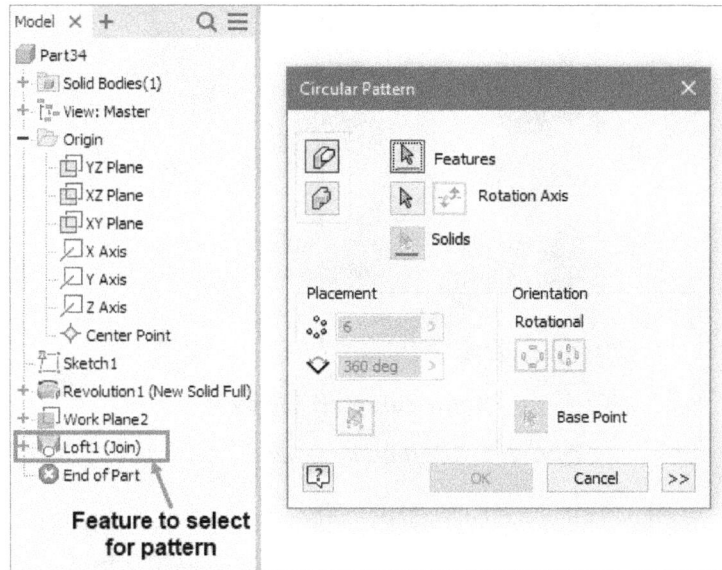

Figure-155. Circular pattern dialog box

- Select the loft feature created earlier and click on the **Rotation Axis** selection button from the dialog box; refer to Figure-156. You will be asked to select the rotation axis.

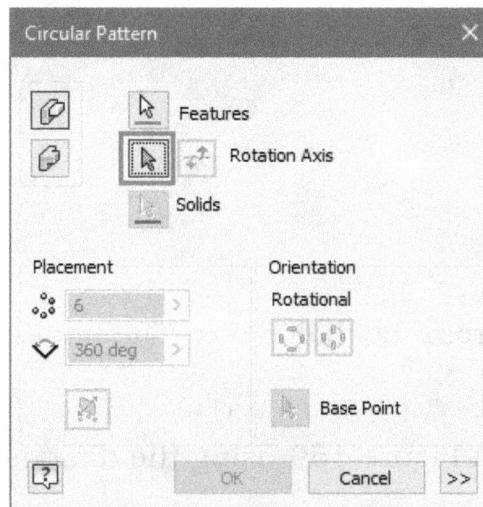

Figure-156. Rotation Axis button

- Select the circular face of the model as shown in Figure-157. Preview of the pattern will be displayed; refer to Figure-158.

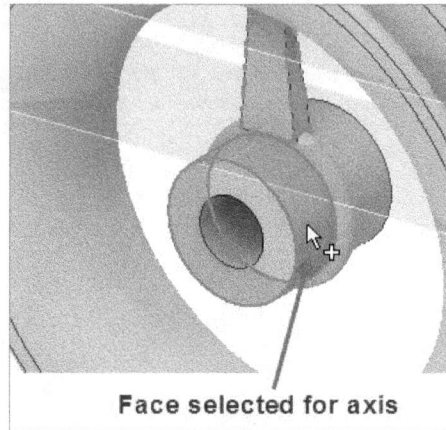

Figure-157. Selecting face for axis

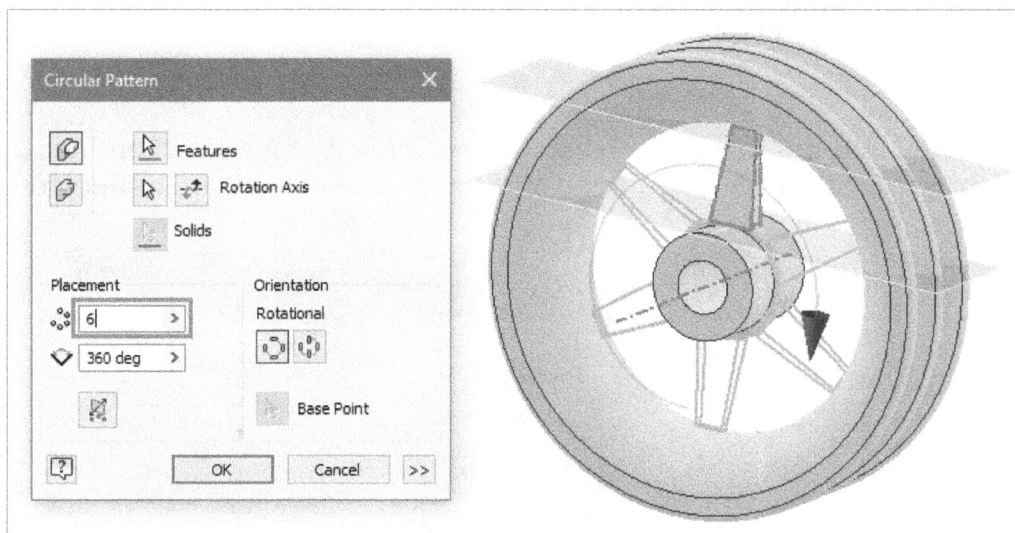

Figure-158. Preview of pattern

* Click on the **OK** button from the dialog box. The final model will be displayed.

PRACTICAL 3

Create a model as shown in Figure-159 using the dimensions given in Figure-160.

Figure-159. Model for Practical 3

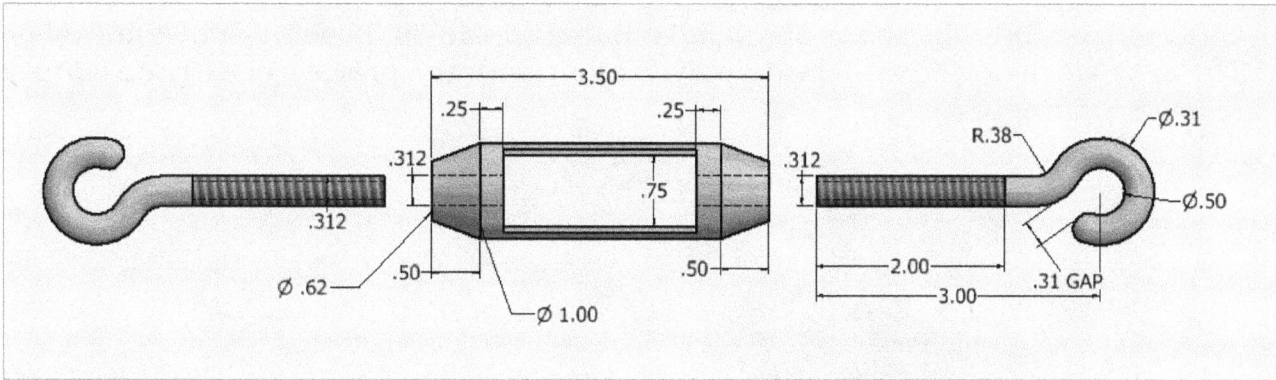

Figure-160. Practical 3 drawing view

Strategy for Creating model

1. Looking at the isometric view and orthographic views, we can find the sketch for extrude feature should be created on the Front plane (YZ Plane).
2. Using the **Extrude** and **Loft** tool, we will create the solid base of the model.
3. Using the **Plane** tools, we will create offset planes at a distance of .50mm to the vertical of loft feature created.
4. Using the **Thread** and **Sweep** tool, we will create the pulley to be joined to the base feature.

Starting Part File

- Start Autodesk Inventor by double-clicking on the Autodesk Inventor Professional 2024 icon from the desktop. (If not started yet.)
- Click on the **New** button from the **Quick Access Toolbar**. The **Create New File** dialog box will be displayed.
- Double-click on **Standard(mm).ipt** icon from the **Metric** templates. The Part environment of Autodesk Inventor will be displayed.

Creating Sketch

- Click on the **Start 2D Sketch** button from **Sketch** panel in the **3D Model** tab of the **Ribbon**. You will be asked to select a sketching plane.
- Select the **XY Plane** (Front Plane) as sketching plane. The sketching environment will become active.
- Create the sketch as shown in Figure-161.

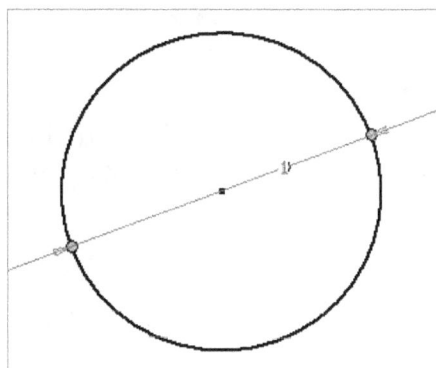

Figure-161. Sketch for Practical 3

- Click on the **Finish Sketch** button from **Exit** panel in the **Ribbon**.

Creating Extrude Feature

- Click on the **Extrude** tool from **Create** panel in the **3D Model** tab of the **Ribbon**. The **Extrude** dialog box will be displayed along with the preview of extrude feature; refer to Figure-162.

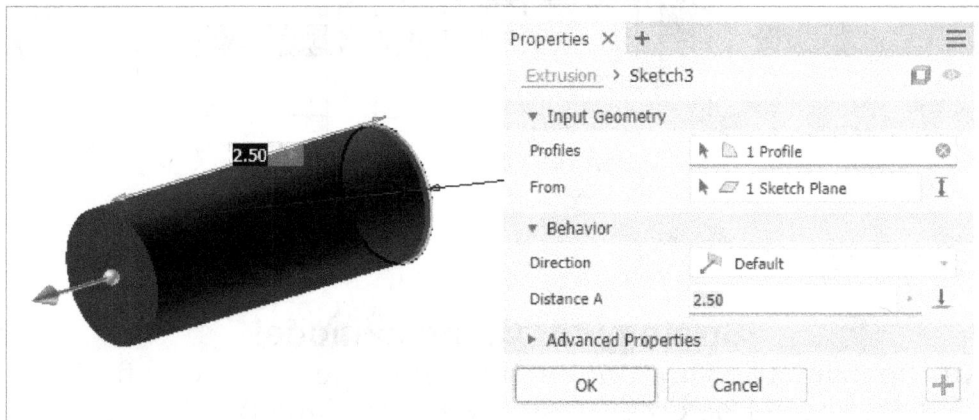

Figure-162. Preview of extrude

- Enter the extrusion value as **2.50** in the **Distance A** edit box in the dialog box.
- Click on the **OK** button from the dialog box to create the feature.

Creating planes for sketches of Loft feature

- Click on the **Offset from Plane** tool from **Plane** drop-down in the **Work Features** panel of the **Ribbon**. You will be asked to select the surface or plane.
- Select the surface as shown in Figure-163. The plane will be created.

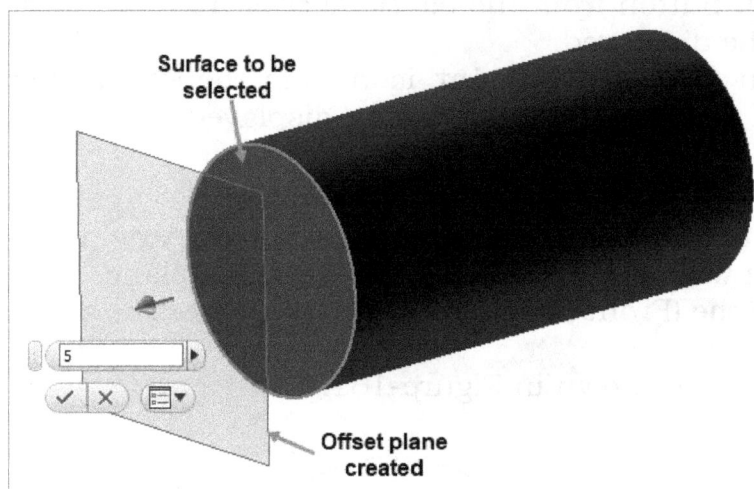

Figure-163. Offset plane creation

- Enter the Offset distance value as **.50** in the edit box displayed.
- Similarly, create another offset plane on the opposite surface of this side; refer to Figure-164.

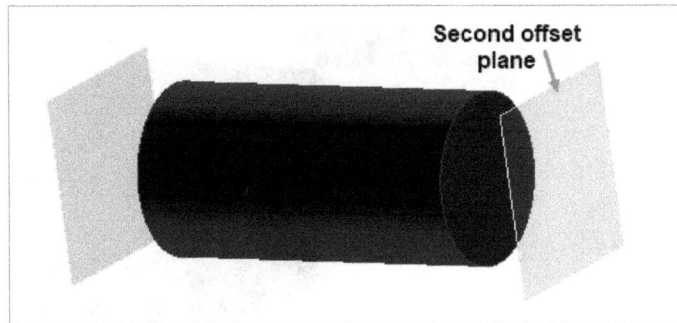

Figure-164. Second offset plane

- Click on the **OK** button to exit the tool.

Creating sketches on planes

- Click on the **Start 2D Sketch** tool from **Sketch** panel in the **3D Model** tab of the **Ribbon**. You will be asked to select a sketching plane.
- Select the newly created planes and create sketch on both the planes as shown in Figure-165.

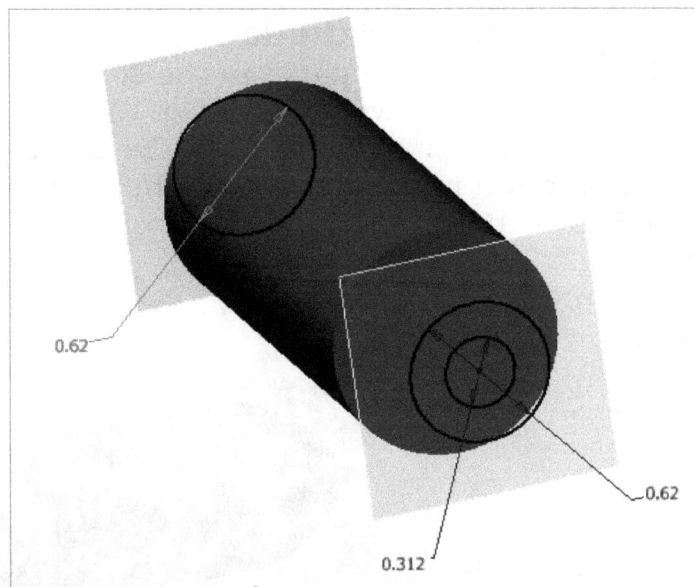

Figure-165. Sketches created for creating loft and hole

- Click on the **Finish Sketch** button from **Exit** panel in the **Ribbon**.

Creating Loft feature

- Click on the **Loft** tool from **Create** panel in the **3D Model** tab of the **Ribbon**. The **Loft** dialog box will be displayed and you will be asked to select the sketches.
- Select the newly created sketch of diameter value **0.62** and the surface on the model upto which the loft feature will be created. Preview of the loft feature will be displayed; refer to Figure-166.

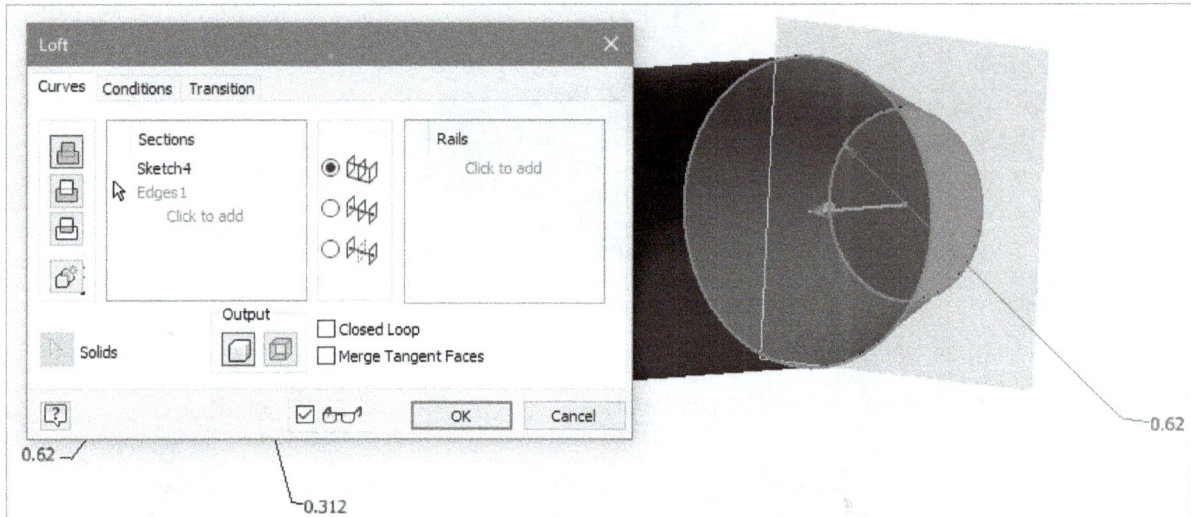

Figure-166. Preview of loft feature

- Click on the **OK** button from the dialog box to create the feature.
- Similarly, create the loft feature on the other side of the model.

Creating Extrude cut

- Click on the **Extrude** tool from **Create** panel in the **3D Model** tab of the **Ribbon**. The **Extrude** dialog box will be displayed asking you to select the profile.
- Select the sketch having diameter of **0.312** from the model and select the **Cut** option from **Boolean** drop-down in the dialog box. The preview of extrude cut will be displayed; refer to Figure-167.

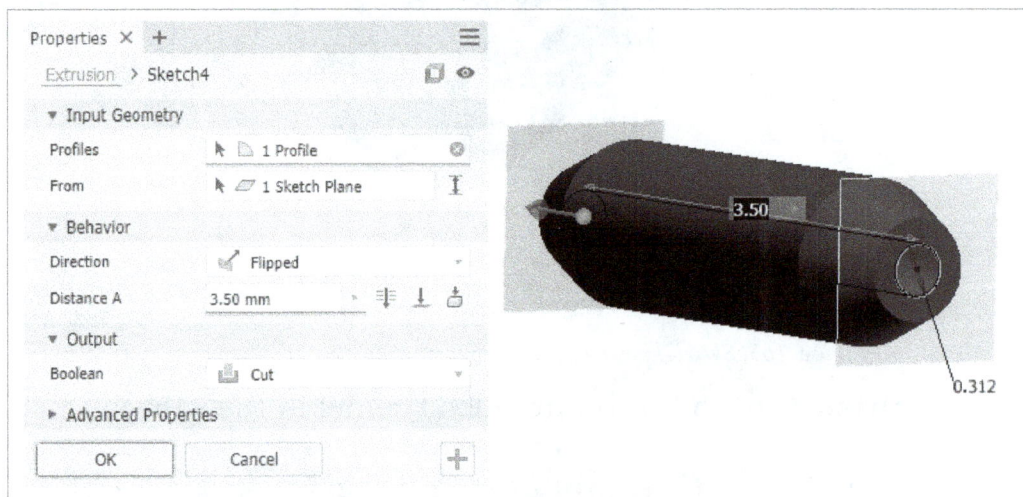

Figure-167. Preview of extrude cut

- Click on the **OK** button from the dialog box to create the feature.

Creating Offset Planes

- Click on the **Offset from Plane** tool from **Plane** drop down in the **Work Features** panel of the **Ribbon**. You will be asked to select the surface or plane.
- Select the surface from the model as shown in Figure-168. The preview of offset plane will be displayed.

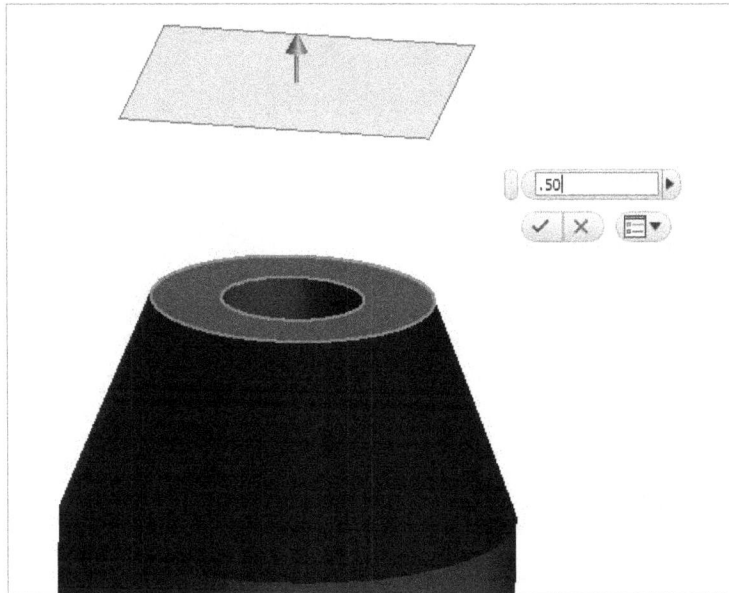

Figure-168. Offset plane created

- Enter the Offset distance value as **.50** in the edit box and click on the **OK** button to create the plane.

Creating sketch on plane

- Click on the **Start 2D Sketch** button from **Sketch** panel in the **3D Model** tab of the **Ribbon**. You will be asked to select the sketching plane.
- Select the newly created plane and create a sketch as shown in Figure-169.

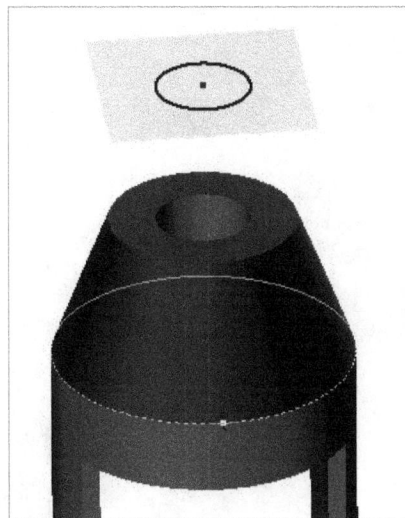

Figure-169. Sketch created on plane

- Click on the **Finish Sketch** button from **Exit** panel in the **Ribbon**.

Creating Extrude and Thread feature

- Click on the **Extrude** tool from **Create** panel in the **3D Model** tab of the **Ribbon**. The **Extrude** dialog box will be displayed along with the preview of extrusion; refer to Figure-170.

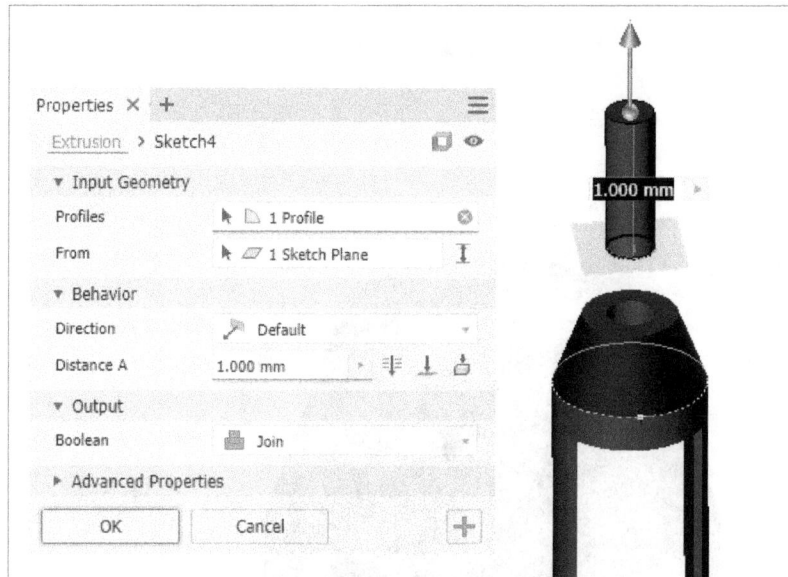

Figure-170. Preview of extrusion

- Enter the value of extrusion as **2** in the **Distance A** edit box of the dialog box.
- Click on the **OK** button from the dialog box to create the feature.
- Click on the **Thread** tool from **Modify** panel in the **3D Model** tab of the **Ribbon**. The **Thread** dialog box will be displayed and you will be asked to select the cylindrical or conical face to thread.
- Select the face to create the thread as shown in Figure-171. The preview of thread will be displayed.

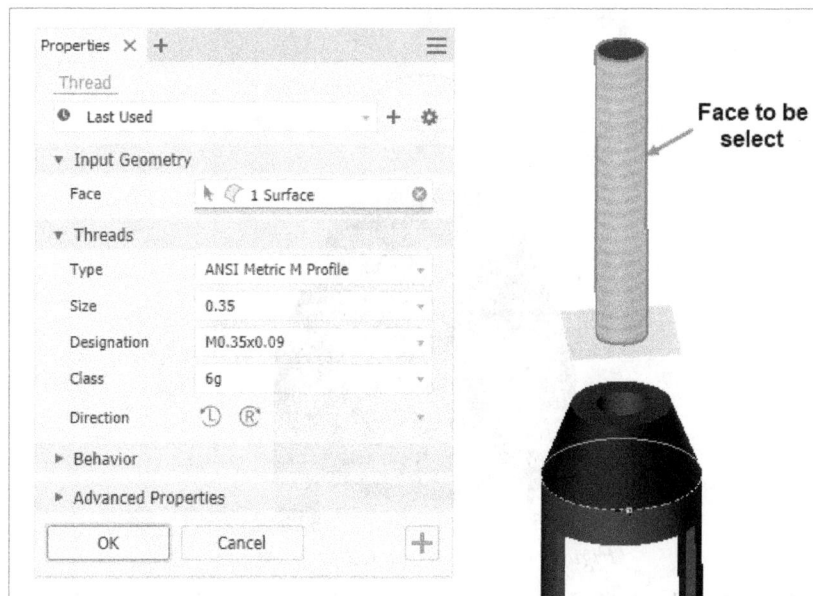

Figure-171. Preview of thread

- Click on the **OK** button from the dialog box to create the feature.

Creating sketch and plane for sweep feature

- Click on the **Start 2D Sketch** button from **Sketch** panel in the **3D Model** tab of the **Ribbon**. You will be asked to select the sketching plane.
- Click on the **XZ plane** as sketching plane. The sketching environment will become active.

- Create the sketch as shown in Figure-172.

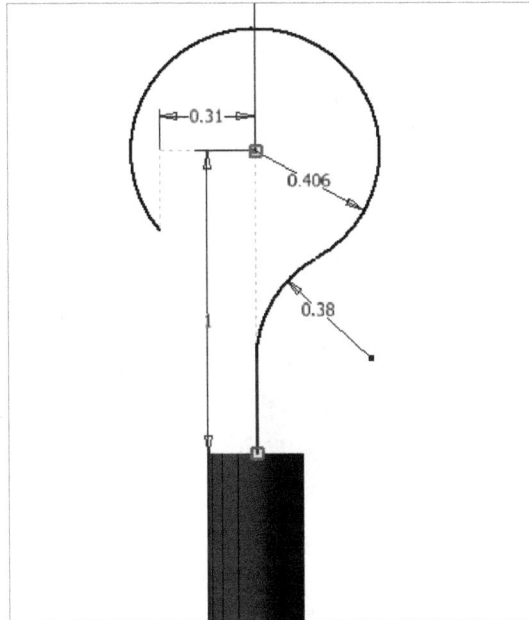

Figure-172. Creating sketch

- Click on the **Finish Sketch** button from **Exit** panel in the **Ribbon**.
- Click on the **Plane** tool from **Plane** drop-down in the **Work Features** panel from the **3D Model** tab of the **Ribbon**. You will be asked to select the geometry.
- Select the point on the newly created sketch as shown in Figure-173. The plane will be created.

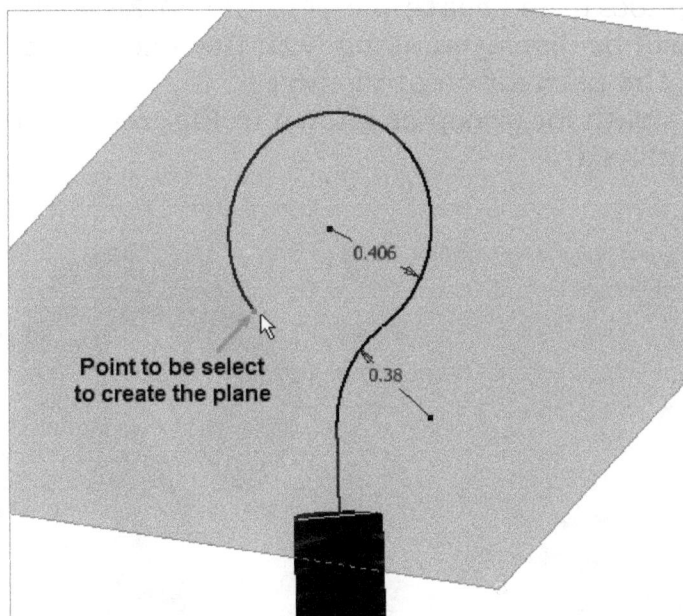

Figure-173. Plane created

- Click on the **Start 2D Sketch** button from **Sketch** panel in the **3D Model** tab of the **Ribbon**. You will be asked to select the plane.
- Select the newly created plane as sketching plane. The sketching environment will become active.
- Create the sketch as shown in Figure-174.

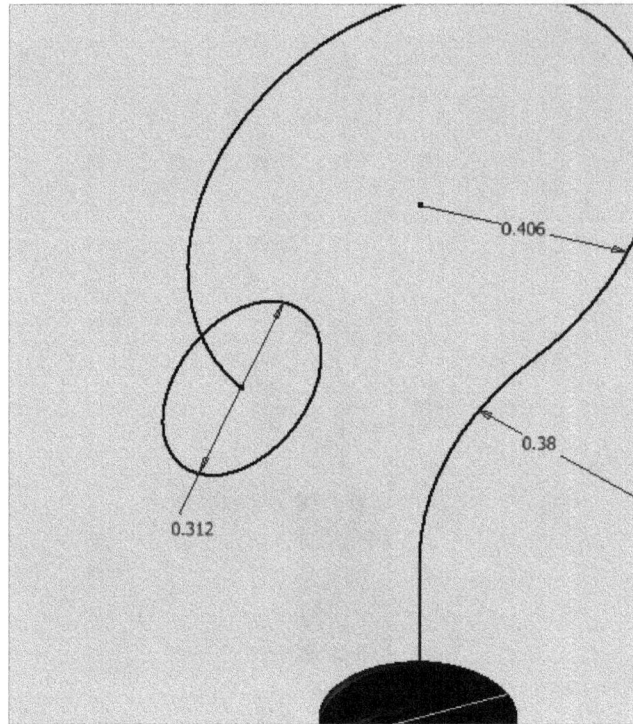

Figure-174. Sketch created

*   Click on the **Finish Sketch** button from **Exit** panel in the **Ribbon**.

Creating Sweep and Fillet feature

*   Click on the **Sweep** tool from **Create** panel in the **3D Model** tab of the **Ribbon**. The **Sweep** dialog box will be displayed along with the selection of profile and you will be asked to select the path for creating sweep.
*   Select the curve as path for sweep as shown in Figure-175. The preview of sweep feature will be displayed.

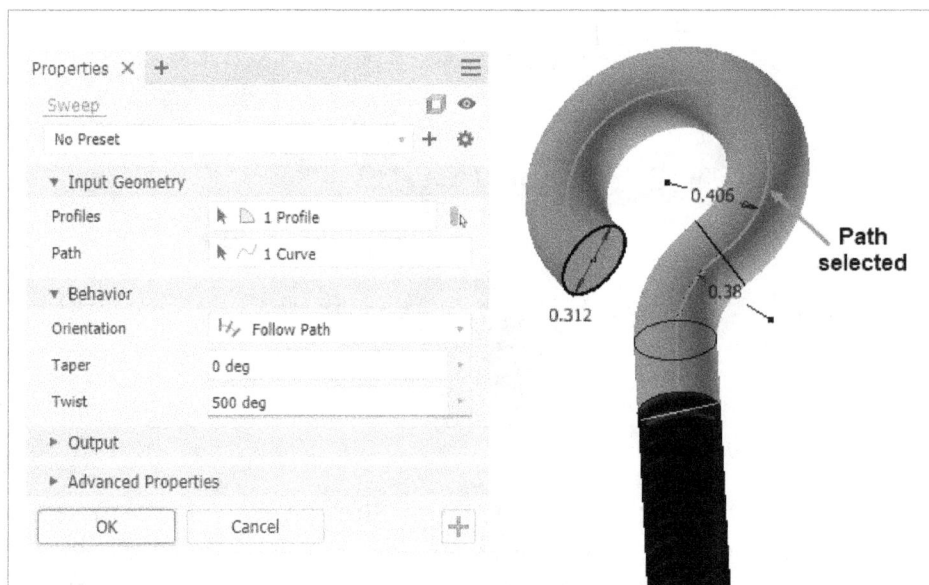

Figure-175. Sweep feature created

*   Click on the **OK** button from the dialog box to create the feature.

- Click on the **Fillet** tool from **Modify** panel in the **3D Model** tab of the **Ribbon**. The **Fillet** dialog box will be displayed and you will be asked to select the edge to be fillet.
- Select the edge as shown in Figure-176. The preview of fillet will be displayed.

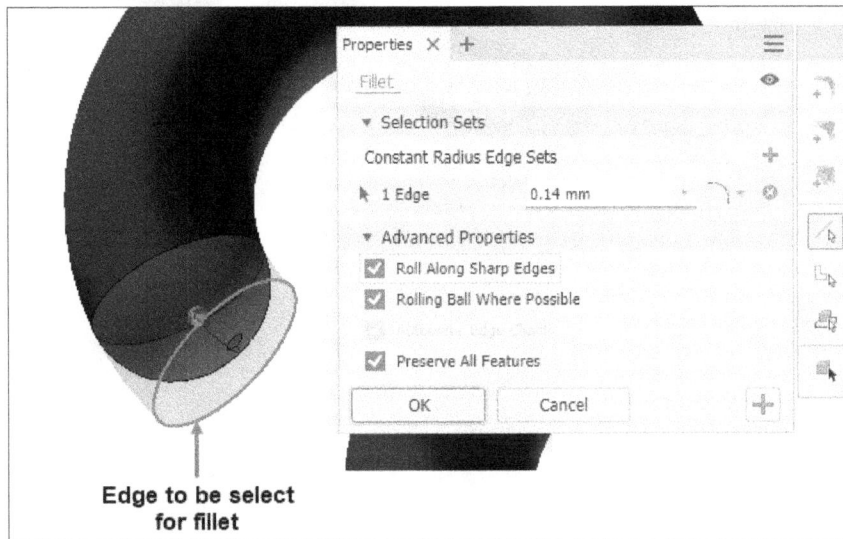

Figure-176. Preview of fillet

- Enter the radius value as **0.14** in the **Radius** edit box of the dialog box.
- Click on the **OK** button from the dialog box to create the feature.

Similarly, create the solid part as shown in Figure-177 on the opposite side of the model. The final model will be displayed; refer to Figure-178.

Figure-177. Solid part

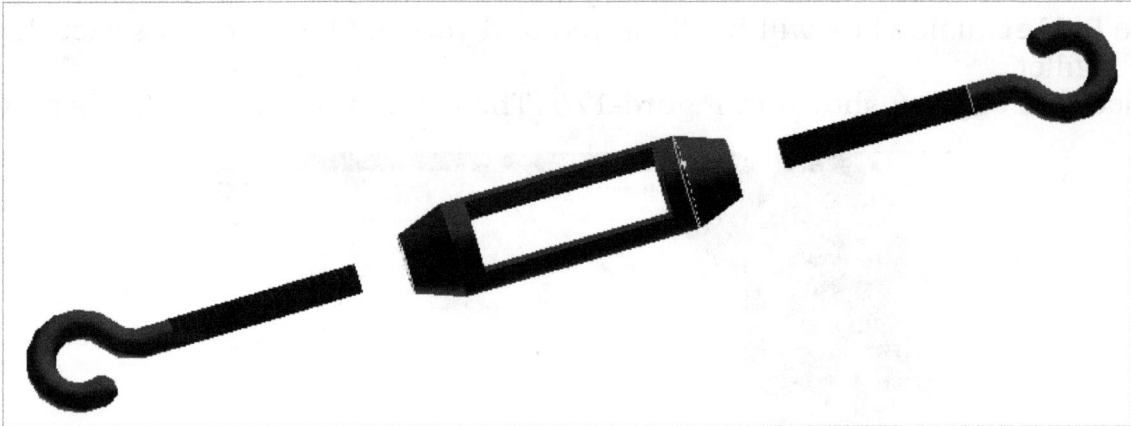

Figure-178. Final model

SELF ASSESSMENT

Q1. Which of the following options should you select to control the outer shape of the sweep feature?

a) Fixed
b) Follow Path
c) Orientation
d) Guide

Q2. In **Loft** feature, which option needs the condition that the curve to be selected for loft must be intersect with all the sections one time?

a) Rail
b) Center Line
c) Area Loft
d) None of the Above

Q3. Which of the following options can be used to create coil by specifying pitch and number of revolution?

a) Revolution and Height
b) Pitch and Height
c) Spiral
d) None of the Above

Q4. By selecting which tool, you can imprint a profile on the selected face?

a) Decal
b) Rib
c) Emboss
d) All of the above

Q5. Which of the following types of hole allow you to choose standard holes from the library?

a) Simple Hole
b) Clearance Hole
c) Tapped Hole
d) Taper Tapped Hole

Q6. Which of the following tools is used to divide the selected entity with the help of trimming reference?

a) Shell
b) Draft
c) Offset
d) Split

Q7. The **Chamfer** tool is used to chisel the sharp points of the model. (True/False)

Q8. Setbacks are created at the corners where three fillets coincide at a point. (True/False)

Q9. The .... tool is used to apply radius at sharp edges.

Q10. The .... tool is used to flattened the faces that cannot be flattened with the Unfold or sheetmetal flat pattern command.

Chapter 6

Advanced Modeling Tools and Practical

Topics Covered

The major topics covered in this chapter are:

- *Direct Editing*
- *Bend Tool*
- *Practical 1 to 5*
- *Practice*

INTRODUCTION

In this chapter, we will discuss the remaining tools in Advanced Solid Modeling. We will also practice on the tools and techniques discussed in previous chapters.

DIRECT TOOL

The **Direct** tool is used to edit imported parametric features of the model by using simple drag and drop operations. The tool is available in the **Modify** panel of the **3D Model** tab in the **Ribbon**. The procedure to use this tool is given next.

*   Make sure you have an imported model or solid model already created in the viewport.
*   Click on the **Direct** tool from **Modify** panel in the **3D Model** tab of the **Ribbon**. The **Direct Editing** toolbar will be displayed; refer to Figure-1.

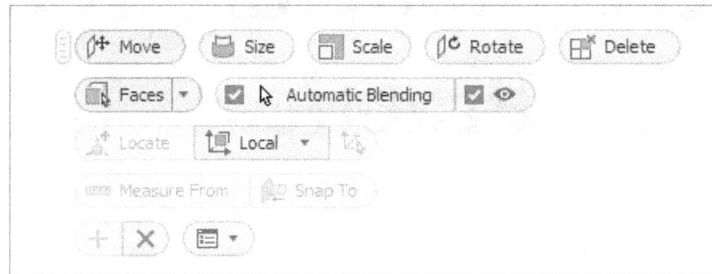

Figure-1. Direct Editing toolbar

There are five options in toolbar to perform direct editing; Move, Size, Scale, Rotate, and Delete. We will discuss each of the option one by one.

Moving Faces by Direct Editing

*   Click on the **Move** button from the toolbar. The toolbar will be displayed as shown in Figure-2.

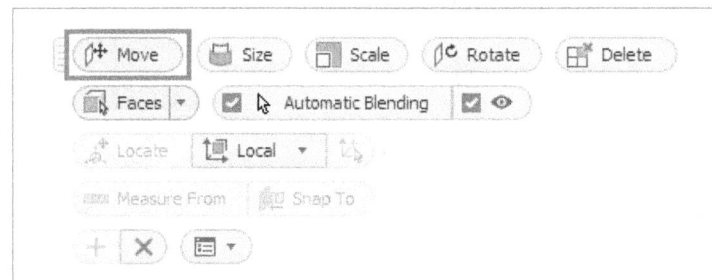

Figure-2. Move by direct editing

*   Click on the face that you want to move; refer to Figure-3.
*   Select the **Preview** check box to display the preview of face while modifying by direct editing.

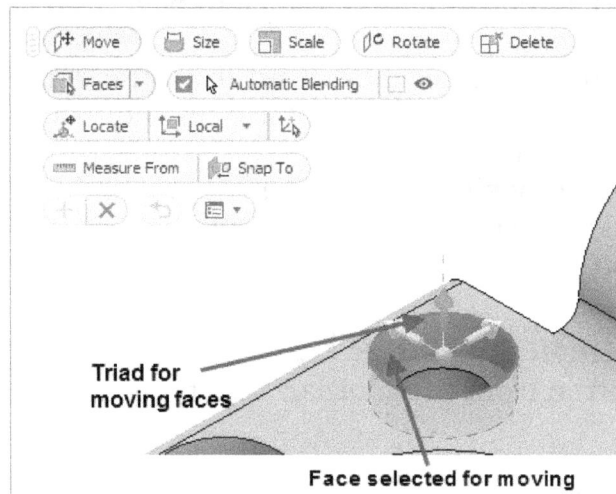

Figure-3. Selecting face for moving

- If you want to select multiple faces then you can do so by holding the **CTRL** key while selecting faces. If you want to remove a face from selection then you can do so by holding the **SHIFT** key while selecting the face.
- To change the location of triad, click on the **Locate** button from the toolbar and click at desired location to place the triad.
- By default, the triad is oriented as per the selected faces but if you want to orient the triad as per the World Coordinate system then you can do so by selecting the **World** option from the drop-down as shown in Figure-4.

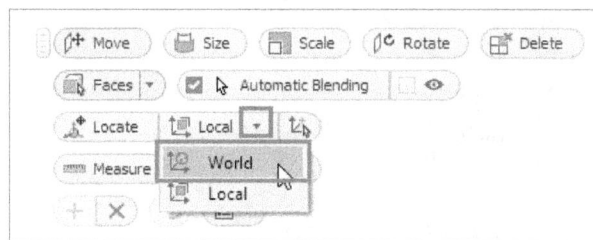

Figure-4. World orientation of triad

- Click on the arrow in desired direction and drag the arrow to move the selected face/faces; refer to Figure-5.

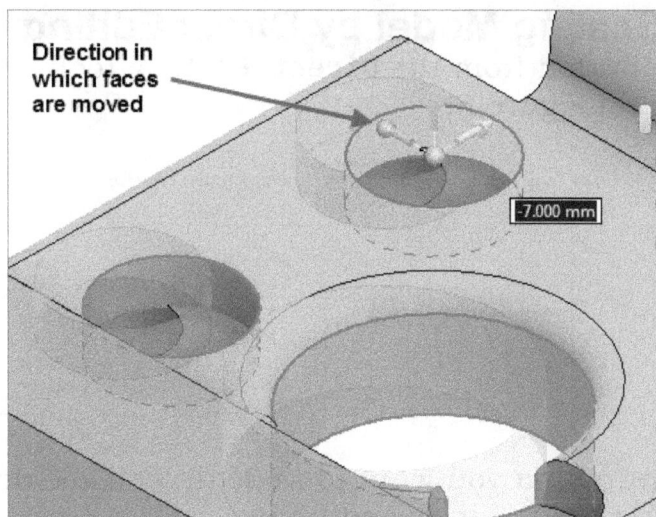

Figure-5. Direction arrow used to move faces

- You can also specify desired value of distance to be moved in the edit box displayed on dragging arrow.
- By default, the distance value is measured from the center of first selected face but you can select other entities by clicking on the **Measure From** button in the toolbar and selecting desired entity.
- You can select the target position for movement by clicking on the **Snap To** button in the toolbar and selecting the point to which you want to move the selected faces.
- Click on the **+** button from the toolbar to apply the change.

Changing Size by Direct Editing

- Click on the **Size** button from the toolbar. The toolbar will be displayed as shown in Figure-6.

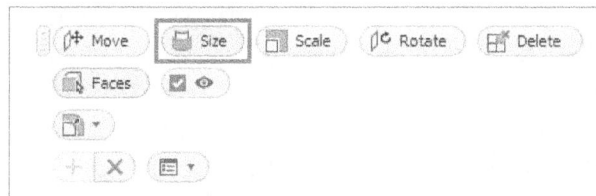

Figure-6. Size button in Direct Editing toolbar

- Select the face/faces for changing size. Triad to change the size will be displayed.
- Click on desired option from the dimension type drop-down; refer to Figure-7.

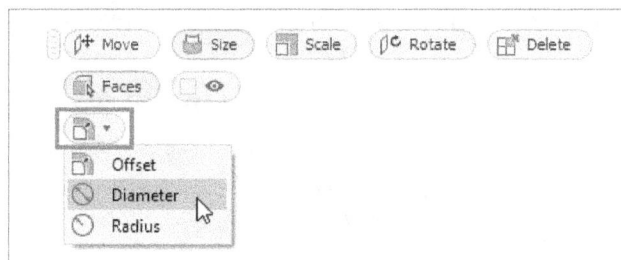

Figure-7. Options for changing size

- Specify desired value in the edit box or drag the handle in desired direction to change the value.
- Click on the **+** button to apply change.

Scaling Model by Direct Editing

- Click on the **Scale** button from the **Direct Editing** toolbar. The toolbar will be displayed as shown in Figure-8.

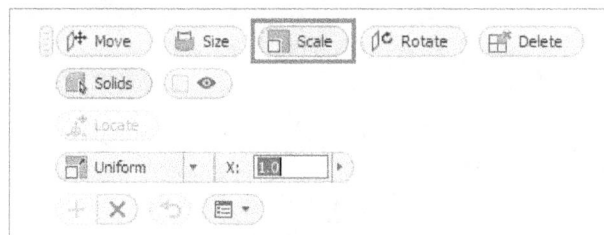

Figure-8. Scale button in Direct Editing toolbar

- Select the solid feature that you want to scale up or scale down.
- Specify desired value of scale in the edit box.
- Click on the **+** button from the toolbar to apply the changes.

Rotating Faces by Direct Editing

- Click on the **Rotate** button from the **Direct Editing** toolbar. The toolbar will be displayed as shown in Figure-9.

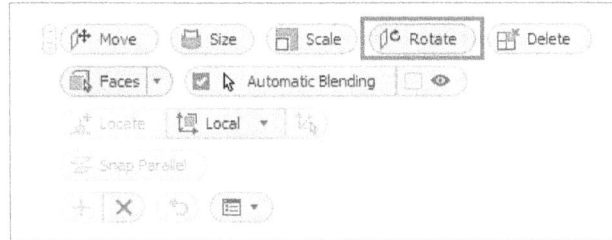

Figure-9. Rotate button in Direct Editing toolbar

- Select the face that you want to rotate. Drag handles for rotation will be displayed; refer to Figure-10.

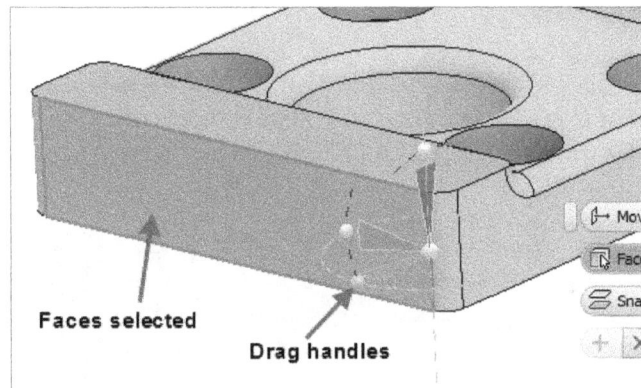

Figure-10. Drag handle for rotating face

- Click on the ball of drag handle and drag to rotate the face.
- Click on the **+** button to apply the change.

Deleting Faces by Direct Editing

- Click on the **Delete** button from the **Direct Editing** toolbar. The toolbar will be displayed as shown in Figure-11.

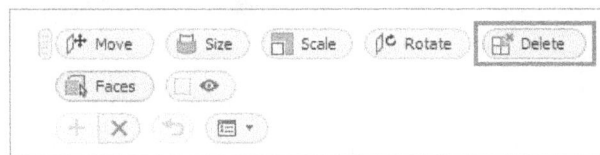

Figure-11. Delete button in Direct Editing toolbar

- Click on the face/faces to be deleted. Preview of the deleted faces will be displayed if feasible; refer to Figure-12.

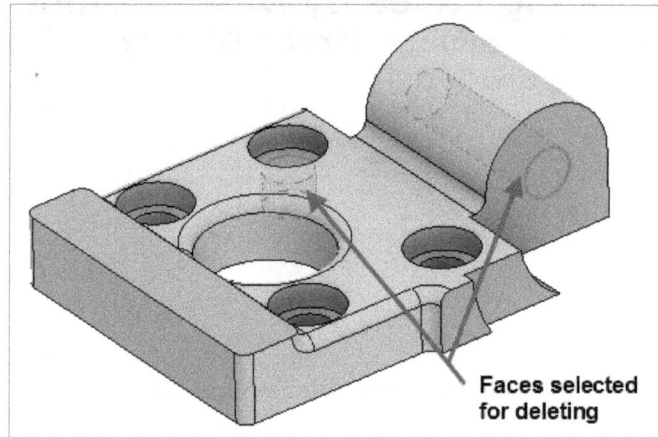

Figure-12. Faces selected for deleting

- Click on the **Apply** button to delete the faces.

BEND PART TOOL

The **Bend Part** tool is used to bend part by using sketched line. The tool is available in the expanded **Modify** panel of the **Ribbon**; refer to Figure-13. The procedure to use this tool is given next.

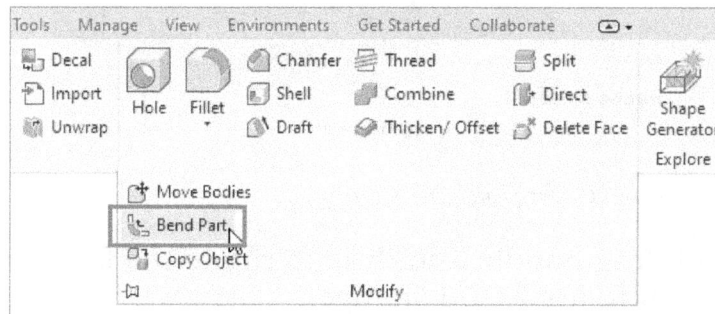

Figure-13. Bend Part tool

- Make sure a bend line is created by using the sketching tools.
- Click on the **Bend Part** tool from **Modify** panel in the **3D Model** tab of the **Ribbon**. The **Bend Part** dialog box will be displayed; refer to Figure-14.

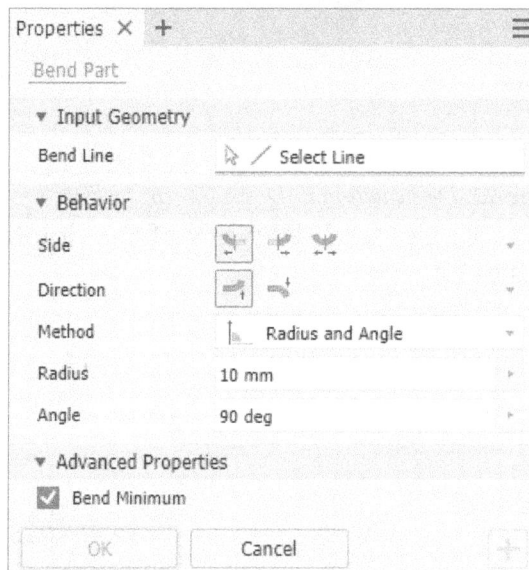

Figure-14. Bend Part dialog box

- Click on the bend line created earlier in the sketch.
- Select desired side on which part bends around the bend line from **Side** area in the **Behavior** rollout of the dialog box.
- Specify desired direction to bend the part from **Direction** area in the **Behavior** rollout.
- Select desired option from **Method** drop-down in the **Behavior** rollout. Select **Radius and Angle** option to create a feature using a radius and angle that you specify. Select **Radius and Arc Length** option to create a feature using a radius and arc length that you specify. Select **Arc Length and Angle** option to create a feature using values you specify for Arc Length and Angle.
- Specify desired radius and angle value in the **Radius** and **Angle** edit boxes, respectively.
- Select **Bend Minimum** check box from **Advanced Properties** rollout to specify which portions to bend.
- Specify desired parameters. Preview of the bend part will be displayed; refer to Figure-15.

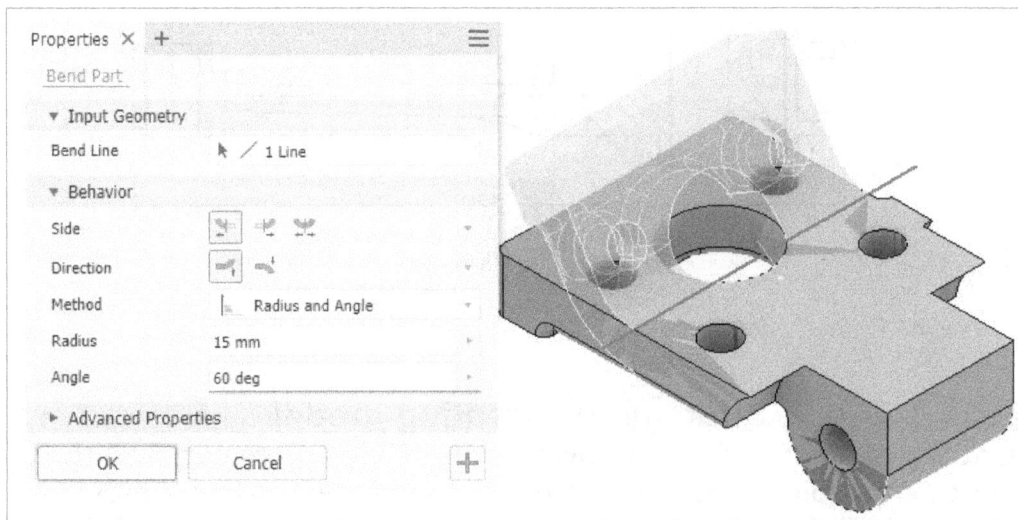

Figure-15. Preview of bended part

- Click on the **OK** button from the dialog box.

PRACTICAL 1

Create the model as shown in Figure-16. The dimensions are given in Figure-17.

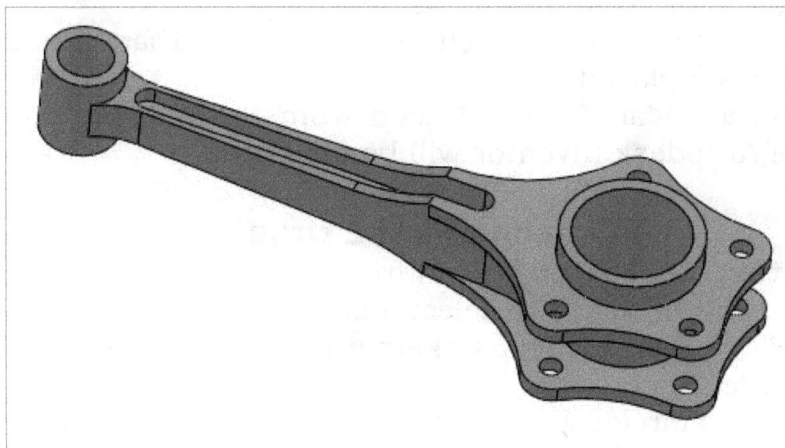

Figure-16. Practical 1

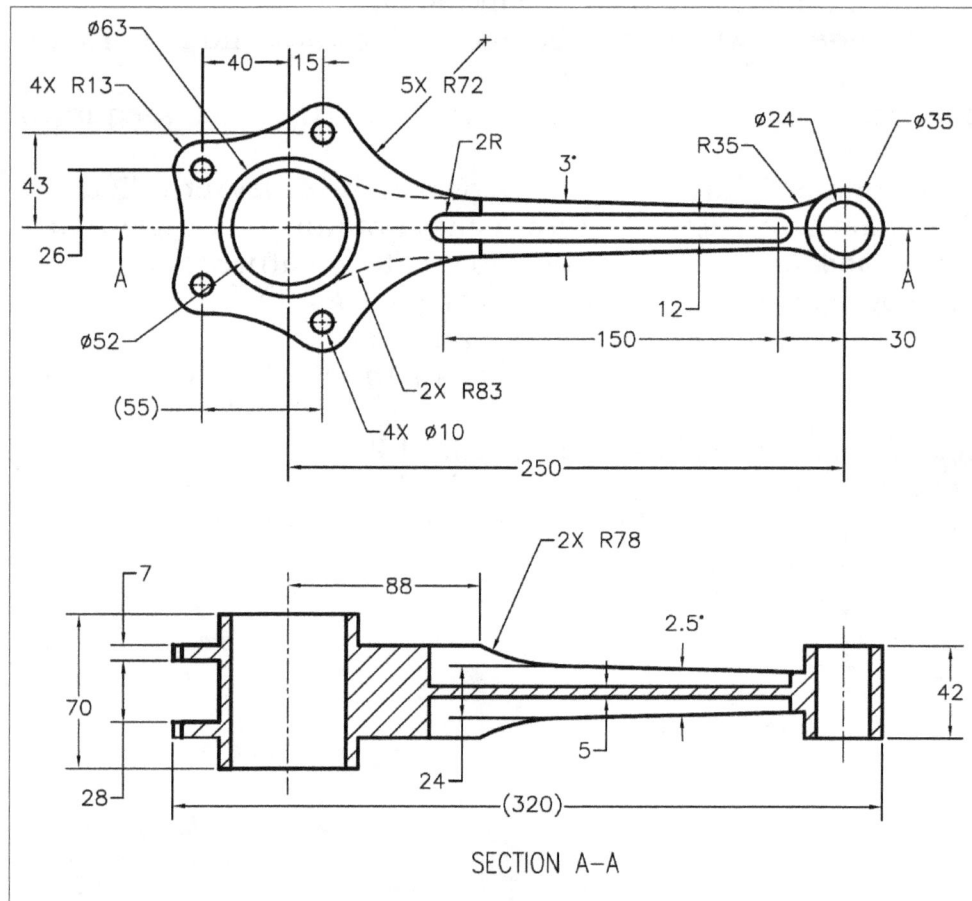

Figure-17. Practical 1 drawing

Strategy for Creating model

1. Looking at the isometric view and orthographic views, we can find that the model is created by multiple extrude features and extrude cut features.
2. First sketch (which is a ring) is to be created on the **XZ** plane (Top plane) and then it is to be extruded.
3. Create the other extrusion features and remove the extra portion from them.

Starting Part File

- Start Autodesk Inventor by double-clicking on the Autodesk Inventor Professional 2024 icon from the desktop. (If not started yet.)
- Click on the **New** button from the **Quick Access Toolbar**. The **Create New File** dialog box will be displayed.
- Double-click on **Standard(mm).ipt** icon from the **Metric** templates. The Part environment of Autodesk Inventor will be displayed.

Creating First Extrusion

- Click on the **Start 2D Sketch** button from **Sketch** panel in the **3D Model** tab of the **Ribbon**. You will be asked to select a sketching plane.
- Select the **XZ** Plane (Top Plane) as sketching plane. The sketching environment will become active.
- Create a ring of two circles as shown in Figure-18.

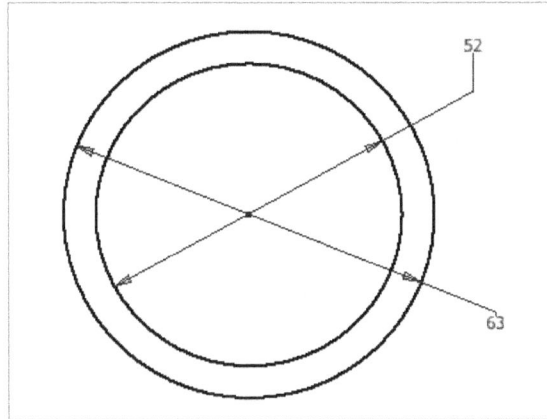

Figure-18. Circles created

- Click on the **Finish Sketch** button from the **Exit** panel in the **Ribbon**.
- Click on the **Extrude** tool from **Create** panel in the **3D Model** tab of the **Ribbon**. You will be asked to select the profile for extrusion.
- Select the sketch created and extrude it to both side by **70**; refer to Figure-19.

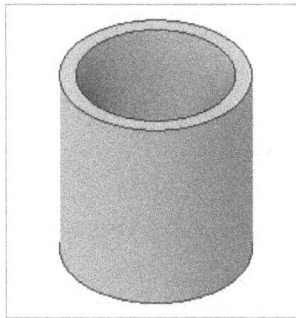

Figure-19. First Extrusion

Creating the Second Extrusion

- Click on the **Start 2D Sketch** button from **Sketch** panel in the **3D Model** tab of the **Ribbon**. You will be asked to select a sketching plane.
- Select the **XZ** Plane (Top Plane) as sketching plane.
- Create the sketch as shown in Figure-20.

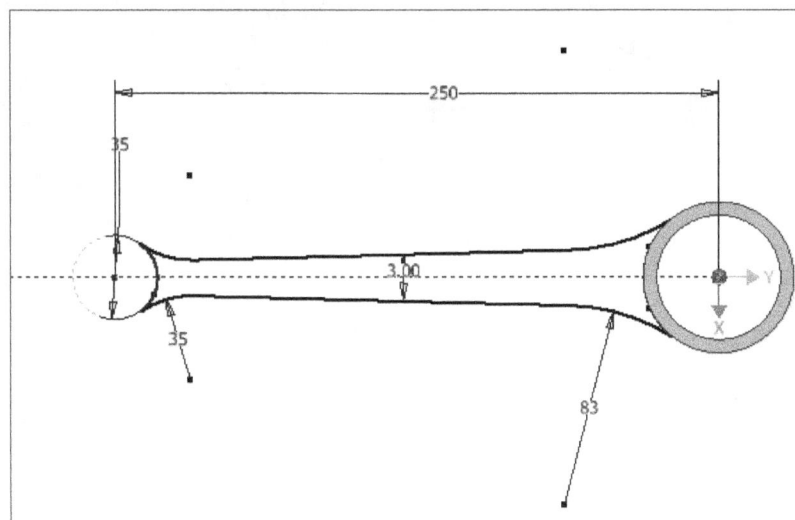

Figure-20. Sketch for second extrusion

- Click on the **Finish Sketch** button from **Exit** panel in the **Ribbon** to exit the sketching environment.
- Extrude the sketch to both side at a distance of **42**; refer to Figure-21.

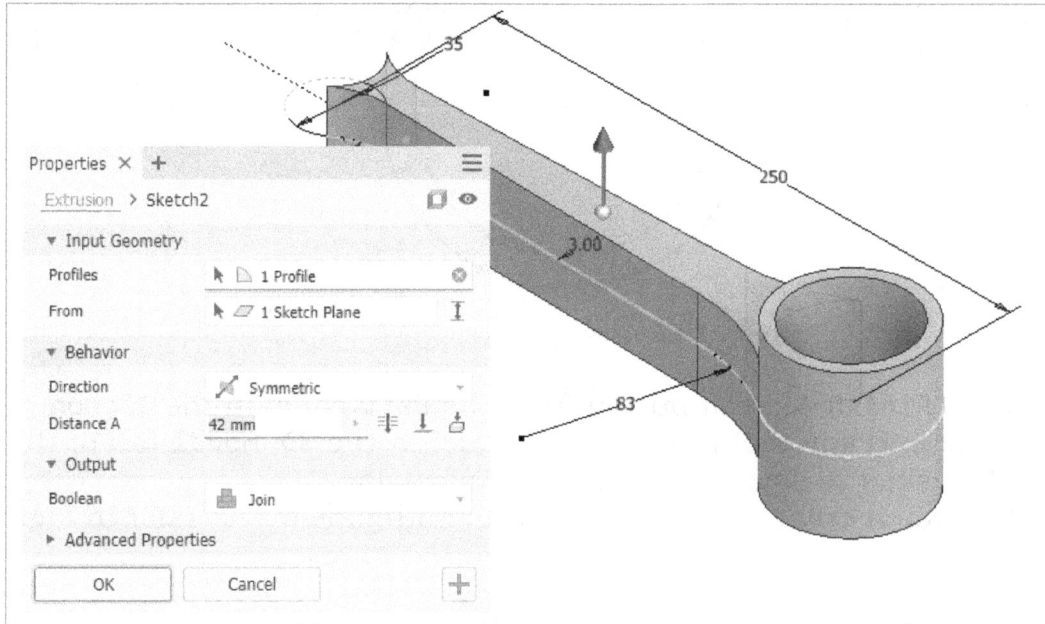

Figure-21. Second Extrusion

- Click on the **OK** button from the dialog box to create the feature.

Creating Third Extrusion Feature

- Click on the **Start 2D Sketch** button from **Sketch** panel in the **3D Model** tab of the **Ribbon**. You will be asked to select the sketching plane or face.
- Select the top face of second extrude feature. The sketch environment will display.
- Create the sketch as shown in Figure-22.

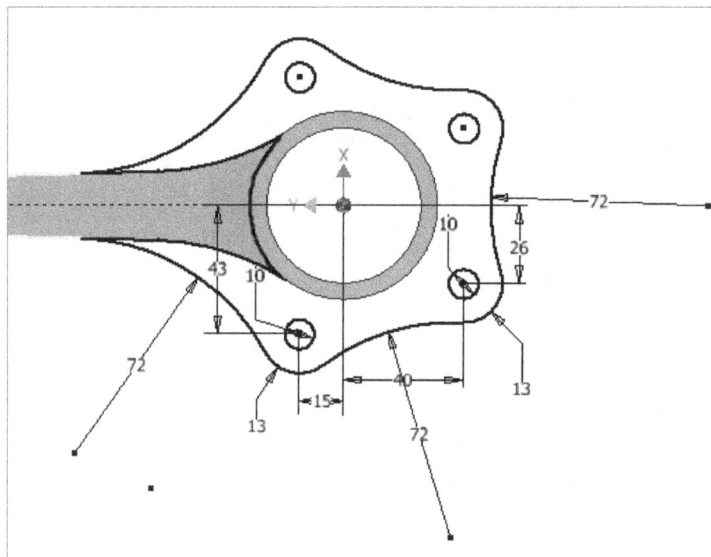

Figure-22. Sketch for third extrusion feature

- Extrude the sketch by **7** mm downward; refer to Figure-23.

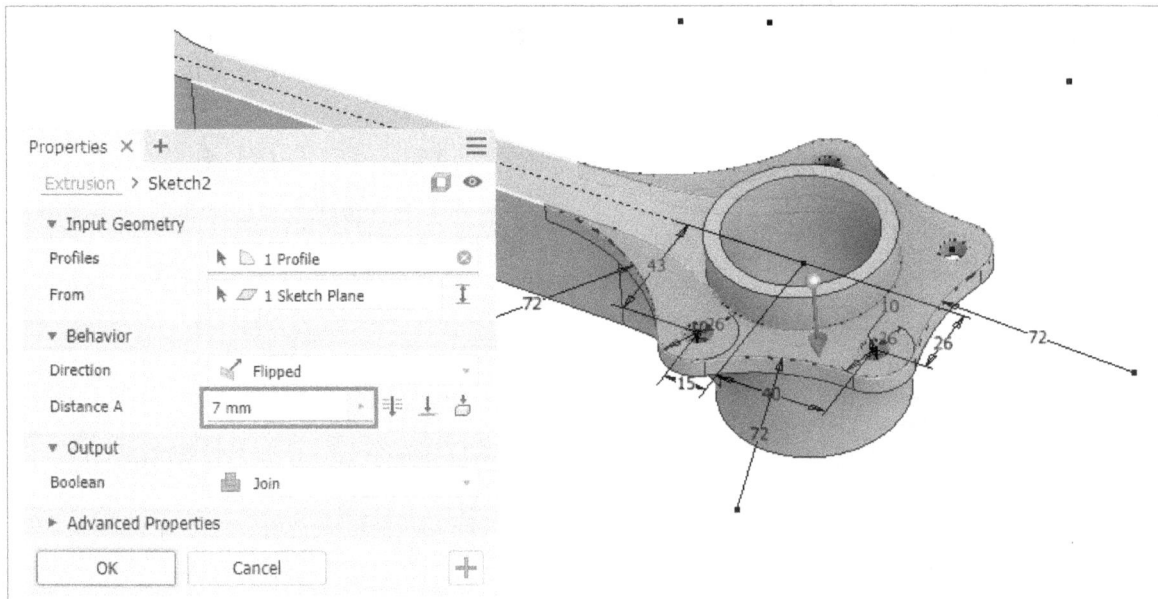

Figure-23. Extruding third feature

- Click on the **OK** button from the dialog box to create the feature.

Creating Mirror of Third Feature

- Click on the **Mirror** tool from **Pattern** panel in the **3D Model** tab of the **Ribbon**. The **Mirror** dialog box will be displayed as shown in Figure-24.

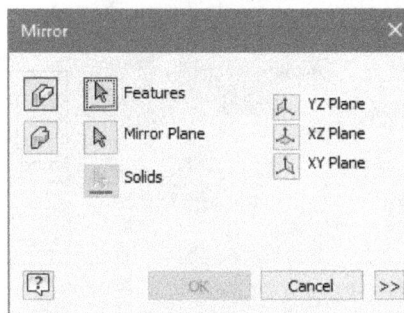

Figure-24. Mirror dialog box

- Select the third extrusion feature from the **Model Browser** bar at the left of the application window; refer to Figure-25.

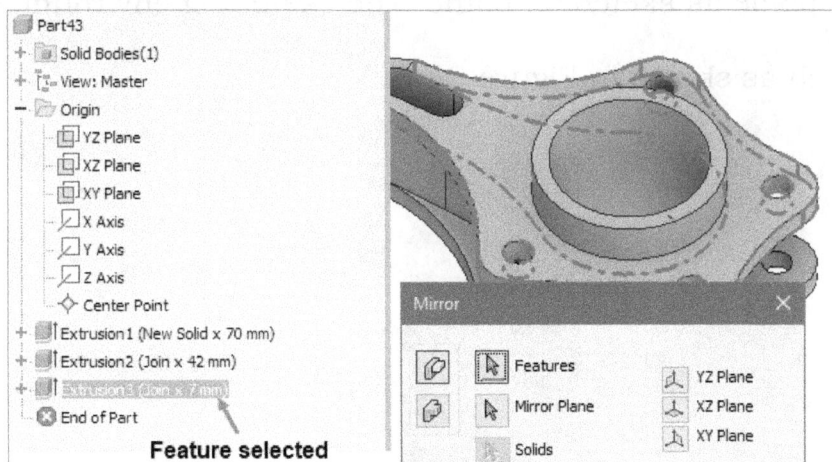

Figure-25. Selecting feature for mirroring

• Click on the **Mirror Plane** selection button from the dialog box and select the **XZ Plane** button as mirror plane from the dialog box or from the **Model Browser** bar; refer to Figure-26. Preview of the mirror feature will be displayed; refer to Figure-27.

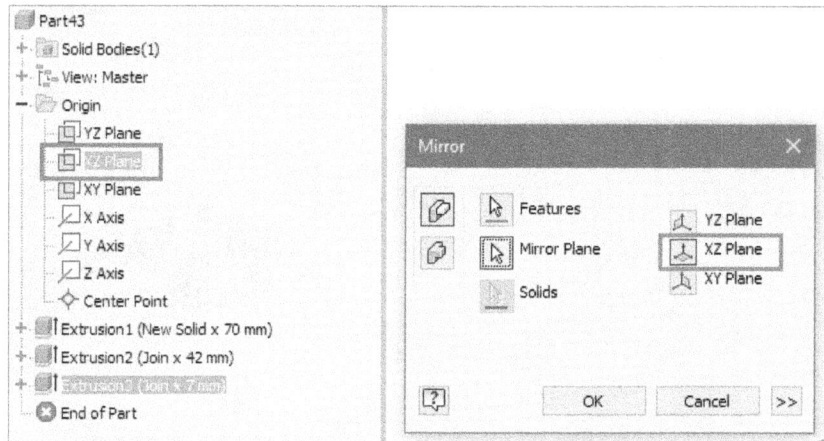

Figure-26. Plane to be selected

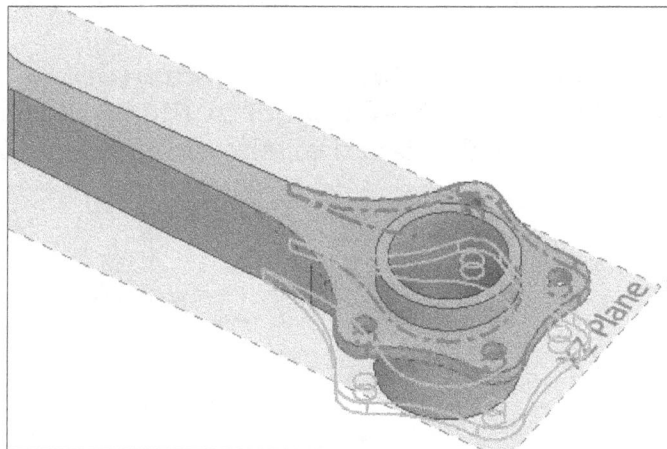

Figure-27. Preview of mirror feature

Creating First Extrude Cut Feature

• Click on the **Start 2D Sketch** button from **Sketch** panel in the **3D Model** tab of the **Ribbon**. You will be asked to select a sketching plane.
• Select the **YZ** plane as sketching plane. The sketching environment will become active.
• Create a sketch as shown in Figure-28.

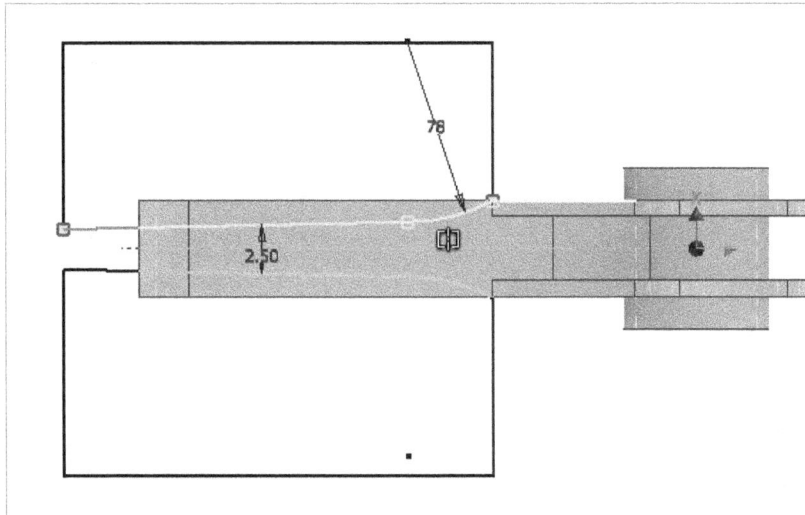

Figure-28. Sketch for extrude cut feature

- Click on the **Finish Sketch** button from **Exit** panel in the **Ribbon**.
- Click on the **Extrude** tool from **Create** panel in the **3D Model** tab of the **Ribbon**. The **Extrude** dialog box will be displayed asking you to select the profile.
- Select the newly created sketch and extrude both the sides with **Cut** option selected from **Boolean** area of the dialog box; refer to Figure-29.

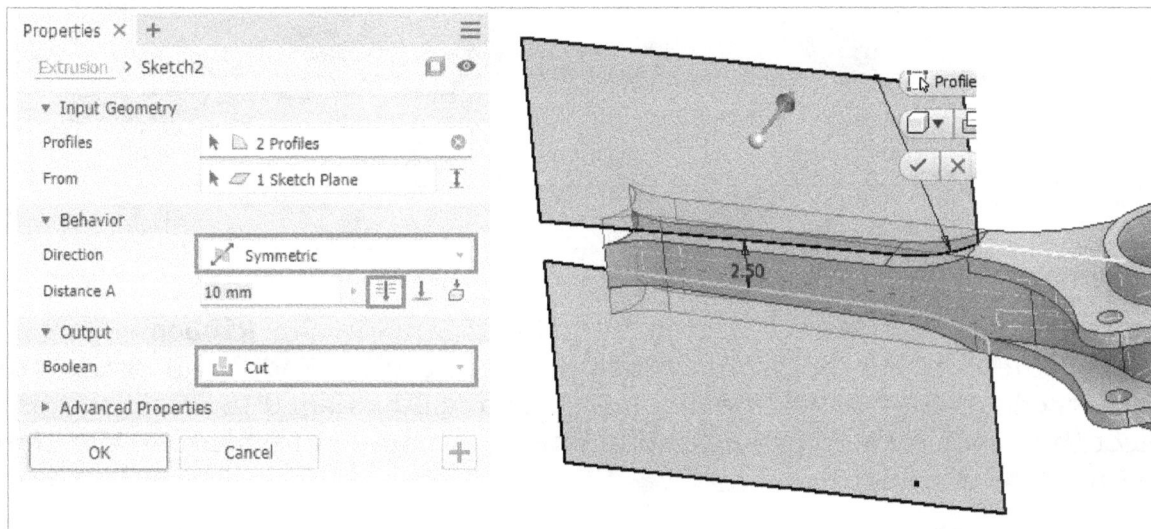

Figure-29. Creating cut feature

- Click on the **OK** button from the dialog box to create the feature.
- After creating extrude cut feature, the model will be displayed as shown in Figure-30.

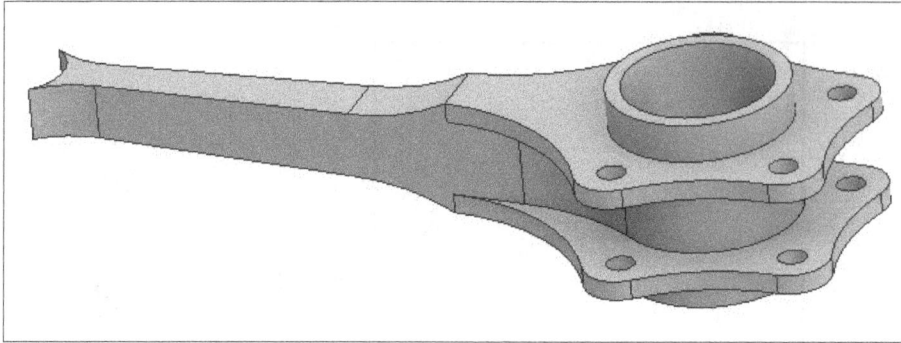

Figure-30. After creating extrude cut feature

Creating the Second and Third Extrude Cut Feature

- Click on the **Start 2D Sketch** button from **Sketch** panel in the **3D Model** tab of the **Ribbon**. You will be asked to select the sketching plane or face.
- Select the top face of the third extrusion feature created earlier as sketching face.
- Create a sketch as shown in Figure-31.

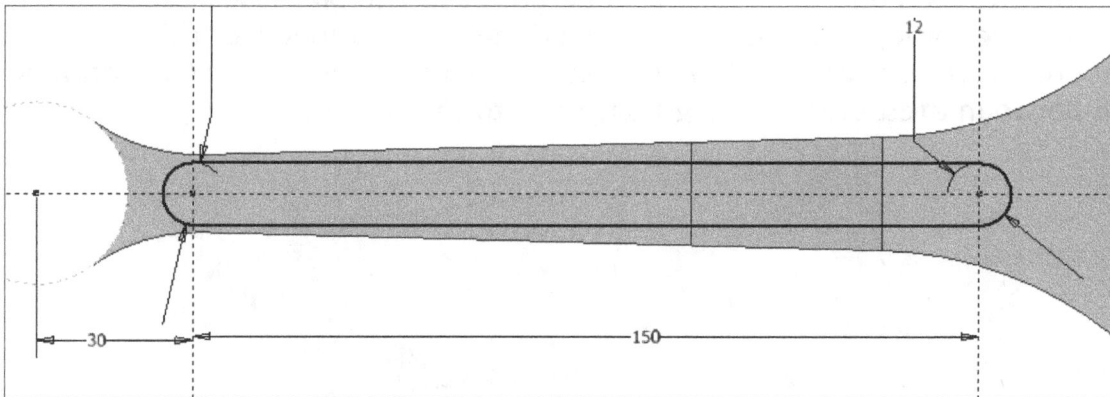

Figure-31. Sketch for second cut feature

- Click on the **Finish Sketch** button from **Exit** panel in the **Ribbon**.
- Click on the **Extrude** tool from **Create** panel in the **3D Model** tab of the **Ribbon**. The **Extrude** dialog box will be displayed. You will be asked to select the profile.
- Select the newly created profile and create an extrude cut feature to the depth of **18.5** as shown in Figure-32.

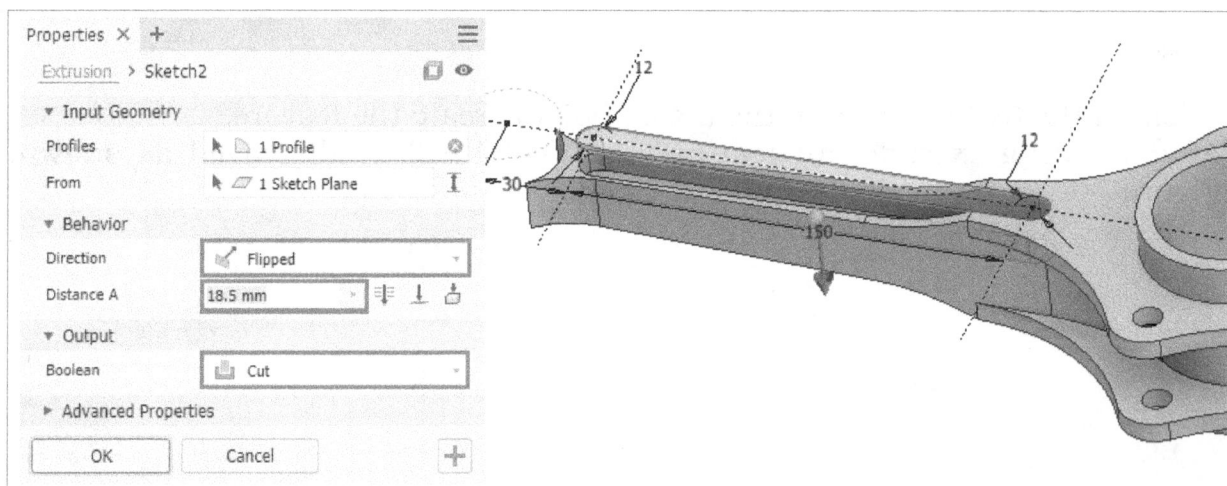

Figure-32. Second extrude cut feature

- Click on the **OK** button from the dialog box to create the feature.
- Click on the **Mirror** tool from **Pattern** panel in the **3D Model** tab of the **Ribbon**. The **Mirror** dialog box will be displayed asking you to select the feature.
- Select the newly created extrude feature to be mirrored to the other side and select the **XZ Plane** button as **Mirror Plane** from the dialog box or **Model Browser** bar. The preview of mirror feature will be displayed; refer to Figure-33.

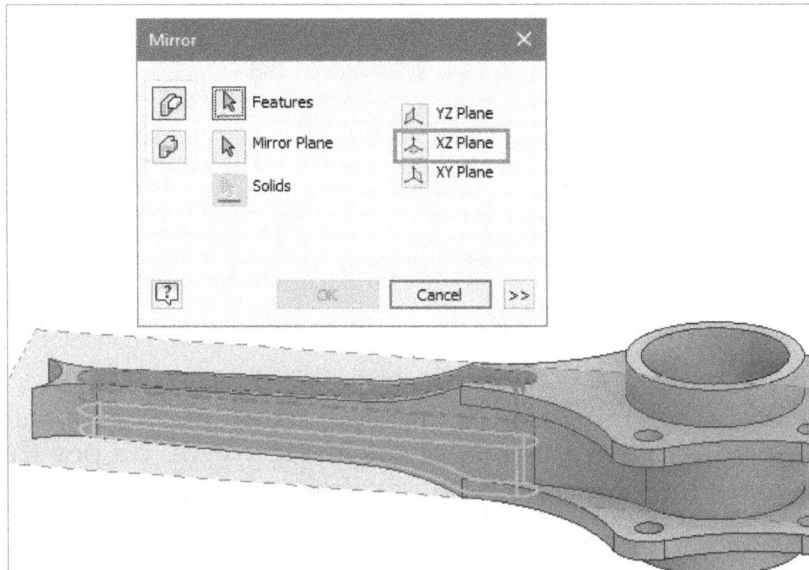

Figure-33. Mirror feature created

- Click on the **OK** button from the dialog box to create the feature.

Creating the last extrude feature

- Click on the **Start 2D Sketch** button from **Sketch** panel in the **3D Model** tab of the **Ribbon**. You will be asked to select the sketching plane.
- Select the **XZ** plane as sketching plane and create the sketch as shown in Figure-34.

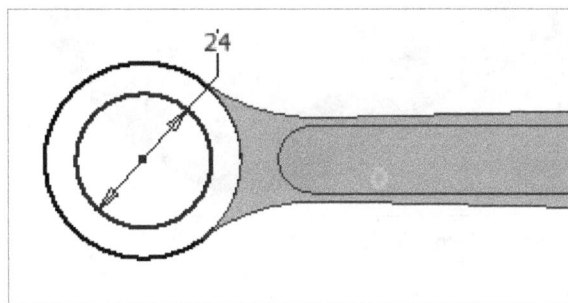

Figure-34. Sketch for last extrude feature

- Click on the **Finish Sketch** button from **Exit** panel in the **Ribbon**.
- Click on the **Extrude** tool from **Create** panel in the **3D Model** tab of the **Ribbon**. The **Extrude** dialog box will be displayed asking you to select the profile.
- Select the newly created sketch and extrude it to both the sides of distance **42** mm. The preview of extruded feature will be displayed; refer to Figure-35.

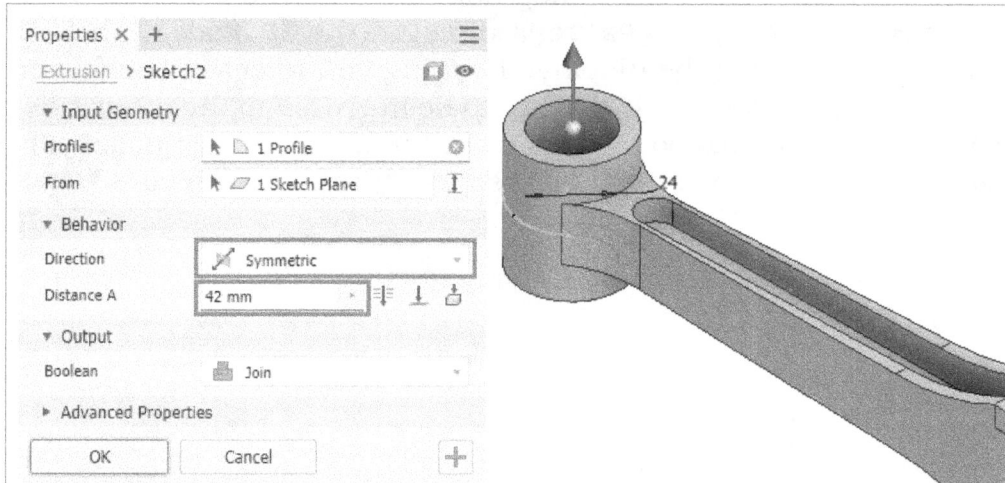

Figure-35. Extrude feature created

- Click on the **OK** button from the dialog box to create the feature.

PRACTICAL 2

In this practical, we will create a model given in Figure-36 as per the drawing given in Figure-37. The model is of a pipe with varying section and it will require the use of datum curve.

Figure-36. Practical 2 model

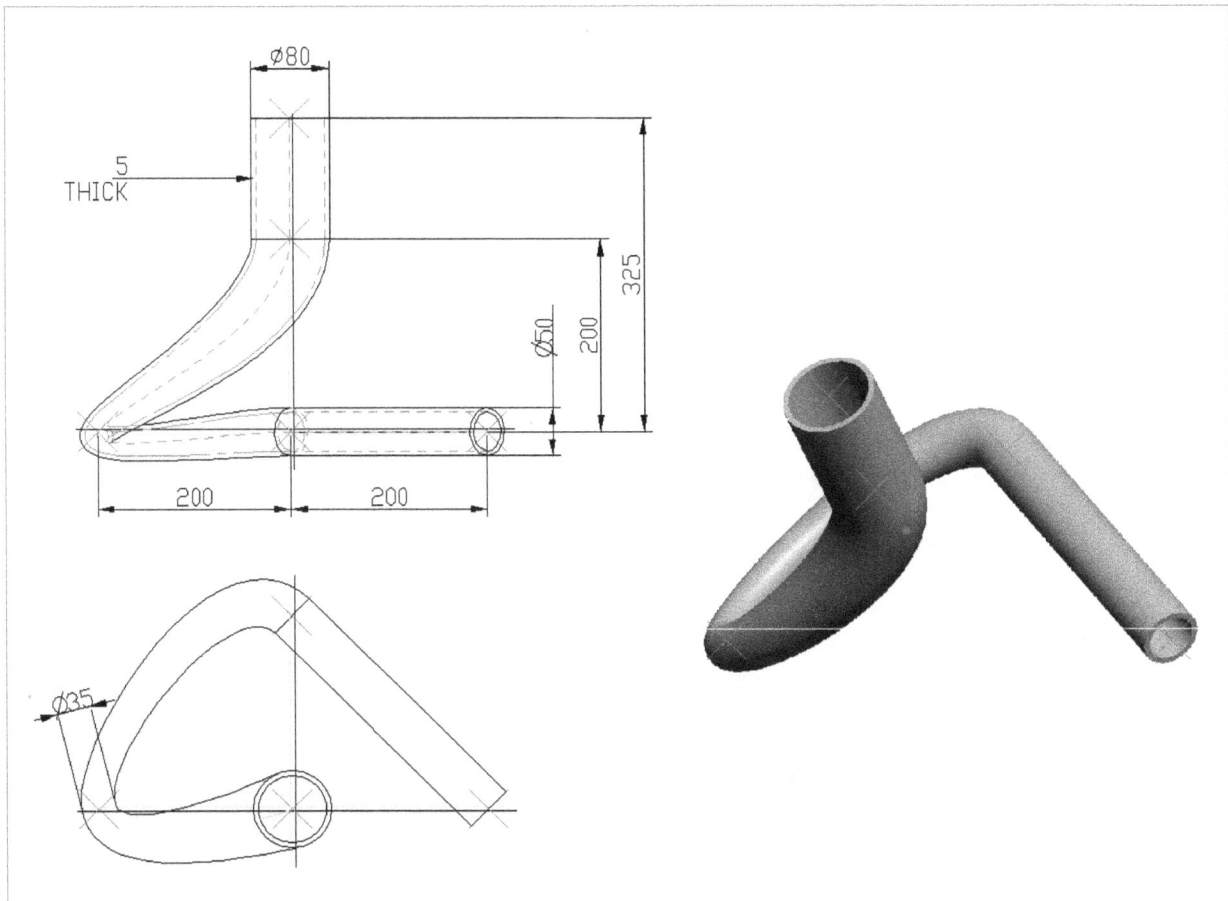

Figure-37. Practical 2

Strategy for Creating model

1. Looking at the isometric view and orthographic views, we can find that the model is created by using a spline passing through specified points.
2. First create the required points on XZ and YZ planes.
3. Create a 3D spline passing through the specified points.

Starting Part File

- Start Autodesk Inventor by double-clicking on the Autodesk Inventor Professional 2024 icon from the desktop. (If not started yet.)
- Click on the **New** button from the **Quick Access Toolbar**. The **Create New File** dialog box will be displayed.
- Double-click on **Standard(mm).ipt** icon from the **Metric** templates. The Part environment of Autodesk Inventor will be displayed.

Creating Sketch for Profile

- Click on the **Start 2D Sketch** button from the **Sketch** panel in the **3D Model** tab of the **Ribbon**. You will be asked to select a sketching plane.
- Select the **XZ** Plane (Top Plane) as sketching plane. The sketching environment will be displayed.
- Create three points on the sketching plane as shown in Figure-38.

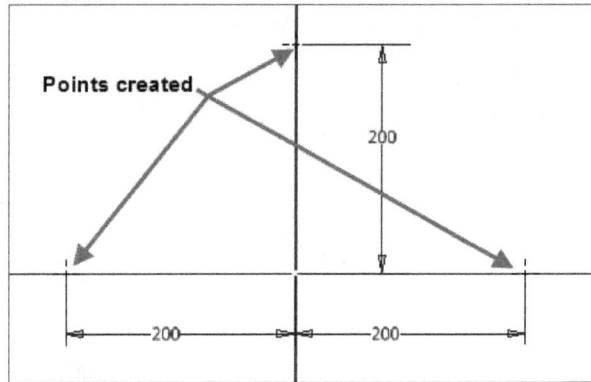

Figure-38. Points to be created

- Click on the **Finish Sketch** button from **Exit** panel in the **Ribbon**.
- Click on the **Start 2D Sketch** button again and select **YZ** Plane (Right Plane) as sketching plane. The sketching environment will be displayed.
- Draw two points as shown in Figure-39.

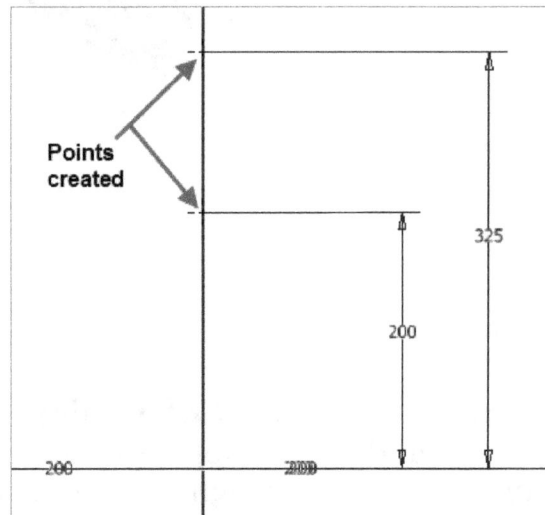

Figure-39. Points created in second sketch

- Click on the **Finish Sketch** button from **Exit** panel in the **Ribbon**.
- Click on the **Start 3D Sketch** button from **Sketch** panel in the **Sketch** tab of **Ribbon**. The tools related to 3D sketching will be displayed in the **Ribbon**.
- Click on the **Spline Interpolation** tool from the **Spline** drop-down in the **Draw** panel of **3D Sketch** tab in the **Ribbon**; refer to Figure-40. You will be asked to select starting point of spline.

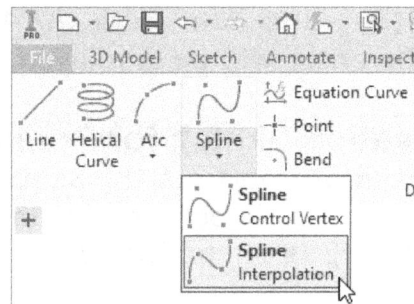

Figure-40. Spline Interpolation tool

- Create a spline passing through the earlier created points as shown in Figure-41.

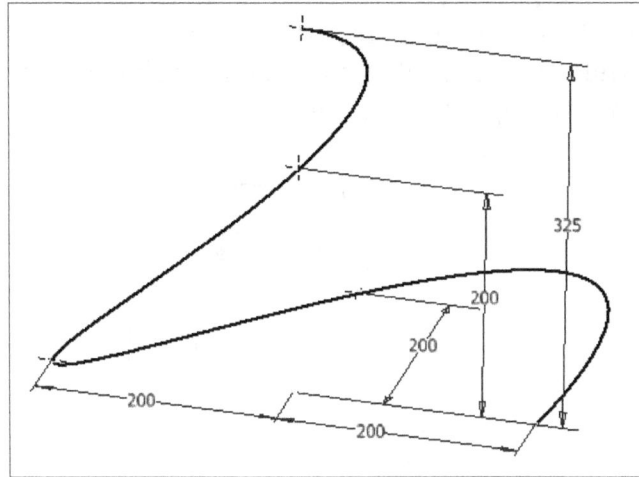

Figure-41. Spline passing through selected points

- Right-click on the spline and click on the **AutoCAD** option from the **Fit Method** cascading menu in the shortcut menu; refer to Figure-42, to change the shape of spline as per **AutoCAD** style; refer to Figure-43.

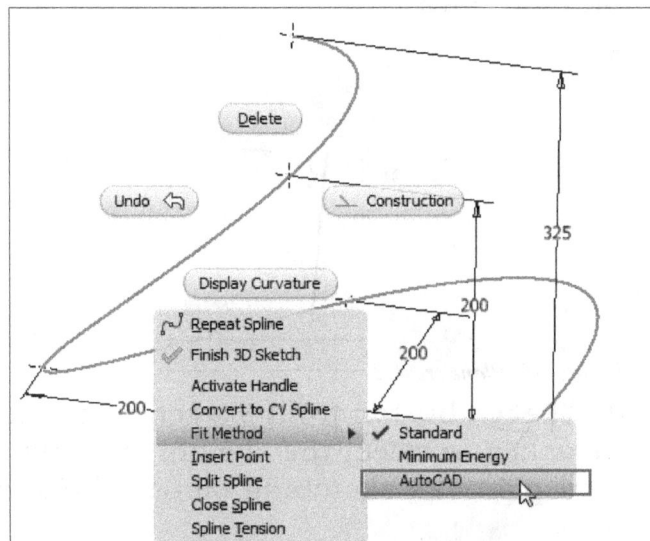

Figure-42. Changing fit method for spline

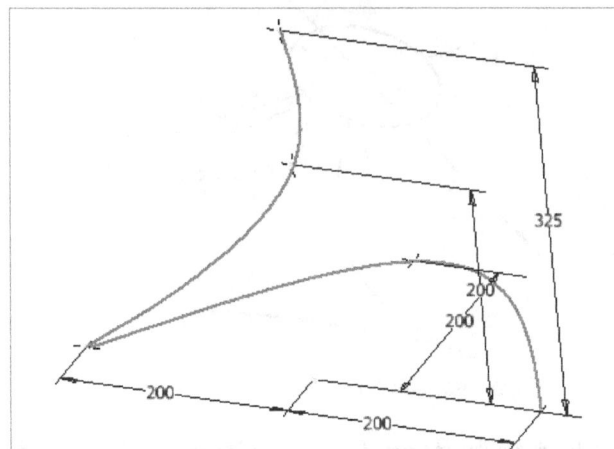

Figure-43. Spline after changing fit method

- Click on the **Finish Sketch** button from the **Exit** panel in the **Ribbon**.

Creating Section Sketches

- Click on the **Plane** tool from **Plane** drop-down in the **Work Features** panel of the **Ribbon**. You will be asked to select geometry as reference for plane.
- Select the curve near the top point as shown in Figure-44.

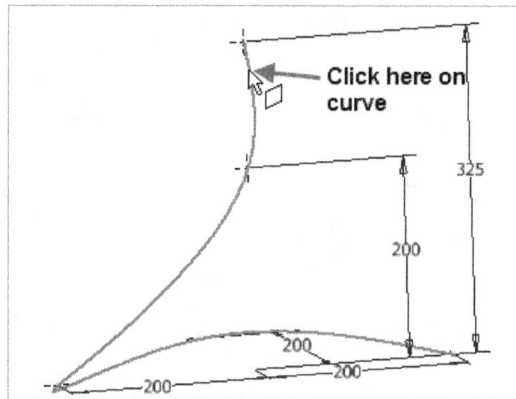

Figure-44. Selecting curve for plane creation

- Now, click on the top point to specify second reference for plane. The plane will be created as shown in Figure-45.

Figure-45. Plane created

- Click on the **Start 2D Sketch** button from **Sketch** panel in the **3D Model** tab of the **Ribbon**. You will be asked to select the sketching plane.
- Select the newly created plane as sketching plane. The sketching environment will be displayed.
- Create a 2D sketch as shown in Figure-46.

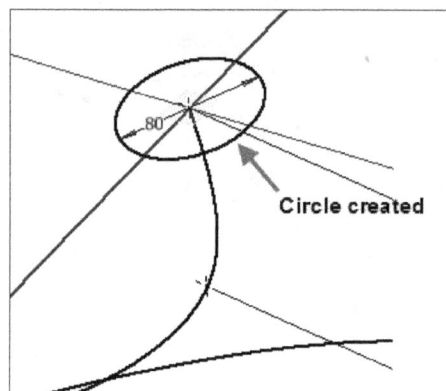

Figure-46. Circle created

- Click on the **Finish Sketch** button from **Exit** panel in the **Ribbon**.
- Similarly, create planes at other points and create sketches as shown in Figure-47.

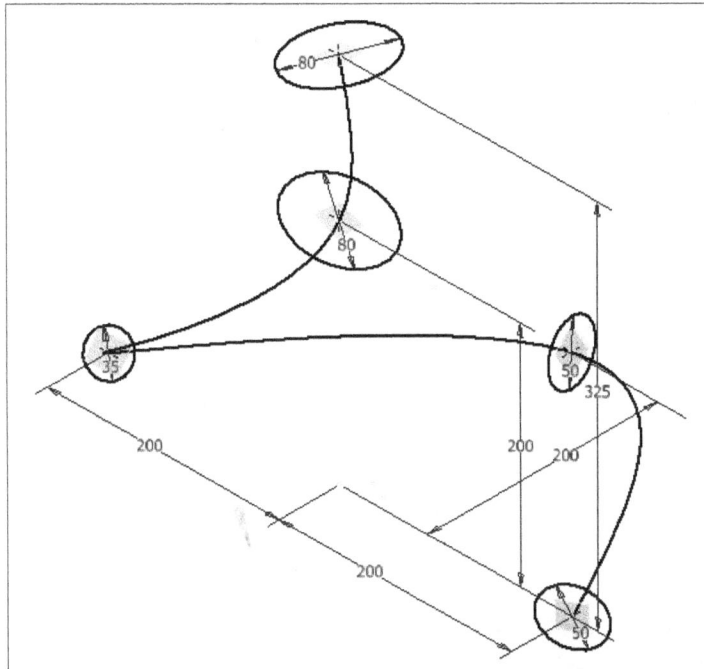

Figure-47. Sketches created on planes

Creating Loft Feature

- Click on the **Loft** tool from **Create** panel in the **3D Model** tab of the **Ribbon**. The **Loft** dialog box will be displayed as shown in Figure-48.

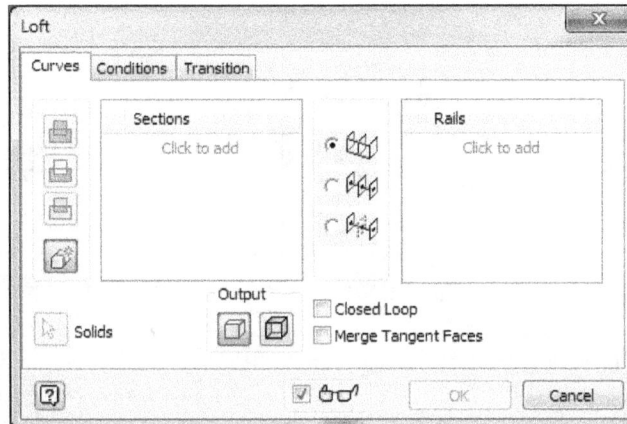

Figure-48. Loft dialog box

- Select the **Center Line** radio button from the dialog box and click in the **Center Line** selection box; refer to Figure-49.

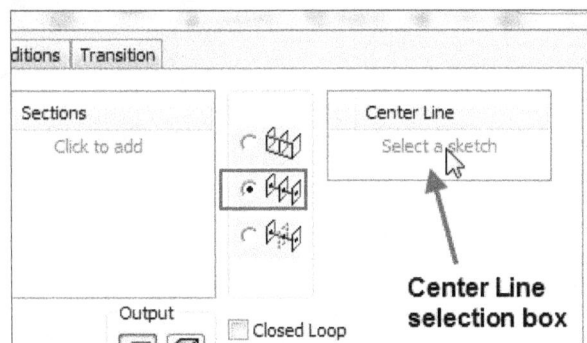

Figure-49. Centerline selection box

- Select the 3D spline created earlier; refer to Figure-50.

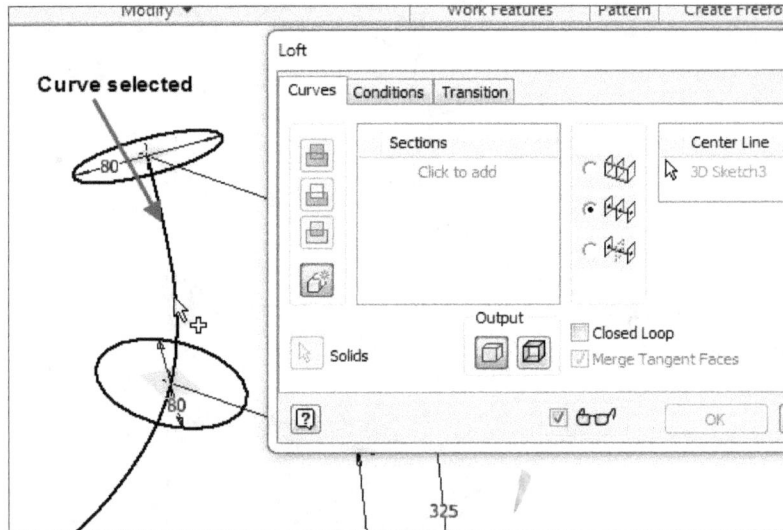

Figure-50. Curve to be selected as centerline

- Click in the **Sections** selection box and one by one select the section sketches as per the flow of loft feature; refer to Figure-51.

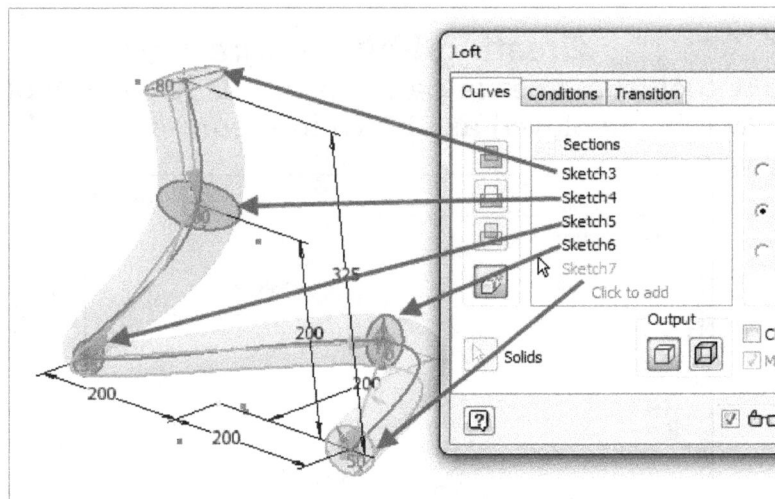

Figure-51. Sections selected

- Click on the **OK** button from the dialog box to create the feature. The loft feature will be displayed as shown in Figure-52.

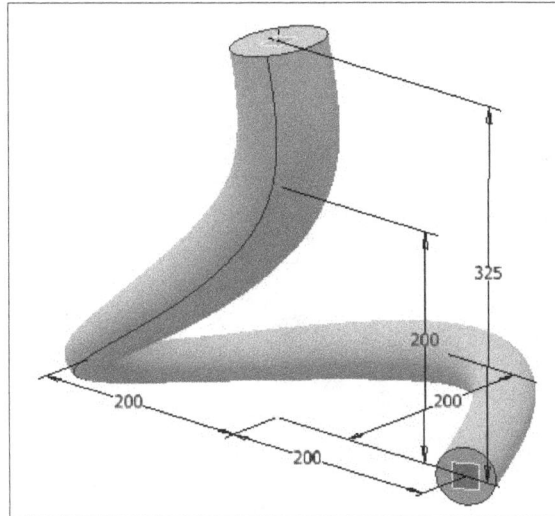

Figure-52. Loft feature created

Creating Shell feature

- Click on the **Shell** tool from **Modify** panel in the **3D Model** tab of the **Ribbon**. The **Shell** dialog box will be displayed as shown in Figure-53.

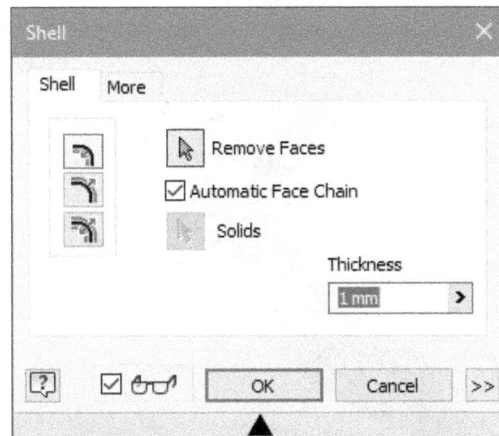

Figure-53. Shell dialog box

- Click in the **Thickness** edit box and specify the value as **5** mm.
- Select both the end faces of the loft feature to remove them; refer to Figure-54.

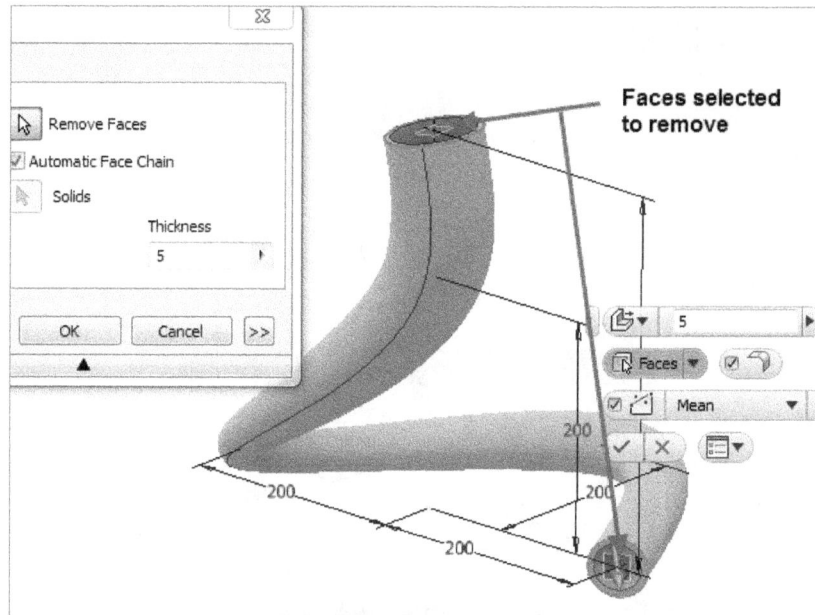

Figure-54. Faces to remove while shelling

- Click on the **OK** button from the dialog box. The feature will be created; refer to Figure-55.

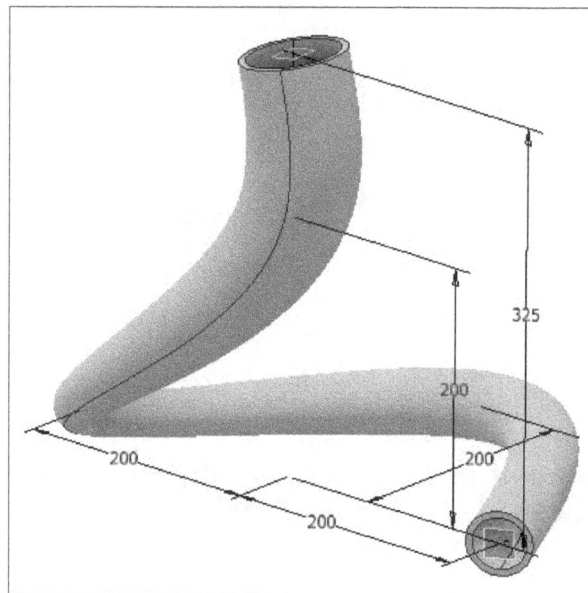

Figure-55. Model after applying shell feature

Now, select all the sketches and planes which are not required to be displayed in viewport and select the **Visibility** option from the shortcut menu displayed in the **Model Browser** bar; refer to Figure-56. The extra geometry will get hidden.

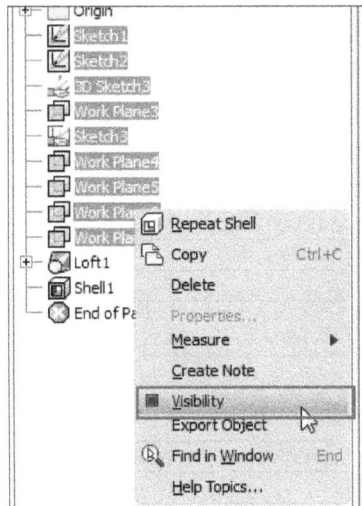

Figure-56. Visibility option

PRACTICAL 3

In this practical, we will create a model of spring as shown in Figure-57 with parameters as :

Wire Diameter	:	5 mm
Pitch	:	10 mm
Spring Diameter	:	60 mm
Height	:	150 mm

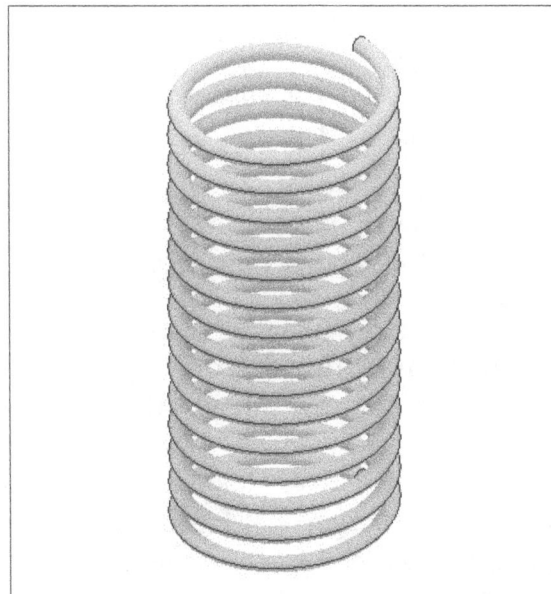

Figure-57. Model of spring

Strategy for Creating model

1. We have a direct tool to create springs in Autodesk Inventor but for using this tool, we must have a profile and an axis for the spring.
2. Create a sketch of profile with an axis.
3. Select the **Coil** tool and set the parameters of the spring.

Starting Part File

- Start Autodesk Inventor by double-clicking on the Autodesk Inventor Professional 2024 icon from the desktop. (If not started yet.)
- Click on the **New** button from the **Quick Access Toolbar**. The **Create New File** dialog box will be displayed.
- Double-click on **Standard(mm).ipt** icon from the **Metric** templates. The Part environment of Autodesk Inventor will be displayed.

Creating Sketch for Profile and axis

- Click on the **Start 2D Sketch** button from **Sketch** panel in the **3D Model** tab of the **Ribbon**. You will be asked to select a sketching plane.
- Select the **YZ** Plane (Right Plane) to create sketch.
- Create a circle and a line as shown in Figure-58.

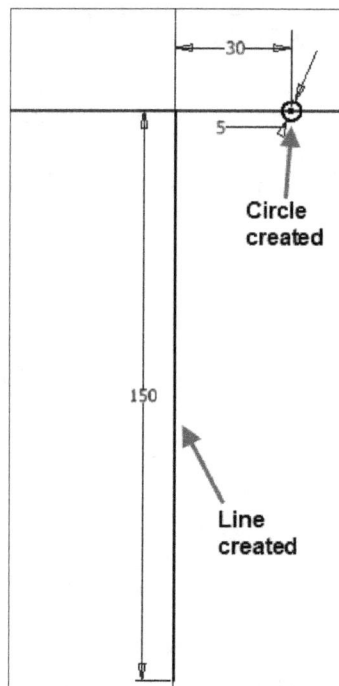

Figure-58. Sketch for spring

- Click on the **Finish Sketch** button from **Exit** panel in the **Ribbon**.
- Note that diameter of circle in this sketch represents diameter of wire and dimension of 30 represents the spring radius. Also, the line of **150** mm represents length of spring.

Creating Coil Feature

- Click on the **Coil** tool from **Create** panel in the **3D Model** tab of the **Ribbon**. The **Coil** dialog box will be displayed with profile selected automatically; refer to Figure-59. Also, you will be asked to select axis.

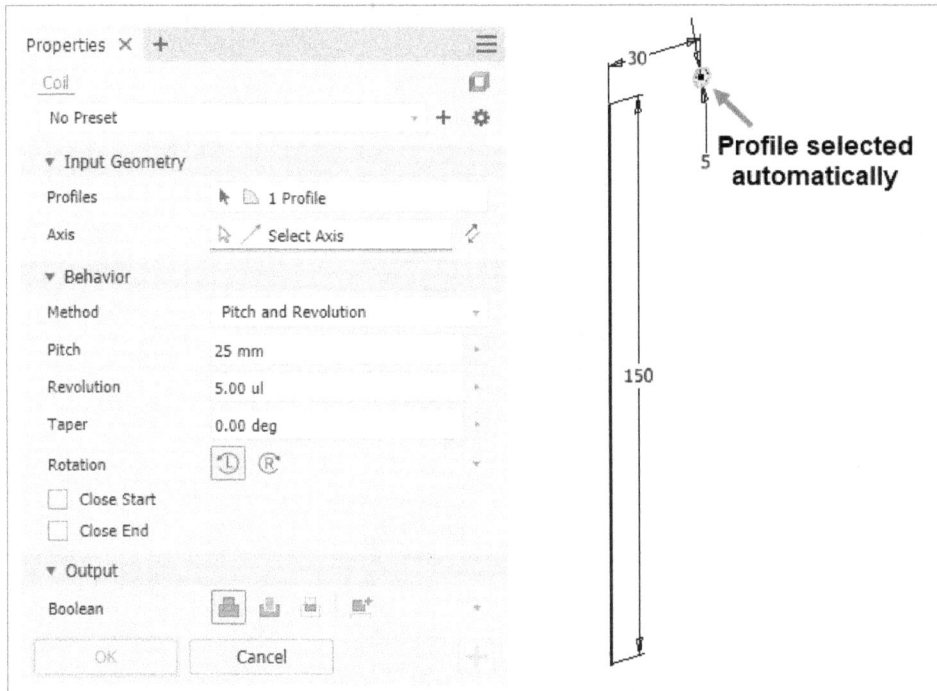

Figure-59. Coil dialog box with profile selected

- Select the line. The preview of coil will be displayed; refer to Figure-60.

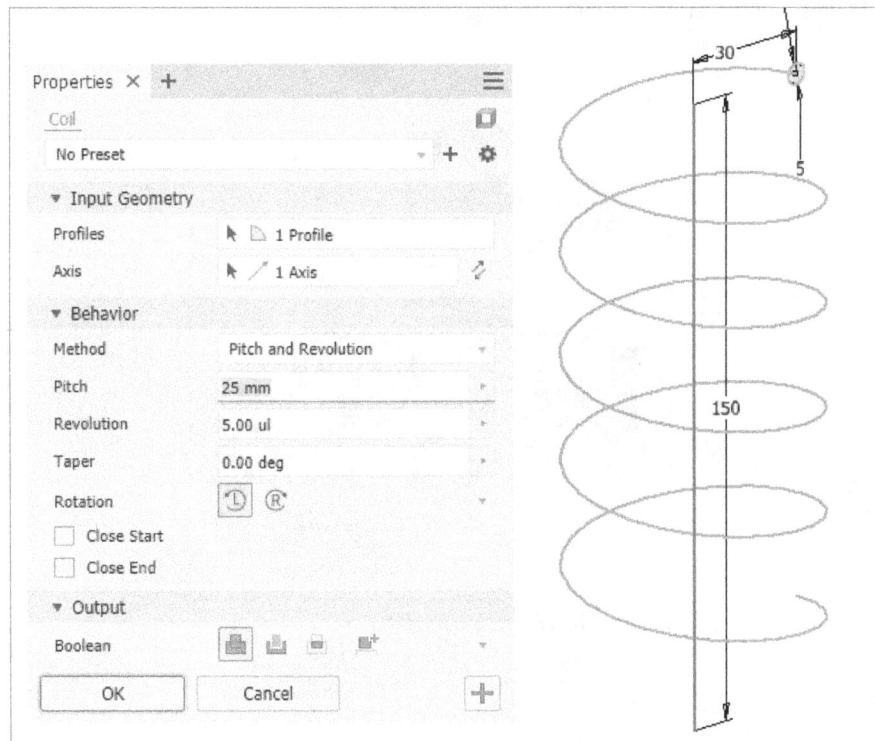

Figure-60. Preview of coil

- Click in the **Method** drop-down and select the **Pitch and Height** option from the drop-down. The **Pitch** and **Height** edit boxes will become active.
- Specify the value of pitch as **10** and height as **150** in the respective edit boxes.
- Click on the **OK** button from the dialog box. The spring will be created as shown in Figure-61.

Figure-61. Model of spring

PRACTICAL 4

In this practical, we will create a model of support bracket as shown in Figure-62. Dimensions of the model are given in Figure-63.

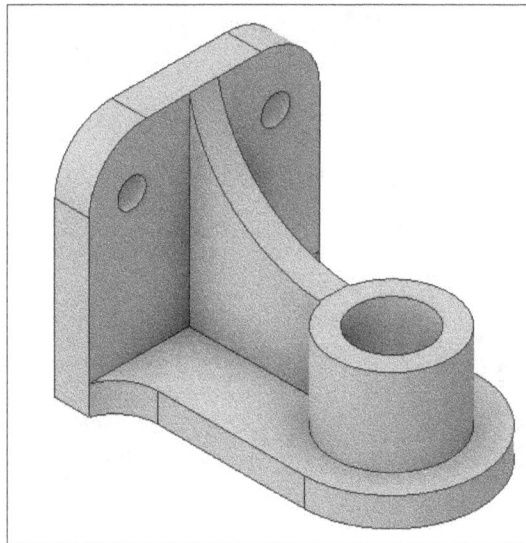

Figure-62. Model for Practical 4

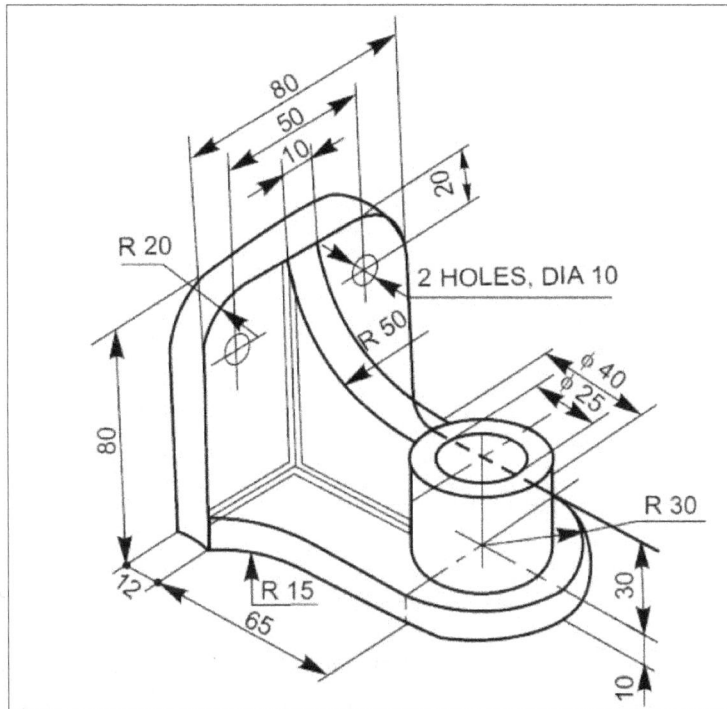

Figure-63. Dimensions for Practical 4

Strategy for Creating model

1. There are three extrude featured in the model; left side wall, bottom plate, and small tube. We will use the **Extrude** tool to create these features.
2. There is one rib feature which we will create by using the **Rib** tool.

Starting Part File

- Start Autodesk Inventor by double-clicking on the Autodesk Inventor Professional 2024 icon from the desktop. (If not started yet.)
- Click on the **New** button from the **Quick Access Toolbar**. The **Create New File** dialog box will be displayed.
- Double-click on **Standard(mm).ipt** icon from the **Metric** templates. The Part environment of Autodesk Inventor will be displayed.

Creating First Extrude Feature

- Click on the **Start 2D Sketch** button from **Sketch** panel in the **3D Model** tab of the **Ribbon**. You will be asked to select a plane.
- Select the **YZ** Plane (Right Plane) and create a sketch as shown in Figure-64.

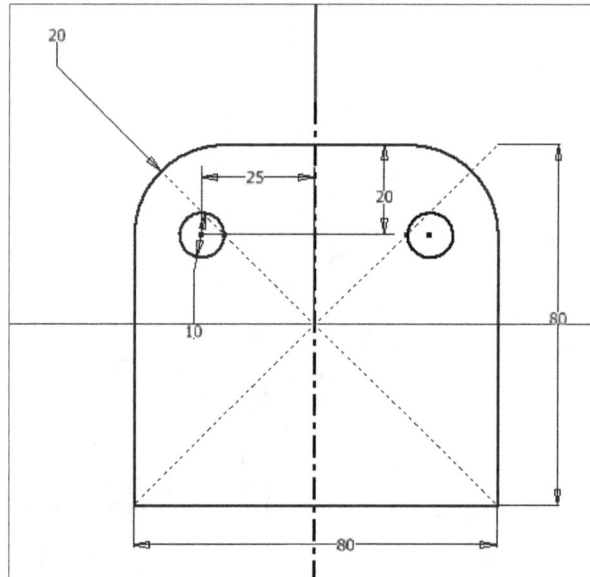

Figure-64. Sketch for base feature

- Click on the **Finish Sketch** button from **Exit** panel in the **Ribbon**.
- Click on the **Extrude** tool from **Create** panel in the **3D Model** tab of the **Ribbon**. You will be asked to select the section for extrusion.
- Select the inner loop of sketch and specify the extrusion distance as **12** in the **Distance A** edit box; refer to Figure-65. Preview of extrusion will be displayed.

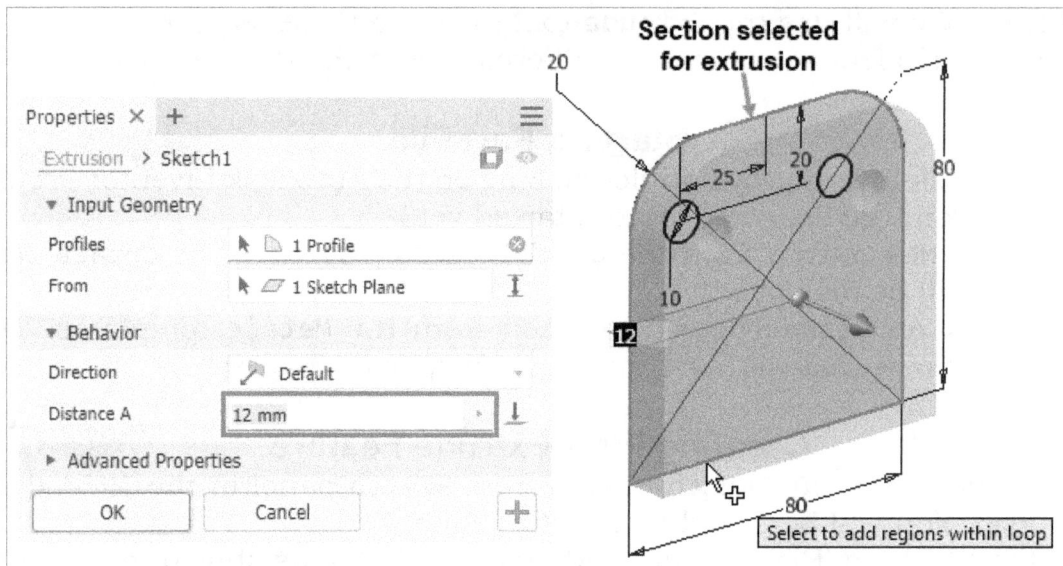

Figure-65. Section selected for extrusion

- Click on the **OK** button to create extrude feature.

Creating Second Extrude Feature

- Click on the **Start 2D Sketch** tool from **Sketch** panel in the **3D Model** tab of the **Ribbon**. You will be asked to select a sketching plane.
- Select the bottom face of extrude feature created earlier as sketching plane; refer to Figure-66.

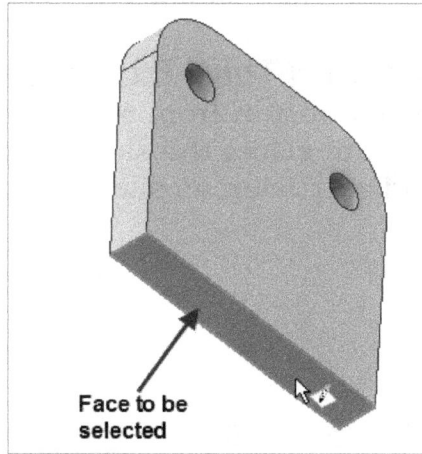

Figure-66. Face to select for sketching

- Create a sketch as shown in Figure-67.

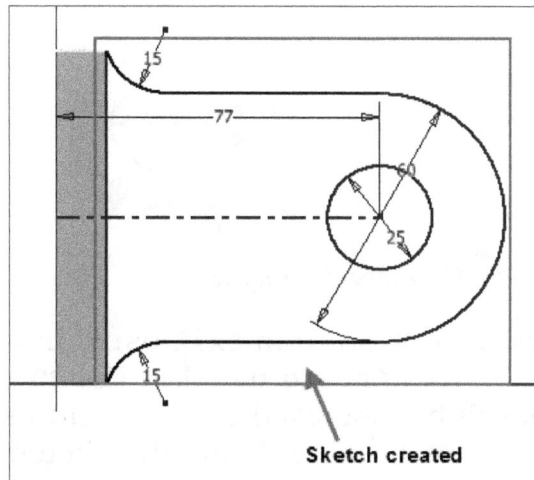

Figure-67. Sketch created for second extrusion

- Click on the **Finish Sketch** button from **Exit** panel of the **Ribbon**.
- Click on the **Extrude** tool from **Create** panel in the **3D Model** tab of the **Ribbon**. The **Extrude** dialog box will be displayed asking you to select the profile.
- Select the newly created sketch and extrude this sketch at a height of **10** mm; refer to Figure-68.

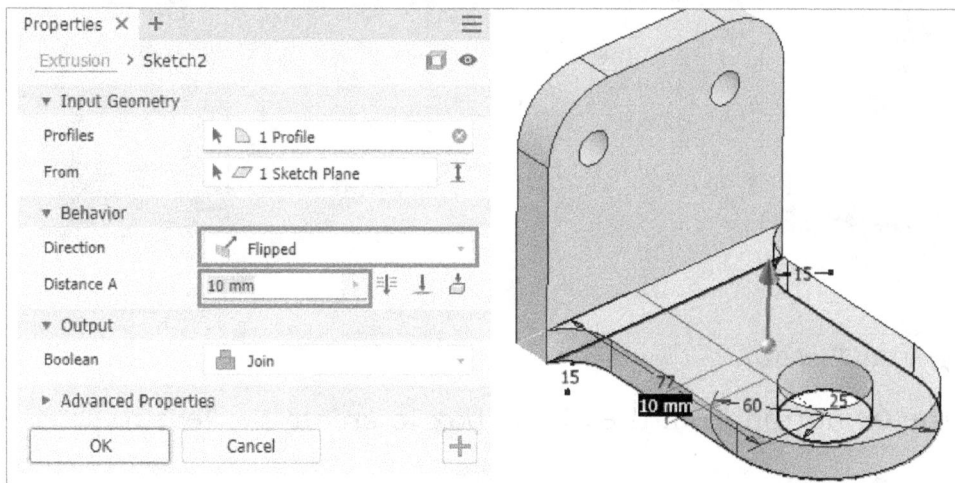

Figure-68. Extruding second sketch

- Click on the **OK** button from **Extrude** dialog box to create the feature.

Creating Third Extrusion

- Click on the **Start 2D Sketch** button from **Sketch** panel in the **3D Model** tab of the **Ribbon**. You will be asked to select the sketching plane or face.
- Select the flat face of second extrusion as sketching plane and create the sketch as shown in Figure-69.

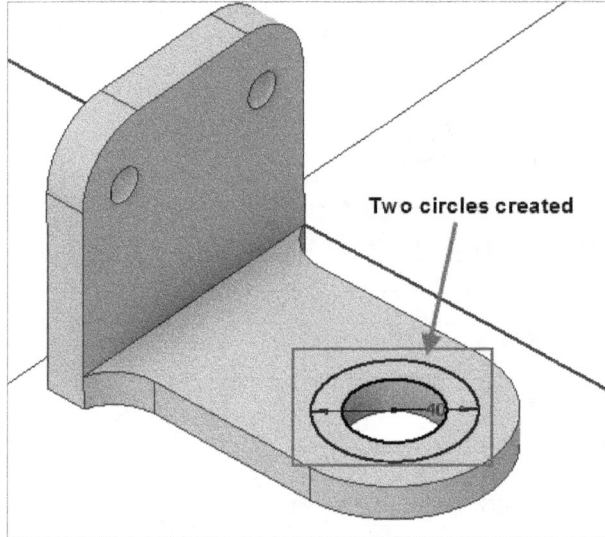

Figure-69. Sketch created on flat face

- Click on the **Finish Sketch** button from **Exit** panel in the **Ribbon**.
- Click on the **Extrude** tool from **Create** panel in the **3D Model** tab of the **Ribbon**. The **Extrude** dialog box will be displayed asking you to select the profile.
- Select the newly created sketch and extrude the sketch to the height of **30** mm; refer to Figure-70.

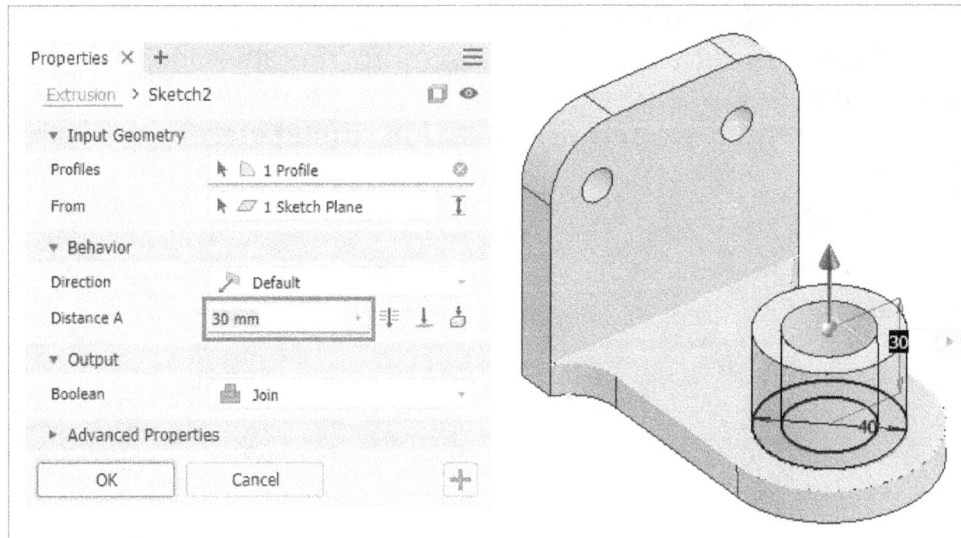

Figure-70. Third extrude feature creation

- Click on the **OK** button from the dialog box to create the feature.

Creating Rib Feature

- Click on the **Start 2D Sketch** button from **Sketch** panel in the **3D Model** tab of the **Ribbon**. You will be asked to select the sketching plane.
- Select the **XY** plane as sketching plane from the expanded **Origin** node in **Model Browser** bar. The sketching environment will become active; refer to Figure-71.

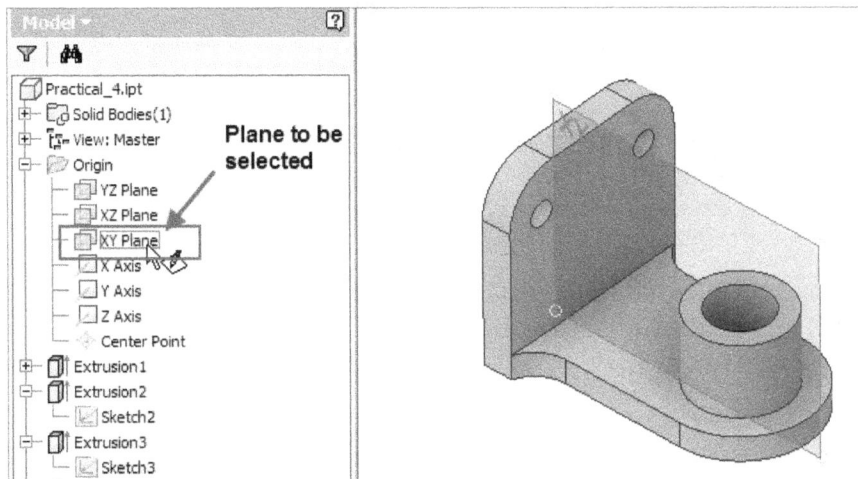

Figure-71. Selecting plane for rib feature sketch

- Create an arc of **50 mm** radius as shown in Figure-72. Note that end points of arc are coincident to faces of previous extrude features.

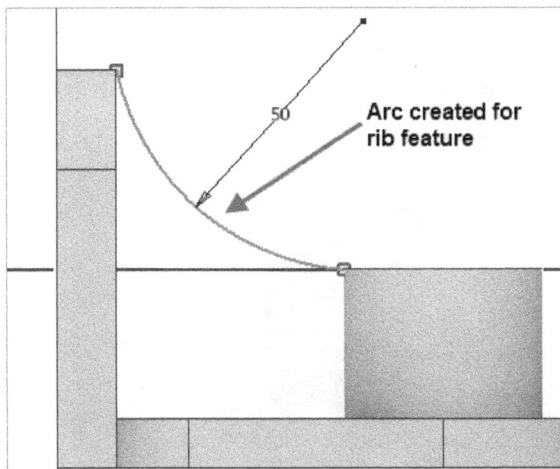

Figure-72. Sketch for rib feature

- Click on the **Finish Sketch** button from **Exit** panel in the **Ribbon**.
- Click on the **Rib** tool from **Create** panel in the **3D Model** tab of the **Ribbon**. The **Rib** dialog box will be displayed; refer to Figure-73.

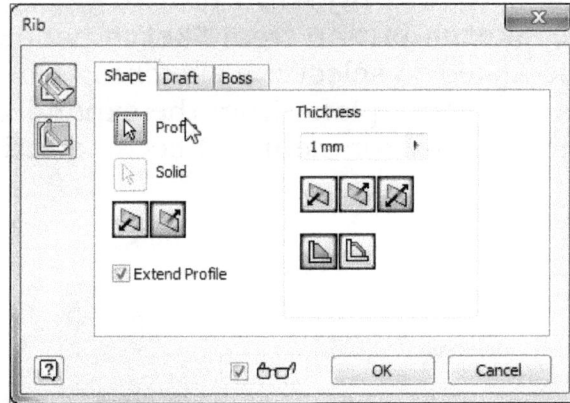

Figure-73. Rib dialog box

- Specify the thickness as **10** mm and select the **Parallel to Sketch Plane** button from the left area in the dialog box.
- Select the other buttons in the dialog box as shown in Figure-74.

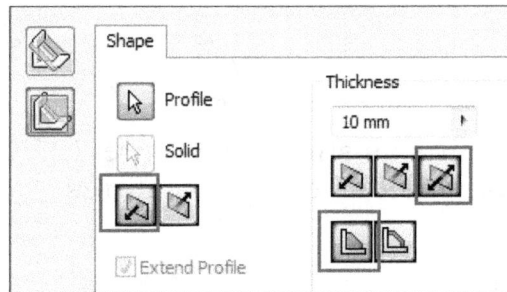

Figure-74. Buttons selected in Rib dialog box

- Click on the **OK** button from the dialog box. The rib feature will be created as shown in Figure-75.

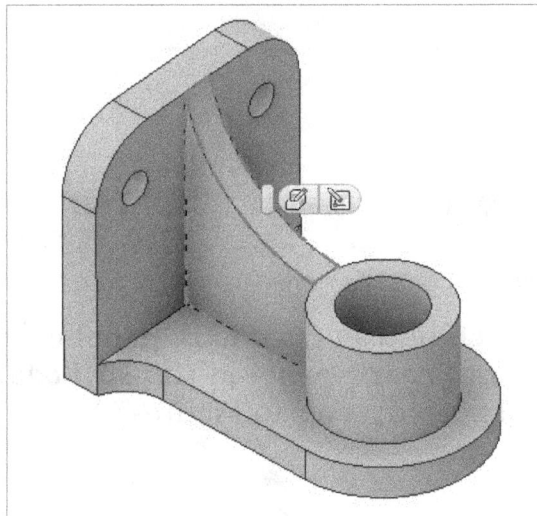

Figure-75. Rib feature created

PRACTICAL 5

In this practical, we will create a model as shown in Figure-76. Dimensions of the model are given in Figure-77.

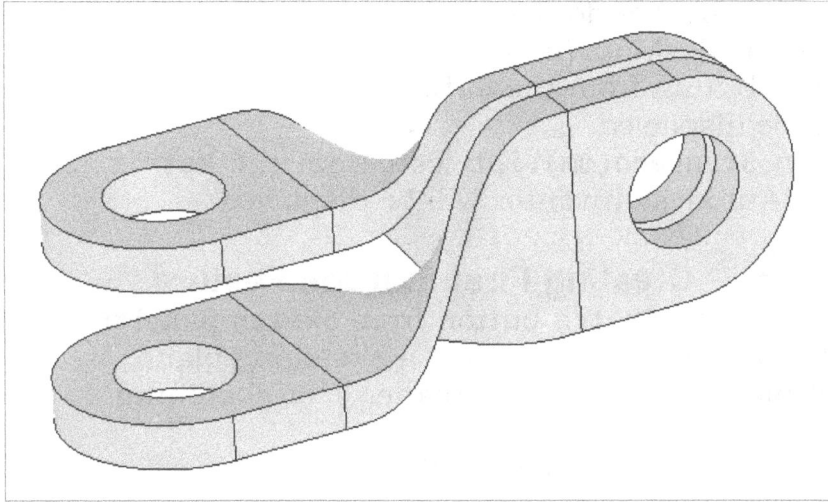

Figure-76. Model for Practical 5

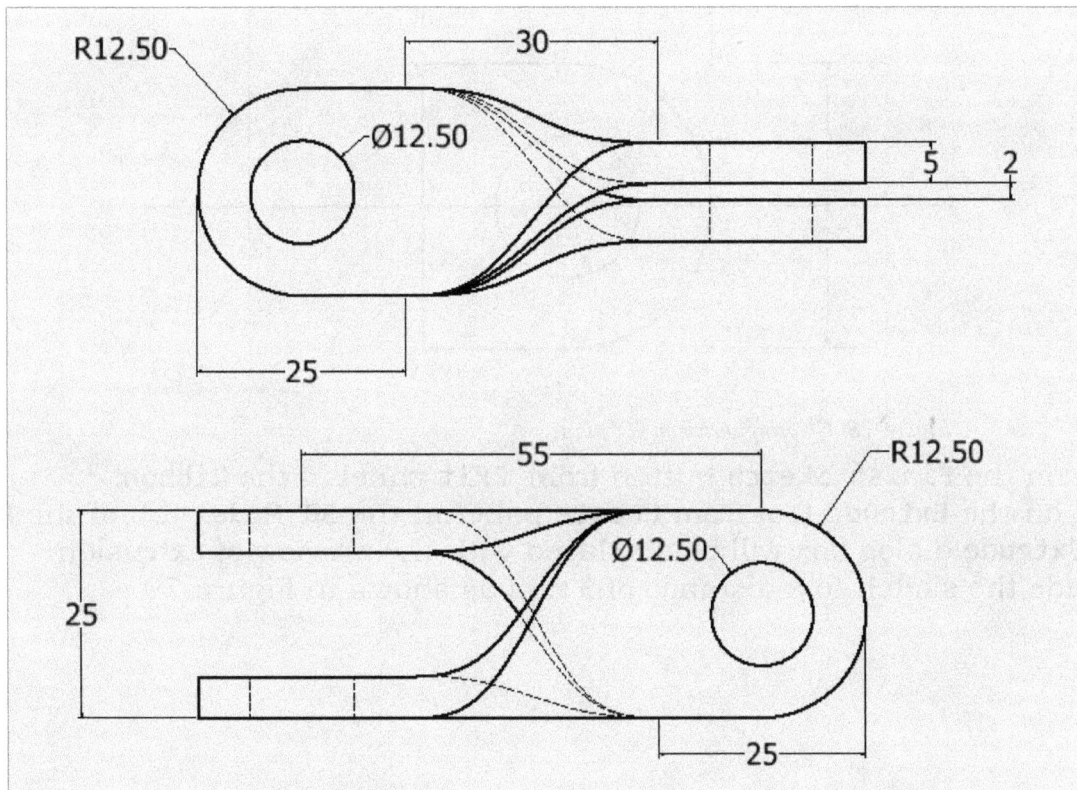

Figure-77. Dimensions for Practical 5

Strategy for Creating model

1. We need to create an extrude feature and then create rectangular pattern of that feature.
2. Similarly, we will create another extrude feature on the created offset plane and create a mirror copy of that extruded feature.
3. Using **Loft** tool, we will create loft feature between the extruded features.

4. Then, we will change the position of transition points in the **Transition** tab of the **Loft** dialog box.

Starting Part File

- Start Autodesk Inventor by double-clicking on the Autodesk Inventor Professional 2024 icon from the desktop. (If not started yet.)
- Click on the **New** button from the **Quick Access Toolbar**. The **Create New File** dialog box will be displayed.
- Double-click on **Standard(mm).ipt** icon from the **Metric** templates. The Part environment of Autodesk Inventor will be displayed.

Creating First Extrude Feature

- Click on the **Start 2D Sketch** button from **Sketch** panel in the **3D Model** tab of the **Ribbon**. You will be asked to select a sketching plane.
- Select the **XY** plane as sketching plane. The sketching environment will be displayed.
- Create a sketch as shown in Figure-78.

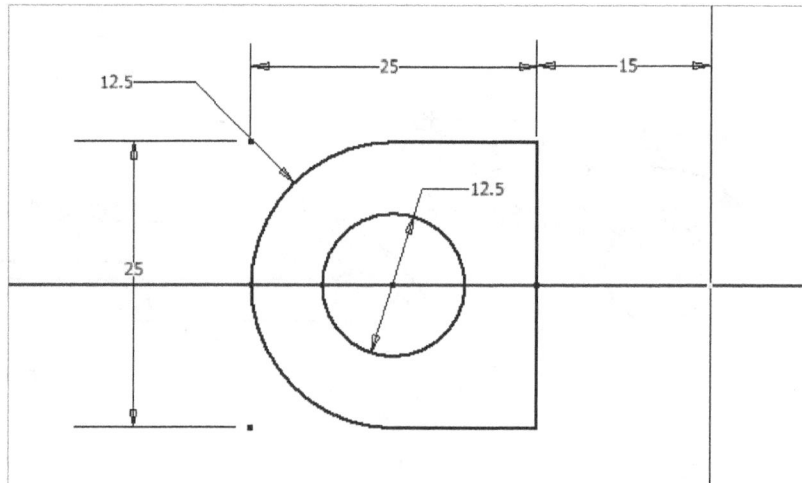

Figure-78. Creating sketch on XY plane

- Click on the **Finish Sketch** button from **Exit** panel in the **Ribbon**.
- Click on the **Extrude** tool from **Create** panel in the **3D Model** tab of the **Ribbon**. The **Extrude** dialog box will be displayed with the preview of extrusion.
- Extrude the sketch to a distance of **5** mm as shown in Figure-79.

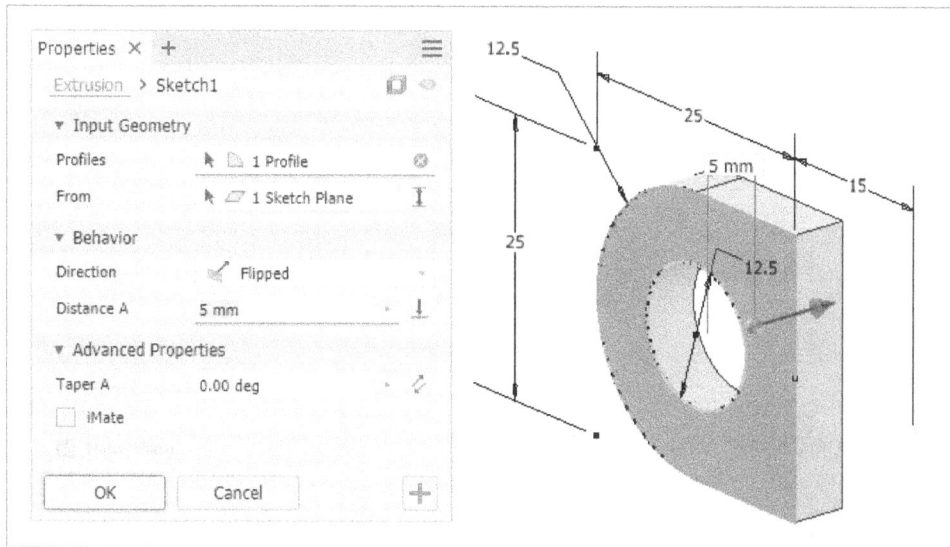

Figure-79. Extruding the sketch

- Click on the **OK** button from the **Extrude** dialog box to create the feature.

Creating pattern of the feature

- Click on the **Rectangular Pattern** tool from **Pattern** panel in the **3D Model** tab of **Ribbon**. The **Rectangular Pattern** dialog box will be displayed. You will be asked to select the feature.
- Select the recently created extrude feature to be patterned; refer to Figure-80.

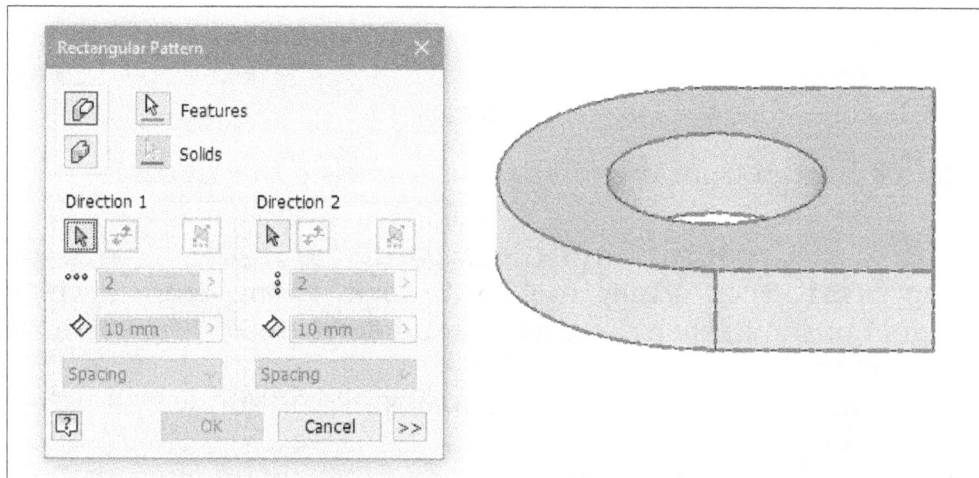

Figure-80. Selecting the feature to be patterned

- Click on the selection button of **Direction 1** area of the dialog box and select the edge as shown in Figure-81.

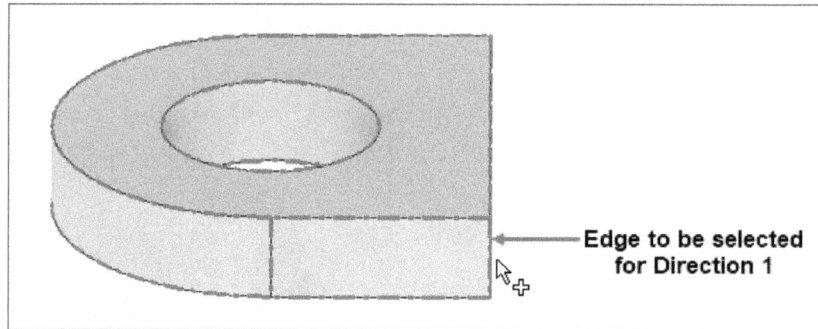

Figure-81. Edge selected for direction reference

- Specify the value of distance as **25** and number of instances as **2** in the respective edit boxes in **Direction 1** area of the dialog box; refer to Figure-82.

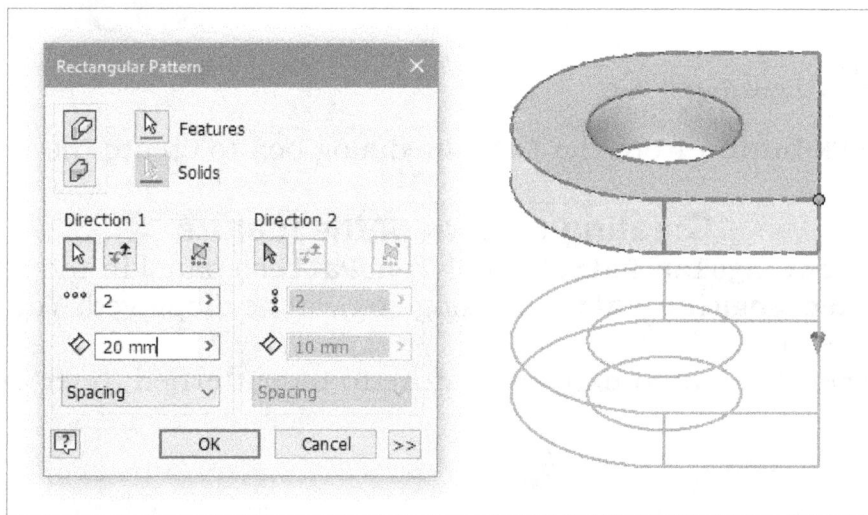

Figure-82. Specifying parameters to create the pattern

- Click on the **OK** button from the dialog box to create the feature.

Creating Second Extrude Feature

- Click on the **Offset from Plane** tool from **Work Plane** drop down in the **Work Features** panel of the **Ribbon**. You will be asked to select the plane or face as a reference to create the plane.
- Select **XZ** plane as a reference from **Model Browser** bar. You will be asked to specify the offset distance.
- Specify offset distance value as **1** mm in the edit box of the mini toolbar displayed and click on **OK** button from the mini toolbar to create the plane; refer to Figure-83.

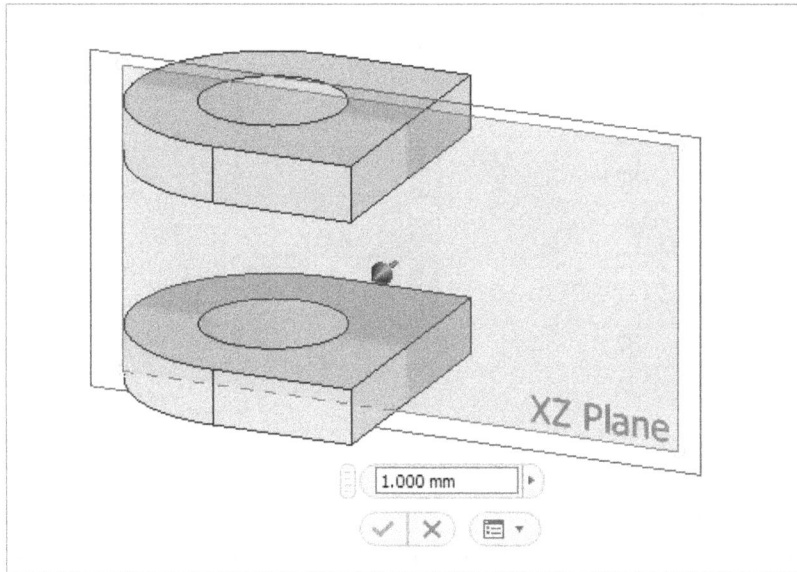

Figure-83. Specifying offset distance for plane

- Click on the **Start 2D Sketch** button from **Sketch** panel in the **3D Model** tab of the **Ribbon**. You will be asked to select a sketching plane.
- Select newly created offset plane as sketching plane. The sketching environment will be displayed.
- Create the sketch as shown in Figure-84.

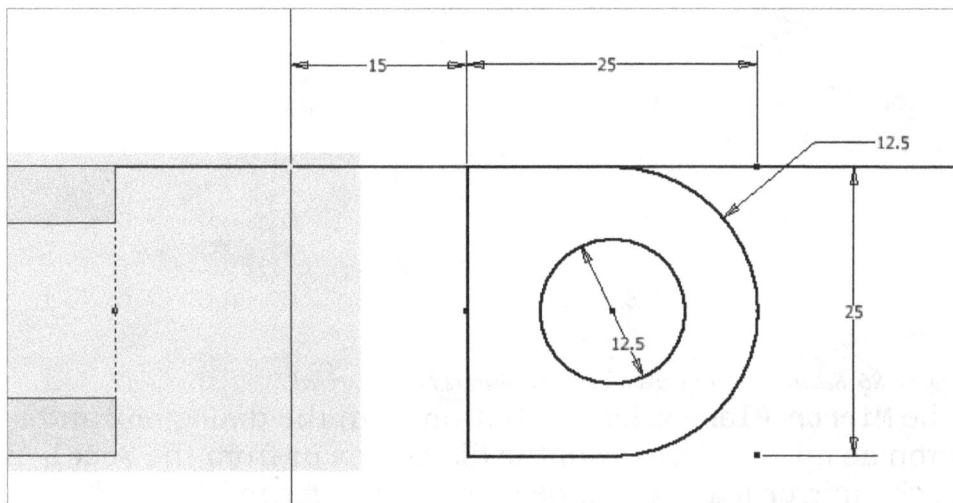

Figure-84. Creating sketch on offset plane

- Click on the **Finish Sketch** button from **Exit** panel in the **Ribbon**.
- Click on the **Extrude** tool from **Create** panel in the **3D Model** tab of the **Ribbon**. The **Extrude** dialog box will be displayed with the preview of extrusion.
- Extrude the sketch to a distance of **5** mm as shown in Figure-85.

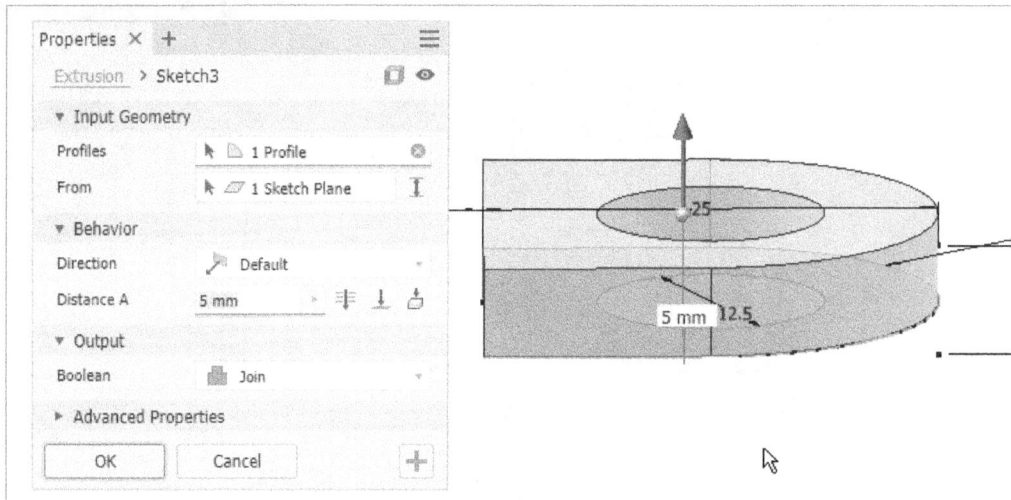

Figure-85. Extruding the second sketch

- Click on the **OK** button from the **Extrude** dialog box to create the feature.

Creating Mirror Feature

- Click on the **Mirror** tool from **Pattern** panel in the **3D Model** tab of the **Ribbon**. The **Mirror** dialog box will be displayed. You will be asked to select the feature to be mirrored.
- Select the recently created extrude feature as shown in Figure-86.

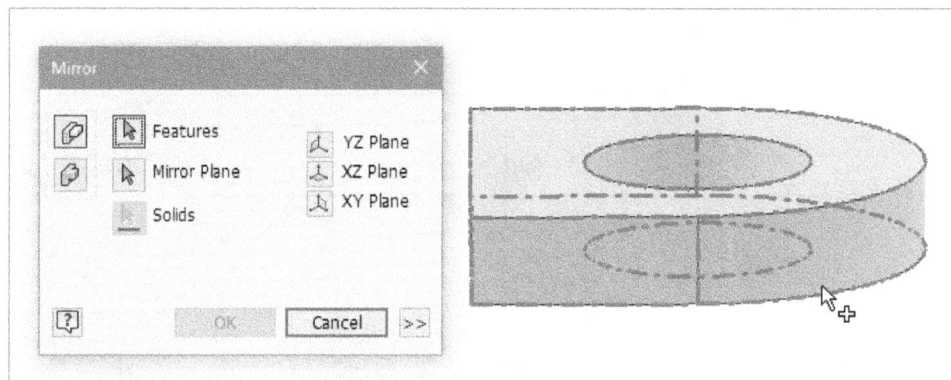

Figure-86. Selecting recently created extrude feature to be mirrored

- Click on the **Mirror Plane** selection button from the dialog box and select the **XZ Plane** button as mirror plane from the dialog box or from the **Model Browser** bar. Preview of the mirror feature will be displayed; refer to Figure-87.

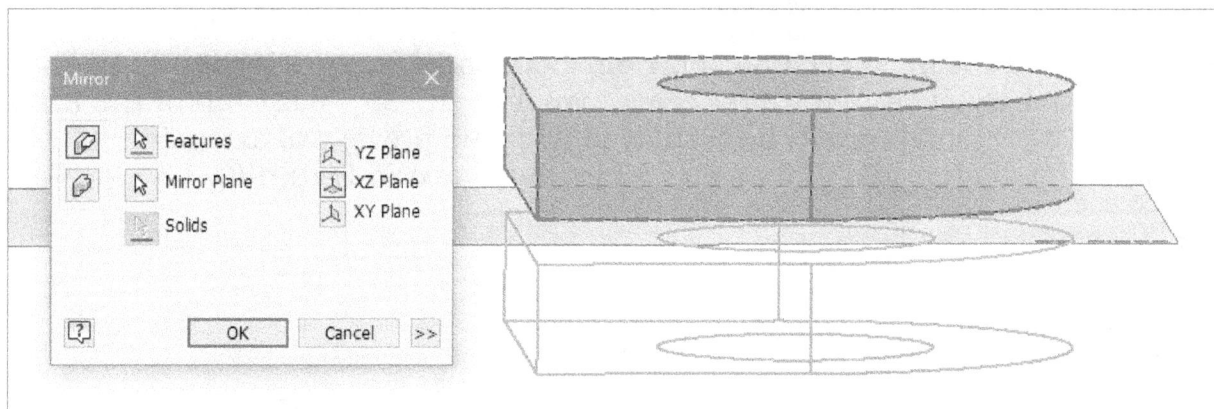

Figure-87. Selecting the mirror plane

- Click on the **OK** button from the dialog box to create the mirror feature.

Creating Loft Feature

- Click on the **Loft** tool from **Create** panel in the **3D Model** tab of the **Ribbon**. The **Loft** dialog box will be displayed. You will be asked to select the profiles to loft.
- One by one select the profiles between which you want to create the loft feature. Preview of the loft feature will be displayed; refer to Figure-88.

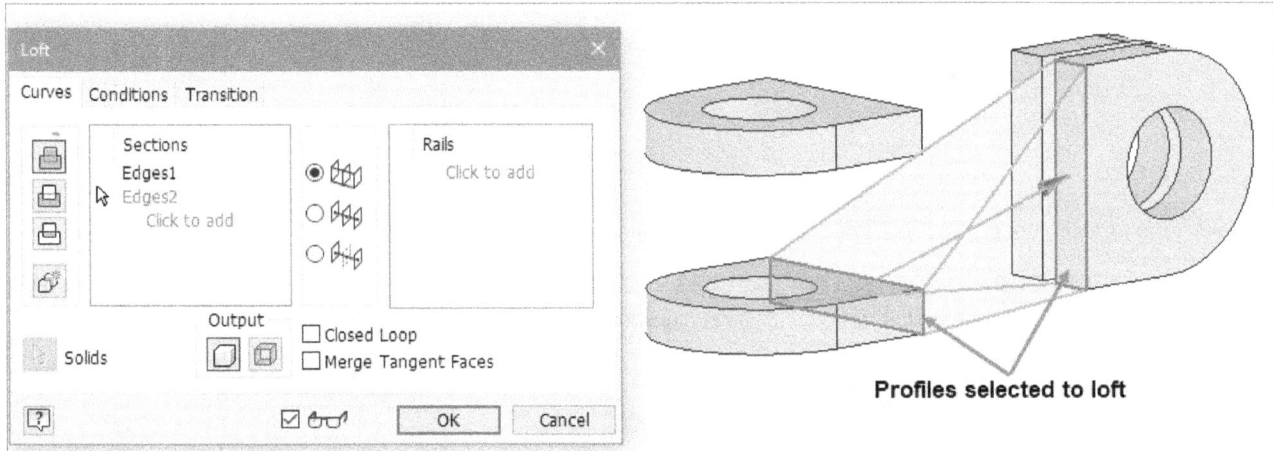

Figure-88. Preview of loft feature

- Click on the **Transition** tab of the **Loft** dialog box and clear the **Automatic Mapping** check box. The options to change position of transition points will be displayed; refer to Figure-89.

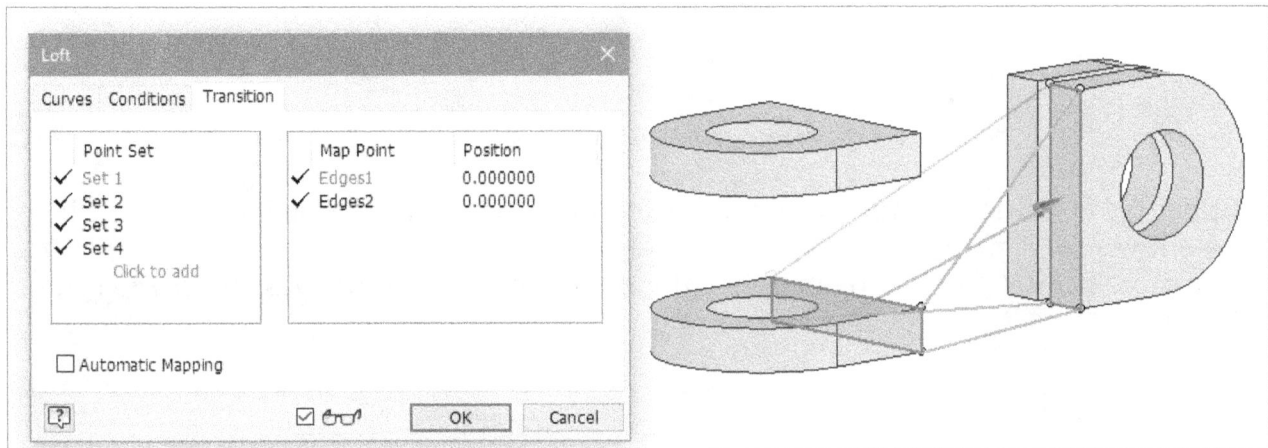

Figure-89. Options to change position of transition points

- Drag the keypoints to change the transition between the profiles as shown in Figure-90.

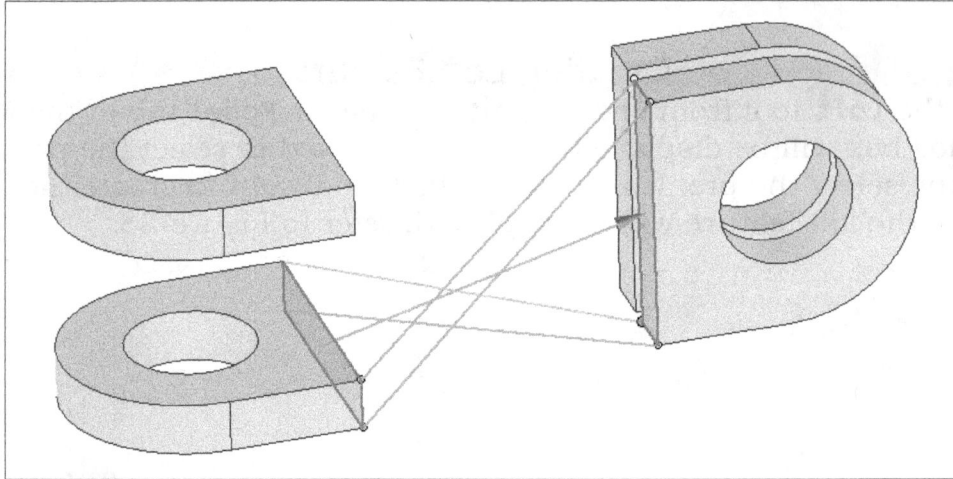

Figure-90. Changing the transition points

- Click on the **Conditions** tab of the **Loft** dialog box and select **Smooth (G2) Condition** option from **Conditions** drop-down of both the sections; refer to Figure-91.

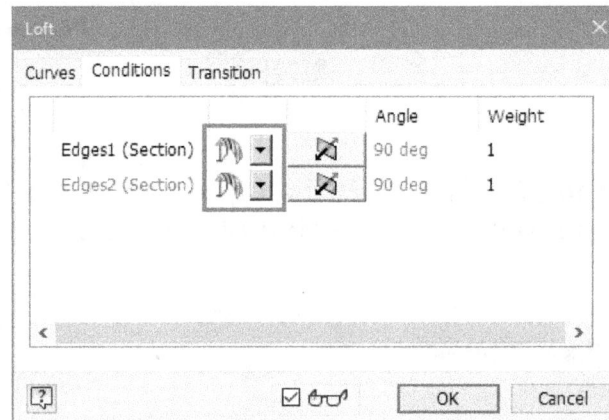

Figure-91. Specifying conditions for loft feature

- Specify the value of angle and weight for loft feature in the respective edit boxes as shown in Figure-91.
- Click on the **OK** button from the **Loft** dialog box to create the feature; refer to Figure-92.

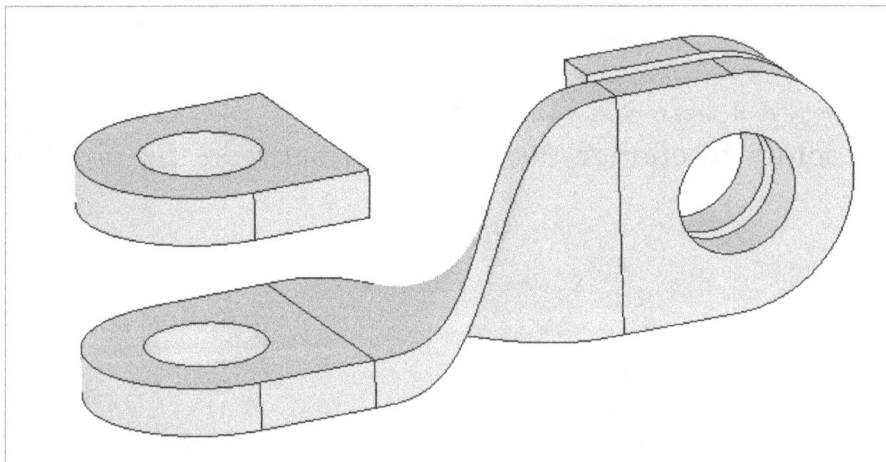

Figure-92. Creating the loft feature

- Similarly, create the second loft feature between the other profiles. The final model will be created as shown in Figure-93.

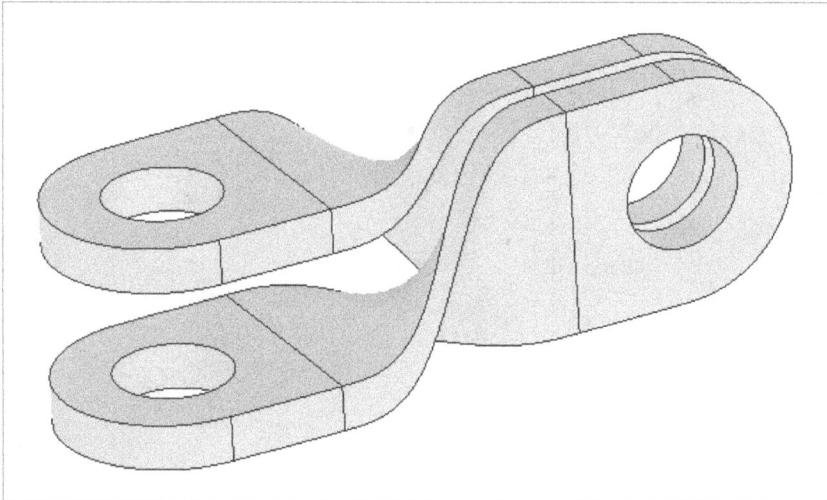

Figure-93. Final model of loft feature created

PRACTICAL 6

In this practical, we will create a model as shown in Figure-94. Dimensions of the model are given in Figure-95.

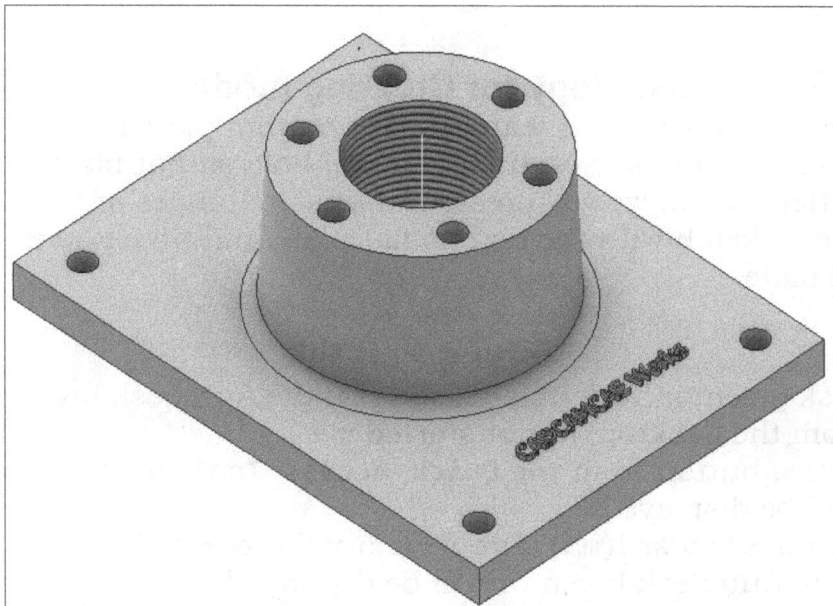

Figure-94. Practical 6 model

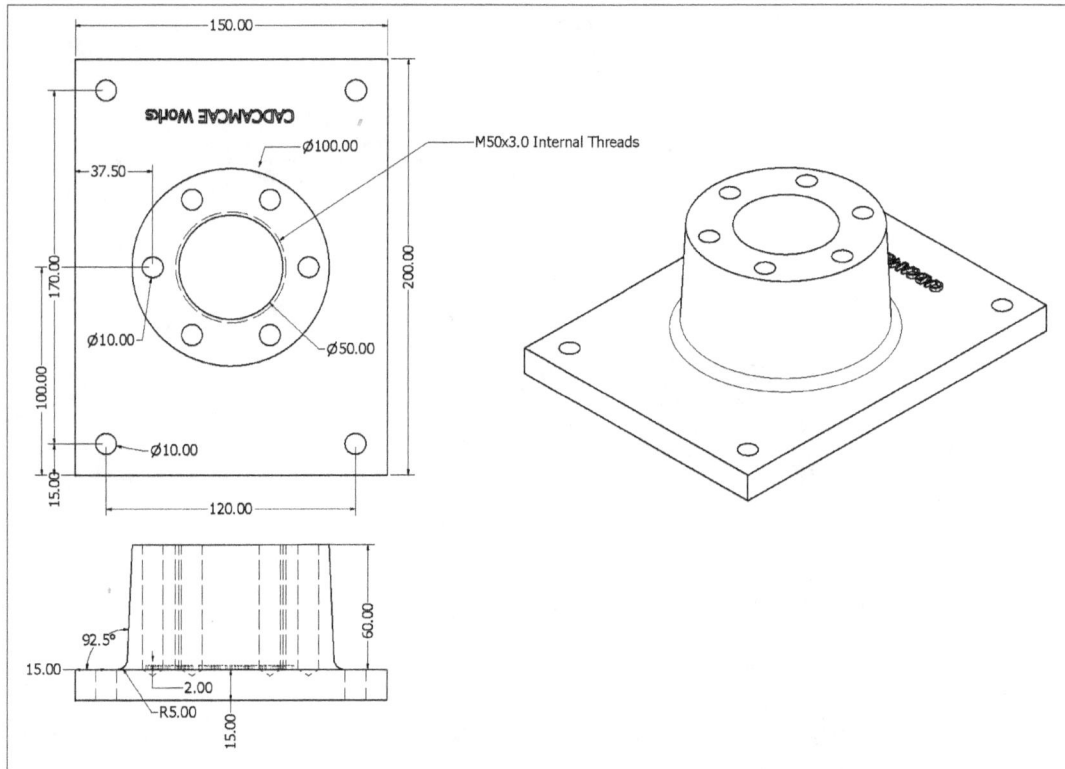

Figure-95. Practical 6 drawing

Strategy for Creating model

1. We need to create two extrude features; flat bottom plate and round tube on it.
2. We will then create one hole on tube and one hole on flat plate.
3. Using the Pattern tools, we will create multiple instances of hole.
4. We will create a sketch with the text on flat plate and then use the **Emboss** tool to emboss it on plate.

Starting Part File

- Start Autodesk Inventor by double-clicking on the Autodesk Inventor Professional 2024 icon from the desktop. (If not started yet.)
- Click on the **New** button from the **Quick Access Toolbar**. The **Create New File** dialog box will be displayed.
- Double-click on **Standard(mm).ipt** icon from the **Metric** templates. The Part environment of Autodesk Inventor will be displayed.

Creating First Extrude Feature

- Click on the **Start 2D Sketch** button from **Sketch** panel in the **3D Model** tab of the **Ribbon**. You will be asked to select a sketching plane.
- Select the **XZ** plane as sketching plane. The sketching environment will be displayed.
- Create a sketch as shown in Figure-96.

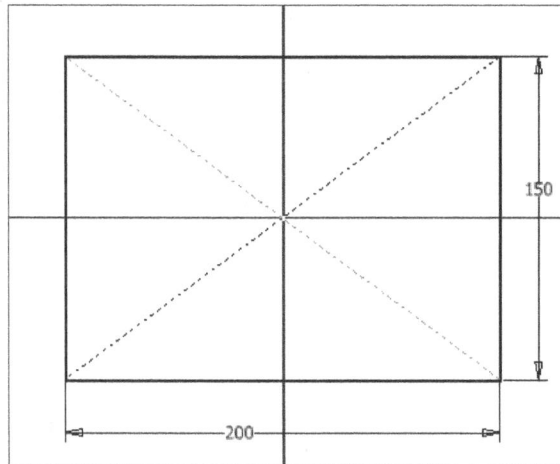

Figure-96. Sketch on XZ plane

- Click on the **Finish Sketch** button from **Exit** panel in the **Ribbon**.
- Click on the **Extrude** tool from **Create** panel in the **3D Model** tab of the **Ribbon**. The **Extrude** dialog box will be displayed with the preview of extrusion.
- Extrude the sketch to a height of **15** mm; refer to Figure-97.

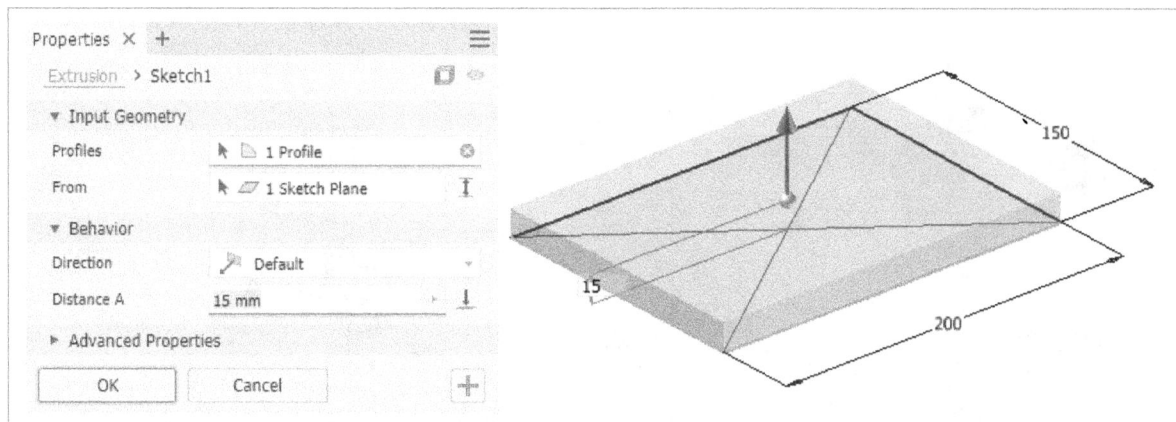

Figure-97. Extruded sketch

- Click on the **OK** button from the **Extrude** dialog box to create the feature.

Creating Second Extrude Feature

- Click on the **Start 2D Sketch** button from **Sketch** panel in the **3D Model** tab of the **Ribbon**. You will be asked to select a sketching plane.
- Select the top face of the plate as sketching plane. The sketching environment will be displayed.
- Create a sketch as shown in Figure-98.

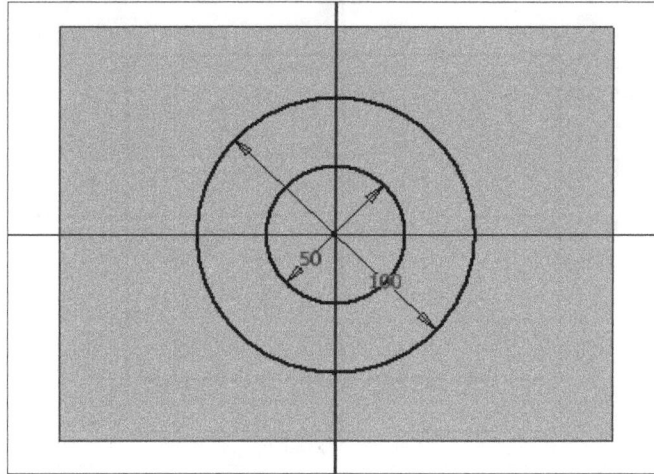

Figure-98. Sketch created on plate

- Click on the **Finish Sketch** button from **Exit** panel in the **Ribbon**.
- Click on the **Extrude** tool from **Create** panel in the **3D Model** tab of the **Ribbon**. The **Extrude** dialog box will be displayed asking you to select the profile.
- Select the newly created sketch and extrude the sketch to a height of **60** mm; refer to Figure-99.

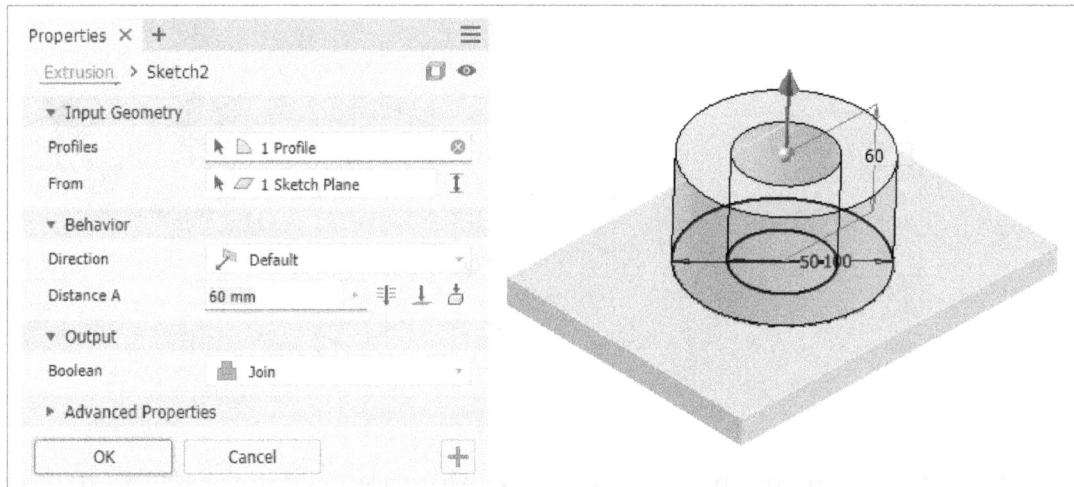

Figure-99. Extruding second feature

- Click on the **OK** button from the dialog box to create the feature.

Creating Holes

- Click on the **Hole** tool from **Modify** panel in the **3D Model** tab of the **Ribbon**. The **Hole** dialog box will be displayed; refer to Figure-100.

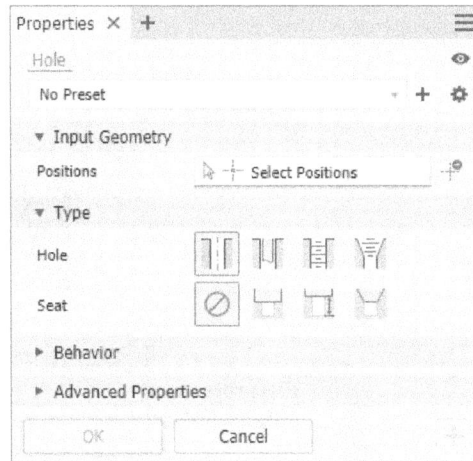

Figure-100. Hole dialog box

- Select the top face of the second extrude feature; refer to Figure-101.

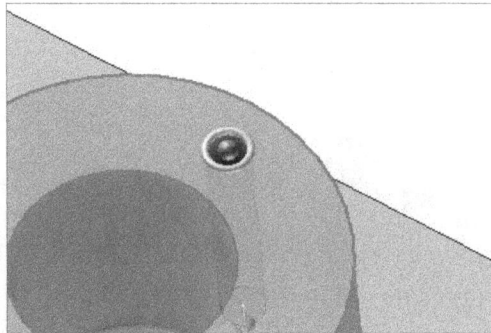

Figure-101. Position selected on the top face of extrude feature

- Specify the diameter of the hole as **10** mm in the corresponding edit box; refer to Figure-102.

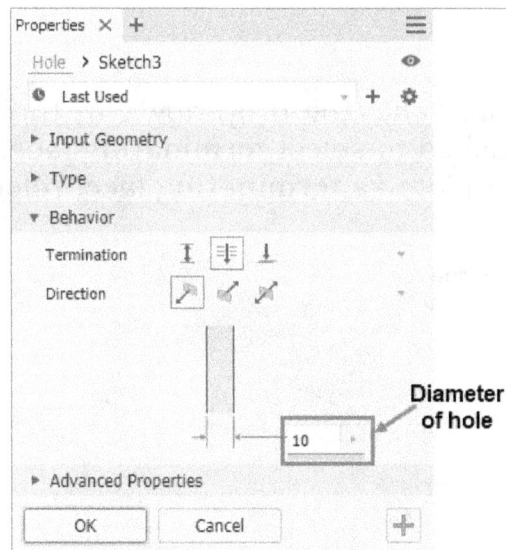

Figure-102. Edit box for specifying hole diameter

- Select the first reference linear edge to create the dimension and set the distance as **100** mm in the corresponding edit box; refer to Figure-103.

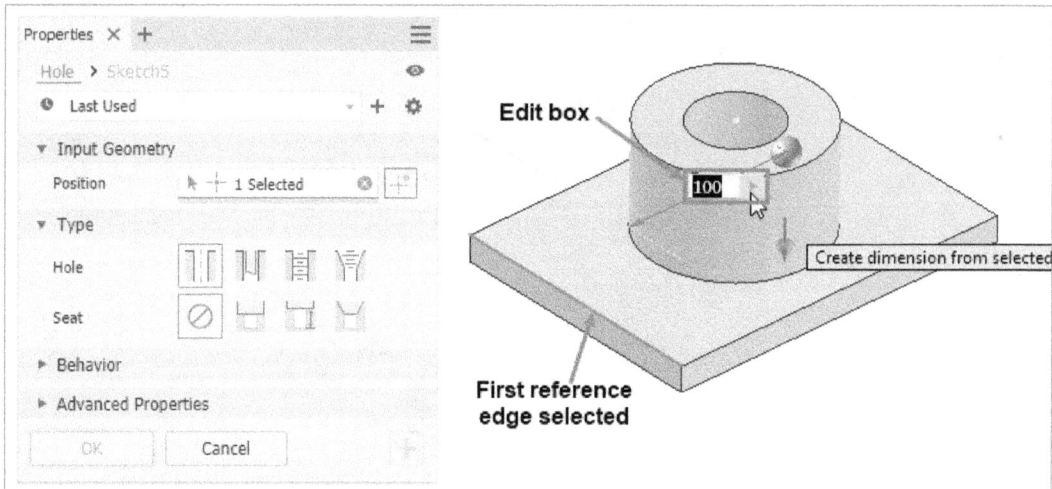

Figure-103. Selecting first reference edge

- Select the second reference linear edge to create the dimension and specify the distance as **37.5** in the corresponding edit box displayed; refer to Figure-104.

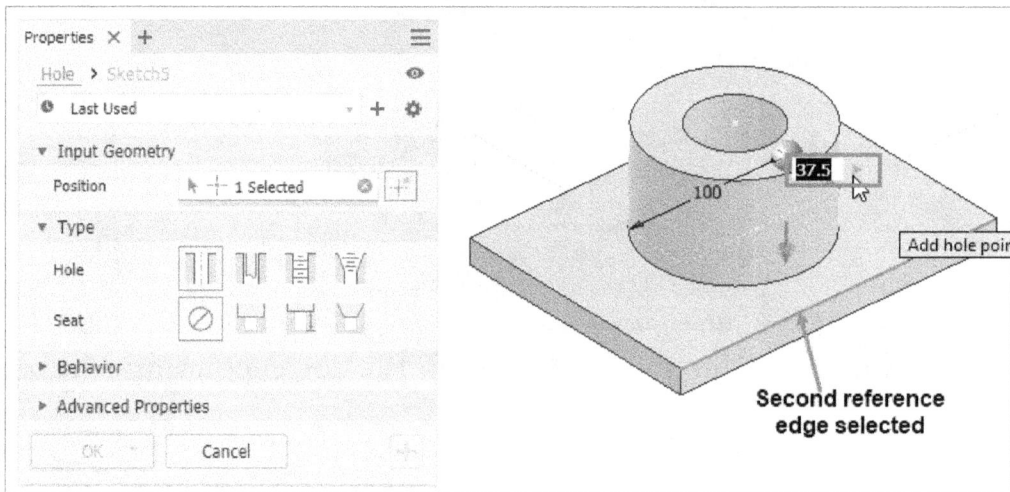

Figure-104. Selecting second reference edge

- Select the **To** button from **Behavior** drop-down in the **Termination** area of the dialog box. You will be asked to select terminating face.
- Select the top face of flat plate as terminating face; refer to Figure-105.

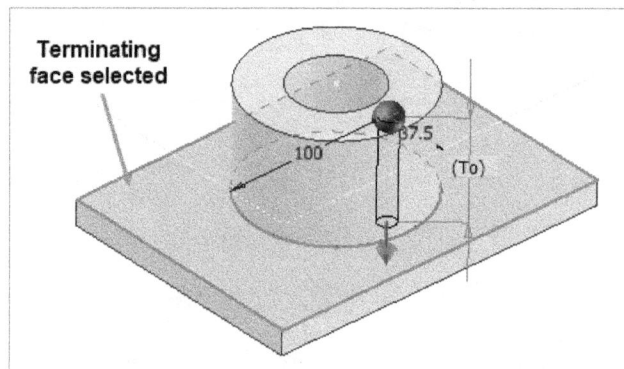

Figure-105. Terminating face selected

- Select the **Terminate feature by extending the face** button since the selected face is not intersecting with the path of hole.
- Click on the **OK** button from the dialog box to create the feature.

- Similarly, create a hole on the flat plate at a distance of **15** mm from the both edge, diameter **10** mm, and termination as through all; refer to Figure-106.

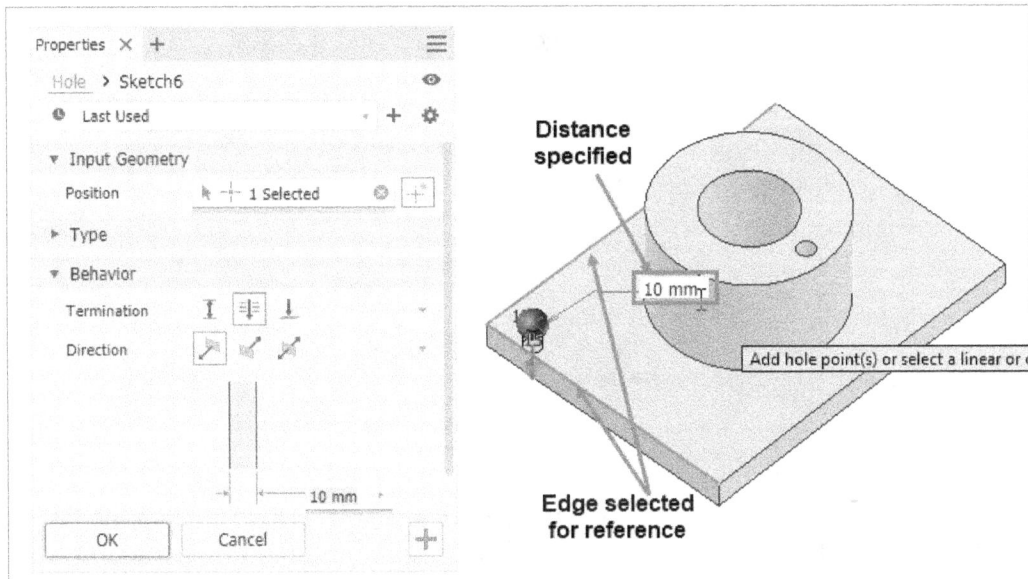

Figure-106. Creating second hole

Creating Patterns of Holes

- Click on the **Rectangular Pattern** tool from **Pattern** panel in the **3D Model** tab of **Ribbon**. The **Rectangular Pattern** dialog box will be displayed; refer to Figure-107.

Figure-107. Rectangular Pattern dialog box

- Select the hole created on flat plate.
- Click on the selection button for first direction from the **Direction 1** area of the dialog box and select the edge as shown in Figure-108.

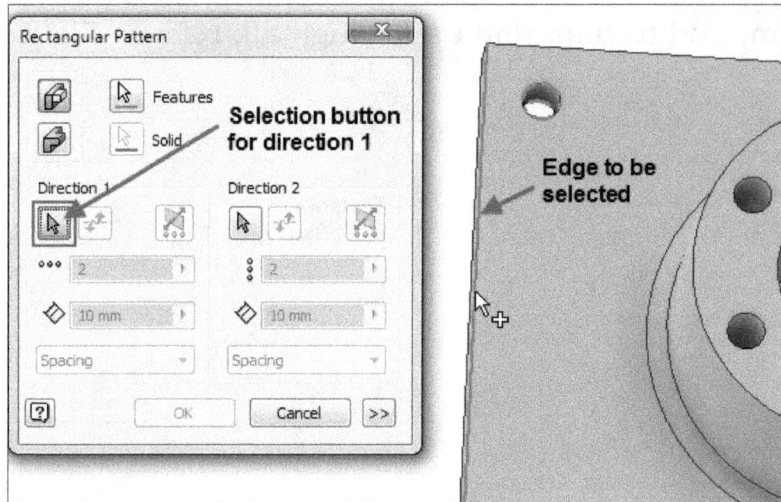

Figure-108. Edge to be selected

- Specify the value of distance as **120** and number of instances as **2** in the respective edit boxes in **Direction 1** area of the dialog box; refer to Figure-109.

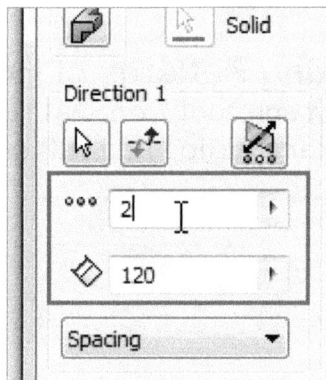

Figure-109. Parameters specified for direction 1

- Similarly, specify the distance in second direction as **170** and number of instances as **2**; refer to Figure-110.

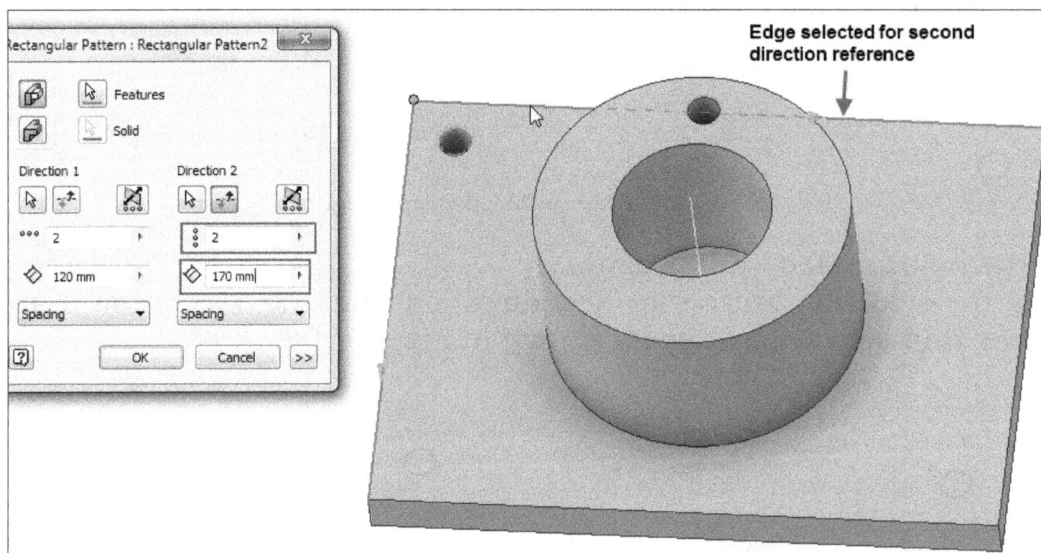

Figure-110. Parameters specified for direction 2

- Click on the **OK** button from the dialog box.

- Click on the **Axis** button from **Work Features** panel in the **3D Model** tab of the **Ribbon**. You will be asked to select geometry for creating axis.
- Select the internal round face of the second extrude feature; refer to Figure-111. (Note that the outer round face will become tapered after applying draft, so we are selecting the internal round face.)

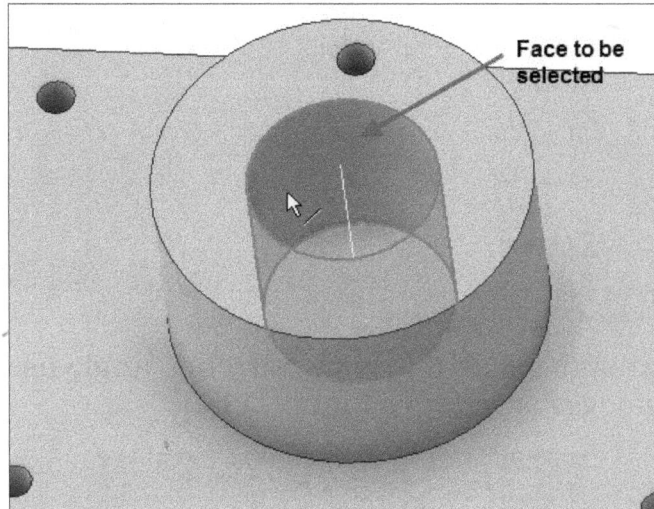

Figure-111. Selecting round face for axis

- Click on the **Circular Pattern** tool from **Pattern** panel in the **3D Model** tab of the **Ribbon**. The **Circular Pattern** dialog box will be displayed; refer to Figure-112. Also, you will be asked to select features to be patterned.

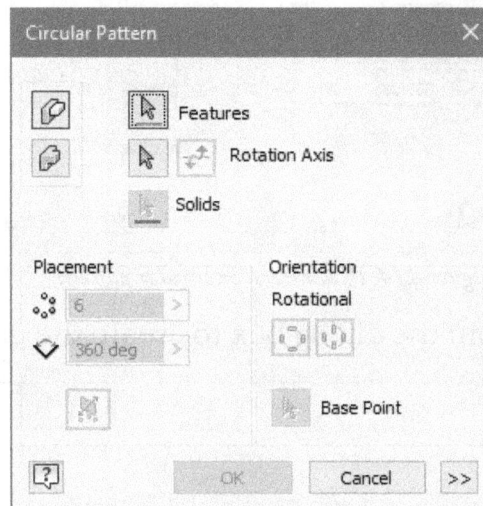

Figure-112. Circular pattern dialog box

- Select the hole created on the face of tube.
- Click on the selection button for **Rotation Axis** in the dialog box and select the axis we have created earlier; refer to Figure-113.

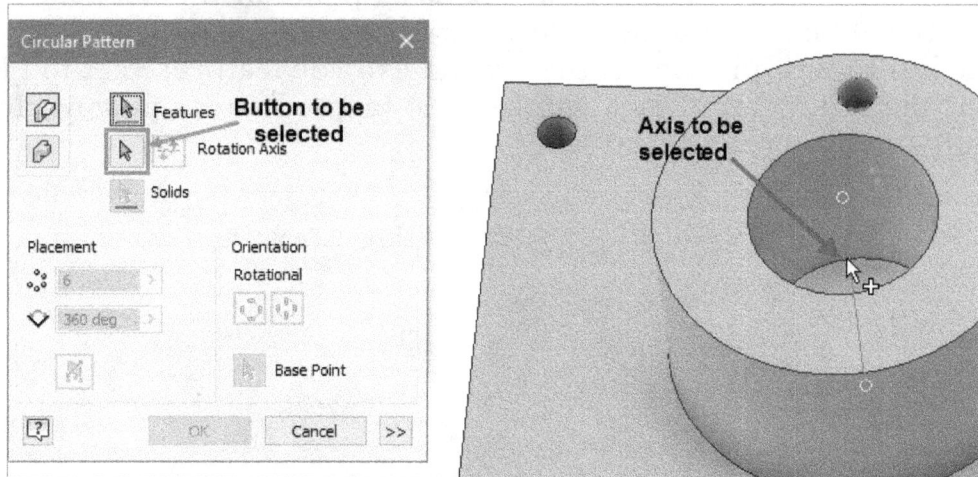

Figure-113. Axis selected for circular pattern

- Set the number of instances as **6** and Occurrence angle as **360** in the respective edit boxes in the dialog box; refer to Figure-114.

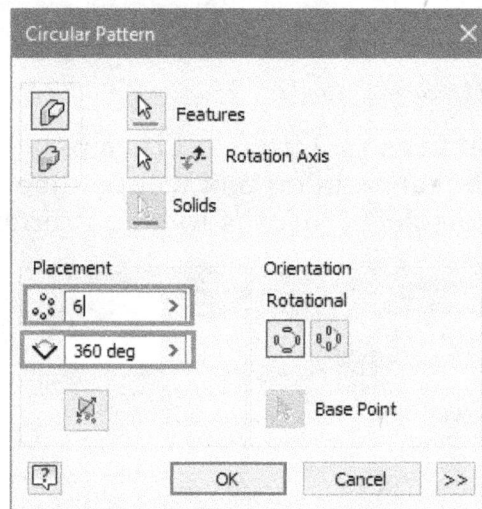

Figure-114. Parameters for circular pattern

- Click on the **OK** button from the dialog box to create the pattern; refer to Figure-115.

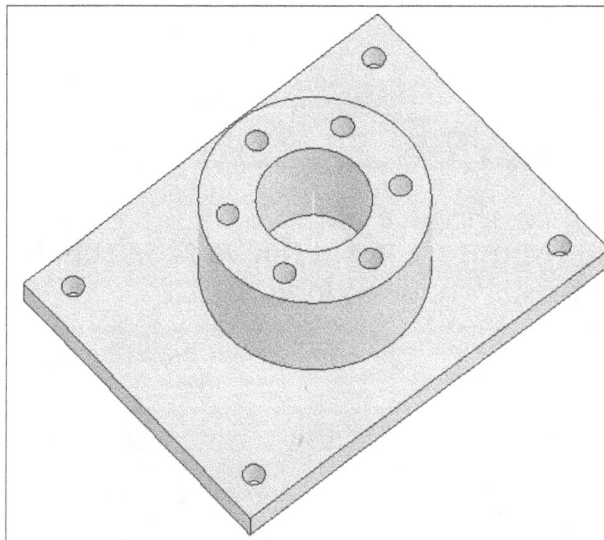

Figure-115. Model after pattern creation

Applying Draft to Round Face

- Click on the **Draft** tool from **Modify** panel in the **3D Model** tab of the **Ribbon**. The **Face Draft** dialog box will be displayed; refer to Figure-116.

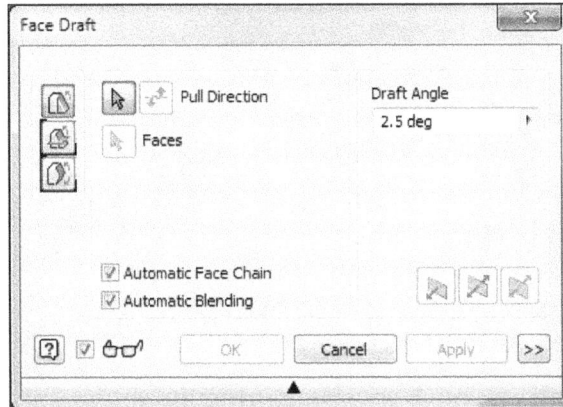

Figure-116. Face Draft dialog box

- Select the top face of the extruded tube feature as pull direction face. You will be asked to select the faces on which you want to apply draft.
- Select the round face of the tube feature; refer to Figure-117.

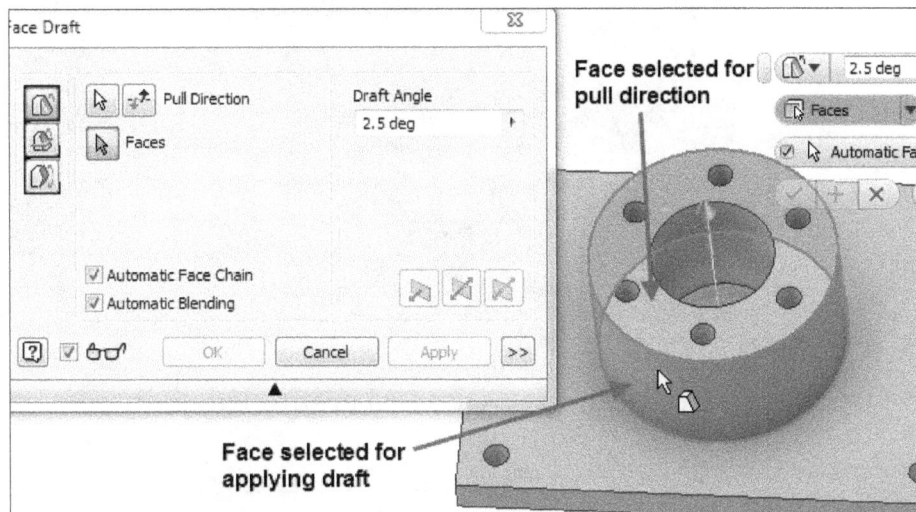

Figure-117. Faces selected for draft

- Specify the angle value as **2.5** deg in the **Draft Angle** edit box of the dialog box.
- Click on the **OK** button to apply draft.

Applying Fillet

- Click on the **Fillet** tool from **Modify** panel in the **3D Model** tab of the **Ribbon**. The **Fillet** dialog box will be displayed; refer to Figure-118. Also, you will be asked to select edges to apply fillet.

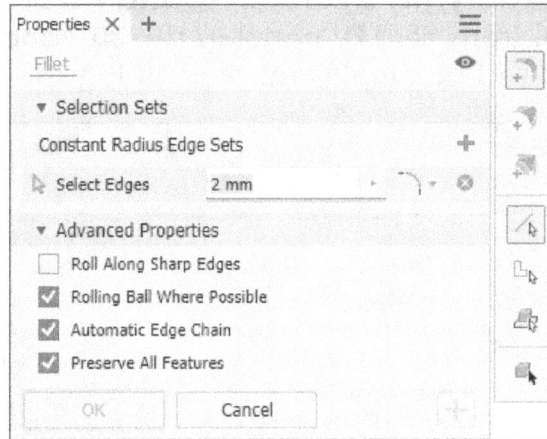

Figure-118. Fillet dialog box

- Specify the value of radius as **5** in the dialog box.
- Select the round edge of the extruded tube feature as shown in Figure-119. Preview of the fillet will be displayed.

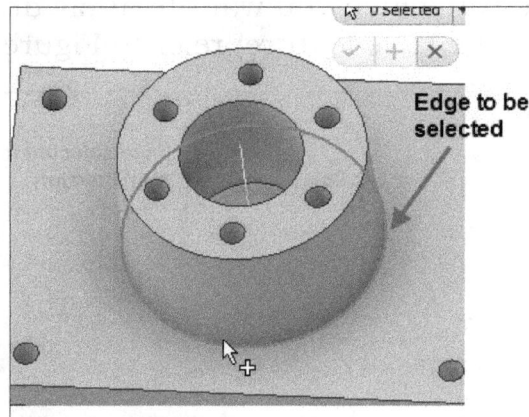

Figure-119. Edge selected for fillet

- Click on the **OK** button from the dialog box to create the feature.

Creating Emboss Feature

- Click on the **Start 2D Sketch** tool from **Sketch** panel in the **3D Model** tab of the **Ribbon**. You will be asked to select sketching plane.
- Select the top face of flat plate. The sketching environment will become active.
- Position the model as shown in Figure-120. Note that you may need to use the rotation arrows of **ViewCube** to change orientation of model; refer to Figure-121.

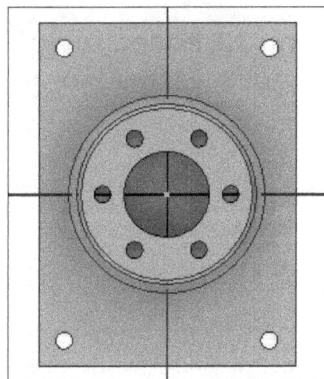

Figure-120. Model after changing orientation

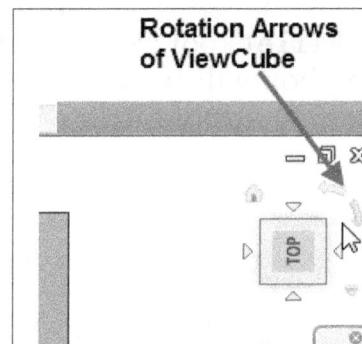

Figure-121. Rotation arrows of ViewCube

- Click on the **Text** tool from **Create** panel in the **Sketch** tab of the **Ribbon**. You will be asked to specify the location of the text box.
- Click at the location approximately as shown in Figure-122. The **Format Text** dialog box will be displayed; refer to Figure-123.

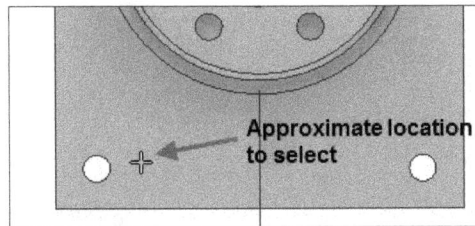

Figure-122. Approximate location to select

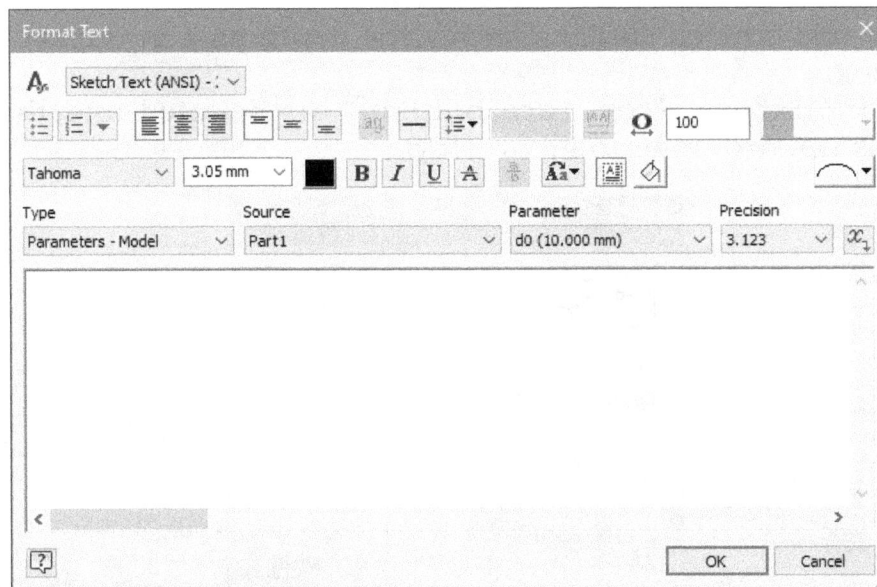

Figure-123. Format Text dialog box

- Select the font size as **6.10** from the size drop-down; refer to Figure-124.

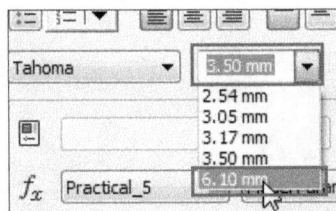

Figure-124. Font size drop-down

- Type the text as **CADCAMCAE Works** in the space provided for it.
- Click on the **OK** button from the dialog box.
- Press **ESC** from the keyboard to exit the tool.
- Click on the text created and drag it to desired position.
- Click on the **Finish Sketch** tool from **Exit** panel in the **Ribbon**.
- Click on the **Emboss** tool from **Create** panel in the **3D Model** tab of the **Ribbon**. The **Emboss** dialog box will be displayed; refer to Figure-125. Also, you will be asked to select the profile for embossing.

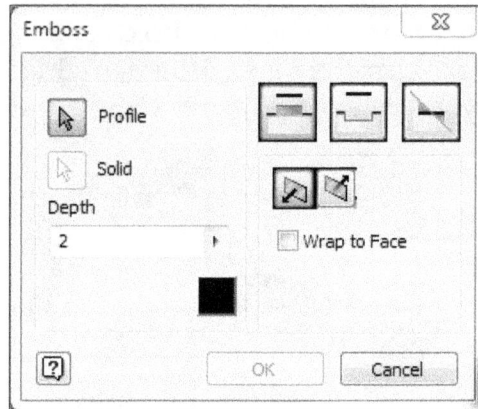

Figure-125. Emboss dialog box

- Click on the text created earlier. Specify the depth as **2** in the **Depth** edit box in the dialog box.
- Make sure the **Emboss from Face** button is selected and the direction of embossing is upward; refer to Figure-126.

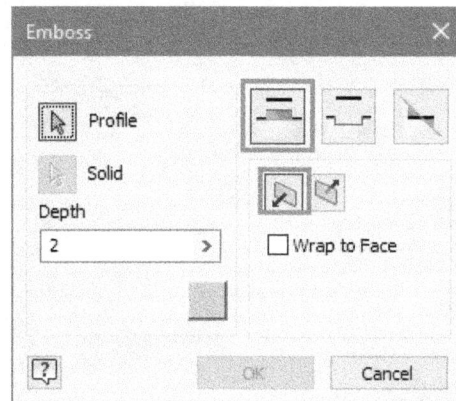

Figure-126. Option to be select for embossing

- Click on the **OK** button from the dialog box. The embossing will be created; refer to Figure-127.

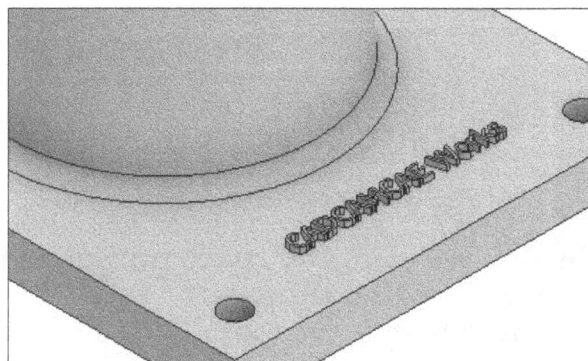

Figure-127. Embossing created

Creating Thread

- Click on the **Thread** tool from **Modify** panel in the **3D Model** tab of the **Ribbon**. The **Thread** dialog box will be displayed; refer to Figure-128.

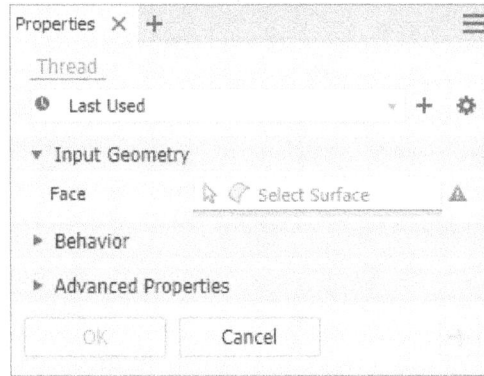

Figure-128. Thread dialog box

- Select the internal face of the extruded tube feature; refer to Figure-129. The updated **Thread** dialog box will be displayed; refer to Figure-130.

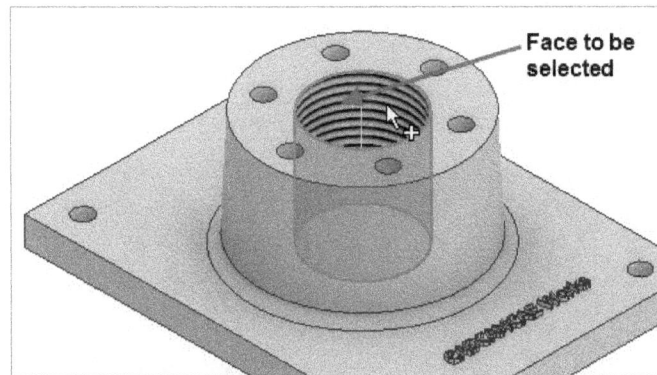

Figure-129. Face selected for thread

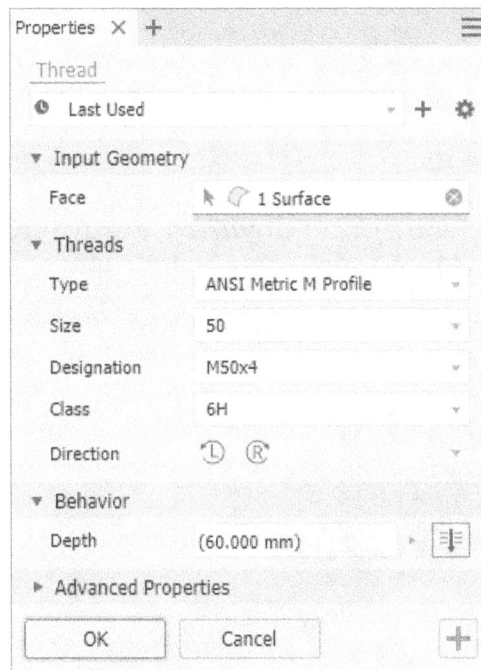

Figure-130. Updated thread dialog box

- Select the size as **50**, class as **6H**, and then click in the **Designation** drop-down and select the **M50x3** option from the list; refer to Figure-131.

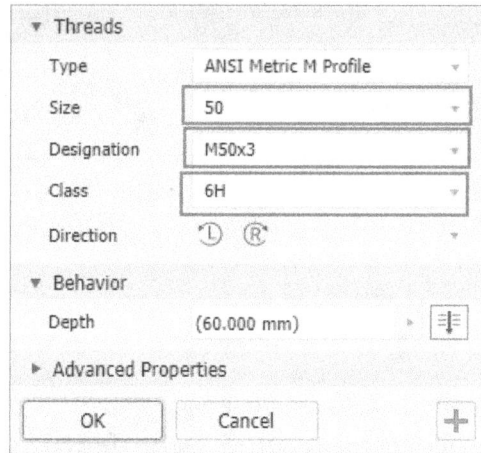
Figure-131. Parameters for thread

- Click on the **OK** button from the dialog box to create the feature; refer to Figure-132.

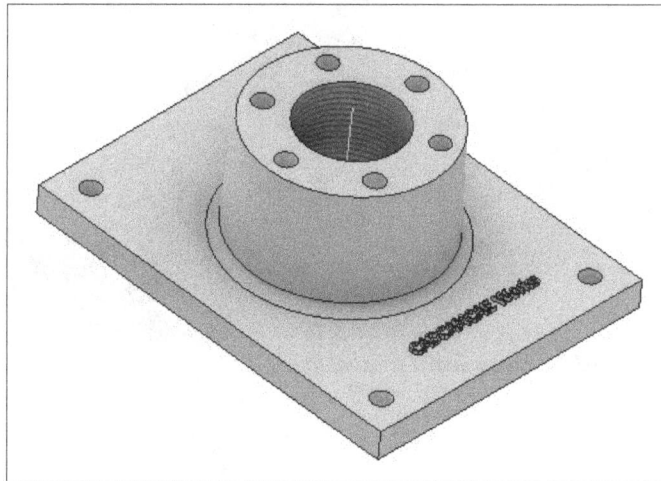
Figure-132. Thread feature created

PRACTICE 1

Create the model (isometric view) as shown in Figure-133. The dimensions of the model are given in Figure-134.

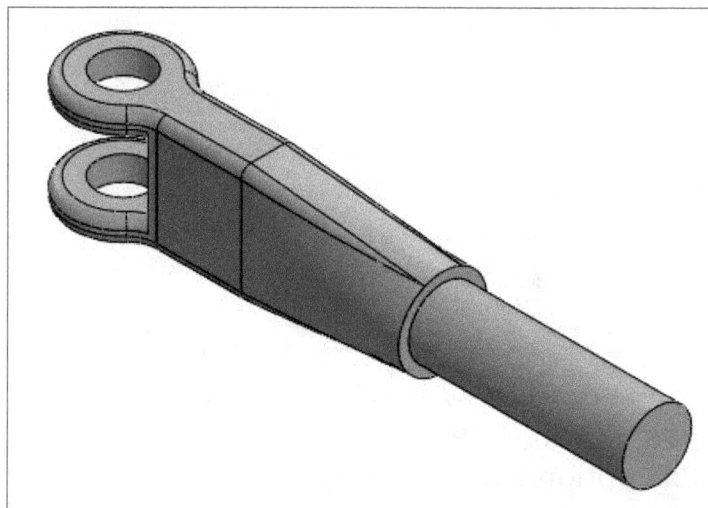
Figure-133. Model for Practice 1

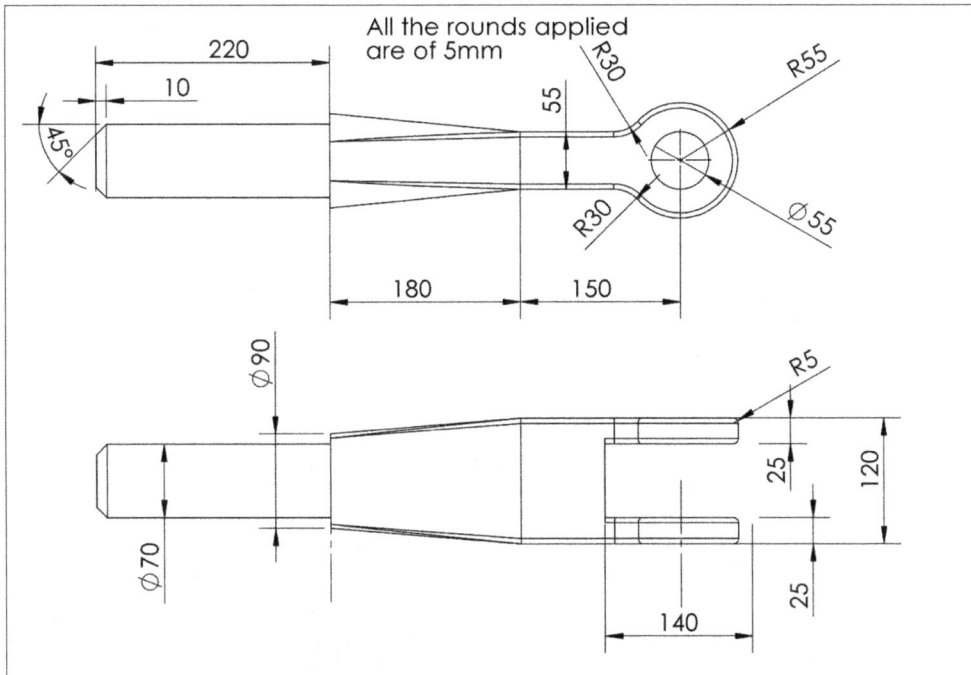

Figure-134. Practice 1 drawing views

PRACTICE 2

Create the model as shown in Figure-135. Dimensions are given in Figure-136. Assume the missing dimensions.

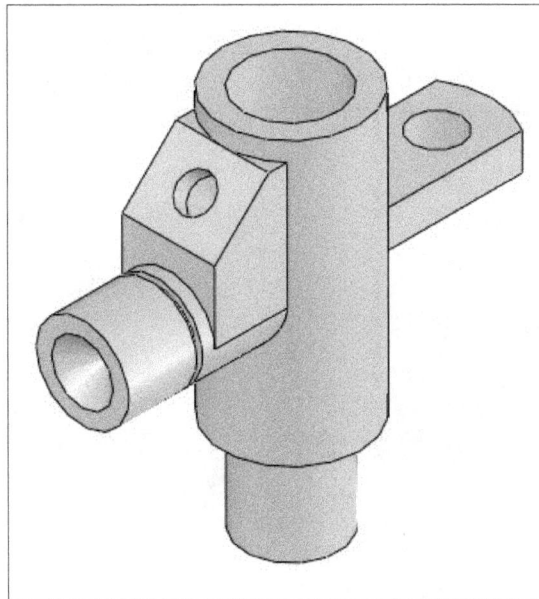

Figure-135. Practice2 model

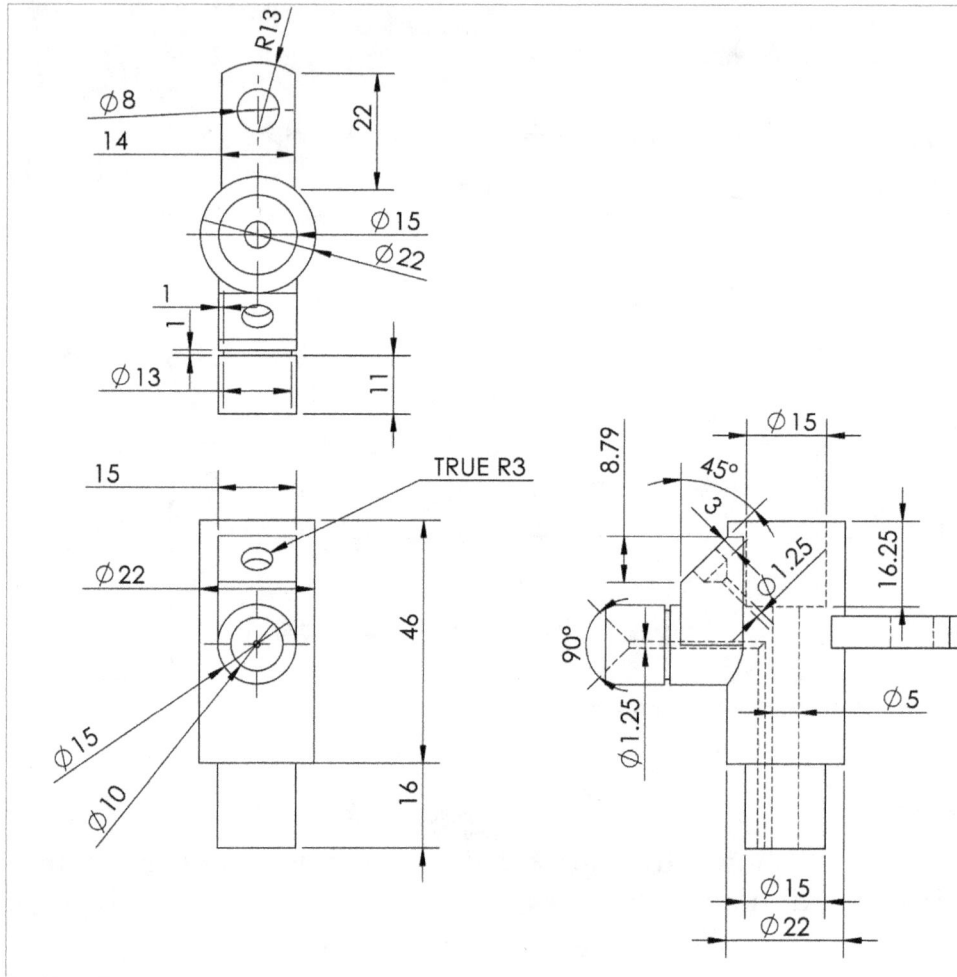

Figure-136. Practice 2

PRACTICE 3

Create the model by using the dimensions given in Figure-137.

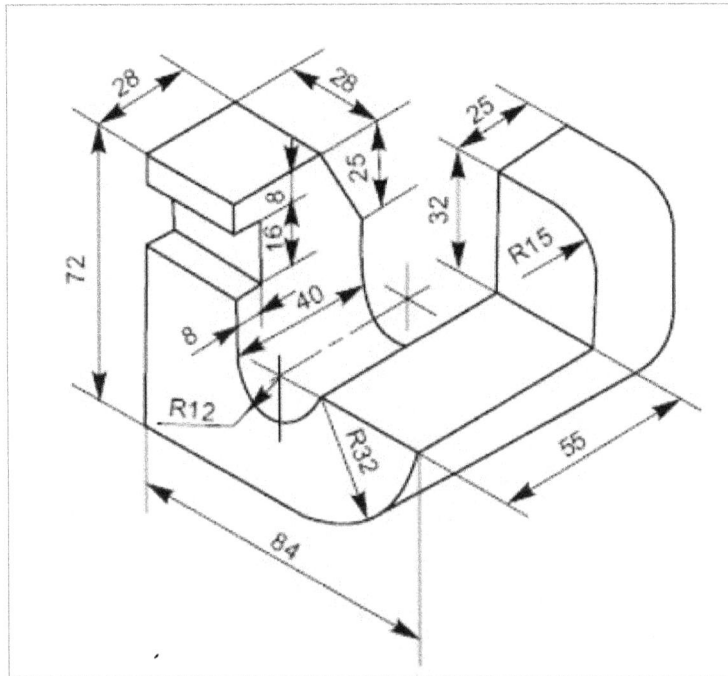

Figure-137. Practice 3

PRACTICE 4

Create the model by using the dimensions given in Figure-138.

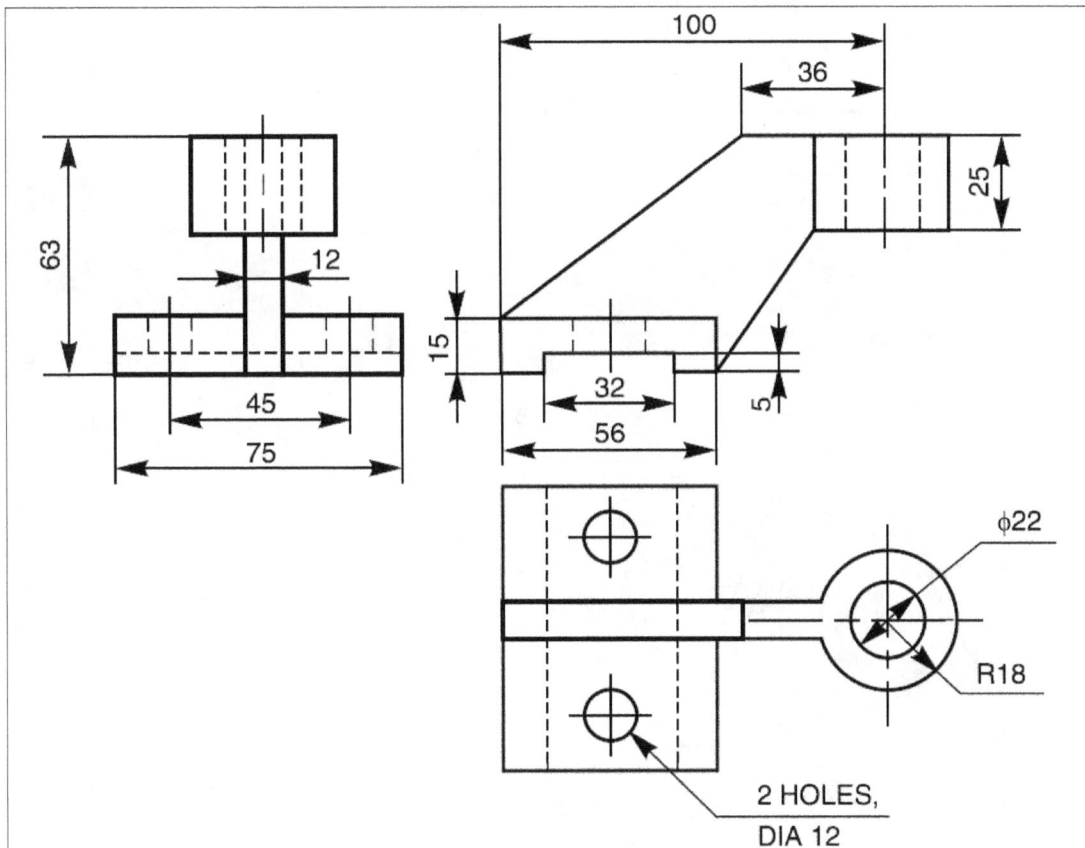

Figure-138. Practice 4

PRACTICE 5

Create the model as shown in Figure-139 by using the dimensions given in Figure-140.

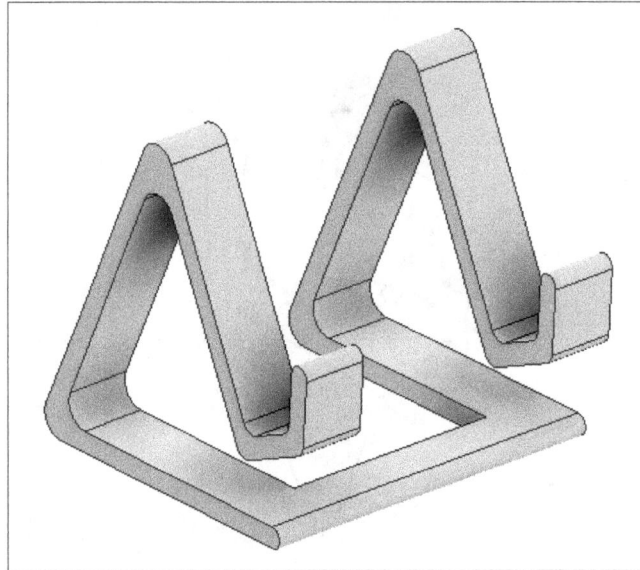

Figure-139. Practice 5

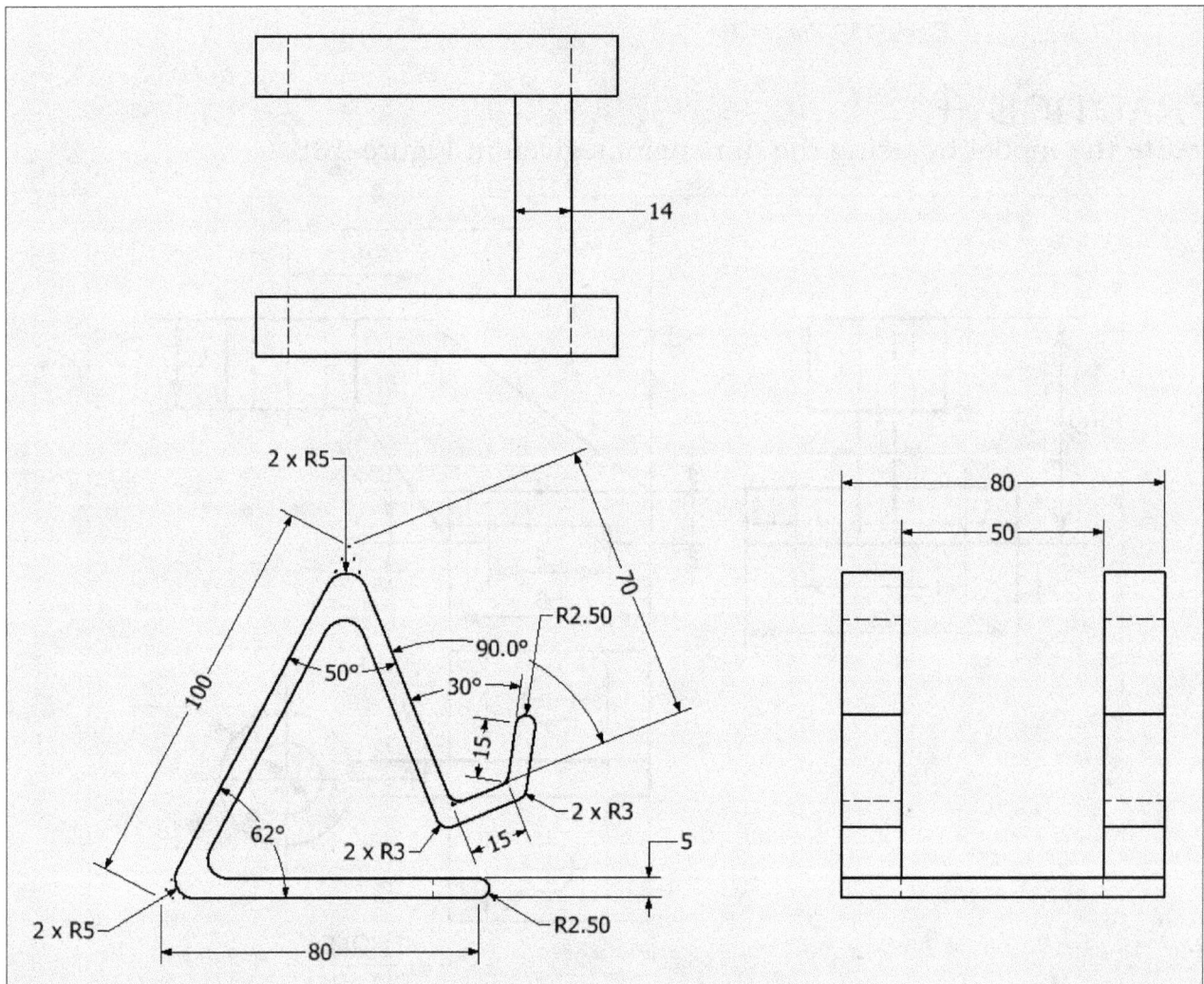

Figure-140. Practice 5 drawing

PRACTICE 6

Create the model as shown in Figure-141 by using the dimensions given in Figure-142.

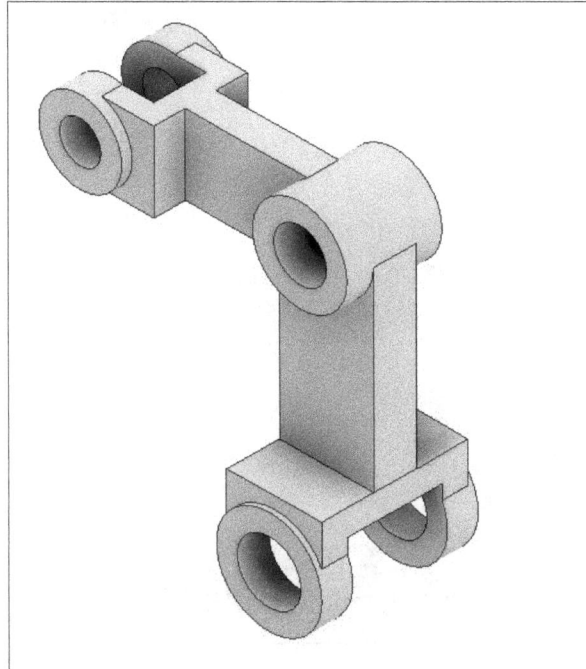

Figure-141. Practice 6

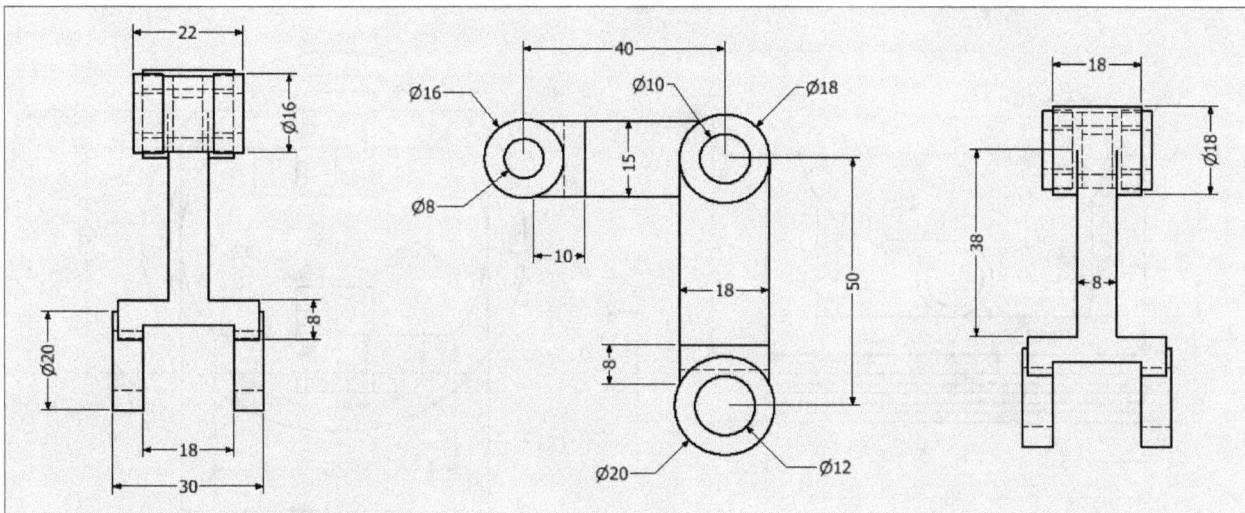

Figure-142. Practice 6 drawing

PRACTICE 7

Create the model as shown in Figure-143 by using the dimensions given in Figure-144.

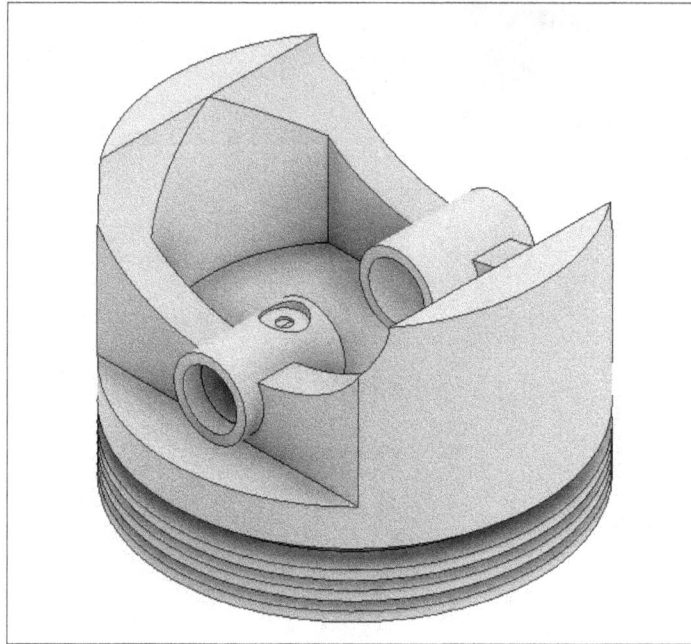

Figure-143. Practice 7

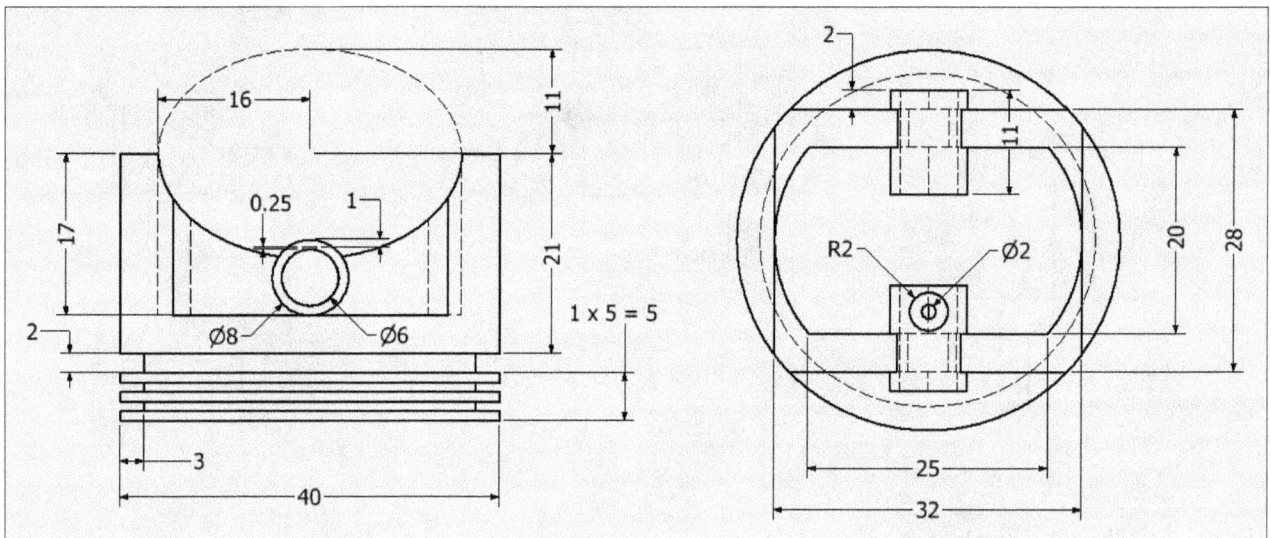

Figure-144. Practice 7 drawing

PRACTICE 8

Create the model as shown in Figure-145 by using the dimensions given in Figure-146.

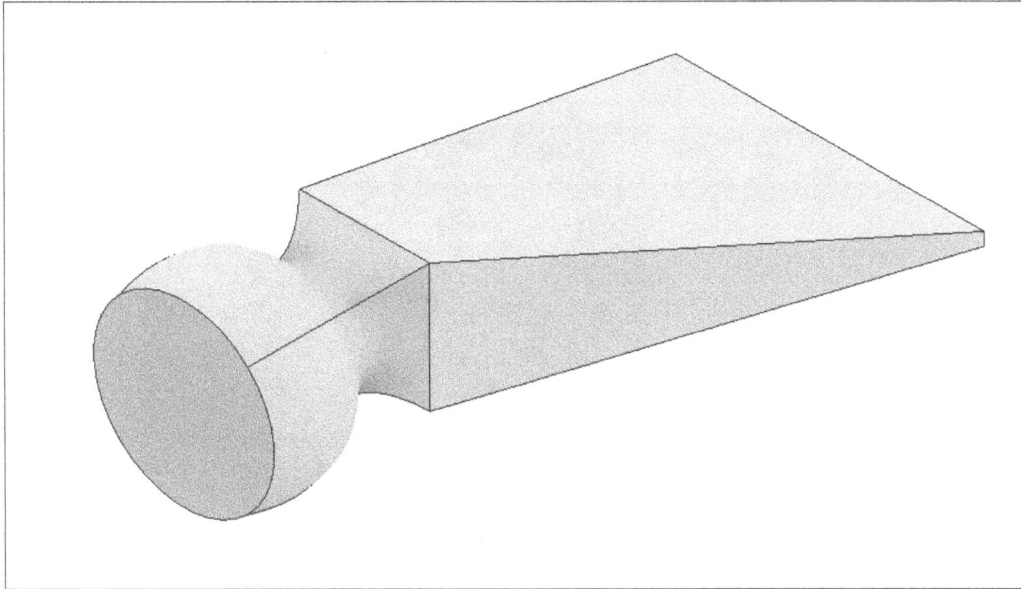

Figure-145. Practice 8

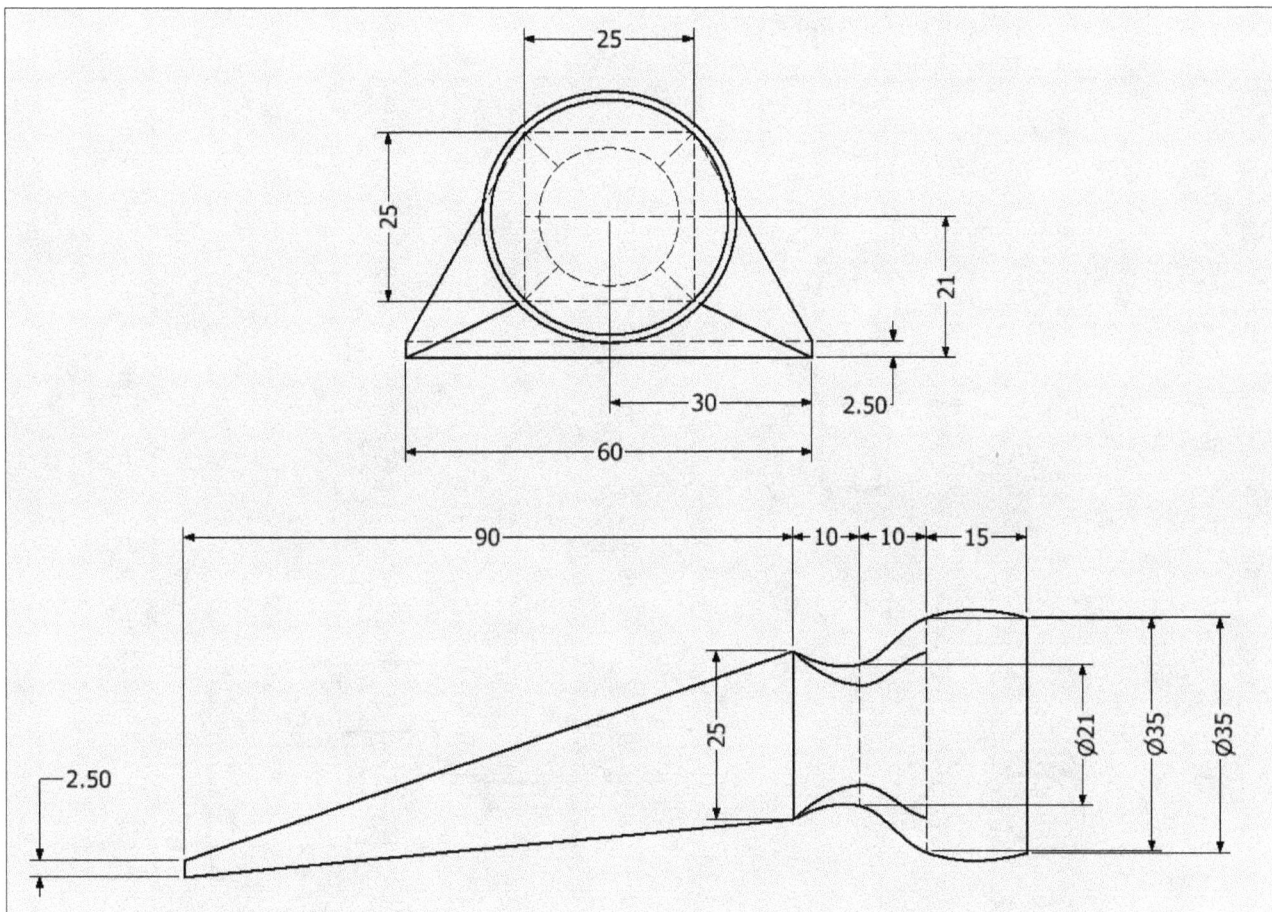

Figure-146. Practice 8 drawing

PRACTICE 9

Create the model as shown in Figure-147 by using the dimensions given in Figure-148.

Figure-147. Practice 9

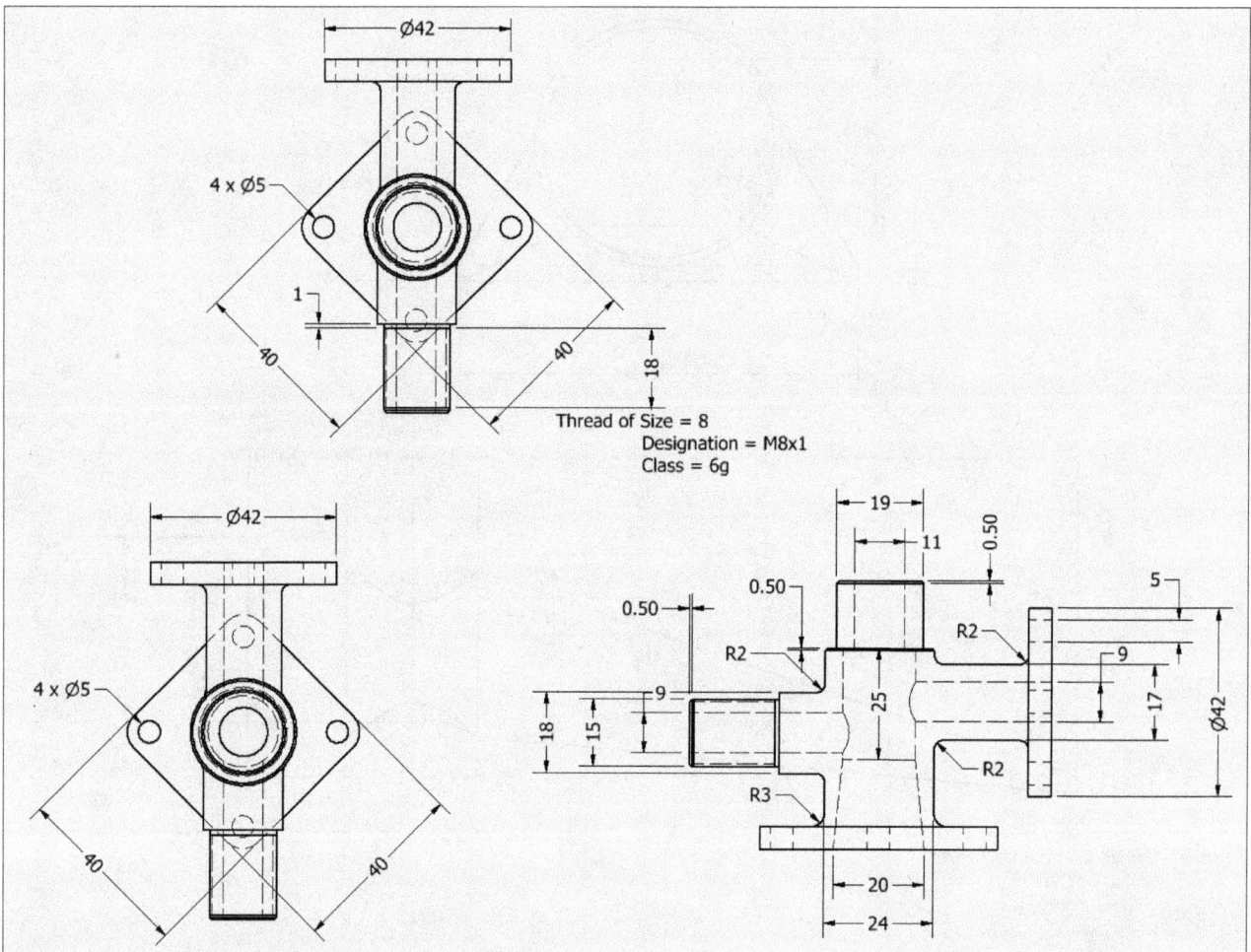

Figure-148. Practice 9 drawing

PRACTICE 10

Create the model as shown in Figure-149 by using the dimensions given in Figure-150.

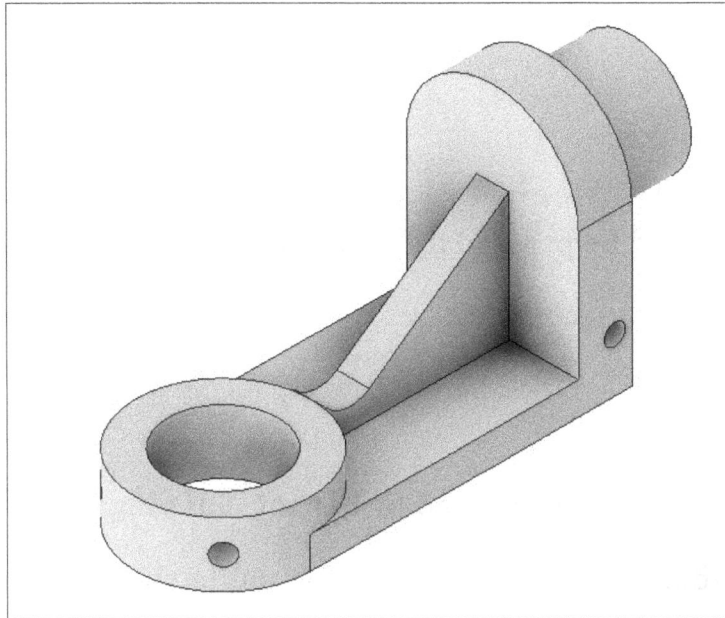

Figure-149. Practice 10

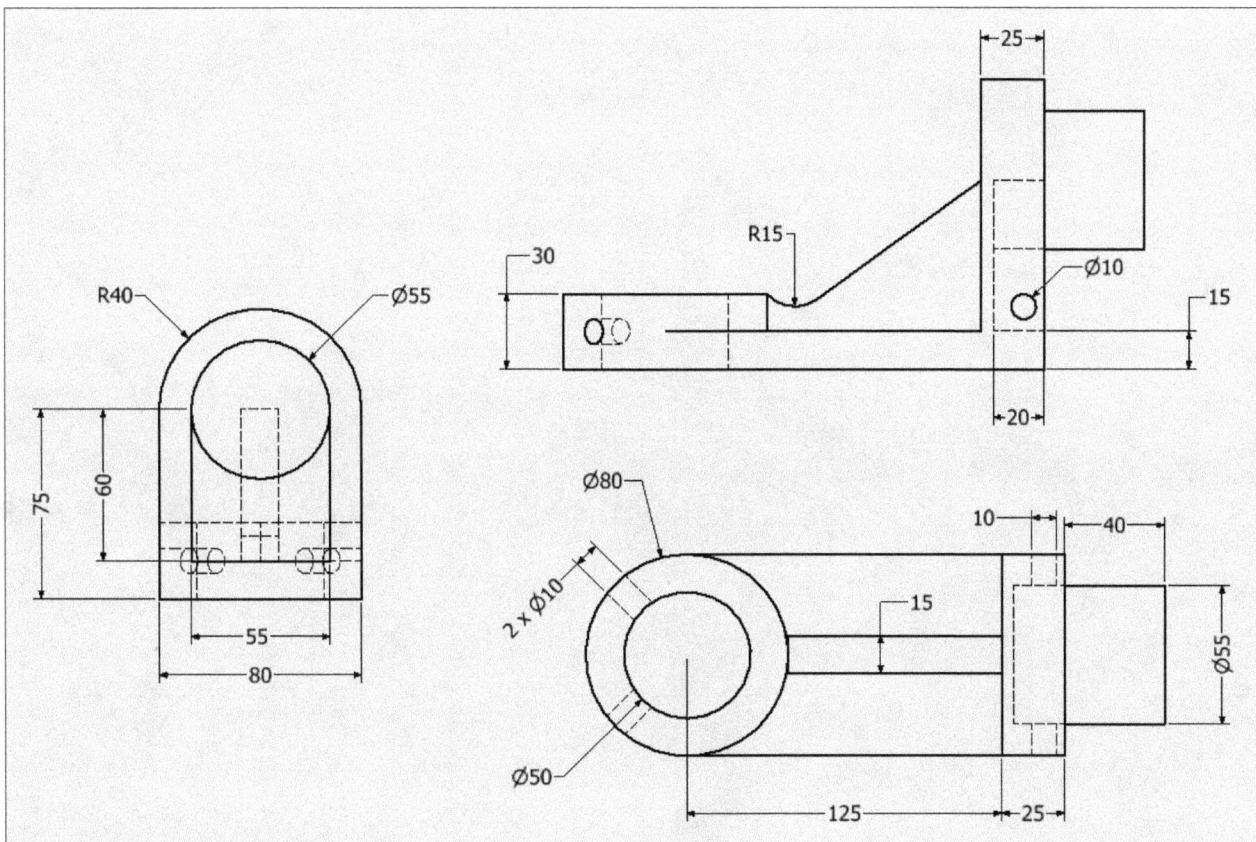

Figure-150. Practice 10 drawing

SELF ASSESSMENT

Q1. Which of the following options are available in **Direct Editing Toolbar**?

a) Move
b) Scale
c) Delete
d) All of the Above

Q2. Which button should be pressed and held while removing a face from selection?

a) Alt
b) Ctrl
c) Shift
d) Tab

Q3. The size of faces cannot be change by **Direct Editing Toolbar**. (True/False)

Q4. The **Bend Part** tool is used to bend part by using sketched line. (True/False)

Q5. The .... tool is used to edit imported parts of parametric features of the model by using simple drag and drop operations.

Chapter 7

Assembly Design

Topics Covered

The major topics covered in this chapter are:

- *Starting Assembly Design*
- *Placing Components in Assembly*
- *Replacing Components*
- *Top Down and Bottom Up Assembly approach*
- *Creating Components in Assembly*
- *Positioning Tools*
- *Constraints*
- *Joints*
- *Contact Sets*
- *Bill of Materials*
- *Driving Constraints for animation/motion study*
- *Introduction to Presentation*

INTRODUCTION

In engineer's language, assembly is the combination of two or more components and these components are constrained to each other in a specified manner called assembly constraint. In Autodesk Inventor, there is a separate environment to create assembly of parts.

STARTING ASSEMBLY ENVIRONMENT

- Start Autodesk Inventor using icon on desktop or in Start menu.
- Click on the **File Menu** button and select the **Assembly** option from **New** cascading menu; refer to Figure-1.

Or

- Click on the **Assembly** option from the **New** drop-down in **Quick Access Toolbar**; refer to Figure-2.

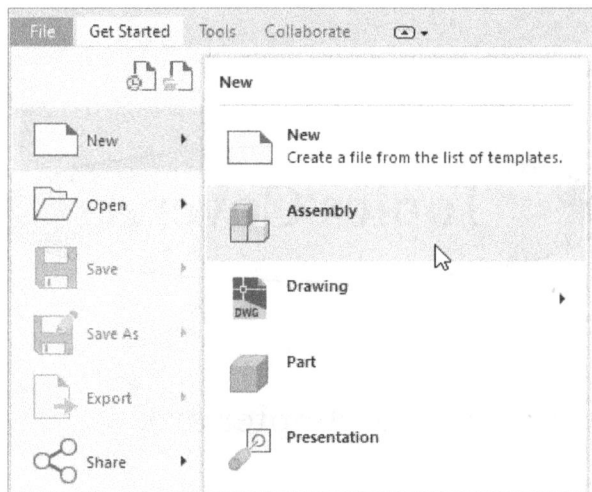

Figure-1. Assembly option

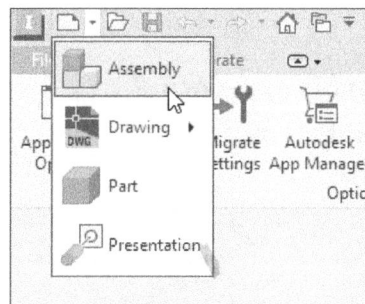

Figure-2. Assembly option in Quick Access toolbar

- The assembly environment will be displayed as shown in Figure-3.

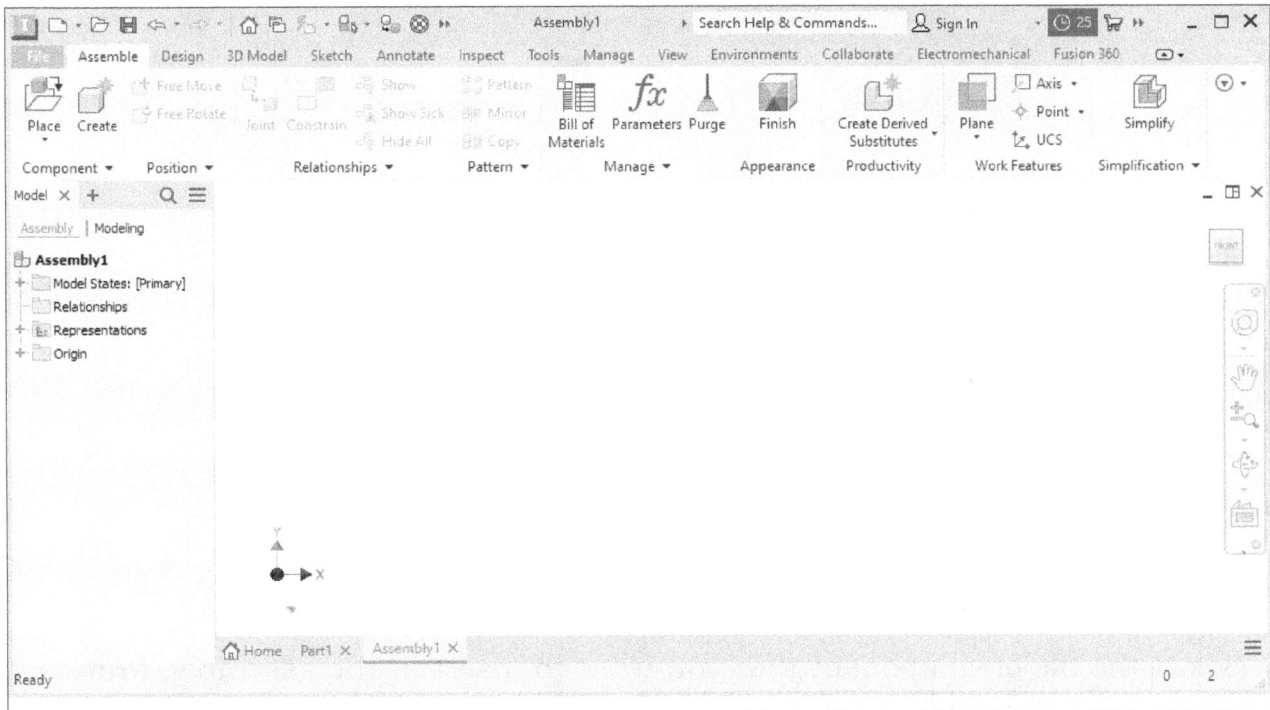

Figure-3. Assembly environment

PLACING COMPONENTS IN ASSEMBLY

There are six ways to insert components in an assembly in Autodesk Inventor.

- Placing Component from Local Storage
- Placing Component from Content Center
- Placing Imported CAD Files
- Placing iLogic Component
- Placing Electrical Catalog Browser

These methods of inserting components are discussed next.

Placing Component from Local Storage

Using the **Place** tool, you can insert the components in the assembly which are stored in the local drive of your computer. These components are the part files that we have created earlier. The procedure to insert components by this method is discussed next.

- Click on the **Place** tool from **Component** panel in the **Assemble** tab of the **Ribbon**. The **Place Component** dialog box will be displayed; refer to Figure-4.

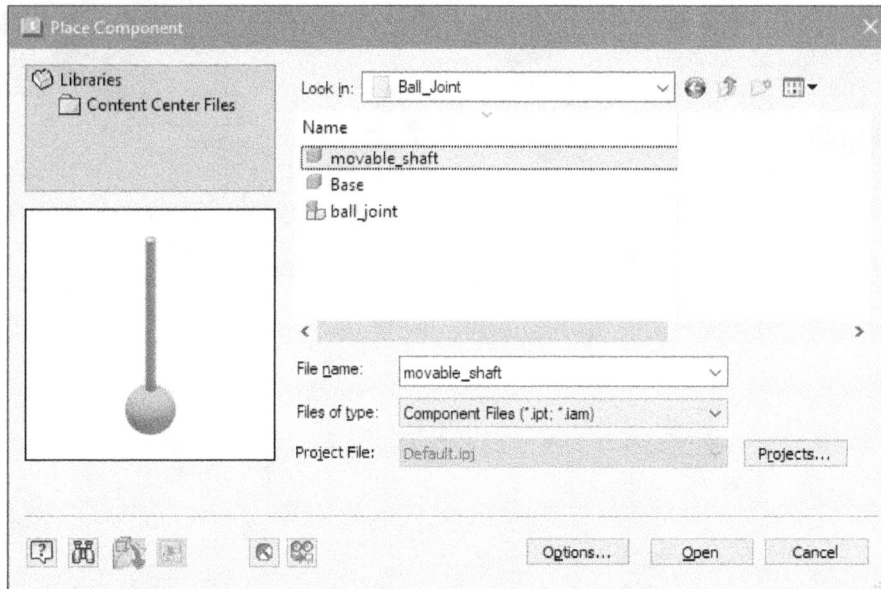

Figure-4. Place Component dialog box

- Select the file of component that you want to insert in the assembly. Preview of the component will be displayed.
- You can select desired file format from the **Files of type** drop-down if you wish to import CAD files of other software; refer to Figure-5.

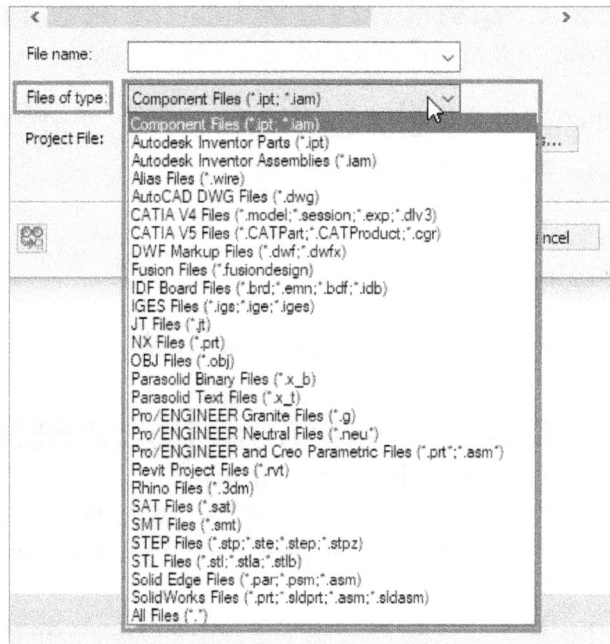

Figure-5. Files of type drop-down

- After selecting the file, click on the **Open** button from the dialog box. The component will get attached to the cursor; refer to Figure-6.

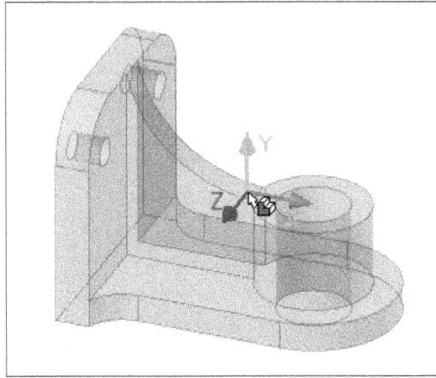

Figure-6. Component attached to cursor

- Click in the viewport to place the component.
- If you want to insert more copies of the component then click again in the viewport.
- Press **ESC** from keyboard to exit inserting more components.

Place from Content Center

- Make sure you have a base component already available in the viewport (inserted by previous method).
- Click on the **Place from Content Center** tool from the **Place** drop-down in the **Component** panel of the **Ribbon**. The **Place from Content Center** window will be displayed; refer to Figure-7.

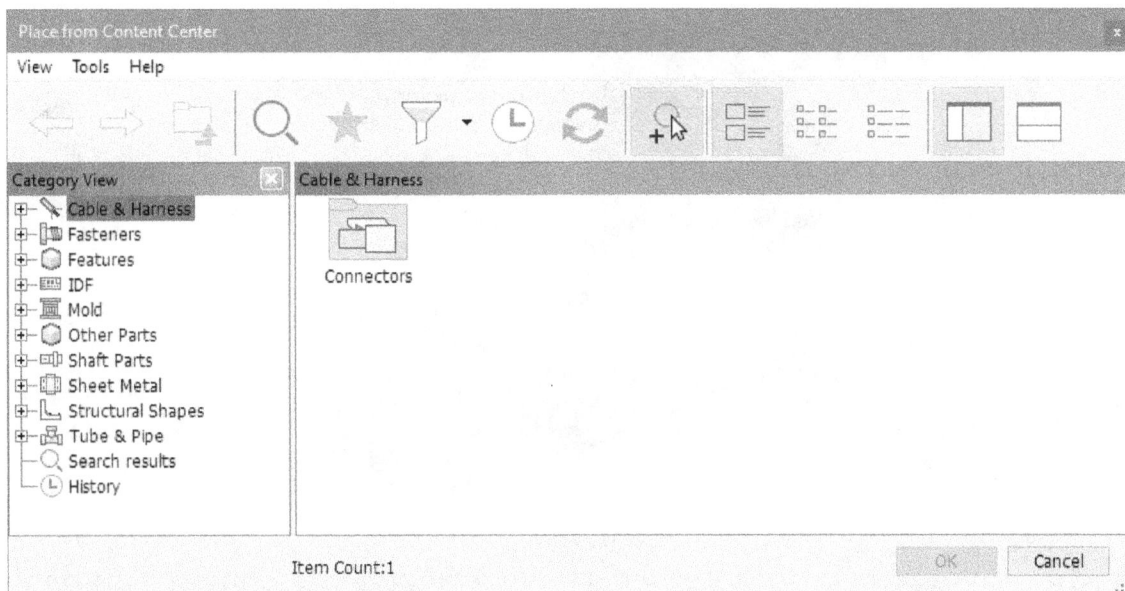

Figure-7. Place from Content Center window

- Click on desired category, components in that category will be displayed in the right; refer to Figure-8.

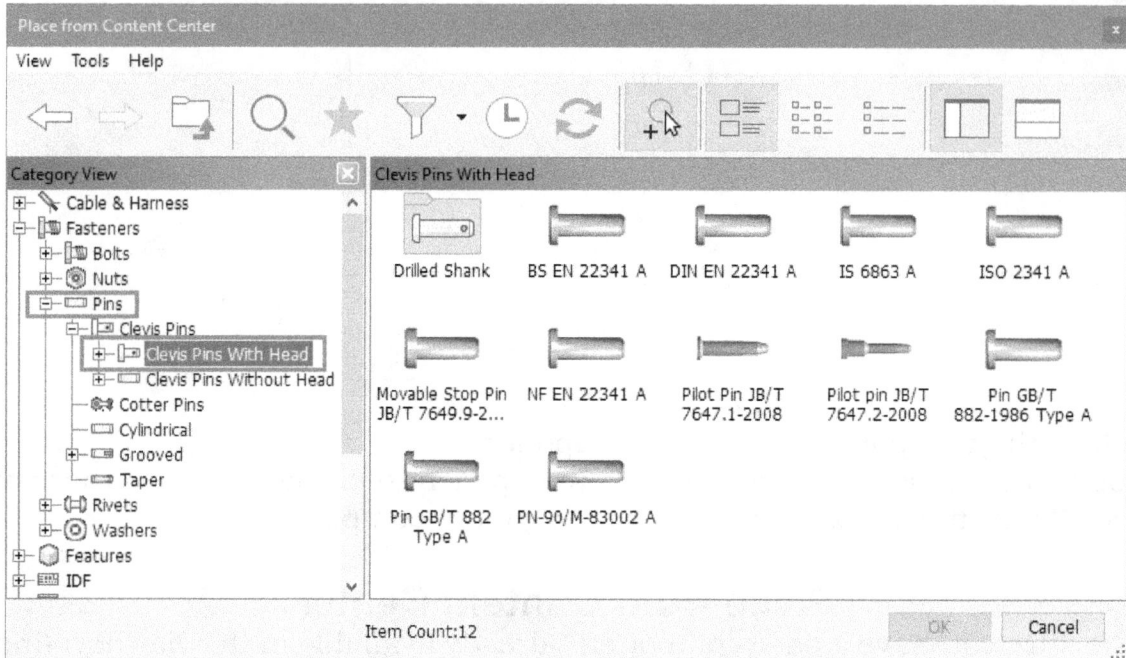

Figure-8. Categories of components

- Double-click on desired component from the right area of the window.
- Preview of the component will be displayed if you hover the cursor on the related entity; refer to Figure-9.

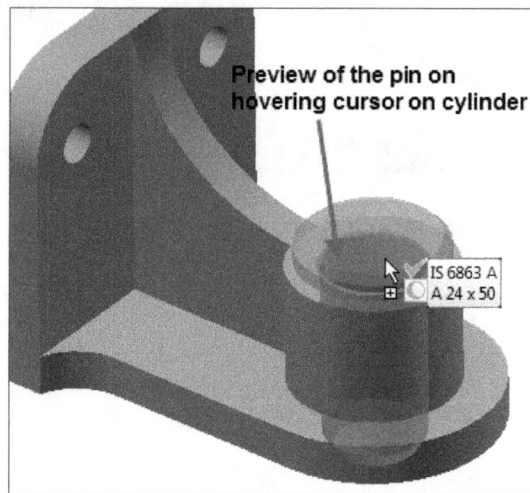

Figure-9. Preview of the pin being inserted

- Note that the size of pin and other similar components changes automatically based on the selected base component, once you place the component at desired location.
- If you want to place the component then click **RMB** in the modeling area and click on the **Place** button from the options displayed; refer to Figure-10. The component will be placed.

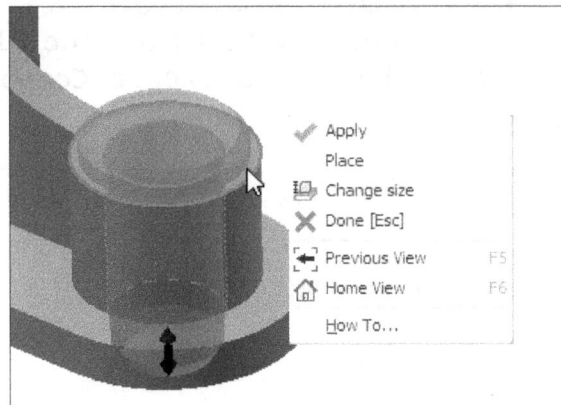

Figure-10. Options displayed on right clicking

- If you want to insert more components of same type then click on the **Apply** button in place of **Place** button.
- If you want to manually change the size of component then you can click on the **Change size** button. A dialog box with various size options will be displayed; refer to Figure-11.

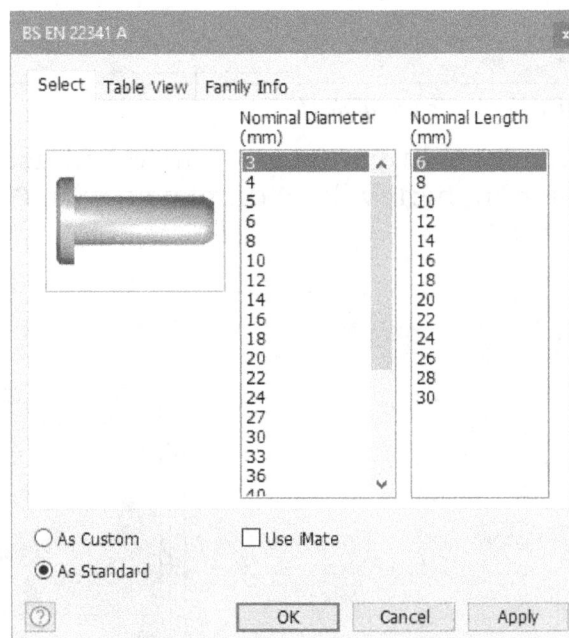

Figure-11. Size options in the dialog box

- Select desired size and click on the **OK** button.
- Note that you can change the size of components inserted from content center anytime by right-click on them in **Model Browser**.

Place Imported CAD files

- Click on the **Place Imported CAD Files** tool from the **Place** drop-down in the **Component** panel of the **Ribbon**. The **Place Component** dialog box will be displayed as discussed earlier.
- Note that the **All Models** option is selected in the **Files of type** drop-down in the dialog box, so you can select the model file of any CAD software.

Place iLogic Component

*   Click on the **Place iLogic Component** tool from the **Place** drop-down in the **Component** panel of the **Ribbon**. The **Place iLogic Component** dialog box will be displayed; refer to Figure-12.

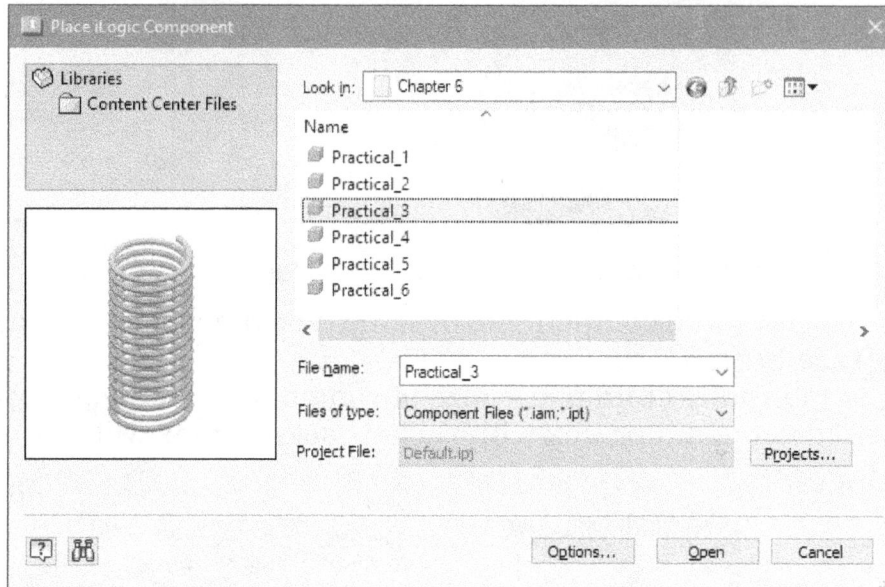

Figure-12. Place iLogic Component dialog box

*   Select the file of component that you want to insert in assembly and click on the **Open** button. Selected component will be displayed with **Place iLogic Component** window; refer to Figure-13.

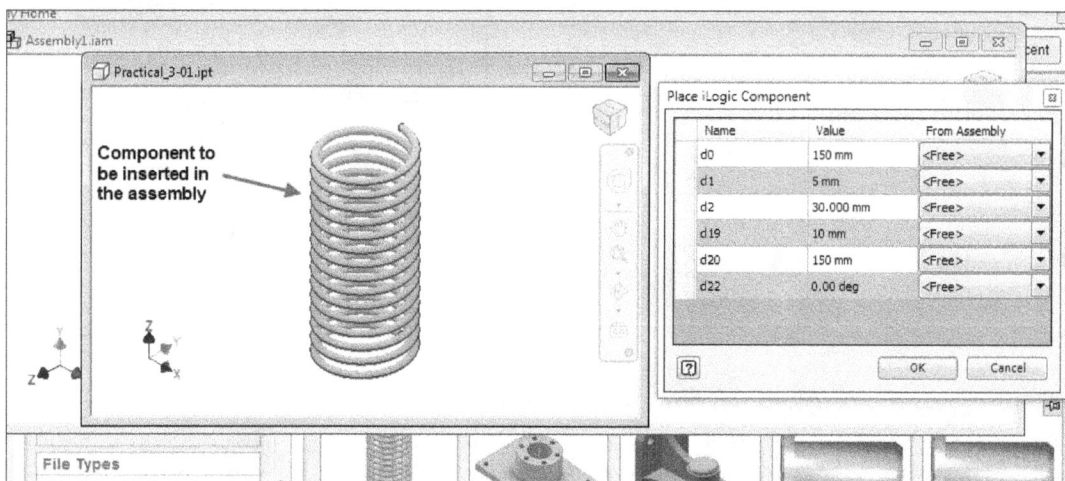

Figure-13. Place iLogic Component window

*   Set desired values of dimensions in the dialog box and click on the **OK** button. The component will get attached to the cursor.
*   Click in the viewport to place the component.
*   Press **ESC** to exit inserting more component.

Electrical Catalog Browser

Using this option, you can insert the electrical components in Autodesk Inventor. This option is useful when you are working with electrical assemblies. The procedure to use this option is same as discussed earlier.

REPLACING COMPONENT

Sometimes, we need to replace an already existing component with a new component. You can simply delete the earlier inserted component and insert a new component but in this case, you have to specify the assembly constraints again. To solve this problem, we use **Replace** tool. The procedure is given next.

- Click on the **Replace** tool from the **Replace** drop-down in the expanded **Component** panel in the **Ribbon**; refer to Figure-14. You will be asked to select the component which you want to replace.

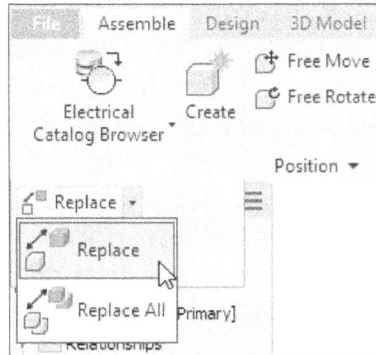

Figure-14. Replace tool

- Click on the component that you want to replace and press **ENTER** from the keyboard.
- The **Place Component** dialog box will be displayed as discussed earlier; refer to Figure-15.

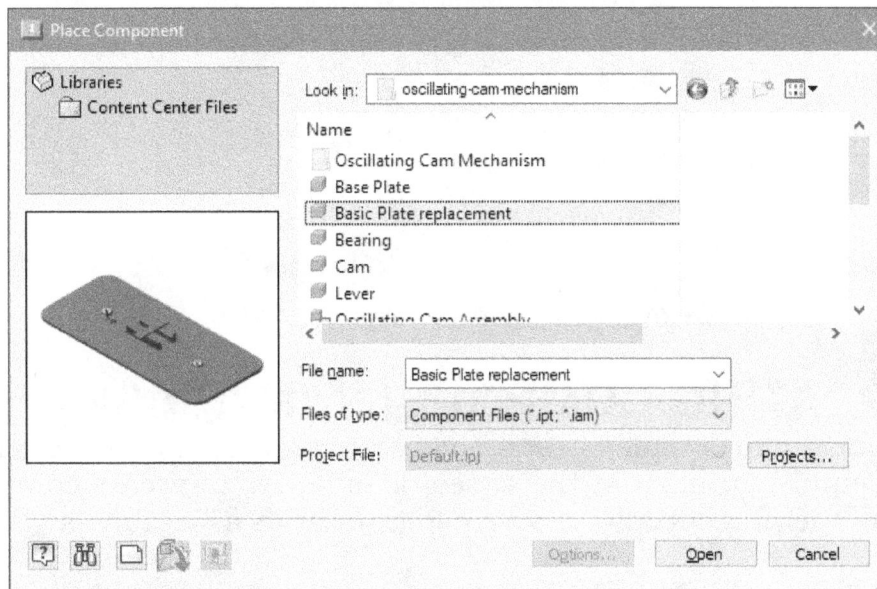

Figure-15. Place Component dialog box for replacing component

- Click on the **Open** button from the dialog box. The component will get replaced automatically if the locations of faces and other references are same.
- If you do not have the equal number of references in same orientation then an error message will be displayed.

TOP-DOWN AND BOTTOM-UP APPROACH FOR ASSEMBLY

There are two approaches in creating assembly, Top-Down approach and Bottom-Up approach. In Top-Down approach, the component are created in the assembly and constraints can be applied before creating the components. In Bottom-Up approach, the components are created in Part environment first. After creating, these components are brought in the assembly environment and are assembled by constraints.

Earlier, we have inserted the components which were already created in Part environment which means we have used the Bottom-Up approach. Now, we will create the components in assembly itself which means we will use the Top-Down approach. Now, the question arises why do we need Top-Down approach if we can easily work with Bottom-Up approach. Answer to this question lies in practical use. Sometimes, we need to design a component which fits exactly to the space left in assembly. For example, we have inserted components in an assembly as shown in Figure-16. At one stage of designing, we need a component to join the holes of discs as shown in Figure-16. In such cases, it is easy to create component in the assembly which means it is easy to use Top-Down approach in such cases.

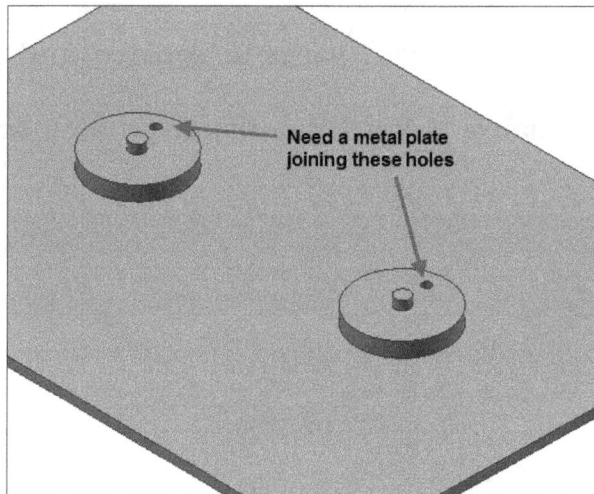

Figure-16. Example for top-down approach

CREATING COMPONENT IN ASSEMBLY

When working on some complex assemblies, you may need a new component to be created for assembly. In most of the cases, it is better to create component in the assembly itself. The procedure to create component in assembly is given next.

*   Click on the **Create** tool from the **Component** panel in the **Assemble** tab of the **Ribbon**. The **Create In-Place Component** dialog box will be displayed; refer to Figure-17.

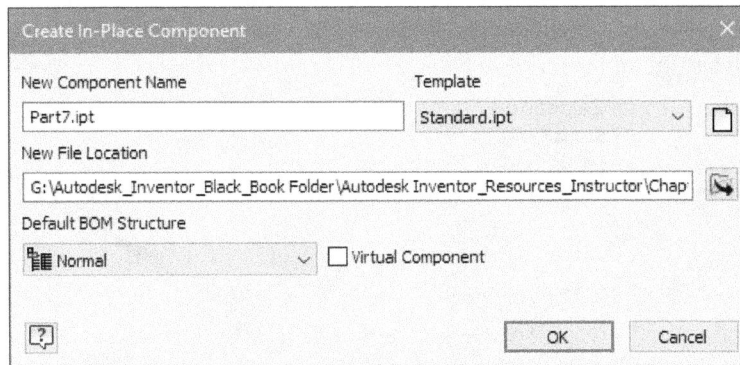

Figure-17. Create In-Place Component dialog box

- Specify desired name of the component in the **New Component Name** edit box.
- Select desired template from the **Template** drop-down. If you want to select a template from broader list then click on the **Browse Template** button. The **Open Template** dialog box will be displayed; refer to Figure-18.

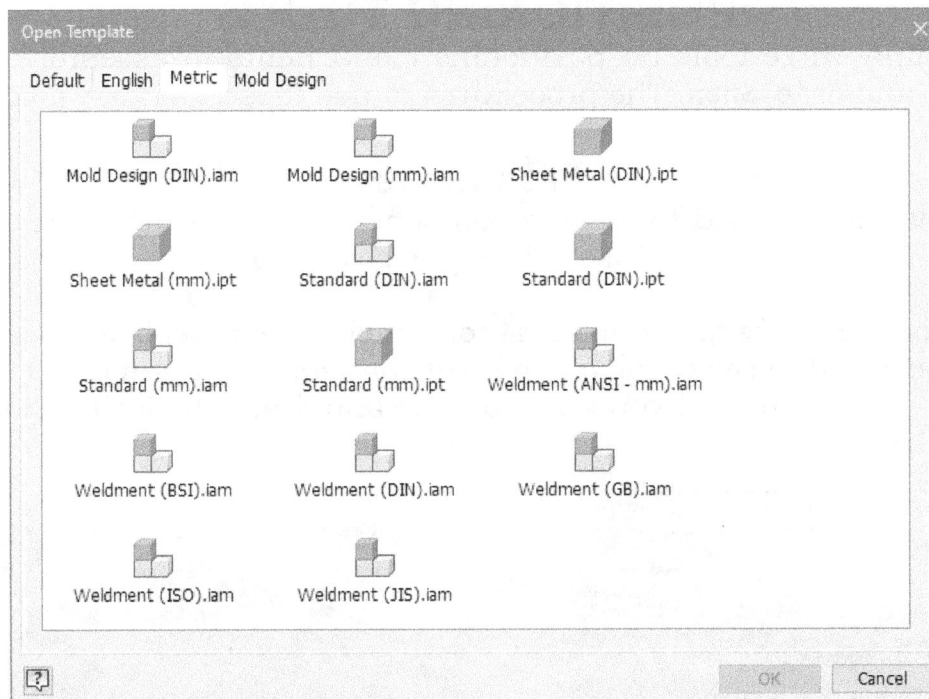

Figure-18. Open Template dialog box

- Double-click on desired template from the dialog box to select the template.
- Set desired location to save the component file in the **New File Location** edit box.
- Clear the **Constrain sketch plane to selected face or plane** check box if you do not want the sketch plane to constrained to selected plane or face.
- Click on the **OK** button from the dialog box. You will be asked to select a sketching plane/face.
- Select desired sketching plane/face; refer to Figure-19. The tools for 3D Modeling will be activated.
- Create desired component by using 3D Modeling tools as discussed earlier; refer to Figure-20.

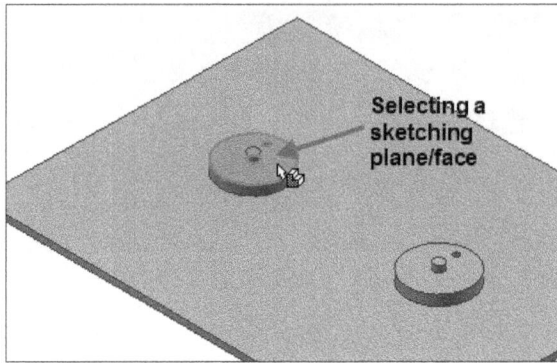

Figure-19. Selecting sketching face

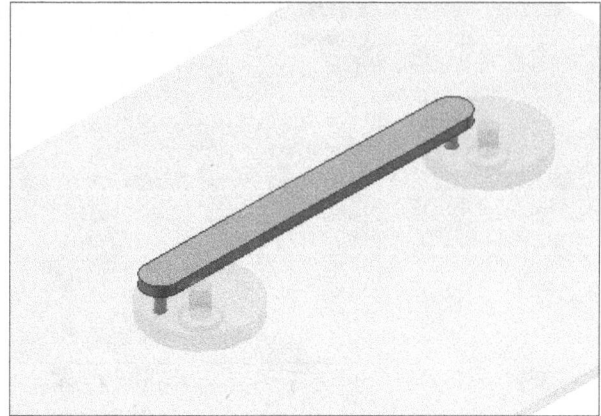

Figure-20. Part created in assembly

*   Click on the **Return** tool from the **Return** panel in the **Ribbon** to return to assembly environment.

POSITIONING TOOLS

There are mainly three tools for positioning components in assembly; **Free Move**, **Free Rotate**, and **Grid Snap**. The procedures to use these tools are discussed next.

Free Move

The **Free Move** tool is used to move a component freely in 3D space of assembly environment. The procedure to use this tool is given next.

*   Click on the **Free Move** tool from **Position** panel in the **Assemble** tab of the **Ribbon**. You will be asked to select the component that you want to move.
*   Select the component that you want to move and drag it to desired position; refer to Figure-21.

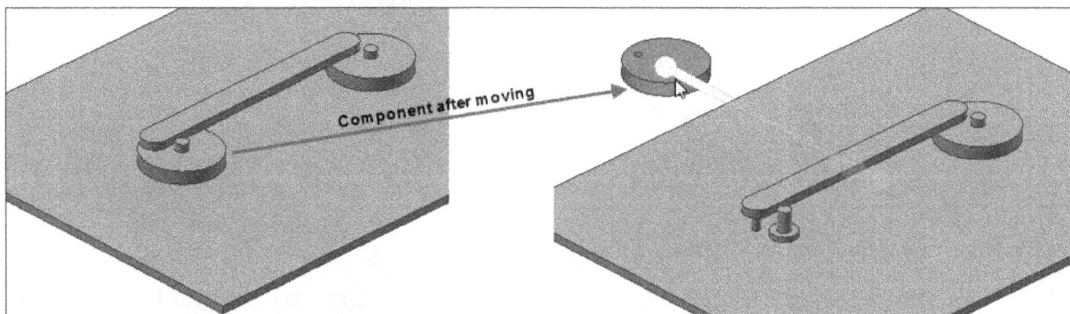

Figure-21. Freely moving component

*   Note that you can move only one component at a time by using this tool.

Free Rotate

The **Free Rotate** tool works similar to **Free Move** tool but in place of moving the object is rotated freely. The procedure to use this tool is given next.

*   Click on the **Free Rotate** tool from **Position** panel in the **Assemble** tab of the **Ribbon**. You will be asked to select component to be rotated.
*   Select the component that you want to rotate. A window will be displayed to rotate the component; refer to Figure-22.

- Drag the component in desired direction to rotate.
- Click in the viewport to exit the tool.

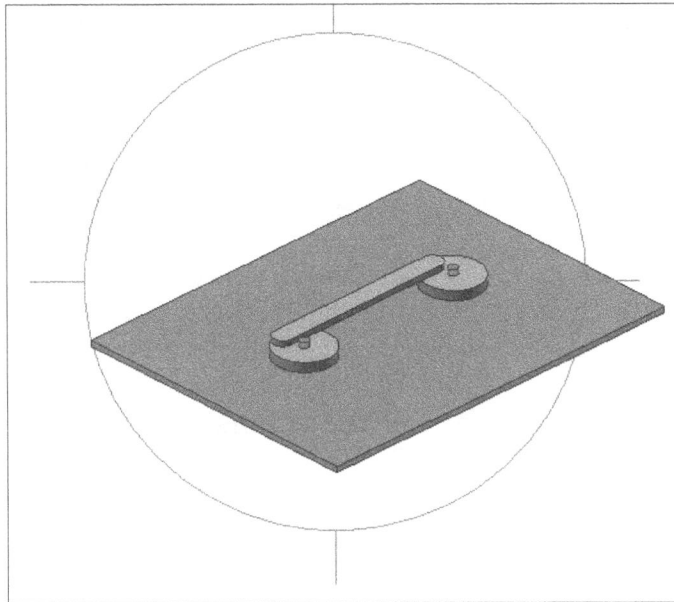

Figure-22. Free rotation orbit

Grid Snap

The **Grid Snap** tool is used to move the component in constrained environment. For example, you can make the component move in planar direction only or you can make the component move upward/downward only. The procedure to use this tool is given next.

- Click on the **Grid Snap** tool from the expanded **Position** panel in the **Assemble** tab of the **Ribbon**. You will be asked to click on a geometry of the component.
- Select desired geometry of component. Note that the selection of geometry can change the buttons displayed for movement in the toolbar; refer to Figure-23.

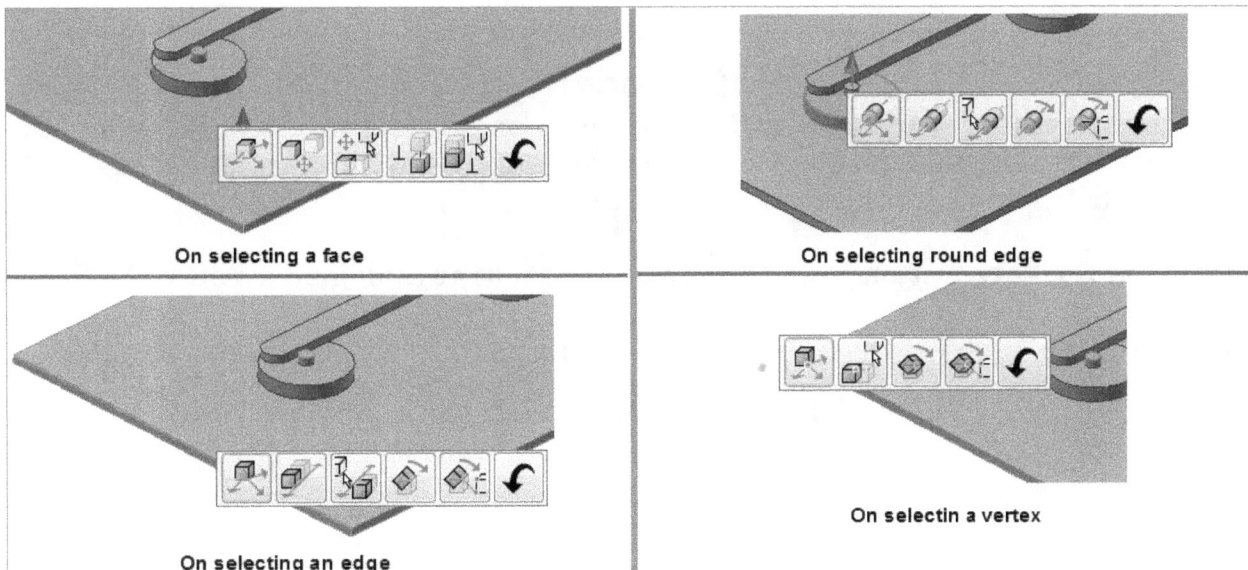

Figure-23. Buttons for movement of component

- Click on desired button to move/rotate the component.
- Move/rotate the component in the direction corresponding to the selected button.

CONSTRAINS

The Constraints are used to restrict the movement of objects. It is very important to apply constraints to the components in assembly because without them every component is free to move in any direction. Later in the book, you will perform various studies on the assembly like stress analysis, motion study, and so on. If the components are not properly constrained then results of these studies will be ambiguous. You can apply various constraints in Autodesk Inventor by using the **Constrain** tool. The procedure to apply constrain is given next.

- Click on the **Constrain** tool from the **Relationships** panel in the **Assemble** tab of the **Ribbon**. The **Place Constraint** dialog box will be displayed; refer to Figure-24.

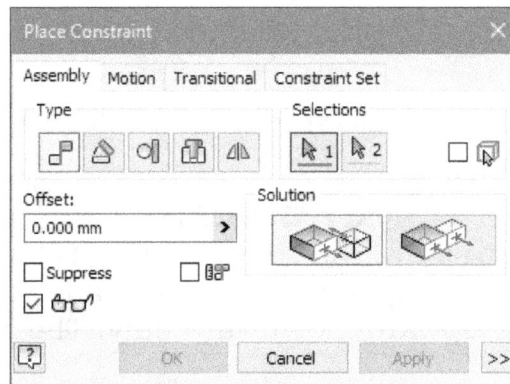

Figure-24. Place Constraint dialog box

- There are various buttons in the dialog box to apply different type of constraints.

Various constraints that can be applied in an assembly in Autodesk Inventor are discussed next.

Mate Constrain

The Mate constrain is used to make two faces or edges share the same location. This constrain can also be used to fix the distance between two faces/edges. The procedure to apply this constrain is given next.

- Click on the **Mate** button from the **Place Constraint** dialog box. You will be asked to select the first geometry.
- Click on the first geometry (Face/edge/vertex). You will be asked to select the second geometry.
- Click on the second geometry; refer to Figure-25.

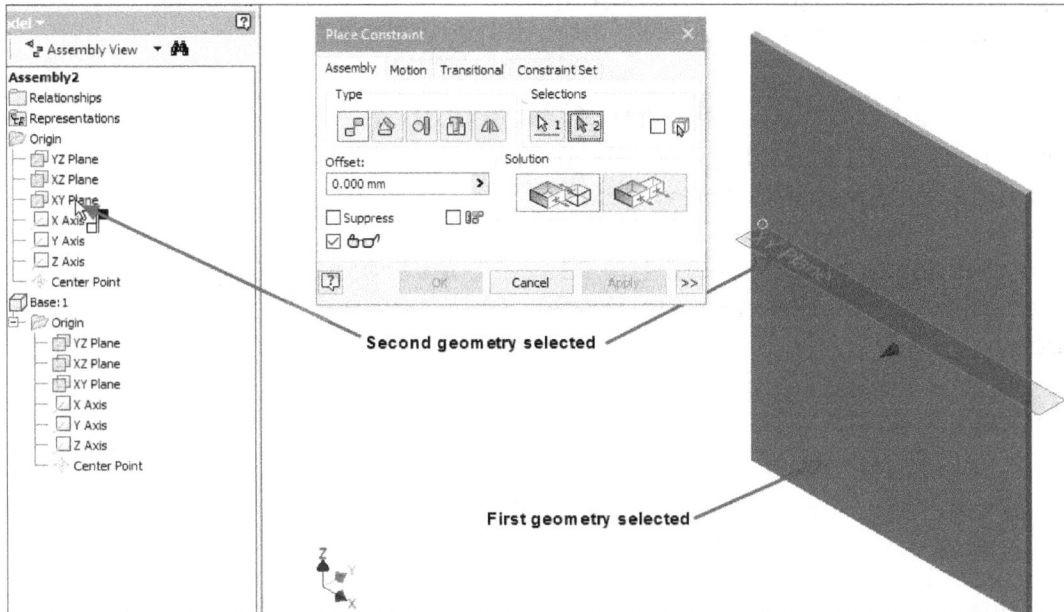

Figure-25. Selecting geometry for mate constraint

- If you want to specify an offset distance between the selected geometries then specify desired value in the **Offset** edit box.
- Select the **Suppress** check box to suppress the constraints.
- If you want to change the side of mate constraint, click on the **Flush** button in the **Solution** area of the dialog box.
- You can also specify the limit for movement of selected geometry by using the options in expanded dialog box; refer to Figure-26. To expand the dialog box, click on the **>>** button from the dialog box.

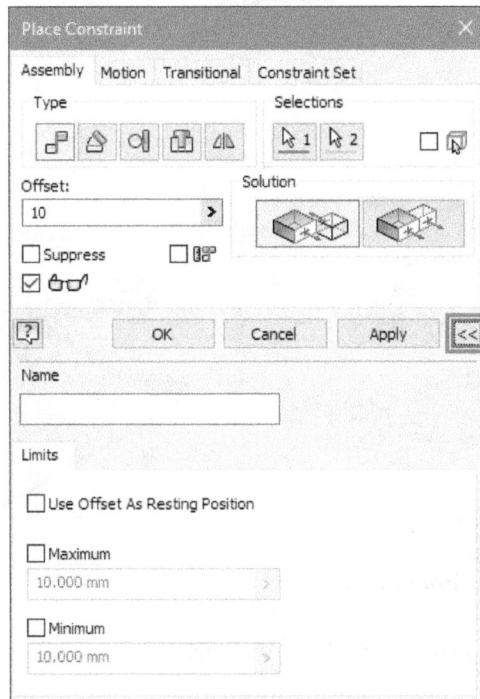

Figure-26. Expanded Place Constraint dialog box

- Click on the **Maximum** check box and specify the maximum distance limit between the selected geometries in the edit box displayed.

- Click on the **Minimum** check box and specify the minimum distance limit between the selected geometries in the edit box displayed below.

Angle Constraint

The **Angle** constraint is used to set angle between two components. The procedure to specify angle constraint is given next.

- Click on the **Angle** button from the **Place Constraint** dialog box. The dialog box will be displayed as shown in Figure-27. Also, you will be asked to select the first geometry.

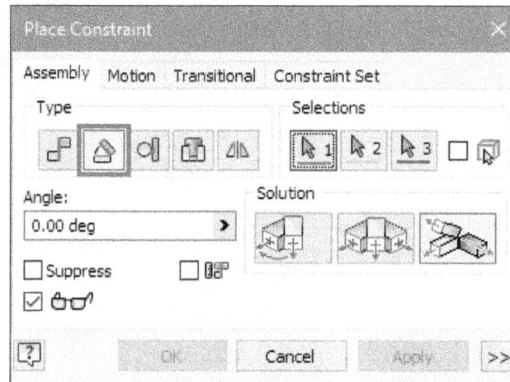

Figure-27. Place Constraint dialog box for Angle constraint

- Select the first geometry. You will be asked to select the second geometry.
- Select the second geometry. Preview of the angle constraint will be displayed; refer to Figure-28.

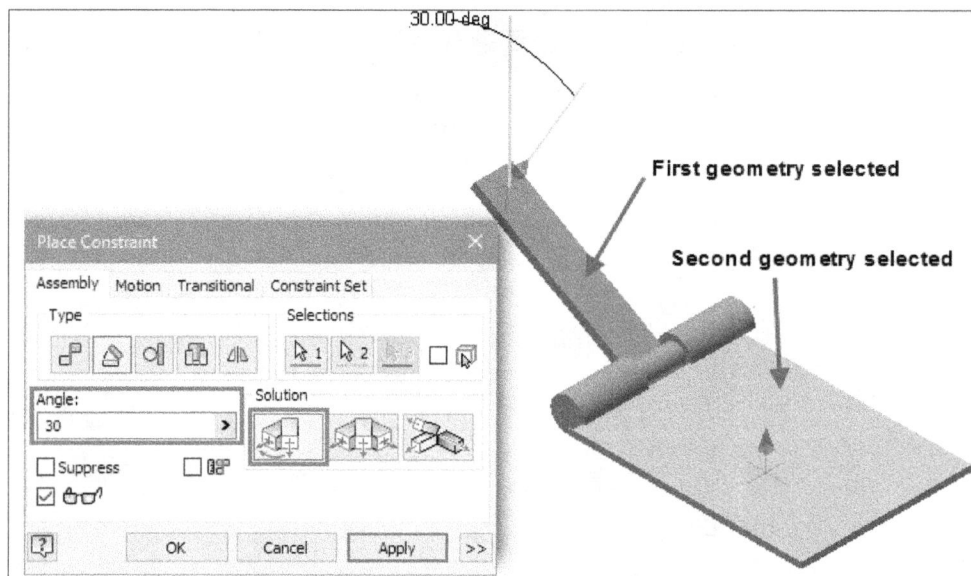

Figure-28. Preview of angle constraint

- Specify desired value in the **Angle** edit box.

- Note that there are three buttons in the **Solution** area of the dialog box; **Directed Angle**, **Undirected Angle**, and **Explicit Reference Vector**. Select the **Directed Angle** button if you want to use right-hand rule in specifying the angle. Select the **Undirected Angle** button if you want to use the left-hand rule in specifying angle between the geometries. You can select the reference vector for specifying angle direction by using the **Explicit Reference Vector** button.

- If you have selected the **Explicit Reference Vector** button then three selection buttons will be available in the **Selections** area of the dialog box; refer to Figure-29.

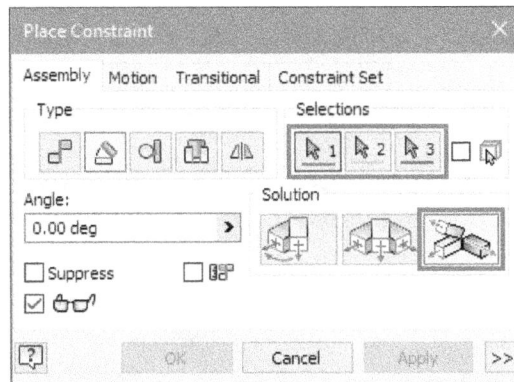

Figure-29. Explicit Reference Vector button selected

- Select the two geometries between which the angle is to be specified and then select a face/edge/axis to specify the direction for angle vector; refer to Figure-30.

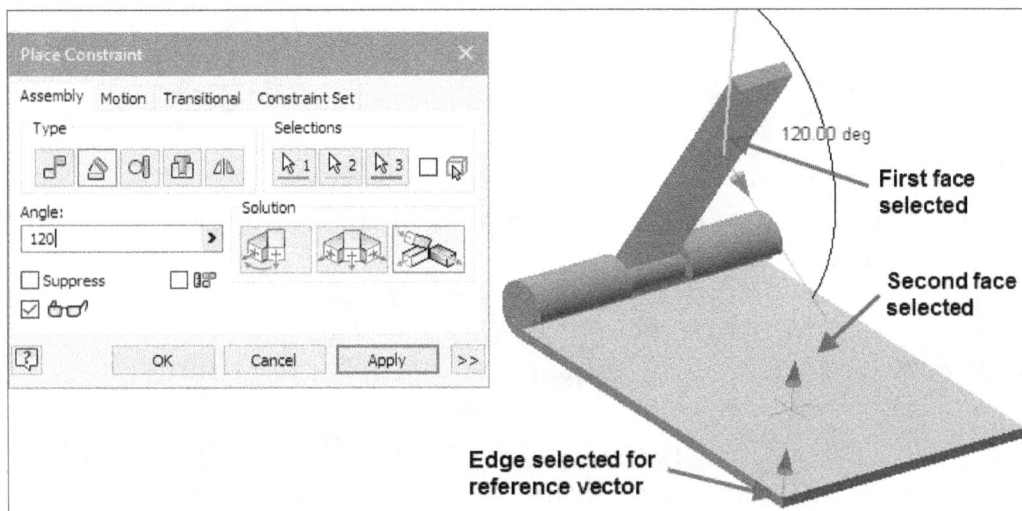

Figure-30. Preview of angle constraint with explicit vector

- Click on the **OK** button to apply the constraint.

Tangent Constraint

The **Tangent** constraint is used to make a selected geometry tangent to the other geometry. The procedure to apply this constraint is given next.

- Click on the **Tangent** button from the **Place Constraint** dialog box. The dialog box will be displayed as shown in Figure-31. Also, you will be asked to select the first geometry.

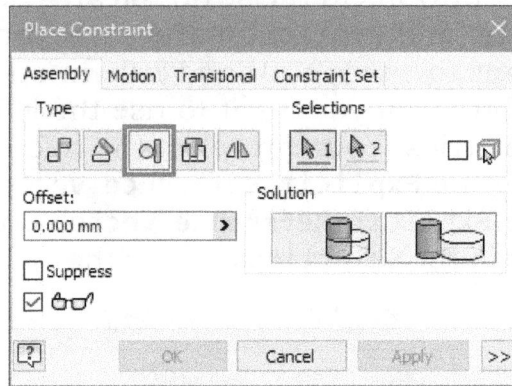

Figure-31. Place Constraint dialog box for Tangent constraint

- Select the first geometry. You will be asked to select the second geometry.
- Select the second geometry. Preview of the tangent constraint will be displayed; refer to Figure-32.

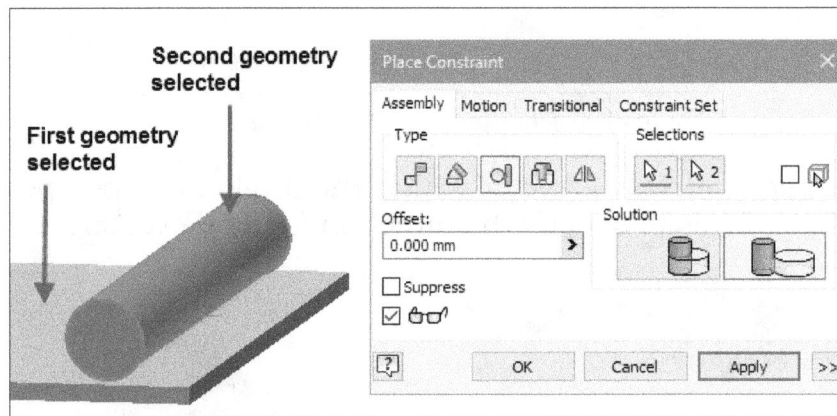

Figure-32. Preview of tangent constraint

- Select desired button from the **Solution** area of the dialog box to change the side of tangent constraint.
- If you want to specify distance between the two selected geometries then specify desired offset value in the **Offset** edit box in the dialog box.
- Click on the **OK** button from the dialog box to apply constraint.

Insert Constraint

The **Insert** constraint is used to insert pin like geometries inside hole like geometries. The procedure to use this constraint is given next.

- Click on the **Insert** button from the **Place Constraint** dialog box. The dialog box will be displayed as shown in Figure-33.

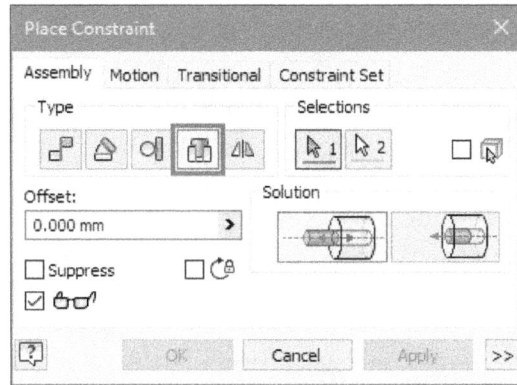

Figure-33. Place Constraint dialog box for Insert constraint

- Select the round edge of first entity. You will be asked to select the second entity.
- Select the second entity. Preview of the insert constraint will be displayed; refer to Figure-34.

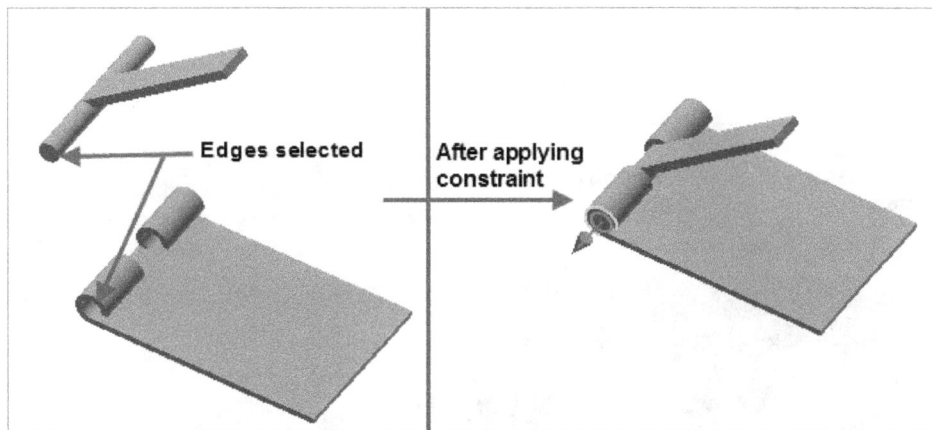

Figure-34. Preview of insert constraint

- Note that if we break down the insert constraint then it does two things, making axis of the selected geometries coincident and making selected round edges aligned in one plane.
- To change the direction of placement, select desired button from the **Solution** area of the dialog box.
- Click on the **OK** button from the dialog box to apply the constraint.

Symmetry Constraint

The **Symmetry** constraint is used to make the selected two components symmetric about the selected plane. The procedure to apply this constraint is given next.

- Click on the **Symmetry** button from the **Place Constraint** dialog box. The dialog box will be displayed as shown in Figure-35. Also, you will be asked to select a geometry.

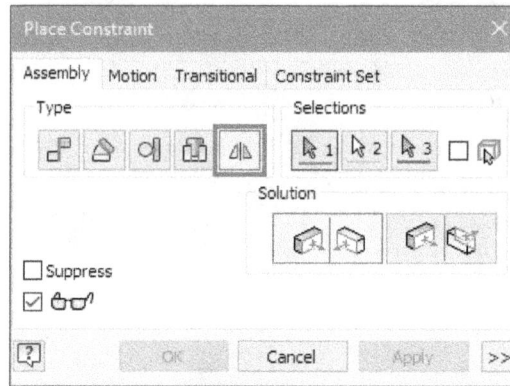

Figure-35. Place Constraint dialog box for Symmetry constraint

- Select the face/edge/axis of the first component. You will be asked to select the second geometry.
- Select the second geometry. You will be asked to select a plane for defining symmetry reference.
- Select a plane about which you want the components to be symmetric. Preview of the symmetry will be displayed; refer to Figure-36.

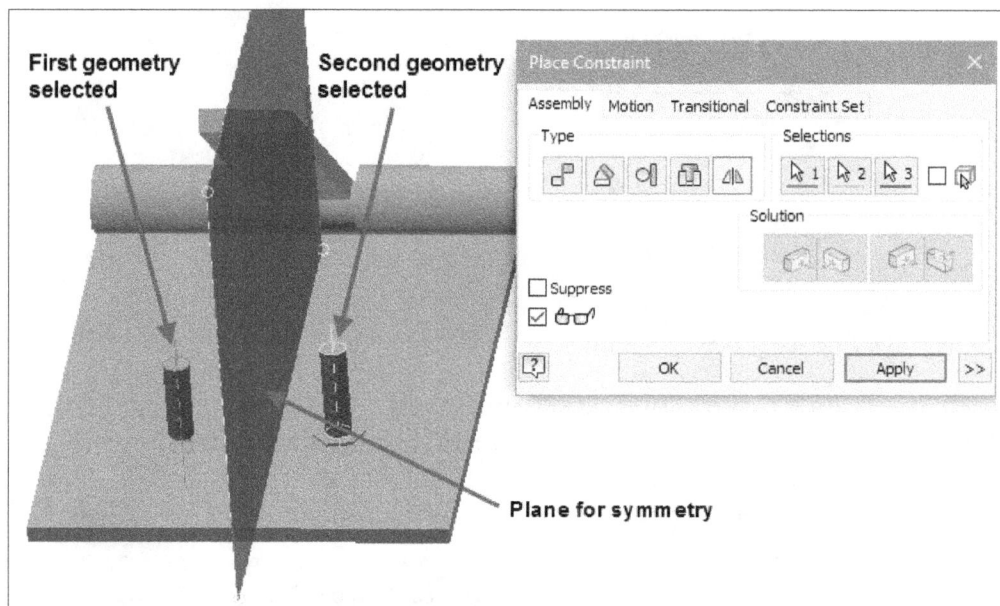

Figure-36. Preview of symmetry constraint

- Click on the **OK** button from the dialog box to apply the constraint.

Motion Constraint

There are two type of constraints for setting motion of objects in Autodesk Inventor assembly, **Rotation** and **Rotation-Translation**. Using the **Rotation** constraint, you can specify the ratio of rotation between the two objects like in gears, bearings, etc. Using the **Rotation-Translation** constraint, you can make one component translate on rotation of other component or you can make one component rotate on translation of other component. Examples can be screw jack motion, rack and pinion design, etc. The procedures to apply these constraints are discussed next.

Rotation Constraint

- Click on the **Motion** tab in the **Place Constraint** dialog box. The options in the dialog box will be displayed as shown in Figure-37.

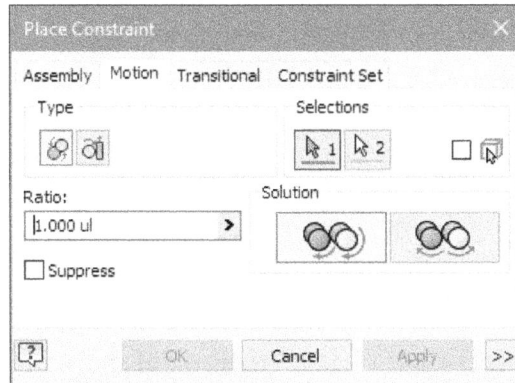

Figure-37. Place Constraint dialog box with Motion tab selected

- Click on the **Rotation** button ⊗ from the **Type** area of the dialog box. You will be asked to select the first geometry.
- Select the round face of first object. You will be asked to select the second geometry.
- Select the round face of the second object.
- Specify desired ratio for rotation; refer to Figure-38.

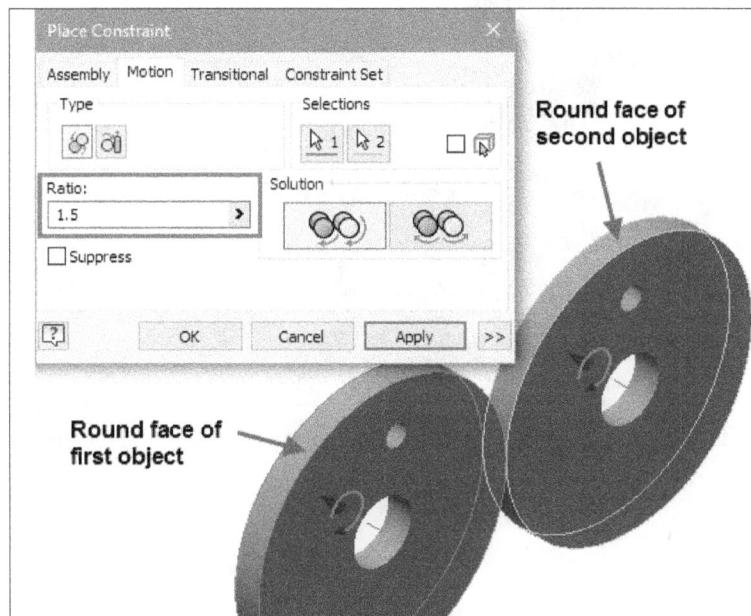

Figure-38. Options for rotation constraint

- To change the direction of rotation of the objects, click on desired button from the **Solution** area of the dialog box.
- Click on the **OK** button from the dialog box.

Rotation-Translation Constraint

- Click on the **Motion** tab in the **Place Constraint** dialog box and select the **Rotation-Translation** button ⊙. The options in the dialog box will be displayed as shown in Figure-39. Also, you will be asked to select the first geometry.

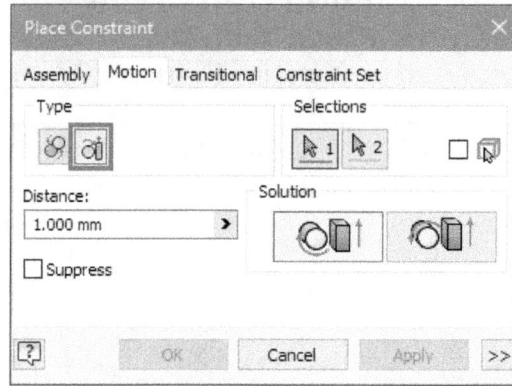

Figure-39. Place Constraint dialog box for rotation translation option

- Select the round face of the first object (disc); refer to Figure-40. You will be asked to select the second geometry.

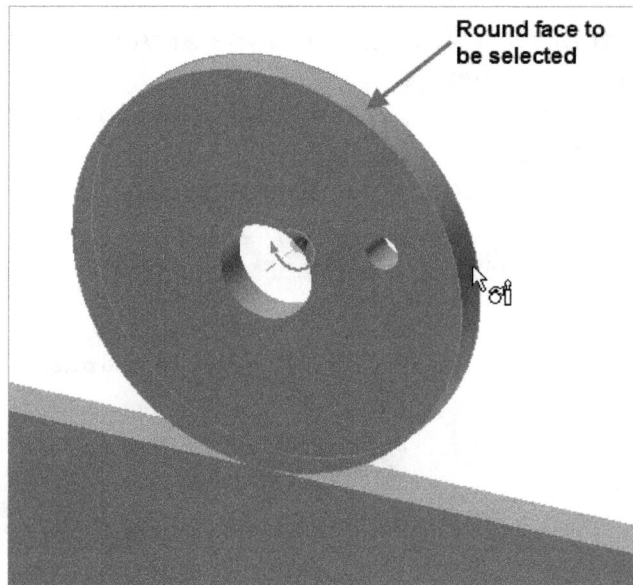

Figure-40. Round face to be selected

- Select the edge of the second object (rail) coinciding with the round face; refer to Figure-41.

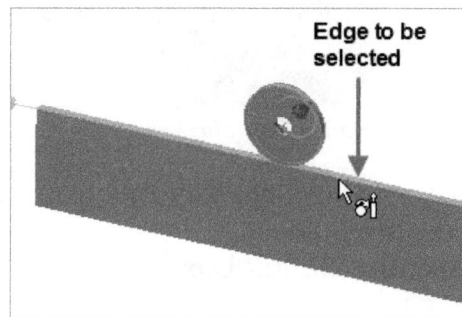

Figure-41. Edge selected for constraint

- Specify the distance moved by rail (second geometry) on one rotation of the disc object (first geometry) in the **Distance** edit box of the dialog box.
- Click on the **OK** button from the dialog box to create the constraint. Note that if you rotate the disc by dragging then the rail will move by specified distance.

Transitional Constraint

Transitional constraint makes a component follow the path of other component. You can use this constraint to make cam-follower or slot-follower mechanism. The procedure to use this constraint is given next.

- Click on the **Transitional** tab in the **Place Constraint** dialog box. The dialog box will be displayed as shown in Figure-42. Also, you will be asked to select the first geometry.

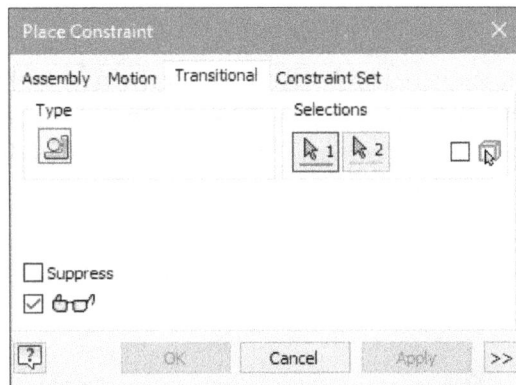

Figure-42. Place Constraint dialog box with Transitional tab selected

- Select the face of first component. You will be asked to select the second geometry.
- Select the face of slot/cam (second object). Preview of the constraint will be displayed; refer to Figure-43.

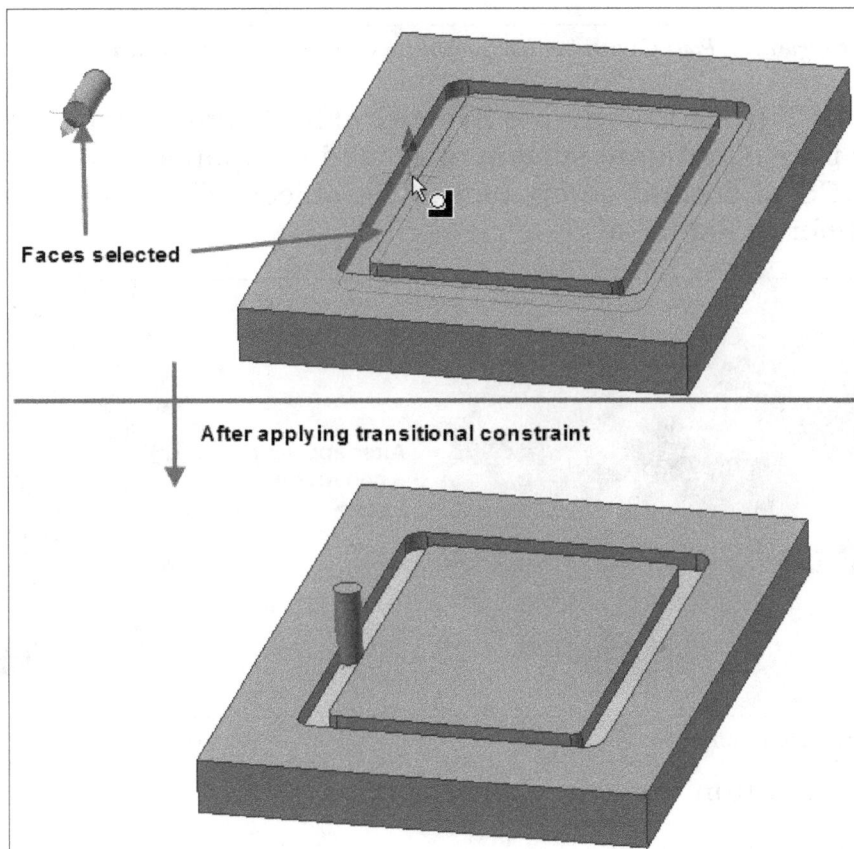

Figure-43. Preview of transitional constraint

- Click on the **OK** button from the dialog box to apply the constraint. Note that in the above example, you would have to apply the transitional constraint also on walls of the slot and follower. After applying the constraint, if you move the follower by dragging then it will move only inside the slot.

Constraint Set

The **Constraint Set** tool is used to constrain two components in such a way that coordinate systems of the two components are coincident. The procedure to apply this constraint is given next.

- Make sure that UCS are created for the components before using this constraint.
- Click on the **Constraint Set** tab in the **Place Constraint** dialog box. The options in the dialog box will be displayed as shown in Figure-44. Also, you will be asked to select the UCS of first object.

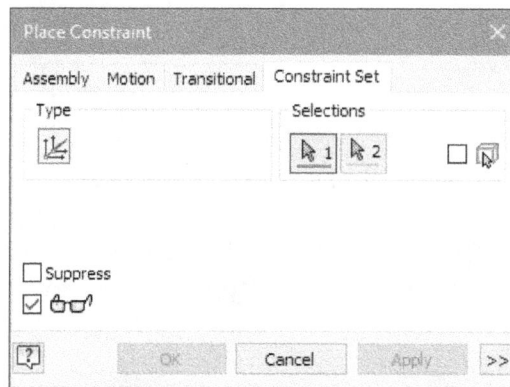

Figure-44. Place Constraint dialog box with Constraint Set tab selected

- Click on the UCS (User Coordinate System) of first component. You will be asked to select the user coordinate system of second component.
- Select the UCS of second component. The object will be placed as both UCS coincident; refer to Figure-45.

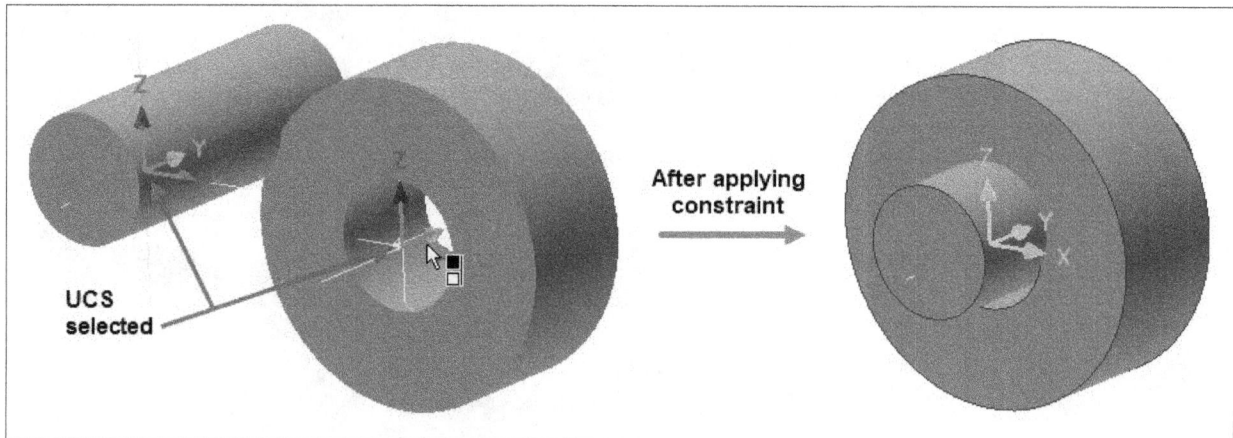

Figure-45. Preview of Constraint set

- Click on the **OK** button from the dialog box.

JOINTS

Joints are similar to constraints we have discussed earlier in terms of applying them. But, joints also provide motion along with placing the components at desired location. Various joints and procedures to apply them are discussed next.

Rigid Joint

The **Rigid joint** is used to fix a component at desired location. This type of joints are used in placing bolts, keys, pins, etc. The procedure to use this joint is given next.

- Click on the **Joint** tool from **Relationships** panel in the **Assemble** tab of the **Ribbon**. The **Place Joint** dialog box will be displayed; refer to Figure-46.

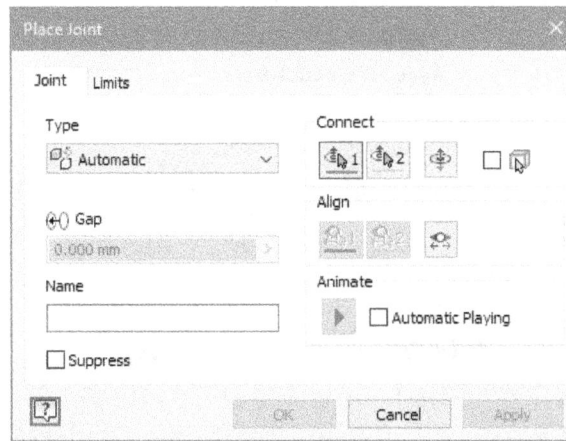

Figure-46. Place Joint dialog box

- Select the **Rigid** option from **Type** drop-down in the dialog box; refer to Figure-47. You will be asked to place origin on the first object.

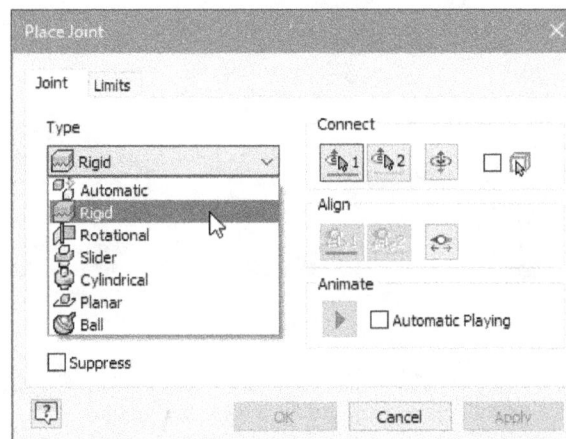

Figure-47. Rigid option in Type drop down

- Click on desired location on first object. You will be asked to place origin on the next object. Note that the location denoted by ring and point will be the origin of that component while creating joint.
- Select the location of second object; refer to Figure-48. The rigid joint will be created.

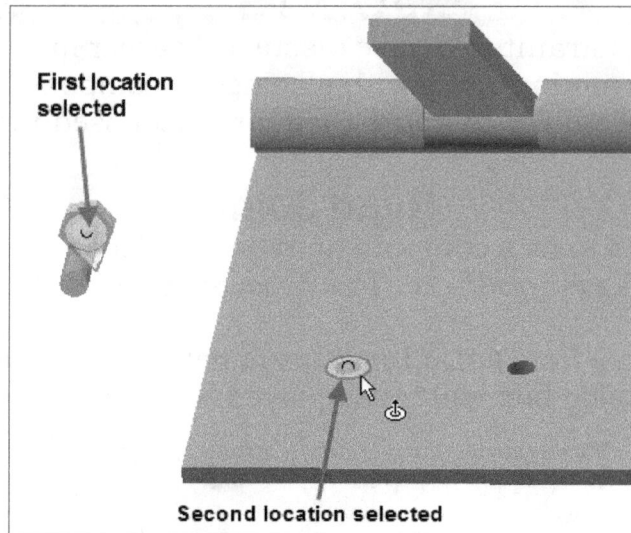

Figure–48. Selecting locations on components

- Click in the **Gap** edit box in the dialog box and specify the gap between selected faces, if required.
- Click on the **OK** button to create joint.

Rotational Joint

- Click on the **Joint** tool from **Relationships** panel in the **Assemble** tab of the **Ribbon**. The **Place Joint** dialog box will be displayed as discussed earlier.
- Select the **Rotational** option from **Type** drop-down in the dialog box; refer to Figure-49. You will be asked to place origin on the first component.

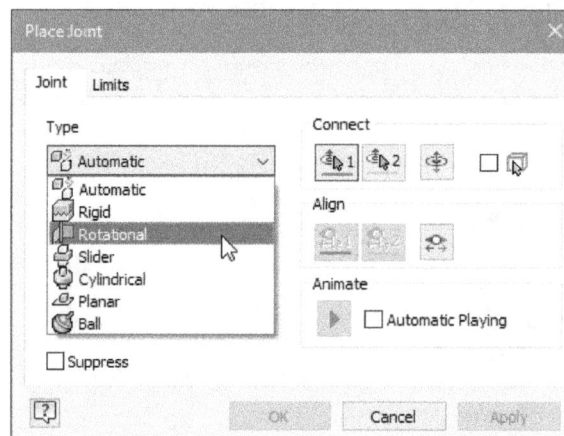

Figure–49. Rotational option in Type drop down

- Click at desired location on the first component. You will be asked to select a location on the second component.
- Click at desired location on the second component; refer to Figure-50. Preview of the rotational joint will be displayed.

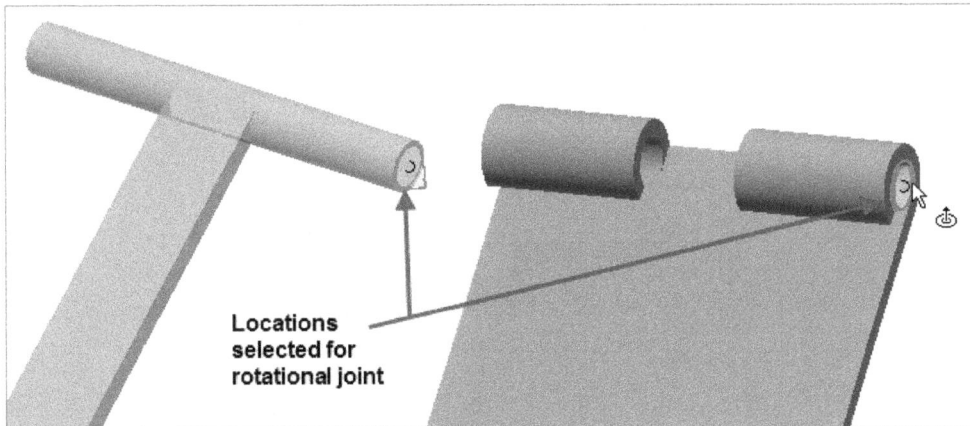

Figure-50. Locations selected for rotational joint

- If you want to set the maximum and minimum limit of rotation then click on the **Limits** tab in the dialog box. The dialog box will be displayed as shown in Figure-51.

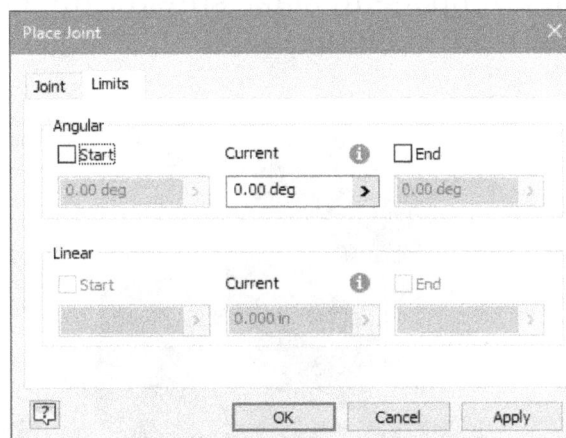

Figure-51. Limits tab in Place Joint dialog box

- Select the **Start** check box and specify the starting point of limit.
- Similarly, specify the end point of limit by using the **End** check box.
- To specify the current position of the component, click in the **Current** edit box and specify desired value of angle.
- Click on the **OK** button from the dialog box.

Slider Joint

The **Slider** joint is used to create joint when you need sliding motion of a component. For example, piston in cylinder of engine. The procedure to apply this joint is given next.

- Click on the **Joint** tool from **Relationships** panel in the **Assemble** tab of the **Ribbon**. The **Place Joint** dialog box will be displayed.
- Select the **Slider** option from **Type** drop-down in the dialog box; refer to Figure-52. You will be asked to select the first origin.

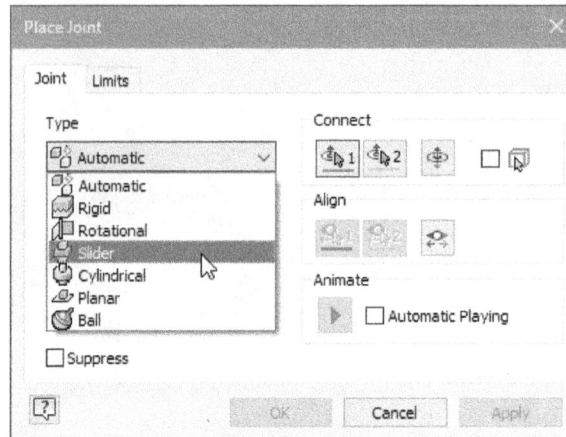

Figure-52. Slider option in Type drop down

- Select the face of first component (slider). You will be asked to select origin of the other component.
- Select the face of second component. Make sure the direction of motion is aligned properly after selection; refer to Figure-53.

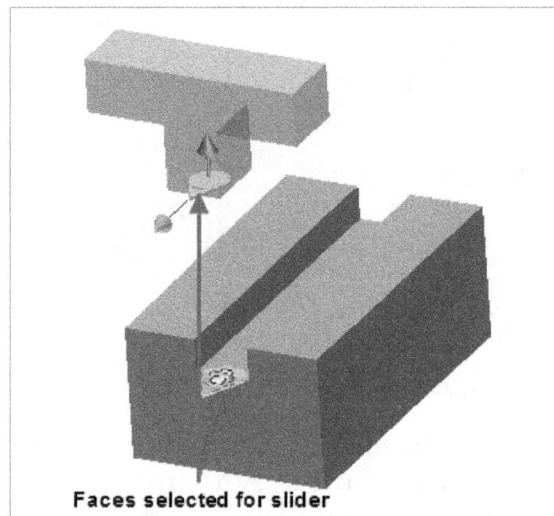

Faces selected for slider

Figure-53. Faces selected for slider joint

- Set desired gap and distance limits.
- Click on the **OK** button to create the joint.

Cylindrical Joint

The **Cylindrical** joint is used to create joint when you need sliding and rotating motion of a shaft. For example, motion of a shaft in a tube. The procedure to apply this joint is given next.

- Click on the **Joint** tool from **Relationships** panel in the **Assemble** tab of the **Ribbon**. The **Place Joint** dialog box will be displayed.
- Select the **Cylindrical** option from **Type** drop-down in the dialog box; refer to Figure-54. You will be asked to select the first origin.

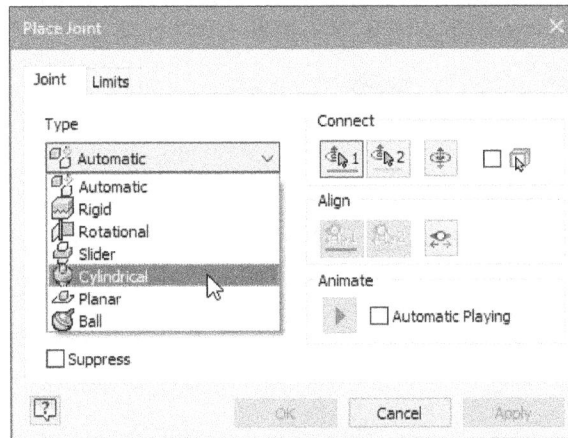

Figure-54. Cylindrical option in Type drop down

- Click on the round face of the first component. You will be asked to select origin of next component.
- Click on the round face of the second component. Preview of the joint will be displayed; refer to Figure-55.

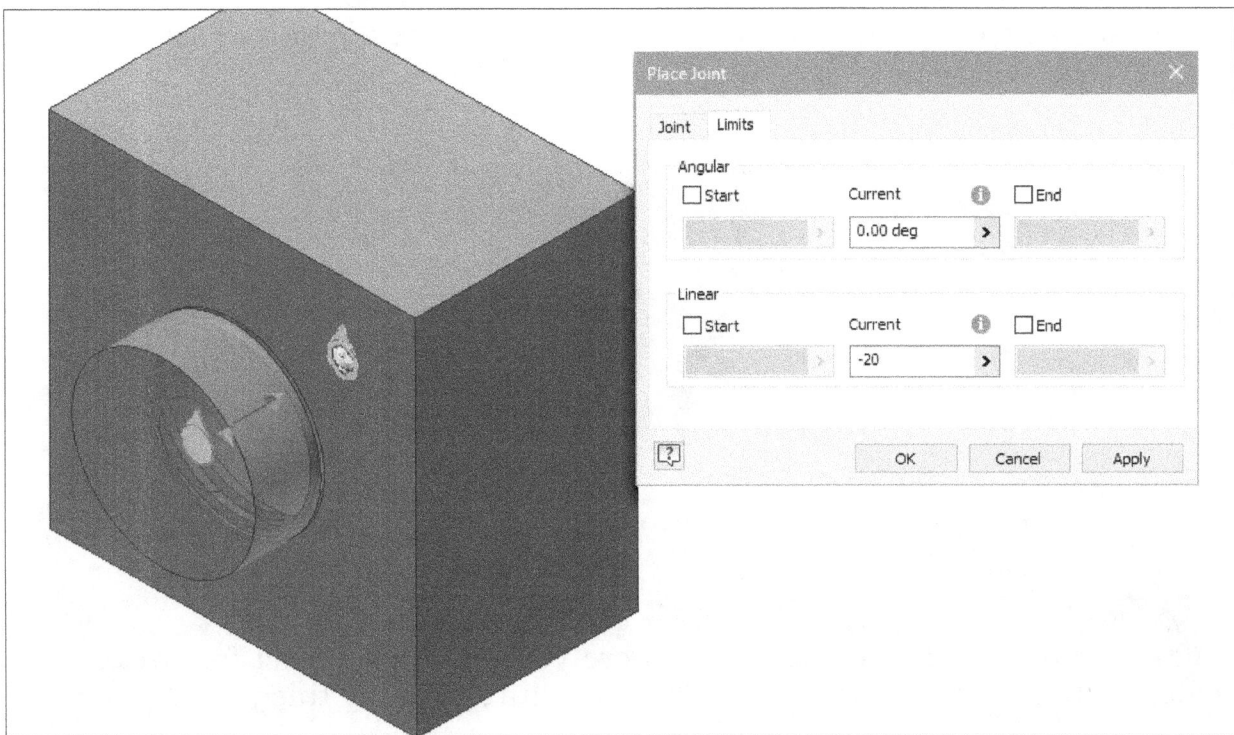

Figure-55. Preview of cylindrical joint

- Set desired limits for the component.
- Click on the **OK** button from the dialog box.

Planar Joint

The **Planar** joint is used to create joint when you need component to move in a specified plane. Note that one rotational degree of freedom will also be free of the component on applying this joint. The procedure to apply this constraint is given next.

- Click on the **Joint** tool from **Relationships** panel in the **Assemble** tab of the **Ribbon**. The **Place Joint** dialog box will be displayed.

- Select the **Planar** option from **Type** drop-down in the dialog box; refer to Figure-56. You will be asked to select the first origin.

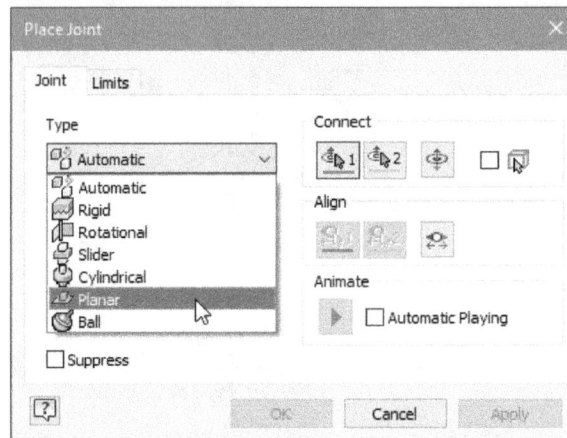

Figure-56. Planar option in Type drop down

- Click on the face of first component. You will be asked to place origin on next component.
- Click on the face of second component. Preview of the joint will be displayed.
- Set desired gap and click on the **OK** button from the dialog box to create joint; refer to Figure-57.

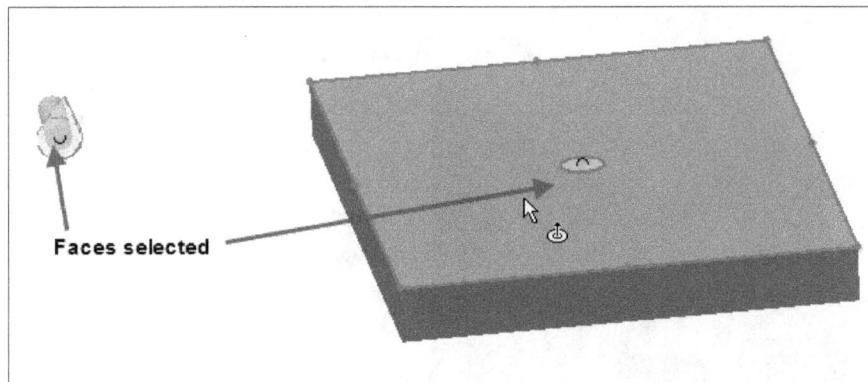

Figure-57. Faces selected for planar joint

Ball Joint

The **Ball** joint is used to create joint where you need the object free in rotational dimensions but fixed in translation. The procedure to apply this joint is given next.

- Click on the **Joint** tool from **Relationships** panel in the **Assemble** tab of the **Ribbon**. The **Place Joint** dialog box will be displayed.

- Select the **Ball** option from **Type** drop-down in the dialog box; refer to Figure-58. You will be asked to select the first origin.

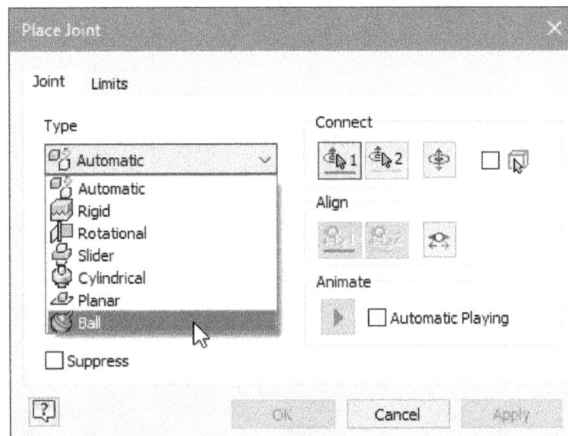

Figure-58. Ball option in Type drop down

- Select the round face of first component. You will be asked to select origin of next component.
- Select the round face of second component. The preview of ball joint will be displayed; refer to Figure-59.

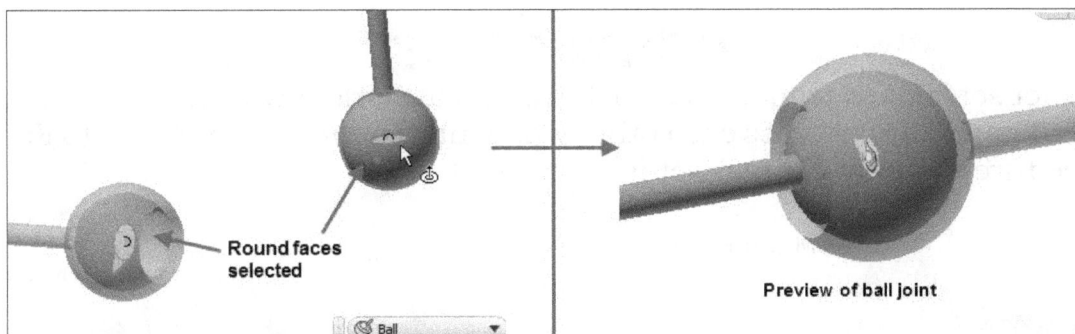

Figure-59. Creating ball joint

- Click on the **OK** button from the dialog box to create the joint.

GROUNDED CONSTRAINT

As the name suggests, the Grounded constraint is used to fix the component at its location. This constraint is generally used to fix the base component so that other components based on it does not move automatically while applying mechanism on them. The procedure to ground a component is given next.

- Right-click on the component that you want to be grounded (fixed) from the **Model Browser** bar. A shortcut menu will be displayed; refer to Figure-60.

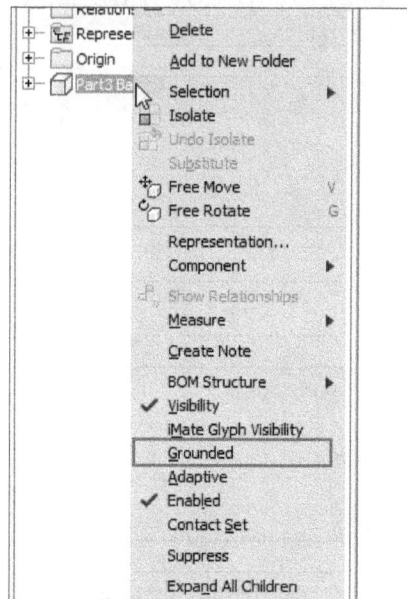

Figure-60. Shortcut menu for component in assembly

• Click on the **Grounded** option from the shortcut menu. The component will be fixed at its current position.

CONTACT SET

The Contact set is used to make part behave in real world environment. It specifically means that parts cannot cross each other when interference occurs; refer to Figure-61. The procedure to apply contact set is given next.

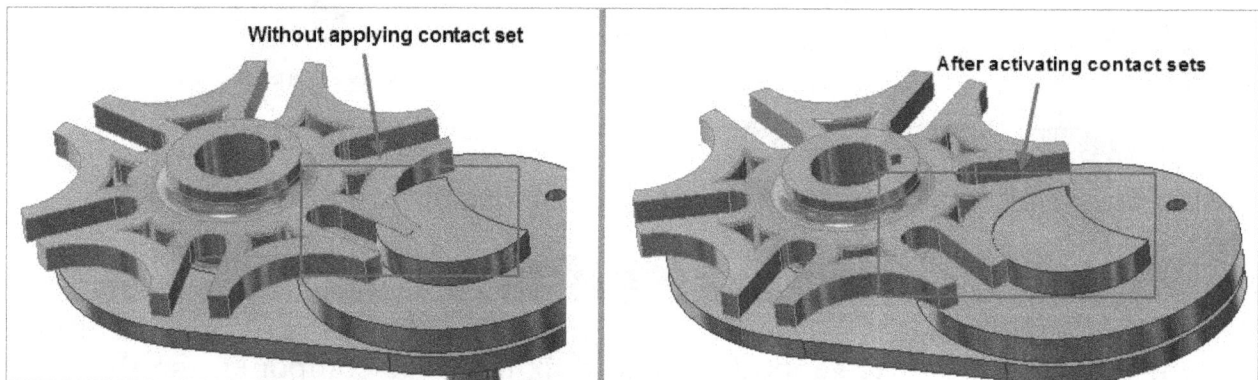

Figure-61. Interference without activating contact sets

• Right-click on the component which you want to use in contact set from the **Model Browser** bar. A shortcut menu will be displayed.
• Select the **Contact Set** option from the shortcut menu to apply contact set; refer to Figure-62. Note that the icon of component in **Model Browser** bar will change as shown in Figure-63.

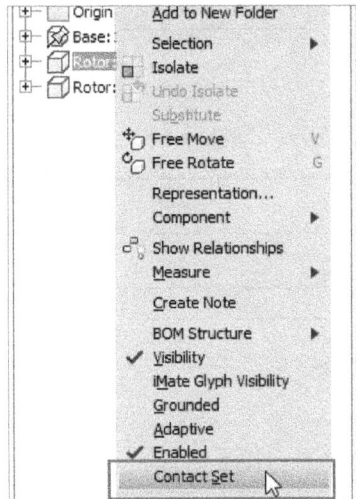

Figure-62. Contact Set option

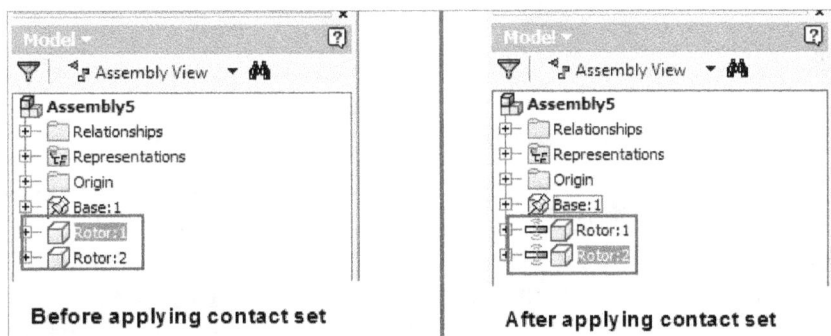

Figure-63. Applying contact set

- We have applied the contact sets but we need to activate them before making use in assembly. To do so, click on the **Inspect** tab in the **Ribbon** and select **Activate Contact Solver** button from the **Interference** panel in the **Ribbon**; refer to Figure-64. Contact sets will be activated.

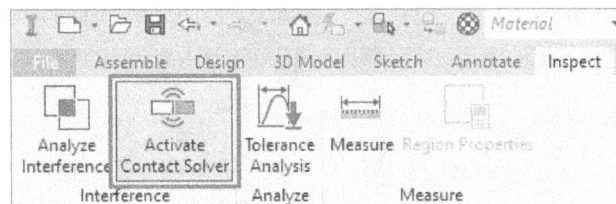

Figure-64. Activate Contact Solver button

BILL OF MATERIALS

The Bill of Materials (in short BOM) is used to form a table of components that are used in the assembly. In this table, we have name of components and their respective quantity. It becomes very important for purchasing the components from market or manage inventory of components required in making assembly. The procedure to create BOM in Autodesk Inventor is given next.

- Click on the **Bill of Materials** tool from the **Manage** panel in the **Assemble** tab of the **Ribbon**. The **Bill of Materials** dialog box will be displayed; refer to Figure-65.

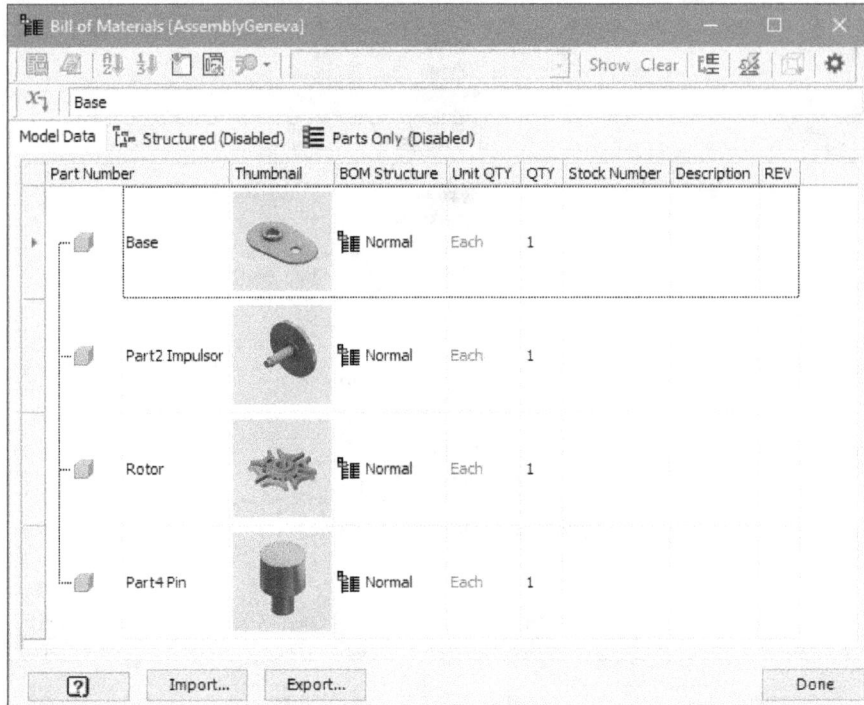

Figure-65. Bill of Materials dialog box

- Click in desired field to change the value of that item. For example, you can change the quantity from 1 to 5 in the fields under **QTY** column, you can change the name of part by clicking in the respective field, and so on.
- Enter the other details in respective fields of the dialog box.
- If you want to export the bill of material to other software in xml format then click on the **Export** button at the bottom in the dialog box. Similarly, you can import the bill of material in xml format from other software by using the **Import** button.
- Click on the **Done** button to exit the dialog box.

DRIVING CONSTRAINT

You can apply motor to a constraint by using the **Drive** option. The drive can be applied to both rotational as well as translational motion. The procedure to apply the **Drive** option is given next.

- Right-click on desired constraint from the **Relationships** category in the **Model Browser** bar. A shortcut menu will be displayed; refer to Figure-66. Note that the selected constraint should have one or more degree of freedom.

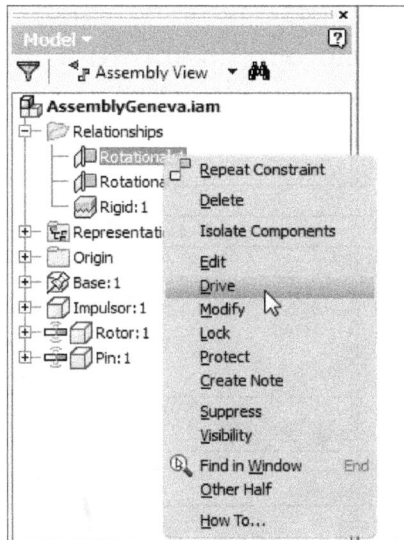

Figure-66. Shortcut menu for constraint

- Click on the **Drive** option from the shortcut menu. The **Drive** dialog box will be displayed; refer to Figure-67.

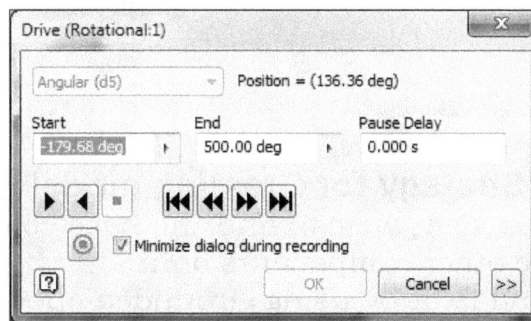

Figure-67. Drive dialog box

- Specify desired starting and end value in the respective edit boxes.
- You can specify desired amount of delay in the **Pause Delay** edit box.
- Click on the **Forward** and **Reverse** buttons from the dialog box to check the motion of assembly.
- Click on the **OK** button to apply drive.

PRACTICAL 1

Create assembly of Geneva mechanism as shown in Figure-68. The components of assembly are available in the resource kit.

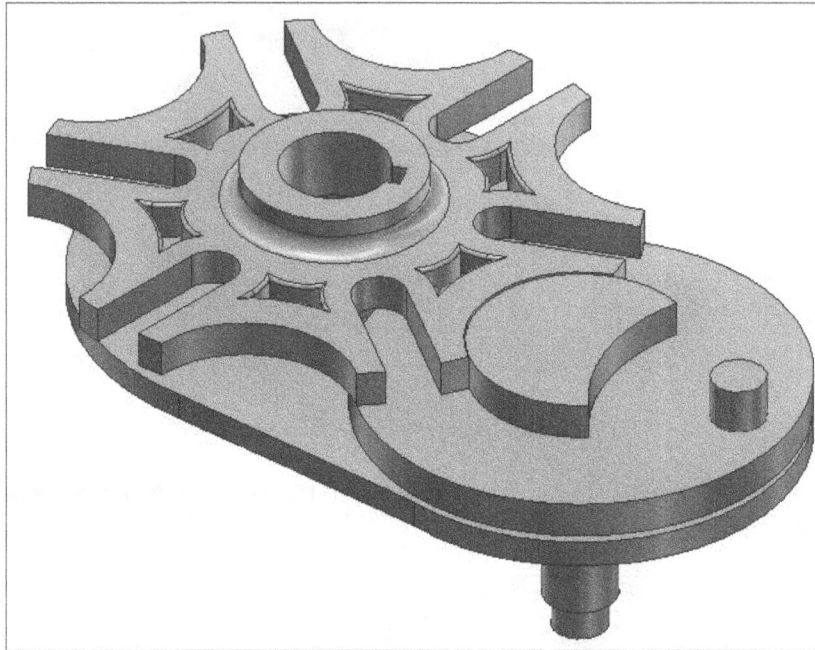

Figure-68. Geneva mechanism

Strategy for Creating model

1. Looking at the isometric view, we can find the base component which is a base plate supporting all the other components on it.
2. Insert the base plate and fix it by using Grounded option.
3. One by one insert the other components and apply constraints which provide motion also.

Starting Assembly File

- Start Autodesk Inventor by double-clicking on the Autodesk Inventor Professional 2024 icon from the desktop. (If not started yet.)
- Click on the **New** button from the **Quick Access Toolbar**. The **Create New File** dialog box will be displayed.
- Double-click on **Standard(mm).iam** icon from the Assembly category in **Metric** templates; refer to Figure-69. The Assembly environment of Autodesk Inventor will be displayed.

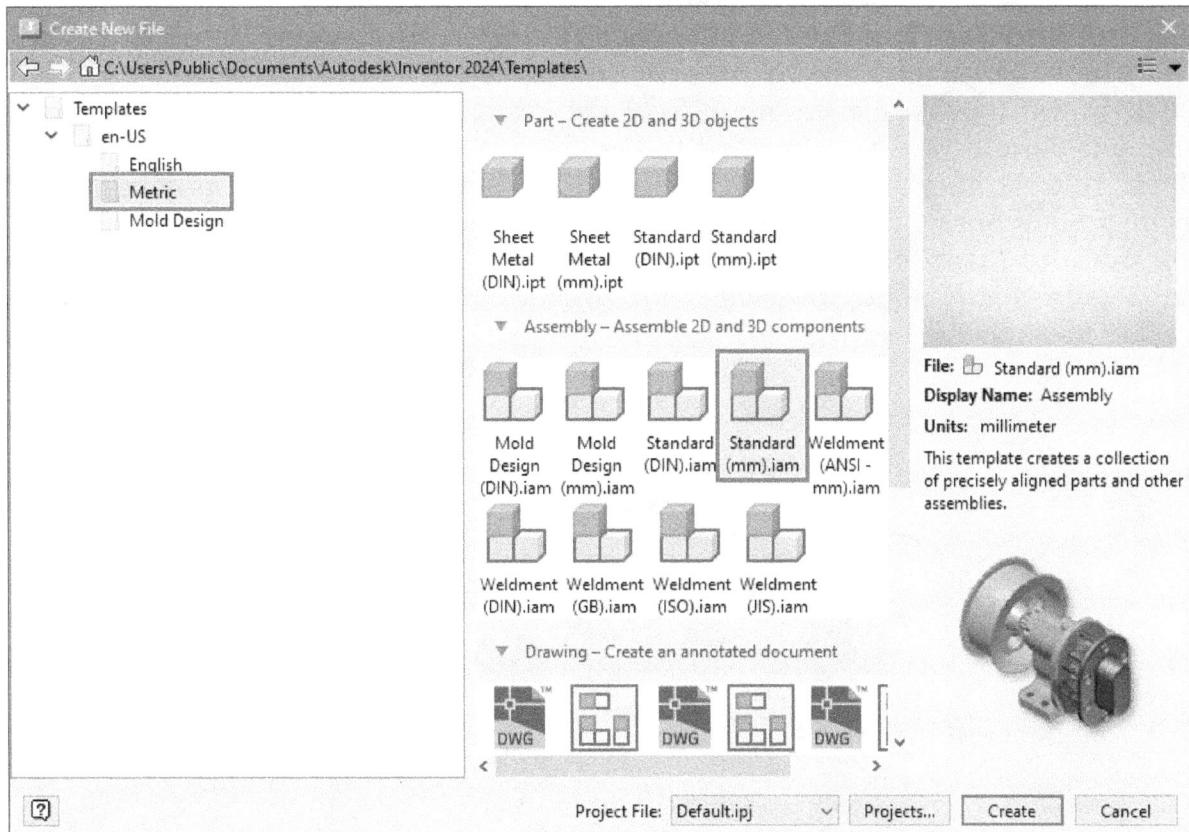
Figure-69. Assembly option in Create New File dialog box

Inserting First Component

- Click on the **Place** button from the **Component** panel in the **Assemble** tab of the **Ribbon**. The **Place Component** dialog box will be displayed.
- Select the **Base.ipt** from the dialog box and click on the **Open** button. The component will get attached to the cursor.
- Click in the viewport to place the component. A copy of the component still get attached to cursor. Press **ESC** from the keyboard to exit inserting more copies of the component.
- Right-click on the name of component from the **Model Browser** bar. A shortcut menu will be displayed; refer to Figure-70.

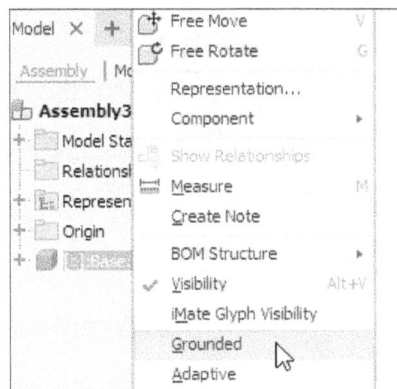
Figure-70. Shortcut menu for grounding component

- Click on the **Grounded** option from it. The component will get fixed.

Inserting Impulsor component

*   Click on the **Place** tool from the **Component** panel in the **Assemble** tab of the **Ribbon**. The **Place Component** dialog box will be displayed.
*   Click on the **Impulsor** component from the dialog box and click on the **Open** button. The component will get attached to the cursor.
*   Click in the empty area of the viewport. The component will be placed. Press **ESC** from the keyboard to exit inserting more copies of the component.
*   Click on the **Joint** tool from **Relationships** panel in the **Assemble** tab of the **Ribbon**. The **Place Joint** dialog box will be displayed; refer to Figure-71.

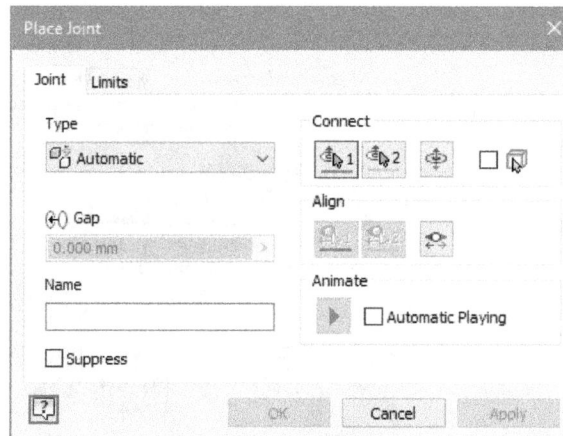

Figure-71. Place Joint dialog box

*   Select the **Rotational** option from **Type** drop-down in the dialog box. You will be asked to place origin on the first component.
*   Click on the face of first component; refer to Figure-72. You will be asked to place origin on the second component.
*   Select the edge of second component; refer to Figure-72. The component will be placed accordingly.

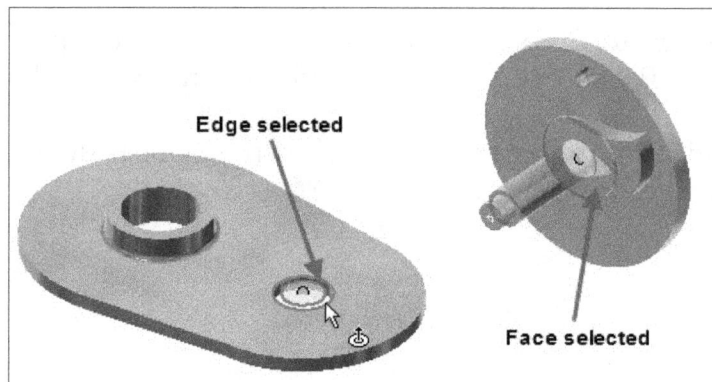

Figure-72. Entities selected for rotational joint

*   Flip the side of the component if required by using the **Flip component** button from the dialog box.
*   Click on the **OK** button from the dialog box to create the joint.

Inserting Rotor Component

*   Click on the **Place** tool from the **Component** panel in the **Assemble** tab of the **Ribbon**. The **Place Component** dialog box will be displayed.

- Click on the **Rotor** component from the dialog box and click on the **Open** button. The component will get attached to the cursor.
- Click in the empty area of the viewport. The component will be placed. Press **ESC** from the keyboard to exit inserting more copies of the component.
- Click on the **Joint** tool from the **Relationships** panel in the **Assemble** tab of the **Ribbon**. The **Place Joint** dialog box will be displayed.
- Select the **Rotational** option from **Type** drop-down in the dialog box. You will be asked to place origin on the first component.
- Click on the edge of first component; refer to Figure-73. You will be asked to place origin on the second component.
- Select the edge of second component; refer to Figure-73. The joint will be created.

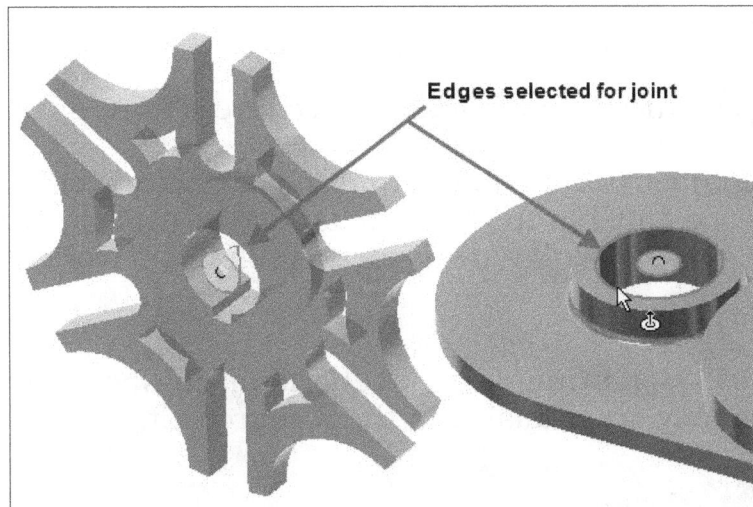

Figure-73. Edges selected for rotational joint

- Click on the **OK** button from the dialog box.
- Right-click on the **Rotor** component from the **Model Browser** bar and select the **Contact Set** option from the shortcut menu displayed.

Inserting Pin

- Click on the **Place** tool from the **Component** panel in the **Assemble** tab of the **Ribbon**. The **Place Component** dialog box will be displayed.
- Click on the **Pin** component from the dialog box and click on the **Open** button. The component will get attached to the cursor.
- Click in the empty area of the viewport. The component will be placed. Press **ESC** from the keyboard to exit inserting more copies of the component.
- Click on the **Joint** tool from the **Relationships** panel in the **Assemble** tab of the **Ribbon**. The **Place Joint** dialog box will be displayed.
- Select the **Rigid** option from **Type** drop-down in the dialog box. You will be asked to place origin on the first component.
- Click on the face of first component; refer to Figure-74. You will be asked to place origin on the next component.
- Click on the edge of second component as shown in Figure-74. The rigid joint will be created.

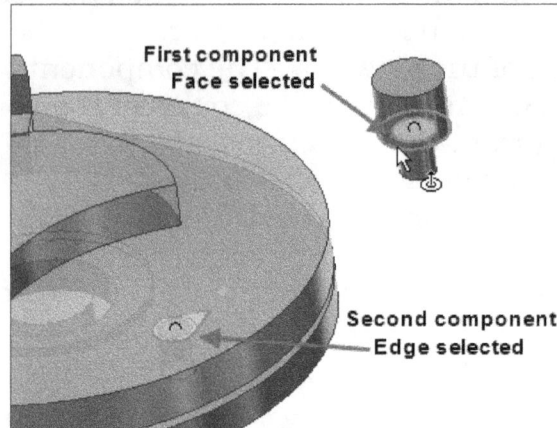

Figure-74. Entities selected for rigid joint

• Apply contact set on the **Pin** component. Procedure is same as discussed for **Rotor**.

Now, activate the contact solver by selecting the **Activate Contact Solver** button from the **Interference** panel in the **Inspect** tab of the **Ribbon**. Rotate the **Impulsor** component by dragging to check if the geneva mechanism works as expected.

PRACTICAL 2

Create assembly of Oscillating Cam Mechanism as shown in Figure-75. The components of assembly are available in the resource kit.

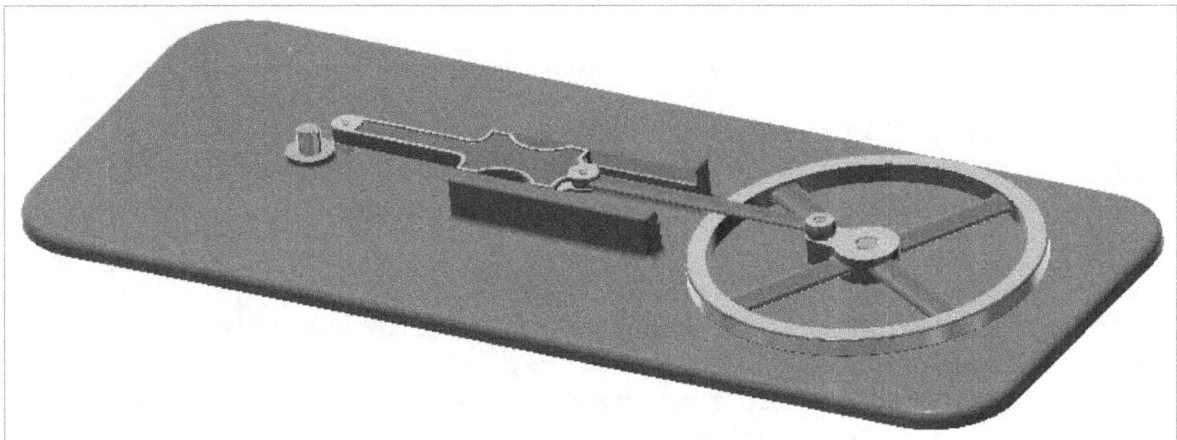

Figure-75. Partial assembly of oscillating cam mechanism

Strategy for Creating model

1. Looking at the isometric view, we can find the base component which is a base plate supporting all the other components on it.
2. Insert the base plate and fix it by using **Grounded** option.
3. One by one insert the other components and apply constraints which provide motion also.
4. Run the mechanism by dragging the wheel.

Starting Assembly File

• Start Autodesk Inventor by double-clicking on the Autodesk Inventor Professional 2024 icon from the desktop. (If not started yet.)

- Click on the **New** button from the **Quick Access Toolbar**. The **Create New File** dialog box will be displayed.
- Double-click on **Standard(mm).iam** icon from the **Assembly** category in **Metric** templates. The Assembly environment of Autodesk Inventor will be displayed.

Inserting the Base Component

- Click on the **Place** tool from the **Component** panel in the **Assemble** tab of the **Ribbon**. The **Place Component** dialog box will be displayed.
- Click on the **Base Plate** component from the dialog box and click on the **Open** button. The component will get attached to the cursor.
- Click in the empty area of the viewport. The component will be placed. Press **ESC** from the keyboard to exit inserting more copies of the component.
- Right-click on the component name in the **Model Browser** bar and select the **Grounded** option from the shortcut menu displayed. The plate will get fixed.

Inserting Oscillating Wheel Component

- Click on the **Place** tool from the **Component** panel in the **Assemble** tab of the **Ribbon**. The **Place Component** dialog box will be displayed.
- Click on the **Oscillating Wheel** component from the dialog box and click on the **Open** button. The component will get attached to the cursor.
- Click in the empty area of the viewport. The component will be placed. Press **ESC** from the keyboard to exit inserting more copies of the component.
- Click on the **Constrain** tool from **Relationships** panel in the **Assemble** tab of the **Ribbon**. The **Place Constraint** dialog box will be displayed.
- Click on the **Insert** button from the **Type** area of the dialog box and select the edges of the base component and wheel as shown in Figure-76.

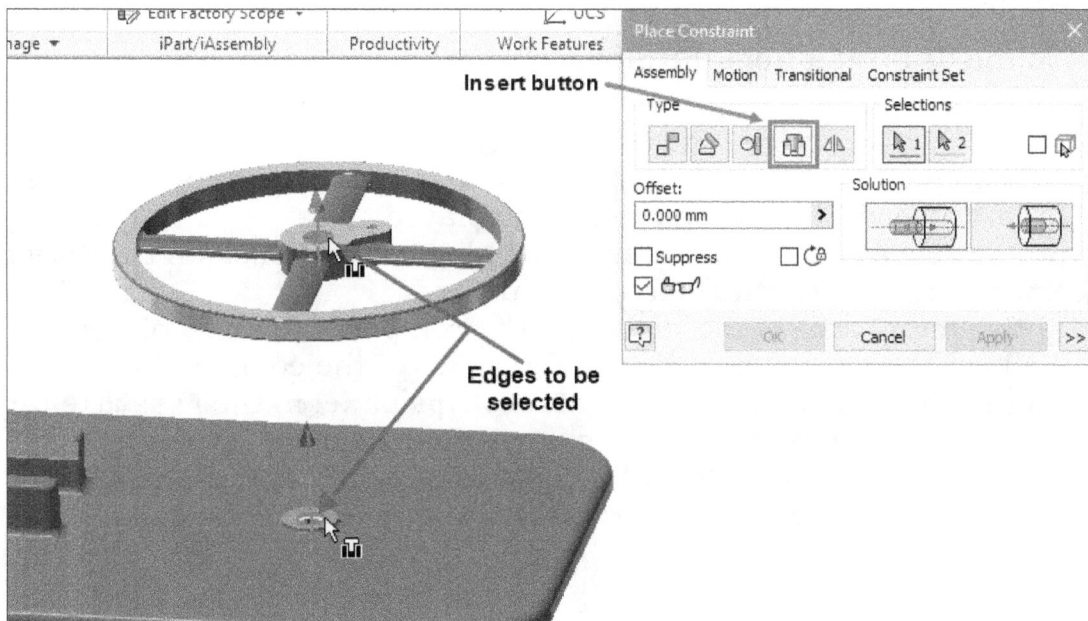

Figure-76. Inserting Oscillating wheel

- Flip the component if required and then click on the **OK** button from the dialog box.

Inserting Lever

- Click on the **Place** tool from the **Component** panel in the **Assemble** tab of the **Ribbon**. The **Place Component** dialog box will be displayed.
- Click on the **Lever** component from the dialog box and click on the **Open** button. The component will get attached to the cursor.
- Click in the empty area of the viewport. The component will be placed. Press **ESC** from the keyboard to exit inserting more copies of the component.
- Click on the **Constrain** tool from **Relationships** panel in the **Assemble** tab of the **Ribbon**. The **Place Constraint** dialog box will be displayed.
- Click on the **Insert** button from the **Type** area of the dialog box and select the edges of oscillating wheel and lever; refer to Figure-77.

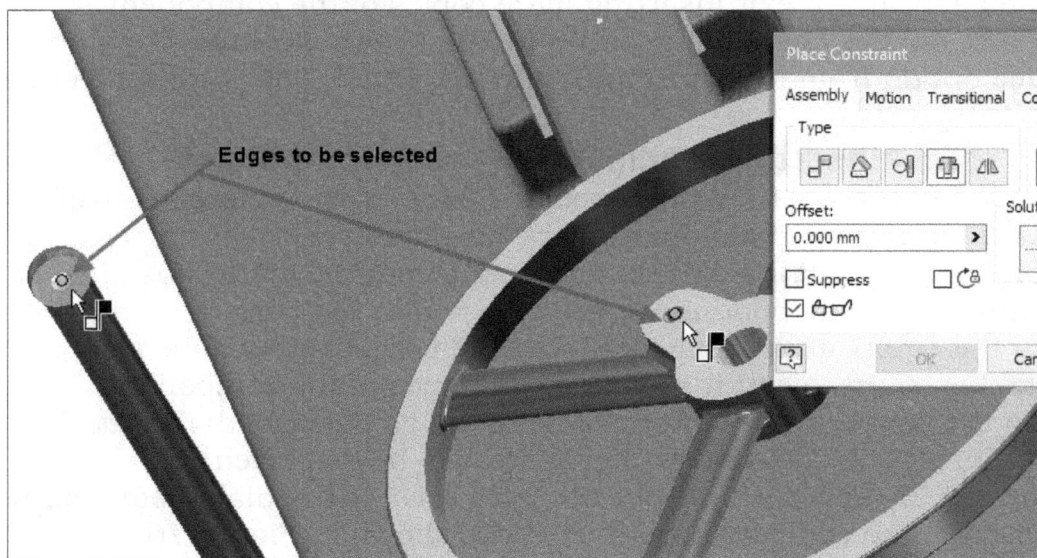

Figure-77. Edges selected for insert constraint

- Click on the **OK** button from the dialog box.

Inserting Tappet

- Click on the **Place** tool from the **Component** panel in the **Assemble** tab of the **Ribbon**. The **Place Component** dialog box will be displayed.
- Click on the **Tappet** component from the dialog box and click on the **Open** button. The component will get attached to the cursor.
- Click in the empty area of the viewport. The component will be placed. Press **ESC** from the keyboard to exit inserting more copies of the component.
- Click on the **Constrain** tool from **Relationships** panel in the **Assemble** tab of the **Ribbon**. The **Place Constraint** dialog box will be displayed.
- Click on the **Insert** button from the **Type** area of the dialog box and select the edges of oscillating wheel and lever; refer to Figure-78.

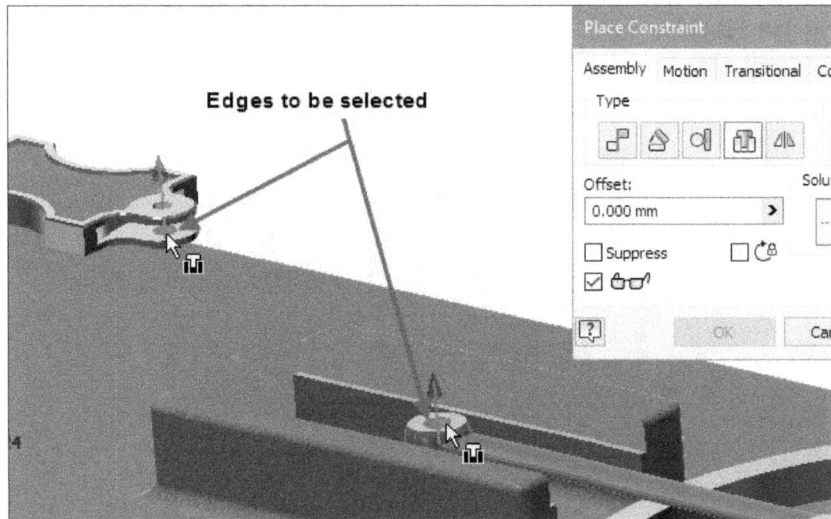

Figure-78. Connecting tappet with lever

- Click on the **Apply** button from the dialog box.
- Drag the tappet outside so that you can see the side faces of tappet; refer to Figure-79.

Figure-79. Tappet dragged outside

- Click on the **Mate** button from the **Type** area of the dialog box and select the side face of tappet; refer to Figure-80.

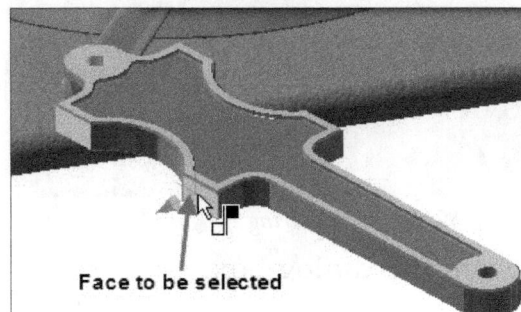

Figure-80. Face of tappet to be selected

- Select the side face of the guide in base plate; refer to Figure-81. Preview of mate constraint will be displayed.

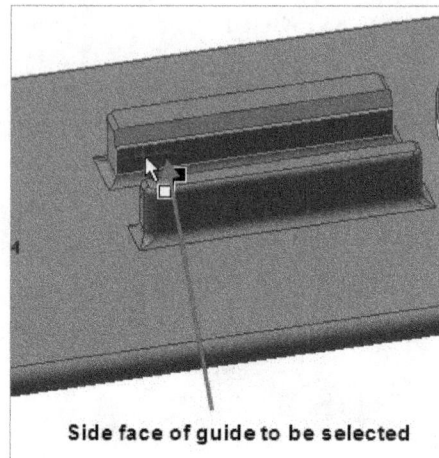

Figure-81. Side faces of guide to be selected

- Click on the **OK** button from the dialog box.

Inserting bearing

- Click on the **Place** tool from the **Component** panel in the **Assemble** tab of the **Ribbon**. The **Place Component** dialog box will be displayed.
- Click on the **Bearing** component from the dialog box and click on the **Open** button. The component will get attached to the cursor.
- Click in the empty area of the viewport. The component will be placed. Press **ESC** from the keyboard to exit inserting more copies of the component.
- Click on the **Constrain** tool from **Relationships** panel in the **Assemble** tab of the **Ribbon**. The **Place Constraint** dialog box will be displayed.
- Click on the **Insert** button from the **Type** area of the dialog box and select the edges of bearing and tappet as shown in Figure-82.

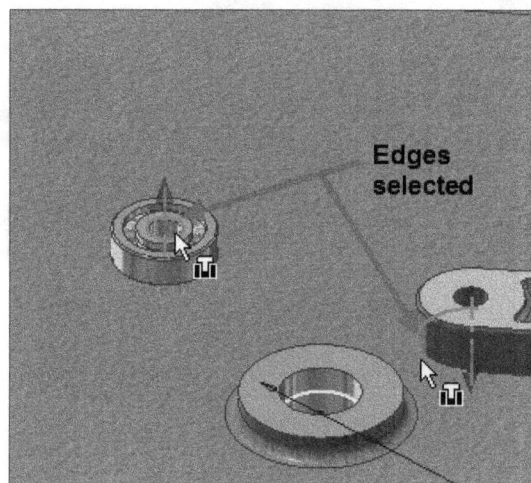

Figure-82. Inserting bearing

- Click on the **OK** button from the dialog box.

Inserting Pins

There are four different size pins to be placed at five places in the assembly. Insert these pins using the **Insert** constraint as shown in Figure-83.

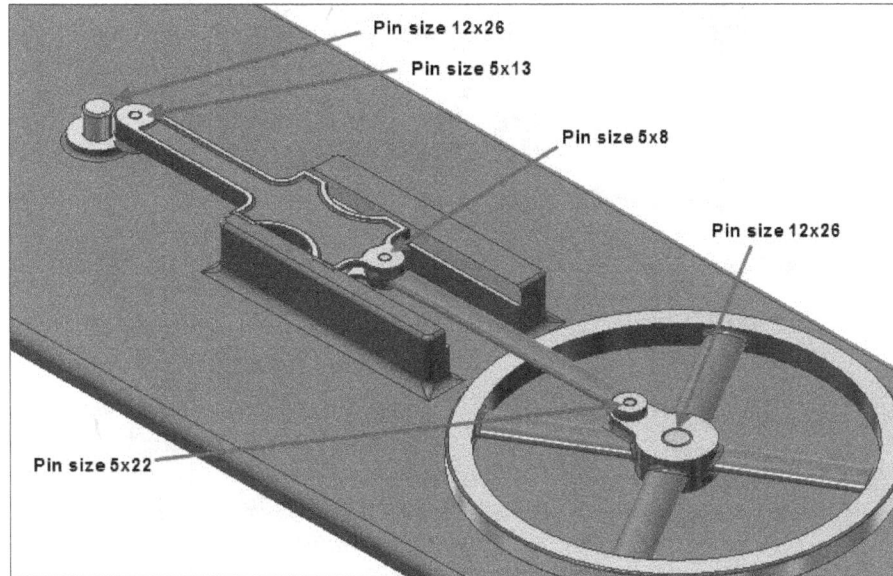

Figure-83. Pins inserted in assembly

Inserting Cam

- Click on the **Place** tool from the **Component** panel in the **Assemble** tab of the **Ribbon**. The **Place Component** dialog box will be displayed.
- Click on the **Cam** component from the dialog box and click on the **Open** button. The component will get attached to the cursor.
- Click in the empty area of the viewport. The component will be placed. Press **ESC** from the keyboard to exit inserting more copies of the component.
- Click on the **Constrain** tool from **Relationships** panel in the **Assemble** tab of the **Ribbon**. The **Place Constraint** dialog box will be displayed.
- Click on the **Insert** button from **Type** area of the dialog box and select the edges as shown in Figure-84.

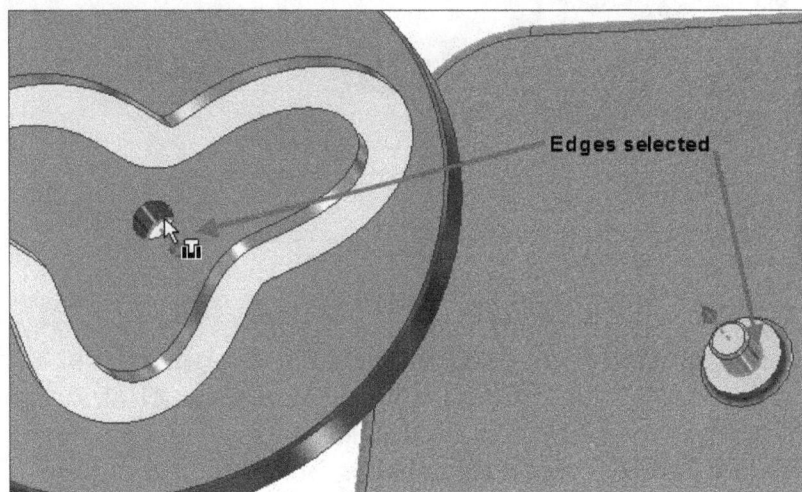

Figure-84. Edges to be selected

- Click on the **Apply** button from the dialog box.
- Click on the **Transitional** tab of the dialog box. You will be asked to select the entities to be constrained.

- Select the round face of bearing and inner face of the groove; refer to Figure-85.

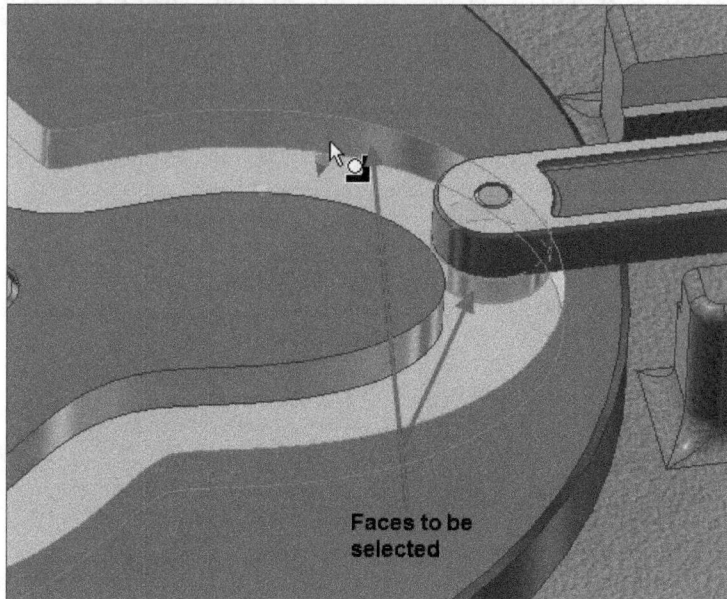

Figure-85. Faces selected for translational constraint

- Click on the **OK** button from the dialog box to apply the constraint.

PRACTICE 1

Create the working assembly of a simple vice as shown in Figure-86. The components are available in the resource kit.

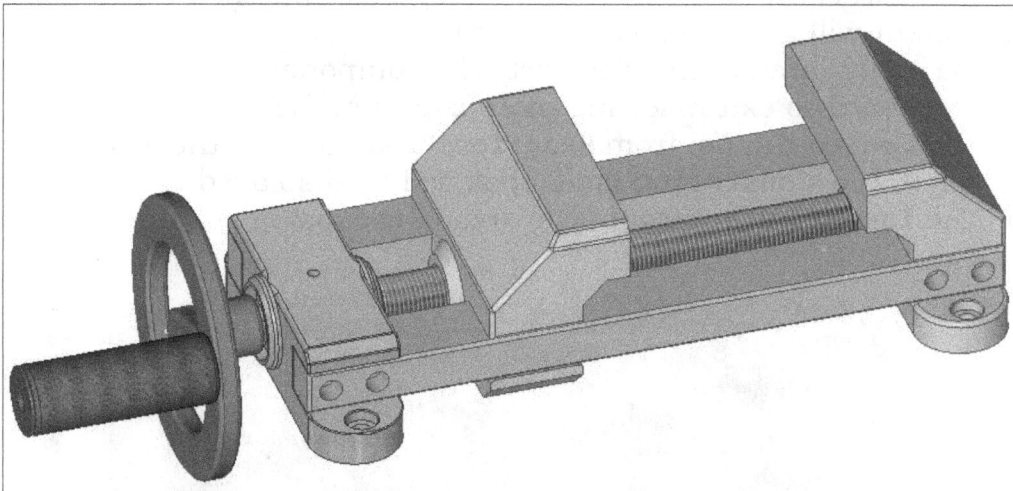

Figure-86. Simple Vice

INTRODUCTION TO PRESENTATION

The Presentation environment is used to generate snapshots and video clips of animation of the assembly. To activate this environment, click on the **Presentation** option from the **New** cascading menu of the **File** menu. The tools of Presentation environment will be displayed with **Insert** dialog box; refer to Figure-87. Select desired assembly file to be used in presentation and click on the **Open** button. Interface of presentation environment will be displayed; refer to Figure-88.

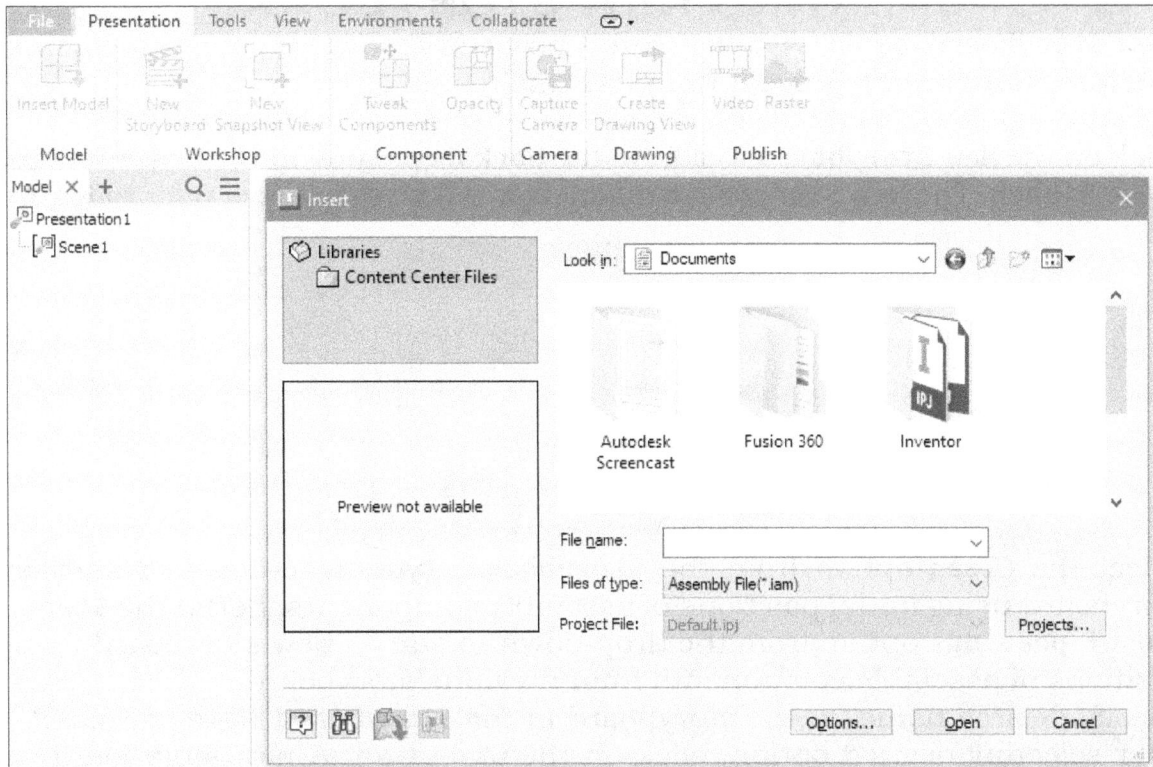

Figure-87. Presentation environment

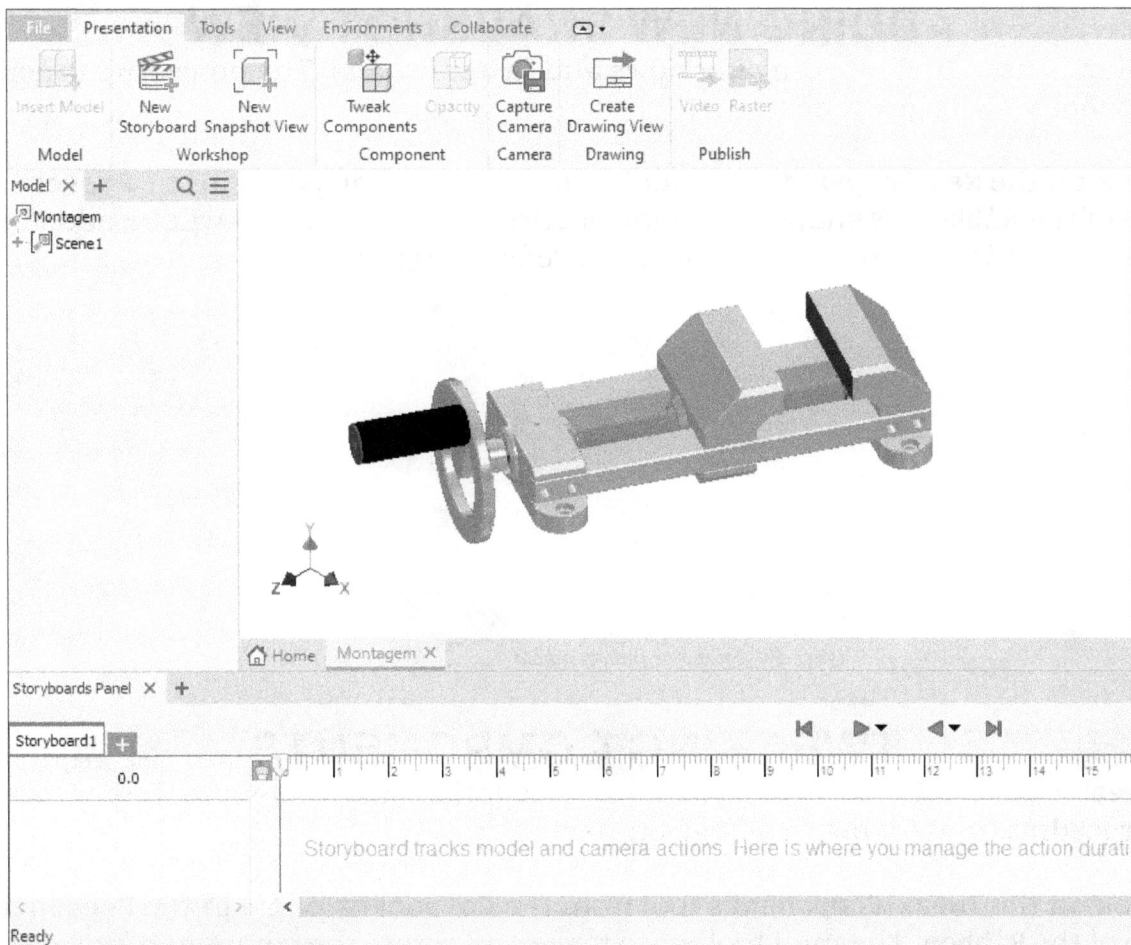

Figure-88. Presentation interface

CREATING NEW STORYBOARD

The **New Storyboard** tool is used to create a new storyboard for generating animation of assembly motion. The procedure to use this tool is given next.

- Click on the **New Storyboard** tool from the **Workshop** panel in the **Presentation** tab of the **Ribbon**. The **New Storyboard** dialog box will be displayed; refer to Figure-89.

Figure-89. New Storyboard dialog box

- Select the **Clean** option from the **Storyboard Type** drop-down to start a new storyboard using initial positions of assembly components. Select the **Start from end of previous** option from the drop-down to start a new storyboard using the positions of assembly components from previous storyboard.
- Specify desired name for the storyboard in the **Storyboard Name** edit box.
- After selecting desired option, click on the **OK** button. A new storyboard will be added in the presentation.

ADDING NEW SNAPSHOT VIEW

Snapshot is used to create drawing views and raster images. The procedure to generate a snapshot view is given next.

- Click on the **New Snapshot View** tool from the **Workshop** panel in the **Presentation** tab of the **Ribbon**. A snapshot image of current view of model will be added in the **Snapshot Views** panel of the interface; refer to Figure-90.

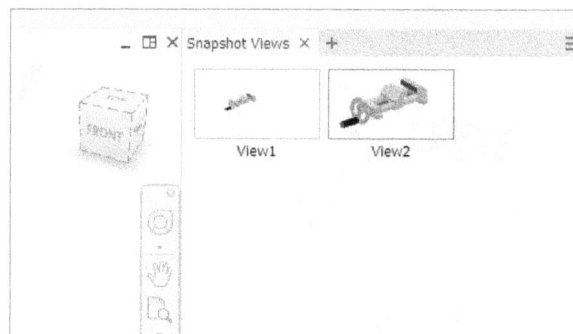

Figure-90. Snapshot images added

TWEAKING COMPONENTS

The **Tweak Components** tool is used to apply translation and rotation to the components. The procedure to use this tool is given next.

- Click on the **Tweak Components** tool from the **Component** panel in the **Presentation** tab of the **Ribbon**. The mini toolbar to move and rotate the components of assembly will be displayed; refer to Figure-91.

- Set desired value in **Duration** edit box of mini toolbar to define time in seconds within which the component will complete applied translation or rotation.
- Select desired component from the assembly and then select the **Move** or **Rotate** toggle button to define whether you want to move the component or rotate it. After selecting the toggle button, use the drag handles on the component to create motion; refer to Figure-92.

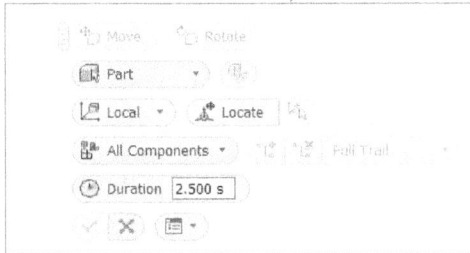

Figure-91. Mini toolbar for tweaking components

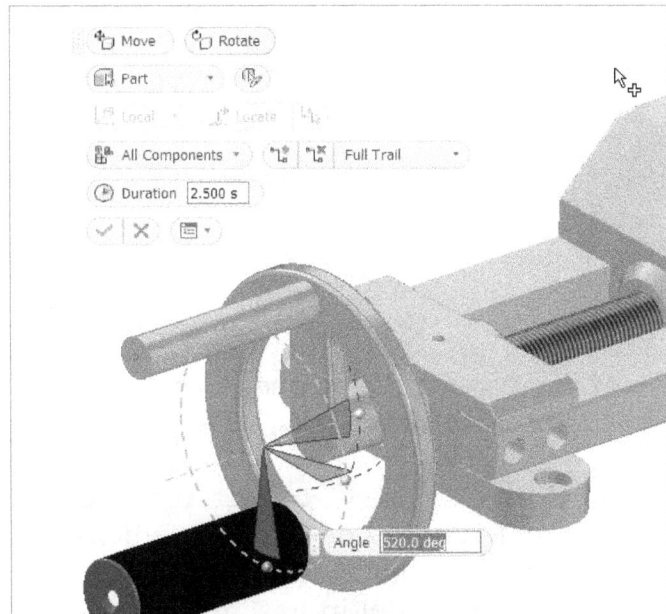

Figure-92. Rotation applied to model

- After setting desired parameters, click on the **OK** button from the mini toolbar. The tweak will be created and added in the **Model Browser**. Also, the time bar of motion will be added in the **Storyboards Panel**; refer to Figure-93.

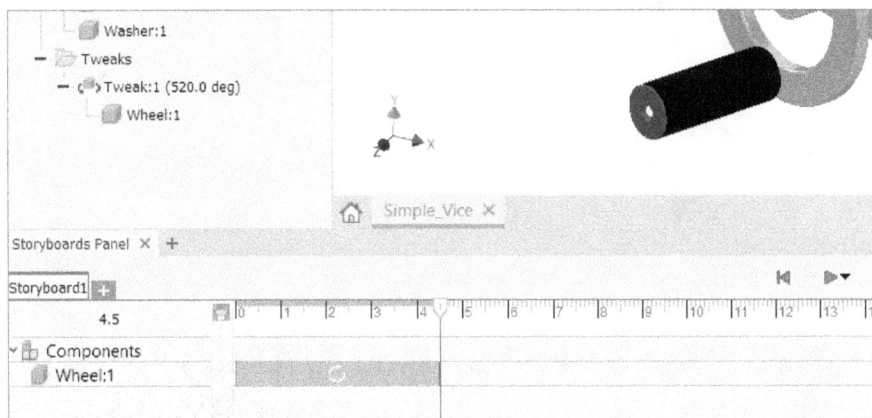

Figure-93. Tweak motion added

You can drag the motion tiles for various components in **Storyboards Panel** to desired time values. Use the standard media buttons in Storyboards Panel to check animation of motion.

SETTING OPACITY OF COMPONENTS

The **Opacity** tool is used to set transparency of selected component. The procedure to use this tool is given next.

- Select desired component from the graphics area for which you want to set opacity and click on the **Opacity** tool from the **Component** panel in the **Presentation** tab of the **Ribbon**. The **Opacity** mini toolbar will be displayed; refer to Figure-94.

Figure-94. Opacity mini toolbar

- Use the slider at the top in the mini toolbar to set transparency of selected component. After setting desired parameters, click on the **OK** button from the mini toolbar.

SETTING CAMERA POSITION

The **Capture Camera** tool is used to set position and orientation of camera for defining point of view for animation. The procedure to use this tool is given next.

- Click on the **Capture Camera** tool from the **Camera** panel in the **Presentation** tab of the **Ribbon**. The camera feature will be added in the **Storyboard Panel**; refer to Figure-95. Note that model view will reach to current status from default orientation of model in assembly file.

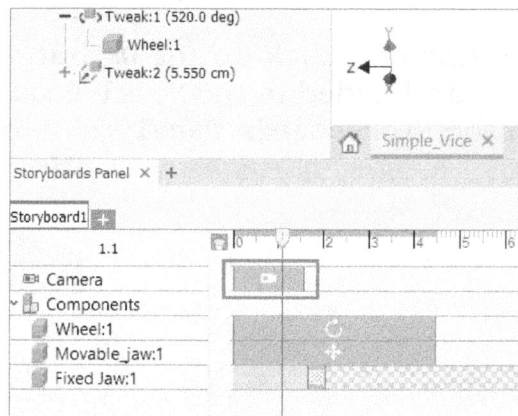

Figure-95. Camera added to storyboard

CREATING DRAWING VIEW USING SNAPSHOT

The **Create Drawing View** tool is used to generate a drawing view using selected snapshot view from the **Snapshot Views** panel in the interface. After selecting the snapshot, click on the **Create Drawing View** tool from the **Drawing** panel in the **Presentation** tab of the **Ribbon**. The drawing view will be placed in a new drawing or drawing linked with the current model.

PUBLISHING VIDEOS AND IMAGES

The **Video** tool in **Publish** panel of **Presentation** tab is used to generate video of the storyboards created in the environment. The procedure to use this tool is given next.

- Click on the **Video** tool from the **Publish** panel in the **Presentation** tab of the **Ribbon**. The **Publish to Video** dialog box will be displayed; refer to Figure-96.

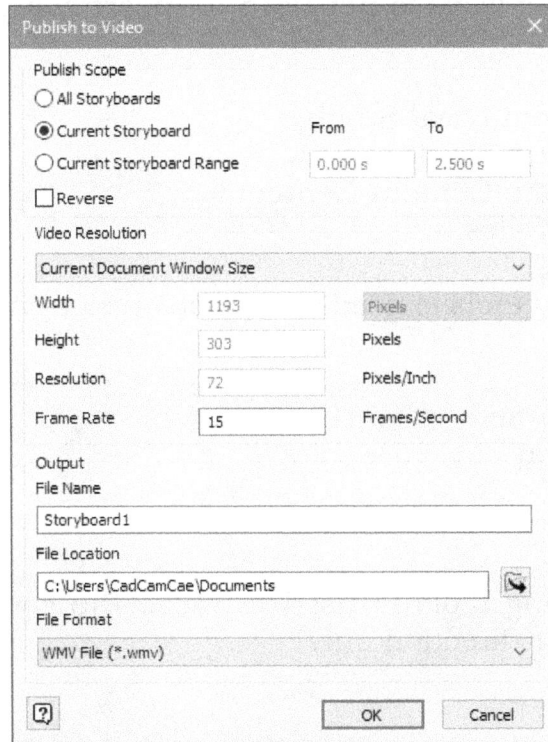

Figure-96. Publish to Video dialog box

- Specify the parameters as discussed earlier and click on the **OK** button. The **Video Compression** dialog box will be displayed; refer to Figure-97.

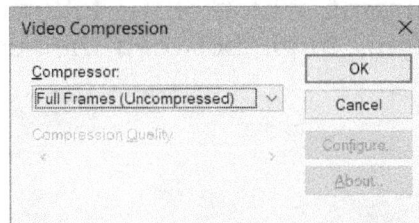

Figure-97. Video Compression dialog box

- Select desired compression option from the **Compressor** drop-down and click on the **OK** button. Video clip of animation will be generated.

Similarly, you can use the **Raster** tool from the **Publish** panel to generate image file.

SELF ASSESSMENT

Q1. Which of the following ways is not used to insert components in an Assembly in Autodesk Inventor?

a) Placing Component from Local Storage
b) Placing Component from Vault
c) Placing Component from Content Center
d) Placing Exported CAM files

Q2. Which of the following tools is used to replace an existing component with a new component?

a) Electrical Catalog Browser
b) Place iLogic Component
c) Both a and b
d) None of the Above

Q3. Which of the following tools is used to move the component in constrained environment i.e. in planar direction only?

a) Grid Snap
b) Free Move
c) Free Rotate
d) None of the Above

Q4. When placing Angle constraint, which of the following options should be select to use left-hand rule in specifying angle between the geometries?

a) Directed Angle
b) Undirected Angle
c) Explicit Reference Vector
d) Both a and b

Q5. Which of the following constraints is used to make cam follower or slot follower mechanism?

a) Rotation Constraint
b) Rotation-Translation Constraint
c) Transitional Constraint
d) Mate Constraint

Q6. What is the full form of UCS?

a) Unified Combined System
b) User Combined System
c) Unified Coordinate System
d) User Coordinate System

Q7. Which of the following joints is used when you need the object free in rotational dimensions but fixed in translation?

a) Cylindrical
b) Rotational
c) Ball
d) Slider

Q8. The **Insert** Constraint is used to insert pin like geometries inside hole like geometries. (True/False)

Q9. **Driving** Constraint can only be applied to rotational motion. (True/False)

Q10. The .... is used to form a table of components that are used in the Assembly.

FOR STUDENT NOTES

Chapter 8

Advanced Assembly and Design

Topics Covered

The major topics covered in this chapter are:

- *Sub-Assembly*
- *Pattern in Assembly*
- *Mirror and Copy in Assembly*
- *Bolted Connections*
- *Pin Joints*
- *Shaft Design*
- *Gear Design*
- *Belt Design*
- *Bearing Design*
- *Key Design*
- *Cam Design*

INTRODUCTION

In previous chapter, we learnt to insert various components in an assembly and apply different type of constraints between them. In this chapter, we will do some advanced operations on the assemblies. We will learn to use sub-assemblies. We will perform tasks related to reusability of assembly/assembly's components. We will also learn about various design components that can be directly used in our assemblies.

SUB-ASSEMBLIES

As the name suggests, sub-assemblies are the small sections of big assemblies. When we are working on an assembly where a group of components are repeating then it is a good idea to make a sub-assembly of those components and use it in the main assembly. There is no special tool to create sub-assemblies. They are created in the same way as assemblies. An example of using sub-assembly is discussed next in the form of tutorial.

Use of Sub-Assembly

We need to create an assembly as shown in Figure-1. Note that the wheel assembly is repeating in the main assembly, so we can make a sub-assembly of wheel and use it in the main assembly. The steps to do so are given next.

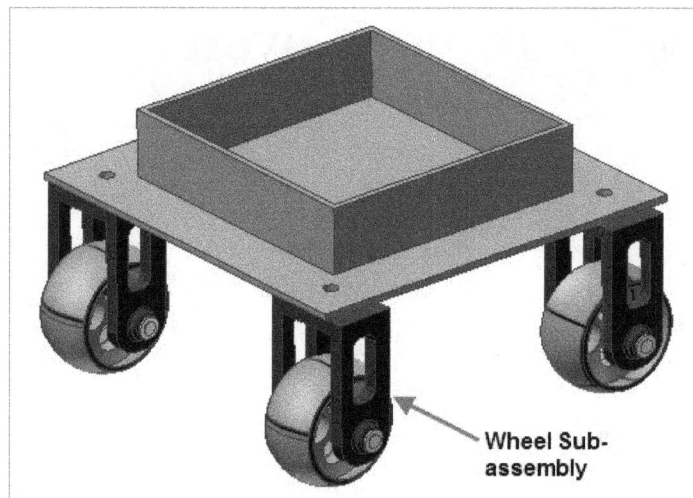

Figure-1. Trolley Assembly

*   Create an assembly of the components used in the wheel and save it by the name Wheel sub-assembly; refer to Figure-2.

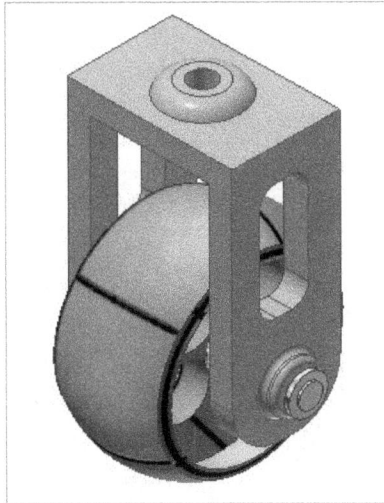

Figure-2. Wheel sub-assembly

- Start a new assembly and click on the **Place** button from the **Component** panel in the **Assemble** tab of the **Ribbon**. The **Place Component** dialog box will be displayed.
- Double-click on the Base component file from the downloaded resource folder using the **Place Component** dialog box and place it as grounded in the viewport.
- Double-click on the Wheel sub-assembly file from the dialog box. The component get attached to cursor.
- Click in the viewport to place the component.
- Move the cursor over the top round edge of the Wheel sub-assembly and click on it; refer to Figure-3. The sub-assembly get attached to the cursor.

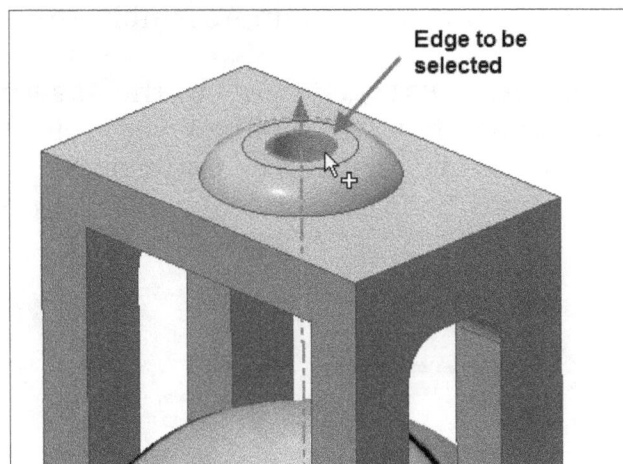

Figure-3. Edge of sub-assembly to be selected

- Move the cursor on the round edge of holes in the base. Preview of the **Insert** constraint will be displayed; refer to Figure-4. Click at this location to place the sub-assembly.

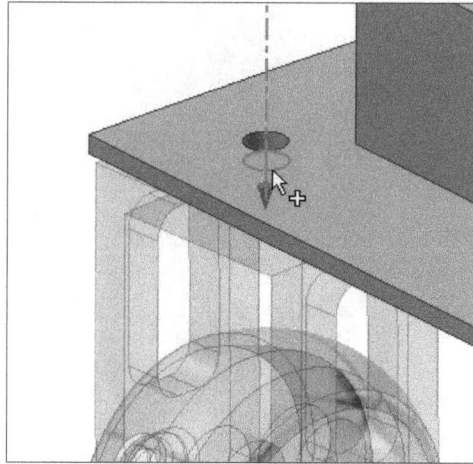

Figure-4. Preview of insert constrain

- Click on the **OK** button from the toolbar displayed. Repeat the procedure for rest of the wheel sub-assemblies.
- Press **ESC** when you have inserted all the wheel sub-assemblies.

From this small tutorial, you can understand the use of sub-assemblies in the main assembly.

PATTERN IN ASSEMBLY

Pattern tool is used to create multiple copies of a component/sub-assembly. The **Pattern** tool in Assembly works in similar way to the tool discussed in Modeling environment. The procedure to use this tool in Assembly environment is given next.

- Click on the **Pattern** tool from **Pattern** panel in the **Assemble** tab of the **Ribbon**. The **Pattern Component** dialog box will be displayed; refer to Figure-5. Also, you will be asked to select the components to be patterned.

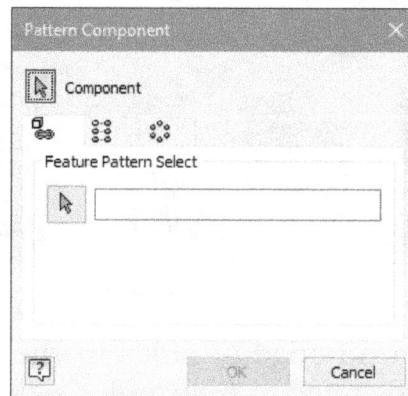

Figure-5. Pattern Component dialog box

- Click on the component that you want to be patterned and then click on the **Rectangular** tab or **Circular** tab as required. If you select the **Rectangular** tab then options will be displayed to create rectangular pattern (in the form of rows and columns); refer to Figure-6. If you select the **Circular** tab then options will be displayed to create circular pattern; refer to Figure-7.

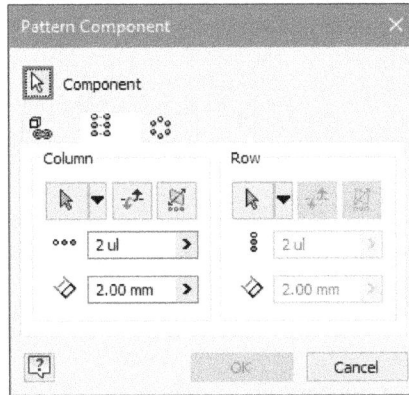

Figure-6. Pattern Component dialog box with Rectangular tab selected

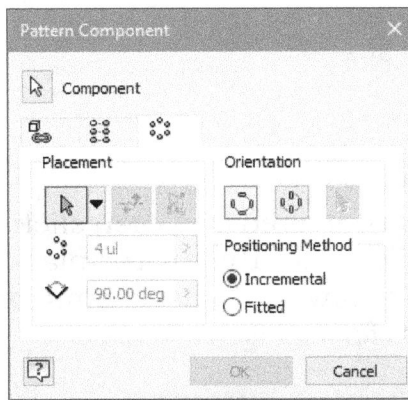

Figure-7. Pattern Component dialog box with Circular tab selected

- Rest of the procedure is same as discussed in chapter related to 3D Modeling.

MIRROR AND COPY

The **Mirror** tool and **Copy** tool in Assembly environment work in the same way as they do in 3D Modeling environment. We will discuss the procedure to use **Mirror** tool. You can use similar steps to use **Copy** tool.

- Click on the **Mirror** tool from **Pattern** panel in the **Assemble** tab of the **Ribbon**. The **Mirror Components** dialog box will be displayed; refer to Figure-8. Also, you will be asked to select the components to be mirrored.

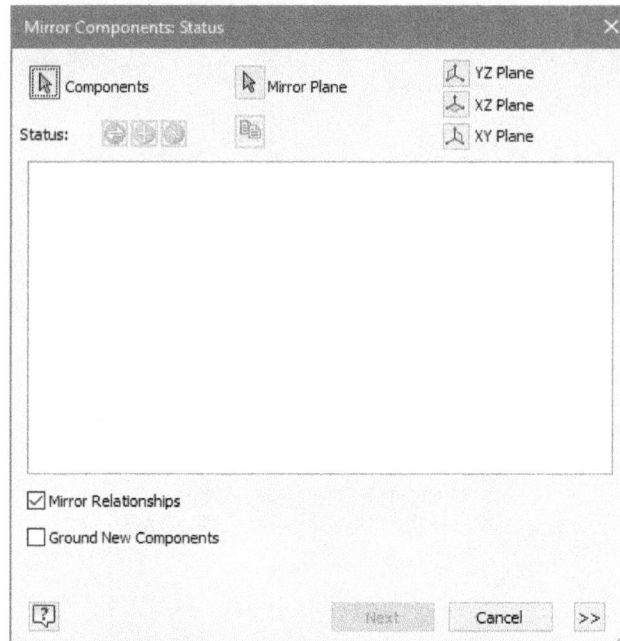

Figure-8. Mirror Components dialog box

- Select the components to be mirrored and then click on the **Mirror Plane** selection button. You will be asked to select the mirror plane.
- Click on a plane/face that you want to use as mirror plane. Preview of the mirror feature will be displayed; refer to Figure-9.

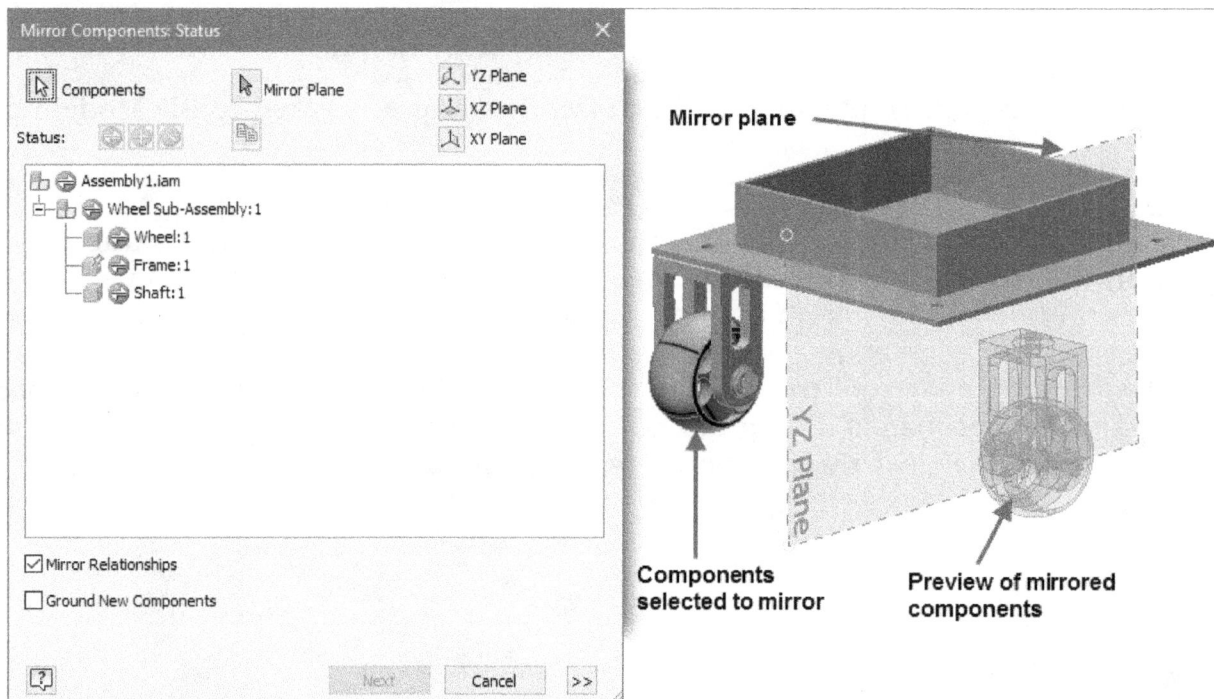

Figure-9. Preview of the mirror feature

- Exclude desired components by selecting them and then clicking on the **Exclude** button [img] from the **Status** area of the dialog box. If you want to create a new file by copying selected components then select the **Mirror** button [img] and if you want to just reuse the mirrored components but do not want a new file to be generated by mirroring then select the **Reuse** button [img].

- Click on the **Next** button after performing desired changes. The **Mirror Components** dialog box will be displayed with list of components to be mirrored; refer to Figure-10.

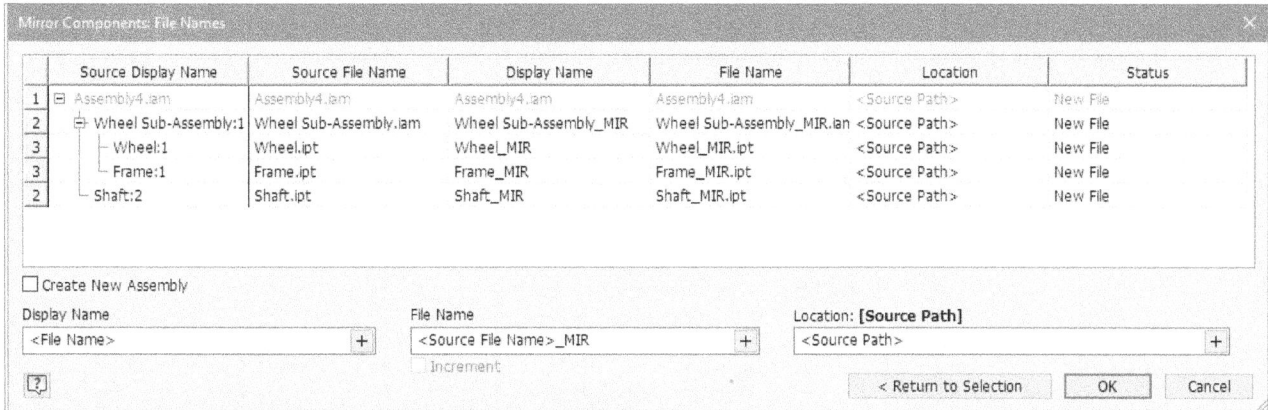

		Source Display Name	Source File Name	Display Name	File Name	Location	Status
1	⊟	Assembly4.iam	Assembly4.iam	Assembly4.iam	Assembly4.iam	<Source Path>	New File
2	⊕	Wheel Sub-Assembly:1	Wheel Sub-Assembly.iam	Wheel Sub-Assembly_MIR	Wheel Sub-Assembly_MIR.iam	<Source Path>	New File
3		Wheel:1	Wheel.ipt	Wheel_MIR	Wheel_MIR.ipt	<Source Path>	New File
3		Frame:1	Frame.ipt	Frame_MIR	Frame_MIR.ipt	<Source Path>	New File
2		Shaft:2	Shaft.ipt	Shaft_MIR	Shaft_MIR.ipt	<Source Path>	New File

☐ Create New Assembly

Display Name: `<File Name>` File Name: `<Source File Name>_MIR` Location: **[Source Path]** `<Source Path>`

Increment

< Return to Selection   OK   Cancel

Figure-10. Mirror Components dialog box with list of components

- Click on the **OK** button from the dialog box to create the mirror copy.

DESIGN ACCELERATOR COMPONENTS

The Design Accelerator components are mechanical standard components that change shape and size as per the specified parameters. For example, bolt, cam, gears, etc. All the Design Accelerator components are divided into four categories; Fasten, Frame, Power Transmission, and Spring. The components in each category are discussed next.

COMPONENTS FOR FASTENING

There are two type of fastening components available as Design Accelerator components: Bolt and Pin. The procedure to use each of the component is given next.

Bolted Connection

The Bolted connection component is used to represent the bolted connection in assembly with real-world parameters. The procedure to apply bolted connection is given next.

- Click on the **Bolted Connection** tool from **Fasten** panel in the **Design** tab of the **Ribbon**. The **Bolted Connection Component Generator** dialog box will be displayed; refer to Figure-11.

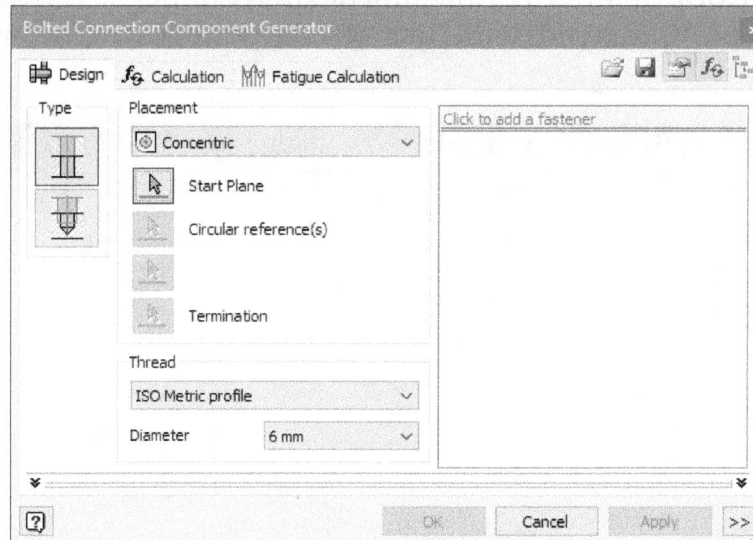

Figure-11. Bolted Connection Component Generator dialog box

- Select desired type of hole by selecting **Through All** or **Blind** connection type buttons from the left area of the dialog box. If you select the **Through All** button then the bolt created will be have nut on the other end. If you select the blind connection then bolt created will terminate with blind hole (Like a foundation bolt).
- On selecting desired button, you will be asked to select a face/plane to specify starting point of the bolted connection.
- Click on desired face/plane. You will be asked to select circular reference to specify axis of the bolted connection.
- Select the circular face/edge of the hole for which you want to create bolted connection. You will be asked to select plane/face to specify termination or blind start plane depending on the connection type button selected.
- Select desired face/plane. Preview of connection will be displayed with dashed line; refer to Figure-12.

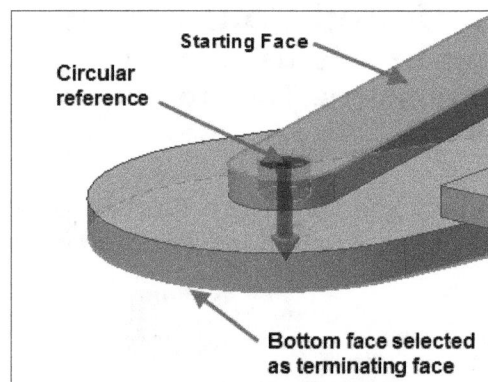

Figure-12. Preview of bolted connection

- Note that we have selected **Concentric** option from the drop-down in **Placement** area of the dialog box, you can select different option and specify references for bolted connection accordingly.
- Click on the **Click to add a fastener** option from the right area of the dialog box. A selection box will be displayed to select fastener; refer to Figure-13.

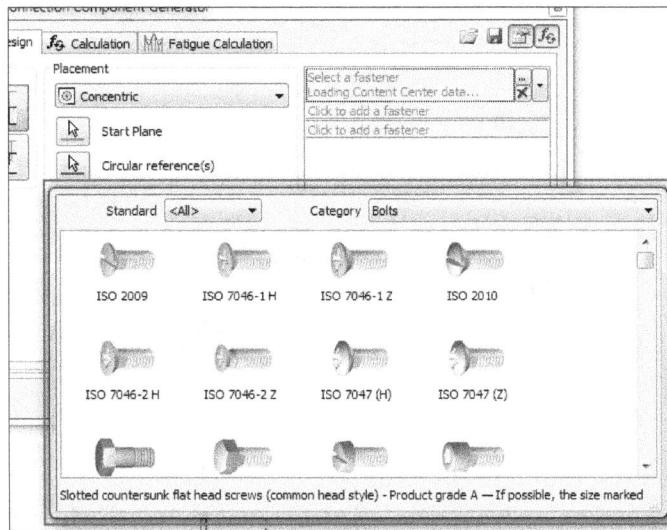

Figure-13. Selection box for fasteners

- Click in the **Category** drop-down in selection box and select desired category. The bolts in selected category will be displayed; refer to Figure-14.

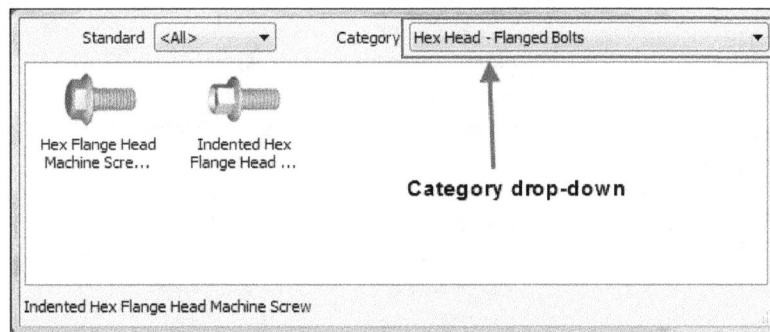

Figure-14. Category drop-down in selection box

- Click on desired fastener. Preview of the fastener will be displayed in the viewport; refer to Figure-15.

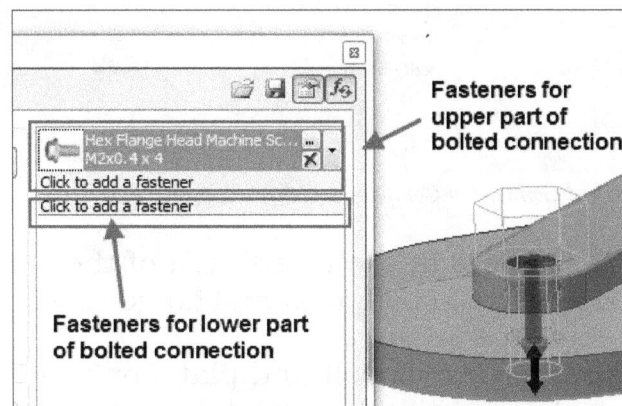

Figure-15. Preview of fastener

- Click again on the **Click to add a fastener** option in the dialog box to add more parts of the bolted connection and apply other fasteners of the bolted connection. Note that the options above dividing line are for upper part and options below dividing line are for lower part of bolted connection; refer to Figure-15. Figure-16 shows preview of a bolted connection after adding washers and nut.

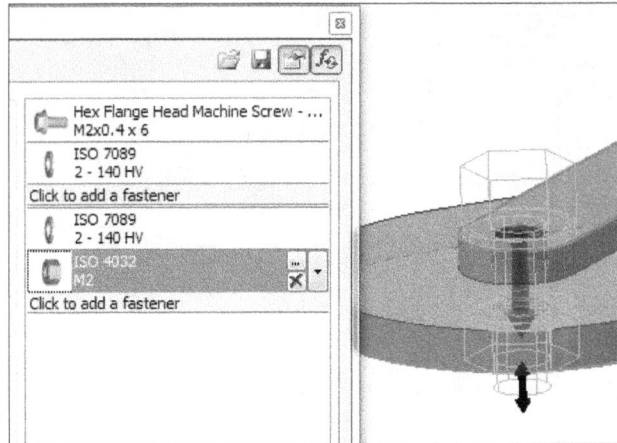

Figure-16. Preview of bolted connection with fasteners

- Click on the **Calculation** tab in the dialog box. The dialog box will be displayed as shown in Figure-17.

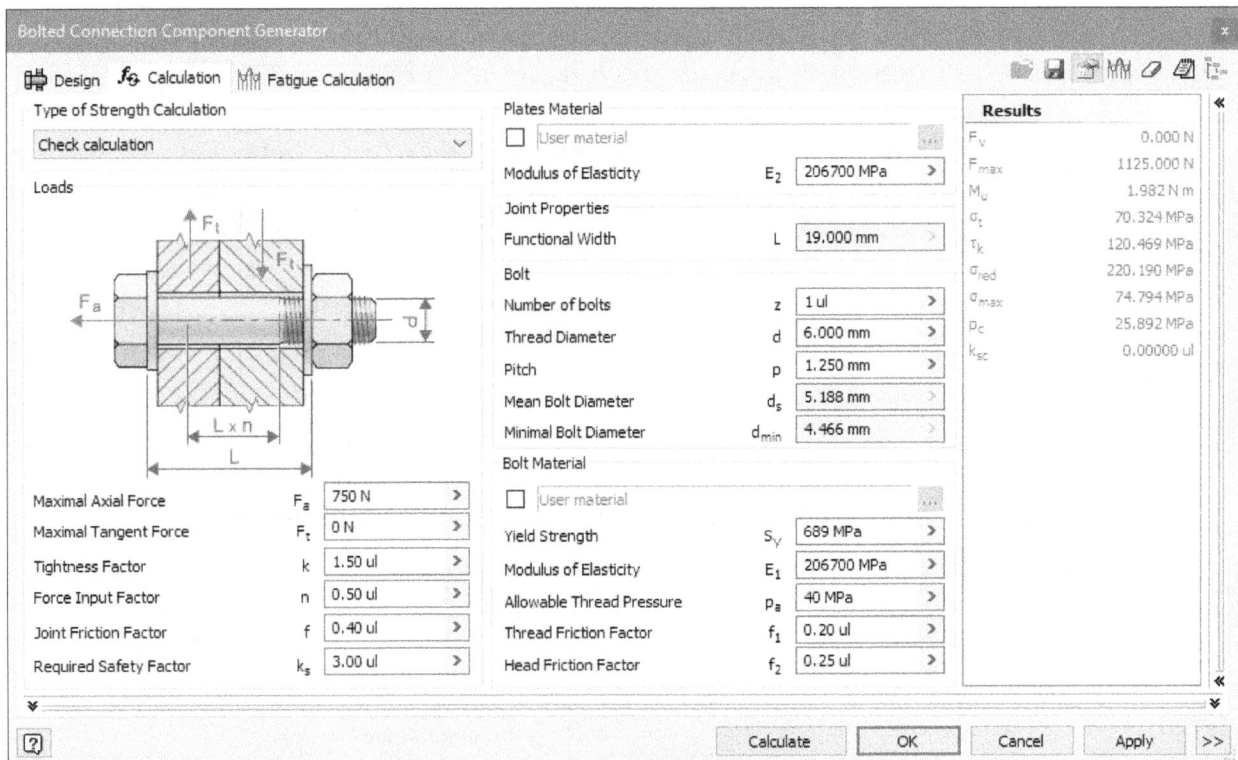

Figure-17. Calculation tab in Bolted Connection Component Generator dialog box

- Specify the design parameters in the **Loads** area of the dialog box like, maximal axial force that can be applied on bolt, maximal tangential force that can be applied on bolt, Factor of Safety, and so on.
- Specify the material properties for bolt and plate, or you can select the material available in the library by selecting the check box in the respective area.
- After specify the values, click on the **Calculate** button to check whether the use of current bolt is feasible; refer to Figure-18. Note that factor of safety is much lower than the standard value, so we need to modify the bolt parameters. There are many ways to change parameters of bolt to make it feasible like, change **Thread Diameter** from 2 to 4, change the material to more tough material, or increase the number of bolts from 1 to 5.

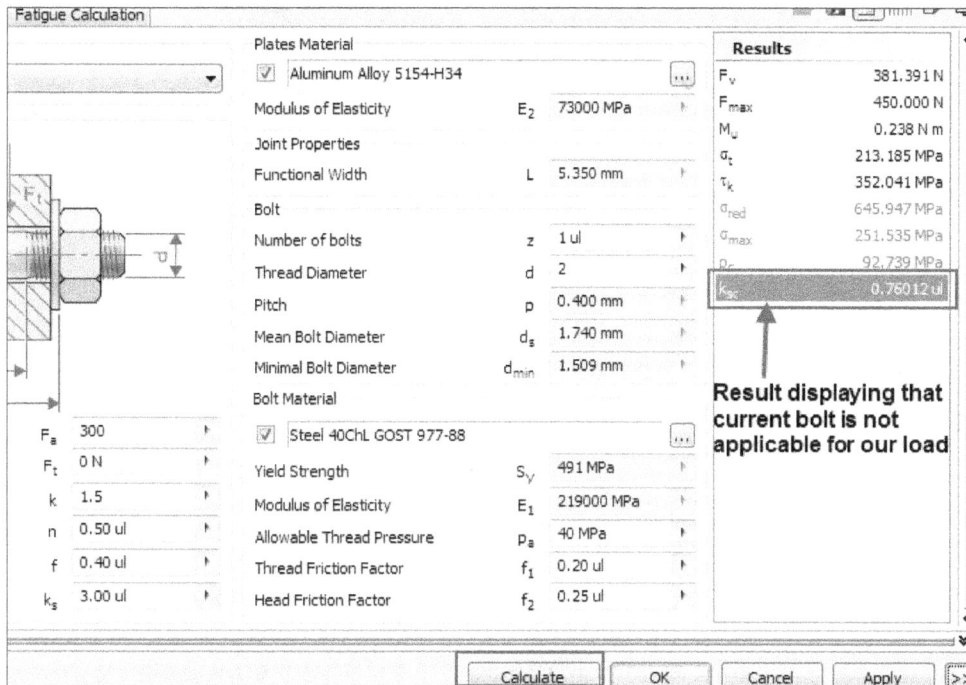

Figure-18. Result of bolt loading calculation

• Click on the **OK** button from the dialog box to create bolted connection.

Clevis Pin

Clevis Pin is a type of fastener that will allow the rotation or swivel of the connected parts about the axis of the pin linkage. A clevis pin, sometimes referred to as a link pin or hinge pin, consists of a head, shank, and cross drilled hole. When using a fastening, such as a clevis pin, the hole which is at the opposite end of the pin to the head is inserted through the items to be linked and then a cotter pin, R clip, or similar retaining fastener is inserted through the hole to fix the clevis pin in place.

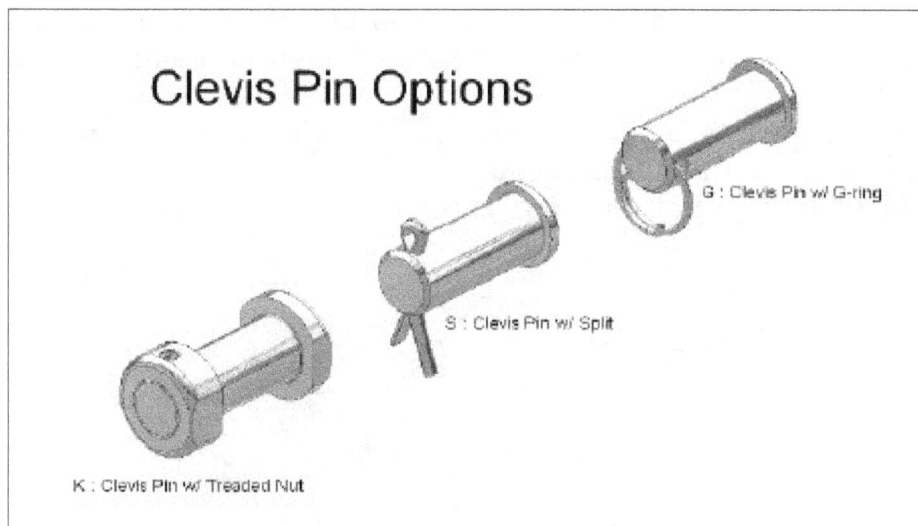

The procedure to create clevis pin component is given next.

• Click on the **Clevis Pin** tool from **Fasten** panel in the **Design** tab of the **Ribbon**. The **Clevis Pin Component Generator** dialog box will be displayed; refer to Figure-19.

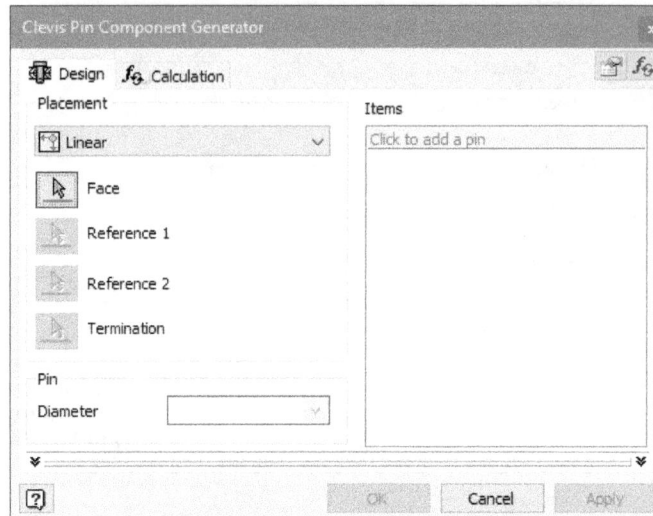

Figure-19. Clevis Pin Component Generator dialog box

- Click on the **Linear** option from **Placement** drop-down in the dialog box. Various placement options will be displayed in the drop-down; refer to Figure-20. Click on desired option for placement. (We have selected **Concentric** option because we want to make the pin placed concentric to hole.)

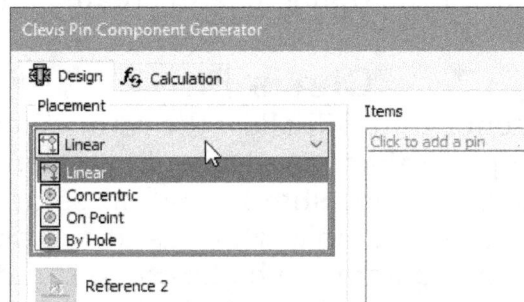

Figure-20. Drop-down for placement options

- On selecting the **Concentric** option, you will be asked to select face to be used as starting face for pin. Select desired face. You will be asked to select circular reference.
- Select the circular edge/face as reference for defining the diameter of pin. You will be asked to select the terminating face.
- Select the face for terminating the pin; refer to Figure-21.

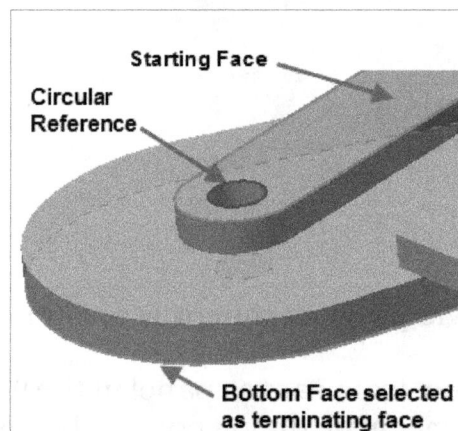

Figure-21. References selected for Clevis pin

- Click on the **Click to add a pin** option from the right area of the dialog box. A selection box will be displayed with different types of clevis pins; refer to Figure-22.

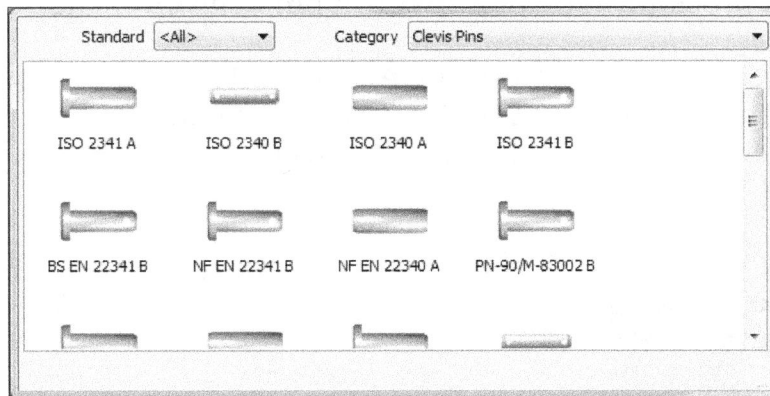

Figure-22. Selection box for clevis pin

- Select desired type of pin from the selection box. Preview will be displayed.
- Click on the **Calculation** tab from the dialog box. The dialog box will be displayed as shown in Figure-23.

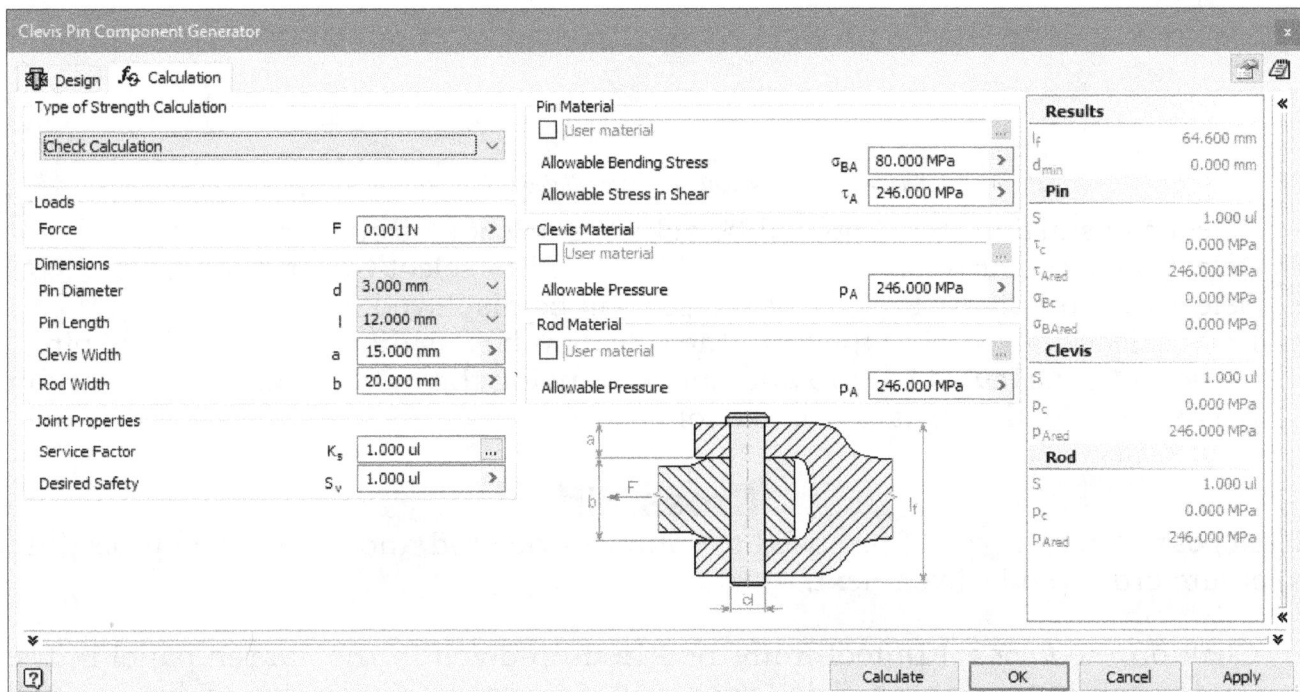

Figure-23. Calculation tab in Clevis Pin Component Generator dialog box

- Specify the loads, dimensions, and joint properties in the respective edit boxes to input the actual parameters of your joint.
- Similarly, specify the material properties of the pin, clevis, and rod.
- Click on the **Calculate** button to check whether the pin is feasible for specified parameters or not. If the pin is not feasible (i.e. one of the parameter in results is displayed as red) then perform the necessary modifications.
- Click on the **OK** button from the dialog box to create the pin.
- Click on the **OK** button again to accept the name and location of component.

Secure Pin

The **Secure Pin** is fastener used to bind together two rotating parts along their common axis. The procedure to create secure pin is given next.

- Click on the **Secure Pin** tool from the **Pin** drop-down in the **Fasten** panel in the **Design** tab of the **Ribbon**. The **Secure Pin Component Generator** dialog box will be displayed; refer to Figure-24. Also, you will be asked to select face to specify location of pin.

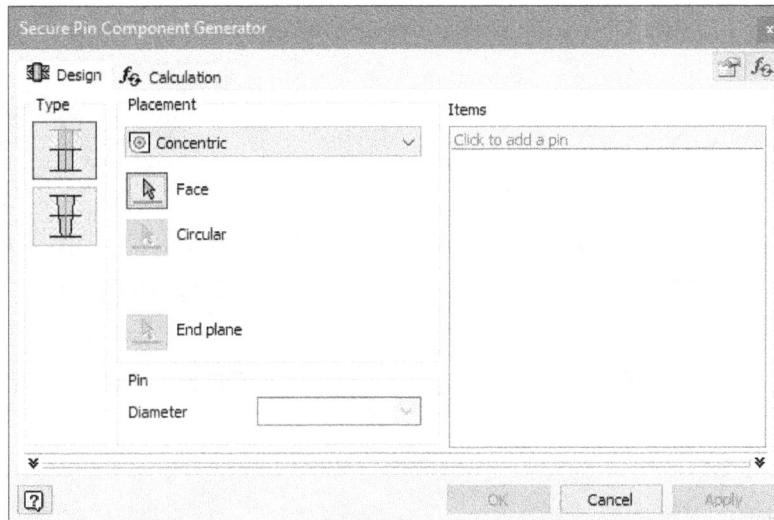

Figure-24. Secure Pin Component Generator dialog box

- Select the starting face. You will be asked to select a circular reference.
- Select the circular reference. You will be asked to select the end plane.
- Select the end plane. Preview of the pin will be displayed.
- Now, select desired pin type from the selection box as discussed for clevis pin.
- Click on the **Calculation** tab and check by using the options in tab whether pin needs modification or it is feasible for use.

Cross Pin

The **Cross Pin** is a type of pin used to connect draw rod and sleeve. The procedure to create cross pin is given next.

- Click on the **Cross Pin** tool from the **Pin** drop-down in the **Fasten** panel in the **Design** tab of the **Ribbon**. The **Cross Pin Component Generator** dialog box will be displayed; refer to Figure-25. Also, you will be asked to select face to specify location of pin.

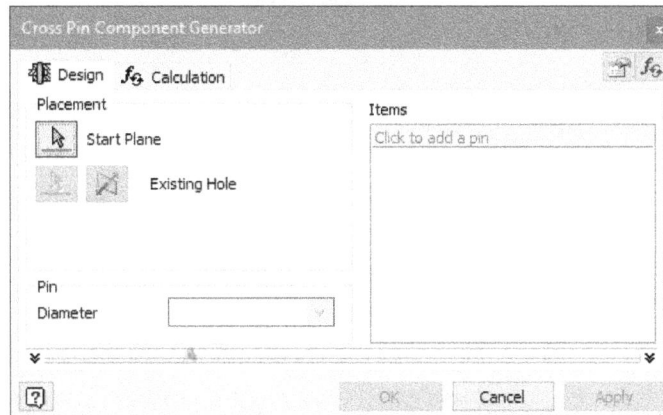

Figure-25. Cross Pin Component Generator dialog box

- Select the starting face. You will be asked to select a hole.
- Select the hole to be used as circular reference for pin.
- Now, select desired pin type from the selection box as discussed for clevis pin.
- Click on the **Calculation** tab and check by using the options in tab whether pin needs modification or it is feasible for use.

Joint Pin

The **Joint Pin** tool is used to design and calculate pins loaded with torque. The procedure to create joint pin is given next.

- Click on the **Joint Pin** tool from the **Pin** drop-down in the **Fasten** panel in the **Design** tab of the **Ribbon**. The **Joint Pin Component Generator** dialog box will be displayed; refer to Figure-26. Also, you will be asked to select face to specify location of pin.

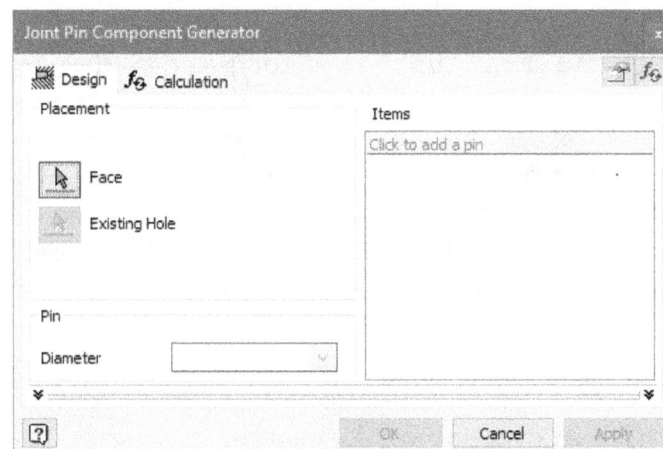

Figure-26. Joint Pin Component Generator dialog box

- Select the starting face. You will be asked to select a hole reference.
- Select the hole to be used as circular reference for pin.
- Now, select desired pin type from the selection box as discussed for cross pin.
- Click on the **Calculation** tab and check by using the options in tab whether pin needs modification or it is feasible for use.

Radial Pin

The procedure to create radial pin is same as discussed for cross pin.

* Click on the **Radial Pin** tool from the **Pin** drop-down in the **Fasten** panel in the **Design** tab of the **Ribbon**. The **Radial Pin Component Generator** dialog box will be displayed. Also, you will be asked to select face to specify location of pin.
* Do the same as discussed earlier for **Cross pin**.

COMPONENTS FOR POWER TRANSMISSION

There are various components for transmission like shaft, gears, bearing, belts, cams, and so on. The procedures to create various power transmission components are given next.

Shaft

A shaft is a rotating machine element used to transmit power from one location to other. Power is delivered to the shaft by some tangential force and resultant torque generated in the shaft makes the transfer of power to other linked elements. The standard size of transmission shafts are given as:

25 mm to 60 mm diameter with 5 mm steps
60 mm to 110 mm diameter with 10 mm steps
110 mm to 140 mm diameter with 15 mm steps
140 mm to 500 mm diameter with 20 mm steps

Standard lengths of shafts are 5 m, 6 m, and 7 m.

While designing a shaft, we need to take care of following stresses:

1. Shear stresses due to the transmission of torque (i.e. due to torsional load).
2. Bending stresses (tensile or compressive) due to the forces acting upon machine elements like gears, pulleys, etc. as well as due to the weight of the shaft itself.
3. Stresses due to combined torsional and bending loads.

According to American Society of Mechanical Engineers (ASME) code for the design of transmission shafts, the maximum permissible working stresses in tension or compression may be taken as,

(a) 112 MPa for shafts without allowance for keyways.
(b) 84 MPa for shafts with allowance for keyways.

The maximum permissible shear stress may be taken as
(a) 56 MPa for shafts without allowance for key ways.
(b) 42 MPa for shafts with allowance for keyways.

In Autodesk Inventor Assembly environment, the **Shaft** tool is used to create shaft in an assembly to transmit power. The procedure to create shaft is given next.

- Click on the **Shaft** tool from the **Power Transmission** panel in the **Design** tab of the **Ribbon**. The **Shaft Component Generator** dialog box will be displayed; refer to Figure-27. Also, preview of the shaft will be displayed attached to the cursor; refer to Figure-27.

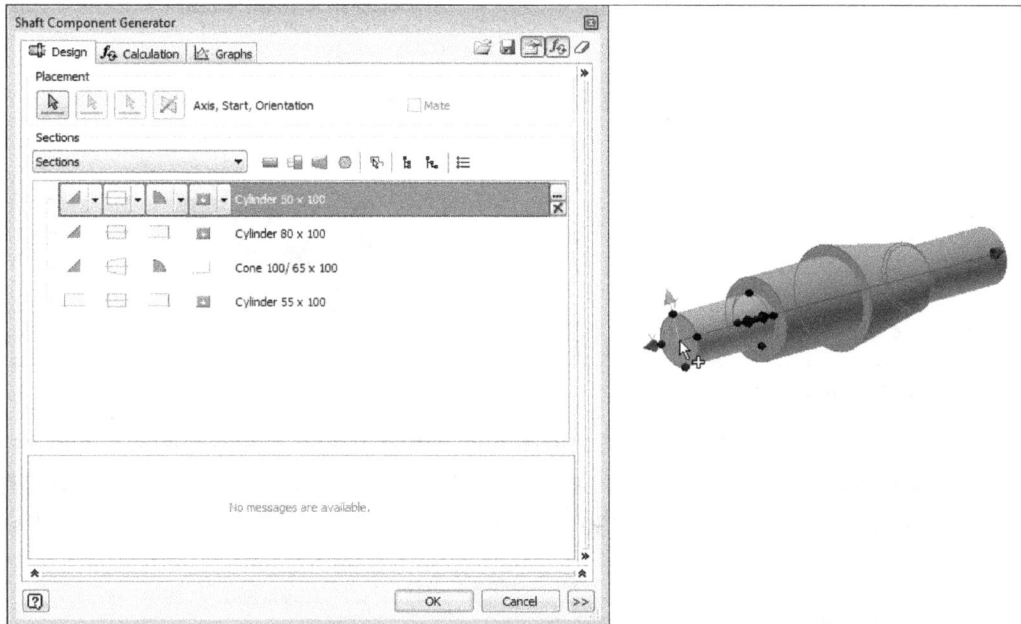

Figure-27. Shaft Component Generator dialog box

- Change the sections of the shaft as required. Each shaft section has four features to change; First edge feature, Section type, Second edge feature, and Section feature like keyways, etc. To change the shape of section, click on the drop-down for desired feature. Various options will be displayed; refer to Figure-28.

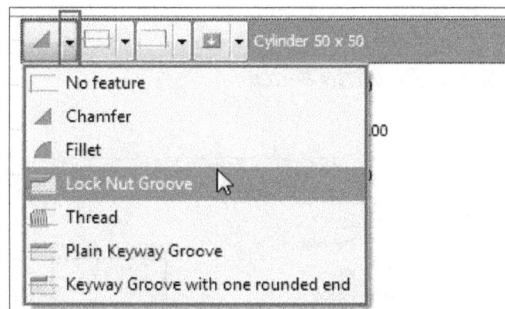

Figure-28. First Edge Feature drop-down

- Select desired option from the drop-down. The related dialog box will be displayed to specify parameter. Specify the parameter and click on the **OK** button.
- Similarly, you can change other features of selected section.
- To add a new section in the shaft, select the shaft section from the list or from the model preview after which you want to add the new section and select desired button from the dialog box; refer to Figure-29.

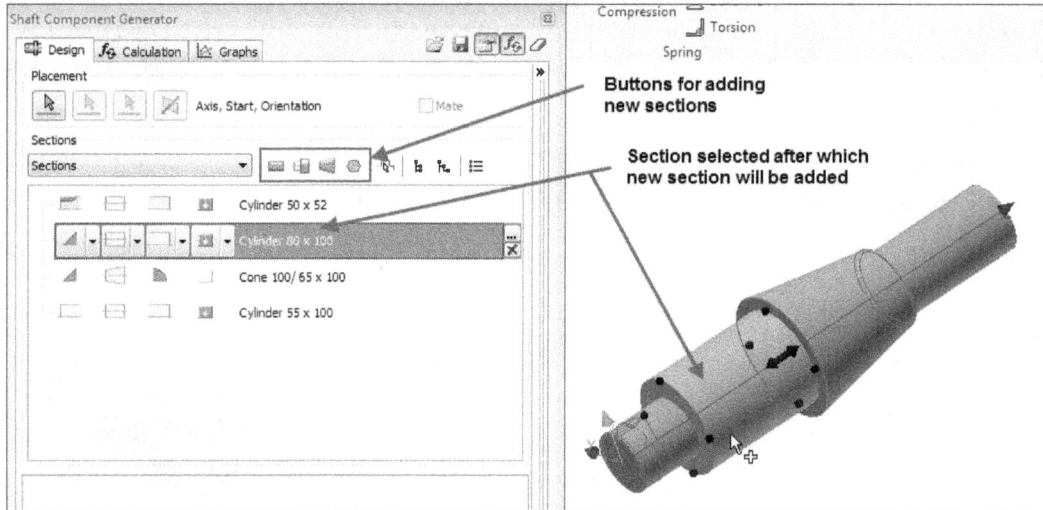

Figure-29. Adding new sections in shaft

- On selecting the button, a new section will be added in the shaft. To change the parameters of shaft section, click on the **Section Properties** button next to the selected section; refer to Figure-30. The related dialog box will be displayed; refer to Figure-31.

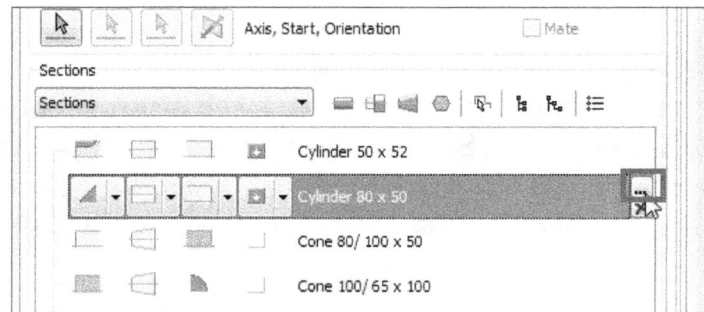

Figure-30. Section Properties button

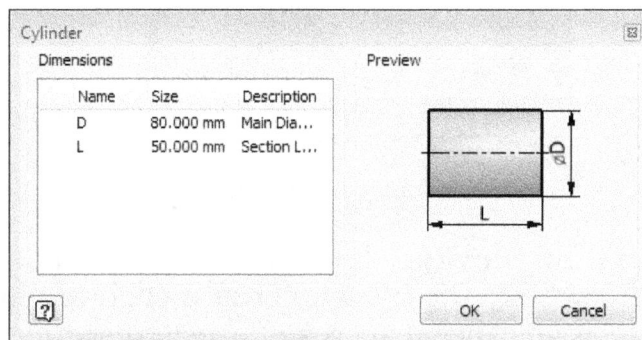

Figure-31. Cylinder dialog box

- Click on the dimension value in the dialog box that you want to change and specify desired value. Click on the **OK** button to exit.
- After making desired shape, click on the **Calculation** tab in the dialog box. The dialog box will be displayed as shown in Figure-32.

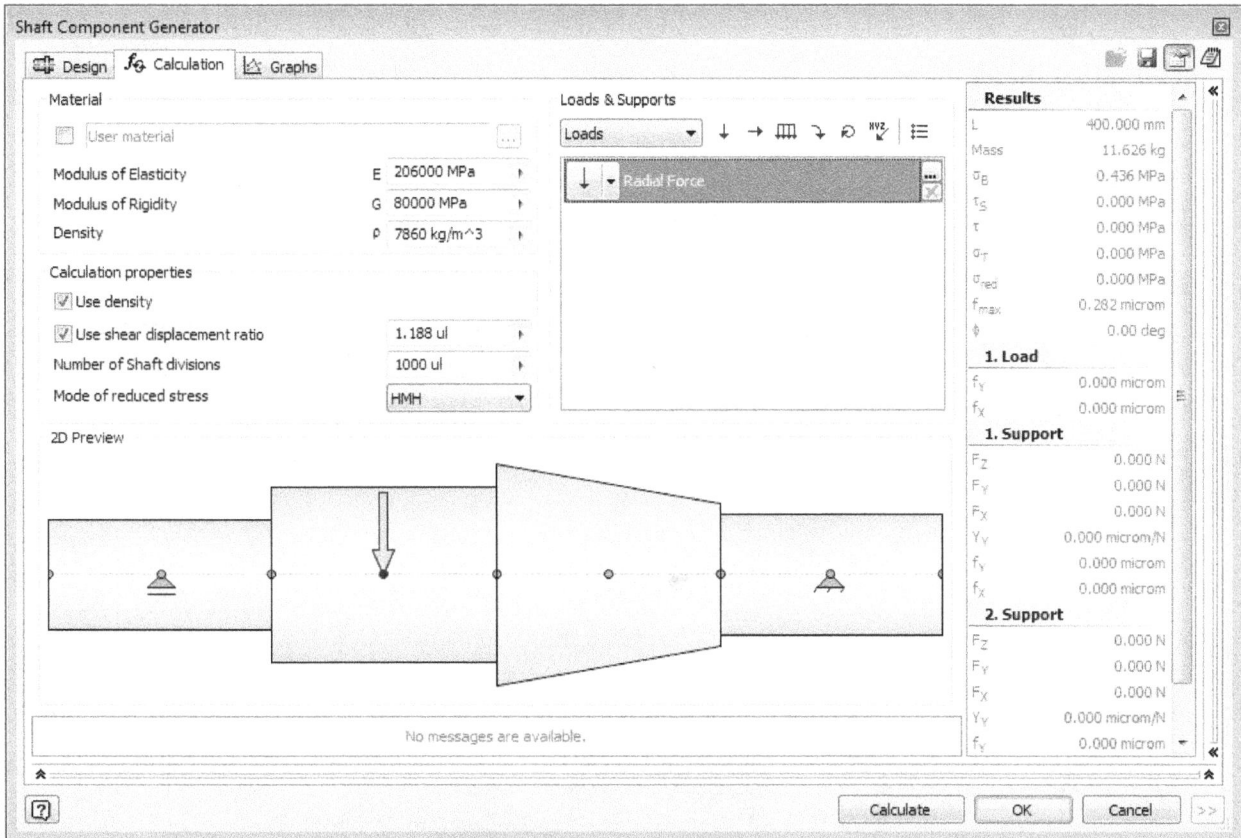

Figure-32. Shaft Component Generator dialog box with Calculation tab

- Click on the check box for material in the **Material** area of the dialog box and select desired material.
- If you want to specify the material properties specifically then clear the check box and specify desired values in the edit box of the **Material** area.
- Select the **Use Density** check box to include the mass in the calculation. If you clear the check box then Density is not included in the calculation.
- Select the **Use shear displacement ratio** check box if the shaft profile is thick/rigid. Default value of ratio for cylindrical profile is 1.18. Note that if the shaft has thin profile then effect of shear force is not counted.
- Specify the number of shaft divisions for calculations in the **Number of Shaft divisions** edit box.
- Click on the drop-down for **Mode of reduced stress** in the **Calculation Properties** area of the dialog box and select either **HMH** or **Tresca-Guest** mode of reduced stress calculation.

Formula for reduced stress is given by,
Reduced stress:

$$\sigma_{red} = \sqrt{(\sigma_B + \sigma_T)^2 + \alpha * (\tau^2 + \tau_S{}^2)}$$

where:

σ_B bending stress
σ_T tension stress
τ torsion stress
τ_S shear stress
a constant $a = 3$ for HMH
                                 $a = 4$ for Tresca-Guest

- Now, we need to apply loads and supports for our calculations. To apply a load, select the **Loads** option from the drop-down in the **Loads & Supports** area of the dialog box; refer to Figure-33. The buttons in the **Loads & Supports** area are displayed as shown in Figure-34.

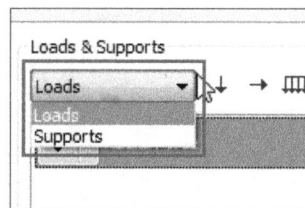

Figure-33. Drop-down in Loads & Supports area

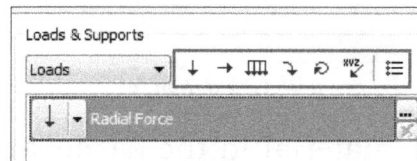

Figure-34. Buttons in the Loads & Supports area

- Click on desired point on which you want to apply load from the **2D Preview** area of the dialog box; refer to Figure-35.

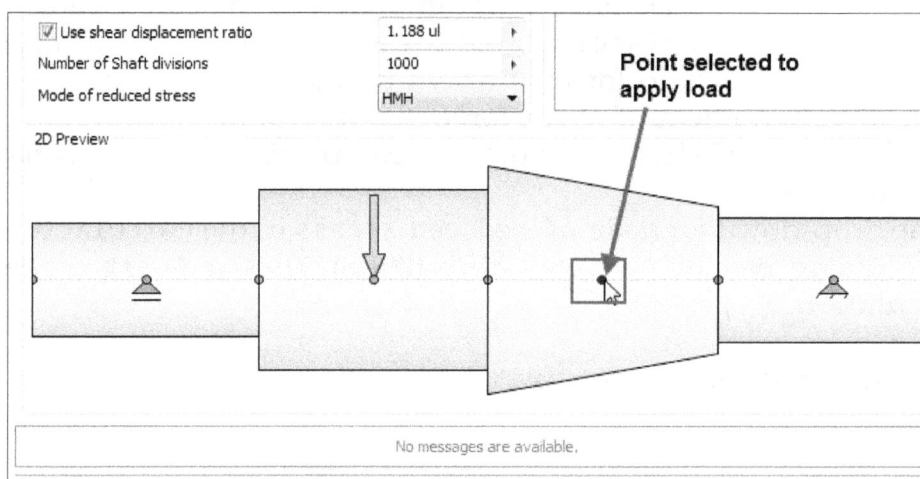

Figure-35. Selecting point to apply load

- Click on desired button to apply respective force. There are six types of loads that can be applied; Radial Force, Axial Force, Continuous Load, Bending Moment, Torque, and Common Load. On clicking at the button, respective dialog box will be displayed. Like, we have selected **Add Axial Force** button from the **Loads & Supports** area. The **Axial Force** dialog box will be displayed; refer to Figure-36.

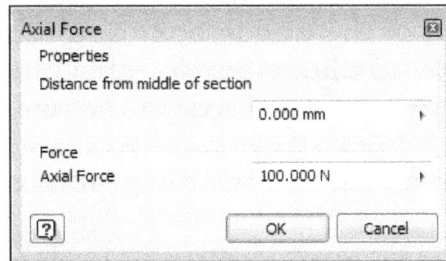

Figure-36. Axial Force dialog box

- Specify desired parameters in the dialog box and click on the **OK** button to exit. The load with specified parameters will be applied. Note that if you are applying Torque then you must apply equal and opposite torque at other point of shaft to form the equilibrium equation for stress calculations.
- To add support to the shaft, click on the drop-down in the **Loads & Supports** area and select the **Supports** option. The buttons in the area will be displayed as shown in Figure-37.

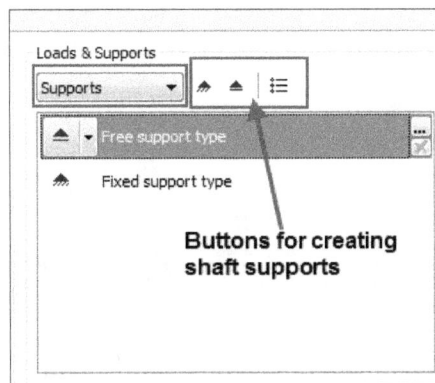

Figure-37. Creating shaft supports

- There are two buttons to apply supports; **Add Fixed Support** and **Add Free Support**. Select the **Add Fixed Support** button if you want the shaft to be fixed at desired point. If you want the shaft to be supported via bearing which means free to rotate then select the **Add Free Support** button.
- On selecting a button to add support, the respective dialog box will be displayed; refer to Figure-38.

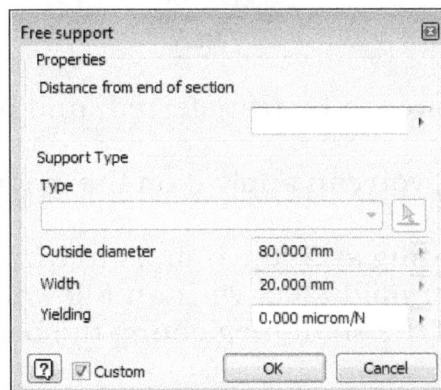

Figure-38. Free Support dialog box

- Specify desired parameters in the dialog box and click on the **OK** button. The support will be added to the shaft.

- Click on the **Calculate** button from the dialog box. The related results will be displayed in the right area of the dialog box. Note that you may need to double-click on the splitter to expand dialog box for checking results.
- Change the parameters if there is any error in the results then perform the related modification in shaft.
- Click on the **Graphs** tab to check the results graphically; refer to Figure-39.

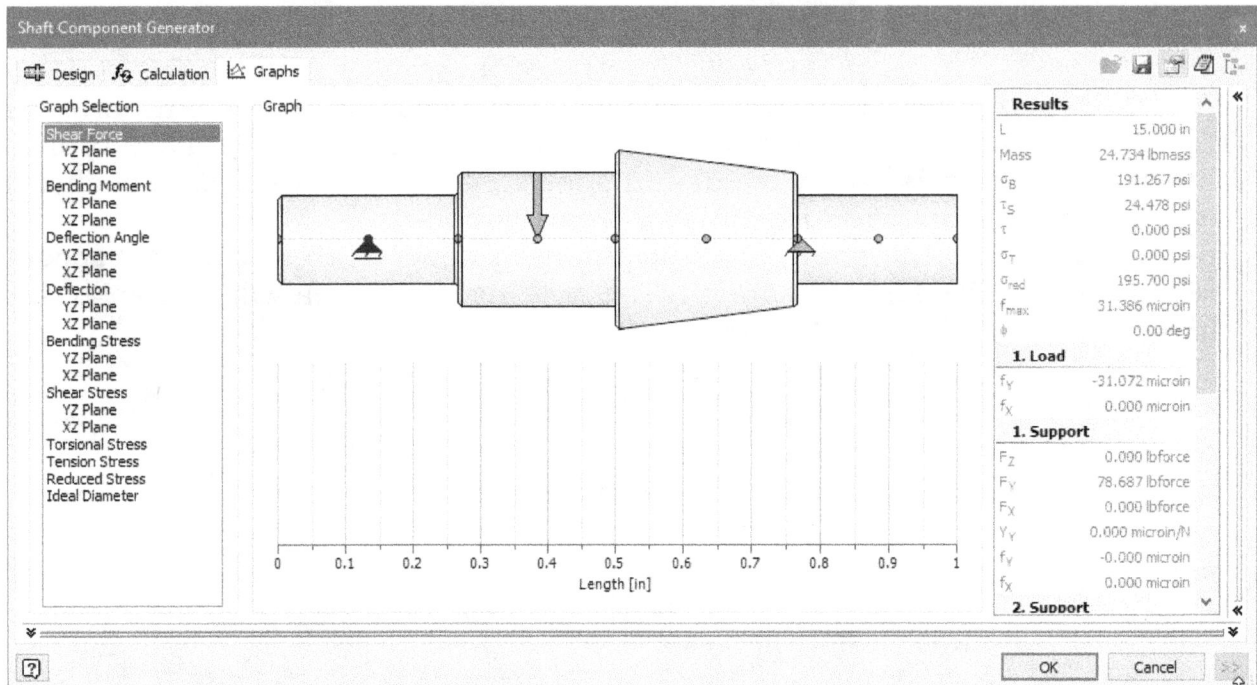

Figure-39. Graphs tab in the dialog box

- Click on the **OK** button from the dialog box to create the shaft.

Finding Torque

Most of the time in real world problems, you will get the load in Watt and RPM like, a shaft need to transfer 20KW load at 200 RPM. In such cases, you need to convert this value into torque. The formula is given next.

$$T = \frac{P \times 60}{2 \pi N}$$

Here, T is torque, P is the power to be transferred, and N is the RPM.

After getting the torque value, you can apply it on the model to test the failure of shaft.

Example, A line shaft rotating at 200 r.p.m. is to transmit 20 kW. The shaft may be assumed to be made of mild steel with an allowable shear stress of 42 MPa. Determine the diameter of the shaft, neglecting the bending moment on the shaft.

Finding Bending Moment

Sometimes, you will get the problems like a load of 50kN is working at a distance of 100 mm outside the wheelbase. In such cases, you need to find out the bending moment by using the formula,

$$M = W \times L$$

Here, M is the bending moment, W is the load, and L is the distance from wheelbase.

Example, A pair of wheels of a railway wagon carries a load of 50 kN on each axle box, acting at a distance of 100 mm outside the wheel base. The gauge of the rails is 1.4 m. Find the diameter of the axle between the wheels, if the stress is not to exceed 100 MPa.

Spur Gear

For any kind of gear design, there are a few requirements of designers. In designing a gear drive, following data is usually given :

1. The power to be transmitted,
2. The speed of the driving gear,
3. The speed of the driven gear or the velocity ratio, and
4. The centre distance.

The following requirements must be met in the design of a gear drive :

(a) The gear teeth should have sufficient strength so that they will not fail under static loading or dynamic loading during normal running conditions.
(b) The gear teeth should have wear characteristics so that their life is satisfactory.
(c) The use of space and material should be economical.
(d) The alignment of the gears and deflections of the shafts must be considered because they effect on the performance of the gears.
(e) The lubrication of the gears must be satisfactory.

The different modes of failure of gear teeth and their possible remedies to avoid the failure, are as follows :

1. **Bending failure**. Every gear tooth acts as a cantilever. If the total repetitive dynamic load acting on the gear tooth is greater than the beam strength of the gear tooth then the gear tooth will fail in bending, i.e. the gear tooth will break. In order to avoid such failure, the module and face width of the gear is adjusted so that the beam strength is greater than the dynamic load.

2. **Pitting**. It is the surface fatigue failure which occurs due to many repetition of Hertz contact stresses. The failure occurs when the surface contact stresses are higher than the endurance limit of the material. The failure starts with the formation of pits which continue to grow resulting in the rupture of the tooth surface. In order to avoid the pitting, the dynamic load between the gear tooth should be less than the wear strength of the gear tooth.

3. Scoring. The excessive heat is generated when there is an excessive surface pressure, high speed, or supply of lubricant fails. It is a stick-slip phenomenon in which alternate shearing and welding takes place rapidly at high spots. This type of failure can be avoided by properly designing the parameters such as speed, pressure, and proper flow of the lubricant, so that the temperature at the rubbing faces is within the permissible limits.

4. Abrasive wear. The foreign particles in the lubricants such as dirt, dust, or burr enter between the tooth and damage the form of tooth. This type of failure can be avoided by providing filters for the lubricating oil or by using high viscosity lubricant oil which enables the formation of thicker oil film and hence permits easy passage of such particles without damaging the gear surface.

5. Corrosive wear. The corrosion of the tooth surfaces is mainly caused due to the presence of corrosive elements such as additives present in the lubricating oils. In order to avoid this type of wear, proper anti-corrosive additives should be used.

Designing Spur Gear

In order to design spur gears, the following procedure may be followed :

The design tangential tooth load is obtained from the power transmitted and the pitch line velocity by using the following relation :

$$W_T = \frac{P}{v} \times C_S$$

where W_T = Permissible tangential tooth load in newtons,

P = Power transmitted in watts,

v = Pitch line velocity in m / s = (πDN)/60

D = Pitch circle diameter in metres,

We know that circular pitch,

p_c = π D / T = π m

D = m.T

Thus, the pitch line velocity may also be obtained by using the following relation, i.e.

$$v = \frac{\pi D.N}{60} = \frac{\pi m.T.N}{60} = \frac{p_c.T.N}{60}$$

where,   m = Module in metres, and

T = Number of teeth.

Now, we will learn the procedure to create spur gears in Autodesk Inventor.

- Click on the **Spur Gear** tool from the **Gear** drop-down in the **Power Transmission** panel of **Design** tab in the **Ribbon**. The **Spur Gears Component Generator** dialog box will be displayed; refer to Figure-40.

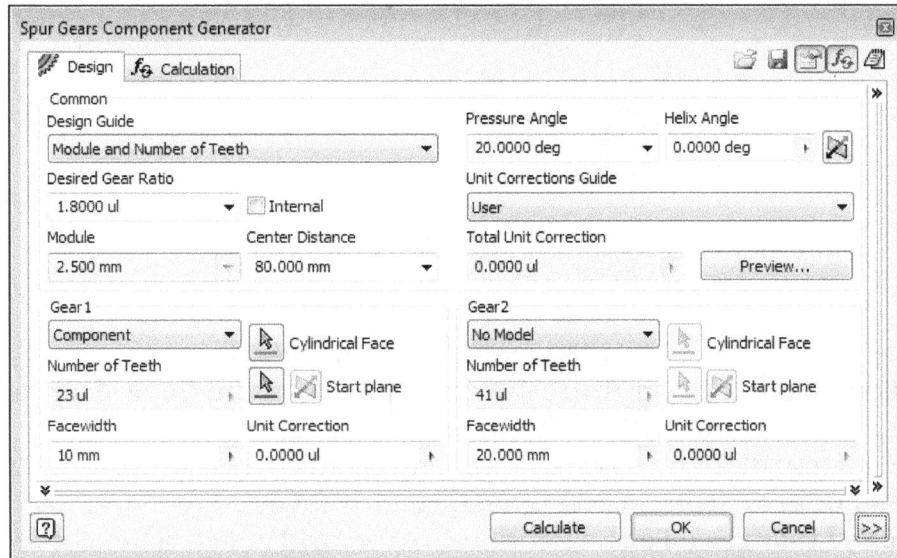

Figure-40. Spur Gear Component Generator dialog box

- Click in the **Design Guide** drop-down in the **Common** area of **Design** tab in the dialog box. The options in the drop-down will be displayed as shown in Figure-41.

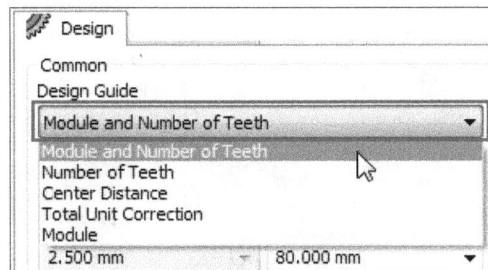

Figure-41. Design Guide drop-down

- Select desired option from the drop-down. The parameters in the dialog box will be activated accordingly. Select only that option from the drop-down which you want to make variable. For example, if you select the **Module and Number of Teeth** option from drop-down then system will find out the module and number of teeth based on the parameters specified by you. If you know all the parameters of gear then you should select the **Total Unit Correction** option from the drop-down.

- Specify desired value of gear ratio, module, and center distance in the edit boxes. In engineering equations, gear ratio is given by $G = T_2/T_1$ where T_1 is number of teeth on first gear and T_2 is number of teeth on second gear. Module is given by m = (Pitch Circle Dia.)/ Number of teeth. The **Center Distance** is distance between centers of two mating gears.

- Click on the drop-down in **Gear 1** area of the dialog box and select desired option; refer to Figure-42. If you select the **Component** option from the drop-down then a new component for gear will be created. If you select the **Feature** option then gear will be created as featured on existing disc. If you selected **No Model** option then no model will be created.

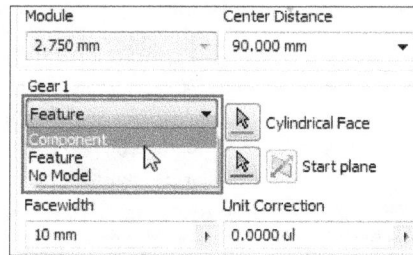

Figure-42. Drop-down in Gear 1 area

Creating Gear Component

- Select the **Component** option from the drop-down in the **Gear 1** area of the dialog box. The related options will be displayed in the area.
- Click in the **Facewidth** edit box and specify desired value. Note that facewidth is the thickness of gear from starting face to end face.
- Click on the **Cylindrical Face** selection button and select a face or axis to position the gear.
- Now, click on the **Start Plane** selection button and select the start face/plane for gear. Preview of the gear will be displayed; refer to Figure-43.

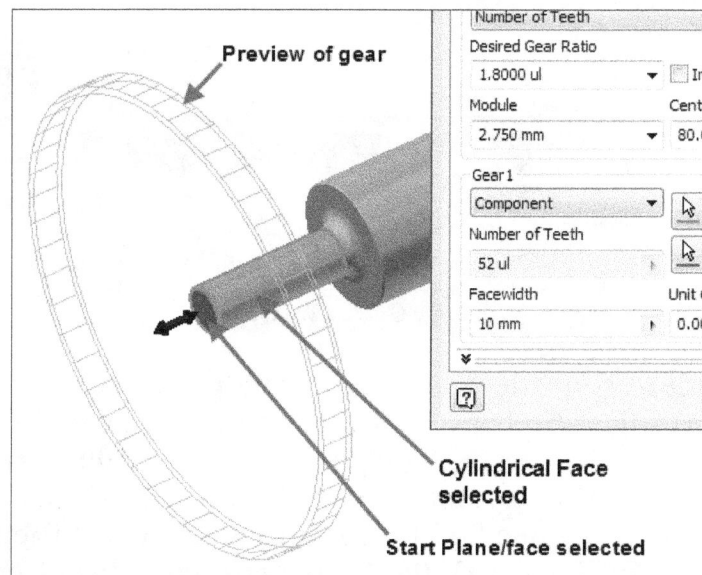

Figure-43. Preview of gear

- Specify the other parameters like number of teeth, center distance, module, etc. based on option selected in the **Design Guide** drop-down in the dialog box.

Creating Gear Feature

- Select the **Feature** option from the drop-down in the **Gear 1** area of the dialog box. The related options will be displayed in the area.
- Click in the **Facewidth** edit box and specify desired value.
- Click on the **Cylindrical Face** selection button and select a face or axis to position the gear.
- Now, click on the **Start Plane** selection button and select the start face/plane for gear. Preview of the gear will be displayed; refer to Figure-44.

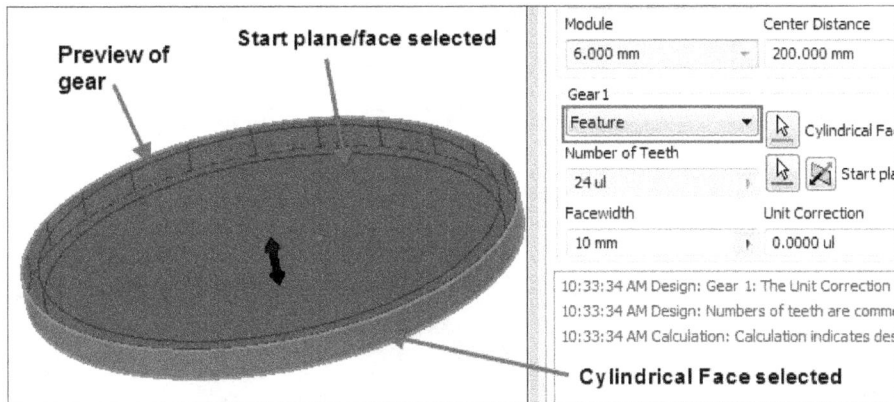

Figure-44. Creating gear feature

- Specify the other parameters like number of teeth, center distance, module, etc. based on option selected in the **Design Guide** drop-down in the dialog box.
- Note that you can also cut partial teeth on the disc by taking lesser facewidth in this method; refer to Figure-45.

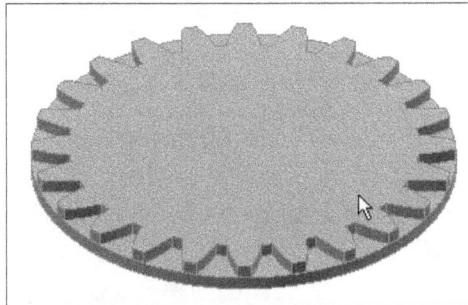

Figure-45. Gear feature generated by lesser face width

- Similarly, you can generate the other mating gear using the options in the **Gear 2** area of the dialog box.
- Now, click on the **Calculate** tab in the dialog box to check the feasibility of gear in real environment. The options in the dialog box will be displayed as shown in Figure-46.

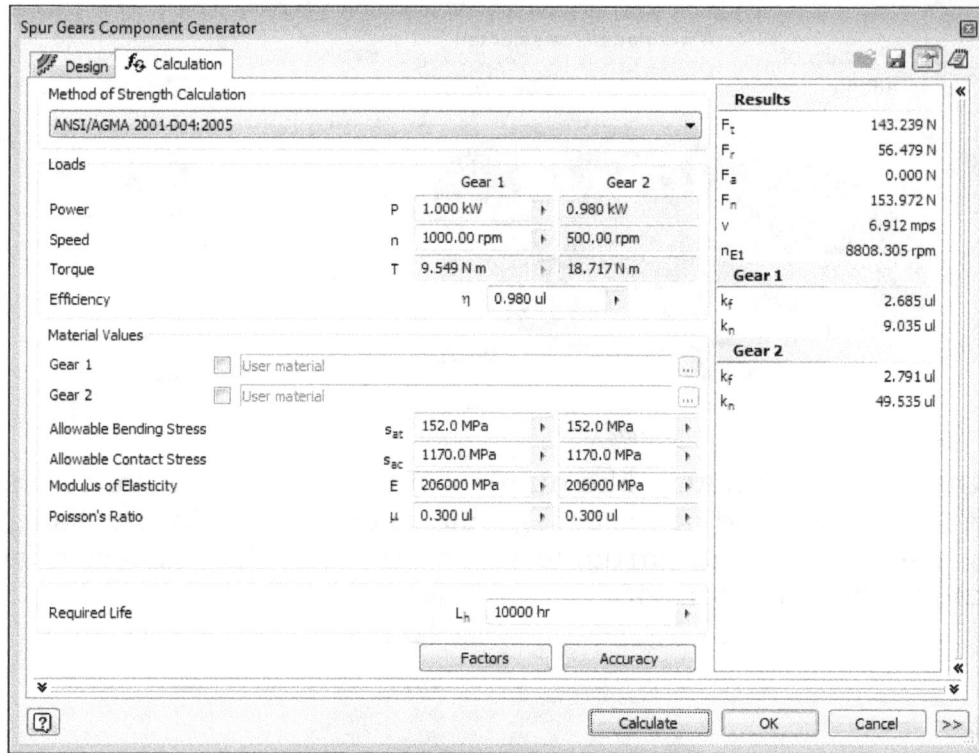

Figure-46. Calculation tab in Spur Gear Component Generator dialog box

- Select desired method of strength calculation from the **Method of Strength Calculation** drop-down; refer to Figure-47. The options in the dialog box will be modified accordingly.

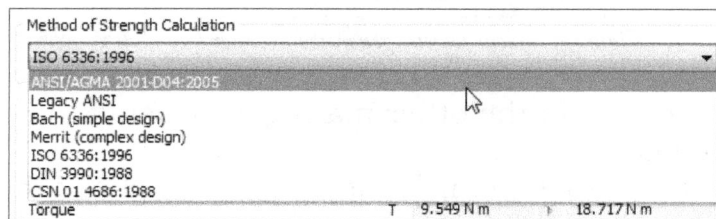

Figure-47. Method of Strength Calculation drop-down

- Specify the parameters related to load in the **Loads** area and parameters related to material in the **Material Values** area.
- Click on the **Calculate** button from the dialog box to check the feasibility of gear under specified load.
- Modify the gear if problems are displayed in the **Results** pane of dialog box and click on the **OK** button to create the gear.
- If you want to modify the minimum Factor of Safety and type of loading calculations then expand the dialog box by clicking on the **More options** button and specify desired values; refer to Figure-48.

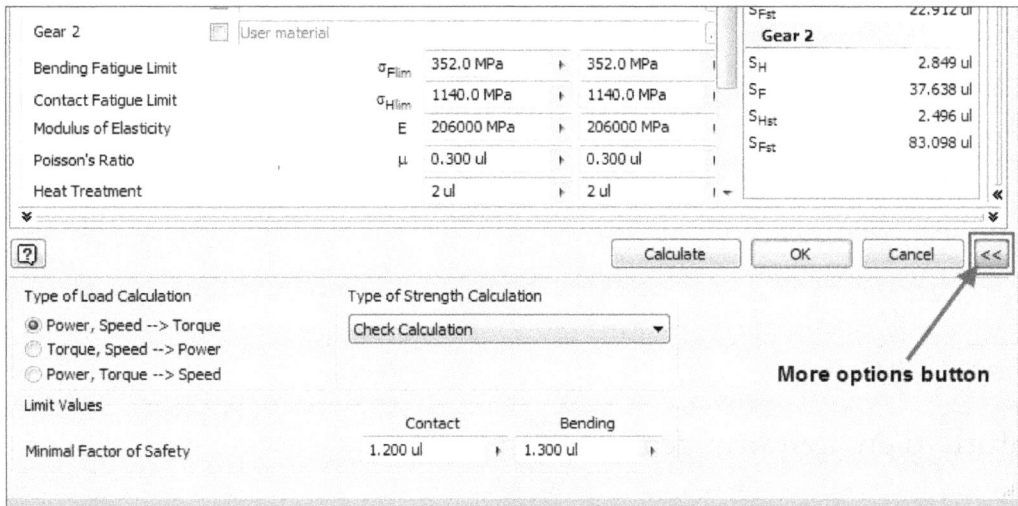

Figure-48. Expanded Spur Gear Component Generator dialog box

Designing Worm Gear

The Worm gear arrangement is used when large speed reduction is required. It is common for worm gears to have reductions of 20:1 and even up to 300:1 or greater. In a worm gear arrangement, there is one worm created on shaft and a worm gear. The axis of rotation of worm and worm gear is generally perpendicular to each other; refer to Figure-49.

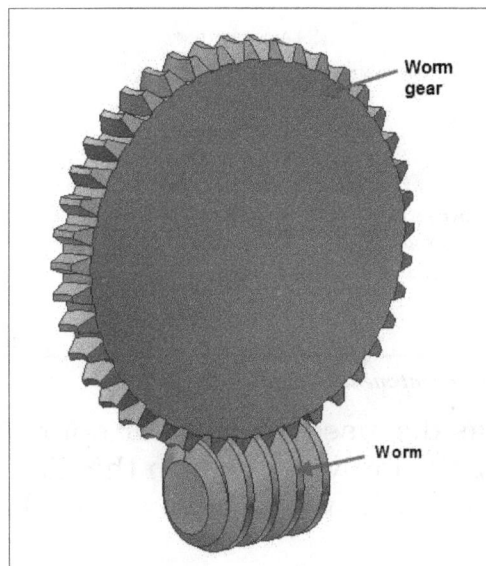

Figure-49. Worm gear arrangement

The worm gearing is classified as non-interchangeable, because a worm wheel cut with a hob of one diameter will not operate satisfactorily with a worm of different diameter, even if the thread pitch is same. The following are two types of worms :

1. Cylindrical or straight worm and
2. Cone or double enveloping worm. Refer to Figure-50.

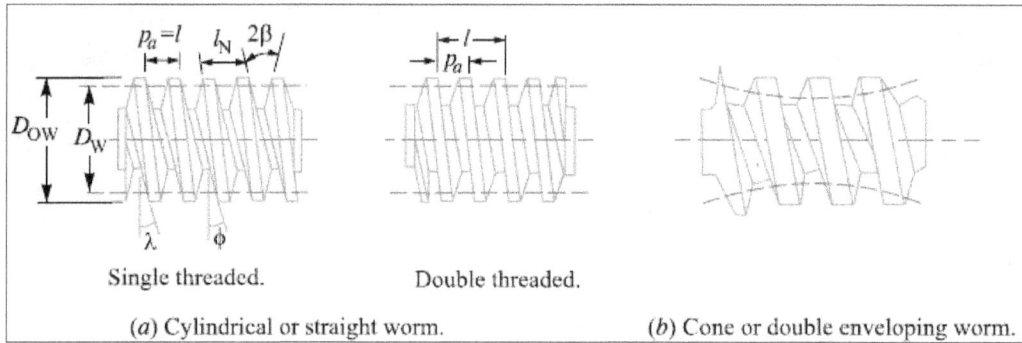

Figure-50. Type of worms

The procedure to create worm gear is given next.

*   Click on the **Worm Gear** tool from the **Gear** drop-down in the **Power Transmission** panel of the **Design** tab in the **Ribbon**. The **Worm Gears Component Generator** dialog box will be displayed; refer to Figure-51.

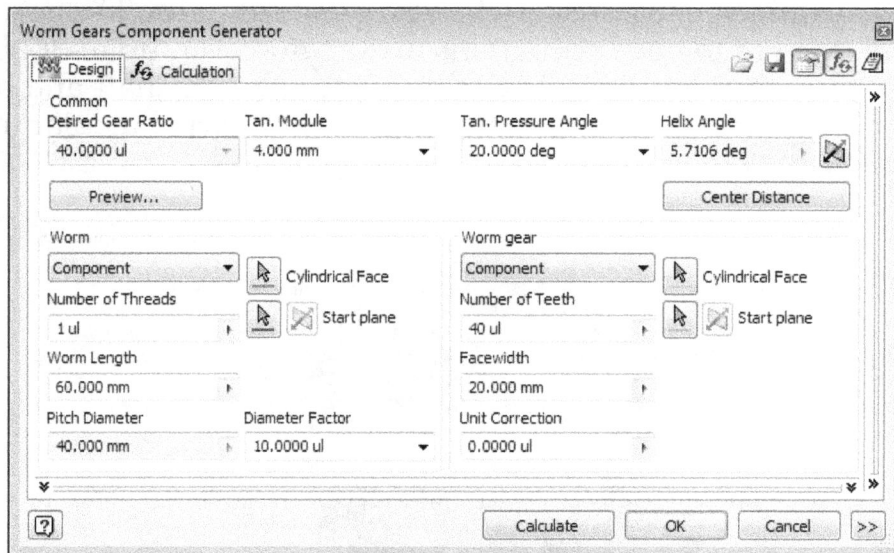

Figure-51. Worm Gears Component Generator dialog box

*   Specify the parameters as discussed earlier for spur gear. Note that you cannot specify the gear ratio directly for worm gear in the dialog box but you can change the value of number of teeth in worm gear or number of threads on worm to change the gear ratio.
*   Click on the **Calculation** tab to modify the load parameters or material parameters.
*   Click on the **Calculate** button from the dialog box. If errors are displayed then change the parameters accordingly.
*   Click on the **OK** button to create the worm gear system.

Designing Bevel Gear

The bevel gears are used for transmitting power at a constant velocity ratio between two shafts whose axes intersect at a certain angle. The pitch surfaces for the bevel gear are frustums of cones; refer to Figure-52.

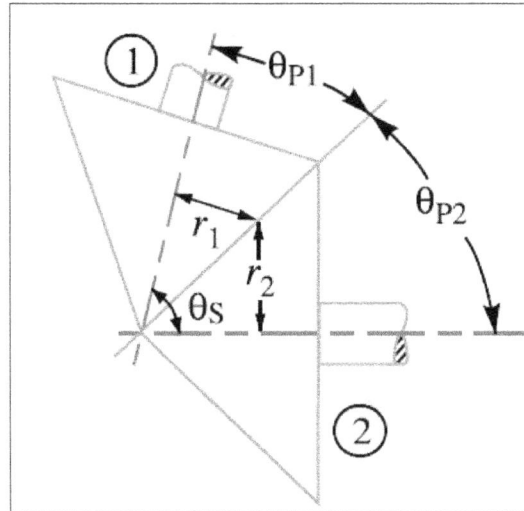

Figure-52. Bevel gear arrangement

Here, r_1 and r_2 are radii of gear 1 and gear 2, respectively.
Θ_{P1} and Θ_{P2} are cone angle for gears
and, Θ_S is shaft angle.

The procedure to create bevel gears is given next.

- Click on the **Bevel Gear** tool from the **Gear** drop-down in the **Power Transmission** panel of the **Design** tab in the **Ribbon**. The **Bevel Gears Component Generator** dialog box will be displayed; refer to Figure-53.

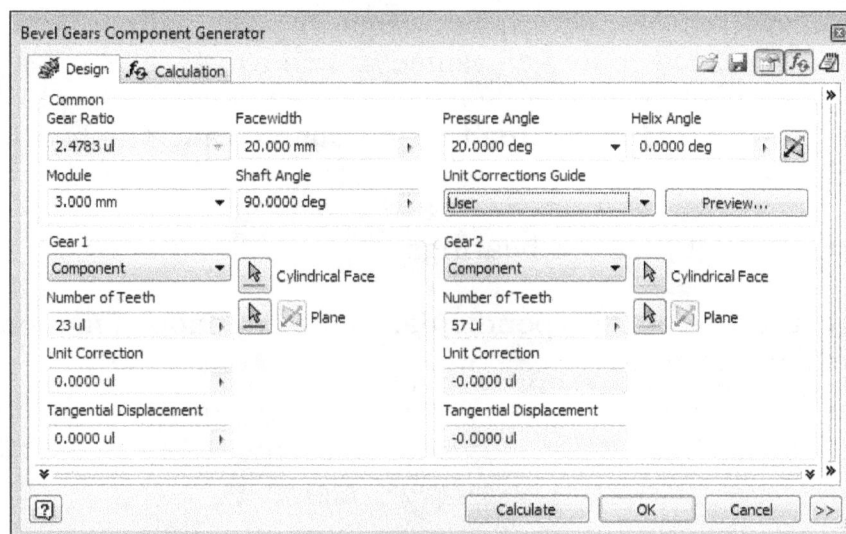

Figure-53. Bevel Gears Component Generator dialog box

- Specify the module, facewidth, and shaft angle in the respective edit boxes in the dialog box.
- Set the number of teeth for both the gears. Note that gear ratio will be calculated based on the specified number of teeth.
- Select the cylindrical face and start plane for gear 1 and gear 2 if required. These options have already been discussed in previous topics.
- Click on the **Calculation** tab and set the loading & material parameters.

• Click on the **Calculate** button. If results are fine then click on the **OK** button otherwise make the changes as per the results. The bevel gear arrangement will be created; refer to Figure-54.

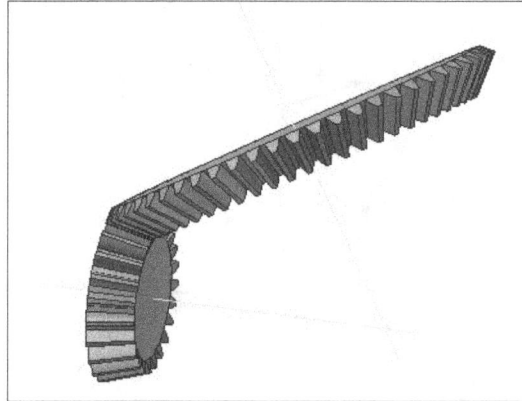

Figure-54. Bevel gear

Designing Bearing

A bearing is a machine element which support another moving machine element (known as journal). It permits a relative motion between the contact surfaces of the members, while carrying the load. A little consideration will show that due to the relative motion between the contact surfaces, a certain amount of power is wasted in overcoming frictional resistance and if the rubbing surfaces are in direct contact, there will be rapid wear. In order to reduce frictional resistance and wear and in some cases to carry away the heat generated, a layer of fluid (known as lubricant) may be provided. The lubricant used to separate the journal and bearing is usually a mineral oil refined from petroleum, but vegetable oils, silicon oils, greases, etc., may be used. Bearings are broadly classified in two ways:

Depending upon the direction of load to be supported. The bearings under this group are classified as:

(a) Radial bearings, and (b) Thrust bearings.

In radial bearings, the load acts perpendicular to the direction of motion of the moving element as shown in Figure-55.

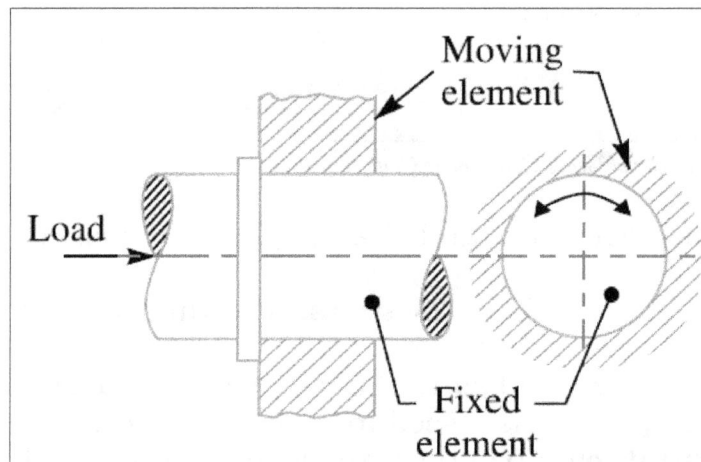

Figure-55. Radial Bearing

In thrust bearings, the load acts along the axis of rotation as shown in Figure-56.

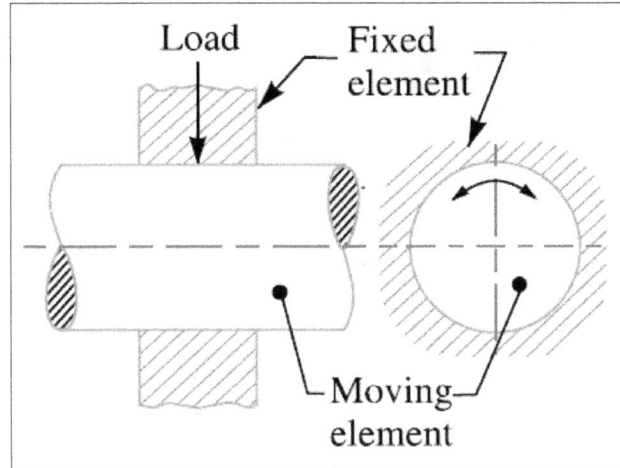

Figure-56. Thrust Bearing

Depending upon the nature of contact. The bearings under this group are classified as :
(a) Sliding contact bearings, and (b) Rolling contact bearings.

In sliding contact bearings, as shown in Figure-57, the sliding takes place along the surfaces of contact between the moving element and the fixed element. The sliding contact bearings are also known as plain bearings.

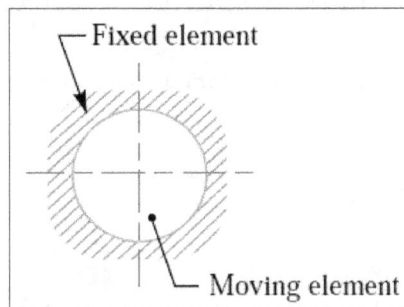

Figure-57. Sliding Contact Bearing

In rolling contact bearings, as shown in Figure-58, the steel balls or rollers, are interposed between the moving and fixed elements. The balls offer rolling friction at two points for each ball or roller.

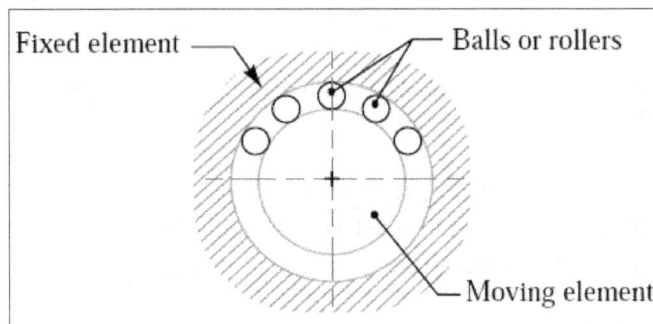

Figure-58. Rolling Contact Bearing

The procedure to design various bearings in Autodesk Inventor is given next.

- Click on the **Bearing** tool from the **Power Transmission** panel in the **Design** tab of the **Ribbon**. The **Bearing Generator** dialog box will be displayed; refer to Figure-59.

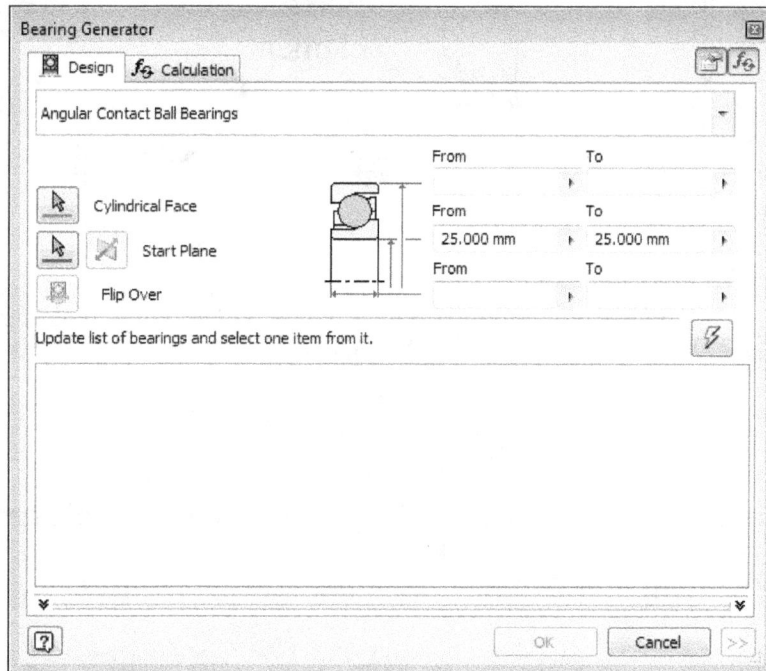

Figure-59. Bearing Generator dialog box

- Click on the **Browse for bearing** drop-down at the top in the dialog box. The Content center for bearing will be displayed; refer to Figure-60.

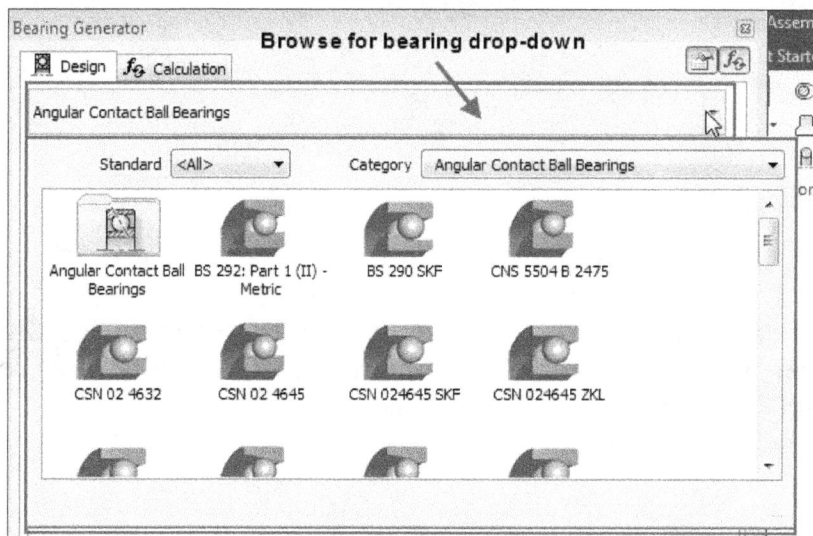

Figure-60. Content Center for Bearing

- Select desired standard and category of bearings from the **Standard** and **Category** drop-down in the Content center. List of short listed bearings will be displayed.
- Select the folder of bearing from the Content center; refer to Figure-61.

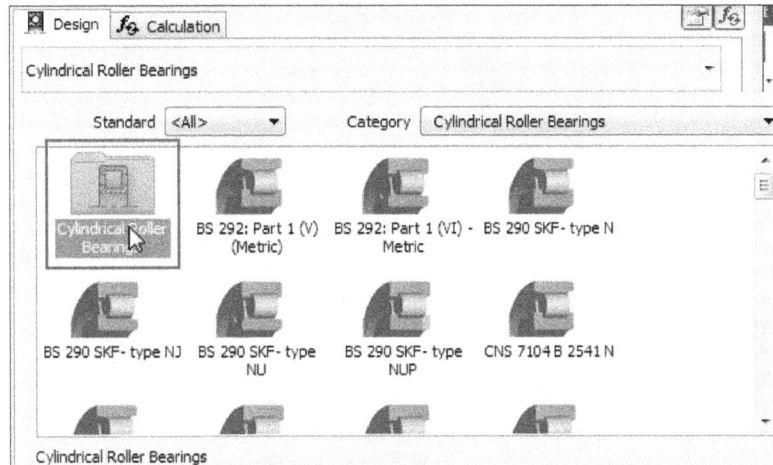

Figure-61. Folder for Bearings

- Specify the parameters for sizing of bearing in the edit boxes in the dialog box; refer to Figure-62.

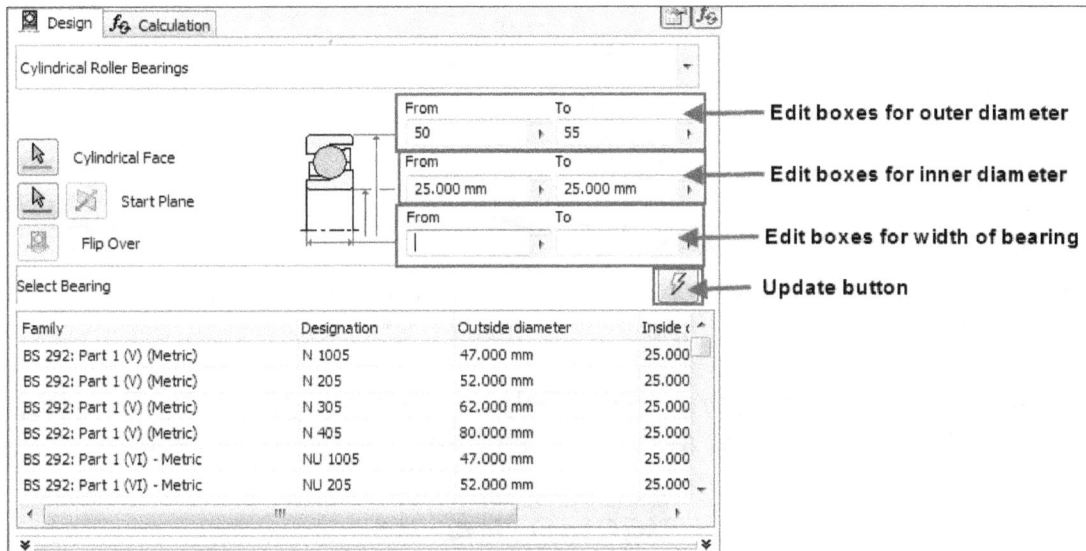

Figure-62. Parameters for sizing bearing

- Click on the **Update** button to update the list of bearings available in the Content center.
- Click on the **Calculation** tab in the dialog box to check the feasibility of bearing under load. The dialog box will be displayed as shown in Figure-63.

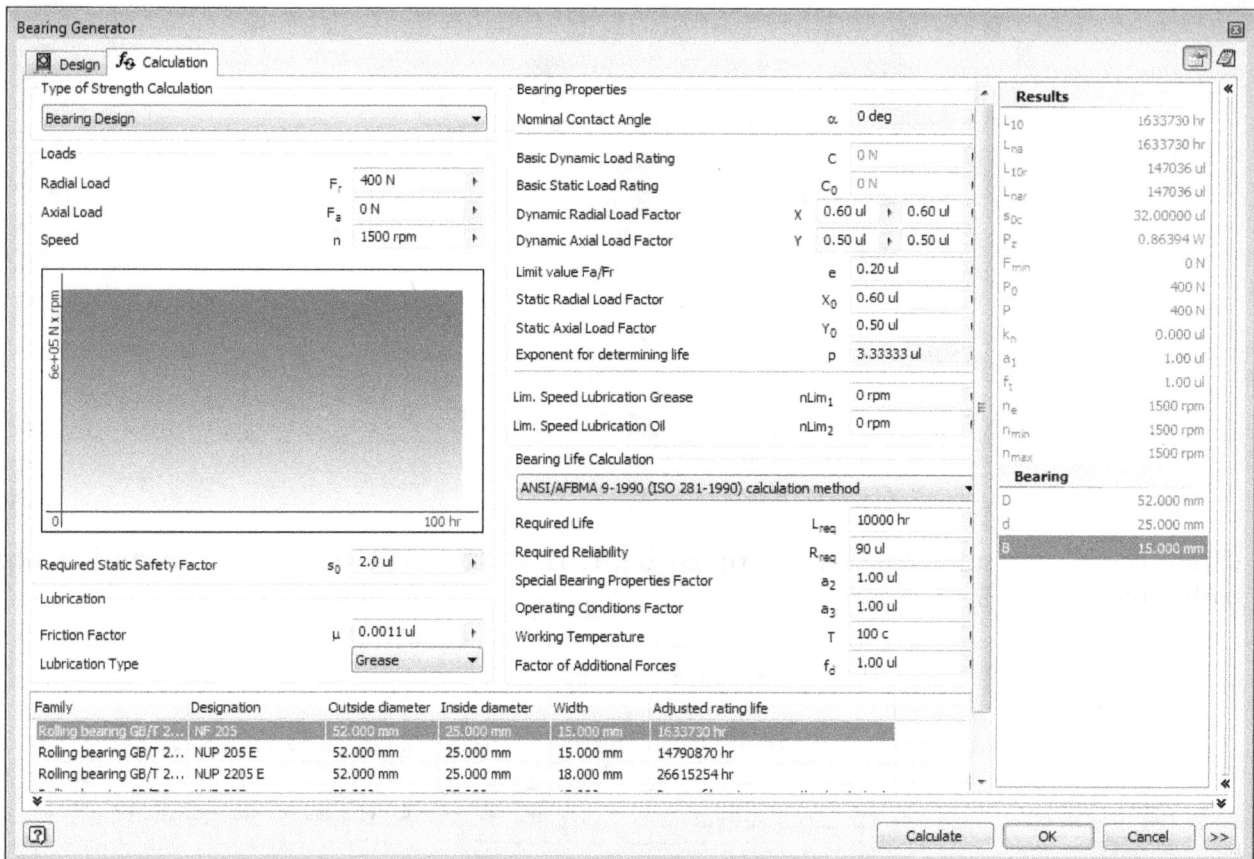

Figure-63. Calculation tab in Bearing Generator dialog box

- Specify the loading parameters and then click on the **Calculate** button. List of bearings qualifying the loading conditions will be displayed in the list at bottom in the dialog box.
- Select the suitable bearing and then click on the **OK** button. The bearing will get attached to the cursor. Click at desired location to place the bearing.

Designing V-Belts

The pulleys are used to transmit power from one shaft to another by means of flat belts, V-belts, or ropes. Since the velocity ratio is the inverse ratio of the diameters of driving and driven pulleys, therefore the pulley diameters should be carefully selected in order to have a desired velocity ratio. The pulleys must be in perfect alignment in order to allow the belt to travel in a line normal to the pulley faces. The pulleys may be made of cast iron, cast steel or pressed steel, wood, and paper. The cast materials should have good friction and wear characteristics. The pulleys made of pressed steel are lighter than cast pulleys, but in many cases they have lower friction and may produce excessive wear.

The V-belts are made of fabric and cords moulded in rubber and covered with fabric and rubber. These belts are moulded to a trapezoidal shape and are made endless. These are particularly suitable for short drives. The included angle for the V-belt is usually from 30° to 40°. The power is transmitted by the wedging action between the belt and the V-groove in the pulley or sheave. The wedging action of the V-belt in the groove of the pulley results in higher forces of friction. A little consideration will show that the wedging action and the transmitted torque will be more if the groove angle of the pulley is small. But a small groove angle will require more force to pull the belt out of the groove which will result in loss of power and excessive belt wear due to friction and heat.

Hence selecting groove angle is a compromise between the two. Usually, the groove angles of 32° to 38° are used.

In Autodesk Inventor, you can design both V-belt and pulley together. The procedure to create V-belt with pulley is given next.

- Click on the **V-Belts** tool from the **Belts** drop-down in the **Power Transmission** panel of the **Design** tab in the **Ribbon**. The **V-Belts Component Generator** dialog box will be displayed as shown in Figure-64.

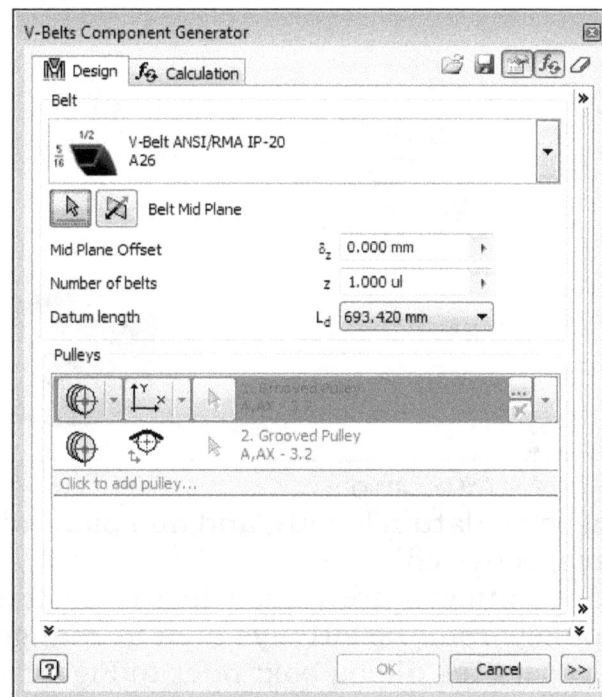

Figure-64. V-Belts Component Generator dialog box

- Click in the **Browse for belt type** drop-down and select desired belt type; refer to Figure-65.

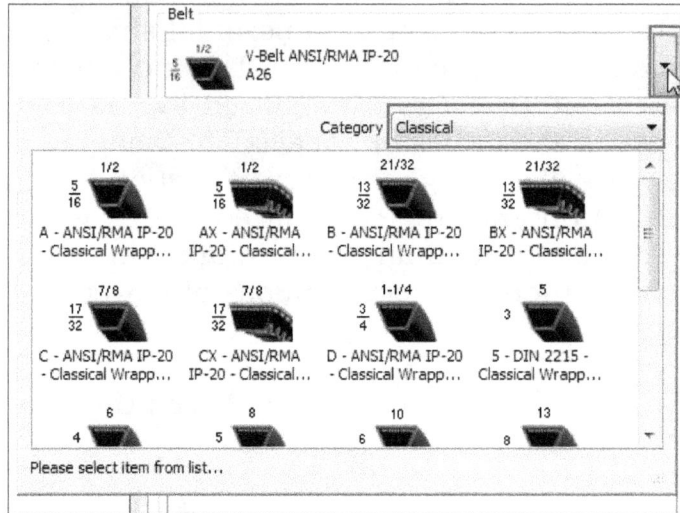

Figure-65. Browse for belt type drop-down

- Click on the selection button for **Belt Mid Plane** and select the middle plane/ face for belt and pulley creation. Preview of the belt and pulley will be displayed; refer to Figure-66.

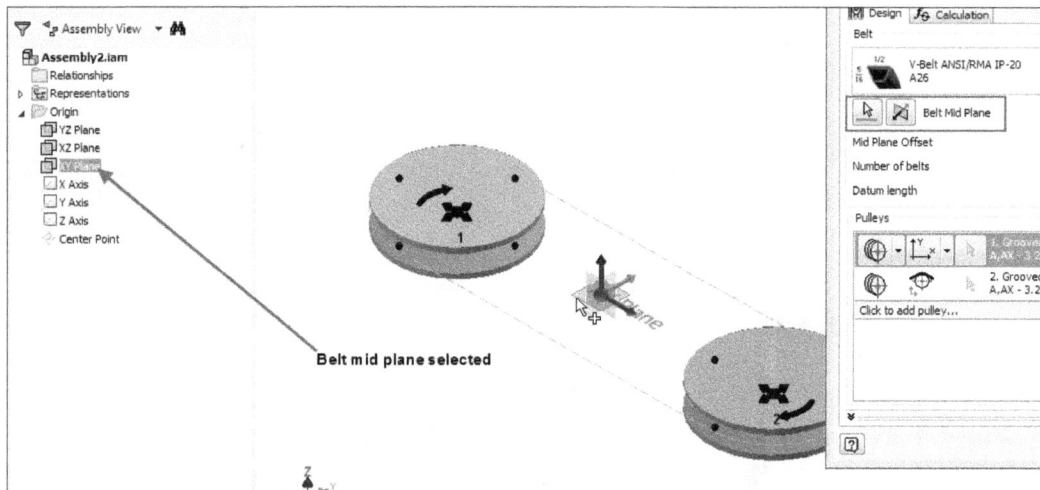

Figure-66. Mid plane selected for belt and pulley

- Specify the number of belts, datum length, and mid plane offset for the belt-pulley arrangement in the respective edit boxes.
- By default, two grooved pulleys are added in the system; refer to Figure-67. Double-click on the pulley from the **Pulleys** area to change its parameters using the **Groove pulley properties** dialog box; refer to Figure-68.

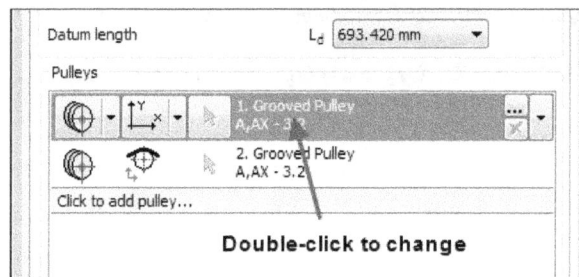

Figure-67. Pulleys

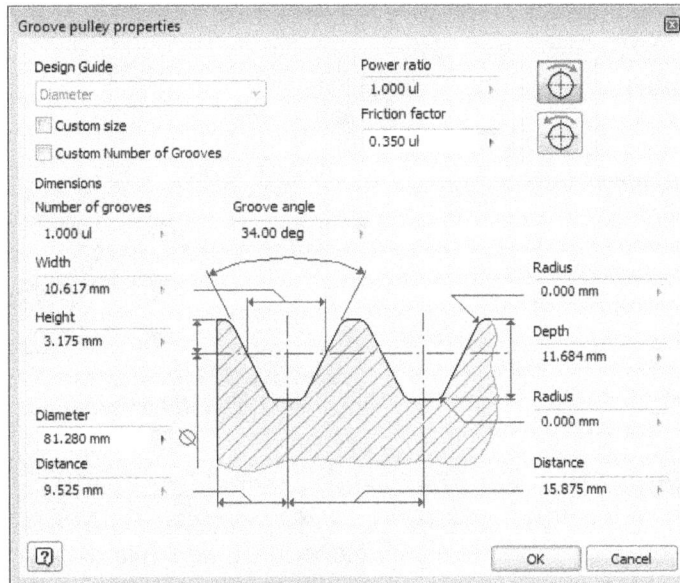

Figure-68. Groove pulley properties dialog box

- By default, you can change only the diameter of pulley and friction factor. If you want to change the other parameters then select the **Custom size** and **Custom Number of Grooves** check boxes. After changing the parameters, click on the **OK** button to exit the dialog box.
- If you want to add more pulleys in the system then click on the **Click to add pulley** option in the **Pulleys** area of the dialog box. A flyout with various pulley options will be displayed; refer to Figure-69.

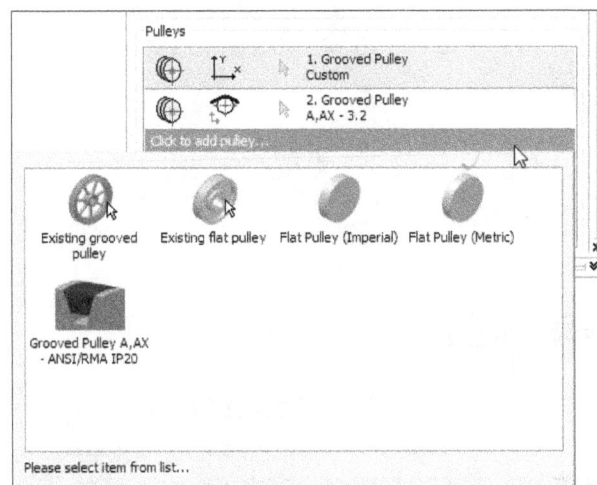

Figure-69. Flyout with various pulley options

- Select the **Grooved Pulley** option because we are creating pulley for V-belt. A new pulley will be added in the system. You can change the parameters of pulley as discussed earlier.
- To change the position of pulley, we have five options in the **Pulley placement guide** drop-down; refer to Figure-70.
- Select the **Fixed position by coordinates** option to change the position of pulley by giving coordinates. After selecting this option from drop-down, double-click on the move handles on pulley. The **Coordinates** dialog box will be displayed; refer to Figure-71. Specify desired values and click on the **OK** button to change the position of pulley.

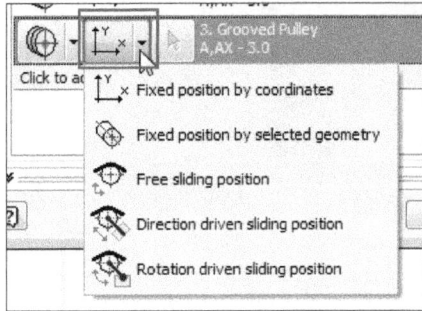

Figure-70. Pulley placement guide drop-down

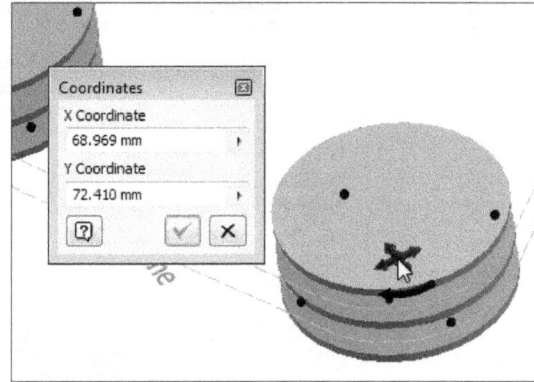

Figure-71. Coordinates dialog box

- Select the **Fixed position by selected geometry** option to place the pulley on selected geometry. You will be asked to select shaft axis, cylindrical or conical face, vertex, work point or work axis. Select desired geometry. The pulley will be placed accordingly.
- Select the **Free sliding position** option to freely move the pulley to desired location. On selecting this option, drag the pulley by using the move handles on the pulley; refer to Figure-72.

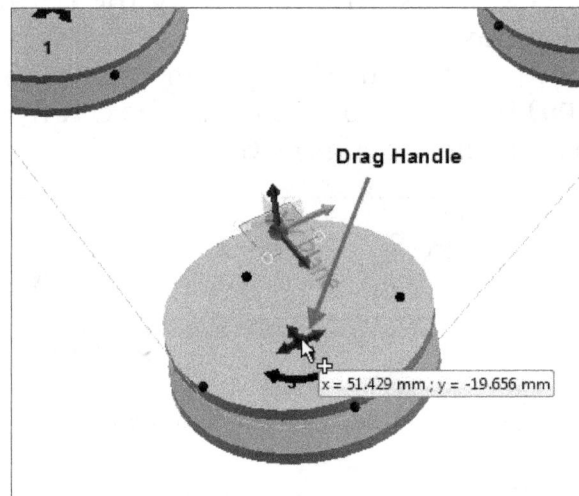

Figure-72. Drag handle on pulley

- Select the **Direction Driven Sliding Position** option to freely move the pulley along the selected plane or planar face. On selecting this option, you will be asked to select a sliding plane or planar face. Select a plane/planar face to be used as sliding plane. Drag handles will be displayed on the pulley. Move the pulley at desired location; refer to Figure-73. Note that pulley will automatically snap to the locations where belt lengths are available.

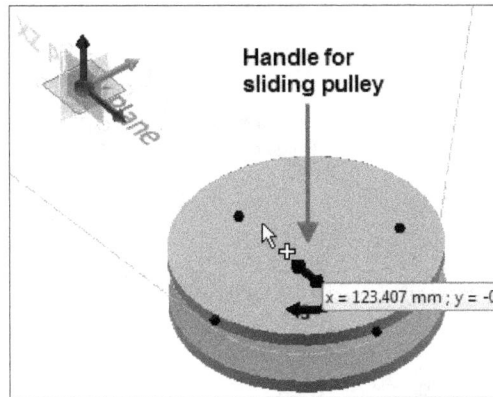

Figure-73. Handle for sliding along plane

- Similarly, select the **Rotation Driven Sliding Position** option to freely move the pulley around the selected axis.
- Click on the **Calculation** tab to check the feasibility of belt and pulley under specified load. The dialog box will be displayed; refer to Figure-74.

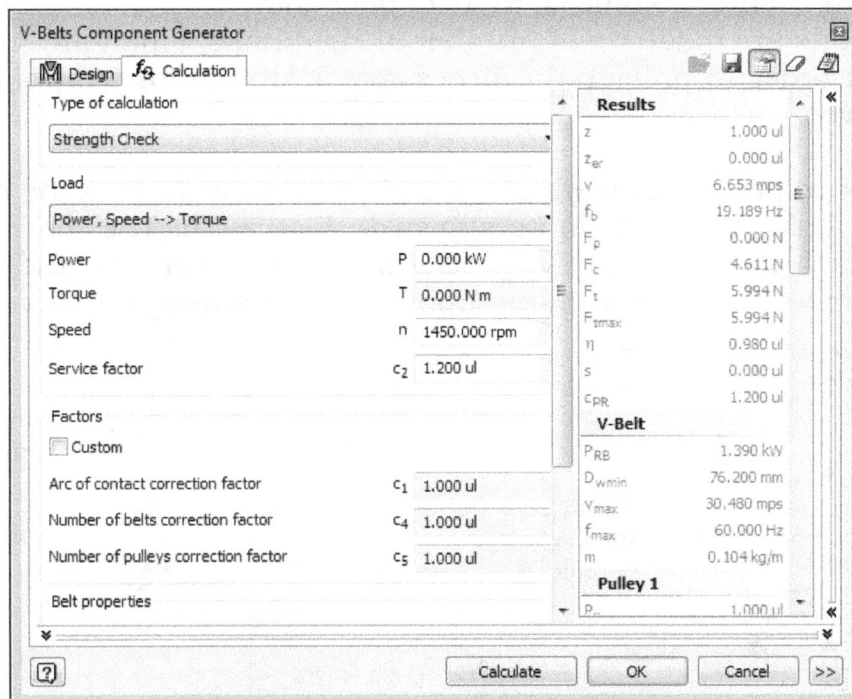

Figure-74. Calculation tab in V-Belts Component Generator dialog box

- Specify various parameters related to loading and belt properties.
- Click on the **Calculate** button. Modify the belt and pulley if error is displayed and then click on the **OK** button to create the V-belt system.

Designing Synchronous Belts

The synchronous belt, also called timing belt, toothed belt, cogged belt or cog belt is used as a non-slipping mechanical drive belt. It is made as a flexible belt with teeth moulded onto its inner surface. It runs over matching toothed pulleys or sprockets; refer to Figure-75. The procedure to create synchronous belt transmission system is same as discussed for V-Belt.

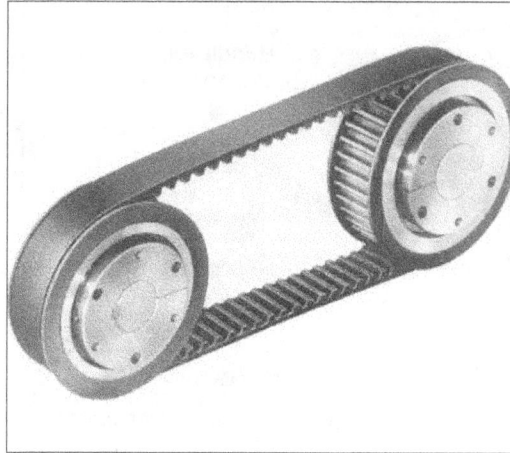

Figure-75. Synchoronous Belt

Designing Roller Chains

There is always some slippage in belt and rope transmission system. To avoid this slippage at shorter center distances, we use roller chain transmission system. In this system, the chains are made up of number of rigid links which are hinged together by pin joints in order to provide the necessary flexibility for wrapping around the driving and driven wheels.

These wheels have projecting teeth of special profile and fit into the corresponding recesses in the links of the chain as shown in Figure-76. The toothed wheels are known as sprocket wheels or simply sprockets. The sprockets and the chain are thus constrained to move together without slipping and ensures perfect velocity ratio.

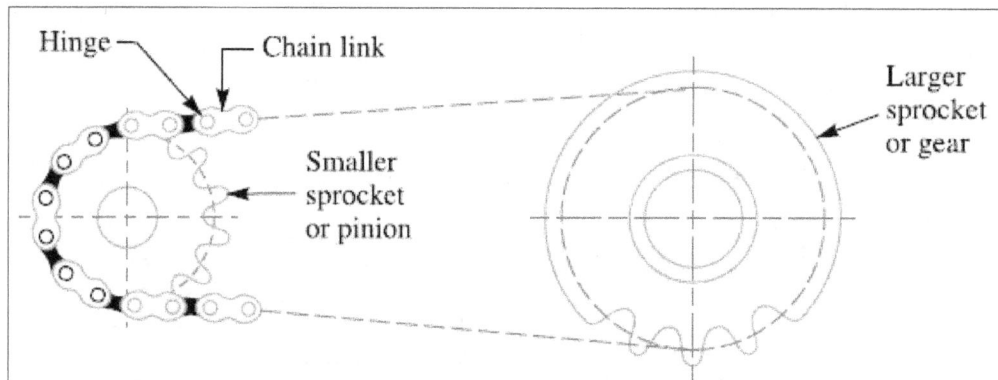

Figure-76. Roller Chain system

The chains are mostly used to transmit motion and power from one shaft to another, when the centre distance between their shafts is short such as in bicycles, motor cycles, agricultural machinery, conveyors, rolling mills, road rollers, etc. The chains may also be used for long centre distance of upto 8 metres. The chains are used for velocities up to 25 m/s and for power upto 110 kW. In some cases, higher power transmission is also possible.

The procedure to design roller chain system in Autodesk Inventor is given next.

•   Click on the **Roller Chains** tool from the **Belts** drop-down in the **Power Transmission** panel of the **Design** tab in the **Ribbon**. The **Roller Chains Generator** dialog box will be displayed; refer to Figure-77.

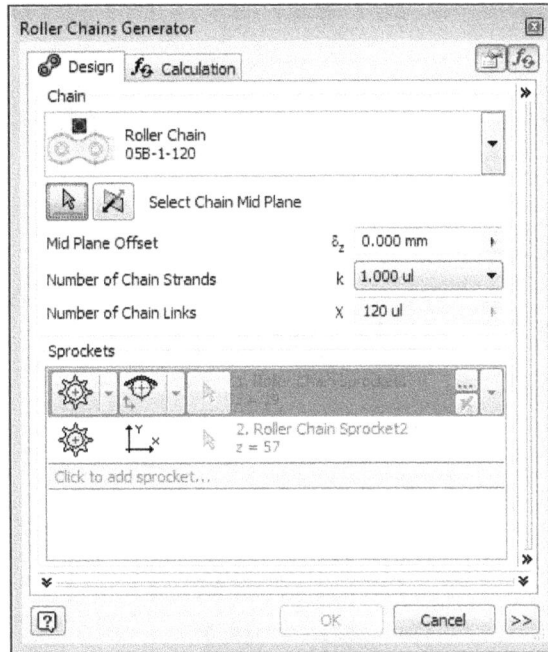

Figure-77. Roller Chains Generator dialog box

- Click on the **Browse for a chain** drop-down in the **Chain** area of the dialog box and select desired chain from the selection box displayed; refer to Figure-78.

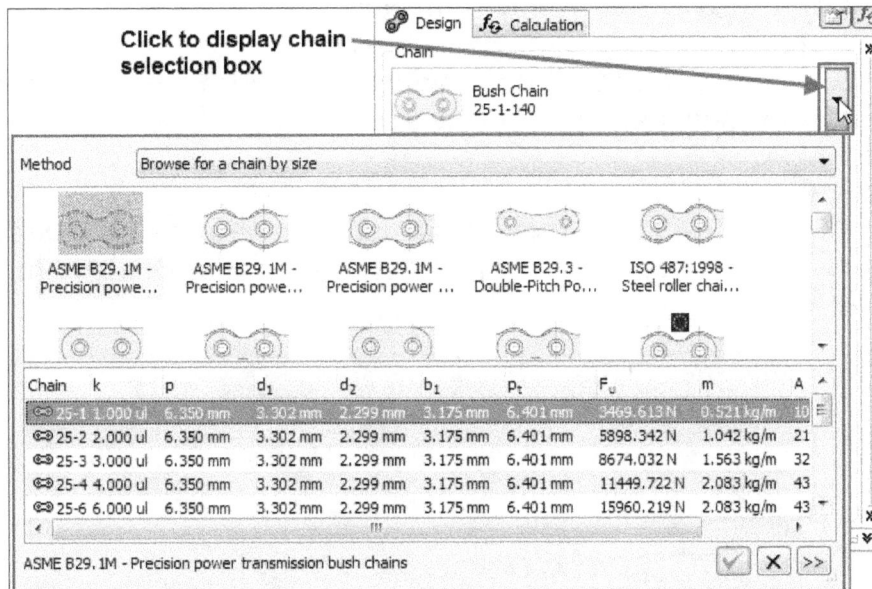

Figure-78. Chain selection box

- If you want to check the preview and maximum power transmission capacity of the selected chain then click on the **More options** button from the selection box and click on the related tab; refer to Figure-79.

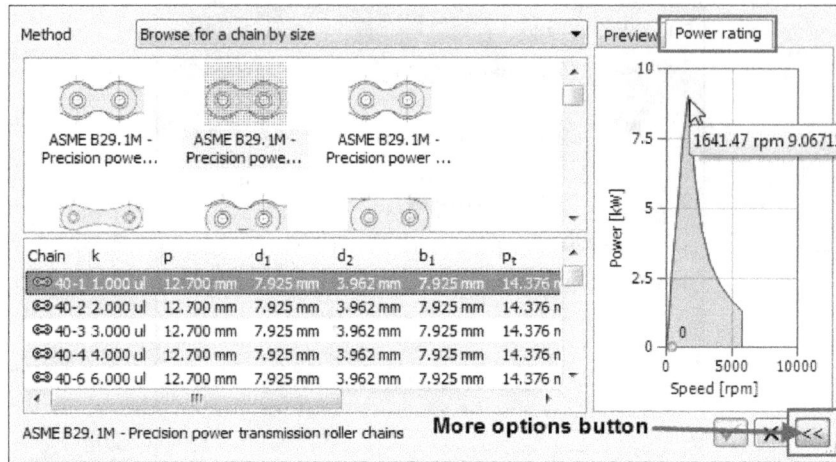

Figure-79. Power rating of selected chain

- Rest of the procedure to create chain system is same as discussed for V-Belt.

Designing Key

A key is a piece of mild steel inserted between the shaft and hub or boss of the pulley to connect these together in order to prevent relative motion between them. It is always inserted parallel to the axis of the shaft. Keys are used as temporary fastenings and are subjected to considerable crushing and shearing stresses. A keyway is a slot or recess in a shaft and hub of the pulley to accommodate a key.

The **Key** tool in Autodesk Inventor is used to generate both key and keyway based on the load requirements specified. The procedure to design key and keyway in Autodesk Inventor is given next.

- Click on the **Key** tool from the **Power Transmission** panel of the **Design** tab in the **Ribbon**. The **Parallel Key Connection Generator** dialog box will be displayed; refer to Figure-80.

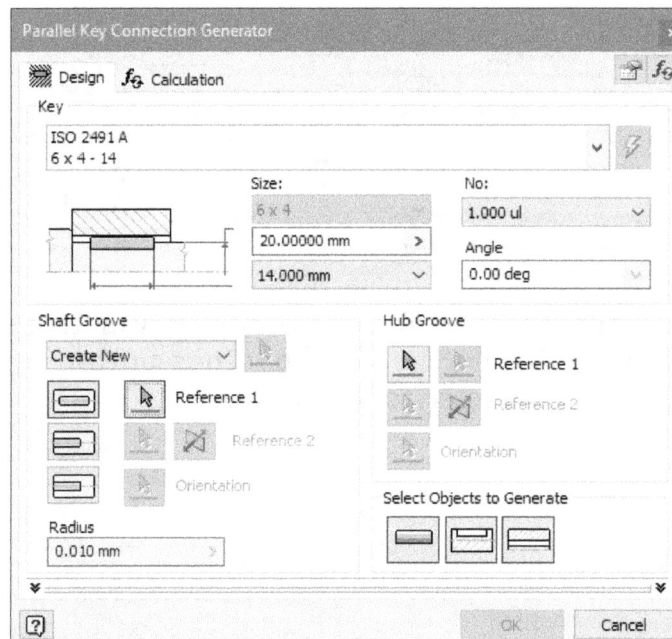

Figure-80. Parallel Key Connection Generator dialog box

- Click in the **Browse for key** drop-down in the **Key** area of the dialog box. Various options to select key are displayed; refer to Figure-81.

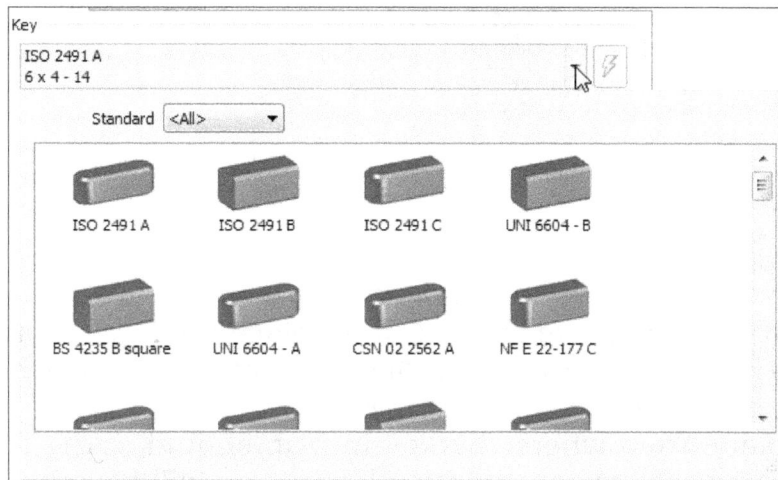

Figure-81. Browse for key drop-down

- Specify the size of key in the edit boxes available in the **Key** area of the dialog box.
- If you have a key groove in the shaft already then select the **Select Existing** option from the drop-down in the **Shaft Groove** area of the dialog box. You will be asked to select the existing groove. Select the groove.
- If you do not have groove created in the shaft then select the **Create New** option from the drop-down in the **Shaft Groove** area of the dialog box. You will be asked to select a cylindrical face for placing the groove (If not asked then click on the **Reference 1** selection button in the **Shaft Groove** area). Select the cylindrical face. You will be asked to select the starting face/plane for key groove. Select the face/plane; refer to Figure-82. Preview of the groove will be displayed.

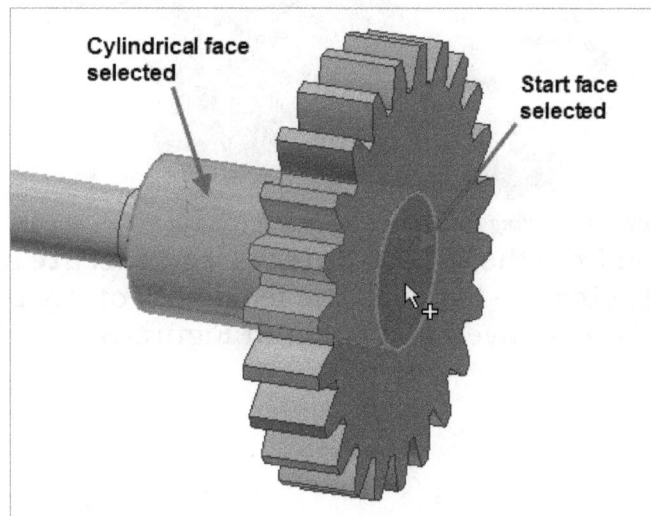

Figure-82. Faces selected for creating groove

- Click on desired button for groove shape from the **Shaft Groove** area of the dialog box; refer to Figure-83.

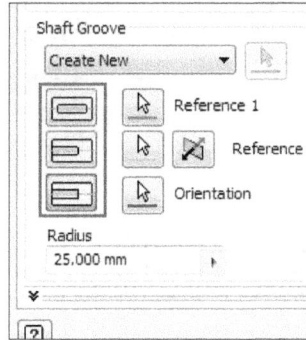

Figure-83. Buttons for shaping grooves

- Now, we need to create groove in the hub to place the key. Click on the selection button from **Reference 1** in the **Hub Groove** area of the dialog box. You will be asked to select a plane/face as starting reference.
- Select a face/plane from where the key in hub should start. The next selection button will become selected automatically and you will be asked to select a work point as center point for hub.
- Select circumferential edge of the hub or center point of the hub; refer to Figure-84. Preview of the groove will be displayed.

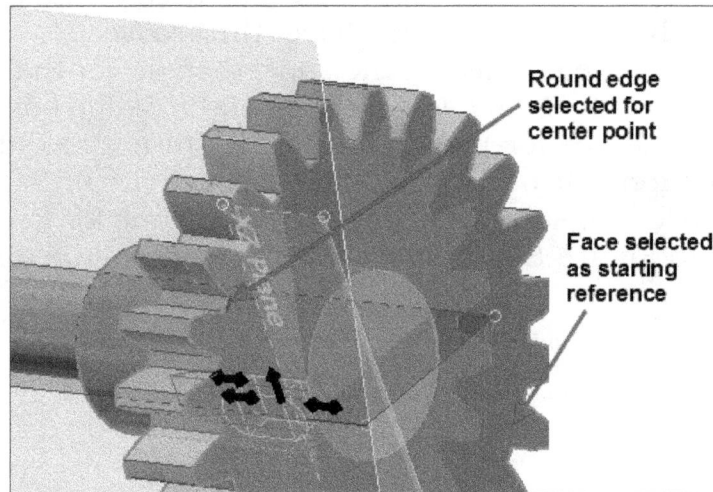

Figure-84. Selecting faces for groove in hub

- Select desired button from the **Select Objects to Generate** area of the dialog box.
- Click on the **Calculation** tab to check the feasibility of key under specified loads. The dialog box will be displayed as shown in Figure-85.

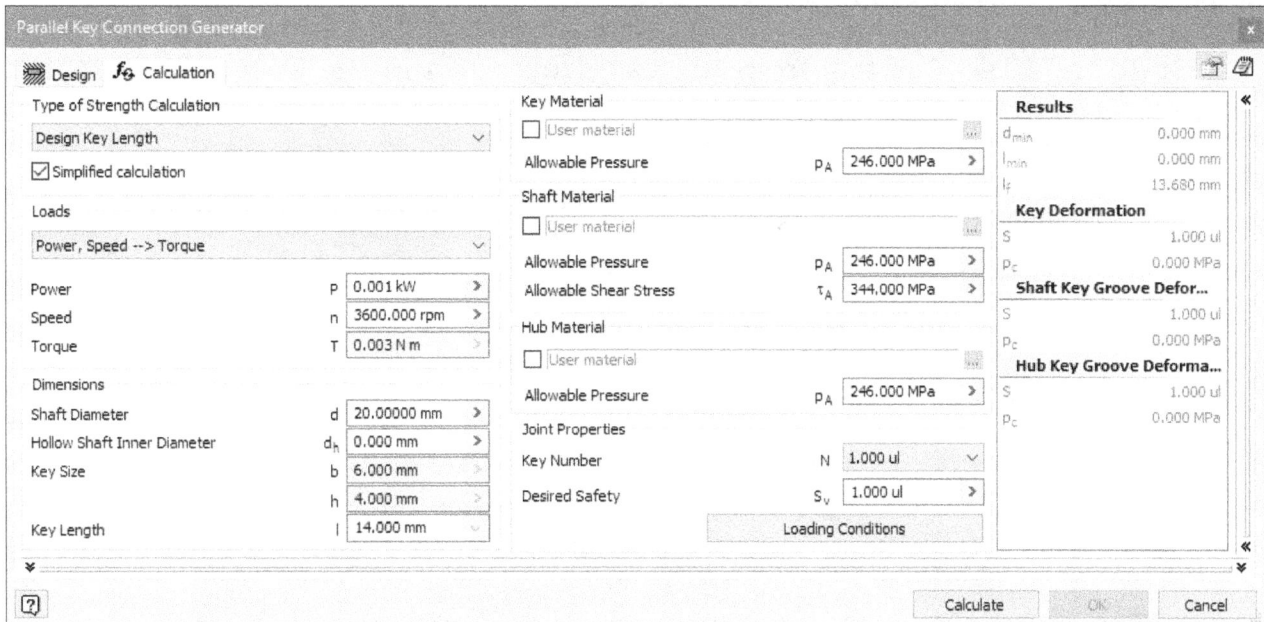

Figure-85. Calculation tab in Parallel Key Connection Generator dialog box

• Specify desired parameters and click on the **Calculate** button to check the results. Modify the key and grooves if error is displayed and then click on the **OK** button to create the key and grooves.

Designing Disc Cam

The transformation of one of the simple motions, such as rotation, into any other motions is often conveniently accomplished by means of a cam mechanism. A cam mechanism usually consists of two moving elements, the cam and the follower, mounted on a fixed frame. Cam devices are versatile, and almost any arbitrarily-specified motion can be obtained. In some instances, they offer the simplest and most compact way to transform motions. In Plate cam or disk cam, the follower moves in a plane perpendicular to the axis of rotation of the camshaft. A translating or a swing arm follower must be constrained to maintain contact with the cam profile. Figure-86 shows the nomenclature of a disc cam.

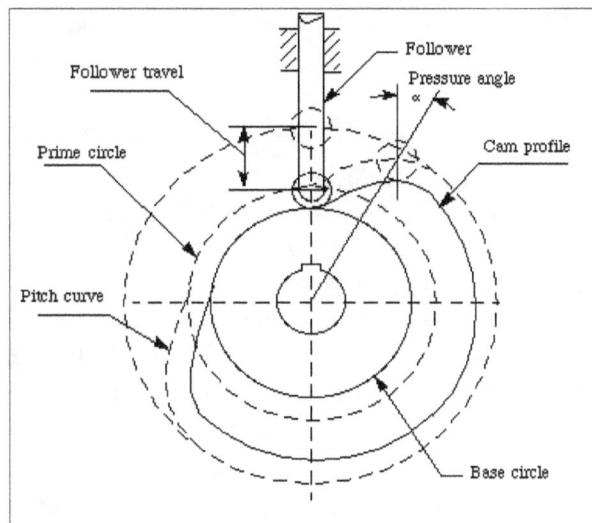

Figure-86. Disc cam nomenclature

The procedure to design disc cam in Autodesk Inventor is given next.

- Click on the **Disc Cam** tool from the **Cam** drop-down in the **Power Transmission** panel in the **Design** tab of the **Ribbon**. The **Disc Cam Component Generator** dialog box will be displayed; refer to Figure-87.

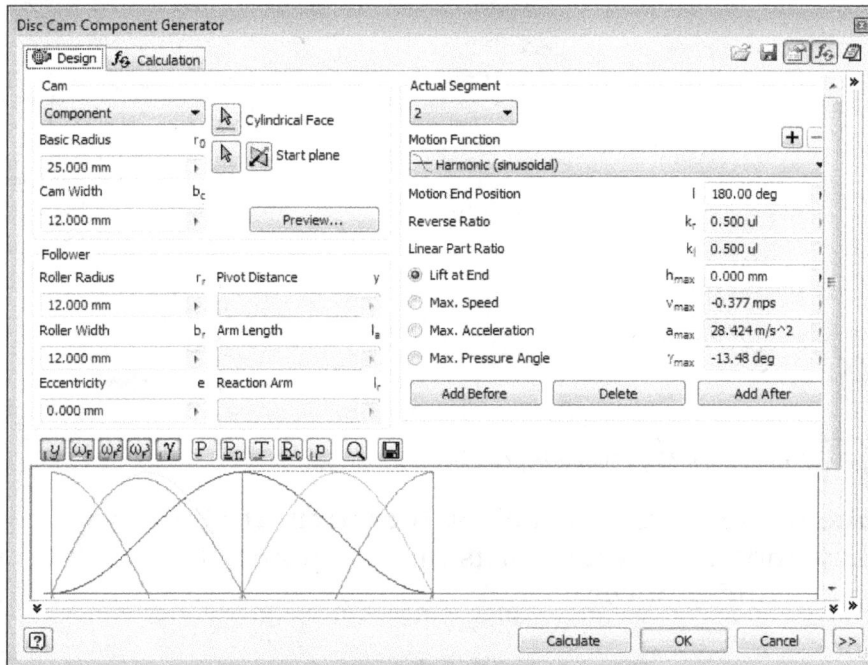

Figure-87. Disc Cam Component Generator dialog box

- Specify the base radius and width of cam in the **Basic Radius** and **Cam Width** edit boxes, respectively.
- Click on the **Preview** button if you want to check the preview of disc cam.
- Similarly, specify the roller radius, roller width, and eccentricity of roller in the edit boxes of the **Follower** area in the dialog box.
- Now, we will design profile of the cam. In Autodesk Inventor, cam profile is designed by segments like, for 0 to 120 degree, the total lift by cam will be 0 to 10 mm, after that the lift will move to 0 from 10 mm till 240 degree and then it will remain 0. To do so, click on the **Actual Segment** drop-down and select **1** in it, if not selected by default; refer to Figure-88.

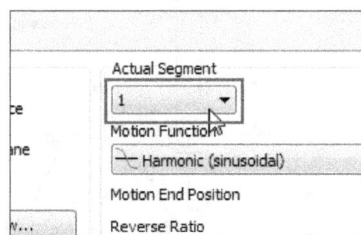

Figure-88. Actual Segment drop-down

- Click in the **Motion Function** drop-down and select desired motion type.
- Click in the **Motion End Position** edit box and specify the total angle span for current segment which is **120** for our current example.
- Select the **Lift at End** radio button to specify the lift in follower by the end of current segment which is **10** mm for current example. You can specify maximum speed, maximum acceleration, or maximum pressure angle in place of lift by using the respective radio button in the dialog box; refer to Figure-89.

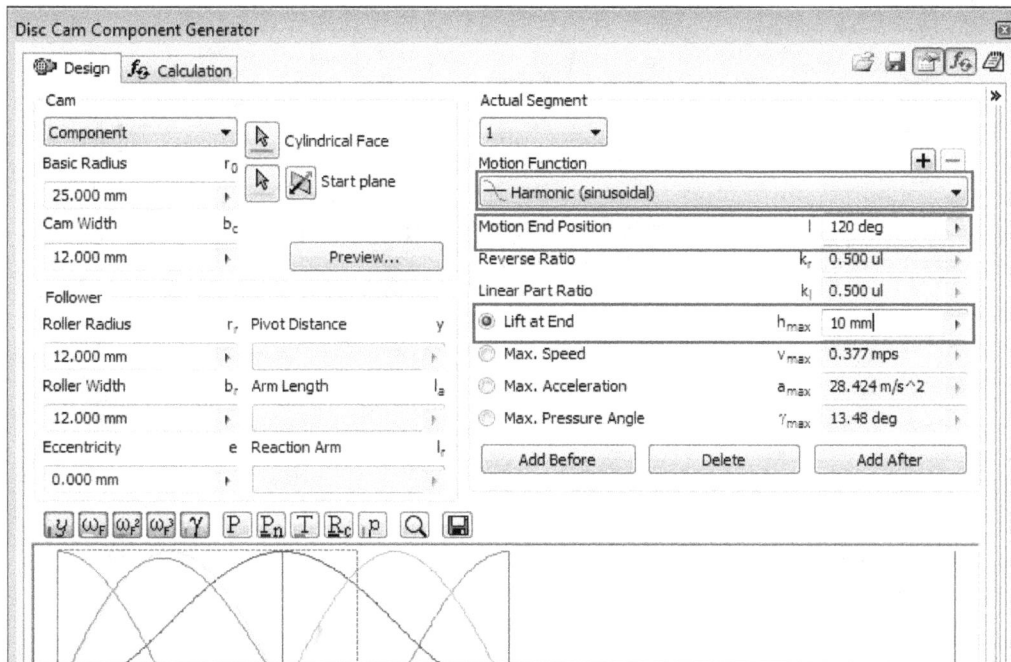

Figure-89. Cam profile parameters for segment 1

- Click in the **Actual Segment** drop-down and select the next segment. Specify the parameters in the same way.
- Note that the total motion of cam is **360** degree and you can make as many sections as you want in this total motion. To add a new motion segment, click on the **Add Before** or **Add After** button. A new segment will be added before or after the current segment as per the button selected.
- After performing the changes, click on the calculate button to check the preview of cam profile in the preview area; refer to Figure-90.

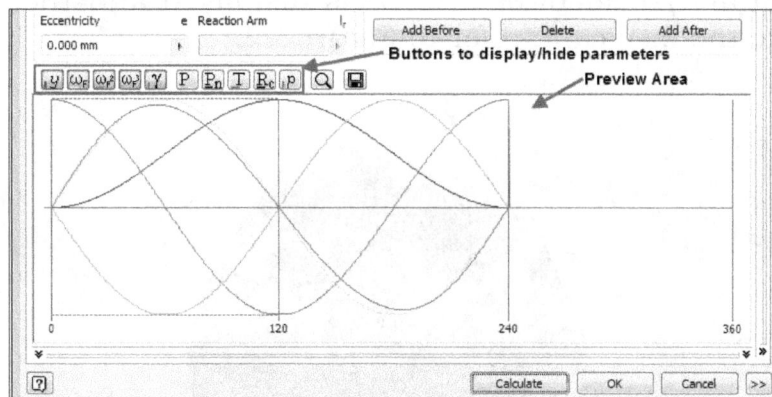

Figure-90. Preview of cam profile

- You can display or hide any parameter in the preview by using buttons overhead. You can save the graphical preview in the form of tab delimited text by using the **Save** button in the preview area.
- Click on the **Calculation** tab to check the feasibility of cam in real working environment. The dialog box will be displayed as shown in Figure-91.

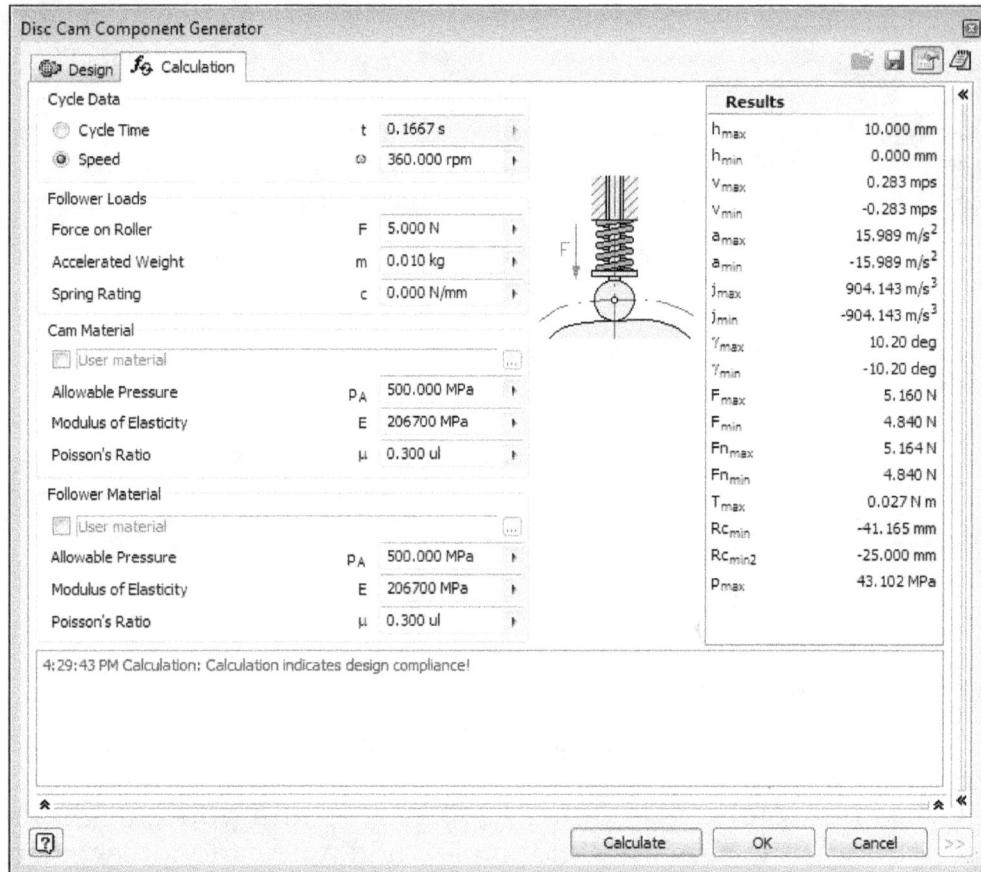

Figure-91. Calculation tab in Disc Cam Component Generator dialog box

- Specify the parameters like, speed of cam, loads on follower, material properties etc. and click on the **Calculate** button to check the report.
- Modify the cam and/or follower if error is displayed and then click on the **OK** button to create the disc cam; refer to Figure-92.

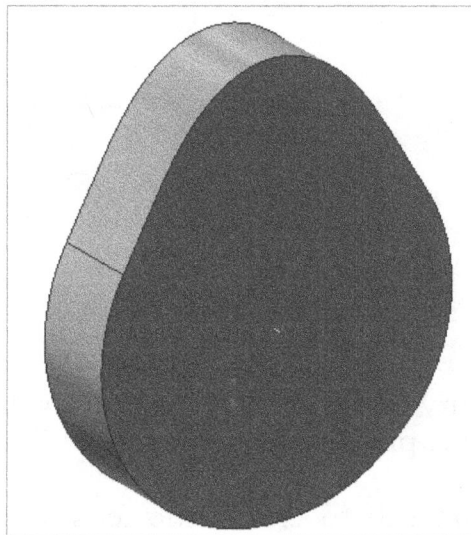

Figure-92. Disc cam created

Till this point, we have discussed various tools related to advanced assembly and design accelerator features/components. In the next chapter, we will learn more about the design accelerator features/components.

PRACTICE 1

In this practice problem, you need to create an assembly of vice as shown in Figure-93. The exploded view of the assembly is given in Figure-94. You can find the components of assembly in the resource kit of the book.

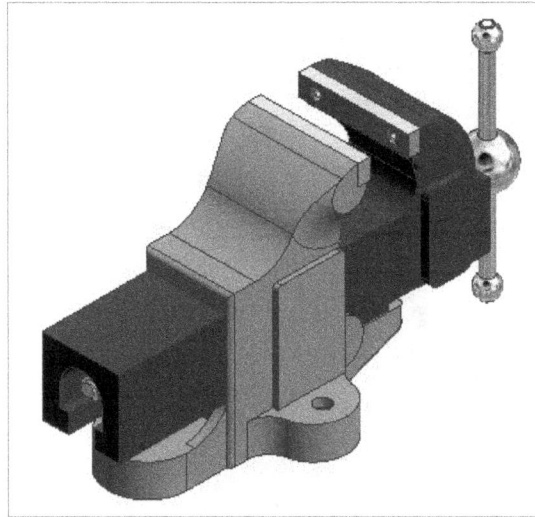

Figure-93. Vice

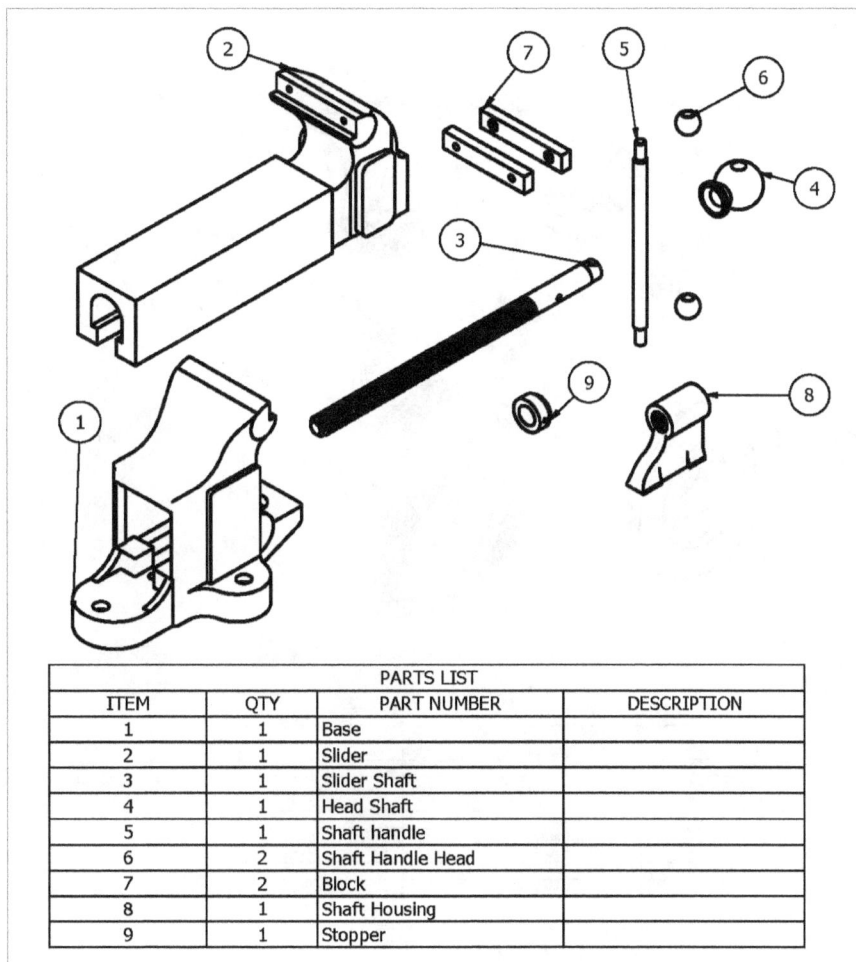

PARTS LIST			
ITEM	QTY	PART NUMBER	DESCRIPTION
1	1	Base	
2	1	Slider	
3	1	Slider Shaft	
4	1	Head Shaft	
5	1	Shaft handle	
6	2	Shaft Handle Head	
7	2	Block	
8	1	Shaft Housing	
9	1	Stopper	

Figure-94. Exploded view of vice

PRACTICE 2

In this practice problem, we will create an assembly of 3 Jaw Chuck as shown in Figure-95. Exploded view of the assembly is given in Figure-96. Note that you will find the assembly parts in resource kit for this book which can be asked at **cadcamcaeworks@gmail.com**.

Figure-95. 3 Jaw Chuck Assembly

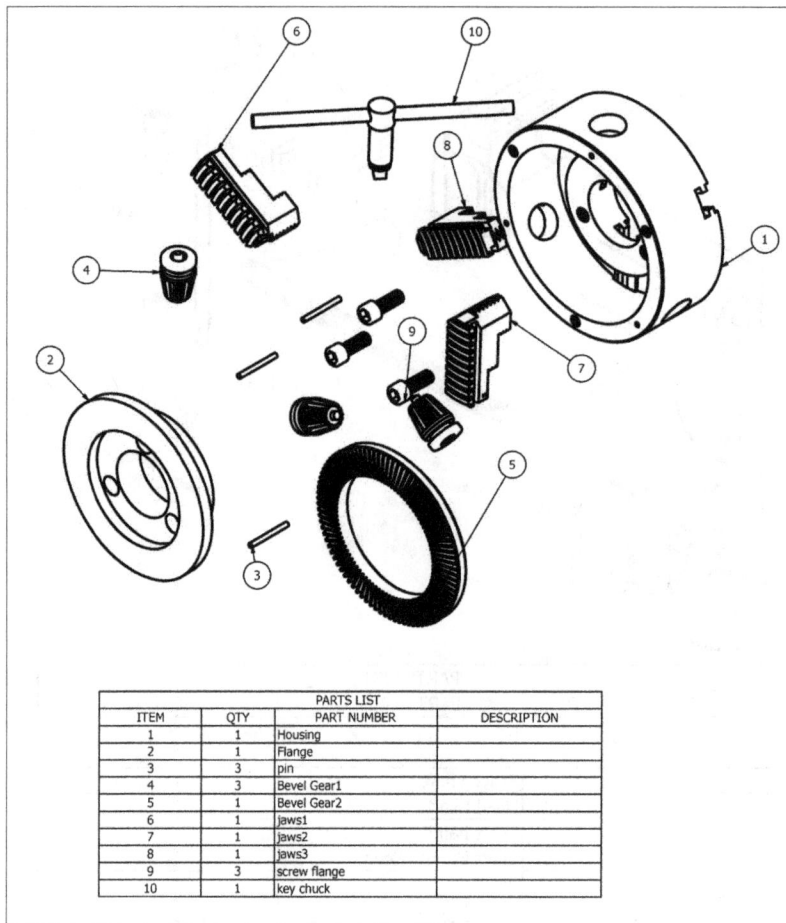

PARTS LIST			
ITEM	QTY	PART NUMBER	DESCRIPTION
1	1	Housing	
2	1	Flange	
3	3	pin	
4	3	Bevel Gear1	
5	1	Bevel Gear2	
6	1	jaws1	
7	1	jaws2	
8	1	jaws3	
9	3	screw flange	
10	1	key chuck	

Figure-96. Exploded view of 3 jaw chuck

PRACTICE 3

Design a cam, with a minimum radius of 25 mm, rotating clockwise at a uniform speed to give a roller follower, at the end of a valve rod, motion described below :

1. To raise the valve through 50 mm during 120° rotation of the cam ;
2. To keep the valve fully raised through next 30°;
3. To lower the valve during next 60°; and
4. To keep the valve closed during rest of the revolution i.e. 150°;

The diameter of the roller is 20 mm and the diameter of the cam shaft is 25 mm.

PRACTICE 4

A gear drive is required to transmit a maximum power of 22.5 kW. The velocity ratio is 1:2 and r.p.m. of the pinion is 200. The approximate centre distance between the shafts may be taken as 600 mm. The teeth has 20° stub involute profiles. Find the module, face width, and number of teeth on each gear. Take Factor of safety as 1.4.

PRACTICE 5

A worm drive transmits 15 kW at 2000 r.p.m. to a machine carriage at 75 r.p.m. The worm is triple threaded and has 65 mm pitch diameter. The worm gear has 90 teeth of 6 mm module. The tooth form is to be 20° full depth involute. The coefficient of friction between the mating teeth may be taken as 0.10. Find out : 1. tangential force acting on the worm and 2. axial thrust and separating force on worm.

SELF ASSESSMENT

Q1. Which of the following options should be selected from the **Bolted Connection Component Generator** dialog box so that the bolt created will be have nut on the other end?

a) Through All
b) Blind
c) Concentric
d) None of the Above

Q2. By selecting which of the following options, the **Clevis pin** can be placed in the assembly?

a) Concentric
b) By hole
c) Linear
d) All of the Above

Q3. Which type of pin is used to design and calculate pins loaded with torque?

a) Joint Pin
b) Cross Pin
c) Secure Pin
d) Radial Pin

Q4. What are the standard lengths of shafts?

a) 5m, 6m, and 7m
b) 3m, 5m, and 6m
c) 4m, 7m, and 8m
d) 6m, 7m, and 8m

Q5. What are the maximum permissible working stresses that can be taken for shafts in tension or compression?

a) 96 MPa without allowance and 86 MPa with allowance
b) 112 MPa without allowance and 84 MPa with allowance
c) 124 MPa without allowance and 96 MPa with allowance
d) 118 MPa without allowance and 88 MPa with allowance

Q6. What are the maximum permissible shear stresses that can be taken for shafts?

a) 66 MPa without allowance and 24 MPa with allowance
b) 72 MPa without allowance and 40 MPa with allowance
c) 56 MPa without allowance and 42 MPa with allowance
d) 48 MPa without allowance and 32 MPa with allowance

Q7. Which of the following loads cannot be applied in the **Shaft Component Generator** dialog box?

a) Radial Force
b) Tension Force
c) Axial Force
d) Bending Moment

Q8. Which of the following failures of gear teeth has a stick-slip phenomena in designing gear drive?

a) Bending Failure
b) Pitting
c) Scoring
d) Abrasive wear

Q9. Which of the following is the correct gear ratio in designing spur gears according to engineering equations?

a) T1 = G / T2
b) T2 = G / T1
c) G = T1 / T2
d) G = T2 / T1

Q10. The Synchronous Belt is also called as-

a) Timing Belt
b) Toothed Belt
c) Cogged Belt
d) All of the Above

FOR STUDENT NOTES

Chapter 9

Advanced Assembly and Design-II

Topics Covered

The major topics covered in this chapter are:

- *Designing Linear and Cylindrical Cam*
- *Designing Splines*
- *Performing Design Related Calculations*
- *Designing Frames*
- *Measurement tools*

INTRODUCTION

In previous chapter, we learnt about some advanced tools for assembly editing. We have also learned about various design accelerator components. In this chapter, we will continue discussing about various design accelerator components used in machineries.

POWER TRANSMISSION COMPONENTS

We have discussed about various power transmission components like shaft, gear, bearing, belts, keys, and so on. We will now continue discussing about rest of the power transmission components.

Designing Linear Cam

Linear Cam is also called flat plate cam. In this system, follower moves up and down as the linear cam moves left and right; refer to Figure-1. The procedure to design linear cam in Autodesk Inventor is given next.

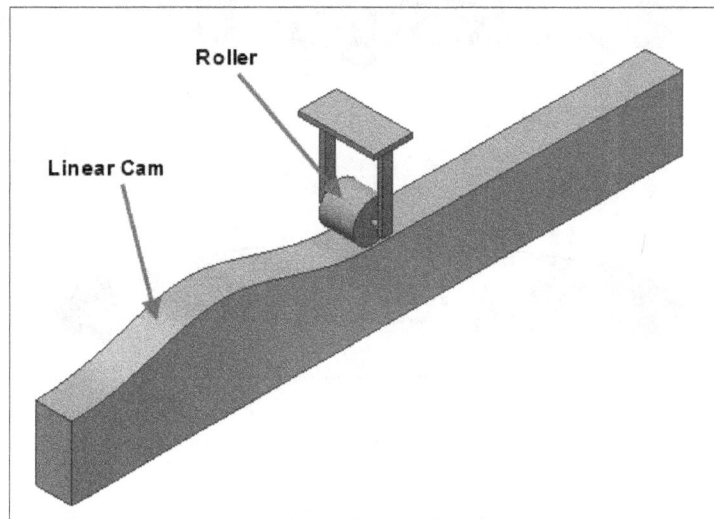

Figure-1. Linear cam

- Click on the **Linear Cam** tool from **Cam** drop-down in the **Power Transmission** panel of the **Design** tab in the **Ribbon**. The **Linear Cam Component Generator** dialog box will be displayed; refer to Figure-2.
- Specify the cam width and length in the **Cam Width** and **Motion Length** edit boxes, respectively.
- Similarly, specify the parameters for follower in the **Follower** area of the dialog box.
- Create the motion path of cam as created in previous chapter under **Disc Cam** topic.
- Click on the **More options** button at the bottom right corner of the dialog box to expand the dialog box; refer to Figure-3.

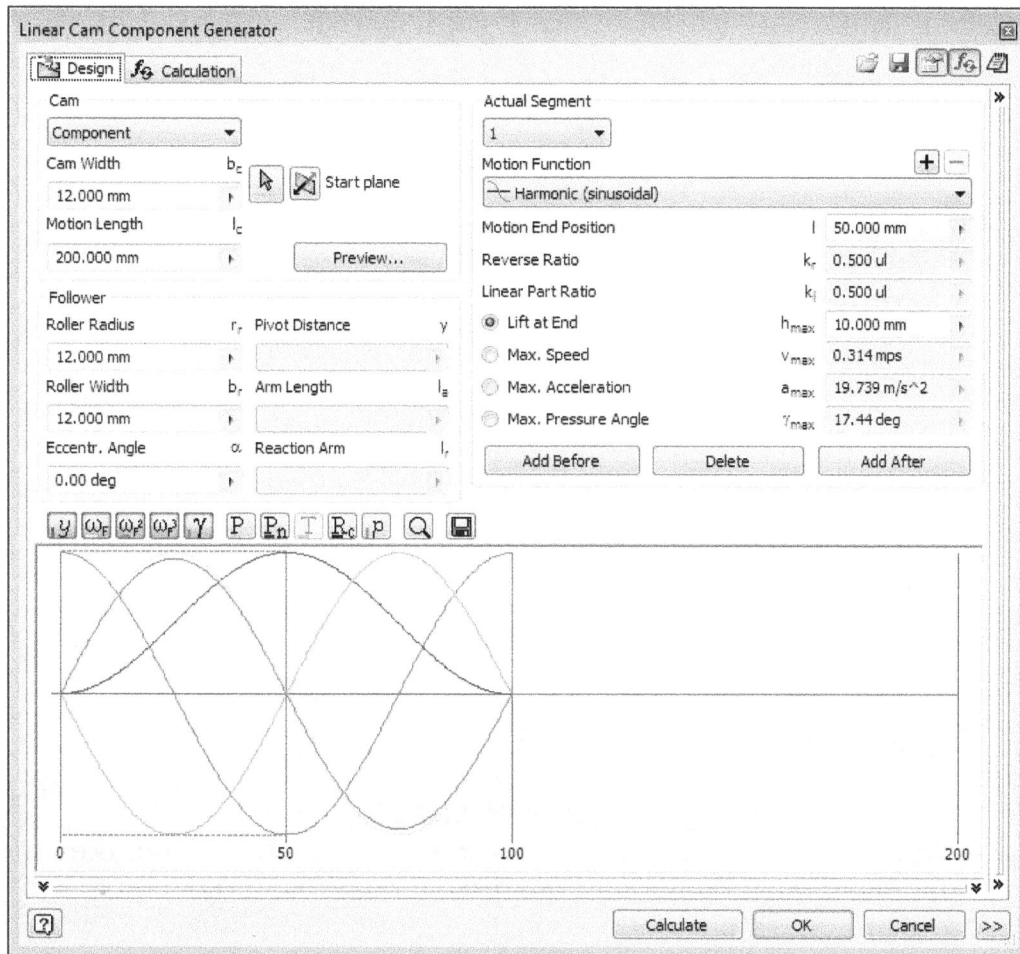

Figure-2. Linear Cam Component Generator dialog box

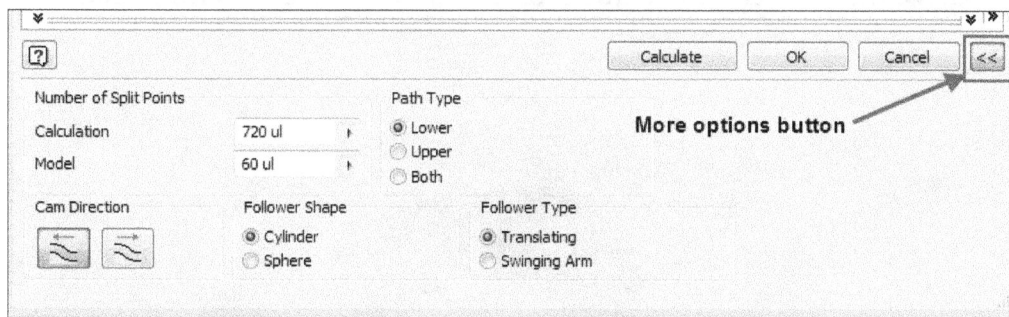

Figure-3. Expanded dialog box

- Set the advanced parameters for cam in the expanded dialog box and click on the **Calculation** tab. The dialog box will be displayed as shown in Figure-4.
- Set the parameters for calculation in the edit boxes of the dialog box and click on the **Calculate** button. Modify the cam and follower if error occurs otherwise click on the **OK** button to create the cam.

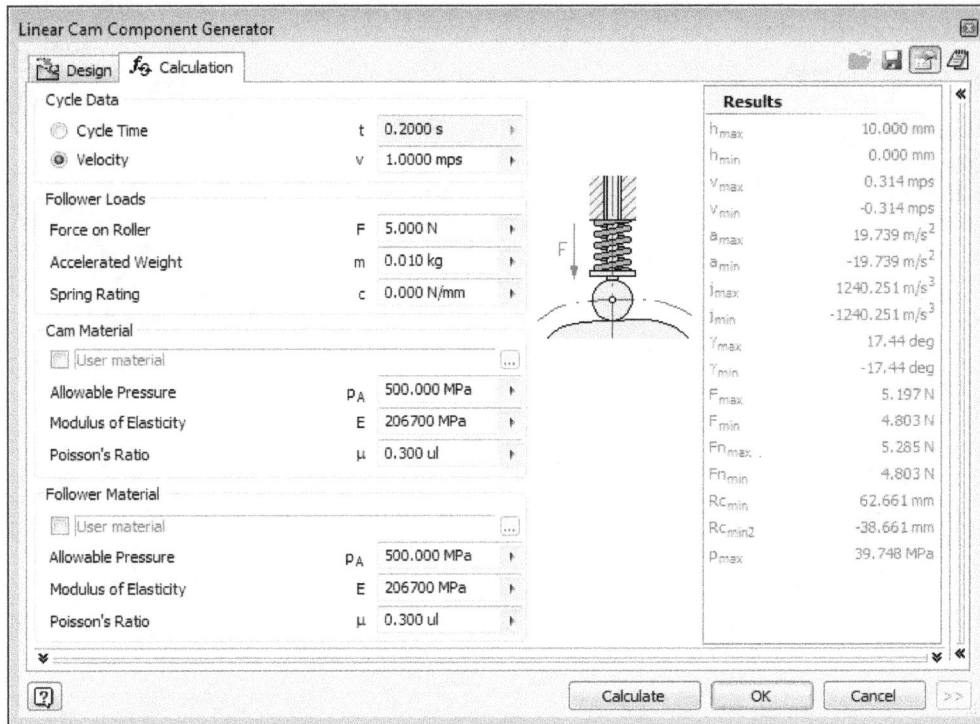

Figure-4. Calculation tab in Linear Cam Component Generator dialog box

Designing Cylindrical Cam

In Cylindrical Cam/Barrel Cam, a cylinder cam profile rotates and makes the follower moves upwards/downward; refer to Figure-5. These unusual cams are normally composed of a cylinder which has a groove cut out of its surface and it is in this that the follower runs up and down; refer to Figure-6. This type of cam can be seen in some old clock mechanisms and still in modern sewing machines. Machines that perform repetitive movements may use a cylinder cam profile. The procedure to design cylindrical cam is same as disc cam.

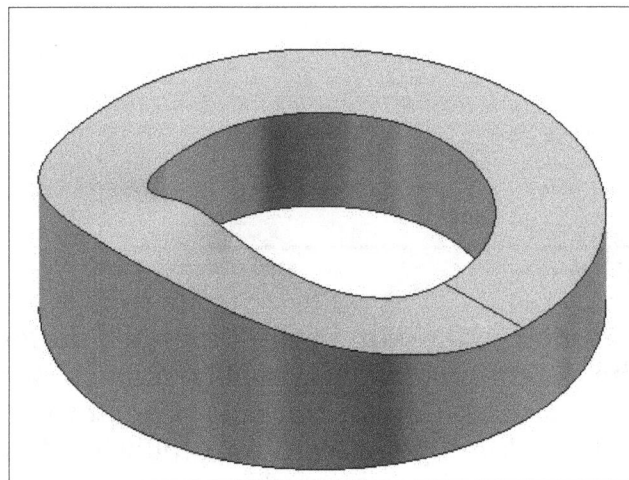

Figure-5. Cylindrical Cam

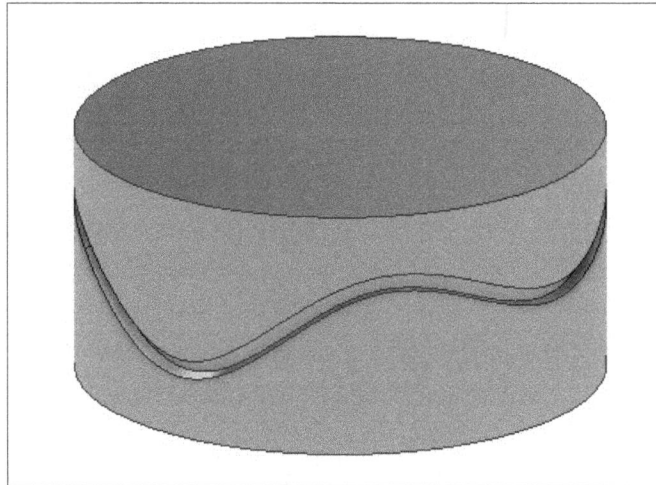

Figure-6. Cylindrical cam

Designing Parallel Spline

Splines are ridges or teeth on a drive shaft that mesh with grooves in a mating piece and transfer torque to it, maintaining the angular correspondence between them. The procedure to design parallel spline is given next. Note that to use this tool, you must have a shaft and a hub already in the drawing area.

- Click on the **Parallel Splines** tool from the **Spline** drop-down in the **Power Transmission** panel in the **Design** tab of the **Ribbon**. The **Parallel Splines Connection Generator** dialog box will be displayed as shown in Figure-7.

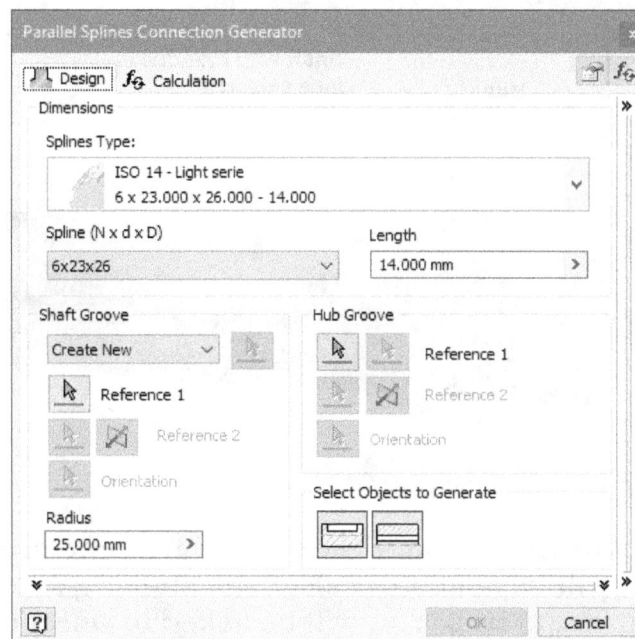

Figure-7. Parallel Splines Connection Generator dialog box

- Click in the **Splines Type** drop-down and select desired type of spline; refer to Figure-8.

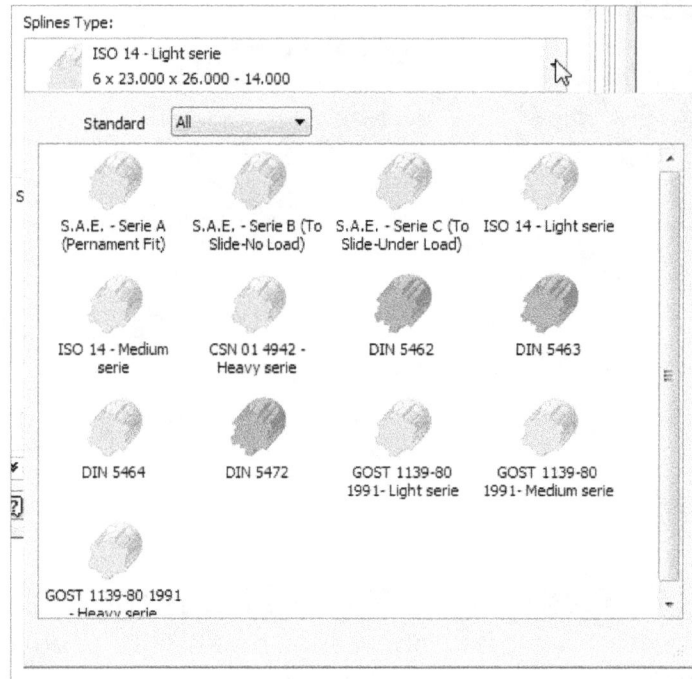

Figure-8. Splines Type drop-down

- Click in the **Spline (N x d x D)** drop-down and select desired size of spline.
- Select the **Create New** option from the drop-down in the **Shaft Groove** area of the dialog box to create spline grooves on the shaft. You will be asked to select the cylindrical face of the shaft.
- Select the cylindrical face of shaft in the drawing area; refer to Figure-9. You will be asked to select starting face for the spline.

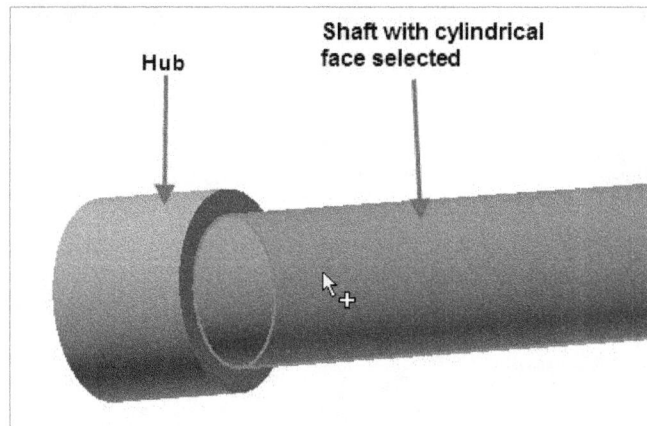

Figure-9. Cylindrical face of shaft selected

- Select the start face for the spline. Preview of the spline grooves on shaft will be displayed. Also, you will be asked to select the starting face of hub.
- Click on the starting face of hub. You will be asked to select center point for spline on hub.
- Select a work point or edge of a circular face to select its center. Preview of the spline connection will be displayed; refer to Figure-10.

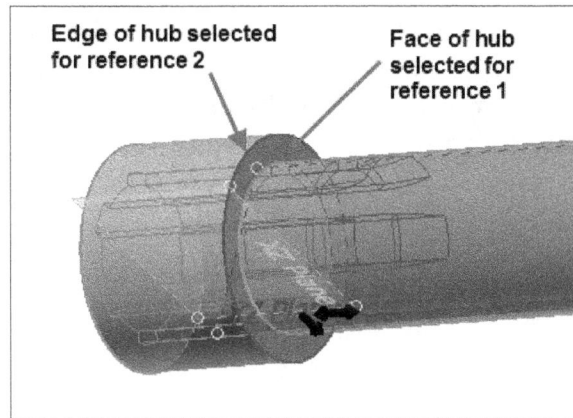

Figure-10. Preview of spline connection

- Specify desired value of runout radius in the **Radius** edit box of the dialog box. Run out radius should not be confused with the run out used in GD&T for deviation in geometric center and actual center of shaft, it is the radius provided at the ends of spline grooves for soft locking of hub and shaft.
- Specify the length of spline grooves on the shaft in **Length** edit box.
- Click on the **Calculation** tab to check the feasibility of the spline connection. The dialog box will be displayed as shown in Figure-11.

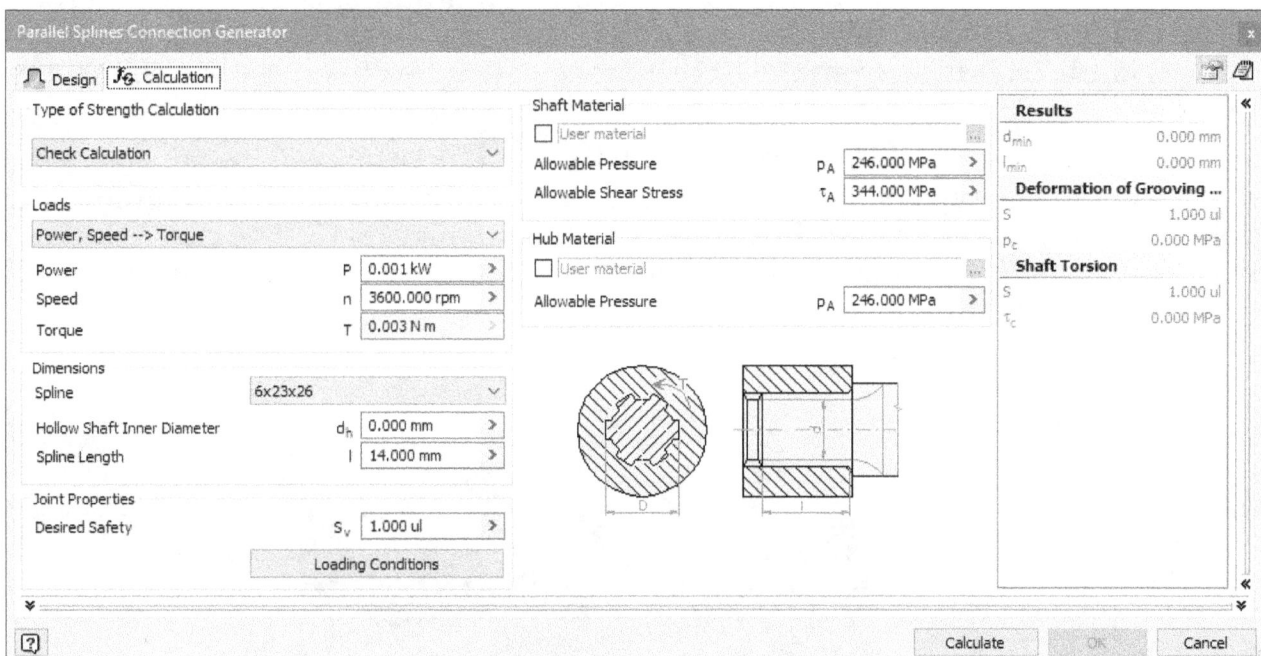

Figure-11. Calculation tab in Parallel Splines Connection Generator dialog box

- Select desired option from the **Type of Strength Calculation** drop-down. There are three options in the drop-down; **Check Calculation**, **Length Design**, and **Diameter Design**. Select the **Check Calculation** option if you know all the parameters of spline connection and just want to check whether the connection will do the job or not. Select the **Length Design** option if you want the length of spline to be decided by system based on the load calculations. Select the **Diameter Design** option if you want the diameter of spline to be decided by system based on the load calculations.
- Select desired option from the **Loads** drop-down and specify the parameters in the edit boxes of **Loads** area.

- Similarly, apply the material properties to shaft and hub by using the options in the **Shaft Material** and **Hub Material** area of the dialog box.
- Click on the **Calculate** button and check the results. If there is some error then modify the shaft and hub splines accordingly.
- Click on the **OK** button to create the spline connection. The **File Naming** dialog box will be displayed; refer to Figure-12.

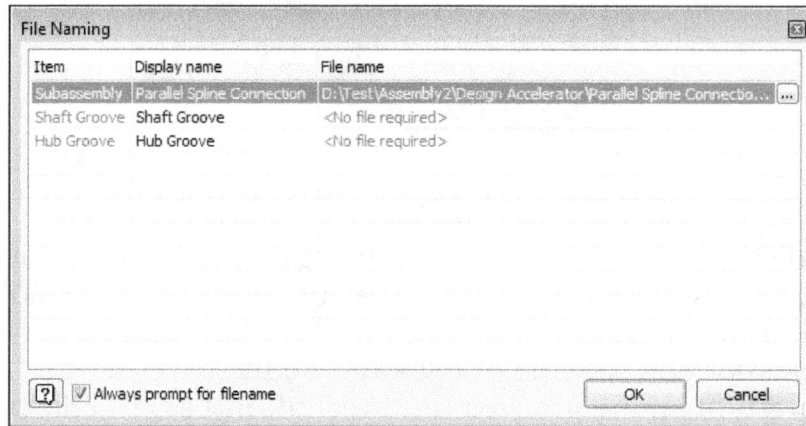

Figure-12. File Naming dialog box

- Double-click on the field under **File name** column and specify the location and name of the sub-assembly file.
- Click on the **OK** button from the **File Naming** dialog box to create the connection.

Designing Involute Spline

Involute spline is a type of spline where the sides of the equally spaced grooves are involute, as in an involute gear, but not as tall; refer to Figure-13. The curves increase strength by decreasing stress concentrations. The procedure to create involute spline is same as for Parallel spline.

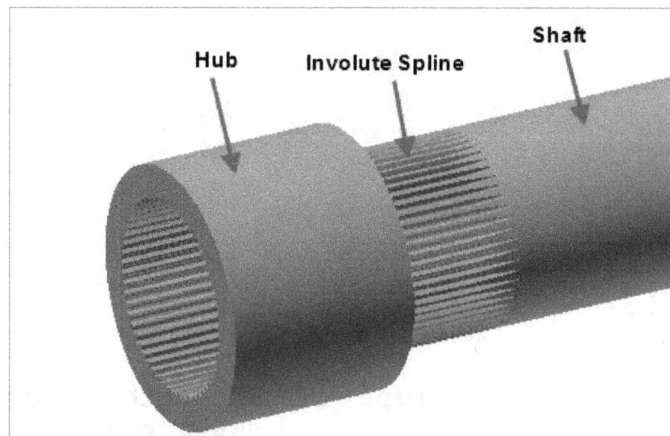

Figure-13. Involute spline created

Generating O-Ring

An O-ring, also known as a packing, or a toric joint, is a mechanical gasket in the shape of a torus; it is a loop of elastomer with a round cross-section, designed to be seated in a groove and compressed during assembly between two or more parts, creating a seal at the interface. The O-ring may be used in static applications or in dynamic applications where there is relative motion between the parts and the O-ring. Dynamic examples include rotating pump shafts and hydraulic cylinder pistons.

The procedure to generate O-ring in Inventor is given next.

*   Click on the **O-Ring** tool from **Power Transmission** panel in the **Design** tab of the **Ribbon**. The **O-Ring Component Generator** dialog box will be displayed; refer to Figure-14. Also, you will be asked to select a cylindrical surface, planar face, or work plane.

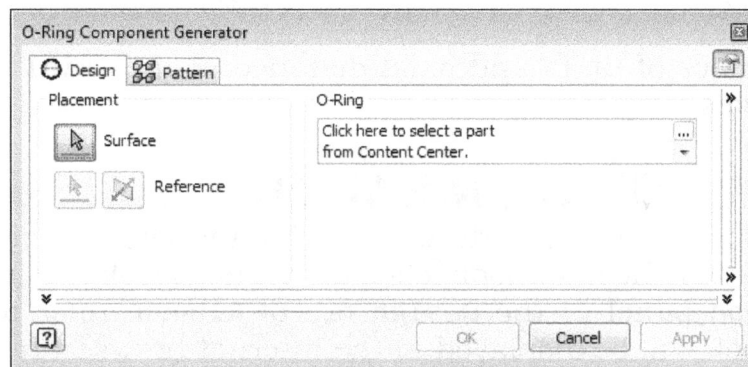

Figure-14. O-Ring Component Generator dialog box

*   Select desired reference, you will be asked to specify positional reference for the O-ring.
*   Select the positional reference. Note that if you have selected cylindrical surface earlier then you will be prompted to select planar face/work plane for positional reference and if you have selected planar face/work plane earlier then you will be prompted to select edge, point, or axis for positional reference.
*   After selecting the references, click in the **Browse for o-ring** drop-down in the **O-Ring** area of the dialog box. The options in the drop-down will be displayed as shown in Figure-15.

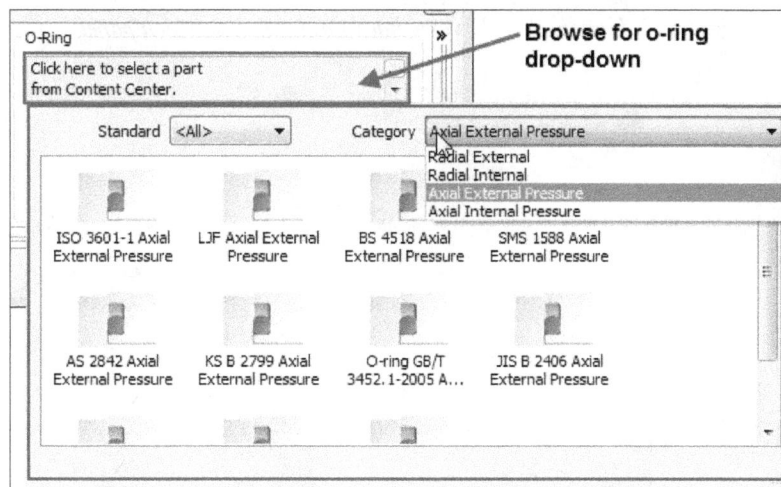

Figure-15. O-Ring drop-down

- Click in the **Category** drop-down and select desired category of rings and then select the o-ring from list displayed. Preview of the ring will be displayed.
- If you want to create more than one rings then click on the **Pattern** tab and select the **Axial** radio button from the dialog box; refer to Figure-16.

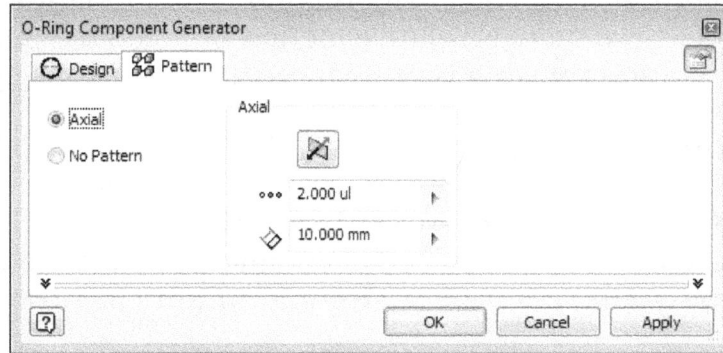

Figure-16. Pattern tab in O-Ring Component Generator dialog box

- Specify the number of units and axial distance in the **Axial count** and **Axial Spacing** edit boxes, respectively.

DESIGN HANDBOOK

The **Handbook** tool is used to display the engineer's handbook for common mechanical components. To display the handbook, click on the **Handbook** tool from the expanded **Power Transmission** panel in the **Design** tab of **Ribbon**; refer to Figure-17. The Engineer's Handbook will be displayed in the internet browser; refer to Figure-18.

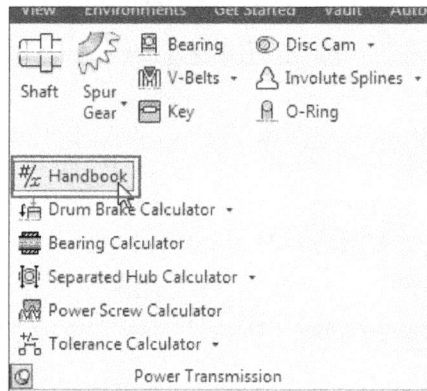

Figure-17. Handbook tool in expanded Power Transmission panel

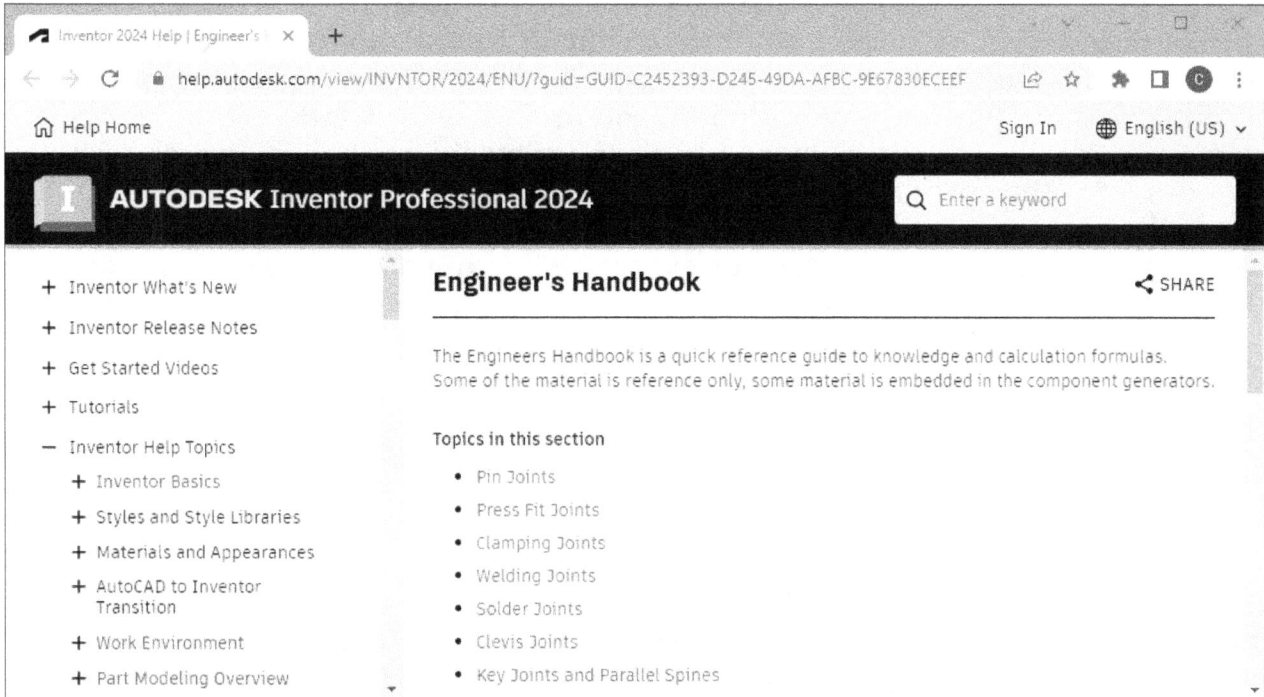

Figure-18. Engineer's Handbook in browser

Click on desired topic in the browser. Related calculation formulae will be displayed; refer to Figure-19.

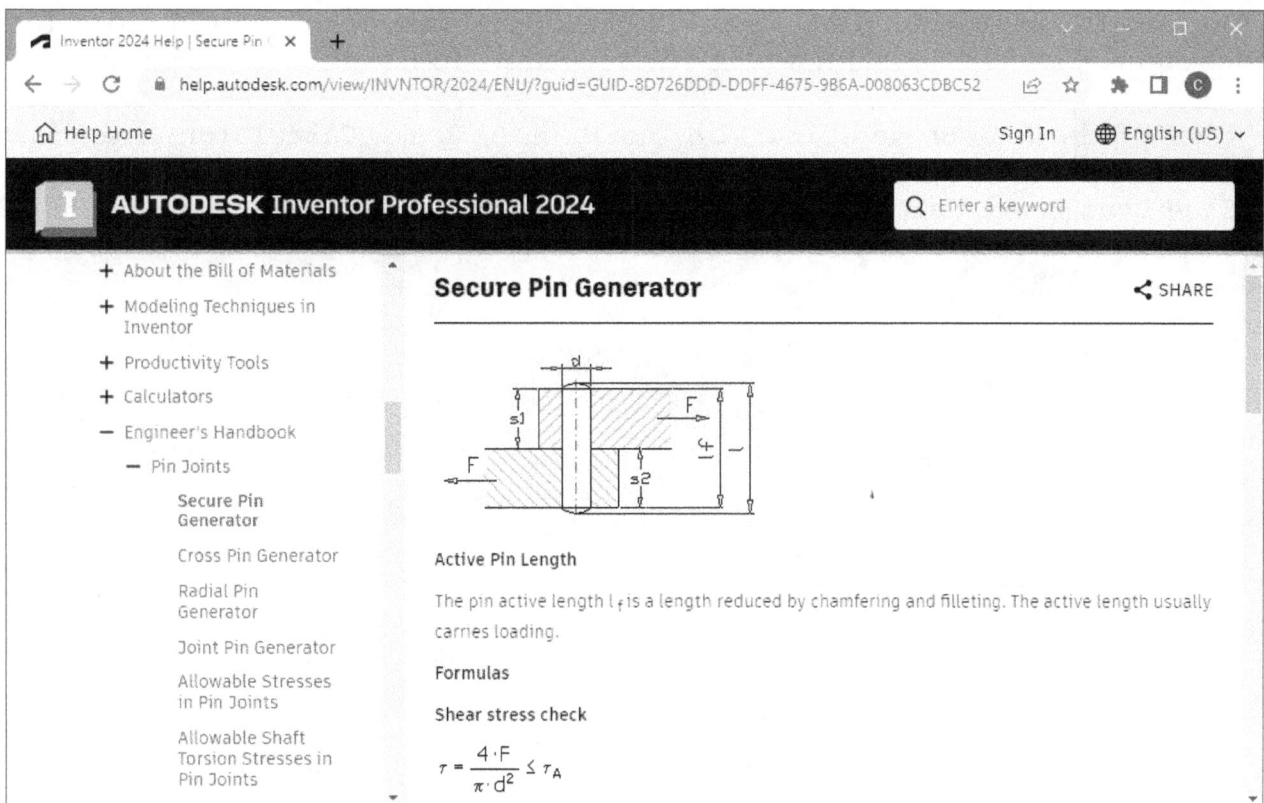

Figure-19. Secure Pin Generator calculations

BRAKE CALCULATORS

A brake is a device by means of which artificial frictional resistance is applied to a moving machine member, in order to retard or stop the motion of a machine. In the process of performing this function, the brake absorbs either kinetic energy of the moving member or potential energy given up by objects being lowered by hoists, elevators, etc. The energy absorbed by brakes is dissipated in the form of heat. This heat is dissipated in the surrounding air (or water which is circulated through the passages in the brake drum), so that excessive heating of the brake lining does not take place. The design or capacity of a brake depends upon the following factors :

1. The unit pressure between the braking surfaces,
2. The coefficient of friction between the braking surfaces,
3. The peripheral velocity of the brake drum,
4. The projected area of the friction surfaces, and
5. The ability of the brake to dissipate heat equivalent to the energy being absorbed.

There are mainly four type of brakes used in mechanical engineering; Drum brake, Disc brake, Band brake, and Cone brake. In Autodesk Inventor, we have tools to calculate various parameters related to these brakes. These tools and their related calculations are discussed next.

Drum Brake Calculator

The **Drum Brake Calculator** tool is used to calculate various parameters related to drum braking like friction force, moment of inertia, brake revolution number, total time required for braking, and so on. The procedure to use this tool is given next.

- Click on the **Drum Brake Calculator** tool from the **Brake Calculators** drop-down in the expanded **Power Transmission** panel in the **Design** tab of the **Ribbon**. The **Shoe Drum Brake Calculator** dialog box will be displayed as shown in Figure-20.

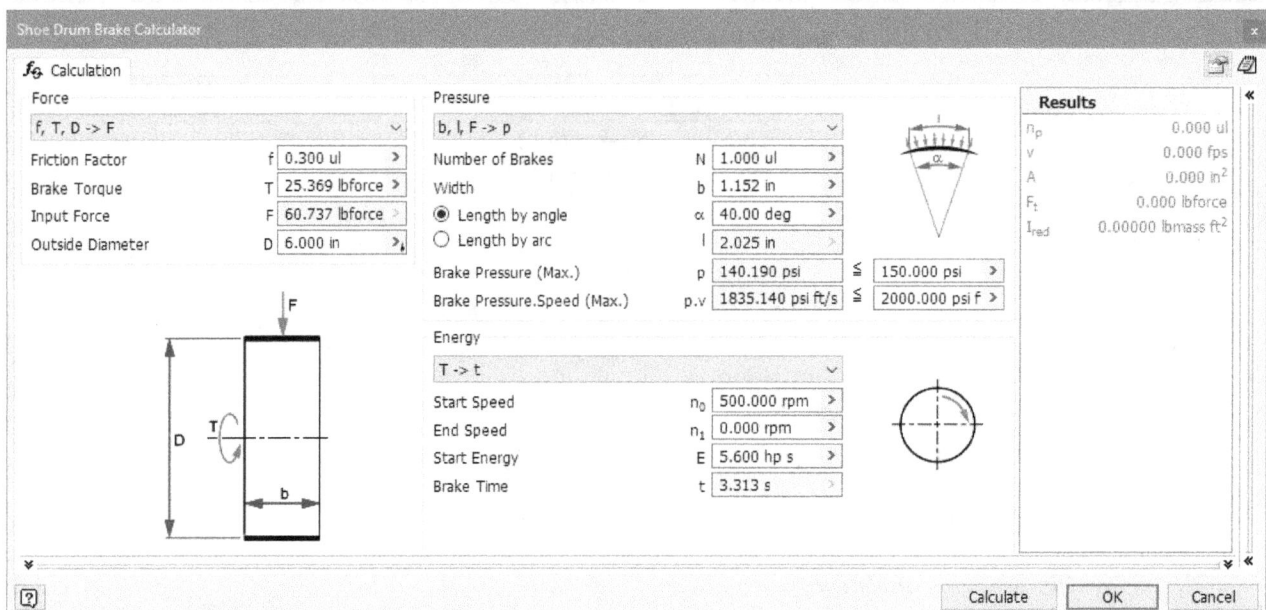

Figure-20. Shoe Drum Brake Calculator dialog box

- Click in the **Force** drop-down and select desired option. There are four options in the drop-down; **f,T,D->F**, **T,D,F->f**, **f,D,F->T**, and **f,T,F->D**. Note that the parameters which are on the left side of arrow in these options are specified by user and the parameter on the other side of arrow is calculated by system based on the inputs.
- Specify desired parameters in the **Force** area of the dialog box.
- Similarly, specify the parameters in **Pressure** and **Energy** areas of the dialog box.
- Click on the **Calculate** button and check the results in **Results** area and driven edit boxes; refer to Figure-21.

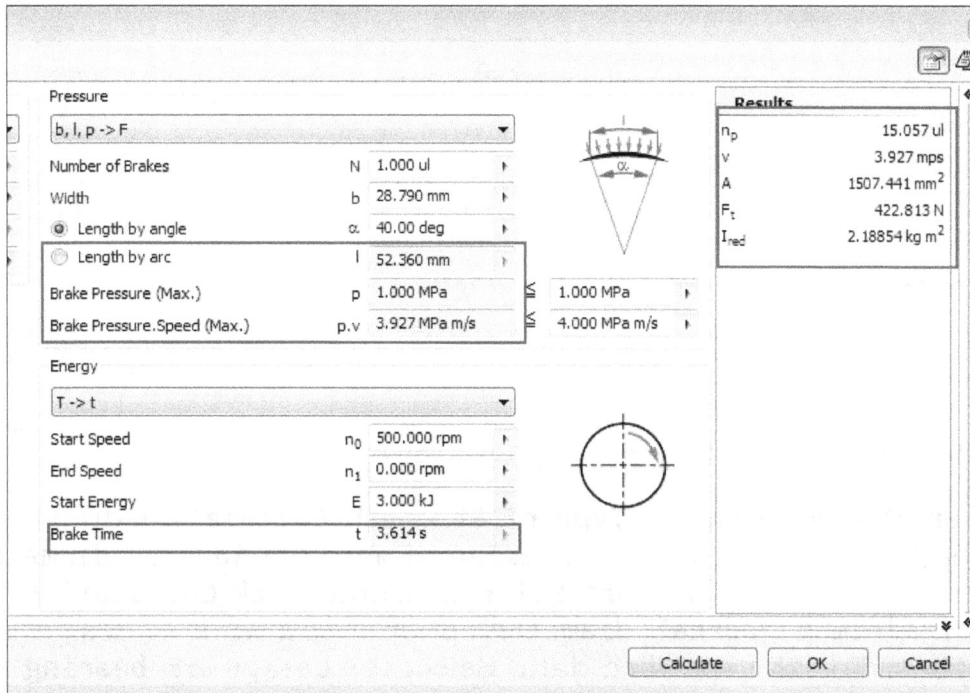

Figure-21. Results of shoe drum brake calculator

- Click on the **OK** button to add calculations of shoe drum brake in the **Model Browser Bar**.

Disc Brake, Band Brake, and Cone Brake Calculator

The **Disc Brake Calculator**, **Band Brake Calculator**, and **Cone Brake Calculator** tools are used to calculate parameter related to disc brake, band brake, and cone brake. The procedure to use these tools is similar to the procedure discussed in previous topic.

BEARING CALCULATOR

The **Bearing Calculator** tool is used to calculate parameters related to journal bearing like, minimum journal diameter, bearing clearance design, lubricant selection and so on. The procedure to use the **Bearing Calculator** tool is given next.

- Click on the **Bearing Calculator** tool from the expanded **Power Transmission** panel in the **Design** tab of the **Ribbon**. The **Plain Bearing Calculator (SI units)** dialog box will be displayed; refer to Figure-22.

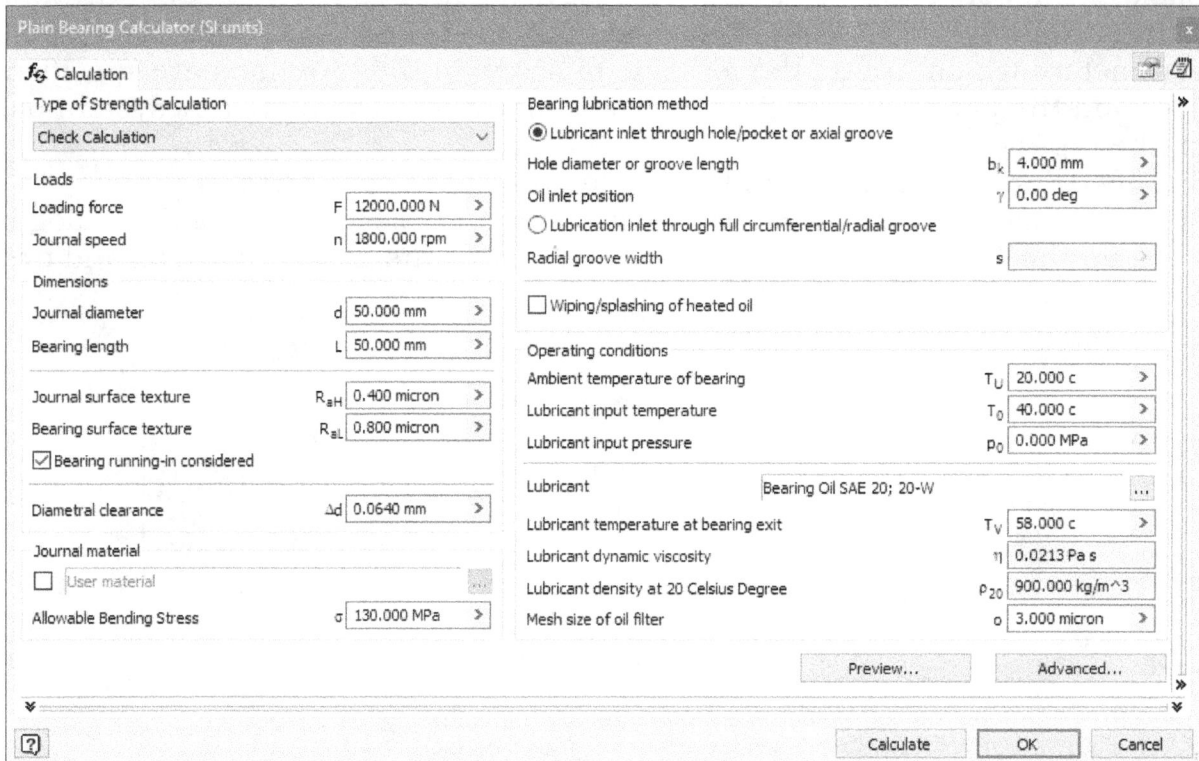

Figure-22. Plain Bearing Calculator dialog box

- Select desired option from the **Type of Strength Calculation** drop-down. There are four options in this drop-down; **Design of minimum journal diameter**, **Design of bearing clearance**, **Lubricant Selection**, and **Check Calculation**. Select the **Design of minimum journal diameter** option if you want to design the journal diameter based on the specified data. Select the **Design of bearing clearance** option if you want to design the bearing clearance based on the parameters specified in calculator. Select the **Lubricant Selection** option if lubricant type is to be selected based on the specified parameters. Select the **Check Calculation** option from the drop-down if you want to check the feasibility of journal bearing under current loading conditions.
- Specify the parameters based on the selected option and click on the **Calculate** button. If an error occurs then modify the parameters accordingly.
- Click on the **OK** button to add calculation in **Model Browser Bar**. The **File Naming** dialog box will be displayed. Specify the location for file and then click on the **OK** button.

SEPARATED HUB CALCULATOR

Hubs are used to transmit mechanical power from a drive motor by coupling it to an output device such as a wheel or an arm. A separated hub is made of two halves joined by bolts. The **Separated Hub Calculator** tool in Autodesk Inventor is used to check the feasibility of designed hub under specified loading conditions. The procedure to use this tool is given next.

- Click on the **Separated Hub Calculator** tool from the **Hub Calculator** drop-down in the expanded **Power Transmission** panel in the **Design** tab of the **Ribbon**. The **Separated Hub Joint Calculator** dialog box will be displayed; refer to Figure-23.

- Select desired option from the **Type of Strength Calculation** drop-down in the dialog box. There are three options in this drop-down; **Hub Length Design**, **Shaft Diameter Design**, and **Check Calculation**. Select the **Hub Length Design** option if you want to design the length of hub based on the loads specified. Select the **Shaft Diameter Design** if you want to modify the shaft diameter based specified loading conditions. If you want to check the feasibility of separated hub for the specified loading conditions and parameter then select the **Check Calculation** option.

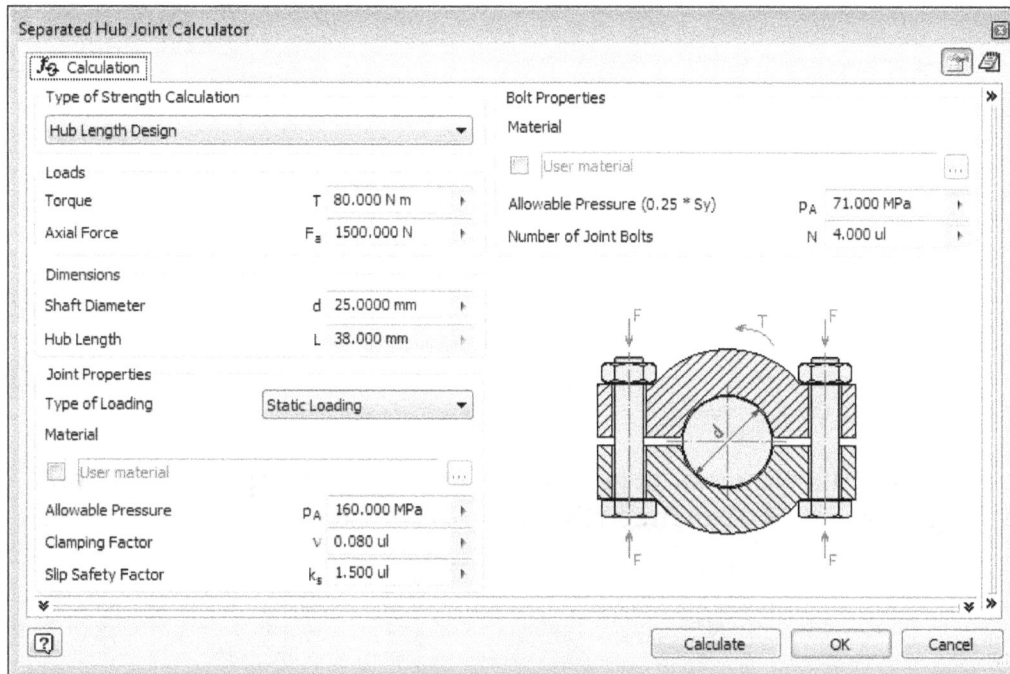

Figure-23. Separated Hub Joint Calculator dialog box

- Specify the load values and dimensions in the **Loads** and **Dimensions** areas, respectively.
- Select desired type of loading from the **Type of Loading** drop-down in the **Joint Properties** area of the dialog box. There are three options in this drop-down; **Static Loading**, **Repeated Loading**, and **Alternating Loading**. If the load to be applied is stable then select the **Static Loading** option. If the loading is repeated in small intervals then select the **Repeated Loading** option. If the loading inverse after small intervals then select the **Alternating Loading** option.
- Specify the other parameters and click on the **Calculate** button. Check the results if there is any error then modify the parameters accordingly.
- Click on the **OK** button to add the calculation to **Model Browser Bar**. Click **OK** in the **File Naming** dialog box displayed.

In the same way, you can use the **Slotted Hub Calculator** and **Cone Joint Calculator** tools available in the **Hub** drop-down and **Power Screw Calculator** tool in the expanded **Power Transmission** panel.

TOLERANCE CALCULATOR

The **Tolerance Calculator** tool is used to find out the tolerance of a dimension based on the other related dimensions. The procedure to use this tool is given next.

- Click on the **Tolerance Calculator** tool from the **Fit/Tolerance Calculator** drop-down in the expanded **Power Transmission** panel in the **Design** tab of the **Ribbon**. The **Tolerance Calculator** dialog box will be displayed as shown in Figure-24.

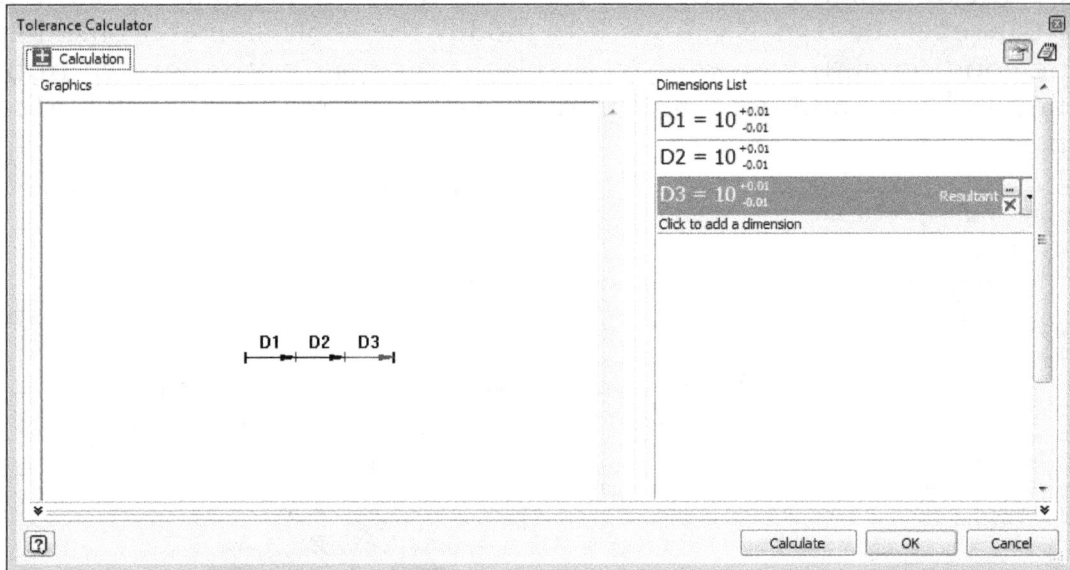

Figure-24. Tolerance Calculator dialog box

- Double-click on the dimension you want to change in the **Dimensions List** area of the dialog box. The **Tolerance** dialog box will be displayed; refer to Figure-25.

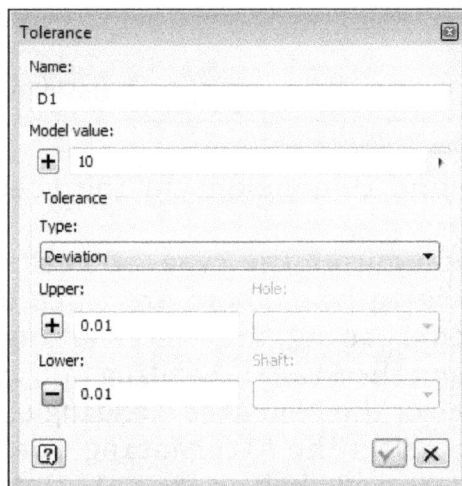

Figure-25. Tolerance dialog box

- Specify desired model value and tolerance values in the related edit boxes. Click on the **OK** button to apply the changes.
- Click on the **Click to add a dimension** option from the dialog box to add more dimensions. Note that the dimension value with cyan color background is the resultant dimension, so you need to drag all the dimensions above it to calculate their resultant.
- Now, click on the **Calculate** button to find out the resultant dimension with equal tolerance. The value of resultant dimension will be displayed in the **Dimensions List** area and graphical form of dimension will be displayed in the **Graphics** area of the dialog box.

LIMITS/FITS CALCULATOR

The **Limits/Fits Calculator** tool is used to check the appropriate limit/fit values for the shaft and hole. The procedure to use this tool is given next.

- Click on the **Limits/Fits Calculator** tool from the **Fit/Tolerance Calculator** drop-down in the expanded **Power Transmission** panel of the **Design** tab in the **Ribbon**. The **Limits and Fits Mechanical Calculator** dialog box will be displayed; refer to Figure-26.

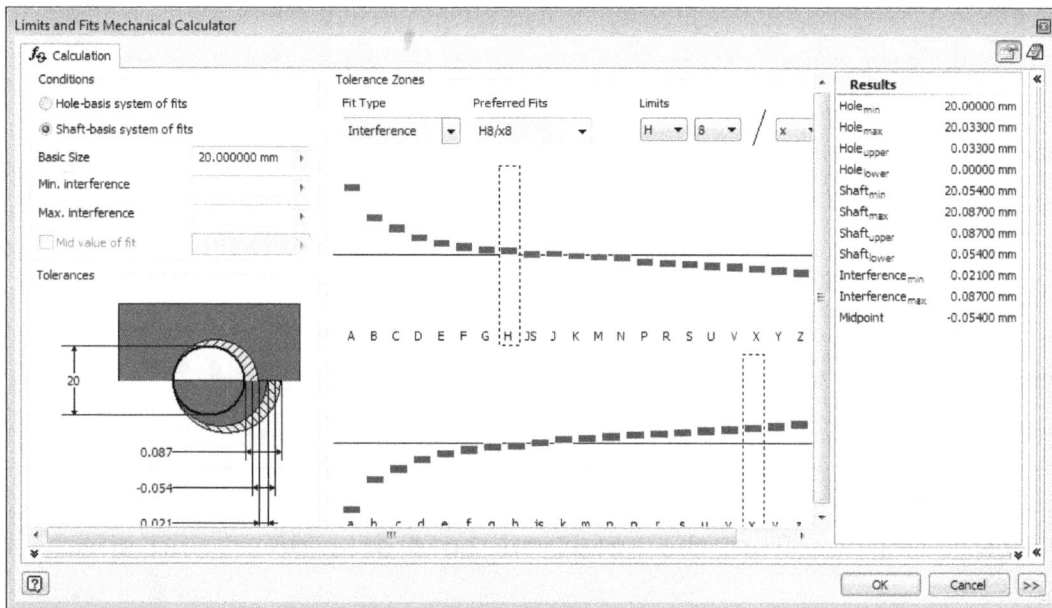

Figure-26. Limits and Fits Mechanical Calculator dialog box

- Select desired radio button from the **Conditions** area. There are two radio buttons; **Hole-basis system of fits** and **Shaft-basis system of fits**. Select the **Hole-basis system of fits** radio button if you want to design limits/fits based on hole. If you want to design limits/fits based on shaft then select the other radio button.
- Specify the diameter of shaft or hole, based on radio button selected, in the **Basic Size** edit box.
- Specify the minimum interference and maximum interference in the respective edit boxes or you can define the mid value of fit by selecting the **Mid value of fit** check box and specifying the value in related edit box.
- Select desired option from the **Fit Type** drop-down, the most relevant fits and limits will be selected automatically in the **Preferred Fits** and **Limits** drop-downs.
- Click on the **OK** button to add the Limits and Fits calculations in the **Model Browser Bar**. The **File Naming** dialog box will be displayed. Click on the **OK** button.

PRESS FIT CALCULATOR

The **Press Fit Calculator** tool is used to check the parameters related to press fitting of shaft in the hole. Press fitting of shaft is based on the thermal expansion of metals. When we heat up the hole and insert shaft in it then after gradual cooling, the hole contracts and tightly holds the shaft. The procedure to use this tool is given next.

- Click on the **Press Fit Calculator** tool from the **Fit/Tolerance** drop-down in the expanded **Power Transmission** panel of the **Design** tab in **Ribbon**. The **Press Fit Calculator** dialog box will be displayed; refer to Figure-27.

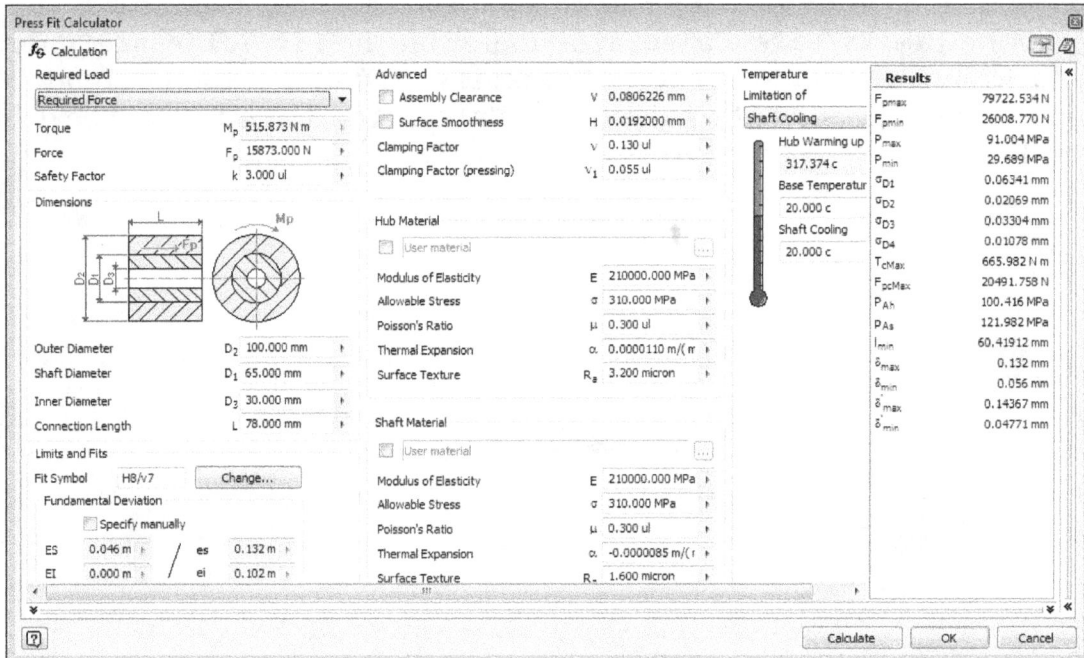

Figure-27. Press Fit Calculator dialog box

- Select desired option from the **Required Load** drop-down and specify the load parameters.
- Specify the basic diameters of shaft and hole, and the connection length in the **Dimensions** area of the dialog box.
- To change the limit/fit of shaft-hole connection, click on the **Change** button in the **Limits and Fits** area of the dialog box. The **Limits and Fits Mechanical Calculator** dialog box will be displayed as discussed earlier. Change the parameters as required and click on the **OK** button.
- Specify the other parameters related to material and temperature.
- Click on the **Calculate** button. If error is displayed then modify the parameters accordingly.
- Click on the **OK** button to add calculation in **Model Browser Bar**.

SPRINGS

A spring is defined as an elastic body, whose function is to distort when loaded and to recover its original shape when the load is removed. The various important applications of springs are as follows :

1. To cushion, absorb, or control energy due to either shock or vibration as in car springs, railway buffers, air-craft landing gears, shock absorbers, and vibration dampers.
2. To apply forces, as in brakes, clutches, and spring loaded valves.
3. To control motion by maintaining contact between two elements as in cams and followers.
4. To measure forces, as in spring balances and engine indicators.
5. To store energy, as in watches, toys, etc.

There are four tools in Autodesk Inventor to create springs viz. **Compression, Extension, Belleville**, and **Torsion**. The procedure to use these tools are discussed next.

Designing Compression Spring

The **Compression** tool in Autodesk Inventor is used to design compression springs. The procedure to do so is given next.

- Click on the **Compression** tool from the **Spring** panel in the **Design** tab of the **Ribbon**. The **Compression Spring Component Generator** dialog box will be displayed; refer to Figure-28.

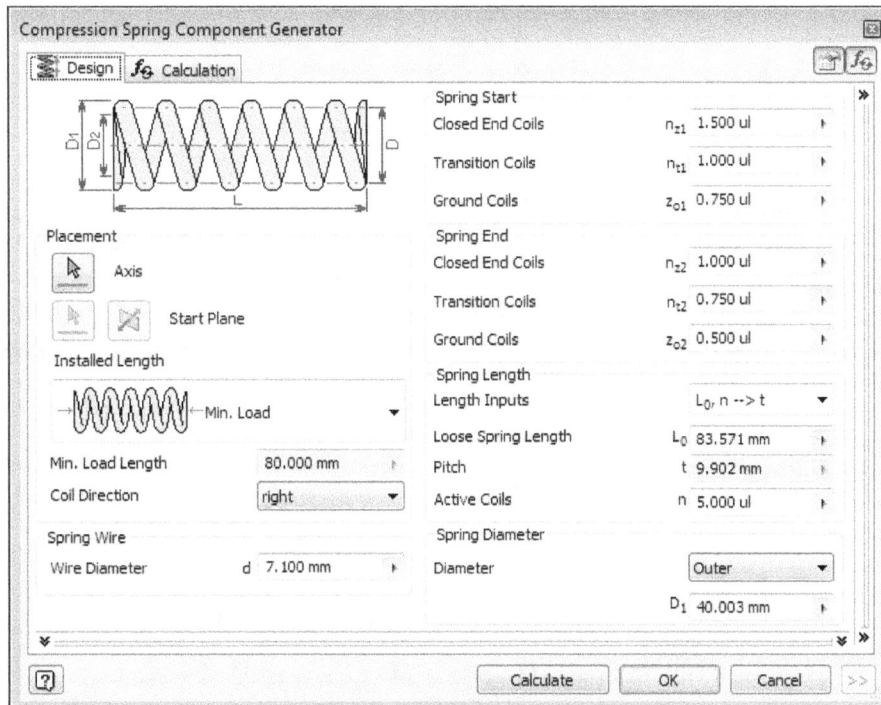

Figure-28. Compression Spring Component Generator dialog box

- Click on the **Axis** selection button from the **Placement** area of the dialog box. You will be asked to select the center axis for the spring.
- Select an axis. You will be asked to select the starting plane.
- Click on the plane/planar face you want to use as starting plane. Preview of the spring will be displayed in the drawing area; refer to Figure-29.

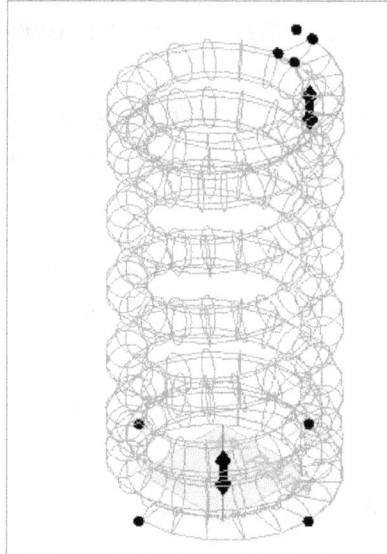

Figure-29. Preview of spring

- You can flip the direction of spring coils by using the **Flip Side** button for **Start Plane** in the **Placement** area.
- Click in the **Installed Length** drop-down and select desired option. There are four options in this drop-down; **Min. Load**, **Working Load**, **Max. Load**, and **Custom**; refer to Figure-30.

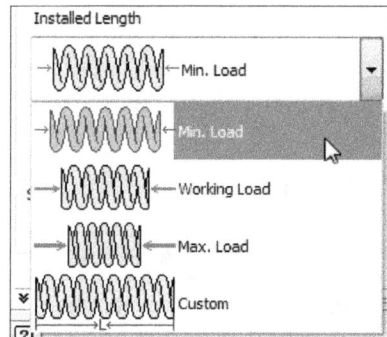

Figure-30. Installed Length drop-down

- Select the **Min. Load** option to define installed length of spring at minimum load. Select the **Working Load** option to define installed length of spring at working load. Select the **Max. Load** option to define installed length of spring at maximum load. Note that on selecting these options, the installed length is automatically calculated based on inputs given in **Spring Length** area of the dialog box. Select the **Custom** option to define the installed length manually. Note that the specified value must be less than loose spring length and greater than maximum load length.
- Specify the other parameters related to spring start coil, spring end coil, spring length, and spring diameter in the relative areas of the dialog box.
- Click in the **Spring Strength Calculation** drop-down and select desired option. There are three options in this drop-down viz. **Compression Spring Design**, **Spring Check Calculation**, and **Work Forces Calculation**. Select the **Compression Spring Design** option to design compression spring based on the specified parameters. Select the **Spring Check Calculation** option to check the feasibility of spring for specified loading conditions. Select the **Work Forces Calculation** option to find out the minimum, maximum, and working load of spring.

- Select desired option from the **Design Type** drop-down if you have selected the **Compression Spring Design** option from the **Spring Strength Calculation** drop-down. If you have selected the **F,D-->d,L0,n,Assembly Dimension** from the **Design Type** drop-down then you can select desired option from the **Design of Assembly Dimensions** drop-down to specify the assembly parameters.

- The options in the **Method of Stress Curvature Correction** drop-down are used to apply correction in stress curvature of the calculation.

- Specify the other parameters based on the options selected in the drop-downs.

- Click on the **Calculate** button. The driven parameters will be calculated automatically based on your inputs. If there is any error displayed then change the parameters accordingly.

- Click on the **OK** button to create the spring. The **File Naming** dialog box will be displayed.

- Set desired location and file names and then click on the **OK** button. The spring will get attached to the cursor.

- Click in the drawing area to place it.

Designing Extension Spring

Extension springs, also known as a tension spring, are helical wound coils, wrapped tightly together to create tension. Extension springs usually have hooks, loops, or end coils that are pulled out and formed from each end of the body. The function of an extension spring is to provide extended force when the spring is pulled apart from its original length. In Autodesk Inventor, the **Extension** tool is used to design extension springs. The procedure to do so is given next.

- Click on the **Extension** tool from the **Spring** panel in the **Design** tab of the **Ribbon**. The **Extension Spring Component Generator** dialog box will be displayed; refer to Figure-31.

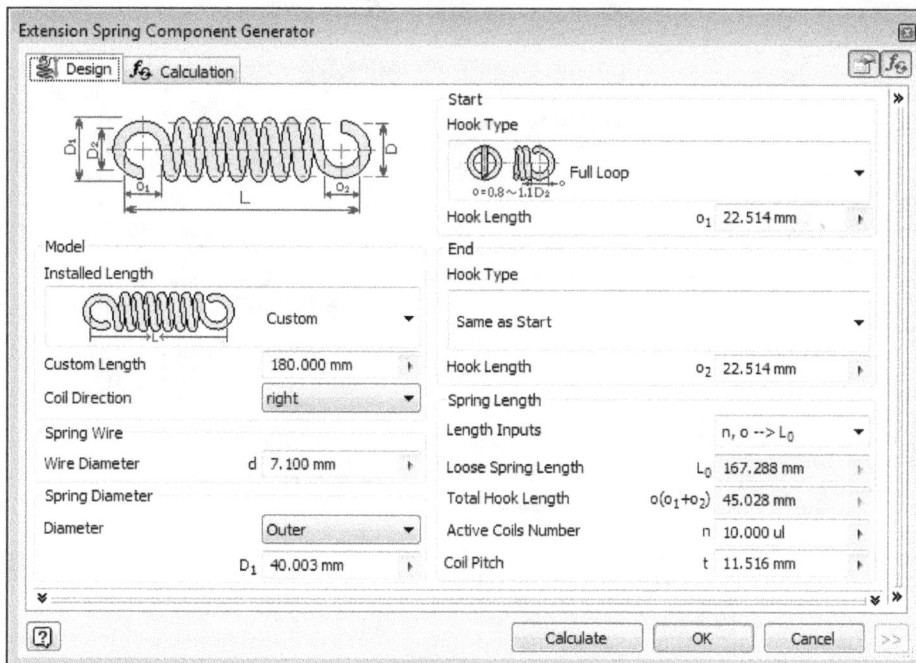

Figure-31. Extension Spring Component Generator dialog box

- Click in the **Installed Length** drop-down of **Model** area of the dialog box. Four options viz. **Min. Load**, **Working Load**, **Max. Load**, and **Custom** will be displayed. Select desired option. Use of each option has already been discussed in previous topic.
- Specify spring wire diameter and spring diameter in the respective edit boxes.
- Click in the **Hook Type** drop-down for both start and end, and select desired type of hook; refer to Figure-32.

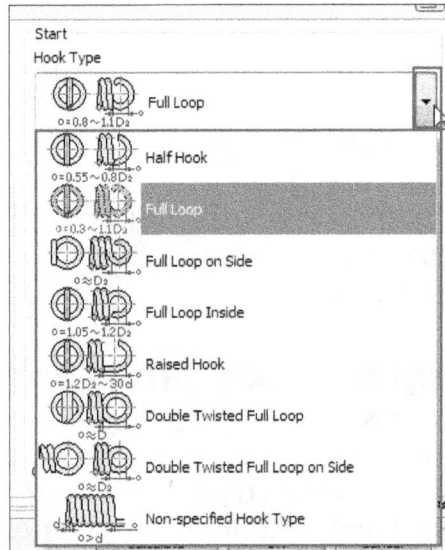

Figure-32. Hook Type drop-down

- Specify the other parameters related to spring length and hook length.
- Click on the **Calculate** button to check the validity of parameters specified so far.
- Click on the **Calculation** tab to check the feasibility of spring under specified load. The dialog box will be displayed as shown in Figure-33.

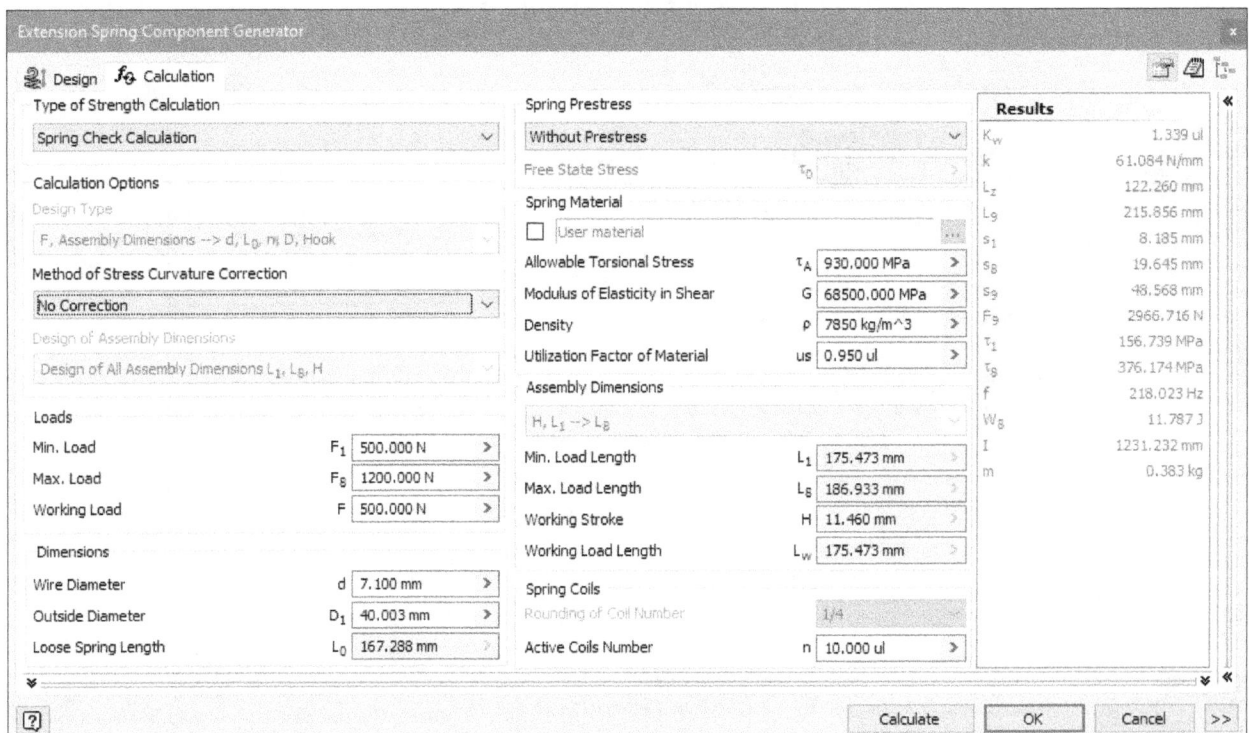

Figure-33. Calculation tab in Spring Component Generator dialog box

- Specify the parameters as discussed in previous topic and click on the **Calculate** button. If an error is displayed then modify the spring accordingly.
- Click on the **OK** button to create the spring. The **File Naming** dialog box will be displayed.
- Specify desired name and location and then click on the **OK** button to create the spring. The spring will get attached to the cursor.
- Click at desired location in the drawing area to place the spring; refer to Figure-34.

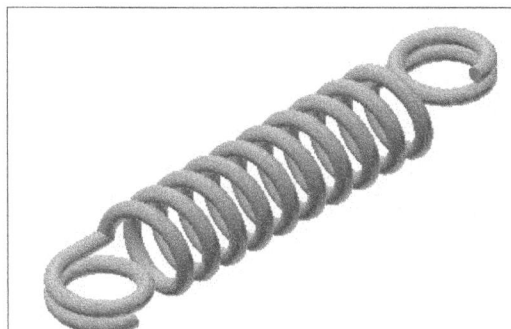

Figure-34. Spring created

Similarly, you can design Belleville and Torsion springs by using the **Belleville** and **Torsion** tools in the **Spring** panel of **Design** tab in **Ribbon**.

FRAME DESIGNING

In Autodesk Inventor, there is a set of tools specifically grouped to design frames in the **Frame** panel of **Design** tab in the **Ribbon**; refer to Figure-35.

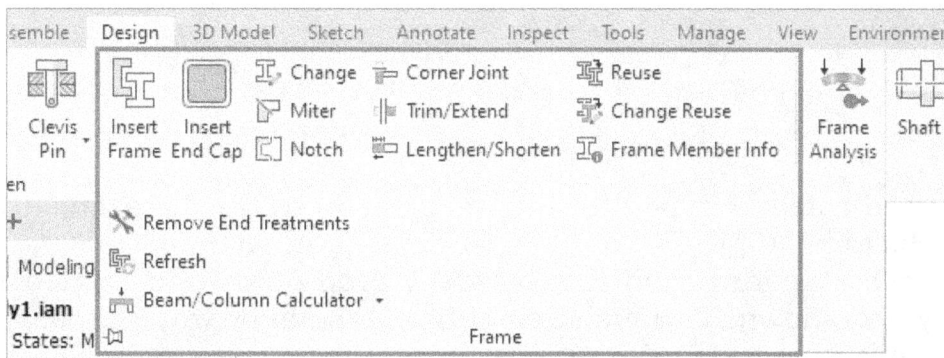

Figure-35. Frame panel

The **Insert Frame** tool in the **Frame** panel is used to insert the base members of frame and the other tools are used to perform various modifications in the frame. The procedures to use these tools are discussed next.

Inserting Frame

The **Insert Frame** tool is used to insert the frame members along the selected edges or point to point. The procedure to use this tool is given next.

- Click on the **Insert Frame** tool from the **Frame** panel in the **Design** tab of the **Ribbon**. The **Insert Frame** dialog box will be displayed; refer to Figure-36.

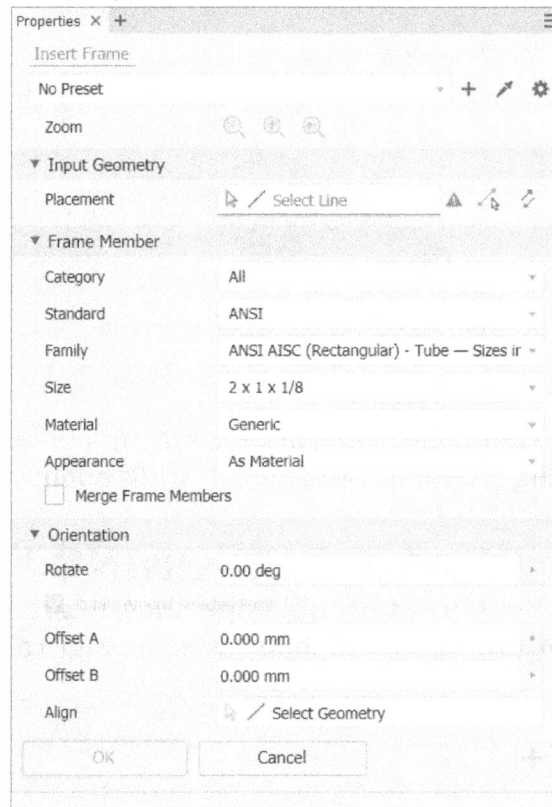

Figure-36. Insert Frame dialog box

- Select desired category type of frame member from the **Category** drop-down in the dialog box like Angles, Channels, I-Beams, and so on.
- Select desired standard from the **Standard** drop-down and related shape from the **Family** drop-down. Various sizes of frame member will become available for selected family in the **Size** drop-down.
- Select desired size and material from the **Size** and **Material** drop-downs, respectively.
- Select desired appearance from the **Appearance** drop-down if you want appearance different from the selected material.
- Select the edges/sketch entities along which you want to create the frame members. Preview of the frame will be displayed; refer to Figure-37. Note that you can select the sketch entities that are created in Part mode. The sketches created in the Assembly environment can not be used to create frame.
- If the frame members are intersecting at corner points then select the **Merge Frame Members** check box to merge them at intersections.

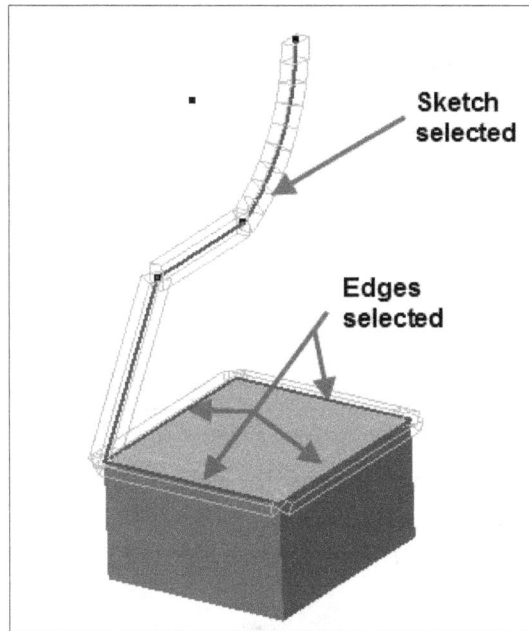

Figure-37. Preview of Frame members

- If you want to select start and end points in place of edges/lines then click on the **Insert Members between Points** button from the **Input Geometry** area of the dialog box.
- Click on the **Flip Direction** button in the **Input Geometry** area to reverse the direction of frame member along selected edge/line/points.
- Specify the offset values in the **Vertical Offset** and **Horizontal Offset** edit boxes if required.
- Click in the **Rotate** edit box and specify desired angle value if you want to rotate the frame member about its center line.
- If the frame members are not aligned as desired then click in the **Align** selection box of **Orientation** area. You will be asked to select a geometry to which the frame members are to be aligned. Select desired geometry, the frame members will be aligned to it; refer to Figure-38.

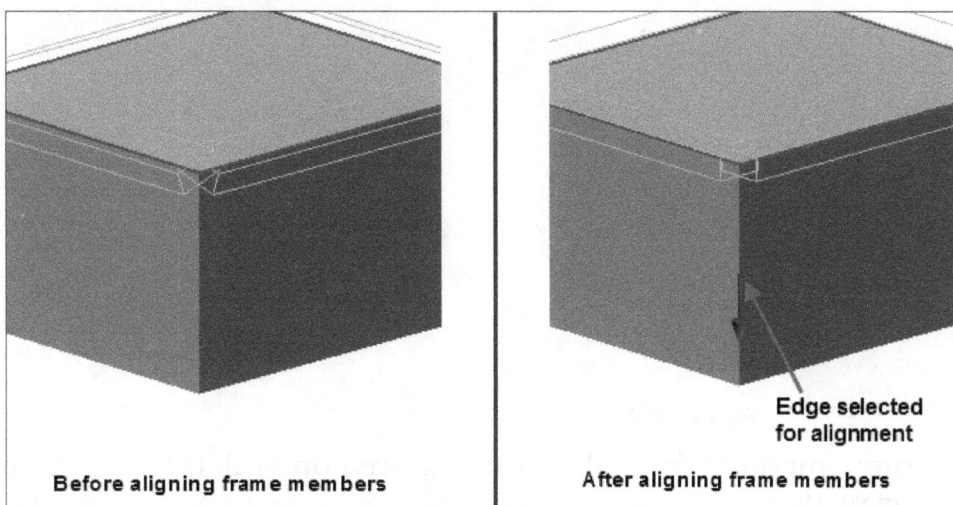

Figure-38. Aligning frame members

- Click on the **OK** button to create the frame members. The **Create New Frame** dialog box will be displayed; refer to Figure-39.

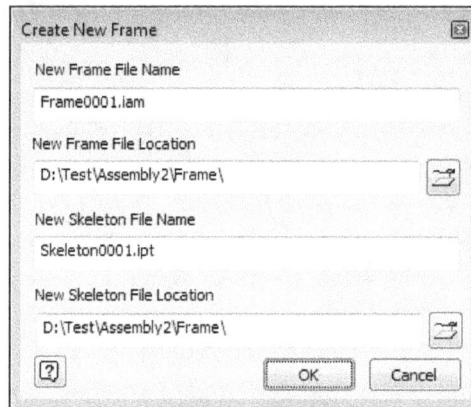

Figure-39. Create New Frame dialog box

- Specify desired frame file name, frame location, skeleton file name, and skeleton file location in the respective edit boxes.
- Click on the **OK** button to create the frame. You can hide the base object now.

Inserting End Cap

The **Insert End Cap** tool is used to place end caps on selected frame members. The procedure to use this tool is discussed next.

- Click on the **Insert End Cap** tool from the **Frame** panel in the **Design** tab of the **Ribbon**. The **End Caps** dialog box will be displayed; refer to Figure-40.

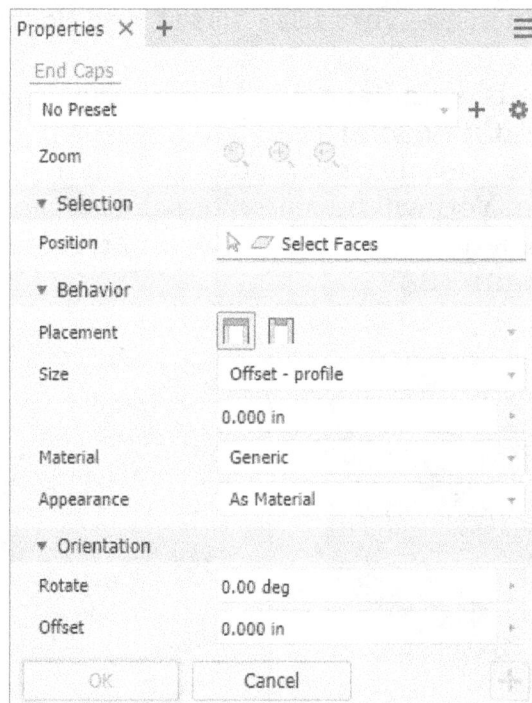

Figure-40. End Caps dialog box

- Select the frame member from the drawing area on which you want to place the end caps. Parameters of the selected member will be displayed in the dialog box.
- Select the frame member with desired parameters by using the options in the dialog box. The options in the dialog box are same as discussed in previous topic.
- Click on the **OK** button to apply the changes.

Changing Frame Members

The **Change** tool in the **Frame** panel in the **Design** tab of the **Ribbon** is used to change the parameters related to frames. The procedure to use this tool is discussed next.

- Click on the **Change** tool from the **Frame** panel in the **Design** tab of the **Ribbon**. The **Change Frame** dialog box will be displayed; refer to Figure-41.

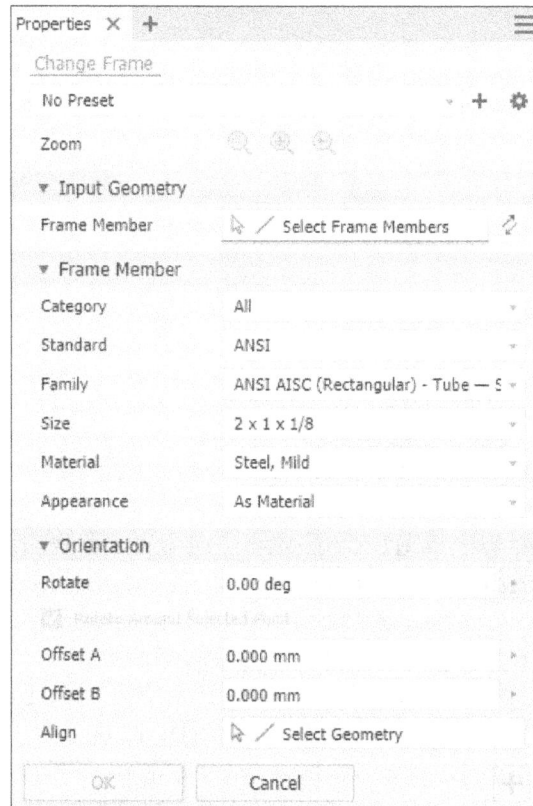

Figure-41. Change Frame dialog box

- Select the frame member that you want to change from the drawing area. Parameters of the selected member will be displayed in the dialog box.
- Select the frame member with desired parameters by using the options in the dialog box. The options in the dialog box are same as discussed in previous topic.
- Click on the **OK** button to apply the changes.

Mitering Corners

A miter joint (mitre in British English), sometimes shortened to miter, is a joint made by beveling each of two parts to be joined, usually at a 45° angle, to form a corner. The procedure to apply miter to a joint is discussed next.

- Click on the **Miter** tool from the **Frame** panel in the **Design** tab of the **Ribbon**. The **Miter** dialog box will be displayed; refer to Figure-42. Also, you will be asked to select the frame member to be cut.

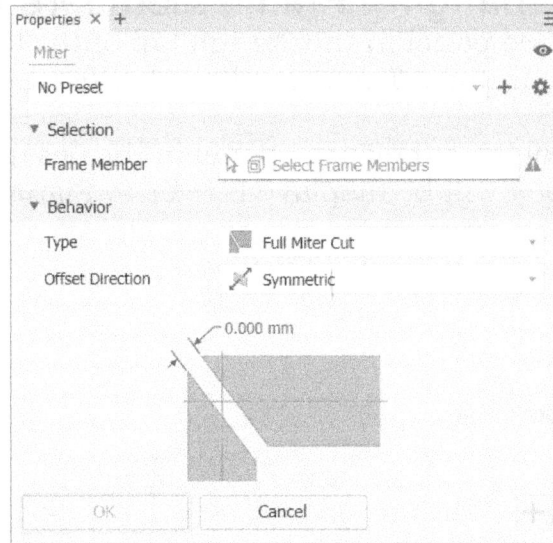

Figure–42. Miter dialog box

- Select a frame member you want to be cut. You will be asked to select the other member to be cut.
- Select the other intersecting frame member. Preview of the miter cut will be displayed.
- By default, the **Full Miter Cut** option is selected from **Type** drop-down in the **Behavior** area of the dialog box to match full edges of both the frame members after miter cut. If you want frame member to be joined at their bisector then select the **Bisect Miter Cut** option from **Type** drop-down.
- Select desired offset type from **Offset** drop-down in the dialog box and specify desired offset distance in the **Miter Offset** edit box.
- Click on the **Apply** button to create the current miter cut and continue to create the next or click on the **OK** button to create the miter cut and exit the tool. The miter cut will be created; refer to Figure-43.

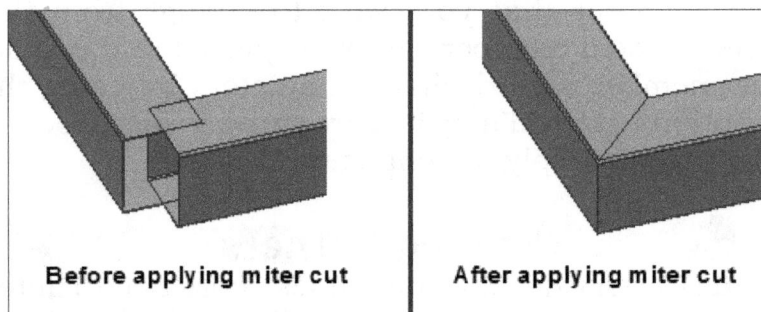

Figure–43. Applying Miter cut

Applying Notch Cut

The notch cutting is used to cut the ends of two frame members so that they can be joined. Sometimes notching is done to make other joints like miter joints; refer to Figure-44.

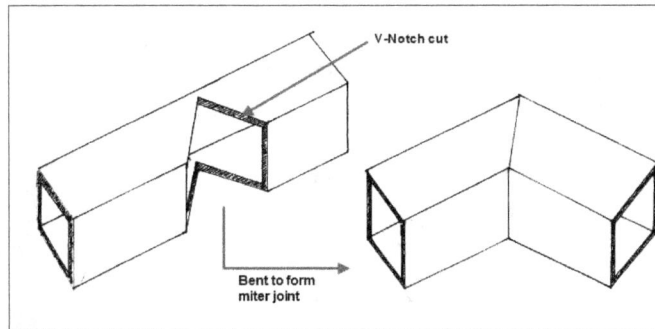

Figure-44. Notched tube

The procedure to apply notch cut is given next.

- Click on the **Notch** tool from the **Frame** panel in the **Design** tab of the **Ribbon**. The **Notch** dialog box will be displayed; refer to Figure-45. Also, you will be asked to select the frame member to be cut.

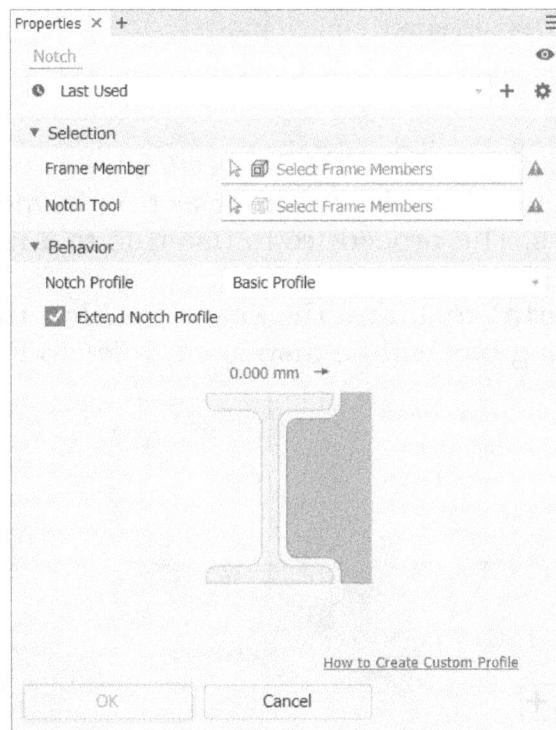

Figure-45. Notch dialog box

- Select the frame member that you want to be cut. You will be asked to select the notch tool.
- Select the notch tool to be used as cutting tool reference.
- Select desired option from **Notch Profile** drop-down in the **Behavior** area of the dialog box. There are three options in this drop-down, viz. **Basic Profile**, **Custom Profile**, and **Custom I template**. Select the **Basic Profile** option to set a single offset value for the notch. Select the **Custom Profile** option, if available, to use the custom notch profile in the Content Center family. Or select the **Custom I template** option, if available, to set multiple offset values.
- Select the **Extend Notch Profile** check box, if available, to extend the profile to an intersection.

- Select the **Perpendicular Cut** check box, if available, to create precise intersections such as a laser cut notch for a pipe.
- Specify desired value in the **Notch Offset** edit box in the dialog box.
- Click on the **OK** button, the notch cut will be applied; refer to Figure-46.

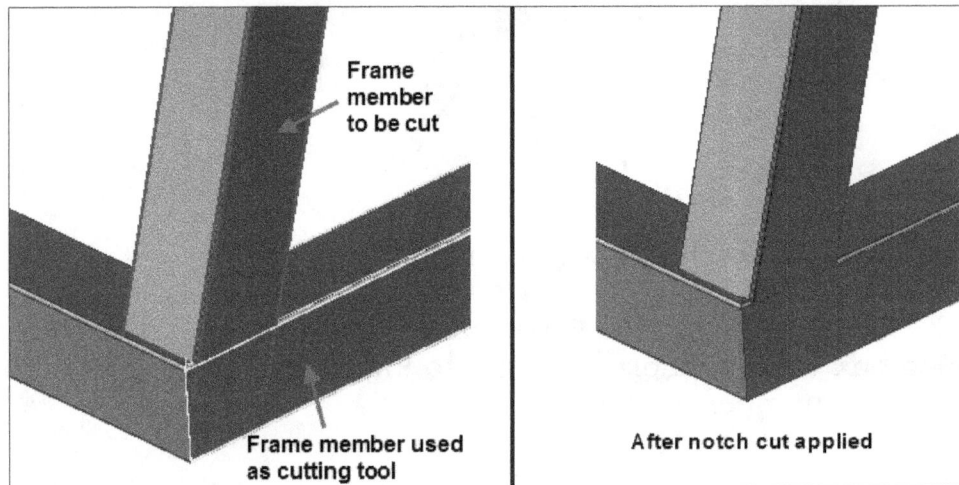

Figure-46. Applying notch cut

Corner Joint

The **Corner Joint** tool is used to join two intersecting frame members by trimming or extending the members. The procedure to use this tool is given next.

- Click on the **Corner Joint** tool from the **Frame** panel in the **Design** tab of **Ribbon**. The **Corner Joint** dialog box will be displayed; refer to Figure-47.

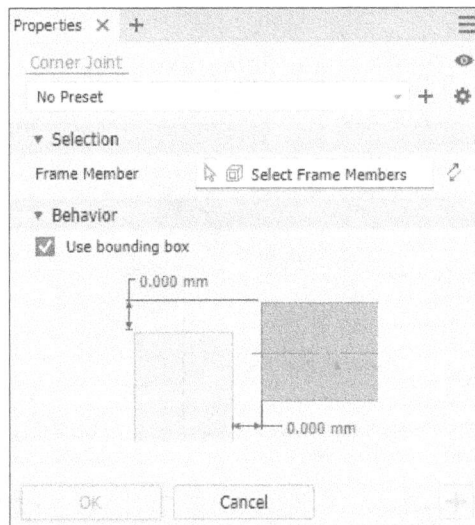

Figure-47. Corner Joint dialog box

- Select the two intersecting frame members where you want to create corner joint. Preview of corner joint will be displayed; refer to Figure-48.
- Specify desired distance values in the edit boxes of **Behavior** area in dialog box and click on the **OK** button to create the joint.

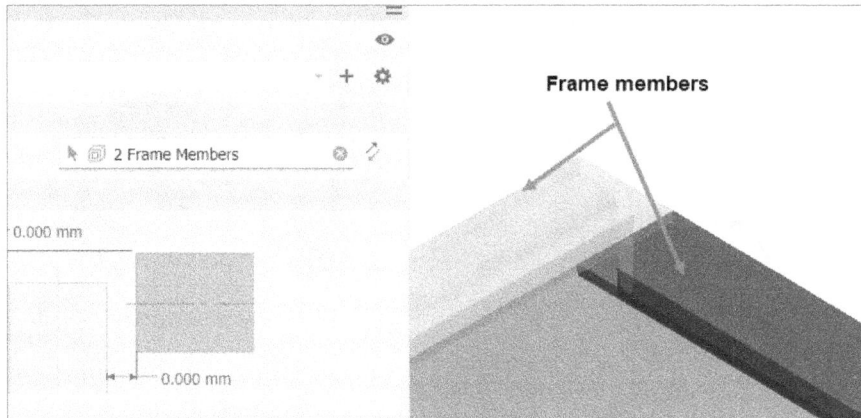

Figure-48. Corner joint preview

Trim/Extend

The **Trim/Extend** tool is used to trim the selected frame member at the other intersecting frame member. The procedure to use this tool is given next.

- Click on the **Trim/Extend** tool from the **Frame** panel in the **Design** tab of the **Ribbon**. The **Trim/Extend** dialog box will be displayed; refer to Figure-49.

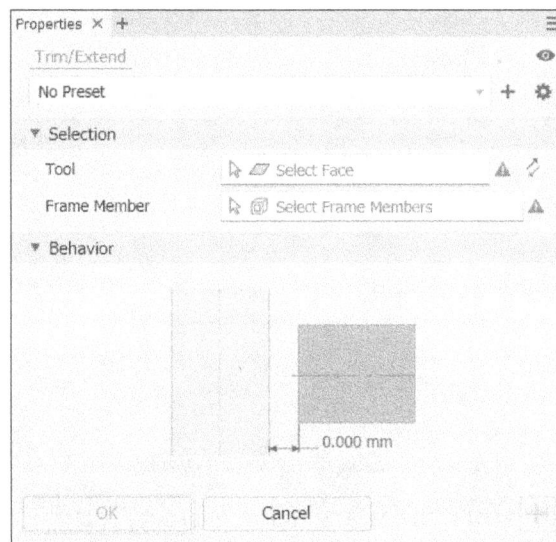

Figure-49. Trim Extend dialog box

- Select desired face that you want to be used as tool for trimming. Note that selected face will be used as reference for trimming or extension. On selecting the face, you will be asked to select the other frame member to be trimmed or extended.
- Select the frame member intersecting/approaching the first one. Preview of feature will be displayed.
- Specify the gap values in the **Behavior** area of the dialog box, if required.
- Click on the **Apply** button to create the current trimming and continue to create the next or click on the **OK** button to create the trimmed parts and exit the tool. The trimmed frame members will be created; refer to Figure-50.

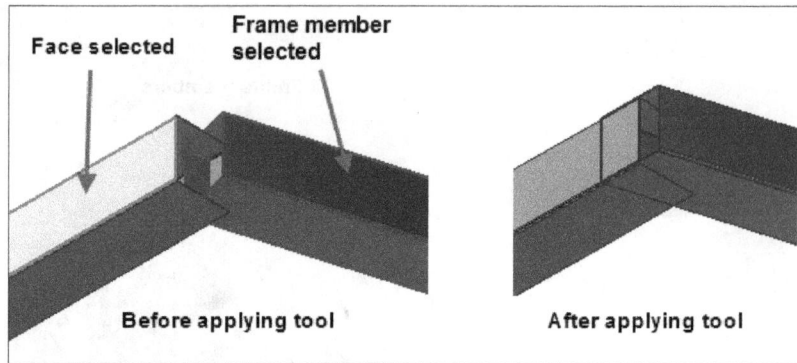

Figure-50. Trimming-Extending frame member

Lengthen/Shorten Frame Member

The **Lengthen/Shorten** tool is used to lengthen/shorten the frame member by specified value. The procedure to use this tool is given next.

*   Click on the **Lengthen/Shorten** tool from the **Frame** panel in the **Design** tab of the **Ribbon**. The **Lengthen/Shorten** dialog box will be displayed; refer to Figure-51.

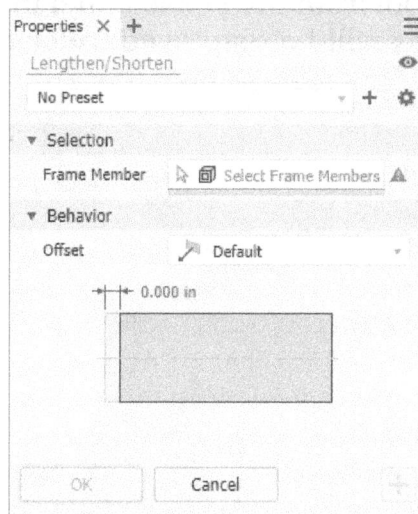

Figure-51. Lengthen Shorten dialog box

*   Select the frame member(s) you want to shorten/lengthen.
*   Select desired option from the **Offset** drop-down in the **Behavior** area of the dialog box.
*   Specify desired value in the **Lengthen/Shorten** edit box of the dialog box.
*   Click on the **OK** button to apply the changes and exit the tool.

Similarly, you can use the other editing tools like **Reuse** tool, **Change Reuse** tool, **Remove End Treatments** (in expanded **Frame** panel) etc. available in the **Frame** panel.

Beam/Column Calculations

After creating the frame, next step is to check whether it will sustain the specified load or not. The procedure to perform these calculations is given next.

- Click on the **Beam/Column Calculator** tool from the **Calculators** drop-down in the expanded **Frame** panel of the **Design** tab in the **Ribbon**; refer to Figure-52. The **Beam and Column Calculator** dialog box will be displayed; refer to Figure-53. Also, you are asked to select a beam or column object.

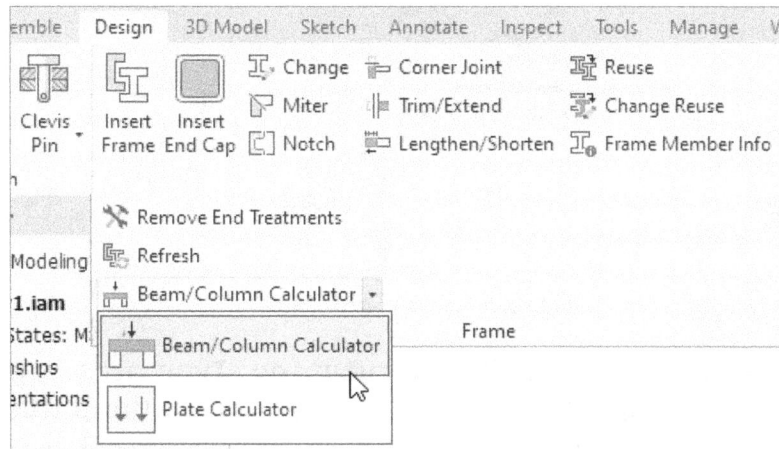

Figure-52. Beam-Column Calculator tool

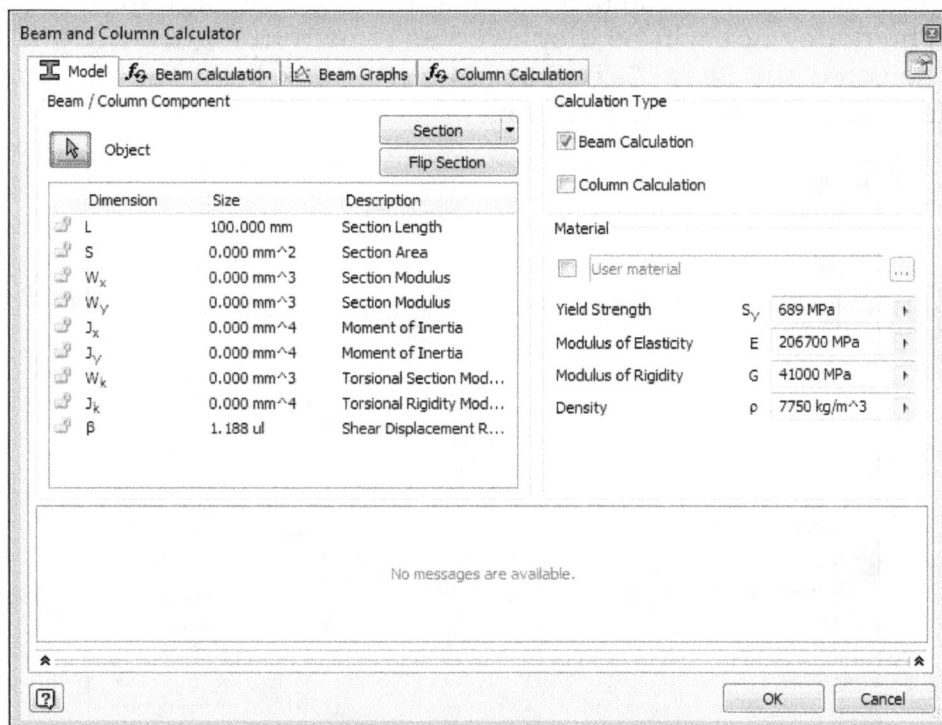

Figure-53. Beam and Column Calculator dialog box

- Select a frame member on which you want to perform the loading test.
- Click on the **Section** button and select the shape that closely match to your frame member; refer to Figure-54. Specify the dimensions of section in the dialog box displayed and click **OK** button.

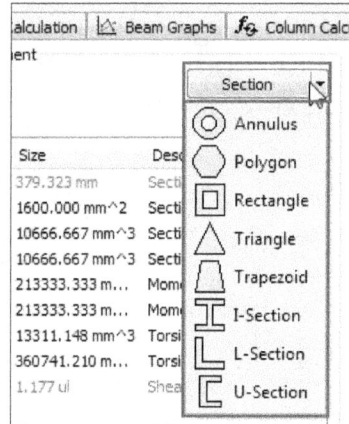

Figure-54. Section drop-down

- Select desired check box from the **Calculation Type** area to mark it as beam or column. If you want to perform both type of calculations then select both the check boxes in the area.
- Specify the material properties in the **Material** area of the dialog box.
- Click on the **Beam Calculation** tab to check the effect of load on the selected frame member. The dialog box will be displayed as shown in Figure-55. Make sure that you have selected the **Beam Calculation** check box from the **Calculation Type** area of the **Model** tab in the dialog box.

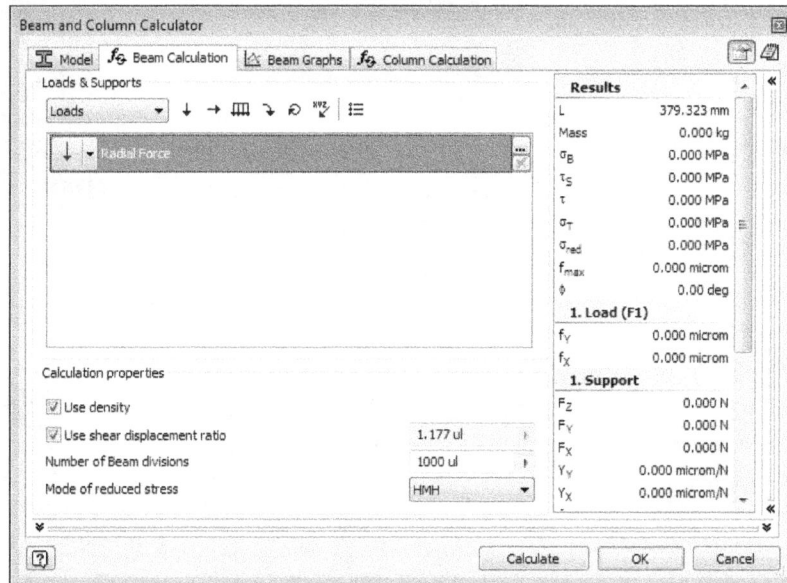

Figure-55. Beam Calculation tab in Beam and Column Calculator dialog box

- Specify desired loads as discussed in previous chapter and click on the **Calculate** button. The results will be displayed in the **Results** area of dialog box. Modify the frame member, if error occurs.
- Similarly, you can perform the column calculations and check the beam graph.
- After performing calculations, click on the **OK** button to exit the dialog box.

MEASUREMENT TOOLS

There are various measurement tools to quickly measure the components in Autodesk Inventor. These tools are available in the **Inspect** tab as well as **Design** tab. To display the measurement tools in **Design** tab, click on the down button in **Design** tab. A shortcut menu will be displayed as shown in Figure-56. Select the **Measure** check box and click in drawing area. The **Measure** panel will be added in the **Ribbon**.

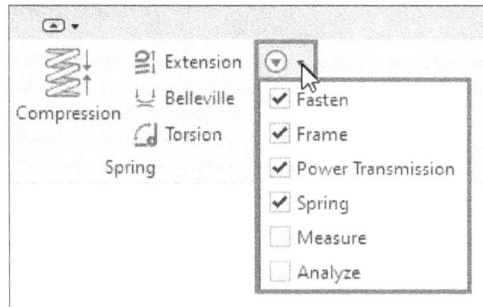

Figure-56. Option to add measurement tool

The tools in the **Measure** panel are discussed next.

Measure Tool

The **Measure** tool is used to measure distance between two entities like points, lines, edges, faces, etc. The procedure to use this tool is given next.

- Click on the **Measure** tool from the **Measure** panel in the **Design** tab of the **Ribbon**. The **Measure** dialog box will be displayed; refer to Figure-57.

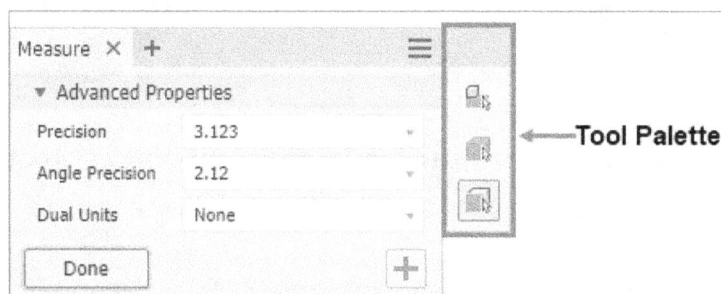

Figure-57. Measure dialog box

- The tool palette displays the selection priority options; refer to Figure-57. Select desired selection priority option from the **Component Priority** , **Part Priority** , or **Select Faces and Edges** .
- Select an edge or line to measure its length. Select an arc to measure its radius. Select a circle to measure its diameter. Select a face to measure its perimeter and area.
- Click on the **Restart Measure** button from the dialog box before measuring next entity; refer to Figure-58.

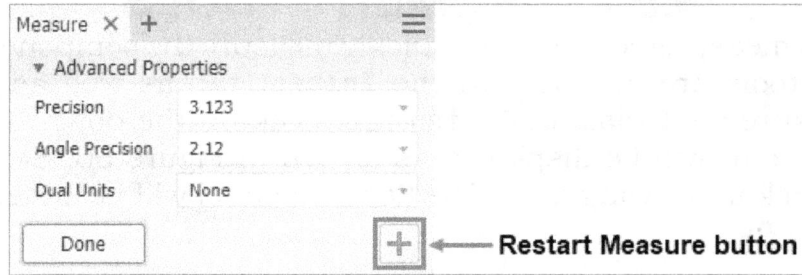

Figure-58. Restart Measure button in Measure dialog box

- To measure distance between two lines, edges, or faces, select the first entity and then hold the **CTRL** key while selecting the next entity; refer to Figure-59.

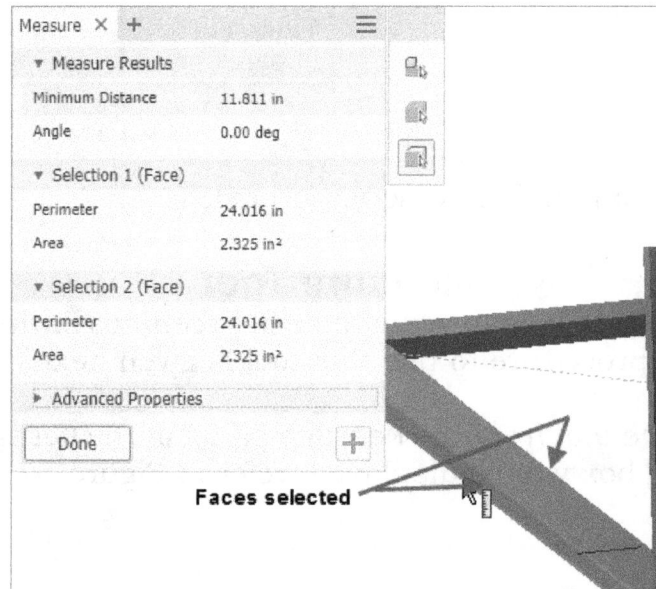

Figure-59. Faces selected for measurement

- Similarly, you can use the other options in **Measure** dialog box to measure various parameters.

PROBLEM 1

A compression coil spring made of an alloy steel is having the following specifications :

Mean diameter of coil = 50 mm ; Wire diameter = 5 mm ; Number of active coils = 20.

If this spring is subjected to an axial load of 500 N; Design the spring.

PROBLEM 2

Design and draw a valve spring of a petrol engine for the following operating conditions :

Spring load when the valve is open = 400 N
Spring load when the valve is closed = 250 N
Maximum inside diameter of spring = 25 mm
Length of the spring when the valve is open = 40 mm
Length of the spring when the valve is closed = 50 mm
Maximum permissible shear stress = 400 MPa

PROBLEM 3

The splined ends and gears attached to the A-36 steel shaft are subjected to the torques shown in Figure-60. Determine the angle of twist of end B with respect to end A. The shaft has a diameter of 40 mm. Properties of A-36 steel should be checked online.

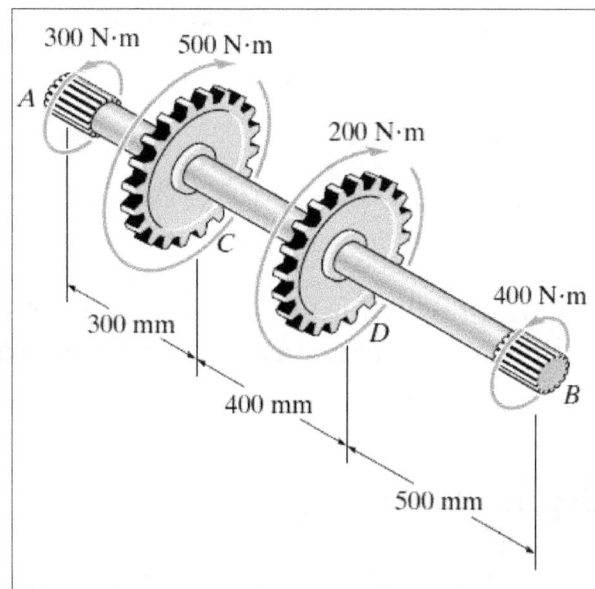

Figure-60. Gear spline system under load

SELF ASSESSMENT

Q1. Which of the following options is also known as **Toric Joint** which is a power transmission component?

a) O-Ring
b) Involute Spline
c) Parallel Spline
d) Linear Cam

Q2. Which of the following factors are responsible for design or capacity of a **Brake Calculator**?

a) The unit pressure between the braking surfaces
b) The projected area of the friction surfaces
c) Both a and b
d) None of the Above

Q3. Which of the following types of **Brake Calculator** is not used in Mechanical Engineering?

a) Drum Brake Calculator
b) Disc Brake Calculator
c) Dynamic Brake Calculator
d) Cone Brake Calculator

Q4. Which of the following options should be selected in the **Bearing Calculator** dialog box to check the feasibility of journal bearing under current loading conditions?

a) Lubricant Selection
b) Check Calculation
c) Design of minimum Journal Diameter
d) None of the Above

Q5. Which of the following statements is incorrect?

a) Springs are used to control or absorb energy due to either shock or vibration as in car springs, railway buffers, etc.

b) Springs are used to apply forces as in brakes, clutches, etc.

c) Springs are used to control motion by maintaining contact between two elements as in cams and followers.

d) Springs are used to measure distance as in spring balances and engine indicators.

Q6. Which of the following tools is not used in creating springs?

a) Centrifugal
b) Compression
c) Extension
d) Belleville

Q7. Splines are ridges or teeth on a drive shaft that mesh with grooves in mating piece. (True/False)

Q8. A brake is a device by means of which artificial speed is applied to a static machine member. (True/False)

Q9. .... also known as .... are helical wound coils, wrapped tightly together to create tension.

Q10. The dimension value with .... color background is the resultant dimension in the **Tolerance Calculator** dialog box.

FOR STUDENT NOTES

Chapter 10

Sheetmetal Design

Topics Covered

The major topics covered in this chapter are:

- *Creating Sheet Metal part files*
- *Setting Sheet Metal Defaults*
- *Industrial Terms related to Sheet Metal Design*
- *Sheet Metal Creation and Modification Tools*
- *Creating Flat Patterns*
- *Practical on Sheet Metal Design*
- *Practice and Problems*

INTRODUCTION

Sheet metal work is an important aspect of Mechanical engineering. Many parts around us are manufactured via sheetmetal processes. For example, car body, vents in houses, Air-conditioner ducts, spoon, metal bowls, and so on. The sheetmetal parts generally have thickness ranging from fraction of millimeter to 12.5 millimeters i.e. up to half inch. Like welding and machining, sheetmetal also has its own processes like, bending, punching, stamping, spinning, rolling, and so on. In this chapter, we will discuss about the tools available in Autodesk Inventor related to sheetmetal designing.

STARTING SHEETMETAL ENVIRONMENT

In Autodesk Inventor, there are two ways to start sheetmetal environment:

1. Make a part in the Standard Part environment and then convert it to Sheetmetal.
2. Start a new file in the Sheetmetal environment by using Sheetmetal template.

You will learn about converting the part in to sheetmetal later in this chapter. We will now start a new file in sheetmetal environment.

Starting A New File in Sheetmetal Environment

The procedure to start a new file in sheetmetal environment is given next.

- Start Autodesk Inventor and click on the **New** button from **New** cascading menu in the **File** menu of the **Ribbon**. The **Create New File** dialog box will be displayed; refer to Figure-1.

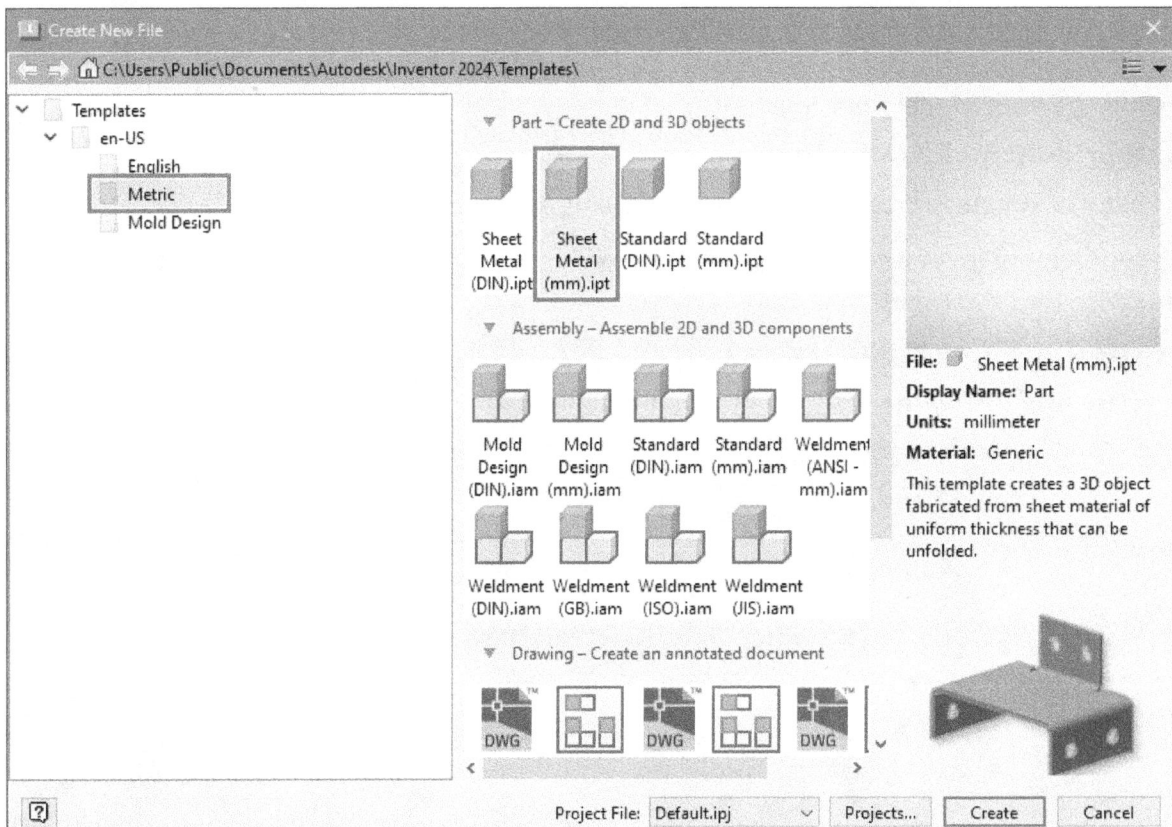

Figure-1. Create New File dialog box

- Select the Sheet Metal(mm).ipt template or SheetMetal(DIN).ipt template to start a new file in sheetmetal environment using metric units. Click on the **Create** button from the dialog box. A new file will be created in sheetmetal environment; refer to Figure-2.

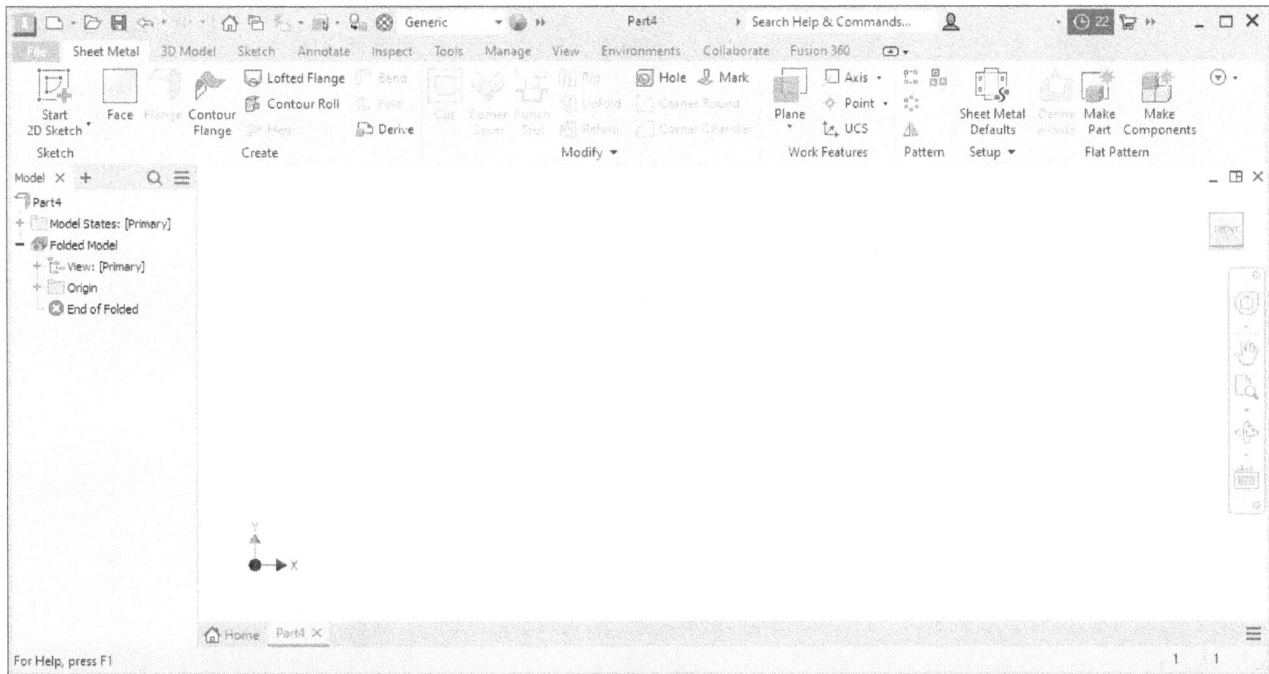

Figure-2. Sheetmetal Environment

Now, we have entered in the Sheetmetal environment; the next step is to set sheetmetal parameters.

SHEET METAL DEFAULTS SETTING

The sheetmetal defaults are the default values of various parameters related to sheetmetal designing. Once you have set the default values then they will be applied on the sheetmetal project automatically. The procedure to change the sheetmetal defaults is given next.

- Click on the **Sheet Metal Defaults** tool from the **Setup** panel in the **Sheet Metal** tab of the **Ribbon**. The **Sheet Metal Defaults** dialog box will be displayed; refer to Figure-3.

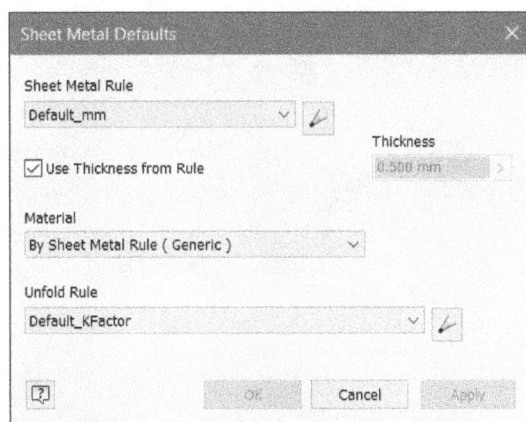

Figure-3. Sheet Metal Defaults dialog box

- Select desired template from the **Sheet Metal Rule** drop-down. There are two templates by default; **Default** and **Default mm**. The **Default** template works with imperial unit system and **Default mm** template works on metric unit system.
- Click on the **Edit Sheet Metal Rule** button next to **Sheet Metal Rule** drop-down. The **Style and Standard Editor** dialog box will be displayed; refer to Figure-4.

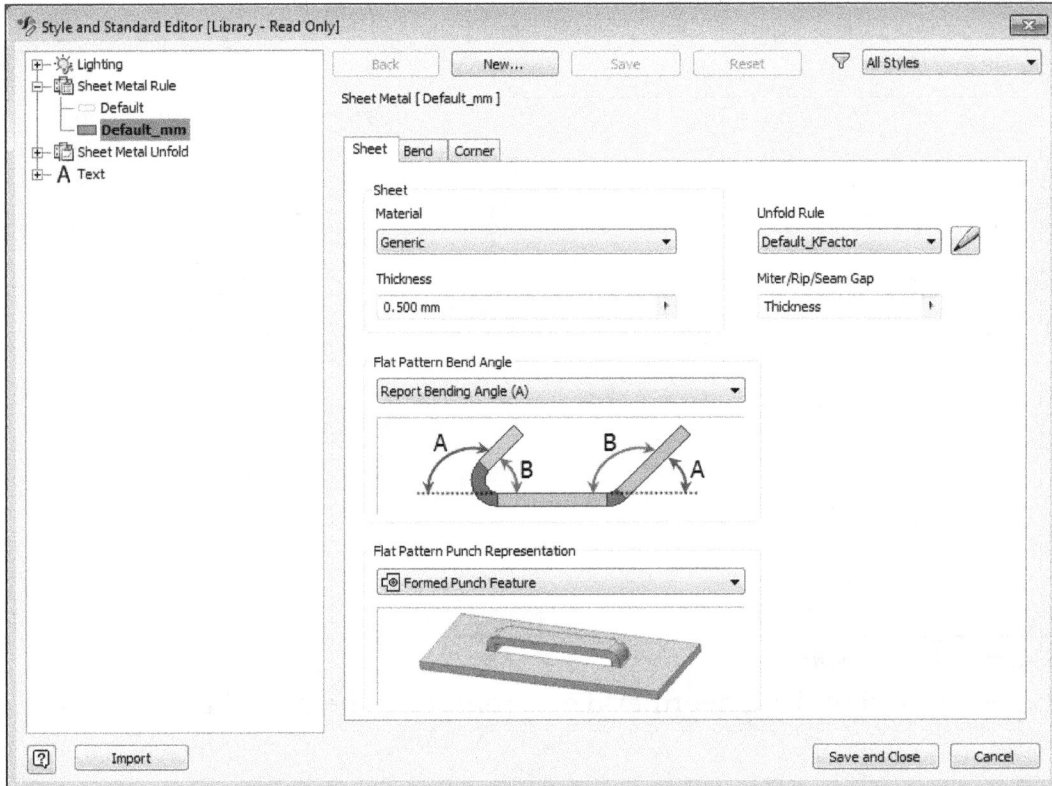

Figure-4. Style and Standard Editor dialog box

- Click in the **Material** drop-down of **Sheet** area in **Sheet** tab of the dialog box. A list of materials stored in library will be displayed; refer to Figure-5.

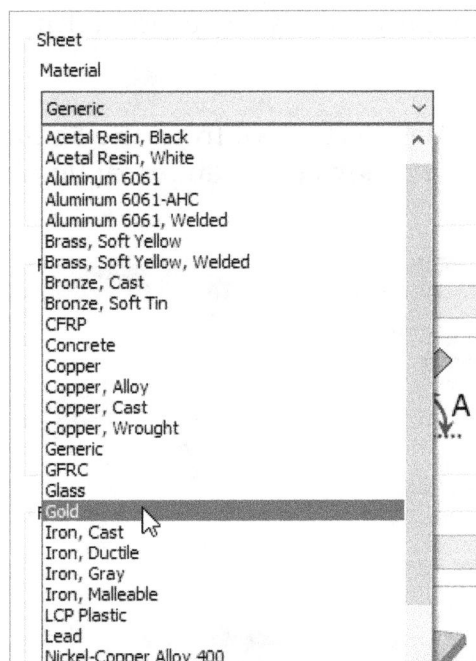

Figure-5. Material drop down

- Select desired material from the drop-down.
- Click in the **Thickness** edit box and specify desired value of the thickness.
- Click in the **Miter/Rip/Seam Gap** edit box and specify desired value of gap. By default, the **Thickness** value is specified in the edit box which means the gap will be equal to thickness of sheetmetal.
- Click in the **Unfold Rule** drop-down and select desired rule for unfolding of sheetmetal object. There are two options in this drop-down; **BendCompensation** and **Default KFactor**. Select the **BendCompensation** option if you want to compensate for the length of sheet required in creating bend by using the custom formulas based on bending angle and thickness of sheet. Select the **Default KFactor** option if you want to compensate for bend length by using the K factor.
- Similarly, you can set the other parameters in the **Sheet** tab like punch representation in flat pattern and flat pattern bending angle.
- Click on the **Bend** tab from the dialog box. The dialog box will be displayed; refer to Figure-6.

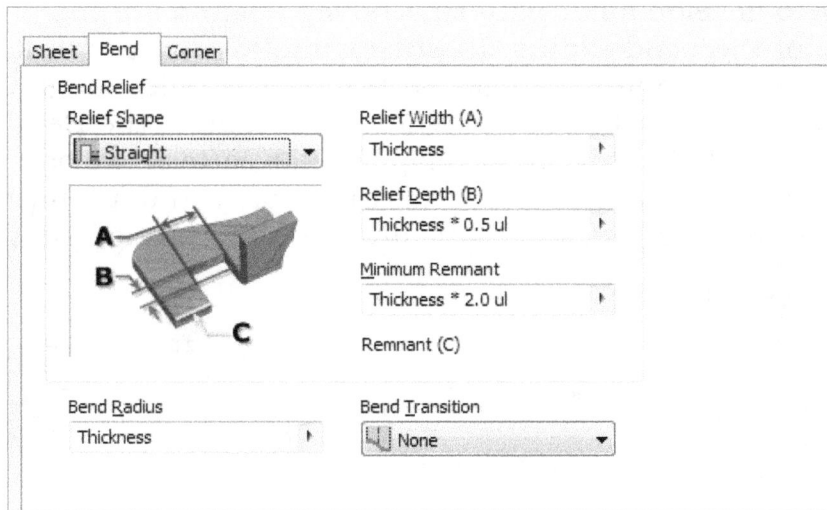

Figure-6. Bend tab in the Style and Standard Editor dialog box

- Select desired shape of bend relief from the **Relief Shape** drop-down in the **Bend Relief** area of the dialog box.
- Similarly, select desired option from the **Bend Transition** drop-down.
- Click in the **Bend Radius**, **Relief Width**, **Relief Depth**, and **Minimum Remnant** edit boxes and specify desired values.
- Similarly, click on the **Corner** tab in the dialog box and specify the relief parameters for corners in sheetmetal parts.
- Click on the **Save and Close** button to save and apply the parameters specified.

SHEET METAL DESIGN TERMS

While going forward in the chapter, you will come across some technical terms which are used in sheet metal industry. Here, we will discuss about these technical terms and their effects on production.

Sheet Metal Definition
Metal that has been rolled into a sheet having a thickness between foil and plate. The thickness of metal can vary from fraction of millimeters to 12.5 mm, in general.
Gauge

Gauge is a traditional measurement unit of sheet thickness commonly used in USA, India, and many other parts of world. It is a non-linear unit. Higher the gauge number, the thinner is the sheet metal. Like gauge 0000000 means 12.7 mm thickness and gauge 38 means 0.16 mm thickness. You can find a table on gauge to mm conversion easily in local stores. Use of Gauge to designate sheet metal thickness is discouraged by numerous international standards organizations. For Example, ASTM states in specification ASTM A480-10a "The use of gauge number is discouraged as being an archaic term of limited usefulness not having general agreement on meaning."

Bend Allowance

When the sheet metal is put through the process of bending, the metal around the bend is deformed and stretched. As this happens, you gain a small amount of total length in your part. Likewise, when you are trying to develop a flat pattern, you will have to make a deduction from your desired part size to get the correct flat size. The **Bend Allowance** is defined as the material you will add to the actual leg lengths of the part in order to develop a flat pattern. The leg lengths are the part of the flange which is outside of the bend radius. In our example, a part with flange lengths of 2" and 3" with an inside radius of .250" at 90° will have leg lengths of 1.625" and 2.625", respectively; refer to Figure-7. When we calculate the Bend Allowance, we find that it equals .457". In order to develop the flat pattern, we add .457" to 1.625" and 2.625" to arrive at 4.707". In Autodesk Inventor, you don't need to specify the value of bend allowance as it is automatically calculated based on K factor.

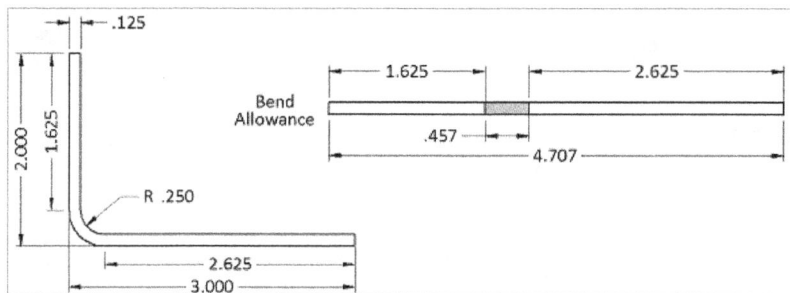

Figure-7. Bend Allowance example

K-Factor

The K-Factor in sheet metal designing is the ratio of the neutral axis to the material thickness. When metal is bent, the upper section is going to compress and the lower section is going to stretch. The line where the transition from compression to stretching occurs is called the neutral axis. The location of the neutral axis varies and is based on the material's physical properties and its thickness. The K-Factor is the ratio of the Neutral Axis' Offset (t) and the Material Thickness (MT). Figure-8 shows how the top of the bend is compressed and the bottom is stretched.

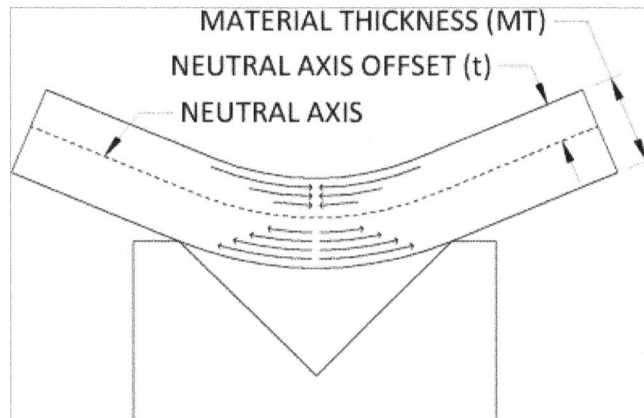

Figure-8. Neutral axis of bent model

$$K = t/MT$$

Generally, K is taken as 0.33 for soft materials and 0.4 for hard materials as a thumb rule. Since, K factor is not just mathematical term, you need experiments to find exact value of K for your material situations. We have given the general steps to find out K-factor by experiment.

Calculating the K-Factor

Since, the K-Factor is based on the property of the metal and its thickness, there is no simple way to calculate it ahead of the first bend. Typically, the K-Factor is going to be between 0 and .5. In order to find the K-Factor, you will need to bend a sample piece and deduce the Bend Allowance. The Bend Allowance is then plugged into the equation to find the K-Factor.

• Begin by preparing sample blanks which are of equal and known sizes. The blanks should be at least a foot long to ensure an even bend, and a few inches deep to make sure you can sit them against the back stops. For our example, let's take a piece that is 14 Gauge, .075", 4" Wide and 12" Long. The length of the piece won't be used in our calculations. Preparing at least 3 samples and taking the average measurements from each will help.
• Set up your press brake with desired tooling, you'll be using to fabricate this metal thickness and place a 90° bend in the center of the piece. For our example, this means a bend at the 2" mark.
• Once you've bent your sample pieces carefully, measure the flange lengths of each piece. Record each length and take the average of lengths. The length should be something over half the original length. For our example, the average flange length is 2.073".
• Second, measure the inside radius formed during the bending. A set of radius gauges will get you fairly close to finding the correct measurement. However to get an exact measurement, an optical comparator will give you the most accurate reading. For our example, the inside radius is measured at .105".
• Now that you have your measurements, we'll determine the Bend Allowance. To do this, first determine your leg length by subtracting the material thickness and inside radius from the flange length. (Note this equation only works for 90° bends because the leg length is from the tangent point.) For our example, the leg length will be 2.073 − .105 − .075 = 1.893.
• Subtract twice the leg length from the initial length to determine the Bend Allowance. 4 − 1.893 * 2 = .214.

- Plug the Bend Allowance (BA), the Bend Angle (A), Inside Radius (R), and Material Thickness (T) into the below equation to determine the K-Factor (K).

$$K = \frac{-R + \dfrac{BA}{\pi A/180}}{T}$$

Spline Factor

The **Spline factor** is a special term used in Autodesk Inventor for sheet metal work. Since, we can create Contour Flanges, Contour Rolls, or Lofted Flanges with elliptical or spline segments within the feature profile, there is some extra length created in flat pattern of these features. To compensate for this extra length in these special cases, you specify the value of Spline factor in Sheet Metal Defaults. By default, the value is 0.5.

CREATION TOOLS

The tools to create components in sheetmetal are available in the **Create** panel of the **Sheet Metal** tab of the **Ribbon**; refer to Figure-9.

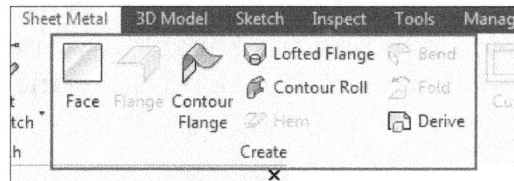

Figure-9. Create panel in Sheet Metal tab

The procedures of using the tools in this panel are discussed next.

Using Face Tool

The **Face** tool is used to create a sheet metal body by extruding closed sketch profile. The procedure to use this tool is given next.

- Click on the **Face** tool from **Create** panel in the **Sheet Metal** tab of the **Ribbon**. The **Face** dialog box will be displayed; refer to Figure-10.

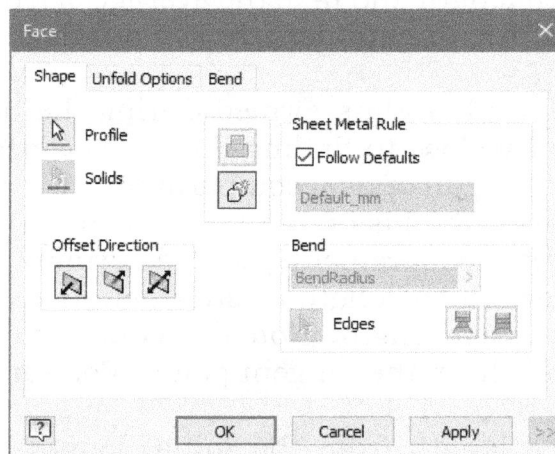

Figure-10. Face dialog box

- If there is single sketch in the drawing area then it will get automatically selected otherwise you will be asked to select a sketch using which you want to create a face feature; refer to Figure-11.

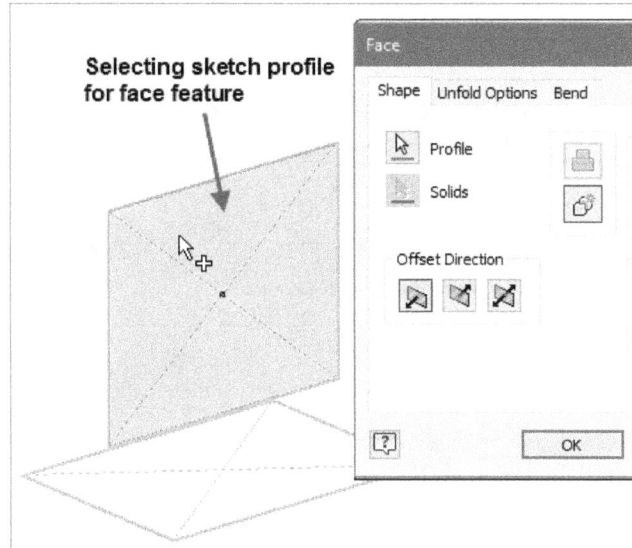

Figure-11. Sketch profile selected for face feature

- On selecting the profile, preview of the sheetmetal face feature will be displayed. Note that the default thickness will be added to the profile.
- Select desired button from the **Offset Direction** area in the **Shape** tab of the dialog box to define the side of thickness with respect to selected profile.
- By default, the Sheet Metal Default values are applied for unfolding and bending but you can change the values by using the options in the **Unfold Options** and **Bend** tab. The options in these tabs are same as discussed earlier.
- Click on the **Apply** button and select the other profile to create face feature or click on the **OK** button to create the current face feature and exit the tool.

Using Flange Tool

The **Flange** tool is used to create a face of sheet metal at specified angle to the selected edge; refer to Figure-12.

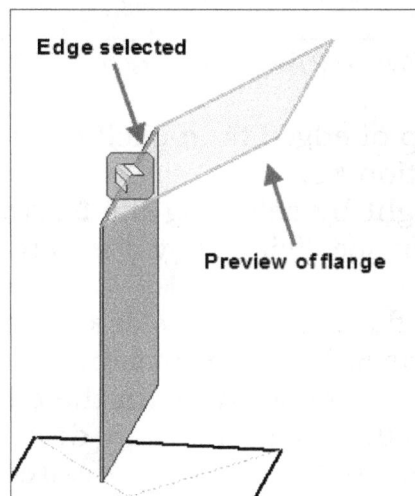

Figure-12. Preview of flange

The procedure to use this tool is given next.

- Click on the **Flange** tool from the **Create** panel in the **Sheet Metal** tab of the **Ribbon**. The **Flange** dialog box will be displayed; refer to Figure-13 and you will be asked to select an edge.

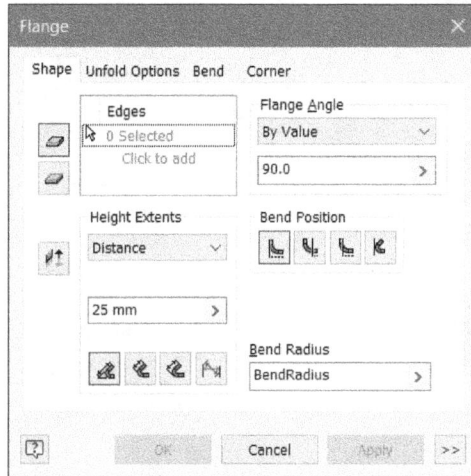

Figure-13. Flange dialog box

- Select the edge/edges on which you want to create the flange. Preview of the flange/flanges will be displayed; refer to Figure-14.

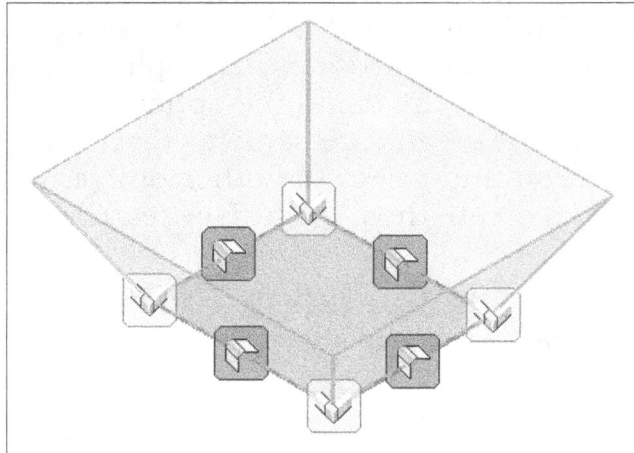

Figure-14. Preview of flanges

- If you want to select a loop of edges then click on the **Loop Select Mode** button adjacent to **Edges** selection box.
- Apply desired value of height by selecting the **Distance** option from the **Height Extents** drop-down and then specifying the value in the **Distance** edit box displayed below the drop-down.
- If you want to extend the flanges up to selected reference vertex or point then select the **To** option from the **Height Extents** drop-down and select the reference point. You can also specify the offset value in the edit box displayed.
- Click in the **Flange Angle** edit box and specify desired value of flange angle.
- Click in the **Bend Radius** edit box and specify desired value of bend radius at the selected edges. By default, the value set in **Sheet Metal Defaults** is applied.
- Change the reference for measuring height of flange by selecting desired button from the **Height Datum** area.

- You can change the bend position by selecting desired button from the **Bend Position** area of the dialog box.
- If you want to flip the direction of flanges then click on the **Flip Direction** button adjacent to **Height Extents** area.

Mitered Flange

Mitered Flanges are created in similar way to creating edge flange in Autodesk Inventor with the only difference that the **Apply Auto-Mitering** check box is selected in the **Corner** tab of the dialog box; refer to Figure-15. Figure-16 shows mitered flanges and edge flanges. Note that the **Apply Auto-Mitering** check box is selected by default in the dialog box.

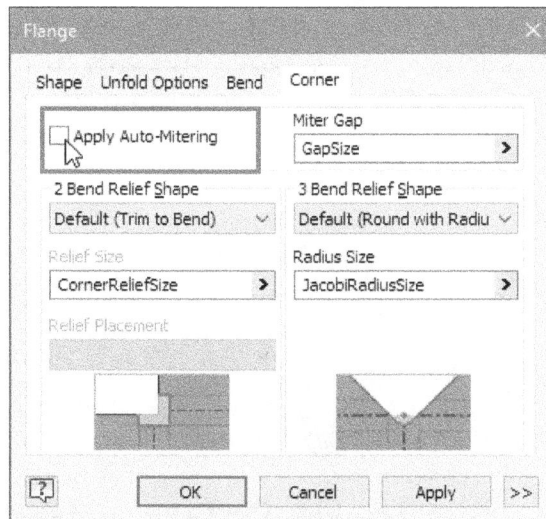

Figure-15. Apply Auto Mitering check box

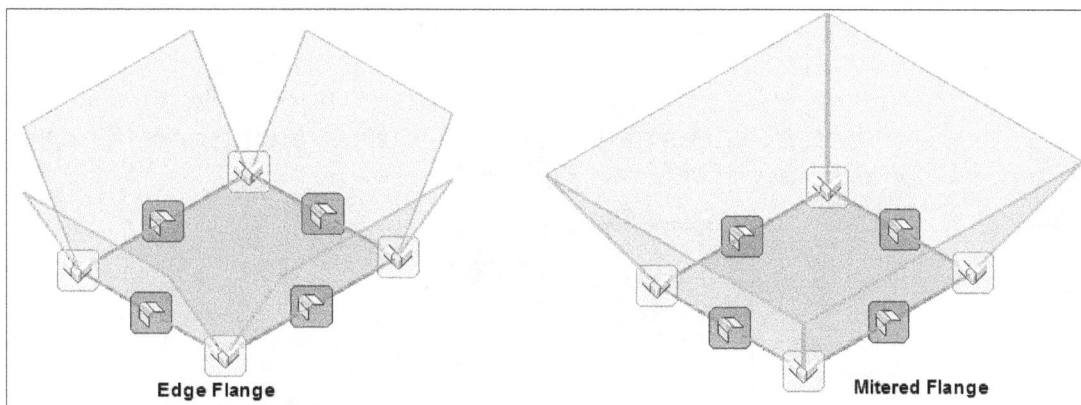

Figure-16. Edge Flange and Mitered Flange

Dynamic Editing of Flange

If you look at the preview of flange in drawing area then two type of buttons are displayed on corners and edges viz. **Corner Edit** button and **Bend Edit** button. The procedure to use these buttons is given next.

- Click on the **Corner Edit** button to edit the parameters related to respective corner. The **Corner Edit** dialog box will be displayed; refer to Figure-17.

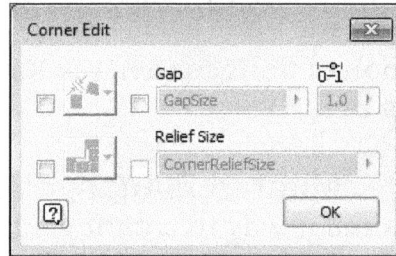

Figure-17. Corner Edit dialog box

- Select the upper left check box to change the parameters related to gap. The drop-down next to the check box will become active; refer to Figure-18. There are four options in this drop-down; **Symmetric Gap**, **Overlap**, **Reverse Overlap**, and **No Seam**. Select the **Symmetric Gap** option if you want the gap to be equal on both sides of dividing line at corner. If you want the gap to be on one side of dividing line then select the **Overlap** or **Reverse Overlap** option. If you want to apply no seam and want the flanges created with default gap then select the **No Seam** option.

Figure-18. Gap drop-down

- If you have selected the **Symmetric Gap**, **Overlap**, or **Reverse Overlap** option then you can select the **Gap** check box and specify desired value of gap in the edit box.
- Similarly, select the **Trim to Bend** check box and specify desired options related to trimming at bends of sheet for making flange.
- Click on the **OK** button from the **Corner Edit** dialog box to apply the changes.
- If you want to edit extents of bends in the flange then click on the **Bend Edit** button displayed at the edges of the flanges in the drawing area. The **Bend Edit** dialog box will be displayed; refer to Figure-19.

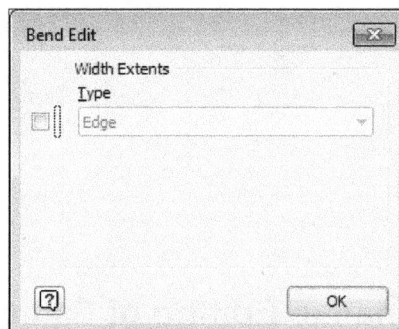

Figure-19. Bend Edit dialog box

- Select the check box from the dialog box to activate the options in the drop-down.
- By default, the **Edge** option is selected in the drop-down which means the flange will be created up to the full length of the selected edge.

- If you want to specify the extents of flange measured from the center line then select the **Width** option from the drop-down. The dialog box will be displayed as shown in Figure-20 along with the preview of flange. Select the **Centered** radio button and specify desired value for width.

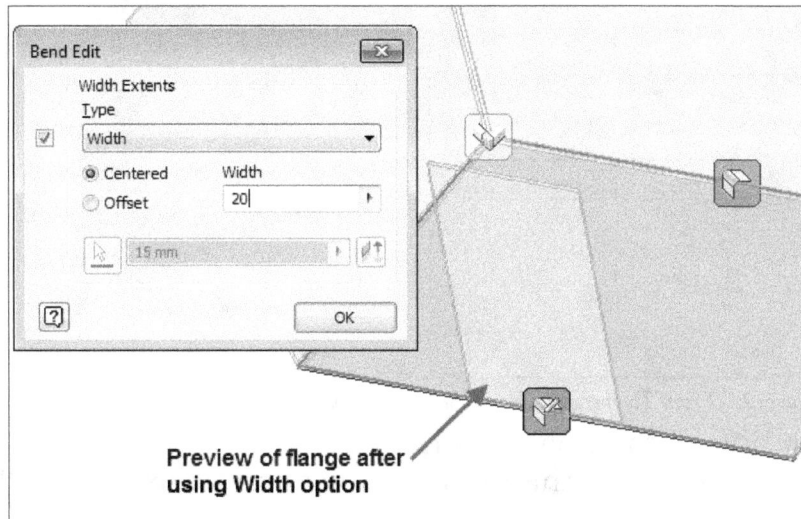

Figure-20. Preview of flange with Width option selected

- If you want to move the flange to desired location over the edge then select the **Offset** radio button. You will be asked to specify the starting point of the flange. Click on the side edge to specify starting point and specify desired value of offset in the related edit box. Preview of the flange will be displayed; refer to Figure-21. If you want the flange to be created between the specified points then select the **From To** option from the drop-down and select the points. Preview of the flange will be displayed; refer to Figure-22.

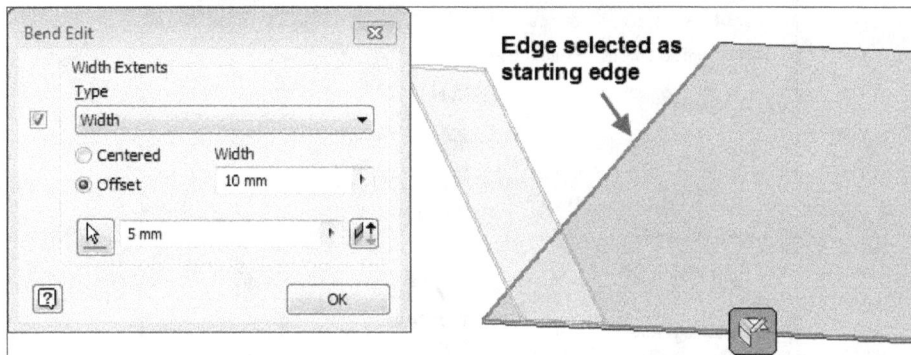

Figure-21. Preview of offset flange

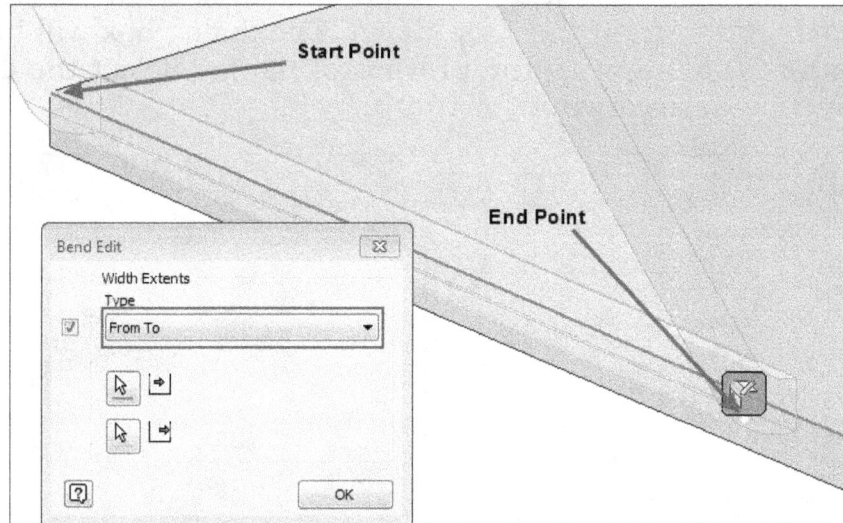

Figure-22. From To option in Bend Edit dialog box

- Click on the **OK** button to apply the changes.
- Click on the **OK** button from the **Flange** dialog box to create the flange.

Using Contour Flange

The **Contour Flange** tool is used to create flange with specified shape (contour). The procedure to use this tool is given next.

- Click on the **Contour Flange** tool from the **Create** panel in the **Sheet Metal** tab of the **Ribbon**. The **Contour Flange** dialog box will be displayed; refer to Figure-23. Make sure that you have an open sketch in the drawing area.

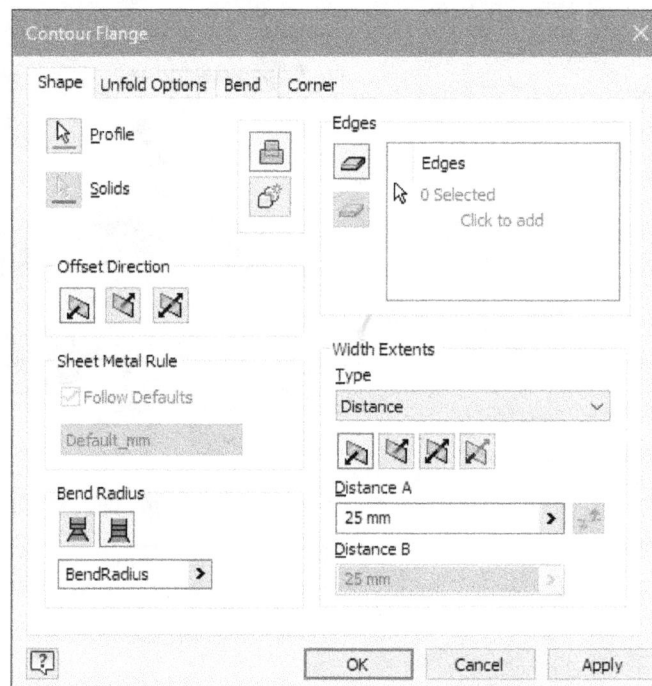

Figure-23. Contour Flange dialog box

- If there is only one profile sketch then it will be selected automatically. If there are two or more profile sketches then first created profile will be selected. To select a different profile, click on the **Profile** button and select desired profile sketch; refer to Figure-24.

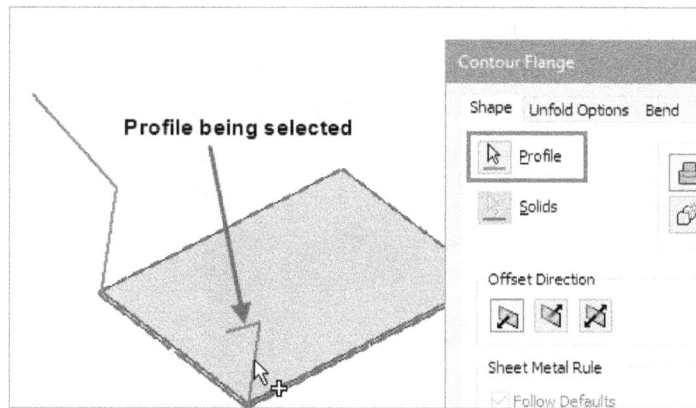

Figure-24. Profile to be selected

- On selecting the profile, you will be asked to select the connected edge. Select an edge and preview of the contour flange will be displayed; refer to Figure-25. Select more edges if required, preview will be displayed accordingly.

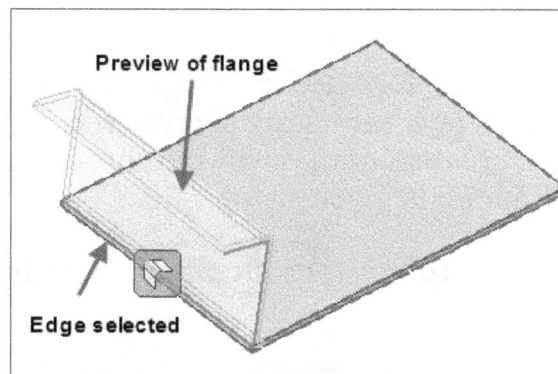

Figure-25. Preview of contour flange

- Most of the options in this dialog box are same as discussed for **Flange** tool except **Distance** option in the **Type** drop-down of the **Width Extents** area in the **Contour Flange** dialog box; refer to Figure-26.

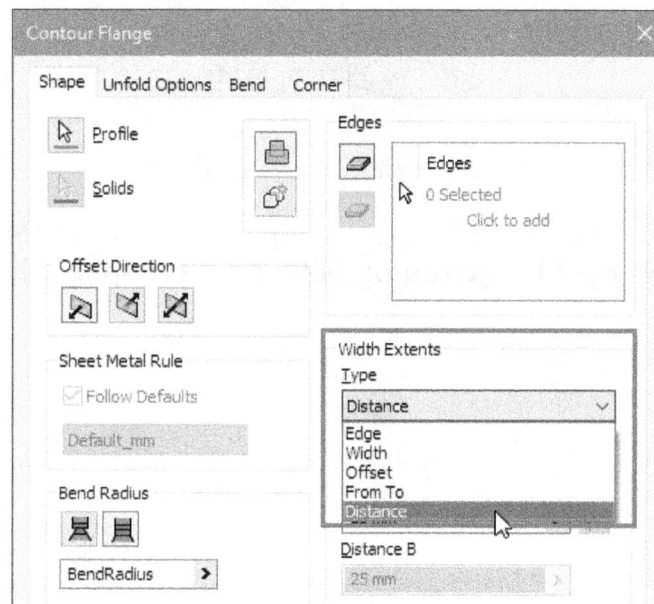

Figure-26. Width Extents drop-down

- If you do not have edges as path reference and want to create a contour flange at straight path of specified distance then select the **Distance** option from the **Type** drop-down in the **Width Extents** area of the dialog box. The options to specify distance and direction will be displayed in the dialog box; refer to Figure-27.

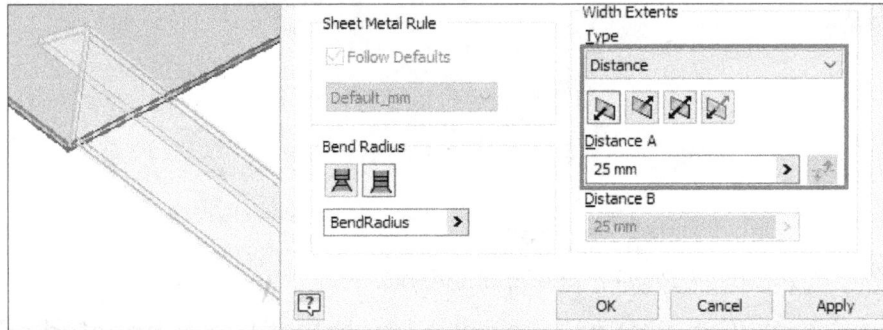

Figure-27. Options to specify distance and direction

- Specify desired value of distance and select desired direction.
- Click on the **OK** button to create the flange.

Using Lofted Flange Tool

The **Lofted Flange** tool is used to create flange between two selected profile sketches. Note that the edges of the created flange will be shaped as selected profiles. The procedure to create lofted flange is given next.

- Create the two sketches of profiles for lofted flanges; refer to Figure-28.

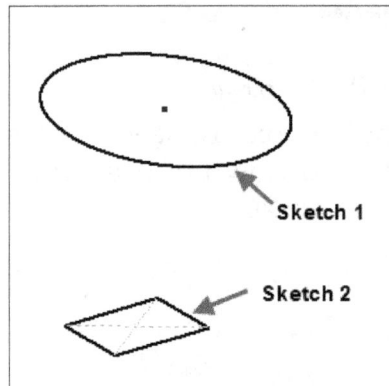

Figure-28. Profile sketches for lofted flange

- Click on the **Lofted Flange** tool from the **Create** panel in the **Sheet Metal** tab of the **Ribbon**. The **Lofted Flange** dialog box will be displayed; refer to Figure-29.

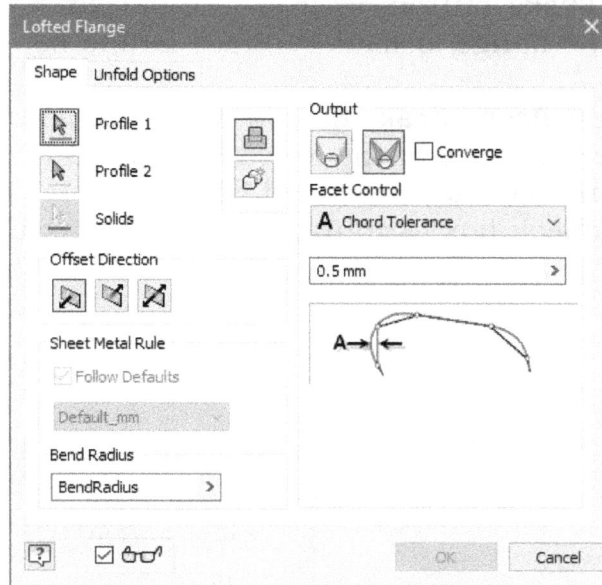

Figure-29. Lofted Flange dialog box

- Select the first profile from the drawing area. You will be asked to select the second profile.
- Click on the second profile from the drawing area. Preview of the lofted flange will be displayed; refer to Figure-30.

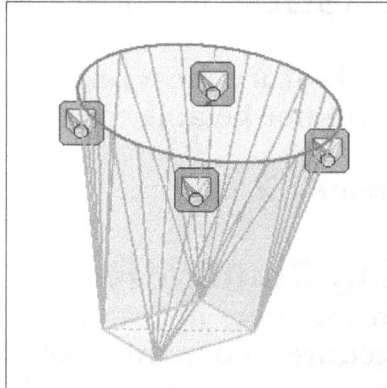

Figure-30. Preview of the lofted flange

- Click on the **Die Formed** button from the **Output** area of the dialog box if you want the flange created as die formed; refer to Figure-31.

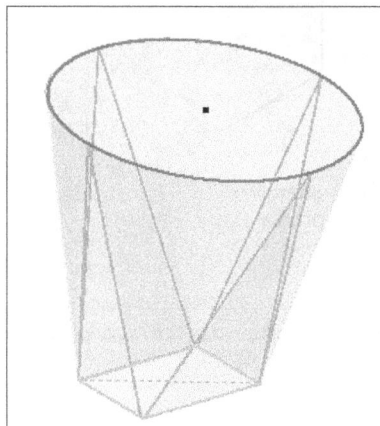

Figure-31. Preview of die formed lofted flange

- Click on the **Press Brake** button ⬚ from the **Output** area of the dialog box if you want to create the flange formed by Press-Brake bending machine; refer to Figure-30.
- If you have selected the **Press Brake** button then you can set the chord tolerance, facet angle, and facet distance of the lofted flange. All the three parameters are used to modify the surface shape of lofted flange.
- Select the **Converge** check box to converge all the break lines at one point; refer to Figure-32.

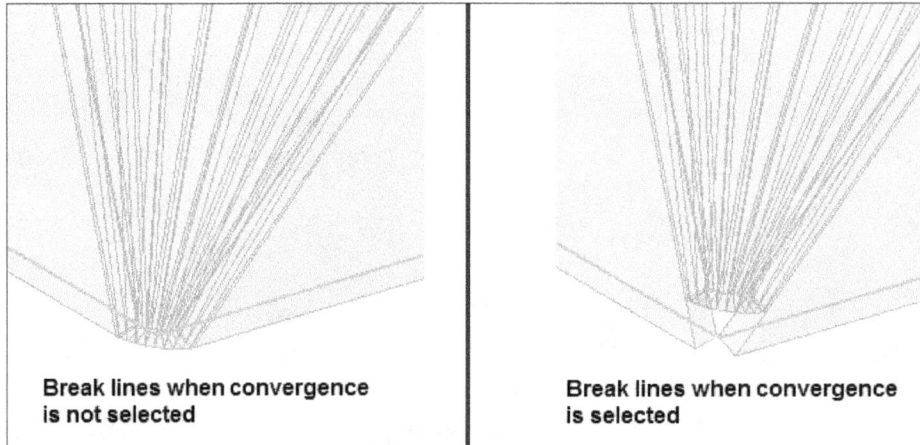

Break lines when convergence is not selected

Break lines when convergence is selected

Figure-32. Effect of Convergence check box

- The options in the **Unfold Options** tab of the dialog box have already been discussed.
- Note that if **Enable/Disable feature preview** check box ☑ 👓 is not selected in the dialog box then you will not be able to see the preview while creating the feature.
- Click on the **OK** button to create the feature.

Using Contour Roll Tool

The **Contour Roll** tool is used to create sheet metal flange by rolling the selected profile about an axis. The procedure to use this tool is given next.

- Create a sketch with an axis similar to shown in Figure-33.

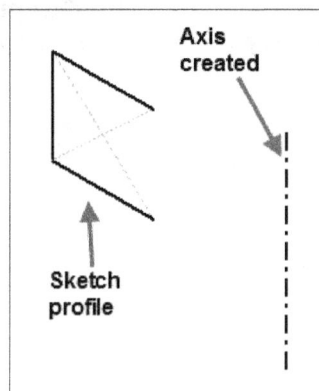

Axis created

Sketch profile

Figure-33. Sketch created for contour roll feature

- Click on the **Contour Roll** tool from the **Create** panel in the **Sheet Metal** tab of the **Ribbon**. The **Contour Roll** dialog box will be displayed; refer to Figure-34. If you have only a profile sketch and axis then preview will be displayed with profile sketch and axis selected automatically; refer to Figure-35.

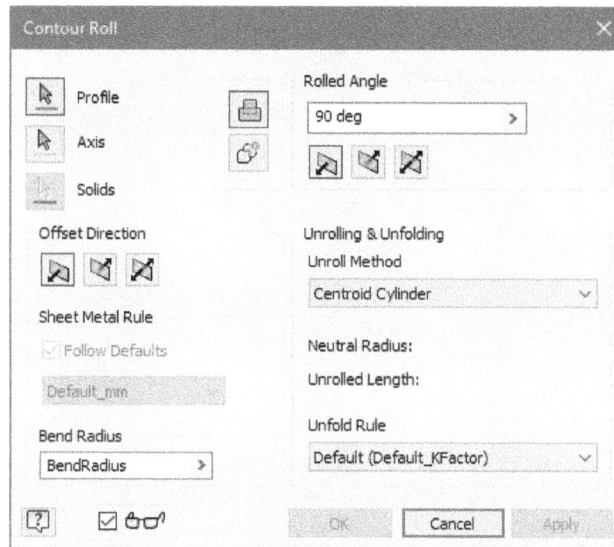

Figure-34. Contour Roll dialog box

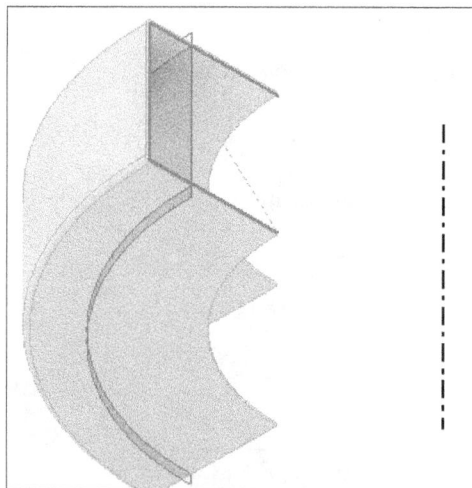

Figure-35. Preview of the contour rolled flange

- Set desired angle value of roll in the **Rolled Angle** edit box.
- The three buttons ⬚ ⬚ ⬚ in the **Offset Direction** area of the dialog box are used to flip the thicken side of flange with respect to profile sketch.
- The three buttons ⬚ ⬚ ⬚ in the **Rolled Angle** area of the dialog box are used to flip the direction of rolling.
- Set desired unfolding and unrolling options and then click on the **OK** button from the dialog box to create the feature.

Using Hem Tool

Hemming is a forming operation in which sheet is folded on its edge or folded over another part in order to achieve a tight fit. Normally, hemming operations are used to connect parts together, to improve the appearance of a part and to reinforce part edges. In Autodesk Inventor, we use **Hem** tool to create hemming in sheet metal part. The procedure to use this tool is given next.

- Click on the **Hem** tool from the **Create** panel in the **Sheet Metal** tab of the **Ribbon**. The **Hem** dialog box will be displayed; refer to Figure-36. Also, you will be asked to select an edge to create hem. Note that the **Hem** tool will be active in **Create** panel only when you have a sheet metal base part created in the drawing area.

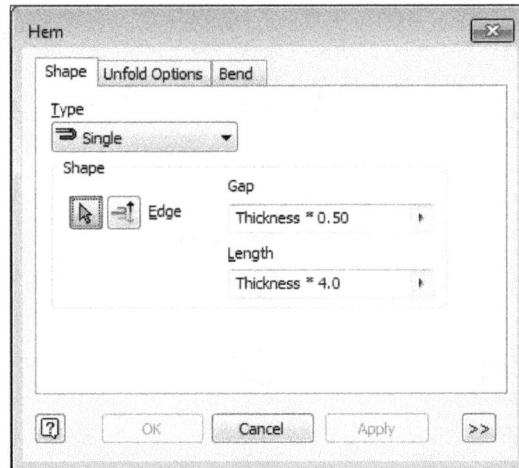

Figure-36. Hem dialog box

- Select an edge through which you want to create hem. Preview of the hem will be displayed; refer to Figure-37.

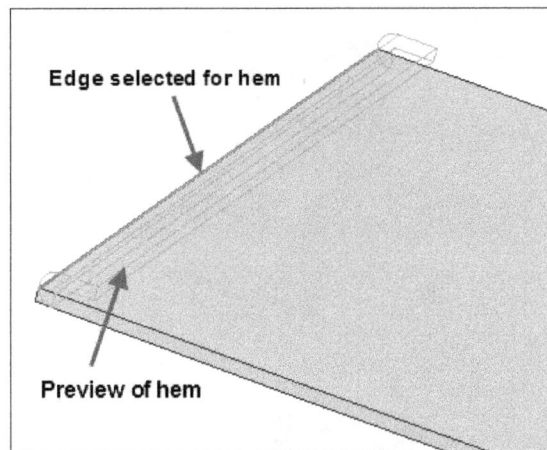

Figure-37. Preview of hem

- Specify desired value of **Gap** and **Length** in the respective edit boxes in the dialog box.
- If you want to create **Teardrop**, **Rolled**, or **Double** hem then select the respective option from the **Type** drop-down. Specify desired parameters based on your selection.
- Expand the dialog box to modify the options related to **Width Extents**; refer to Figure-38.

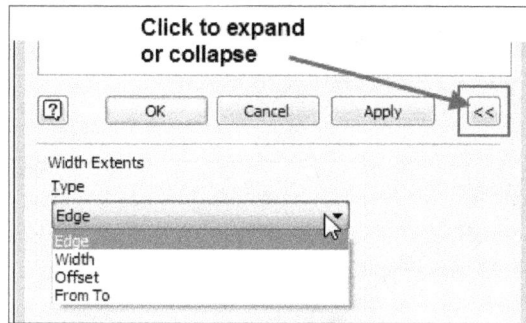
Figure-38. Expanded Hem dialog box

• Specify the other required parameters as discussed earlier in the chapter. Click on the **OK** button to create the feature; refer to Figure-39.

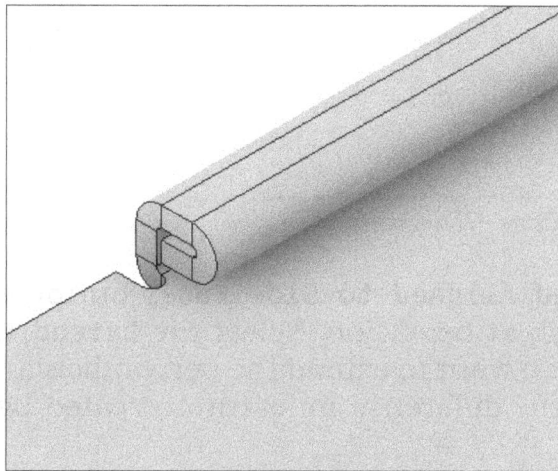
Figure-39. Double hem with width specified and round relief

Using Bend Tool

The **Bend** tool is used to create a bend by joining two sheet metal parts. You will be asked to select edge of each part that should be joined while creating the bend. The procedure to use this tool is given next.

• Click on the **Bend** tool from the **Create** panel in the **Sheet Metal** tab of the **Ribbon**. The **Bend** dialog box will be displayed; refer to Figure-40. Also, you will be asked to select edges of the planar sheet metal faces.

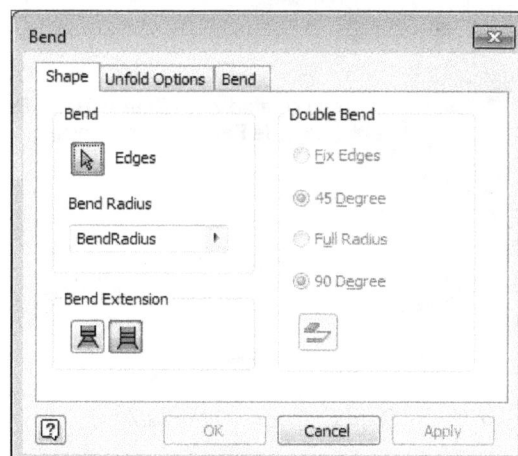
Figure-40. Bend dialog box

- Select the edges of the planar faces of the two sheetmetal parts that are to be joined by bend; refer to Figure-41.

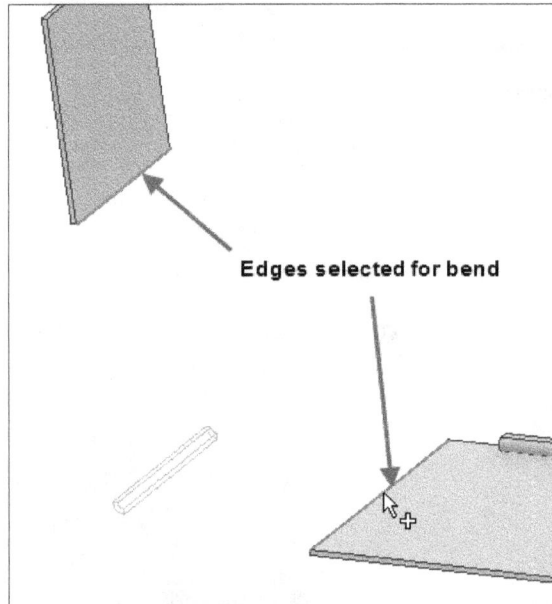

Figure–41. Edges selected for bend

- Select the **Extend Bend Aligned to Side Faces** button 🔲 if you want to extend the side faces to match at bend line. Select the **Extend Bend Perpendicular to Side Faces** button if you want to extend the perpendicular face up to the side face. Figure-42 illustrates the difference in features created by selecting two button.

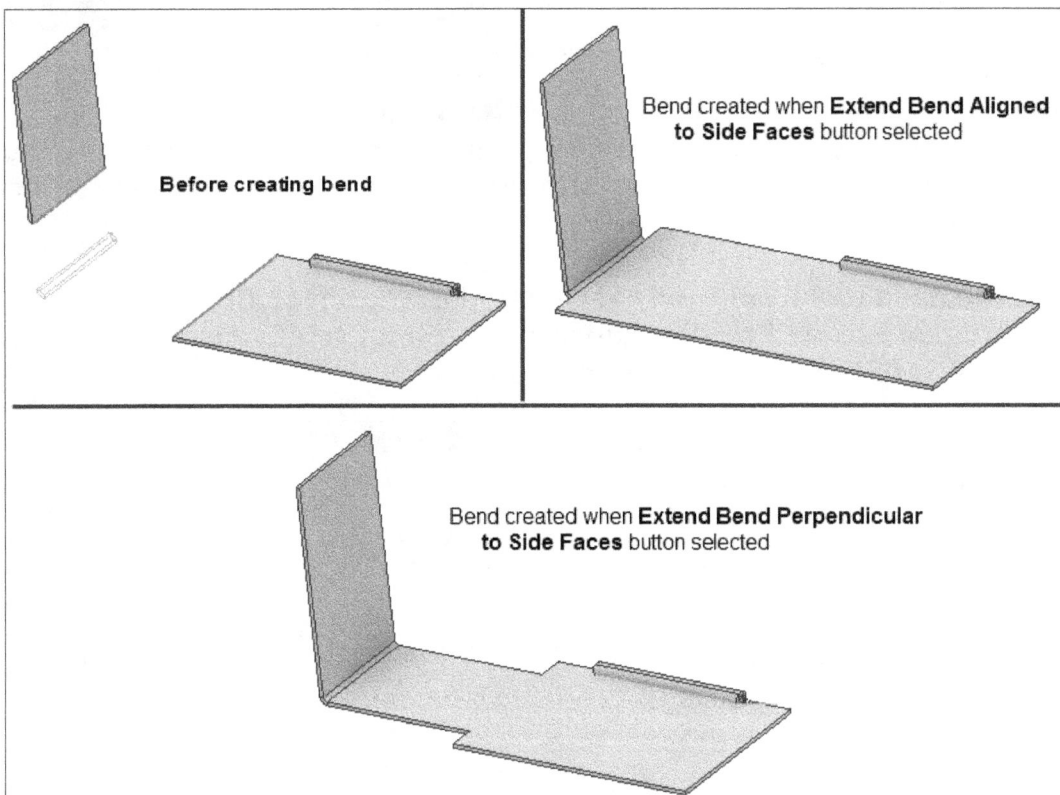

Figure–42. Bend created using different Bend Extension options selected

- Specify desired value of bend radius in the **Bend Radius** edit box.

- If you have two parallel sheet metal planar faces not in the same plane to join by bend then options in the **Double Bend** area of the dialog box will become active on selecting the edges. Figure-43 shows the preview of bend using **Fix Edges** and **45 Degree** radio buttons. Figure-44 shows the preview of bend using the **Full Radius** and **90 Degree** radio buttons.

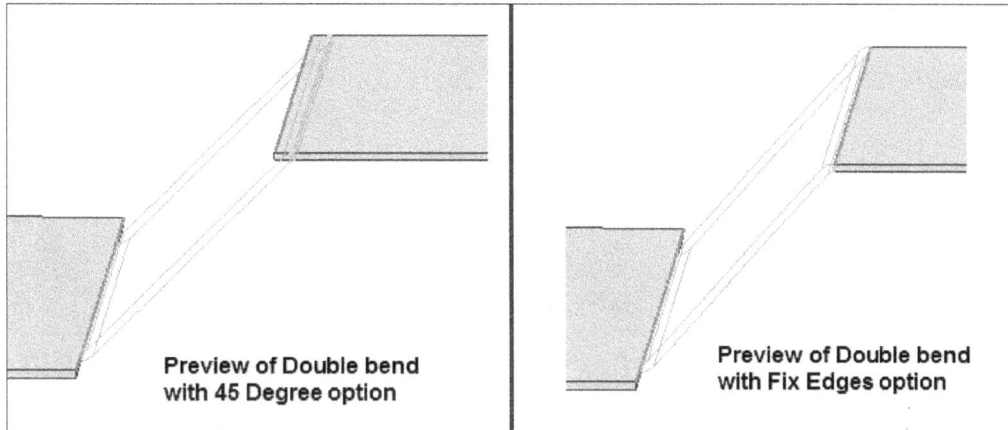

Preview of Double bend with 45 Degree option

Preview of Double bend with Fix Edges option

Figure-43. Preview of double bend with 45 Degree and Fix Edges options

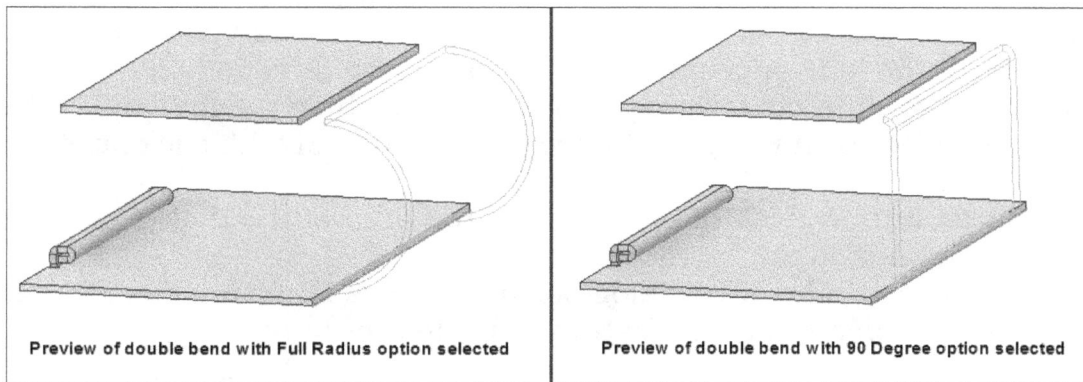

Preview of double bend with Full Radius option selected

Preview of double bend with 90 Degree option selected

Figure-44. Preview of double bend with Full Radius and 90 Degree options

- Click on the **Flip Fixed Edge** button 🔁 to flip the fixed edge between the two edges selected for bend.
- Click on the **OK** button from the dialog box to create the bend.

Using Fold Tool

The **Fold** tool is used to fold the selected sheet metal face using a bend line. Note that the end points of bend line should end at the edges of face selected. The procedure to use this tool is given next.

- Click on the **Fold** tool from the **Create** panel in the **Sheet Metal** tab of the **Ribbon**. The **Fold** dialog box will be displayed as shown in Figure-45. Also, you will be asked to select the bend line.

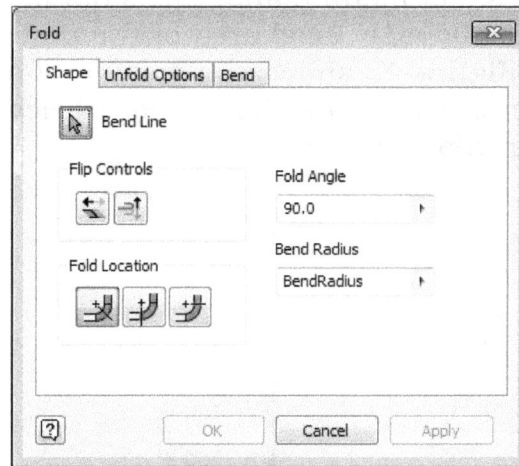

Figure-45. Fold dialog box

- Select the bend line created on the sheet metal face which you want to fold. Note that the bend line is a single sketched line created on the sheet metal face to be folded. The end points of bend line always end at the edges of the sheet metal face to be folded.
- On selecting the bend line, two arrows are displayed on the bend line which denote the direction of fold. The horizontal arrow directs towards the side that will be folded (the side opposite to horizontal arrow direction will remain fixed). The vertical arrow directs toward the direction in which the face will be folded, it can be upward or downward. To change the direction of fold, click on the buttons in the **Fold Controls** area of the dialog box.
- Select desired button from the **Fold Location** area of the dialog box to change the start location of fold.
- The other options in the **Fold** dialog box are same as discussed earlier. Click on the **OK** button to create the fold feature; refer to Figure-46.

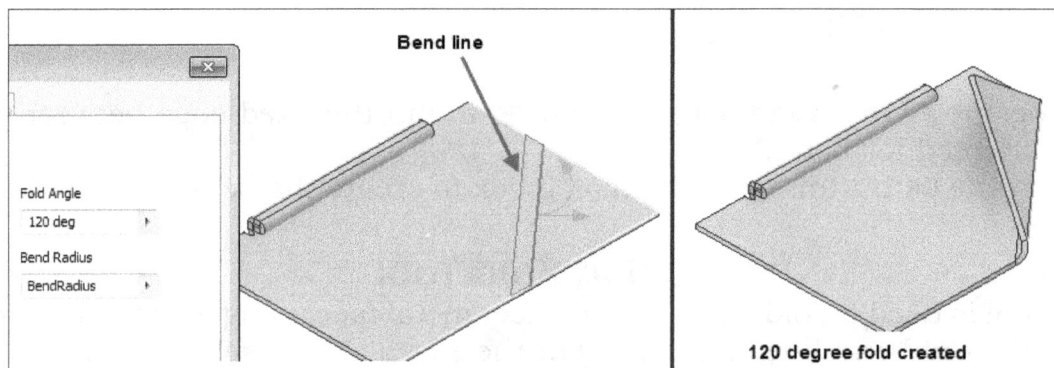

Figure-46. Fold feature creation

Using Derive Tool

The **Derive** tool is used to derive solid parts or assemblies into the current environment. The procedure to use this tool is given next.

- Click on the **Derive** tool from the **Create** panel in the **Sheet Metal** tab of the **Ribbon**. The **Open** dialog box will be displayed as shown in Figure-47.

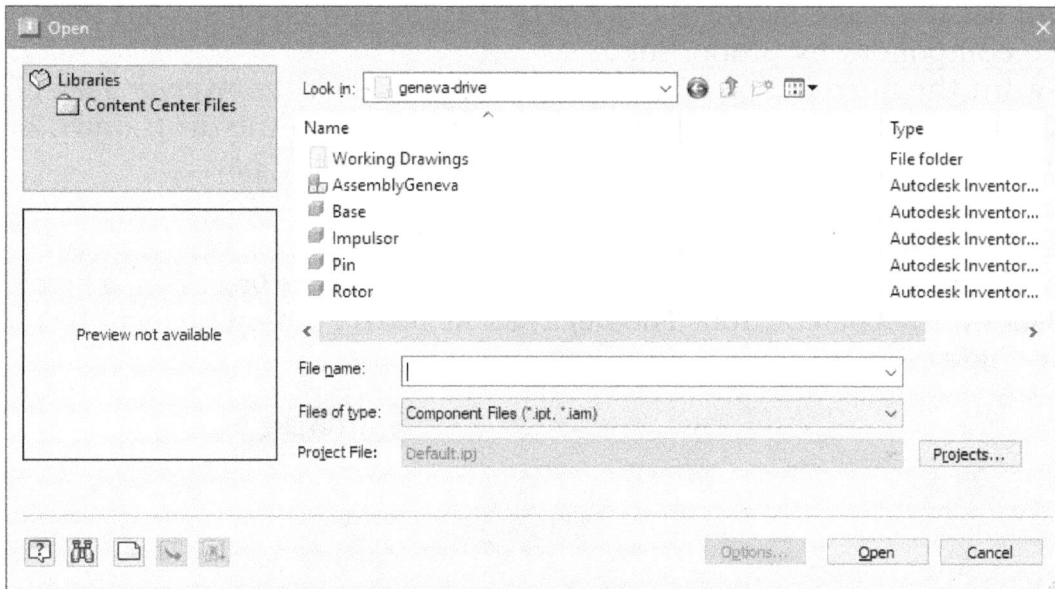

Figure-47. Open dialog box for deriving parts or assemblies

Deriving part

• Select a part file from the dialog box and click on the **Open** button. The **Derived Part** dialog box will be displayed as shown in Figure-48.

Figure-48. Derived Part dialog box

• Select desired derive style from the **Derive Style** options. There are four buttons to change derive style in the dialog box; **Single solid body merging out seams between planar faces**, **Solid body keep seams between planar faces**, **Maintain each solid as a solid body**, and **Body as Work Surface**. These buttons perform the task as per their names.

• Click on the plus sign before the component to include/exclude it from deriving. Similarly, you can include/exclude other properties of the part by using the plus sign.

- You can set the scale factor in the **Scale factor** edit box from **Options** tab to re-size the component by proportion.
- If you want the mirror copy of derived part then select the **Mirror part** check box from **Options** tab and select desired mirror plane from the drop-down below it.
- Click on the **OK** button from the dialog box to insert the part.

Deriving Assembly

- After clicking on the **Derive** tool, select an assembly in the **Open** dialog box displayed and click on the **Open** button. The **Derived Assembly** dialog box will be displayed; refer to Figure-49.

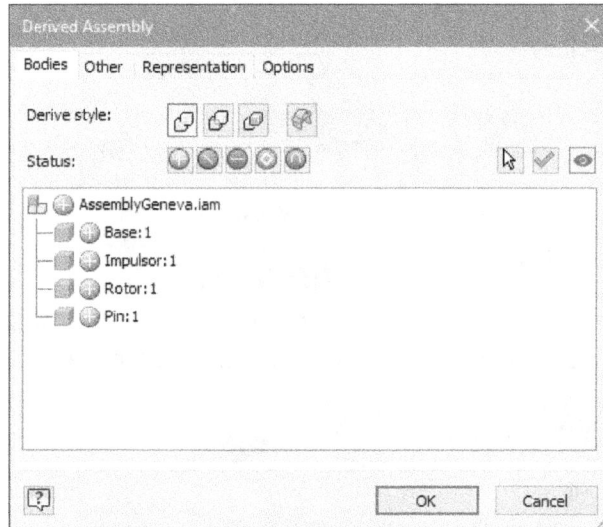

Figure-49. Derived Assembly dialog box

- The options for derive style are same as discussed earlier.
- Click on the plus sign before the component to switch between different modes of deriving like includes the selected components ⊙, excludes the selected components ⊙, subtracts the selected components ⊙, includes bounding boxes of the selected component ⊙, or intersects the selected components with the derived result ⊙.
- Click on the **Other** tab in the dialog box and select the component to be included/excluded in deriving. The options in this tab have already been discussed.
- The options in the **Representation** tab are used to change the orientation, presentation, and level of detail for deriving assembly components.
- The options in the **Options** tab are used to modify miscellaneous details of deriving assembly components like scale factor, hole patching, mirroring assembly, and so on.
- After setting desired values, click on the **OK** button to derive the selected components.

Till this point, we have discussed about the sheet metal feature creation tool. Now, we will discuss about the tools used for modification in sheet metal parts.

MODIFICATION TOOLS

The modification tools are available in the **Modify** panel of the **Sheet Metal** tab in the **Ribbon**; refer to Figure-50. The procedures to use these tools are discussed next.

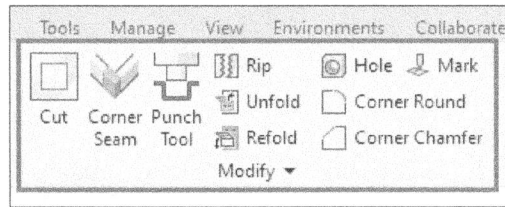
Figure-50. Modify panel in Sheet Metal tab

Using Cut Tool

The **Cut** tool, as the name suggests is used to cut material from the sheet metal face. The procedure to use this tool is given next.

- Click on the **Cut** tool from the **Modify** panel in the **Sheet Metal** tab of the **Ribbon**. The **Cut** dialog box will be displayed; refer to Figure-51. Also, you will be asked to select a close profile to make a cut.

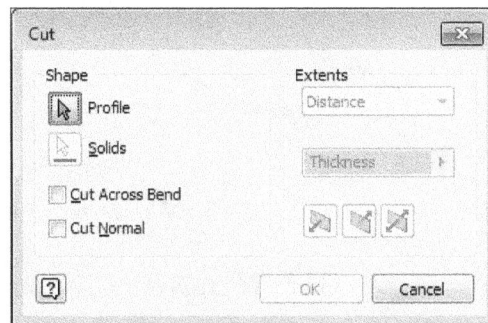
Figure-51. Cut dialog box

- Select the profile sketch from the drawing area. Preview of the cut will be displayed; refer to Figure-52.

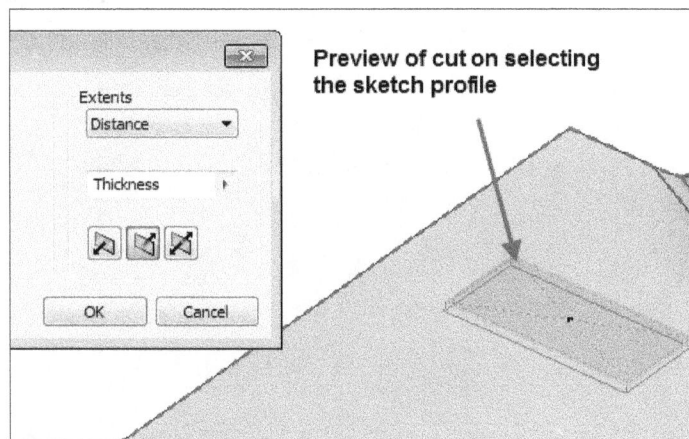
Figure-52. Preview of cut

- Select desired extent option from the drop-down in the **Extents** area and specify the related parameters. Most of the time, we select the **To Next** option from the drop-down because generally we are cutting one layer of sheet metal at a time.
- If you are cutting a sheet metal at bend then you have two options to select:- **Cut Across Bend** and **Cut Normal**. Select the **Cut Across Bend** check box if you want the complete selected profile to be cut from base material; refer to Figure-53.

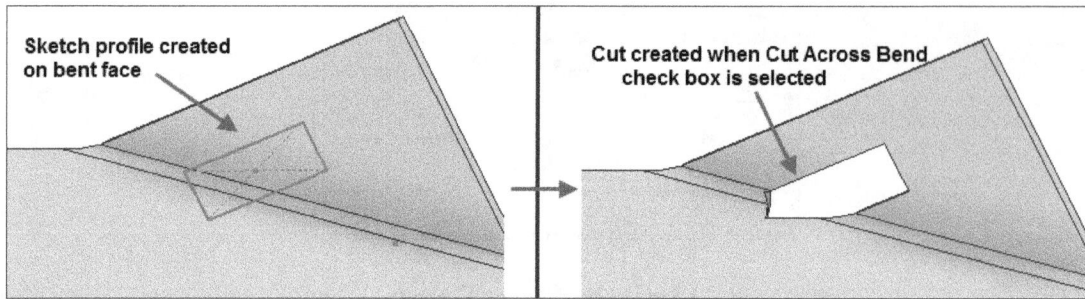

Figure-53. Cut created using Cut Across Bend check box

- If you have selected the **Cut Normal** check box in place of **Cut Across Bend** in the same example then the cut would be created as shown in Figure-54.

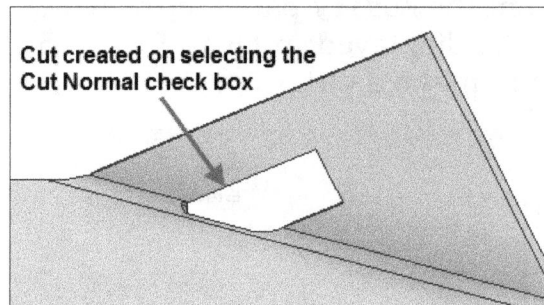

Figure-54. Cut created using Cut Normal check box

Using Corner Seam Tool

The **Corner Seam** tool is used to define or create gap between the adjoining faces of flange walls. The procedure to use this tool is given next.

- Click on the **Corner Seam** tool from the **Modify** panel in the **Sheet Metal** tab of the **Ribbon**. The **Corner Seam** dialog box will be displayed; refer to Figure-55.

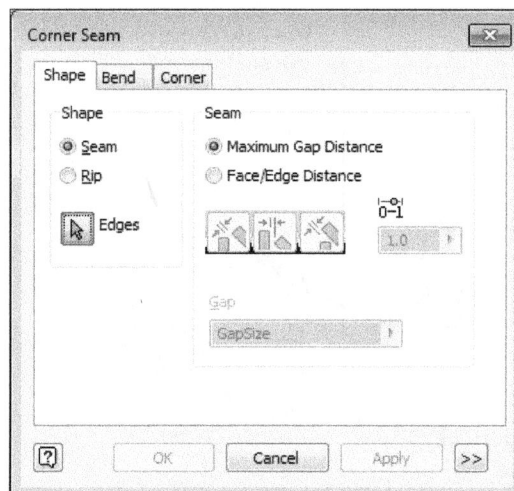

Figure-55. Corner Seam dialog box

Using Seam Option

- Select the **Seam** radio button to modify bend/corner already created. You will be asked to select the edges to be seamed.
- Select two adjacent edges of bend faces. Preview of the seam will be displayed; refer to Figure-56.

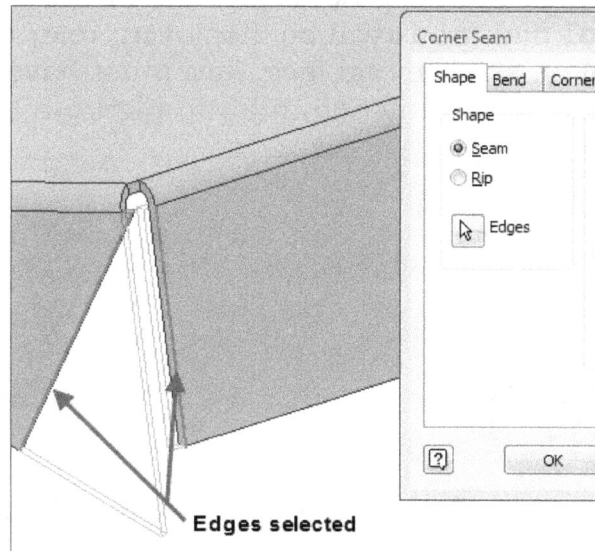

Figure-56. Edges selected for corner seam

- Select the **Maximum Gap Distance** or **Face/Edge Distance** radio button to specify whether you want to specify the maximum gap distance or face/edge distance.
- Depending on the radio button selected, you can select the **Symmetric Gap**, **Overlap**, **No Overlap**, or **Reverse Overlap** button to modify the gap between two faces.
- Specify desired gap value and click on the **OK** button to create the seam.

Using Rip Option

- Select the **Rip** radio button from the dialog box if you want to control gap created by **Rip** tool or gap created during conversion from solid to sheetmetal part. On selecting the **Rip** radio button, you will be asked to select the edge of ripped section.
- Select the edge of ripped corner; refer to Figure-57.
- Rest of the procedure is same as discussed for **Seam** radio button.

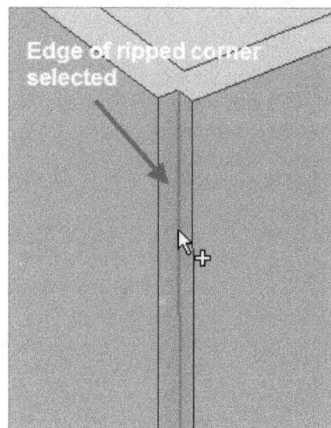

Figure-57. Edge of ripped corner selected

Using Punch Tool (Modified)

The **Punch Tool** is used to make indent mark of the shape carved on punch on to the sheet metal face. The procedure to use this tool is given next.

- Click on the **Punch Tool** from the **Modify** panel in the **Sheet Metal** tab of the **Ribbon**. The **PunchTool** dialog box will be displayed; refer to Figure-58. Note that before using this tool on sheet metal face, you must have a sketch point on the face of sheet metal part to be punched; refer to Figure-59.

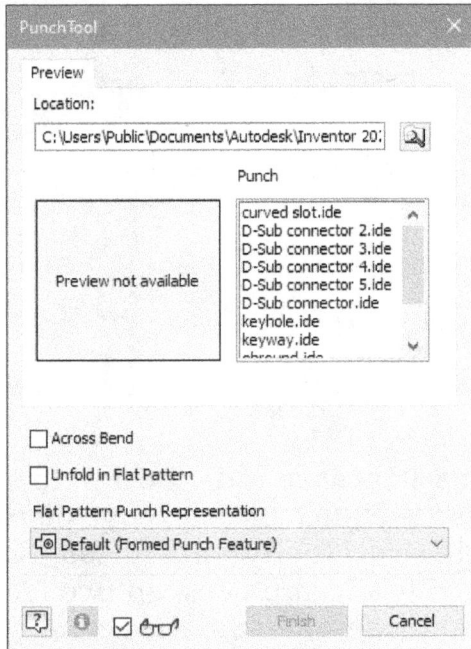

Figure-58. PunchTool dialog box

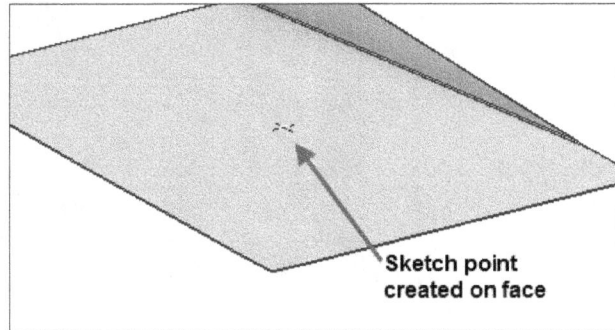

Figure-59. Sketch point created on face

- Click on the **Select PunchTool Library Folder** button from the dialg box. The **Punch Tool Directory** dialog box will be displayed; refer to Figure-60.

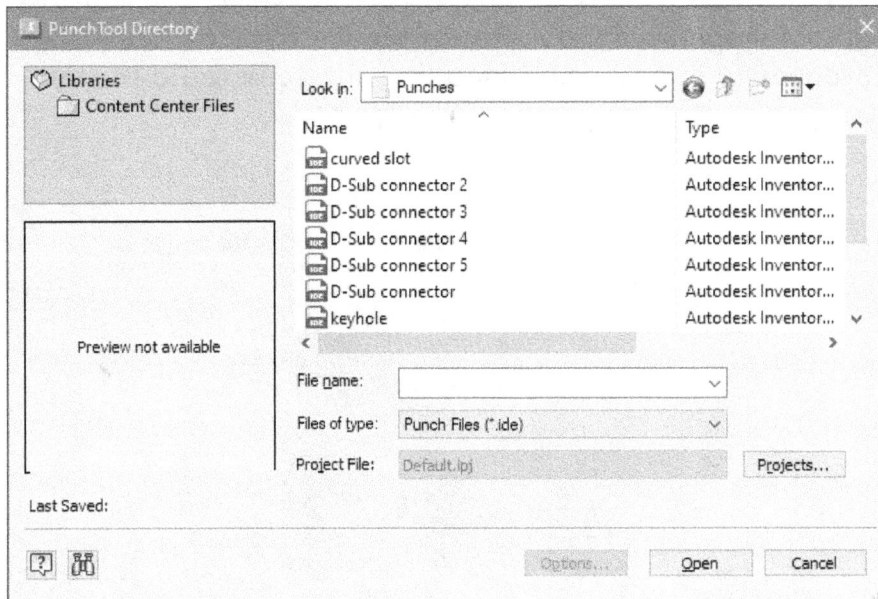

Figure-60. PunchTool Directory dialog box

- Select desired punch file from the dialog box and click on the **Open** button. The **Punch Tool** dialog box will be displayed along with preview sketch of punch; refer to Figure-61.

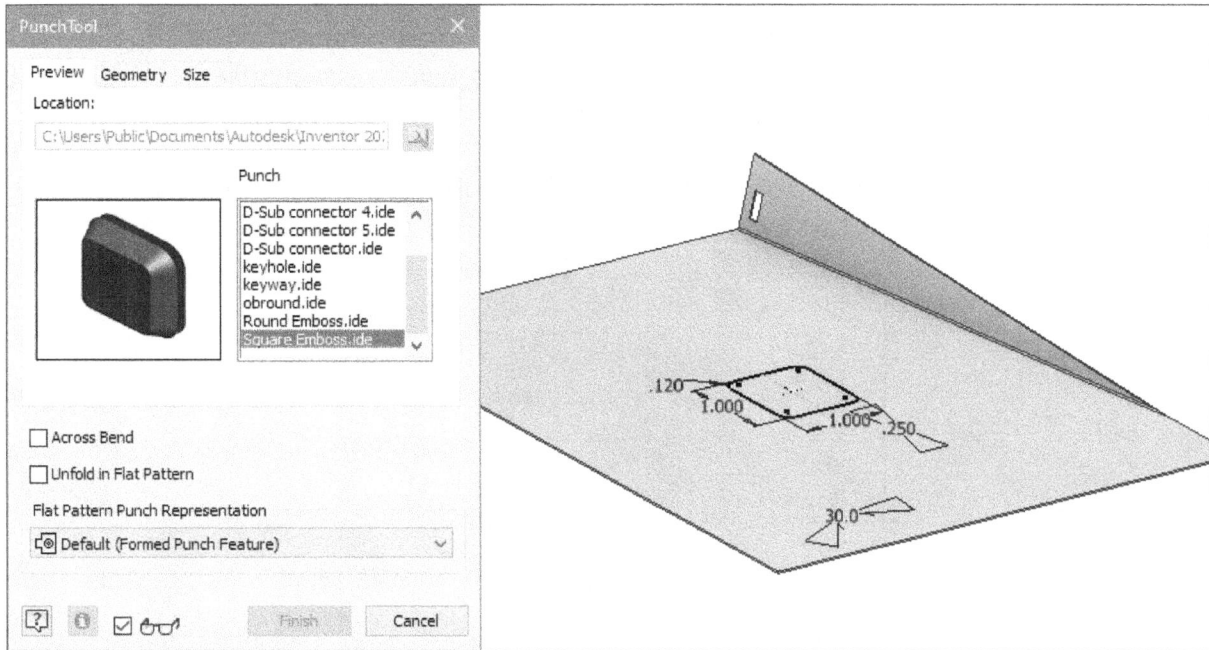

Figure-61. Punch Tool dialog box with sketch preview

- Select the **Across Bend** check box if you want the imprint of punch across the bends. If this check box is not selected then punch will stop at bends.
- Select the **Unfold in Flat Pattern** check box if you want to unfold the indent made by punch in the flat pattern.
- Click on the **Geometry** tab in the dialog box and specify desired value of angle in the **Angle** edit box; refer to Figure-62.

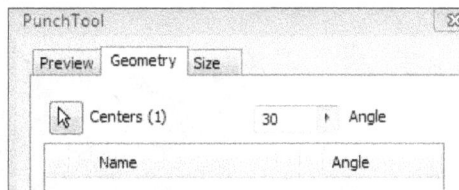

Figure-62. Angle edit box in PunchTool dialog box

- Click on the **Size** tab and specify the parameters of the punch in the table.
- Select desired representation of punch in flat pattern by clicking on the **Flat Pattern Punch Representation** drop-down.
- Click on the **Finish** button from the dialog box. The punch indent will be created; refer to Figure-63.

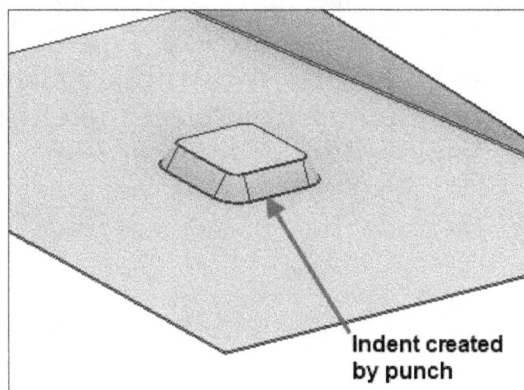

Figure-63. Indent created by punch

Using Rip Tool

The **Rip** tool is used to provide relief at the corners and edges for folding/unfolding of sheet metal components. This tool is most commonly used after creating lofted sheet metal feature. The procedure of using this tool is given next.

* Click on the **Rip** tool from the **Modify** panel in the **Sheet Metal** tab of the **Ribbon**. The **Rip** dialog box will be displayed as shown in Figure-64.

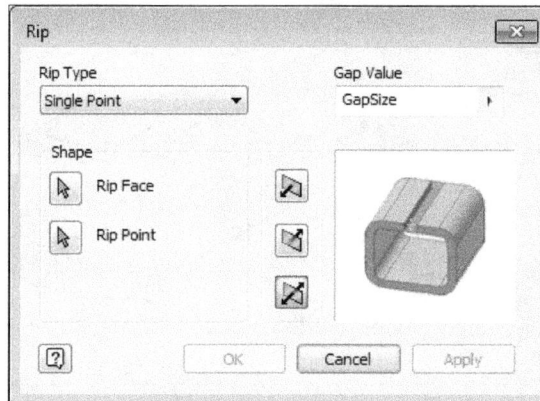

Figure-64. Rip dialog box

* There are three option in the **Rip Type** drop-down to rip the face(s). Select the **Single Point** option if you want to rip the face by the straight line passing through the selected point. You will be asked to select rip face and then rip point. Select the entities, preview of the rip feature will be displayed; refer to Figure-65.

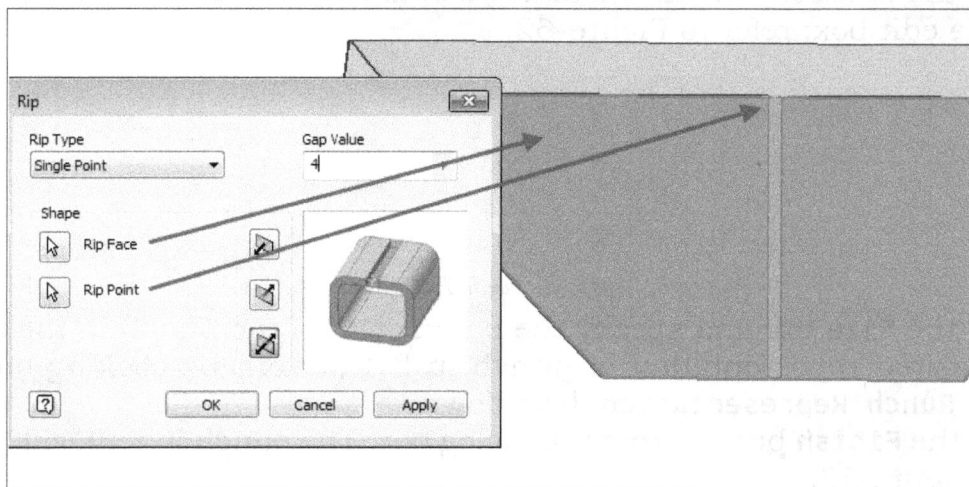

Figure-65. Preview of Rip with Single Point Rip Type option selected

* Select the **Point to Point** option from the **Rip Type** drop-down if you want to rip face between the two selected points. You will be asked to select rip face, start point, and then end point of rip feature. Select the entities accordingly. Preview of the rip feature will be displayed; refer to Figure-66.

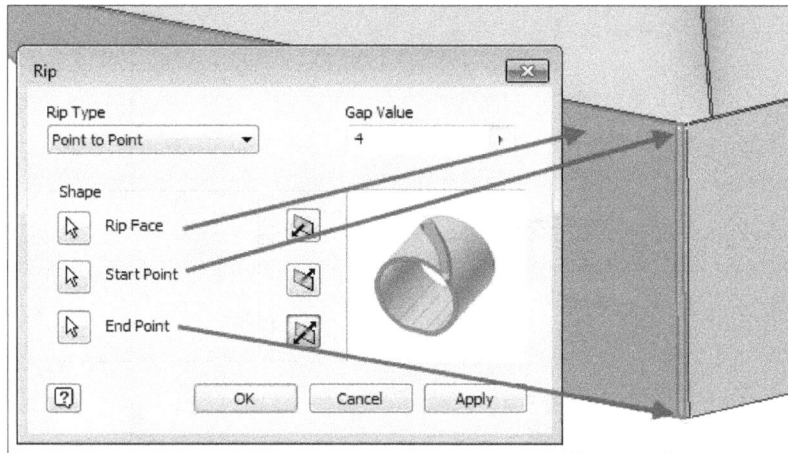

Figure-66. Preview of Rip with Point to Point Rip Type option selected

- Select the **Face Extents** option from the **Rip Type** drop-down if you want to rip the selected face completely. You will be asked to select rip face. Select the face you want to rip off. Preview of the rip feature will be displayed; refer to Figure-67.

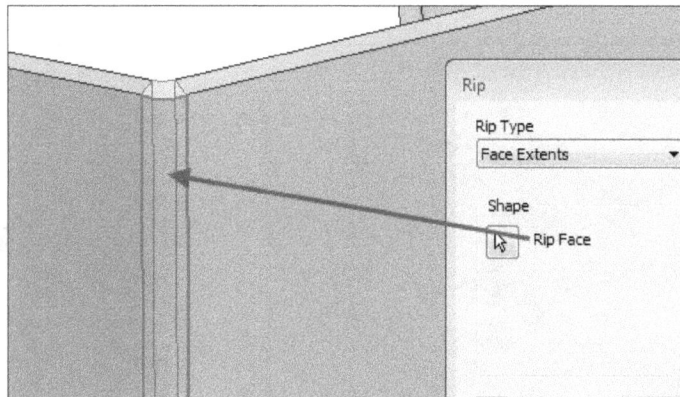

Figure-67. Preview of Rip with Face Extents Rip Type option selected

- Select the **Flip Side** buttons to change the side of rip and specify desired gap value in the **Gap Value** edit box. Click on the **OK** button to rip the face.

Using Unfold Tool

The **Unfold** tool, as the name suggests, is used to unfold the bends while making selected face stationary. The procedure to use this tool is given next.

- Click on the **Unfold** tool from the **Modify** panel in the **Sheet Metal** tab of the **Ribbon**. The **Unfold** dialog box will be displayed; refer to Figure-68. Also, you will be asked to select a stationary face.

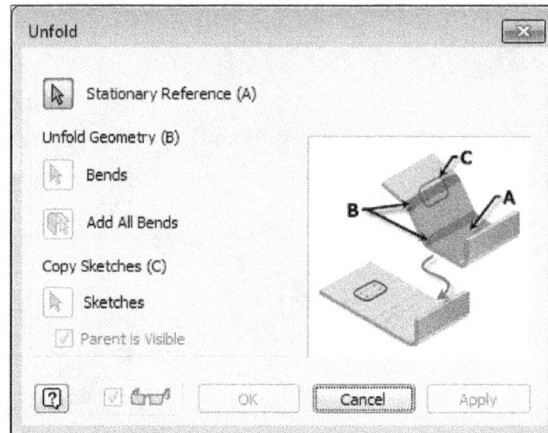

Figure-68. Unfold dialog box

- Select a stationary face of the sheet metal part to be unfolded. You will be asked to select bends to be unfolded.
- Select desired bend/bends. Preview of the unfold feature will be displayed; refer to Figure-69.

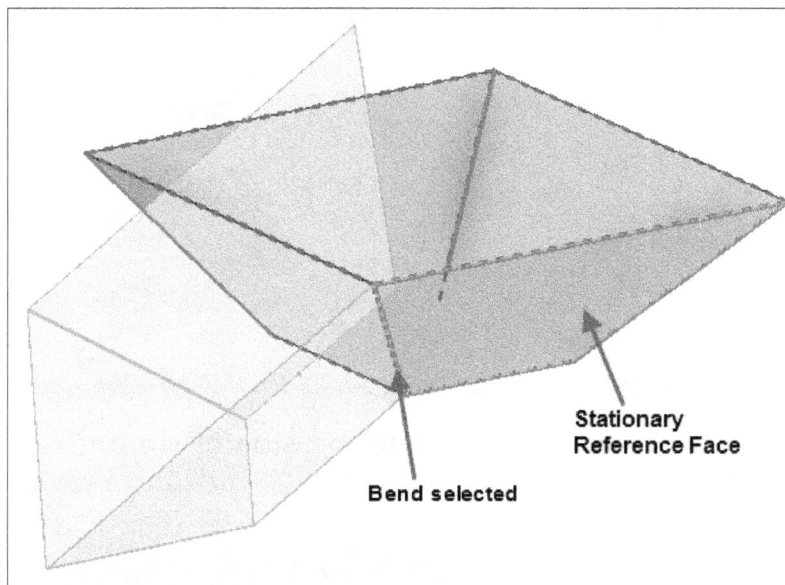

Figure-69. Preview of unfold feature

- If you want to unfold all the bends in the part then click on the **Add All Bends** button from the **Unfold Geometry** area of the dialog box.
- If you have a sketch on the face to be unfolded then you can select it after clicking on the **Sketches** selection button in the **Copy Sketches** area of the dialog box. The preview of unfolded feature with copied sketch will be displayed; refer to Figure-70.

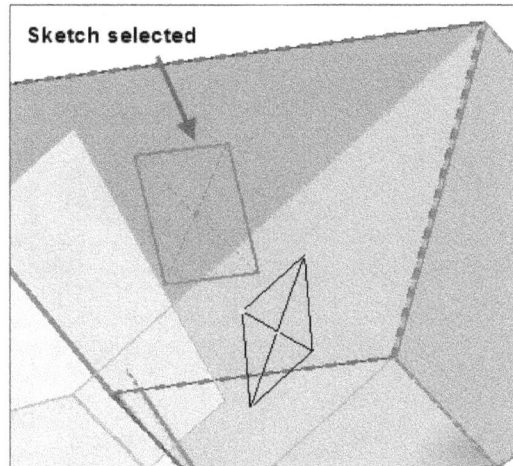

Figure-70. Preview of unfold feature with copied sketch

- Click on the **OK** button from the dialog box to create the unfold feature.

Using Refold Tool

The **Refold** tool is used to refold the unfolded bends. In simple words, it does the reverse of **Unfold** tool. The procedure to use this tool is given next.

- Click on the **Refold** tool from the **Modify** panel in the **Sheet Metal** tab of the **Ribbon**. The **Refold** dialog box will be displayed; refer to Figure-71. Also, you will be asked to select stationary reference.

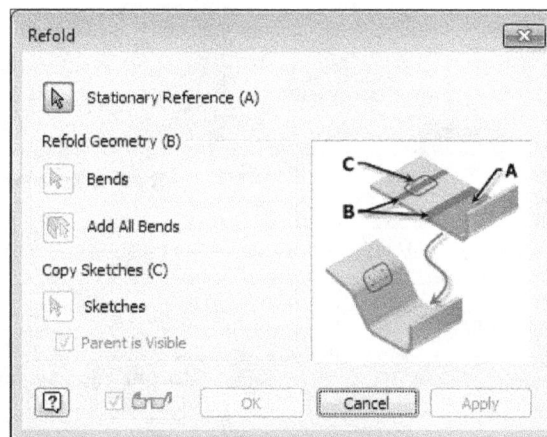

Figure-71. Refold dialog box

- Select the face of unfolded part that you want to be stationary. You will be asked to select the bends.
- Select the unfolded bend. Preview of the refold feature will be displayed; refer to Figure-72. Rest of the procedure is same as discussed for **Unfold** tool.

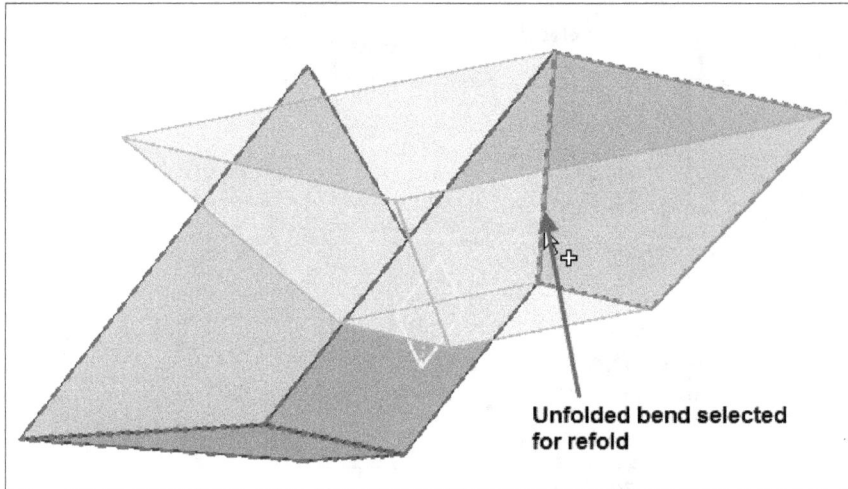

Figure-72. Preview of refold feature

The **Hole** tool in **Sheet Metal** tab works in the same way as **Hole** tool in **3D Model** tab which has been discussed already. The **Corner Chamfer** and **Corner Round** tools work in the same way as **Chamfer** tool in **3D Model** tab but in case of sheet metal, you can select only corner edges on choosing these tools; refer to Figure-73 and Figure-74.

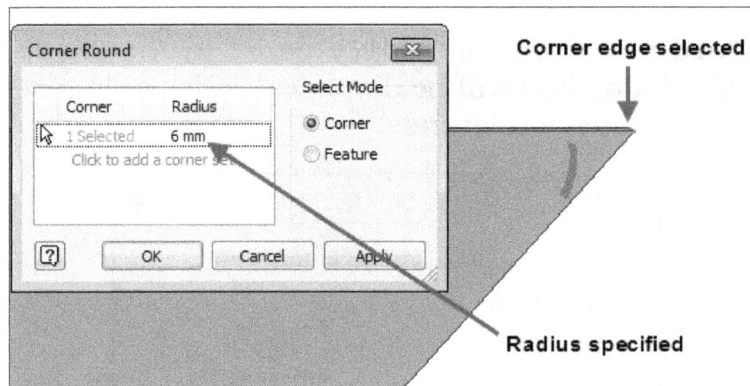

Figure-73. Corner Round preview

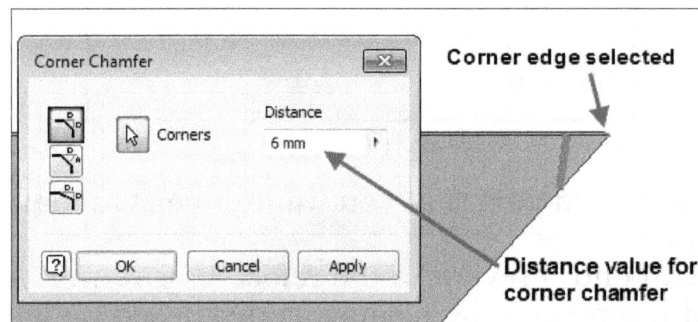

Figure-74. Preview of corner chamfer

CREATE FLAT PATTERN

The last step after creating a sheet metal part is creating the flat pattern of the part. Based on this flat pattern, metal sheets are cut and bend to form the required sheet metal part. The **Create Flat Pattern** tool is used to create flat pattern of sheet metal part. The procedure is given next.

- Click on the **Create Flat Pattern** tool from the **Flat Pattern** panel in the **Sheet Metal** tab of the **Ribbon**. The flat pattern of the sheet metal model will be displayed; refer to Figure-75.

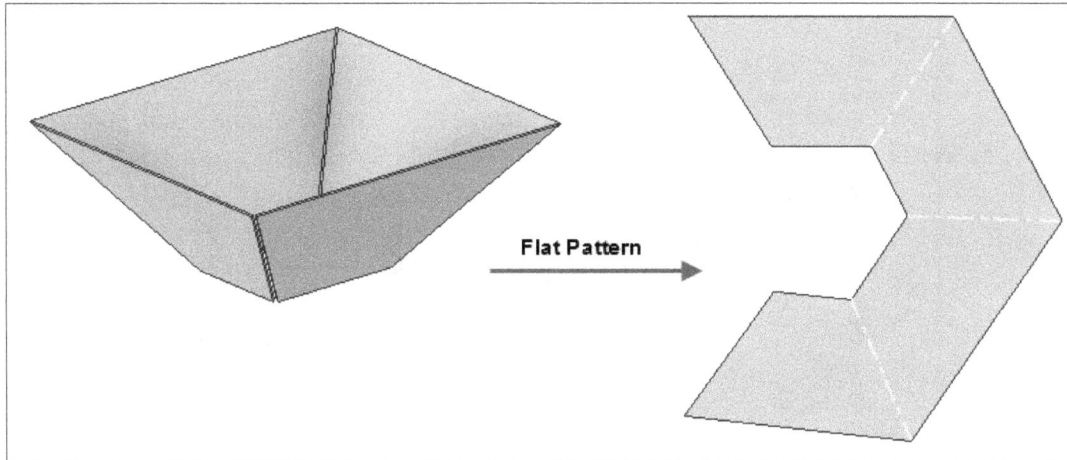

Figure-75. Flat pattern created

- To display the bending order of flat pattern, click on the **Bend Order Annotation** tool from the **Manage** panel in the **Flat Pattern** tab after using **Flat Pattern** tool. The bending order will be annotated on the flat pattern; refer to Figure-76.

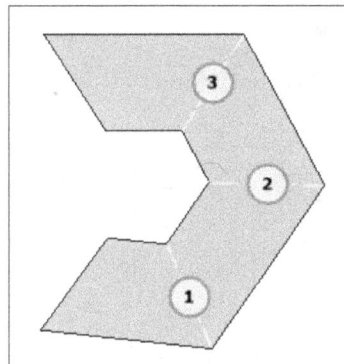

Figure-76. Bending order annotated on flat pattern

- To revert back to folded part, click on the **Go to Folded Part** tool from the **Folded Part** panel in the **Flat Pattern** tab of the **Ribbon**.

PRACTICAL 1

Create the sheet metal model as shown in Figure-77. The drawing is given in Figure-78.

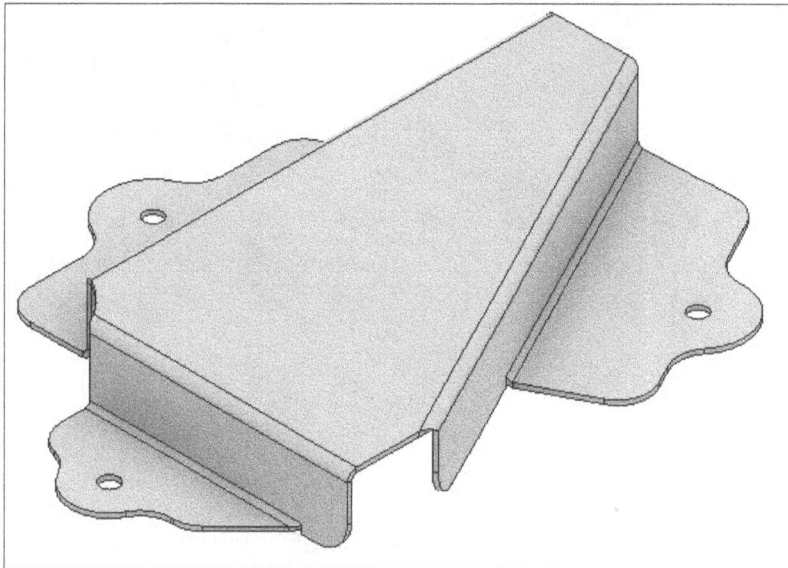

Figure-77. Model for Practical 1

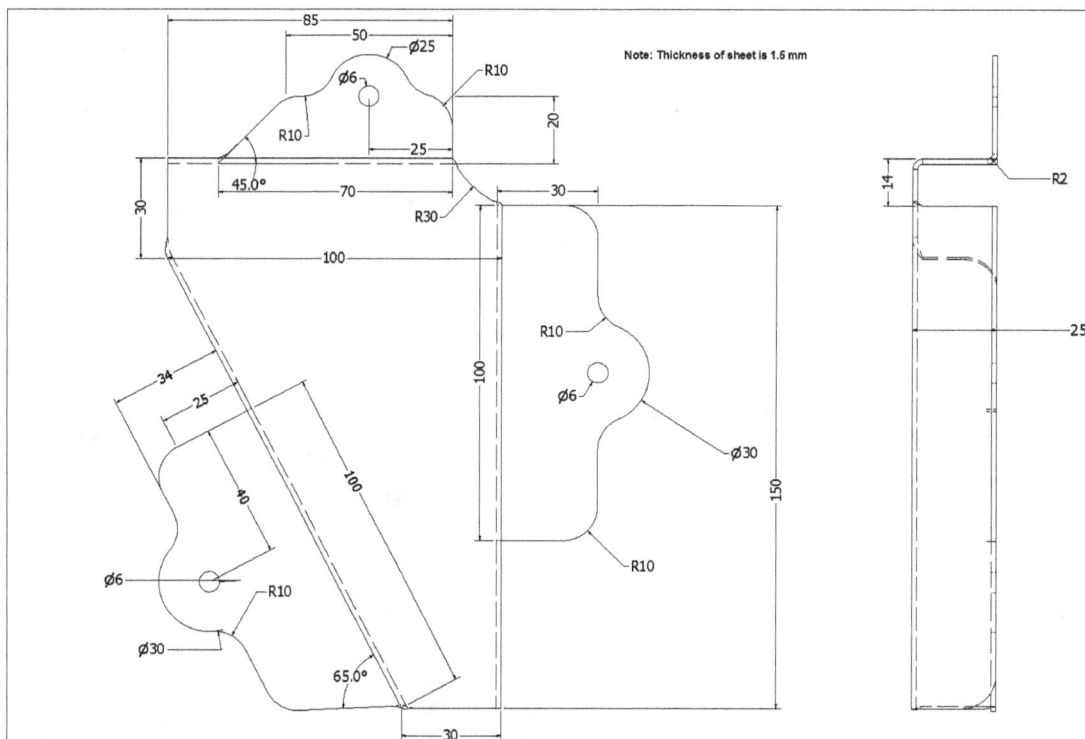

Figure-78. Drawing for Practical 1

Although, the drawing may look terrible at first look but if you read the drawing in pieces then you will find that there is a base of triangular shape with very clear dimensions then there are flanges on the edges and then there are flat faces on each flange with clear dimensions. We will create this sheet metal part in steps as given next.

Start a New Sheet Metal Part

- Start **Autodesk Inventor** by using **Start** menu or icon on the Desktop of your computer (If not started yet).
- Click on the **New** button from **New** cascading menu in the **File** menu of the **Ribbon**. The **Create New File** dialog box will be displayed.
- Click on the **Metric** folder under **Template** category in the left of the dialog box and double click on the **Sheet Metal (mm).ipt** template from the **Part** area of the dialog box; refer to Figure-79.

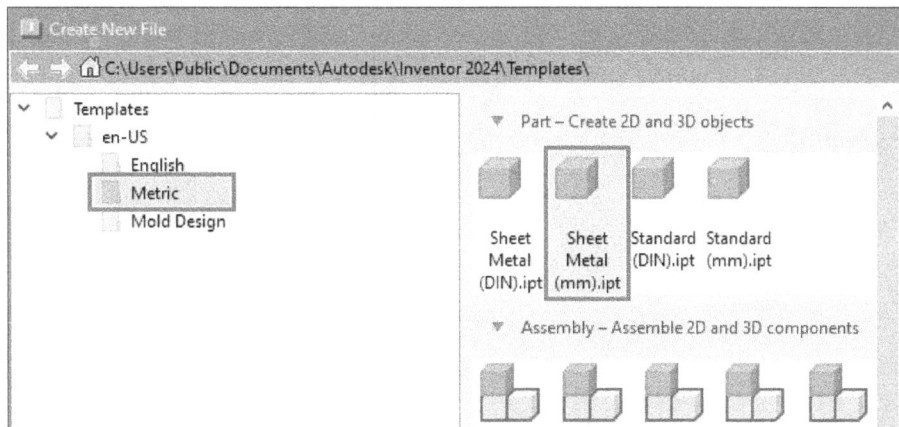

Figure-79. Starting new sheet metal part

Setting Default Thickness

- Click on the **Sheet Metal Defaults** button from the **Setup** panel in the **Sheet Metal** tab of **Ribbon**. The **Sheet Metal Defaults** dialog box will be displayed; refer to Figure-80.

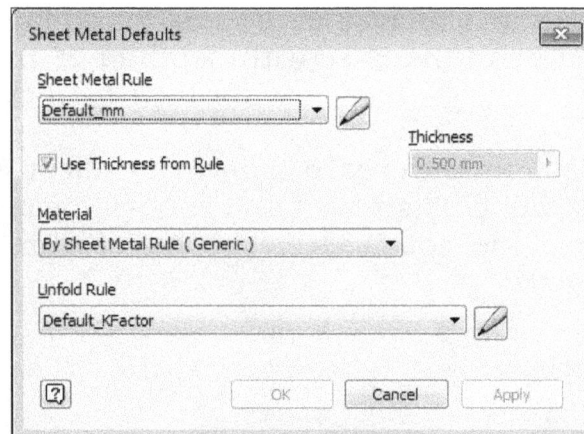

Figure-80. Sheet Metal Defaults dialog box

- Clear the **Use Thickness from Rule** check box and specify the sheet thickness as **1.5** mm in the **Thickness** edit box.
- Click on the **OK** button to apply the settings.

Creating Base

- Click on the **2D Sketch** button from the **Sketch** panel in the **Sheet Metal** tab of the **Ribbon**. You will be asked to select a sketching plane.
- Select the **XZ** Plane as sketching plane. The **Sketch** tab in **Ribbon** will become active and tools related to sketching will be displayed.

- Create the sketch as shown in Figure-81.

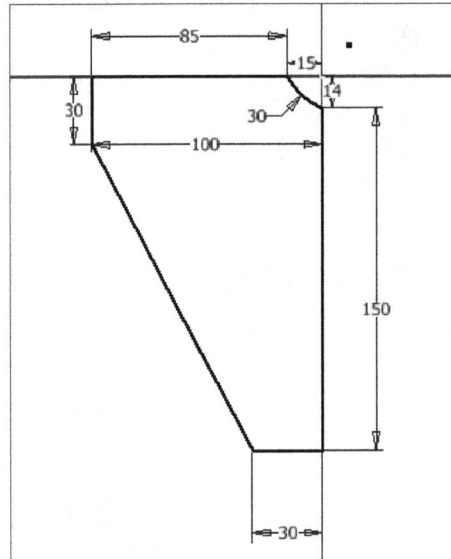

Figure-81. Sketch for base

- After creating the sketch, click on the **Finish Sketch** button from the **Exit** panel in the **Sketch** tab of the **Ribbon**.
- Click on the **Face** tool from the **Create** panel in the **Sheet Metal** tab of the **Ribbon**. The **Face** dialog box will be displayed. Also, the sketch will be selected automatically.
- Specify the parameters as per the drawing and click on the **OK** button from the dialog box. The face will be created.

Creating Flanges

As per the model and drawing displayed in Figure-77 and Figure-78, there are three flanges created on the edges of the base with depth of **25** mm. The steps to create flanges are given next.

- Click on the **Flange** tool from the **Create** panel in the **Sheet Metal** tab of the **Ribbon**. The **Flange** dialog box will be displayed and you will be asked to select edges.
- Select the lower edges of the side faces as shown in Figure-82. The preview of the flanges will be displayed.

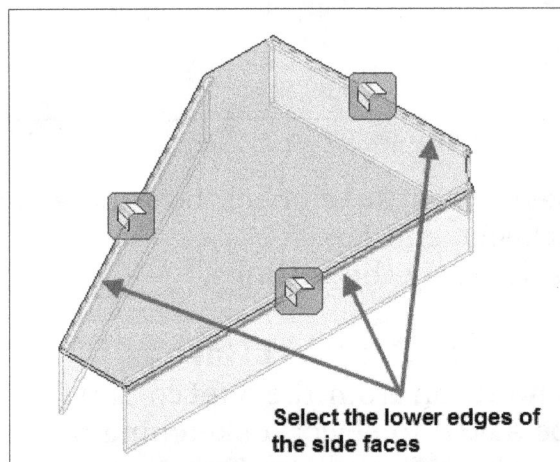

Figure-82. Edges selected for flange creation

- Click in the **Distance** edit box of **Height Extents** area of the **Flange** dialog box and specify the value as **25** mm. Make sure the other parameters is same as given in Figure-83.

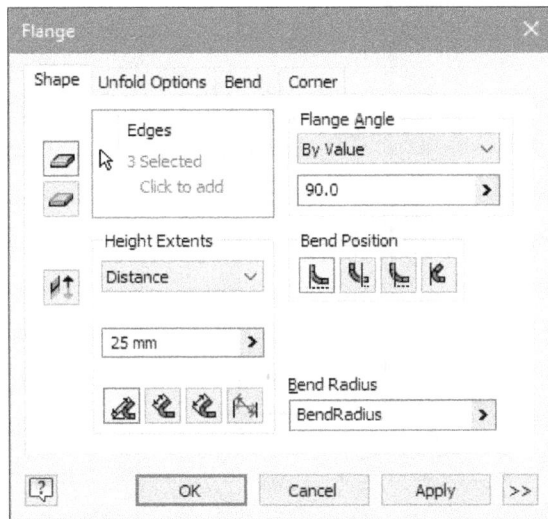

Figure-83. Flange dialog box with values specified

- Click on the **OK** button from the dialog box. The flanges will be created.

Creating First Face Feature on Flange

As per the model, there are three faces created on three flanges. We will create the face feature on first flange near round cut. The procedure is given next.

- Click on the **2D Sketch** button from the **Sketch** panel in the **Sheet Metal** tab of the **Ribbon**. You will be asked to select the sketching plane.
- Select the planar face of flange near round cut as shown in Figure-84. The sketching environment will be activated.

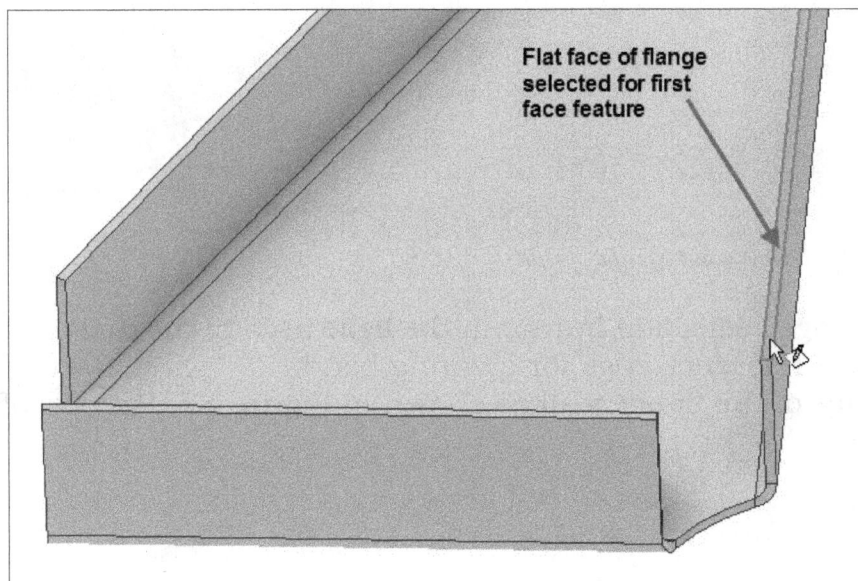

Figure-84. Face of flange selected for first face feature

- Create the sketch as shown in Figure-85.

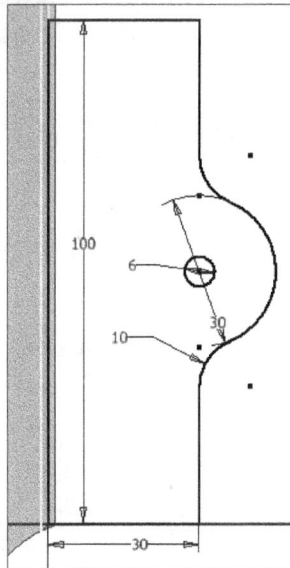

Figure-85. Sketch created for first face feature

- Click on the **Finish Sketch** button and then click on the **Face** tool from the **Create** panel in the **Ribbon**. You will be asked to select a profile.
- Select the newly created sketch by clicking inside the sketch; refer to Figure-86. Preview of the face feature will be displayed.

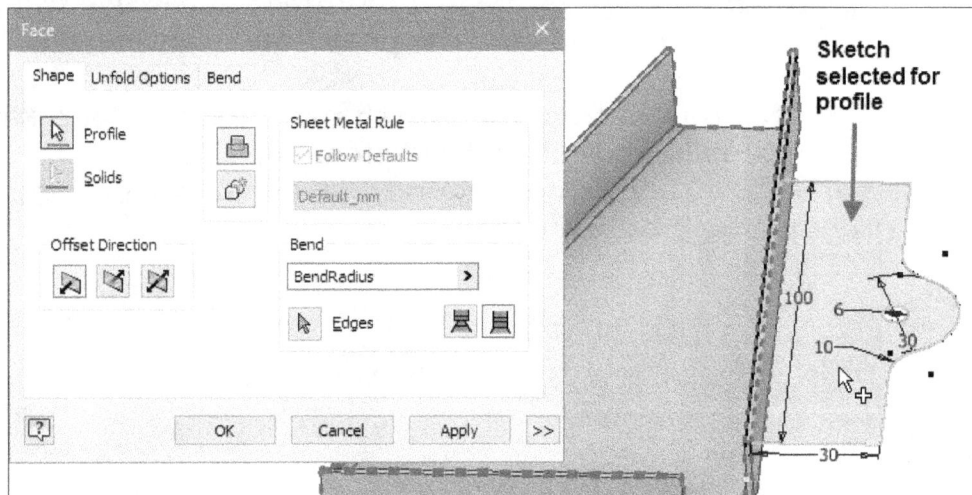

Figure-86. Selecting sketch for face profile

- Click on the **Edges** selection button in the **Bend** area of the dialog box. You will be asked to select reference edge for creating bend.
- Select the edge of the flange wall as shown in Figure-87. Preview of the bend will be displayed.

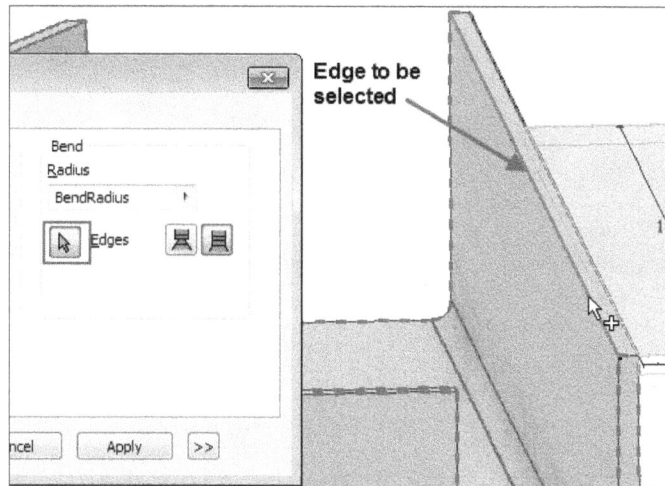

Figure-87. Edge selected for bending

- Click on the **Bend** tab of the dialog box and select the **Tear** option from the **Relief Shape** drop-down; refer to Figure-88.

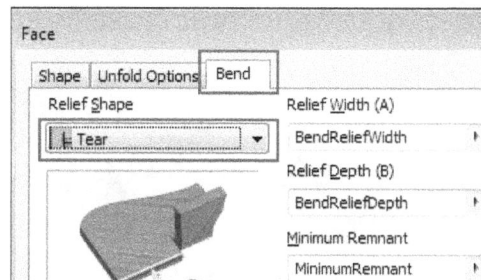

Figure-88. Tear option in Bend tab

- Click on the **OK** button from the dialog box to create the face.

Creating Second Face Feature on Flange

- Click on the **Start 2D Sketch** tool from the **Sketch** panel in the **Sheet Metal** tab of the **Ribbon**. You will be asked to select a sketching face/plane.
- Select the flat face of the smaller flange as shown in Figure-89. The sketching environment will be displayed.

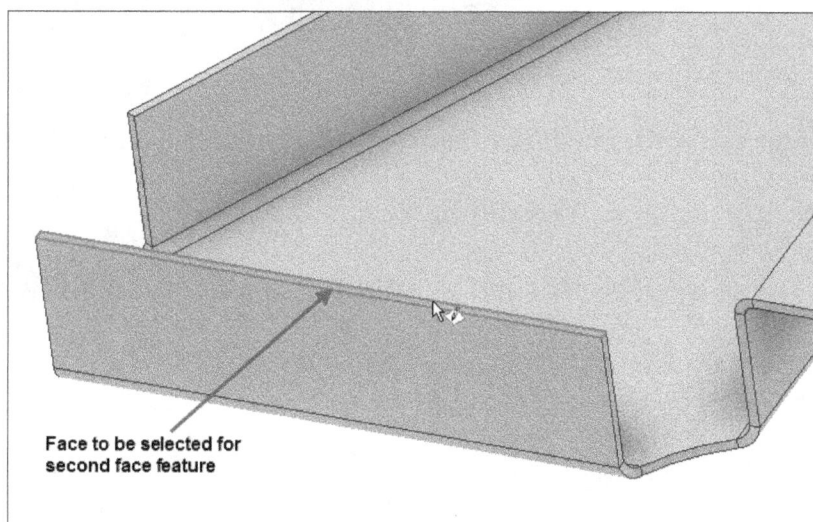

Figure-89. Face of flange selected for second face feature

- Create the sketch on face as shown in Figure-90.

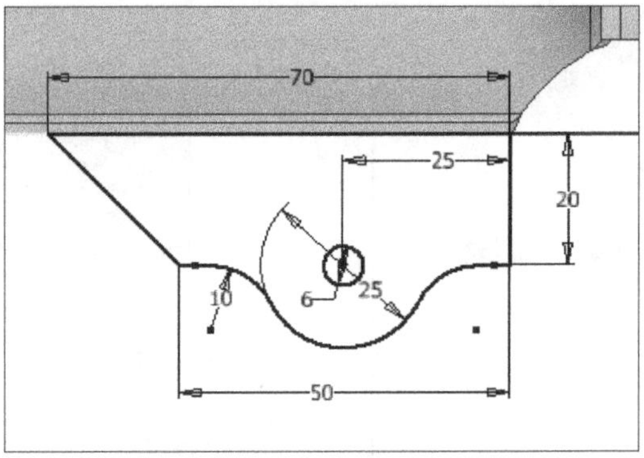

Figure-90. Sketch created for second face feature

- Click on the **Finish Sketch** button from the **Ribbon** and then click on the **Face** tool from the **Create** panel in the **Sheet Metal** tab of the **Ribbon**. The **Face** dialog box will be displayed.
- Select the newly created sketch by clicking inside it. Preview of the face feature will be displayed.
- Click on the **Edges** selection button and select the bend edge as shown in Figure-91.

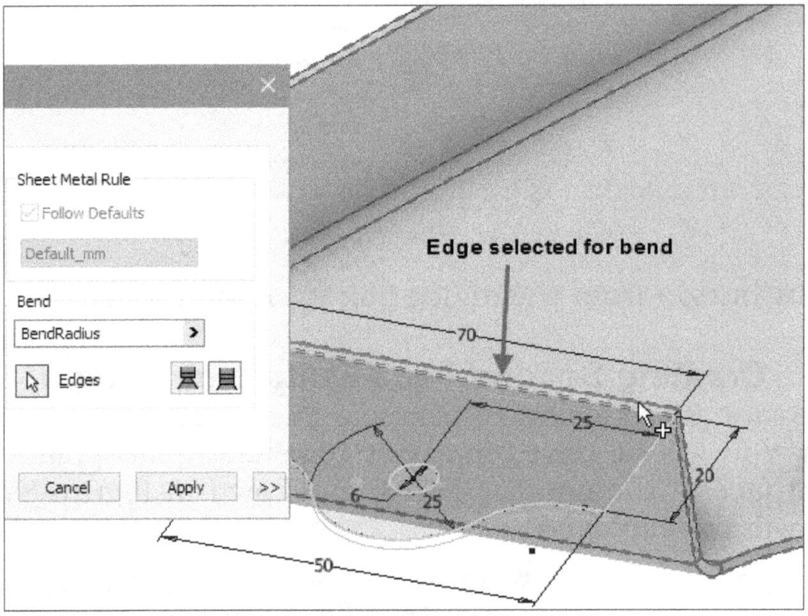

Figure-91. Bend edge selected for second feature

- Click on the **Bend** tab and select the **Tear** option from the **Relief Shape** drop-down as discussed earlier.
- Click on the **OK** button from the dialog box.

Similarly, create the third face feature by using the sketch as shown in Figure-92.

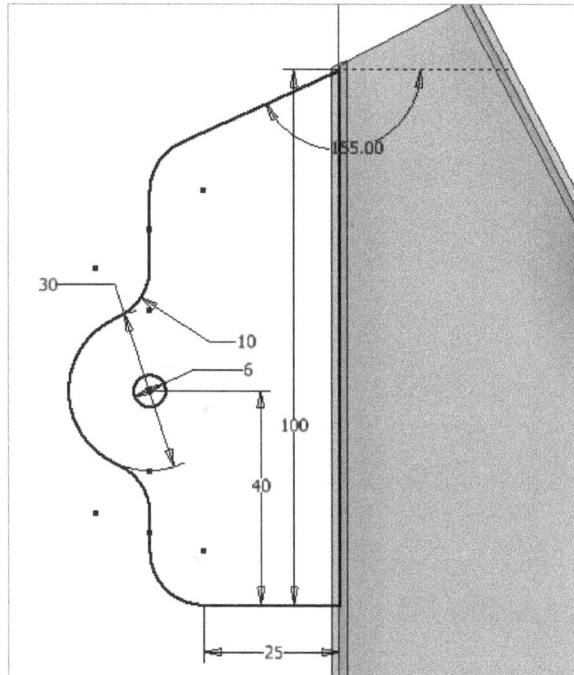

Figure-92. Sketch for third face feature

Applying Corner Rounds

- Click on the **Corner Round** tool from the **Modify** panel in the **Sheet Metal** tab of the **Ribbon**. The **Corner Round** dialog box will be displayed; refer to Figure-93.

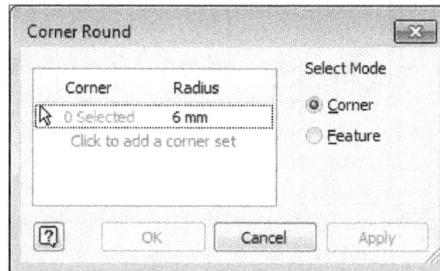

Figure-93. Corner Round dialog box

- Click on the **Radius** value and specify it as **10** mm.
- Click on the **Feature** radio button from the **Select Mode** area of the dialog box. You will be asked to select the feature.
- Click on the flange wall and then click on the **OK** button from the dialog box. Rounds will be applied to corners.

PRACTICAL 2

Create the sheet metal model of lighter cap as shown in Figure-94. The drawing is given in Figure-95.

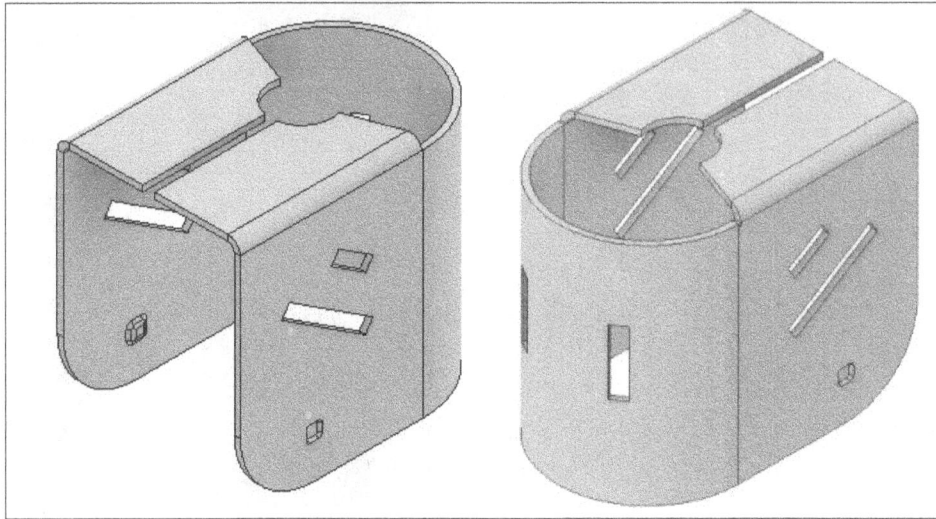

Figure-94. Model of Lighter cap

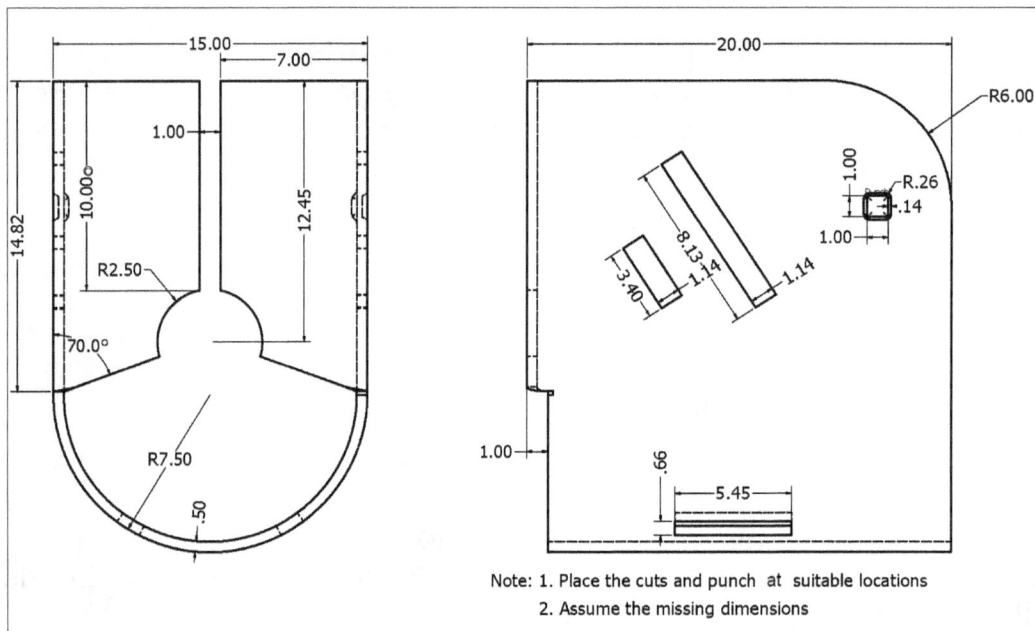

Note: 1. Place the cuts and punch at suitable locations
2. Assume the missing dimensions

Figure-95. Drawing for Practical 2

If you see the model carefully then you will find that there is a round face in the sheet metal model and flanges are attached to it. In Autodesk Inventor, it is not possible to create flange on Contour roll feature. So, we will take a different approach here using Solid modeling tools. The steps to create the part are given next.

Start a New Sheet Metal Part

- Start Autodesk Inventor by using Start menu or icon on the Desktop of your computer (If not started yet).
- Click on the **New** button from **New** cascading menu in the **File** menu of the **Ribbon**. The **Create New File** dialog box will be displayed.

- Click on the **Metric** folder under **Template** category in the left side of the dialog box and double click on the **Sheet Metal (mm).ipt** template from the **Part** area of the dialog box.

Setting Default Thickness

- Click on the **Sheet Metal Defaults** button from the **Setup** panel in the **Sheet Metal** tab of **Ribbon**. The **Sheet Metal Defaults** dialog box will be displayed.
- Clear the **Use Thickness from Rule** check box and specify the sheet thickness as **0.5 mm** in the **Thickness** edit box.
- Click on the **OK** button to apply the settings.

Creating Extrusion

- Click on the **2D Sketch** button from the **Sketch** panel in the **Sheet Metal** tab of the **Ribbon**. You will be asked to select a sketching plane.
- Select the **XZ** plane from the Model Tree. The sketching tools will become active.
- Create the sketch as shown in Figure-96. Click on the **Finish Sketch** button from the **Ribbon** to exit.

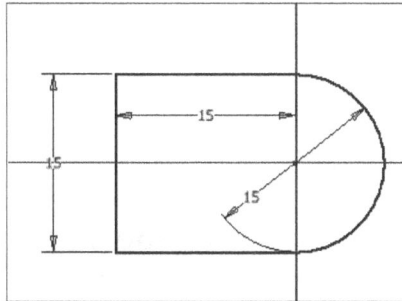

Figure-96. Sketch for extrusion

- Click on the **Extrude** tool from the **Create** panel in the **3D Model** tab of the **Ribbon**. The **Extrude** dialog box will be displayed along with preview of extrusion.
- Specify the distance value as **20** mm in the **Distance** edit box; refer to Figure-97.

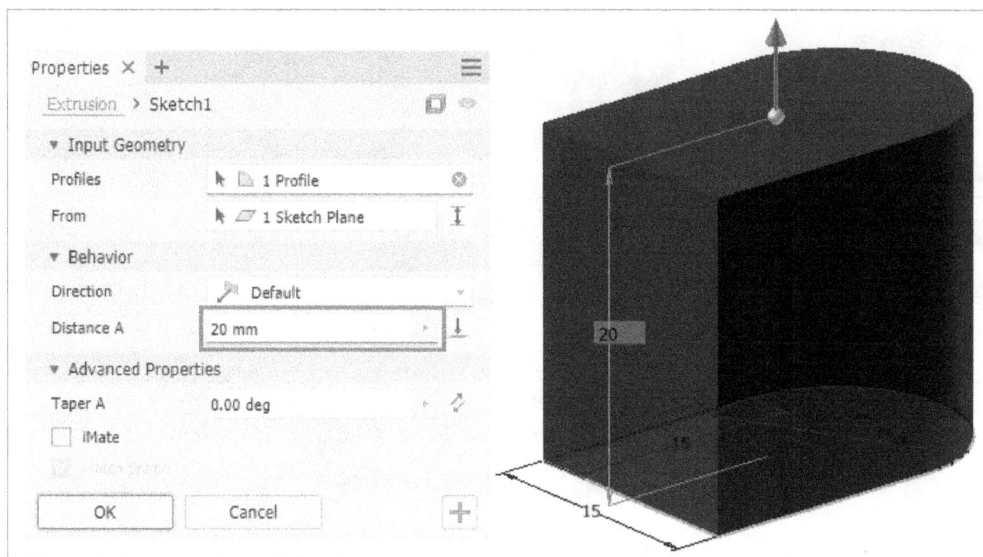

Figure-97. Distance edit box

- Click on the **OK** button from the dialog box.

Applying Shell Feature

We have the outer shape of the sheet metal model but we have the solid model not suitable for sheet metal operations. We will scoop out the extra material to form the part as sheet metal part. The steps are given next.

- Click on the **Shell** tool from the **Modify** panel in the **3D Model** tab of the **Ribbon**. The **Shell** dialog box will be displayed along with the preview of shell; refer to Figure-98.

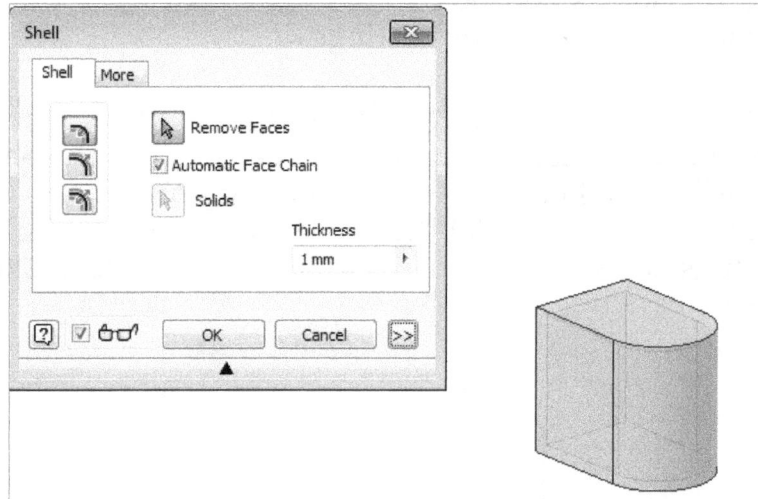

Figure-98. Shell dialog box

- Specify the thickness value as **0.5** mm. You will be asked to select the faces to be removed.
- Select the back face and bottom face of the extrude feature as shown in Figure-99.

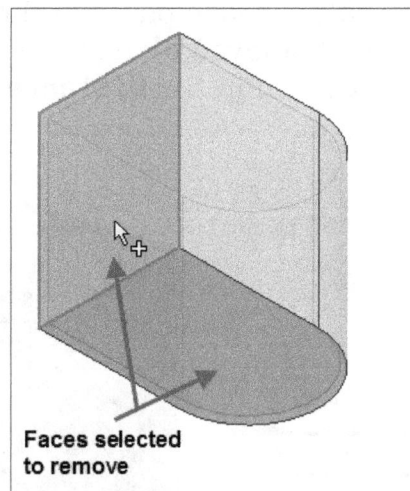

Figure-99. Faces to be selected for removing

- Click on the **OK** button from the dialog box to create the feature. The model will be displayed as shown in Figure-100.

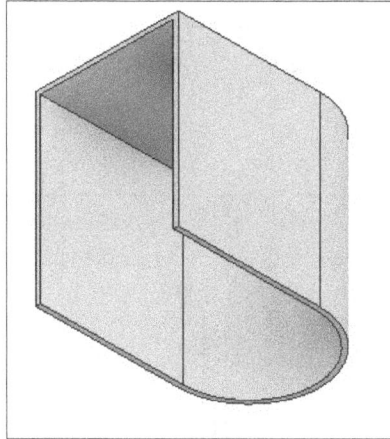

Figure-100. Model after applying shell tool

Creating Cut Feature on Top Face

- Click on the **2D Sketch** tool from the **Sketch** panel in the **Ribbon** and select the top face of the model as sketching plane; refer to Figure-101.

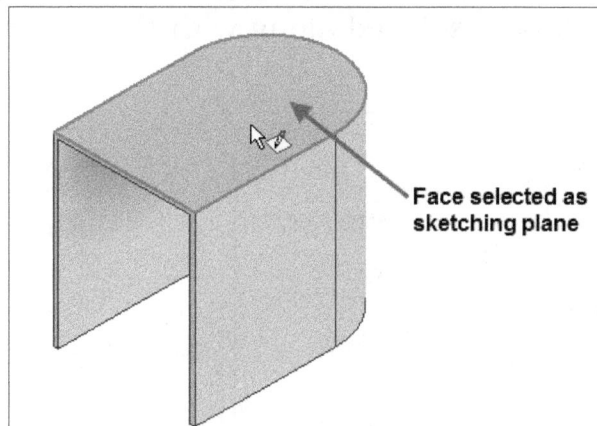

Face selected as
sketching plane

Figure-101. Face selected as sketching plane

- Create the sketch as shown in Figure-102. Click on the **Finish Sketch** button from the **Ribbon** to exit.

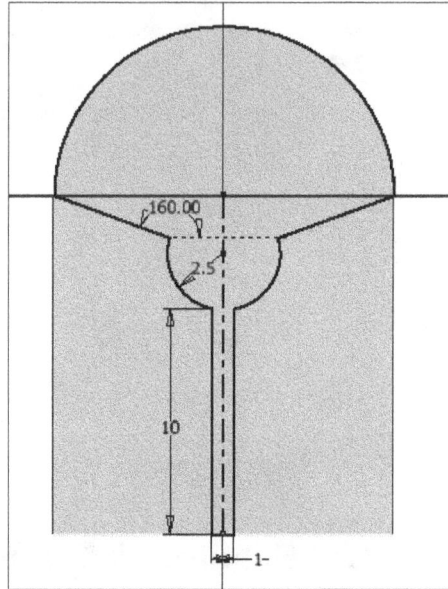

Figure-102. Sketch for cut feature

- Click on the **Cut** tool from the **Modify** panel in the **Sheet Metal** tab of the **Ribbon**. The **Cut** dialog box will be displayed along with the preview of cut feature; refer to Figure-103.

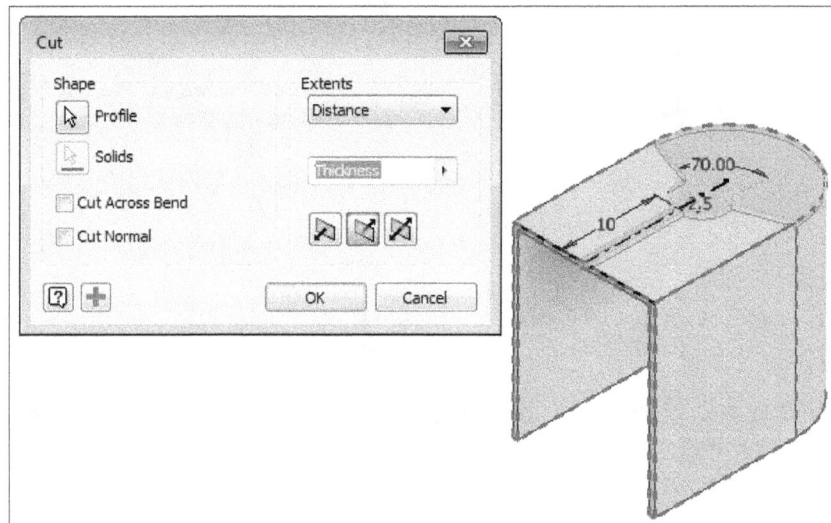

Figure-103. Cut dialog box with cut feature

Converting Edges into Bends

Since, we started creating model using the solid modeling tools, so we will not be able to create flat pattern automatically. To facilitate flat pattern, we need to convert the sharp edges into bends. The steps are given next.

- Click on the **Bend** tool from the **Create** panel in the **Sheet Metal** tab of the **Ribbon**. The **Bend** dialog box will be displayed; refer to Figure-104. Also, you will be asked to select edges.

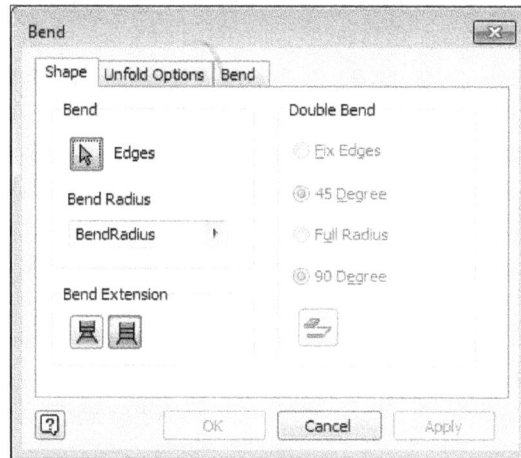

Figure-104. Bend dialog box

- Select the sharp edge of the model. Preview of the bend will be displayed; refer to Figure-105.

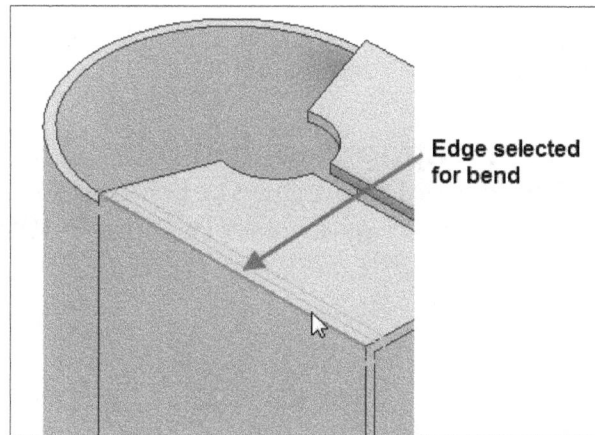

Figure-105. Edge selected for bend

- Click on the **Apply** button to create the bend.
- Similarly, create the bend at the other edge of the model; refer to Figure-106.

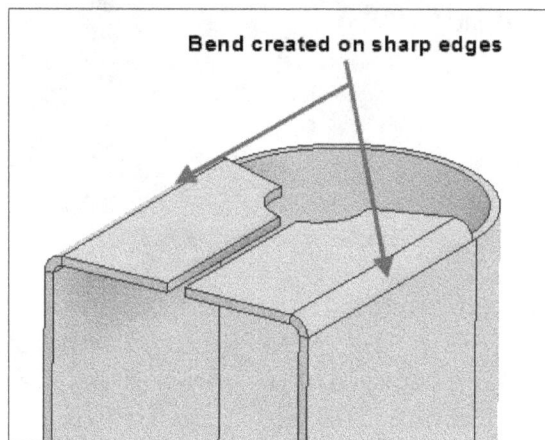

Figure-106. Bend created on sharp edges

Applying Corner Rounds

- Click on the **Corner Round** tool from the **Modify** panel in the **Sheet Metal** tab of the **Ribbon**. The **Corner Round** dialog box will be displayed.
- Select the **Feature** radio button from the **Select Mode** area of the dialog box and select the model. Preview of the corner round will be displayed.
- Make sure the radius value is specified as **6** mm and then click on the **OK** button to create corner round; refer to Figure-107.

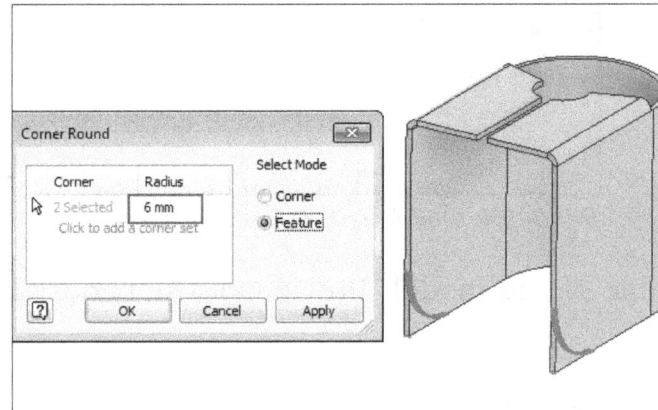

Figure-107. Preview of corner round

Creating Punch Marks

- Click on the **Start 2D Sketch** tool from the **Sketch** panel in the **Ribbon** and create a point on the inner face of model as shown in Figure-108.

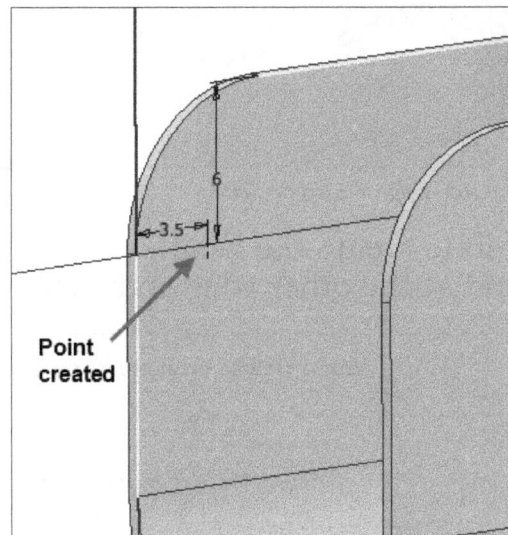

Figure-108. Point created on inner face

- Click on the **Finish Sketch** button from the **Ribbon** to exit.
- Click on the **Punch Tool** from the **Modify** panel in the **Sheet Metal** tab of the **Ribbon**. The **PunchTool Directory** dialog box will be displayed.
- Double-click on **Square Emboss.ide** file from the dialog box. Preview of punch with **PunchTool** dialog box will be displayed.
- Click on the **Size** tab in the dialog box and specify the value as shown in Figure-109.

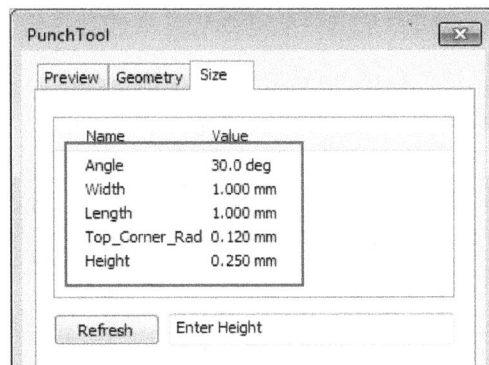

Figure-109. Values specified in mm

- Click on the **Finish** button from the dialog box to create the feature.
- Click on the **Mirror** tool from the **Pattern** panel in the **Sheet Metal** tab of the **Ribbon**. The **Mirror** dialog box will be displayed. Also, you will be asked to select the feature to be mirrored.
- Select the punch mark created and then click on the **Mirror Plane** selection button from the dialog box; refer to Figure-110. You will be asked to select the mirror plane.

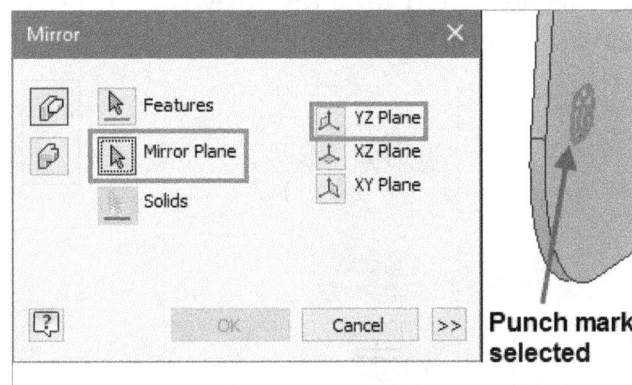

Figure-110. Punch mark selected and Mirror Plane selection button

- Select the **YZ** Plane as mirror plane from the dialog box and then click on the **OK** button from the dialog box.

Creating Cuts

There are some cuts to be made on the lighter cap to facilitate heat escape. We can not create cuts on round faces directly. To do so, we will flatten the sheet metal part and then perform the operation.

- Click on the **Unfold** tool from the **Modify** panel in the **Sheet Metal** tab of **Ribbon**. The **Unfold** dialog box will be displayed and you will be asked to select a stationary reference.
- Select the face as shown in Figure-111. You will be asked to select the bends.

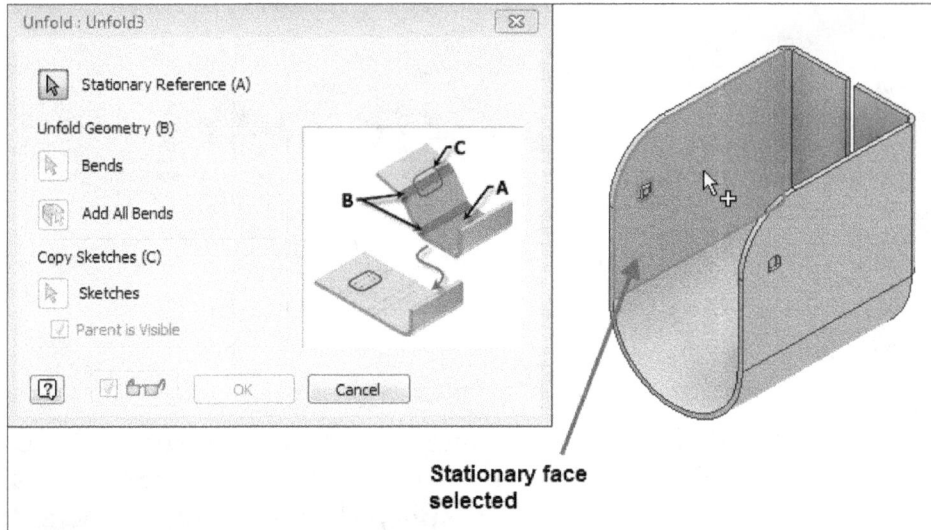

Figure-111. Stationary reference to be selected

- Click on the **Add All Bends** button from the dialog box and then click on the **OK** button. The unfold feature will be created.
- Create 2D sketch on the flattened face as shown in Figure-112. Dimensions are as per the drawing given in Figure-95.

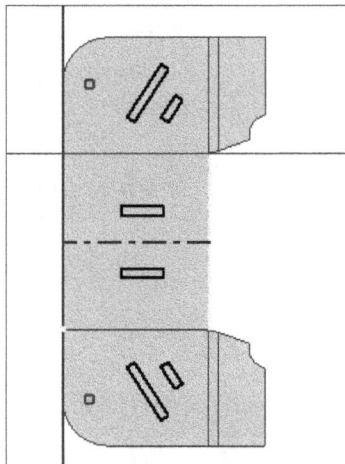

Figure-112. Sketch created on flattened face

- Click on the **Cut** tool from the **Modify** panel in the **Sheet Metal** tab of the **Ribbon**. You will be asked to select the profile sketch.
- Select all the close loops in the sketch and also select the **Cut Across Bend** check box from the dialog box; refer to Figure-113.

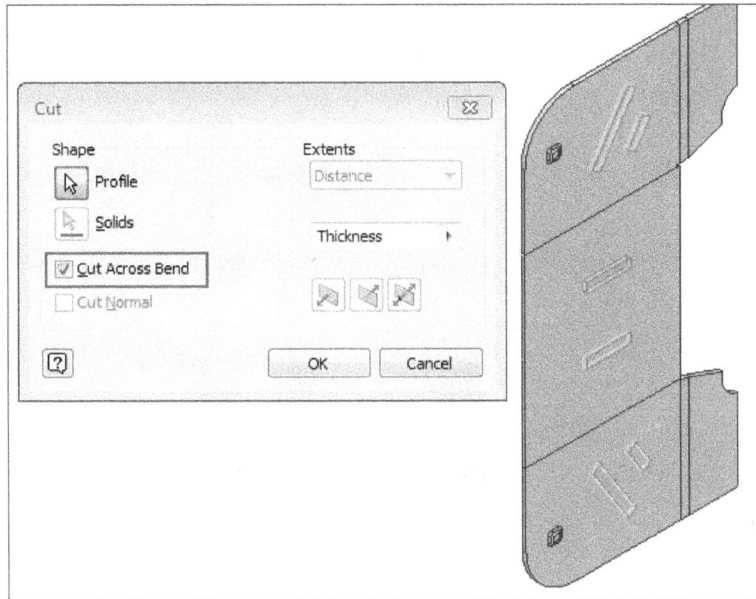

Figure-113. Profile selected for cut feature

- Click on the **OK** button to create the cut feature.
- Click on the **Refold** tool from the **Modify** panel in the **Ribbon**. The **Refold** dialog box will be displayed and you will be asked to select the stationary reference.
- Select the side face of the model as shown in Figure-114.

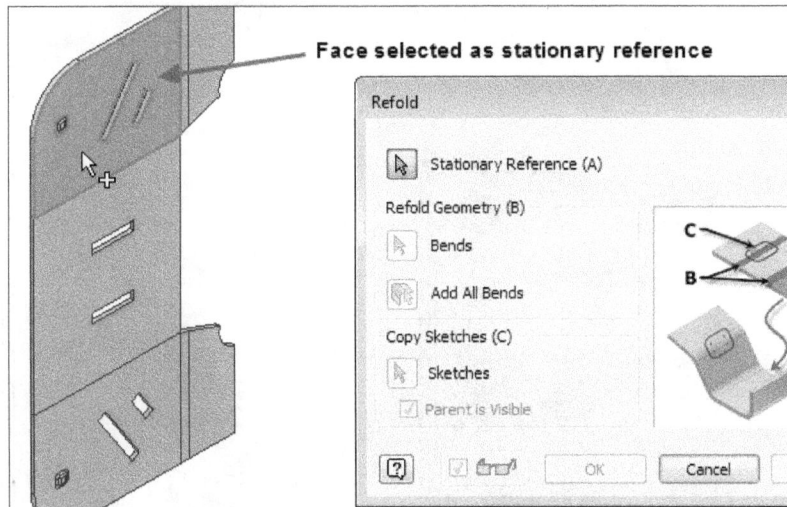

Figure-114. Face selected for refold stationary reference

- Click on the **Add All Bends** button from the dialog box and then click on the **OK** button. The part will be created.

In this chapter, we have created sheet metal models but things are not manufactured barely on the basis of sheet metal models. In industry, we need flat patterns with all the dimensions and bending orders applied on them. In the next chapter, we will learn about creating drawings for shop floor prints. In next chapter, we will also work on creating flat patterns of sheet metal models with necessary dimensions.

Problem 1

Create the sheet metal model of door hand as shown in Figure-115. Drawing of the model is given in Figure-116.

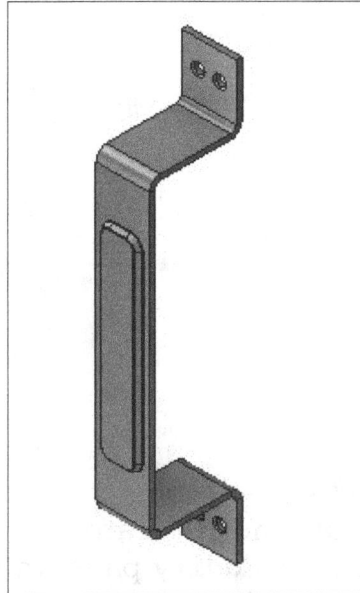

Figure-115. Practice 1 Model

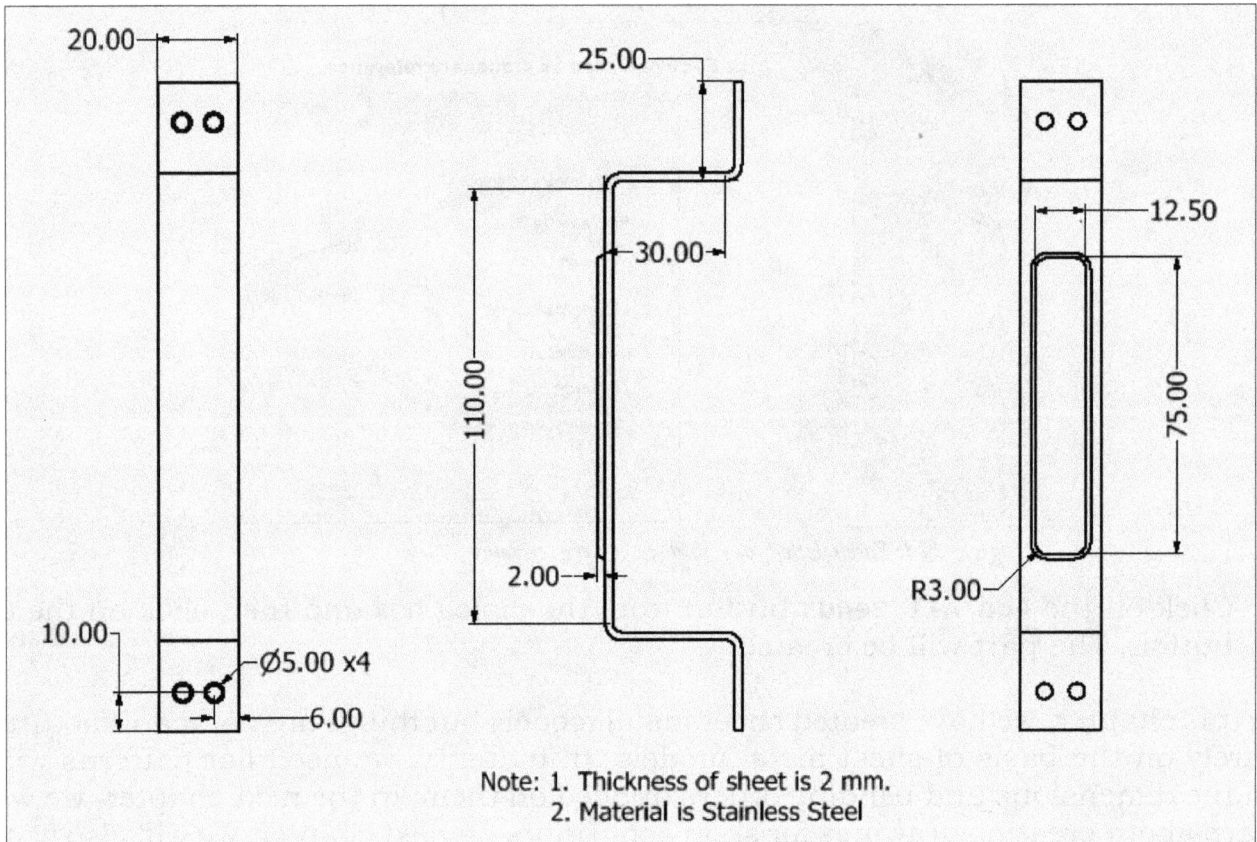

Note: 1. Thickness of sheet is 2 mm.
2. Material is Stainless Steel

Figure-116. Practice 1 Drawing

PROBLEM 2

Create the sheet metal model of hexagonal cover as shown in Figure-117. The dimensions of the model are given in Figure-118.

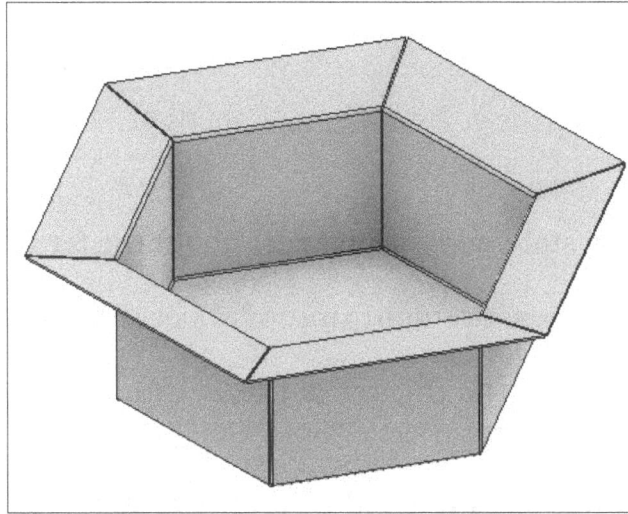

Figure-117. Model for Practice 2

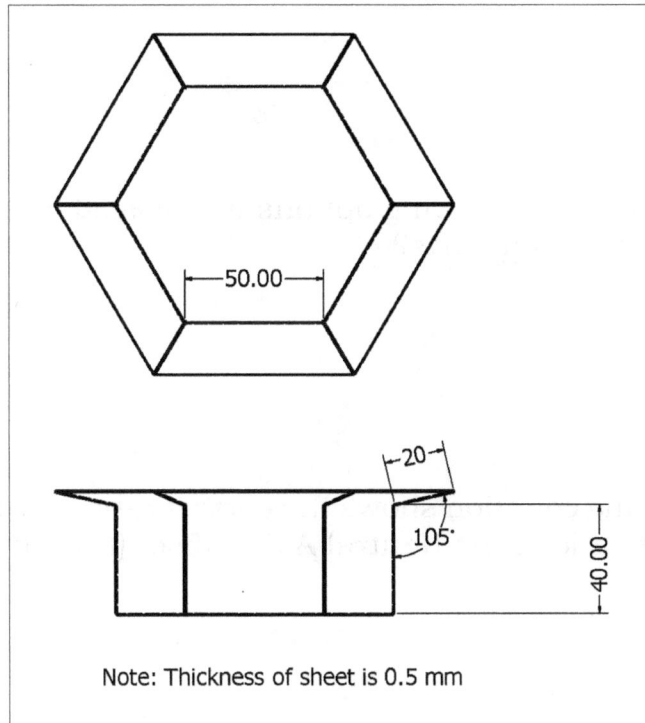

Note: Thickness of sheet is 0.5 mm

Figure-118. Drawing for Practice 2

SELF ASSESSMENT

Q1. Which of the following templates should be selected to start a new file in sheetmetal environment?

a) SheetMetal(mm).igt
b) SheetMetal(DIM).ipt
c) Both a and b
d) None of the Above

Q2. Which of the following statement is incorrect about the term 'Gauge' in sheetmetal?

a) Gauge is a traditional measurement unit of sheet thickness
b) Gauge is a non-linear unit
c) Higher the gauge number, the thicker is the sheetmetal
d) None of the Above

Q3. Which of the following tools is used to create a face of sheetmetal at specified angle to the selected edge?

a) Face
b) Flange
c) Contour Flange
d) Lofted Flange

Q4. By selecting which of the following options in the **Bend** dialog box, the bend will not be create in the sheetmetal parts?

a) Half Radius
b) 45 Degree
c) Full Radius
d) 90 Degree

Q5. Which of the following equation shows the correct ratio of K-Factor in sheet metal designing which is the ratio of the Neutral Axis' Offset (t) to the Material Thickness (MT)?

a) MT = t / K
b) t = K / MT
c) K = t / MT
d) K = MT / t

Q6. The **Fold** tool is used to fold the selected sheet metal face using a straight line. (True/False)

Q7. The **Corner Seam** tool is used to define or create gap between the adjoining faces of flange walls. (True/False)

Q8. The **Refold** tool is used to refold the folded bends. (True/False)

Q9. The .... tool is used to provide relief at the corners and edges for folding or unfolding of sheetmetal components.

Q10. The value of **Spline Factor** in **Sheet Metal Defaults** is ....

FOR STUDENT NOTES

Chapter 11

Weldment Assembly

Topics Covered

The major topics covered in this chapter are:

- *Welding Symbols and Representations*
- *Dimensioning of Weld Bead*
- *Starting Weldment Assembly*
- *Creating Fillet Weld and Fillet Weld Dimension*
- *Creating Groove/Plug Weld and Weld Dimension*
- *Creating Cosmetic Welds*
- *Calculating Weld Bead size by Weld Calculators*
- *Practical on Weldment Assembly*
- *Practice and Problems*

INTRODUCTION

Welding is known to almost every engineer and designer. Like every other dimension, welding symbols are also included in the engineering drawings. In Autodesk Inventor, we have a separate environment to apply weldment representations and symbols to the assembly which later can be derived in engineering drawing. In this chapter, we will discuss the tools related to weldments. But, before we start using Autodesk Inventor for weldments, it's important to revise some basics of welding.

WELDING SYMBOLS AND REPRESENTATION IN DRAWING

The symbols to represent various type of welds are given next.

Butt/Groove Weld Symbols

Various symbols that come under this category are given next. Refer to Figure-1 and Figure-2.

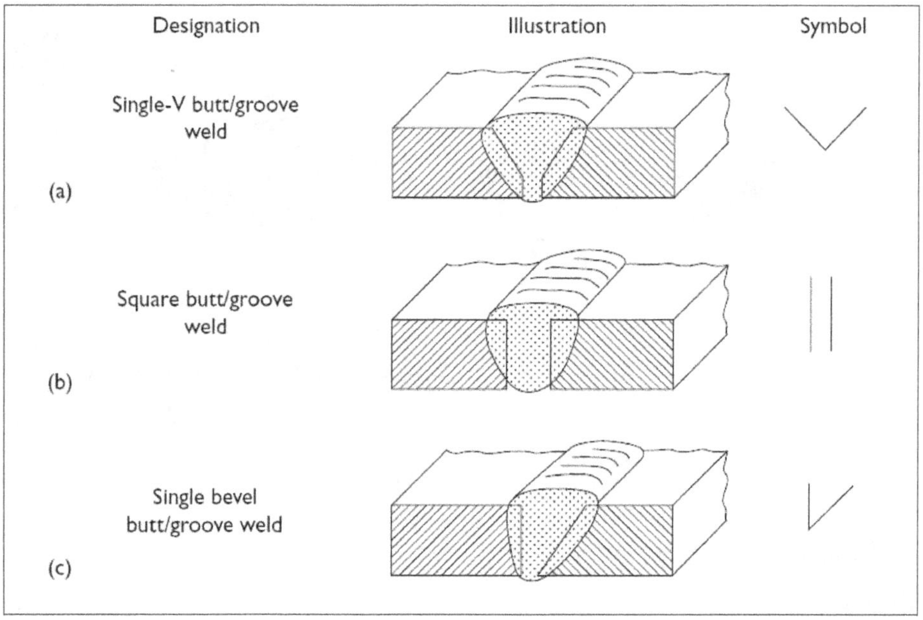

Figure-1. Welding symbols list 1

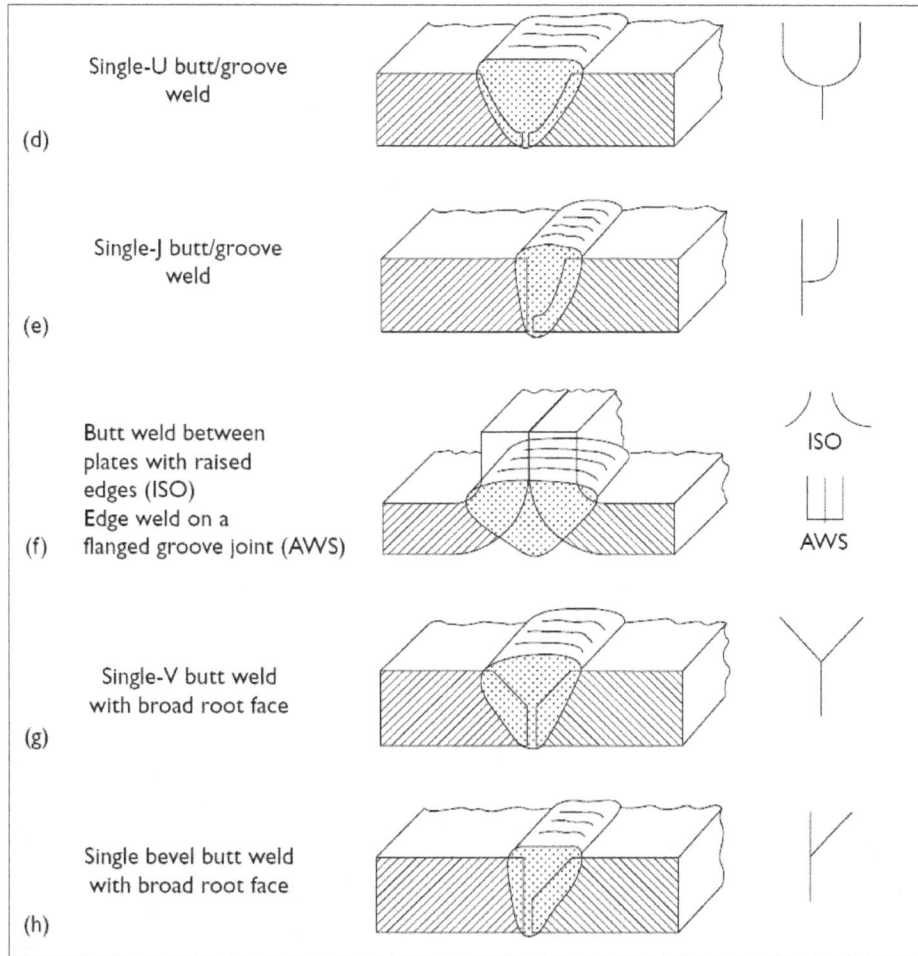

Figure-2. Welding symbols list 2

Fillet and Edge Weld Symbols

Various symbols that come under this category are given next. Refer to Figure-3.

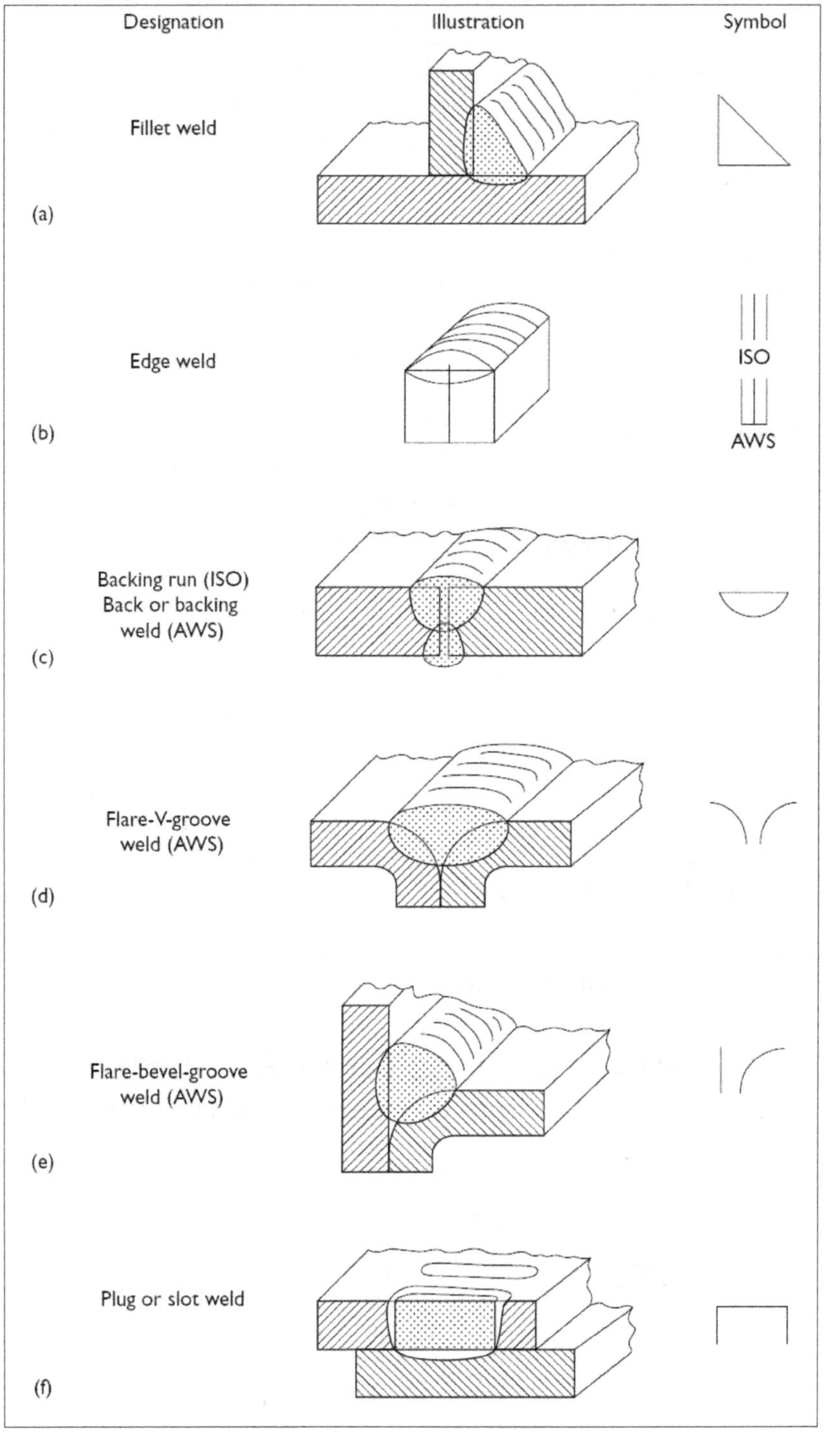

Figure-3. Welding symbols list 3

Miscellaneous Weld Symbols

Various symbols that come under this category are given next. Refer to Figure-4.

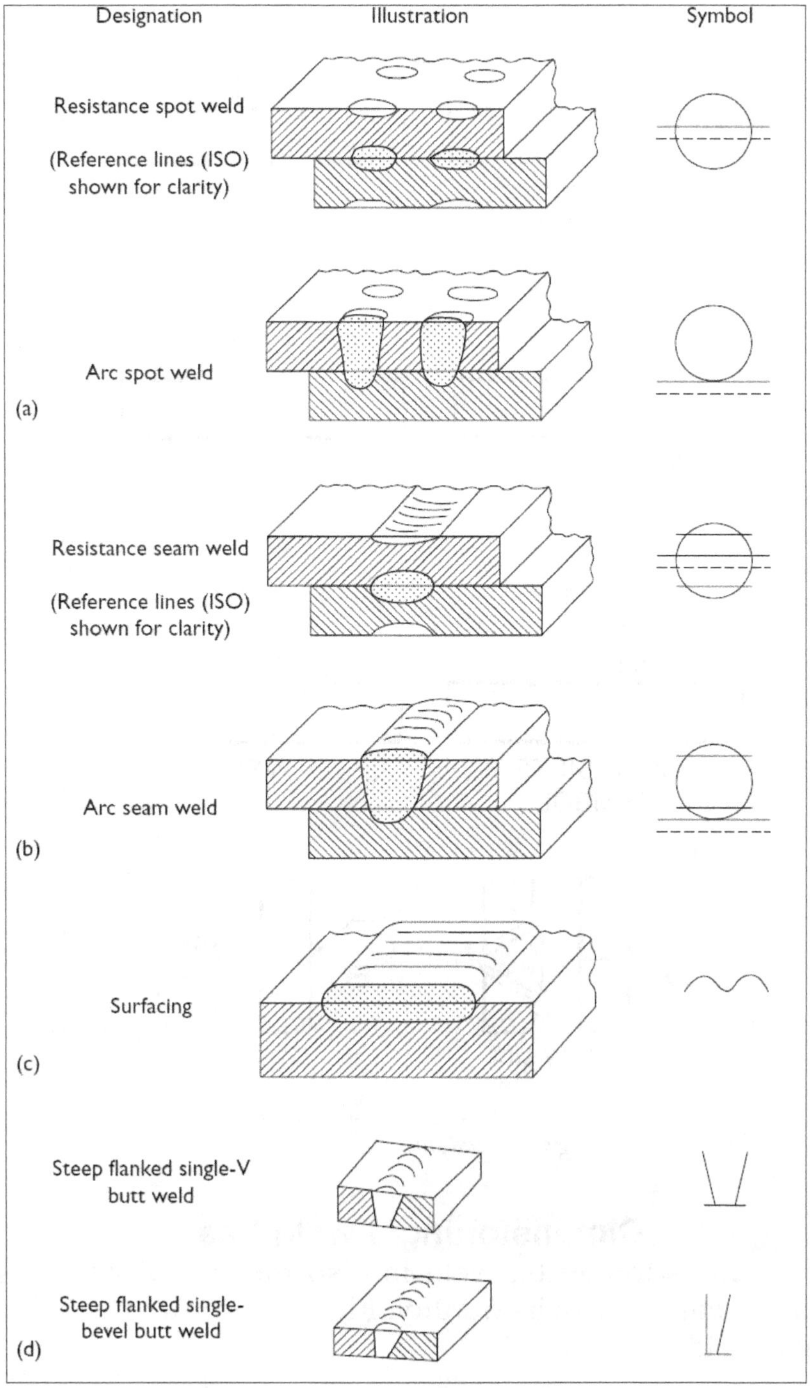

Designation	Illustration	Symbol

Resistance spot weld

(Reference lines (ISO) shown for clarity)

Arc spot weld

(a)

Resistance seam weld

(Reference lines (ISO) shown for clarity)

Arc seam weld

(b)

Surfacing

(c)

Steep flanked single-V butt weld

Steep flanked single-bevel butt weld

(d)

Figure-4. Welding symbols list 4

Now, we know various symbols used in welding drawings but keep a note that placement of welding symbol along the arrow decides the side on which the welding will be done on the object; refer to Figure-5.

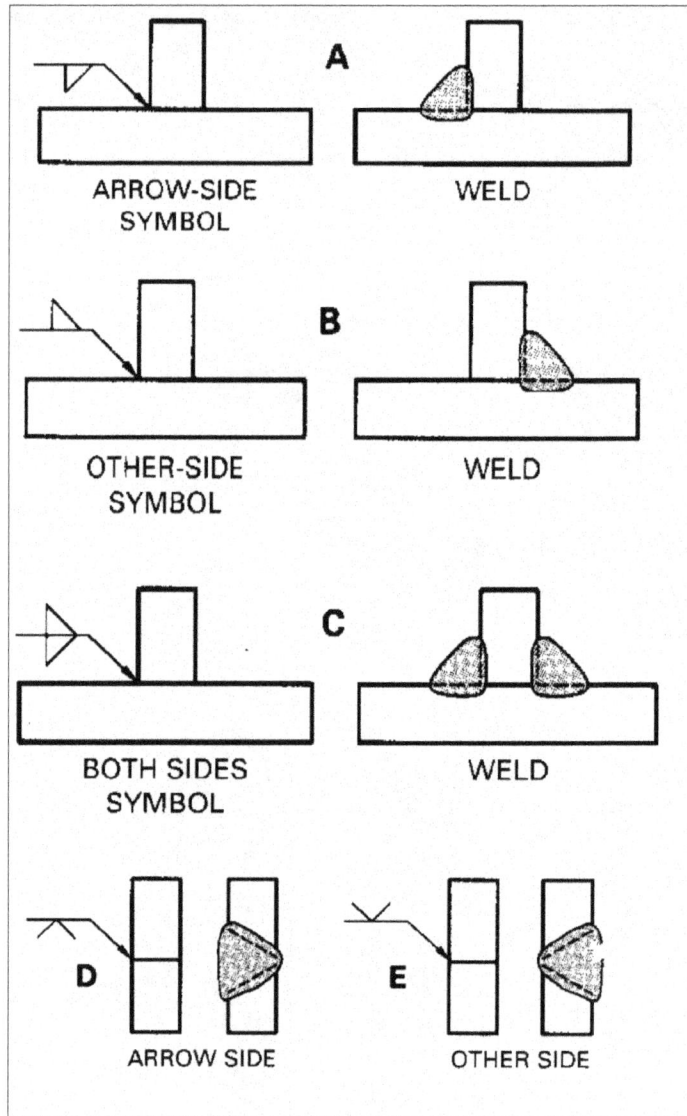

Figure-5. Deciding weld bead side

Dimensioning a weld bead

Just like the other measurements, weld is also measured with respect to various references, so that we can control its quality. Figure-6 shows the information required for dimensioning a weld bead.

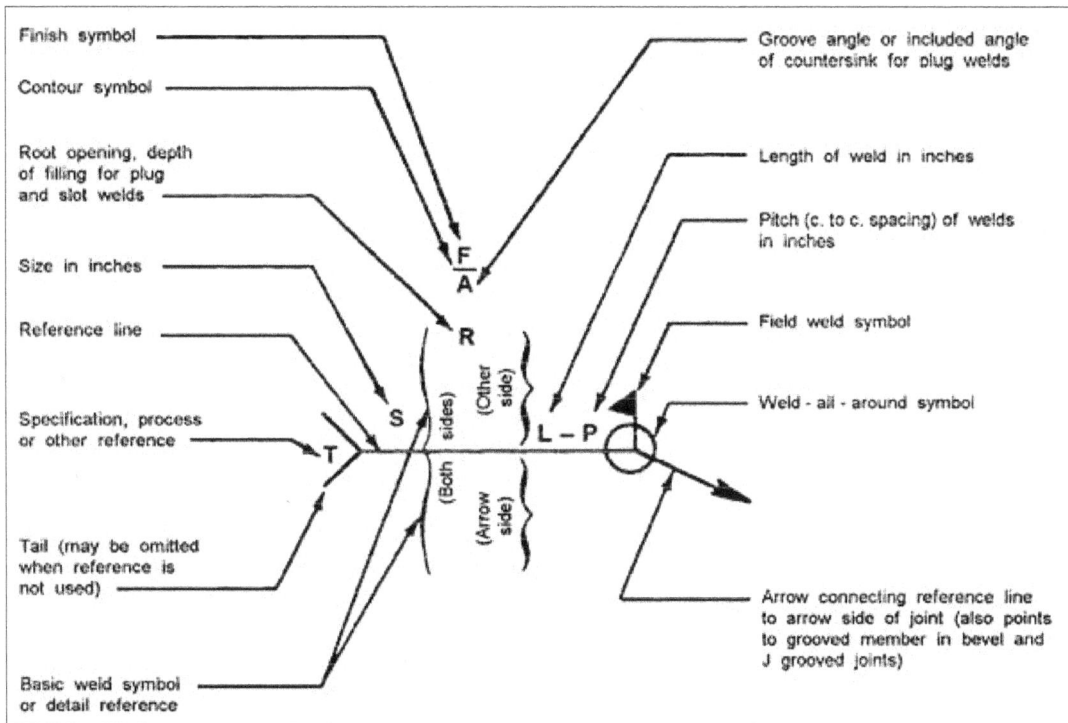

Figure-6. Welding dimension

Till this point, we have learned the basics of weld symbol representations in drawings. So, we are ready to dive into Autodesk Inventor for creating welding representations.

STARTING WELDMENT ASSEMBLY

- Start Autodesk Inventor if not started yet. Click on the **New** button from **New** cascading menu in the **File** menu of the **Ribbon**. The **Create New File** dialog box will be displayed.
- Double-click on the **Weldment (ANSI-mm).iam** template or any other required template of weldment in the **Assembly** area of the dialog box; refer to Figure-7. The assembly environment will be displayed with **Weld** tab selected; refer to Figure-8.

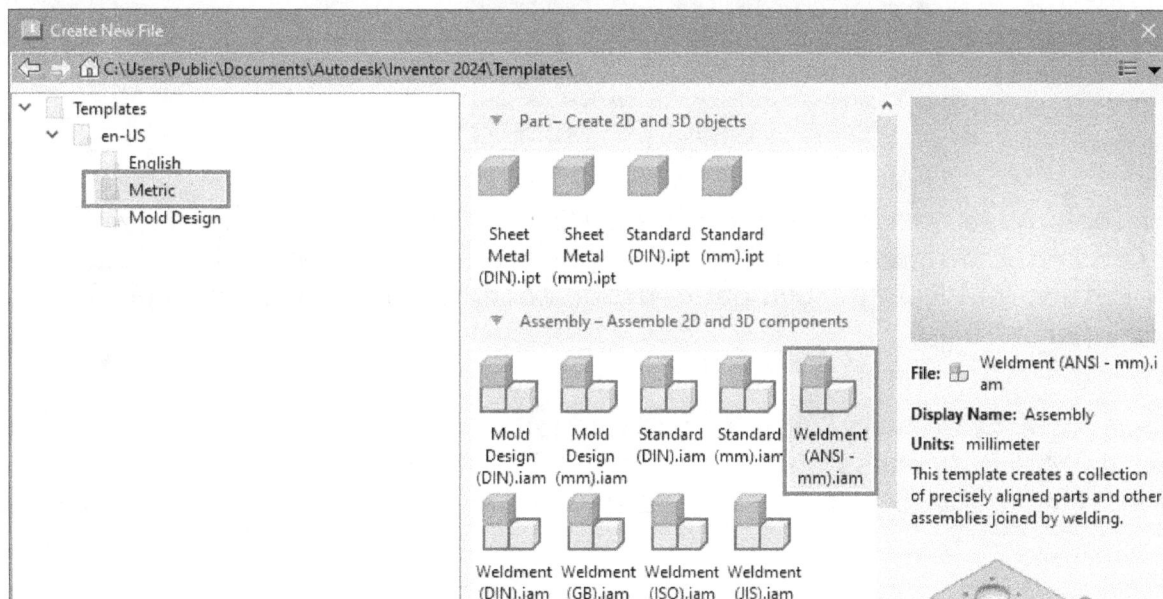

Figure-7. Weldment template in Create New File dialog box

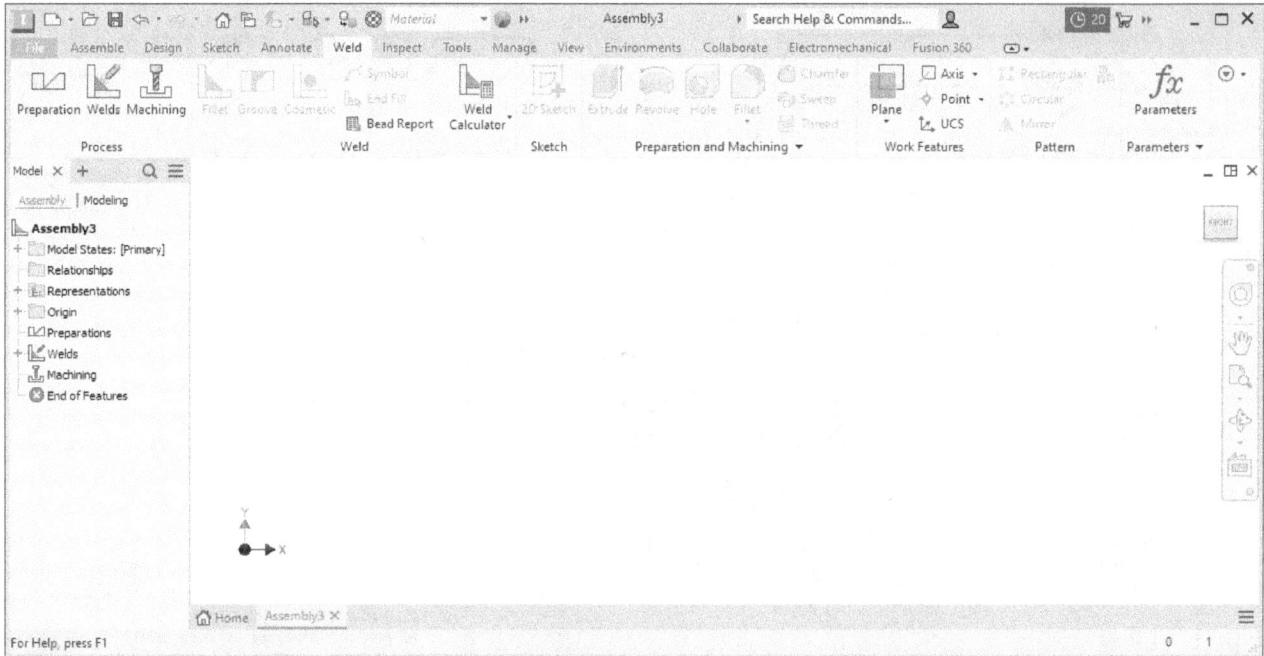

Figure-8. Assembly environment with Weld tab selected

In Autodesk Inventor, the weldment parts are prepared in three processes; Part Preparation, Machining, and Welding. The tools related to these processes discussed next.

PREPARATION

Before we apply welding representations to the model, it is necessary to prepare the part for welding. The tools to perform preparation of model are activated on clicking **Preparation** tool from the **Process** panel in the **Weld** tab of the **Ribbon**; refer to Figure-9.

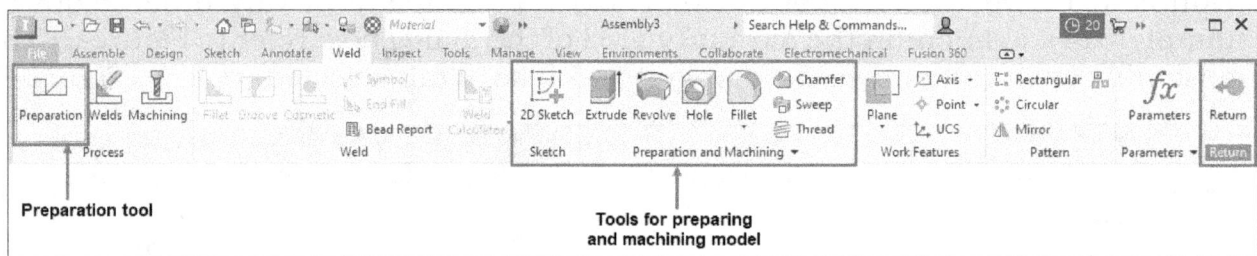

Figure-9. Tools for preparing model

All the tools have already been discussed in previous chapter. Here, you can use these tools only to remove material. After performing the preparation of model, click on the **Return** tool from **Return** panel in the **Weld** tab of the **Ribbon** to perform other welding operation.

WELDING

The **Welds** tool in **Process** panel of **Weld** tab is used to activate the tools related to welding; refer to Figure-10. The procedures of using these tools are discussed next.

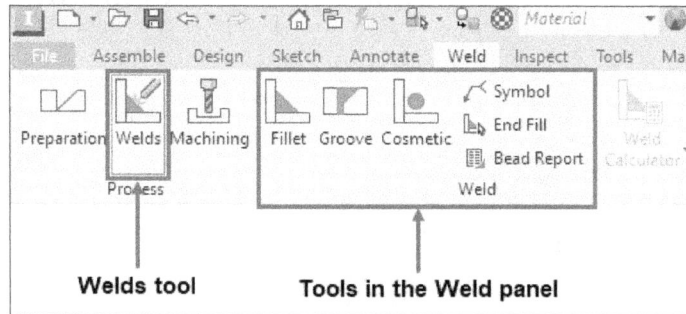

Figure-10. Welding tools

Using Fillet Weld Tool

The **Fillet Weld** tool is used to create fillet shaped weld. Various types of fillet welds have been shown in Figure-3. The procedure to apply fillet weld is given next.

- Click on the **Fillet Weld** tool from the **Weld** panel in the **Weld** tab of the **Ribbon**. The **Fillet Weld** dialog box will be displayed; refer to Figure-11. Also, you will be asked to select the faces to be welded by fillet weld.

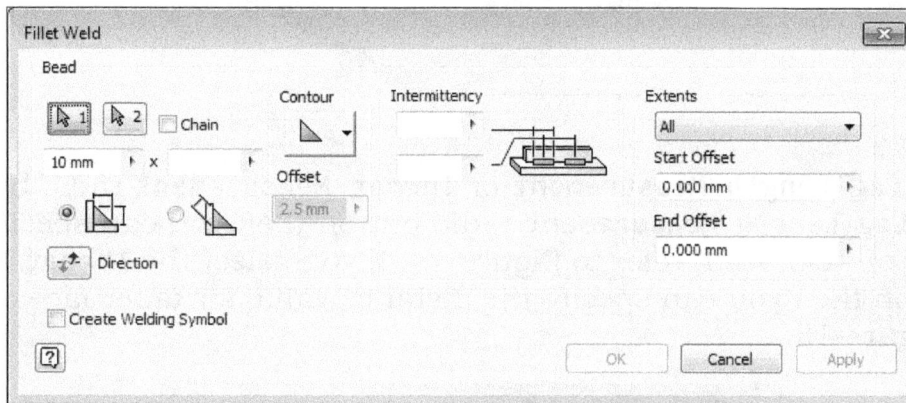

Figure-11. Fillet Weld dialog box

- Select the face/faces of the first face group. Refer to Figure-12.

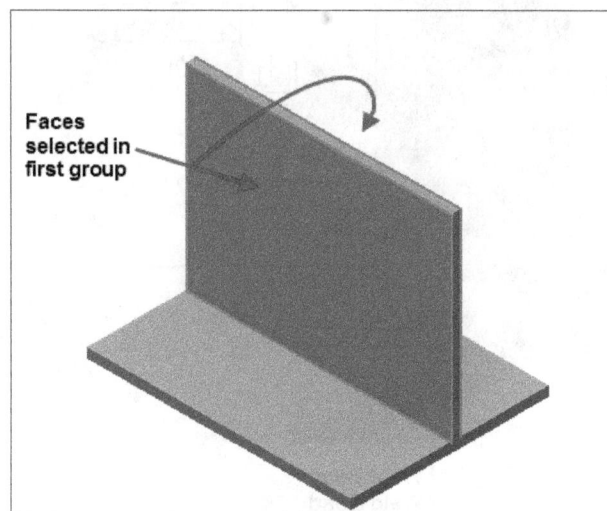

Figure-12. Faces selected in first group for weld

- Click on the **2** selection button from the **Bead** area of the dialog box. You will be asked to select the second face.

- Select the second set of faces. Preview of the fillet weld will be displayed; refer to Figure-13.

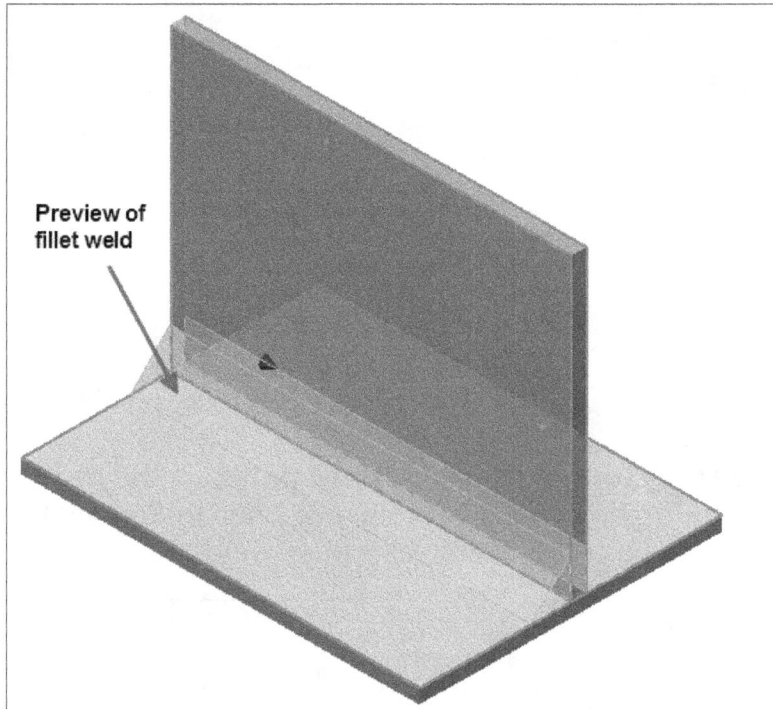

Figure-13. Preview of fillet weld

- Select the **Leg Length Measurement** or **Throat Measurement** radio button. If you select the **Leg Length Measurement** radio button then you can specify the length of each leg of weld bead; refer to Figure-14. If you select the **Throat Measurement** radio button then you can specify the distance value for taper face of weld bead; refer to Figure-15.

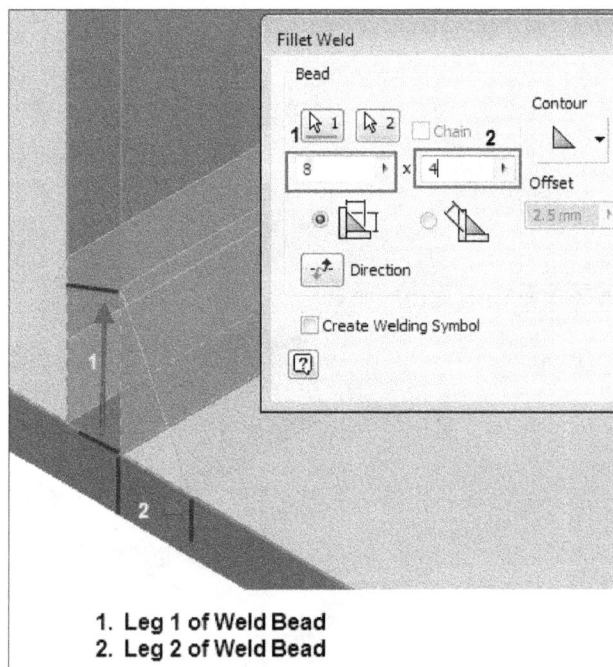

1. Leg 1 of Weld Bead
2. Leg 2 of Weld Bead

Figure-14. Fillet Bead on selecting the Leg Length Measurement radio button

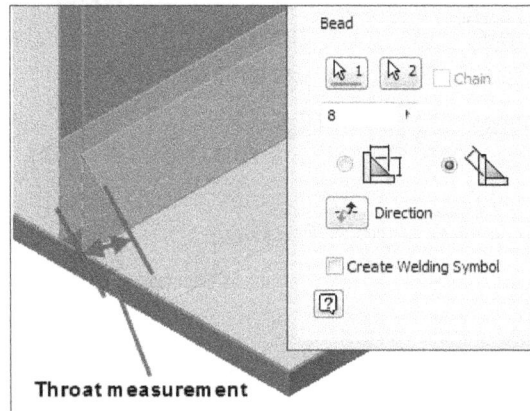

Figure-15. Preview of fillet weld with Throat Measurement radio button selected

- Select desired contour of weld bead from the **Contour** drop-down in the **Contour** area of the dialog box; refer to Figure-16.

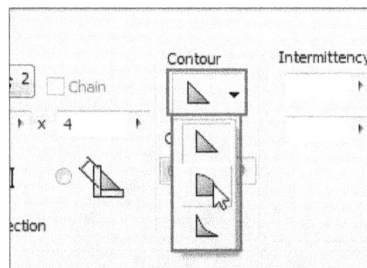

Figure-16. Contour drop-down

- If you have selected the **Convex** or **Concave** contour button then you can specify the contour offset value in the **Offset** edit box to change the shape of contour.
- If you want to create intermittent weld then specify the intermittent length and pitch in the respective edit boxes in the **Intermittency** area of the dialog box.
- Similarly, you can set the starting or end offset value in the **Extents** area of the dialog box to modify starting and end position of the welding bead; refer to Figure-17.

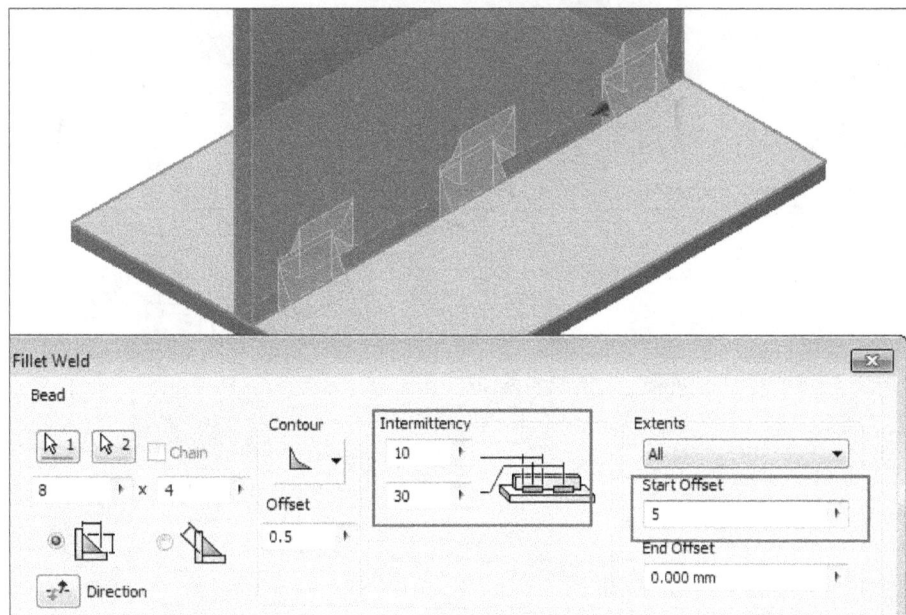

Figure-17. Intermittent welding bead with start offset

Creating Welding Symbol

- Select the **Create Welding Symbol** check box at the bottom of the dialog box. The **Fillet Weld** dialog box will expand as shown in Figure-18.

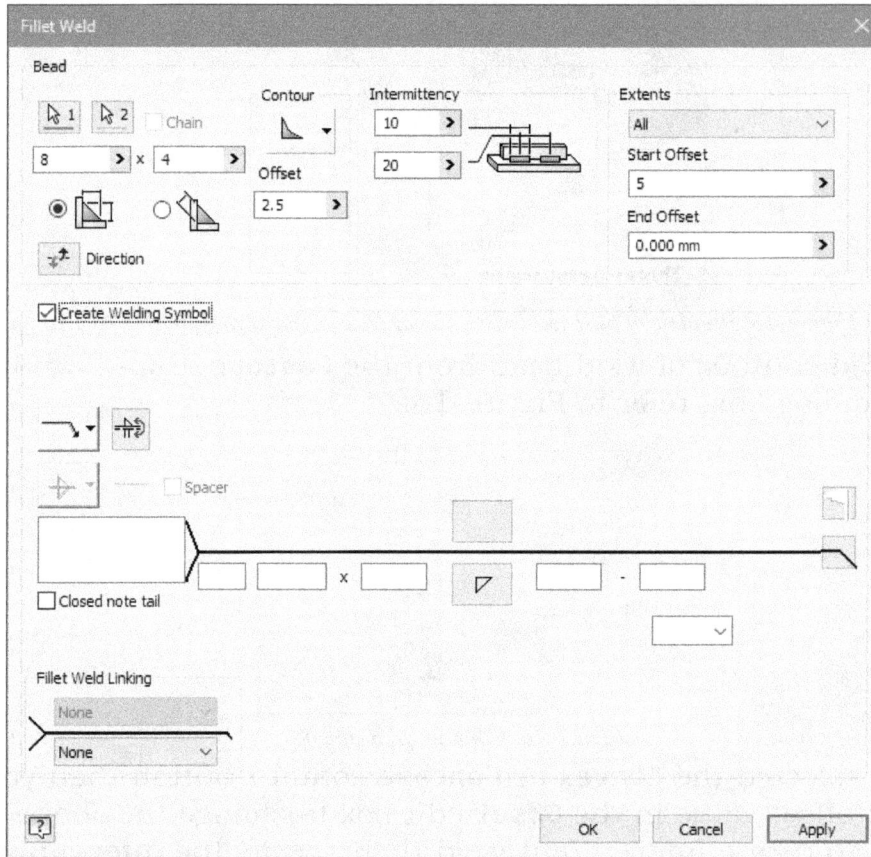

Figure-18. Expanded Filled Weld dialog box

- The edit boxes and drop-downs linked to the weld symbol are used to specify various parameters of weld. We have numbered each option of weld symbol in Figure-19 and use of each option is discussed in table next.

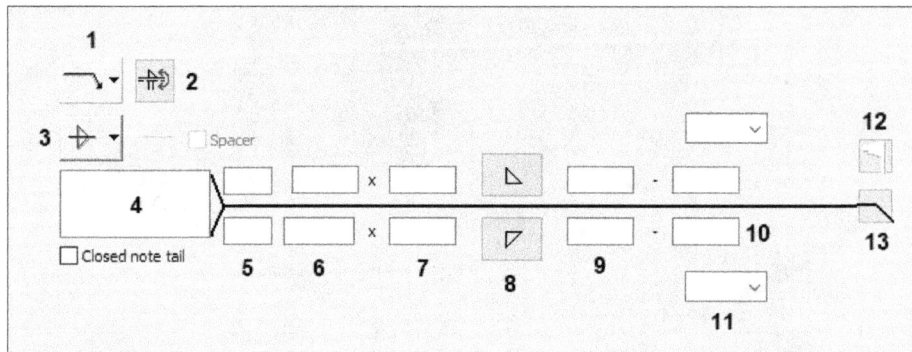

Figure-19. Welding symbol options numbered

Option Number in figure	Option Name	Use in symbol
1	Identification Line drop-down	In welding symbol, identification line is used to denote that weld is to be done at a position far from this symbol location. If the welds are symmetrical on both sides of the plate then identification line is omitted. If the identification line is above the full line then the symbol for the nearside weld is drawn below the reference line and the symbol for the far side weld is above the dashed line.
2	Swap Arrow/ Other Symbols button	The **Swap Arrow/Other Symbols** button is used to swap the parameters above the symbol line with parameters below the symbol line.
3	Stagger drop-down	The options in this drop-down are active only when you are creating weld symbol for both weld of model face. There are four options to create staggered weld.
4	Tail Note Box	This is an edit box used to specify notes on weld like welding class or standard used. Select the **abc** check box to enclose note in a box.
5	Prefix edit box	The **Prefix** edit box is used to specify prefix for value specified in **Leg 1** edit box.
6	Leg 1	This edit box is used to specify the dimension for first leg of weld bead.
7	Leg 2	This edit box is used to specify the dimension for second leg of weld bead.
8	Weld Symbol	This button is used to select the desired symbol for weld.
9	Length	This edit box is used to specify the length of weld bead.
10	Pitch	This edit box is used to specify the gap distance between intermittent weld.

11	Contour drop-down	The options in this drop-down are used to specify the contour of weld bead. On selecting the contour type, the Method drop-down gets displayed next to it. Select the desired option from the drop-down to specify the method of contouring. C means Chipping, G means Grinding, R means Rolling, H means Hammering, M means Machining, and U means Unspecified Mechanical means.
12	Field Welding Symbol	Select this button if you want to tell manufacturer that welding will be done on the assembly line.
13	All Around Symbol	Select this button if you want the weld to be done all around the joint.

- Set desired parameters and click on the **OK** button to create the weld symbol and representation; refer to Figure-20.

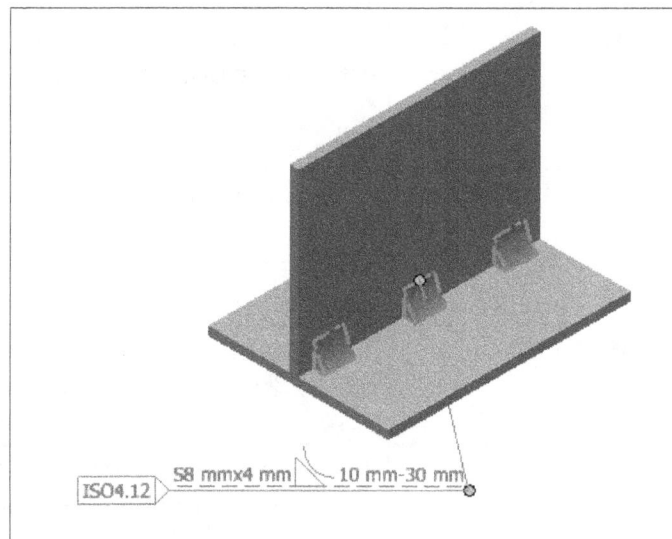

Figure-20. Fillet weld with respective drawing symbol

Using Groove Weld Tool

The **Groove Weld** tool is used to create weld bead to fill groove between two components. With the help of groove, more metal is filled and hence a stronger bond is created. The procedure of using this tool is given next.

- Click on the **Groove Weld** tool from the **Weld** panel in the **Weld** tab of the **Ribbon**. The **Groove Weld** dialog box will be displayed; refer to Figure-21. Also, you will be asked to select the first set of faces.

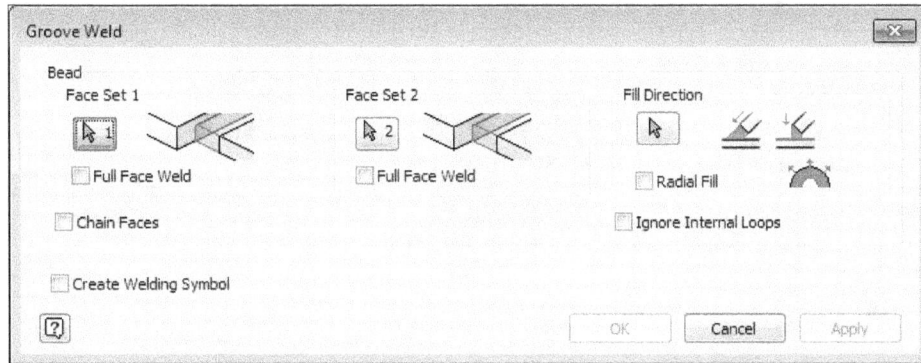

Figure-21. Groove Weld dialog box

- Click on the first face/set of faces.
- Click on the **2** selection button in the **Face Set 2** area of the dialog box. You will be asked to select the second face/set of faces.
- Select the second set of faces; refer to Figure-22.

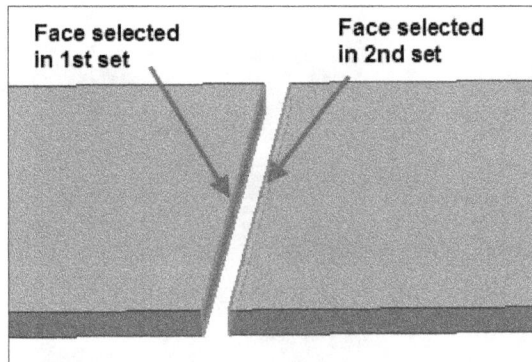

Figure-22. Faces selected for groove weld

- Click on the **Select Fill Direction** selection button in the **Fill Direction** area of the dialog box. You will be asked to select the direction reference for fill direction.
- Select desired face/plane/edge/axis to specify direction of weld filling. Preview of the weld bead will be displayed; refer to Figure-23.

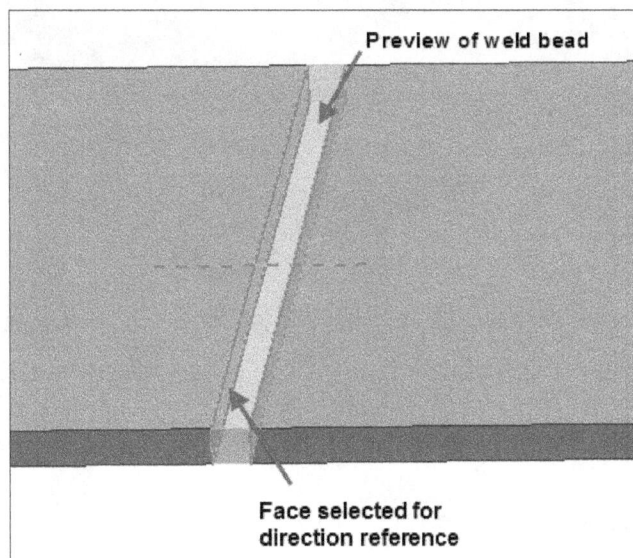

Figure-23. Preview of groove weld bead

- Select the **Full Face Weld** check box when a face is shorter than the joining face but you want the faces to be completely welded; refer to Figure-24.

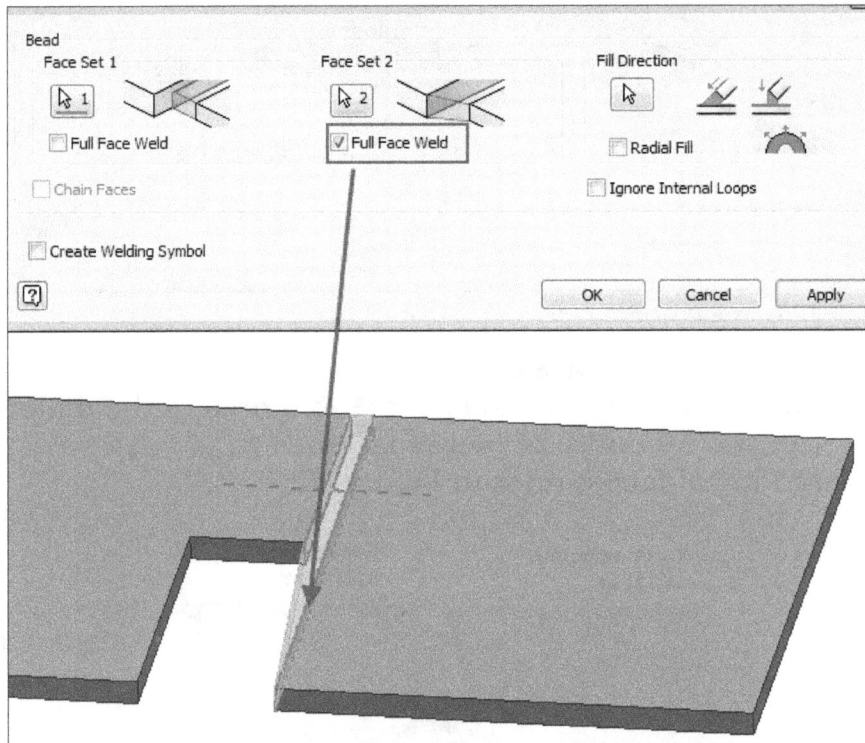

Figure-24. Preview of full face groove weld

- Select the **Radial Fill** check box if you have round faces to be welded; refer to Figure-25.

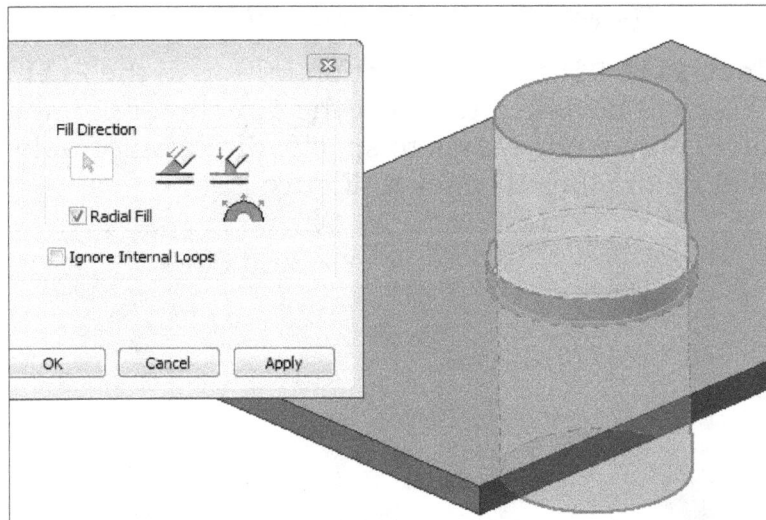

Figure-25. Preview of groove weld with radial fill

- Select the **Ignore Internal Loops** check box if you want to make sure that internal faces are not getting welded automatically.
- Select the **Create Welding Symbol** check box to create the welding symbol as discussed earlier.
- Click on the **OK** button to create the weld bead.

Using Cosmetic Weld Tool

Cosmetic The **Cosmetic Weld** tool is used to assign welding symbol to desired edge(s)/loop. If you apply cosmetic weld then graphical weld will not be displayed. The procedure of using this tool is given next.

- Click on the **Cosmetic Weld** tool from the **Weld** panel in the **Weld** tab of the **Ribbon**. The **Cosmetic Weld** dialog box will be displayed; refer to Figure-26.

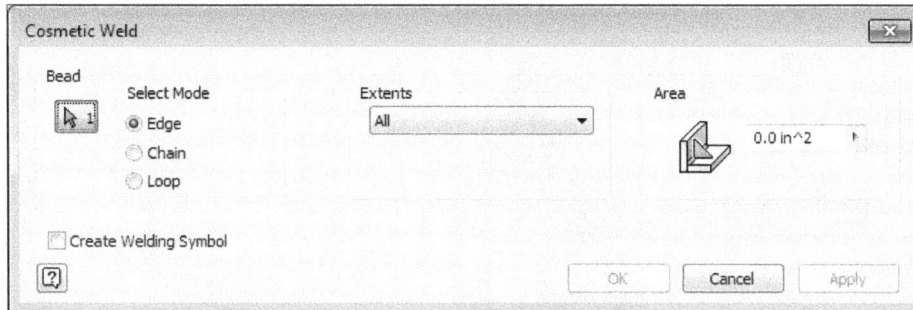

Figure-26. Cosmetic Weld dialog box

- Select desired radio button from the **Select Mode** area of the dialog box. There are three options; **Edge**, **Chain**, and **Loop**.
- Select the entities based on the radio button selected.
- Click in the edit box of **Area** area and specify desired value of cross-section area of weld bead.
- Select the **Create Welding Symbol** check box and specify the values for welding symbol as discussed earlier.
- Click on the **OK** button to create the cosmetic weld.

Using Welding Symbol Tool

Symbol The **Welding Symbol** tool is used to assign welding symbol to welding beads for which welding symbols have not been created. Before using this tool, make sure that welding beads without welding symbol are available in the drawing area. The procedure of using this tool is given next.

- Click on the **Welding Symbol** tool from the **Weld** panel in the **Weld** tab of the **Ribbon**. The **Welding Symbol** dialog box will be displayed; refer to Figure-27. Also, you will be asked to select the welding bead.

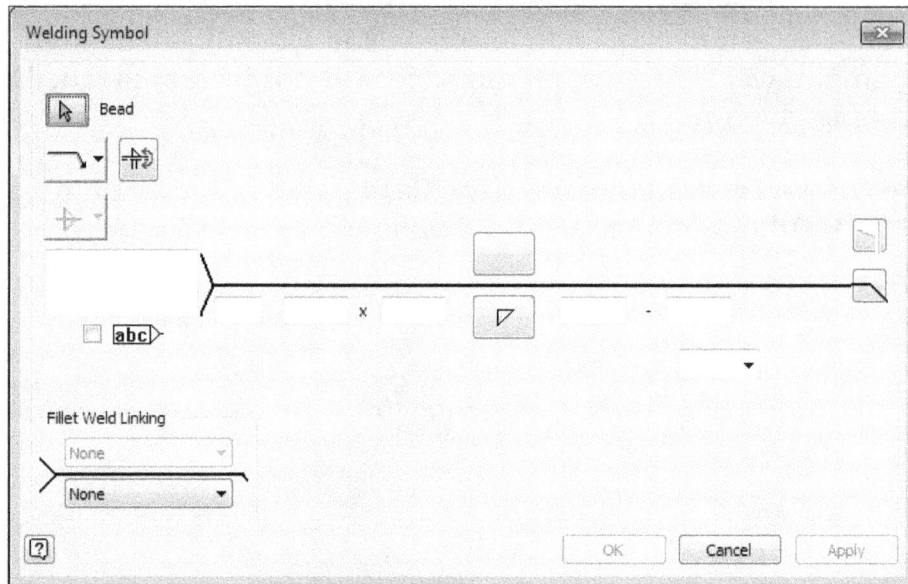

Figure-27. Welding Symbol dialog box

- Select the welding bead to which the welding symbol will be attached.
- Specify the parameters of welding symbol as discussed earlier.
- Click on the **OK** button to create the weld symbol.

Using End Fill

End Fill The **End Fill** tool is used to assign end to the weld bead. Fillet welds get end fill automatically assigned but groove welds are not assigned end fills and hence, we need to assign end fills to the faces of groove weld beads. The procedure of using this tool is given next.

- Click on the **End Fill** tool from the **Weld** panel in the **Weld** tab of the **Ribbon**. You will be asked to select the faces to be recognized as end fills.
- Select the faces of the weld bead to be recognized as end fills; refer to Figure-28.
- Right-click in the drawing area and select the **OK** button to create the feature; refer to Figure-29.

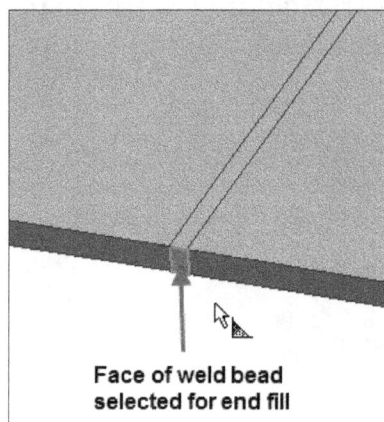

Figure-28. Face selected for end fill

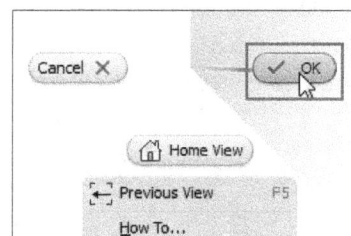

Figure-29. Right-click menu

Using Bead Report

Bead Report The **Bead Report** tool is used to create weld bead report in Microsoft Excel format or other tabulated data formats. The procedure of using this tool is given next.

- Click on the **Bead Report** tool from the **Weld** panel in the **Weld** tab of the **Ribbon**. The **Weld Bead Report** dialog box will be displayed; refer to Figure-30.

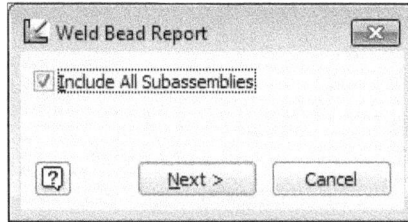

Figure-30. Weld Bead Report dialog box

- Select the **Include All Subassemblies** check box if you want to include weld beads from all the sub-assemblies otherwise the weld beads of current assembly file will only be included.
- Click on the **Next** button from the dialog box. The **Report Location** dialog box will be displayed; refer to Figure-31.

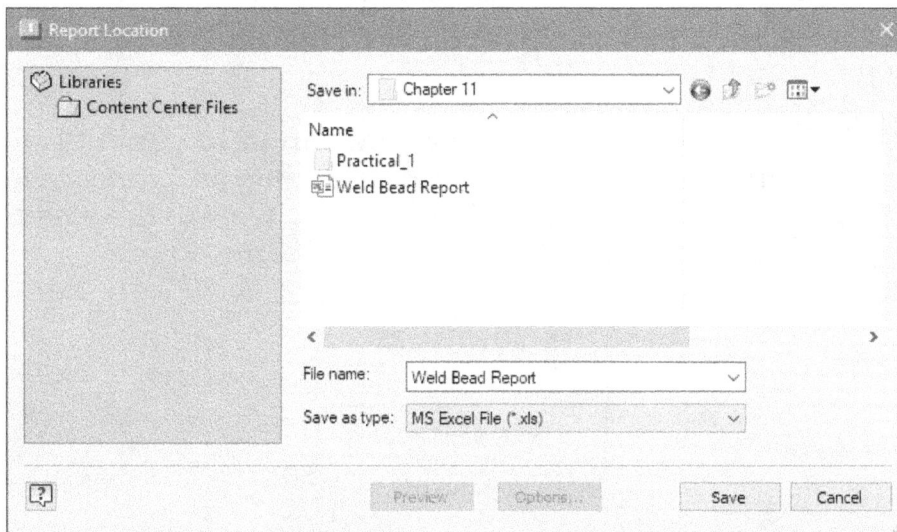

Figure-31. Report Location dialog box

- Specify the name of file and desired location in the dialog box. Click on the **Save** button to save the report. The report file will be created; refer to Figure-32.

Document	ID	Type	Length	UoM	Mass	UoM	Area	UoM	Volume	UoM	L
Weld_assembly.iam	Groove Weld 1	Groove	N/A		0.001	kg	812	mm^2	480	mm^3	
	Fillet Weld 2	Fillet	320	mm	0.022	kg	7.93E+03	mm^2	8.00E+03	mm^3	
	Fillet Weld 3	Fillet	56.549	mm	0.002	kg	955.644	mm^2	588.91	mm^3	
	Fillet Weld 4	Fillet	56.549	mm	0.002	kg	955.644	mm^2	588.91	mm^3	

Figure-32. Weld Bead Report

MACHINING

The tools for the machining process are active when you click on the **Machining** tool from the **Process** panel in the **Weld** tab of the **Ribbon**. Note that the tools for preparation and machining are same; refer to Figure-33. There is only one difference between use of preparation and machining tools, you cannot use the preparation tools after creating weld beads.

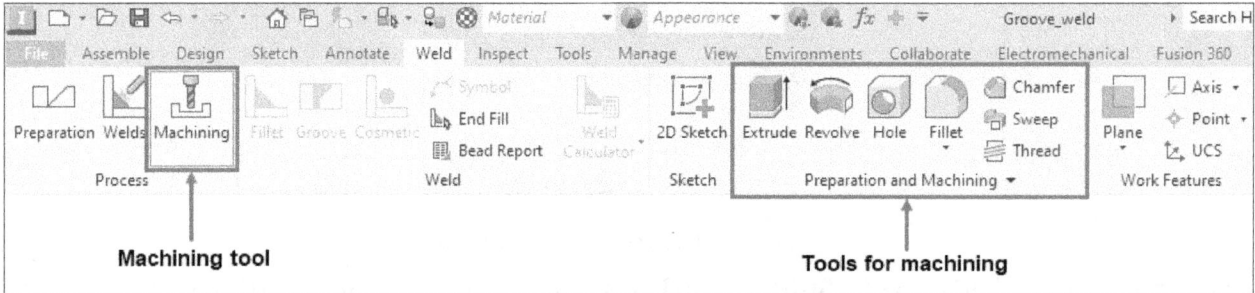

Figure-33. Tools for machining

WELD CALCULATOR

Yes! **Weld Calculator** should be discussed before creating weld beads but some of my book readers are very new to welding, so I explained the weld creation tools earlier. In this way, they know shapes of different types of welds which will be helpful in weld calculations.

There are **10** tools in **Weld Calculator** drop-down in **Weld** panel of the **Weld** tab to make calculations related to welding bead (and Soldering also); refer to Figure-34. The procedure of using all the calculator is almost same. We will discuss the procedure of using one weld calculator and you can apply the same other.

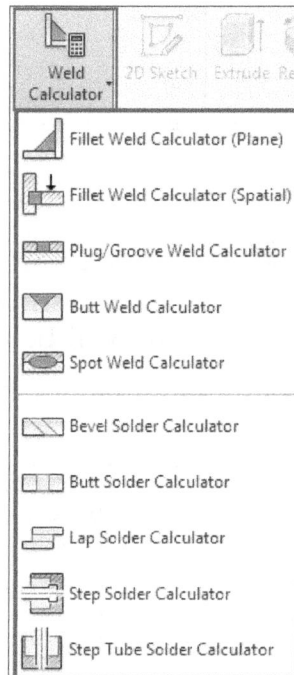

Figure-34. Weld Calculator dialog box

Fillet Weld Calculator (Plane)

The **Fillet Weld Calculator (Plane)** tool in the **Weld Calculator** drop-down is used to calculate whether the fillet weld bead will be able to sustain the given load conditions or not. The procedure of using this tool is given next.

- Click on the **Fillet Weld Calculator (Plane)** tool from the **Weld Calculator** drop-down in the **Weld** panel of the **Weld** tab in the **Ribbon**. The **Fillet Weld (Connection Plane Load) Calculator** dialog box will be displayed; refer to Figure-35.

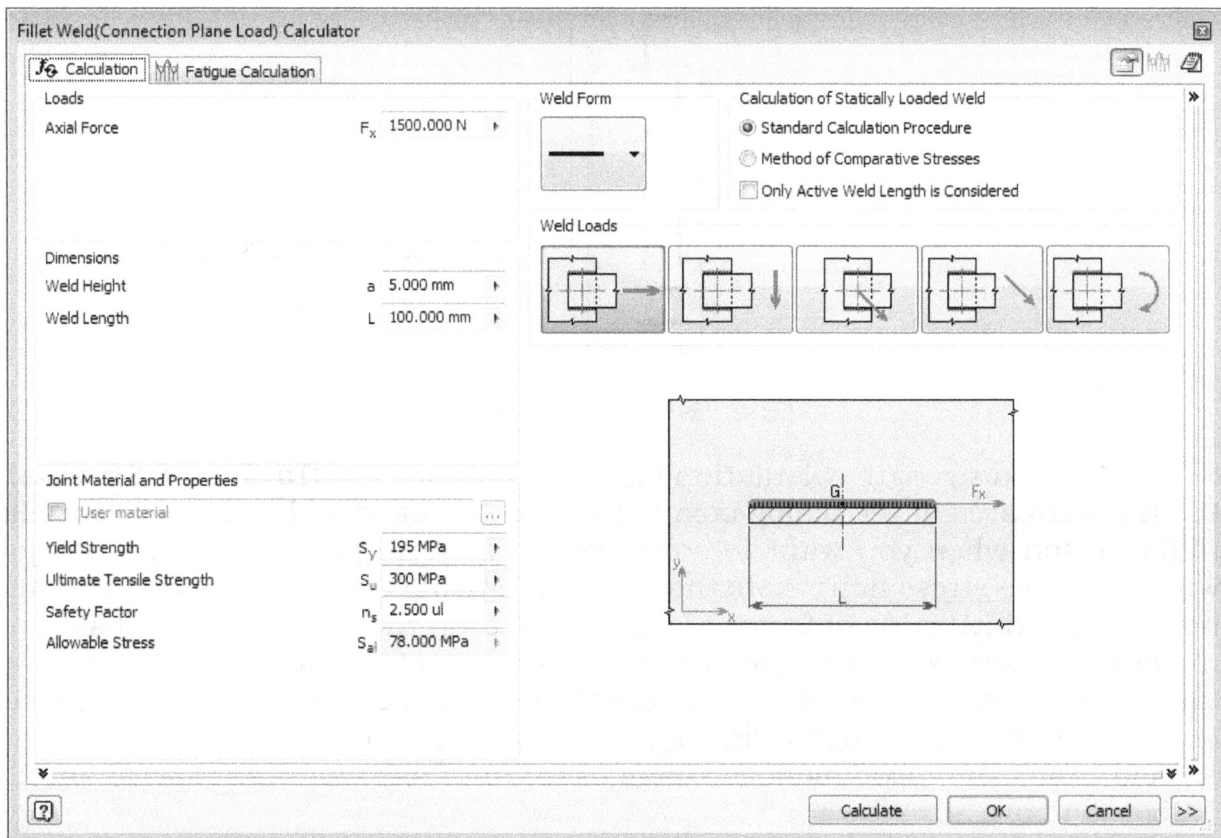

Figure-35. Fillet Weld (Connection Plane Load) Calculator dialog box

- Select desired weld form from the **Weld Form** drop-down; refer to Figure-36. The options in the **Weld Form** drop-down represent the shape of welding bead around the welded objects.

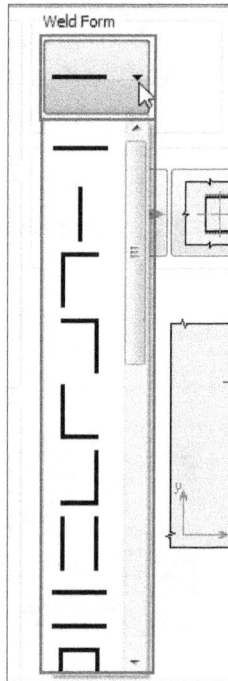

Figure-36. Weld Form drop-down

- Select the weld strength calculation method from the **Calculation of Statically Loaded Weld** area of the dialog box. Select the **Standard Calculation Procedure** radio button when you want to compare the <u>total stress induced</u> in weld joint with <u>allowable stress limit</u>. Note that if you are using this method then you must have appropriate value of factor of safety based on the weld type, weld size, and material properties. The recommended values of factor of safety are given in the table next. Select the **Method of Comparative Stresses** radio button if you want to compare <u>allowable stress</u> with <u>partial stresses</u> induced at the ends and sides of weld bead. The recommended range of factor of safety is less than or equal to 1.25 and above 2.

Weld type, loading	n_s
Butt welds loaded with traction	1.6 to 2.2
Butt welds loaded with bend	1.5 to 2.0
Butt welds loaded with shear	2.0 to 3.0
Butt welds loaded with loading	1.4 to 2.7
Fillet welds loaded in the plane of joining the part	2.0 to 3.0
Fillet welds loaded spatially	1.4 to 2.7
Plug and groove welds	2.0 to 3.0
Plug (resistant) welds loaded with shear	1.6 to 2.2
Plug (resistant) welds loaded with tearing	2.5 to 3.3

- Select the **Only Active Weld Length is Considered** check box to make sure that the active weld length is considered in calculations.

- Select desired load from the **Weld Loads** area of the dialog box based on the selected weld form. There are five buttons in the **Weld Loads** area named **Axial force, Bending Force parallel with the neutral axis of the weld group, Common force acting in the center of gravity of the weld group, Common force acting outside the center of gravity of the weld group,** and **Bending Moment**. On selecting the load, the graphical preview of load action will be displayed in the **Weld Loads** area; refer to Figure-37.

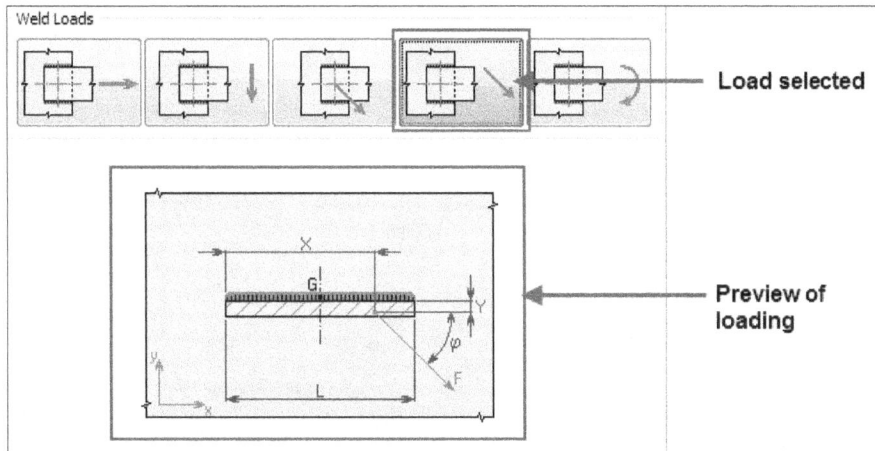

Figure-37. Preview of load on weld bead

- Specify the parameters related to load in the edit boxes of **Loads** area of the dialog box.
- Similarly, set the weld bead size in the **Dimensions** area of the dialog box.
- Set the material properties in the **Joint Material and Properties** area of the dialog box. Specify the Safety Factor value carefully as it can make huge impact on calculation and usability of component made using the design.
- Click on the **More options** button at the bottom right corner of the dialog box to expand the dialog box; refer to Figure-38.

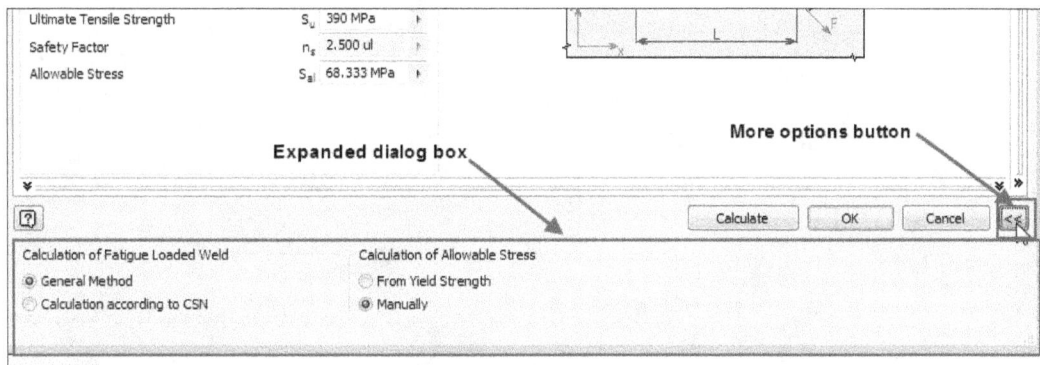

Figure-38. Expanded Fillet Weld (Connection Plane Load) Calculator dialog box

- Select the **Manually** radio button from the **Calculation of Allowable Stress** area of the expanded dialog box to manually specify the allowable stress.
- Select the **Calculation according to CSN** radio button from the **Calculation of Fatigue Loaded Weld** area in the expanded dialog box to change the method of fatigue stress calculations.

Fatigue Calculation

- Click on the **Enable/disables fatigue calculation** button from the top-right corner of the dialog box. The **Fatigue Calculation** tab in the dialog box will become active. Note that the **Enable/disables fatigue calculation** button will not be available for single straight weld bead weld form.

- Click on the **Fatigue Calculation** tab from the dialog box. The dialog box will be displayed as shown in Figure-39.

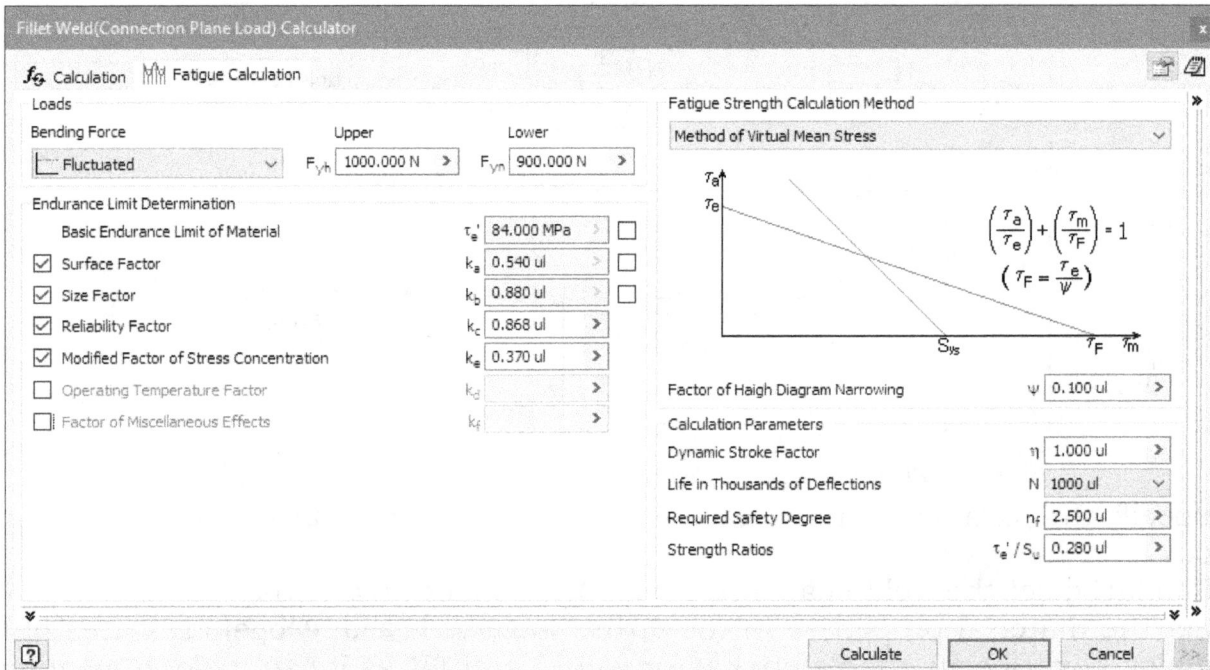

Figure-39. Fatigue Calculation tab in Fillet Weld (Connection Plane Load) Calculator dialog box

- Select the method of load repetition from the drop-down in the **Loads** area of the dialog box; refer to Figure-40.

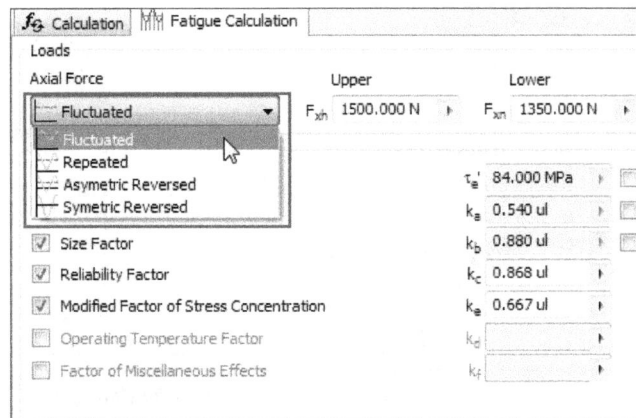

Figure-40. Drop-down in Loads area

- Specify the upper and lower load value in the **Upper** and **Lower** edit boxes.
- Set the other parameters for calculations and endurance limit.
- Click on the **Calculate** button. The result will be calculated and displayed in right pane of the dialog box; refer to Figure-41.

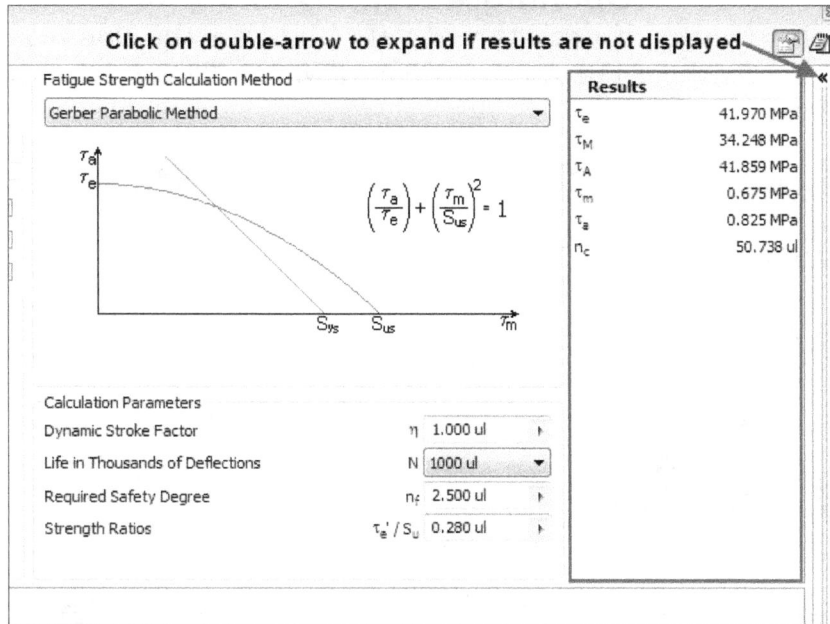

Figure–41. Results of fatigue calculation for fillet weld

• If the factor of safety is above required value then the weld bead can sustain the specified load. Based on the results, you can create weld bead in the model as discussed earlier in the chapter.

In the same way, you can use the calculators in the **Weld Calculator** drop-down.

PRACTICAL 1

In this practical, we will apply plug weld on the assembly model as shown in Figure-42.

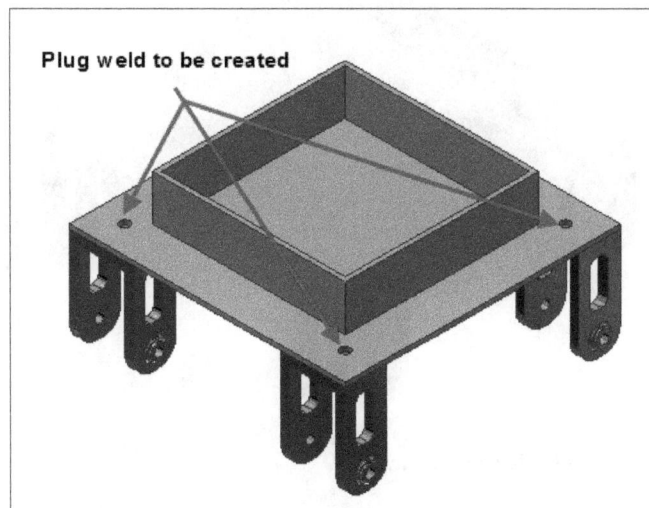

Figure–42. Plug weld in model for Practical1

When we create a weldment assembly in Autodesk Inventor, the first step is to find out appropriate size of weld for desired load conditions. Once, we get the right size of weld, we can use welding tools to model the weld and apply welding symbols. The steps to create weldment model for Practical 1 are given next.

Opening Assembly File

- Start Autodesk Inventor if not started yet. Click on the **Open** button from **Open** cascading menu in the **File** menu of the **Ribbon**. The **Open** dialog box will be displayed; refer to Figure-43.

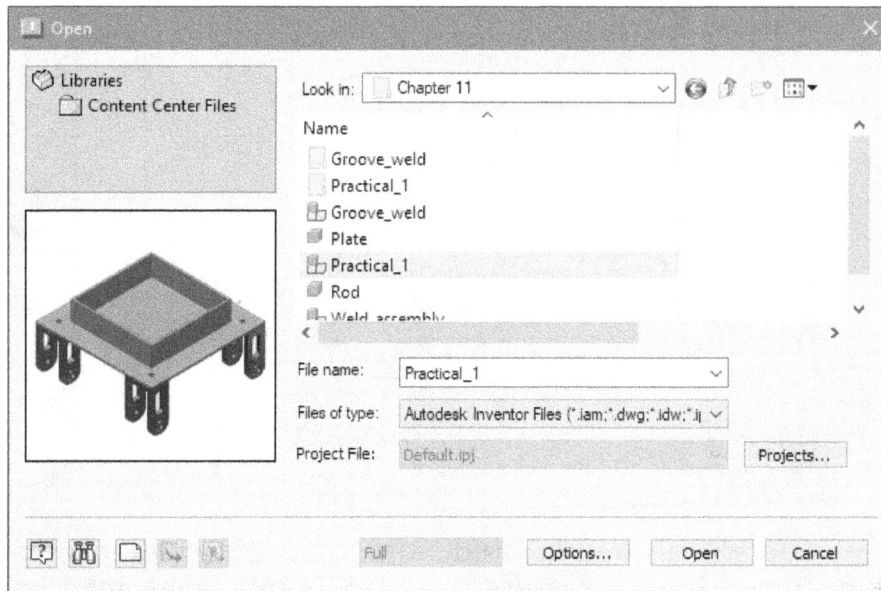

Figure-43. Open dialog box

- Double-click on the **Practical 1** file in **Chapter 11** folder of resource kit. The model will be displayed as shown in Figure-44. (Note that you need to write us an e-mail at cadcamcaeworks@gmail.com to get resource kit of the book.)

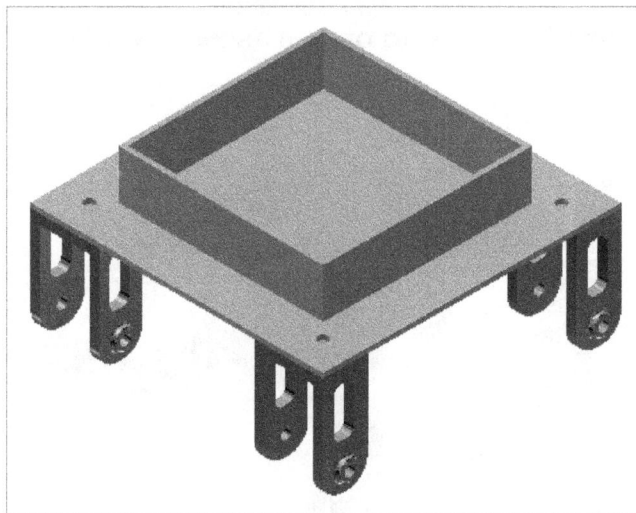

Figure-44. Model for Practical 1

Calculating Plug Weld Size

- Click on the **Plug/Groove Weld Calculator** tool from the **Weld Calculator** dropdown in the **Weld** panel of the **Weld** tab in the **Ribbon**. The **Plug and Groove Weld Calculator** dialog box will be displayed; refer to Figure-45.

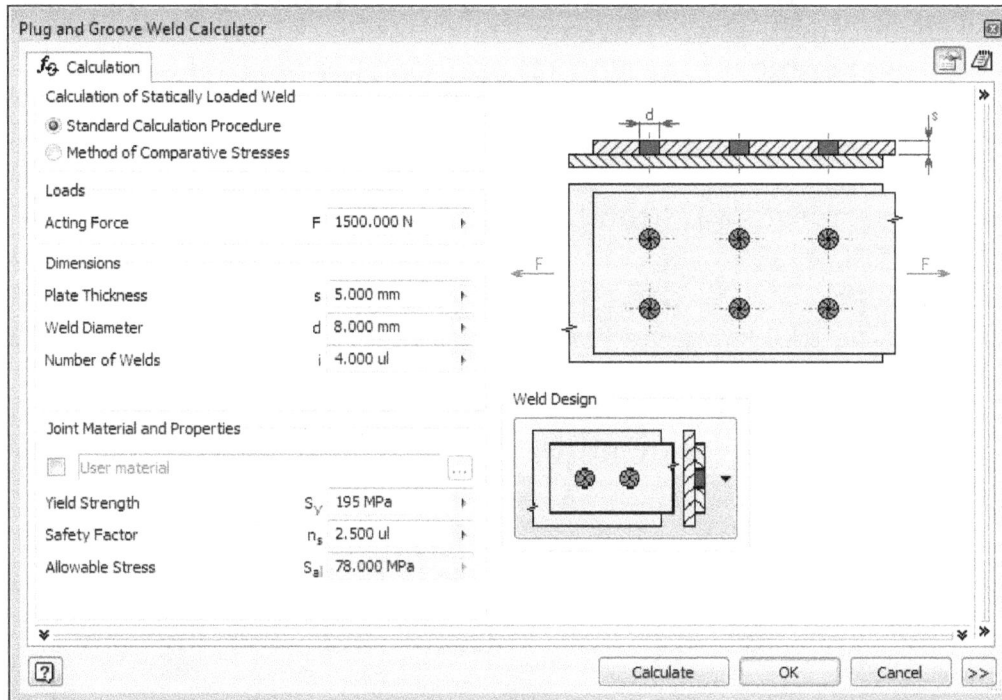

Figure-45. Plug and Groove Weld Calculator dialog box

- Specify the value of force as **15000 N** in the **Acting Force** edit box in the **Loads** area of the dialog box.
- Specify the plate thickness as **5** mm, Weld diameter as **11** mm, and number of welds as **1**. These values are taken after measuring the plate thickness and diameter of hole.
- Click on the **User material** edit box in the **Joint Material and Properties** area of the dialog box. The **Joint Material** dialog box will be displayed; refer to Figure-46.

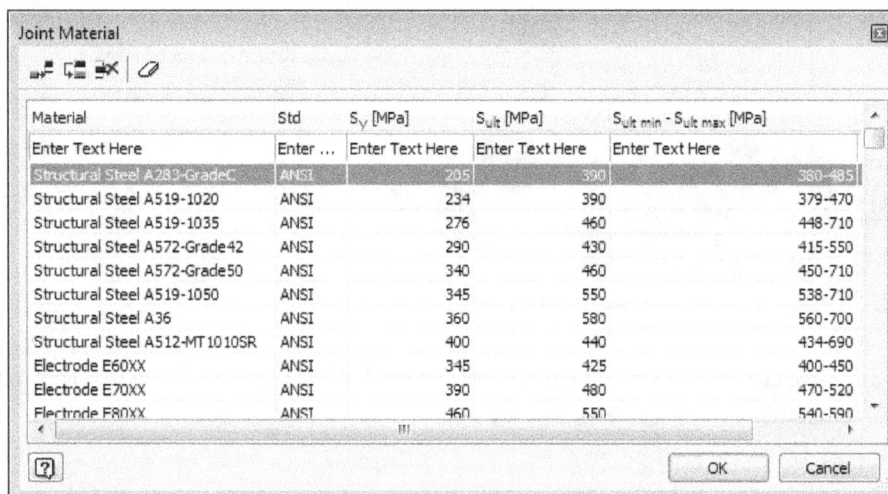

Figure-46. Joint Material dialog box

- Select the **Electrode E120XX** material from the dialog box and click on the **OK** button.

- Click on the **Calculate** button from the **Plug and Groove Weld Calculator** dialog box to test the strength of weld. We get the message "Calculation indicates design compliance!" which means we can create the plug weld of size specified in the **Dimensions** area of the dialog box. If you get any error in calculation results then you need to increase the plug diameter, increase number of welds or weld material.
- Click on the **OK** button from the dialog box. The **File Naming** dialog box will be displayed. Click on the **OK** button to create the calculation feature.

Creating Plug Weld

- Click on the **Welds** tool from the **Process** panel in the **Weld** tab of the **Ribbon**. The tools related to weld will become active.
- Click on the **Groove** tool from the **Weld** panel in the **Ribbon**. The **Groove Weld** dialog box will be displayed; refer to Figure-47.

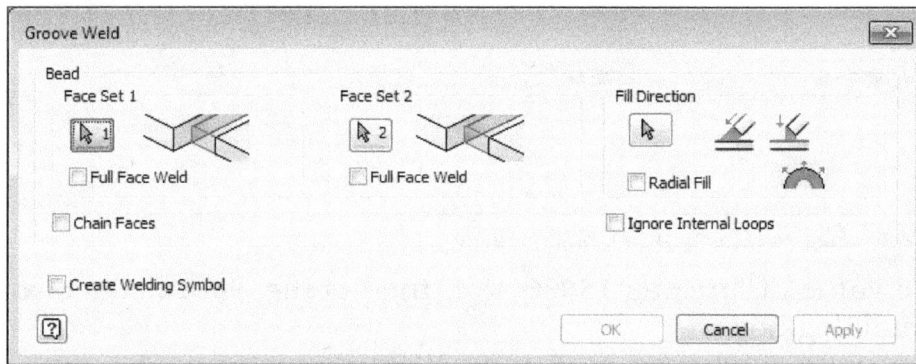

Figure-47. Groove Weld dialog box

- Select the round face of the hole as shown in Figure-48.

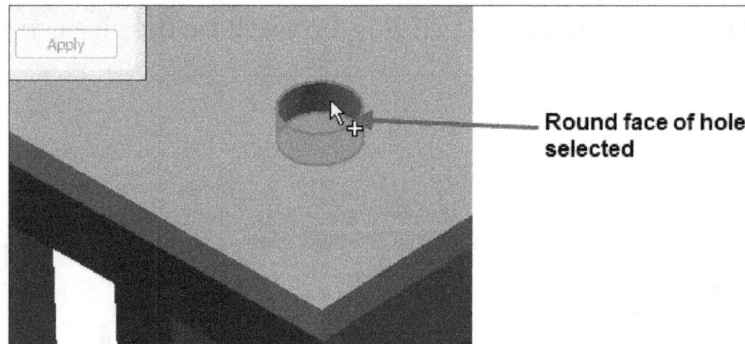

Figure-48. Round face of hole selected

- Click on the selection button in the **Face Set 2** area of dialog box and select the face below the hole; refer to Figure-49.

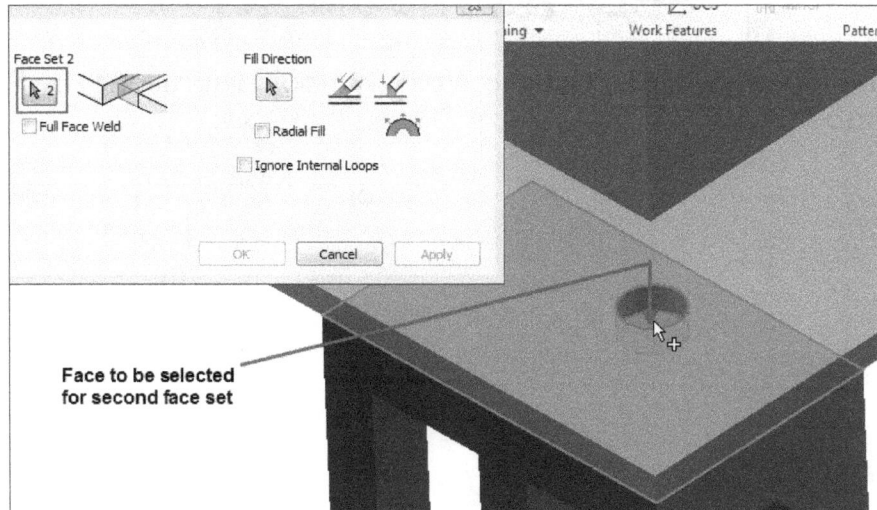

Figure-49. Face selected for Face Set 2

- Select the **Radial Fill** check box. Preview of the plug weld will be displayed.
- Select the **Create Weld Symbol** check box. The dialog box will get expanded.
- Set the parameters of welding symbol as shown in Figure-50.

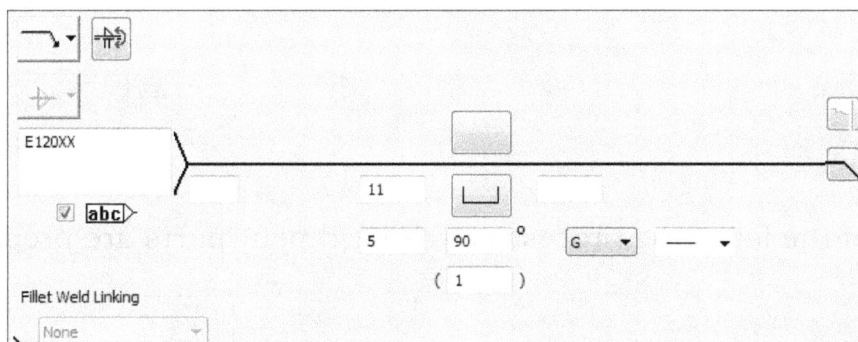

Figure-50. Parameters for welding symbol

- Click on the **OK** button to create the weld bead and symbol. Repeat the same procedure for other holes.

Till this chapter, you have learned about modeling and assembly tools as well as weldment assembly. Now, I want you to open up your imaginations and make the components around you in CAD. Grab a scale and measuring tape, and start making assemblies (for this reason we have not given any exercise for this chapter!). Use the Student Notes area next to note down the drawings of component you have found around you.

SELF ASSESSMENT

Q1. Which of the following Butt/Groove Weld Symbols represent the Single-U butt/groove weld symbol?

a)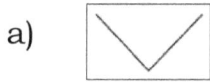

c)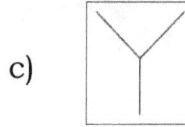

b)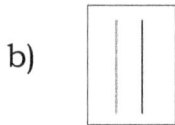

d)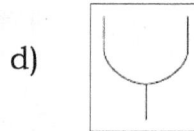

Q2. Which of the following weld symbols represent the Arc Seam Weld symbol?

a)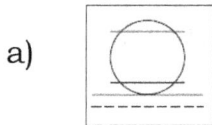

c)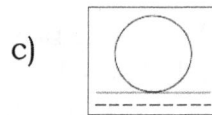

b)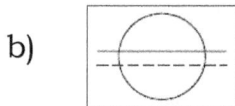

d)

Q3. By which of the following processes, the weldment parts are prepared?

a) Machining
b) Welding
c) Both a and b
d) None of the Above

Q4. Which of the following options is incorrect definition of contour type for fillet weld symbol?

a) C means Chipping
b) G means Grooming
c) R means Rendering
d) H means Hammering

Q5. By using which of the following weld tools, more metal is filled and hence a stronger bond is created?

a) Groove Weld
b) Fillet Weld
c) Cosmetic Weld
d) Both a and c

Q6. Which of the following welds required End Fill to be assigned to the faces of weld beads?

a) Fillet Weld
b) Cosmetic Weld
c) Groove Weld
d) Both a and c

Q7. Which of the following weld loads is not present in the **Fillet Weld Calculator** dialog box?

a) Axial Force
b) Bending Moment
c) Common force acting inside the CoG of weld group
d) Common force acting beside the CoG of weld group

Q8. Which of the following values is a result value of analysis rather than properties of material?

a) Yield Strength
b) Ultimate Tensile Strength
c) Safety Factor
d) Allowable Stress

Q9. The **Bead Report** tool is used to create weld bead report in Microsoft Excel format. (True/False)

Q10. With the help of Groove weld, less metal is filled and hence a weaker bond is created. (True/False)

FOR STUDENT NOTES

Chapter 12

Mold Design

Topics Covered

The major topics covered in this chapter are:

- *Guidelines for preparing plastic part for molding*
- *Plastic Part Preparation Tools*
- *Starting Mold Assembly*
- *Importing Plastic Part*
- *Setting gate location*
- *Practical issues with pressure and temperature settings*
- *Workpiece, Patch surface, and Runoff surface creation*
- *Designing Runner, Cooling Channel, Gate, and Secondary Sprue*
- *Mold Base assembly and component insertion*
- *2D Drawing creation*

INTRODUCTION TO MOLD DESIGN

Mold Designing is the engineering of creating mold dies for manufacturing plastic parts. There are two major parts of mold; core and cavity. Both core and cavity are combined together and molten plastic is filled to create the component. The most important part of mold designing is preparing part and checking mold-ability. Below are some guidelines for preparing part for mold designing.

Designing Wall Thickness

Wall thickness strongly influences many key part characteristics, including mechanical performance and feel, cosmetic appearance, mold-ability, and economy. The optimum thickness is often a balance between opposing tendencies, such as strength versus weight reduction or durability versus cost. Give wall thickness careful consideration in the design stage to avoid expensive mold modifications and molding problems in production.

In simple, flat-wall sections, each 10% increase in wall thickness provides approximately a 33% increase in stiffness. Increasing wall thickness also adds to part weight, cycle times, and material cost. Consider using geometric features—such as ribs, curves, and corrugations — to stiffen parts. These features can add sufficient strength, with very little increase in weight, cycle time, or cost. Some materials, polycarbonate for example, lose impact strength if the thickness exceeds a limit known as the critical thickness. Above the critical thickness parts made of polycarbonate can show a marked decrease in impact performance. Walls with thickness greater than the critical thickness may undergo brittle, rather than ductile, failure during impact. The critical thickness reduces with lowering temperature and molecular weight. The critical thickness for medium-viscosity polycarbonate at room temperature is approximately 3/16 inch.

- Avoid designs with thin areas surrounded by thick perimeter sections as they are prone to gas entrapment problems,
- Maintain uniform nominal wall thickness; and
- Avoid wall thickness variations that result in filling from thin to thick sections; refer to Figure-1.

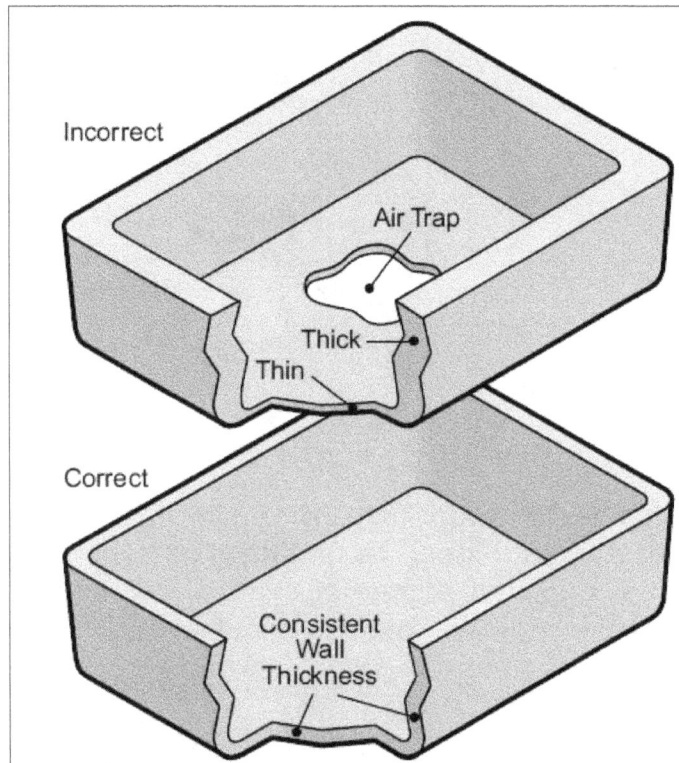

Figure-1. Problem with non uniform thickness of basic wall

Thin-walled parts — those with main walls that are less than 1.5 mm thick or those with wall thicknesses greater than 2 mm can also be considered as thin-walled parts if their flow-length to thickness ratios are too high for conventional molding.

Designing Ribs

Ribs provide a means to economically augment stiffness and strength in molded parts without increasing overall wall thickness. Other uses for ribs include:

• Locating and captivating components of an assembly;
• Providing alignment in mating parts; and
• Acting as stops or guides for mechanisms.

Proper rib design involves five main issues: thickness, height, location, quantity, and mold-ability. Consider these issues carefully when designing ribs.

Rib Thickness and Size

Many factors go into determining the appropriate rib thickness. Thick ribs often cause sink and cosmetic problems on the opposite surface of the wall to which they are attached; refer to Figure-2. On parts with wall thicknesses that are 1.0 mm or less, the rib thickness should be equal to the wall thickness. Rib thickness also directly affects mold-ability. Very thin ribs can be difficult to fill. Because of flow restrictions, thin ribs near the gate can sometimes be more difficult to fill than those farther away.

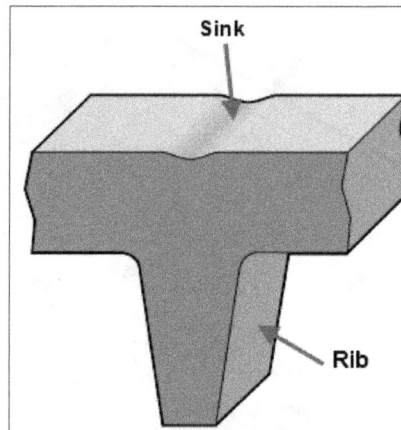

Figure-2. Sink opposite to rib

Ribs usually project from the main wall in the mold-opening direction and are formed in blind holes in the mold steel. To facilitate part ejection from the mold, ribs generally require at least one-half degree of draft per side; refer to Figure-3. More than one degree of draft per side can lead to excessive rib thickness reduction and filling problems in tall ribs.

Generally, taller ribs provide greater support. To avoid mold filling, venting, and ejection problems, standard rules of thumb limit rib height to approximately three times the rib-base thickness. Because of the required draft for ejection, the tops of tall ribs may become too thin to fill easily.

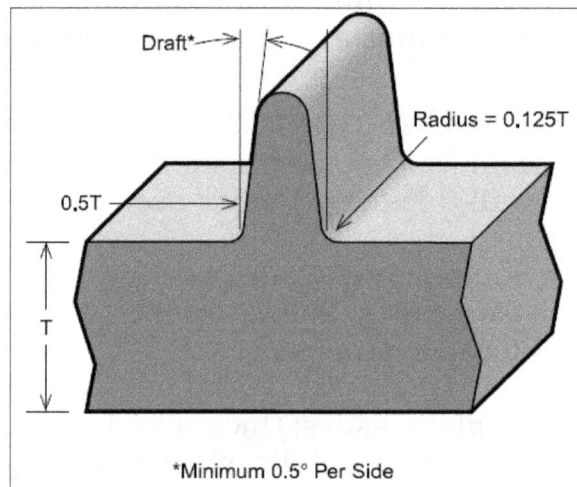

Figure-3. Rib thickness design

Rib Location and Numbers

Carefully, consider the location and quantity of ribs to avoid worsening problems the ribs were intended to correct. For example, ribs added to increase part strength and prevent breakage might actually reduce the ability of the part to absorb impacts without failure. Likewise, a grid of ribs added to ensure part flatness may lead to mold-cooling difficulties and warpage. Typically much easier to add than remove, ribs should be applied sparingly in the original design and added as needed to fine tune performance. Maintain enough space between ribs for adequate mold cooling: for short ribs allow at least two times the wall thickness. Replace the large problematic ribs with multiple shorter ribs; refer to Figure-4.

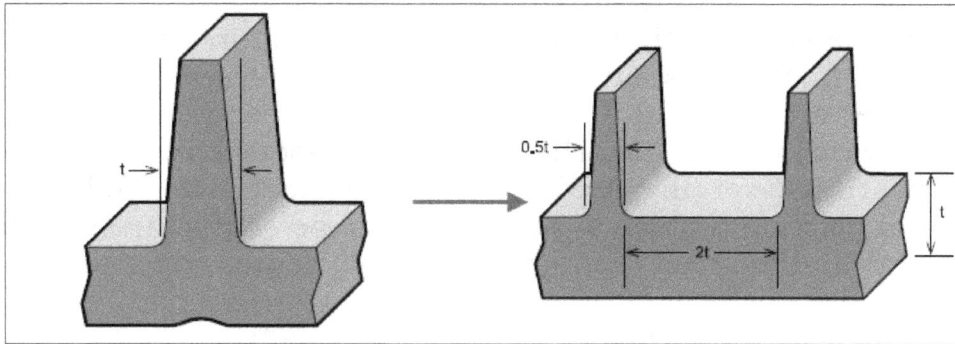

Figure-4. Replacing large rib with shorter ribs

Designing Bosses

Bosses find use in many part designs as points for attachment and assembly. The most common variety consists of cylindrical projections with holes designed to receive screws, threaded inserts, or other types of fastening hardware. As a rule of thumb, the outside diameter of bosses should remain within 2.0 to 2.4 times the outside diameter of the screw or insert; refer to Figure-5. Long core in bosses should be replaced as shown in Figure-6.

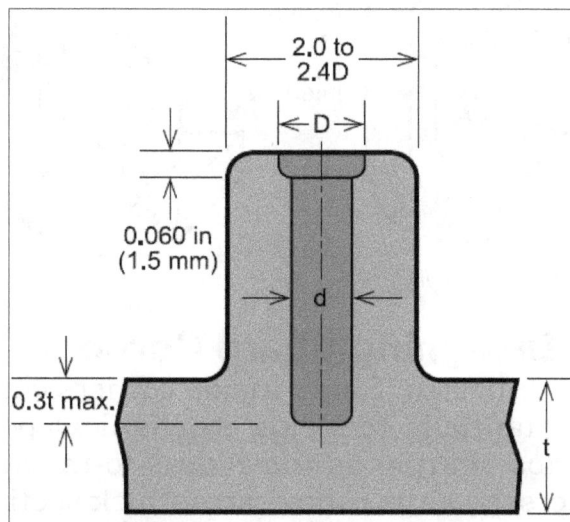

Figure-5. Boss Design

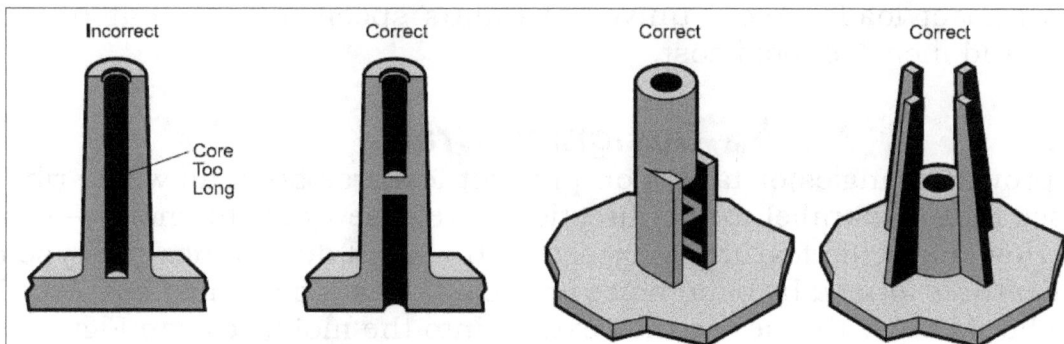

Figure-6. Long core replacement in boss

Designing Gussets

Gussets are rib-like features that add support to structures such as bosses, ribs, and walls. Gusset thickness should be one-half to two-thirds the thickness of the walls to which they are attached if sink is a concern. Specify proper draft and draw polishing to help with mold release. The location of gussets in the mold steel generally prevents practical direct venting. Avoid designing gussets that could trap gasses and cause filling and packing problems. Adjust the shape or thickness to push gasses out of the gussets and to areas that are more easily vented; refer to Figure-7.

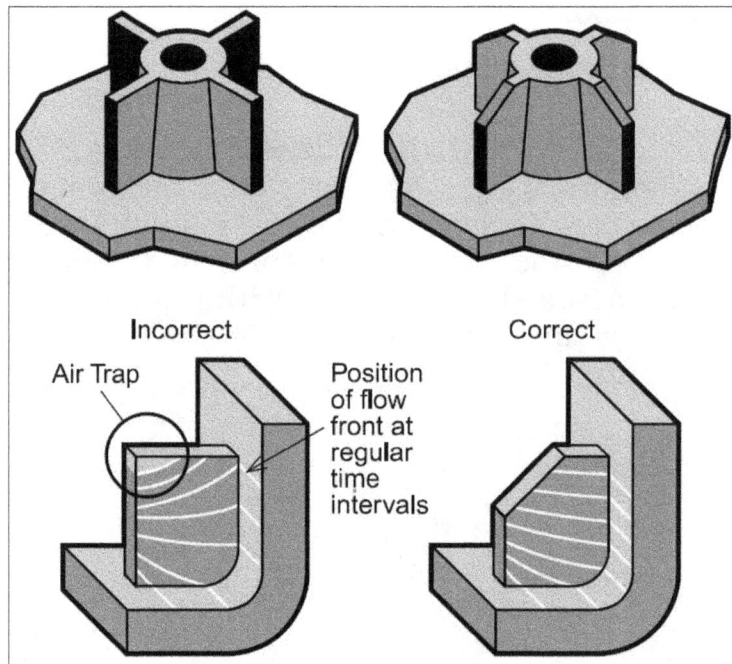

Figure-7. Gusset Designing

Designing Sharp Corners

Avoid sharp corners in your design. Sharp inside corners concentrate stresses from mechanical loading, substantially reducing mechanical performance. The stress concentration factor climbs sharply as the radius-to-thickness ratio drops below approximately 0.2. Conversely, large ratios cause thick sections, leading to sinks or voids. A radius-to-thickness ratio of approximately 0.15 provides a good compromise between performance and appearance for most applications subjected to light to moderate impact loads. Avoid universal radius specifications that round edges needlessly and increase mold cost.

Designing Draft

Draft — providing angles or tapers on product features such as walls, ribs, posts, and bosses that lie parallel to the direction of release from the mold — eases part ejection. How a specific feature is formed in the mold determines the type of draft needed. Features formed by blind holes or pockets — such as most bosses, ribs, and posts — should taper thinner as they extend into the mold; refer to Figure-8.

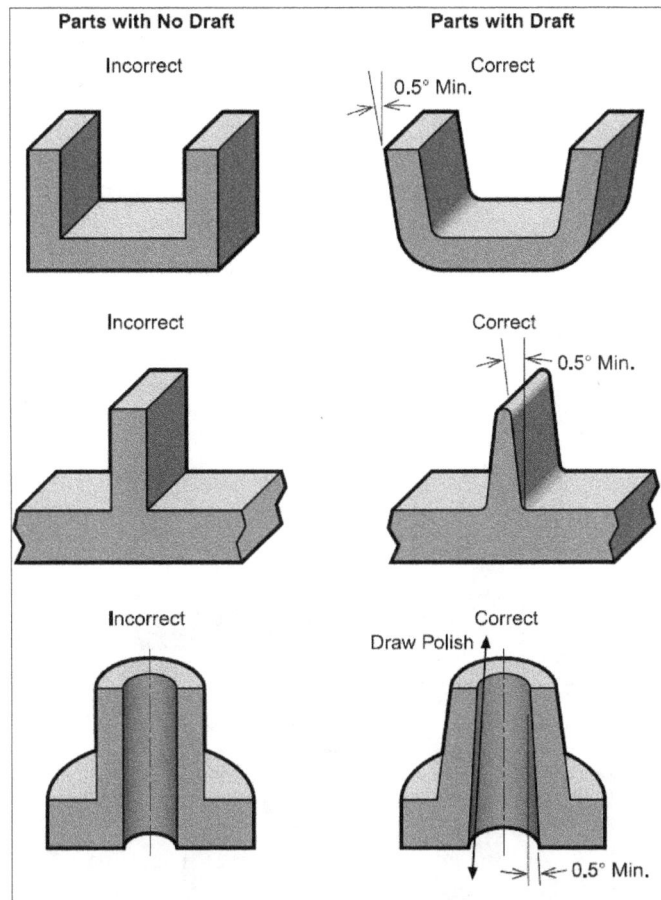

Figure-8. Draft Design

Surfaces formed by slides may not need draft if the steel separates from the surface before ejection. Other rules of thumb for designing draft include:

• Draft all surfaces parallel to the direction of steel separation;
• Angle walls and other features that are formed in both mold halves to facilitate ejection and maintain uniform wall thickness;
• Use the standard one degree of draft plus one additional degree of draft for every 0.001 inch of texture depth as a rule of thumb; and
• Use a draft angle of at least one-half degree for most materials. Design permitting, use one degree of draft for easy part ejection. SAN resins typically require one to two degrees of draft.

The mold finish, resin, part geometry, and mold ejection system determine the amount of draft needed. Generally, polished mold surfaces require less draft than surfaces with machined finishes. An exception is thermoplastic polyurethane resin, which tends to eject easier from frosted mold surfaces. Parts with many cores may need a higher amount of draft.

Designing Holes and Cores

Generally, the depth-to-diameter ratio for blind holes should not exceed 3:1. Ratios up to 5:1 are feasible if filling progresses symmetrically around the unsupported hole core or if the core is in an area of slow-moving flow. Consider alternative part designs that avoid the need for long delicate cores; refer to Figure-9. If the core is supported on both ends, the guidelines for length-to-diameter ratio double: typically 6:1 but up to 10:1 if the filling around the core is symmetrical. The level of support on the core ends determines the maximum suggested ratio.

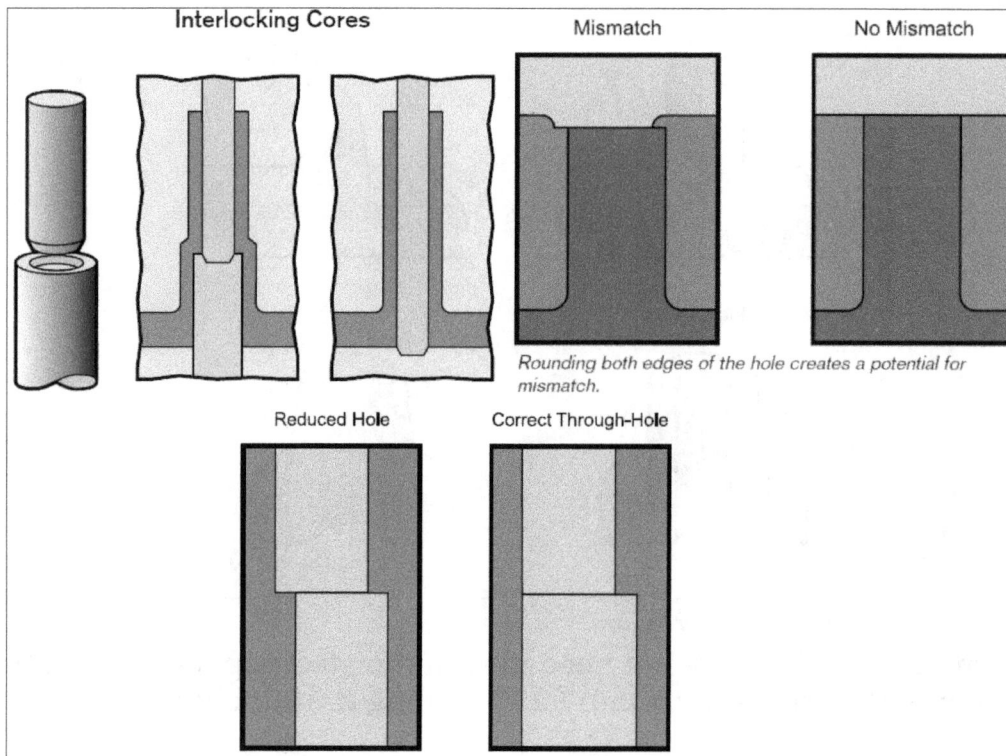

Figure-9. Designing Holes and cores

Designing Undercuts

Some design features, because of their orientation, place portions of the mold in the way of the ejecting plastic part.

Called "undercuts," these elements can be difficult to redesign. Sometimes, the part can flex enough to strip from the mold during ejection, depending upon the undercut's depth and shape and the resin's flexibility; refer to Figure-10. Undercuts can only be stripped if they are located away from stiffening features such as corners and ribs. Generally, avoid stripping undercuts in parts made of stiff resins such as polycarbonate, polycarbonate blends, and reinforced grades of polyamide 6. Undercuts up to 2% are possible in parts made of these resins, if the walls are flexible and the leading edges are rounded or angled for easy ejection. Typically, parts made of flexible resins, such as unfilled polyamide 6 or thermoplastic polyurethane elastomer, can tolerate 5% undercuts. Under ideal conditions, they may tolerate up to 10% undercuts. Most undercuts cannot strip from the mold, needing an additional mechanism in the mold to move certain components prior to ejection.

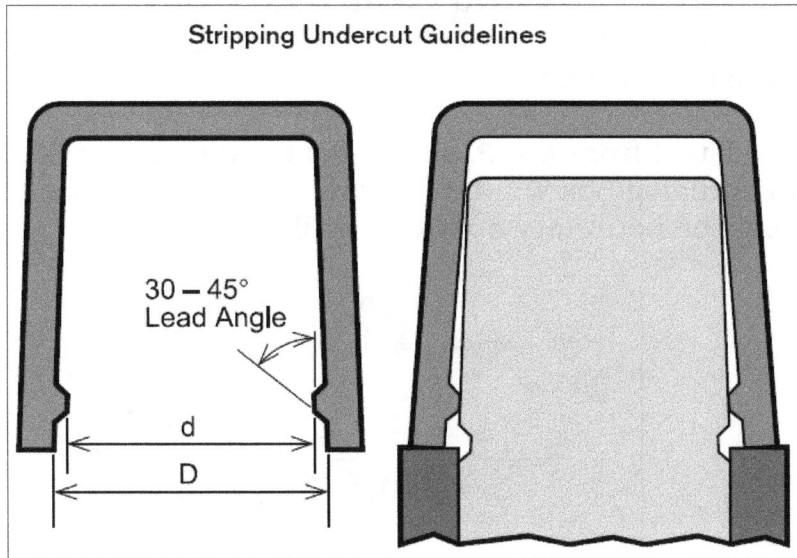

Figure-10. Undercut design

Till now in this chapter, we have discussed some basic rules for preparing part for mold design. Now, we will discuss the tools in Autodesk Inventor for making plastic part.

PLASTIC PART PREPARATION TOOLS

The tools to prepare plastic part are available in the Part environment which has been discussed next. To display tools for plastic part preparation, follow the steps given next.

* Start Autodesk Inventor and create a new part file with **Standard (mm).ipt** template.
* Click on the **Show Panels** button at the right corner in the **Ribbon**. The list of panels will be displayed; refer to Figure-11.
* Select the **Plastic Part** check box from the list. The **Plastic Part** panel will be added in the **Ribbon**; refer to Figure-12.

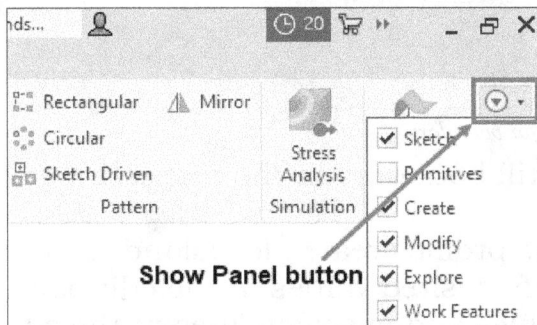

Show Panel button

Figure-11. List of Panels

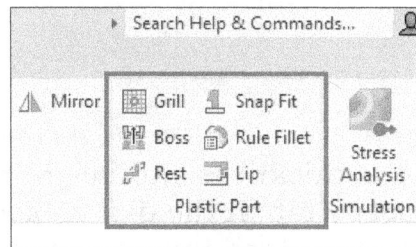

Figure-12. Plastic Part panel

In previous chapters, you have learned the use of 3D modeling tools, so we will not repeat the use of those tools here. Now, we will discuss the use of tools in **Plastic Part** panel.

Using Grill Tool

The **Grill** tool, as the name suggests, is used to create grill on the selected part. The procedure to use this tool is given next.

- Click on the **Grill** tool from the **Plastic Part** panel in the **3D Model** tab of the **Ribbon**. The **Grill** dialog box will be displayed; refer to Figure-13. Also, you will be asked to select the boundary curve for grill.

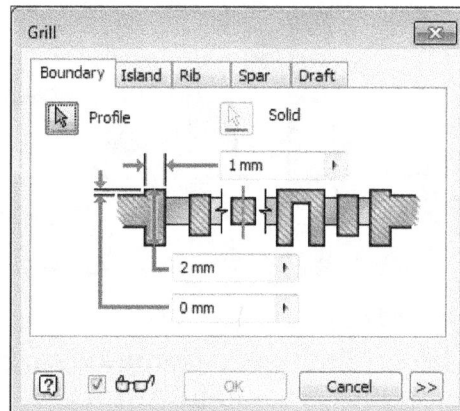

Figure-13. Grill dialog box

- Select the boundary lines of the sketch created for grill; refer to Figure-14.

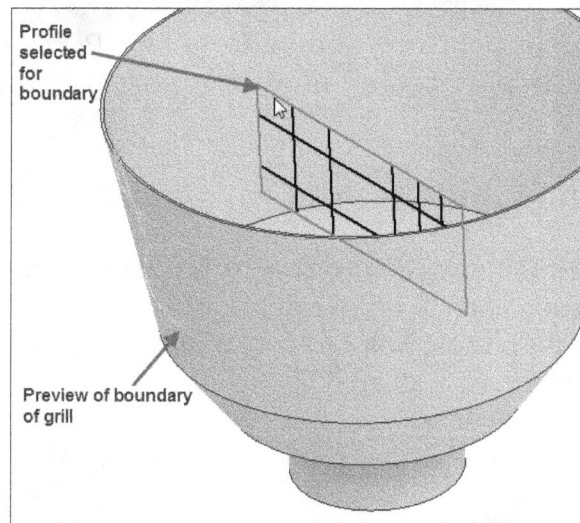

Figure-14. Profile selected for boundary of grill

- Specify the thickness and height of grill boundary in the respective edit boxes of the **Boundary** tab in the dialog box.
- Click on the **Island** tab and select the profile created for island feature. Preview of the feature will be displayed. Specify desired values in the edit boxes.
- Similarly, select the rib and spar profiles and specify values in the edit boxes to create the grill; refer to Figure-15.

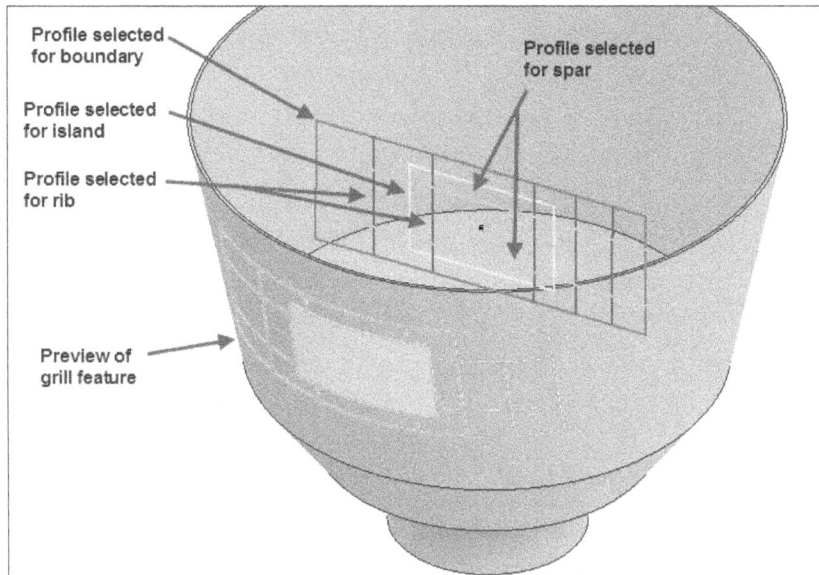

Figure-15. Profiles selected for grill feature

- Click on the **Draft** tab and specify desired value of draft angle in the edit box. If the grill fall on parting line then select the **Parting Element** check box and specify the offset distance.
- Click on the **OK** button to create the grill; refer to Figure-16.

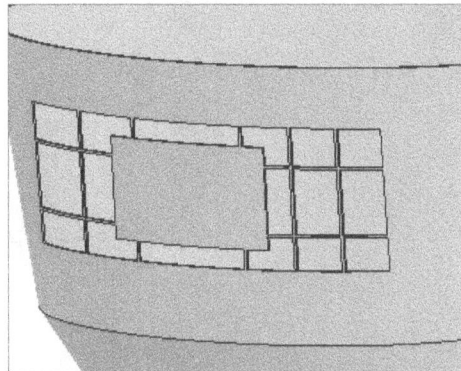

Figure-16. Grill created on the round face

Using Boss Tool

The **Boss** tool in **Plastic Part** panel is used to create boss feature at the specified location with parameters set in dialog box. The procedure of using this tool is given next.

- Click on the **Boss** tool from the **Plastic Part** panel in the **3D Model** tab of the **Ribbon**. The **Boss** dialog box will be displayed; refer to Figure-17.

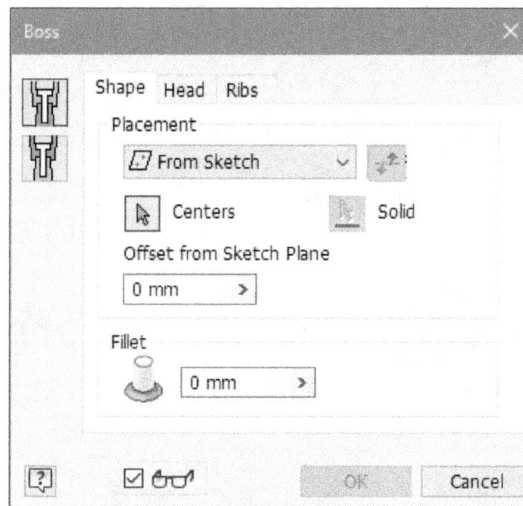

Figure-17. Boss dialog box

- Select the work points/sketch points on which you want to create the boss features. Preview of the feature will be displayed; refer to Figure-18.

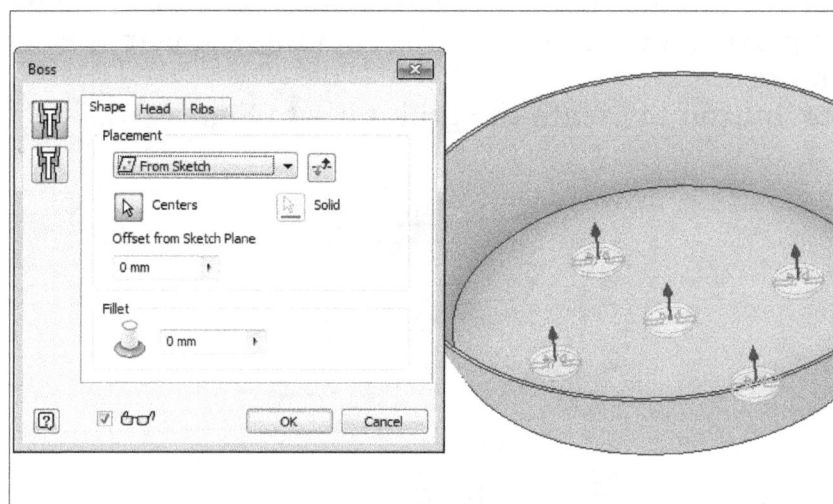

Figure-18. Preview of boss feature

- Set desired value of offset in the **Offset from Sketch Plane** edit box.
- Specify desired value of fillet radius in the edit box of **Fillet** area in the dialog box.
- Select the **Head** or **Thread** button from the left area in the dialog box. Select the **Head** button 🔩 if you want to create the female part of the boss fastener. Select the **Thread** button 🔩 from the dialog box if you want to create male part of the boss fastener.
- Click on the **Head** or **Thread** tab depending on the button selected. The dialog box will be displayed as shown in Figure-19.

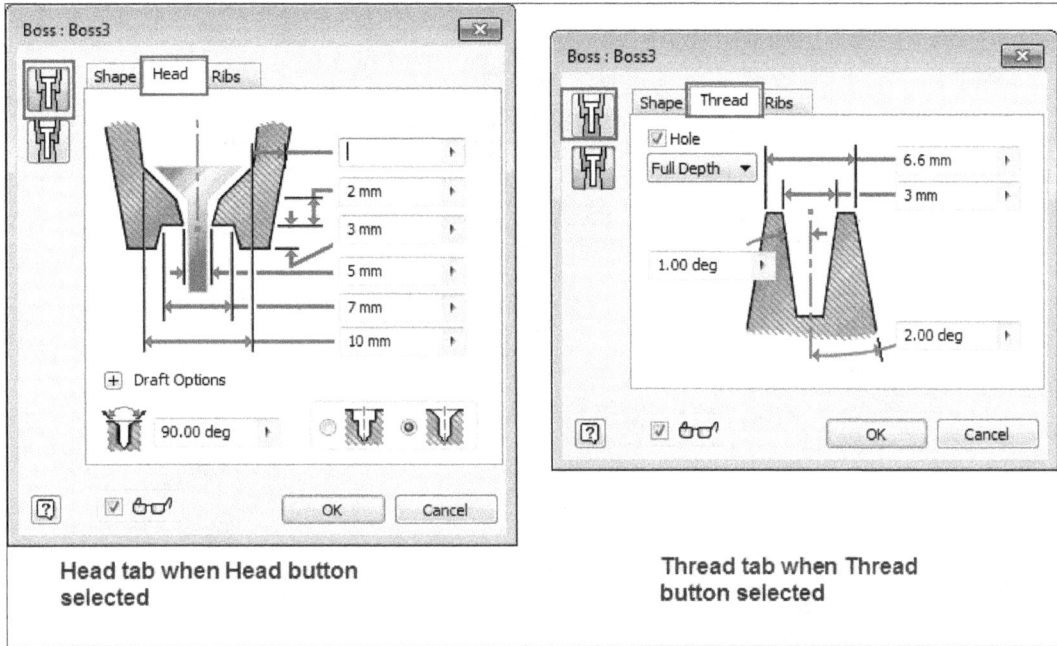

Figure-19. Head tab and Thread tab in Boss dialog box

- Specify desired parameters in the dialog box and then click on the **Ribs** tab. The **Boss** dialog box will be displayed as shown in Figure-20.
- Select the **Stiffening Ribs** check box to add ribs to the boss feature. Preview of ribs will be displayed; refer to Figure-21.

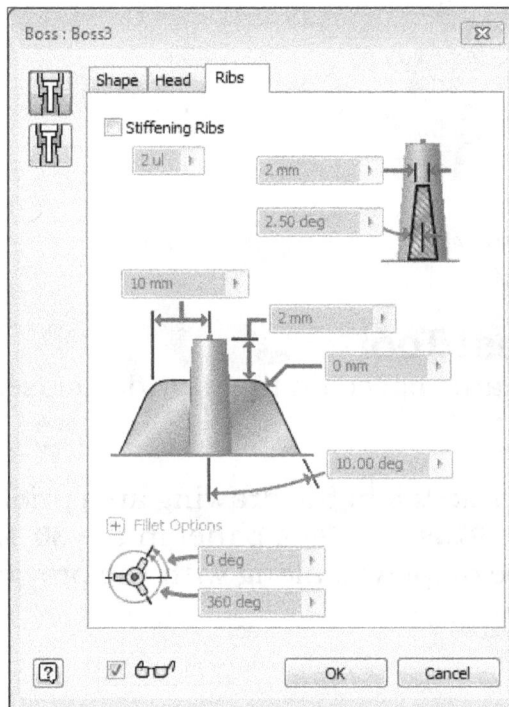

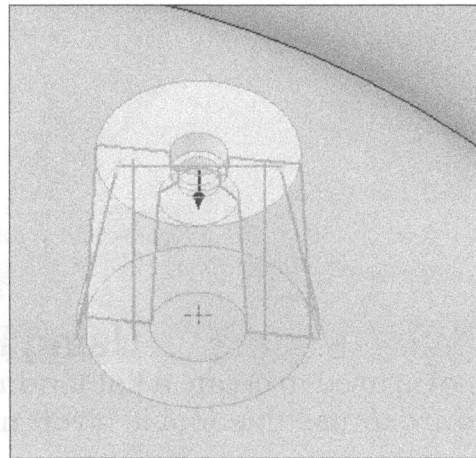

Figure-21. Preview of rib in head feature

Figure-20. Boss dialog box with Rib tab selected

- Set desired parameters for the rib in the dialog box; refer to Figure-22. Click on the **OK** button to create the boss feature; refer to Figure-23.

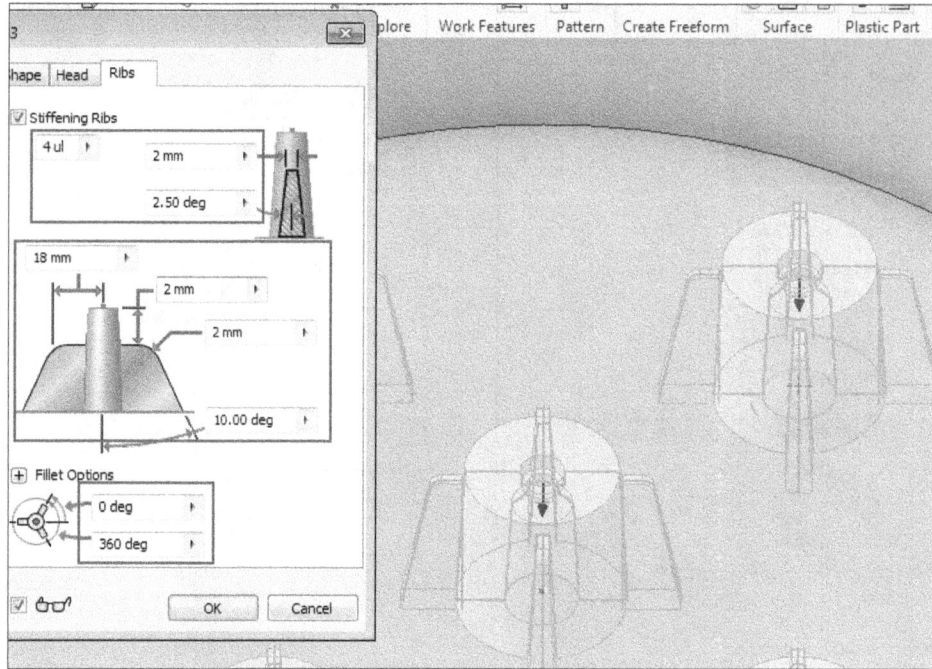

Figure-22. Parameters specified for rib

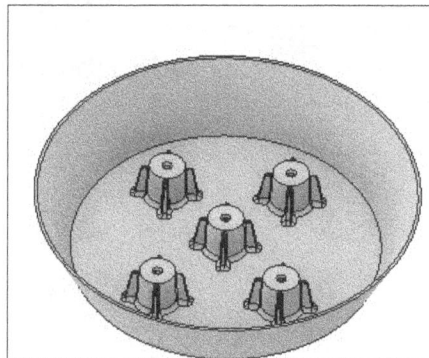

Figure-23. Boss feature created

Using Rest Tool

The **Rest** tool is used to create a flat landing area based on the provided closed sketch. The procedure to use this tool is given next.

• Make sure you have created a closed loop sketch in the drawing area prior to using this tool. Click on the **Rest** tool from the **Plastic Part** panel in the **3D Model** tab of the **Ribbon**. The **Rest** dialog box will be displayed along with the preview of rest feature; refer to Figure-24.

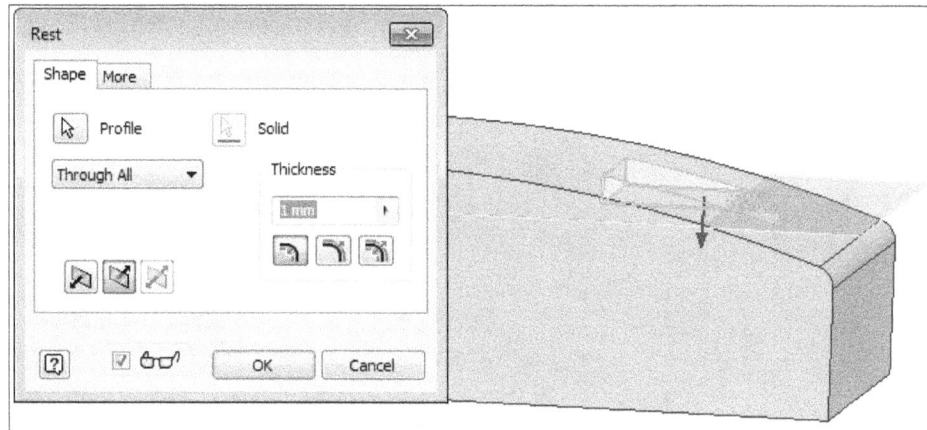

Figure-24. Rest dialog box with preview of rest feature

- Select desired option from the drop-down to specify depth of the rest feature; refer to Figure-25.

Figure-25. Drop-down for depth options

- Set desired depth and direction. Similarly, specify the thickness of rest feature in the **Thickness** edit box and set the direction of thickness.
- Click on the **More** tab to specify advanced options for rest feature. The dialog box will be displayed as shown in Figure-26.

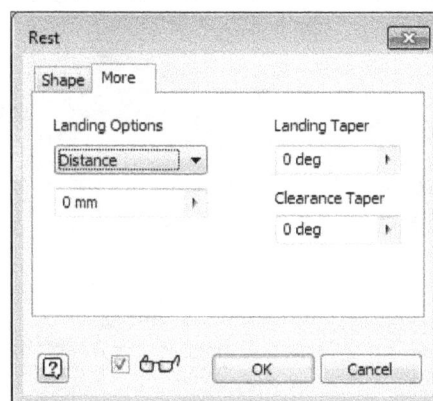

Figure-26. More tab in the Rest dialog box

- Set the landing distance value in the **Distance** edit box or select the **To surface** option from the **Landing Options** drop-down and select desired limiting surface.
- Specify landing taper and clearance taper angle values in the **Landing Taper** and **Clearance Taper** edit boxes, respectively.
- Click on the **OK** button to create the feature; refer to Figure-27.

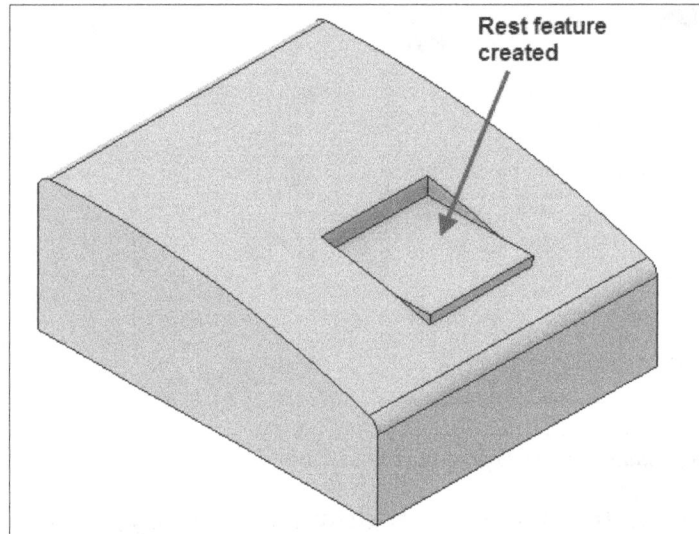

Figure-27. Rest feature created

Using Snap Fit Tool

The **Snap Fit** tool is used to create snap fit hook or loop, so that the two components can be joined without the use of screws, glues, clips, or other joining elements. Make sure you have created points at the locations required for snap fits before using this tool. The procedure to use this tool is given next.

- Click on the **Snap Fit** tool from the **Plastic Part** panel in the **3D Model** tab of the **Ribbon**. The **Snap Fit** dialog box will be displayed. Also, you will be asked to select the sketch or points to create the snap fit.
- Select the sketch or points on the part on which you want to create the snap fit. The preview of snap fit will be displayed as shown in Figure-28.

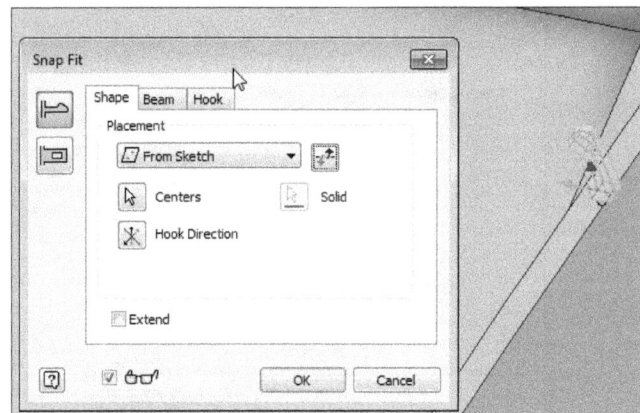

Figure-28. Snap Fit dialog box

- Click on the **Hook Direction** button to set the direction of hook. On clicking the button, you will be asked to select a yellow arrow of desired direction displayed with the preview. Select desired arrow to set hook direction.
- Click on the **Cantilever Snap Fit Hook** or **Cantilever Snap Fit Loop** button from the left area of the dialog box to create snap fit hook or loop.

- Specify the parameters in **Beam** and **Hook** tabs of the dialog box if you have selected the **Cantilever Snap Fit Hook** button. Similarly, specify the parameters in **Clip** and **Catch** tabs of the dialog box if you have selected the **Cantilever Snap Fit Loop** button.
- Click on the **OK** button to create the snap fit feature; refer to Figure-29.

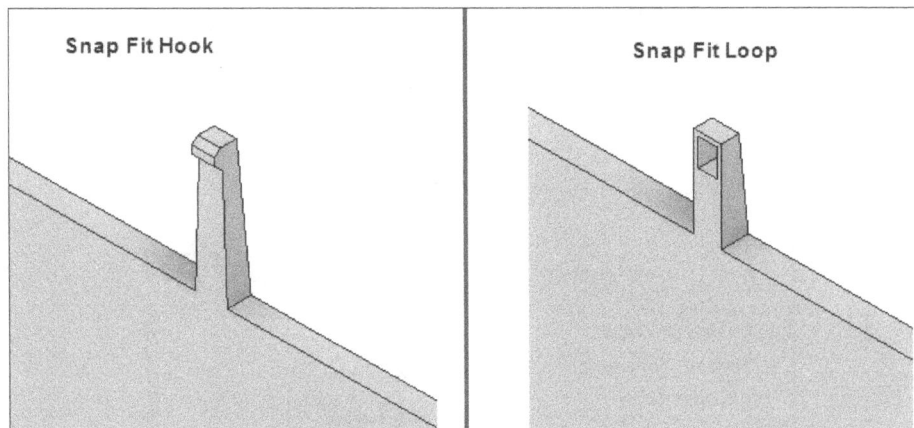

Figure-29. Snap Fit Hook and Loop created

Using Rule Fillet Tool

The **Rule Fillet** tool is used to fillet the edges based on specified rules. The steps to use this tool are given next.

- Click on the **Rule Fillet** tool from the **Plastic Part** panel in the **3D Model** tab of the **Ribbon**. The **Rule Fillet** dialog box will be displayed as shown in Figure-30.
- Click on the field under **Source** column and select **Feature** or **Face** option as required from the drop-down displayed; refer to Figure-31.

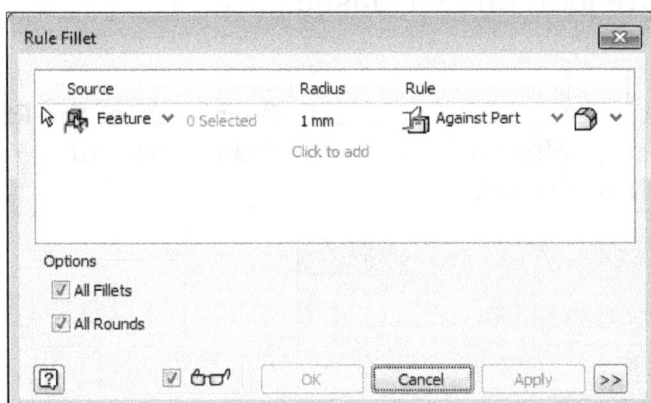

Figure-31. Drop-down in Source column

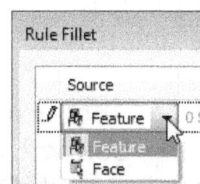

Figure-30. Rule Fillet dialog box

- Click in the field under **Radius** column and specify desired value of radius for fillet.
- Click in the field under **Rule** column and select desired rule from the drop-down. If you have selected the **Against Part** option then only the edges formed by the faces of the features and the faces of the part body are filleted. If you have selected the **Against Features** option then only the edges that are generated by intersection of selected features will be filleted. If you have selected the **Free Edges** option from the drop-down then all the edges formed by faces of the selected feature will be filleted. If you have selected the **All Edges** option then all the edges of the selected features will be filleted.

- To change the advanced parameters of the fillet, click on the **More** button at the bottom-right in the dialog box. The dialog box will expand as shown in Figure-32.

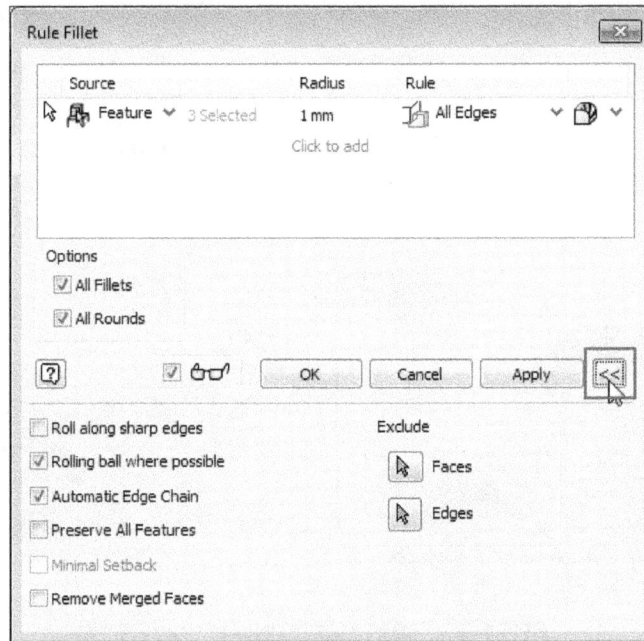

Figure-32. Expanded Rule Fillet dialog box

- Select desired check boxes from the expanded dialog box to modify fillet. Click on the **Faces** or **Edges** selection button from the **Exclude** area and select the entities that you do not want to be filleted.
- Click on the **OK** button to fillet the model based on specified rules.

Using Lip Tool

The **Lip** tool is used to create lip-groove joint on two plastic parts. The procedure to use this tool is given next.

- Click on the **Lip** tool from the **Plastic Part** panel in the **3D Model** tab of the **Ribbon**. The **Lip** dialog box will be displayed; refer to Figure-33. Also, you will be asked to select an edge to define the path of lip/groove.

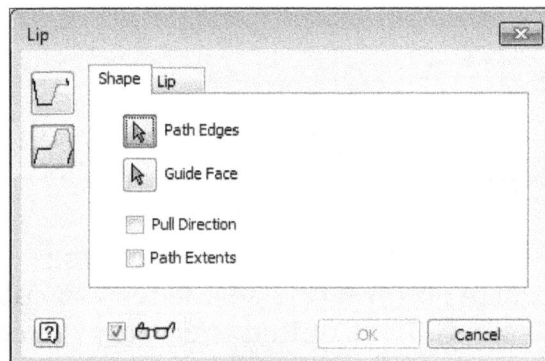

Figure-33. Lip dialog box

- Select the edge of plastic part on which you want to create the lip/groove.
- Click on the **Guide Face** selection button and select the face connected to earlier selected edge. Preview of the lip/groove will be displayed; refer to Figure-34.

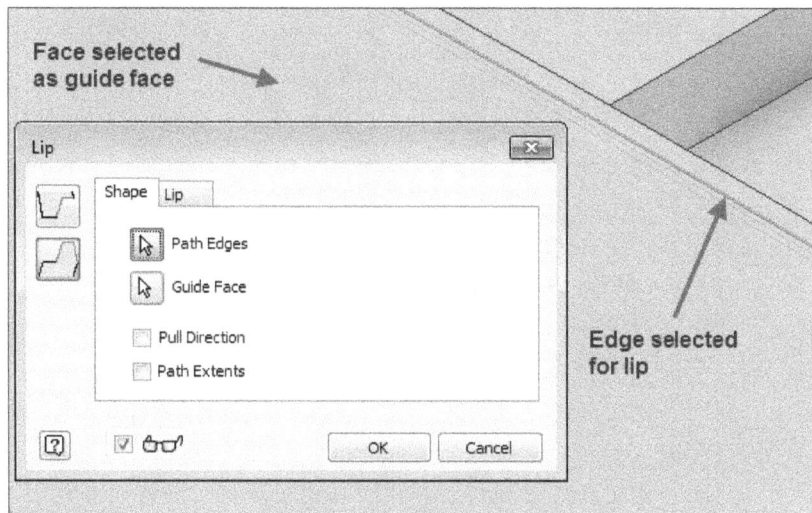

Figure-34. Lip preview at side face of model

- Click on the **Groove** ⌣ or **Lip** ⌐ button from the left area of the dialog box to create groove or lip, respectively. The **Groove** tab will be added in the dialog box if you have selected the **Groove** button and the **Lip** tab will be added in the dialog box if you have selected the **Lip** button from the left area.
- Click on the tab and specify desired dimensions for lip/groove; refer to Figure-35.

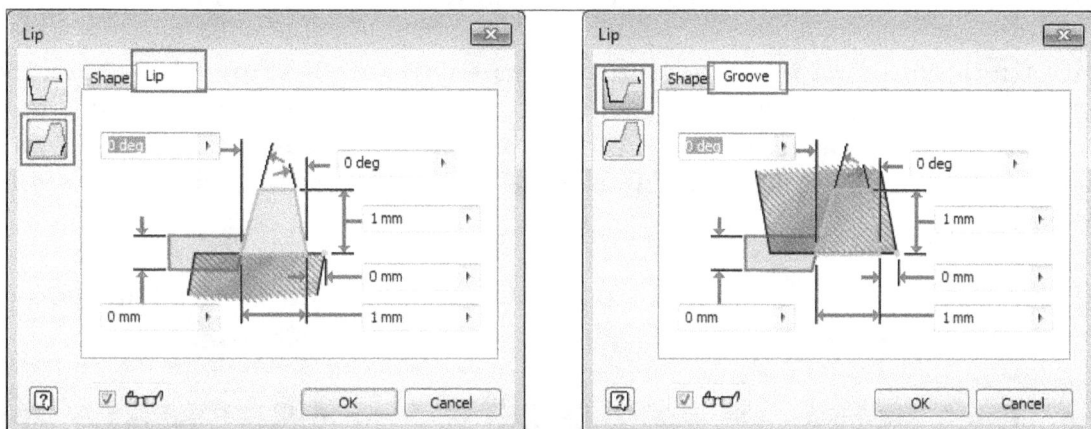

Figure-35. Lip and Groove tabs in the Lip dialog box

- After specifying the dimensions, click on the **OK** button to create the feature.

Till now in this chapter, we have learned about the tools specific to plastic part making. Now, we will discuss the tools in assembly environment which are used in mold making.

STARTING MOLD DESIGN ASSEMBLY

- Start Autodesk Inventor (if not started yet). If you have a file opened in Autodesk Inventor then save the file and close it. Next, click on the **New** button from **New** cascading menu in the **File** menu of the **Ribbon**. The **Create New File** dialog box will be displayed.
- Double-click on the **Mold Design (mm).iam** in the **Assembly** area of **Metric** folder in the dialog box. The **Create Mold Design** dialog box will be displayed with Assembly environment opened; refer to Figure-36.

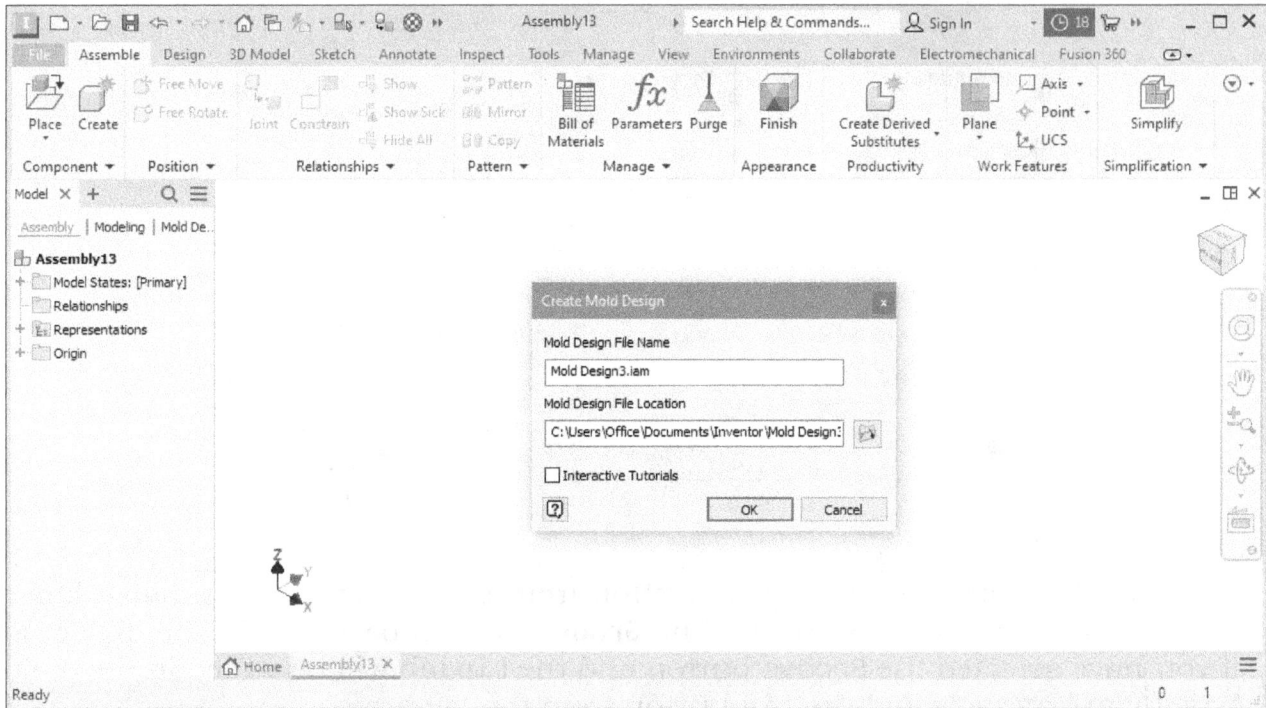

Figure-36. Create Mold Design dialog box

- Specify the name of mold design file in the **Mold Design File Name** edit box of the dialog box.
- Specify the location of mold design files in the **Mold Design File Location** edit box or click on the browse button next to edit box and select the location.
- Click on the **OK** button from the dialog box. The **Mold Layout** and **Mold Assembly** tab will be added to the **Ribbon**. The tools to create mold design are available in these tabs.

Now, we will discuss the use of each tool in these tabs. But before that you need to understand the workflow in Autodesk Inventor for mold design.

WORKFLOW OF MOLD DESIGN ASSEMBLY IN AUTODESK INVENTOR

The workflow of mold design assembly in Autodesk Inventor can be given by Figure-37.

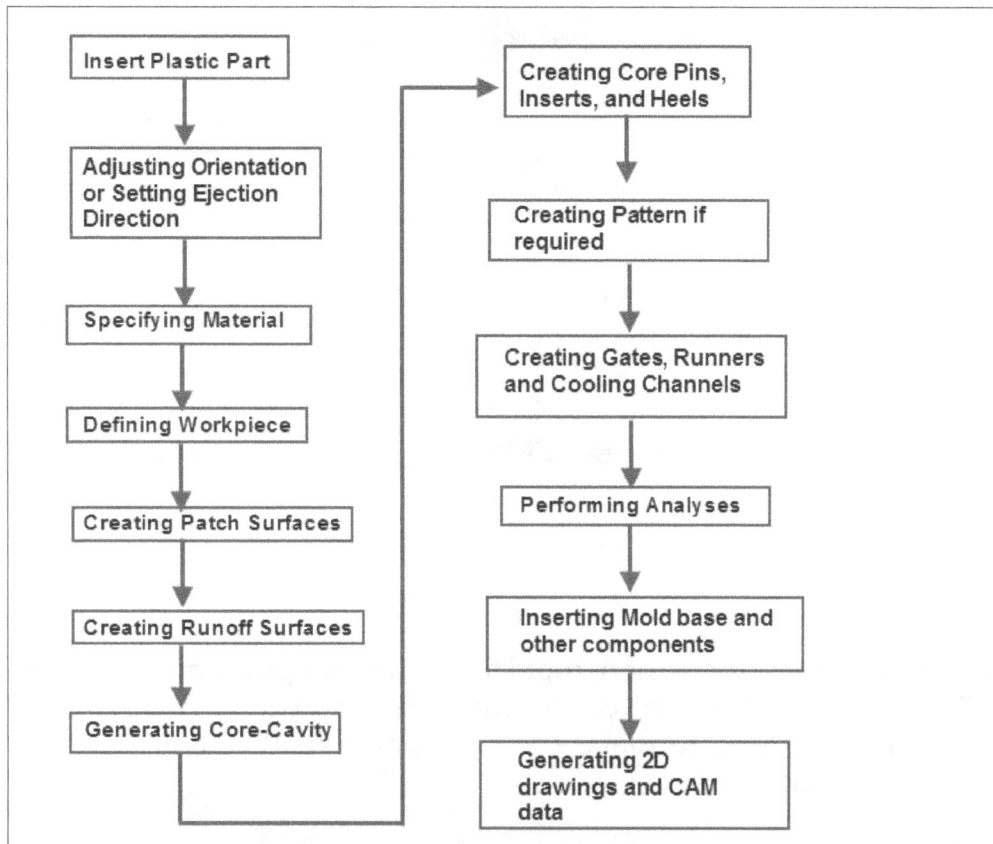

Figure-37. Workflow in Mold design

Now, we will discuss the tools involved in each step of the work flow in detail.

INSERTING PLASTIC PART

The very first step in mold designing is to insert the plastic part for which we are creating the mold. The procedure to insert plastic part is given next.

- Click on the **Plastic Part** tool from the **Mold Layout** panel in the **Mold Layout** tab of the **Ribbon**. The **Plastic Part** dialog box will be displayed as shown in Figure-38.

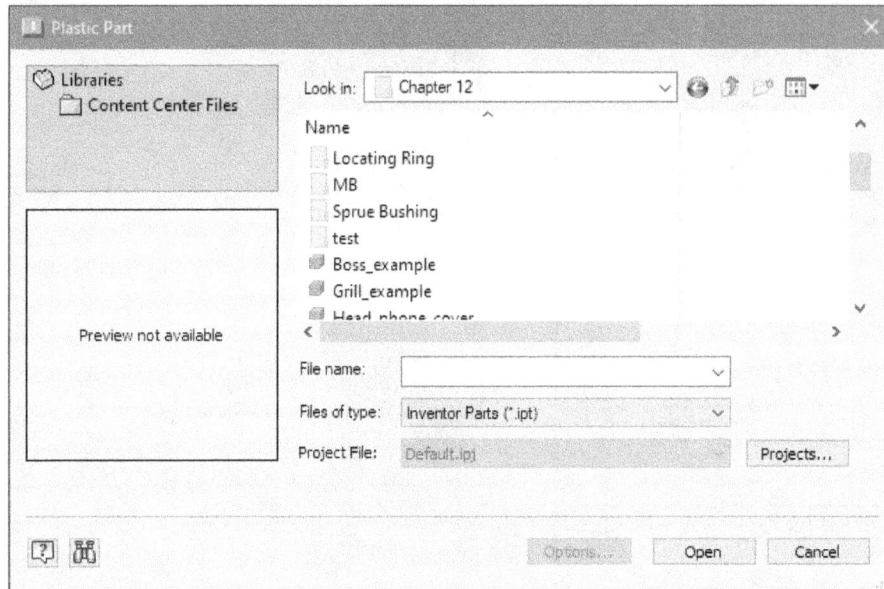

Figure-38. Plastic Part dialog box

- Select the plastic part for which mold is to be designed and click on the **Open** button. The part gets placed at coordinate system automatically.
- Click in the drawing area to place the part and exit the tool. The other tools in the panel will get activated.

ADJUSTING ORIENTATION/POSITION OF PART

The direction of mold ejection is based on the orientation of the part in mold. We will now learn how to orient the part or say decide the ejection direction of mold. The steps are given next.

Adjusting Orientation of Part

- Click on the **Adjust Orientation** tool from the **Mold Layout** panel in the **Mold Layout** tab of the **Ribbon**. Preview of the opening direction will be displayed along with the **Adjust Orientation** dialog box; refer to Figure-39.

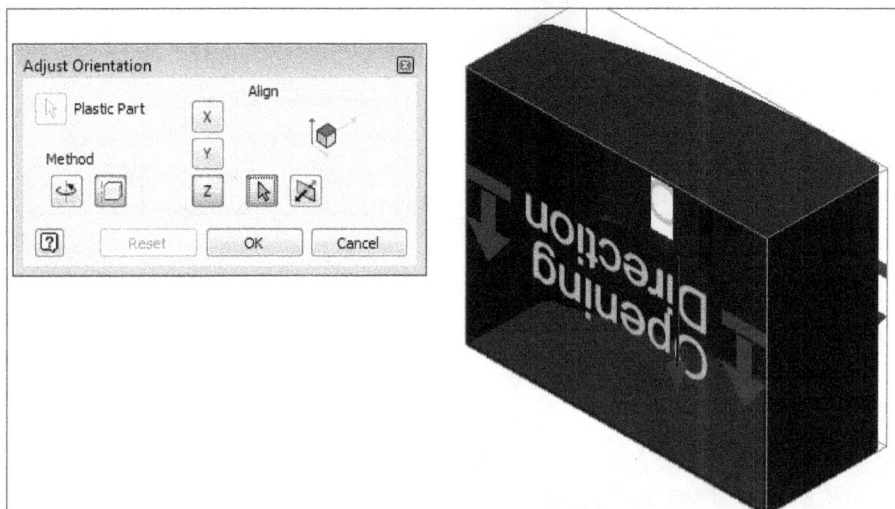

Figure-39. Adjust Orientation dialog box with preview of opening direction

- Make sure **Align with Axis** 🔲 and **Z** buttons are selected in the dialog box. Select the flat face perpendicular to which you want the opening direction. The orientation of the model will change accordingly; refer to Figure-40.

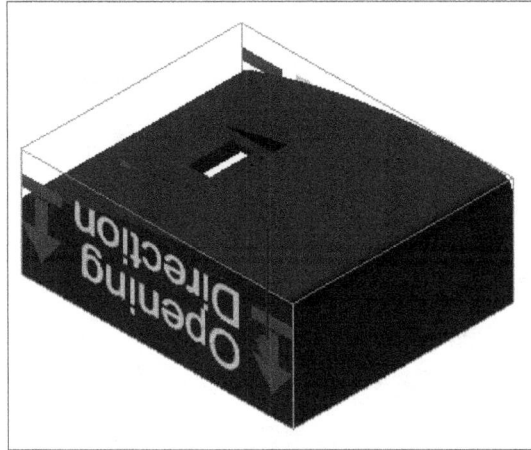

Figure-40. Part after changing direction

- Flip the part if required by using the **Flip moldable part** button. Click on the **OK** button to set the orientation.

Adjusting Position of Part

- Click on **Adjust Position** tool from the **Adjust Orientation** drop-down in the **Mold Layout** panel of the **Ribbon**; refer to Figure-41. The **Adjust Position** dialog box will be displayed; refer to Figure-42. Also, you will be asked to select a flat face in the direction of opening to specify reference for positioning.

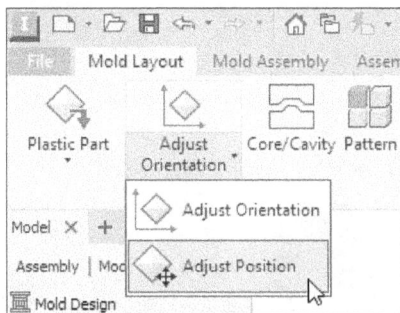

Figure-41. Adjust Position tool

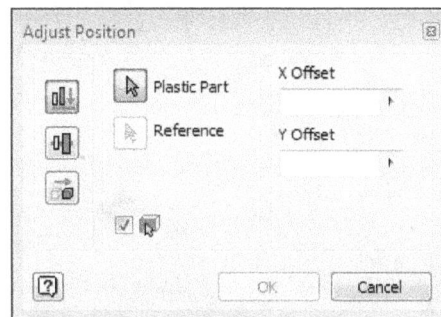

Figure-42. Adjust Position dialog box

- Select desired button from the left area of the dialog box. Select the **Align XY Plane with Reference** button 🔳 if you want to position the part according to the **XY** plane of coordinate system; refer to Figure-43. Select the **Align Center with X/Y Direction** button 🔳 if you want to position the plastic part along the X and Y directions of coordinate system selected; refer to Figure-44. Select the **Free Transform** button 🔳 to position the plastic part freely in X, Y, and Z directions by specifying offset values; refer to Figure-45.

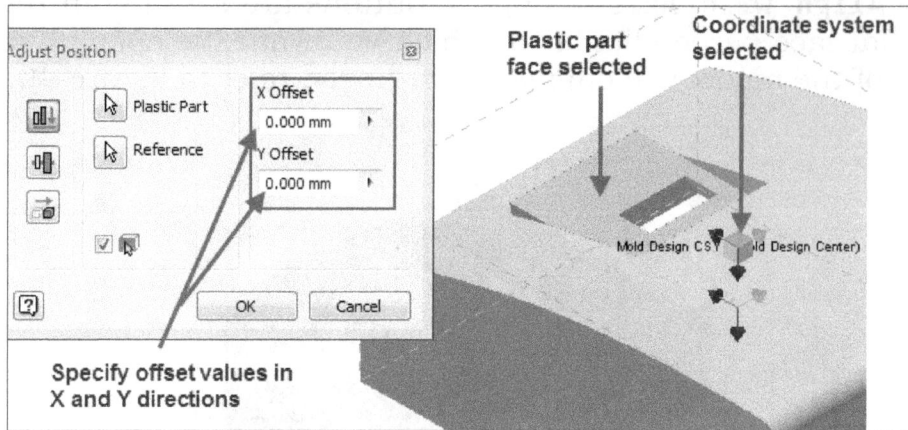

Figure-43. Adjusting position of part in XY plane

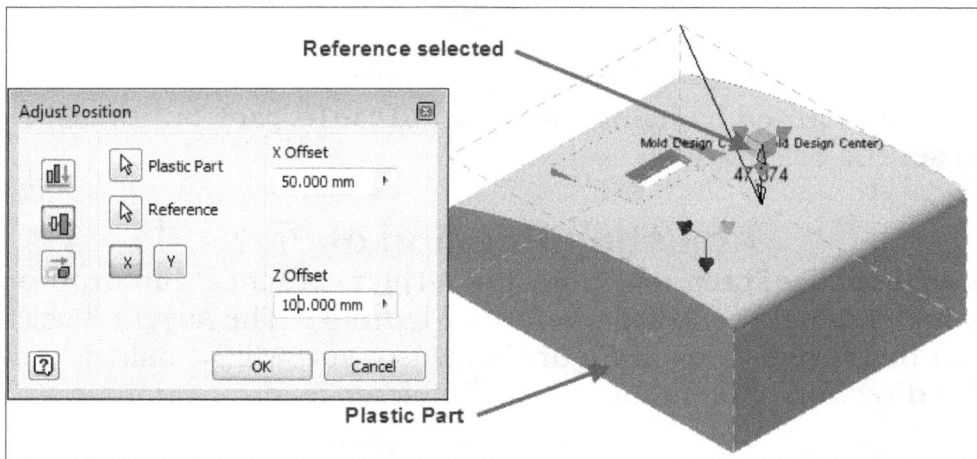

Figure-44. Adjusting position of part in X and Y directions

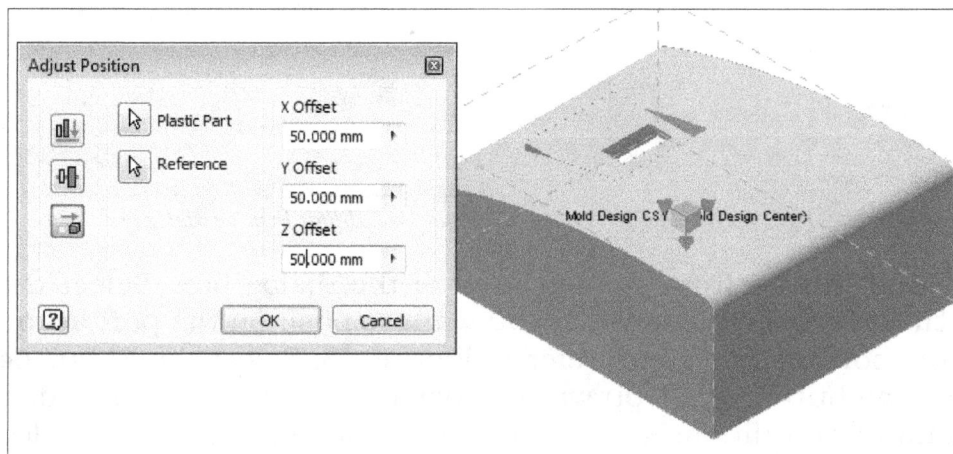

Figure-45. Adjusting position of part freely in XY and Z directions

- After setting desired position, click on the **OK** button from the dialog box.

PREPARING CORE AND CAVITY

The tools to prepare core and cavity are available in a contextual tab named Core/Cavity. The **Core/Cavity** tab is displayed on selecting the **Core/Cavity** tool from the **Mold Layout** panel in the **Mold Layout** tab of the **Ribbon**; refer to Figure-46.

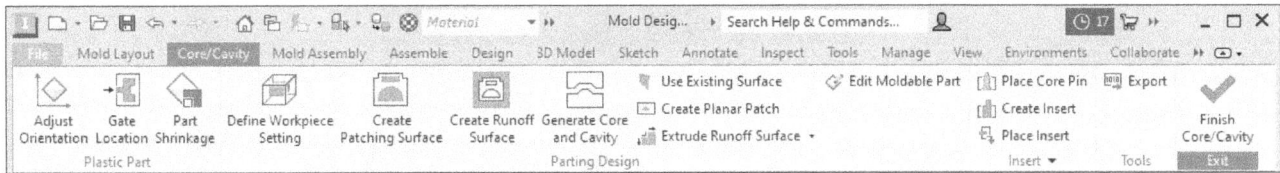

Figure-46. Core-Cavity contextual tab

Various tools in this tab are discussed next in order of their use in mold design.

Part Shrinkage Tool

The **Part Shrinkage** tool is used to provide shrinkage allowance in the plastic model. As the molten plastic cools down in mold, it gets shrunk by certain amount called shrinkage allowance. The procedure to use **Part Shrinkage** tool is given next.

- Click on the **Part Shrinkage** tool from the **Plastic Part** panel in the **Core/Cavity** contextual tab of the **Ribbon**. The **Part Shrinkage** dialog box will be displayed; refer to Figure-47.

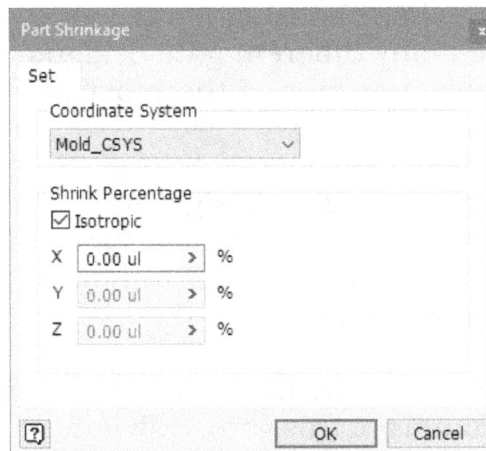

Figure-47. Part Shrinkage dialog box

- Select desired coordinate system from the drop-down in the **Coordinate System** area of the dialog box.
- Select the **Isotropic** check box from **Shrink Percentage** area if the shrinkage is equal in all three directions; X,Y, and Z. If shrinkage is not equal then clear the check box.
- Specify the percentage of shrinkage in the edit boxes of **Shrink Percentage** area.
- Click on the **OK** button to apply the changes.

Gate Location Tool

The **Gate Location** tool is used to specify the location of orifice to be used for filling the molten plastic in the mold. The procedure to use this tool is given next.

- Click on the **Gate Location** tool from the **Plastic Part** panel in the **Core/Cavity** contextual tab in the **Ribbon**. The **Gate Location** dialog box will be displayed; refer to Figure-48.
- Click at desired location on the plastic model to specify gate location.
- Click on the **Apply** button to create the gate location.
- Click on the **Done** button to finish the process.

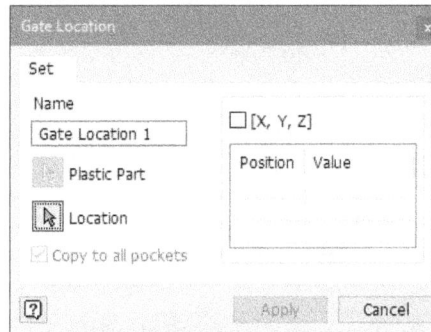

Figure-48. Gate Location dialog box

Gate Design

Although, Autodesk Inventor has a tool to suggest location of gates but there are a few guidelines which will be helpful to reduce rework losses in molding due to gate problems. Here is a short note on gate design:

Common Gate Designs

The largest factor to consider when choosing the proper gate type for your application is the gate design. There are many different gate designs available based on the size and shape of your part. Below are four of the most popular gate designs used by mold makers:

The **Edge Gate** is the most common gate design. As the name indicates, this gate is located on the edge of the part and is best suited for flat parts. Edge gates are ideal for medium and thick sections and can be used on multi-cavity two plate tools. This gate will leave a scar at the parting line.

The **Sub Gate** is the only automatically trimmed gate on the list. Ejector pins will be necessary for automatic trimming of this gate. Sub gates are quite common and have several variations such as banana gate, tunnel gate, and smiley gate to name a few. The sub gate allows you to gate away from the parting line, giving more flexibility to place the gate at an optimum location on the part. This gate leaves a pin sized scar on the part.

The **Hot Tip Gate** is the most common of all hot runner gates. Hot tip gates are typically located at the top of the part rather than on the parting line and are ideal for round or conical shapes where uniform flow is necessary. This gate leaves a small raised nub on the surface of the part. Hot tip gates are only used with hot runner molding systems. This means that, unlike cold runner systems, the plastic is ejected into the mold through a heated nozzle and then cooled to the proper thickness and shape in the mold.

The **Direct or Sprue Gate** is a manually trimmed gate that is used for single cavity molds of large cylindrical parts that require symmetrical filling. Direct gates are the easiest to design and have low cost and maintenance requirements. Direct gated parts are typically lower stressed and provide high strength. This gate leaves a large scar on the part at the point of contact.

Gate Locations

To avoid problems from your gate location, below are some guidelines for choosing the proper gate location(s):

- Place gates at the heaviest cross section to allow for part packing and minimize voids & sink.
- Minimize obstructions in the flow path by placing gates away from cores & pins.
- Be sure that stress from the gate is in an area that will not affect part function or aesthetics.
- If you are using a plastic with a high shrink grade, the part may shrink near the gate causing "gate pucker" if there is high molded-in stress at the gate.
- Be sure to allow for easy manual or automatic de-gating.
- Gate should minimize flow path length to avoid cosmetic flow marks.
- In some cases, it may be necessary to add a second gate to properly fill the parts.
- If filling problems occur with thin walled parts, add flow channels or make wall thickness adjustments to correct the flow.
- Sometimes, two gates are not better than one. When we fill melt through two gates then a weld line is formed at the middle of the gates. If this weld lines falls on highly stresses area of the part then your product will fail in its applications.
- If a part is round and needs to be absolutely round then you need to gate it in the center.
- If a part is long and narrow and needs to be absolutely straight then you need to gate it on the end.
- Try to gate into the thickest area and avoid gating into a thin area. Failure to do so can result in voids (air bubbles).

Practical Issues with Pressure and Temperature Parameters

There are a few issues that can cause problem in mold which are discussed next.

Excessive Injection Fill Speed

The speed and pressure of the melt as it enters the mold determine both density and consistency of melt in packing the mold. If the fill is too fast, the material tends to "slip" over the surface and will "skin" over before the rest of the material solidifies. The slipped skin area does not faithfully reproduce the mold steel surface, as does the material in other areas, because it has not been packed tightly against the steel.

Solution: One solution is to adjust the fill speed rate until the optimum has been achieved. This will help eliminate blushing.

Melt Temperature Too High Or Too Low

Although this may sound contradictory, either condition might cause blushing. If the injection barrel heat is too high, the material will flow too quickly, resulting in slippage of the surface skin, as mentioned above. If the barrel heat is too low, the material may solidify before full packing occurs and the plastic will not be pushed against the mold steel, especially in the gate area because that is the last area to pack.

Solution: Melt temperature must be adjusted to the optimum for a specific material and specific product design.

Low Injection Pressure

The plastic material must be injected into the mold in such a way as to cause proper filling and packing while maintaining consistent solidification of the melt. Injection pressure is one of the main control variables of the machine and must be high enough to pack the plastic molecules against the steel of the mold while the plastic cools. Low pressure will not achieve this packing and the material will appear dull in local areas that do not have enough pressure.

Solution: Increasing the injection pressure forces the material against the mold surface, producing a truer finish that replicates the steel finish.

Low Mold Temperature

A low mold temperature may cause the molten material to slow down and solidify before the mold is packed out. This will cause dull areas where the plastic was not forced against the steel finish.

Solution: Increasing the mold temperature allows the material to flow farther and pack properly. The material temperature could also be raised to accomplish the same effect.

Define Workpiece Setting Tool

The **Define Workpiece Setting** tool is used to create a block of steel from which the core and cavity is to be cut. The procedure to use this tool is given next.

- Click on the **Define Workpiece Setting** tool from the **Parting Design** panel in the **Core/Cavity** contextual tab of the **Ribbon**. The **Define Workpiece Setting** dialog box will be displayed; refer to Figure-49.

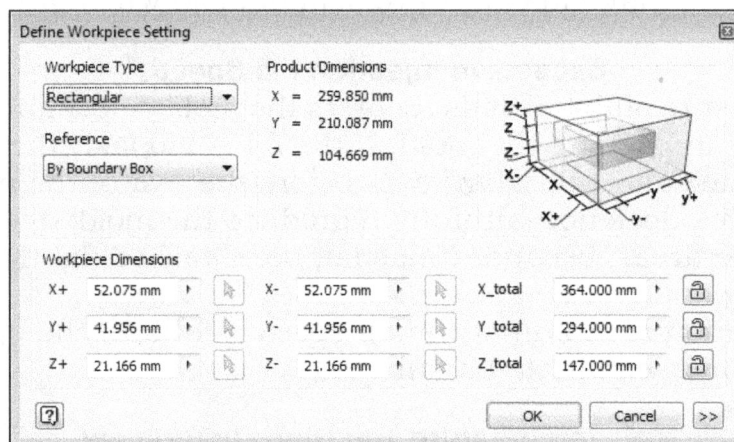

Figure-49. Define Workpiece Setting dialog box

- Select the shape of work piece from the **Workpiece Type** drop-down in the dialog box suitable for your mold base. There are two options in the drop-down; **Rectangular** and **Cylinder** to create rectangular and cylindrical workpiece, respectively.
- Specify the dimensions as per your requirement in the **Workpiece Dimensions** area of the dialog box.
- Click on the **OK** button. A transparent workpiece will be created around your plastic part; refer to Figure-50.

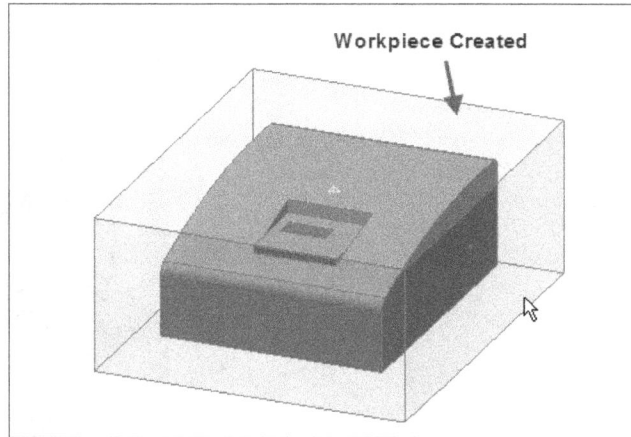

Figure-50. Workpiece created around plastic part

Create Patching Surface Tool

The **Create Patching Surface** tool is used to patch the cuts and holes in the model so that the core and cavity can be separated. The procedure to use this tool is given next.

- Click on the **Create Patching Surface** tool from the **Parting Design** panel in the **Core/Cavity** contextual tab of the **Ribbon**. The **Create Patching Surface** dialog box will be displayed; refer to Figure-51.

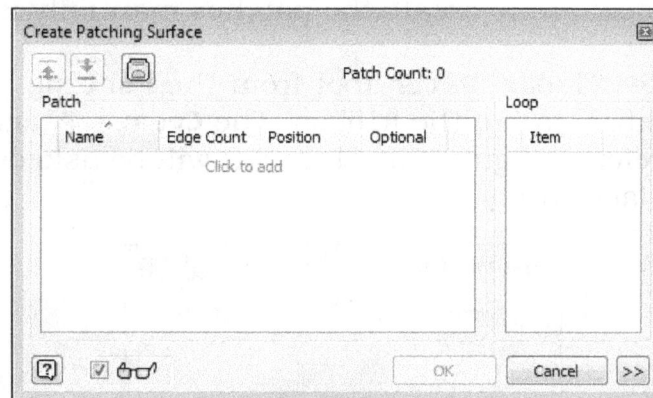

Figure-51. Create Patching Surface dialog box

- Click on the **Auto Detect** button from the dialog box. Preview of the patches will be displayed; refer to Figure-52.

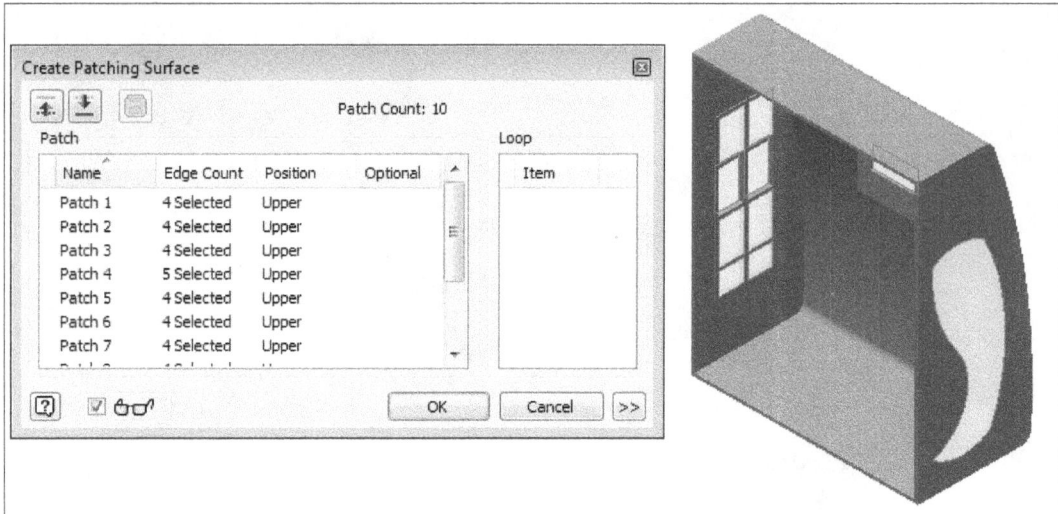

Figure-52. Preview of patches

- If you want to switch between upper and lower position of the patches then click on the **All Upper** ⬆ or **All Lower** ⬇ button from the dialog box.
- Click on the **OK** button from the dialog box to create the patch surface.

Manually Creating Patch Surface

Using the previous tool, you were able to create the patches automatically. But, sometimes it becomes necessary to create the surface patches manually to achieve desired shape. The procedure to create the patches manually is given next.

- Click on the **Create Planar Patch** tool from the **Parting Design** panel in the **Core/Cavity** contextual tab of the **Ribbon**. The **Create Planar Patch** dialog box will be displayed; refer to Figure-53. Also, you will be asked to select connective edges to create surface patch.

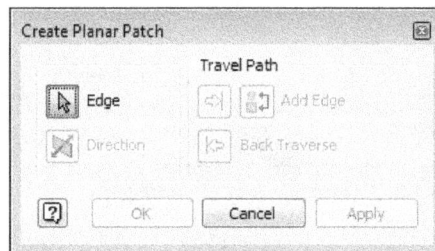

Figure-53. Create Planar Patch dialog box

- Select an edge of cut to be patched. The tools in the dialog box will become active and preview will be displayed; refer to Figure-54.

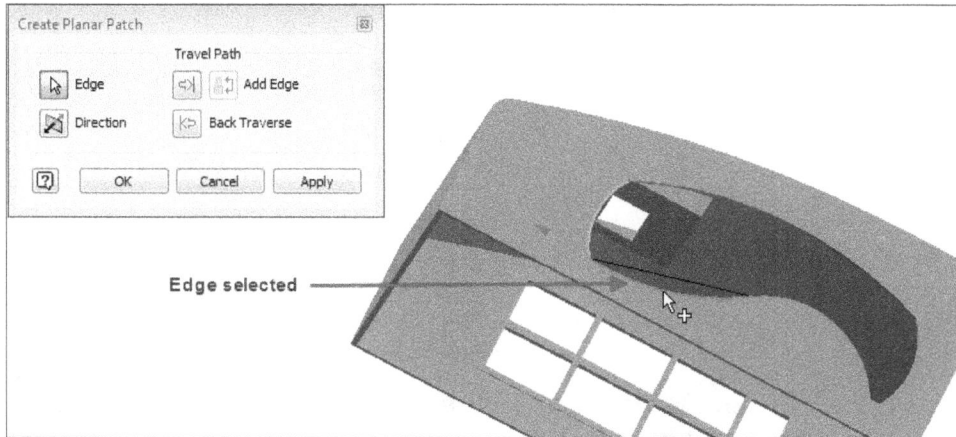

Figure-54. Preview of patch surface

- The orange colored edge in preview is next suggested edge to be selected. The black line in preview closing the selected edge into a loop is generated automatically and if we click on the **Apply** button now then this loop will become surface patch.
- Click on the **Add Edge** button in the **Travel Path** area of the dialog box to select the next suggested edge. Repeat this step until you get desired loop to create surface patch.
- To revert back in edge selection, click on the **Back Traverse** button in the **Travel Path** area of the dialog box.
- Click on the **OK** button to create the surface patch; refer to Figure-55.

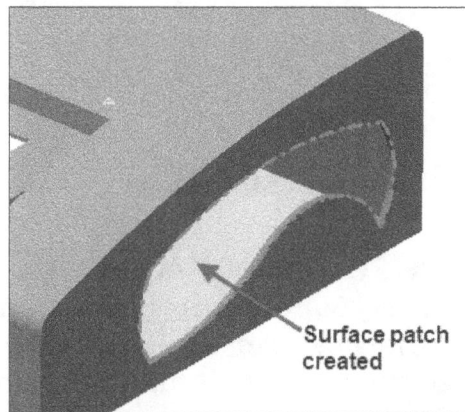

Figure-55. Planar surface patch created

Create Runoff Surface Tool

The **Create Runoff Surface** tool is used to create surface using the outer boundary edges of the model. This surface acts as a splitting tool to split core and cavity steel. The procedure to use this tool is given next.

- Click on the **Create Runoff Surface** tool from the **Parting Design** panel of the **Core/Cavity** contextual tab in the **Ribbon**. The **Create Runoff Surface** dialog box will be displayed; refer to Figure-56.

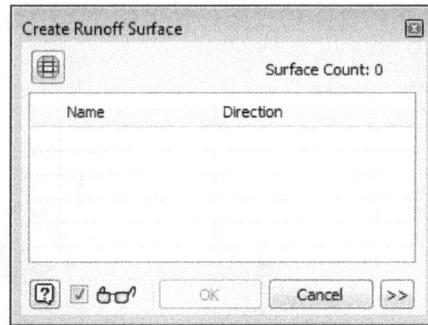

Figure-56. Create Runoff Surface dialog box

- One by one click on the boundary edges of the model to create runoff surface manually or click on the **Auto Detect** button. Preview of the Runoff surface will be displayed; refer to Figure-57.

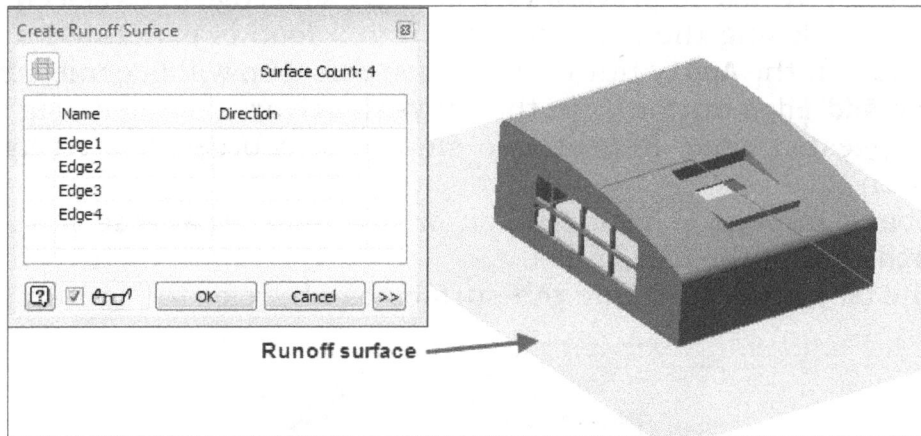

Figure-57. Preview of runoff surface

- Click on the **OK** button to create the surface.

Other Tools to Create Runoff Surfaces

There are four tools in the **Runoff Surface** drop-down in **Parting Design** panel of the **Ribbon** to create runoff surfaces; refer to Figure-58. Besides the **Create Runoff Surface** tool, you can also use these tools to manually create runoff surfaces.

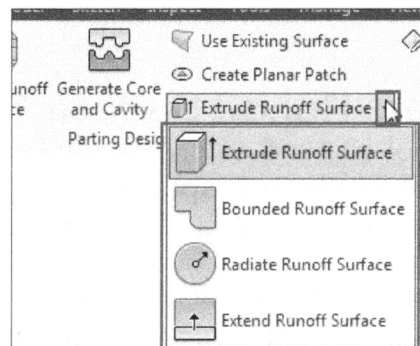

Figure-58. Runoff Surface drop-down

The application of each tool is discussed next.

Extrude Runoff Surface Tool

The **Extrude Runoff Surface** tool is used to create a runoff surface by extruding the selected curve/edge. The procedure to use this tool is given next.

- Click on the **Extrude Runoff Surface** tool from the **Runoff Surface** drop-down in the **Parting Design** panel of **Ribbon**. The **Extrude Runoff Surface** dialog box will be displayed; refer to Figure-59. Also, you will be asked to select the edges for creating extruded runoff surface.

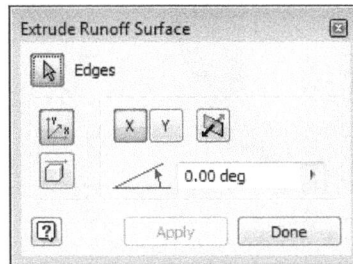

Figure-59. Extrude Runoff Surface dialog box

- Select an edge and then select the **X** or **Y** button to specify the direction of extrusion. Preview of the extruded surface will be displayed; refer to Figure-60. Click on the **Flip** button to change the side of extrusion if required.

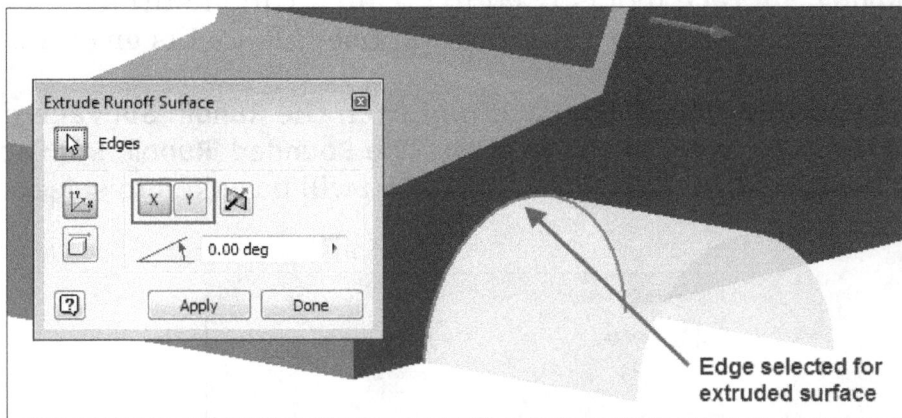

Figure-60. Preview of extruded runoff surface

- If you want to use any axis or edge to specify direction of extrusion then click on the **Align Axis** button from the dialog box and select the reference edge/axis to specify direction; refer to Figure-61.

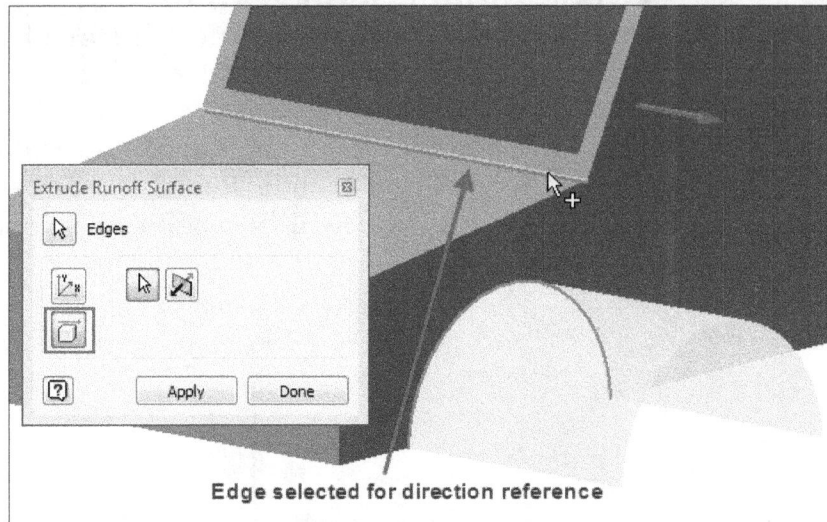

Figure-61. Edge selected for direction reference

- Click on the **Apply** button to create the surface and then click on the **Done** button to exit the dialog box.

Bounded Runoff Surface Tool

The **Bounded Runoff Surface** tool is used to create a runoff surface using the region bounded by selected curves. The procedure to use this tool is given next.

- Click on the **Bounded Runoff Surface** tool from the **Runoff Surface** drop-down in the **Parting Design** panel of the **Ribbon**. The **Bounded Runoff Surface** dialog box will be displayed; refer to Figure-62. Also, you will be asked to select the geometry for creating surface.

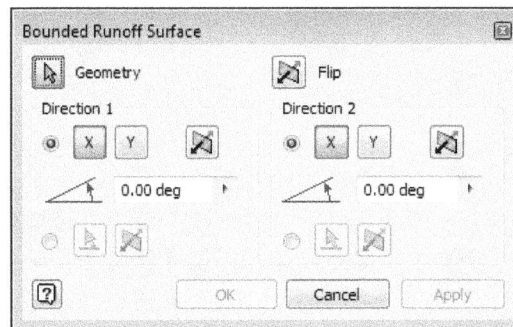

Figure-62. Bounded Runoff Surface dialog box

- Select the curve/edge of the plastic model to be used for creating bounded runoff surface. Preview of the surface will be displayed; refer to Figure-63.

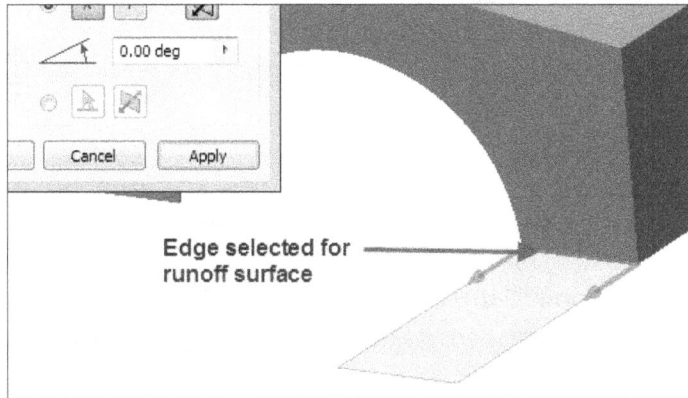

Figure-63. Preview of bounded runoff surface

- Set the other parameters as required and then click on the **OK** button to create the surface.

Radiate Runoff Surface Tool

The **Radiate Runoff Surface** tool is used to create non-planar runoff surfaces based on the selected curves. The procedure to use this tool is given next.

- Click on the **Radiate Runoff Surface** tool from the **Runoff Surface** drop-down in the **Parting Design** panel of the **Ribbon**. The **Radiate Runoff Surface** dialog box will be displayed; refer to Figure-64.

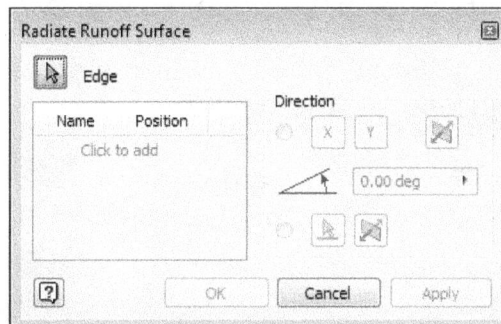

Figure-64. Radiate Runoff Surface dialog box

- Select the curve to create the surface. Preview of the surface will be displayed; refer to Figure-65.

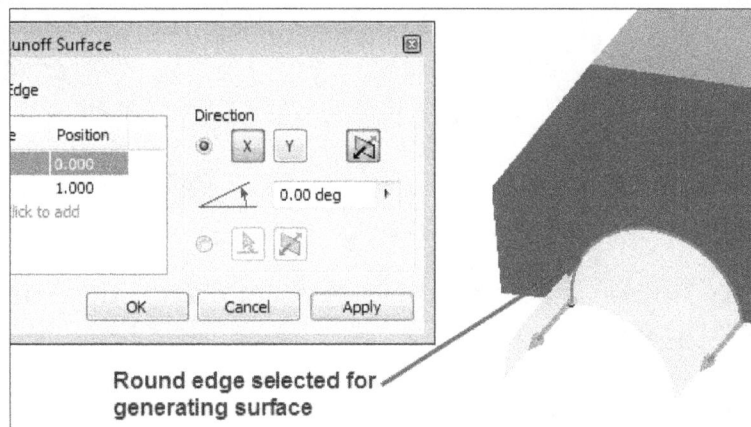

Figure-65. Preview of the radiate runoff surface

- Select desired position from the list box and change the direction using options in the **Direction** area of the dialog box, if needed.
- Click on the **OK** button to create the surface.

Extend Runoff Surface Tool

The **Extend Runoff Surface** tool is used to create surface by extending an edge of selected face. The procedure to use this tool is given next.

- Click on the **Extend Runoff Surface** tool from the **Runoff Surface** drop-down in the **Parting Design** panel of the **Ribbon**. The **Extend Runoff Surface** dialog box will be displayed; refer to Figure-66.

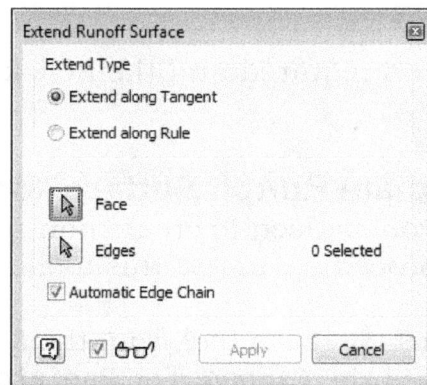

Figure-66. Extend Runoff Surface dialog box

- Select the face of plastic part which you want to extend to create runoff surface. You will be asked to select an edge.
- Select the edge of earlier selected face to define side of extension. Preview of the surface will be displayed; refer to Figure-67.

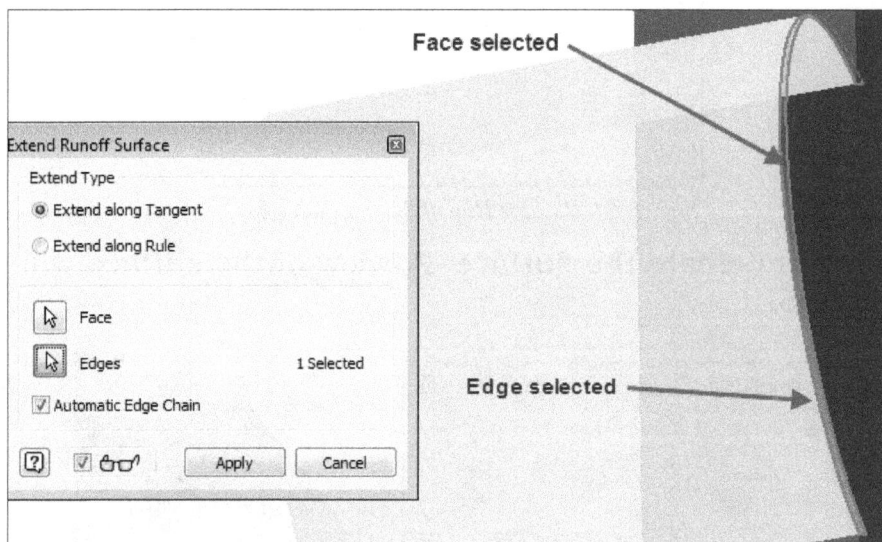

Figure-67. Preview of extend runoff surface

- Click on the **Apply** button from the dialog box to create the surface. Click on the **Done** button to exit the dialog box.

Although, we have learned the necessary steps to create core and cavity but there is one more tool which is important to discuss so that you can create core and cavity with sliders (inserts).

Create Insert Tool

The **Create Insert** tool is used to create sliders or say inserts so that features with undercuts can be created in the plastic part. The procedure to use this tool is given next.

- Click on the **Create Insert** tool from the **Insert** panel in the **Core/Cavity** contextual tab of the **Ribbon**. The **Create Insert** dialog box will be displayed; refer to Figure-68. Also, you will be asked to specify the starting face of the insert.

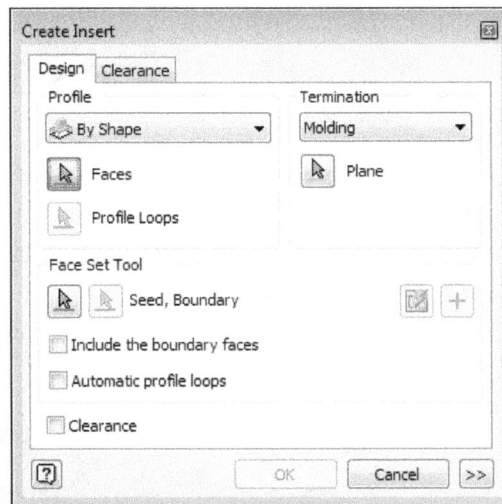

Figure-68. Create Insert dialog box

- Select the face/faces that you want to be head of the insert; refer to Figure-69.

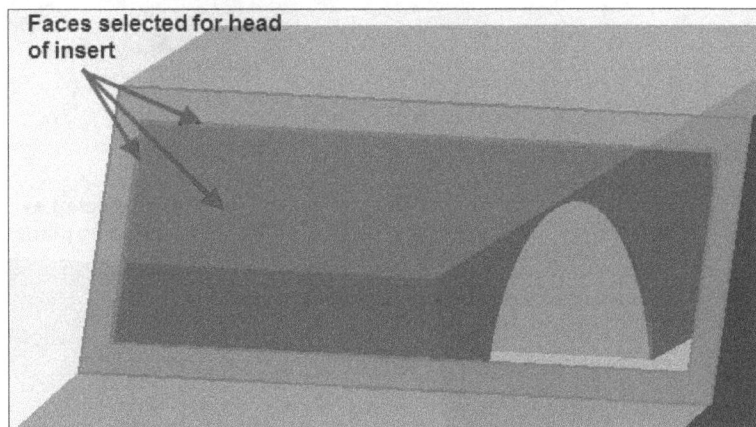

Figure-69. Faces selected for head of insert

- Click on the **Profile Loops** selection button from the dialog box and select the curves to specify outer profile of the insert; refer to Figure-70.

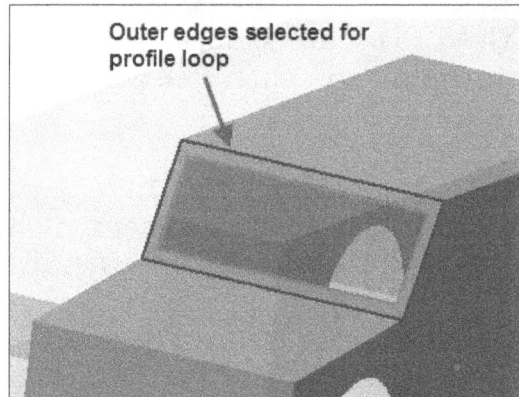

Figure-70. Edges selected for profile loop

- Select the **Molding** option from the drop-down in **Termination** area of the dialog box if you want the terminating face to be on mold. After selecting the **Molding** option, click on the **Plane** selection button and select the face of mold. Preview of the insert will be displayed; refer to Figure-71.

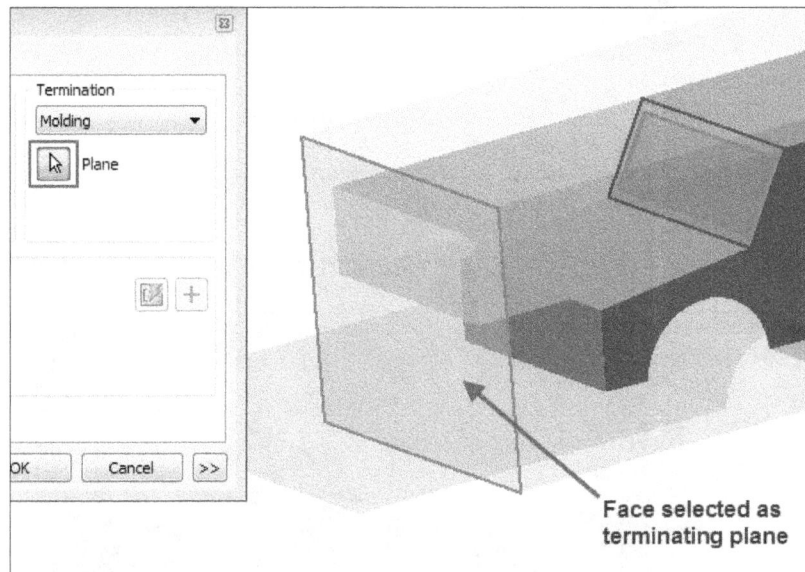

Figure-71. Preview of insert with molding termination

Or

- Select the **Distance** option from the drop-down in the **Termination** area of the dialog box if you want to specify the direction and length of the insert. On selecting the **Distance** option, you will be asked to select a direction reference. Select the face perpendicular to which you want to create the insert and specify desired length in the edit box displayed in **Termination** area of the dialog box. Preview of the insert will be displayed; refer to Figure-72.

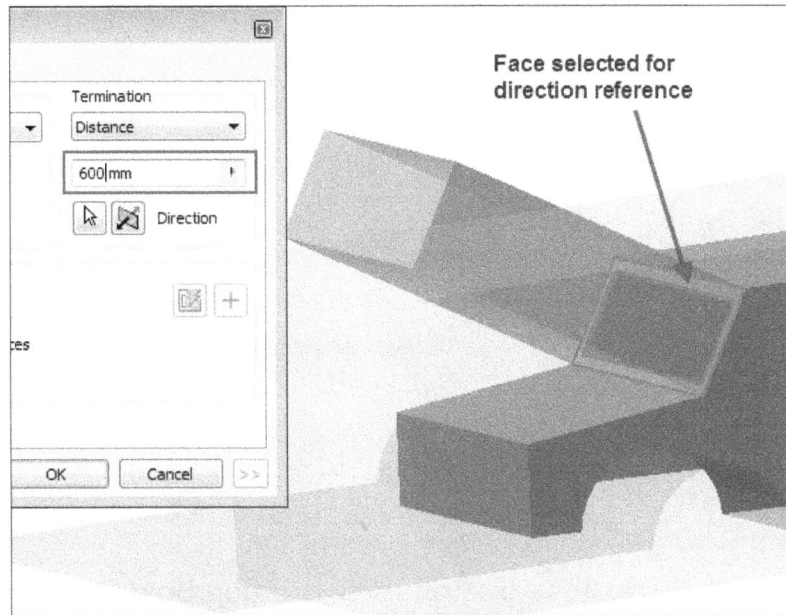

Figure-72. Preview of insert with distance termination

- Using the tools in the **Face Set Tool** area, you can define the shape of side faces of the insert.
- Select the **Clearance** check box and specify the required value of clearance for easy extraction of insert. Expand the dialog box by clicking on the **More** button and specify the clearance value in the edit box displayed; refer to Figure-73.

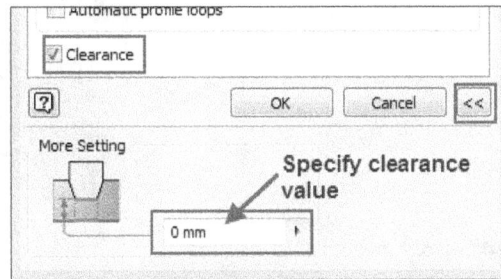

Figure-73. Specifying clearance value for insert

- Click on the **OK** button from the dialog box to create the insert.

Now, we are ready to generate the core and cavity. The tool is discussed next.

Generate Core and Cavity

The **Generate Core and Cavity** tool is used to generate the core and cavity based on specified parameters. The procedure to use this tool is given next.

- Click on the **Generate Core and Cavity** tool from the **Parting Design** panel of the **Core/Cavity** contextual tab in the **Ribbon**. The **Generate Core and Cavity** dialog box will be displayed; refer to Figure-74.

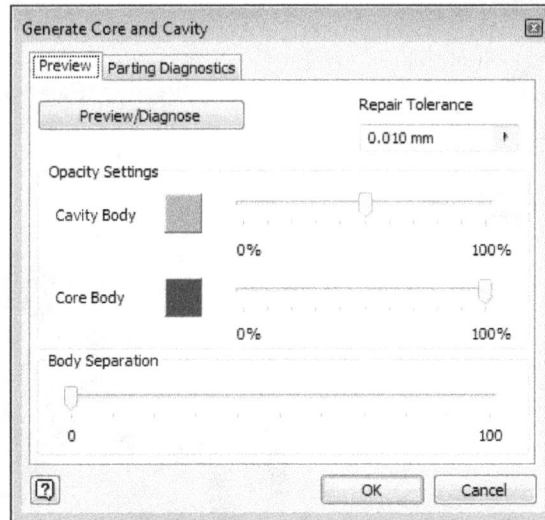

Figure-74. Generate Core and Cavity dialog box

- Click on the **Preview/Diagnose** button in the dialog box. Preview of the core and cavity will be displayed.
- Move the **Body Separation** slider to separate core and cavity; refer to Figure-75.

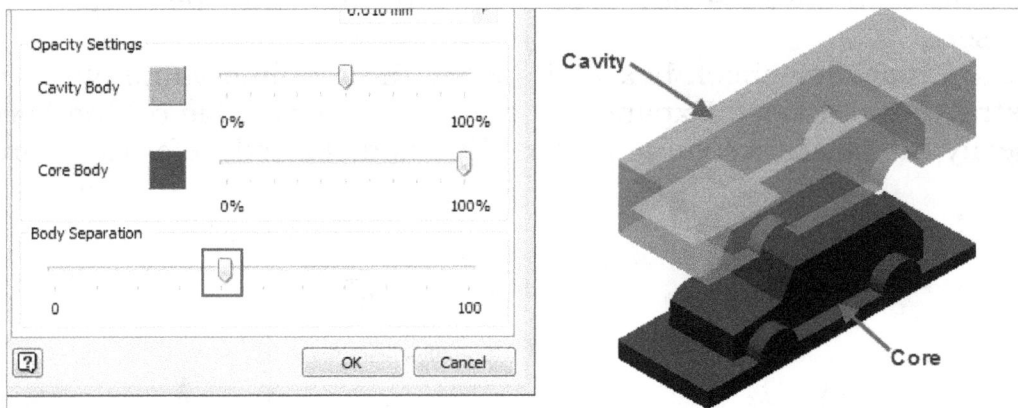

Figure-75. Separated core and cavity preview

- Click on the **Parting Diagnostics** tab in the dialog box to check the list of problems in current core and cavity; refer to Figure-76.

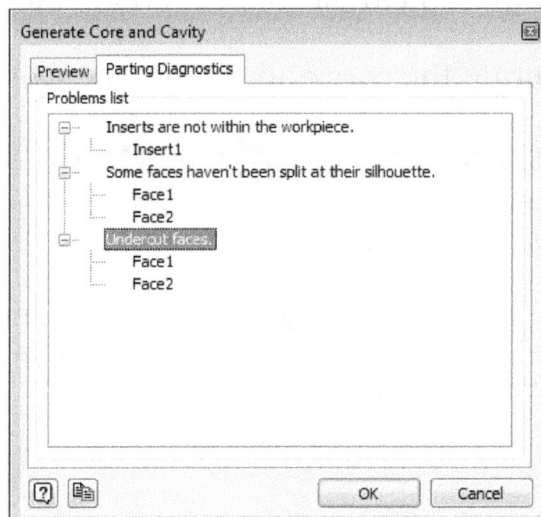

Figure-76. Parting Diagnostics tab in Generate Core and Cavity dialog box

- Click on the **Cancel** button and perform the changes in model as necessary to eradicate the problems.
- If you find that the problems displayed in the dialog box do not affect the molding quality then click on the **OK** button from the dialog box to create core and cavity.

Place Core Pin

The **Place Core Pin** tool is used to place core pins in the core to create hole or cavity in the plastic part. The procedure to use this tool is given next.

- Click on the **Place Core Pin** tool from the **Insert** panel in the **Core/Cavity** contextual tab of the **Ribbon**. The **Place Core Pin** dialog box will be displayed; refer to Figure-77. Also, you will be asked to select a face to specify position of the core pin.

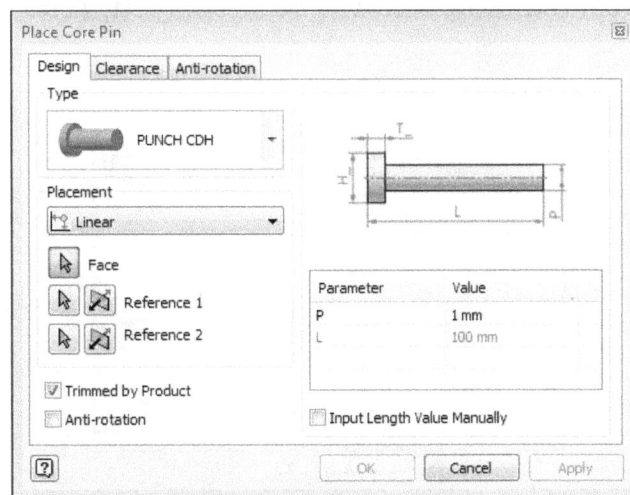

Figure-77. Place Core Pin dialog box

- Select the face on mold. You will be asked to specify first reference. Select an edge of the mold or plastic model in first direction. You will be asked to select second reference.
- Select the other edge of mold/plastic model in second direction. Preview of the core pin will be displayed; refer to Figure-78.
- Set the values of dimensions as required by double-clicking on them.
- Select the **Input Length Value Manually** check box and specify the diameters & length of the pin using the options in the list box.

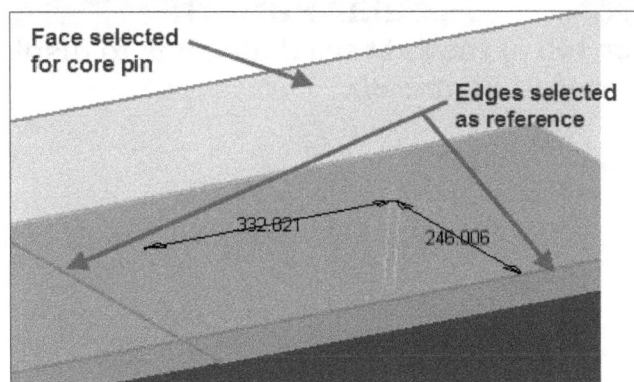

Figure-78. Preview of core pin

- Set the other options as required and then click on the **OK** button. The core pin will be created.

Once you have performed the operations related to core and cavity generation then click on the **Finish Core/Cavity** button from the **Exit** panel in the **Core/Cavity** contextual tab of the **Ribbon** to exit the contextual tab.

CREATING PATTERN OF CORE AND CAVITY

For smaller parts, we create multiple core and cavities in the same mold to get many molded parts at a time. To facilitate this process, we create pattern of core and cavity in Autodesk Inventor. The procedure to do so is given next.

- Click on the **Pattern** tool from the **Mold Layout** panel in the **Mold Layout** tab of the **Ribbon**. The **Pattern** dialog box will be displayed; refer to Figure-79.

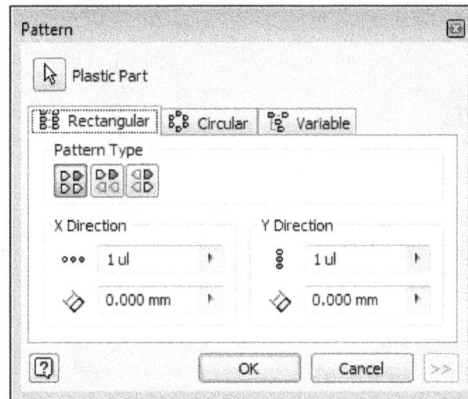

Figure-79. Pattern dialog box

Rectangular Pattern

- By default, the **Rectangular** tab is selected in the dialog box and hence you can create rectangular pattern of the molds.
- Specify desired number of instances and distance between them in the edit boxes of **X Direction** and **Y Direction** areas of the dialog box.
- Select the **Base Pattern** ⊞, **X Balance** ⊞, or **Y Balance** ⊞ button from the **Pattern Type** area of the dialog box to specify the pattern type.
- Click on the **OK** button to create the pattern.

Circular Pattern

- Click on the **Circular** tab of the **Pattern** dialog box to display the options related to circular pattern; refer to Figure-80.

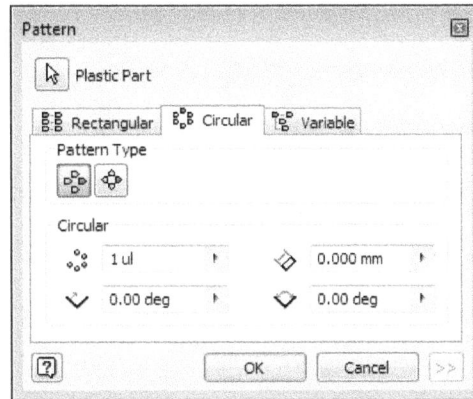
Figure-80. Circular tab in Pattern dialog box

- Specify the number of instances in the **Count** edit box ⠂. Specify the radius of circle in the **Radius** edit box ◇. Specify the rotational angle of instance around its center points in the **Rotational Angle** edit box ⌄. Specify the space angle between the instances in the **Pattern Angle** edit box ◇.
- Select the **Base Pattern** or **Radial Pattern** button from the **Pattern Type** area to specify the type of pattern. Preview of the pattern will be displayed; refer to Figure-81.

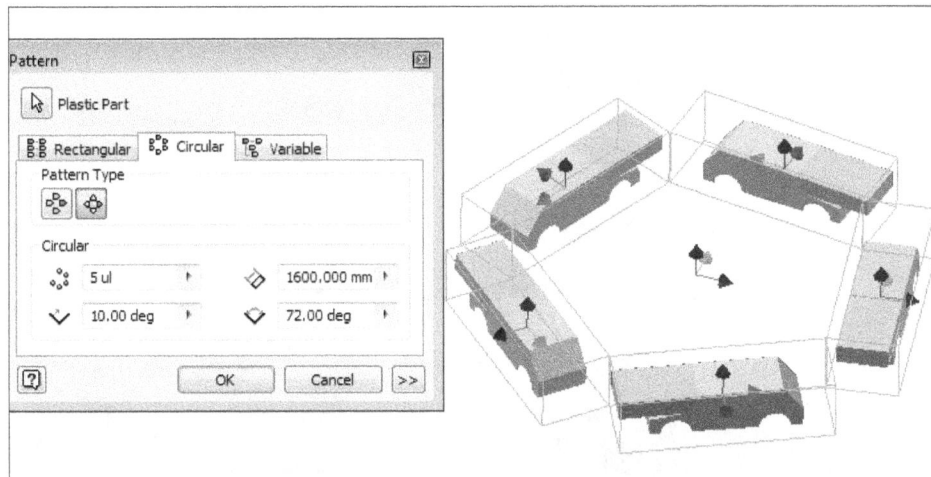
Figure-81. Preview of pattern

- Click on the **OK** button to create the pattern.

Variable Pattern

- Click on the **Variable** tab in the **Pattern** dialog box. The dialog box will be displayed as shown in Figure-82.

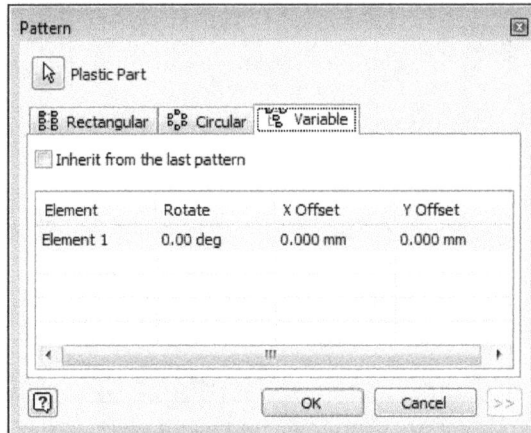

Figure-82. Variable tab in Pattern dialog box

- Right-click in the list box and select the **Add** option from the shortcut menu displayed; refer to Figure-83. A new instance will be added.
- Specify the rotation, x offset, and y offset as required by using the table in the list box.
- Repeat the procedure until you get required number of instances.

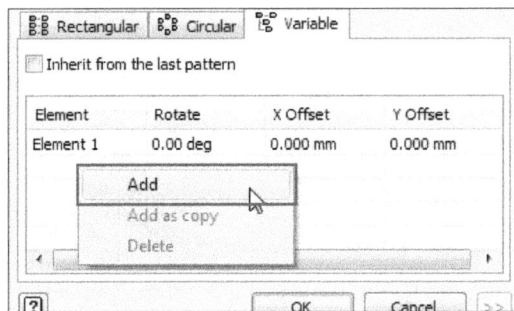

Figure-83. Add option for variable pattern

- Click on the **OK** button to create the pattern.

CREATING GATE

Earlier, we have learned to specify the location of gate. Now, we will learn to define the shape of gate. The procedure to do so is given next.

- Click on the **Gate** tool from the **Runners and Channels** panel of the **Mold Layout** tab in the **Ribbon**. The **Create Gate** dialog box will be displayed; refer to Figure-84.

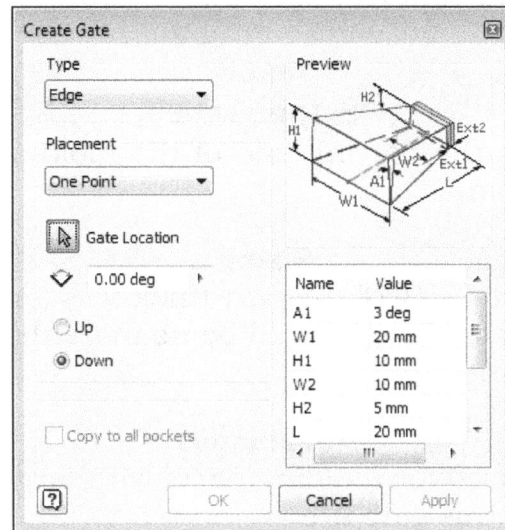

Figure-84. Create Gate dialog box

- Select the type of gate from the **Type** drop-down in the dialog box. You will be asked to select a gate location to place gate. (If you are not prompted to select gate location then click on the **Gate Location** selection button in the dialog box.)
- Click on a gate location created earlier on the plastic model. Preview of the gate will be displayed; refer to Figure-85.

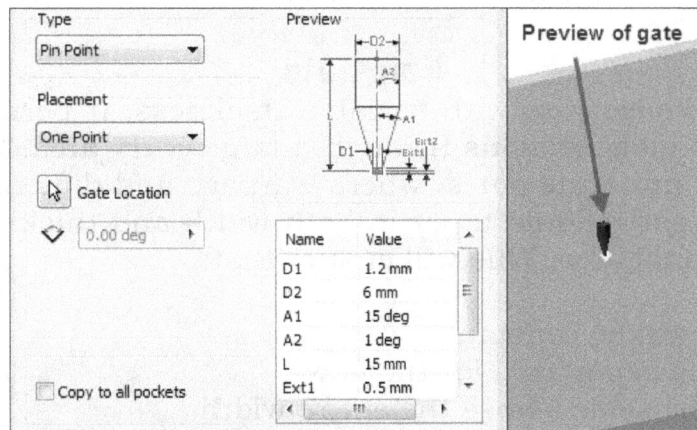

Figure-85. Preview of gate

- If you are using gate other than **Pin**, **Pin Point**, or **Sprue** type then you can select the **Up** or **Down** radio button from the dialog box to specify the orientation of the gate.
- Set the dimensions of the gate in the table at the bottom right in the dialog box.
- If you have multiple gate locations and want the same gate to be created on all the gate locations then select the **Copy to all pockets** check box from the dialog box.
- Click on the **OK** button to create the gate.

Types of Gates and Their Practical Applications

A gate is the connection between the runner system and the molded part. It must permit enough flow to fill the mold cavity, plus additional material to allow for part shrinkage and cooling. The gate type, location, and size has a great effect on the molding process. It affects physical properties, appearance, and size of the part.

Here, we will discuss the advantages and disadvantages of using different type of gates in mold design.

Edge Gate

An edge gate is located on the parting line of the mold and typically fills the part from the side, top, or bottom.

Sizing

The typical gate size is 6% to 75% of the part thickness (or 0.4 to 6.4 mm thick) and 1.6 to 12.7 mm wide. The gate land should be no more than 1.0 mm in length, with 0.5 mm being the optimum.

Advantages

An edge gate is primarily used for molding parts with large surfaces and thin walls. Some of the advantages of this gate are:

- Parallel orientation across the whole width (important for optical parts).
- In each case, uniform shrinkage in the direction of flow and transverse (important for crystalline materials).
- No inconvenient gate mark on the surface.

Disadvantages

This style gate is a problem with low viscosity, long fiber, or bead filled resins.

Fan Gate

A fan gate is a wide edge gate with variable thickness. It permits rapid filling of large parts or fragile mold sections through a large entry area. It is used to create a uniform flow front into wide parts, where warpage and dimensional stability are main concerns. The gate should taper in both width and thickness, to maintain a constant cross sectional area. This will ensure that:

- The melt velocity will be constant
- The entire width is being used for the flow
- The pressure is the same across the entire width.

Sizing

As with other manually trimmed gates, the maximum thickness should be no more than 75% of the part thickness. Typical gate sizes are from 0.25 to 1.6 mm thick. The gate width is typically from 6.4 mm to 25% of the cavity length.

Advantages

Fan gate is mainly applied to the tablet-shaped, shallow shell-shaped, and box-shaped article. Some of the advantages are:

- The injection speed is uniform which can reduce internal stress in the article and the possibility of air entrapment.
- Easy removal of the gate.
- Melt plastic flow from gate into the big cross-sectional area(cavity), so melt flow is good.
- Avoid deformation and maintain dimensional stability.

Disadvantages

The gate is best removed by using a trim fixture, especially on a thick section part. Since it makes a rather large scar, it is advisable to locate the gate at a non-cosmetic/ non-functional area, if possible.

Pin Gate

This type of gate relies on a three-plate mold design, where the runner system is on one mold parting line and the part cavity is in the primary parting line. Reverse taper runners drop through the middle (third) plate, parallel to the direction of the mold opening. As the mold cavity parting line is opened, the small-diameter pin gate is torn from the part. A secondary opening of the runner parting line ejects the runners. Alternatively, the runner parting line opens first. An auxiliary, top-half ejector system extracts the runners from the reverse taper drops, tearing the runners from the parts.

Sizing

The gate diameter like that of all other gates depends on the section thickness of the part and the processed plastic material and is independent of the system. One can generally state that smaller cross-sections facilitate the break-off. Therefore, as high a melt temperature as possible is used in order to keep the gate as small as possible.

Advantages

The design is particularly useful when multiple gates per part are needed to assure symmetric filling or where long flow paths must be reduced to assure packing to all areas of the part.

Disadvantages

This gate type has disadvantage of too much scraps rates as the runner is big.

Pin Point Gate

The Pin Point gate or simply called point gate is similar to Pin gate in shape. But the size of gate is smaller in this case.

Sizing

Typical gate sizes are 0.25 to 1.6 mm in diameter.

Advantages

Since the size of gate is very small so smaller scar is generated on the plastic part. Some designs allow gating into a very small concave dish on the part surface to ensure that the scar break is below a flat surface. In such cases, we can use this type of gate.

Disadvantages

The small cross-sectional opening of the point gate becomes a problem with respect to resin filled plastic. This type of gate can cause a problem with low viscosity, long fiber, or bead filled resins. The small gate restriction could raise the melt temperature and affect heat sensitive resins. Ignoring these factors could result in high mold maintenance.

Submarine Gate

A submarine gate is used in two-plate mold construction. An angled, tapered tunnel is machined from the end of the runner to the cavity, just below the parting line. As the parts and runners are ejected, the gate is sheared at the part. If a large diameter pin is added to a non-functional area of the part, the submarine gate can be built into the pin, avoiding the need of a vertical surface for the gate. If the pin is on a surface that is hidden, it does not have to be removed.

Multiple submarine gates into the interior walls of cylindrical parts can replace a diaphragm gate and allow automatic de-gating. The out-of-round characteristics are not as good as those from a diaphragm gate, but are often acceptable.

Sizing

The typical size is 0.25 to 2.0 mm in diameter. It is tapered to the spherical side of the runner.

Advantages

- Submarine gate can be machined to the exact size, without the fitting problem as to shape.
- During the injection part stripping, it could be automatically removed from the product for full-automation production.
- This type of gate is widely employed for automatically plastic products production.

Disadvantages

The small cross-sectional opening of the submarine gate becomes a problem with respect to resin filled plastic. This style gate could be a problem with low viscosity, long fiber, or bead filled resins. The small gate restriction could raise the melt temperature and affect heat sensitive resins. Ignoring these factors could result in high mold maintenance.

The Tunnel, Flat Bottom Submarine, and Sprue gates share the same properties as of Submarine gate. These gates are used based on the requirement of runner.

CREATING RUNNER

Runner is a channel machines in the mold to supply molten plastic to the gates through which the melt goes into cavity. In Autodesk Inventor, runner creation process can be divided into two sections; runner sketch creation and runner model creation. These sections are discussed next.

Creating Runner Sketch

There are two tools in Autodesk Inventor to create runner; **Auto Runner Sketch** and **Manual Sketch**. If you select the **Auto Runner Sketch** tool then you will find some predefined shapes of runners. The procedure to use both the tools are discussed next.

Using Auto Runner Sketch Tool

- Click on the **Auto Runner Sketch** tool from the **Runner Sketch** drop-down in the **Runners and Channels** panel of the **Mold Layout** tab in the **Ribbon**. The **Auto Runner Sketch** dialog box will be displayed; refer to Figure-86.

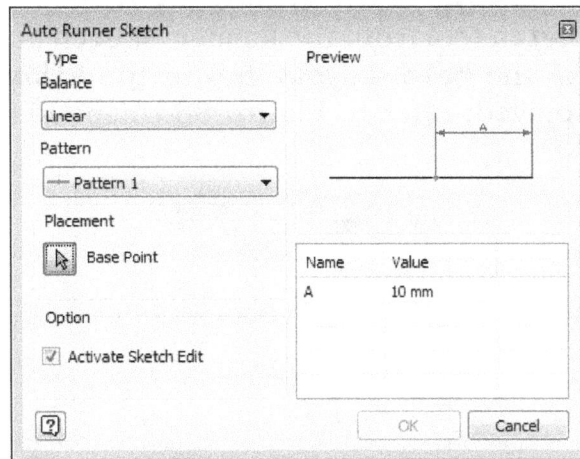

Figure-86. Auto Runner Sketch dialog box

- Select the type of runner from the **Balance** drop-down in the **Type** area of the dialog box and select desired pattern (shape) from the **Pattern** drop-down. You will be asked to specify the base point of the runner.
- Click on the mold at desired location to specify the base point. Preview of the runner sketch will be displayed.
- Set the dimensions for runner in the table given in dialog box. The preview of runner will change accordingly; refer to Figure-87.

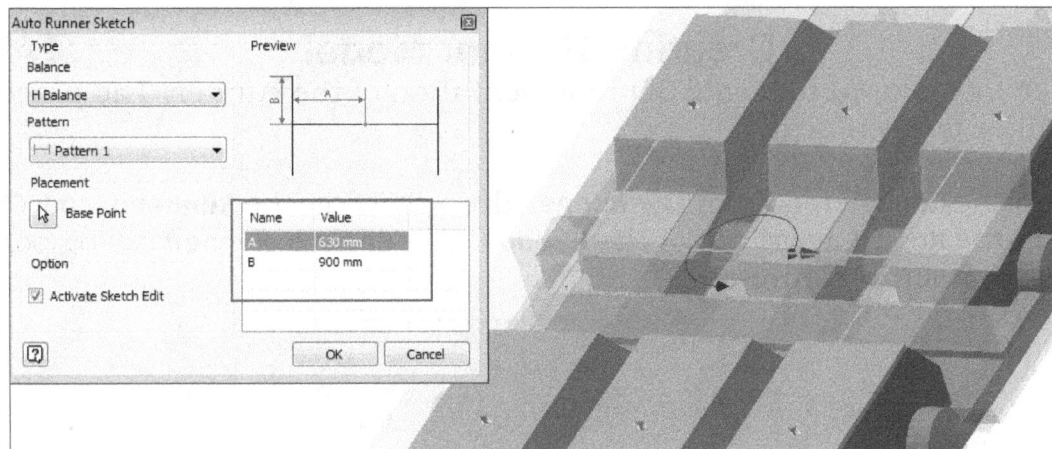

Figure-87. Preview of auto runner sketch

- Move and rotate the sketch as required by using the handles displayed on the preview.
- Clear the **Activate Sketch Edit** check box if you do not want to edit the sketch otherwise keep it selected.
- Click on the **OK** button from the dialog box. The sketch will be created. If the **Activate Sketch Edit** check box is selected then sketching environment will be displayed. Edit the runner sketch as required and then click on the **Finish Sketch** button from the **Exit** panel in the **Sketch** tab of the **Ribbon**. Click on the **Return** button from the **Return** panel in the **3D Model** tab of the **Ribbon** to return back to mold design environment.

Using Manual Sketch Tool

- Click on the **Manual Sketch** tool from the **Runner Sketch** drop-down in the **Runners and Channels** panel of the **Mold Layout** tab in the **Ribbon**. The **Manual Sketch** dialog box will be displayed; refer to Figure-88. Also, you will be asked to select a sketching plane.

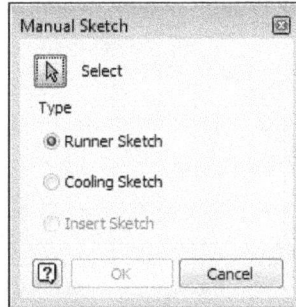

Figure-88. Manual Sketch dialog box

- Make sure the **Runner Sketch** radio button is selected in the dialog box and then click on the flat face of the mold to define sketching plane and click on the **OK** button from the dialog box. The sketching environment will be activated.
- Create the sketch of runner using the lines. Once the sketch is completed, click on the **Finish Sketch** button from the **Exit** panel in the **Ribbon** and then click on the **Return** button from **Return** panel in the **Ribbon**.

Creating Runner Model

The **Runner** tool is used to create 3D representation of the runner. The procedure to create runner is given next.

- Click on the **Runner** tool from the **Runner** drop-down in the **Runners and Channels** panel in the **Mold Layout** tab of the **Ribbon**. The **Create Runner** dialog box will be displayed; refer to Figure-89.

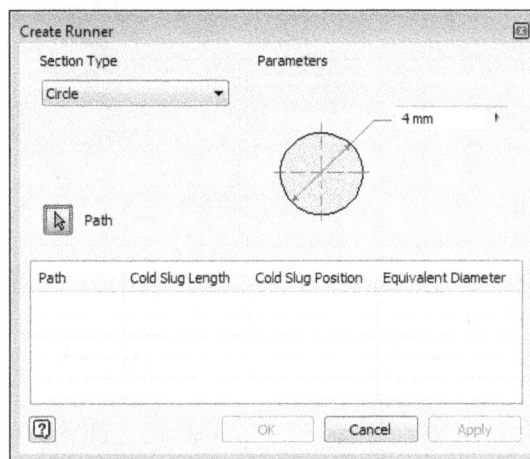

Figure-89. Create Runner dialog box

- Select desired section type from the **Section Type** drop-down in the dialog box. The preview of runner cross-section and related parameters will be displayed in the **Parameters** area of the dialog box.
- Set desired dimensions in the edit boxes of **Parameters** area of the dialog box.
- Select the runner sketch earlier created to draw the runner. Preview of the runner will be displayed; refer to Figure-90.

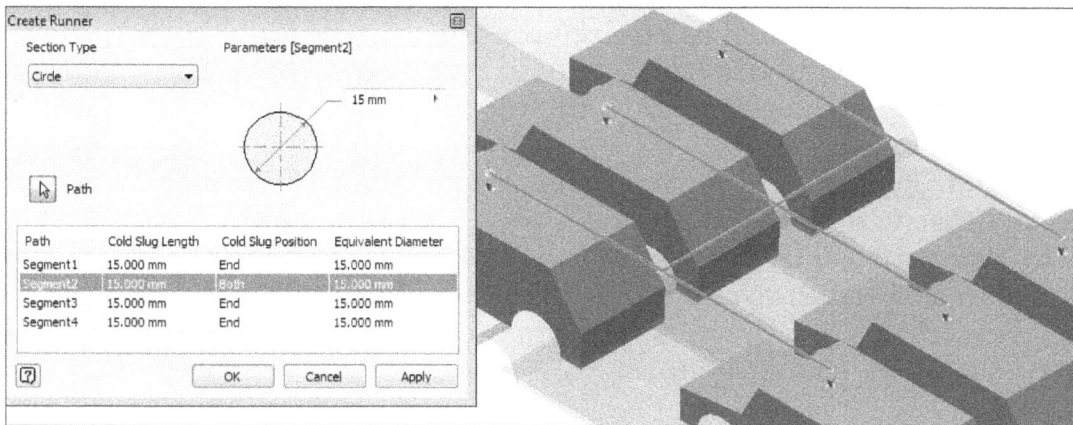

Figure-90. Preview of runner

- Select desired positions of cold slug from the field under **Cold Slug Position** column in the table of dialog box.
- Click on the **OK** button from the dialog box to create the runner.

Runner Design Guidelines

The runner is a channel machined into the mould plate to connect the sprue with the entrance or gate to the impression. In the basic two-plate mould, the runner is positioned on the parting surface while on more complex designs, the runner is positioned below the parting surface. Below are some points to be taken care of while making runner for mold.

- The wall of the runner channel must be smooth to prevent any restriction to flow.
- There must be no machine marks which would tend to retain. In the runner in the mould plate, to ensure this it is desirable for the mould design to specify that the runner is polished "in line of draw".

There are some other considerations for determining the runner.

- The shape of the cross section of the runner
- The size of the runner
- The runner layout

Determining Cross-section Shape of Runner

The cross sectional shape of the runner used in a mould is usually one of the following forms:

1. Fully round
2. Semi Round
3. U Shaped
4. Trapezoidal
5. Modified trapezoidal
6. Hexagonal

1. Full Round Runner: The full-round runner is the best in terms of a maximum volume-to-surface ratio, which minimizes pressure drop and heat loss. However, the tooling cost is generally higher because both halves of the mold must be machined so that the two semi-circular sections are aligned when the mold is closed.

2. Semi Round Runner: The use of semi round shaped runner is generally not recommended because of lower cross-section size of the runner. But when it is not possible to use trapezoidal, hexagonal, or u shaped runner then it is a cheaper option as compared to full round runner. For some plastics materials like Eastman polymers, the semi-round shape is recommended as it cause lesser flow restriction as compared to other runner shapes.

3. U Shaped Runner: The U shaped runner is second most efficient runner shape. The width of U-shape runner is smaller than that of circular runner. With the same efficiency and length (unit length), the volume of U-shape is minimal among the semicircular, ladder, and U-shape runners and next to that of circular runner. Therefore, U-shape is most suitable in the cool-runner mould.

4. Trapezoidal Runner: The trapezoidal runner permits the runner to be designed and cut on one side of the mold. It is commonly used in three-plate molds, where the full-round runner may not be released properly, and at the parting line in molds, where the full-round runner interferes with mold sliding action.

5. Modified Trapezoidal Runner: The modified trapezoidal runner is a combination of both trapezoidal runner and round runner. The bottom of runner is round and upper part is trapezoidal. This type of runner can be used in both hot and cold runners. But the material wastage in case of trapezoidal and modified trapezoidal is higher as compared to the first three runners discussed. Note that the material left in runner is wastage after molding.

6. Hexagonal Runner: The hexagonal runner is basically a double trapezoidal runner, where the two halves of the trapezium meet at parting surface. It is easier to match the two halves of the hexagonal runner compared to that of a round runner. This point applies particularly to runners, which are less than 3 mm in width. The hexagonal runner has minimum flow resistance after full round runner.

Hydraulic Diameter and Flow Resistance

To compare runners of different shapes, you can use the hydraulic diameter, which is an index of flow resistance. The higher the hydraulic diameter, the lower the flow resistance. Hydraulic diameter can be defined as:

$$D_h = \frac{4A}{P}$$

where D_h = hydraulic diameter

      A = cross section area

      P = perimeter

Based on this formula, the hydraulic diameters of different cross-sections are given next.

Cross Section			R=H/2	
Dh	D	0.9523D	0.9116D	0.8862D

Cross Section				
Dh	0.8771D	0.8642D	0.8356D	0.7090D

Determining Size of Runner

Ideally, the size of the runner diameter will take many factors into account — part volume, part flow length, runner length, machine capacity gate size, and cycle time. Generally, runners should have diameters equal to the maximum part thickness, but within the 4 mm to 10 mm diameter range to avoid early freeze-off or excessive cycle time. The runner should be large enough to minimize pressure loss, yet small enough to maintain satisfactory cycle time. Smaller runner diameters have been successfully used as a result of computer flow analysis where the smaller runner diameter increases material shear heat, thereby assisting in maintaining melt temperature and enhancing the polymer flow. Large runners are not economical because of the amount of energy that goes into forming, and then regrinding the material that solidifies within them.

By an empirical formula, calculation of main runner size can be given as:
$$D = (W^{1/2} \times L^{1/4})/3.7$$
Where,
D=runner diameter (mm)
W=weight of moulding (g)
L=height/length of runner (mm)

Theoretically, the cross-sectional area of main runner should be equal to/in excess of the combined cross-sectional areas of the branch runners that is feeding the material.

Determining Runner Layout

The purpose of different type of runner layouts is to balance the flow of melt, so that all the cavities in mold are filled properly without pressure loss. In a balanced layout, the length of runner from sprue to gate is equal for all the mold cavities. Proper runner layout also makes it possible to produce family molds. In family molds, the flow of plastic is maintained by size and length of runner, so that cavities are filled in same time. There are hardly any guidelines for runner layout as it all depends on flow of plastic. Only way to find out best runner layout is to perform plastic flow analysis on the mold.

CREATING SECONDARY SPRUE

The secondary sprue, also called the sprue runner, is used to supply molten melt from one runner to the other runner. It is generally included in 3 plate molds. The procedure to create secondary sprue is given next.

- Click on the **Secondary Sprue** tool from the **Runner** drop-down in the **Runners and Channels** panel in the **Mold Layout** tab of the **Ribbon**. The **Create Secondary Sprue** dialog box will be displayed; refer to Figure-91. Also, you will be asked to select a sketch point to locate secondary sprue.

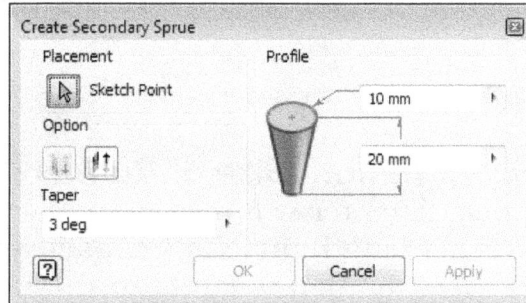

Figure-91. Create Secondary Sprue dialog box

- Select a sketch point created (by using the **Start 2D Sketch** tool) on the runner line. Preview of the secondary sprue will be displayed; refer to Figure-92.

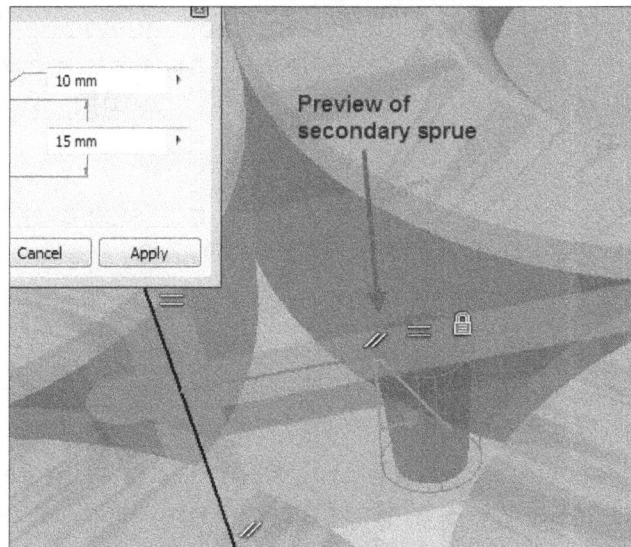

Figure-92. Preview of secondary sprue

- Specify desired parameters and click on the **OK** button to create the sprue.

Most of the designing work has been done till this point. Now, we will learn to add a mold base to the core and cavity so that we can check the practical fitting of core and cavity in mold plates.

ADDING MOLD BASE TO ASSEMBLY

The tools to add and manage mold base are available in the **Mold Assembly** tab of the **Ribbon**; refer to Figure-93.

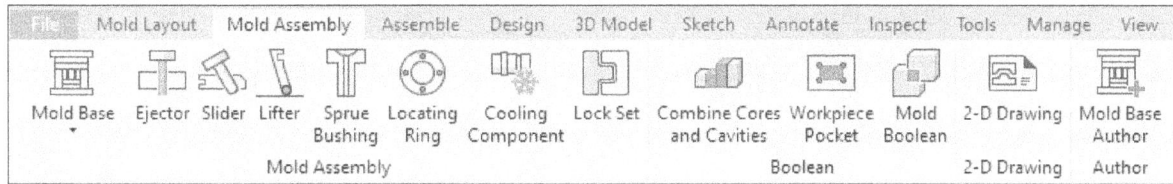

Figure-93. Mold Assembly tab in the Ribbon

The procedure to add mold base to the assembly is given next.

- Click on the **Mold Base** tool from the **Mold Base** drop-down in the **Mold Assembly** panel of **Mold Assembly** tab in the **Ribbon**. The **Mold Base** dialog box will be displayed; refer to Figure-94.

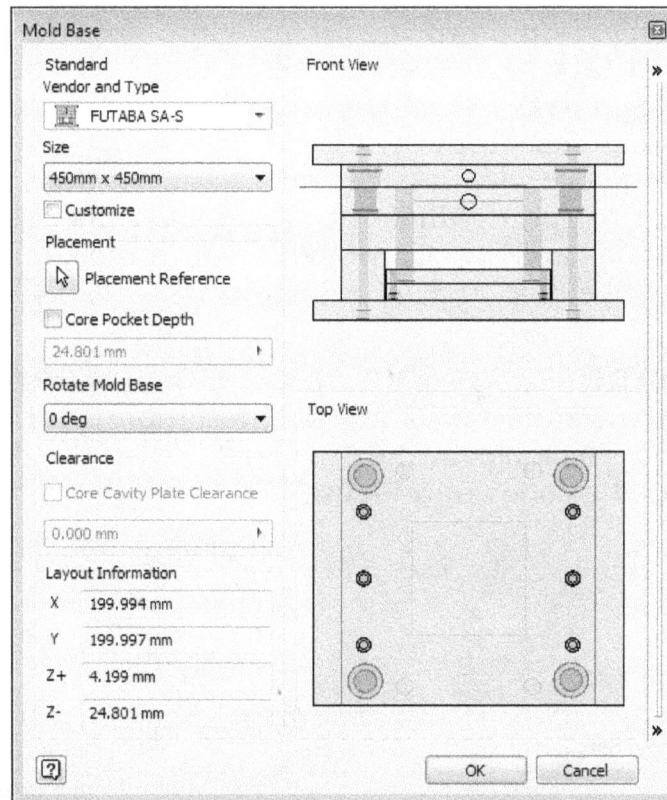

Figure-94. Mold Base dialog box

- Click in the **Vendor and Type** box of **Standard** area of the dialog box. The list of mold base vendor will be displayed in the selection box for selected categories; refer to Figure-95.

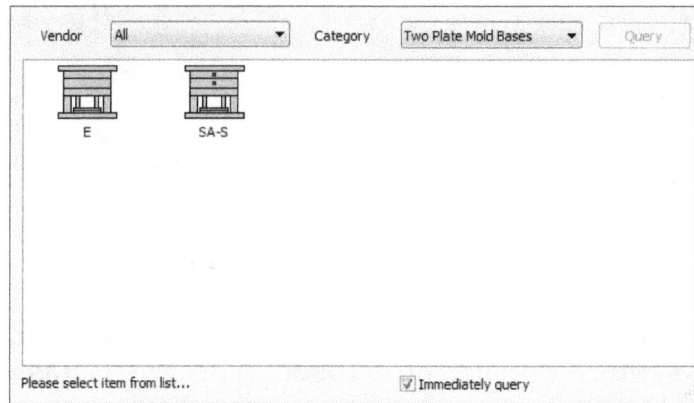

Figure-95. List of mold base vendors

- Click on the **Category** drop-down in the selection box and select the **Two Plate Mold Bases** or **Three Plate Mold Bases** option as required. The corresponding mold bases will be displayed in the selection box. Click on the mold base to select it.
- Select the size of mold base from the **Size** drop-down.
- By default, the core pocket depth is decided based on the core-cavity created by you but if you want to specify different core pocket depth then select the **Core Pocket Depth** check box and specify desired value in the edit box below it. Preview in the dialog box will be modified accordingly.

Customizing Mold Base Components

- If you want to customize the size of any of the component in the mold base then select the **Customize** check box in the **Standard** area of the dialog box. The dialog box will get expanded as shown in Figure-96.

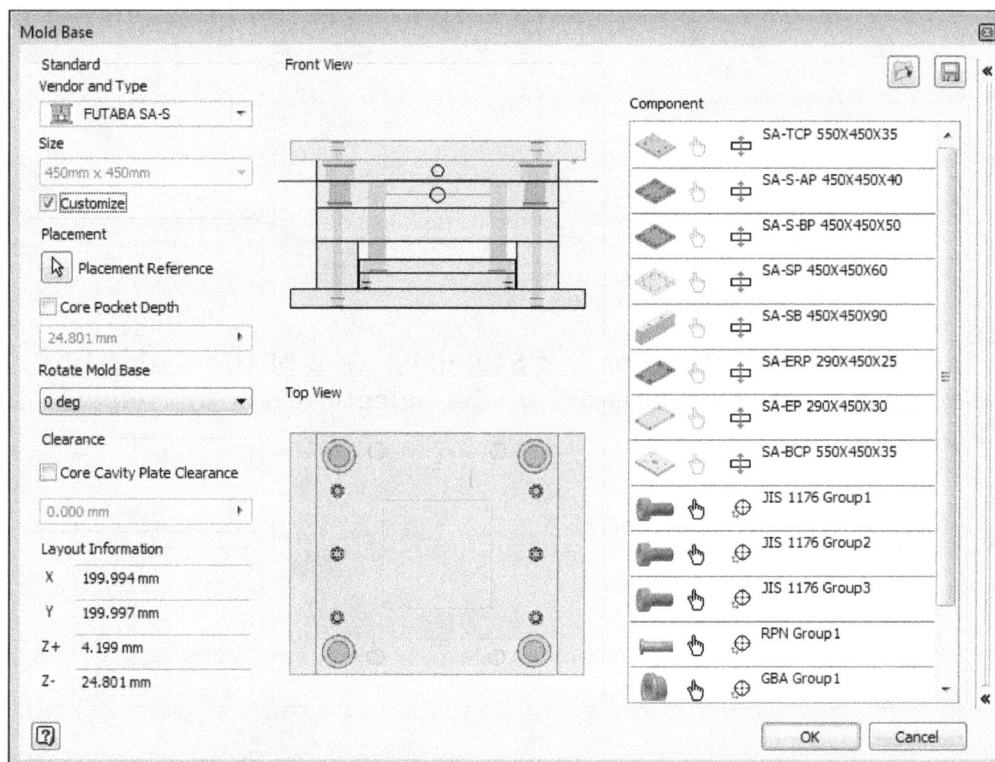

Figure-96. Expanded Mold Base dialog box

- There are four options to modify each component of the mold base; **Position Settings**, **Property Settings**, **Delete**, and **Type Settings**. These options are displayed on selecting the component from the **Component** area of the dialog box; refer to Figure-97.

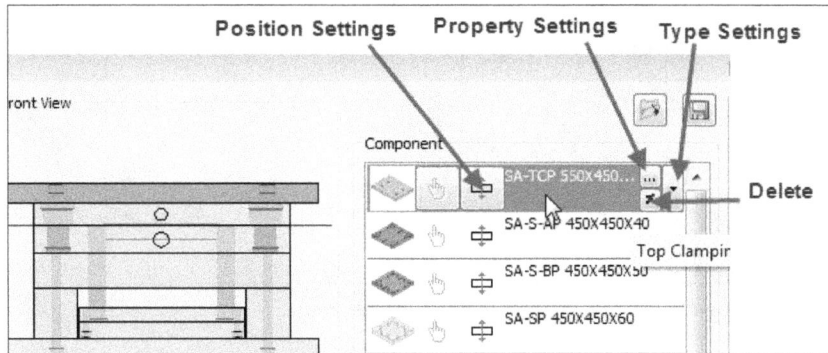

Figure-97. Options for mold base components

- Click on the **Position Settings** button of the component once and then click at desired location in the preview to place the component; refer to Figure-98.

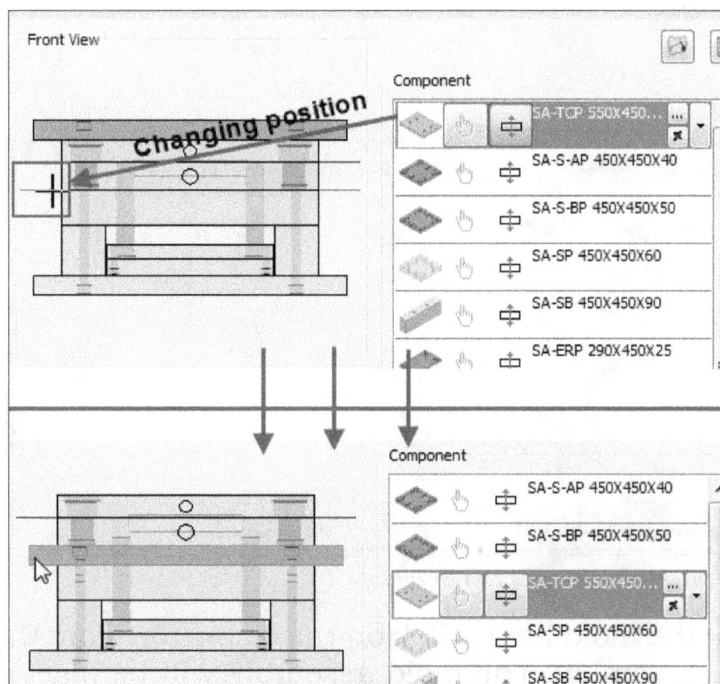

Figure-98. Changing position of mold base component

- Click on the **Property Settings** button to change the dimensions of the selected component. On clicking this button, the dialog box to change dimensions will be displayed; refer to Figure-99. Specify desired dimensions and click on the **OK** button to apply dimensions.

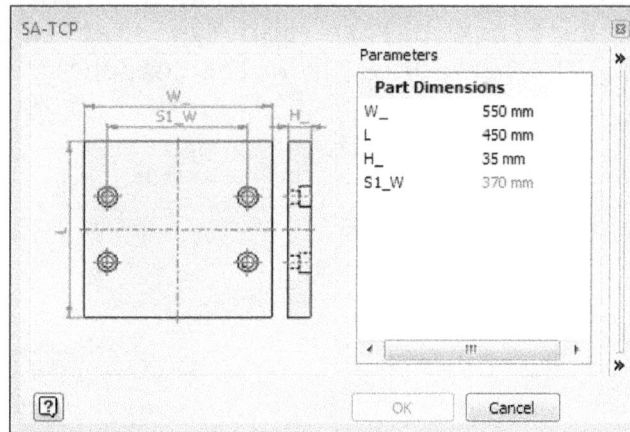

Figure-99. Dialog box to change dimensions

- Click on the **Delete** button to delete the selected component.
- Click on the **Type Settings** button to replace the selected component from the other component in the catalog. On selecting this button, a selection box is displayed with replacement components; refer to Figure-100. Select desired component to replace the current.

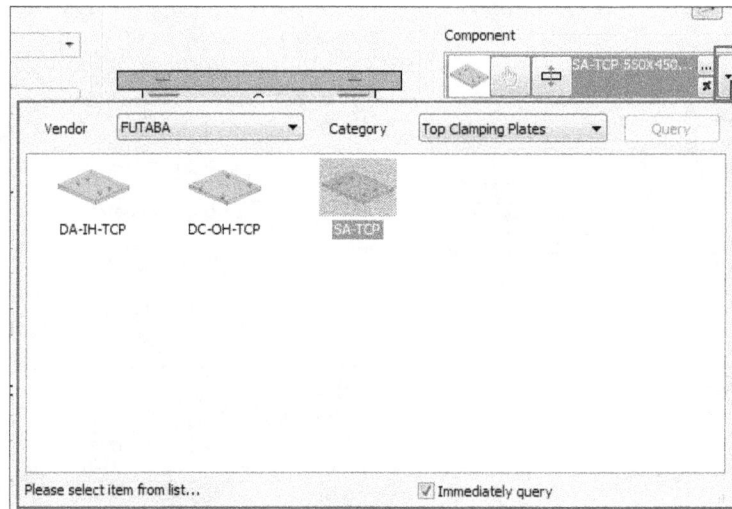

Figure-100. Changing component type

- After customizing the mold base, click on the **OK** button from the dialog box to add mold base to the assembly. The mold assembly will be placed in the mold base automatically based on the specified parameters; refer to Figure-101.

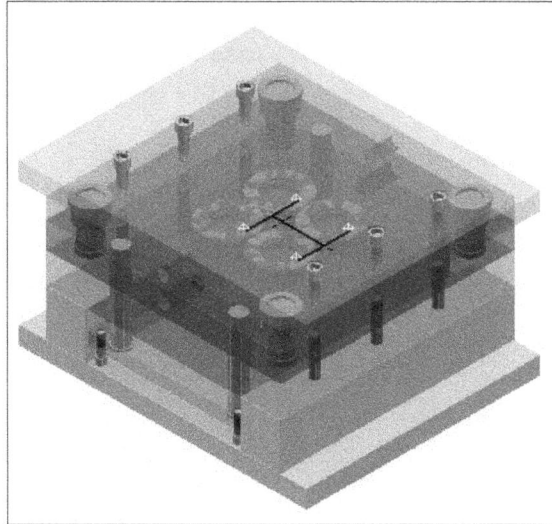

Figure-101. Mold assembly with mold base

COOLING CHANNEL

As runner is important for transportation of plastic melt to the cavity, in the same way cooling channel is important to maintain right temperature in the mold. Most of the people think that the work of cooling channel is to cool down molds but this is half correct. A cooling channel may raise the temperature of mold at some points if necessary. In this way, it gives equal time to all areas of the plastic component to cool down. The procedure to create cooling channel is discussed next.

Creating Cooling Channel

• Click on the **Cooling Channel** tool from the **Runners and Channels** panel in the **Mold Layout** tab of the **Ribbon**. The **Cooling Channel** dialog box will be displayed; refer to Figure-102. Also, you will be asked to select a face to place cooling line.

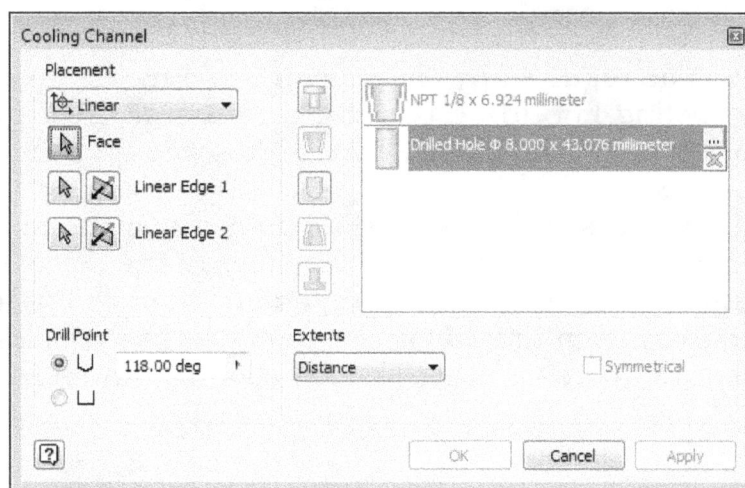

Figure-102. Cooling Channel dialog box

• Select the face of mold base to place the cooling channel line. Preview of the cooling line will be displayed; refer to Figure-103. Also, you will be asked to select linear edge for dimension reference.

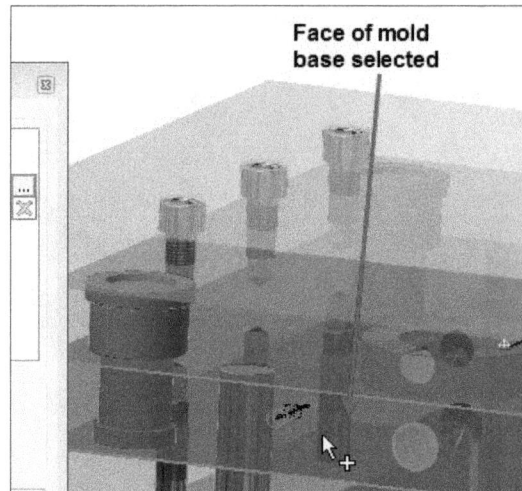

Figure-103. Face selected for cooling channel line

- Select the first reference edge and then second reference edge to place the cooling line fully constrained; refer to Figure-104.

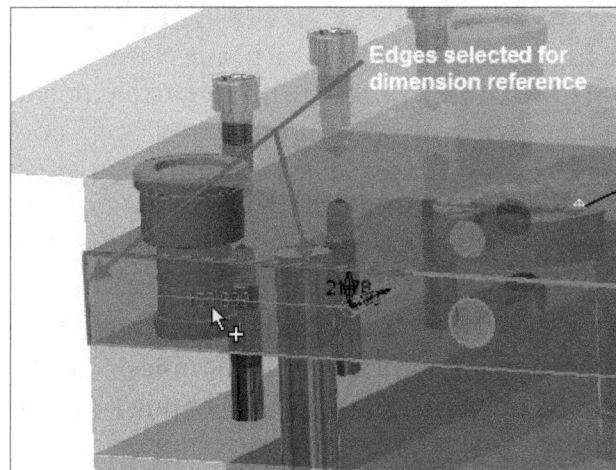

Figure-104. Edges selected for placement of cooling line

- Set the dimensions as required by clicking on them in the preview.
- If you want the cooling line to be passing through all the objects in the mold assembly then select the **Through All** option from the drop-down in the **Extents** area of the dialog box.
- Select desired components of the cooling line using the buttons given in the center of the dialog box. You can change the parameters of the selected components by clicking on the ellipse button for that component; refer to Figure-105.
- Click on the **OK** button from the dialog box to create the cooling line.
- Repeat the procedure to create other lines of cooling channel.

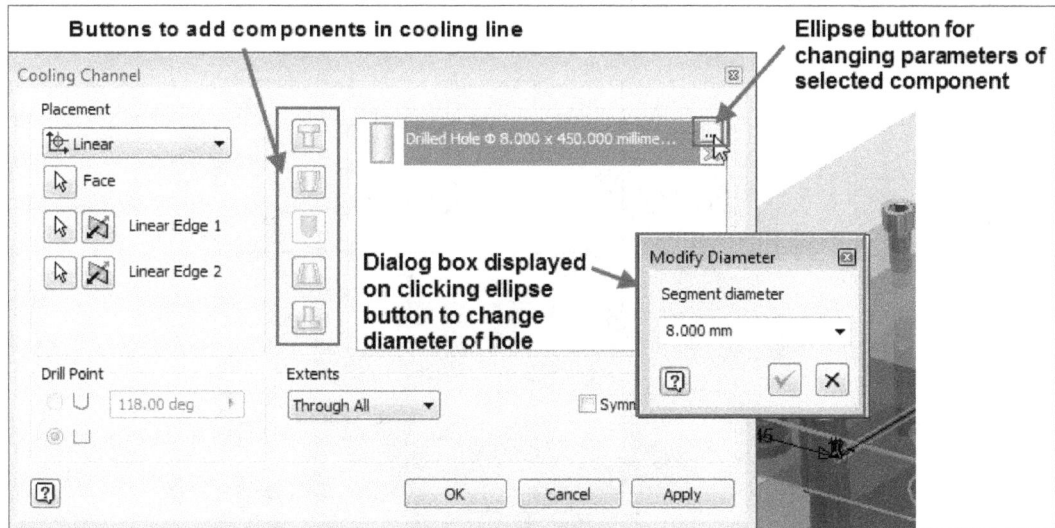

Figure-105. Adding and changing components for cooling line

Sketch Method for Cooling Channel Creation

In this method, we will first create sketched lines for cooling channel. Based on these sketch lines we will create the cooling channel. The procedure is discussed next.

- Create a workplace using the **Plane** tools at the location suitable for cooling channels.
- Click on the **Manual Sketch** button from the **Runner** drop-down in the **Runners and Channels** panel of the **Mold Layout** tab in the **Ribbon**. The **Manual Sketch** dialog box will be displayed as discussed earlier.
- Select the **Cooling Sketch** radio button from the dialog box. You will be asked to select a plane for creating sketch.
- Select the work plane you have created for cooling sketch; refer to Figure-106. Click on the **OK** button from the dialog box. The sketching environment will be activated.

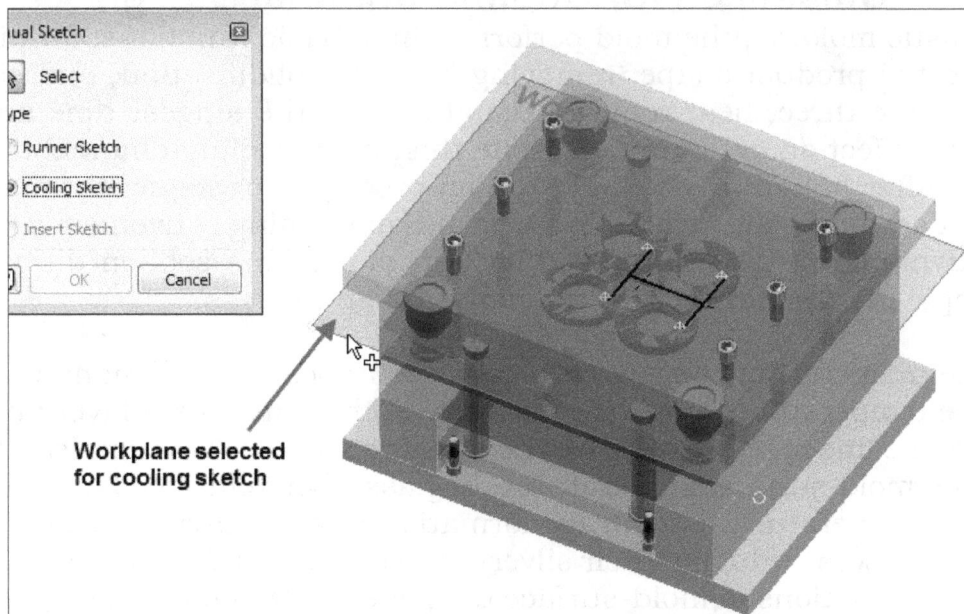

Figure-106. Selecting workplane for cooling sketch

- Create the sketch for cooling channel passing through mold steel; refer to Figure-107.

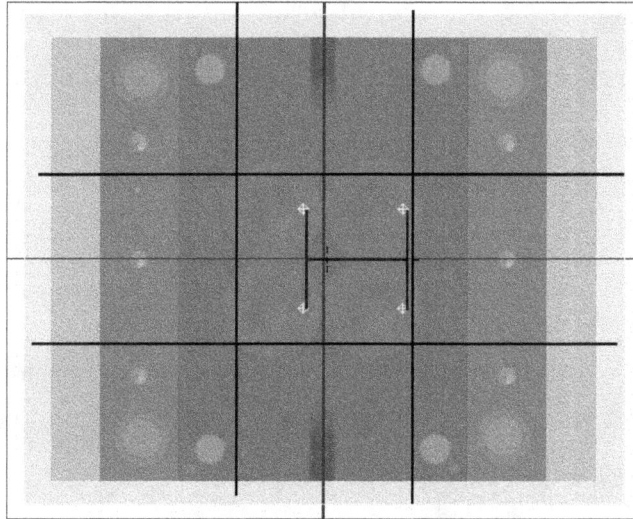

Figure-107. Sketch lines created for cooling channel

- Click on the **Finish Sketch** button from the **Exit** panel in the **Ribbon** and then click on the **Return** button from the **Return** panel in the **Ribbon**.
- Click on the **Cooling Channel** tool from the **Runners and Channels** panel in the **Mold Layout** tab of the **Ribbon**. The **Cooling Channel** dialog box will be displayed as discussed earlier.
- Select the **From Sketch** option from the **Placement** drop-down in the dialog box. You will be asked to select the reference line.
- Select a line of the sketch and add/modify the cooling channel components as discussed earlier.
- Click on the **Apply** button and then select the next line of cooling channel. Repeat the procedure till you have created all the lines of cooling channel.
- Click on the **Done** button to exit the dialog box.

Guidelines for Cooling Channel Design

In thermoplastic molding, the mold performs three basic functions: forming molten material into the product shape, removing heat for solidification, and ejecting the solid part. Of the three, heat removal usually takes the longest time and has the greatest direct effect on cycle time. Despite this, mold cooling-channel design often occurs as an afterthought in the mold-design process. Consequently, many cooling designs must accommodate available space and machining convenience rather than the thermodynamic needs of the product and mold. This section discusses mold cooling, a topic to consider early in the mold-design process.

Mold-surface temperature can affect the surface appearance of many parts. Hotter mold-surface temperatures lower the viscosity of the outer resin layer and enhance replication of the fine micro texture on the molding surface. This can lead to reduced gloss at higher mold-surface temperatures. In glass-fiber-reinforced materials, higher mold-surface temperatures encourage formation of a resin-rich surface skin. This skin covers the fibers, reducing their silvery appearance on the part surface. Uneven cooling causes variations in mold-surface temperature that can lead to non-uniform part-surface appearance.

Before, heat from the melt can be removed from the mold, it must first conduct through the layers of plastic thickness to reach the mold surface. Material thermal conductivity and part wall thickness determine the rate of heat transfer. Generally, good thermal insulators, plastics conduct heat much more slowly than typical mold materials. Cooling time increases as a function of part thickness squared; doubling wall thickness quadruples cooling time.

Below are some points to be taken care of while designing cooling channel:

- Core out thick sections or provide extra cooling in thick areas to minimize the effect on cycle time.
- Avoid low-conductivity mold materials, such as stainless steel, when fast cycles and efficient cooling are important.
- Place cooling-channel center lines approximately 2.5 cooling-channel diameters away from the mold cavity surface.
- As a general rule of thumb, use center-to-center spacing of no more than three cooling-channel diameters.
- Consider using baffles and bubblers to remove heat from deep cores.
- Adjust the bubbler tube or baffle length for optimum cooling. If they are too long, flow can become restricted. If too short, coolant flow may stagnate at the ends of the hole.
- Consider using spiral channels cut into inserts for large cores.

CREATING COLD WELL

The cold well is created in runner line to hold the plastic which is cold compared to the melt in the runner line. The procedure to create cold well is discussed next.

- Click on the **Cold Well** tool from the **Runners and Channels** panel in the **Mold Layout** tab of the **Ribbon**. The **Cold Well** dialog box will be displayed; refer to Figure-108.

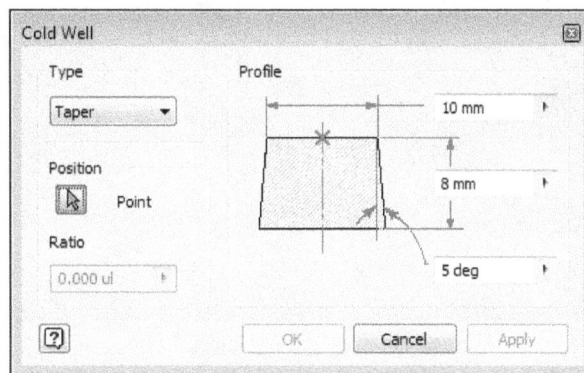

Figure-108. Cold Well dialog box

- Click in the **Type** drop-down and select desired type of cold well. You will be asked to select a point on the runner line.
- Click on the runner line at desired location to place the cold well. Preview of the cold well will be displayed; refer to Figure-109.

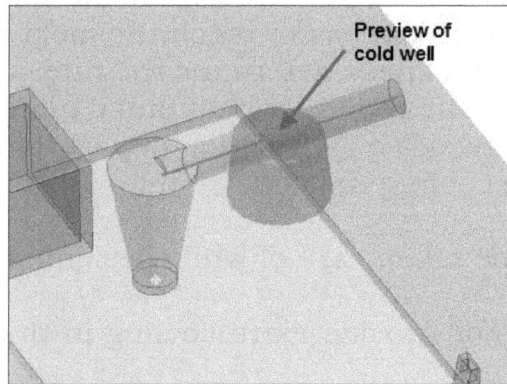

Figure-109. Preview of cold well

- Set the ratio value in **Ratio** edit box to position the cold well on the runner line.
- Set the parameters of cold well as required in the dialog box and click on the **OK** button to create the cold well.

ADDING SPRUE BUSHING

The sprue bushing is used to create pathway for filling molten plastic into the runner from where it will get filled in the cavity. This is a very simple pick and place step in Autodesk Inventor, although it is a tedious job in the workshop. The procedure to add sprue bushing in the assembly is given next.

- Click on the **Sprue Bushing** tool from the **Mold Assembly** panel in the **Mold Assembly** tab of the **Ribbon**. The **Sprue Bushing** dialog box will be displayed; refer to Figure-110.

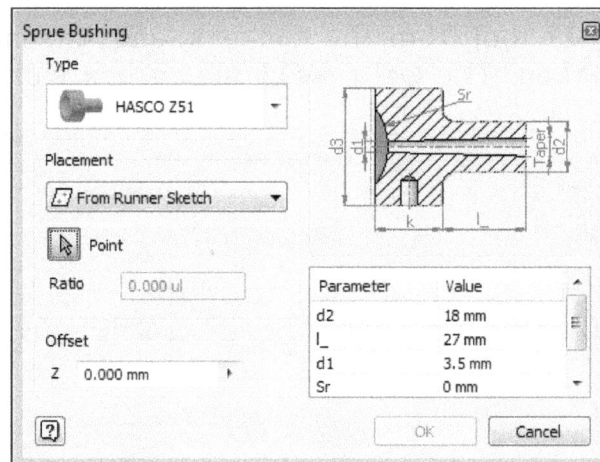

Figure-110. Sprue Bushing dialog box

- Select desired bushing type from the **Type** selection box in the dialog box.
- Select the **From Runner Sketch** option from the **Placement** drop-down to select a point on the runner sketch or select the **Linear** option from the drop-down to select two references to place the sprue. Since, the **From Runner Sketch** option is more widely used by mold makers so we will select this option.
- On selecting the **From Runner Sketch** option, you will be asked to select a point on the runner sketch. Select a point on the runner sketch, preview of the sprue bushing will be displayed; refer to Figure-111.

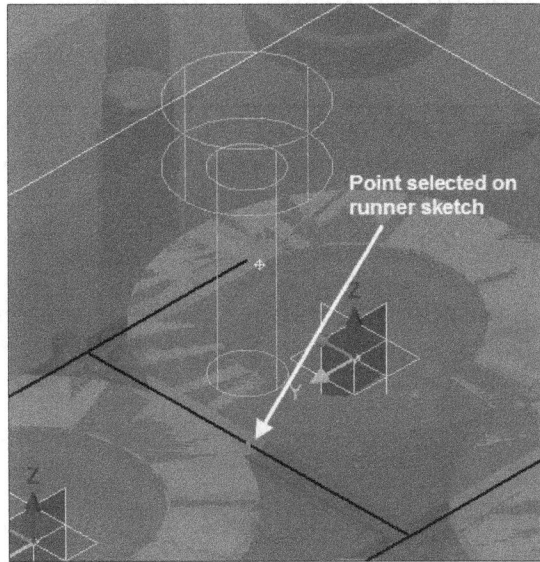

Figure-111. Point selected for sprue bushing

- Set the parameters of sprue bushing as required by using the options in the table.
- Click on the **OK** button to create the sprue bushing.

Now, you can run the mold fill analysis on the assembly of runner, gate, and cavities by using the **Mold Fill Analysis** tool in the **Mold Simulation** panel of **Mold Layout** tab in the **Ribbon**.

PLACING LOCATING RING

Locating Ring is an important part of mold base. This ring helps to align the mold with the plastic injection machine. It is also used to align and fasten the sprue bushing. The procedure to place locating ring in mold base is given next.

- Click on the **Locating Ring** tool from the **Mold Assembly** panel in the **Mold Assembly** tab of the **Ribbon**. The **Locating Ring** dialog box will be displayed and preview of locating ring will be displayed automatically aligned to the sprue bushing; refer to Figure-112.

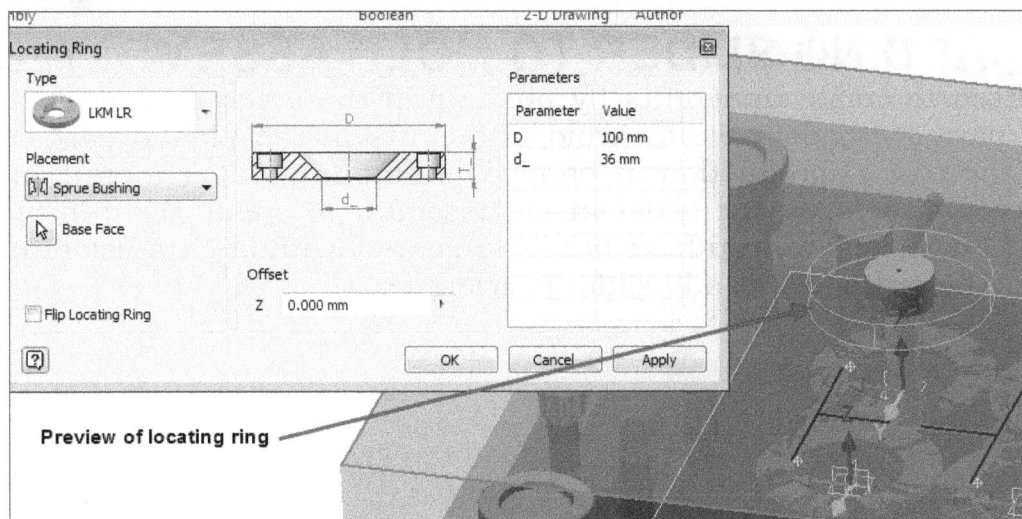

Figure-112. Locating Ring dialog box and preview

- Set the parameters of ring as required and then click on the **OK** button to create the locating ring.

PLACING EJECTOR PIN

Ejector pins are used to eject the plastic part from the cavity steel. The procedure to place ejector pin in mold assembly is given next.

- Click on the **Ejector** tool from the **Mold Assembly** panel in the **Mold Assembly** tab of the **Ribbon**. The **Ejector** dialog box will be displayed with plastic part and runoff surface in drawing area; refer to Figure-113.

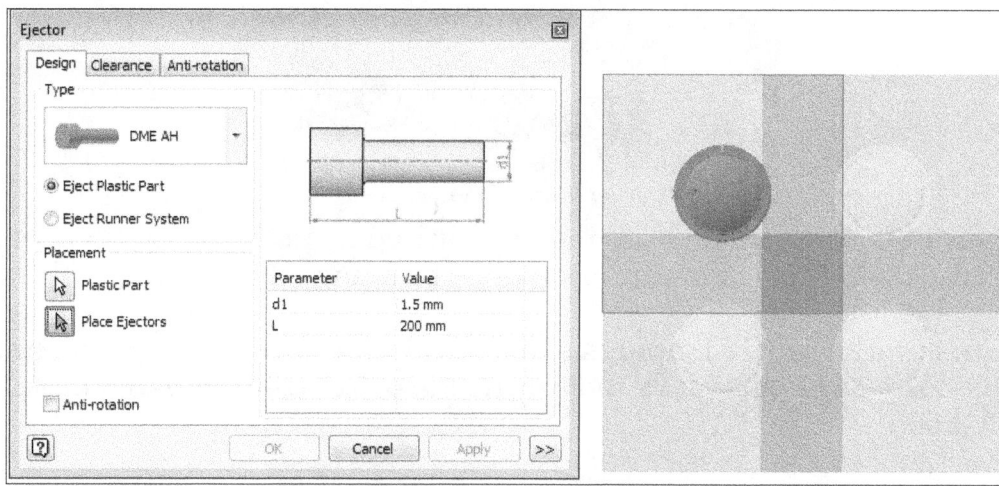

Figure-113. Ejector dialog box with plastic model

- Select desired type of ejector pin from the **Type** selection box.
- Select the **Eject Plastic Part** radio button if you want the ejector pins to be used on plastic part. Select the **Eject Runner System** radio button if you want the ejector to push out the runner.
- Click on the plastic part or runner system at desired locations to create ejector pins at those locations.
- Set the parameters of pins as required.
- Click on the **OK** button from the dialog box to create the ejector pins.

ADDING SLIDER TO MOLD ASSEMBLY

Slider is used to create undercuts by pushing in the insert in the mold. On the designers part, it is advisable to avoid undercuts in the plastic part because it can raise the cost of production in handsome amount. But if there is no escape from undercuts then use the slider insert assembly to create them in mold; refer to Figure-114. Make sure you have created inserts for undercuts before using this procedure. The procedure to add slider is given next.

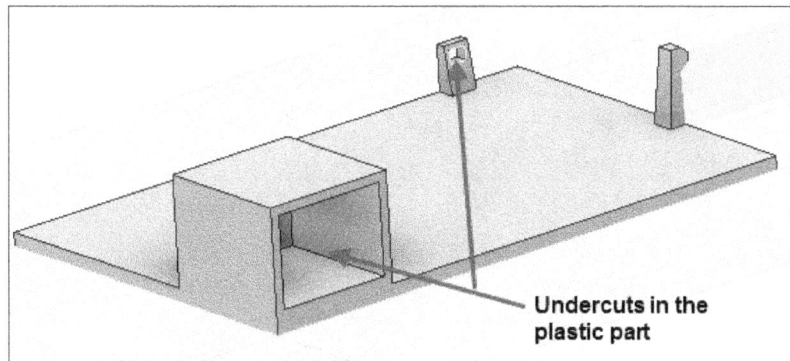

Figure-114. Undercut in plastic part

- Click on the **Slider** tool from the **Mold Assembly** panel in the **Mold Assembly** tab of the **Ribbon**. The **Slider** dialog box will be displayed; refer to Figure-115.

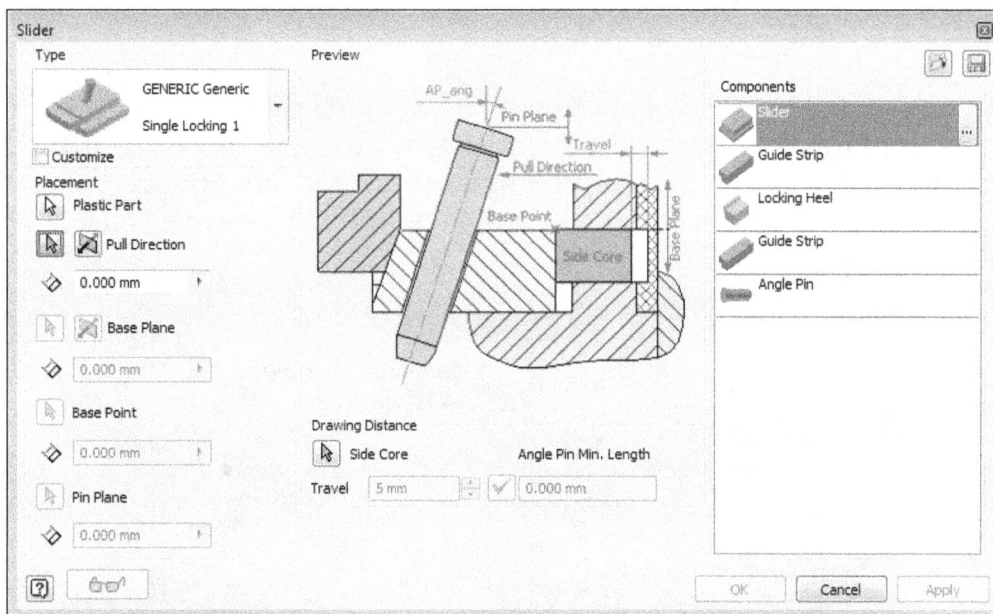

Figure-115. Slider dialog box

- Select desired type of slider from the **Type** selection box in the dialog box. You will be asked to select plane for pull direction.
- Select a face on the insert to specify pull direction. Note that pull direction is the direction opposite to the insert insertion; refer to Figure-116. On selecting the face, you will be asked to select base plane.

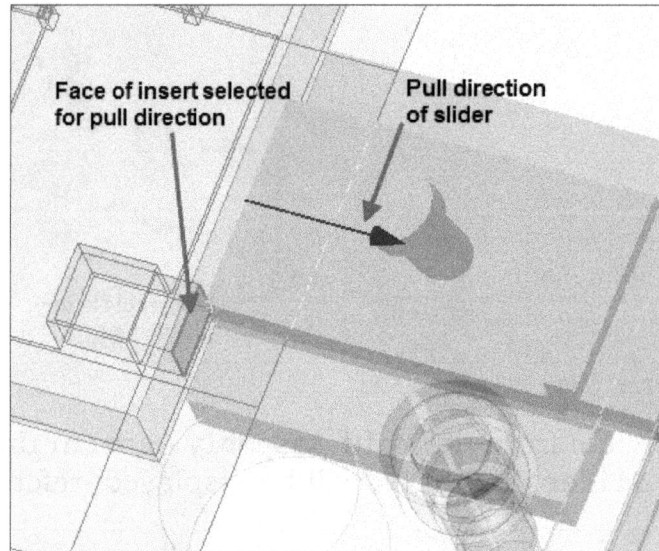

Figure-116. Selecting face for pull direction

- Select the upper face of the insert to specify as base plane; refer to Figure-117. In general, base plane is the plane which aligns with the top face of the slider. On selecting the base plane, you will be asked to select base point.
- Select a point on the insert to specify as base point; refer to Figure-118. You will be asked to specify pin plane.

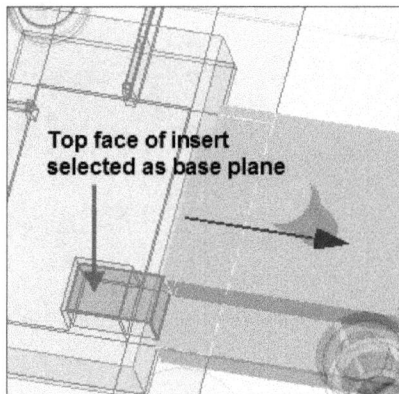

Figure-117. Face selected for base plane

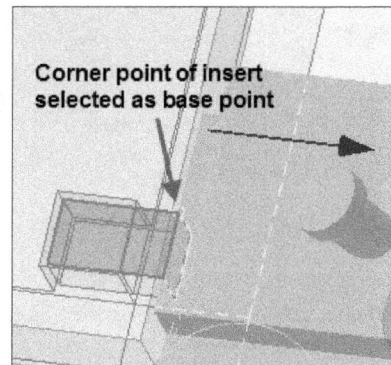

Figure-118. Selecting base point of slider

- Select a plane/face at which you want the pin to be when mold assembly is closed; refer to Figure-119.

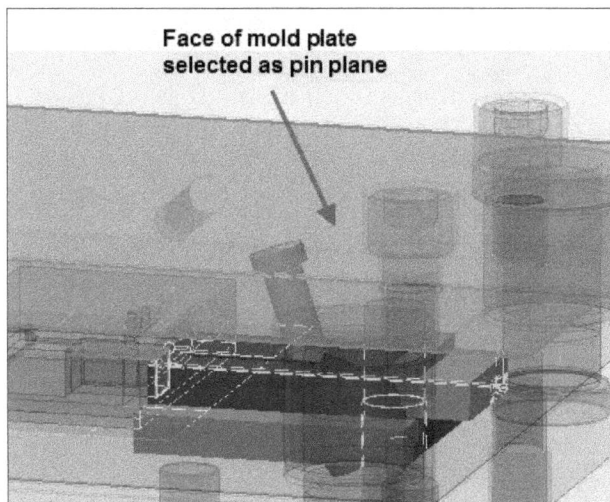

Figure-119. Face selected as pin plane

- Click on the **Side core** selection button from the **Drawing Distance** area of the dialog box. You will be asked to select the insert.
- Select the insert created earlier for slider; refer to Figure-120.

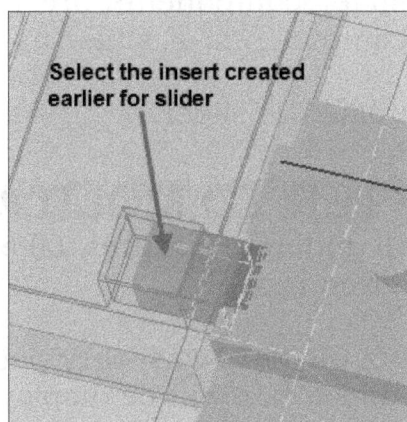

Figure-120. Selecting insert for slider

- Specify the total travel of slider along with insert when the mold is open in the **Travel** edit box in the **Drawing Distance** area of the dialog box. This should be enough travel so that the insert is completely out of core/cavity block; refer to Figure-121.

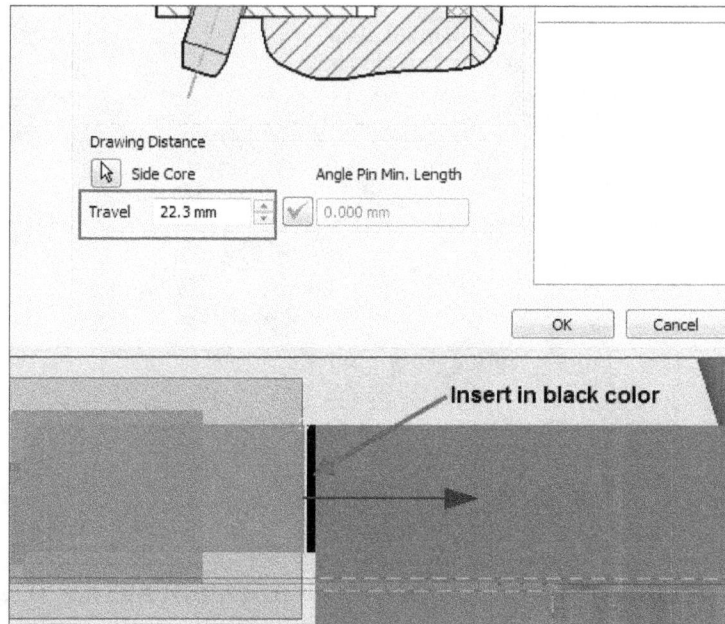

Figure-121. Travel of slider specified

- Now, you may find that slider components are bigger or smaller than what is required. So, edit the components of slider as we did for cooling channel components.
- Click on the **OK** button to create the slider.

ADDING LIFTER TO MOLD ASSEMBLY

Lifter is used to create the small undercut features. Like snap fits on the plastic part. The procedure to add lifter is given next.

- Click on the **Lifter** tool from the **Mold Assembly** panel in the **Mold Assembly** tab of the **Ribbon**. The **Lifter** dialog box will be displayed as shown in Figure-122. Also, you will be asked to select a face for pull direction.

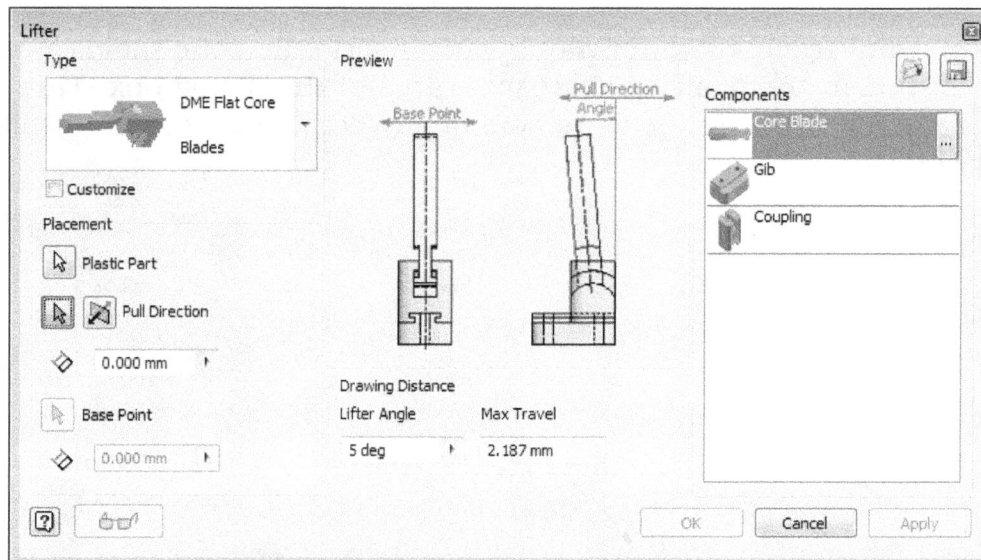

Figure-122. Lifter dialog box

- Select the flat face of insert created for lifter to specify pull direction; refer to Figure-123. You will be asked to select the base point.

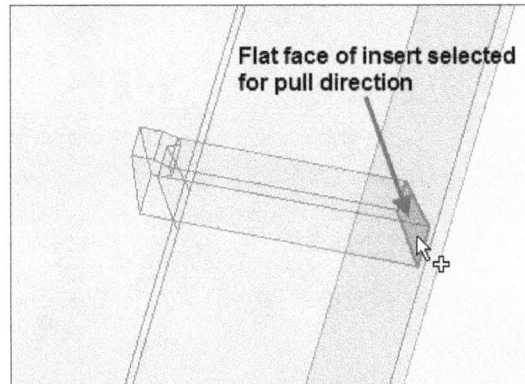

Figure-123. Face selected for pull direction

- Click at a point on lifter insert. Note that the selected point will align to the mid plane of the lifter assembly.
- Set the pull direction offset distance and base point offset distance in their related edit boxes.
- Specify the lifter angle in the **Lifter Angle** edit box and click on the **OK** button to create the lifter assembly. The lifter assembly will be added to the mold assembly; refer to Figure-124.

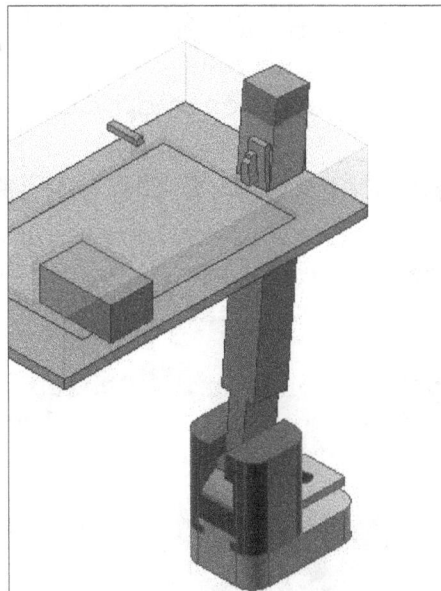

Figure-124. Lifter assembly created

COMBINING CORES AND CAVITIES

The **Combine Cores and Cavities** tool is used to combine the selected cores and cavities in one set of core and cavity plates. The procedure to use this tool is given next.

- Click on the **Combine Cores and Cavities** tool from the **Boolean** panel in the **Mold Assembly** tab of the **Ribbon**. The **Combine Cores/Cavities** dialog box will be displayed along with the cores and cavities in the assembly; refer to Figure-125.

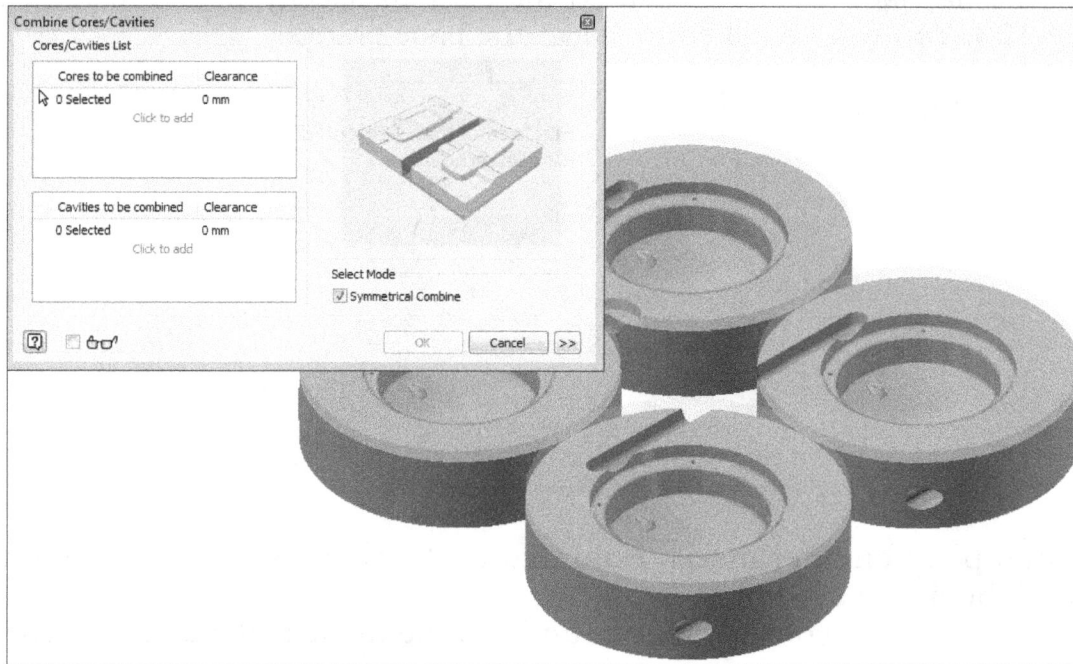

Figure-125. Combine Cores/Cavities dialog box

- One by one select the models displayed in the drawing area.
- Select the **Symmetrical Combine** check box if you want to combine the core-cavities in symmetry.
- Click on the more button to expand the dialog box and set desired corner conditions.
- Click on the **OK** button to combine cores and cavities.

CREATING POCKET IN WORKPIECE

In Autodesk Inventor, pockets for core and cavity are not created automatically in the mold plates. So, this work need to be performed manually. The procedure to create the pockets in mold plates is discussed next.

- Click on the **Workpiece Pocket** tool from the **Boolean** panel in the **Mold Assembly** tab of the **Ribbon**. The **Workpiece Pocket** dialog box will be displayed; refer to Figure-126.

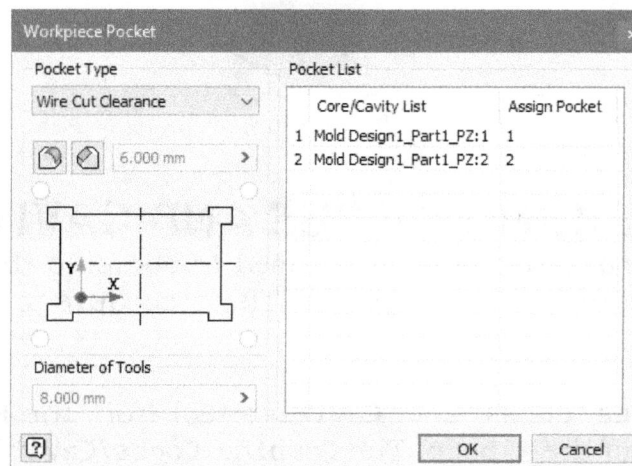

Figure-126. Workpiece Pocket dialog box

- Select desired pocket type and click on the **OK** button. The pockets will be created in the workpiece accordingly.

MOLD BOOLEAN

There are many components in the mold assembly which are not part of mold base but need to be added or subtracted from the mold base plates. Like the cooling channel and runner are not subtracted from mold base plates automatically, inserts are not added to the slider or lifter automatically. To perform such operations, we use **Mold Boolean** tool in Autodesk Inventor. The procedure to use this tool is given next.

- Click on the **Mold Boolean** tool from the **Boolean** panel in the **Mold Assembly** tab of the **Ribbon**. The **Mold Boolean** dialog box will be displayed; refer to Figure-127.

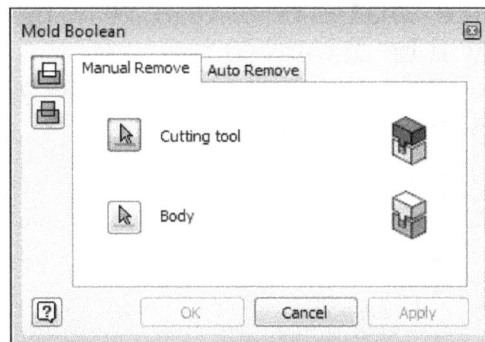

Figure-127. Mold Boolean dialog box

Removing Material

- Make sure the **Remove** button ▣ is selected at the left in the dialog box. Select the cutting tool like cooling channel tube, runner, etc. from the mold assembly. You will be asked to select the body from which material is to be cut.
- Select the plate of mold from which you want to cut the tube, runner, etc.
- Click on the **OK** button to cut the material. Click on the **Apply** button if you want to perform more material removal operations.

You can find the interfering parts and speed up the process of removing material from the mold plates by using the **Auto Remove** tab in the dialog box. The method is discussed next.

- Click on the **Auto Remove** tab in the dialog box. The **Mold Boolean** dialog box will be displayed as shown in Figure-128.

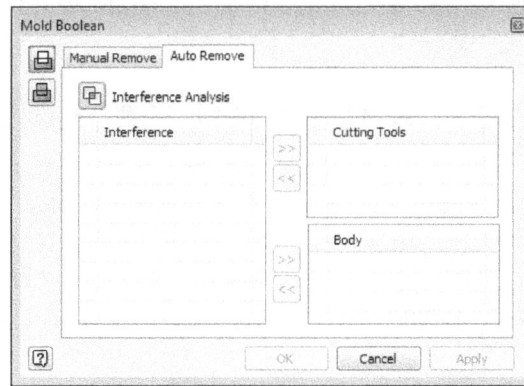

Figure-128. Auto Remove tab in Mold Boolean dialog box

- Click on the **Interference Analysis** button ⊞ from the dialog box. The list of interfering components will be displayed.
- Select the component from the list and click on the **>>** button of **Cutting Tools** or **Body** list as required; refer to Figure-129.

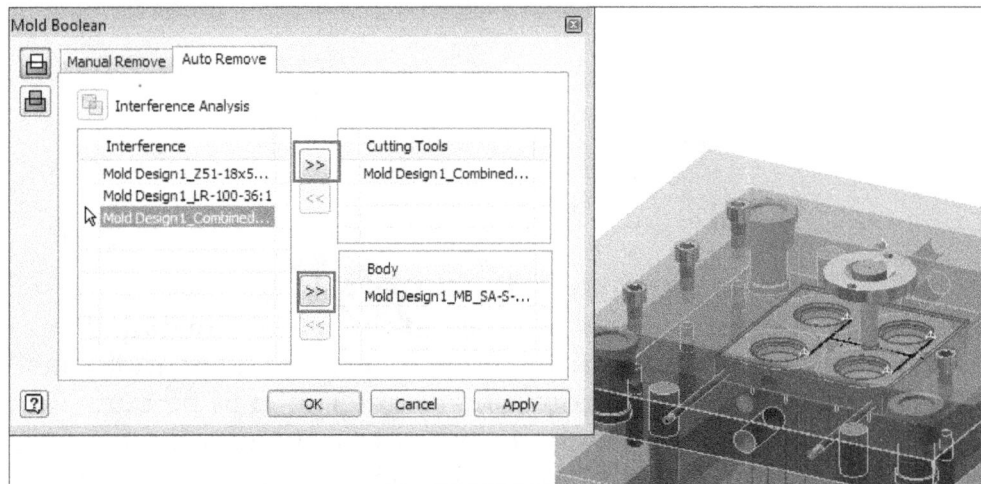

Figure-129. Categorizing interfering components for material removal

- Click on the **OK** or **Apply** button to perform the material removal operation; refer to Figure-130.

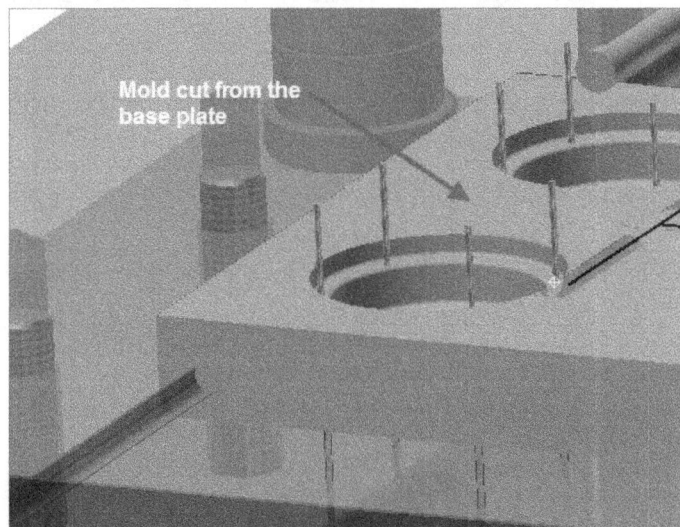

Figure-130. Mold cut from the base plate

Adding Two Bodies

- Click on the **Add** button ⊞ from the left area in the dialog box. The dialog box will be displayed as shown in Figure-131.

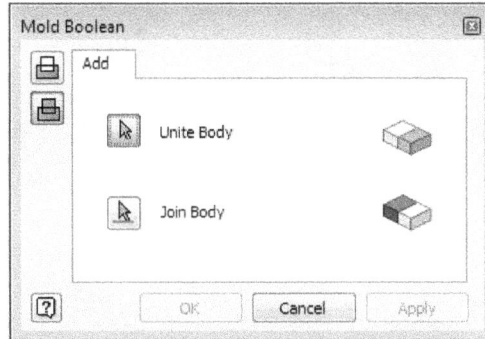

Figure-131. Add tab in Mold Boolean dialog box

- Select the first body (insert) to be united and then the second body (slider or lifter) to be united.
- Click on the **OK** button to unite the bodies.

CHANGING REPRESENTATION OF MOLD

Autodesk Inventor has options to change the representation of mold as open or closed so that you can view the inner area of the mold. The procedure to change the representation is discussed next.

- Click on the **Show tabs** option in the **Browse Bar**. A drop-down will be displayed with options as shown in Figure-132.

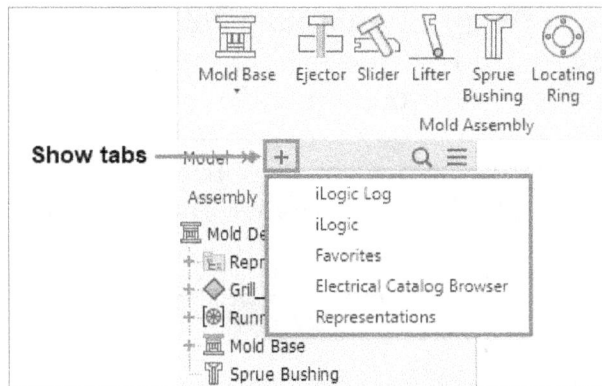

Figure-132. Options on clicking Show tabs

- Click on the **Representations** option from the drop-down. The **Representations Browse Bar** will be displayed; refer to Figure-133.

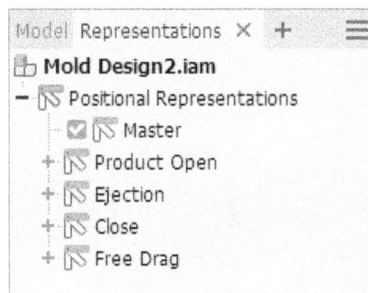

Figure-133. Representations Browse Bar

- Expand the **Positional Representations** node and double-click on the representation that you want to check; refer to Figure-134 which shows open position of mold.

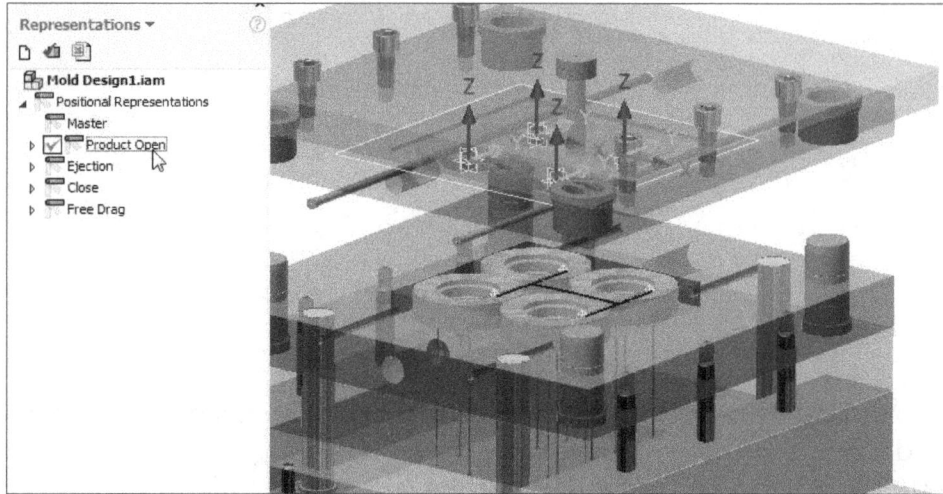
Figure-134. Product Open position of mold

- Remember that if you want to perform any change in the mold assembly then you need to double-click on the **Master** positional representation to return back to original form.
- To return back to **Mold Design Browse Bar**, click on the **Model** tab in the **Browse Bar**.

CREATING 2D DRAWINGS OF MOLD

Creation of 2D drawings is very crucial step for all the manufacturing work. In Autodesk Inventor, there is a direct tool to create 2D drawings for mold assembly. The tool and procedure to create the 2D drawings is given next.

- Click on the **2-D Drawing** tool from the **2-D Drawing** panel in the **Mold Assembly** tab of the **Ribbon**. The **2-D Drawing** dialog box will be displayed as shown in Figure-135.

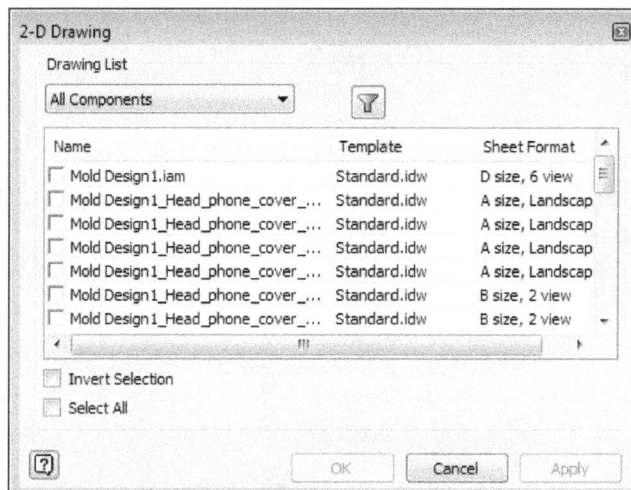
Figure-135. 2-D Drawing dialog box

- Select the check boxes for the component to generate drawings for them.
- Click on the **OK** button from the dialog box to create the selected drawings. The drawing environment will become active and drawings recently created will be displayed.

You will learn about drawing views and their management later in the book.

SELF ASSESSMENT

Q1. Which of the following statements is correct about designing wall thickness in mold design?

a) The critical thickness increases with lowering temperature and molecular weight.
b) The critical thickness increases with increasing temperature and molecular weight.
c) The critical thickness reduces with lowering temperature and molecular weight.
d) The critical thickness reduces with increasing temperature and molecular weight.

Q2. Which of the following issues should be considered when properly designing the ribs?

a) Mold-ability
b) Thickness
c) Location
d) All of the Above

Q3. Which of the following resins is not used in designing undercuts in mold design?

a) Polycarbonate
b) Unfilled Polyamide 6
c) Thermoplastic polyurethane elastomer
d) Polymethylene

Q4. Which of the following options shows the correct workflow of mold design assembly?

a) Insert Plastic Part --> Specifying Material --> Generating Core-Cavity --> Generating 2D drawings and CAM data

b) Insert Plastic Part --> Generating Core Cavity --> Specifying Material --> Generating 2D drawings and CAM data

c) Generating Core-Cavity --> Specifying Material --> Insert Plastic Part --> Generating 2D drawings and CAM data

d) Specifying Material --> Insert Plastic Part --> Generating Core-Cavity --> Generating 2D drawings and CAM data

Q5. In which of the following Gate Designs in preparing core and cavity, the Ejector Pins are necessary for automatic trimming of the gate?

a) Edge Gate
b) Sub Gate
c) Hot Tip Gate
d) Direct or Sprue Gate

Q6. Which of the following Gates is used in two plate mold construction for creating gate in core and cavity?

a) Submarine Gate
b) Fan Gate
c) Edge Gate
d) Pin Gate

Q7. Which of the following issues can cause problems with Pressure and Temperature parameters in creating mold?

a) Excessive Injection Fill Speed
b) Low Injection Pressure
c) Low Mold Temperature
d) All of the Above

Q8. Which of the following cross-sectional shape of the runner is not used in a mould when creating core and cavity?

a) Fully round
b) U-Shaped
c) Hexagonal
d) Pentagonal

Q9. The **Define Workpiece Setting** tool is used to create a block of steel from which the core and cavity is to be join. (True/False)

Q10. The higher the hydraulic diameter of runner used in mold, the lower the flow resistance. (True/False)

Chapter 13

Surface Design and Freeform Creation

Topics Covered

The major topics covered in this chapter are:

- *Introduction to Surface Design*
- *Surfacing Tools*
- *Freeform Designing Tools*
- *Practical and Practice on Surface Design*

INTRODUCTION TO SURFACE DESIGN

Surface design is a very important aspect of CAD software. Surface designing gives the power to design any shape of object which is possible in real-world. There can be many modeling problems which can not be solved by solid modeling. In such situations, the surface modeling comes for rescue. Check the model shown in Figure-1 and ask yourself how much effort will it take to create in solid modeling.

In this chapter, we will learn about the tools related to surface design and Freeform modeling. **The tools for surface design and freeform modeling are available in the Part environment so you need to start a new part file now.**

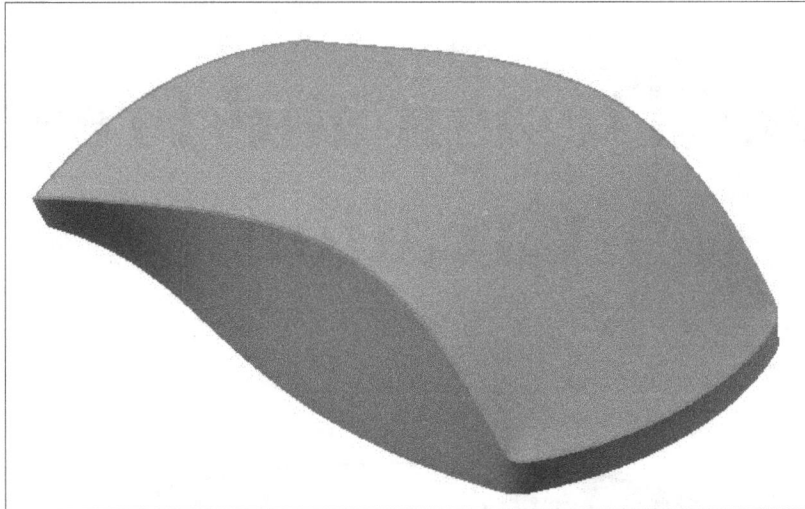

Figure-1. Model of a mouse

Some of the tools discussed in solid modeling can also be used for surface designing. So, we will start discussing these tools and then we will discuss the tools in **Surface** panel.

EXTRUDED SURFACE

An extruded surface can be created by using the same **Extrude** tool that we have discussed for solid modeling. The procedure to do so is given next.

- Click on the **Extrude** tool from the **Create** panel in the **3D Model** tab of the **Ribbon**. If there is no sketch in the drawing area then you will be asked to select a sketching plane to create the sketch.
- Select the sketching plane and create an open or close loop sketch.
- Click on the **Finish Sketch** button from the **Exit** panel in the **Sketch** tab of the **Ribbon**. The **Extrude** dialog box will be displayed along with the sketch extruded as solid model; refer to Figure-2.
- Click on the button in the upper right side of the dialog box to toggle for surface mode; refer to Figure-3.
- Set the other parameters as required and then click on the **OK** button.

In the same way, you can use the **Revolve**, **Sweep**, **Loft**, and **Coil** tools to create surfaces.

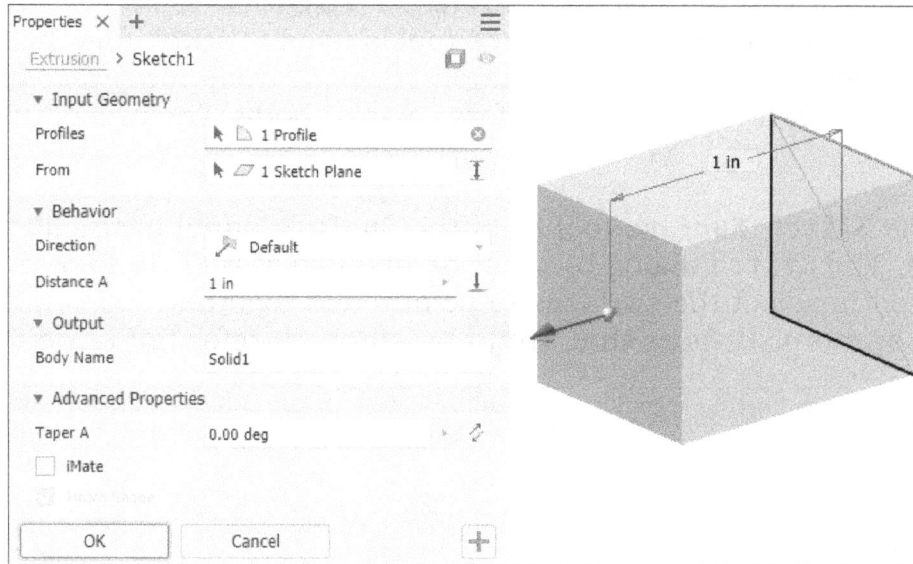

Figure-2. Extrude dialog box in solid mode

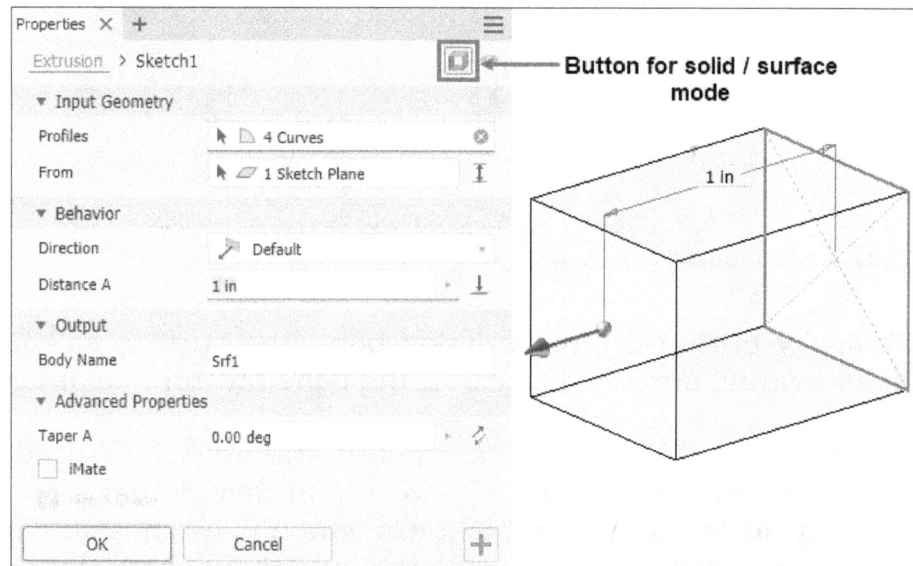

Figure-3. Extrude dialog box in surface mode

SURFACING TOOLS

The tools to perform surfacing operations are available in the **Surface** panel in the **3D Model** tab of the **Ribbon**; refer to Figure-4.

Figure-4. Surface panel

The applications of these tools are discussed next.

Stitch Surface Tool

As the name suggests, the **Stitch Surface** tool is used to stitch together (or say join) the surfaces, which are meeting at common edges. If the joining faces form a close boundary then you can use this tool to form a solid body. The procedure to use this tool is given next.

- Click on the **Stitch Surface** tool from the **Surface** panel in the **3D Model** tab of the **Ribbon**. The **Stitch** dialog box will be displayed; refer to Figure-5. Also, you will be asked to select the surfaces to be stitched together.
- Select two or more surfaces that are to be stitched; refer to Figure-6.

Figure-5. Stitch dialog box

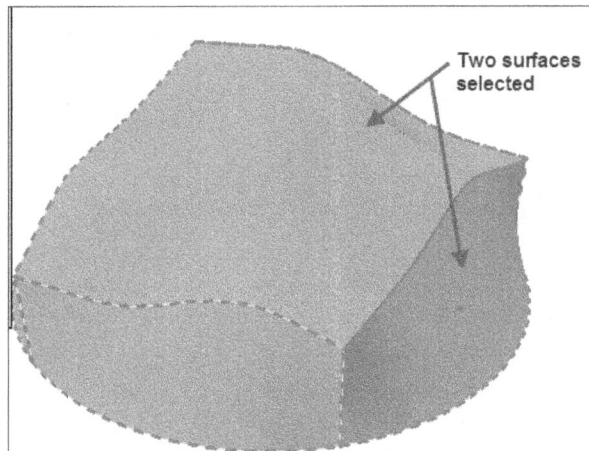

Figure-6. Selecting surfaces for stitching

- Click on the **Apply** button from the dialog box. One surface will be created by stitching the individual surfaces. Click on the **Done** button to exit the dialog box.

A question arises here, Why should we care about stitching the surface? Answer of this question lies with **Thicken** tool. In real-world, surface do not exist as an entity because theoretically surfaces have 0 thickness which is not possible in real world. So, we apply some thickness to the surfaces to make them solid. If we do not stitch the surfaces before applying **Thicken** tool then there will be gap between individual surfaces which makes the manufacturing impossible.

Boundary Patch Tool

The **Boundary Patch** is used to create planar or 3D surface patch based on the selected edges and guidelines. The procedure to use this tool is given next.

- Click on the **Boundary Patch** tool from the **Surface** panel in the **3D Model** tab of the **Ribbon**. The **Boundary Patch** dialog box will be displayed; refer to Figure-7. Also, you will be asked to select an edge or sketch curve.

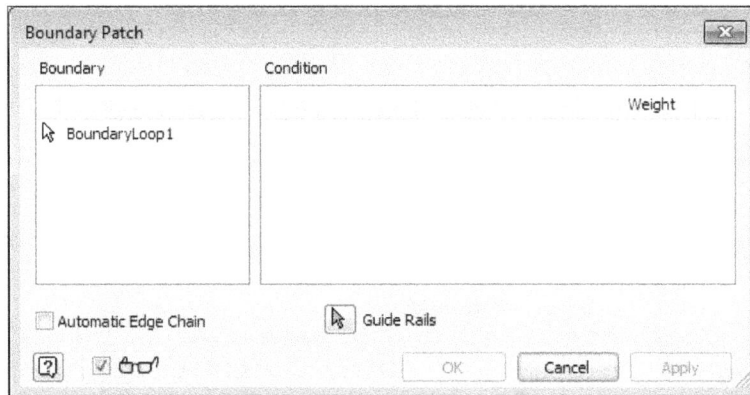

Figure-7. Boundary Patch dialog box

- Select the edges to form a closed loop to create patch. Preview of the boundary patch will be displayed; refer to Figure-8.

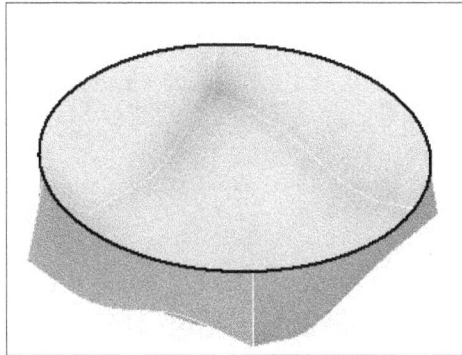

Figure-8. Preview of boundary patch

- Select the **Automatic Edge Chain** check box if you want to select the connecting edges along with the selected edge to form a closed loop.
- Click on the **Guide Rails** selection button to select the guide rail for boundary patch; refer to Figure-9.

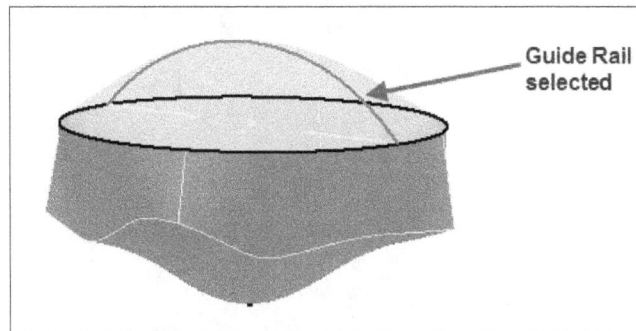

Figure-9. Boundary Patch surface with guide rail

- Click on the **OK** button to create the patch.

Sculpt Tool

The **Sculpt** tool is used to create a solid region bounded by selected surfaces. The procedure to use this tool is given next.

- Click on the **Sculpt** tool from the **Surface** panel in the **3D Model** tab of the **Ribbon**. The **Sculpt** dialog box will be displayed; refer to Figure-10.

Figure-10. Sculpt dialog box

- Select the surfaces to create sculpt feature. The close region bounded by surfaces will be displayed as preview of sculpt feature; refer to Figure-11.

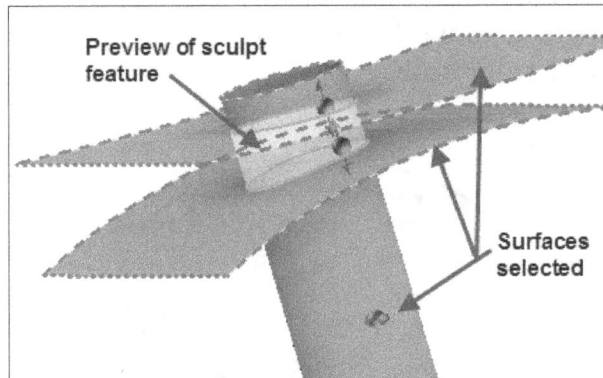

Figure-11. Preview of sculpt feature

- If there is any solid feature passing through the sculpt feature then you can add or subtract the sculpt feature from the solid feature by using the **Add** or **Remove** button from the dialog box; refer to Figure-12.

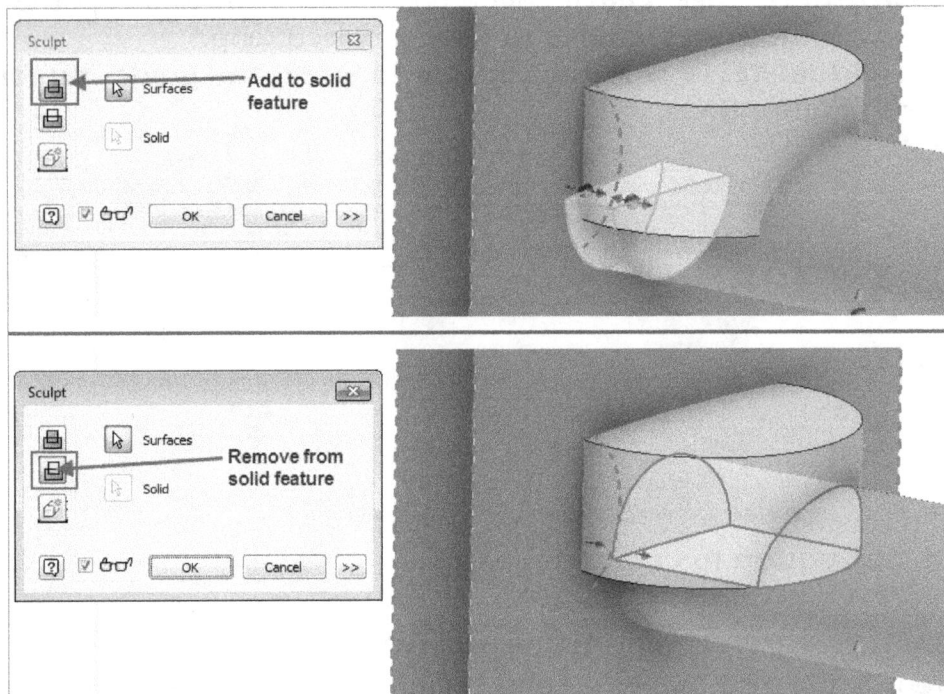

Figure-12. Add or remove sculpt feature

- Click on the **OK** button from the dialog box to create the feature.

Ruled Surface Tool

The **Ruled Surface** tool is used to create surface(s) normal or tangent to the faces of selected edges. The procedure to use this tool is given next.

- Click on the **Ruled Surface** tool from the **Surface** panel in the **3D Model** tab of the **Ribbon**. The **Ruled Surface** dialog box will be displayed as shown in Figure-13.

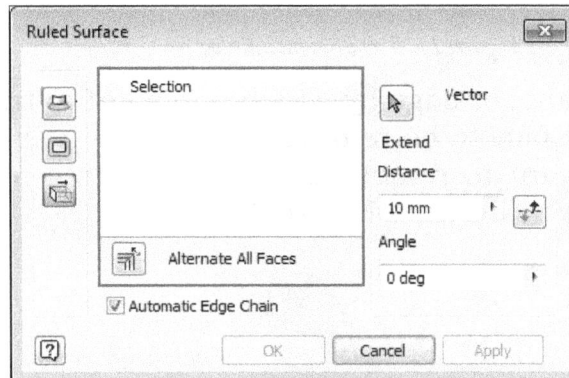

Figure-13. Ruled Surface dialog box

- Select the edge(s) for which you want to create the ruled surface. Preview of the ruled surface will be displayed; refer to Figure-14.

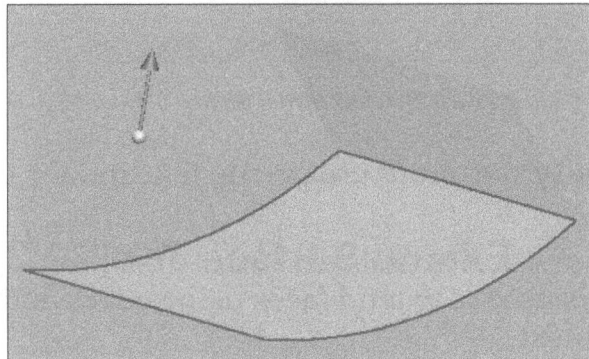

Figure-14. Preview of ruled surface

- Set desired angle in the **Angle** edit box of the dialog box.
- Note that by default, the **Normal** button is selected in the right side of the dialog box and hence the surfaces are created perpendicular to the selected edges. Select the **Tangent** button if you want to create the tangent surfaces. Select the **Vector** button and select desired direction reference to create surfaces in that direction.
- Click on the **OK** button to create the surface.

Trim Surface Tool

As the name suggests, the **Trim Surface** tool is used to trim the surfaces using the other surface, work plane, or sketch. The procedure to use this tool is given next.

- Click on the **Trim Surface** tool from the **Surface** panel in the **3D Model** tab of the **Ribbon**. The **Trim Surface** dialog box will be displayed; refer to Figure-15.

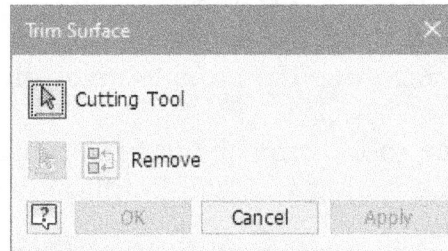

Figure-15. Trim Surface dialog box

- Select the surface, plane, or edge by which you want to trim the surface. You will be asked to select the face to be removed.
- Select the face you want to remove using the cutting tool. Preview of the trim feature will be displayed; refer to Figure-16.

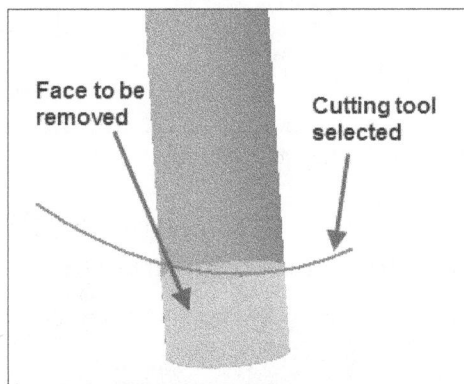

Figure-16. Preview of trim feature

- Click on the **OK** or **Apply** button to create the feature.

Extend Surface Tool

The **Extend Surface** tool is used to extend faces using edges of the faces. The procedure to use this tool is given next.

- Click on the **Extend Surface** tool from the **Surface** panel in the **3D Model** tab of the **Ribbon**. The **Extend Surface** dialog box will be displayed; refer to Figure-17.

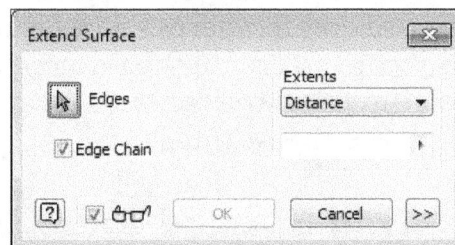

Figure-17. Extend Surface dialog box

- Select the edge(s) that you want to use for extending the corresponding face. Preview of the extended surface will be displayed; refer to Figure-18.

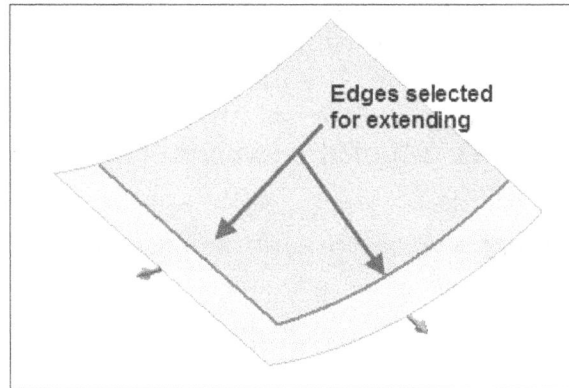

Figure-18. Preview of extend surface

- Set desired distance value in the **Extents** edit box or select the **To** option from the drop-down and select the reference for extension.
- Click on the **OK** button to create the extension surface.

Replace Face Tool

The **Replace Face** tool is used to replace the face of a solid by selecting surface/face. The procedure to use this tool is given next.

- Click on the **Replace Face** tool from the **Surface** panel in the **3D Model** tab of the **Ribbon**. The **Replace Face** dialog box will be displayed; refer to Figure-19.

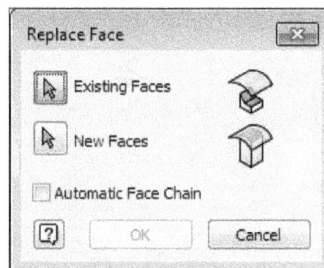

Figure-19. Replace Face dialog box

- Select the face(s) of solid which you want to replaced.
- Click on the **New Faces** selection button from the dialog box and select the face/surface by which you want the face(s) to be replaced; refer to Figure-20.

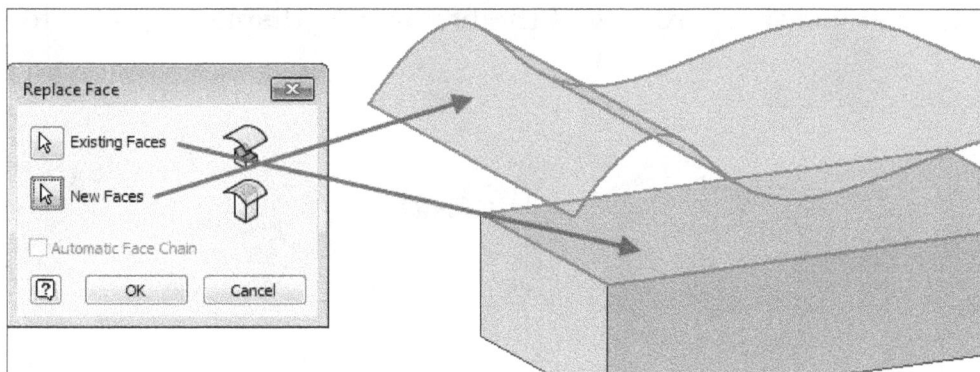

Figure-20. Faces selected for replacement

- Click on the **OK** button from the dialog box to create the feature.

FREEFORM DESIGNING

Freeform designing has many applications in CAD and CAE. Whenever you need an unconventional shape of the object, you can always use the freeform modeling techniques. The tools to perform freeform modeling are available in the **Create Freeform** panel in the **3D Model** tab of the **Ribbon**; refer to Figure-21. Various tools in this panel are discussed next.

Figure-21. Create Freeform panel

BOX TOOL

The **Box** tool is used to create box shapes by free forming. The procedure to use this tool is given next.

*   Click on the **Box** tool from the drop-down in the **Create Freeform** panel of the **Ribbon**. The **Box** dialog box will be displayed as shown in Figure-22. Also, you will be asked to select a face/plane to place the freeform box.

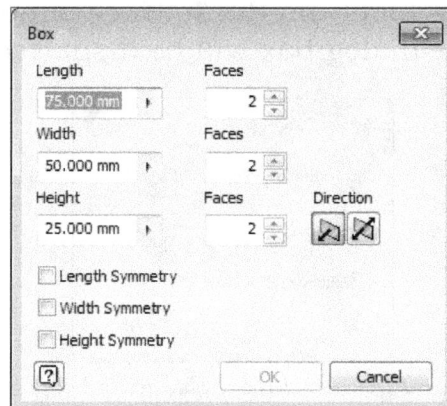
Figure-22. Box dialog box

*   Select the face/plane to place the box. You will be asked to specify the center point of the box.
*   Click at desired location. Preview of the box will be displayed; refer to Figure-23.

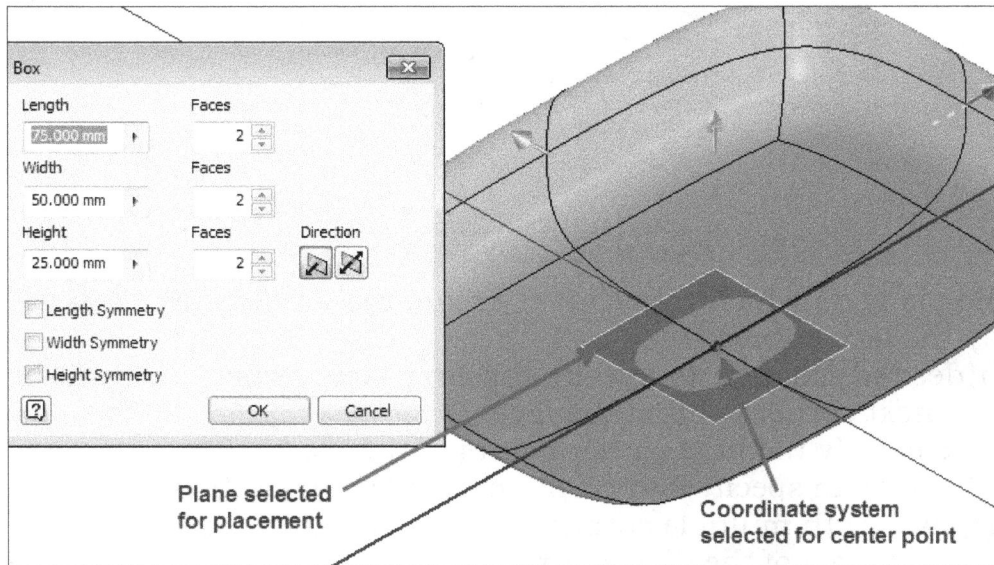

Figure-23. Preview of the freeform box

- Specify the length, width, and height of the box in the respective edit boxes of the dialog box or drag the arrows in preview.
- Select the **Length Symmetry**, **Width Symmetry**, and **Height Symmetry** check boxes to make the box symmetric in length, width, and height with respect to corresponding center planes.
- Set the number of faces on each side of the box by using the **Faces** spinners.
- Click on the **OK** button to activate the **Freeform** contextual tab in the **Ribbon**; refer to Figure-24.

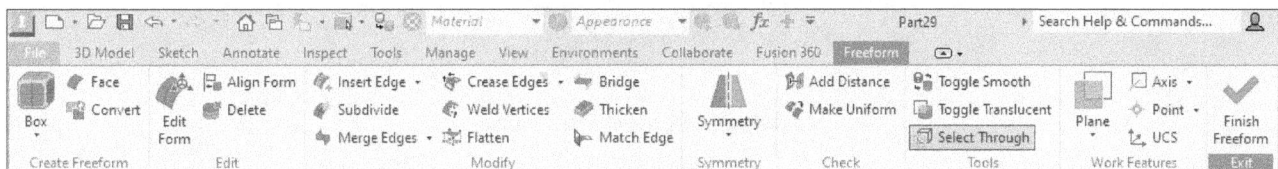

Figure-24. Freeform contextual tab

- The tools in this tab are used to modify the shape of box and other freeform objects. We will discuss the tools in this tab later in this chapter.
- After modifications, click on the **Finish Freeform** button in the **Exit** panel of the **Freeform** contextual tab in the **Ribbon**.

In the same way, you can use the other freeform creation tools like **Plane**, **Cylinder**, **Sphere**, **Torus**, and **Quadball** available in the drop-down in the **Create Freeform** panel of the **Ribbon**.

FACE TOOL

The **Face** tool is used to create individual face based on selected vertices. The procedure to use this tool is given next.

- Click on the **Face** tool from the **Create Freeform** panel in the **3D Model** tab of the **Ribbon**. The **Face** dialog box will be displayed and you will be asked to select a plane or face for placement of freeform face; refer to Figure-25.

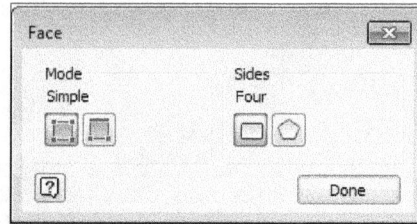
Figure-25. Face dialog box

- Select the plane on which you want to create the face feature. You will be asked to specify the face corner.
- Click on desired location to specify the first face corner. You will be asked to specify the next corner point of the face.
- One by one specify the next corner points of the face.
- By default, you can specify four corner points of the face but if you want to create a freeform face with multiple corner points then click on the **Multiple** button from the **Sides** area of the dialog box.
- After specifying desired points for multiple corner face, press **ENTER** from the keyboard. The freeform face will be created; refer to Figure-26. Perform the freeform operations and click on the **Finish Freeform** button from the **Exit** panel in the contextual tab of the **Ribbon**.

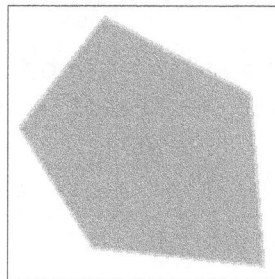
Figure-26. Freeform face created

CONVERT TOOL

The **Convert** tool is used to convert the selected solid face or surface to freeform face. The procedure to use this tool is given next.

- Click on the **Convert** tool from the **Create Freeform** panel in the **3D Model** tab of the **Ribbon**. The **Convert to Freeform** dialog box will be displayed; refer to Figure-27.

Figure-27. Convert to Freeform dialog box

- Select the face that you want to be converted to freeform face. Preview of the freeform face will be displayed; refer to Figure-28.

Figure-28. Preview of freeform face

- Set the number of faces across length and width in the respective edit boxes in the dialog box.
- Click on the **OK** button. The **Freeform** contextual tab will be displayed.
- Perform the modifications on face and click on the **Finish Freeform** tool from the **Exit** panel.

FREEFORM CONTEXTUAL TAB

The tools in the **Freeform** contextual tab are used to modify the shape of selected freeform object. Various tools in this tab are discussed next.

Edit Form Tool

The **Edit Form** tool is used to edit the shape of freeform object. The procedure to use this tool is given next.

- Click on the **Edit Form** tool from the **Edit** panel in the **Freeform** contextual tab of the **Ribbon**. The **Edit Form** dialog box will be displayed; refer to Figure-29.

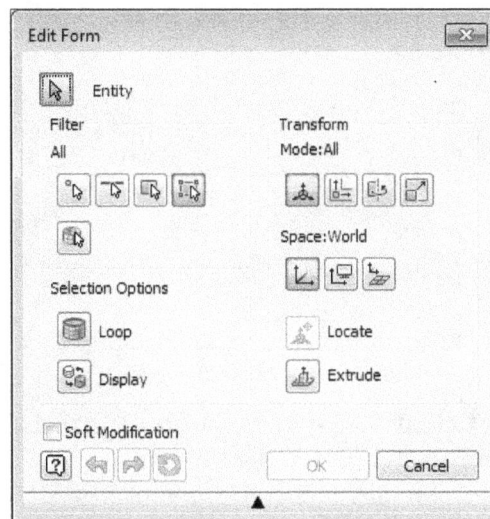

Figure-29. Edit Form dialog box

- Make sure **All** buttons are selected in the **Filter** and **Transform** area of the dialog box so that you can edit every type of geometry of freeform object.
- Select desired coordinate system for modification in the **Space** area.
- Click on the edge/vertex/face of the object to modify it. The transformation handles will be displayed on the selected entity; refer to Figure-30.

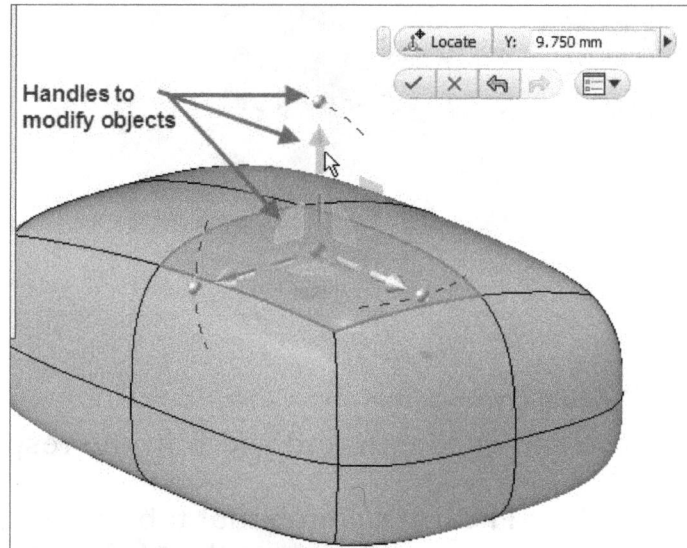

Figure-30. Handles to modify freeform object

- If you do not want the other entities of the freeform to be affected by modification then select the **Extrude** button from the dialog box and then perform the modifications.
- You can use the **Undo**, **Redo**, and **Reset** buttons in the dialog box to undo, redo, and reset the model, respectively.
- Click on the **OK** button from the dialog box, once you have performed the modifications.

Align Form Tool

The **Align Form** tool is used to align the vertex of freeform object to the selected plane. The procedure to use this tool is given next.

- Click on the **Align Form** tool from the **Edit** panel in the **Freeform** contextual tab of the **Ribbon**. The **Align Form** dialog box will be displayed; refer to Figure-31.

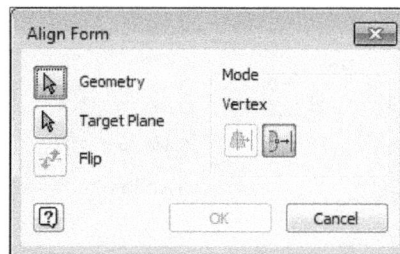

Figure-31. Align Form dialog box

- Select the vertex of object that you want to be aligned. You will be asked to select the plane to which you want to align the vertex.
- Select desired plane or planar face. The object will be aligned at the selected plane/ face through the vertex; refer to Figure-32.

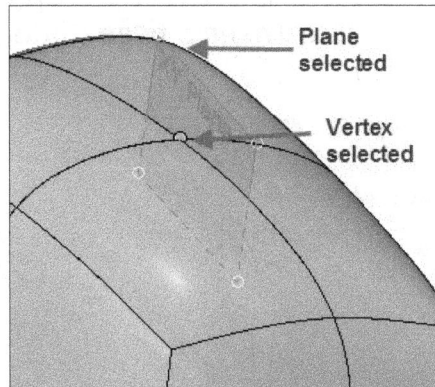

Figure-32. Aligning object

- Click on the **OK** button from the dialog box.

Delete Tool

The **Delete** tool is used to delete the selected entity of the freeform object. The procedure to use this tool is given next.

- Click on the **Delete** tool from the **Edit** panel in the **Freeform** contextual tab of the **Ribbon**. The **Delete** dialog box will be displayed as shown in Figure-33.

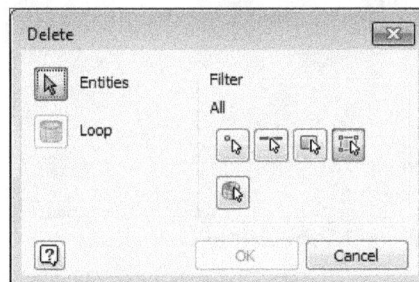

Figure-33. Delete dialog box

- Select the entity that you want to delete and click on the **OK** button. The model after deleting the selected entity will be displayed; refer to Figure-34.

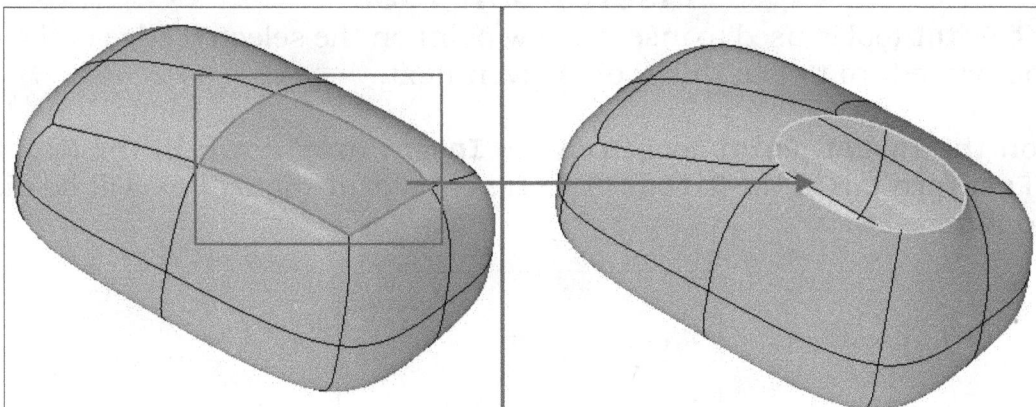

Figure-34. Deleting face of freeform object

Insert Edge Tool

The **Insert Edge** tool is used to insert new edge(s) to the sides of selected edge. The procedure to use this tool is given next.

- Click on the **Insert Edge** tool from the **Insert** drop-down in the **Modify** panel of the **Ribbon**; refer to Figure-35. The **Insert Edge** dialog box will be displayed; refer to Figure-36.

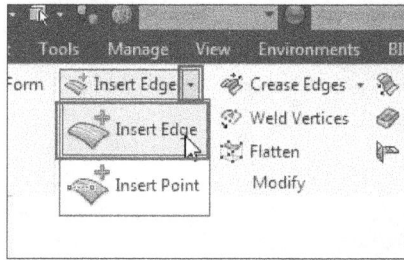

Figure-35. Insert Edge tool

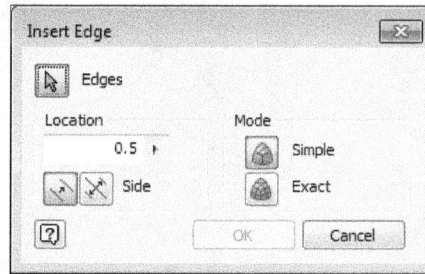

Figure-36. Insert Edge dialog box

- Select the reference edge on sides of which you want to create the new edges. Preview of the edges will be displayed; refer to Figure-37.

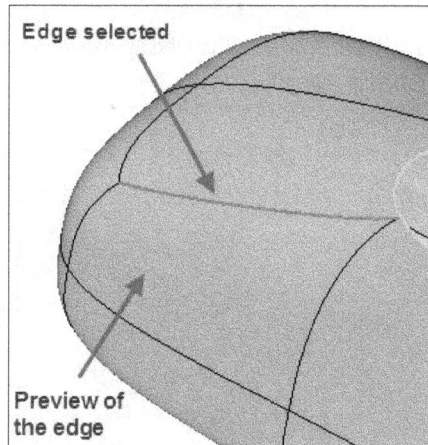

Figure-37. Preview of inserted edge

- Select the **Both** button to create edge on both the sides of the selected edge.
- Select the location ratio in the **Location** edit box and click on the **OK** button to create the edges.

Insert Point Tool

The **Insert Point** tool is used to insert a new point on the selected edge of the freeform object. The procedure to use this tool is given next.

- Click on the **Insert Point** tool from the **Insert** drop-down in the **Modify** panel of the **Freeform** tab in the **Ribbon**. The **Insert Point** dialog box will be displayed; refer to Figure-38.

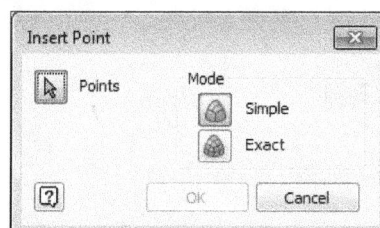

Figure-38. Insert Point dialog box

- Select points on edges of the freeform object and click on the **OK** button to create the points.

Subdivide Tool

The **Subdivide** tool is used to divide the selected face into sub-divisions. The procedure to use this tool is given next.

- Click on the **Subdivide** tool from the **Modify** panel in the **Freeform** contextual tab of the **Ribbon**. The **Subdivide** dialog box will be displayed; refer to Figure-39.

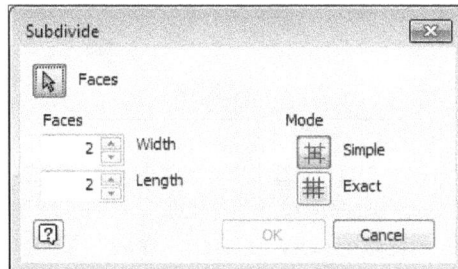

Figure-39. Subdivide dialog box

- Set the number of faces along width and length in the respective spinners in the dialog box.
- Select desired mode of division and click on the face of freeform object. Preview of the sub-division will be displayed; refer to Figure-40.

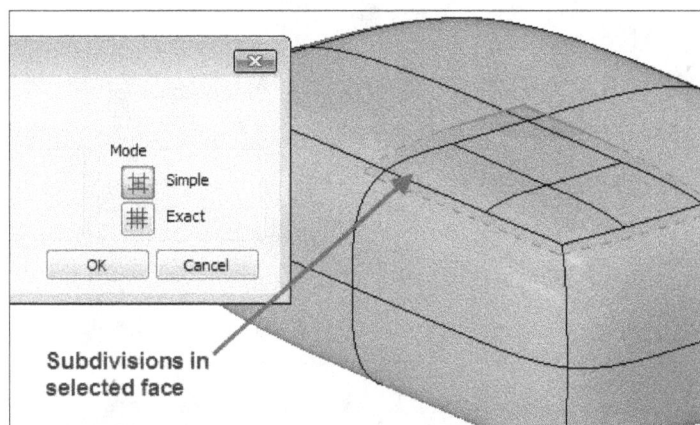

Figure-40. Preview of subdivisions

- Click on the **OK** button to create the sub-division faces.

Merge Edges Tool

The **Merge Edges** tool is used to merge open edges of the two freeform bodies. The procedure to use this tool is given next.

- Click on the **Merge Edges** tool from the **Edges** drop-down in the **Modify** panel of the **Freeform** contextual tab in **Ribbon**. The **Merge Edge** dialog box will be displayed; refer to Figure-41.

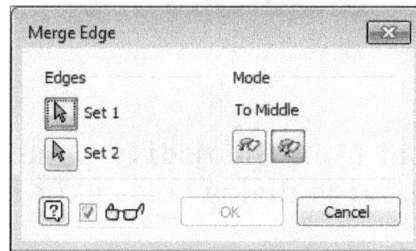

Figure-41. Merge Edges dialog box

- Select the open edges of first freeform object. Note that you can select only adjacent edges of the first selected edge.
- Click on the **Set 2** selection button from the **Edges** area of the dialog box and select the open edges of the second freeform object; refer to Figure-42. Preview of the edge merge will be displayed if possible. The **Preview** check box ☑ 👓 is selected in the dialog box and still preview is not displayed on your selection then you need to change your selection or geometry of edges selected.

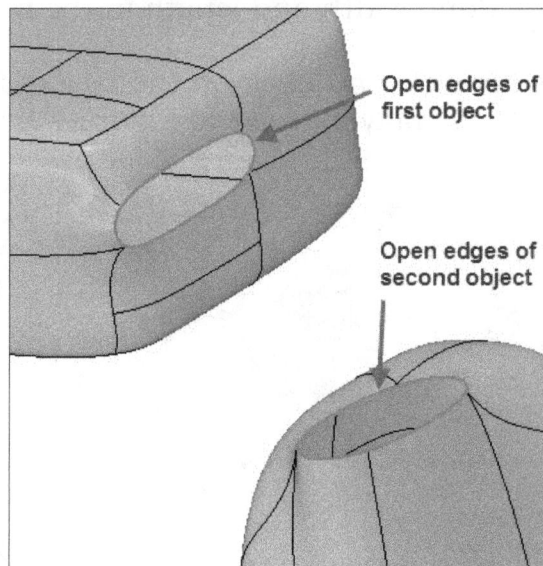

Open edges of
first object

Open edges of
second object

Figure-42. Edges selection for merging

- Click on the **OK** button from the dialog box to merge the edges.

Unweld Edges Tool

The **Unweld Edges** tool is used to break the merged edges. After selecting this tool from the **Modify** panel, select the edges earlier merged and click on the **OK** button from the dialog box.

Crease Edges Tool

The **Crease Edges** tool is used to straighten the curve of selected edges of the freeform object. The procedure to use this tool is given next.

- Click on the **Crease Edges** tool from the **Edges** drop-down in the **Modify** panel in the **Freeform** contextual tab of the **Ribbon**. The **Crease** dialog box will be displayed; refer to Figure-43.

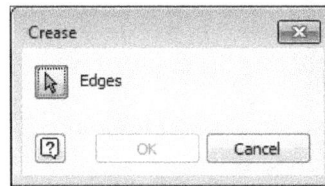

Figure-43. Crease dialog box

- Select the edges to which you want to apply the crease. Preview of the crease feature will be displayed; refer to Figure-44. Click on the **OK** button from the dialog box to create the feature.

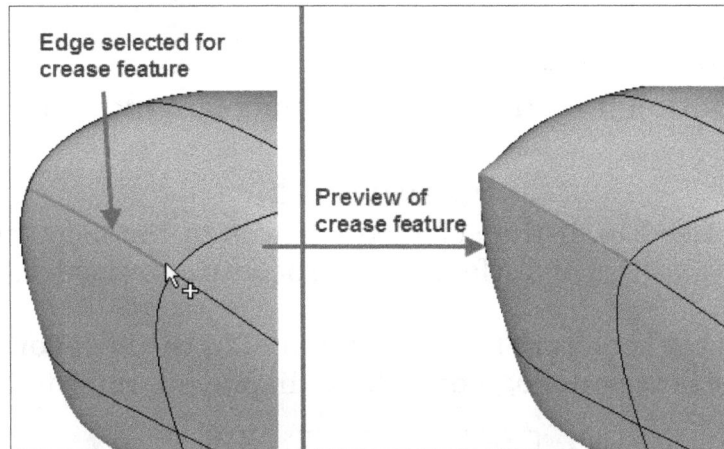

Figure-44. Preview of crease feature

Uncrease Edges Tool

The **Uncrease Edges** tool is used to undo the crease created by **Crease Edges** tool. The procedure of using this tool is similar to **Crease Edges** tool.

Weld Vertices Tool

The **Weld Vertices** tool is used to join two vertices of a freeform object. The procedure to use this tool is given next.

- Click on the **Weld Vertices** tool from the **Modify** panel in the **Freeform** contextual tab of the **Ribbon**. The **Weld Vertices** dialog box will be displayed; refer to Figure-45.

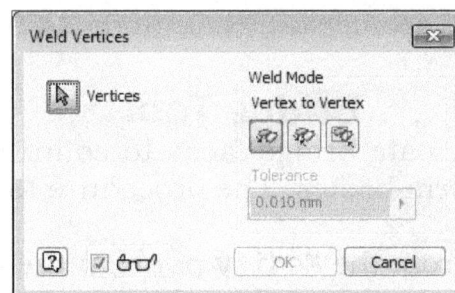

Figure-45. Weld Vertices dialog box

- Select desired button to set mode of joining vertices from the **Weld Mode** area of the dialog box.
- Select the two vertices that you want to be joined. Preview of the welded vertices will be displayed; refer to Figure-46.

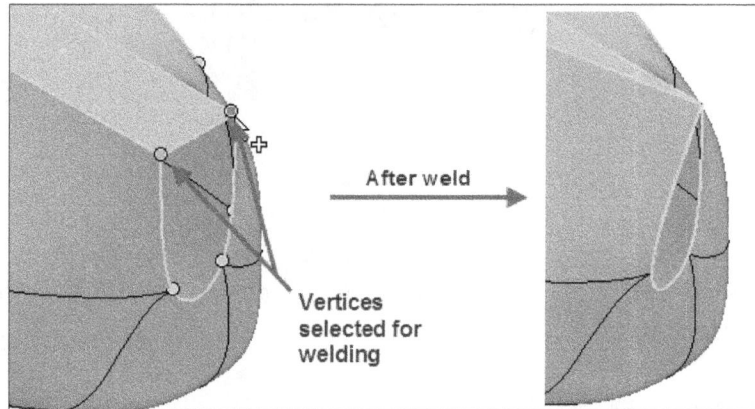

Figure-46. Vertices weld feature

- Click on the **OK** button from the dialog box to create the feature.

Flatten Tool

The **Flatten** tool is used to fit the selected vertices to the selected plane or in best position based on curvature of surface. The procedure to use this tool is given next.

- Click on the **Flatten** tool from the **Modify** panel in the **Freeform** contextual tab of the **Ribbon**. The **Flatten** dialog box will be displayed; refer to Figure-47.

Figure-47. Flatten dialog box

- Select the vertices to be flatten. Preview of flatten feature will be displayed.
- Select the **Plane** or **Parallel Plane** button from the **Direction** area of the dialog box to flatten the vertices on selected plane or parallel to selected plane.
- After selecting the required geometries, click on the **OK** button to flatten the vertices.

Bridge Tool

The **Bridge** tool is used to create bridge faces to connect two faces/open-edges of same body or different freeform bodies. The procedure to use this tool is given next.

- Click on the **Bridge** tool from the **Modify** panel in the **Freeform** contextual tab of the **Ribbon**. The **Bridge** dialog box will be displayed; refer to Figure-48.

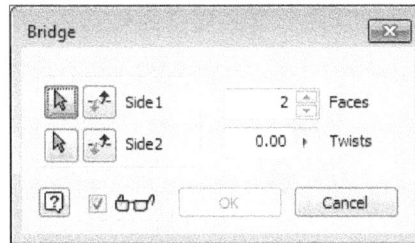

Figure-48. Bridge dialog box

- Select the face or open edge of first side. You will be asked to select the face/edge on the other side.
- Select the edge/face on the other side. Preview of the bridge surface will be displayed; refer to Figure-49.

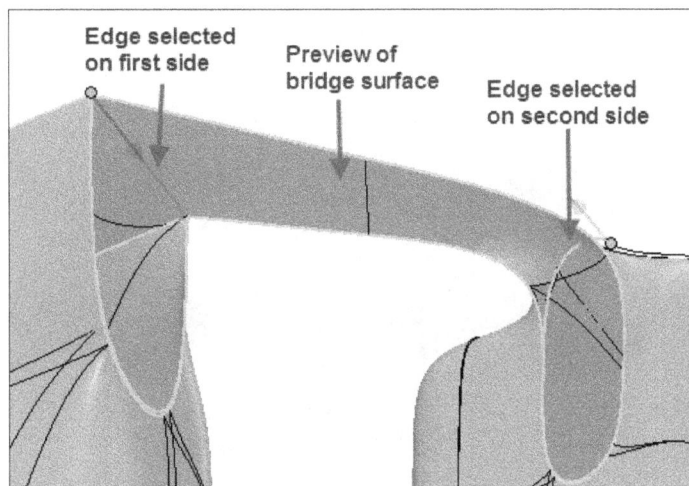

Figure-49. Preview of bridge surface

- You can keep on selecting the edges/faces from side 1 and side 2 to form more surfaces.
- Click on the **OK** button to create the bridge surfaces.

Thicken Tool

The **Thicken** tool is used to thicken the surfaces or offset the faces of solid bodies. The procedure to use this tool is given next.

- Click on the **Thicken** tool from the **Modify** panel in the **Freeform** contextual tab of the **Ribbon**. The **Thicken** dialog box will be displayed; refer to Figure-50.

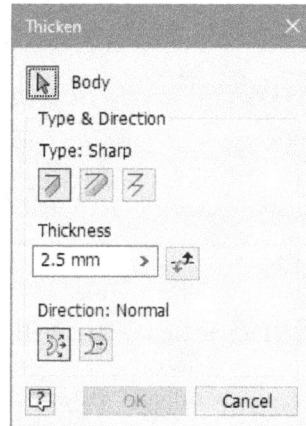

Figure-50. Thicken dialog box

- Select the body to be thicken. Preview of the thicken feature will be displayed; refer to Figure-51.

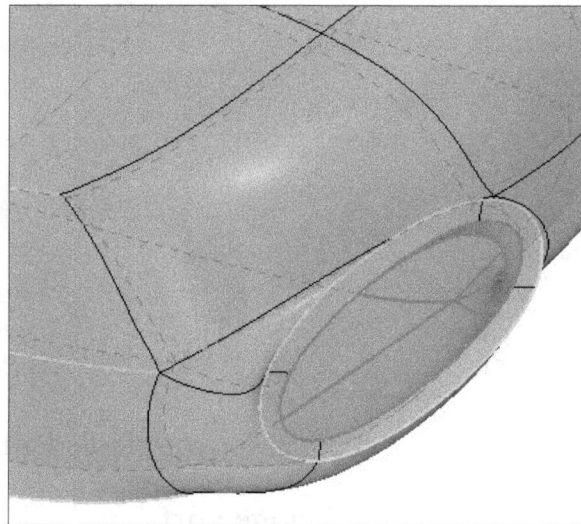

Figure-51. Preview of thicken feature

- By default, the **Sharp** button is selected in the **Type & Direction** area of the dialog box. Hence, the joints and edges in the thicken feature are sharp.
- Select the **Soft** button from the **Type & Direction** area to soften the thicken feature. Select the **No Edge** button to create the surface offset.
- Click on the **OK** button from the dialog box to create the feature.

Match Edge Tool

The **Match Edge** tool is used to match the selected edges to the sketch section or curve. The procedure to use this tool is given next.

- Click on the **Match Edge** tool from the **Modify** panel in the **Freeform** contextual tab of the **Ribbon**. The **Match Edge** dialog box will be displayed; refer to Figure-52.

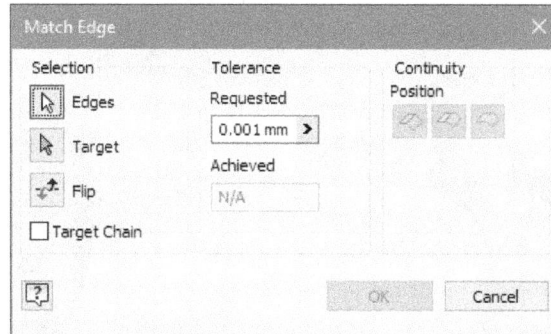

Figure-52. Match Edge dialog box

- Select the edges of the freeform body that are to be matched with sketch section.
- Click on the **Target** selection button and select the sketch curve to be matched with selected edges. Preview of match edge feature will be displayed; refer to Figure-53.

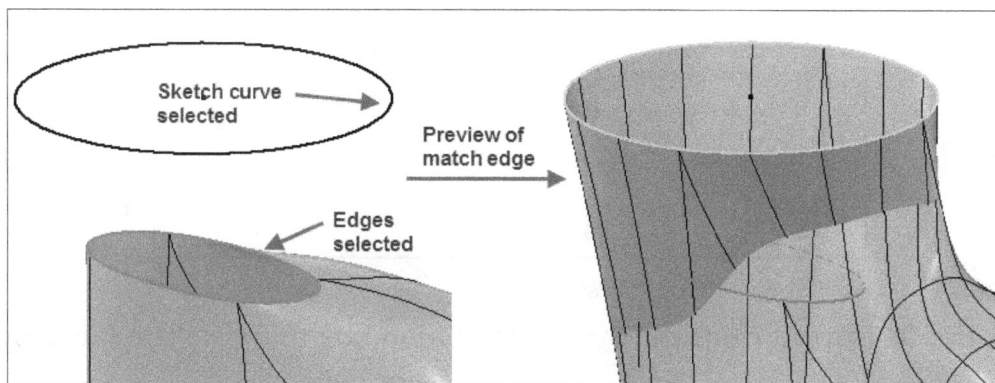

Figure-53. Preview of match edge feature

- Set desired tolerance value in the edit box of **Tolerance** area in the dialog box.
- If the edges are matched opposite to what you need then click on the **Flip** button from the dialog box.
- Click on the **OK** button from the dialog box to create the feature.

Symmetry Tool

The **Symmetry** tool is used to make two faces of the freeform bodies symmetric. Once the faces are symmetric then any change made in one face will also be reflected in other symmetric face. The procedure to use this tool is given next.

- Click on the **Symmetry** tool from the **Symmetry** drop-down in the **Symmetry** panel of the **Ribbon**. The **Symmetry** dialog box will be displayed; refer to Figure-54.

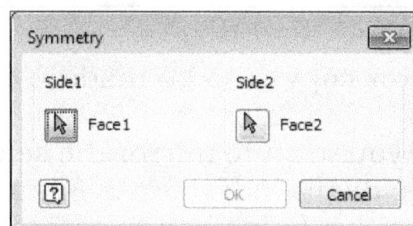

Figure-54. Symmetry dialog box

- Select the face on first side and then a face on other side to make the two faces symmetric. If there is possibility of symmetry in two faces then the **OK** button will become active in dialog box; refer to Figure-55. Otherwise, a warning message will be displayed; refer to Figure-56.

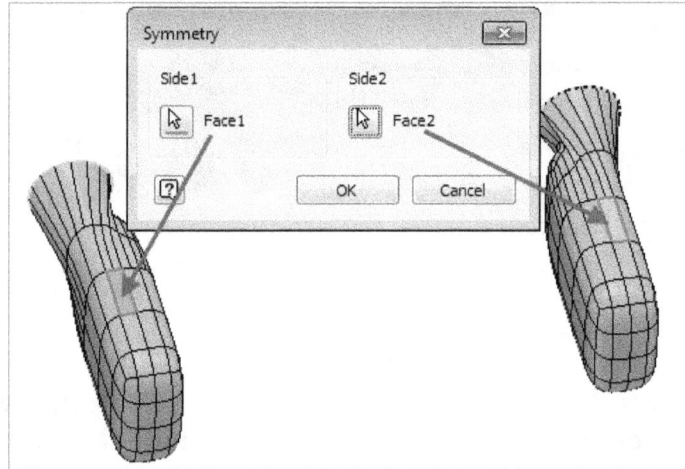

Figure-55. Faces selected for symmetry

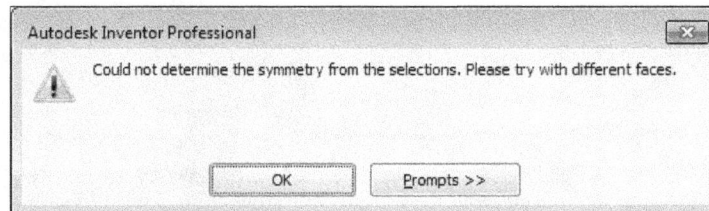

Figure-56. Warning message

- Click on the **OK** button from the **Symmetry** dialog box to make the faces symmetric.

Mirror Tool

The **Mirror** tool is used to create mirror copy of the selected freeform object. The procedure to use this tool is given next.

- Click on the **Mirror** tool from the **Symmetry** drop-down in the **Symmetry** panel of the **Ribbon**. The **Mirror** dialog box will be displayed; refer to Figure-57.

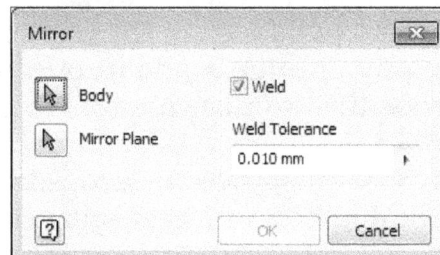

Figure-57. Mirror dialog box

- Select the body whose mirror copy is to be made. You will be asked to select the mirror plane.
- Select a plane about which you want to mirror the selected freeform body. Preview of the mirror feature will be displayed; refer to Figure-58.
- Click on the **OK** button from the dialog box to create the feature.

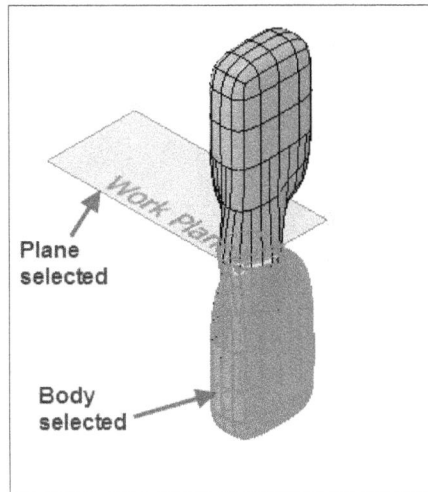

Figure-58. Preview of mirror feature

In the same way, you can use the other tools in the **Freeform** contextual tab to modify the freeform bodies.

Once you have performed all the modifications, click on the **Finish Freeform** tool from the **Exit** panel in the **Freeform** contextual tab of the **Ribbon** to return to modeling environment.

If you have created a closed freeform body then you will get a solid freeform object as output on clicking the **Finish Freeform** tool otherwise, you will get a surface body.

PRACTICAL

Create the model of helmet glass as shown in Figure-59. The dimensions of the model are given in Figure-60.

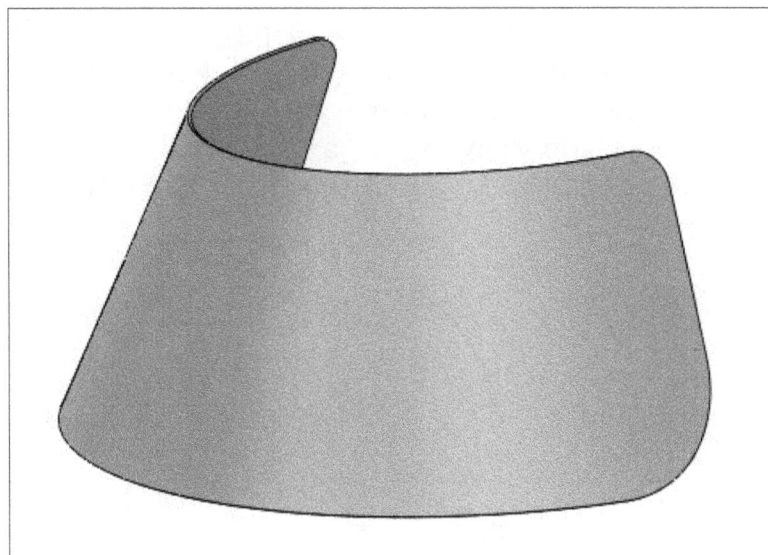

Figure-59. Practical model

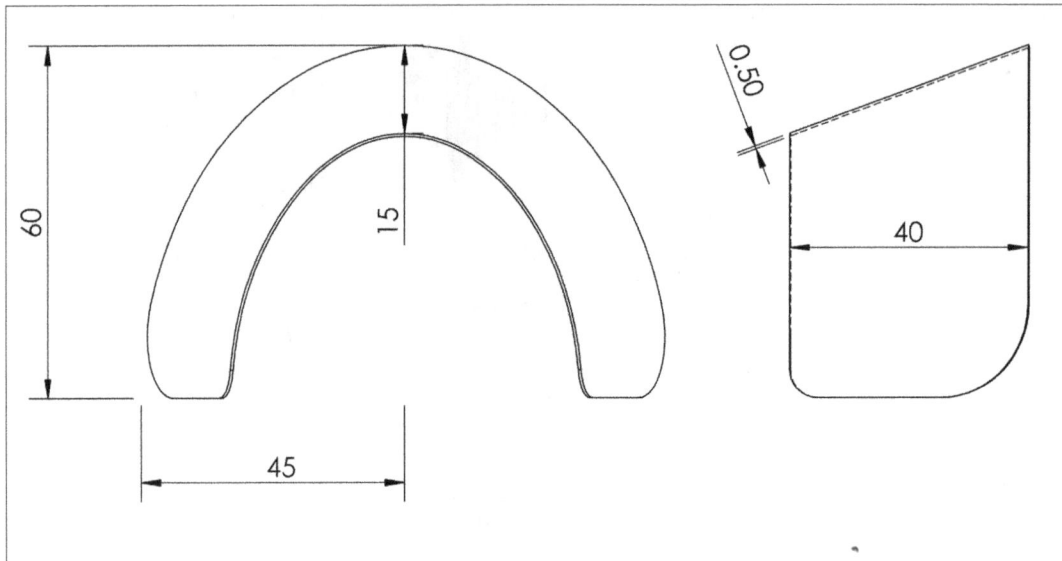

Figure-60. Practical drawing

The model displayed is having very low thickness and its having complex 3D shape. So, it is a good idea to use surfacing in this case.

We can create this model by loft surface easily. For using the **Loft** tool, we need two sketches.

Starting Part file

- Start Autodesk Inventor, if not started yet. Click on the **New** button from **New** cascading menu in the **File** menu of the **Ribbon**. The **Create New File** dialog box will be displayed.
- Double-click on the **Standard(mm).ipt** part file template from the **Metric** templates in the dialog box. The part modeling environment will be displayed.

Creating Sketches

- Click on the **Start 2D Sketch** tool from the **Sketch** drop-down in the **Sketch** panel of the **3D Model** tab in the **Ribbon**. You will be asked to select a sketching plane.
- Select the **XZ** Plane from the drawing area and create the sketch as shown in Figure-61. Click on the **Finish Sketch** button from the **Exit** panel in the **Ribbon**.

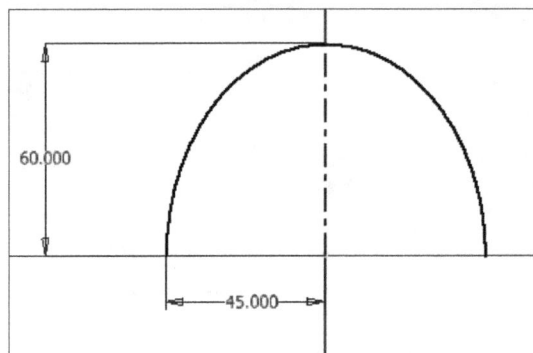

Figure-61. Semi ellipse created as sketch

- Create a plane at an offset distance of **40** above the **XZ** plane; refer to Figure-62.
- Click on the **Start 2D Sketch** tool again and select the newly created plane as sketching plane.

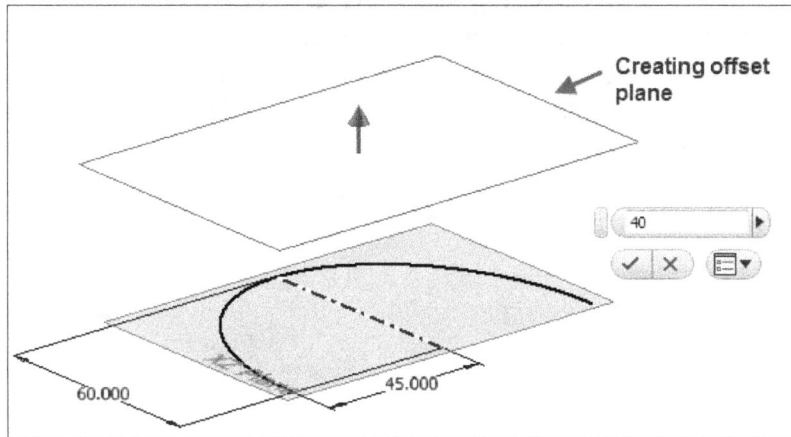

Figure-62. Offset plane to be created

- Click on the **Project Geometry** tool from the **Create** panel in the **Sketch** contextual tab and select the sketch earlier created. A yellow curve will be created coinciding with the sketch; refer to Figure-63.

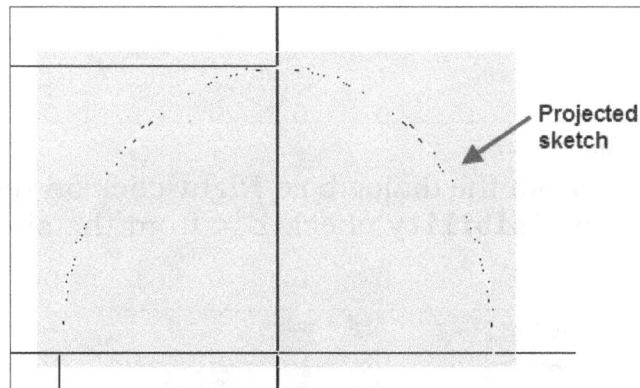

Figure-63. Projected sketch curve

- Click on the **Offset** tool from the **Modify** panel in the **Sketch** contextual tab and select the projected sketch curve. You will be asked to specify the offset distance.
- Move the cursor inside the curve and enter the value as **15** in the edit box displayed attached to cursor. The offset feature will be created; refer to Figure-64.

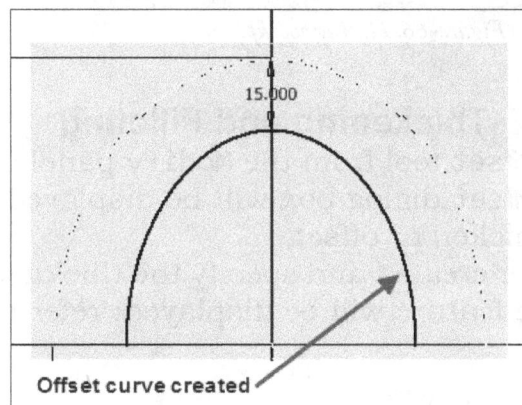

Figure-64. Offset curve created

- Click on the **Finish Sketch** button from the **Exit** panel in the **Ribbon**.

Creating Lofted Surface

- Click on the **Loft** tool from the **Create** panel in the **3D Model** tab of the **Ribbon**. The **Loft** dialog box will be displayed and you will be asked to select the sketches.
- Select the sketch curves created earlier. The preview of lofted surface will be displayed; refer to Figure-65.

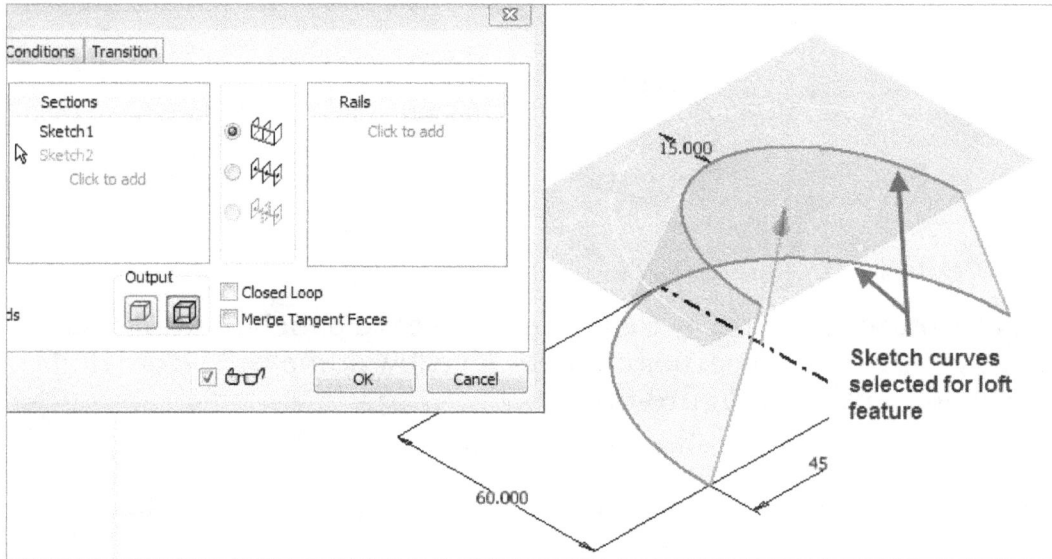

Figure-65. Preview of lofted surface

- Click on the **OK** button from the dialog box. Right-click on the work plane created and hide it by clearing **Visibility** check box from the shortcut menu; refer to Figure-66.

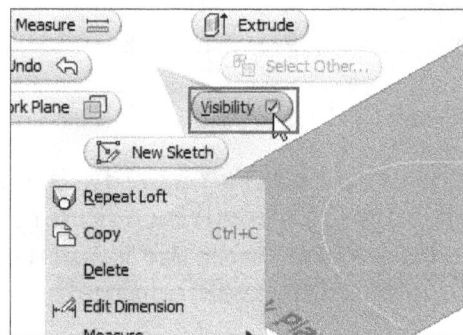

Figure-66. Hiding workplane

Thickening and Filleting

- Click on the **Thicken/Offset** tool from the **Modify** panel in the **3D Model** tab of the **Ribbon**. The **Thicken/Offset** dialog box will be displayed. Also, you will be asked to select the faces to thicken or offset.
- Select the surface earlier created and specify the thickness as **0.5** in the **Distance** edit box. Preview of the feature will be displayed; refer to Figure-67.

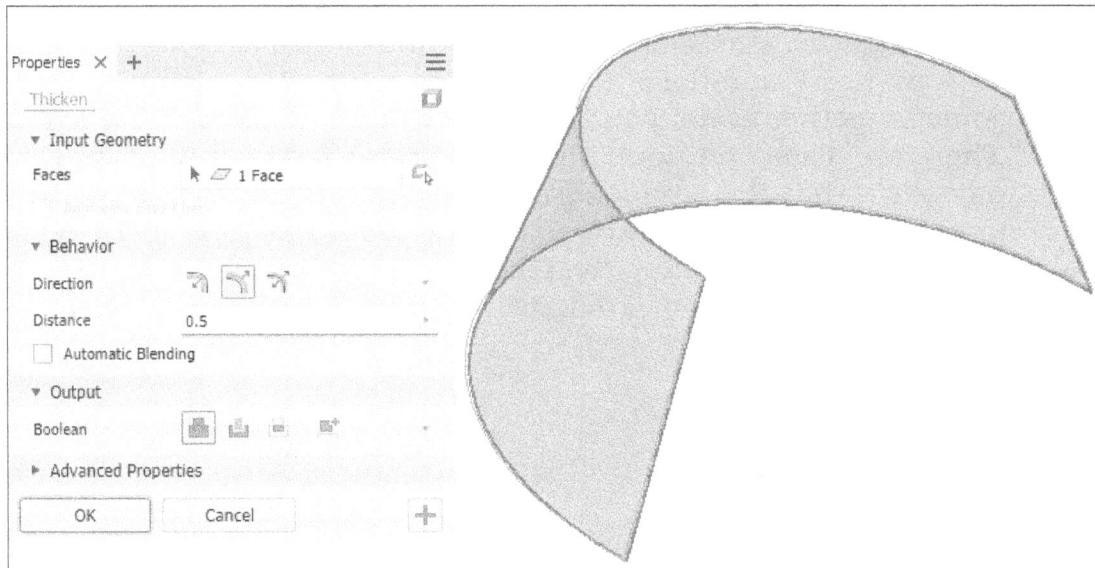

Figure-67. Preview of thicken feature

- Click on the **OK** button from the dialog box. Hide the surface earlier created by using the right-click shortcut menu.
- Click on the **Fillet** tool from the **Modify** panel in the **3D Model** tab of the **Ribbon**. The **Fillet** dialog box will be displayed.
- Select the corner edges of the thicken feature and specify suitable value of radius in the edit box. Preview of fillet will be displayed; refer to Figure-68.

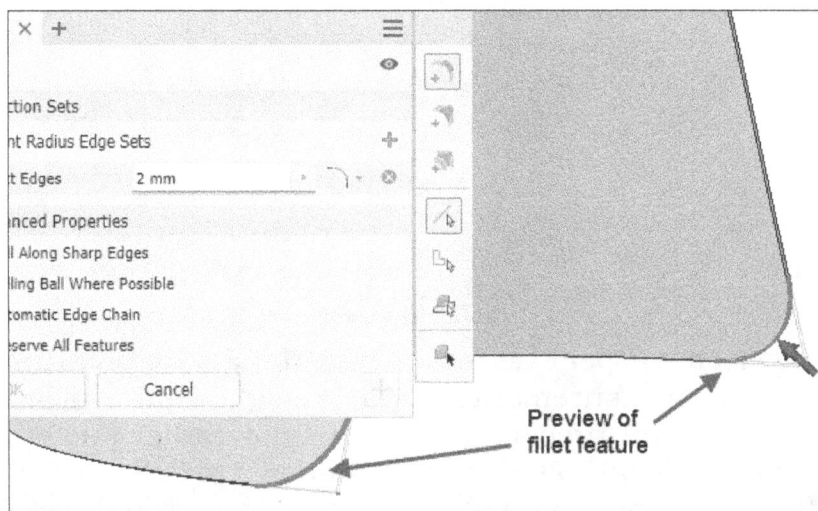

Figure-68. Preview of fillet feature

- Click on the **OK** button from the dialog box to create the feature.

MEASUREMENT TOOLS

When you are working with surfaces, there are various aspects of surface which can not be measured by using standard dimensioning tools. For example, if there are two surfaces joined together at a curve then how will we measured the smoothness of that joint. There are also conditions in mold design that require measurement of taper angle to confirm that the component can be ejected from mold safely. These type of measurements are performed by using the tools available in **Inspect** tab of the **Ribbon**; refer to Figure-69 and Figure-70. We will first discuss the measurement tools for part design and then we will discuss the tools for assembly design.

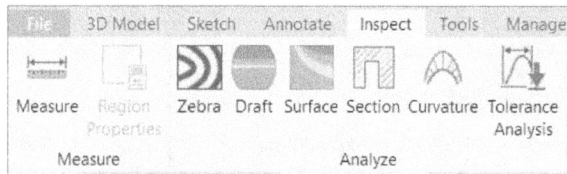

Figure-69. Inspect tab for Part modeling

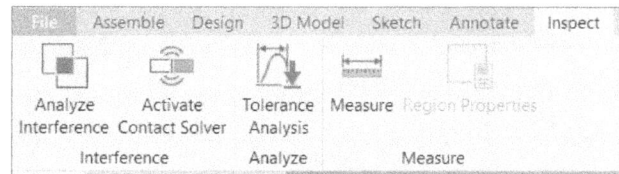

Figure-70. Inspect tab for Assembly modeling

Measuring Entities

The **Measure** tool is used to measure various common parameters of selected entities. The results of measure tool vary depending on selected objects. The procedure to use this tool is given next.

- Click on the **Measure** tool from the **Measure** panel in the **Inspect** tab of **Ribbon**. The **Measure** dialog box will be displayed; refer to Figure-71.

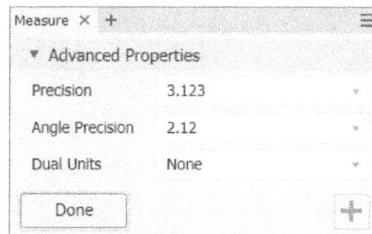

Figure-71. Measure dialog box

- Expand the **Advanced Properties** node from the dialog box to set precision for general and angular measurements. Select desired option from the **Precision** drop-down to set general precision of linear and radial measurements. Select desired option from the **Angle Precision** drop-down to define precision for angular measurements. If you want to display dimensions in two different unit systems then select desired unit option from the **Dial Units** drop-down; refer to Figure-72. Note that default unit set for the model will be inactive in the drop-down as it is already selected.

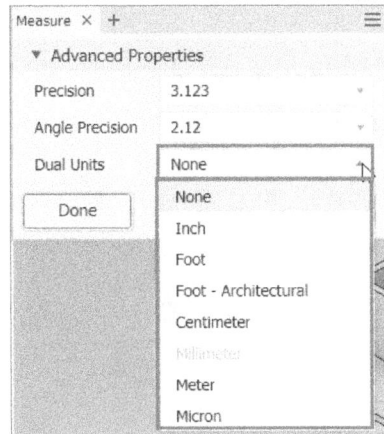

Figure-72. Dual Units drop-down

Based on selected entities, you will get different measurement results which are discussed next.

Measuring Points

- Select desired point from the model to find out its location with respect to origin.
- Select two points from the model to check their location as well as distance between the two; refer to Figure-73. Click on the **Restart Measure** button ⊞ at the bottom in the dialog box to select next set of entities.

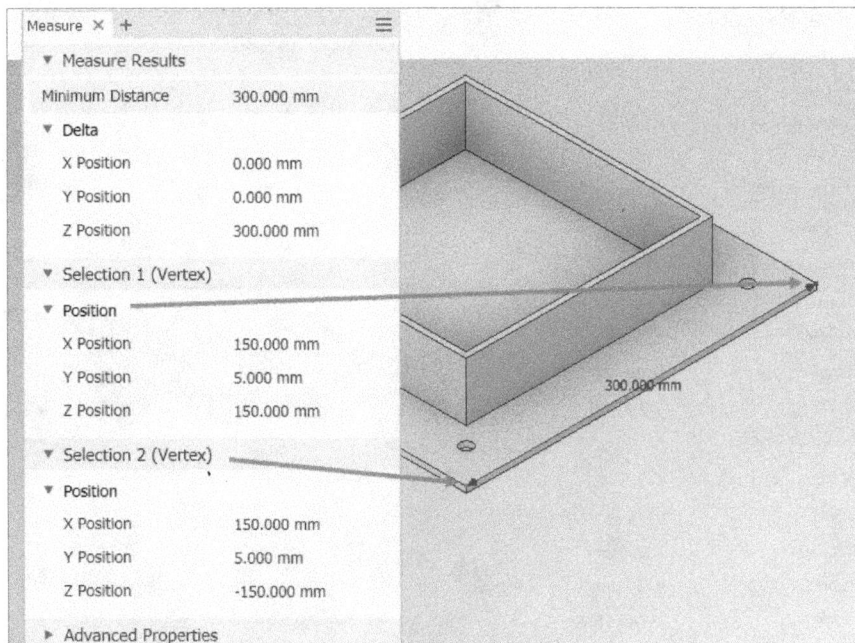

Figure-73. Measuring points

Measuring Lines

- Select a line or edge from the model to check its length. Select another line/edge if you want to check distance between the two lines as well as angle between them; refer to Figure-74.

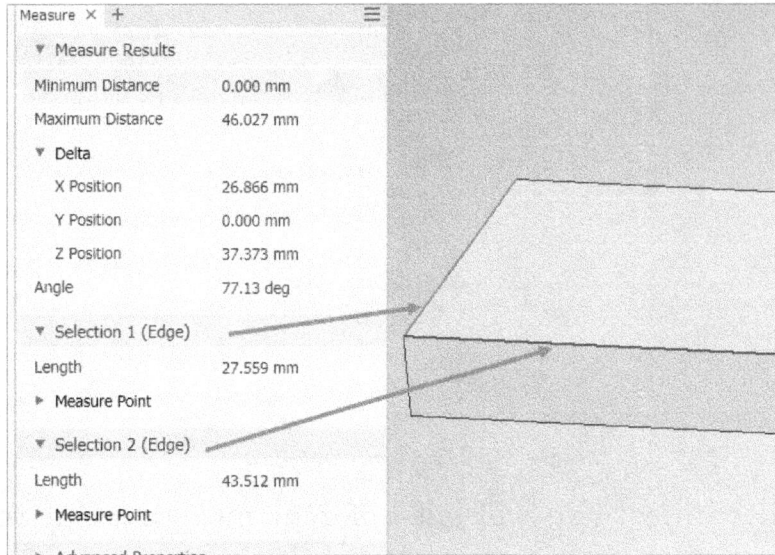

Figure-74. Measuring lines/edges

Measuring Circle/Arc

• Select a circle or arc curve/edge from the model to check diameter, radius, length, and swap angle of selected curves. If two circular curves/edges are selected then distance between center of two curves/edges will be displayed; refer to Figure-75.

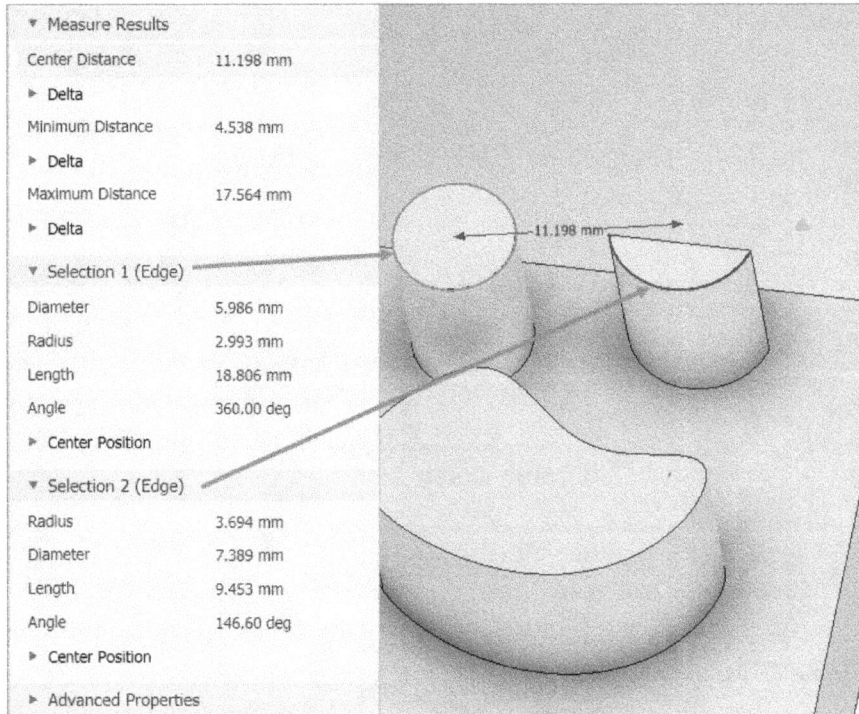

Figure-75. Measuring circle and arc

Similarly, you can measure length of the spline by selecting it. If you select a face then its perimeter, total loop length, and area will be displayed. If two faces selected then distance between the faces will be displayed as well.

Inspecting Region Properties

The **Region Properties** tool is used to check region properties of selected closed sketch loop. Note that this tool is active only when you are in sketching environment. The procedure to use this tool is given next.

- Click on the **Region Properties** tool from the **Measure** panel in the **Inspect** tab of the **Ribbon**. The **Region Properties** dialog box will be displayed; refer to Figure-76.

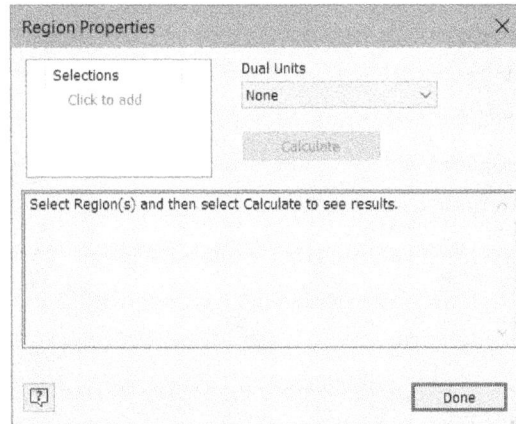

Figure-76. Region Properties dialog box

- Click inside the closed region of desired sketch loop to check its properties and click on the **Calculate** button. The properties of region will be displayed; refer to Figure-77.

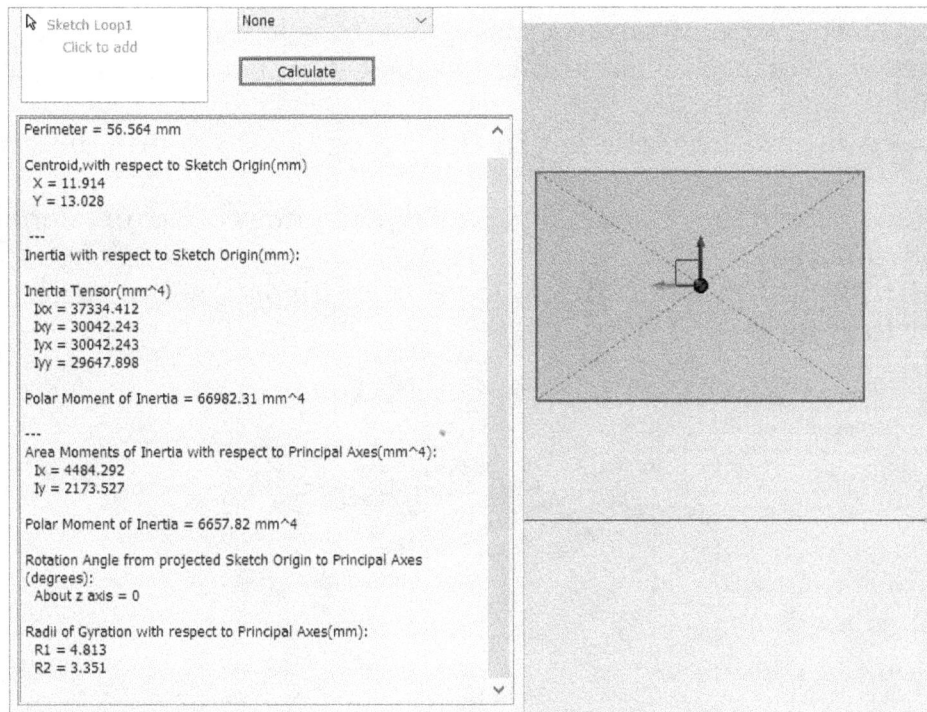

Figure-77. Region properties

- Click on the **Done** button after checking result to exit the dialog box.

Analyzing Surface Continuity using Zebra Pattern

The **Zebra** tool is used to display zebra strips on selected surfaces to check continuity in the surfaces. The procedure to use this tool is given next.

*   Click on the **Zebra** tool from the **Analyze** panel in the **Inspect** tab of the **Ribbon**. The **Zebra Analysis** dialog box will be displayed; refer to Figure-78.

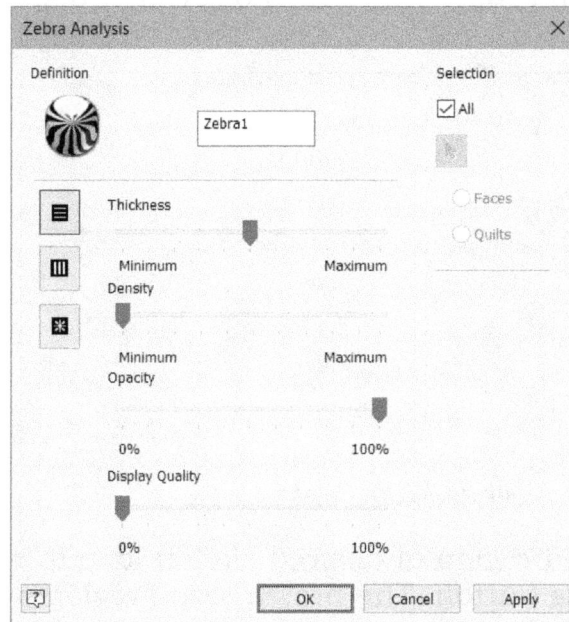

Figure-78. Zebra Analysis dialog box

*   Select the **All** check box from the dialog box to select all the faces of the model to display zebra surface. Clear the check box if you want to select faces/quilts individually.
*   Select desired button from the left in the dialog box to define direction of zebra strips.
*   Set desired parameters in the dialog box like thickness of strips, density, opacity, and display quality.
*   Click on the **OK** button from the dialog box to display zebra strips on selected faces; refer to Figure-79.

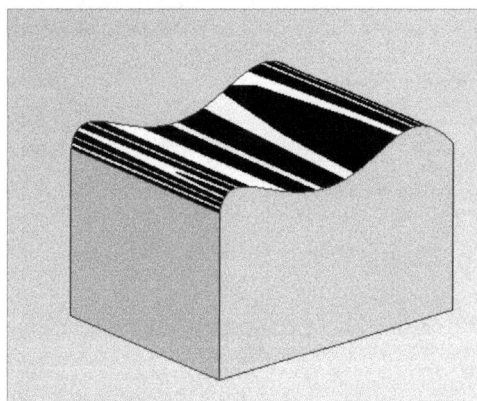

Figure-79. Zebra strips placed on faces

*   To hide the result of analysis, right-click on the **Analysis: Zebra** folder from the **Model Browser** and clear the **Analysis Visibility** option; refer to Figure-80. Expand the **Analysis: Zebra** folder and delete the zebra by pressing **DELETE** key.

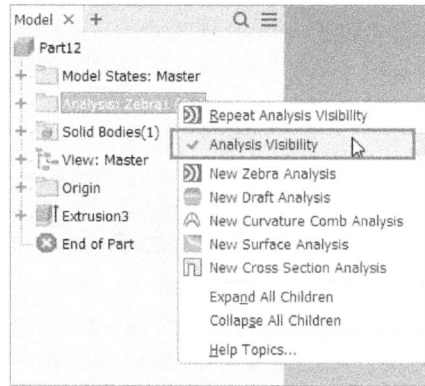

Figure-80. Analysis Visibility option

Performing Draft Analysis

The **Draft** tool is used to check draft angle of selected faces with respect to pull face. The procedure to use this tool is given next.

- Click on the **Draft** tool from the **Analyze** panel in the **Inspect** tab of the **Ribbon**. The **Draft Analysis** dialog box will be displayed; refer to Figure-81.

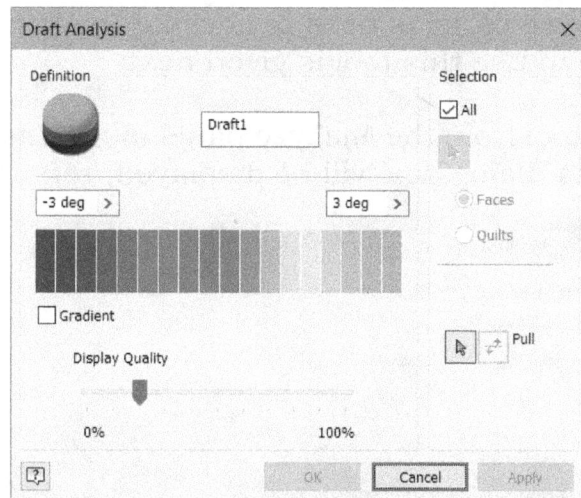

Figure-81. Draft Analysis dialog box

- Set desired values in the **Draft start angle** and **Draft end angle** edit boxes to define angle range for performing draft analysis.
- If **All** check box is selected then all the faces of model will be selected for draft analysis. Clear the check box and select desired faces for performing analysis.
- Click on the **Pull** selection button and select the face to be used as reference face for checking face angles; refer to Figure-82.

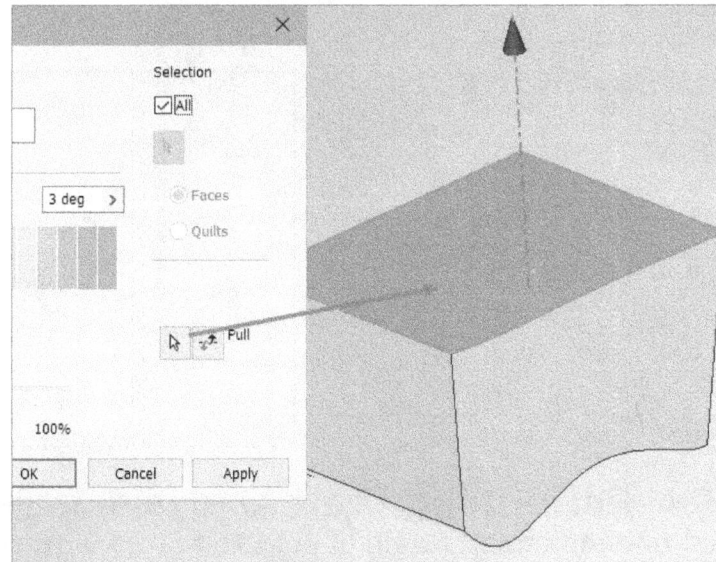

Figure-82. Pull direction face

- Click on the **OK** button from the dialog box to check results of analysis.

Surface Analysis

The **Surface** tool in **Analyze** panel is used to check curvature of surface using color gradient. The procedure to use this tool is given next.

- Click on the **Surface** tool from the **Analyze** panel in the **Inspect** tab of the **Ribbon**. The **Surface Analysis** dialog box will be displayed; refer to Figure-83.

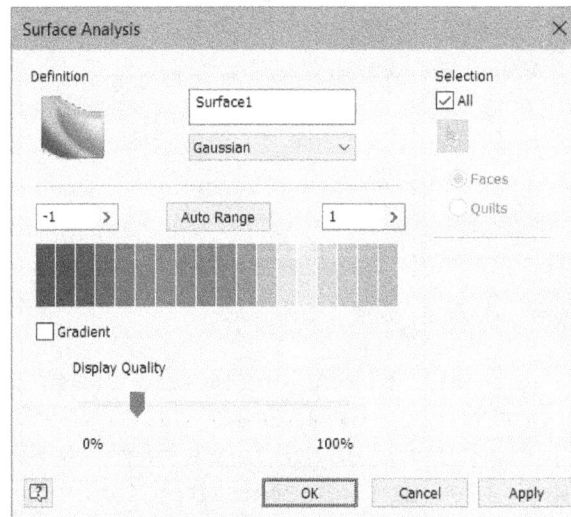

Figure-83. Surface Analysis dialog box

- Select desired method for surface curvature analysis from the **Definition** drop-down.
- Set the other parameters as discussed earlier and click on the **OK** button. The result of surface analysis will be displayed on the model.

Sections

The **Section** tool of **Analyze** panel is used to check cross-section of the model by cutting it using a plane. The procedure to use this tool is given next.

- Click on the **Section** tool from the **Analyze** panel in **Inspect** tab of the **Ribbon**. The **Cross Section Analysis** dialog box will be displayed; refer to Figure-84.

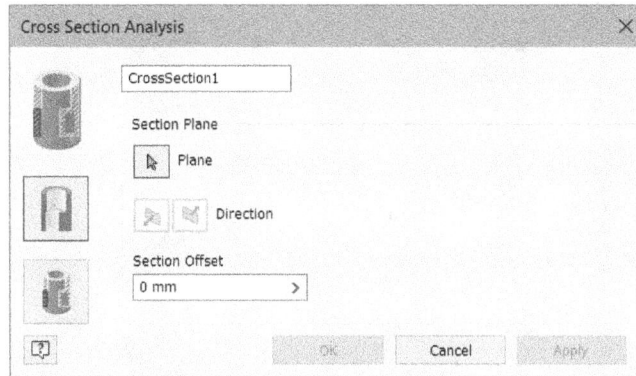

Figure-84. Cross Section Analysis dialog box

- Select desired face/plane to be used as section plane for cutting the model. The **Direction** buttons will become active.
- Select desired direction button to define direction in which section will be performed. The direction of section should be towards the model.
- Set desired value in the **Section offset** edit box to move the section plane in earlier specified direction. You can also use the arrow handle to move section plane; refer to Figure-85.

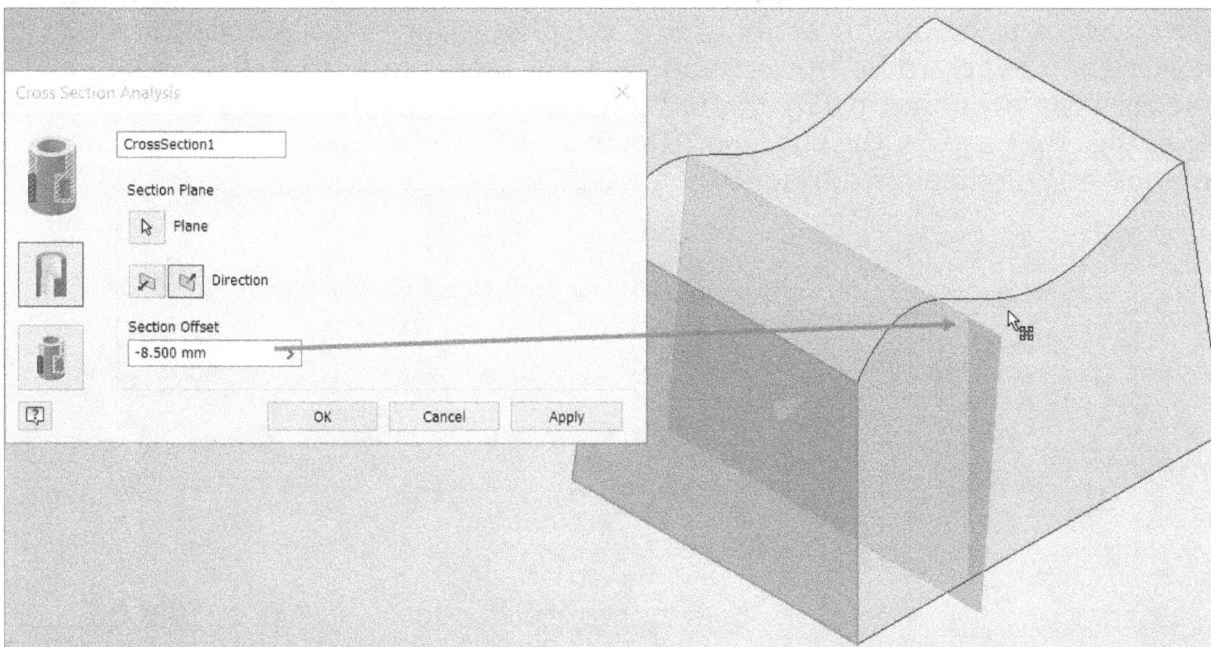

Figure-85. Moving section plane

- Click on the **Apply** button to create section.

Applying Advanced Sections

The options in the **Advanced Cross Section Analysis** dialog box are used to create multiple sections in the model. Click on the **Advanced** button from the left side in the dialog box. The options in the dialog box will be displayed as shown in Figure-86.

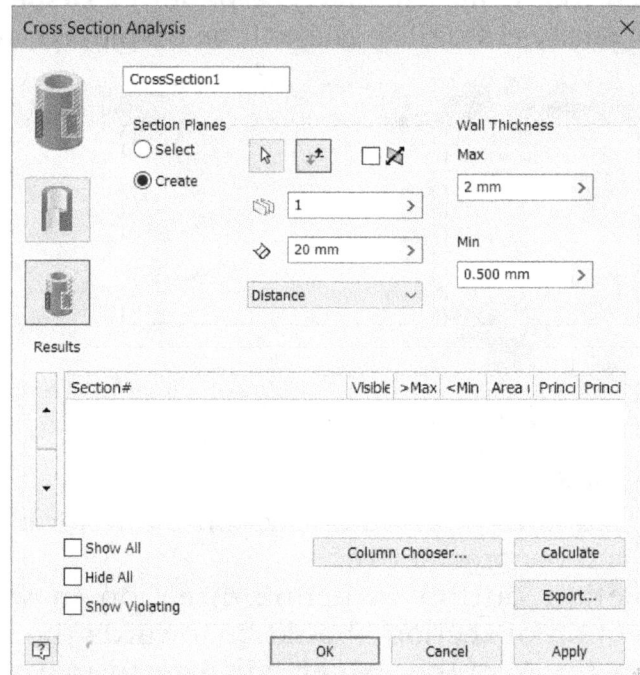

Figure-86. Advanced options

- Select the **Select** radio button from the **Section Planes** area of dialog box and select desired planes from the drawing area to be defined as section planes.
- If you want to create an array of section planes for cross section analysis then select the **Create** radio button and select a face/plane to define starting plane for creating section. Specify related parameters in **Section Number** and **Section Spacing** edit boxes of the **Section Planes** area in the dialog box. Preview of section planes will be displayed; refer to Figure-87.

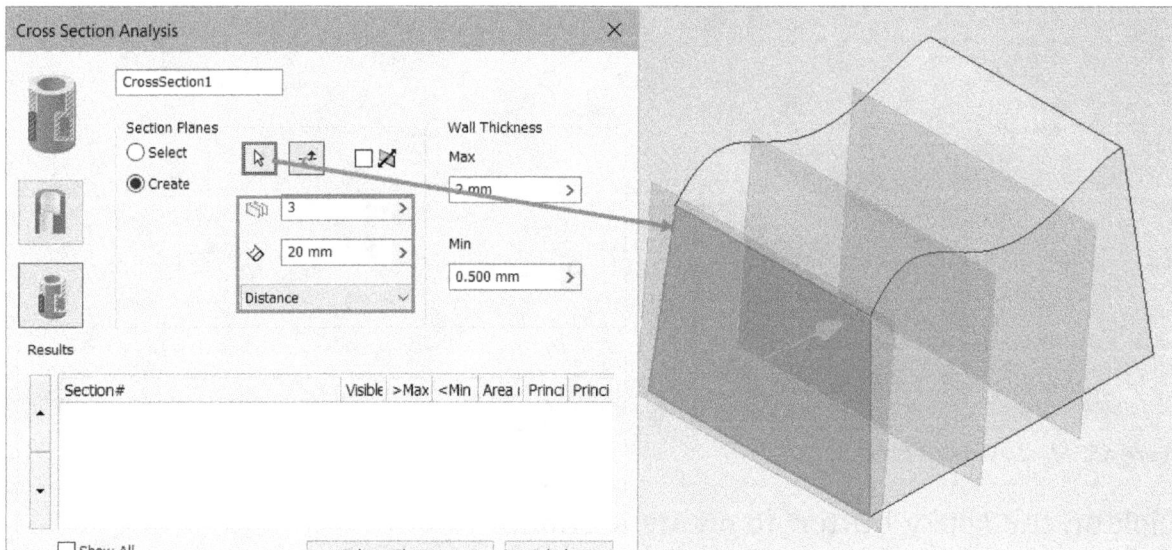

Figure-87. Section planes created

- Set desired parameters in **Wall Thickness** area as discussed earlier.
- After setting desired parameters, click on the **Calculate** button from the dialog box. The sections will be created and their results will be displayed in Results area; refer to Figure-88.

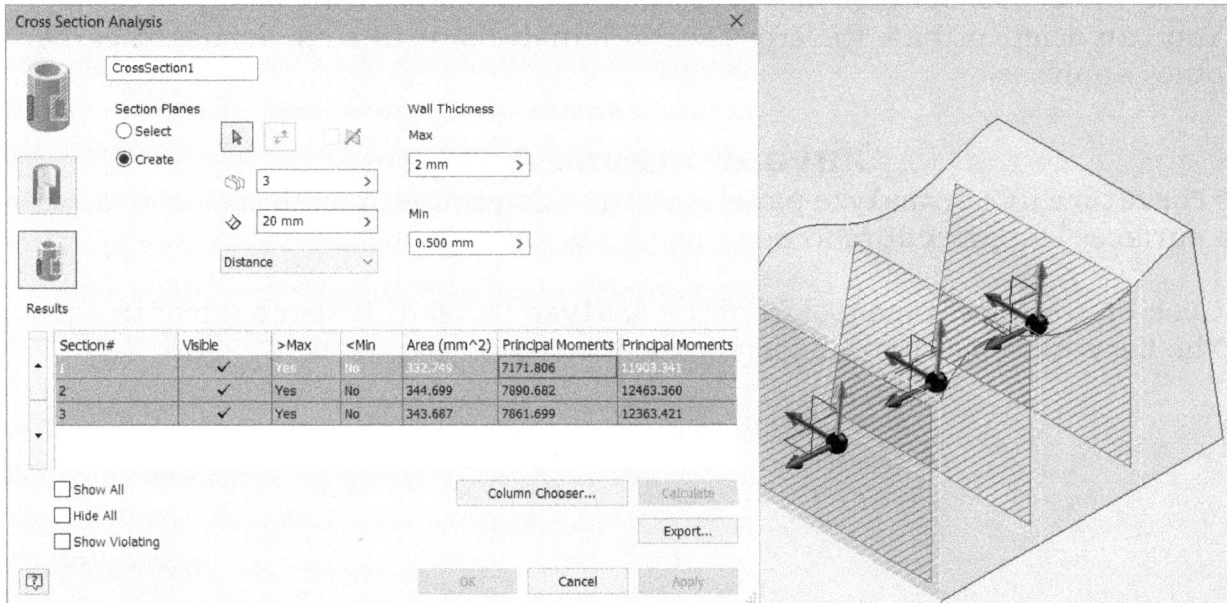

Figure-88. Multiple sections created

- Select the **Show All** check box to display all the sections in model. Select the **Hide All** check box to hide all the sections from model. Select the **Show Violating** check box to display only those sections which exceed the minimum and maximum wall thickness range.
- Click on the **Column Chooser** button if you want to add more result parameters in the table. The **Customization** dialog box will be displayed with list of available parameters.
- Drag desired parameter from the **Customization** dialog box to columns in the table; refer to Figure-89.

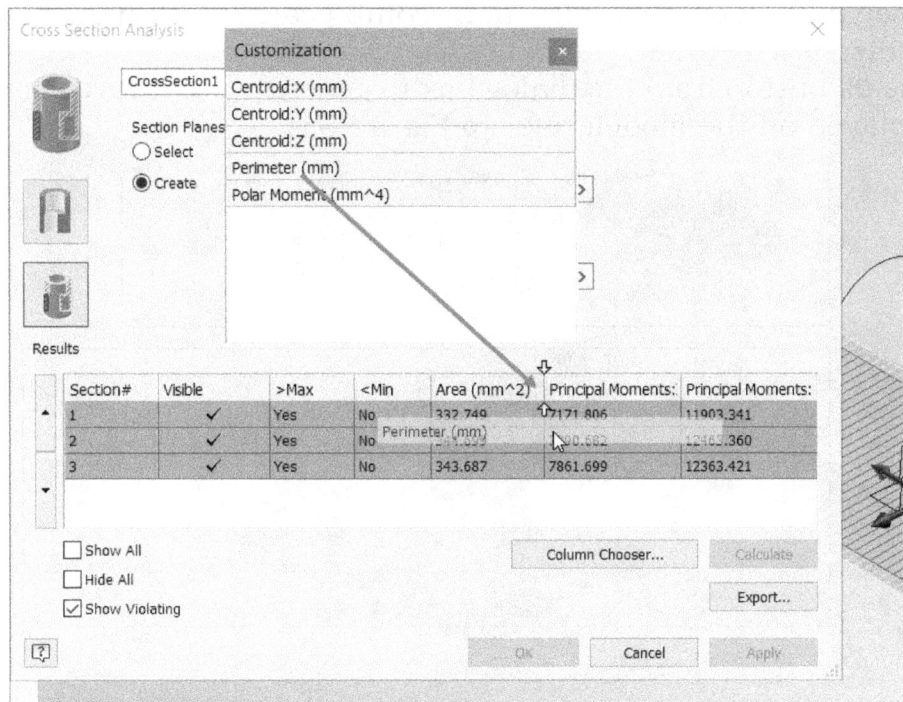

Figure-89. Adding parameters in columns of table

- Close the **Customization** dialog box after performing desired changes.
- Click on the **Export** button to save the data as text file.

- Close the **Cross Section Analysis** dialog box after setting desired parameters. You can delete or hide the cross section analysis results as discussed earlier for other analyses.

Curvature Comb Analysis

The **Curvature** tool in **Analyze** panel is used to determine smoothness and curvature of a surface. The procedure to use this tool is given next.

- Click on the **Curvature** tool from the **Analyze** panel in **Inspect** tab of the **Ribbon**. The **Curvature Analysis** dialog box will be displayed; refer to Figure-90.

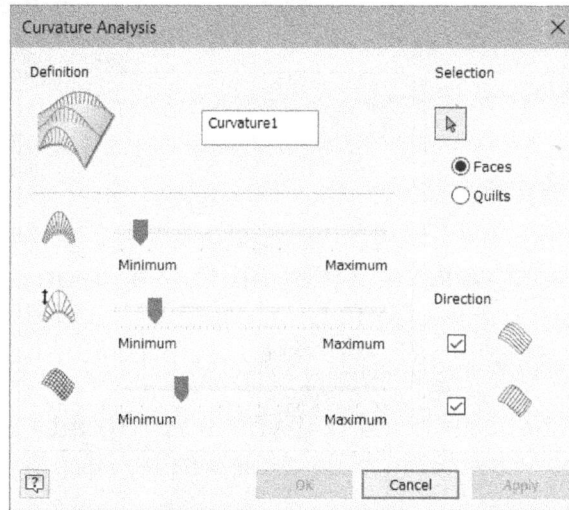

Figure-90. Curvature Analysis dialog box

- Select the face(s) of model on which you want to display curvature comb.
- Set the other parameters like direction, comb scale, comb density, and surface density as discussed earlier.
- Click on the **OK** button from the dialog box to generate curvature comb. The comb will be displayed on the model; refer to Figure-91.

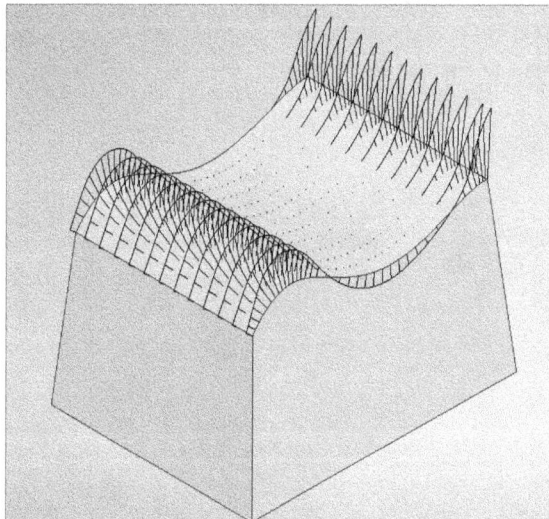

Figure-91. Curvature comb

PRACTICE 1

Create the model of flower vase as shown in Figure-92. The dimensions of the model are given in Figure-93.

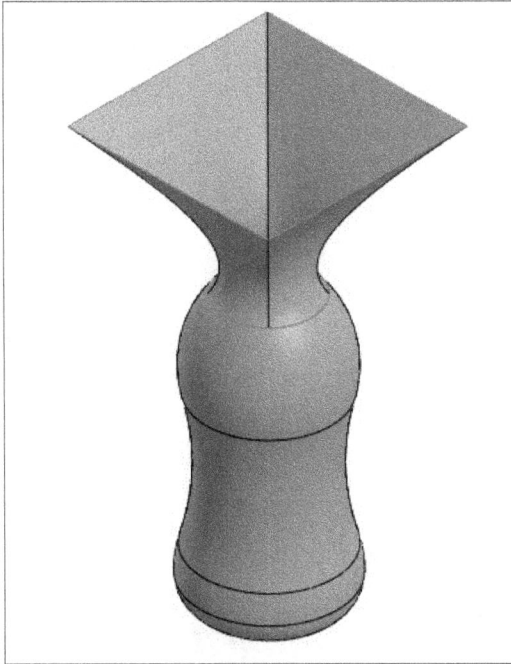

Figure-92. Flower vase model

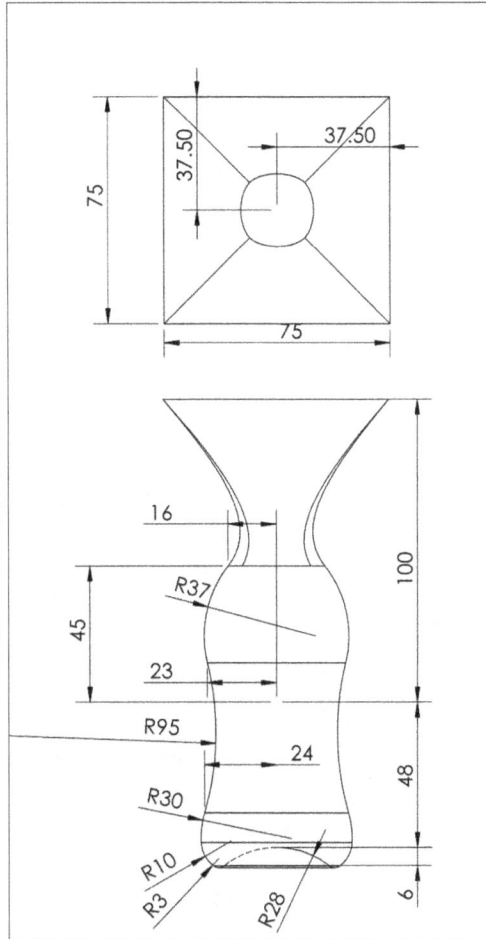

Figure-93. Practice 1 Drawing

PRACTICE 2

Create the surface model of tank as shown in Figure-94. The dimensions of the model are given in Figure-95.

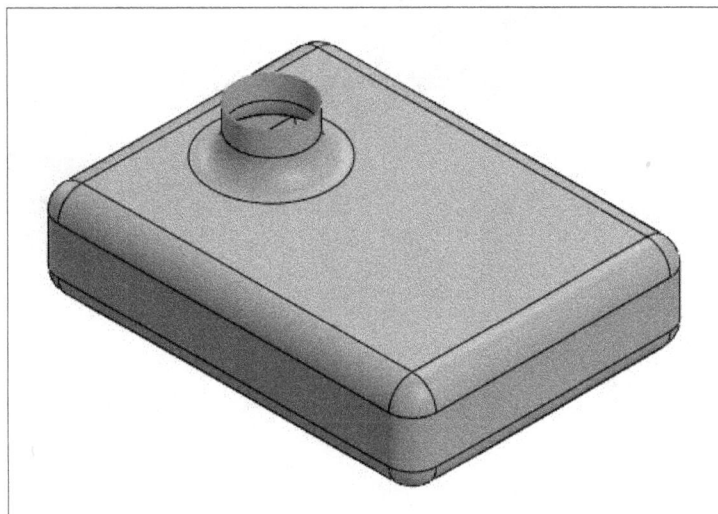

Figure-94. Practice 2 model

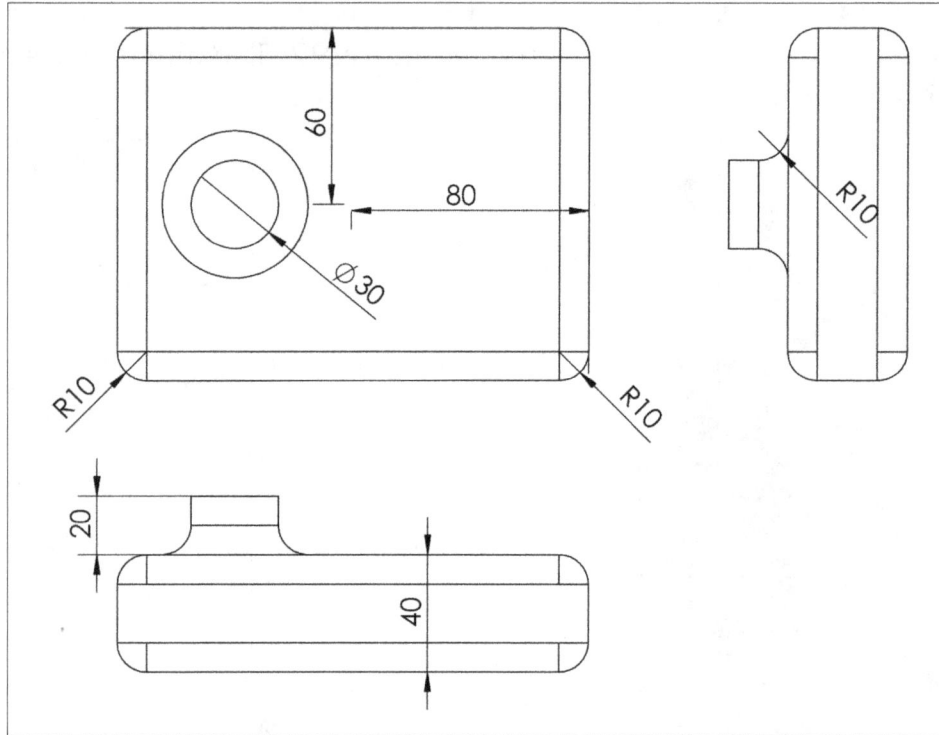

Figure-95. Practice 2 drawing

PRACTICE 3

Create the surface model of car bumper as shown in Figure-96. The dimensions of the model are given in Figure-97. **Assume the missing dimensions**.

Figure-96. Practice 3 model

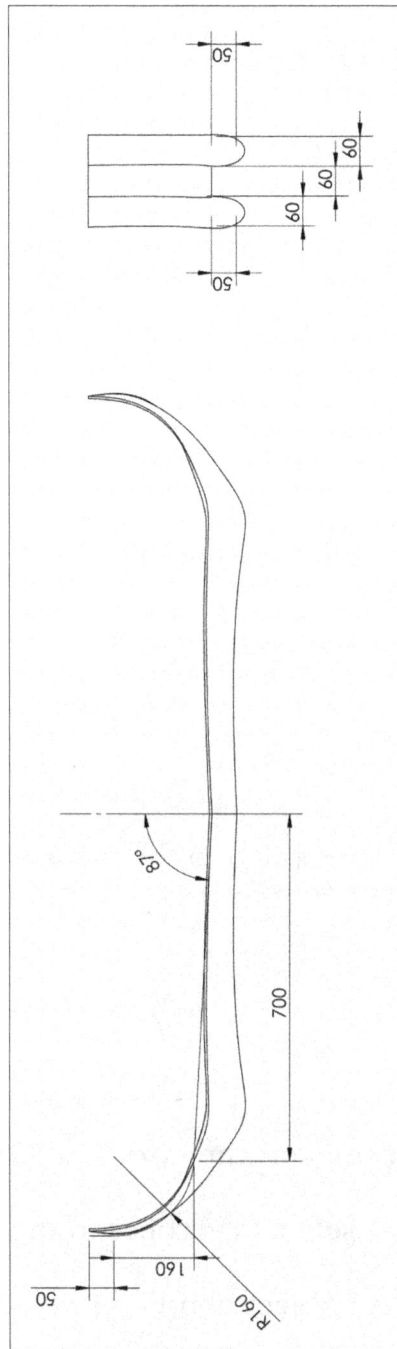

Figure-97. Car bumper

SELF ASSESSMENT

Q1. Which of the following tools is used to join the surfaces, meeting at common edges to form a solid body?

a) Extend Surface
b) Ruled Surface
c) Sculpt
d) None of the Above

Q2. Which of the following tools is the Freeform modeling tool?

a) Box
b) Sculpt
c) Boundary Patch
d) All of the Above

Q3. Which of the following categories consists of tools like Sphere, Torus, Quadball, etc.?

a) Freeform Designing
b) Surface Designing
c) Pattern Designing
d) None of the Above

Q4. Which of the following tools is used to break the merged edges in modifying the freeform objects?

a) Delete
b) Subdivide
c) Unweld Edges
d) Crease Edges

Q5. Which of the following tools is applied to the faces of freeform bodies so that any change made in one face will also be reflected in other face?

a) Mirror
b) Symmetry
c) Bridge
d) Both a and b

Q6. The **Thicken** tool is used to thicken the surfaces. (True/False)

Q7. The **Flatten** tool is used to fit selected vertices to the selected plane. (True/False)

Q8. The two vertices of a freeform object can be breaked by **Weld Vertices** tool. (True/False)

Q9. The **Merge Edges** tool is used to merge .... of the two freeform bodies.

Q10. Whenever you need an unconventional shape of the object, you can always use the .... modeling techniques.

Chapter 14

Drawing Creation

Topics Covered

The major topics covered in this chapter are:

- *Elements of Engineering Drawing*
- *Starting Drawing File and managing drawing Sheets*
- *Views in Engineering Drawing and their placement in Autodesk Inventor*
- *Applying annotations to drawing views*
- *Engineering Drawing Symbols*
- *Inserting symbols in Autodesk Inventor Drawings*
- *GD&T and Bill of Materials*
- *Practical and Practice on Drawing Creation*

INTRODUCTION

Drawing is a very important part of daily life of engineers in workshop. If you are manufacturing a component then there are various steps at which you will need engineering drawing of component. We need engineering drawing while,

- Programming CNC machines for machining component.
- Manufacturing dies for casting, molding, and sheetmetal work.
- Performing quality check on prepared component.
- Preparing costing and budget for manufacturing component and many other things.

In Autodesk Inventor, there is a separate environment to handle drawing. Note that the Drawing environment is well synchronized with Modeling environment and assembly environment. So, any change made in modeling environment or assembly environment is also reflected in drawing environment and vice-versa. This phenomena is called bidirectional associativity.

Before we move on to drawing creation tools in Autodesk Inventor, it is important to understand some basic concepts of engineering drawings.

ELEMENTS OF ENGINEERING DRAWING

There are various important concepts to be known to Design Engineer before he/she converts the model/assembly to drawing on paper. These concepts are discussed next.

Types of Engineering Drawings

There are mainly two types of engineering drawings for mechanical components; Part Drawing and Assembly Drawing. The part drawings are further classified as Machine Drawing and Production Drawings; refer to Figure-1.

There are various types of Assembly Drawings like Design Assembly Drawing, Installation Assembly Drawing, Catalog Drawing, and so on; refer to Figure-2.

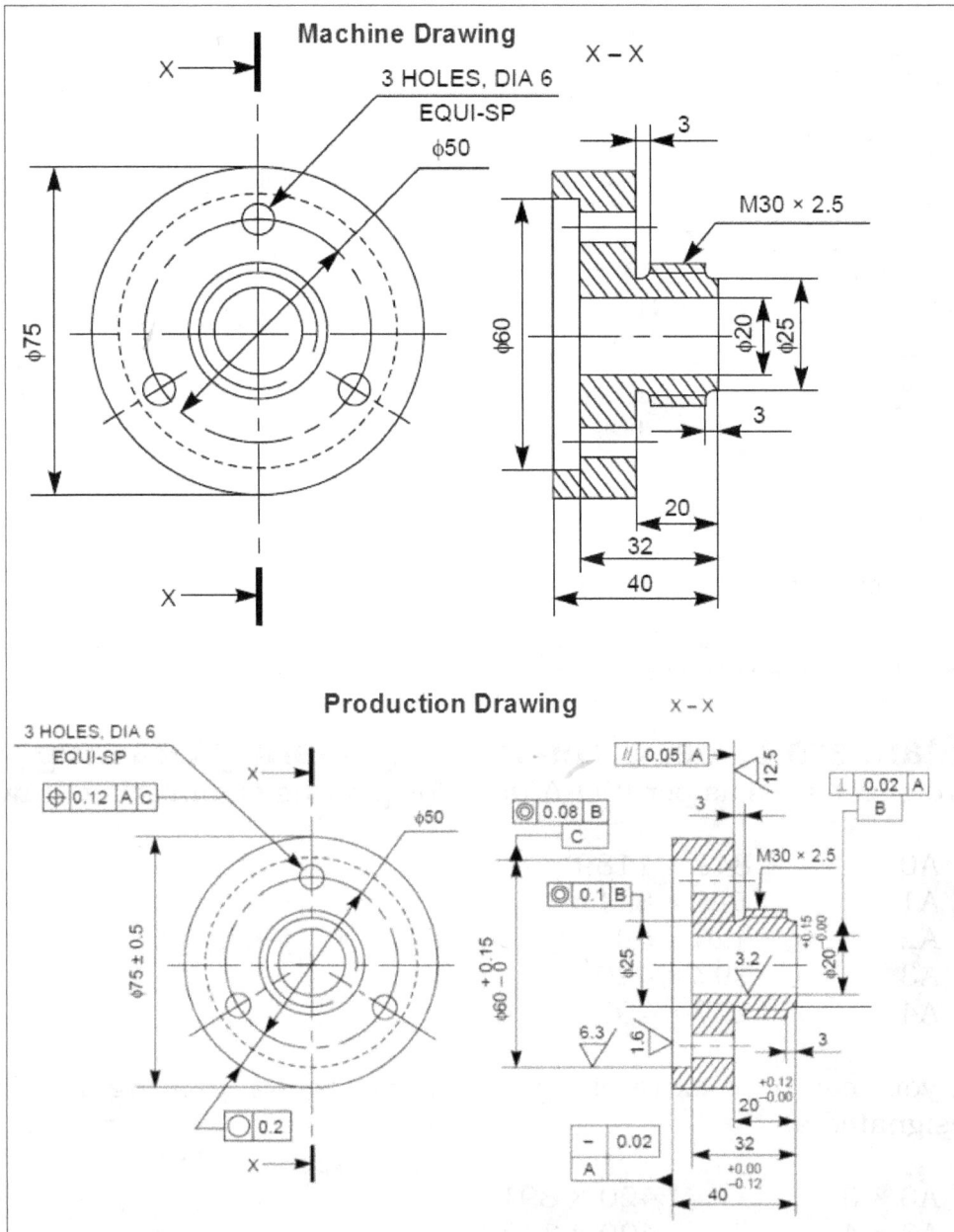

Figure-1. Classification of Part Drawings

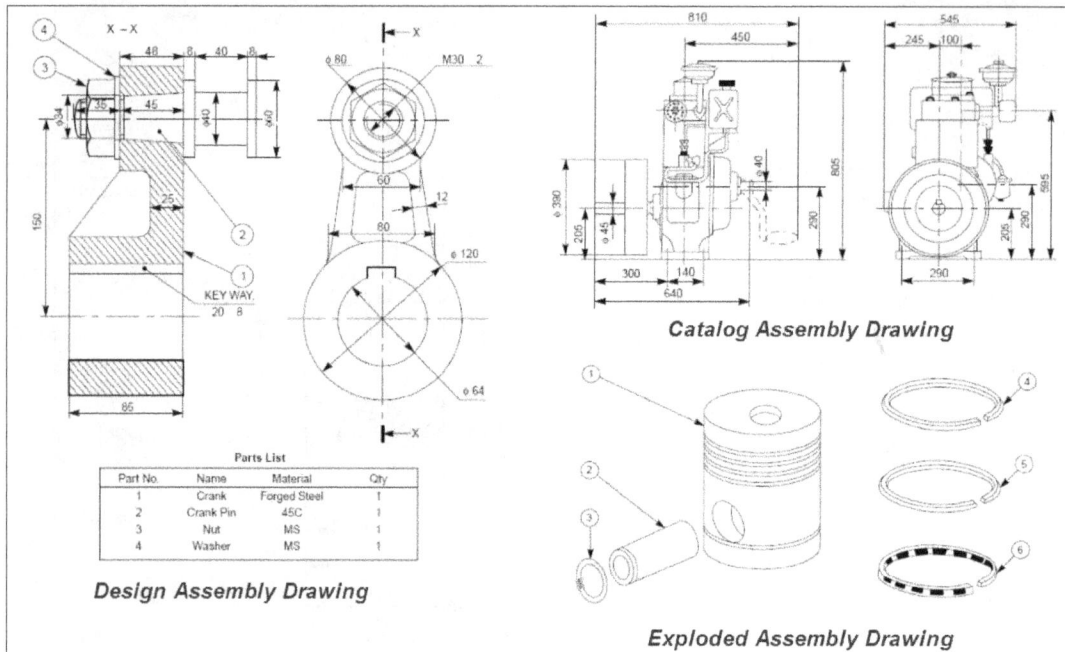

Figure-2. Types of Assembly Drawings

Standard Sheet Sizes for Engineering Drawings

The standard sheet sizes as per ISO-A used for plotting engineering drawings are:

A0	841 × 1189
A1	594 × 841
A2	420 × 594
A3	297 × 420
A4	210 × 297

Sometimes, you may need extra elongated paper size for plotting drawings. These sizes are designated as:

A3 × 3	420 × 891
A3 × 4	420 × 1188
A4 × 3	297 × 630
A4 × 4	297 × 840
A4 × 5	297 × 1050

Title Block

The title block should lie within the drawing space such that, the location of it, containing the identification of the drawing, is at the bottom right hand corner. This must be followed, both for sheets positioned horizontally or vertically; refer to Figure-3.

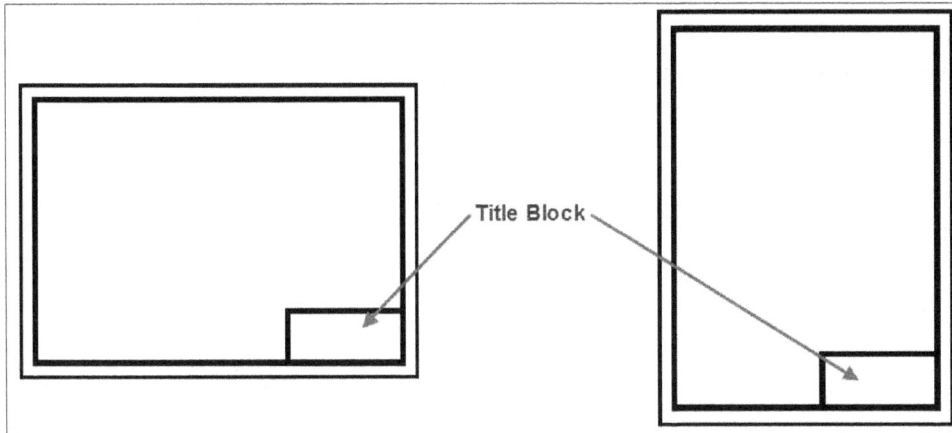

Figure-3. Title block position

The direction of viewing of the title block should correspond in general with that of the drawing. The title block can have a maximum length of **170** mm. Figure-4 shows a typical title block, providing the following information:

1. Title of the drawing
2. Sheet number
3. Scale
4. Symbol, denoting the method of projection
5. Name of the firm
6. Initials of staff drawn, checked, and approved.

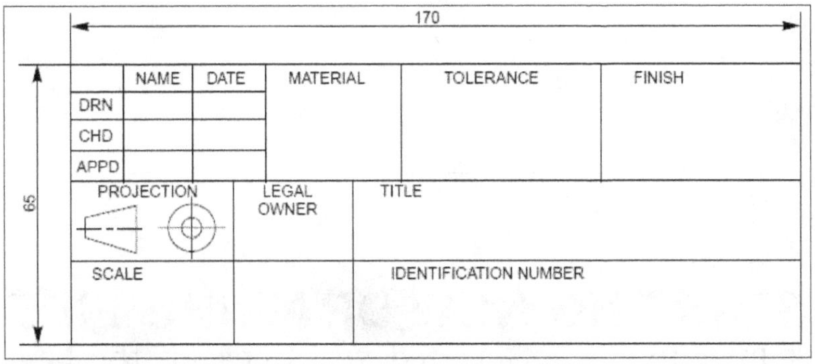

Figure-4. Typical title block

Type of Lines Used in Engineering Drawings

Lines of different types and thicknesses are used for graphical representation of objects. The types of lines and their applications are shown in Figure-5.

These were some of the basics that you need to understand before we start using drawing creation tools in Autodesk Inventor. Note that we will keep on discussing the engineering drawings concepts where ever need in this chapter.

Line	Description	General Applications
A ▬▬▬▬▬	Continuous thick	A1 Visible outlines
B ────────	Continuous thin (straight or curved)	B1 Imaginary lines of intersection B2 Dimension lines B3 Projection lines B4 Leader lines B5 Hatching lines B6 Outlines of revolved sections in place B7 Short centre lines
C ∿∿∿∿	Continuous thin, free-hand	C1 Limits of partial or interrupted views and sections, if the limit is not a chain thin
D ──⋀─⋀─⋀──	Continuous thin (straight) with zigzags	D1 Line (see Fig. 2.5)
E ── ── ── ──	Dashed thick	E1 Hidden outlines
G ──── ── ────	Chain thin	G1 Centre lines G2 Lines of symmetry G3 Trajectories
H	Chain thin, thick at ends and changes of direction	H1 Cutting planes
J ▬▬ ▬ ▬▬ ▬	Chain thick	J1 Indication of lines or surfaces to which a special requirement applies
K ──── ── ──── ──	Chain thin, double-dashed	K1 Outlines of adjacent parts K2 Alternative and extreme positions of movable parts K3 Centroidal lines

Figure-5. Lines in engineering drawings

STARTING NEW DRAWING FILE

- Start Autodesk Inventor, if not started yet. Click on the **New** button from **New** cascading menu in the **File** menu of the **Ribbon**. The **Create New File** dialog box will be displayed.
- Select the **English** or **Metric** template from the **Templates** category in the dialog box and move the slider down in the dialog box; refer to Figure-6.

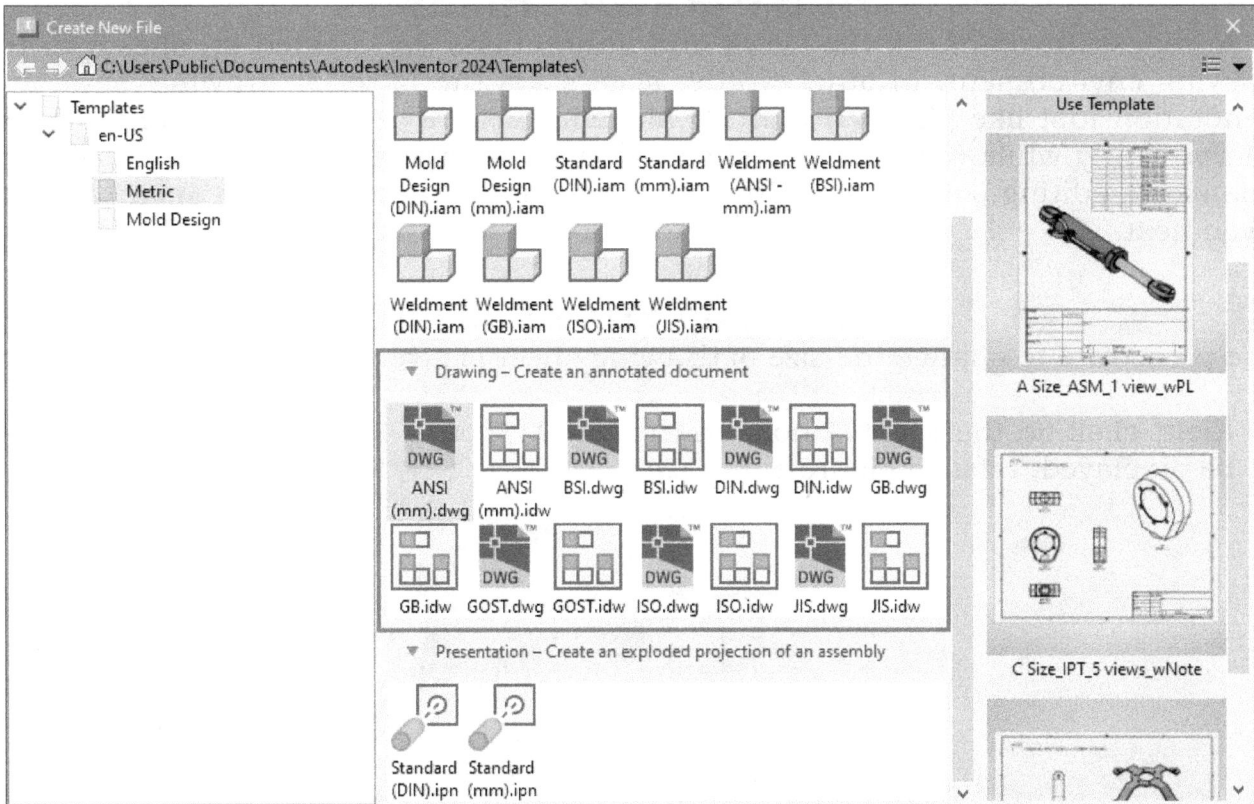

Figure-6. Create New File dialog box

- Select desired template from the **Drawing** area of the dialog box and click on the **Create** button. The drawing environment will be displayed with new file opened; refer to Figure-7 (we have selected **ANSI(mm).idw** template).

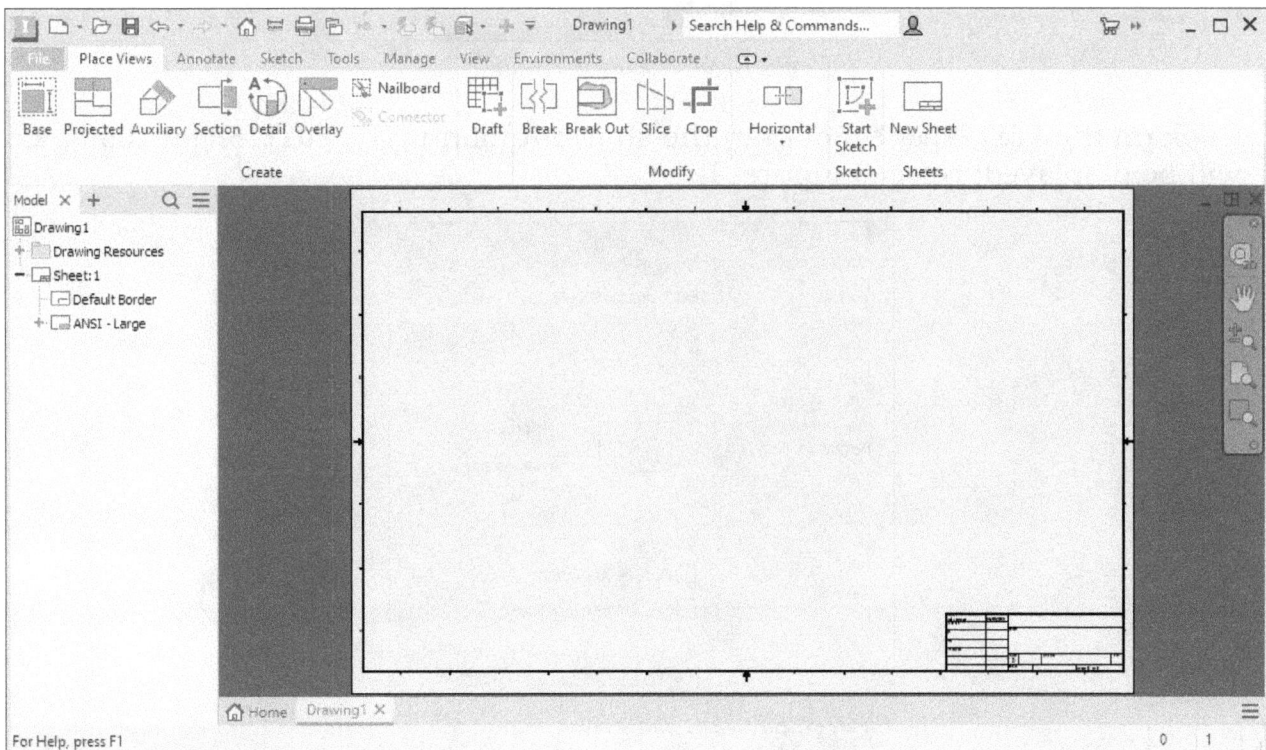

Figure-7. Drawing environment in Autodesk Inventor

SHEET MANAGEMENT

In Autodesk Inventor, all the views and annotations are created on sheets in the drawing environment. In other words, sheets are the base for drawing views and annotation just like planes are in 3D Modeling. There are few operations that we do frequently while working with sheets of drawing like changing size, positioning title block, editing borders, etc. The procedures to do some common operations are given next.

Changing Size of Sheet

The procedure to change the size of drawing sheet is given next.

• Right-click on the **Sheet:1** node in the **Model Browse Bar**. A shortcut menu will be displayed; refer to Figure-8.

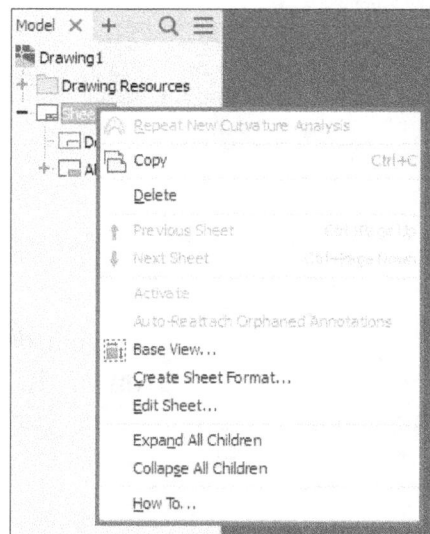

Figure-8. Right-click shortcut menu

• Click on the **Edit Sheet** tool from the shortcut menu. The **Edit Sheet** dialog box will be displayed; refer to Figure-9.

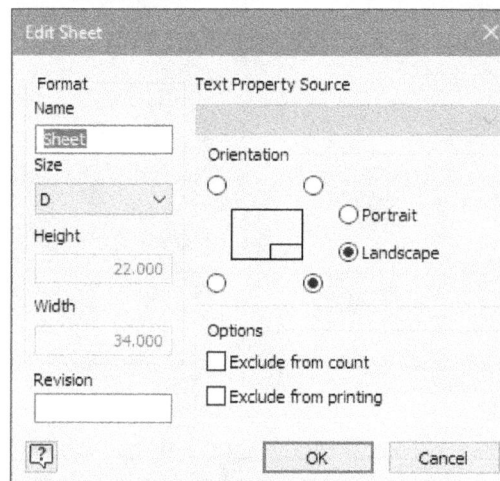

Figure-9. Edit Sheet dialog box

• Specify desired name of sheet in the **Name** edit box.

- Click in the **Size** drop-down and select desired size from the list.
- Set desired revision number in the **Revision** edit box.
- From the **Orientation** area, select radio button to define location of title block. Also, select the **Portrait** or **Landscape** radio button to define the orientation of sheet.
- If you want to exclude the current sheet from counting or printing then select the related check box from the **Options** area of the dialog box.
- Click on the **OK** button to apply the changes. The drawing sheet will be modified accordingly; refer to Figure-10.

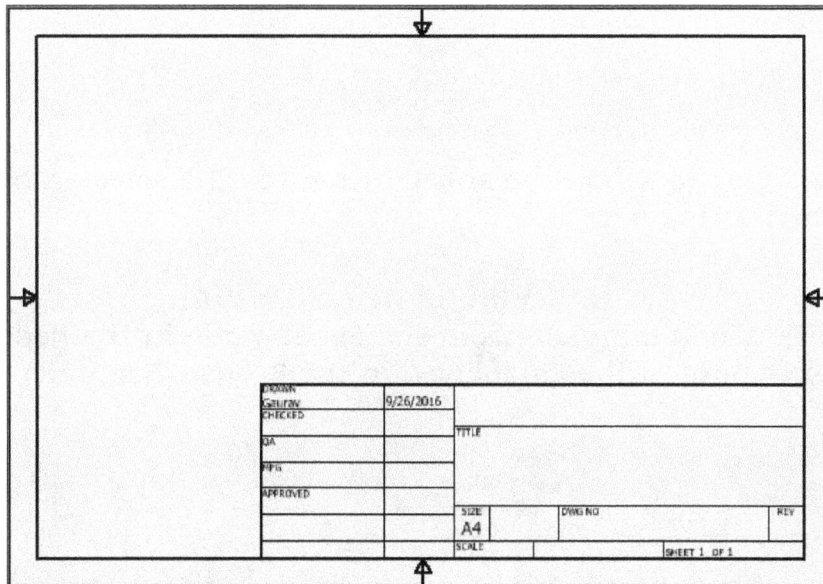

Figure-10. Drawing sheet of A4 size

Note that the size of title block has not changed as per the sheet size. We need to do this work manually.

Changing Title Block

There are two ways in which we generally modify the title block of drawing sheet; changing size of title block and changing parameters in the title block. The methods to do both the operations are discussed next.

Changing Size of Title Block

- Select the title block template from the **Sheet:1** node in the **Model Browse Bar** and press **Delete** from keyboard to delete it; refer to Figure-11.

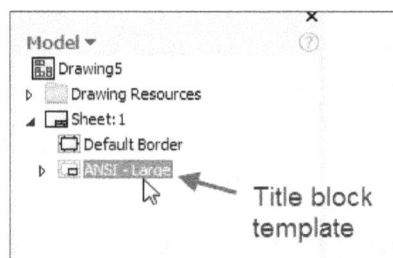

Figure-11. Title block template to delete

- Expand the **Drawing Resources** node and then **Title Blocks** node in the **Model Browse Bar**. The list of title blocks available for current sheet size will be displayed.

- Select desired size from the list and right-click to display the shortcut menu; refer to Figure-12.

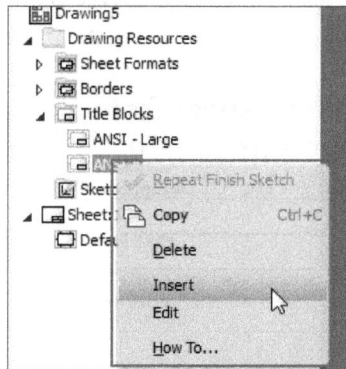

Figure-12. Shortcut menu for title block

- Select the **Insert** option from the shortcut menu. The selected title block will be inserted in the drawing sheet.

Editing Values in Title Block

- Expand the title block template from the **Sheet** node in the **Model Browse Bar**. The **Field Text** option will be displayed in the **Browse Bar**; refer to Figure-13.

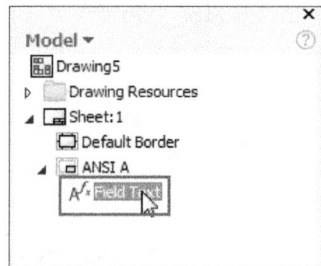

Figure-13. Field Text option

- Double-click on the option. The **Edit Property Fields** dialog box will be displayed; refer to Figure-14.

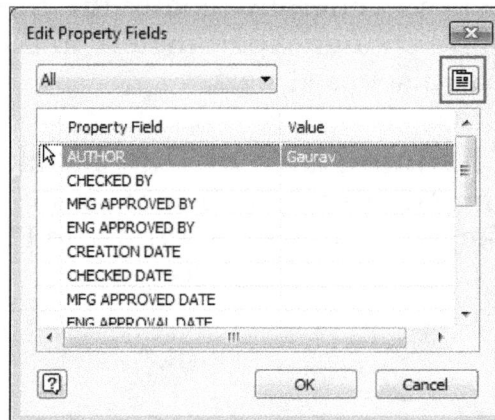

Figure-14. Edit Property Fields dialog box

- Click on the **iProperties** button at the top-right corner of the dialog box. The **Drawing iProperties** dialog box will be displayed; refer to Figure-15.

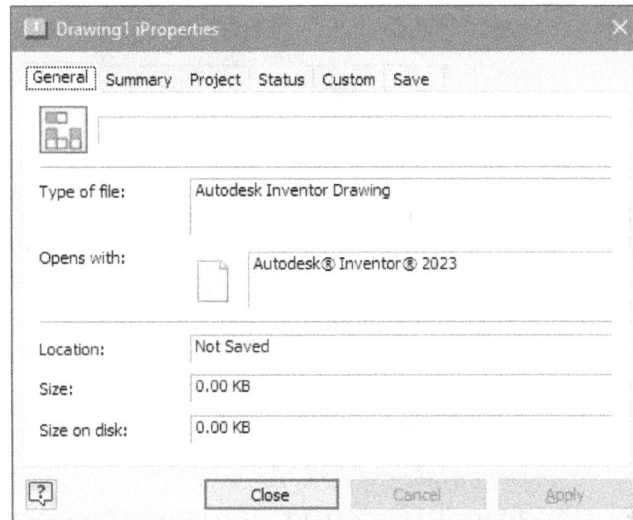

Figure-15. Drawing iProperties dialog box

- Click on the **Summary** tab in the dialog box and specify the parameters of title block like, Title, subject, author of drawing, and so on.
- Click on the **Project** tab in the dialog box and specify the parameters like part number, description, revision number, etc.
- Similarly, specify the other parameters in the dialog box and click on the **Apply** and then **Close** button from the dialog box.
- Click on the **OK** button from the **Edit Property Fields** dialog box. The changes will be reflected in the title block.

Inserting Sheets

Sometimes, while making drawings, you may need more sheets to insert views of the drawings like, in case of machine drawing for assembly. The procedure to insert sheets in drawing is given next.

- Click on the **New Sheet** button from the **Sheets** panel in the **Place Views** tab of the **Ribbon**. A new sheet will be added with the properties of earlier sheets.
- Keep on clicking **New Sheet** button to add as many sheets as required.

Activating and Deleting Sheet

- Right-click on the sheet in the **Model Browse Bar** that you want to activate or delete. A shortcut menu will be displayed; refer to Figure-16.

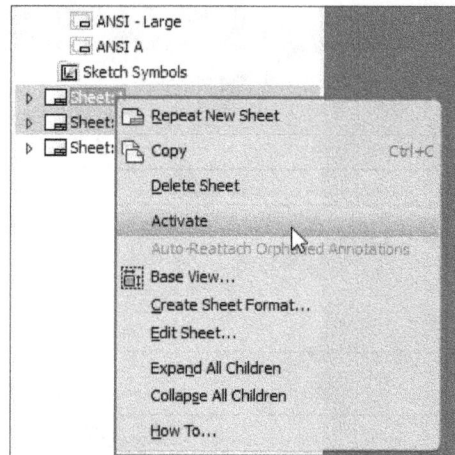

Figure-16. Shortcut menu for sheets

- Select the **Activate** button from the menu to activate the current sheet for inserting views and manipulating them.
- Select the **Delete** button from the menu to delete the current sheet.

VIEWS IN ENGINEERING DRAWING

In Engineering Drawing, 3D objects are represented on paper with the help of various views. A drawing view is projection of 3D object on paper in the direction of view. The most general drawing views used in engineering drawings are:

- Isometric View
- Orthographic View
- Section View
- Detail View
- Auxiliary View

We will now discuss these views in short.

Isometric View or Isometric Projection

Grab any object in front of you. A pencil, remote control, computer mouse, bobble-head doll, coffee cup, etc. Look at it. Now turn it upside down (unless it's full of your favorite beverage!). Turn it around. Turn it sideways. Each time you move it, you're looking at a different view of that object. You can see depth and edges. You can tell that it's not just a picture, but a real thing. Your eyes give you many clues to the objects in your world that are three-dimensional (have height, width, and depth).

A piece of paper has only two dimensions -- it's flat. If you try to draw an object on a piece of paper, you'll notice that it's not easy to make the drawn object look like it has depth. One way is to use an isometric view, which is derived from the Greek words Iso means "equal" and metric means "measurement". When using an isometric view, you line up the drawing along three axes, visible or invisible guidelines that establish directions for measurement, that are separated by 120-degree angles from each other, as shown in Figure-17.

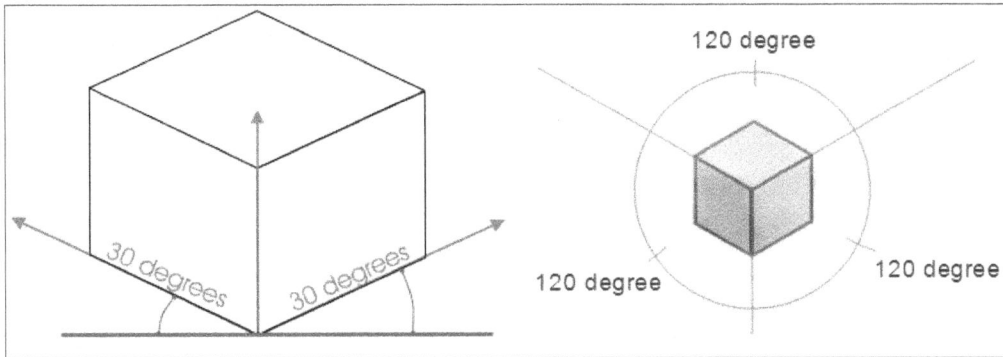

Figure-17. Isometric projection of cube

Orthographic Views or Orthographic Projection

Orthographic (ortho) views are two-dimensional drawings used to represent or describe a three-dimensional object. The ortho views represent the exact shape of an object seen from one side at a time as you are looking perpendicularly to it without showing any depth to the object.

Primarily, three ortho views (top, front, and right) adequately depict the necessary information to illustrate the object; refer to Figure-18. Sometimes, only two ortho views are needed as in a cylinder. The diameter of the cylinder and its length are the only dimension information needed to complete the drawing. A sphere only needs the diameter. It is the same from all angles and remains a perfect circle in the iso drawing.

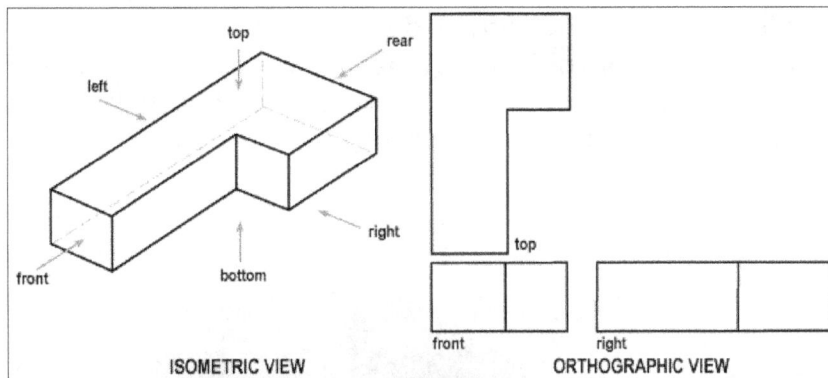

Figure-18. Isometric orthographic representation

The "six" side method is a process of making six primary ortho views that represent the entire image. This method gives you all the information to create complex object.

There are two projection methods to create orthographic views in drawings; First Angle Projection and Third Angle Projection.

First Angle Projection

In First Angle Orthographic Projection, the front, top, and side views are placed opposite to the position from where you are looking at the object. Refer to Figure-19.

In most of the countries except USA, Canada, Japan, and Australia, the First Angle projection is in use rather than Third Angle Projection. You can identify the First Angle Projection by finding symbol as shown in Figure-20.

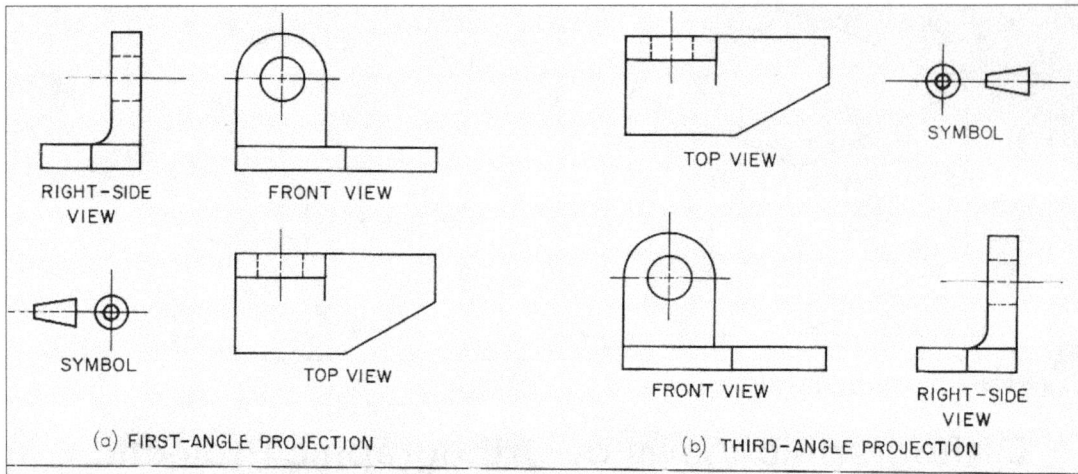

Figure-19. Example of projection

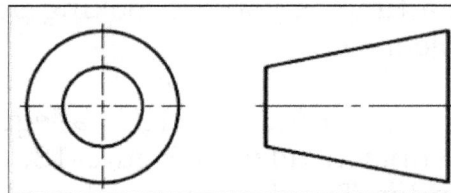

Figure-20. First Angle Projection Symbol

Third Angle Projection

In Third Angle Orthographic Projection, the top view is placed at the top of front view and right side view is placed on the right side of the front view; refer to Figure-18 and Figure-19. If you are still in confusion why have we placed the views like this in first angle and third angle projections then check Figure-21. In First Angle Projection, the object is placed in first quadrant whereas in Third Angle Projection, the object is placed in third quadrant.

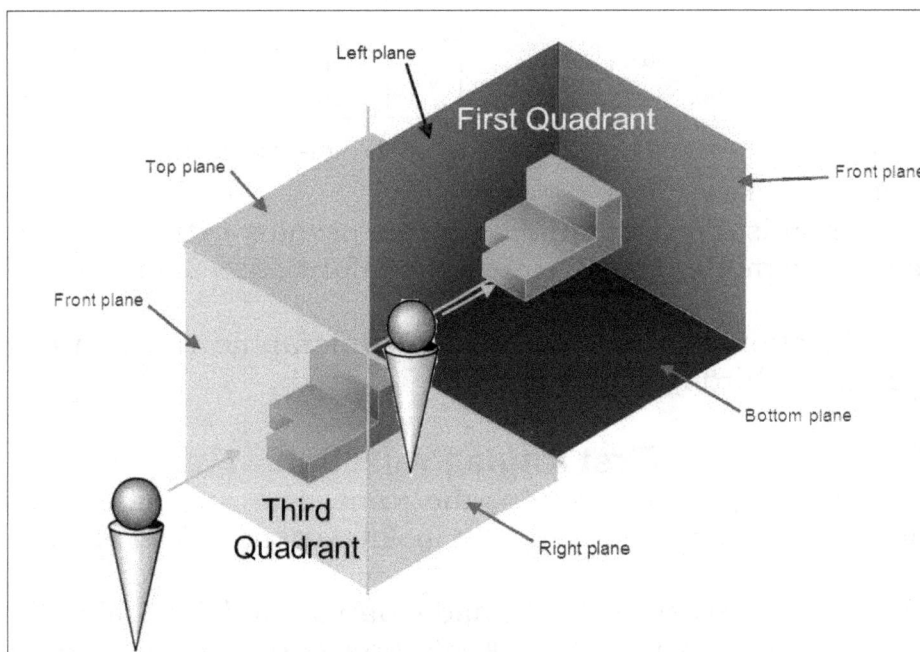

Figure-21. Placement of object in quadrants

Now, we will learn to create these projections in Autodesk Inventor drawing environment.

INSERTING BASE VIEW

In Autodesk Inventor Drawing environment, first drawing view placed is called base view and other views are generally projections. The procedure to insert base view is given next.

- Click on the **Base** tool from the **Create** panel in the **Place Views** tab in the **Ribbon**. The **Drawing View** dialog box will be displayed; refer to Figure-22.

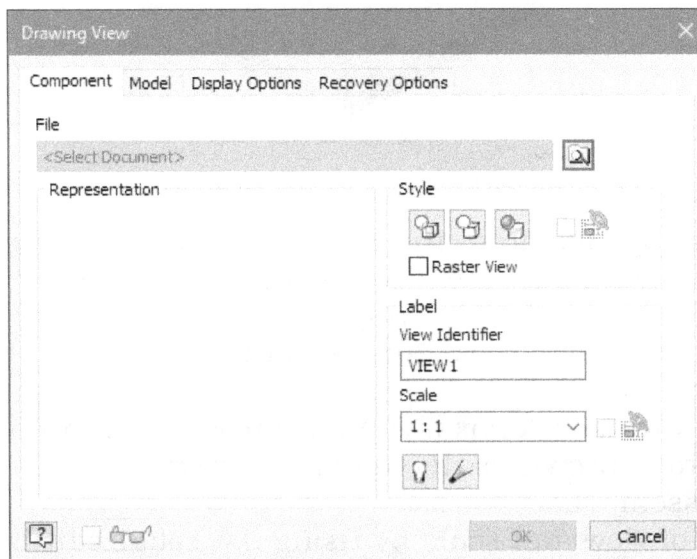

Figure-22. Drawing View dialog box

- Click on the **Open an existing file** button from the dialog box. The **Open** dialog box will be displayed.
- Select the part file or assembly file for which you want to create the drawing views and click on the **Open** button from the dialog box. The selected file will be linked in the drawing file and preview of the drawing view will be displayed on the sheet; refer to Figure-23.

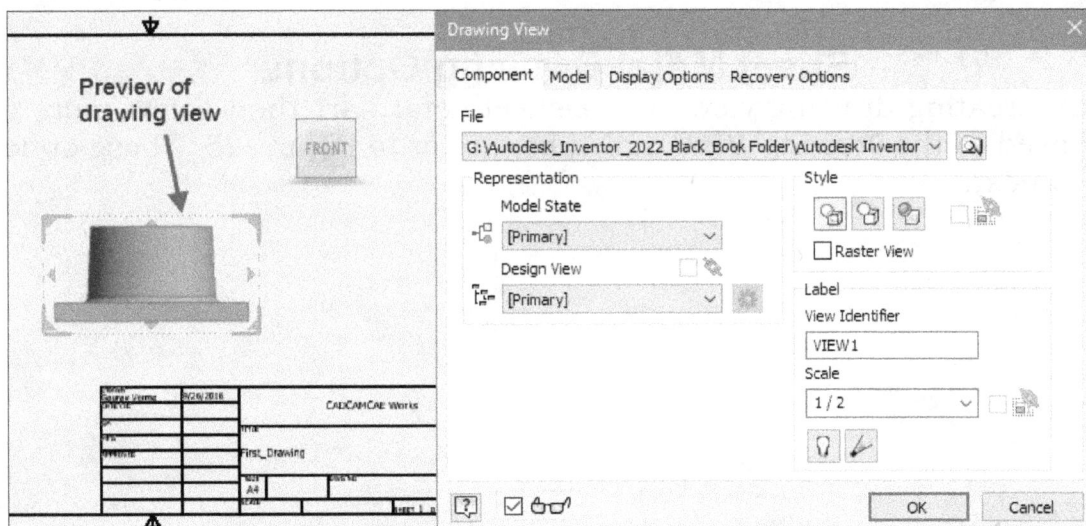

Figure-23. Preview of drawing view

- Click at desired location on the **View Cube** to orient the component to front view, top view, right view, or isometric view; refer to Figure-24.

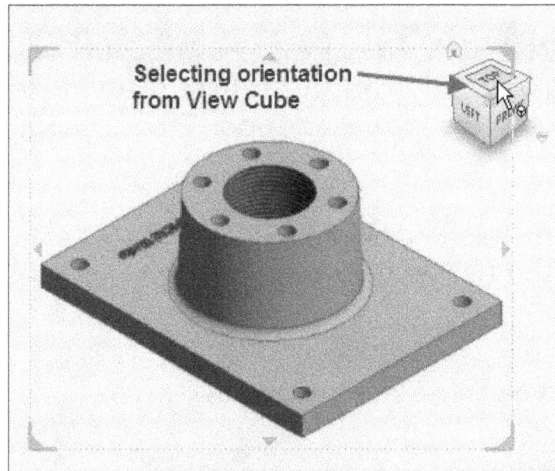

Figure-24. Changing drawing view using View Cube

- Select desired display style from the **Style** area of the dialog box. Like, you can select the **Hidden Line** button to display hidden lines in the view, you can select the **Shaded** button to display drawing view as shaded rather than wireframe or you can select the **Hidden Lines Removed** button to not display hidden lines in the drawing view.
- Select the **Raster View** check box if you want to generate pixel based views rather than accurate parametric views. Raster views are generated very fast so they are useful for large assemblies.
- Change the label of view and scale by using the **Label** edit box and **Scale** drop-down, respectively.
- If there are more than one model states like in case of iAssembly, iParts, or mold design assemblies then select desired model state from the drop-down in the **Model State** tab of the dialog box.
- Set desired display options from the **Display Options** tab in the dialog box.
- Click on the **OK** button to create the view.
- By default, the base view is placed at the center of sheet. Hover the cursor over the borders of view and drag it to desired location for changing placement.

Sheet Metal Drawing Options

If you are creating drawing views for a sheetmetal part then a few more options are displayed in the **Drawing View** dialog box; refer to Figure-25. These options are discussed next.

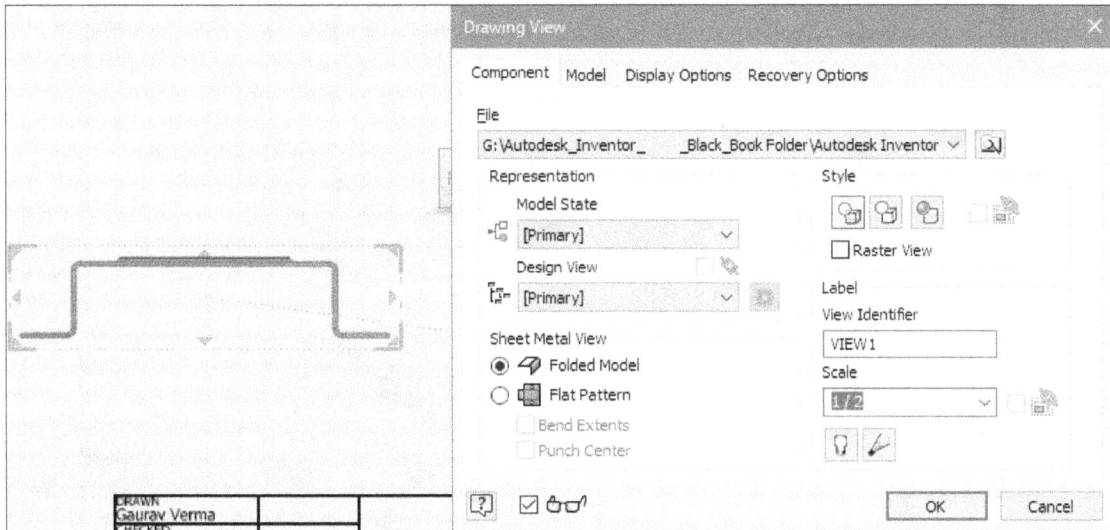

Figure-25. Sheet Metal View area in Drawing View dialog box

- By default, the **Folded Model** radio button is selected in the **Sheet Metal View** area of the dialog box. So, the model being inserted in drawing views is folded.
- Select the **Flat Pattern** radio button if you want to insert the flat pattern drawing view of the model.
- Select the **Bend Extents** check box and **Punch Center** check box if you want to display bend extends and punch center mark in the drawing view.

Similarly, you can use assembly and presentation files for view insertion. Note that exploded view of assembly are created in Presentation environment of Autodesk Inventor.

After placing the base view, next step is to insert projection views to display all details of 3D model. Before we start inserting the projection views, it is important to understand how we can change the parameters of drawing like 1st angle to 3rd angle projection, line width, drawing units, etc.

CHANGING STANDARDS AND STYLES

There are various parameters of drawing which keep on changing based on the Standards used in the company. The procedure to change the standards and styles is given next.

- Click on the **Styles Editor** tool from the **Styles and Standards** panel in the **Manage** tab of the **Ribbon**. The **Style and Standard Editor** dialog box will be displayed; refer to Figure-26.

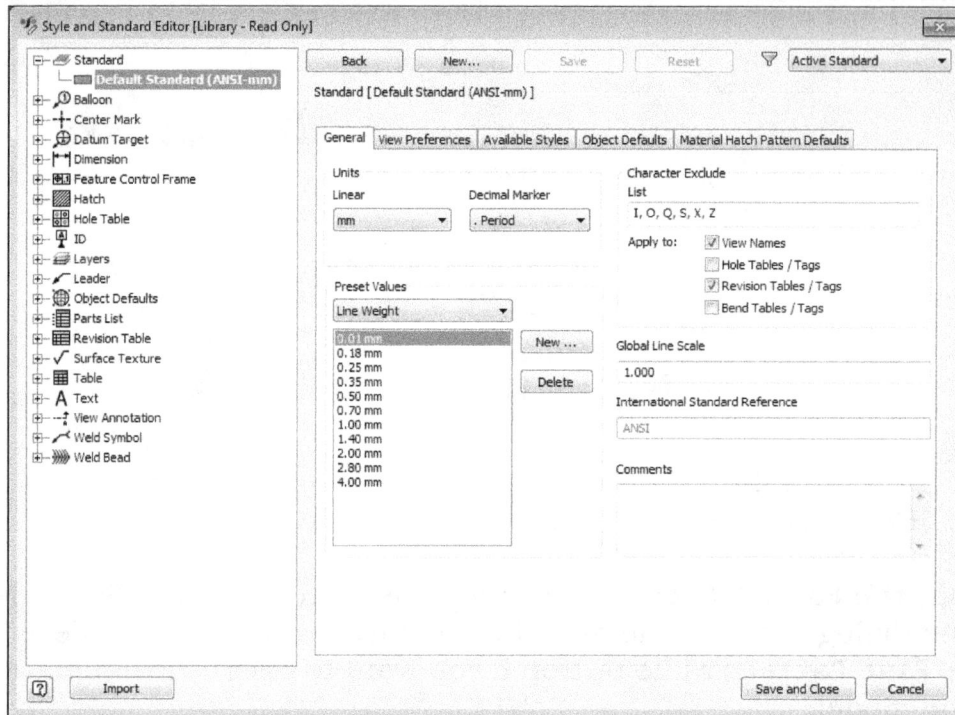

Figure-26. Style and Standard Editor dialog box

- Select desired unit from the **Linear** drop-down in the **Units** area of the dialog box. Similarly, set desired decimal marker from the **Decimal Marker** drop-down in the **Units** area.
- Set the standard line scale using the **Global Line Scale** edit box.
- After performing the changes, click on the **Save** button at the top in the dialog box.

Changing Projection Type and Other View Related Parameters

- Click on the **View Preferences** tab in the dialog box to display view related options; refer to Figure-27.

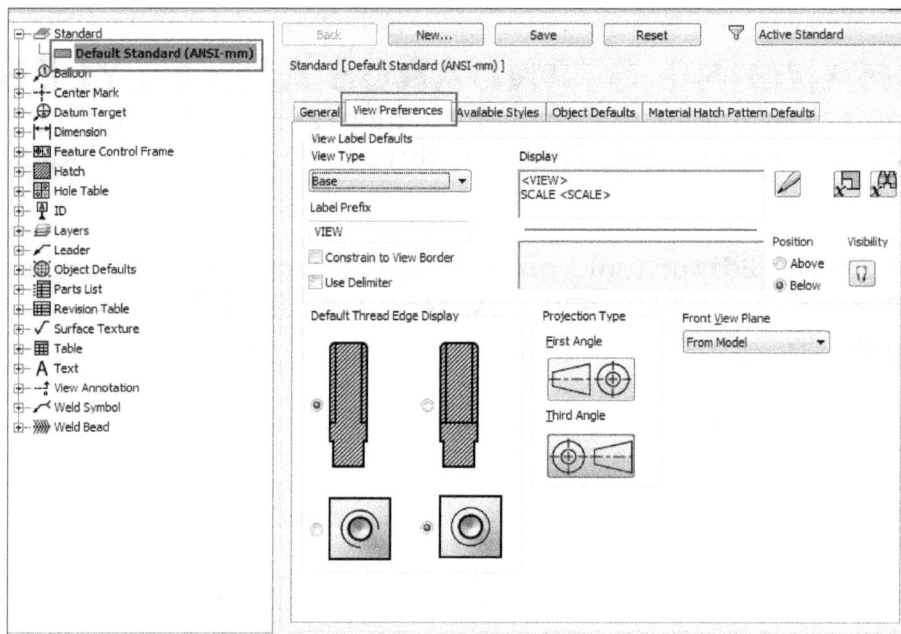

Figure-27. View Preferences tab in the Style and Standard Editor dialog box

- Select desired projection type from the **Projection Type** area of the dialog box. There are two buttons in this area; **First Angle** button to use First Angle Projection in drawing views and **Third Angle** button to use Third Angle Projection in drawing views.
- Select desired radio button in the **Default Thread Edge Display** area of the dialog box to represent threads as required in drawings.
- Similarly, you can change the default label for drawing views and toggle the visibility of labels by using the options in the **Display** area of the tab.
- After performing the changes, click on the **Save** button at the top in the dialog box.

Note that you can modify the drawing view parameters before creating the base view and projections. After creating the views, the modifications done in the dialog box will not be reflected in the drawing.

Modifying Layers of Drawing Objects

- Expand the **Layers** node in the left area of the dialog box. The list of various layers will be displayed; refer to Figure-28.

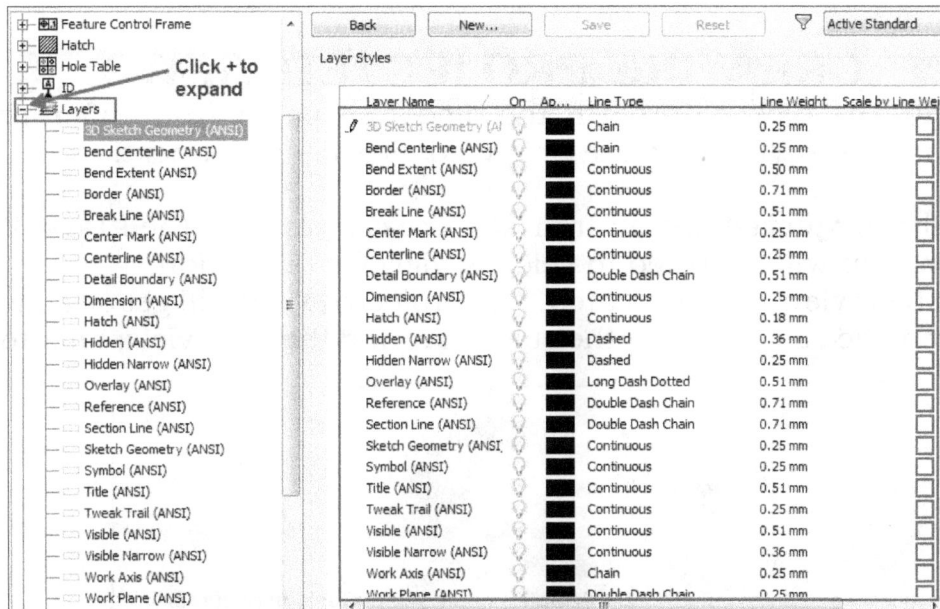

Figure-28. Expanded list of layers

- Select the layer of object and click on the value of the layer to modify it in the table. The option to change it will be displayed. Like, if you want to change the list type for **Bend Centerline (ANSI)** then click on the **Chain** option for it in the table. The selected value will change to a drop-down; refer to Figure-29.

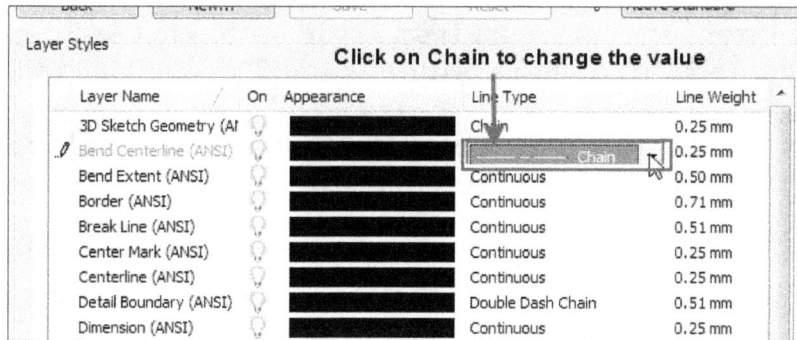

Figure-29. Changing parameters for layer

- Click again on the value to expand the drop-down and select desired value. Similarly, you can change the other values in the table.
- After performing the changes, click on the **Save** button at the top in the dialog box.

There are various other parameters which will be used later in this chapter. You will learn to modify them later in this chapter. For now, click on the **Save and Close** button from the dialog box to exit.

CREATING PROJECTED VIEWS

Projected views are used to present the model on paper from different sides and angles. The procedure to create the projected views is given next.

- Click on the **Projected** button from the **Create** panel in the **Place Views** tab of the **Ribbon**. You will be asked to select a view for projection.
- Select the base view earlier created. The projected view will get attached to cursor and gets modified as you move the cursor around the base view; refer to Figure-30.

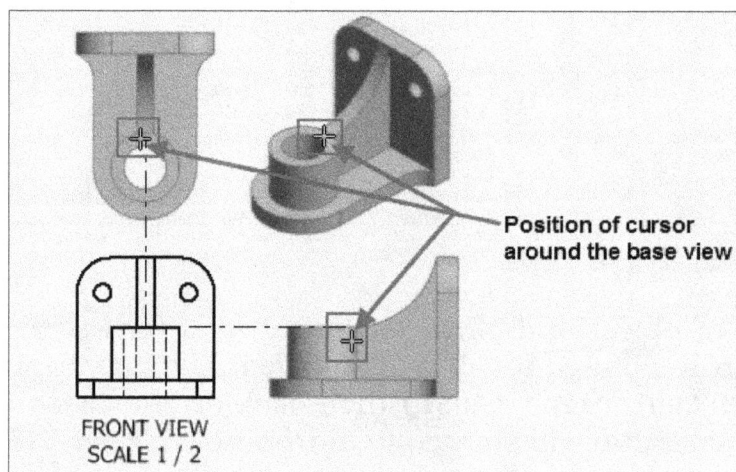

Figure-30. Preview of Projected view

- Click in desired location to place the projection view. Note that the projection view is automatically created based on the Projection Type selected in the **Style and Standard Editor** dialog box.
- You can click at other locations to place more projected views of the model.
- After placing desired number of views, right-click in the drawing area and select the **Create** button from the interactive shortcut menu displayed. The projections views will be created; refer to Figure-31.

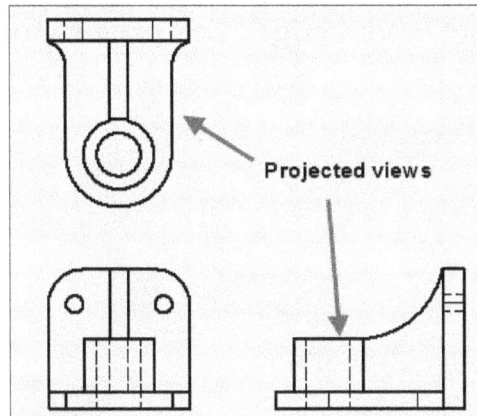

Figure-31. Projected views created

Note that the properties of base view are automatically applied to the projected views. Like, if base view is not shaded then projected views will also be not shaded by default.

CREATING AUXILIARY VIEW

Auxiliary view is an orthographic view taken in such a manner that the lines of sight are not parallel to the principal projection planes (frontal, horizontal, or profile). There are an infinite number of possible auxiliary views of any given object. When creating engineering drawings, it is often necessary to show features in a view where they appear true size so that they can be dimensioned. The object is normally positioned such that the major surfaces and features are either parallel or perpendicular to the principal planes; refer to Figure-32.

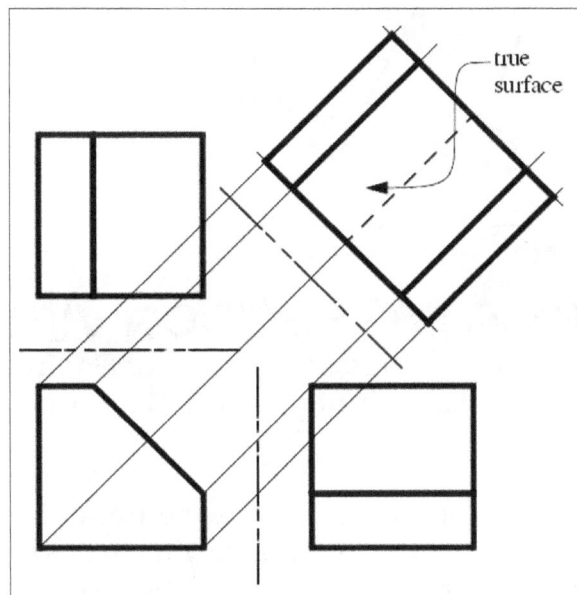

Figure-32. Auxiliary view example

The procedure to create auxiliary view is discussed next.

• Click on the **Auxiliary** tool from the **Create** panel in the **Place Views** tab of the **Ribbon**. You will be asked to select a view.
• Select the view using which you want to create the auxiliary view. The **Auxiliary View** dialog box will be displayed and you will be asked to select the linear model edge to which the auxiliary view plane will be perpendicular.

- Set the view identifier and scale using the respective options in the dialog box and then select the linear model edge. The preview of the auxiliary view will be displayed; refer to Figure-33.

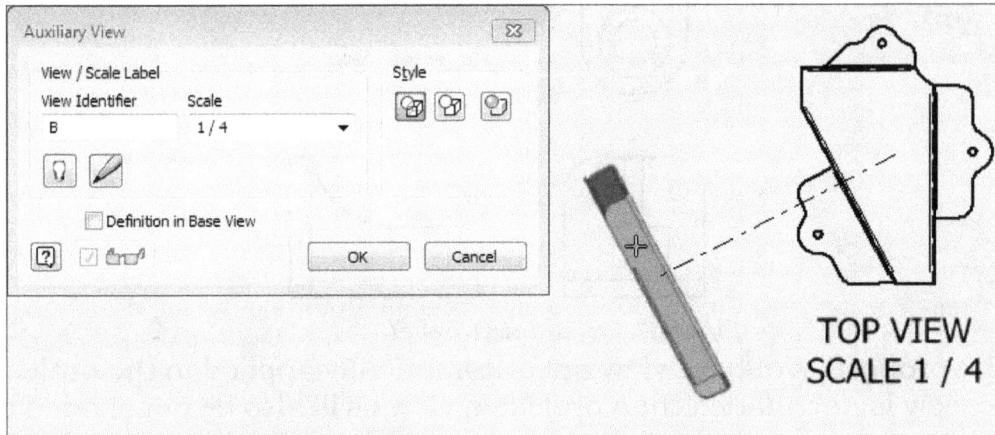

Figure-33. Preview of auxiliary view

- Click in the drawing area to place the view; refer to Figure-34.

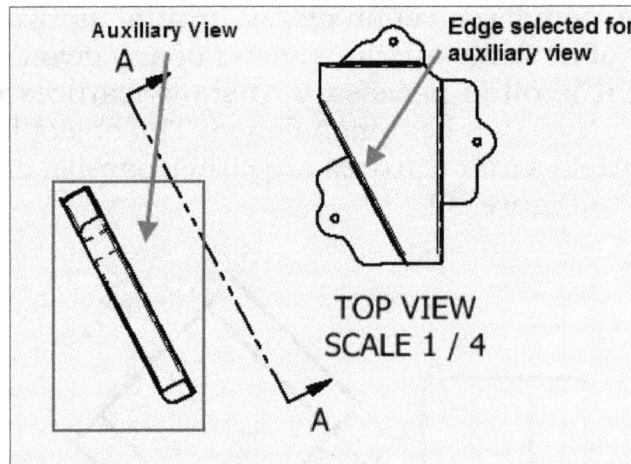

Figure-34. Auxiliary view created

CREATING SECTION VIEW

The **Section View** tool in Autodesk Inventor Drawing Environment is used to create section of the solid model to display inner details. The procedure to use this tool is given next.

- Click on the **Section view** tool from the **Create** panel in the **Place Views** tab of the **Ribbon**. You will be asked to select a base view whose section is to be created.
- Select a view from the drawing area. You will be asked to specify end points of the section line.
- Click to specify the first end point (or say starting point). You will be asked to specify next point of the line segment.
- Click on desired locations to create the section line; refer to Figure-35. Note that the section line is multiline entity.

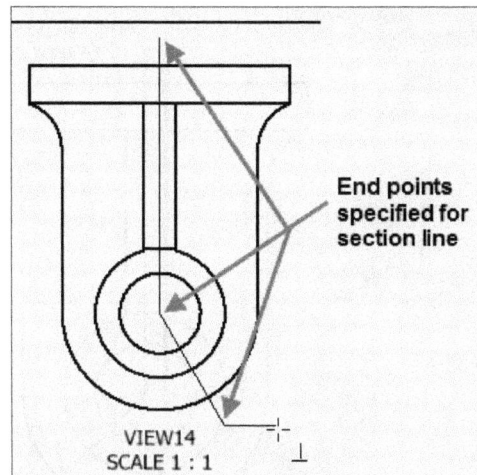

Figure-35. End points of section-line

- Right-click in the drawing area and click on **Continue** button from the shortcut menu. Preview of the section view will be displayed along with **Section View** dialog box; refer to Figure-36.

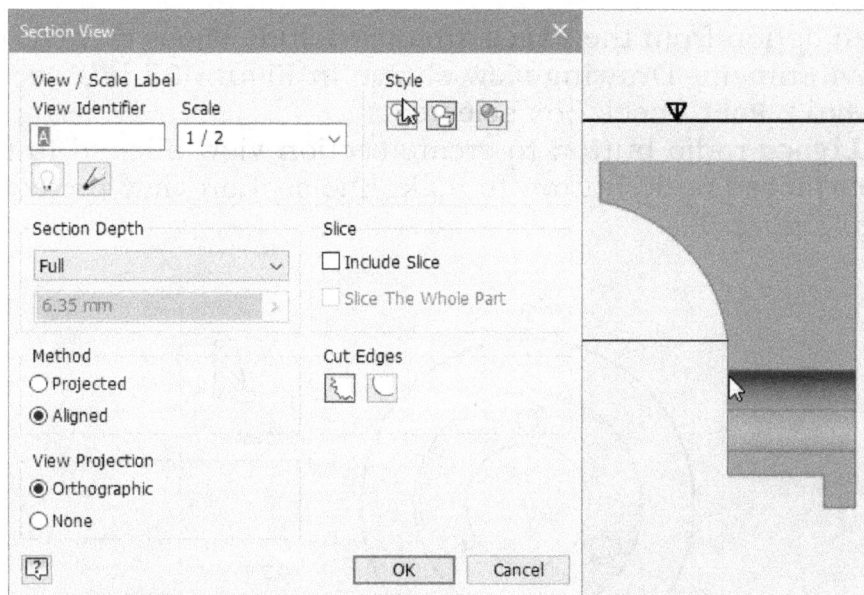

Figure-36. Preview of section view with Section View dialog box

- Specify the alphabet for view identifier in the **View Identifier** edit box.
- Select desired scale from the **Scale** drop-down in the dialog box.
- If you want to create section up to specified depth then select the **Distance** option from the drop-down in the **Section Depth** area and specify the distance value in the edit box below drop-down. Preview of the section will be displayed; refer to Figure-37.

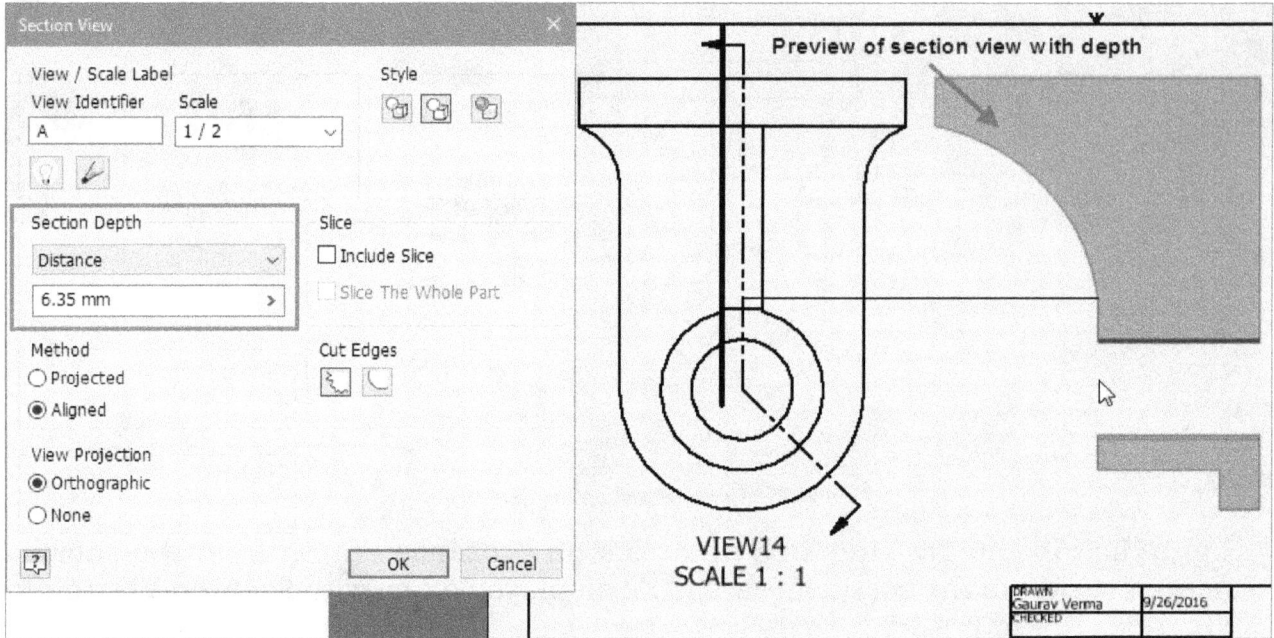

Figure-37. Preview of section view with depth

- Select desired option from the **Slice** area to include slices created for the part in modeling environment. Drawing view shown in Figure-38 is a section view with **Slice The Whole Part** check box selected.
- Select the **Aligned** radio button to create section view aligned to the base view. Select the **Projected** radio button to make the section view projection of section lines; refer to Figure-39.

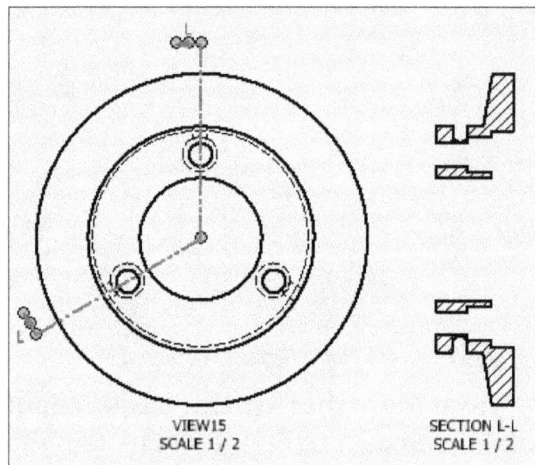

Figure-38. Section view with slice the whole part check box selected

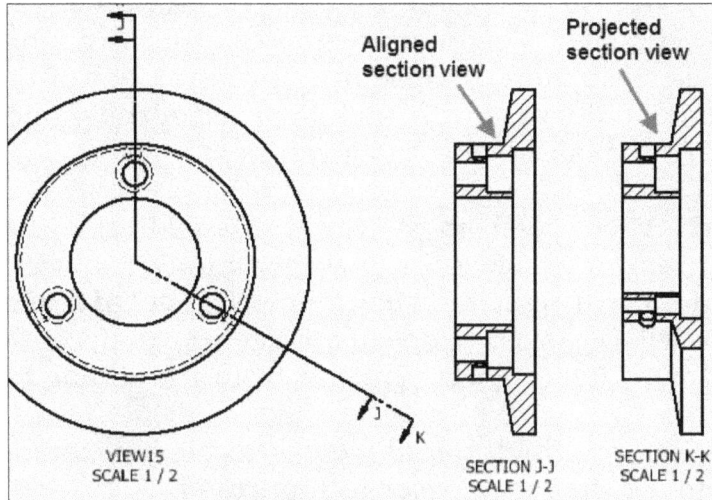

Figure-39. Section views created

- Click in the drawing area to place the view.

Modifying Hatching in Section View

Sometimes, you need to change the hatch pattern in section view to represent different type of materials. The procedure to change hatching is given next.

- Double-click on the hatching in section view. The **Edit Hatch Pattern** dialog box will be displayed; refer to Figure-40.

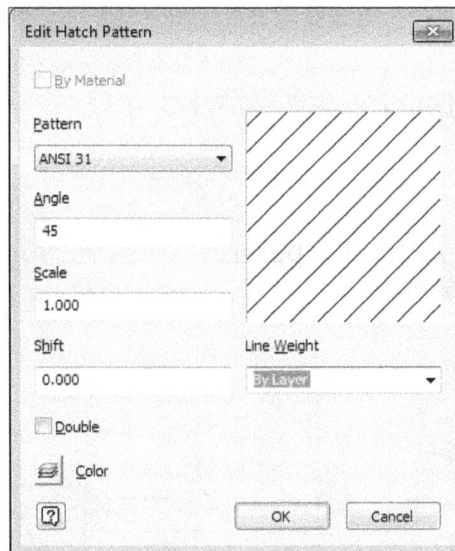

Figure-40. Edit Hatch Pattern dialog box

- Select the type of hatch pattern from the **Pattern** drop-down in the dialog box. Preview will be displayed on the right in the dialog box.
- Specify the other parameters as required and click on the **OK** button to apply the changes.

CREATING DETAIL VIEW

The detail view is an enlarged view of a portion of other drawing view. Detail views are used to provide clearer, more precise annotation. Most of the time detail views are provided to dimension curves at the small corner of the model. The procedure to create detail view in Autodesk Inventor is given next.

*   Click on the **Detail View** tool from the **Create** panel in the **Place Views** tab of the **Ribbon**. You will be asked to select the view for which detail view is to be generated.
*   Click on the view for creating detail view fence. The **Detail View** dialog box will be displayed and you will be asked to specify the center point of the detail view fence.
*   Select the **Circular** or **Rectangular** button from the dialog box to set the shape of detail view fence.
*   Click on the location for which detail view is to be generated. You will be asked to specify end point of the fence; refer to Figure-41.

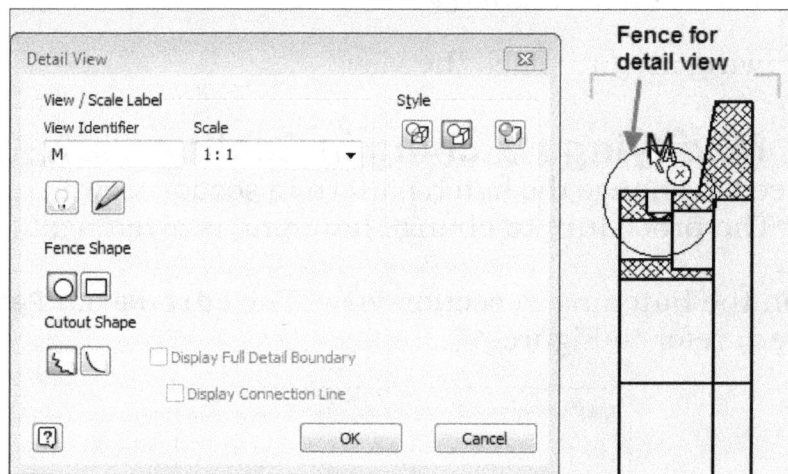

Figure-41. Detail View dialog box with detail view fence

*   Click to specify the span of detail view fence. The detail view get attached to the cursor.
*   Specify desired parameters in the **Detail View** dialog box and click to place the detail view; refer to Figure-42.

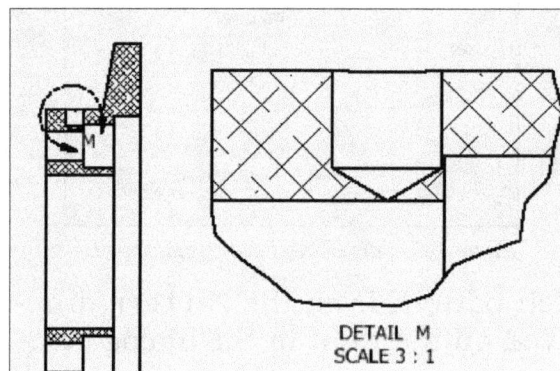

Figure-42. Detail view created

CREATING OVERLAY VIEW

The Overlay view, also called multi-positional view, is used to represent various positions of components of assembly in single view; refer to Figure-43. The procedure to create overlay view is given next.

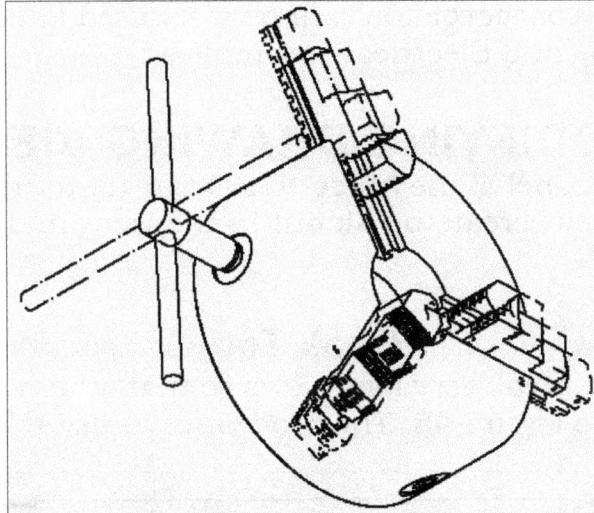

Figure-43. Overlay view example

- Click on the **Overlay** button from the **Create** panel in the **Place Views** tab of the **Ribbon**. You will be asked to select the drawing view for which overlay views are to be created.
- Select the drawing view. The **Overlay View** dialog box will be displayed; refer to Figure-44.

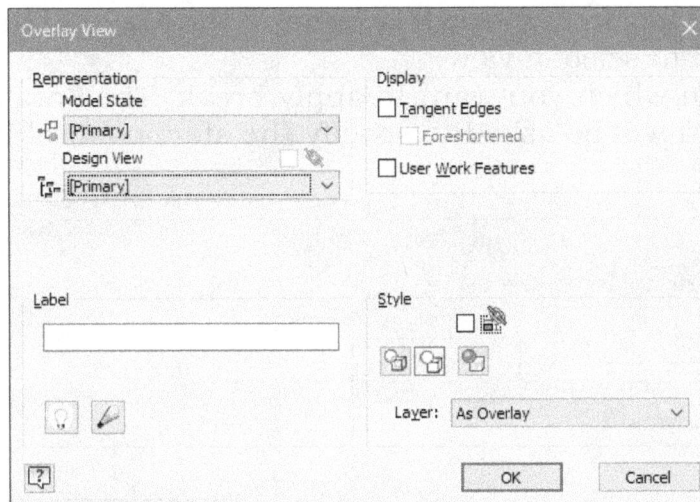

Figure-44. Overlay View dialog box

- Select the position of assembly that you want to overlay on the current drawing view from the **Model State** and **Design View** drop-downs in the **Representation** area of the dialog box.
- Set the other parameters in the dialog box as required and click on the **OK** button. The overlay feature will be created.

Note that the assembly used for overlay drawing view must have more than one positional views. You have learned about creating positional representations in Chapter 12 of this book under the heading "Changing Representation of Mold". You can apply the same method to any assembly.

The **Nailboard** tool and **Connector** tool will be discussed later in this book when we will be discussing Piping and Electrical assemblies.

MODIFYING DRAWING VIEW

The tools in the **Modify** panel of the **Place Views** tab are used to modify the drawing views like break long view, create break out in the view, and so on. These tools are discussed next.

Break Tool

The **Break** tool is used to break very long views so that other views can accommodate in smaller sheet; refer to Figure-45. The procedure to use this tool is given next.

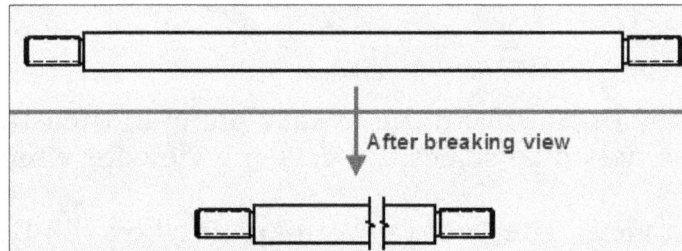

Figure-45. Example of breaking view

- Click on the **Break** tool from the **Modify** panel in the **Place Views** tab of the **Ribbon**. You will be asked to select a view.
- Select the view on which you want to apply break. The **Break** dialog box will be displayed and you will be asked to specify the start point of break section; refer to Figure-46.

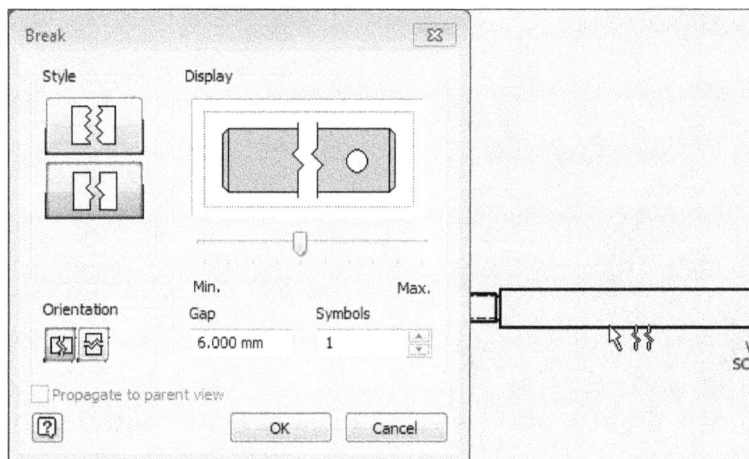

Figure-46. Break dialog box

- Select the style and orientation of the break feature from the **Style** and **Orientation** area of the dialog box.
- Set the gap of break symbol and number of symbols using the **Gap** and **Symbols** edit boxes, respectively.

- Click on the drawing view to specify starting and then end point; refer to Figure-47. The break feature will be applied to the view.

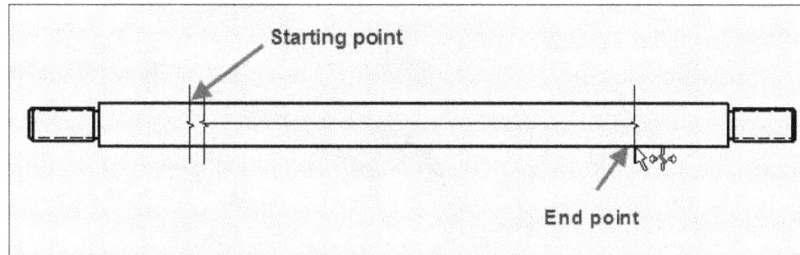

Figure-47. Specifying starting and end point for break

Creating Break Out View

The break out views are used to display inner details of smaller sections of the component. The procedure to create break out view is given next.

- Click on the **Start Sketch** tool from the **Sketch** panel in the **Place Views** tab of the **Ribbon**. You will be asked to select the view to create sketch section for break out view.
- Select the drawing view. Tools of the sketching environment will be activated.
- Create a closed section at desired location on the drawing view and click on the **Finish Sketch** button from the **Exit** panel in the **Ribbon**.
- Now, click on the **Break Out** tool from the **Modify** panel in the **Place Views** tab of the **Ribbon**. You will be asked to select the view with sketch created earlier.
- Select the drawing view. The sketch section will get selected automatically and you will be asked to specify a point for depth reference. Also, the **Break Out** dialog box will be displayed; refer to Figure-48.

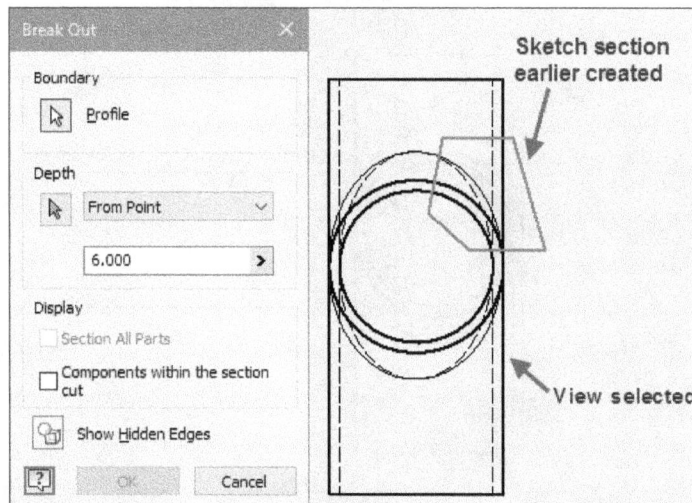

Figure-48. Break Out dialog box with drawing view selected

- Click at desired location inside the sketch created to specify the reference for depth.
- Specify desired value of depth of break out section and click on the **OK** button. The break out view will be created. You can create a projected view of the break out view to give better view of the inside of component; refer to Figure-49.

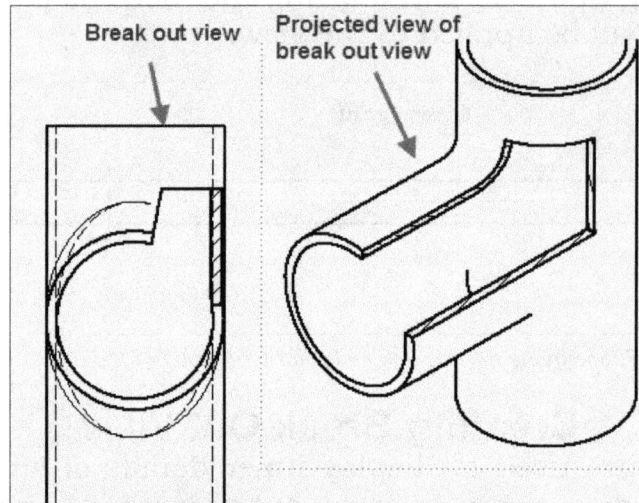

Figure-49. Break out view with its projected view

Creating Slice of Drawing View

The Slice tool is used to create 2D slices of the selected drawing view in its projected representation. The procedure to use this tool is given next.

*   Click on the **Start Sketch** tool from the **Sketch** panel in the **Place Views** tab of the **Ribbon**. You will be asked to select the view to create sketch section for break out view.
*   Select the drawing view. Tools of the sketching environment will be activated.
*   Create an open loop sketch at desired location on the drawing view and click on the **Finish Sketch** button from the **Exit** panel in the **Ribbon**; refer to Figure-50.

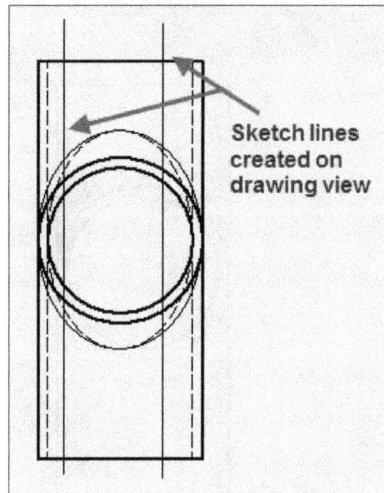

Figure-50. Sketch created for slice

*   Click on the **Projected** tool from the **Create** panel in the **Place Views** tab of the **Ribbon** and create a projected view of base view; refer to Figure-51.

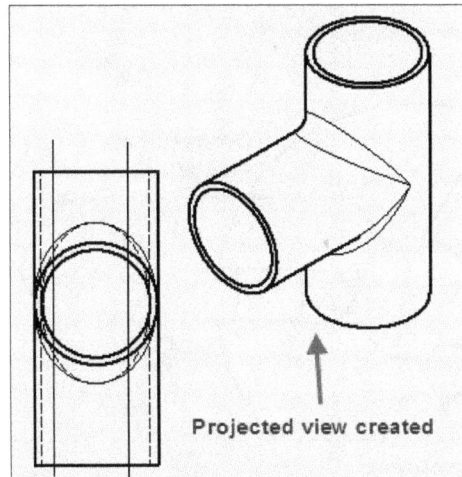

Figure-51. Projected view created

- Click on the **Slice** tool from the **Modify** panel in the **Place Views** tab of the **Ribbon**. You will be asked to select the view to be sliced.
- Select the projected view. The **Slice** dialog box will be displayed and you will be asked to select the sketch for slicing.
- Select the sketch earlier created for slicing. The **OK** button in **Slice** dialog box will become active.
- Select the **Slice All Parts** check box to slice the complete part in view and click on the **OK** button from the **Slice** dialog box displayed. The sliced view will be displayed; refer to Figure-52.

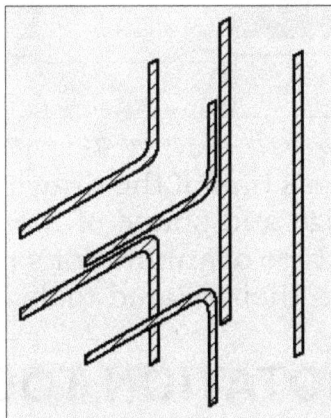

Figure-52. Sliced view

Cropping View

The **Crop** tool is used to crop the drawing view so that only small portion of the drawing can be displayed. The procedure to use this tool is given next.

- Click on the **Crop** tool from the **Modify** panel in the **Place Views** tab of the **Ribbon**. You will be asked to select the view to be cropped.
- Select the drawing view to be cropped. You will be asked to select the first rectangle corner.
- Click at desired location on the view. You will be asked to select second corner of the crop rectangle.
- Click at desired location. The drawing inside the rectangle will remain and rest will be removed; refer to Figure-53.

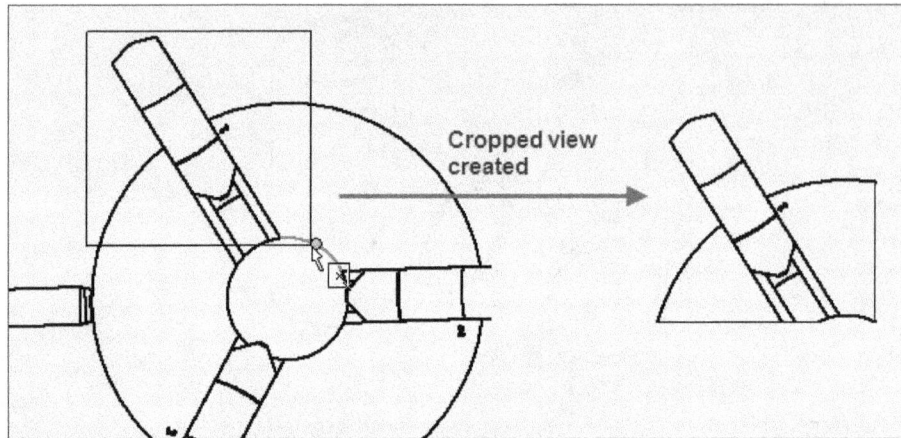

Figure-53. Cropped view created

The alignment tools available in the **Break Alignment** drop-down of the **Modify** panel are used to break or modify the alignment between base view and other child views; refer to Figure-54. I hope you can work on them by yourself. If you get any doubt, please E-mail at **cadcamcaeworks@gmail.com**

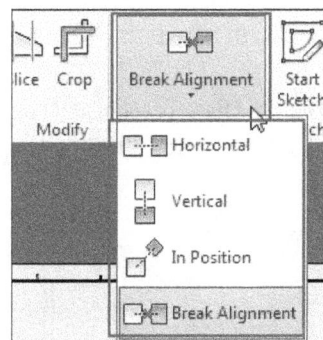

Figure-54. Alignment drop-down

Placing the views in drawing is less half of the drawing work done. The actual use of drawing is to express the real size and shape of the object. To express the size and shape of object, we use different type of annotations in drawing. Now, we will discuss different type of annotations and their related tools in Autodesk Inventor.

ANNOTATION TOOLS

As discussed earlier, the annotations are used to express the real size and shape of object to be manufactured. In Engineering field, annotations are group of different type of dimensions and symbols permitted by national and international standards of drafting. Each type of dimension and symbol has unique and clear meaning in drawing. The tools to apply annotations to drawing views are available in the **Annotate** tab in the **Ribbon**; refer to Figure-55.

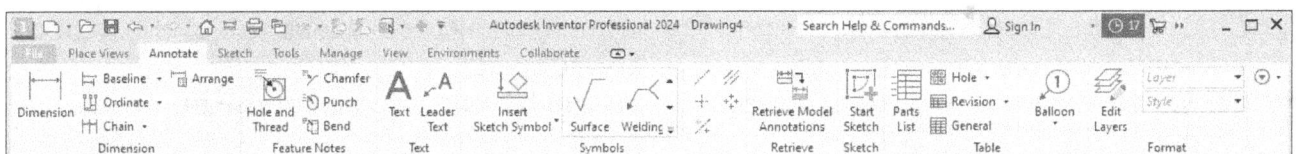

Figure-55. Annotate tab in Ribbon

The tools in the tab are discussed next.

Retrieve Model Annotations Tool

⊞ Retrieve The **Retrieve Model Annotations** tool is used to retrieve all the dimensions applied to the model in the 3D modeling environment. If you have not applied any dimension to the model then you will retrieve nothing by using this tool. Note that while working in Industry, you will apply this tool after placing the views to get all the modeling dimension. The procedure to use this tool is given next.

- Click on the **Retrieve Model Annotations** tool from the **Retrieve** panel in the **Annotate** tab of the **Ribbon**. The **Retrieve Model Annotation** dialog box will be displayed; refer to Figure-56. Also, you will be asked to select the view whose dimensions are to be retrieved.

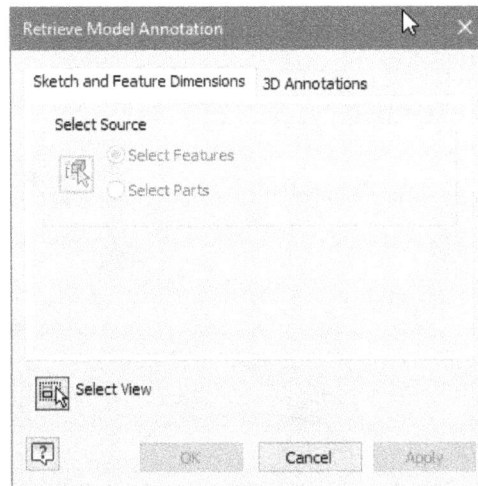

Figure-56. Retrieve Model Annotation dialog box

- Select the drawing view from the drawing area, the preview of dimensions will be displayed; refer to Figure-57. You will be asked to select the features of the model or the dimensions to retrieve.

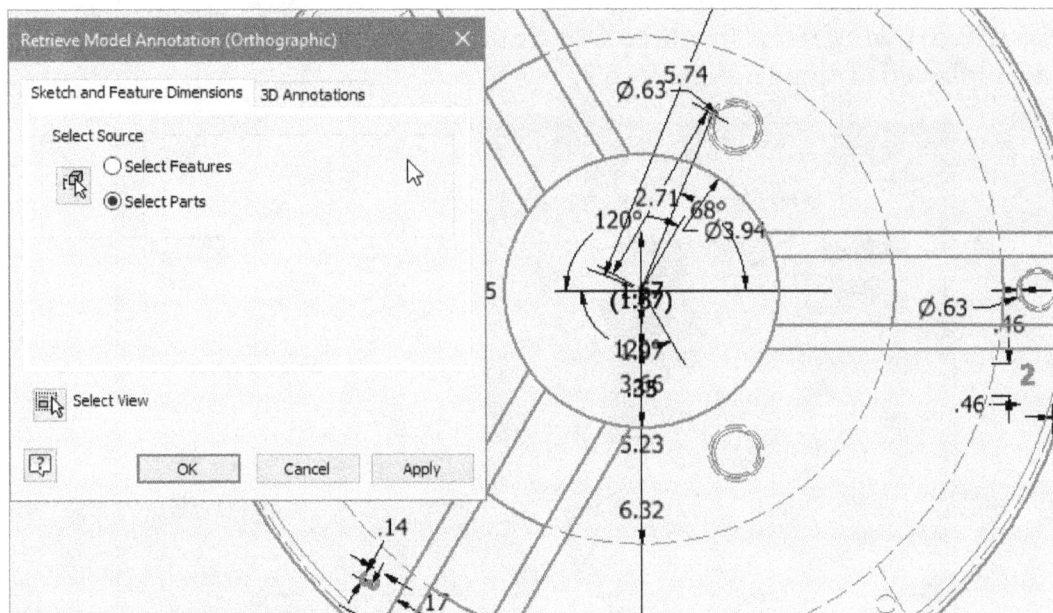

Figure-57. Preview of dimensions retrieved

- Select desired features or the dimensions. If you want to select the parts then select the **Select Parts** radio button from the **Select Source** area in the **Sketch and Feature Dimensions** tab of the dialog box. Note that you can make window selections to select multiple dimensions at one time to retrieve.
- Click on the **OK** button from the dialog box.

Dimension Tool

The **Dimension** tool is used to apply dimension to the selected entity in the drawing view. The procedure to use this tool is given next.

- Click on the **Dimension** tool from the **Dimension** panel in the **Annotate** tab of the **Ribbon**. You will be asked to select the geometry to be dimensioned.
- Select the entity/entities to be dimensioned. The dimension will get attached to cursor; refer to Figure-58. Note that the pattern of selecting entities for dimensioning is same as discussed in 3D Modeling environment.

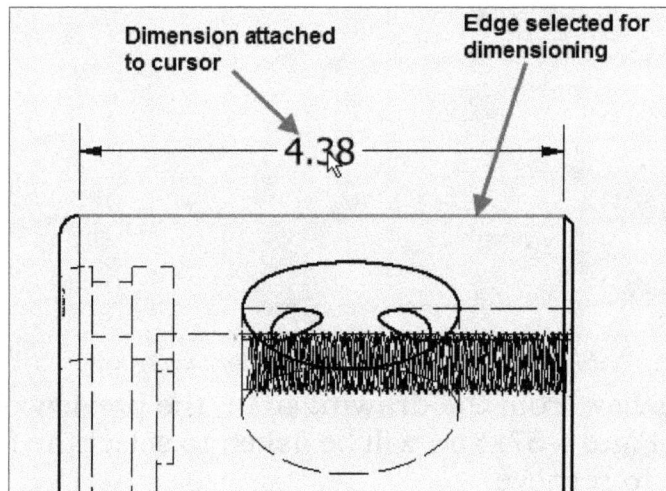

Figure-58. Dimensioning entities

- Click in the drawing area to place the dimension. The **Edit dimension** dialog box will be displayed; refer to Figure-59.

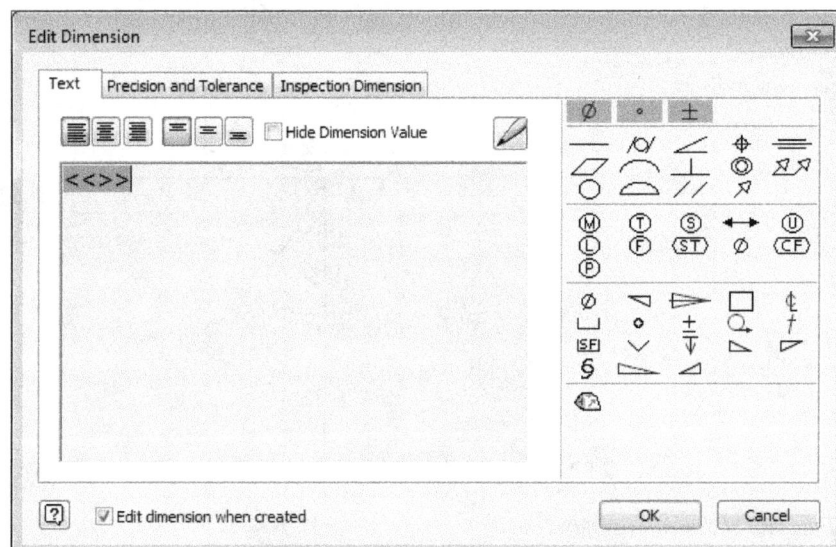

Figure-59. Edit Dimension dialog box

- Specify desired parameters in the dialog box and click on the **OK** button to create the dimension.

The options in the **Edit Dimension** dialog box are discussed next.

Text Options

The options to edit text of dimension are available in the **Text** tab of the dialog box. The major options in this tab are discussed next.

- Type desired text in the text box to add text to the dimension.
- Set the alignment of the text using justification buttons above the text box in the dialog box.
- Select the **Hide Dimension Value** check box to hide the dimension value and display only user defined text.
- You can insert the symbols in the dimension text by using the panel in the right of dialog box; refer to Figure-60.

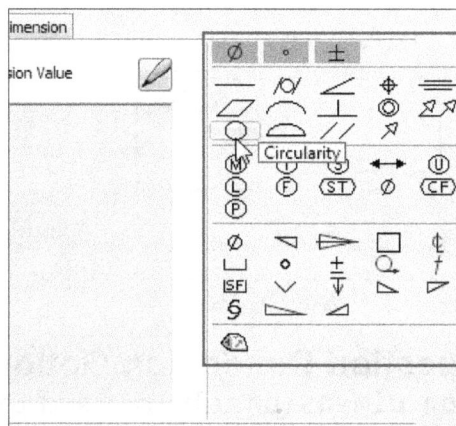

Figure-60. Panel for inserting symbols in the dimension

- If you want to change the formatting of text like changing font, boldface, font size, etc. then click on the **Launch Text Editor** button from the dialog box. The **Format Text** dialog box will be displayed; refer to Figure-61.

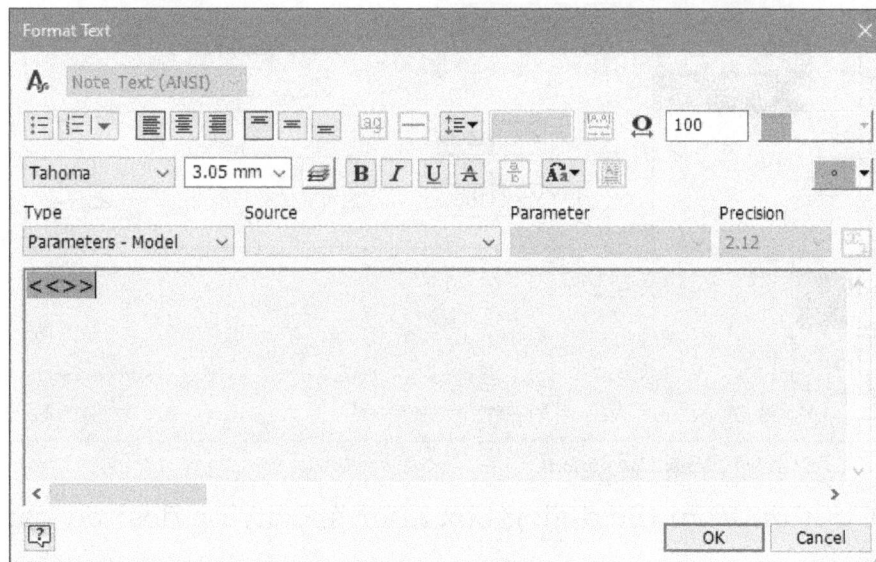

Figure-61. Format Text dialog box

- Using the options in the dialog box, change the formatting of text and then click on the **OK** button.

Precision and Tolerance Options

The options in the **Precision and Tolerance** tab are used to specify the dimensioning tolerance and limits of the component. Select desired tolerance method from the **Tolerance Method** area of the dialog box and set the precision of tolerance from the **Precision** area of the dialog box; refer to Figure-62. Set the value of tolerances in the edit boxes below **Tolerance Method** selection list in the dialog box.

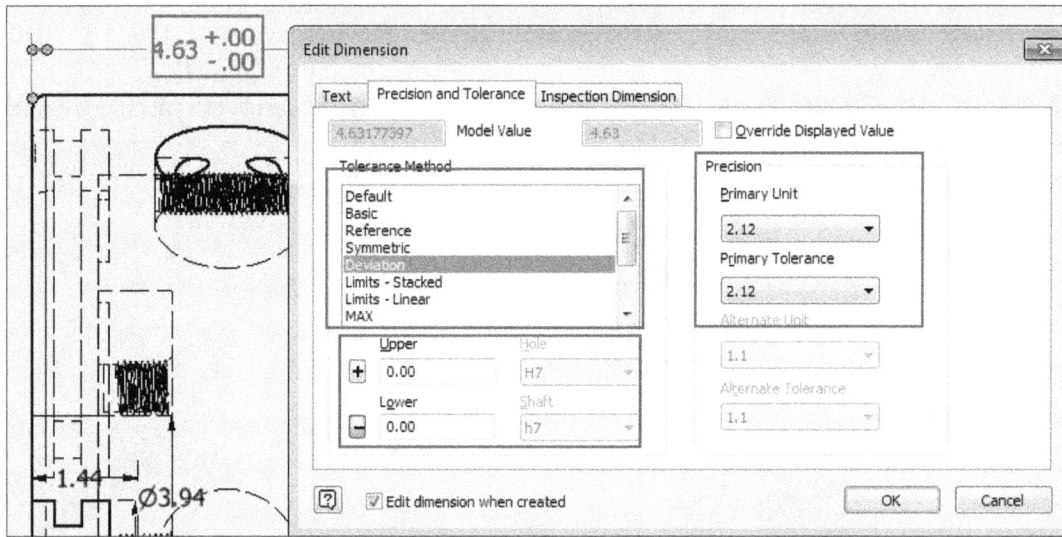

Figure-62. Precision and Tolerance tab in the dialog box

Inspection Dimension Options

The options in the **Inspection Dimension** tab are used to set the inspection criteria for the current dimension. Select the **Inspection Dimension** check box and specify desired values for inspection rate of current dimension; refer to Figure-63.

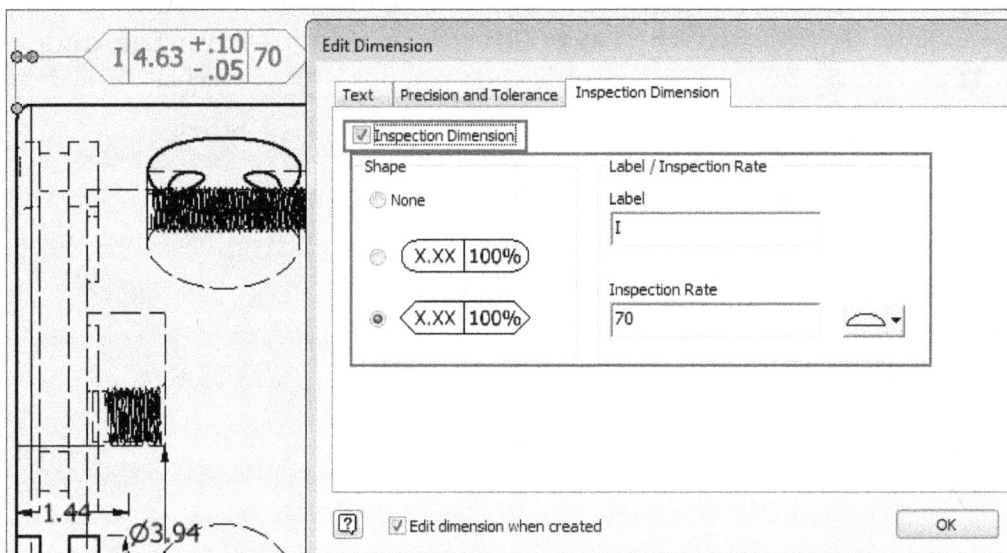

Figure-63. Inspection Dimension options

Click on the **OK** button from the dialog box after specifying desired parameters.

Baseline Dimensioning

There are two tools to create baseline dimensions, **Baseline** and **Baseline Set**. The method of using both the tools is same as discussed next.

*   Click on the **Baseline** or **Baseline Set** tool from the **Baseline Dimension** drop-down in the **Dimension** panel of the **Annotate** tab in the **Ribbon**. You will be asked to select the entity on model which you want to be base for all other dimensions.
*   Select an edge or curve from the drawing view. You will be asked to select geometry to be dimensioned.
*   Select the entities and then right-click in the drawing area. A shortcut menu will be displayed.
*   Click on the **Continue** button from the shortcut menu. The dimensions will get attached to the cursor.
*   Move the cursor along X or Y direction and click to place the dimensions; refer to Figure-64.

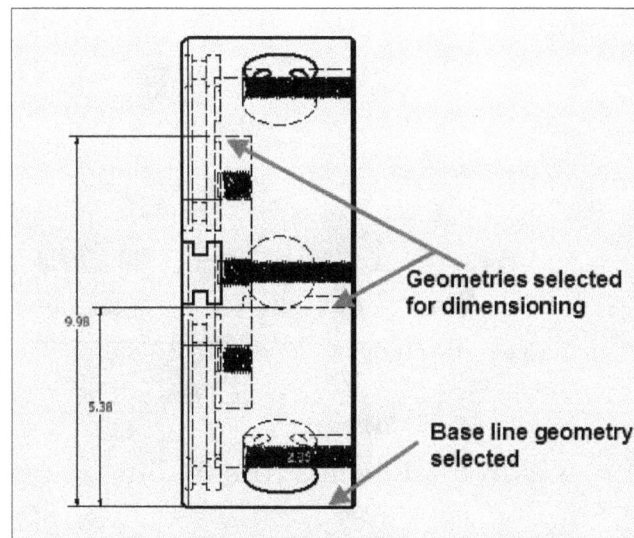

Figure-64. Baseline dimensioning

*   Right-click in the drawing area and select the **Create** option from the shortcut menu.
*   Note that if you use the **Baseline Set** tool then all the dimensions created will be part of a group. If you select one dimension of the baseline set then all dimension will get selected.

Ordinate Dimensioning

The Ordinate dimensioning is used to display the dimensions in the form of coordinate. This type of dimensioning is useful in manual CAM programming the component or when there are lots of dimension in small section of the drawing. The are two tools to create ordinate dimensions; **Ordinate** and **Ordinate Set**. The method of using both the tools is same as discussed next.

*   Click on the **Ordinate** or **Ordinate Set** tool from the **Ordinate Dimension** drop-down in the **Dimension** panel of the **Annotate** tab in **Ribbon**. You will be asked to select the drawing view to be dimensioned.
*   Select the drawing view. You will be asked to specify the origin location.

- Click at desired location in the drawing view to specify **0** level of ordinate dimension. You will be asked to select the geometries to be dimensioned.
- Select the entities in the drawing view and right-click in the drawing area. A shortcut menu will be displayed.
- Select the **Continue** button from the shortcut menu. You will be asked to specify location for placing dimensions.
- Click at desired location in the drawing area to place the dimension; refer to Figure-65.

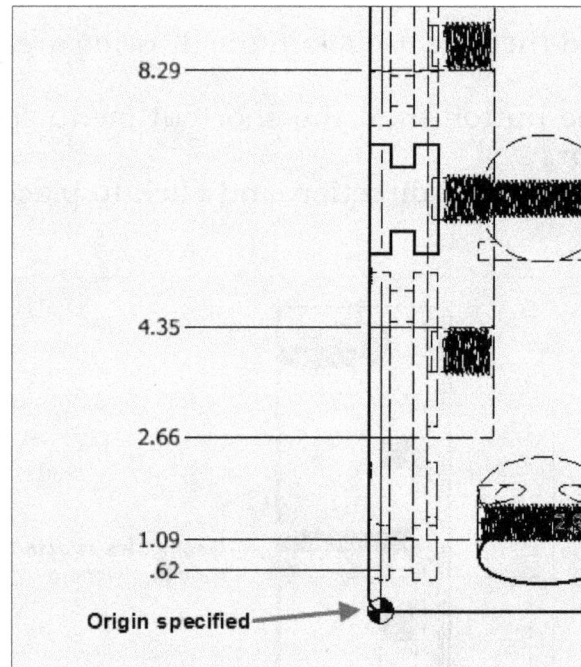

Figure-65. Ordinate dimension

- Right-click in the drawing area and select the **OK** button from the shortcut menu displayed.
- Note that if you use the **Ordinate Set** tool then all the dimensions created will be part of a group. If you select one dimension of the ordinate set then all dimensions will get selected.

Chain Dimensioning

The chain dimensioning is used to create dimensions linked one after the other. There are two tools to create chain dimensions; **Chain** and **Chain Set**. The method of using both the tools is same as discussed next.

- Click on the **Chain** or **Chain Set** tool from the **Chain Dimension** drop-down in the **Dimension** panel of the **Annotate** tab in the **Ribbon**. You will be asked to select entity for base of chain dimensioning.
- Select the geometry from the drawing view. You will be asked to select geometries to be dimensioned.
- Select the entities and right-click in the drawing area. A shortcut menu will be displayed.
- Click on the **Continue** button from the menu. The dimensions will get attached to cursor.
- Click in the drawing area to place the dimensions; refer to Figure-66.

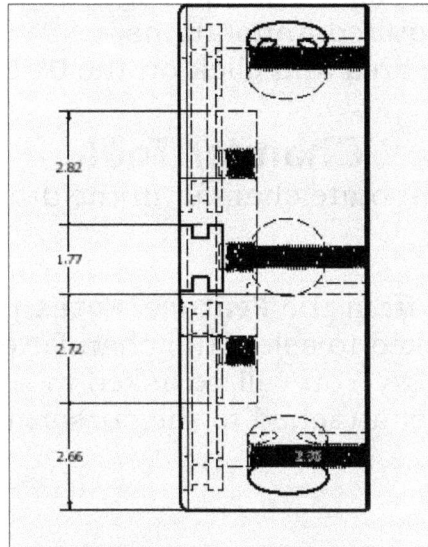

Figure-66. Chain dimensioning

- Right-click in the drawing area and click on the **Create** button.
- Note that if you use the **Ordinate Set** tool then all the dimensions created will be part of a group. If you select one dimension of the ordinate set then all dimension will get selected.

Hole and Thread Tool

The **Hole and Thread** tool is used to annotate holes and threads. The procedure to use this tool is given next.

- Click on the **Hole and Thread** tool from the **Feature Notes** panel in the **Annotate** tab of the **Ribbon**. You will be asked to select hole or thread edges in the drawing view.
- Select the edge of the hole/thread. The annotation will get attached to cursor.
- Click at desired location in the drawing to place the annotation; refer to Figure-67.

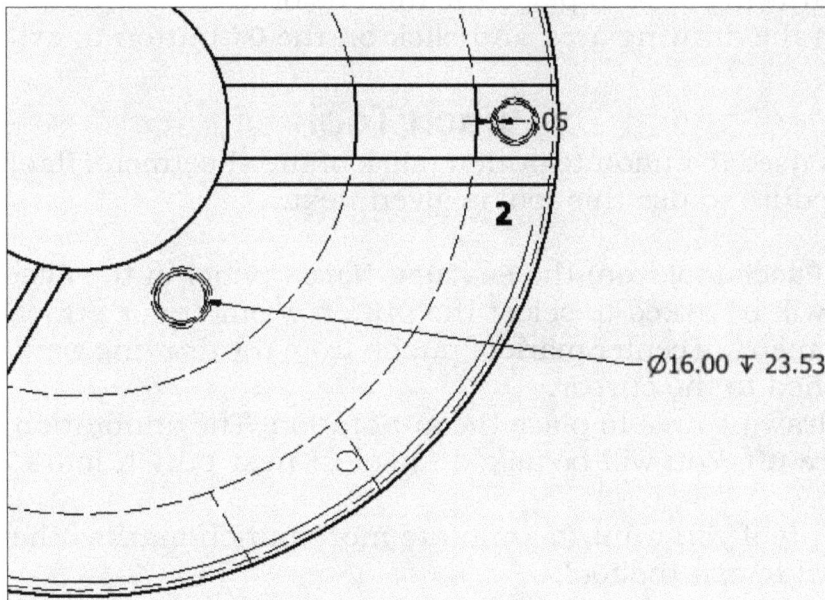

Figure-67. Hole annotation

- You will be asked to select hole or thread edge. Click on the edge and repeat the procedure till you have required annotations.
- Right-click in the drawing area and click on the **OK** button to exit the tool.

Chamfer Tool

The **Chamfer** tool is used to annotate chamfer in the drawing view. The procedure to use this tool is given next.

- Click on the **Chamfer** tool from the **Feature Notes** panel in the **Annotate** tab of the **Ribbon**. You will be asked to select the chamfer edge.
- Click on the chamfered edge. You will be asked to select the reference edge. The chamfer annotation will get attached to the cursor; refer to Figure-68.

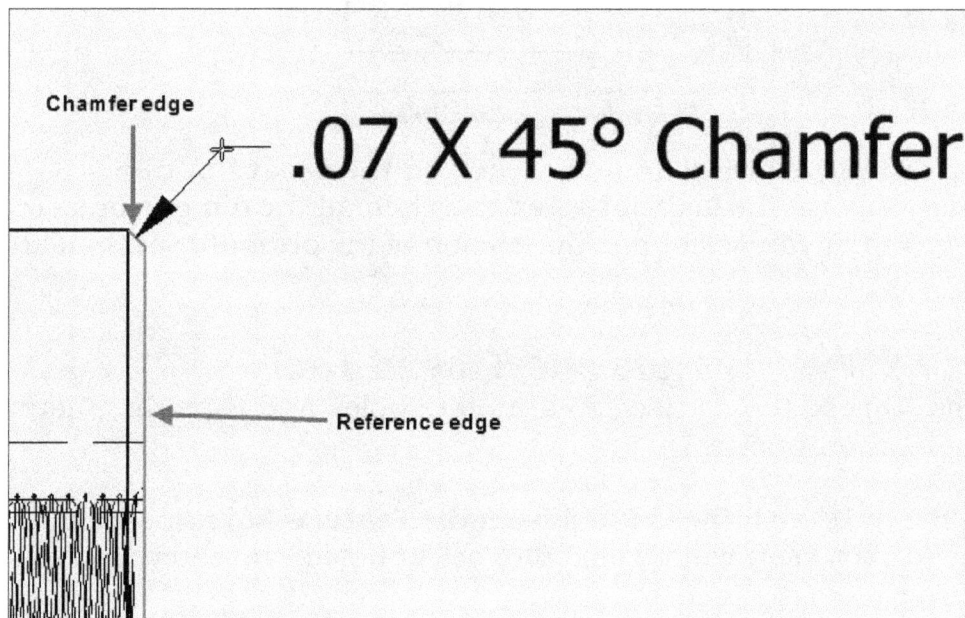

Figure-68. Chamfer dimension

- Click in the drawing area to place the dimension.
- Right-click in the drawing area and click on the **OK** button to exit the tool.

Punch Tool

The **Punch** tool is used to annotate punch mark of the sheetmetal flat pattern drawing views. The procedure to use this tool is given next.

- Click on the **Punch** tool from the **Feature Notes** panel in the **Annotate** tab of the **Ribbon**. You will be asked to select the punch geometry or punch center mark.
- Select the geometry of center mark of punch from the drawing view. The annotation will get attached to the cursor.
- Click in the drawing area to place the annotation. The annotation will be created; refer to Figure-69. You will be asked to select next punch mark or geometry for annotation.
- Select the entity if you want to annotate more punch marks otherwise press **ESC** from keyboard to exit the tool.

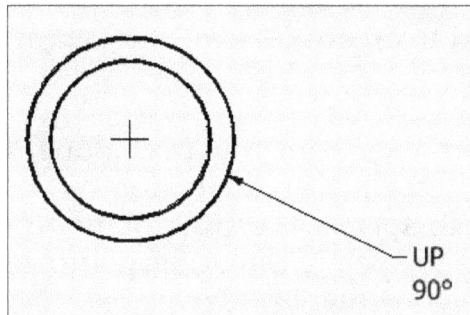

Figure-69. Punch mark annotation

Bend Tool

The **Bend** tool in the **Feature Notes** panel is used to create annotations for sheet metal bends. The procedure to use this tool is given next.

- Click on the **Bend** tool from the **Feature Notes** panel in the **Annotate** tab of the **Ribbon**. You will be asked to select a bend center line.
- Select the center line of bend created in flat pattern drawing view. The bend annotation will be created; refer to Figure-70.

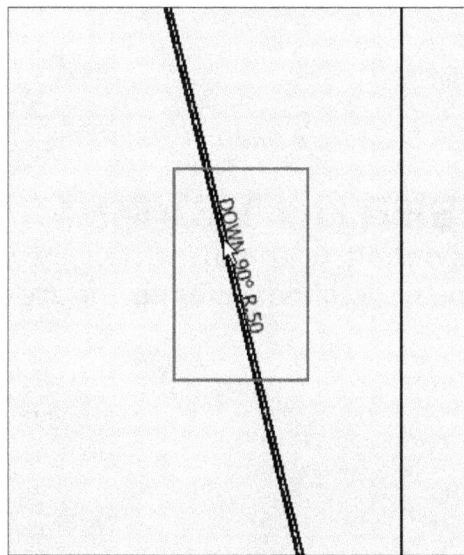

Figure-70. Bend annotation created

- Select the other bend lines to annotate them or press **ESC** to exit the tool.

Text Tool

The **Text** tool is used to write desired text in the drawing like, notes for manufacturer or machinist. The procedure to use this tool is given next.

- Click on the **Text** tool from the **Text** panel in the **Annotate** tab of the **Ribbon**. You will be asked to specify the location of the text.
- Click at desired location to place the text. The **Format Text** dialog box will be displayed as discussed earlier.
- Write down desired text in the text box and set the formatting as required.
- Click on the **OK** button from the dialog box to create the text.
- Press **ESC** to exit the tool.

Leader Text

The **Leader Text** tool is used to create text with leader attached to it. The procedure to use this tool is given next.

- Click on the **Leader Text** tool from the **Text** panel in the **Annotate** tab of the **Ribbon**. You will be asked to click on a location.
- Click at desired location to specify the starting point of the leader.
- Click at desired locations to specify the other points of the leader and then right-click. A shortcut menu will be displayed.
- Select the **Continue** button from the shortcut menu. The **Format Text** dialog box will be displayed as discussed earlier.
- Set desired text and click on the **OK** button. The leader text will be created; refer to Figure-71.
- Press **ESC** from keyboard to exit the tool.

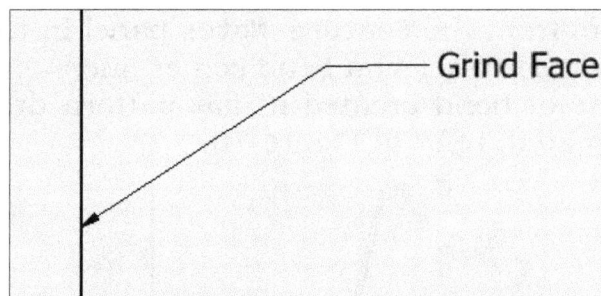

Figure-71. Leader text created

INSERTING SYMBOLS

The tools to insert drawing symbols in the drawing view are available in the **Symbols** panel of the **Annotate** tab in the **Ribbon**; refer to Figure-72. The tools in this panel are discussed next.

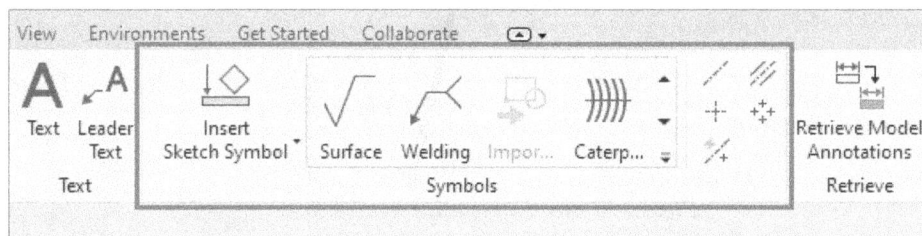

Figure-72. Symbols panel

Insert Sketch Symbol

The **Insert Sketch Symbol** tool is used to insert the symbols saved in local directory of drawing file. Before using this tool, you must have sketched symbols created by using the **Define New Symbol** tool in the **Symbols** panel. The procedure to create and insert symbols is discussed next.

Creating Symbol

- Click on the **Define New Symbol** tool from the **Insert Symbols** drop-down in the **Symbols** panel of the **Annotate** tab in the **Ribbon**; refer to Figure-73. The sketching environment will be displayed and you will be asked to create sketch of the symbol.

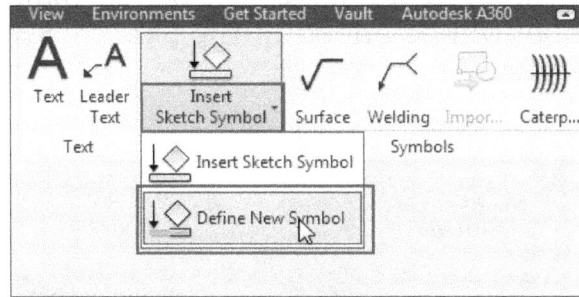

Figure-73. Define New Symbol tool

- If you want to be prompted for specifying text after placing the symbol then click on the **Text** tool from the **Create** panel in the **Sketch** contextual tab. Specify the location of text near the symbol by clicking. The **Format Text** dialog box will be displayed. Select the **Prompted Entry** option from the **Type** drop-down in the dialog box; refer to Figure-74. Set the other parameters and click on the **OK** button from the **Format Text** dialog box. The text prompt will be connected with the symbol; refer to Figure-75. Press **ESC** to exit the tool.

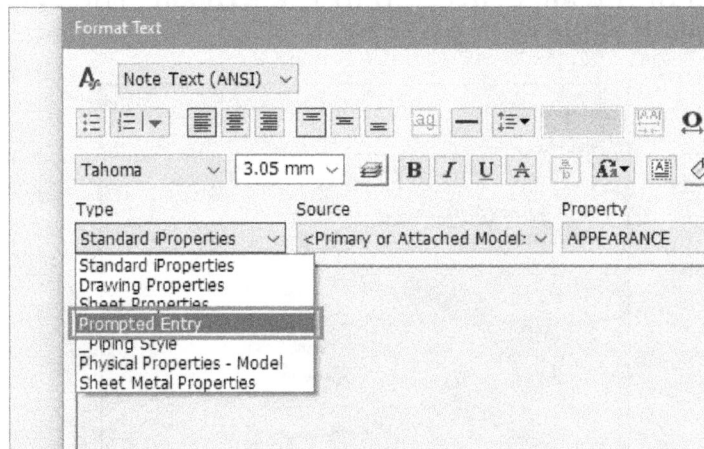

Figure-74. Prompted Entry option

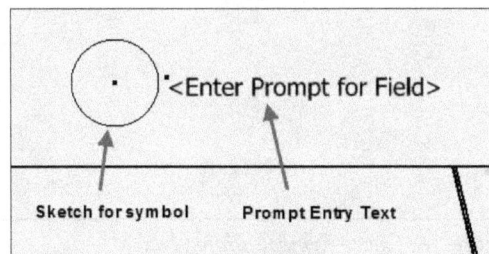

Figure-75. Sketch symbol created

- Click on the **Finish Sketch** button from the **Ribbon**. The **Sketched Symbol** dialog box will be displayed; refer to Figure-76.

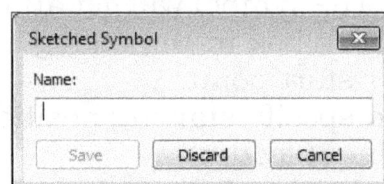

Figure-76. Sketched Symbol dialog box

- Specify the name of symbol in the **Name** edit box of dialog box and click on the **Save** button. The symbol will be added in the **Sketch Symbols** category of **Model Browse Bar**; refer to Figure-77.

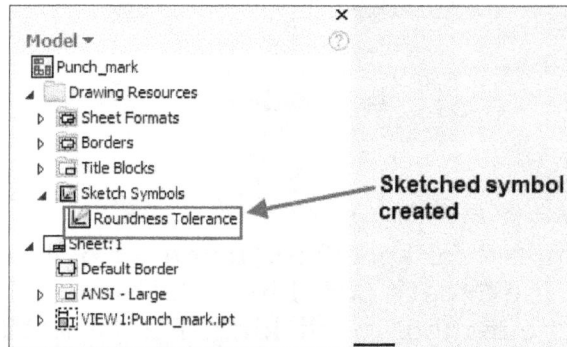

Figure-77. Symbol added in Model Browse Bar

Inserting Sketch Symbol

- Click on the **Insert Sketch Symbol** tool from the **Insert Symbol** drop-down in the **Symbols** panel of the **Annotate** tab in the **Ribbon**. The **Sketch Symbols** dialog box will be displayed; refer to Figure-78.

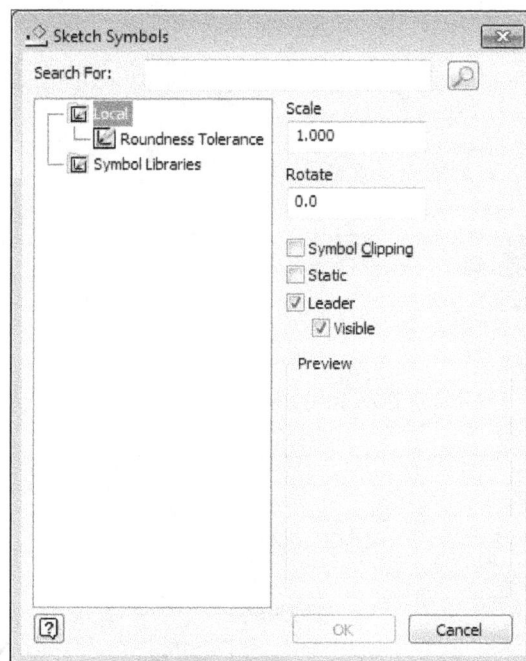

Figure-78. Sketch Symbols dialog box

- Select the symbol from the left area of the dialog box. Preview of the symbol will be displayed.
- Set the parameters like scale, rotation, symbol clipping, etc. in the dialog box and then click on the **OK** button. The symbol will get attached to the cursor and if you have selected the **Leader** check box in the dialog box then you will be asked to specify location of the leader start point.
- Click in the drawing area to specify starting position of symbol leader. You will be asked to specify next point of leader.
- Specify the next point(s) and right-click in the drawing area. A shortcut menu will be displayed.

- Click on the **Continue** button from the shortcut menu. If you have used the **Prompted Entry** option for symbol then **Prompted Texts** dialog box will be displayed; refer to Figure-79.

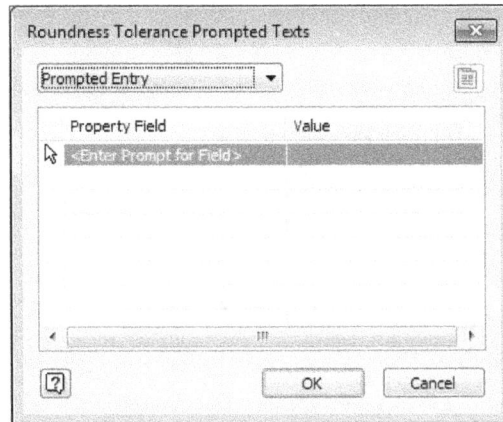

Figure-79. Prompted Texts dialog box

- Click in the **Value** field and specify desired text for symbol.
- Click on the **OK** button from the dialog box. The symbol will be created. Press **ESC** from keyboard to exit the tool.

Inserting Drawing Symbols

There are various symbols that we generally use in engineering drawings like surface finish symbol, Geometric Dimensioning and Tolerance (GD&T) symbols, welding symbols, datum symbols, and so on. In Autodesk Inventor, you do not need to create each and every symbol by using sketch as there are ready to use symbols available in the tool box of **Symbols** panel in the **Annotate** tab of **Ribbon**; refer to Figure-80. The procedure to insert symbols is given next.

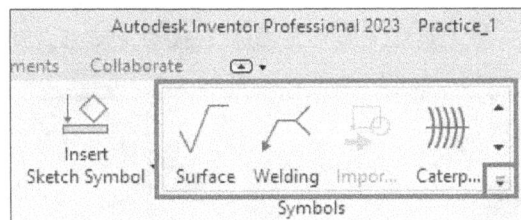

Figure-80. Toolbox in Symbols panel

Inserting Surface Texture Symbol

The Surface Texture symbol is used to express the material condition of surface of metal. The procedure to insert surface texture symbol is given next.

- Click on the **Surface** button in the toolbox of **Symbols** panel in the **Annotate** tab of **Ribbon**. The symbol will get attached to cursor.
- Click at desired location to place the symbol. If you want to create leader with symbol then move the cursor to desired location and click to specify the leader point. Once you have specified desired leader points, right-click and select the **Continue** button from the shortcut menu. If you do not want leader to be created with symbol then right-click after placing the symbol and select the **Continue** button from the shortcut menu. On doing so, the **Surface Texture** dialog box will be displayed; refer to Figure-81.

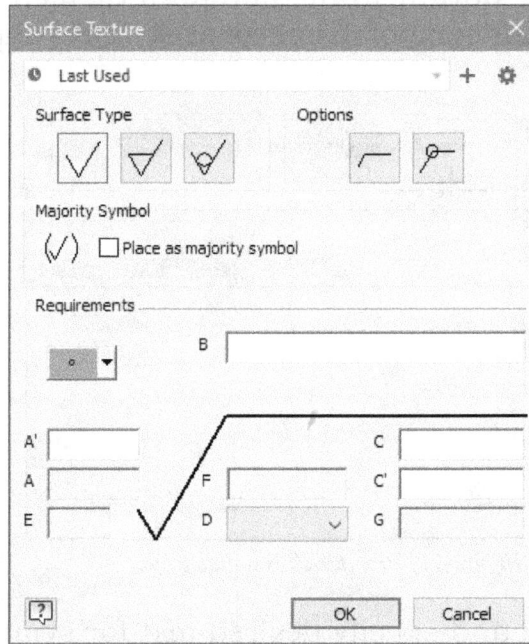

Figure-81. Surface Texture dialog box

- Select desired symbol type from **Surface Type** area of the dialog box.
- Specify the other parameters for the symbol and click on the **OK** button.
- Press **ESC** to exit the tool.

In the same way, you can use the other symbols in the toolbox.

Engineering Drawing Symbols

In this topic, you will learn about the common engineering symbols and their meanings in manufacturing.

Surface Texture Symbols or Surface Roughness Symbols

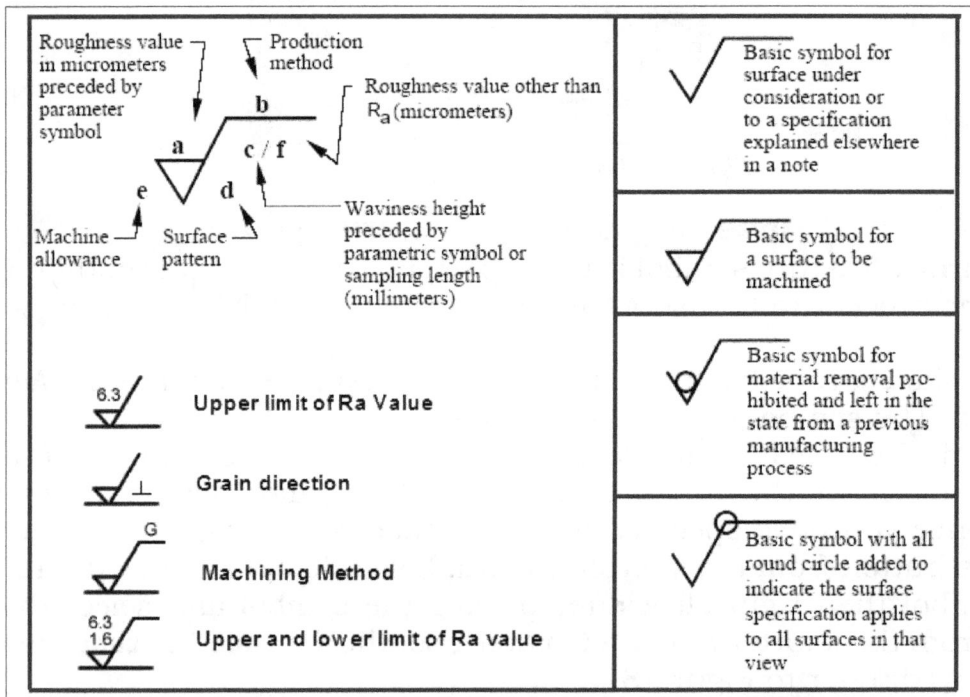

Datum Identifier

A datum is theoretical exact plane, axis, or point location that GD&T or dimensional tolerances are referenced to. You can think of them as an anchor for the entire part; where the other features are referenced from. A datum feature is usually an important functional feature that needs to be controlled during measurement as well.

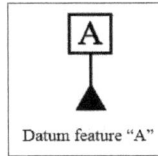

Datum feature "A"

All GD&T symbols except for the form tolerances (straightness, flatness, circularity, and cylindricity) can use datums to help specify what geometrical control is needed on the part. When it comes to GD&T, datum symbols are your starting points where all other features are referenced from.

Placement of Datum identifier is given in Figure-82.

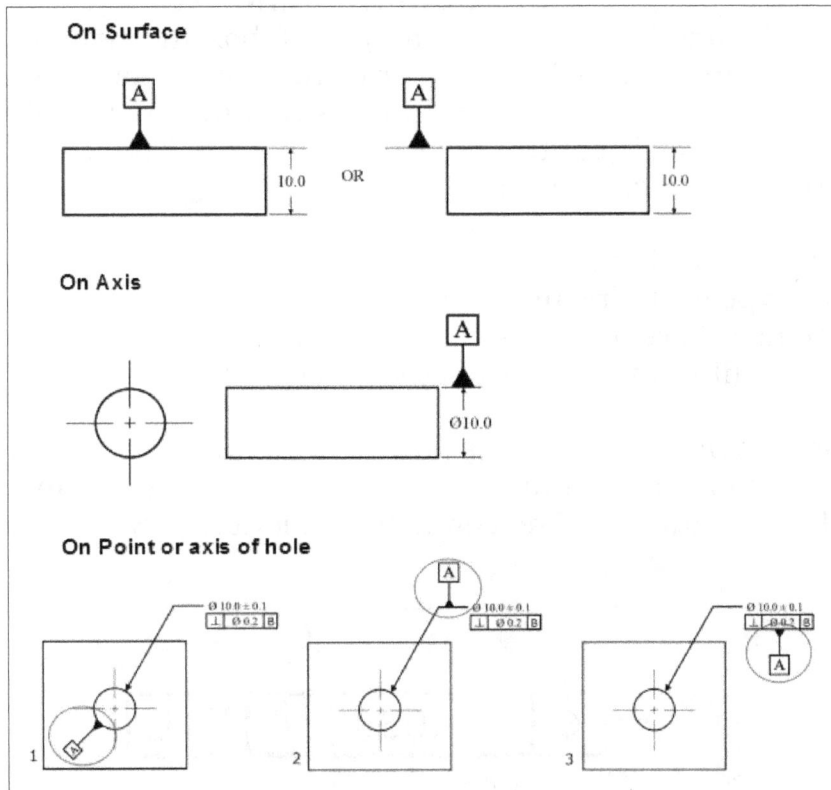

Figure-82. Placement of Datum Identifier

Datum Target Symbols

Some manufacturing processes, such as casting, forging, welding, and heat treating, are likely to produce uneven or irregular surfaces. Datum targets may be used to immobilize parts with such uneven or irregular surfaces. Datum targets may also be used to support irregular-shaped parts that are not easily mounted in a datum reference frame. Datum targets are used only when necessary because, once they are specified, costly manufacturing and inspection tooling is required to process them. The datum target is to be the place of contact (supports of the workpiece) for the manufacturing and inspection equipment.

Note that the upper half of datum target symbol is used only for circular targets and lower half gives Datum identifier with target number; refer to Figure-83. In datum target symbol, A1, A2, A3 are datum target areas and B1, B2 are datum target points and C1 is datum target line.

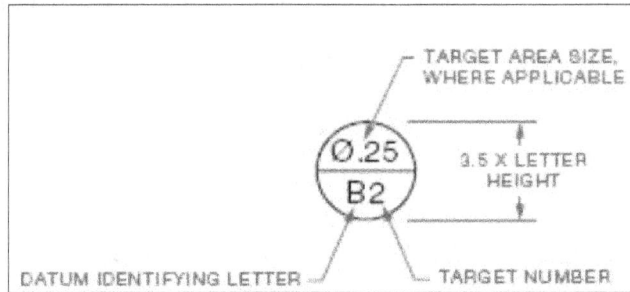

Figure-83. Datum target symbol

In Autodesk Inventor, we have five different tools to insert symbols for datum targets on different geometrical entities.

Feature Control Frame

The feature control frame is also known as GD&T box in laymen's language. The method to insert Feature Control Frame in drawing is same as discussed for Surface Finish symbol. In GD&T, a feature control frame is required to describe the conditions and tolerances of a geometric control on a part's feature. The feature control frame consists of four pieces of information:

1. GD&T symbol or control symbol
2. Tolerance zone type and dimensions
3. Tolerance zone modifiers: features of size, projections...
4. Datum references (if required by the GD&T symbol)

This information provides everything you need to determine what geometrical tolerance needs to be on the part and how to measure or determine if the part is in specification; refer to Figure-84. The common elements of feature control frame are discussed next.

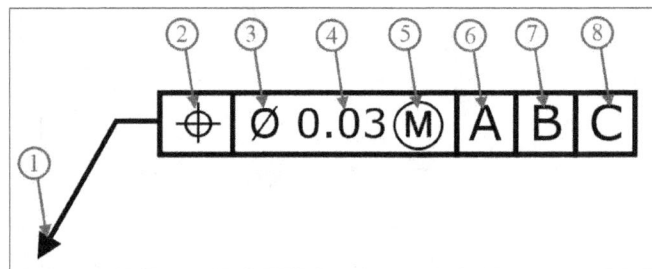

Figure-84. Feature control frame

1. **Leader Arrow** – This arrow points to the feature that the geometric control is placed on. If the arrow points to a surface then the surface is controlled by the GD&T. If it points to a diametric dimension then the axis is controlled by GD&T. The arrow is optional but helps clarify the feature being controlled.
2. **Geometric Symbol** – This is where your geometric control is specified.
3. **Diameter Symbol (if required)** – If the geometric control is a diametrical tolerance then the diameter symbol (Ø) will be in front of the tolerance value.

4. **Tolerance Value** – If the tolerance is a diameter, you will see the Ø symbol next to the dimension signifying a diametric tolerance zone. The tolerance of the GD&T is in same unit of measure that the drawing is written in.

5. **Feature of Size or Tolerance Modifiers (if required)** – This is where you call out max material condition or a projected tolerance in the feature control frame.

6. **Primary Datum (if required)** – If a datum is required, this is the main datum used for the GD&T control. The letter corresponds to a feature somewhere on the part which will be marked with the same letter. This is the datum that must be constrained first when measuring the part. Note: The order of the datum is important for measurement of the part. The primary datum is usually held in three places to fix 3 degrees of freedom.

7. **Secondary Datum (if required)** – If a secondary datum is required, it will be to the right of the primary datum. This letter corresponds to a feature somewhere on the part which will be marked with the same letter. During measurement, this is the datum fixated after the primary datum.

8. **Tertiary Datum (if required)** – If a third datum is required, it will be to the right of the secondary datum. This letter corresponds to a feature somewhere on the part which will be marked with the same letter. During measurement, this is the datum fixated last.

Reading Feature Control Frame

The feature control frame forms a kind of sentence when you read it. Below is how you would read the frame in order to describe the feature.

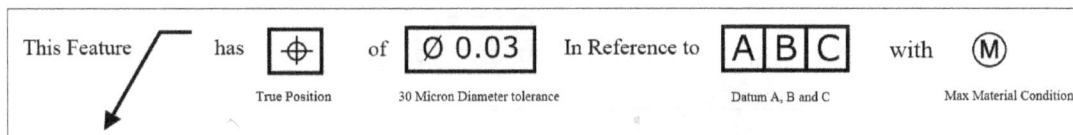

⊕ | Ø 0.03 Ⓜ | A | B | C   gives meaning of

This Feature	has	⊕	of	Ø 0.03	In Reference to	A B C	with	Ⓜ
		True Position		30 Micron Diameter tolerance		Datum A, B and C		Max Material Condition

Meaning of various geometric symbols are given in Figure-85.

SYMBOL	CHARACTERISTICS	CATEGORY
—	Straightness	Form
▱	Flatness	
◯	Circulatity	
⌀	Cylindricity	
⌒	Profile of a Line	Profile
⌓	Profile of Surface	
∠	Angularity	Orientation
⊥	Perpendicularity	
//	Parallelism	
⊕	Position	Location
◎	Concentricity	
≡	Symmetry	
↗	Circular Runout	Runout
↗↗	Total Runout	

Figure-85. Geometric Symbols

Figure-86 and Figure-87 shows the use of geometric tolerances in real-world.

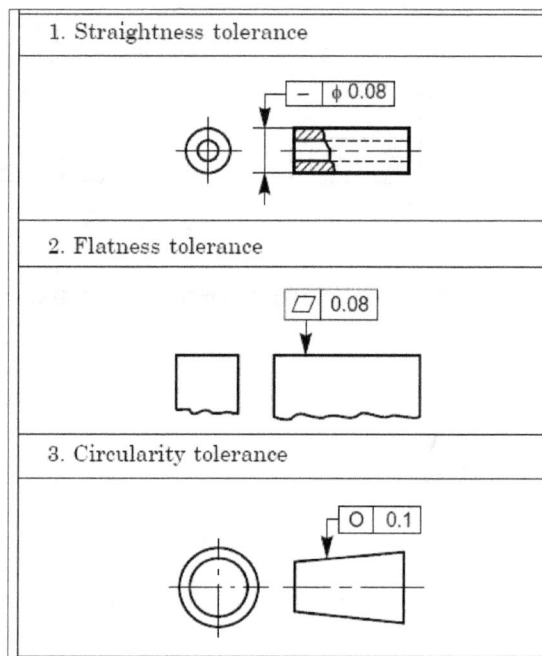

Figure-86. Use of geometric tolerance 1

Note that in applying most of the Geometrical tolerances, you need to define a datum plane like in Perpendicularity, Parallelism, and so on.

There are a few dimensioning symbols also used in geometric dimensioning and tolerances, which are given in Figure-88.

4. Cylindricity tolerance	8. Concentricity and coaxiality tolerance
⌭ 0.1	◎ φ 0.01 A A
5. Parallelism tolerance	
// 0.01 D D	9. Symmetry tolerance
	⌯ 0.08 A A
6. Perpendicularity tolerance	10. Radial run-out
⊥ 0.08 A A	↗ 0.1 D D
7. Angularity tolerance	11. Axial run-out
∠ 0.08 A 40° A	↗ 0.1 D D

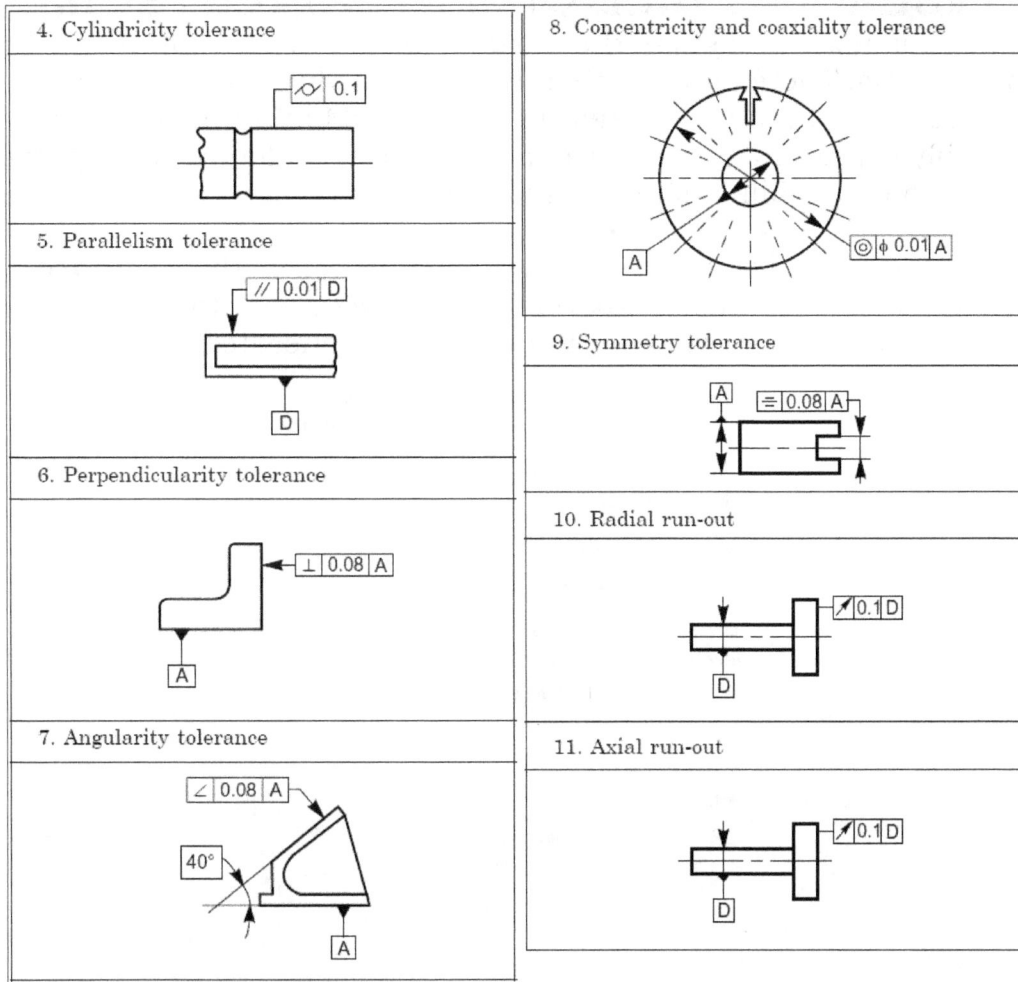

Figure-87. Use of geometric tolerance 2

Symbol	Meaning	Symbol	Meaning
Ⓛ	LMC – Least Material Condition	⟵⊕	Dimension Origin
Ⓜ	MMC – Maximum Material Condition	⊔	Counterbore
Ⓣ	Tangent Plane	∨	Countersink
Ⓟ	Projected Tolerance Zone	�-T	Depth
Ⓕ	Free State	⌀	All Around
∅	Diameter	⟷	Between
R	Radius	✕	Target Point
SR	Spherical Radius	▷	Conical Taper
S∅	Spherical Diameter	◁	Slope
CR	Controlled Radius	☐	Square
Ⓢ̄Ⓣ	Statistical Tolerance		
77	Basic Dimension		
(77)	Reference Dimension		
5X	Places		

Figure-88. Dimensioning symbols

INSERTING BILL OF MATERIALS (BOM)

Bill of Materials is used to display the parts of assembly in a tabulated form in the engineering drawing. There may be other informations like material of parts, number of parts, and so on. Note that we generally insert BOM in a drawing where exploded view of assembly is present so that we can also assign balloons to the parts in BOM. The procedure to insert Bill of Materials is given next.

- Click on the **Parts List** tool from the **Table** panel in the **Annotate** tab of the **Ribbon**. The **Parts List** dialog box will be displayed; refer to Figure-89. Also, you will be asked to select the view for BOM is being generated.

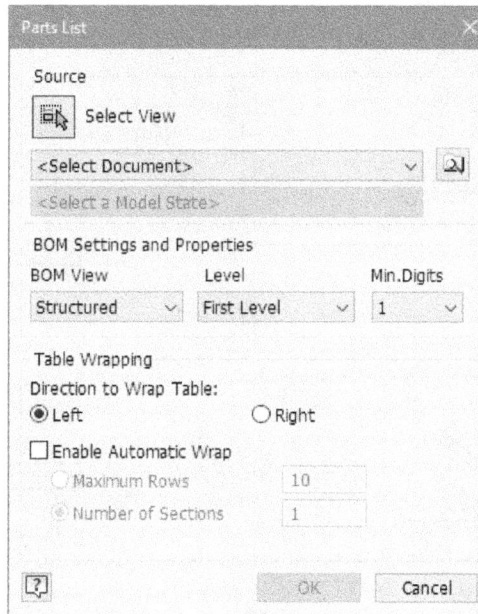

Figure-89. Parts List dialog box

- Select the drawing view from drawing area. The **OK** button will become active in the dialog box.
- Set the parameters in dialog box as required and then click on the **OK** button. The Bill of Material will get attached to cursor.
- Click at desired location in drawing to place the BOM; refer to Figure-90.

ITEM	QTY	PART NUMBER	DESCRIPTION
1	1	Base	
2	1	Slider	
3	1	Slider Shaft	
4	1	Head Shaft	
5	1	Shaft handle	
6	2	Shaft Handle Head	
7	2	Block	
8	1	Shaft Housing	
9	1	Stopper	

Figure-90. Bill of Materials

Editing Bill of Materials

- Double-click on the Bill of Materials created. The **Parts List** dialog box will be displayed as shown in Figure-91.

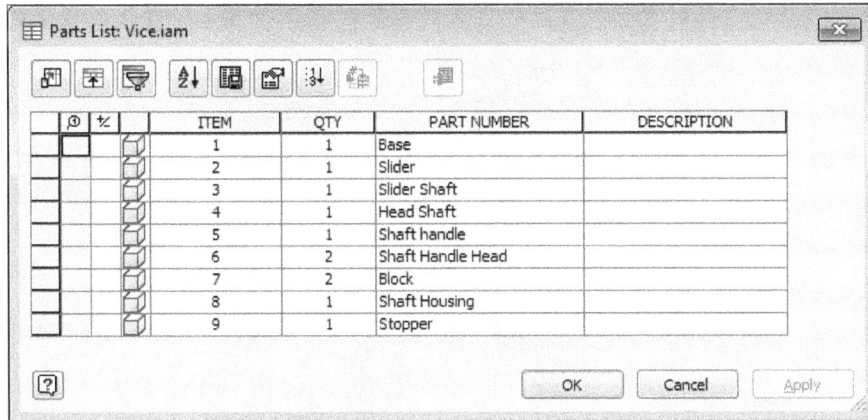

Figure-91. Parts List dialog box

- Click on the **Column Chooser** button from the dialog box. The **Parts List Column Chooser** dialog box will be displayed; refer to Figure-92.

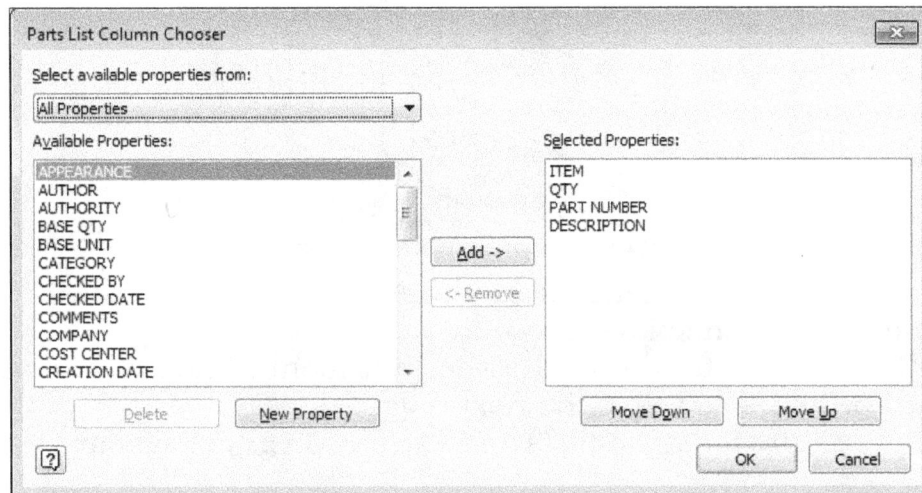

Figure-92. Part List Column Chooser dialog box

- Select the options in the **Available Properties** list box and click on the **Add** button to add column in BOM. To remove a column from BOM, select the option from **Selected Properties** list box and click on the **Remove** button from the dialog box.
- Click on the **OK** button to apply the changes.
- Set the other options in the **Parts List** dialog box as required and then click on the **OK** button.

Creating Balloons

Balloons are used to identify parts in the assembly as per the Bill Of Materials. The procedure to assign balloons to the parts is given next.

- Click on the **Balloon** tool from the **Table** panel in the **Annotate** tab of the **Ribbon**; refer to Figure-93. You will be asked to select a component.

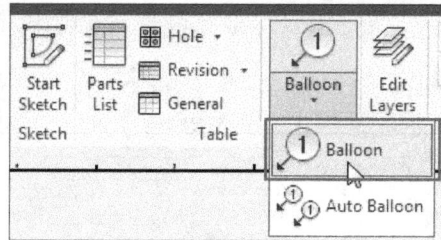

Figure-93. Balloon tool

- Select the component from the assembly. The balloon gets attached to the cursor; refer to Figure-94.

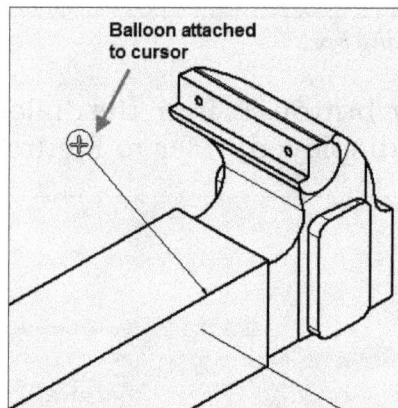

Figure-94. Balloon attached to cursor

- Click to specify the placement location and right-click in the drawing area. A shortcut menu will be displayed.
- Select the **Continue** button from the shortcut menu. The balloon will be placed and you will be asked to select the next part.
- Repeat the procedure for other parts and **Esc** button from keyboard to exit the tool.

You can also use the **Auto Balloon** tool to place balloons automatically.

Note that you can edit the style and standards of balloons, text size, feature control frame, datum identifiers, etc. by using the Styles Editor tool in the Styles and Standards panel of the Manage tab in the Ribbon as discussed earlier.

PRACTICAL

In this practical, you will first create the model of part as per the production drawing given in Figure-95 and then you will create the same production drawing of part using the model.

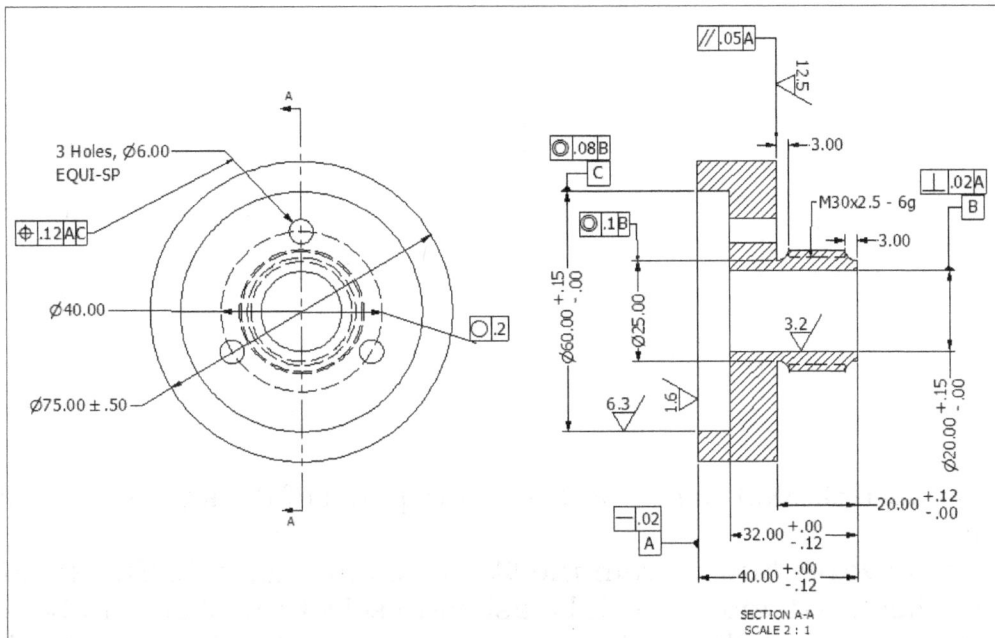

Figure-95. Production drawing for practical 1

If you see the model carefully then you will find that it is single part and can be created in Autodesk Inventor by Part Modeling. The steps to create the part are given next.

Creating a New Part

- Start Autodesk Inventor by using Start menu or icon on the desktop of your computer (If not started yet).
- Click on the **New** button from **New** cascading menu in the **File** menu of the **Ribbon**. The **Create New File** dialog box will be displayed.
- Click on the **Metric** folder under **Template** category in the left of the dialog box and double click on **Standard (mm).ipt**. The Part environment will be displayed.
- Click on the **Revolve** tool from the **Create** panel in the **3D Model** tab of the **Ribbon**. You will be asked to select a sketching plane.
- Select the **YZ Plane** from the drawing area; refer to Figure-96. The Sketching environment will be activated.

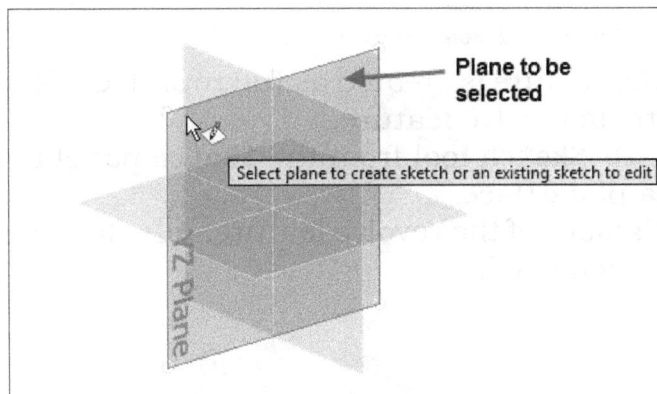

Figure-96. Selecting plane for sketch

- Create the sketch as shown in Figure-97 based on drawing.

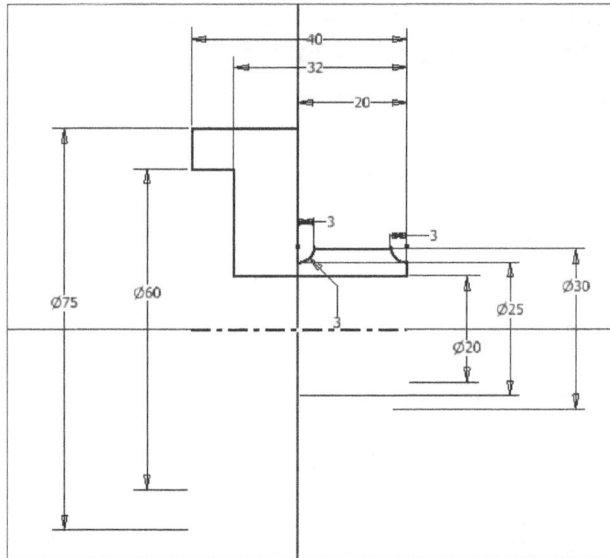

Figure-97. Creating sketch for revolve feature

- Click on the **Finish Sketch** button from **Exit** panel of the **Ribbon**. The sketch will be displayed.
- Click on the **Revolve** button from the **Sketch** contextual tab. The **Revolve** dialog box will be displayed and you will be asked to select the sketch to be revolve.
- Select the newly created sketch and axis to create the feature. Preview of the feature will be displayed; refer to Figure-98.

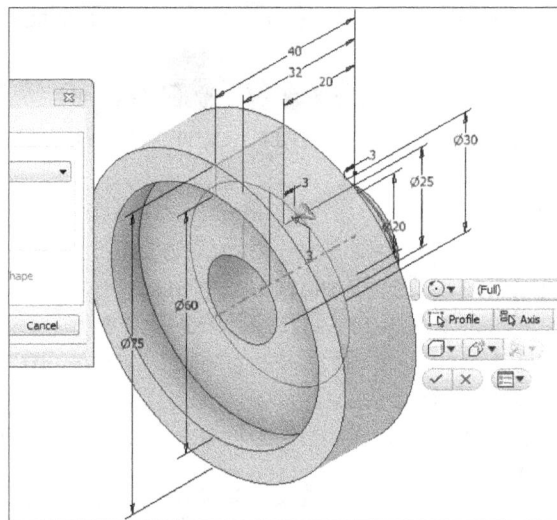

Figure-98. Preview of revolve feature

- Make sure the revolve feature is created as full round. Click on the **OK** button from the dialog box to create revolve feature.
- Click on the **Start 2D Sketch** tool from the **Sketch** panel in the **Ribbon**. You will be asked to select a plane/face.
- Click on the inner flat face of the revolve feature; refer to Figure-99. The sketching environment will be activated.

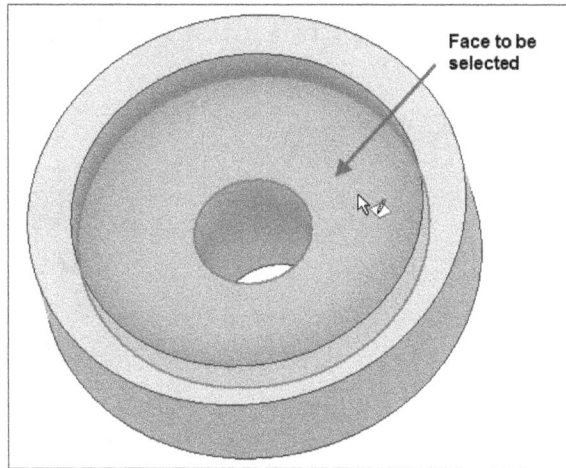

Figure-99. Face selected for creating sketch

- Click on the **Point** tool from the **Create** panel in the **Sketch** contextual tab of **Ribbon**. You will be asked to specify location of point.
- Move the cursor on Y axis in the sketch and specify coordinates as **0** along X and **20** along Y axis as shown in Figure-100. Press **ENTER** to create the point.

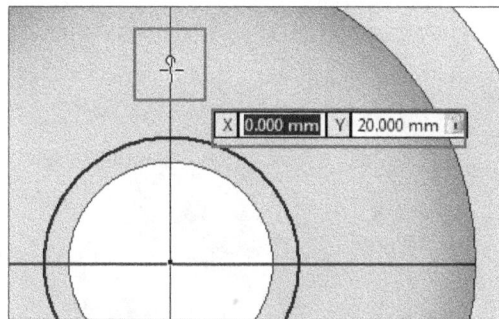

Figure-100. Specifying coordinates of point

- Click on the **Finish Sketch** button to exit the Sketching environment.
- Click on the **Hole** tool from the **Modify** panel in the **3D Model** tab of **Ribbon**. The **Hole** dialog box will be displayed along with the preview of hole located at the sketched point created earlier; refer to Figure-101.

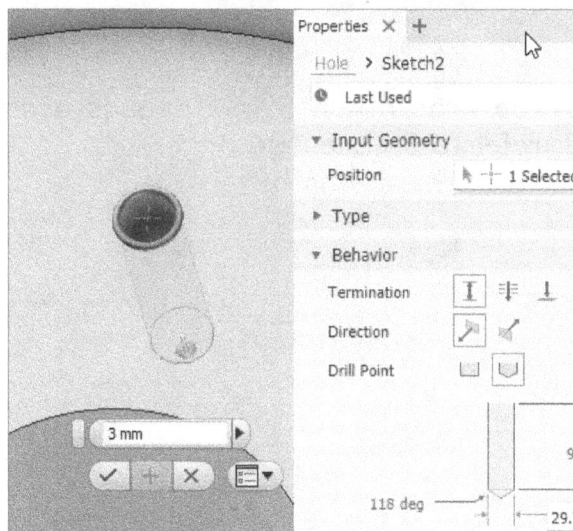

Figure-101. Preview of hole

- Set the hole size as **6** mm and click on the **OK** button from dialog box.
- Create the circular pattern of hole with 3 instances in 360 degree revolution; refer to Figure-102.

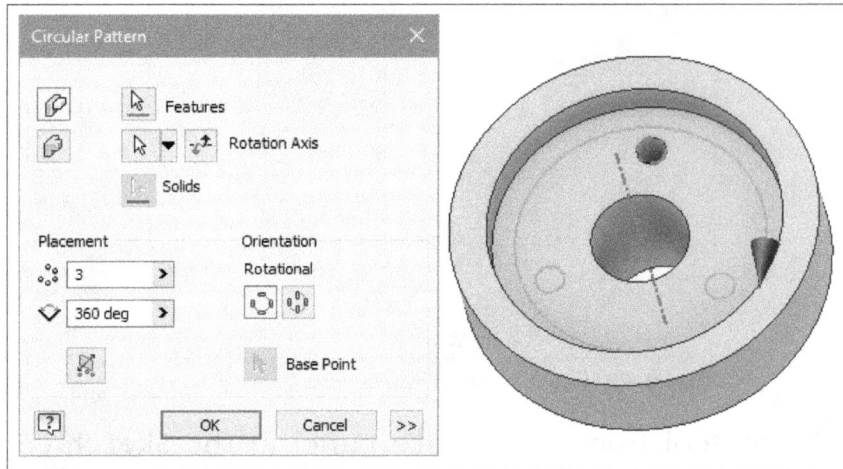

Figure-102. Preview of circular pattern

- Click on the **Thread** tool from the **Modify** panel in the **3D Model** tab of the **Ribbon**. The **Thread** dialog box will be displayed.
- Select the face of hub as shown in Figure-103. Preview of thread will be displayed.

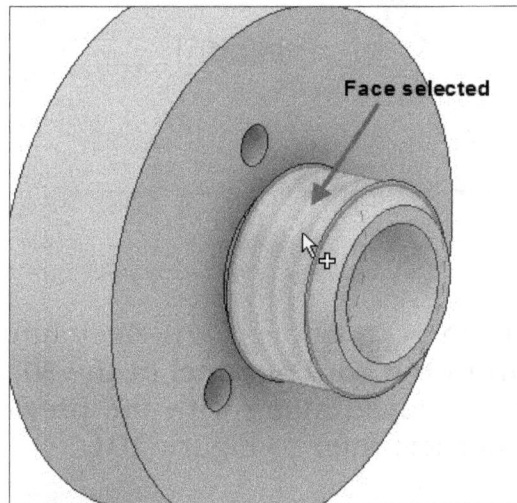

Figure-103. Face selected for Thread tool

- Click on the **Designation** drop-down from **Threads** area of the dialog box and select **M30x2.5** thread; refer to Figure-104.
- Click on the **OK** button from the dialog box.

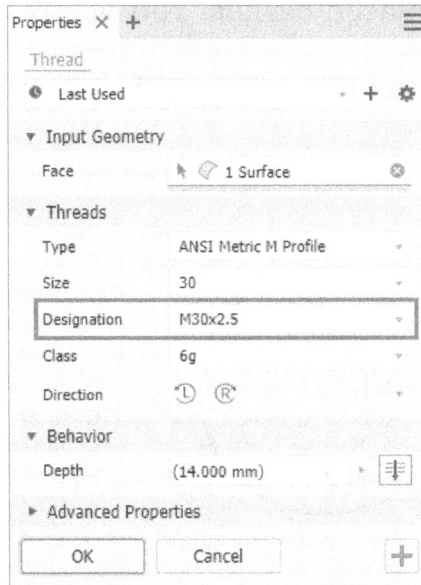
Figure-104. Thread selected in Thread dialog box

Starting A New Drawing

- Click on the **New** button from the **Quick Access Toolbar**. The **Create New File** dialog box will be displayed.
- Double-click on **ANSI(mm).idw** template in the dialog box (because we want the annotations in mm). The drawing environment will be displayed.

Placing Views

- Click on the **Base** tool from the **Create** panel in the **Place Views** tab of the **Ribbon**. The **Drawing View** dialog box will be displayed with preview of base view.
- Set the parameters as required and place the view at the left in the drawing area; refer to Figure-105.

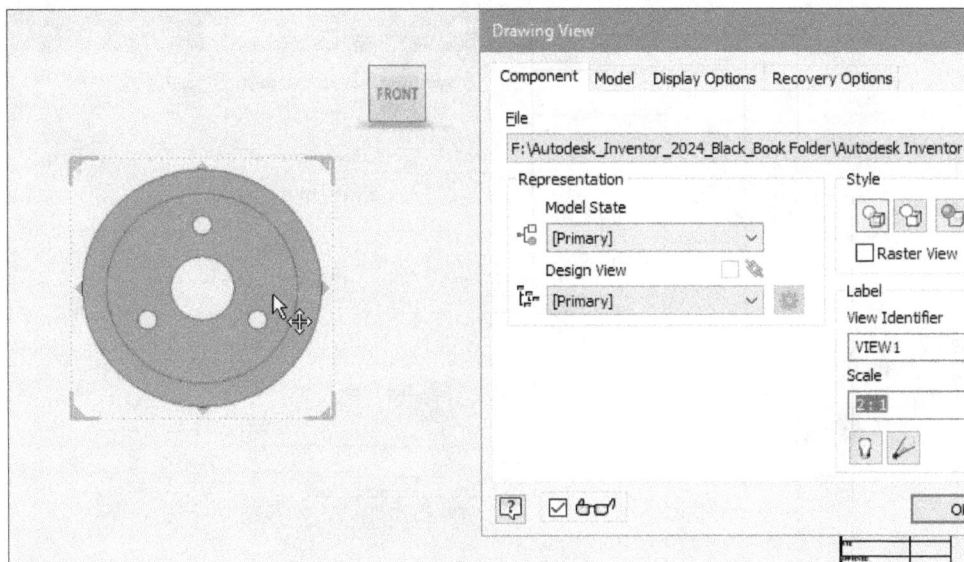
Figure-105. Placing drawing view

- Click on the **OK** button from the dialog box.
- Click on the **Section** tool from the **Create** panel in the **Place Views** tab of the **Ribbon**. You will be asked to select the view for sectioning.

- Select the view you have placed earlier. You will be asked to specify end points of the section line.
- Click at the points shown in Figure-106 and right-click. A shortcut menu will be displayed.

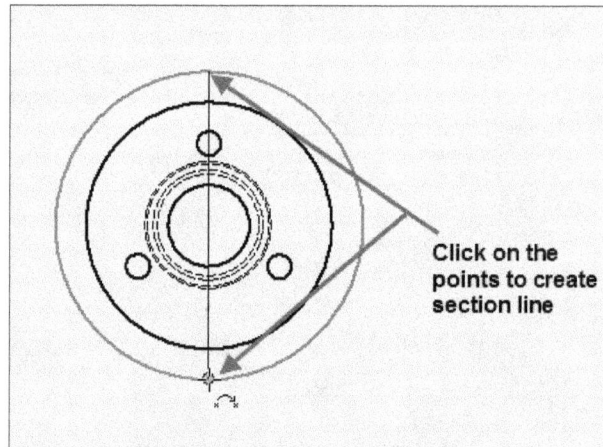

Figure-106. Points selected for section line

- Click on the **Continue** button from the shortcut menu. The preview of section will get attached to cursor.
- Click in the drawing at adequate distance from base view to place the section view.

Applying Annotations

- Click on the **Retrieve Model Annotations** tool from the **Retrieve** panel in the **Annotate** tab of the **Ribbon**. The **Retrieve Model Annotation** dialog box will be displayed. Also, you will be asked to select the view for retrieving dimensions.
- Select the section view created in drawing. The preview of dimensions will be displayed; refer to Figure-107. You will be asked to select the features or the dimensions to be retrieve.

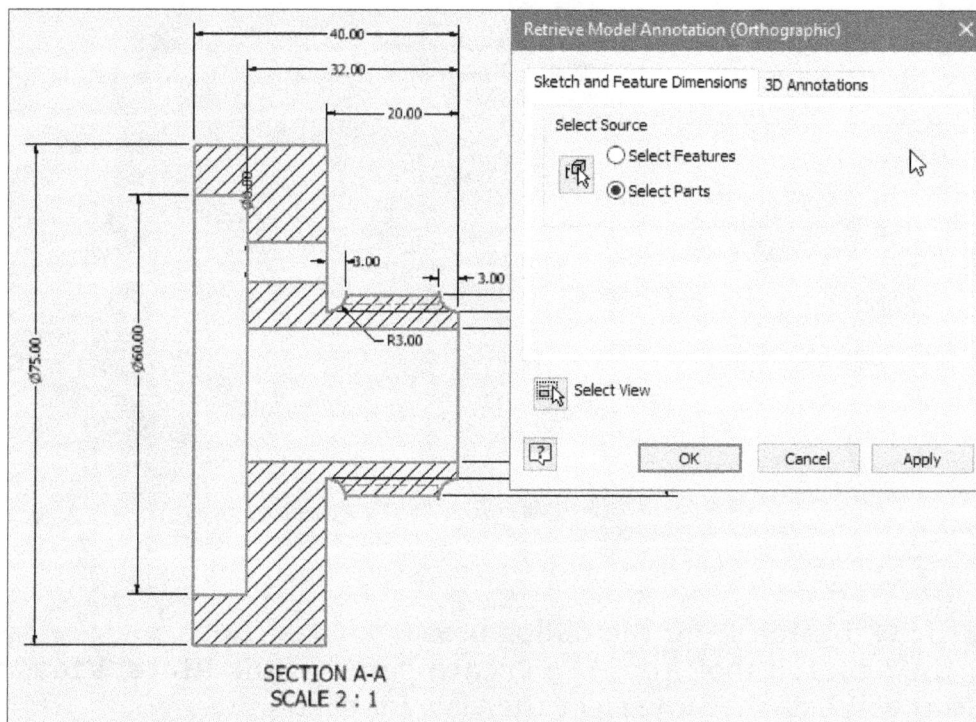

Figure-107. Preview of dimensions

- Select all the dimensions except diameter **6** dimension of hole.
- Click on the **Apply** button to apply dimensions to the view. You will be asked to select the next view for dimensioning.
- Select base view and repeat the procedure to create the dimension in base view.
- Click on the **OK** button to apply dimensions.
- Create the pitch circle for holes in the base view using sketching tools in **Sketch** tab of **Ribbon**; refer to Figure-108.

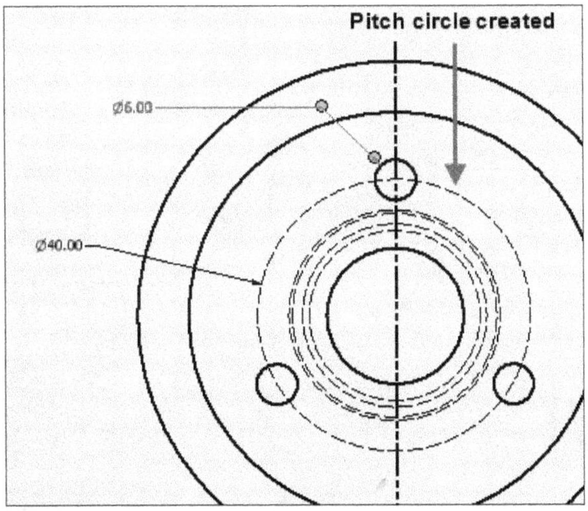

Figure-108. Pitch circle for holes

- Apply the symbols for Datum Identifier, Surface Finish, and Feature Control Frame; refer to Figure-109. Also, double-click on the dimensions and set tolerances as shown in the figure.

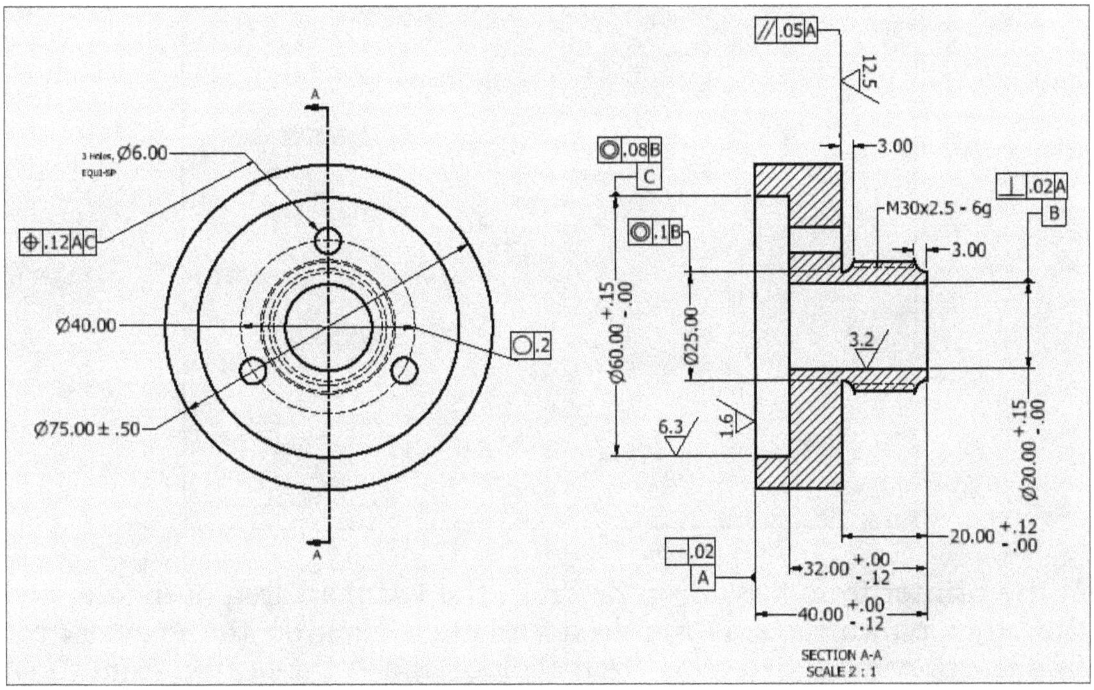

Figure-109. Drawing after applying symbols and tolerances

- Save the drawing using **CTRL+S** at desired location. Yes! It is common sense in Computer works.

Printing Drawing

- Right-click on the sheet from the **Model Browse Bar** and select the **Edit Sheet** from the shortcut menu. The **Edit Sheet** dialog box will be displayed.
- Select the size of sheet which is suitable to accommodate drawing views and annotations; refer to Figure-110.

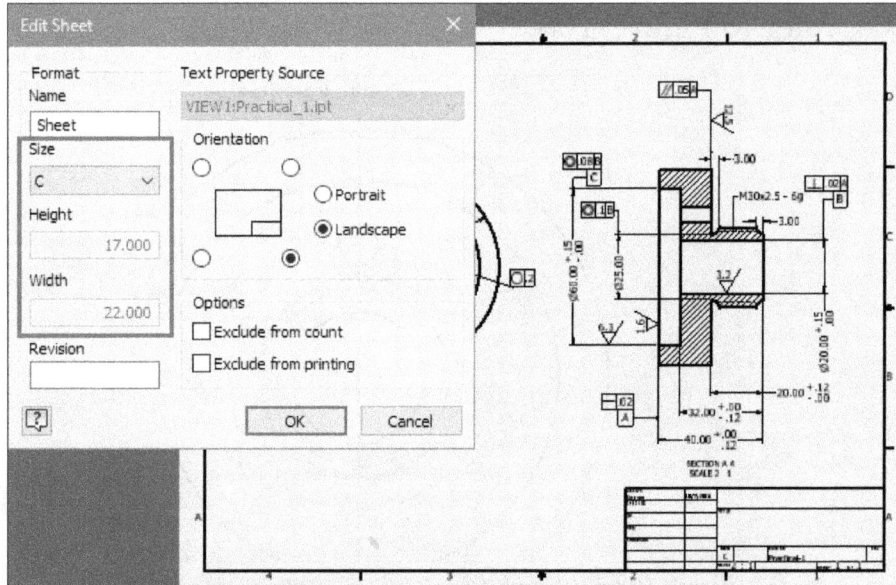

Figure-110. Changing size of sheet

- Press **CTRL+P** from keyboard. The **Print Drawing** dialog box will be displayed; refer to Figure-111.

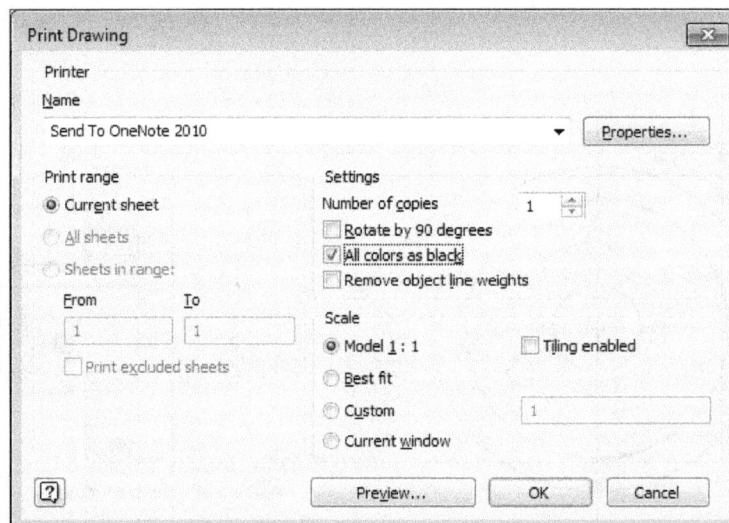

Figure-111. Print Drawing dialog box

- Select the printer from **Name** drop-down in the **Printer** area of the dialog box.
- Set the other parameters as required and then click on the **Preview** button. If you find the preview satisfactory then click on the **Print** button from the **Preview** window and then click on the **OK** button from the dialog box displayed. If you are not satisfied by preview then click on the **Close** button from preview window and make the changes; refer to Figure-112.

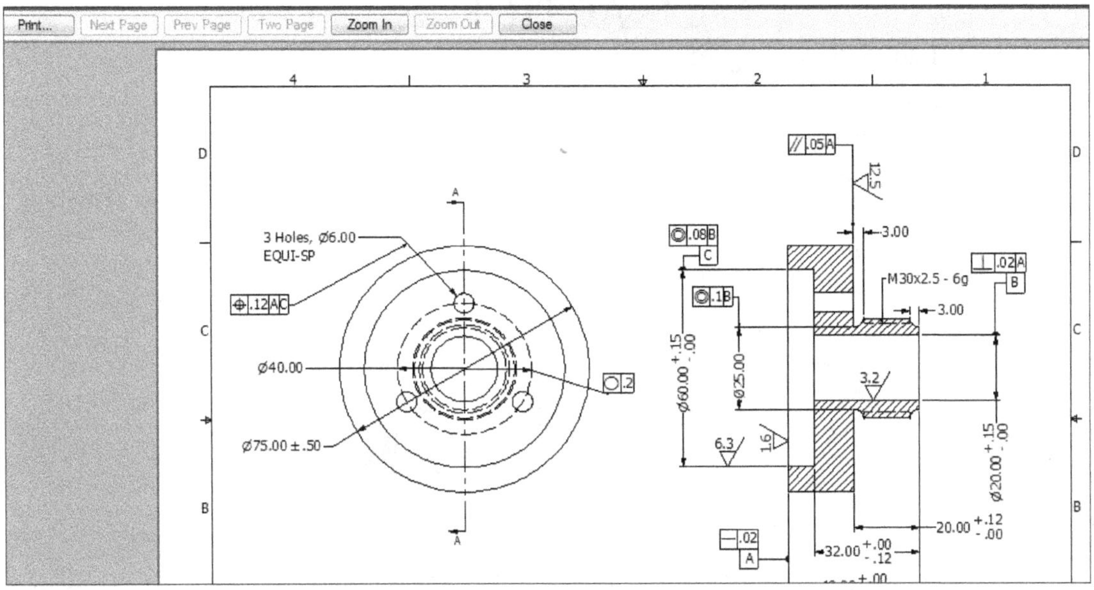

Figure-112. Partial preview of the drawing

PRACTICE 1

Create the model and drawing given in Figure-113.

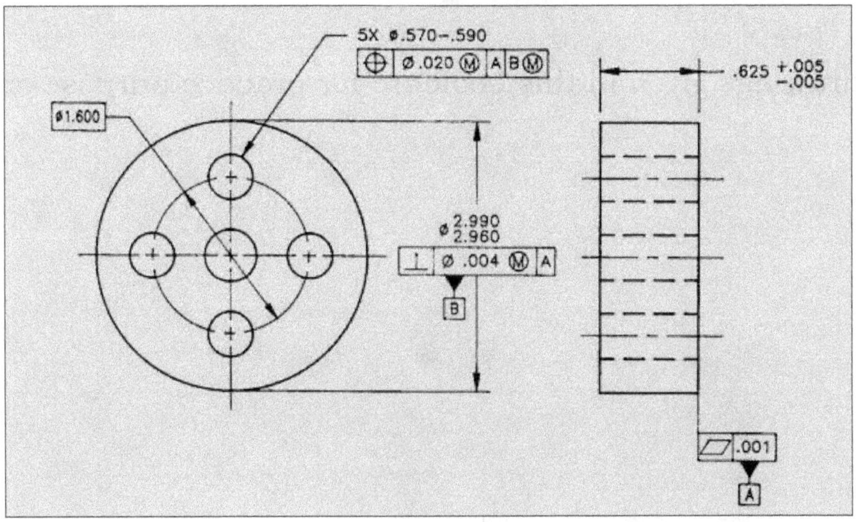

Figure-113. Practice 1

PRACTICE 2

Create the model and drawing given in Figure-114.

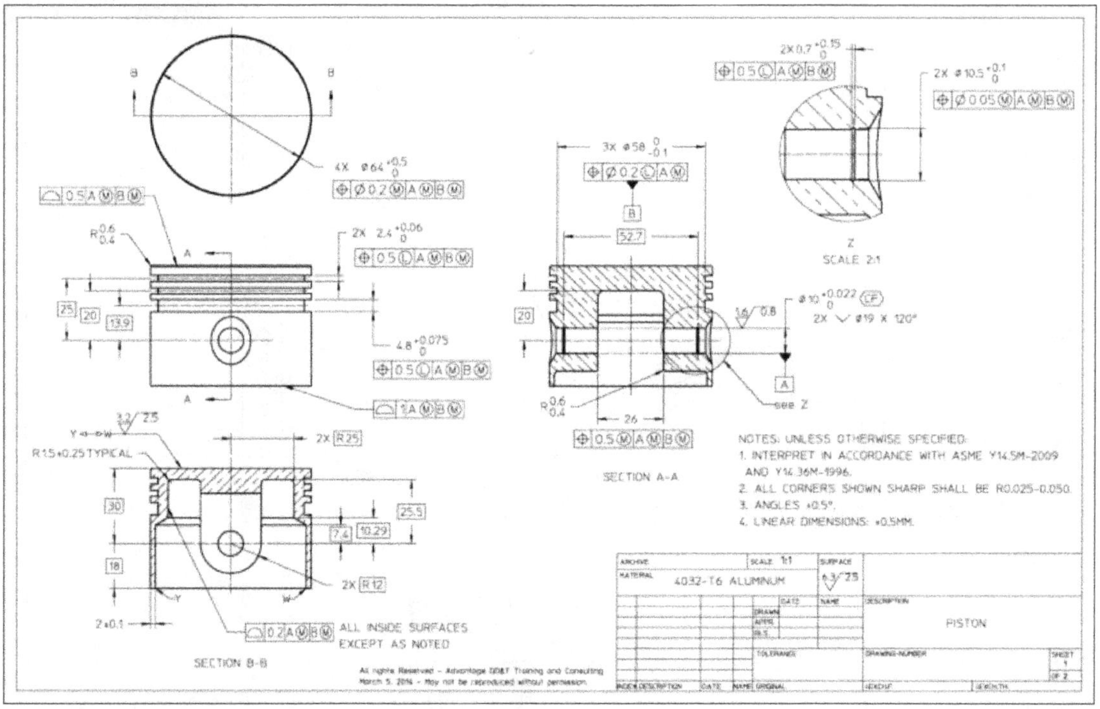

Figure-114. Practice 2

Note that the drawings given in this book are for practice purpose only.

SELF ASSESSMENT

Q1. Which of the following statements is correct about the phenomena of bidirectional associativity?

a) Any change made in modeling environment is also reflected in drawing environment.
b) Any change made in assembly environment is also reflected in drawing environment.
c) Any change made in drawing environment is not reflected in assembly environment.
d) Both a and b

Q2. What is the standard sheet size of A4 as per ISO-A used for plotting engineering drawings?

a) 594 x 841
b) 420 x 594
c) 297 x 420
d) 210 x 297

Q3. Which of the following information is not provided by the title block?

a) Title of the drawing
b) Roll Number
c) Symbol denoting the method of projection
d) Name of the firm

Q4. From which of the following views, the 3D objects are not represented on paper in engineering drawings?

a) Detail view
b) Orthographic view
c) Summary view
d) Both a and c

Q5. Which of the following tools is used to create the view to display the inner details of the solid model?

a) Projected View
b) Auxilliary View
c) Section View
d) Detail View

Q6. Which of the following tools is used to get all the modeling dimensions after placing the views?

a) Retrieve Model Annotations
b) Dimension
c) Chain Set
d) Ordinate Set

Q7. Which of the following dimension tool is used in manual CAM programming of a component?

a) Chain Set
b) Ordinate Set
c) Punch
d) Precision and Tolerance

Q8. Which of the following statements is correct about Datum Target Symbols in Engineering Drawings?

a) A1, A2, A3 are datum target areas and B1, B2 are datum target lines and C1 is datum target point.

b) A1, A2, A3 are datum target areas and B1, B2 are datum target points and C1 is datum target line.

c) A1, A2, A3 are datum target points and B1, B2 are datum target areas and C1 is datum target line.

d) None of the Above

Q9. Which of the following geometrical tolerances represent the cylindricity tolerance?

a)
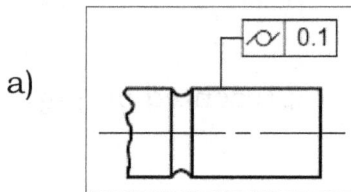

b)
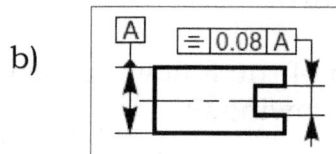

c)
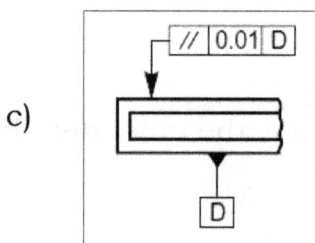

d)
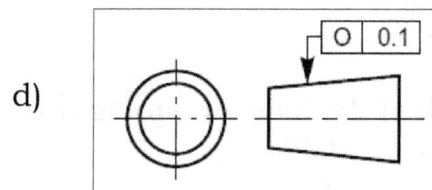

Q10. Bill of Materials is used to display the parts of assembly in a tabulated form in engineering drawing. (True/False)

Chapter 15

Analyses and Simulation

Topics Covered

The major topics covered in this chapter are:

- *Introduction to Stress Analysis*
- *Benefits of Stress Analysis*
- *Starting Analysis in Autodesk Inventor*
- *Performing Static Analysis*
- *Performing Modal Analysis*
- *Performing Shape Generator Study*
- *Report generation*

INTRODUCTION TO STRESS ANALYSIS

Stress analysis is used to find out the stress induced in the selected part under specified loading and contact conditions. In Autodesk Inventor, FEA (Finite Element Analysis) is used to solve the equations for stress analysis. Before we start using Stress Analysis which is based on FEA, we should understand the basics of FEA and its limits.

Basics of FEA

In engineering problems, there are some basic unknowns. If they are found, the behavior of the entire structure can be predicted. The basic unknowns or the Field variables which are encountered in the engineering problems are displacements in solid mechanics, velocities in fluid mechanics, electric and magnetic potentials in electrical engineering, and temperatures in heat flow problems.

In a continuum, these unknowns are infinite. The finite element procedure reduces such unknowns to a finite number by dividing the solution region into small parts called elements and by expressing the unknown field variables in terms of assumed approximating functions (Interpolating functions/Shape functions) within each element. The approximating functions are defined in terms of field variables of specified points called nodes or nodal points. Thus in the finite element analysis, the unknowns are the field variables of the nodal points. Once these are found, the field variables at any point can be found by using interpolation functions.

After selecting elements and nodal unknowns, next step in finite element analysis is to assemble element properties for each element. For example, in solid mechanics, we have to find the force-displacement i.e. stiffness characteristics of each individual element. Mathematically, this relationship is in the form

$$[k]_e \{\delta\}_e = \{F\}_e$$

where $[k]_e$ is element stiffness matrix, $\{\delta\}_e$ is nodal displacement vector of the element, and $\{F\}_e$ is nodal force vector. The element of stiffness matrix k_{ij} represent the force in coordinate direction 'i' due to a unit displacement in coordinate direction 'j'. Four methods are available for formulating these element properties viz. direct approach, variational approach, weighted residual approach, and energy balance approach. Any one of these methods can be used for assembling element properties. In solid mechanics, variational approach is commonly employed to assemble stiffness matrix and nodal force vector (consistent loads).

Element properties are used to assemble global properties/structure properties to get system equations $[k] \{\delta\} = \{F\}$. Then the boundary conditions are imposed. The solution of these simultaneous equations give the nodal unknowns. Using these nodal values, additional calculations are made to get the required values e.g. stresses, strains, moments, etc. in solid mechanics problems.

Thus the various steps involved in the finite element analysis are:

(i)   Select suitable field variables and the elements.
(ii)   Discritise the continua.
(iii)   Select interpolation functions.
(iv)   Find the element properties.
(v)   Assemble element properties to get global properties.
(vi)   Impose the boundary conditions.
(vii)   Solve the system equations to get the nodal unknowns.
(viii)   Make the additional calculations to get the required values.

Assumptions for using FEA

Some assumptions are made in this type of analysis, like:

1. The loads applied does not vary with time.
2. All loads are applied slowly and gradually until they reach to the full magnitude and after reaching the full magnitude, the loads remain constant. Thereby, neglecting impact, inertial, and damping forces.
3. The materials applied to the components satisfy the Hooke's law.
4. The change in stiffness due to loading is neglected.
5. Boundary conditions do not vary during the application of loads. Loads must be constant in magnitude, direction, and distribution.

Geometry Assumptions

1. The part model must represent the required CAD geometry.
2. Only the internal fillets in the area of interest will be included in the study.
3. Shells are created when thickness of the part is small in comparison to its width and length.
4. Thickness of the shell is assumed to be constant.
5. If the dimensions of a particular part are not critical and do not affect the analysis results, some approximations can be made in modeling the particular part.
6. Primary members of structure are long and thin like a beam then idealization is required.
7. Local behavior at the joints of beams or other discontinuities are not of primary interest, so no special modeling of these area is required.
8. Decorative or external features will be assumed insignificant to the stiffness and the performance of the part and will be omitted from the model.

Material Assumptions

1. Material remain in the linear regime. It is understood that either stress levels exceeding yield or excessive displacements will constitute a component failure. That is non linear behavior cannot be accepted.
2. Nominal material properties adequately represent the physical system.
3. Material properties are not affected by load rate.
4. Material properties can be assumed isotropic (Orthotropic) and homogeneous.
5. Part is free of voids or surface imperfections that can produce stress risers and skew local results.
6. Actual non linear behavior of the system can be extrapolated from the linear material results.
7. Weld material and the heat affected zone will be assumed to have same material properties as the base material.

8. Temperature variations may have a significant impact on the properties of the materials used. Change in material properties is neglected.

Boundary Conditions Assumptions

1. Choosing proper BC's require experience.
2. Using BC's to represent parts and effects that are not or cannot be modeled leads to the assumption that the effects of these un-modeled entities can truly be simulated or has no effect on the model being analyzed.
3. For a given situation, there would be many ways of applying boundary conditions. But these various alternatives can be wrong if the user does not understand the assumptions they represent.
4. Symmetry/ anti-symmetry/ reflective symmetry/ cyclic symmetry conditions if exists can be used to minimize the model size and complexity.
5. Displacements may be lower than they would be if the boundary conditions being more appropriate. Stress magnitudes may be higher or lower depending on the constraint used.

Fasteners Assumptions

1. Residual stress due to fabrication, pre-loading of bolts, welding and/or other manufacturing, or assembly processes are neglected.
2. Bolt loading is primarily axial in nature.
3. Bolt head or washer surface torque loading is primarily axial in nature.
4. Surface torque loading due to friction will produce only local effects.
5. Bolts, spot welds, welds, rivets, and/or fasteners which connect two components are considered perfect and acts as rigid joint.
6. Stress relaxation of fasteners or other assembly components will not be considered. Load on threaded portion of the part is evenly distributed on engaged threads.
7. Failure of fasteners will not be reflected in the analysis.

General Assumptions

1. If the results in the particular area are of interest then mesh convergence will be limited to this area.
2. No slippage between interfacing components will be assumed.
3. Any sliding contact interfaces will be assumed frictionless.
4. System damping will be normally small and assumed constant across all frequencies of interest unless otherwise available from published literature or actual tests.
5. Stiffness of bearings in radial or axial directions will be considered infinite.
6. Elements with poor or less than optimal geometry are only allowed in areas that are not of concern and do not affect the overall performance of the model.

Now, we will not go deeper in theoretical area of FEA. We will now start stress analysis in Autodesk Inventor.

STARTING STRESS ANALYSIS

• Open the part file or assembly file on which you want to perform stress analysis in Autodesk Inventor. (C'mon you know it how to open file)

- Click on the **Stress Analysis** tool from the **Begin** panel in the **Environments** tab of the **Ribbon**; refer to Figure-1. The **Analysis** contextual tab will be added in the **Ribbon**; refer to Figure-2.

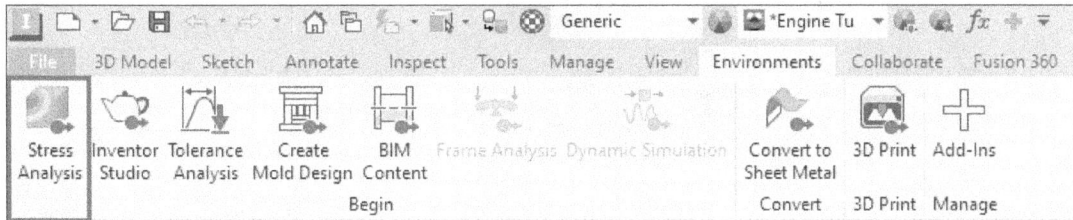

Figure-1. Stress Analysis tool

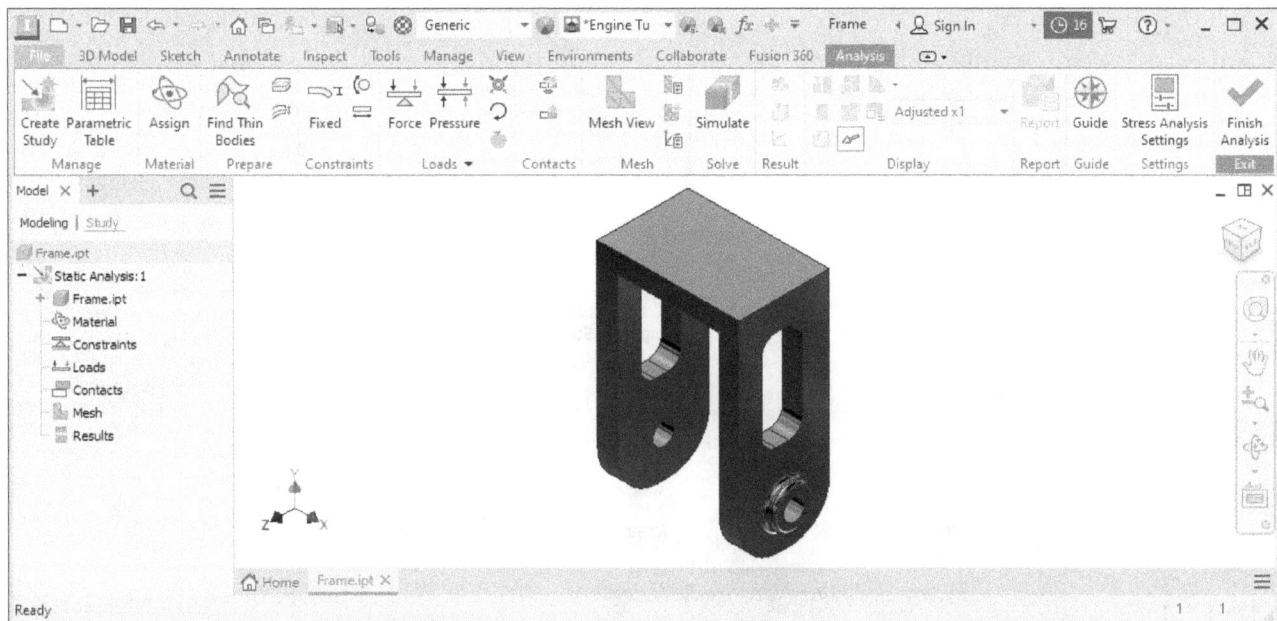

Figure-2. Tools in Analysis contextual tab

Note that only **Create Study** tool is active in the contextual at beginning which means you need to create a simulation study first to use the other tools in tab.

CREATING STUDY

This is the first and very important step in performing stress analysis in Autodesk Inventor. All the results of analysis are directly linked with the options specified here. The procedure to create simulation study is given next.

- Click on the **Create Study** tool from the **Manage** panel in the **Analysis** contextual tab of the **Ribbon**. The **Create New Study** dialog box will be displayed; refer to Figure-3.

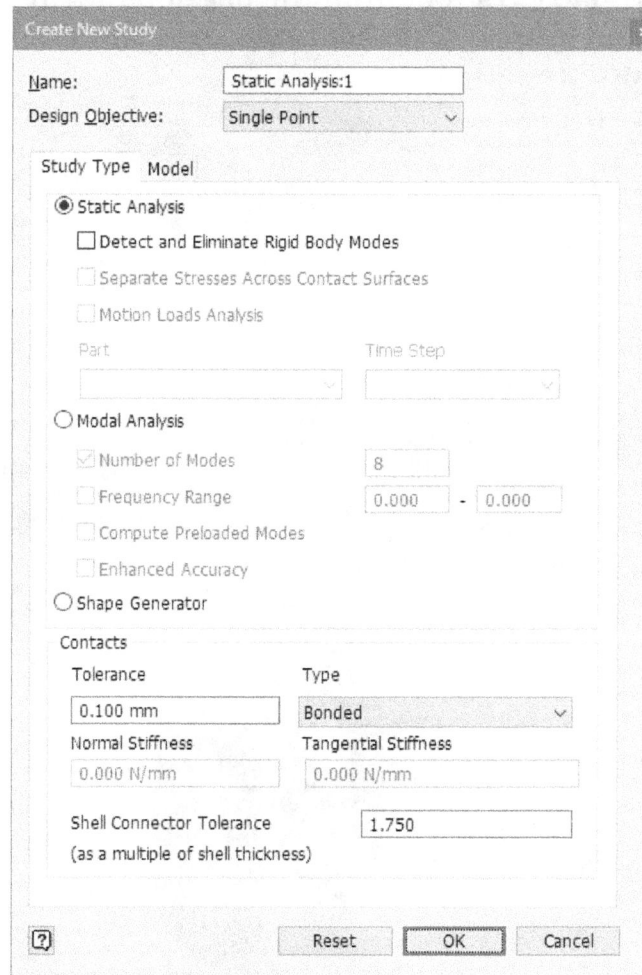

Figure-3. Create New Study dialog box

- Select the type of analysis from the **Study Type** tab of the dialog box. There are three radio buttons to select study type; **Static Analysis**, **Modal Analysis**, and **Shape Generator**. Select the **Static Analysis** radio button if you want to check the stresses induced in part due to applied loads. Select the **Modal Analysis** radio button if you want to find out the natural frequencies and mode shapes along with effect of applied loads on the part. Select the **Shape Generator** radio button if you want to reduce the weight of part for specified loading conditions. This option is available for parts only. Note that **Shape Generator** option will give you the shape mathematically satisfying the weight and loading conditions but it is your duty to find the practicality of shape on our planet!!

- Specify the name of analysis in the **Name** edit box of dialog box.

- Select **Single Point** or **Parametric Dimension** option from the **Design Objective** drop-down. If selected **Single Point** option then analysis result deformation will occur on structure of part. If **Parametric Dimension** option is selected then deformation will obey dimensional stability.

Static Analysis Options

- Select the **Detect and Eliminate Rigid Body Modes** check box if you want to eliminate rigid body modes which are not restricted by providing constraints. In general words, rigid body mode is translational and rotational degree of freedom of complete part. In rigid body, particles of part do not change their position on applying load which means shape of part is constant.

- Select the **Separate Stresses Across Contact Surfaces** check box if you want to discontinue the stress representations of the model at contact surfaces. In simple words, the stress representations at the contact surfaces will be according to their parts rather than mixing of stress shades in results. This option is active for assemblies only.
- Select the **Motion Loads Analysis** check box if you want to run the motion loads analysis on the part. You can run the motion load analysis using the **Dynamic Simulation** tool in the **Environments** tab of **Ribbon** in assembly environment.

Modal Analysis Options

- Select the **Number of Modes** check box and specify the number of modes in the adjacent check box. Number of modes means number of natural frequency for which you want to check the deformation of the part.
- Select the **Frequency Range** check box to define the range of frequency in which natural frequencies should be checked. On selecting this check box, the frequency range edit boxes will become active. Specify desired values in the edit boxes.
- Select the **Compute Preloaded Modes** check box if you want to run a stress analysis first and then run the Modal analysis. In this case, the natural frequencies will be calculated using the pre-stressed model.
- Select the **Enhanced Accuracy** check box if you want to find out natural frequencies up to the order of **10**.

Shape Generator

The **Shape Generator** radio button is used to run analysis to find out the best possible shape of part to survive the applied loads and reduce the mass of part as much as possible. This option is available for parts only.

Contact Options

The options in the **Contacts** area of dialog box are used to specify the parameters related to contacts in analysis. Contact is relation between material structures of two parts in contact. There are various contact type like bonded, spring, and so on. The options in the **Contacts** area of the dialog box are discussed next.

- Select the contact type from the **Type** drop-down in the **Contacts** area of the dialog box which will be automatically applied while working with assemblies in analysis.
- Specify the value of maximum distance/gap in the **Tolerance** edit box up to which the automatic contacts will be created.
- If you have selected the **Spring** option as automatic contact from the **Type** drop-down then you can specify the tangential stiffness and normal stiffness in their respective edit boxes in **Contacts** area of the dialog box.
- Set the representation of assembly in the **Model** tab of the dialog box.
- After specifying all the options as required in the dialog box, click on the **OK** button to create the study.

ASSIGNING MATERIAL

Once you have started an analysis, the very next step is to assign material to the part(s) if not applied earlier. There are various properties that are taken into account while performing analysis like ultimate strength, young's modulus, Poisson's ratio, etc. The procedure to apply material is given next.

*   Click on the **Assign** tool from the **Material** panel in the **Analysis** contextual tab of the **Ribbon**. The **Assign Materials** dialog box will be displayed; refer to Figure-4.

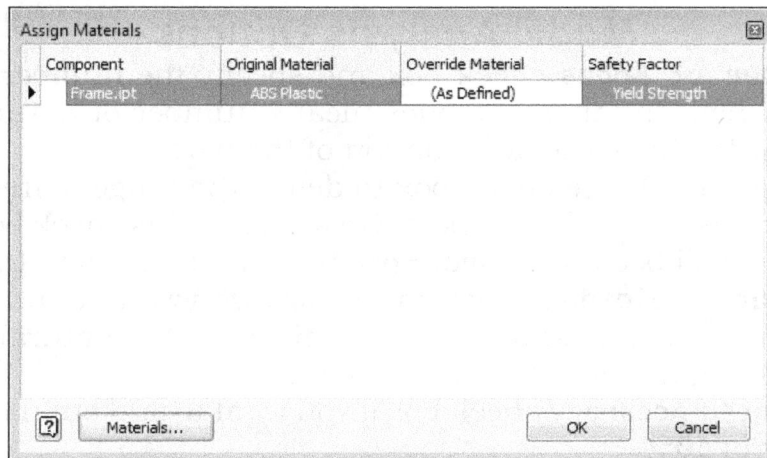

Assign Materials			
Component	Original Material	Override Material	Safety Factor
▶ Frame.ipt	ABS Plastic	(As Defined)	Yield Strength

Figure-4. Assign Materials dialog box

*   If you have applied any material earlier then it will be displayed for the part under **Original Material** column. Click in the field under **Override Material** column in the table. A drop-down will become active.
*   Select desired material from the drop-down.
*   Click on the **OK** button from the dialog box to apply the material. Note that we have discussed about creating custom materials earlier in the book. So, you can create and use the custom materials in the same way.

APPLYING CONSTRAINTS

Constraints means restriction to motion. There are three tools to apply constraints in Stress Analysis environment of Autodesk Inventor. These tools are discussed next.

Fixed Tool

The **Fixed** tool in the **Constraints** panel is used to create fixed constraint. Using this constraint, you can fix the selected portion of part/assembly so that it cannot move in any direction even under the maximum loading conditions. The procedure to use **Fixed** tool is given next.

*   Click on the **Fixed** tool from the **Constraints** panel in the **Analysis** contextual tab of the **Ribbon**. The **Fixed Constraint** dialog box will be displayed; refer to Figure-5. Also, you will be asked to select the entities to be fixed.

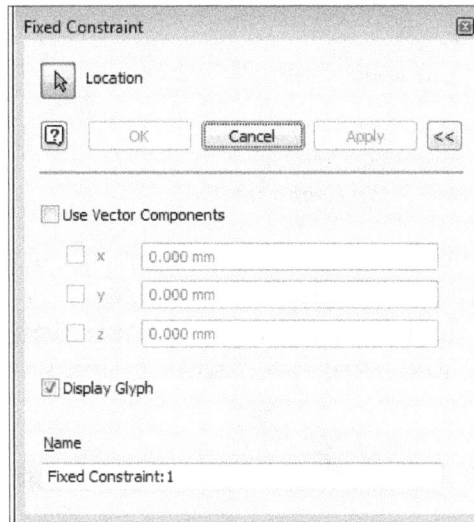

Figure-5. Fixed Constraint dialog box

- Select the entity/entities to be fixed and click on the **OK** button. The selected entities will be fixed for analysis; refer to Figure-6.

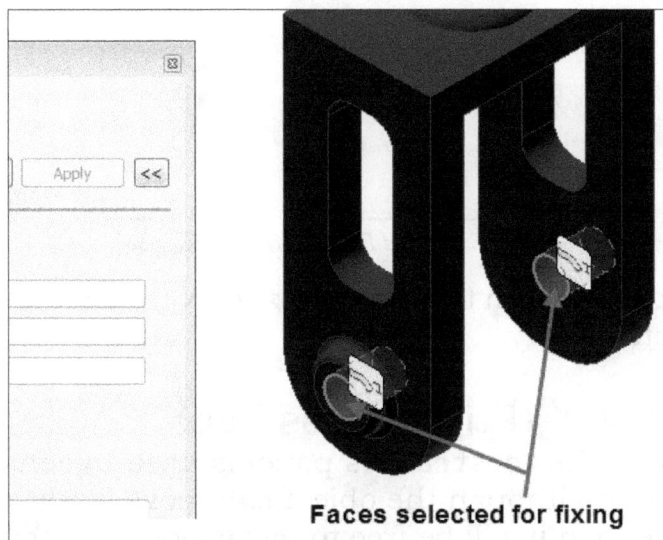

Faces selected for fixing

Figure-6. Faces selected for fixing

Pin Tool

The **Pin** tool in the **Constraints** panel is used to constraint selected entities in such a way that the object is free to rotate but cannot translate in any direction. Note that the constraint represents an imaginary pin at the faces selected. This constraint should be applied to shafts or holes for shafts. The procedure to use **Pin** tool is given next.

- Click on the **Pin** tool from the **Constraints** panel in the **Analysis** contextual tab of the **Ribbon**. The **Pin Constraint** dialog box will be displayed; refer to Figure-7. Also, you will be asked to select the entity/entities on which you want to apply the pin constraint.

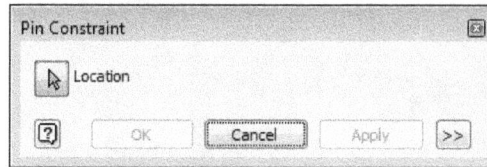

Figure-7. Pin Constraint dialog box

- Select the round face/faces on which you want to apply the pin constraint and then click on the **OK** button from the dialog box.
- To change the advanced options of the pin constraint, click on the **More Options** button in the dialog box. The expanded dialog box will be displayed as shown in Figure-8.

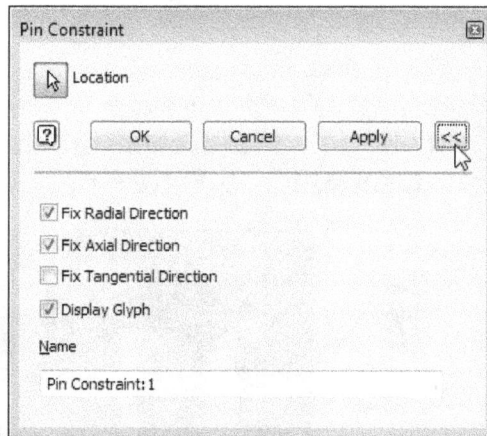

Figure-8. Expanded Pin Constraint dialog box

- Select the fix check boxes from the dialog box to fix the radial, axial, and tangential motion of the selected entity.

Frictionless Tool

The **Frictionless** tool in the **Constraints** panel is used to constraint the motion of object in normal direction although the object can move freely in same plane. If you select a cylindrical face then it will be free to move along the center axis but can not move perpendicular to axis. The procedure to use the tool is given next.

- Click on the **Frictionless** tool from the **Constraints** panel in the **Analysis** tab of the **Ribbon**. The **Frictionless Constraint** dialog box will be displayed; refer to Figure-9. Also, you will be asked to select entities.

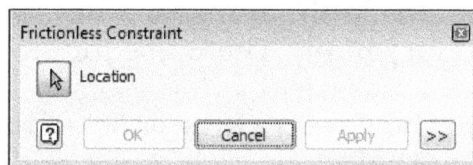

Figure-9. Frictionless Constraint dialog box

- Select the face/faces to which frictionless constraint is to be applied and click on the **OK** button. The constraint will be applied.

APPLYING LOADS

The loads in Analysis environment of Autodesk Inventor represent forces acting on the part/assembly during its practical application. There are seven tools available to apply different type of loads in the **Loads** panel of the **Analysis** contextual tab of **Ribbon**. These tools are discussed next.

Force Tool

The **Force** tool is used to apply force at the selected entity at desired direction. The procedure to use this tool is given next.

*   Click on the **Force** tool from the **Loads** panel in the **Analysis** contextual tab of the **Ribbon**. The **Force** dialog box will be displayed; refer to Figure-10.

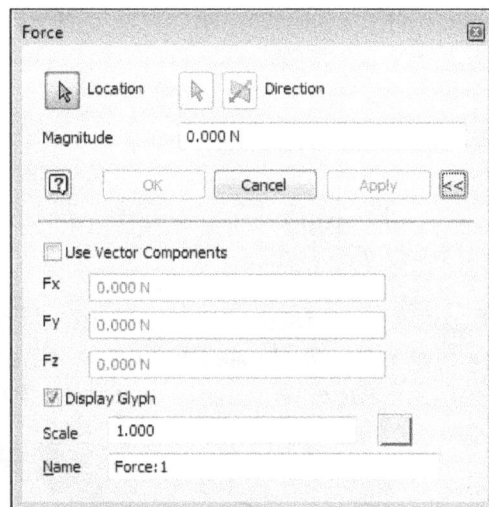

Figure-10. Force dialog box

*   Select the face/edge/vertex on which you want to apply force. Arrow for force direction will be displayed on the selected face/edge/vertex; refer to Figure-11.

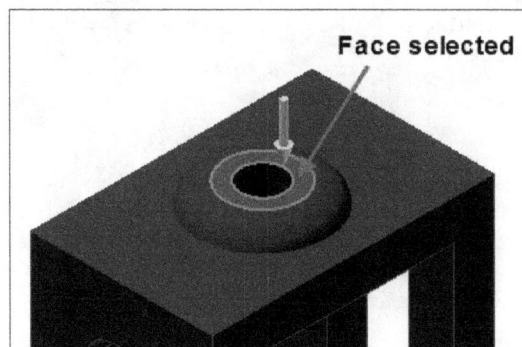

Figure-11. Face selected for applying force

*   Specify desired value of force in the **Magnitude** edit box of the dialog box.
*   By default, force is applied normal to the selected entity. If you want to explicitly specify the force direction then click on the selection button for **Direction** and select the reference for direction of force. Flip the direction if required by using the **Flip** button for **Direction** in the dialog box.

• If you want to specify the X, Y, and Z component of force vector then select the **Use Vector Components** check box in the expanded dialog box and specify desired values in the edit boxes below it.

Pressure Tool

The force applied per unit area is called pressure. Generally, pressure is used to represent the force applied by fluids on the enclosures. The **Pressure** tool is used to apply pressure on the workpiece in Autodesk Inventor Analysis environment. The procedure to use this tool is given next.

• Click on the **Pressure** tool from the **Loads** panel in the **Analysis** contextual tab of the **Ribbon**. The **Pressure** dialog box will be displayed; refer to Figure-12. Also, you will be asked to select the faces on which pressure is to be applied.

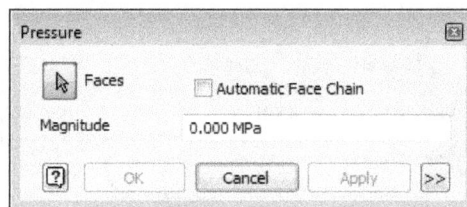

Figure-12. Pressure dialog box

• Select the faces on which pressure is to be applied. If you want to select the faces connected in chain then select the **Automatic Face Chain** check box in the dialog box and then select one of the face in chain; refer to Figure-13.

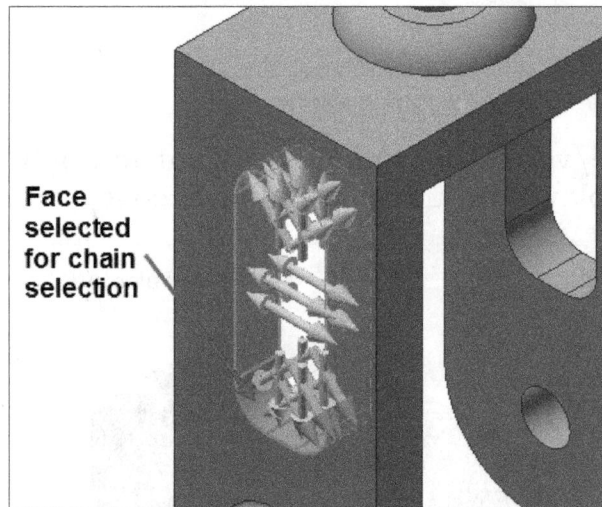

Figure-13. Face selected for applying pressure

• Specify desired value of pressure in the **Magnitude** edit box of the dialog box and click on the **OK** button.

Bearing Load Tool

The **Bearing Load** tool is used to apply the load on part which is sustained by bearing in the actual assembly. The procedure to use **Bearing Load** tool is given next.

• Click on the **Bearing Load** tool from the **Loads** panel in the **Analysis** contextual tab of the **Ribbon**. The **Bearing Load** dialog box will be displayed; refer to Figure-14.

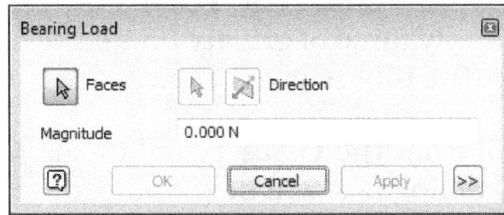

Figure-14. Bearing Load dialog box

- Select the round face on which you want to apply bearing load and specify the value of load in the **Magnitude** edit box.
- Using the options in the expanded dialog box, you can specify the component of the bearing load vector.
- Click on the **OK** button to apply the load.

Moment Tool

The **Moment** tool is used to apply general moment on the selected face. The procedure to use this tool is given next.

- Click on the **Moment** tool from the **Loads** panel in the **Analysis** contextual tab of the **Ribbon**. The **Moment** dialog box will be displayed; refer to Figure-15. Also, you will be asked to select faces.

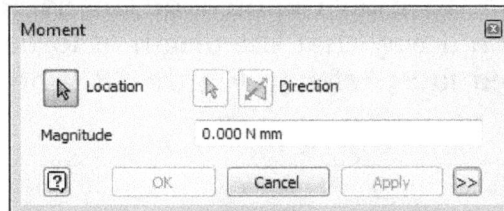

Figure-15. Moment dialog box

- Select the round face on which you want to apply moment and specify desired value in the **Magnitude** edit box of the dialog box; refer to Figure-16.

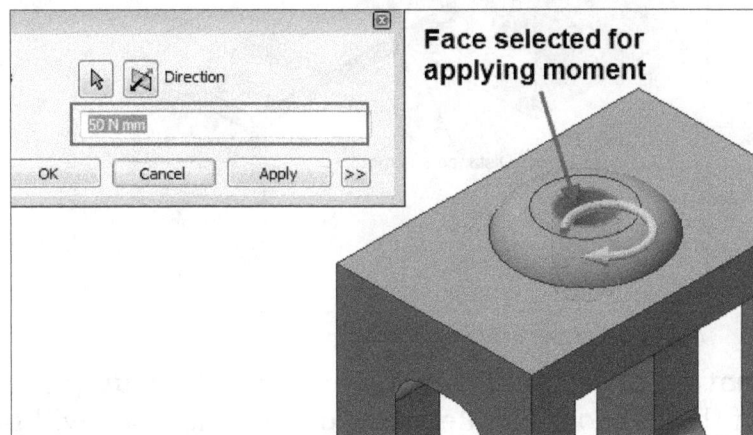

Figure-16. Applying moment

- Click on the **OK** button to apply moment.

Gravity Tool

The **Gravity** tool is used to apply force of gravity on all the components in the drawing area. The procedure to use this tool is given next.

- Click on the **Gravity** tool from the **Loads** panel in the **Analysis** contextual tab of the **Ribbon**. The **Gravity** dialog box will be displayed; refer to Figure-17. Also, you will be asked to select a reference to specify direction of gravity.

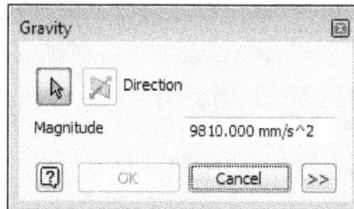

Figure-17. Gravity dialog box

- Select a flat face/edge/axis to specify the direction.
- Specify the magnitude of gravity in the **Magnitude** edit box of the dialog box.
- Use the options in the expanded dialog box to specify components of gravity force.
- Click on the **OK** button to apply the force.

Remote Force Tool

The **Remote Force** tool is used to apply remote load/mass. Remote load/mass is force applied on an object in such a way that the origin of force is somewhere else but it is also affecting the selected face; refer to Figure-18. The procedure to use **Remote Force** tool is given next.

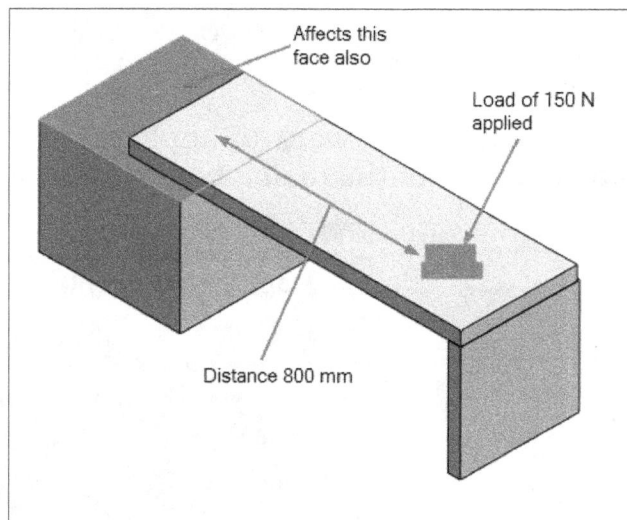

Figure-18. Remote load example

- Click on the **Remote Force** tool from the expanded **Loads** panel in the **Analysis** contextual tab of the **Ribbon**. The **Remote Force** dialog box will be displayed; refer to Figure-19.

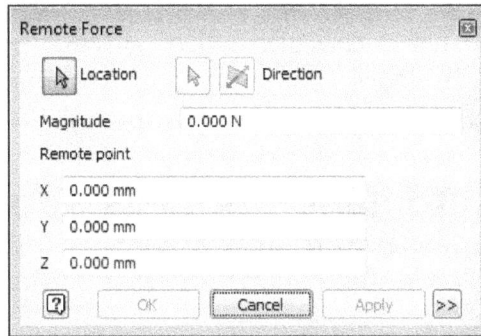

Figure-19. Remote Force dialog box

- Select the face on which you want to apply remote force.
- Click in the **Magnitude** edit box and specify desired value of force.
- Specify the location of remote point in the edit boxes **X**, **Y**, and **Z** given for **Remote point** in the dialog box.
- Click on the **OK** button to create the force.

Body Loads Tool

The **Body Load** tool is used to apply loads causing motion in the body. These loads imitate the effect of forces like centrifugal force, centripetal force, etc. The procedure to apply the body loads is given next.

- Click on the **Body Load** tool from the expanded **Loads** panel in the **Analysis** contextual tab of the **Ribbon**. The **Body Loads** dialog box will be displayed; refer to Figure-20.

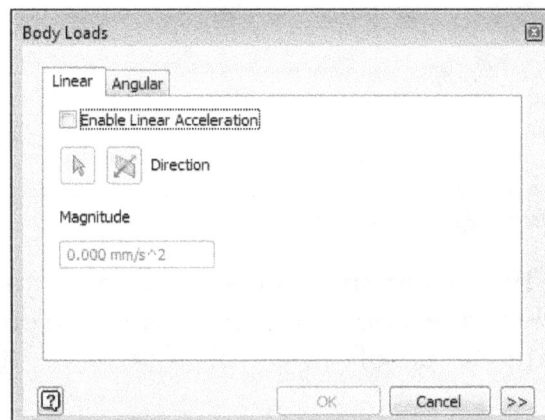

Figure-20. Body Loads dialog box

- Select the **Enable Linear Acceleration** check box from the **Linear** tab if you want to apply linear acceleration. Select the **Enable Angular Velocity and Acceleration** check box from the **Angular** tab if you want to apply angular velocity and acceleration. You can select both check boxes to apply both motions to the selected body.
- Specify the value of velocity and acceleration in their respective **Magnitude** edit boxes in the dialog box.
- Click on the selection button and select the references for body loads.
- Click on the **OK** button to apply the loads.

APPLYING CONTACTS

The contacts in analysis are used to define the contact between two components of assembly at selected faces. The tools to apply contacts are available in the **Contacts** panel in the **Ribbon**. These tools are discussed next.

Automatic Contacts Tool

The **Automatic Contacts** tool is used to apply default contact type to all the possible locations in the assembly. The default contact is specified in the **Create New Study** dialog box discussed earlier.

*   Click on the **Automatic Contacts** tool from the **Contacts** panel in the **Analysis** contextual tab of the **Ribbon**. The **Detecting Automatic Contacts** information box will be displayed for a moment and then the default contacts will be applied automatically.

Manual Contact Tool

The **Manual Contact** tool is used to manually apply contacts in the assembly. The procedure to use this tool is given next.

*   Click on the **Manual Contact** tool from the **Contacts** panel in the **Analysis** contextual tab of the **Ribbon**. The **Manual Contact** dialog box will be displayed; refer to Figure-21.

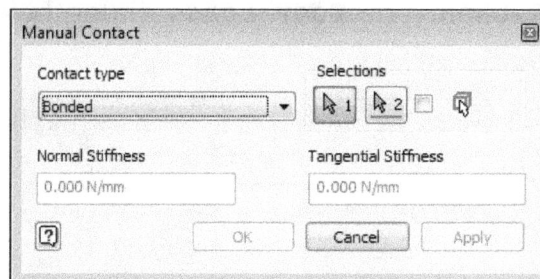

Figure-21. Manual Contact dialog box

*   Select desired contact type from the **Contact Type** drop-down in the dialog box. You will be asked to select reference faces on which the contacts are to be applied.
*   Select the two touching faces of assembly components one by one; refer to Figure-22.

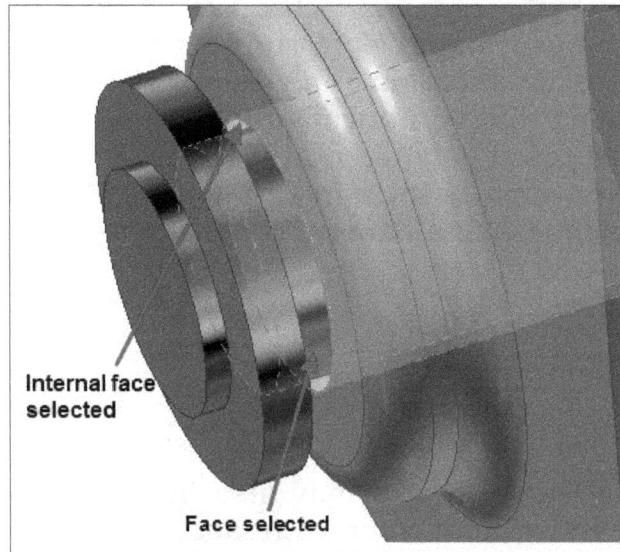

Figure-22. Faces selected for contact

- If you have selected the **Spring** option from the **Contact type** drop-down then specify the stiffness of spring in normal and tangential direction.
- Click on the **Apply** button to apply the contact and start with other faces or click on the **OK** button from the dialog box to apply the contact and exit.

The description of different type of contacts is given next.

Bonded
This creates a rigid bond between selected faces.

Separation
This partially or fully separates selected faces while sliding.

Sliding/No Separation
This creates a normal-to-face direction bond between selected faces while sliding under deformation.

Separation/No Sliding
This partially or fully separates selected faces without them sliding against one another.

Shrink Fit/Sliding
This creates conditions similar to **Separation** but with a negative distance between contact faces, resulting in overlapping parts at the start.

Shrink Fit/No Sliding
This creates conditions of **Separation/No Sliding** but with a negative distance between contact faces, resulting in overlapping parts at the start.

Spring
This creates equivalent springs between the two faces. The **Normal Stiffness** and **Tangential Stiffness** options are available for the **Spring** contact only.

PREPARATION OF PART

It is very important to prepare part before running analysis as there may be many features of part which are not useful in analysis but slow the processing time. There are thousands of iterations that run during analysis, so we should always keep the part model as simple as possible while not destroying the meaning of analysis. In part preparation step, we suppress the non-acting features of part and make sure the part has comparable thickness at all areas. The features that we generally suppress before running analysis are:

*   Fillets and rounds
*   Cosmetic features like threads, taps, etc.
*   Chamfers at edges and so on.

Apart from suppressing features, you may also find the models which have very irregular thickness like sheetmetal component assembled with solid parts or plastic parts. In such models, components that have thin walls compared to the overall size of the model cause problem in stress calculation. We need the tools in **Prepare** panel of the **Analysis** contextual tab to convert such parts to surface bodies to speed up the stress calculations. The tools in the **Prepare** panel are discussed next.

Finding Thin Bodies

In Autodesk Inventor, Shell bodies are considered as Thin bodies. The procedure to find thin bodies and create mid surface is given next.

*   After starting the analysis, click on the **Find Thin Bodies** tool from the **Prepare** panel in the **Analysis** tab of the **Ribbon**. If there are bodies which have high length to thickness radio then a message box will be displayed telling you that thin bodies have found in your model, if you want to generate mid-surfaces then click on the **OK** button.
*   Click on the **OK** button from the dialog box. The **Midsurface** dialog box will be displayed; refer to Figure-23 and the faces of solids to be converted into midsurface will get selected.

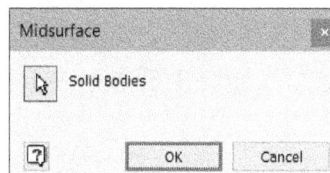

Figure-23. Midsurface dialog box

*   Click on the **OK** button from the dialog box. The mid-surfaces will get created; refer to Figure-24.

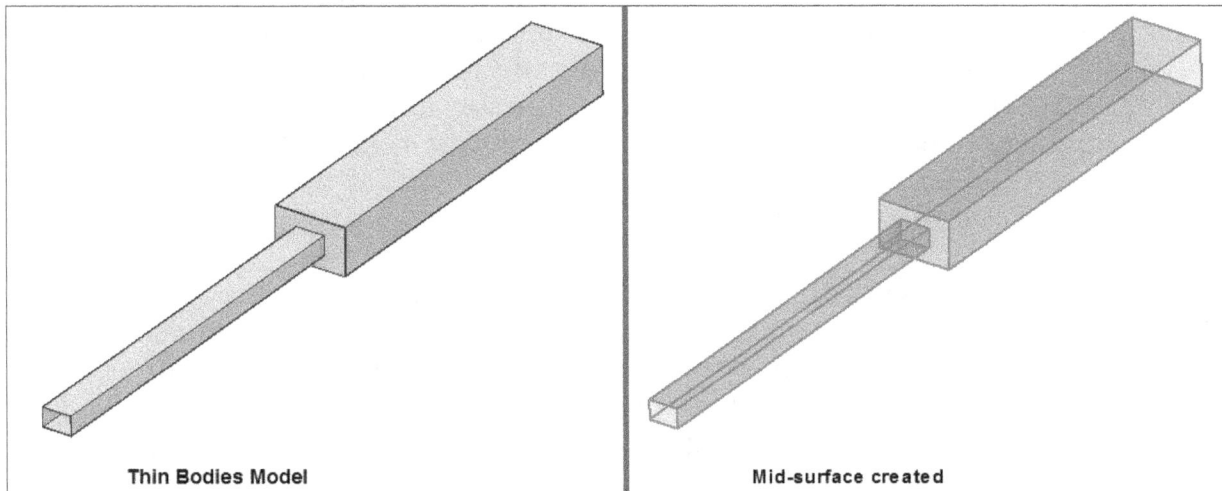

Figure-24. Midsurface created automatically

Similarly, you can use the **Midsurface** tool in the **Prepare** panel to create shell bodies.

Offset

The **Offset** tool in **Prepare** panel is used to create surfaces by offsetting selected faces of the model. The procedure to use this tool is given next.

*   Click on the **Offset** tool from the **Prepare** panel in the **Analysis** tab of the **Ribbon**. The **Offset** dialog box will be displayed; refer to Figure-25. Also, you will be asked to select the faces of the model.

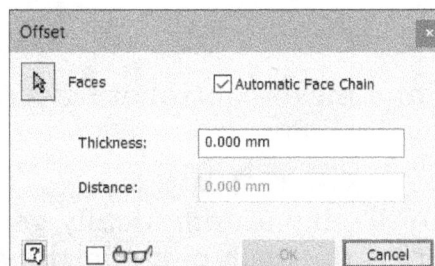

Figure-25. Offset dialog box

*   Select the faces of the model to offset and specify desired thickness in the **Thickness** edit box of the dialog box. The distance value will be half of the thickness specified.
*   Click on the **OK** button to create the surfaces.

MESHING

Meshing of Autodesk Inventor is not advanced so there is a limit on accuracy of analysis to be performed by Autodesk Inventor but still it is useful for less accurate products and educational purposes. Meshing is the base of FEM. Meshing divides the solid/shell models into elements of finite size and shape. These elements are joined at some common points called nodes. These nodes define the load transfer from one element to other element. Meshing is a very crucial step in design analysis. The automatic mesher in the software generates a mesh based on a global element size, tolerance, and local mesh control specifications. Mesh control lets you specify different sizes of elements for components, faces, edges, and vertices.

The software estimates a global element size for the model taking into consideration its volume, surface area, and other geometric details. The size of the generated mesh (number of nodes and elements) depends on the geometry and dimensions of the model, element size, mesh tolerance, mesh control, and contact specifications. In the early stages of design analysis where approximate results may suffice, you can specify a larger element size for a faster solution. For a more accurate solution, a smaller element size may be required.

Meshing generates 3D tetrahedral solid elements, 2D triangular shell elements, and 1D beam elements. A mesh consists of one type of elements unless the mixed mesh type is specified. Solid elements are naturally suitable for bulky models. Shell elements are naturally suitable for modeling thin parts (sheet metals), and beams and trusses are suitable for modeling structural members.

Before you run a simulation, ensure that the mesh is current, and view it in relation to the geometric features of the model. Integrity errors such as small gaps, overlaps, overhangs that are sometimes overlooked in models can cause trouble for mesh creation. In that case, recreate or modify the problematic geometric features. Some spring models mesh when the long helical face is split by a cutting plane containing the axis. If the model is too complex and has geometric singularity, divide it into less complex parts that you can mesh independently. Use a bonded contact between them to make these components behave as a single part.

Use a finer mesh in troublesome areas that you cannot simplify. Decreasing the global mesh size, as well as the local mesh size on certain faces and edges, can help to create a successful mesh.

The tools related to meshing and mesh control are given next.

Mesh View

The **Mesh View** tool is used to display automatically generated mesh of the model. Click on the **Mesh View** tool from the **Mesh** panel in the **Analysis** tab of the **Ribbon**. The mesh model will be displayed; refer to Figure-26.

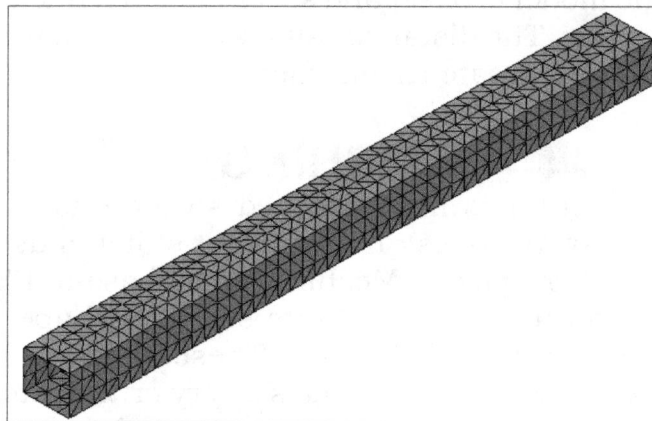

Figure-26. Mesh model

Mesh Settings

The **Mesh Settings** tool is used to change the element size and other parameters for the mesh. The procedure to change mesh settings is given next.

- Click on the **Mesh Settings** tool from the **Mesh** panel in the **Analysis** tab of the **Ribbon**. The **Mesh Settings** dialog box will be displayed; refer to Figure-27.

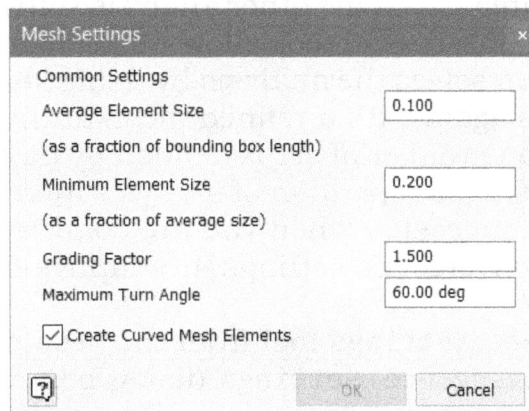

Figure-27. Mesh Settings dialog box

- Specify desired average size and minimum size of elements in the respective edit boxes.
- Specify desired grading factor in the **Grading Factor** edit box. Grading factor is the multiply factor for edge length by which an edge can be bigger than the adjacent edge in case of combination of fine and coarse mesh.
- Set desired maximum turn angle value in the **Maximum Turn Angle** edit box to specify size of element at curves.
- Select the **Create Curved Mesh Elements** check box if you want to create curved elements. Note that it takes more CPU power to generate curved elements. You can use a finer mesh in the stress concentration area around a concave fillet or round to compensate for the lack of curvature if this check box is cleared.
- After specifying desired parameters, click on the **OK** button.

Local Mesh Control

The **Local Mesh Control** tool is used to increase or decrease the density of elements at a specified region. The procedure to use this tool is given next.

- Click on the **Local Mesh Control** tool from the **Mesh** panel in the **Analysis** tab of the **Ribbon**. The **Local Mesh Control** dialog box will be displayed; refer to Figure-28 and you will be asked to select edges/faces to apply local mesh control.

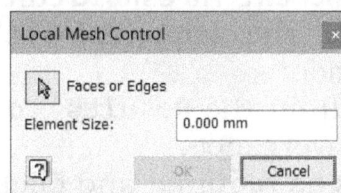

Figure-28. Local Mesh Control dialog box

- Select the faces/edges on which you want to apply local mesh control and specify desired element size in the edit box of the dialog box.
- Click on the **OK** button to apply local mesh control.

Convergence Settings

The **Convergence Settings** tool is used to specify the conditions for refinement of analysis results at stress areas. In some other analysis software, we call this adaptive meshing. In Autodesk Inventor, this tool allows to define h-adaptive meshing. In h-adaptive meshing, system solves the analysis by using standard meshing and then system solves the analysis again with a refined mesh at the locations of strain. This process continues up to the number of steps defined by us or till desired accuracy is achieved from the analysis. This repetition of analysis increases the processing time so this type of meshing is suggested when you are concerned about high accuracy. The procedure to define convergence settings/h-adaptive meshing is given next.

- Click on the **Convergence Settings** tool from the **Mesh** panel in the **Analysis** tab of the **Ribbon**. The **Convergence Settings** dialog box will be displayed; refer to Figure-29.

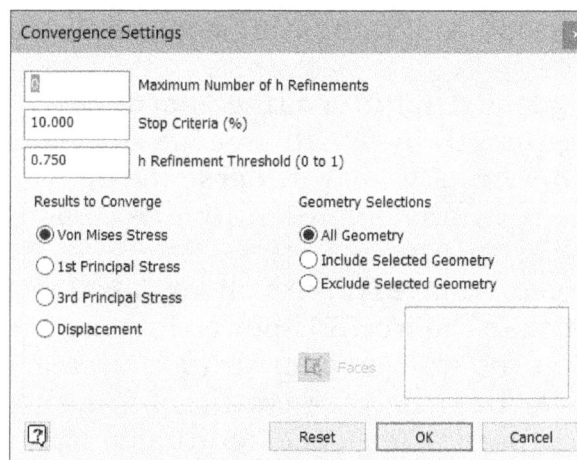

Figure-29. Convergence Settings dialog box

- Specify the number of h-refinements you want to perform in the **Maximum Number of h Refinements** edit box. If you specify value higher than 2 then system will warn you that performance of your system may decrease with such refinements.
- In the **Stop Criteria** edit box, specify the value in percentage to cease refinement. If there is a difference of less than specified value for last two results after refinement then system will cease refinement. For example, if before refinement the stress value was **10.0** and now it becomes **9.0** then system will not refine meshing further.
- Specify desired value in **h Refinement Threshold** edit box. The value can be specified from **0** to **1**. **0** means include all the elements for refinement and **1** means exclude all the elements from refinement.
- Select desired radio button from the **Results to Converge** area to specify the parameter being used for convergence.
- Specify the other parameters as required and click on the **OK** button.

RUNNING STUDY

- After setting all the parameters (applying loads, constraints, and contacts), click on the **Simulate** button from the **Solve** panel in the **Analysis** tab of the **Ribbon**. The **Simulate** dialog box will be displayed; refer to Figure-30.

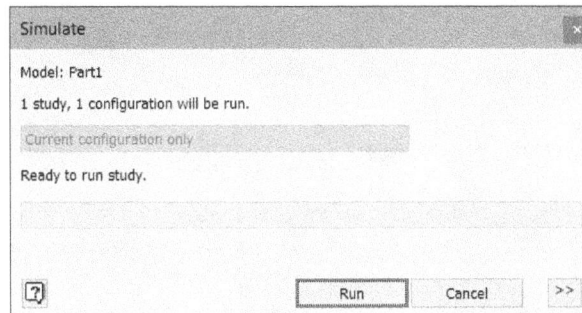

Figure-30. Simulate dialog box

- Click on the **Run** button from the dialog box. The results of analysis will be displayed in the graphics area; refer to Figure-31.

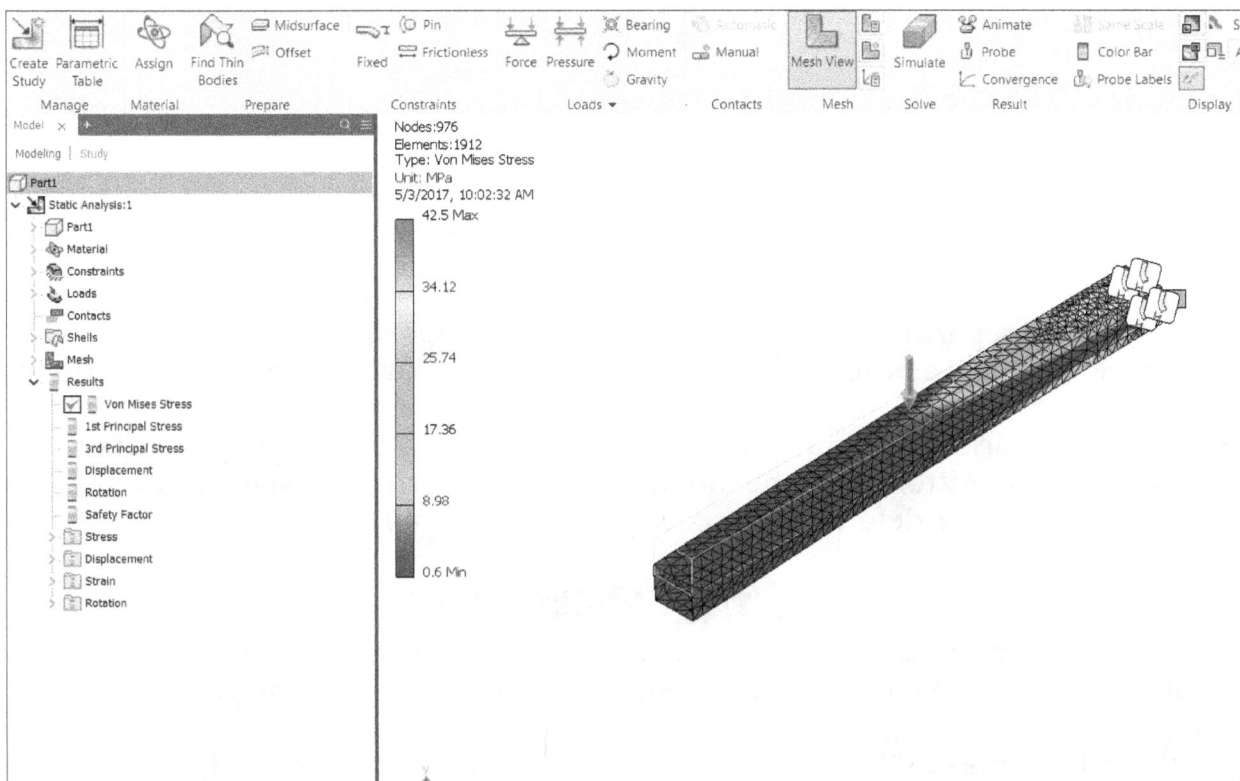

Figure-31. Result of stress analysis

Check the factor of safety and other results to evaluate the feasibility of your component. Generally, the factor of safety should be **3** or higher.

GENERATING REPORTS

After performing the analysis, it is important to properly document the results of analysis. The **Report** tool is used to generate the analysis reports. The procedure to use this tool is given next.

- Click on the **Report** tool from the **Report** panel in the **Analysis** tab of the **Ribbon** after performing the analysis. The **Report** dialog box will be displayed; refer to Figure-32.

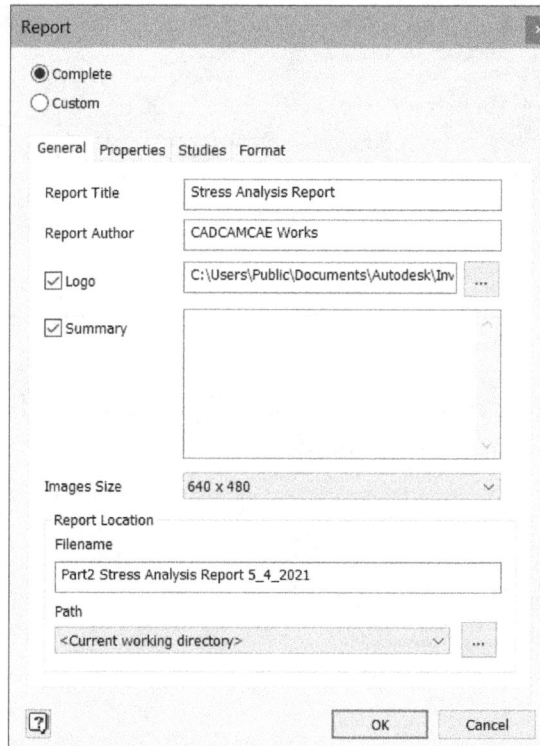

Figure-32. Report dialog box

- Specify desired parameters in the **General** and **Properties** tab.
- Click on the **Studies** tab and select the studies that you want to be included in the report.
- Click on the **Format** tab and select desired file format for report generated.
- Click on the **OK** button from the dialog box. The reports will be generated and will be displayed in the default browser for selected file format.

MODAL ANALYSIS

The Modal analysis is used to find out natural frequencies of the model at which component can deform largely. The procedure to perform modal analysis is given next.

- Click on the **Create Study** tool from the **Manage** panel in the **Analysis** tab of the **Ribbon**. The **Create New Study** dialog box will be displayed as discussed earlier.
- Click on the **Modal Analysis** radio button. The options for the modal analysis will become active.
- Select the **Number of Modes** check box and specify the maximum number of natural frequencies you want to find out.
- Select the **Frequency Range** check box and specify the range in which you are concerned about natural frequencies.
- Set the other parameters as required and click on the **OK** button from the dialog box. The tools to perform modal analysis will become active in the **Analysis** tab.
- Apply constraints and contacts to limit the motion of part. Apply the loads if required.
- Click on the **Simulate** button and then **Run** button from the dialog box displayed. Result of modal analysis will be displayed; refer to Figure-33.

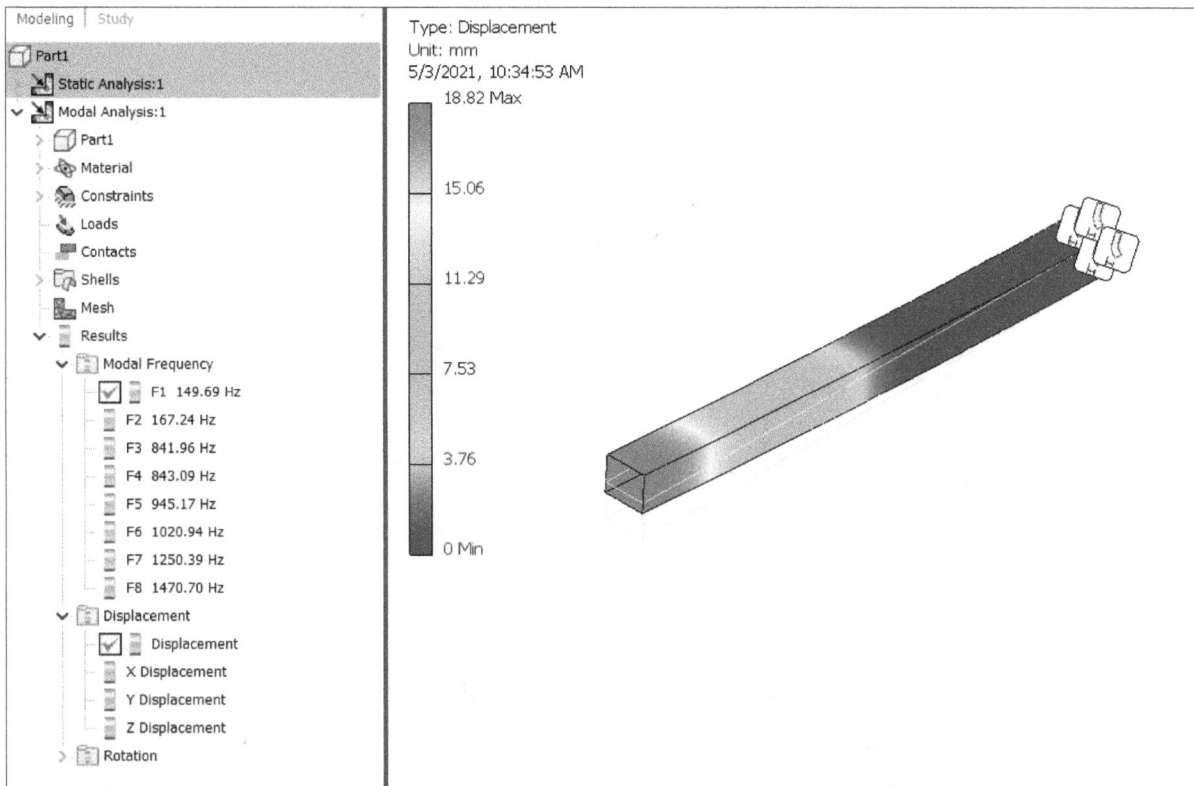

Figure-33. Result of modal analysis

Note that our motive to perform this analysis is to find out the natural frequencies and add/remove material from the part to make sure, the natural frequency do not match with the frequency generated in assembly due to shocks and other working conditions.

SHAPE GENERATOR

The **Shape Generator** tool is used to generate a light weight component based on the load conditions and parameters specified by you. Note that the shape generator analysis can be performed only on parts. You can not perform shape generator study on assembly or surfaces. The procedure to use this tool is given next.

- Click on the **Create Study** tool from the **Manage** panel in the **Analysis** tab of the **Ribbon**. The **Create New Study** dialog box will be displayed as discussed earlier.
- Select the **Shape Generator** radio button from the dialog box and click on the **OK** button. The **Shape Generator** dialog box will be displayed describing the function of tool.
- Click on the **OK** button from the dialog box. The tools related to shape generation will be displayed; refer to Figure-34.

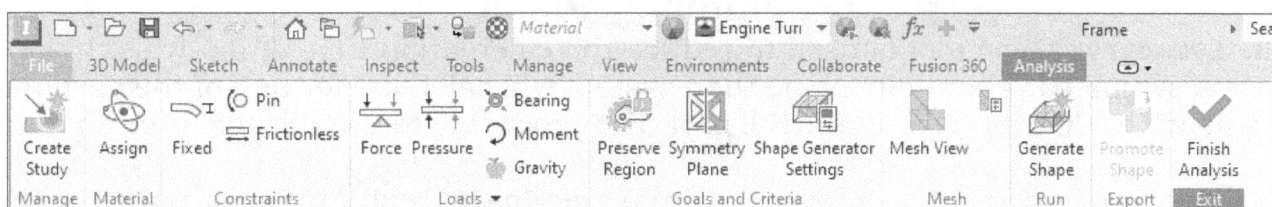

Figure-34. Tools in Analysis tab for shape generation study

• Assign the material, constraints, and load conditions as required; refer to Figure-35.

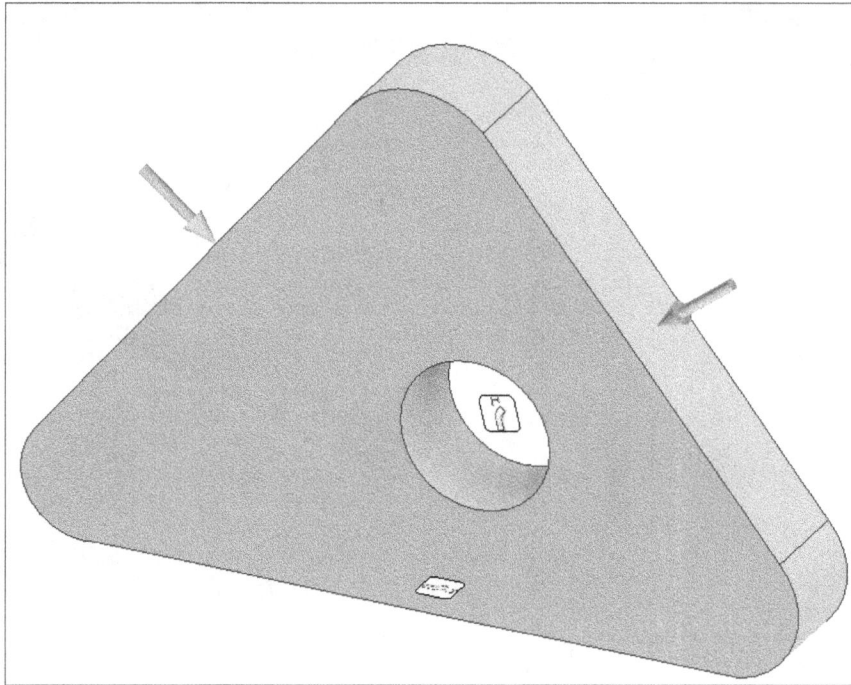

Figure-35. Load and constraint applied to model

Preserve Region

• Click on the **Preserve Region** tool from **Goals and Criteria** panel in the **Analysis** tab of the **Ribbon**; the **Preserve Region** dialog box will be displayed; refer to Figure-36. You will be asked to select the regions to be unchanged after analysis.

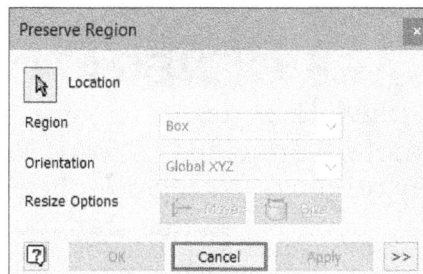

Figure-36. Preserve Region dialog box

• Select the regions like holes, slots, and grooves that you want to be unchanged after shape generation.
• Click on the **Apply** button after selecting one region to select another region or click on the **OK** button if you want to preserve only one region.

Symmetry Plane

The **Symmetry Plane** tool is used to reduce the calculation time of analysis. If your part is symmetric about any plane then system will calculate for half of the part and result will be distributed over the full part. The procedure to use this tool is given next.

• Click on the **Symmetry Plane** tool from the **Goals and Criteria** panel in the **Analysis** tab of the **Ribbon**. The **Symmetry Plane** dialog box will be displayed; refer to Figure-37.

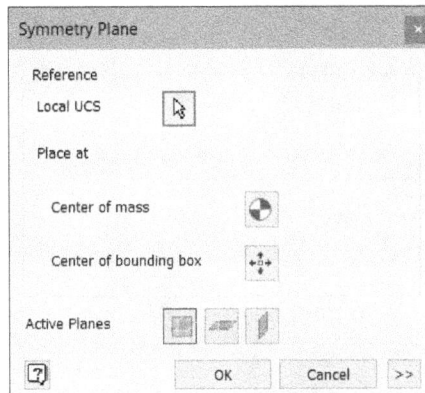

Figure-37. Symmetry Plane dialog box

- Select desired button(s) from the **Active Planes** buttons to use respective planes as symmetry plane; refer to Figure-38.

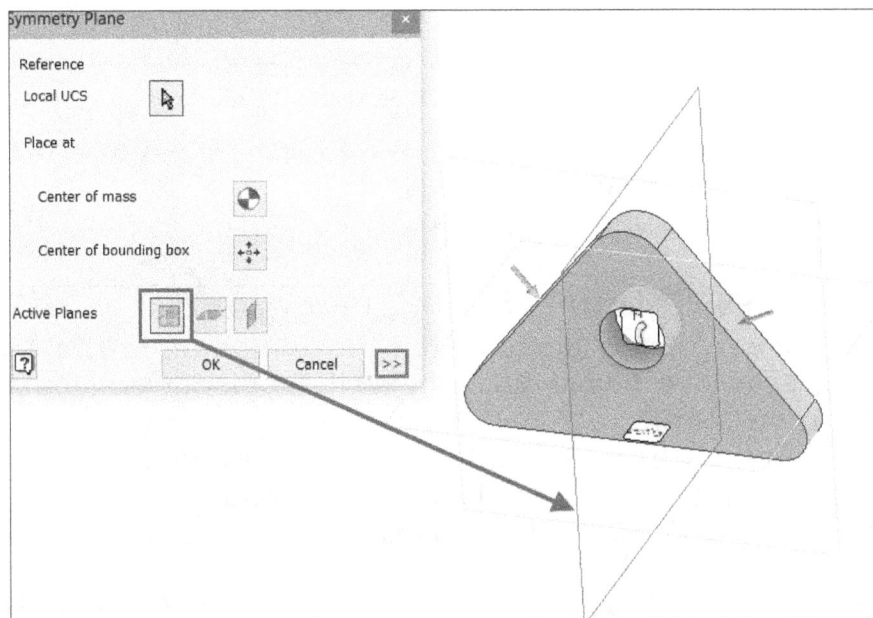

Figure-38. Symmetry plane selected

- Click on the **OK** button from the dialog box.

Shape Generator Settings

The **Shape Generator Settings** tool is used to specify the objectives for shape generation like the percentage of mass to be reduced or the target mass goal. The procedure to use this tool is given next.

- Click on the **Shape Generator Settings** tool from the **Goals and Criteria** panel in the **Analysis** tab of the **Ribbon**. The **Shape Generator Settings** dialog box will be displayed; refer to Figure-39.

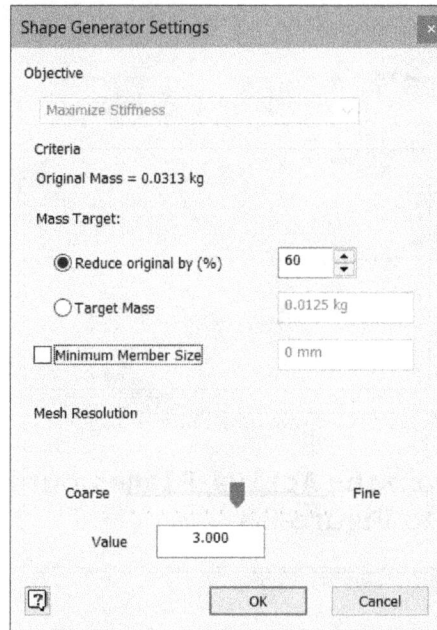

Figure-39. Shape Generator Settings dialog box

- Specify the target mass using the options in the **Mass Target** area of the dialog box.
- Move the slider to make mesh fine or coarse.
- Click on the **OK** button to apply settings.

Performing Shape Generation

- Click on the **Generate Shape** button from the **Run** panel in the **Analysis** tab of the **Ribbon**. The **Generate Shape** dialog box will be displayed similar to **Simulate** dialog box discussed earlier.
- Click on the **Run** button from the dialog box. Depending on computing power of your system, the results will be displayed in a few moments.
- Click on the **Finish Analysis** button to exit the analysis environment.

PRACTICE

Run a static analysis on the model as shown in Figure-40. Find out the factor of safety.

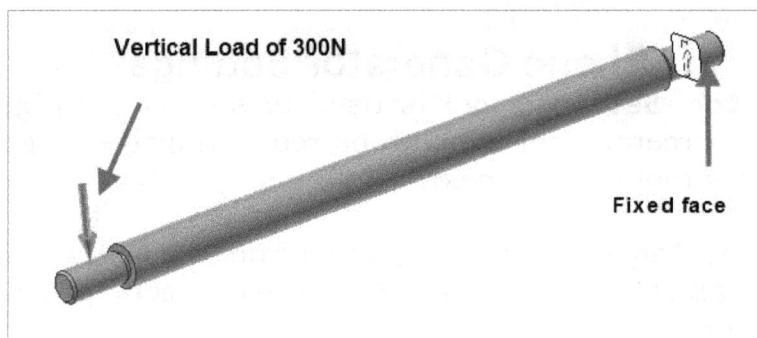

Figure-40. Practice 1

SELF ASSESSMENT

Q1. In Autodesk Inventor, FEA is used to solve the equations for stress analysis? What is the full form of FEA?

a) Fixture Element Analysis
b) Fixture Extract Analysis
c) Finite Element Analysis
d) Finite Extract Analysis

Q2. Which of the following is the correct formula in Finite Element Analysis to find the force-displacement, i.e, stiffness characteristics of each individual element?

a) $\dfrac{[k]_e}{\{F\}_e} = \{\delta\}_e$

b) $[k]_e \{\delta\}_e = \{F\}_e$

b) $\dfrac{\{\delta\}_e}{\{F\}_e} = [k]_e$

d) None of the Above

Q3. Which of the following assumptions is not made using Finite Element Analysis?

a) The loads applied does not vary with time.
b) The change in stiffness due to loading is neglected.
c) The materials applied to the components satisfy the Stefan's Law.
d) Loads must be constant in magnitude, direction, and distribution.

Q4. Which of the following study types is used in **Create New Study** dialog box in Stress Analysis to reduce the weight of part for specified loading conditions and this study type is available for parts only?

a) Shape Generator
b) Modal Analysis
c) Static Analysis
d) All of the Above

Q5. Which of the following tools is used to constraint the selected entity in such a way that the object is free to rotate but cannot translate in any direction?

a) Fixed
b) Pin
c) Frictionless
d) Both b and c

Q6. Which of the following loads is applied on an object when the origin of force is somewhere else but it is affecting the selected face?

a) Body Loads
b) Remote Force
c) Bearing Load
d) Pressure

Q7. Which of the following tools is used to increase or decrease the density of elements at a specified region in meshing of models in Autodesk Inventor?

a) Offset
b) Mesh Settings
c) Local Mesh Control
d) Convergence Settings

Q8. The **Modal Analysis** tool is used to find the natural frequencies of the model at which component can deform largely. (True/False)

Q9. In Autodesk Inventor, Shell bodies are considered as thick bodies. (True/False)

Q10. The Force applied per unit area is called ....

Chapter 16

Model Based Annotations

Topics Covered

The major topics covered in this chapter are:

- *Introduction to Model Based Annotation*
- *Applying Dimensional Annotations to 3D Model*
- *Creating Tolerance Feature*
- *Applying Hole/Thread Notes*
- *Applying Surface Finish Symbol*
- *Creating Leader Note*
- *Creating General Note*
- *Sectioning Model*

INTRODUCTION

In earlier chapters, you have learned to generate drawings after creating solid part or assembly. You have then applied annotations in the drawing to represent the intent of model. In Autodesk Inventor, you can apply the annotations directly to the model in Part or Assembly environment. You can use electronic gadgets at the shop floor to let the machinist find dimensions of his/her interest directly from 3D model, hence saving lot of time which gets wasted in generating 2D drawings (layouts). The scheme by which we apply annotations to 3D model is called **Model Based Annotation**. The tools to apply model based annotations are available in the **Annotate** tab of the **Ribbon**; refer to Figure-1.

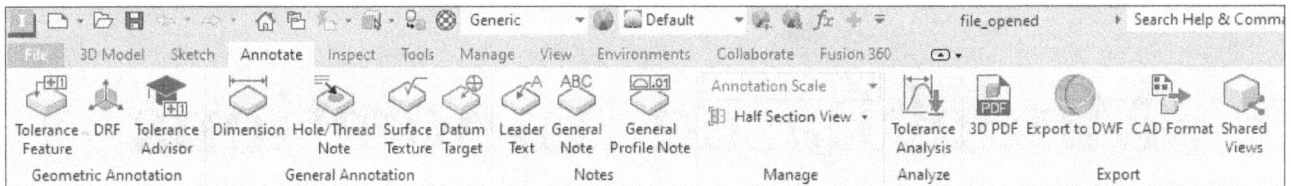

Figure-1. Annotate tab in Ribbon

WORKFLOW FOR MODEL BASED ANNOTATIONS

While performing model based annotations, you can use the tools randomly wherever they are required but if you follow the standard workflow then it can greatly reduce the pain of annotating complex products. Figure-2 shown the workflow for model based annotations. We will discuss the tools of the **Annotate** tab in the same sequence as they are in workflow.

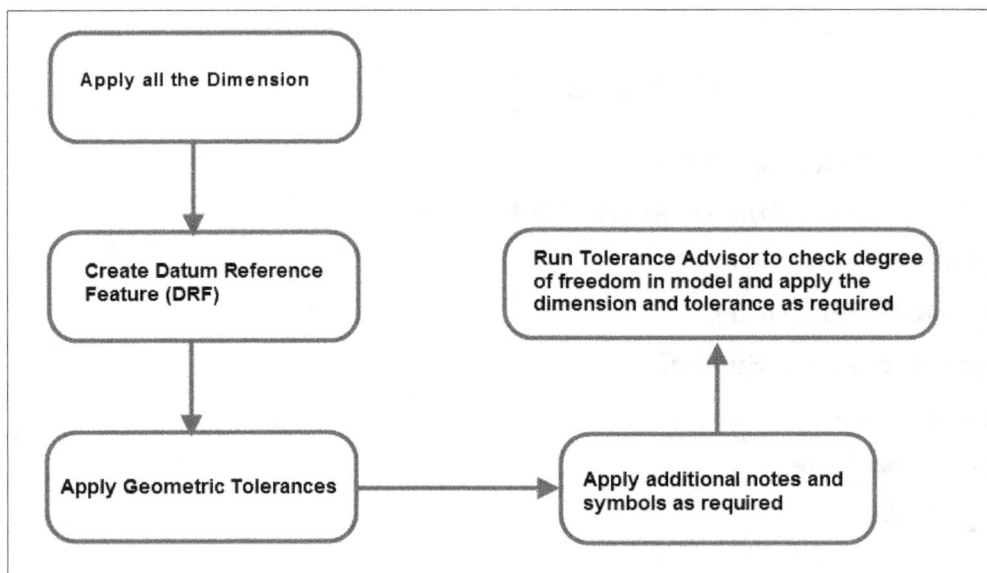

Figure-2. Workflow for Model based annotations

APPLYING DIMENSIONS TO 3D MODEL

The **Dimension** tool in the **Annotate** tab of **Ribbon** in Part/Assembly environment is used to apply dimensions to the 3D model. The procedure to use this tool is given next.

- Click on the **Dimension** tool from the **General Annotation** panel in the **Annotate** tab of the **Ribbon**. The **Standard** dialog box will be displayed; refer to Figure-3 if this is first time you are using this tool.

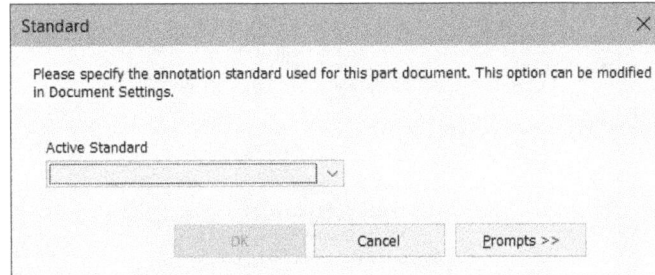

Figure-3. Standard dialog box

- Select desired standard from the **Active Standard** drop-down and click on the **OK** button. You will be asked to select the geometry to be dimensioned.
- Apply the dimensions by selecting the geometries. Like, to create diameter dimension - select the round face of model, to create distance dimension - select the two parallel faces, to apply angle dimension - select the two non-planar faces. Figure-4 shows different type of dimensions applied to the model. Note that you need to click on the **Apply** button from pop-up toolbar after placing each dimension.

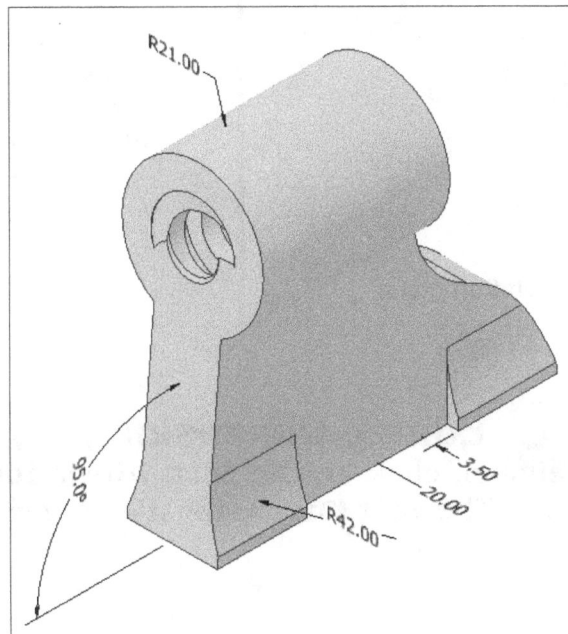

Figure-4. Dimensions applied to model

Changing Annotation Plane

While applying the annotations to 3D model, many times you will feel the need of changing plane for annotations. The procedure to change annotation plane is given next.

- Click on the **Dimension** tool from the **General Annotation** panel in the **Annotate** tab of the **Ribbon**. You will be asked to select geometry to be dimensioned.
- Select the face/edge to apply dimension. The dimension will get attached to cursor. Right-click after moving the dimension away from model. The right-click shortcut menu will be displayed; refer to Figure-5.

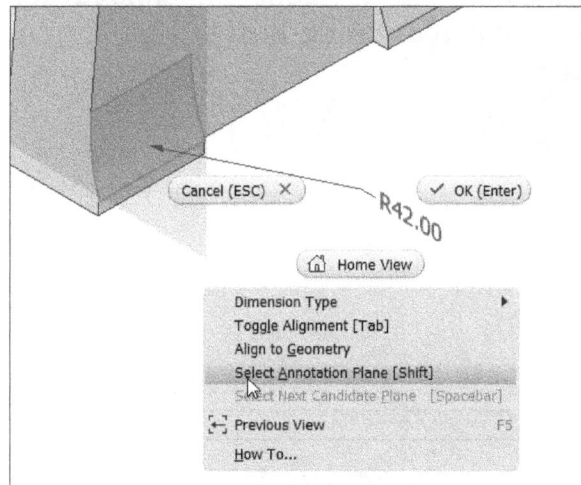

Figure-5. Right-click shortcut menu for annotation

- Click on the **Select Annotation Plane [Shift]** option from the menu. You will be asked to select the face/plane to be used as annotation plane.
- Select desired plane/face. Click to place the dimension; refer to Figure-6.

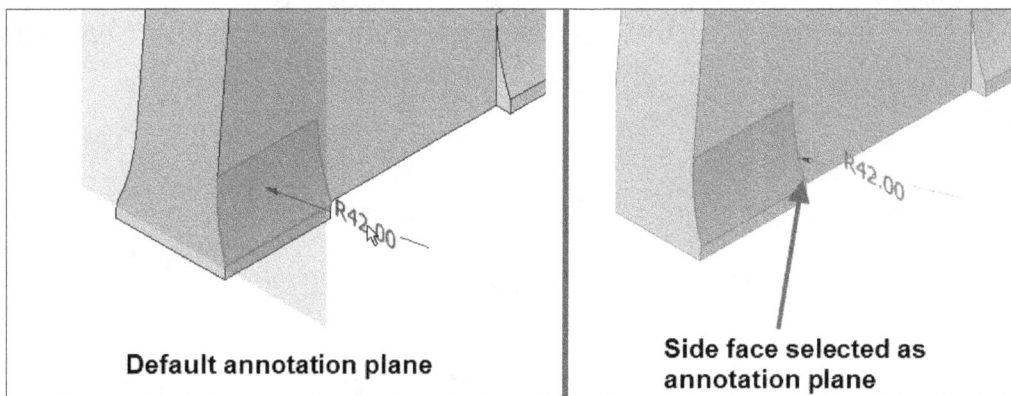

Figure-6. Annotation plane selected

Editing Dimension

- After placing the dimension, click on the **Edit Dimension** tool from the pop-up toolbar; refer to Figure-7. The **Edit Dimension** dialog box will be displayed; refer to Figure-8.

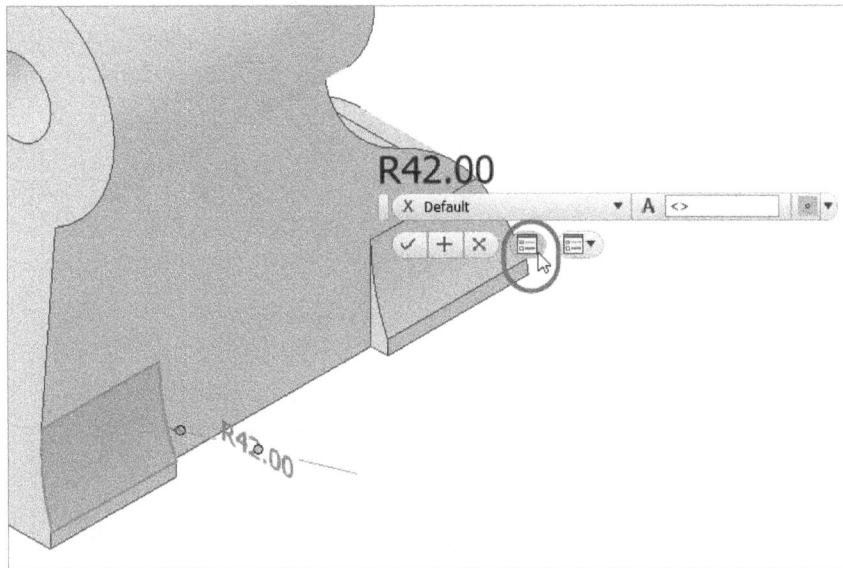

Figure-7. Edit Dimension tool

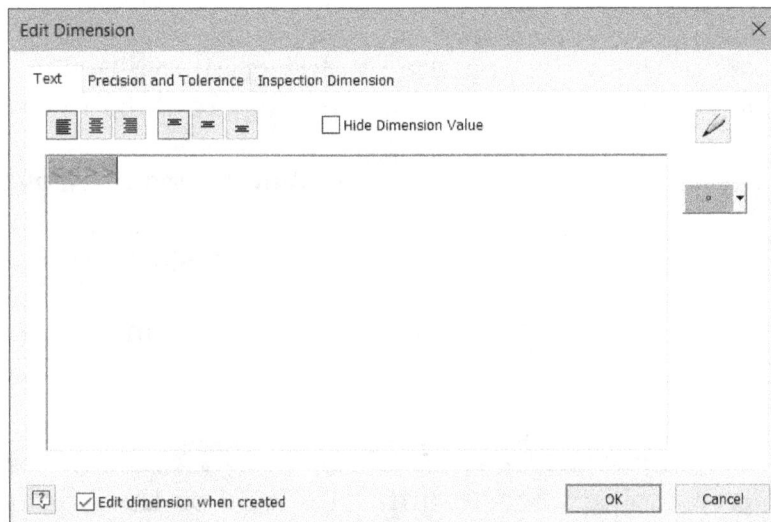

Figure-8. Edit Dimension dialog box

- The options of this dialog box has been discussed in **Chapter 14: Drawing Creation**. After setting desired dimension, click on the **OK** button from the dialog box to apply settings. The pop-up toolbar will be displayed again. Click on the **OK** button to exit the pop-up toolbar.

CREATING TOLERANCE FEATURE

The tolerance feature is used to apply geometric constraints. The procedure to create tolerance feature is given next.

- Click on the **Tolerance Feature** tool from the **Geometric Annotation** panel in the **Annotate** tab of the **Ribbon**. You will be asked to select a face.
- Select the face to which you want to apply geometric tolerance and datum feature and click on the **OK** button from the pop-up toolbar. The tolerance feature will get attached to the cursor. Press **TAB** to change the orientation of tolerance feature.
- Click at desired location to place the feature. The pop-up toolbar will be displayed; refer to Figure-9.

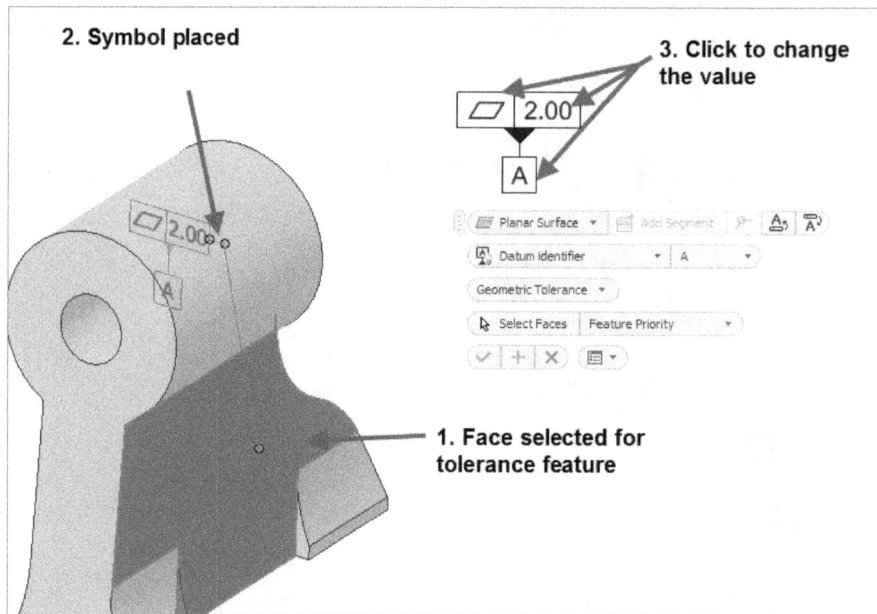

Figure-9. Placing Tolerance feature

- Click on the boxes of tolerance feature to change the values.
- To hide/show the datum feature, click on the **Toggle Datum Feature** button from the pop-up toolbar.
- If you want to add a note then click on the **Add Comment Above** or **Add Comment Below** button from the pop-up toolbar.
- After creating the tolerance feature, click on the **Apply** button to create next geometric tolerance.
- After creating all desired geometric tolerances, click on the **OK** button from the popup toolbar to exit the tool.

HOLE/THREAD NOTES

The **Hole/Thread Note** tool is used to annotate hole or threads in the model. The procedure to use this tool is given next.

- Click on the **Hole/Thread Notes** tool from the **General Annotation** panel in the **Annotate** tab of the **Ribbon**. You will be asked to select the geometry to be annotated.
- Select the thread or hole to annotate. The dimension will get attached to cursor; refer to Figure-10.

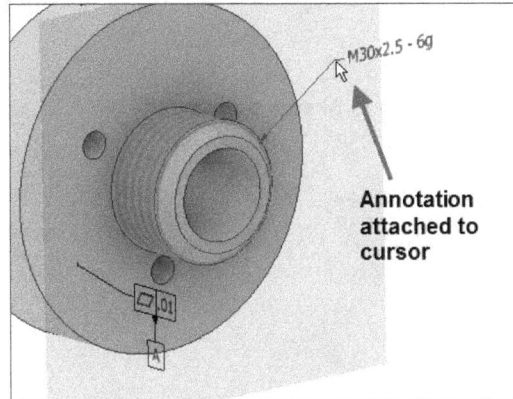

Figure-10. Thread annotation

• Click at desired location to place the dimension.

APPLYING SURFACE TEXTURE SYMBOL

The **Surface Texture** tool is used to add symbol of surface texture to the 3D model. The procedure to use this tool is given next.

• Click on the **Surface Texture** tool from the **General Annotation** panel in the **Annotate** tab of the **Ribbon**. You will be asked to select the face on which surface texture is to be applied.
• Select desired face. Symbol will get attached to the cursor.
• Click at desired location to place the symbol. Preview of the symbol will be displayed with pop-up toolbar; refer to Figure-11.

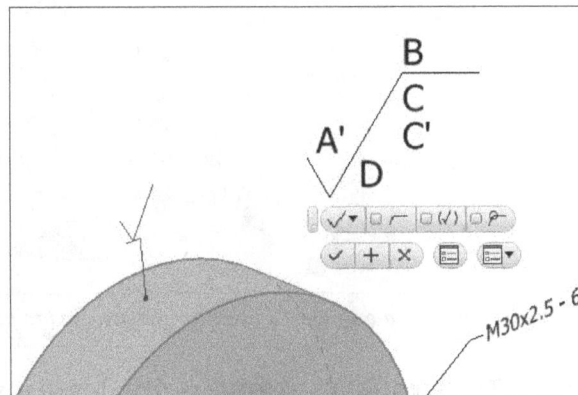

Figure-11. Preview of surface finish symbol

• Select the alphabet to specify surface finish value.
• Use the options in the pop-up toolbar to modify the symbol as required.
• Click on the **Apply** button to apply other surface finish symbols or click on the **OK** button to exit.

APPLYING DATUM TARGET

The **Datum Target** tool is used to add geometric control datum to a plane, axis, or point. The datum provides control for flatness, straightness, cylindricity, and circularity. The procedure to use this tool is given next.

*   Create the sketch points and position them at desired places in the model; refer to Figure-12.

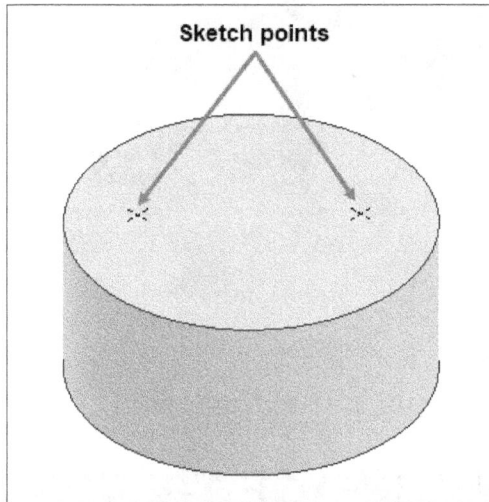

Figure-12. Sketch points created

*   Click on the **Datum Target** tool from the **General Annotation** panel in the **Annotate** tab of the **Ribbon**. The **Datum Target** dialog box will be displayed along with the datum targets attached to the points; refer to Figure-13.

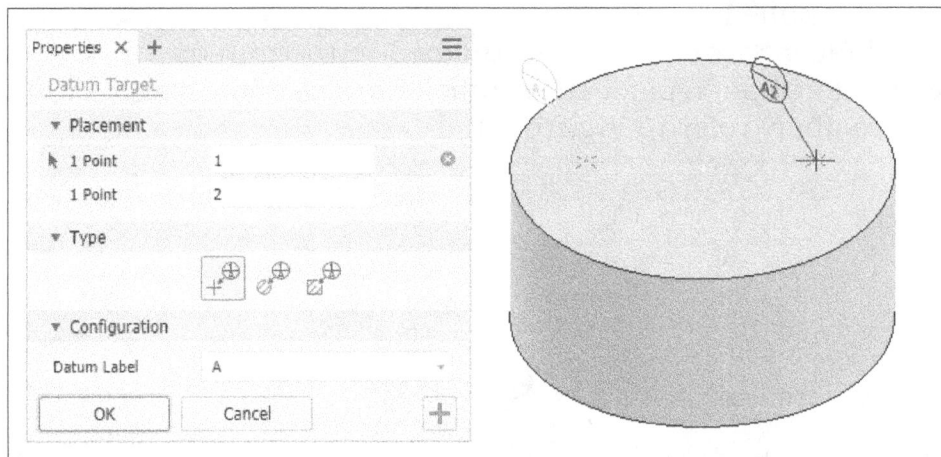

Figure-13. Datum Target dialog box along with datum targets attached to points

*   Select **Point** button from **Type** rollout to use point locations to define the datum target zone. Select **Circle** button to create a circular datum target zone. Select **Rectangle** button to create a rectangular datum target zone.
*   Select desired datum label from **Datum Label** drop-down in the **Configuration** rollout.
*   After specifying desired parameters, click on the **OK** button from the dialog box. The datum targets will be attached.

CREATING LEADER TEXT

Leader text is used to insert text annotation with leader attached to the 3D model. The procedure to create leader text is given next.

- Click on the **Leader Text** tool from the **Notes** panel in the **Annotate** tab of the **Ribbon**. You will be asked to select the geometry to which leader will be attached.
- Select the face, edge, axis, point, or other geometric entity. End point of the leader will get attached to cursor.
- Click at desired location to place the note text. The **Format Text** dialog box will be displayed as discussed in previous chapters.
- Type desired text and set the parameters as required and then click on the **OK** button from the dialog box. The text will be created with leader; refer to Figure-14.

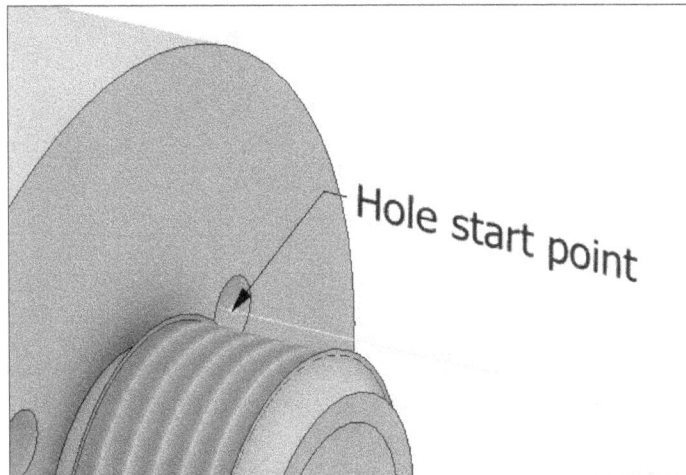

Figure-14. Leader text created

CREATING GENERAL NOTE

The **General Note** tool is used to create notes applicable to the whole part/assembly. The procedure to use this tool is given next.

- Click on the **General Note** tool from the **Notes** panel in the **Annotate** tab of the **Ribbon**. You will be asked to select the quadrant in which note is to be placed.
- Click on desired location to select respective quadrant. The **Format Text** dialog box will be displayed.
- Type desired text and click on the **OK** button after formatting. The note will be created in the selected quadrant.

Note that you can create the **General Profile Note** in the same way.

SECTIONING PART

The tools to perform sectioning of 3D model are available in the **Section View** drop-down from the **Manage** panel in the **Annotate** tab of the **Ribbon**. The procedure to use **Quarter Section View** option is discussed next. You can apply the same procedure to other section options.

- Click on the **Quarter Section View** option from the **Section View** drop-down in the **Manage** panel of the **Annotate** tab in the **Ribbon**. You will be asked to select work plane for sectioning.
- Select the plane from graphics area or from the **Model browser** in the left of the application window. Preview of section by first work plane will be displayed; refer to Figure-15.

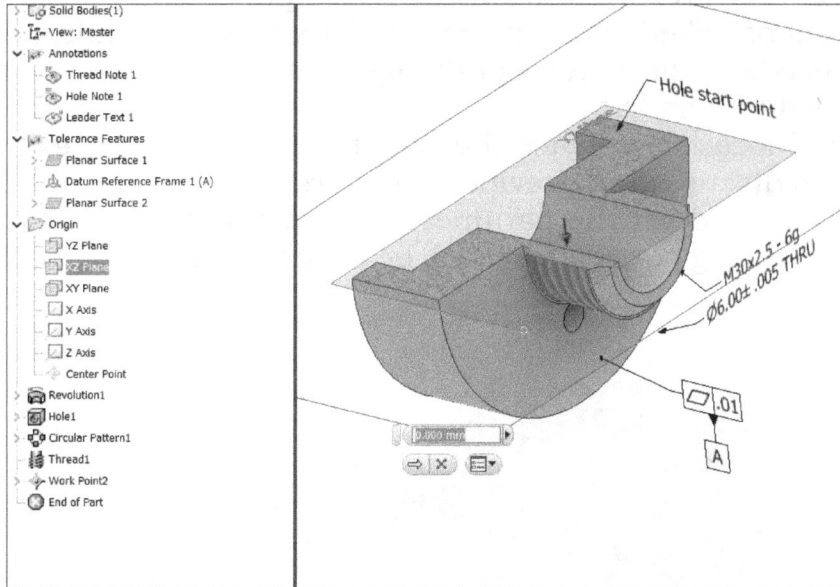

Figure-15. Preview section by first plane

- Specify desired offset value in the edit box of pop-up toolbar and click on the **Continue** button ⇨ from the pop-up toolbar. You will be asked to select the next sectioning plane.
- Select the next plane. Preview of quarter section will be displayed; refer to Figure-16.

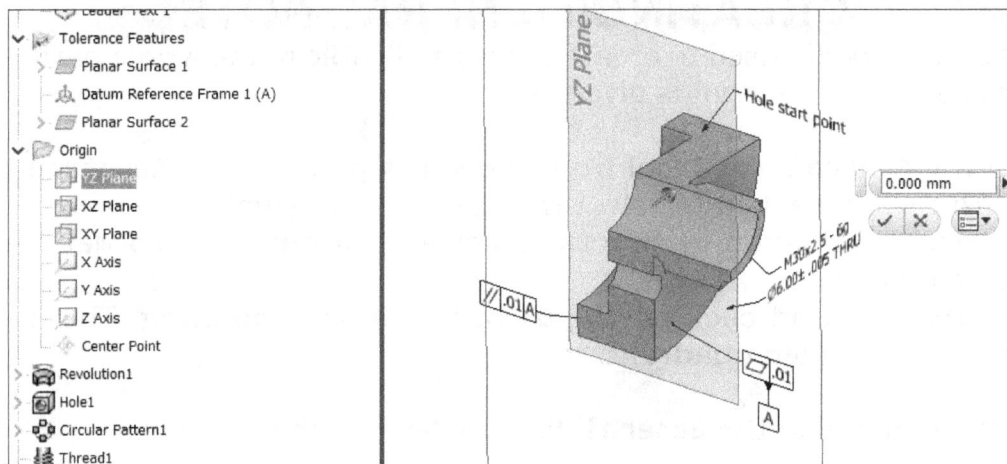

Figure-16. Preview of quarter section

- Set desired parameters and click on the **OK** button from the pop-up toolbar.

The methods for creating 3D PDF and exporting file to other formats have already been discussed in previous chapters.

PRACTICE

Create the model and apply the model based annotations as shown in Figure-17.

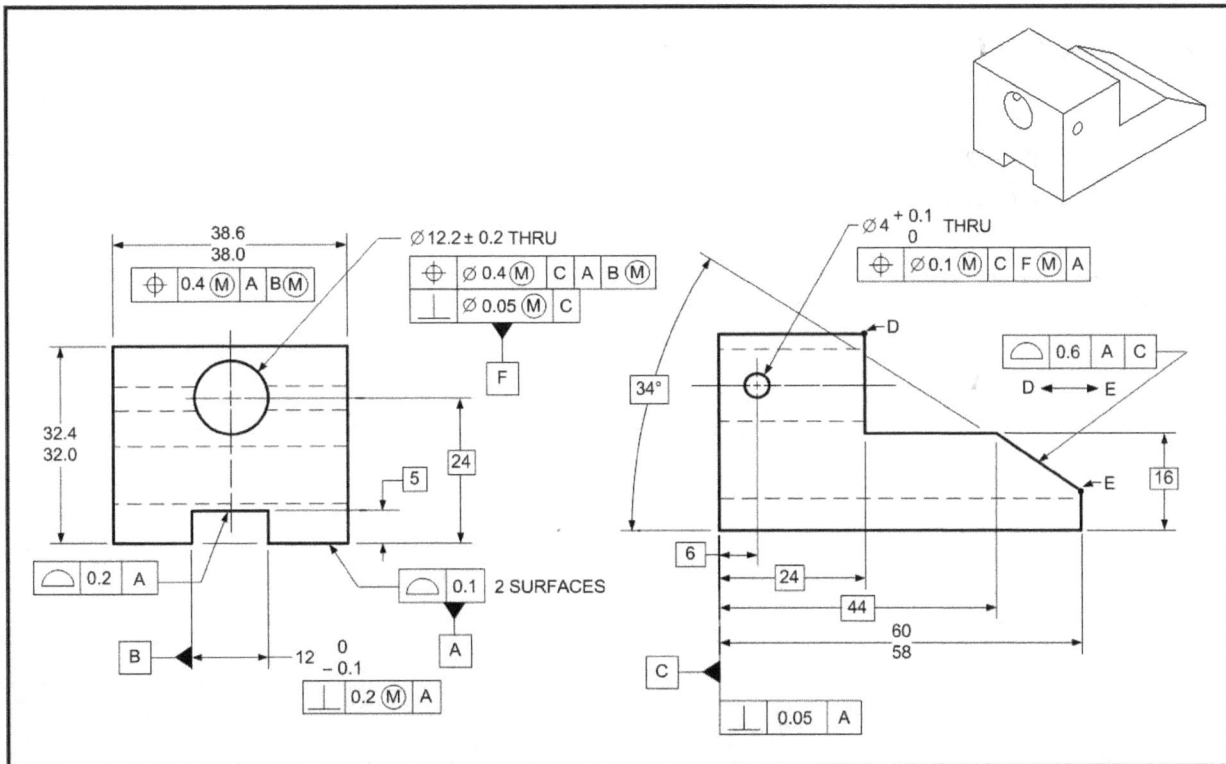

Figure-17. Practice

SELF ASSESSMENT

Q1. Which of the following options is the correct workflow for model based annotations?

a) Apply all the dimension --> Create Datum Reference Feature --> Apply Geometric Tolerances --> Apply additional notes and symbols --> Run Tolerance Advisor to check degree of freedom in model

b) Apply Geometric Tolerances --> Apply all the dimension --> Create Datum Reference Feature --> Apply additional notes and symbols --> Run Tolerance Advisor to check degree of freedom in model

c) Apply all the dimension --> Apply Geometric Tolerances --> Create Datum Reference Feature --> Run Tolerance Advisor to check degree of freedom in model --> Apply additional notes and symbols

d) None of the Above

Q2. Which of the following geometries is to be selected to create distance dimension?

a) Select the round face of model
b) Select the two parallel faces
c) Select the two non-planar faces
d) None of the Above

Q3. Which of the following statement specify the use of Tolerance Feature?

a) To apply Geometric Constraint
b) To change plane for annotations
c) To edit the dimension
d) All of the Above

Q4. The **Leader Text** tool is used to create notes applicable to the whole parts/ assembly. (True/False)

Q5. .... is used to insert text annotation with leader attached to the 3D Model.

FOR STUDENT NOTES

Chapter 17

Application Management

Topics Covered

The major topics covered in this chapter are:

- *Introduction*
- *Managing Materials*
- *Appearance*
- *Application Options*
- *Document Settings*
- *Manage Tab*
- *Styles Editor*
- *Deriving and Importing Objects*
- *Attaching Point Cloud Data*
- *iParts, iMates, and iFeatures*
- *Adding Rules*

INTRODUCTION

In previous chapters, you have learned about the tools used to create various objects and you have also performed various analyses to check the model for production. In this chapter, you will learn about the tools used for managing application level settings. You will also learn to customize various aspects of the software. These tools are available in **Tools**, **Manage**, and **View** tab of the **Ribbon**.

MANAGING MATERIALS

The **Material** tool in **Tools** tab is used to create and manage various materials applied to the models for representing physical properties of the object. The procedure to use this tool is given next.

- Click on the **Material** tool from the **Material and Appearance** panel in the **Tools** tab of the **Ribbon**. The **Material Browser** will be displayed; refer to Figure-1.

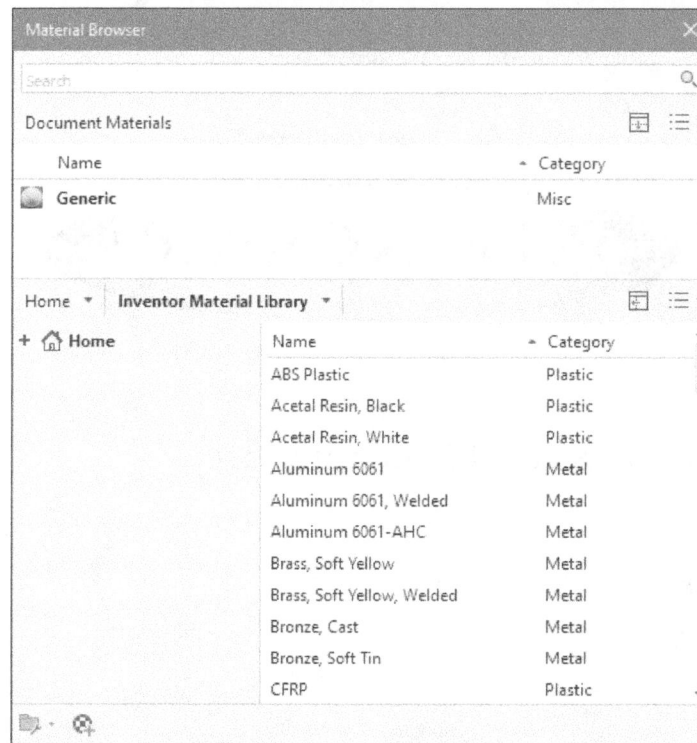

Figure-1. Material Browser

- To apply a material to current part model, select the desired material from the list of materials in the **Material Browser**. By default, the **Inventor Material Library** is selected in the **Browser**. If you want to use another library installed in your system then click on the down button next to **Home** tile and select desired library option; refer to Figure-2.

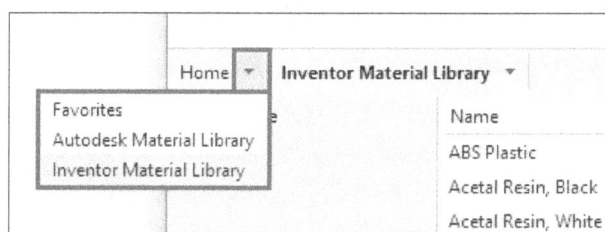

Figure-2. Material library drop-down

- You can filter the list of materials by selecting desired category from the drop-down displayed on clicking down button next to Inventor Material Library tile; refer to Figure-3. For example if you want to use metals in model then select the **Metal** option from the drop-down.

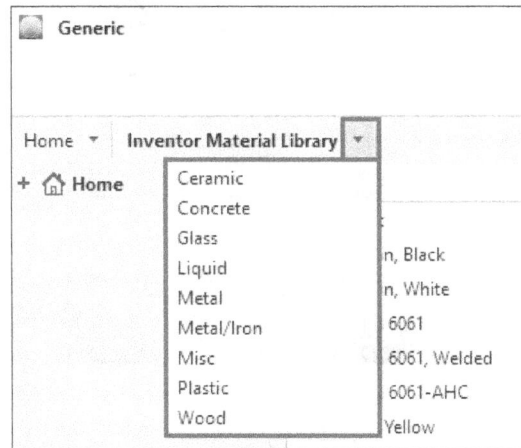

Figure-3. Category drop-down

- If you want to change the style of displaying material in the **Browser** then click on the **View Type** button from the dialog. A drop-down will be displayed with options to modify view style; refer to Figure-4.

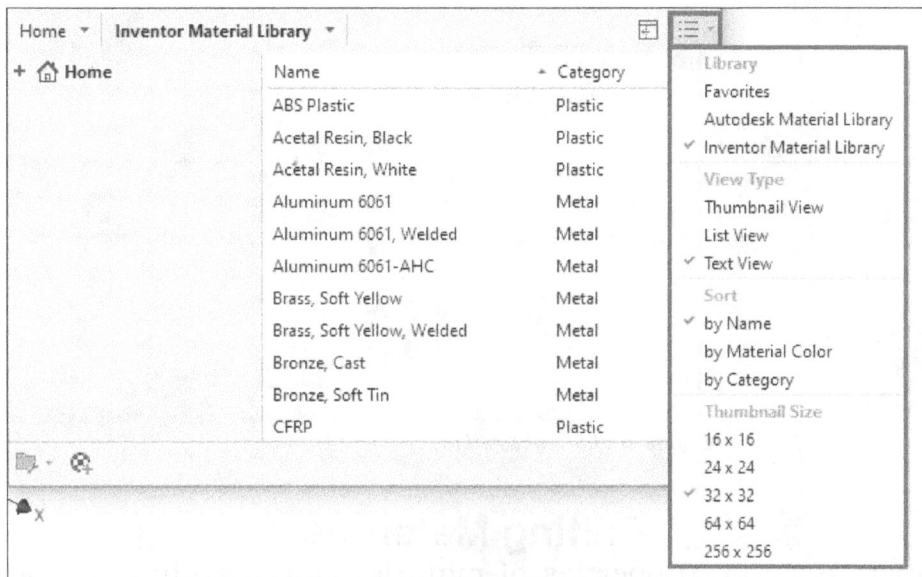

Figure-4. View Type drop-down

- Set desired options in the drop-down. The materials will be displayed accordingly in the **Browser**; refer to Figure-5.

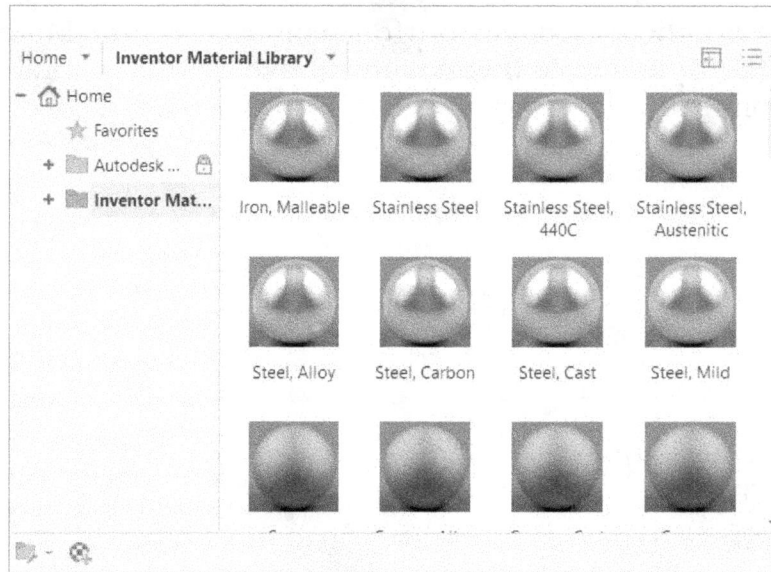

Figure-5. Material displayed in Thumbnail view

- To add materials in the current document, hover the cursor on desired material and click on the **Adds material to document** button; refer to Figure-6. You can add multiple materials to the document for checking different variations of the model.

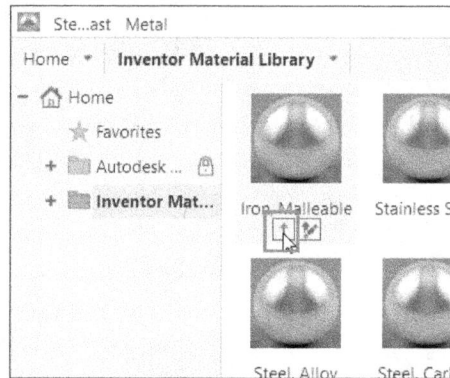

Figure-6. Adds material to document button

Editing Materials

- If you want to edit the properties of material before adding to document then click on the **Adds material to document and displays in editor** button after hovering cursor on desired material in the **Material Browser**; refer to Figure-7. The **Material Editor** dialog box will be displayed; refer to Figure-8.

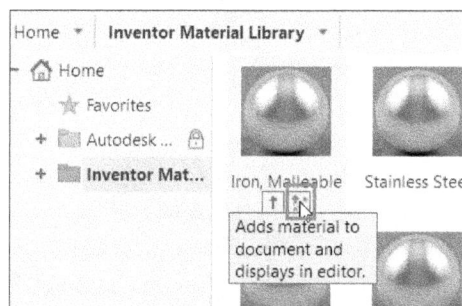

Figure-7. Adds material to document and displays in editor button

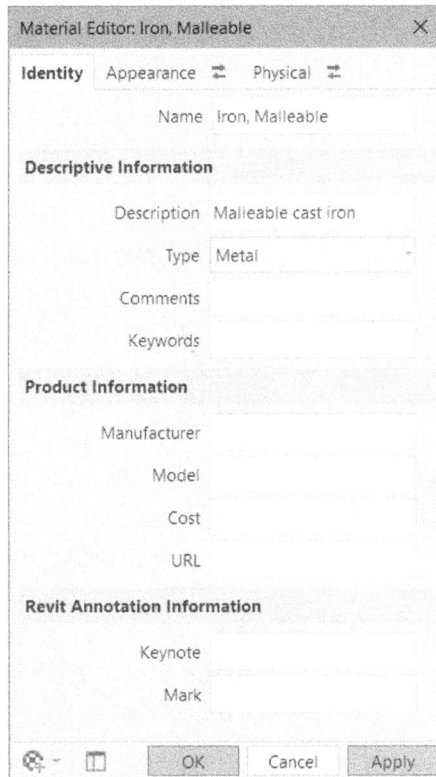

Figure-8. Material Editor dialog box

- The options in various tabs of this dialog box change based on material type. Specify the parameters and click on the **OK** button to modify material and apply to the document. Various parameters related to material are discussed in next topic.

General Parameters of Materials

There are three categories in which properties of materials are defined: Identity, Appearance, and Physical. These categories are available as tabs in the **Material Editor** dialog box when you edit a material or create a new material; refer to Figure-8. The parameters specified in these tabs are discussed next.

Identity Parameters (common for all type of materials)

The parameters of **Identity** tab are used to define description and type of material, manufacturer of material, cost of material and other related data. These parameters are discussed next.

- Specify desired text in the **Description** edit box to add user-define description of material which can be used to identify general use of the material.
- Select desired option from the **Type** drop-down to define type of the material. Note that the type selected here will define the category in which material will be placed in Material Library.
- Specify desired text in the **Comments** edit box to define comments about the material meant to warn or notify the user of material. Like, material is fragile, material is explosive, and so on.
- Specify desired identification keywords in the **Keywords** edit box separated by comma (,) to enable fast filtering of material in the browser based on specified keywords.

- The options in **Product Information** section of this tab are used when generating reports for production or performing cost calculations. Specify desired text in the **Manufacturer** edit box to define the name of vendor from whom your organization purchases the material.
- Specify desired value in the **Model** edit box to define model number of material if provided by your manufacturer.
- Specify desired value in the **Cost** edit box to define cost of material. Note that value specified in this edit box is in the form of text which is not used for any mathematical formula.
- Specify desired value in the **URL** edit box to define website link for the material. If you have Revit installed in your system then you will also find options for **Revit Annotation Information** category. Specify the parameters as discussed earlier.

Any modification in appearance of material does not affect the results of analysis but they can make huge difference when generating rendered images of model. Some important parameters of all five categories of material appearances are discussed next.

Appearance Tab (For metal type materials)

If you are defining parameters of appearance for a metal type material then options in the **Material Editor** will be displayed as shown in Figure-9.

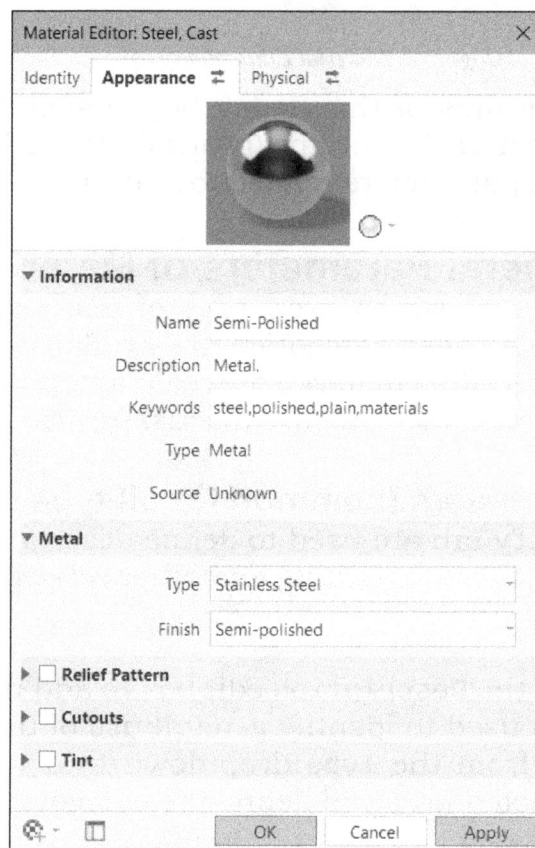

Figure-9. Appearance tab for metal type

- The options in the **Information** section of dialog box are used to define general information about the appearance of material so that user can easily identify the type of material. In this section, you need to specify name, description, and keywords for the material.

- The options in the **Metal** section are used to define color and roughness of surface of material. Click in the field for **Type** option and select desired metallic color. Using the options in **Finish** field, you can define the roughness of surface of material for reflectivity.

- Select the **Relief Pattern** check box to add bumps in the texture of material. Bumps create an illusion of irregularities generally found on surfaces of real objects; refer to Figure-10. Select desired pattern for relief in the **Type** drop-down. Set desired value in **Amount** field to define depth of the bump. Using the **Scale** field, you can define the size of bump. Note that a larger value in **Scale** field means finer pattern. The appearance of material will be modified accordingly.

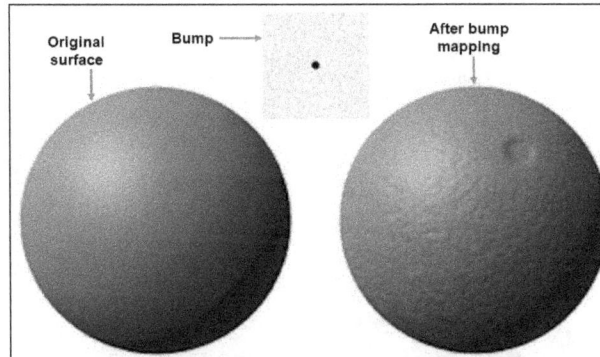

Figure-10. Bump mapping

- Select the **Cutout** check box to make net like texture of material where some portion of material will be transparent and some portion will be opaque; refer to Figure-11. After selecting check box, set desired cutout pattern in the **Type** drop-down and specify related parameters as discussed earlier.

Figure-11. Cutout texture

- Select the **Tint** check box to add shade of selected color to the material surface.

Appearance Tab (Generic)

If you are defining parameters of appearance for generic type material then options in **Material Editor** will be displayed as shown in Figure-12. Most of the options in this tab are same as discussed for metals. The other options of this tab are discussed next.

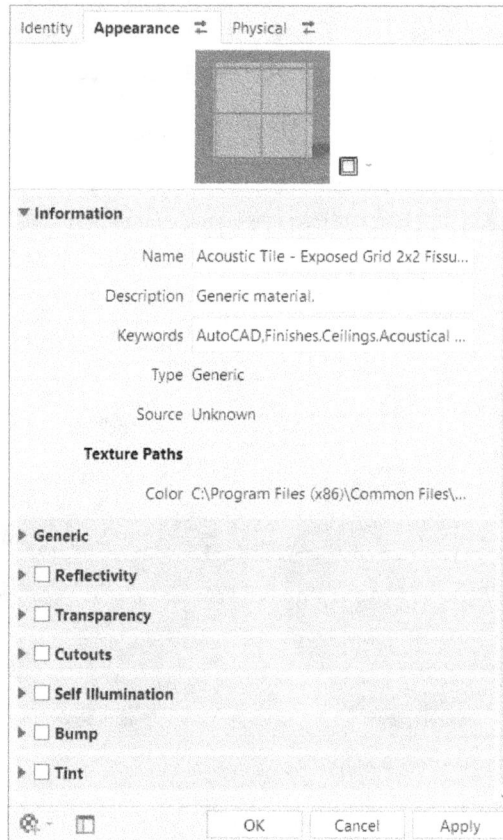

Figure-12. Appearance tab for generic type

- Expand the **Generic** node to modify basic appearance of material. Click in the **Color** field and specify desired color for the model. If you want to use image as appearance texture then click in the **Image** field. The **Texture Editor** window will be displayed; refer to Figure-13. Click in the **Source** field of window and select desired image file. Set the other parameters as desired in the window and close it using button on top-right corner.

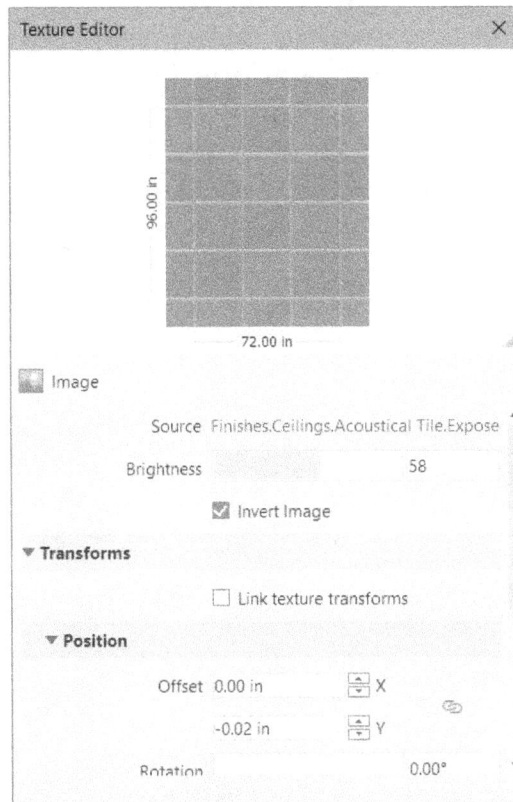

Figure-13. Texture Editor window

- Set desired value in the **Image Fade** field to transparency of image. Similarly, set desired value in the **Glossiness** field to define smoothness of material surface to provide it shine. Select desired option from the **Highlights** drop-down to define whether material surface is metallic or non-metallic.

- Select the **Reflectivity** check box to define how much light will be reflected from the surface of material. Specify desired values in **Direct** and **Oblique** fields to define amount of light that will be reflected when material face is directly in line with light source and when material face is at an angle to the light source, respectively.

- Select the **Transparency** check box to define parameters for free transmission of light through the material. After selecting check box, specify desired value of amount of light which will be freely transmitted though the material using **Amount** slider. Specify desired amount of light to be absorbed by the material surface to give self glow effect in the **Translucency** field. Select desired option from the **Refraction** drop-down to define the medium through which light passes before and after striking the surface of material. The value in edit box next to the drop-down will change based on selected medium. You can specify desired value from 0 to 5 in the edit box to define refraction index of medium.

- Select the **Self Illumination** check box to apply self illumination of material under light. After selecting check box, set desired value using the **Luminance** drop-down to define brightness of light emitting from material in candelas per meter square unit. Set desired value in **Filter Color** selection box to define color of self illuminance. Using the options in **Color Temperature** drop-down, you can define the warmth or coolness of the self illuminance light.

Physical Tab

The options in **Physical** tab are used to define properties that affect the physical data of material used by various simulations (analyses). Parameters like thermal conductivity, yield strength, young's modulus are some example of physical properties. On selecting this tab in **Material Editor** dialog box, the options will be displayed as shown in Figure-14. The options in Information section are same as discussed earlier. The other options are discussed next.

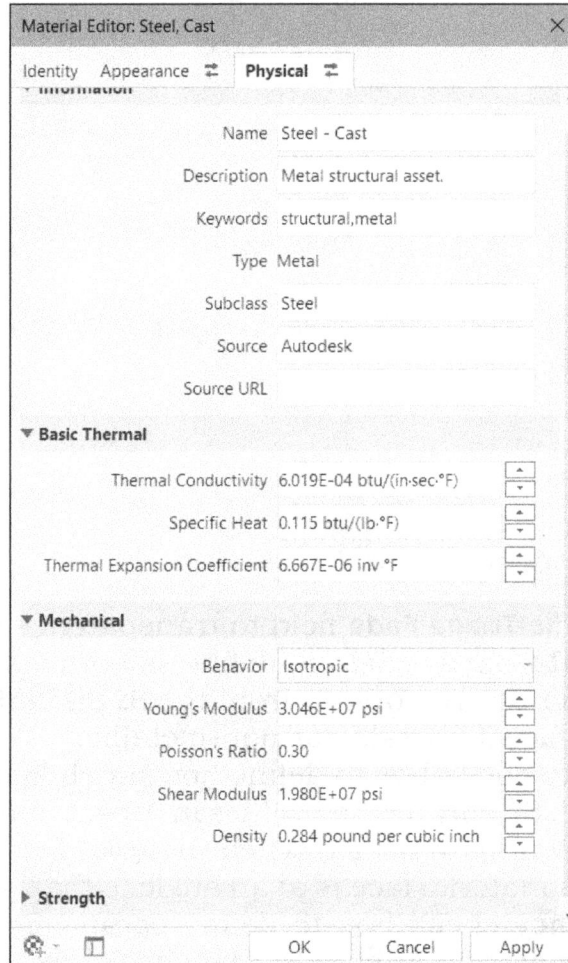

Figure-14. Physical tab for materials

Basic Thermal Properties

*   The options in **Basic Thermal** section are used to define how much heat can be transferred through the material and related parameters. Specify desired value in **Thermal Conductivity** edit box to define the amount of heat in W that can be transferred through the unit length of material at unit temperature. Specify desired value in **Specific Heat** edit box to define the amount of heat energy required by unit mass of material to raise its temperature by 1 unit. Specify desired value in the **Thermal Expansion Coefficient** edit box to define the amount of length increase in material due to unit increase in temperature.

Mechanical Properties

- The options in **Mechanical** section are used to define parameters like Young's modulus, Poisson's ratio, and so on. Specify desired value in **Young's Modulus** edit box to define the amount of stress required for per unit strain in the material. This parameter is also called **Modulus of elasticity** and defines relationship between stress and strain of the material. Specify desired value in **Poisson's Ratio** edit box to define relationship between compression applied on one direction of material causing expansion in other perpendicular direction. You can check the role of this parameter by compression a rectangular piece of sponge. Specify desired value in **Shear Modulus** edit box to define amount of shear stress required for unit shear strain to occur in material. Shear forces cause objects to tilt while their base is fixed. Specify desired value in **Density** edit box to define mass of per unit volume of material. This parameter is used to determine mass of the model. Specify desired value in the **Damping Coefficient** edit box to define a ratio by which oscillations in the material are dissipated. Note that if there is no damping in a spring placed in vacuum then it will keep on oscillating forever once stretched and released. The damping coefficient of material directly affects results of Modal analysis and other frequency related analyses.

- Select the **Isotropic** option from the **Behavior** drop-down to define that properties of the material are same in all the direction. It means you can apply 100 N load in any direction (X, Y, or Z) of material and it will cause same stress in the material because Young's Modulus is same in all directions. Select the **Orthotropic** option from the **Behavior** drop-down to define different physical properties in X, Y, and Z directions for the material. Note that a non-uniform effect of load occurs in these type of materials. Select the **Transverse Isotropic** option from the drop-down if your material has same properties in one plane (transverse plane) and different in direction perpendicular to the plane. Based on selected material behavior, you may get two or three edit boxes for the same property in the dialog box. These edit boxes represent the same property in different directions.

Strength Properties

- The options in **Strength** section are used to define strength parameters of material up to which the material will be useful for mechanical applications. Specify desired value in the **Yield Strength** edit box to define the amount of stress at which permanent deformation will occur in material. Specify desired value in the **Tensile Strength** edit box to define amount of stress required to break off the material. Tensile strength is also called ultimate tensile strength and ultimate strength of material. Select the **Thermally Treated** check box to mark material as thermally treated for drawing representation. Generally, thermal treatment is performed on the material to harden its surface or make the material ductile/brittle.

- After specify desired parameters, click on the **OK** button to apply modifications to material.

Creating a New Material

The **Creates a new material in the document** button is available at the bottom in the **Material Browser**. This button is used to create a new material depending on the user requirements. The procedure to use this button is given next.

- Click on the **Creates a new material in the document** button from the bottom in the **Material Browser**. The **Material Editor** dialog box will be displayed as discussed earlier.
- Specify desired parameters in the dialog box and click on the **OK** button. The material will be created and added in the list of document materials.

Similarly, you can use the options in the **Material library** drop-down to create and manage material libraries; refer to Figure-15. After performing desired operations, close the **Material Browser**.

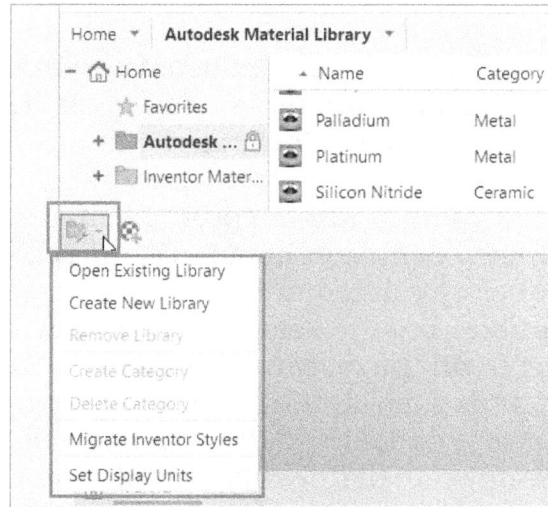

Figure-15. Material library drop-down

APPEARANCE

Appearance represents the way in which material will be displayed when rendering is performed. Applying appearance is similar to applying material without physical properties. You can also use appearance to enhance look of material in graphics area. The **Appearance** tool in **Tools** tab is used to create and manage appearances in Autodesk Inventor. The procedure to use this tool is given next.

- Click on the **Appearance** tool from the **Material and Appearance** panel in the **Tools** tab of the **Ribbon**. The **Appearance Browser** will be displayed; refer to Figure-16.

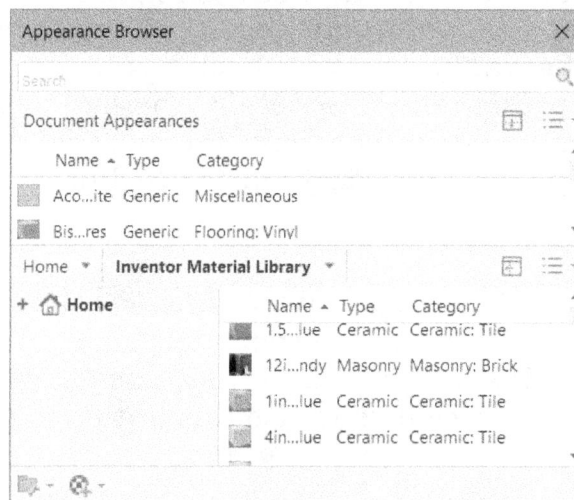

Figure-16. Appearance Browser

- To apply appearance to the model, select desired object from the graphics area and right-click on the appearance to be applied. A shortcut menu will be displayed; refer to Figure-17.

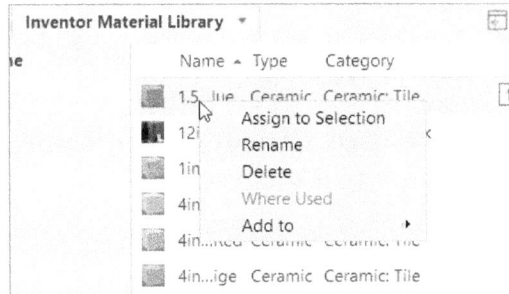

Figure-17. Shortcut menu for appearance

- Select the **Assign to Selection** option from the shortcut menu to apply the appearance to selected objects.

Other options of the **Appearance Browser** are similar to **Material Browser** discussed earlier.

Clearing Appearances

The **Clear** tool in **Tools** tab is used to remove any appearances applied to selected objects. The procedure to use this tool is given next.

- Click on the **Clear** tool from the **Material and Appearance** panel in the **Tools** tab of the **Ribbon**. A mini toolbar will be displayed for managing appearances; refer to Figure-18.

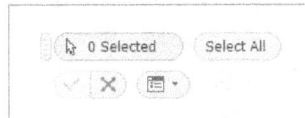

Figure-18. Appearance mini toolbar

- Select desired faces from the model to remove their appearances. If you want to remove appearances from all the faces in the model then click on the **Select All** button.
- Click on the **OK** button from the mini toolbar to remove the appearances.

Adjusting Appearance

The **Adjust** tool in **Tools** tab is used to scale and modify appearance applied to a face. The procedure to use this tool is given next.

- Click on the **Adjust** tool from the **Material and Appearance** panel in the **Tools** tab of the **Ribbon**. The **Adjust Appearance** mini toolbar will be displayed; refer to Figure-19.

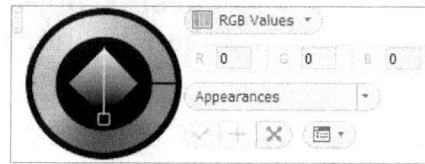

Figure-19. Adjust appearance mini toolbar

• Select the face whose appearance is to be changed. Scale and rotate handles will be displayed on selected face and options in the mini toolbar will be modified according to selected face; refer to Figure-20.

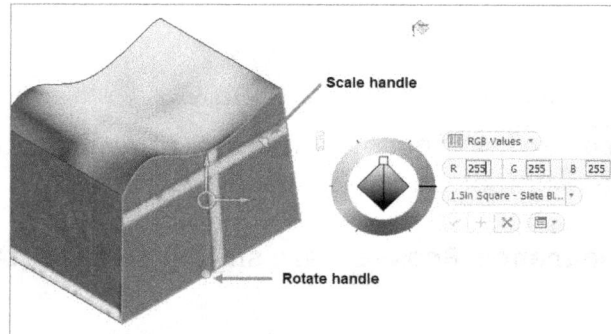

Figure-20. Handles for adjusting appearances

• Drag the handles to scale and rotate appearances. Click on the faces other than earlier selected one to copy the appearance to them.
• Set the other parameters in mini toolbar as discussed earlier and click on the **OK** button to apply adjustments.

APPLICATION OPTIONS

The **Application Options** tool is used to modify common application parameters like interface colors, template file locations, save reminders, display quality, and so on. The procedure to use this tool is given next.

• Click on the **Application Options** tool from the **Options** panel in the **Tools** tab of the **Ribbon**. The **Application Options** dialog box will be displayed; refer to Figure-21.

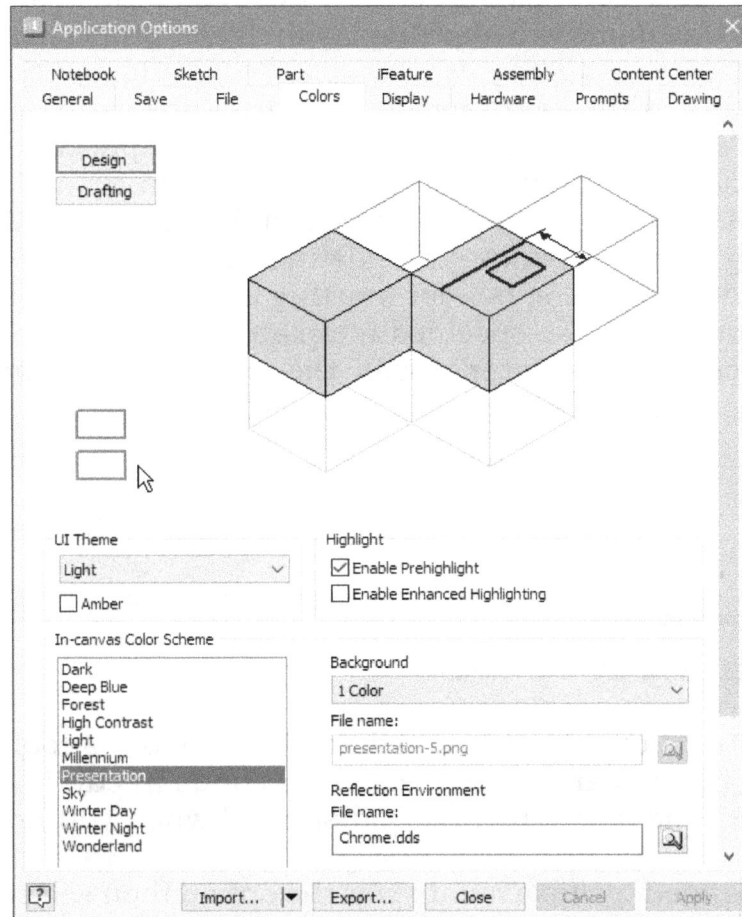

Figure-21. Application Options dialog box

There are enough options in this dialog box that a new chapter can be dedicated to this dialog box. Here, we will given an overview of various tabs in the dialog box and discuss some commonly used options.

General Tab Options

Click on the **General** tab from the dialog box to modify general application parameters; refer to Figure-22. The options in this tab are discussed next.

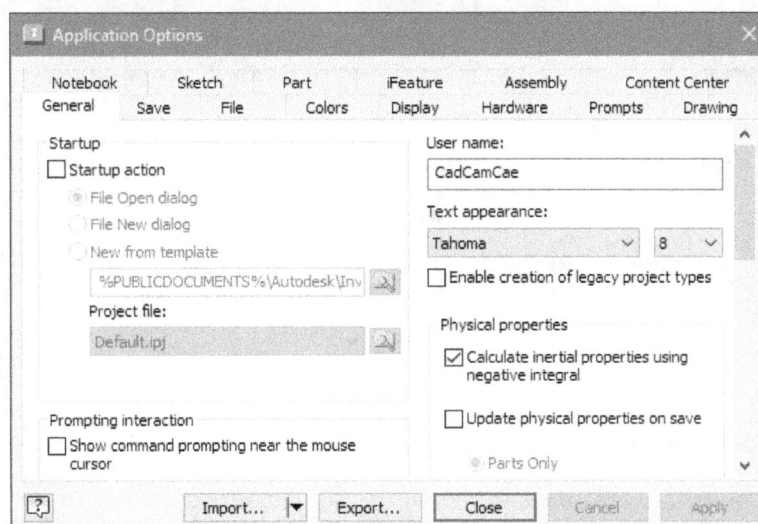

Figure-22. General tab

- Click in the **User name** edit box and specify desired name of software user which is you. Note that this name will also display in model properties and drawing title block.
- Set desired options in the **Text appearance** drop-downs to define text font and text size for text displayed in various dialog boxes of application.
- Select the **Startup action** check box to set action to be performed as soon as software starts. After selecting check box, select the **File Open dialog** radio button to display dialog box for opening an existing file. Select the **File New dialog** radio button to display dialog box for starting a new model. Select the **New from template** radio button to use specified template file and start a new model file.
- Select the **Show command prompting near the mouse cursor** check box to display command prompt near cursor when a tool is active; refer to Figure-23.

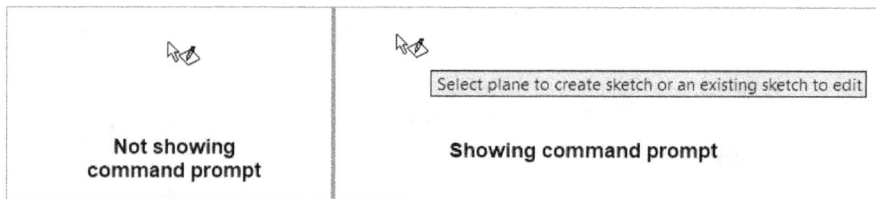

Figure-23. Command prompt near cursor

- Select the **Show Tooltips** check box to display tips about a tool/option when you hover cursor on it. You can set the delay in seconds in edit box below the check box to define the time before tool tip is displayed when you hover cursor on the tool/option.
- If there is an expanded tooltip available for the option then select the **Show second-level Tooltips** check box and specify the delay time as discussed earlier.
- If you want to install local copy of help file then click on the **Download Local Help** link button at the bottom in the dialog box. The download link will open in your default web browser.

Save Tab Options

The options in **Save** tab of **Application Options** dialog box are used to manage parameters related to how software saves the files; refer to Figure-24.

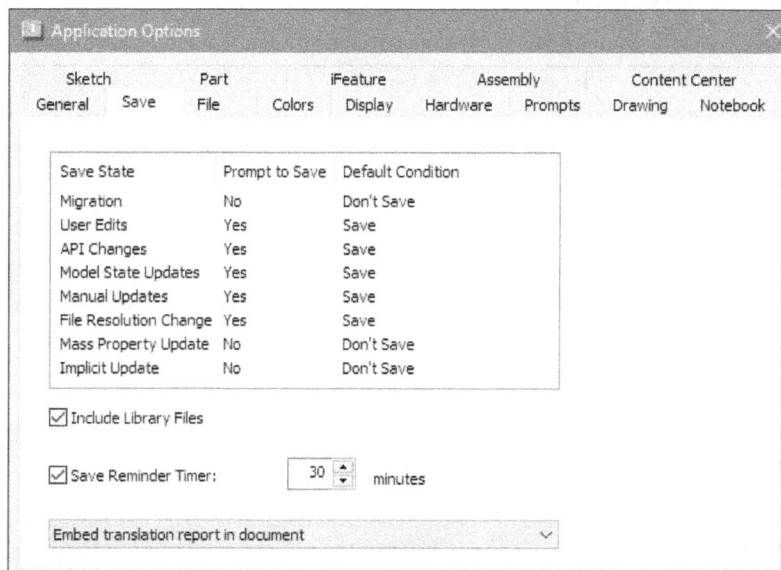

Figure-24. Save tab

- Select the **Include Library Files** check box if you want to save the files used from inventor library with the model file when saving any assembly.
- Select the **Save Reminder Timer** check box and specify desired time duration in edit box next to it so that system prompts you to save any unsaved work after specified time has elapsed.
- If you have performed translation in the model for various texts and annotations then select desired option from the **Translation report** drop-down at the bottom in the dialog box to define how report is saved with the document.

File Tab Options

Select the **File** tab from the dialog box to define options related to file templates and how software opens a file by default; refer to Figure-25.

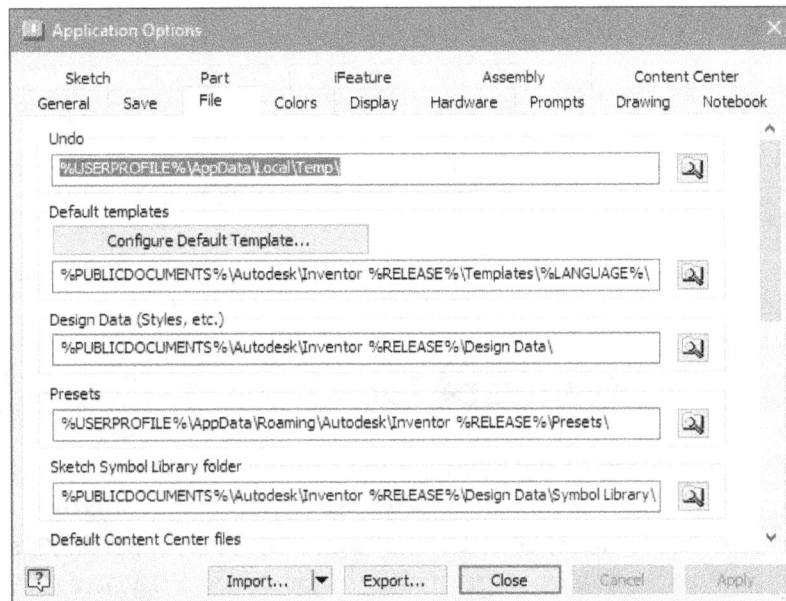

Figure-25. File tab options

- Click on the **Configure Default Template** button from the dialog box to define settings for default templates. The **Configure Default Template** dialog box will be displayed; refer to Figure-26. Select desired radio buttons for default unit and drawing standard to be used in the document. After selecting desired radio buttons, click on the **OK** button from the dialog box. You will be asked whether to overwrite standard template based on specified modifications. Click on the **Overwrite** button to apply changes.

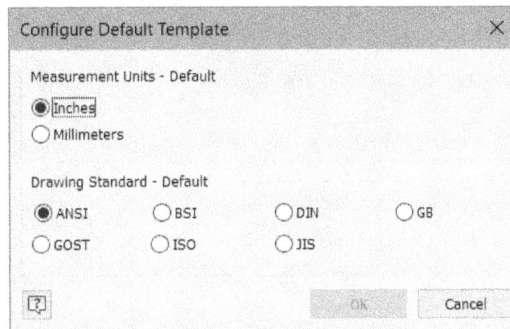

Figure-26. Configure Default Template dialog box

- Click on the **File Open** button from the **Options** area in the dialog box to define the model state and view in which model files will be opened by default by the software.
- Set desired paths for various data files in the fields of this tab.

Colors Tab Options

Using the options in the **Colors** tab of the dialog box, you can set color for background and various objects in the interface as well as drawing area. Select desired option from the **In-canvas Color Scheme** list box in this tab to change the colors using predefined themes.

Display Tab Options

The options in **Display** tab of the dialog box are used to set quality of object display in the drawing area.

- Select desired option from the **Display quality** drop-down to define how smooth the objects will be displayed in the graphics area.
- Using the **Scroll Wheel Sensitivity** slider, you can set the sensitivity of mouse wheel to slower or faster.
- Using the drop-downs in **Middle Mouse Button** area of the dialog box, you can change the functions of various MMB key combinations.

Hardware Tab Options

The options in the **Hardware** tab are used to modify graphic settings for the software. Select the **Quality** radio button to display visualizations in high quality. Select the **Performance** radio button to perform modeling with high performance against the quality of visualizations. Select the **Conservative** radio button to get maximum performance in modeling at the cost of quality of display. If your graphics card is not recognized and you want to use software for visualization performance then select the **Software graphics** check box.

Drawing Tab Options

The options in the **Drawing** tab are used to define settings related to drawing environment; refer to Figure-27.

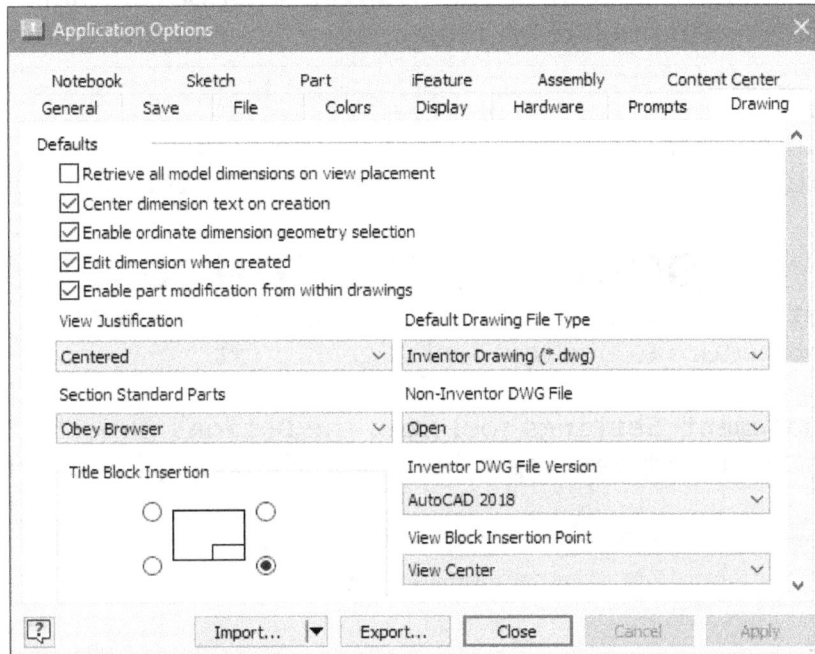

Figure-27. Drawing tab

- Select the **Retrieve all model dimensions on view placement** check box to automatically place model dimensions in the drawing when you place a new view in the drawing.
- Set the other parameters like title block location, default drawing file format, dimension type preferences, and so on in the tab.

Sketch Tab Options

The options in the **Sketch** tab are used to set default settings for sketching environment of Autodesk Inventor; refer to Figure-28.

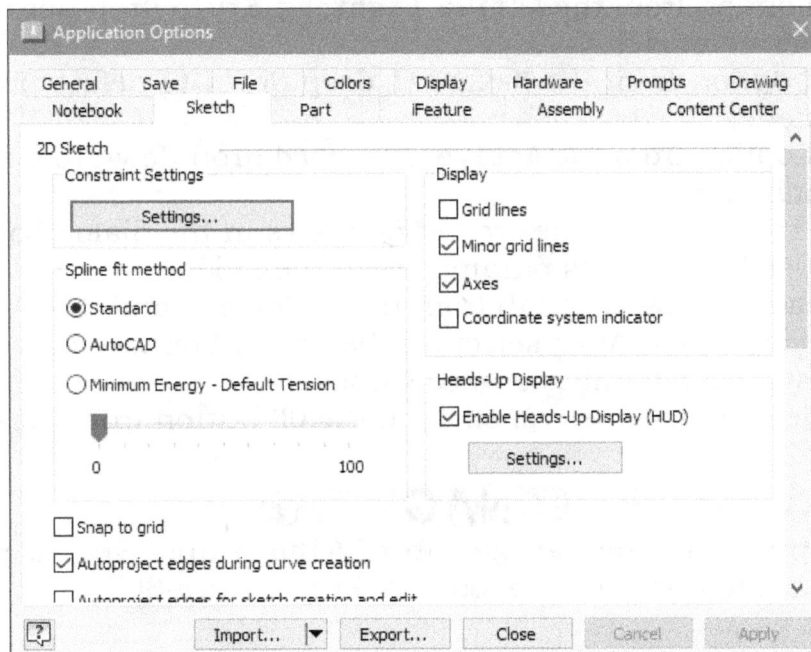

Figure-28. Sketch tab options

- Select desired check boxes from the **Display** area of the dialog box to display various interface elements like grid lines, axes, coordinate system indicator, and so on.

Similarly, you can specify other parameters in the dialog box to define application options. After setting desired parameters, click on the **OK** button from the dialog box.

DOCUMENT SETTINGS

The **Document Settings** tool is used to define default parameters for current open document. The procedure to use this tool is given next.

- Click on the **Document Settings** tool from the **Options** panel in the **Tools** tab of the **Ribbon**. The **Document Settings** dialog box will be displayed; refer to Figure-29.

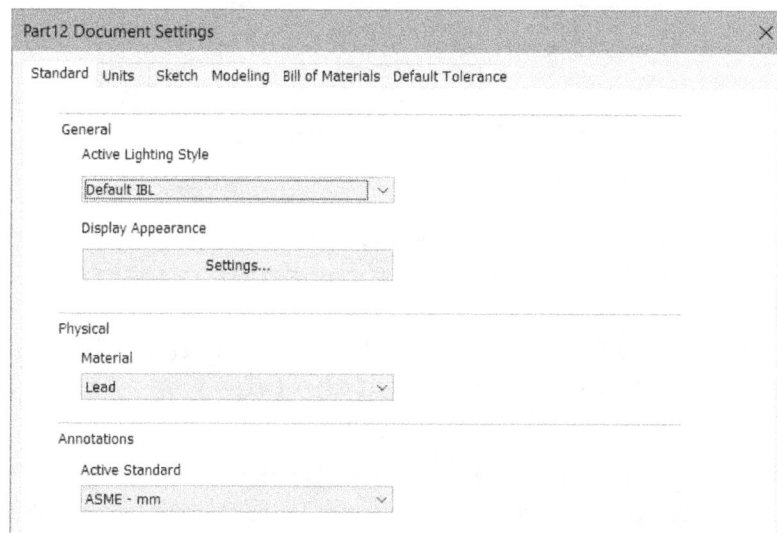

Figure-29. Document Settings dialog box

- Select desired option from the **Active Lighting Style** drop-down to define light setup for rendering model.
- Select desired option from the **Material** drop-down to define physical material used for the model.
- Select desired option from the **Active Standard** drop-down to define units to be used for annotations.
- Select the **Units** tab and set desired parameters in the dialog box to define unit system to be used for various parameters of the model.
- Select the **Default Tolerance** tab from the dialog box and select check boxes to assign tolerance values. After selecting the check box, click in the **Linear** and **Angular** tables to set default tolerance values.
- After setting desired parameters, click on the **OK** button from the dialog box.

MANAGE TAB

The tools and options in the **Manage** tab of **Ribbon** are used to manage styles, parameters, and content related to model; refer to Figure-30.

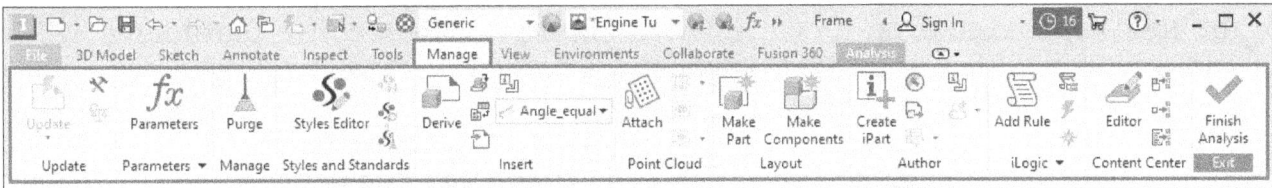

Figure-30. Manage tab

Update Options

The options in the **Update** panel are used to update the changes in the model. Select desired option from the **Update** drop-down to update changes in the assembly, drawing, and modeling environment when you have changed properties of the model but they are not yet reflected in the graphics area. Click on the **Rebuild All** tool from the panel to reflect the changes made in parameters of the model. Click on the **Update Mass** tool from the **Update** panel to update mass of the model based on changes specified in density and other physical properties of the model.

Parameters

The **Parameters** tool is used to create and manage different type of parameters in the model like user parameter and model parameter. The procedure to use this tool is given next.

- Click on the **Parameters** tool from the **Parameters** panel in the **Manage** tab of the **Ribbon**. The **Parameters** dialog box will be displayed showing all the model dimensions as parameters; refer to Figure-31.

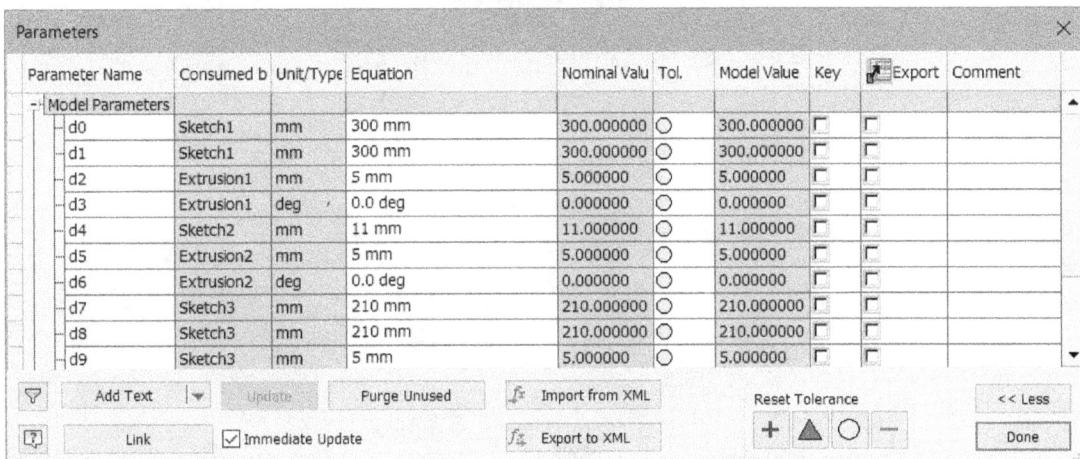

Figure-31. Parameters dialog box

- You can modify the values of all the dimensions by using the fields in the **Equation** column. You can also specify formula for dimension by using the Equation field; refer to Figure-32.

Figure-32. Creating formula using parameter

- If you want to add a new parameter then select desired type of parameter from the **New Parameter** drop-down; refer to Figure-33. Respective parameter will be added in the table under **User Parameters** category.

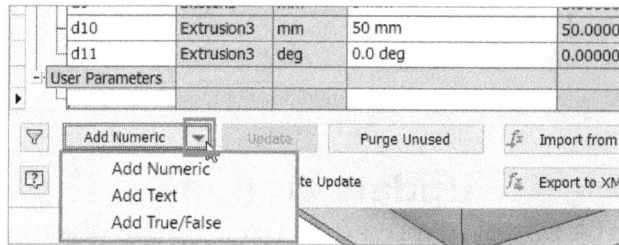

Figure-33. New Parameters drop-down

- Specify desired name for the parameter and define related values in various fields of the dialog box.
- After setting desired parameters, click on the **Done** button. Click on the **Rebuild All** tool from the **Update** panel to update model after changing parameters.

PURGE

The **Purge** tool is used to delete the unused sketches and work features from the active document. The procedure to use this tool is discussed next.

- Click on the **Purge** tool from **Manage** panel in the **Manage** tab of the **Ribbon**. The **Purge Unused** dialog box will be displayed; refer to Figure-34.

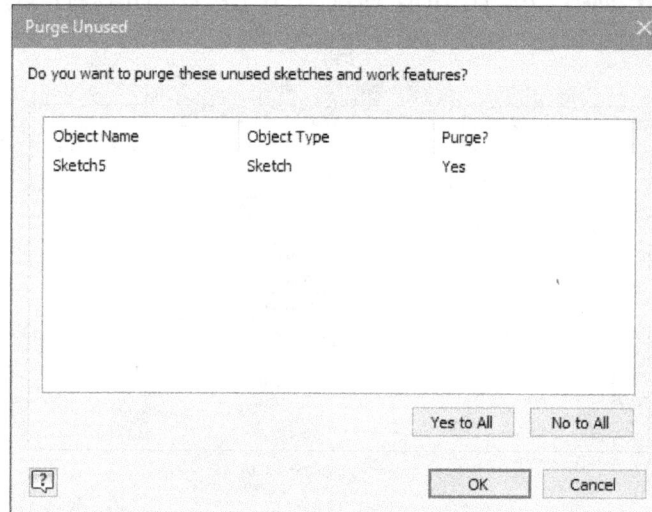

Figure-34. Purge Unused dialog box

- By default, the **Yes** button is selected in the **Purge?** section.
- Click on the **Yes** button from **Purge?** section in the list box to replace with **No** button.
- Click on the **Yes to All** button to purge all the unused sketches or click on the **No to All** button to not to purge any unused sketch.
- After specifying desired parameters, click on the **OK** button from the dialog box.

STYLES EDITOR

The **Styles Editor** tool is used to styles for text and lighting. The procedure to use this tool is given next.

- Click on the **Styles Editor** tool from the **Styles and Standards** panel in the **Manage** tab of the **Ribbon**. The **Styles and Standard Editor** dialog box will be displayed; refer to Figure-35.

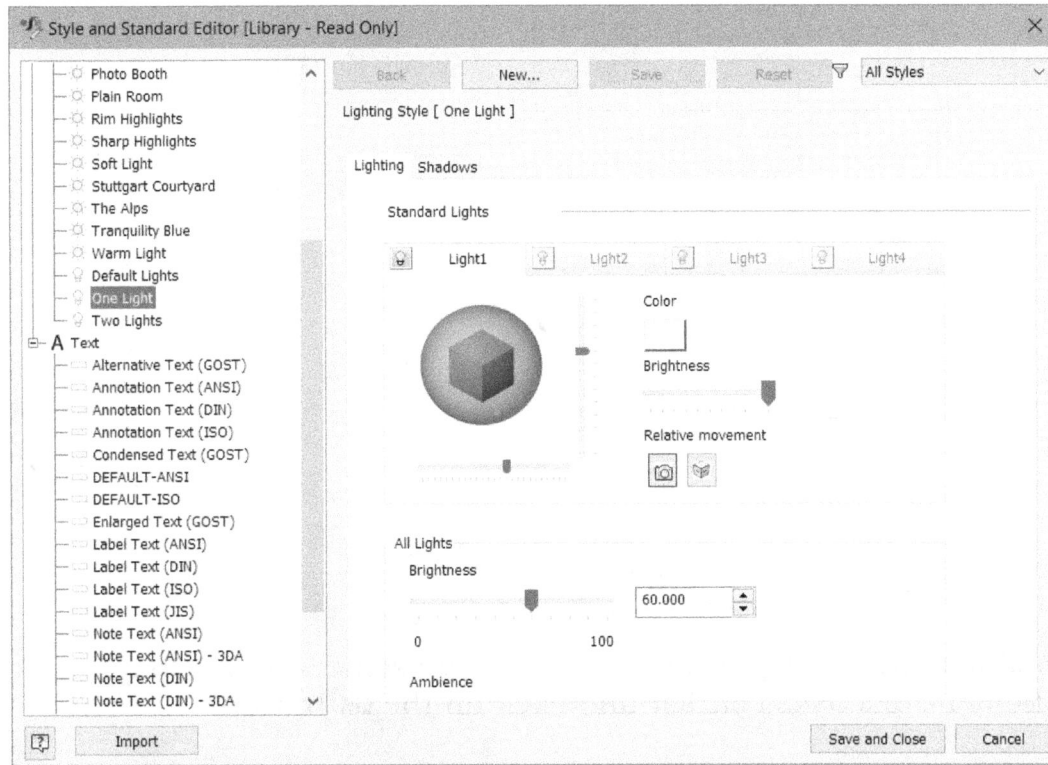

Figure-35. Style and Standard Editor dialog box

- Select desired option from the left area in the dialog box and define the parameters. For example, if you have selected a light then you can define color, brightness, ambience, and location of light.
- The options to modify text style have been discussed earlier.
- After setting desired parameters, click on the **Save and Close** button.

Click on the **Update** tool from the **Styles and Standards** panel to update modified styles in the graphics area.

Click on the **Purge** tool from the **Styles and Standards** panel of **Manage** tab in the **Ribbon** to remove styles from the model which have not been used.

DERIVING COMPONENTS

The **Derive** tool is used to create parts using another part or assembly model as base feature. This tool is useful in creating a multi-body part model. The procedure to use this tool is given next.

- Click on the **Derive** tool from the **Insert** panel in the **Manage** tab of the **Ribbon**. The **Open** dialog box will be displayed and you will be asked to open an existing Autodesk Inventor model file.
- Select desired model file from the dialog box and click on the **Open** button. The **Derived Part** dialog box will be displayed; refer to Figure-36.

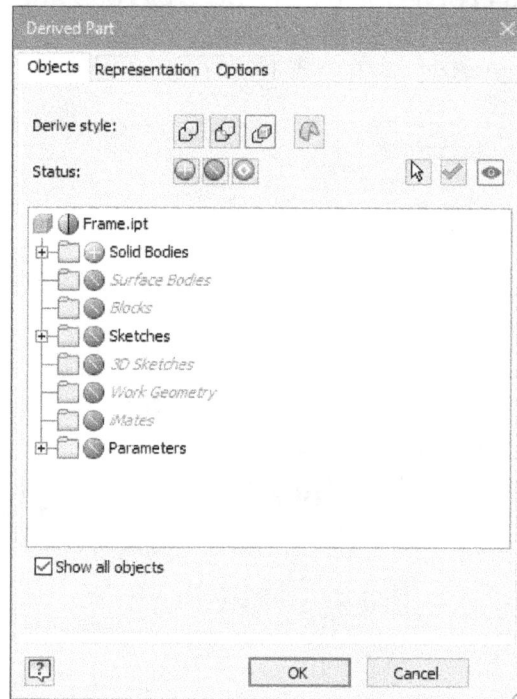

Figure-36. Derived Part dialog box

- The options in this dialog box have been discussed earlier in Chapter 10. Specify the options as discussed earlier and click on the **OK** button.

PLACING FEATURES FROM CONTENT CENTER

The **Feature** tool is used to place objects from content center. The procedure to use this tool is given next.

- Click on the **Feature** tool from the **Insert** panel in the **Manage** tab of the **Ribbon**. The **Place feature from Content Center** dialog box will be displayed; refer to Figure-37.

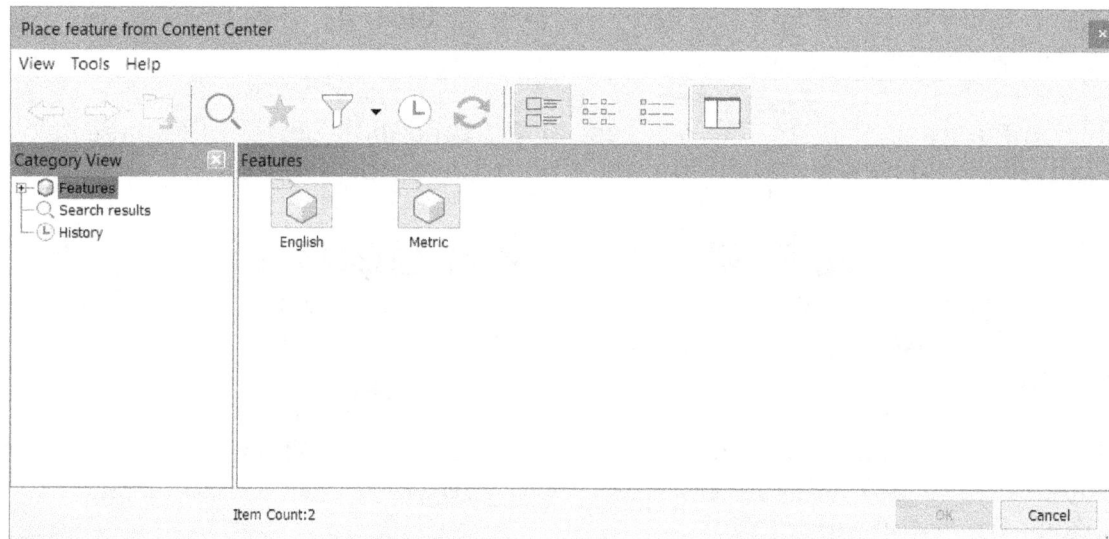

Figure-37. Place feature from Content Center dialog box

- Double-click on the **English** folder to use parts with imperial dimension units or double-click on the **Metric** folder to use parts with metric dimension units. Various categories of parts will be displayed in the dialog box.
- Double-click on the folder of desired category. Related features will be displayed in the dialog box. Note that most of the features will have two options; create a solid protrusion or create a cut using the feature.
- Double-click on desired feature from the dialog box. Related dialog box will be displayed; refer to Figure-38.

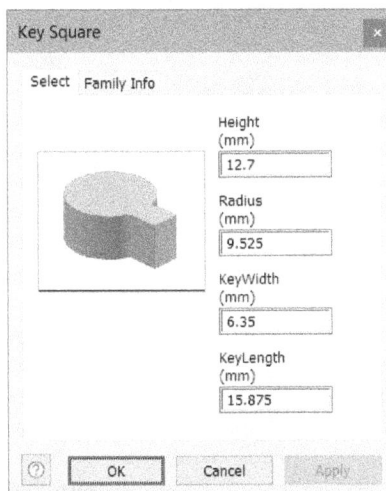

Figure-38. Key Square dialog box

- Set desired parameters in the dialog box and click on the **OK** button. You will be asked to select a face or plane on which the feature will be created.
- Select desired face or plane to be used as base reference for the feature. Preview of the feature will be displayed; refer to Figure-39. Note that when you hover cursor on key points of the model, arrow handles are displayed.
- Drag these arrow handles to modify shape of the model.

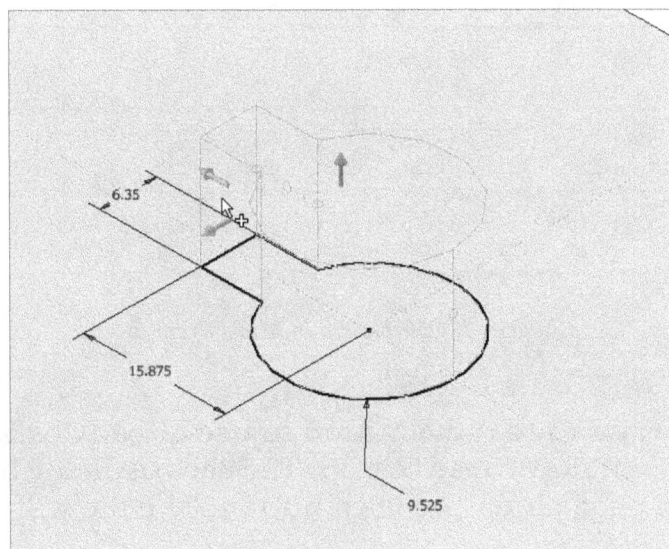

Figure-39. Preview of content center feature

- After setting desired parameters, right-click in the drawing area and select the **Done** option from shortcut menu. The feature will be created.

INSERTING SYSTEM SUPPORTED OBJECTS

The **Insert Object** tool is used to insert objects supported by software installed in your system. For example, if you have Microsoft Excel installed in your system then you can insert excel sheets in the model. The procedure to use this tool is given next.

- Click on the **Insert Object** tool from the **Insert** panel in the **Manage** tab of the **Ribbon**. The **Insert Object** dialog box will be displayed; refer to Figure-40.

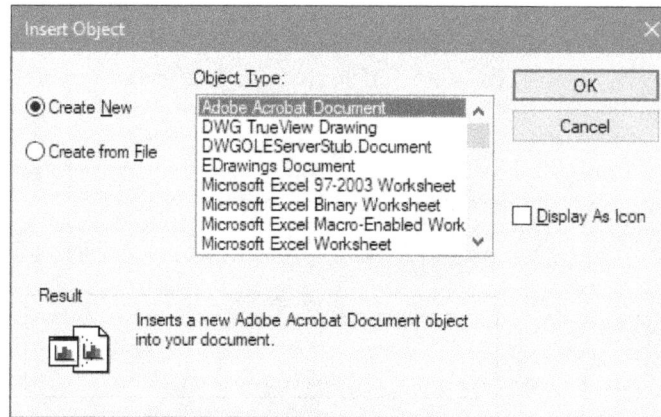

Figure-40. Insert Object dialog box

Creating New File

- Select the **Create New** radio button to create a new object by using the software option available in the **Object Type** area of the dialog box. After selecting radio button, click on the **OK** button. Respective software will open.
- Create the file, save it, and exit the software. Created file will be added in the **3rd Party** node of **Model Browser**; refer to Figure-41. You can edit the object anytime by double-clicking on it.

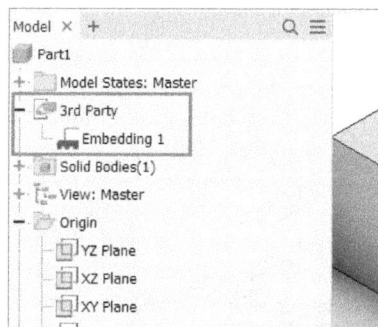

Figure-41. Object added in Model Browser

Using File

- Select the **Create from File** radio button to use already existing object from the local drive and click on the **Browse** button. The **Browse** dialog box will be displayed.
- Select desired file and click on the **Open** button. The file will be added in the edit box.
- Select the **Link** check box to use the file as link in place of saving local copy of the file in project.
- After setting desired parameters, click on the **OK** button from the dialog box. The file will be added in the **Model Browser**.

IMPORTING A CAD FILE MODEL

The **Import** tool is used to import model from a CAD file. The procedure to use this tool is given next.

- Click on the **Import** tool from the **Insert** panel in the **Manage** tab of **Ribbon**. The **Import** dialog box will be displayed. The options of this dialog box have been discussed earlier.

INSERTING iFEATURE

The **Insert iFeature** tool is used to insert iFeature in the part using a plane or planar face. The procedure to use this tool is given next.

- Click on the **Insert iFeature** tool from the **Insert** panel in the **Manage** tab of the **Ribbon**. The **Open** dialog box will be displayed.
- Select desired iFeature file from the dialog box and click on the **Open** button. The **Insert iFeature** dialog box will be displayed with iFeature attached to cursor; refer to Figure-42. The options in the dialog box have been discussed earlier in Chapter 1.

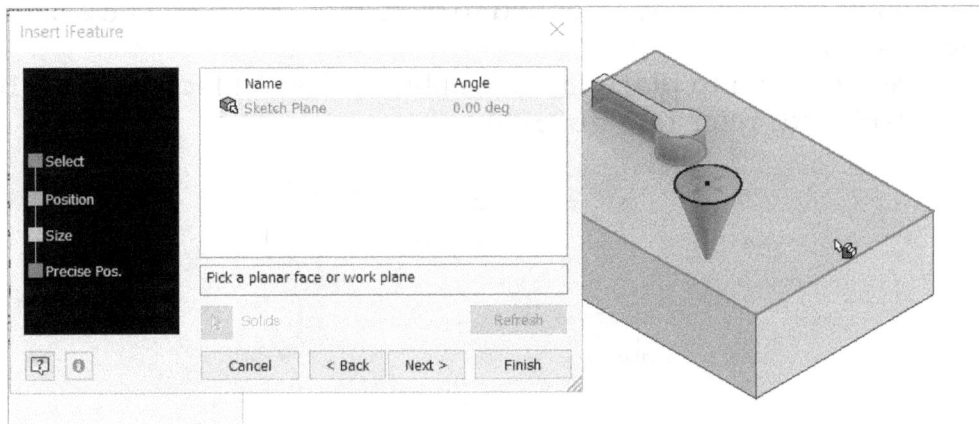

Figure-42. Insert iFeature dialog box and feature attached to cursor

INSERTING FEATURES FROM iFEATURE CATALOG

The options in the **Place iFeature from the iFeature Catalog** drop-down in the **Insert** panel of **Manage** tab in the **Ribbon** are used to insert iFeatures from the catalog of iFeatures. Select desired option from the drop-down and rest of the procedure is same as discussed for Insert iFeature tool. Note that iFeatures saved in C:\Users\Public\Documents\Autodesk\Inventor 2024\Catalog folder (default) will be displayed automatically in the drop-down.

ATTACHING POINT CLOUD

The **Attach** tool is used to insert point cloud data in the model. The procedure to use this tool is given next.

- Click on the **Attach** tool from the **Point Cloud** panel in the **Manage** tab of the **Ribbon**. The **Select Point Cloud File** dialog box will be displayed; refer to Figure-43.

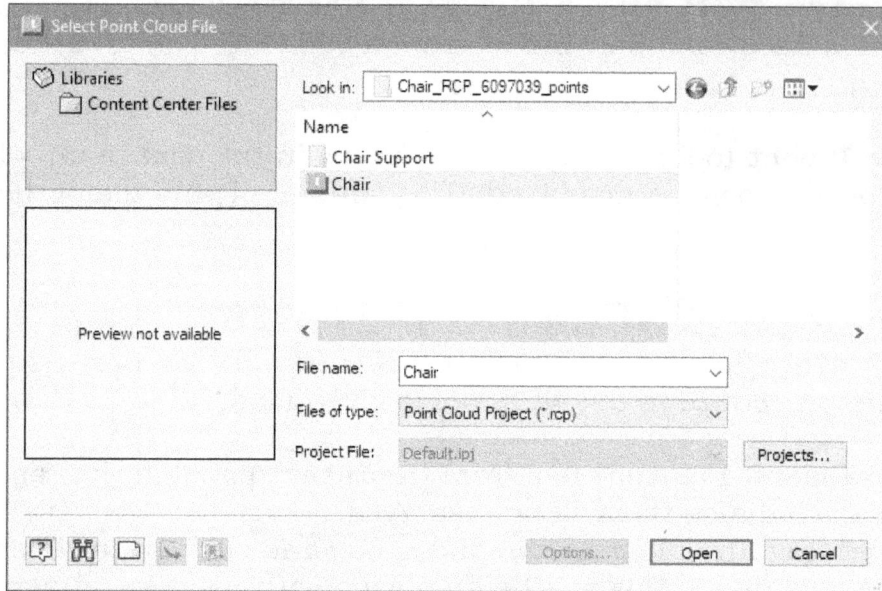

Figure-43. Select Point Cloud File dialog box

- Select desired point cloud data file from the dialog box and click on the **Open** button. The point cloud will be attached to cursor and you will be asked to specify the location to place it.
- Click at desired location to place the data model. The **Attach Point Cloud** dialog box will be displayed; refer to Figure-44.

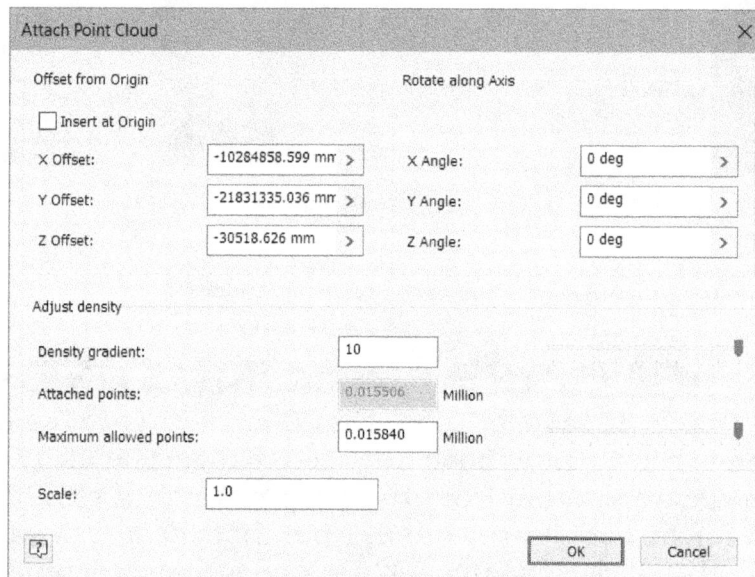

Figure-44. Attach Point Cloud dialog box

- Specify desired parameters in the dialog box like scale value, density, location, and so on.
- Click on the **OK** button to complete the process. The model will be displayed in graphics area; refer to Figure-45.

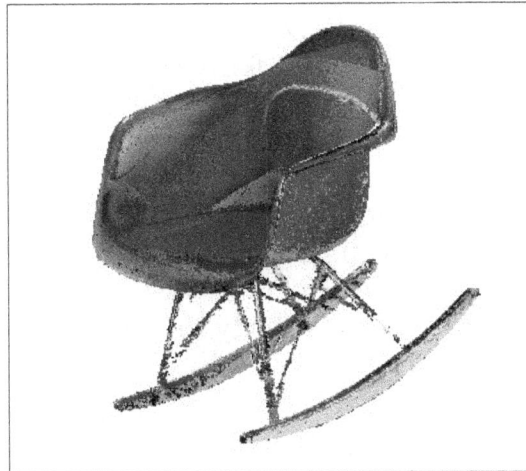
Figure-45. Point cloud model

You can use the other tools in Point Cloud panel to further modify the point cloud data. Note that you can use the point cloud as base for creating your features as points in the point cloud can be easily references for creating features.

MAKING PART USING SELECTED OBJECTS

The **Make Part** tool is used to create a new part model using selected objects as reference. This tool is useful when you have multi-body model file and you want to create individual parts using the bodies so that you can create a functional assembly model. The procedure to use this tool is given next.

- Click on the **Make Part** tool from the **Layout** panel in the **Manage** tab of the **Ribbon**. The **Make Part** dialog box will be displayed with list of all the features in the model; refer to Figure-46.

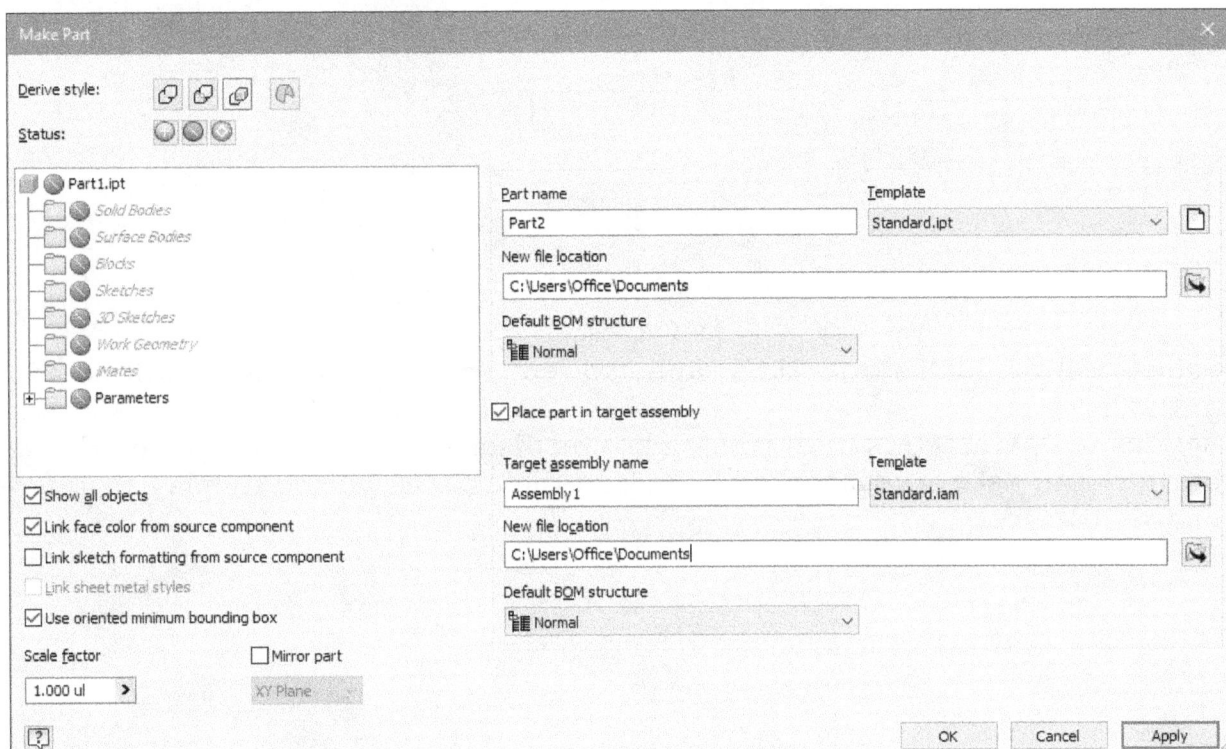
Figure-46. Make Part dialog box

- Double-click on desired bodies, sketches, and features that you want to include in your new part from the list box. Make sure a ⊙ button is displayed against selected feature which means selected feature is derived in the part.
- Click in the **Part name** edit box and specify name of the part file to be created.
- Select desired template from the **Template** drop-down to define basic parameters of model file.
- Click in the **New file location** edit box to define save location for the file.
- You can select the Bill of Materials structure for the model by using options in the **Default BOM structure** drop-down.
- Clear the **Place part in target assembly** check box if you do not want to place the newly created part in an assembly.
- Similarly, set the other parameters and click on the **OK** button. The part file will be created and displayed in the application.

MAKING COMPONENTS

The **Make Components** tool is used to generate components from selected bodies and place them in an assembly file as per their current placements. The procedure to do so is given next.

- Open a multi body part file or create one and then click on the **Make Components** tool from the **Layout** panel in the **Manage** tab of the **Ribbon**. The **Make Components : Selection** dialog box will be displayed; refer to Figure-47.

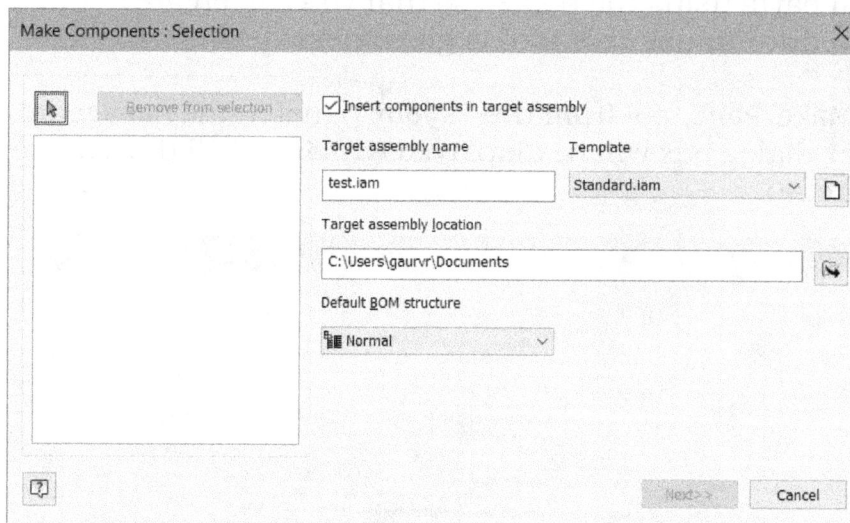

Figure-47. Make Components : Selection dialog box

- Select desired bodies from the graphics area. The selected bodies will be added in the list.
- Set desired parameters in the dialog box as discussed next and click on the **Next** button. The **Make Components : Bodies** dialog box will be displayed; refer to Figure-48.

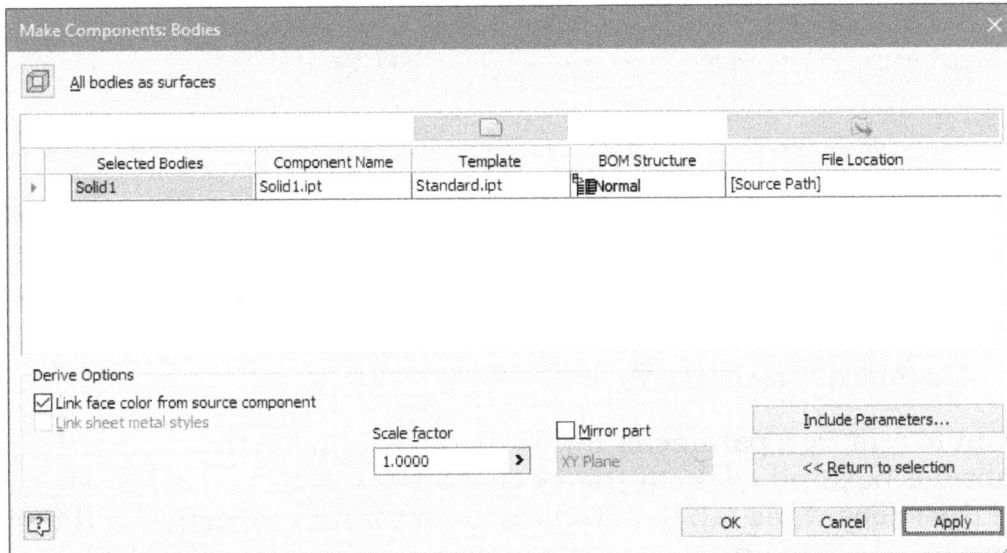

Figure-48. Make Components Bodies dialog box

- You can still modify the parameters as desired by clicking the fields of table. Set the other parameters as discussed earlier and click on the **Apply** button. The assembly file will be created using the bodies as components. Click on the **Cancel** button to exit the dialog box and select the assembly file tile from the bottom bar.

CREATING IPART

The **iPart** tool is used to create a table driven part model in which shape/size of model can be changed by specifying parameters in the table. The procedure to use this tool is given next.

- Click on the **iPart** tool from the **Author** panel in the **Manage** tab of the **Ribbon**. The **iPart Author** dialog box will be displayed; refer to Figure-49.

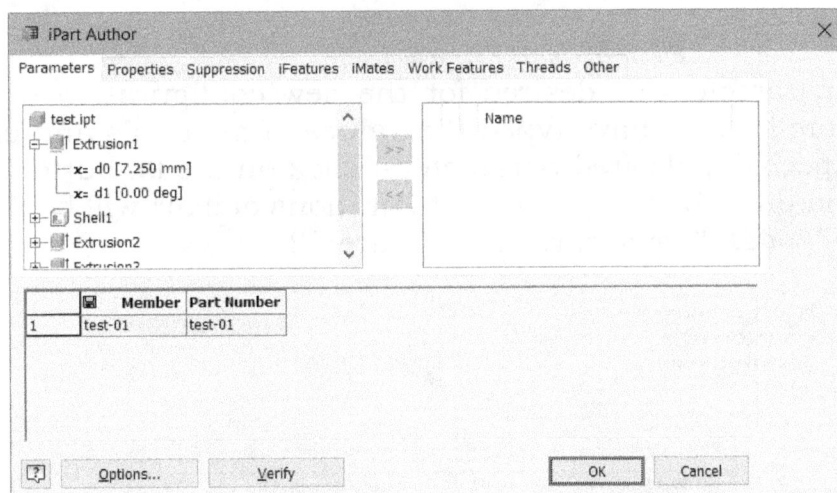

Figure-49. iPart Author dialog box

- Select desired parameter to be added in the iPart from left list box in the dialog box and click on the **>>** button. Selected parameter will be added in the table for customizing part; refer to Figure-50.

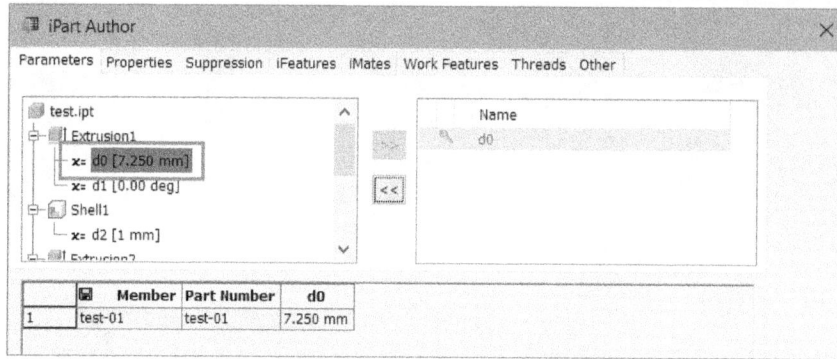

Figure-50. Parameter added for customization

- If you want to remove a parameter from customization list then select the parameter from right list box and click on the **<<** button.
- Click on the **Properties** tab to add parameters from properties of the model. The procedure to add property parameter is same as discussed earlier.
- Click on the **Suppression** tab and select the features to be set as suppressed and unsuppressed in variations of the part.
- Similarly, add other parameters of various tabs in the dialog box to vary the instances of the iPart.
- Right-click in any field of the first row and select the **Insert Row** option from the shortcut menu displayed; refer to Figure-51. A new row will be added in the table.

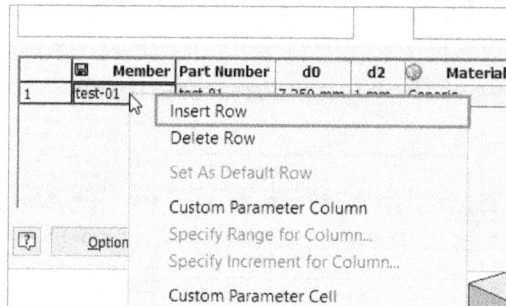

Figure-51. Insert Row option

- Change the parameters as desired for the new configuration of iPart; refer to Figure-52. Note that we have typed the values of materials and compute in the table. After specifying desired parameters, click on the **OK** button. The iPart will be created and table for managing configurations of iPart will be displayed in the **Table** node of **Model Browser**; refer to Figure-53.

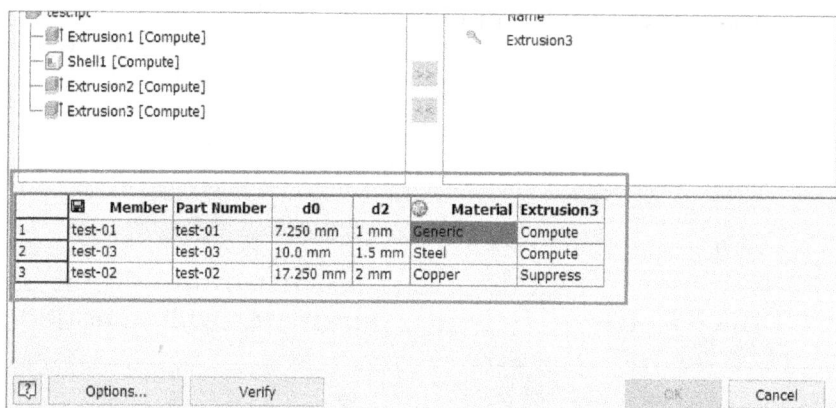

Figure-52. Changing parameters in table

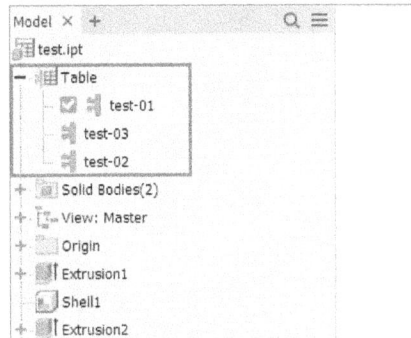
Figure-53. Table of iPart

- Double-click on desired configuration from the **Table** node to activate it in graphics area.
- You can modify the parameters in the dialog box as discussed earlier if needed. Click on the **OK** button from the dialog box to activate the configuration.

DEFINING IMATE

The **iMates** are used to assign default assembly constraints to the part when inserted in an assembly. For example, you can assign Insert constraint to a cylindrical face of part so that when this part is inserted in assembly, it will ask for mating cylindrical face by default. The procedure to define iMate is given next.

- Click on the **iMate** tool from the **Author** panel in the **Manage** tab of the **Ribbon**. The **Create iMate** dialog box will be displayed; refer to Figure-54.

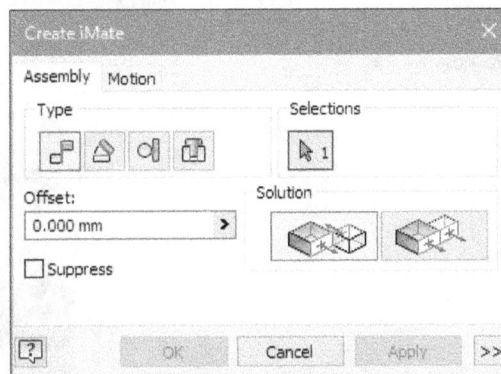
Figure-54. Create iMate dialog box

- Select desired constraint type and set related parameters as discussed earlier in chapter related to Assembly.
- Select desired mate reference (face/edge/axis/point) from the model in graphics area and click on the **OK** button. The iMate will be applied to part.

EXTRACTING IFEATURES

The **Extract iFeature** tool is used to extract features from current model and save them as iFeatures for later use. The procedure to use this tool is given next.

- Click on the **Extract iFeature** tool from the **Author** panel in the **Manage** tab of the **Ribbon**. The **Extract iFeature** dialog box will be displayed; refer to Figure-55.

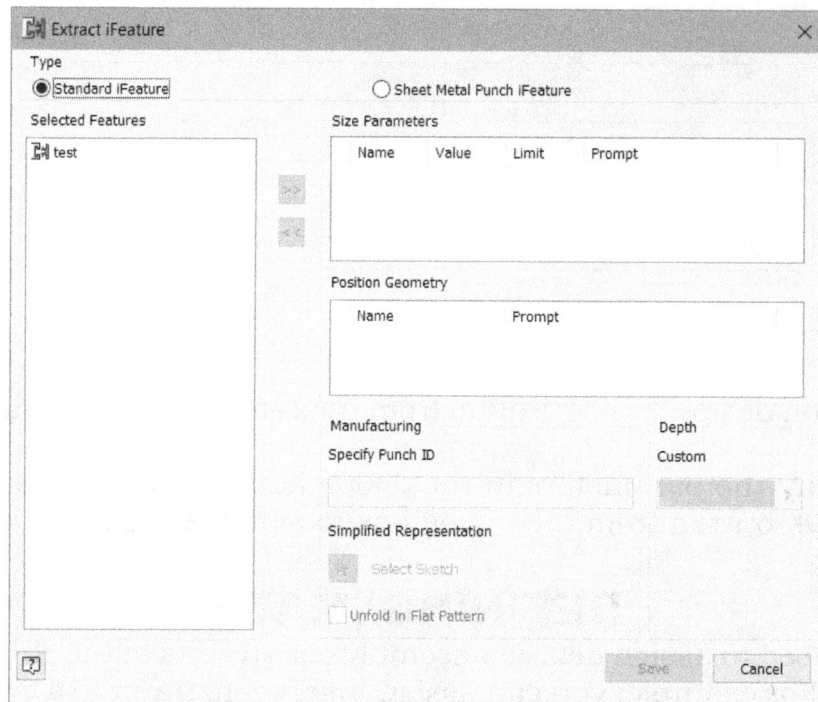

Figure-55. Extract iFeature dialog box

- Select desired feature from the model in graphics area to be used as iFeature. Selected features will be displayed in the **Selected Features** area of the dialog box; refer to Figure-56.

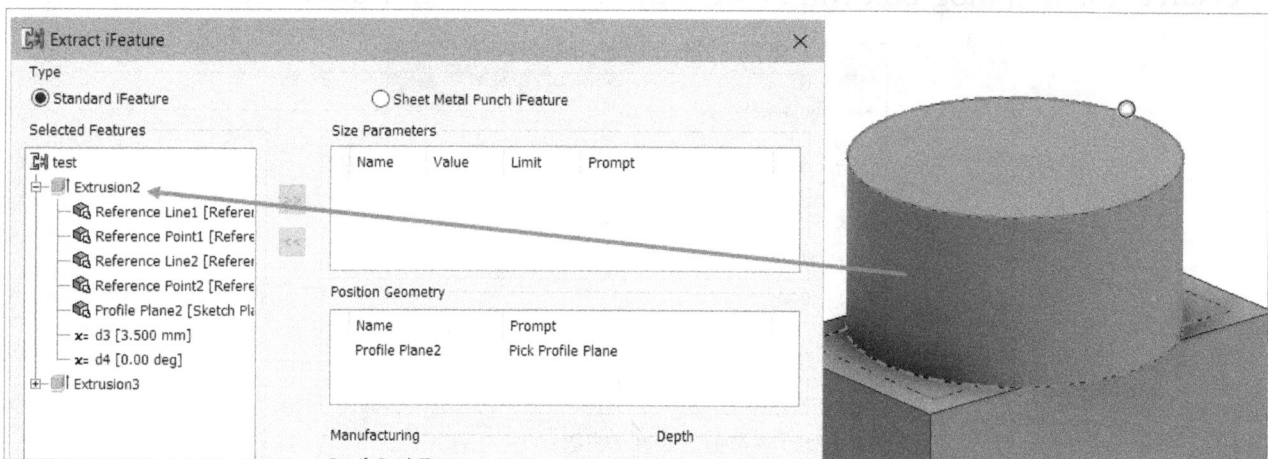

Figure-56. Selected features

- Select desired parameter to be used for size variation from the **Selected Features** area and click on the **>>** button. The parameter will be added in Size Parameters table.
- Similarly, you can add references from the model to be used as positioning geometry; refer to Figure-57.

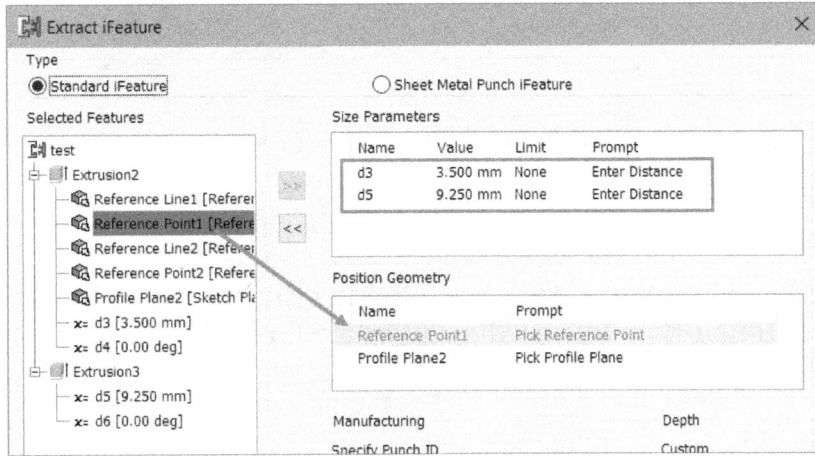

Figure-57. Adding size parameters and position geometry

- After setting desired parameters, click on the **Save** button from the dialog box to save the iFeature. The **Save As** dialog box will be displayed; refer to Figure-58.

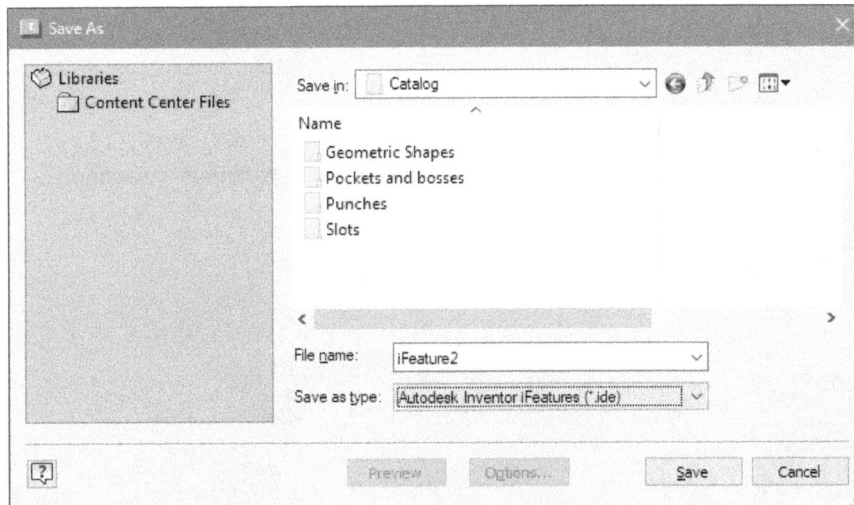

Figure-58. Save As dialog box

- Specify desired name and location of the file, and click on the **Save** button to save iFeature file.

Similarly, you can use the tools in **Component** drop-down of **Author** panel to create iFeature like components related to specific categories.

ADDING RULES TO MODEL

Rules are used to define logic and equations in the model to automate some designing tasks. The procedure to create rules is given next.

- Click on the **Add Rule** tool from the **iLogic** panel in the **Manage** tab of the **Ribbon**. The **Rule Name** dialog box will be displayed; refer to Figure-59.

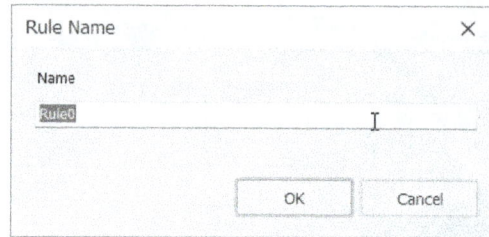

Figure-59. Rule Name dialog box

- Specify desired name for the rule in the **Name** edit box and click on the **OK** button. The **Edit Rule** dialog box will be displayed; refer to Figure-60.

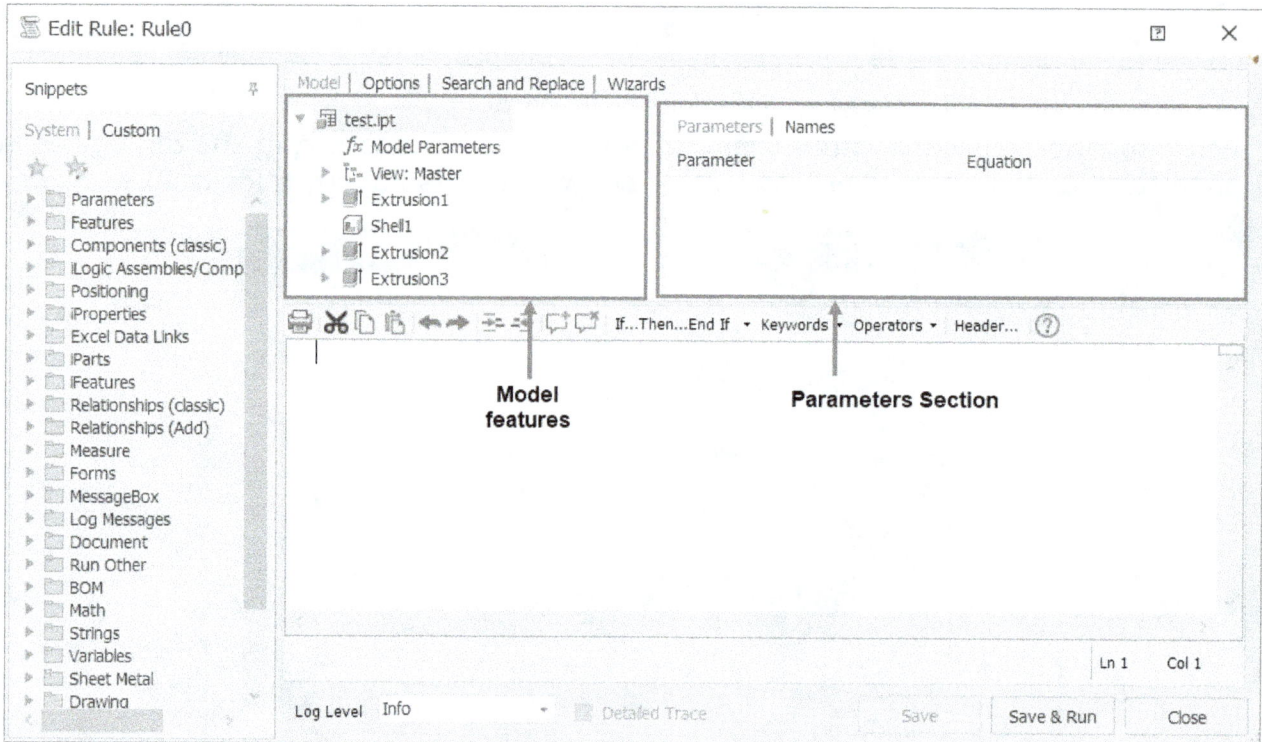

Figure-60. Edit Rule dialog box

- Select desired feature from the **Model** tab to display related parameters in the **Parameters** section of the dialog box.
- Double-click on desired parameter from the **Parameters** section to use it in equation.
- You can create desired arithmetic and logical equations in the equations area of the dialog box; refer to Figure-61.

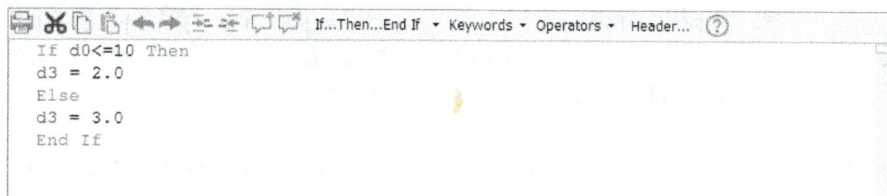

```
If d0<=10 Then
d3 = 2.0
Else
d3 = 3.0
End If
```

Figure-61. Sample equation

- Click on the **Save & Run** button from the dialog box after typing the equation. If there is an error then a message box will be displayed otherwise, model will be modified based on specified rule.

Similarly, you can use the other tools of **Manage** tab in the **Ribbon**.

Index

Ethics of an Engineer

- Engineers shall hold paramount the safety, health and welfare of the public and shall strive to comply with the principles of sustainable development in the performance of their professional duties.

- Engineers shall perform services only in areas of their competence.

- Engineers shall issue public statements only in an objective and truthful manner.

- Engineers shall act in professional manners for each employer or client as faithful agents or trustees, and shall avoid conflicts of interest.

- Engineers shall build their professional reputation on the merit of their services and shall not compete unfairly with others.

- Engineers shall act in such a manner as to uphold and enhance the honor, integrity, and dignity of the engineering profession and shall act with zero-tolerance for bribery, fraud, and corruption.

- Engineers shall continue their professional development throughout their careers, and shall provide opportunities for the professional development of those engineers under their supervision.

www.ingramcontent.com/pod-product-compliance
Lightning Source LLC
Chambersburg PA
CBHW081238220326
41597CB00023BA/3997